Lecture Notes in Artificial Intelligence   4571

Edited by J. G. Carbonell and J. Siekmann

Subseries of Lecture Notes in

Petra Perner (Ed.)

Machine Learning and Data Mining in Pattern Recognition

5th International Conference, MLDM 2007
Leipzig, Germany, July 18-20, 2007
Proceedings

 Springer

Series Editors

Jaime G. Carbonell, Carnegie Mellon University, Pittsburgh, PA, USA
Jörg Siekmann, University of Saarland, Saarbrücken, Germany

Volume Editor

Petra Perner
Institute of Computer Vision and Applied Computer Sciences (IBaI)
Arno-Nitzsche-Str. 43, 04277 Leipzig, Germany
E-mail: pperner@ibai-institut.de

Library of Congress Control Number: 2007930460

CR Subject Classification (1998): I.2, I.5, I.4, F.4.1, H.3

LNCS Sublibrary: SL 7 – Artificial Intelligence

ISSN	0302-9743
ISBN-10	3-540-73498-8 Springer Berlin Heidelberg New York
ISBN-13	978-3-540-73498-7 Springer Berlin Heidelberg New York

Springer is a part of Springer Science+Business Media

springer.com

© Springer-Verlag Berlin Heidelberg 2007
Printed in Germany

Typesetting: Camera-ready by author, data conversion by Scientific Publishing Services, Chennai, India
Printed on acid-free paper   SPIN: 12087358   06/3180   5 4 3 2 1 0

Preface

MLDM / ICDM Medaillie
Meissner Porcellan, the "White Gold" of King
August the Strongest of Saxonia

Gottfried Wilhelm von Leibniz, the great mathematician and son of Leipzig, was watching over us during our event in Machine Learning and Data Mining in Pattern Recognition (MLDM 2007). He can be proud of what we have achieved in this area so far. We had a great research program this year.

This was the fifth MLDM in Pattern Recognition event held in Leipzig (www.mldm.de).

Today, there are many international meetings carrying the title machine learning and data mining, whose topics are text mining, knowledge discovery, and applications. This meeting from the very first event has focused on aspects of machine learning and data mining in pattern recognition problems. We planned to reorganize classical and well-established pattern recognition paradigms from the view points of machine learning and data mining. Although it was a challenging program in the late 1990s, the idea has provided new starting points in pattern recognition and has influenced other areas such as cognitive computer vision.

For this edition, the Program Committee received 258 submissions from 37 countries (see Fig. 1).

To handle this high number of papers was a big challenge for the reviewers. Every paper was thoroughly reviewed and all authors received a detailed report on their submitted work.

After the peer-review process, we accepted 66 high-quality papers for oral presentation, which are included in this proceedings book. The topics range from the classical topics within MLDM such as classification, feature selection and extraction, clustering and support-vector machines, frequent and common item set mining and structural data mining.

China	15.71%	5.04%	USA	10.00%	3.88%	England	7.14%	1.94%
France	4.29%	2.33%	Korea South	4.29%	1.16%	Mexico	4.29%	1.16%
Germany	3.57%	1.94%	Iran	3.57%	0.78%	Italy	3.57%	1.94%
Japan	3.57%	1.55%	Spain	3.57%	1.16%	Turkey	2.86%	1.16%
Lithuania	2.86%	0.78%	Cuba	2.86%	0.78%	Greece	2.14%	1.16%
Poland	2.14%	0.78%	Canada	2.14%	0.78%	Portugal	2.14%	0.39%
Australia	2.14%	0.00%	Switzerland	1.43%	0.78%	Sweden	1.43%	0.78%
Brazil	1.43%	0.39%	Taiwan	1.43%	0.39%	India	1.43%	0.00%
Pakistan	1.43%	0.00%	Chile	0.71%	0.39%	Denmark	0.71%	0.39%
Serbia	0.71%	0.39%	Colombia	0.71%	0.39%	Hungary	0.71%	0.39%
Belgium	0.71%	0.39%	Czech Republic	0.71%	0.39%	Russia	0.71%	0.39%
Netherlands	0.71%	0.39%	Ireland	0.71%	0.39%	Belorussia	0.71%	0.00%
Singapore	0.71%	0.00%						

Fig. 1. Distribution of papers among countries

This year we saw new topics in pattern recognition such as transductive inference and association rule mining. The topics of applied research also increased and cover aspects such as mining spam, newsgroups and blogs, intrusion detection and networks, mining marketing data, medical, biological and environmental data mining, text and document mining. We noted with pleasure an increasing number of papers on special aspects of image mining that are the traditional data in pattern recognition.

24 papers have been selected for poster presentation to be published in the MLDM Poster Proceedings Volume. They cover hot topics like text and document mining, image mining, network mining, support vector machines, feature selection, feature maps, prediction and classification, sequence mining, and sampling methods.

We are pleased to announce that we gave out the best paper award for MLDM for the first time this year.

We also established an MLDM/ICDM/MDA Conference Summary Volume for the first time this year that summarizes the vision of three conferences and the paper presentations and also provides a "Who is Who" in machine learning and data mining by giving each author the chance to present himself.

We also thank members of the Institute of Applied Computer Sciences, Leipzig, Germany (www.ibai-institut.de), who handled the conference. We appreciate the help and understanding of the editorial staff at Springer, and in particular Alfred Hofmann, who supported the publication of these proceedings in the LNAI series.

Last, but not least, we wish to thank all the speakers and participants who contributed to the success of the conference. See you in 2009 again.

July 2007   Petra Perner

International Conference on Machine Learning and Data Mining in Pattern Recognition MLDM 2007

Chair

Petra Perner
Institute of Computer Vision and Applied Computer Sciences IBaI Leipzig, Germany

Program Committee

Agnar Aamodt	NTNU, Norway
Max Bramer	University of Portsmouth, UK
Horst Bunke	University of Bern, Switzerland
Krzysztof Cios	University of Colorado, USA
John Debenham	University of Technology, Australia
Christoph F. Eick	University of Houston, USA
Ana Fred	Technical University of Lisbon, Portugal
Giorgio Giacinto	University of Cagliari, Italy
Howard J. Hamilton	University of Regina, Canada
Makato Haraguchi	Hokkaido University Sapporo, Japan
Tin Kam Ho	Bell Laboratories, USA
Atsushi Imiya	Chiba University, Japan
Horace Ip	City University, Hong Kong
Herbert Jahn	Aero Space Center, Germany
Abraham Kandel	University of South Florida, USA
Dimitrios A. Karras	Chalkis Institute of Technology, Greece
Adam Krzyzak	Concordia University, Montreal, Canada
Lukasz Kurgan	University of Alberta, Canada
Longin Jan Latecki	Temple University Philadelphia, USA
Tao Li	Florida International University, USA
Brian Lovell	University of Queensland, Australia
Ryszard Michalski	George Mason University, USA
Mariofanna Milanova	University of Arkansas at Little Rock, USA
Béatrice Pesquet-Popescu	Ecole Nationale des Télécommunications, France
Petia Radeva	Universitat Autonoma de Barcelona, Spain
Fabio Roli	University of Cagliari, Italy
Gabriella Sanniti di Baja	Instituto di Cibernetica, Italy

Linda Shapiro   University of Washington, USA
Sameer Singh   Loughborough University, UK
Arnold Smeulders   University of Amsterdam, The Netherlands
Patrick Wang   Northeastern University, USA
Harry Wechsler   George Mason University, USA
Sholom Weiss   IBM Yorktown Heights, USA
Djemel Ziou   Université de Sherbrooke, Canada

Additional Reviewers

André Lourenço   Katarzyna Wilamowska
Ashraf Saad   Luca Didaci
Chia-Chi Teng   Mineichi Kuido
Christian Giusti   Natalia Larios Delgado
Chun-Sheng Chen   Oner Ulvi Celepcikay
Claudia Reisz   Orijol Pujol
David Masip   Rachana Parmar
Gary Yngve   Rachsuda Jiamthapthaksin
Gero Szepannek   Roberto Perdisci
Gian Luca Marcialis   Roberto Tronci
Giorgio Fumera   Sara Rolfe
H.S. Wong   Vadeerat Rinsurrongkawong
Helge Langseth   Waclaw Kusnierczyk
Hugo Gamboa   Wei Ding
Igino Corona   Wojciech Stach
Ignazio Pillai   Xingquan Zhu
Indriyati Atmosukarto   Yan Liu
Jordi Vitria

Table of Contents

Invited Talk

Data Clustering: User's Dilemma (Abstract) 1
  Anil K. Jain

Classification

On Concentration of Discrete Distributions with Applications to
Supervised Learning of Classifiers 2
  Magnus Ekdahl and Timo Koski

Comparison of a Novel Combined ECOC Strategy with Different
Multiclass Algorithms Together with Parameter Optimization
Methods 17
  Marco Hülsmann and Christoph M. Friedrich

Multi-source Data Modelling: Integrating Related Data to Improve
Model Performance 32
  Paul R. Trundle, Daniel C. Neagu, and Qasim Chaudhry

An Empirical Comparison of Ideal and Empirical ROC-Based Reject
Rules 47
  Claudio Marrocco, Mario Molinara, and Francesco Tortorella

Outlier Detection with Kernel Density Functions 61
  Longin Jan Latecki, Aleksandar Lazarevic, and Dragoljub Pokrajac

Generic Probability Density Function Reconstruction for Randomization
in Privacy-Preserving Data Mining 76
  Vincent Yan Fu Tan and See-Kiong Ng

An Incremental Fuzzy Decision Tree Classification Method for Mining
Data Streams 91
  Tao Wang, Zhoujun Li, Yuejin Yan, and Huowang Chen

On the Combination of Locally Optimal Pairwise Classifiers 104
  Gero Szepannek, Bernd Bischl, and Claus Weihs

Feature Selection, Extraction and Dimensionality Reduction

An Agent-Based Approach to the Multiple-Objective Selection of
Reference Vectors 117
  Ireneusz Czarnowski and Piotr Jędrzejowicz

On Applying Dimension Reduction for Multi-labeled Problems 131
      Moonhwi Lee and Cheong Hee Park

Nonlinear Feature Selection by Relevance Feature Vector Machine 144
      Haibin Cheng, Haifeng Chen, Guofei Jiang, and Kenji Yoshihira

Affine Feature Extraction: A Generalization of the Fukunaga-Koontz
Transformation 160
      Wenbo Cao and Robert Haralick

Clustering

A Bounded Index for Cluster Validity 174
      Sandro Saitta, Benny Raphael, and Ian F.C. Smith

Varying Density Spatial Clustering Based on a Hierarchical Tree 188
      Xuegang Hu, Dongbo Wang, and Xindong Wu

Kernel MDL to Determine the Number of Clusters 203
      *Ivan O. Kyrgyzov, Olexiy O. Kyrgyzov, Henri Maître, and
      Marine Campedel*

Critical Scale for Unsupervised Cluster Discovery 218
      Tomoya Sakai, Atsushi Imiya, Takuto Komazaki, and Shiomu Hama

Minimum Information Loss Cluster Analysis for Categorical Data 233
      Jiří Grim and Jan Hora

A Clustering Algorithm Based on Generalized Stars 248
      Airel Pérez Suárez and José E. Medina Pagola

Support Vector Machine

Evolving Committees of Support Vector Machines.................... 263
      D. Valincius, A. Verikas, M. Bacauskiene, and A. Gelzinis

Choosing the Kernel Parameters for the Directed Acyclic Graph
Support Vector Machines 276
      Kuo-Ping Wu and Sheng-De Wang

Data Selection Using SASH Trees for Support Vector Machines 286
      Chaofan Sun and Ricardo Vilalta

Dynamic Distance-Based Active Learning with SVM 296
      Jun Jiang and Horace H.S. Ip

Transductive Inference

Off-Line Learning with Transductive Confidence Machines: An
Empirical Evaluation 310
  *Stijn Vanderlooy, Laurens van der Maaten, and
  Ida Sprinkhuizen-Kuyper*

Transductive Learning from Relational Data........................ 324
  *Michelangelo Ceci, Annalisa Appice, Nicola Barile, and
  Donato Malerba*

Association Rule Mining

A Novel Rule Ordering Approach in Classification Association Rule
Mining 339
  Yanbo J. Wang, Qin Xin, and Frans Coenen

Distributed and Shared Memory Algorithm for Parallel Mining of
Association Rules 349
  *J. Hernández Palancar, O. Fraxedas Tormo,
  J. Festón Cárdenas, and R. Hernández León*

Mining Spam, Newsgroups, Blogs

Analyzing the Performance of Spam Filtering Methods When
Dimensionality of Input Vector Changes 364
  J.R. Méndez, B. Corzo, D. Glez-Peña, F. Fdez-Riverola, and F. Díaz

Blog Mining for the Fortune 500 379
  James Geller, Sapankumar Parikh, and Sriram Krishnan

A Link-Based Rank of Postings in Newsgroup 392
  Hongbo Liu, Jiahai Yang, Jiaxin Wang, and Yu Zhang

Intrusion Detection and Networks

A Comparative Study of Unsupervised Machine Learning and Data
Mining Techniques for Intrusion Detection 404
  Reza Sadoddin and Ali A. Ghorbani

Long Tail Attributes of Knowledge Worker Intranet Interactions 419
  Peter Géczy, Noriaki Izumi, Shotaro Akaho, and Kôiti Hasida

A Case-Based Approach to Anomaly Intrusion Detection.............. 434
  Alessandro Micarelli and Giuseppe Sansonetti

Sensing Attacks in Computers Networks with Hidden Markov Models... 449
  Davide Ariu, Giorgio Giacinto, and Roberto Perdisci

Frequent and Common Item Set Mining

FIDS: Monitoring Frequent Items over Distributed Data Streams 464
  Robert Fuller and Mehmed Kantardzic

Mining Maximal Frequent Itemsets in Data Streams Based on
FP-Tree 479
  Fujiang Ao, Yuejin Yan, Jian Huang, and Kedi Huang

CCIC: Consistent Common Itemsets Classifier 490
  Yohji Shidara, Atsuyoshi Nakamura, and Mineichi Kudo

Mining Marketing Data

Development of an Agreement Metric Based Upon the RAND Index
for the Evaluation of Dimensionality Reduction Techniques, with
Applications to Mapping Customer Data............................ 499
  Stephen France and Douglas Carroll

A Sequential Hybrid Forecasting System for Demand Prediction 518
  Luis Aburto and Richard Weber

A Unified View of Objective Interestingness Measures 533
  Céline Hébert and Bruno Crémilleux

Comparing State-of-the-Art Collaborative Filtering Systems 548
  Laurent Candillier, Frank Meyer, and Marc Boullé

Structural Data Mining

Reducing the Dimensionality of Vector Space Embeddings of Graphs ... 563
  Kaspar Riesen, Vivian Kilchherr, and Horst Bunke

PE-PUC: A Graph Based PU-Learning Approach for Text
Classification 574
  Shuang Yu and Chunping Li

Efficient Subsequence Matching Using the Longest Common
Subsequence with a Dual Match Index.............................. 585
  Tae Sik Han, Seung-Kyu Ko, and Jaewoo Kang

A Direct Measure for the Efficacy of Bayesian Network Structures
Learned from Data... 601
  Gary F. Holness

Image Mining

A New Combined Fractal Scale Descriptor for Gait Sequence 616
  Li Cui and Hua Li

Palmprint Recognition by Applying Wavelet Subband Representation
and Kernel PCA 628
Murat Ekinci and Murat Aykut

A Filter-Refinement Scheme for 3D Model Retrieval Based on Sorted
Extended Gaussian Image Histogram 643
Zhiwen Yu, Shaohong Zhang, Hau-San Wong, and Jiqi Zhang

Fast-Maneuvering Target Seeking Based on Double-Action
Q-Learning.. 653
Daniel C.K. Ngai and Nelson H.C. Yung

Mining Frequent Trajectories of Moving Objects for Location
Prediction 667
Mikołaj Morzy

Categorizing Evolved CoreWar Warriors Using EM and Attribute
Evaluation 681
*Doni Pracner, Nenad Tomašev, Miloš Radovanović, and
Mirjana Ivanović*

Restricted Sequential Floating Search Applied to Object Selection...... 694
*J. Arturo Olvera-López, J. Francisco Martínez-Trinidad, and
J. Ariel Carrasco-Ochoa*

Color Reduction Using the Combination of the Kohonen Self-Organized
Feature Map and the Gustafson-Kessel Fuzzy Algorithm 703
Konstantinos Zagoris, Nikos Papamarkos, and Ioannis Koustoudis

A Hybrid Algorithm Based on Evolution Strategies and Instance-Based
Learning, Used in Two-Dimensional Fitting of Brightness Profiles in
Galaxy Images... 716
Juan Carlos Gomez and Olac Fuentes

Gait Recognition by Applying Multiple Projections and Kernel PCA ... 727
Murat Ekinci, Murat Aykut, and Eyup Gedikli

Medical, Biological, and Environmental Data Mining

A Machine Learning Approach to Test Data Generation: A Case Study
in Evaluation of Gene Finders 742
Henning Christiansen and Christina Mackeprang Dahmcke

Discovering Plausible Explanations of Carcinogenecity in Chemical
Compounds 756
Eva Armengol

One Lead ECG Based Personal Identification with Feature Subspace
Ensembles 770
Hugo Silva, Hugo Gamboa, and Ana Fred

Classification of Breast Masses in Mammogram Images Using Ripley's
K Function and Support Vector Machine............................ 784
    Leonardo de Oliveira Martins, Erick Corrêa da Silva,
    Aristófanes Corrêa Silva, Anselmo Cardoso de Paiva, and
    Marcelo Gattass

Selection of Experts for the Design of Multiple Biometric Systems...... 795
    Roberto Tronci, Giorgio Giacinto, and Fabio Roli

Multi-agent System Approach to React to Sudden Environmental
Changes 810
    Sarunas Raudys and Antanas Mitasiunas

Equivalence Learning in Protein Classification 824
    Attila Kertész-Farkas, András Kocsor, and Sándor Pongor

Text and Document Mining

Statistical Identification of Key Phrases for Text Classification......... 838
    Frans Coenen, Paul Leng, Robert Sanderson, and Yanbo J. Wang

Probabilistic Model for Structured Document Mapping 854
    Guillaume Wisniewski, Francis Maes, Ludovic Denoyer, and
    Patrick Gallinari

Application of Fractal Theory for On-Line and Off-Line Farsi Digit
Recognition 868
    Saeed Mozaffari, Karim Faez, and Volker Märgner

Hybrid Learning of Ontology Classes 883
    Jens Lehmann

Discovering Relations Among Entities from XML Documents 899
    Yangyang Wu, Qing Lei, Wei Luo, and Harou Yokota

Author Index 911

Data Clustering: User's Dilemma

Anil K. Jain

Department of Computer Science and Engineering
Michigan State University (USA)
http://www.cse.msu.edu/~jain/

Abstract. Data clustering is a long standing research problem in pattern recognition, computer vision, machine learning, and data mining with applications in a number of diverse disciplines. The goal is to partition a set of n d-dimensional points into k clusters, where k may or may not be known. Most clustering techniques require the definition of a similarity measure between patterns, which is not easy to specify in the absence of any prior knowledge about cluster shapes. While a large number of clustering algorithms exist, there is no optimal algorithm. Each clustering algorithm imposes a specific structure on the data and has its own approach for estimating the number of clusters. No single algorithm can adequately handle various cluster shapes and structures that are encountered in practice. Instead of spending our effort in devising yet another clustering algorithm, there is a need to build upon the existing published techniques. In this talk we will address the following problems: (i) clustering via evidence accumulation, (ii) simultaneous clustering and dimensionality reduction, (iii) clustering under pair-wise constraints, and (iv) clustering with relevance feedback. Experimental results show that these approaches are promising in identifying arbitrary shaped clusters in multidimensional data.

P. Perner (Ed.): MLDM 2007, LNAI 4571, p. 1, 2007.
© Springer-Verlag Berlin Heidelberg 2007

On Concentration of Discrete Distributions with Applications to Supervised Learning of Classifiers

Magnus Ekdahl and Timo Koski

Department of Mathematics
Linköpings University
SE-581 83 Linköping, Sweden

Abstract. Computational procedures using independence assumptions in various forms are popular in machine learning, although checks on empirical data have given inconclusive results about their impact. Some theoretical understanding of when they work is available, but a definite answer seems to be lacking. This paper derives distributions that maximizes the statewise difference to the respective product of marginals. These distributions are, in a sense the worst distribution for predicting an outcome of the data generating mechanism by independence. We also restrict the scope of new theoretical results by showing explicitly that, depending on context, independent ('Naïve') classifiers can be as bad as tossing coins. Regardless of this, independence may beat the generating model in learning supervised classification and we explicitly provide one such scenario.

1   Introduction

Factorization of joint probability distributions in various forms such as conditional independence is widely used in machine learning and statistics in areas such as pattern recognition, learning graphical models, supervised classification and density estimation (see for example [1, 2, 3, 4]). This is not strange since a discrete distribution with independent variables has fewer parameters than a full discrete distribution [5] and often leads to sharper bounds on the performance of learning [6, 7, 8].

While probabilistic independence cannot model all discrete distributions with a preassigned level of accuracy, not much has been published about of what kind of distributions that are worst to model with independence or how much a distribution can differ from the independence factorization.

The main result is a set of distributions that maximizes the absolute probabilistic difference w.r.t. independence, in some sense providing a theoretical counterexample complementing the papers studying this effect on empirical data, c.f. the references in the meta study on the subject available in [4]. This set of distributions also provides a restriction on the extent to which theoretical results concerning independence can be improved. This is coupled with examples in density estimation, classification and supervised classification. Some of the examples

P. Perner (Ed.): MLDM 2007, LNAI 4571, pp. 2–16, 2007.

are ways to demonstrate important points, such as the fact that Naïve Bayes in pattern recognition can be as bad as tossing a coin or that the generating model can be suboptimal in supervised classification.

1.1   Organization

First Section 2 introduces the notation used in the paper then Section 3 presents the main result and discusses implications in density estimation. Section 4 uses the result in Section 3 to show explicitly that independence can yield bad results in pattern recognition and exemplifies that in empirical risk minimization independence outperforms the generating model. All but the shortest proofs can be found in the Appendix.

2   Notation

Let X_i be a discrete random variable, r.v., with the range $\{0, \ldots, r-1\}$ for $r \in [2, \ldots, \infty)$. Random vectors will be written in **bold**, such as $\boldsymbol{X} = (X_i)_{i=1}^d$, that is $\boldsymbol{X}$ is d dimensional, where $d \in [2, \ldots, \infty)$. We denote a state of $\boldsymbol{X}$ with no missing elements as $\boldsymbol{x}$. When referring to the range of $\boldsymbol{X}$, $\mathcal{X} = \bigotimes_{i=1}^d \mathcal{X}_i$ will be used. Note that we will restrict our attention to the multivariate case, where every random element has the same range space, that is, where $\mathcal{X}_i = \{0, \ldots, r-1\}$ for all $i \in [1, \ldots, d]$. The probability that $\boldsymbol{X} = \boldsymbol{x}$, that is $P(\boldsymbol{X} = \boldsymbol{x})$, will be written in the shorter form $P(\boldsymbol{x})$. Let $\mathcal{P}$ denote the set of possible distributions. The individual elements X_i in the random vector $\boldsymbol{X}$ are independent if (and only if) $P(\boldsymbol{x}) = \prod_{i=1}^d P(x_i)$ for all $\boldsymbol{x} \in \mathcal{X}$. As with $P(\boldsymbol{x})$ the short version $P(X_i = x_i) = P(x_i)$ is used for the marginal distribution. Despite of this we are not restricted to the case where all the marginal probabilities for the elements are equal.

3   The Worst Case Deviance

As noted in the introduction the problem of storing probability tables for high dimensional discrete distributions is well known [5]. Independence reduces this problem from $\prod_{i=1}^d r - 1$ to $\sum_{i=1}^d (r - 1)$ required table entries. This does not mean that independence can be introduced without problems however. This section will elaborate on how much a distribution can differ from its independent counterpart, where difference is in a prediction sense.

Definition 1. *A discrete distribution P maximizing the state wise prediction difference w.r.t. its independent counterpart (MSDI distribution for short), is one such that for a reference state $\boldsymbol{x}_j$*

$$P(\boldsymbol{x}) = \begin{cases} d^{\frac{-1}{d-1}} & \boldsymbol{x} = \boldsymbol{x}_j \\ 0 & \boldsymbol{x} \neq \boldsymbol{x}_j \text{ but there exists } i \in [1, \ldots, d] \text{ such that } x_i = x_{ij} \end{cases}$$

In particular a distribution such that

$$
P(\boldsymbol{x}) = \begin{cases} d^{\frac{-1}{d-1}} & x_i = 0 \text{ for all } i \\ 1 - d^{\frac{-1}{d-1}} & x_i = 1 \text{ for all } i \\ 0 & \text{otherwise} \end{cases} \tag{1}
$$

is an MSDI distribution that satisfies $\sum_{\boldsymbol{x} \in \mathcal{X}} P(\boldsymbol{x}) = 1$. Here $0 \leqslant P(\boldsymbol{x}) \leqslant 1$ since

$$
\infty > d > 1 \Rightarrow \infty > d^{\frac{1}{d-1}} > 1 \Rightarrow 0 < d^{\frac{-1}{d-1}} < 1 . \tag{2}
$$

Any distribution satisfying Definition 1 is maximizing the state wise prediction difference of its counterpart, factorized by independence in the following way.

Theorem 1. *For an MSDI distribution $\tilde{P}$*

$$
\max_{P \in \mathcal{P}, \boldsymbol{x} \in \mathcal{X}} \left| P(\boldsymbol{x}) - \prod_{i=1}^{d} P(x_i) \right| = \max_{\boldsymbol{x} \in \mathcal{X}} \left| \tilde{P}(\boldsymbol{x}) - \prod_{i=1}^{d} \tilde{P}(x_i) \right| = d^{\frac{-1}{d-1}} - d^{\frac{-d}{d-1}} . \tag{3}
$$

Proof. For $\tilde{P}$

$$
\left| P(\boldsymbol{x}_j) - \prod_{i=1}^{d} P(x_{ij}) \right| = \left| P(\boldsymbol{x}_j) - \prod_{i=1}^{d} \sum_{\{l | x_{il} = x_{ij}\}} P(\boldsymbol{x}_l) \right| = \left| P(\boldsymbol{x}_j) - \prod_{i=1}^{d} P(\boldsymbol{x}_j) \right|
$$

$$
= \left| d^{\frac{-1}{d-1}} - d^{\frac{-d}{d-1}} \right|, \text{ which is maximal by Theorems 2 and 3 in the appendix A.}
$$

Hence a maximum independent discrete distribution is a distribution that maximizes the absolute worst case probabilistic difference between the joint probability and the product of marginal probabilities (the joint probability under the independence assumption). If $r = 2$ and $d = 2$ (the data is binary and two-dimensional) **the** MSDI distribution is pictured in Table 1. Table 1 clarifies the fact that when $r = 2$ and $d = 2$ the MSDI distribution corresponds to the logical XOR function in the sense that it only has positive probability when x_1 XOR $x_2 = 0$. A similar, but more complicated XOR distribution was constructed in [9]. The maximal state wise difference is

$$
d^{\frac{-1}{d-1}} - d^{\frac{-d}{d-1}} = 2^{\frac{-1}{2-1}} - 2^{\frac{-2}{2-1}} = 2^{-1} - 2^{-2} = 0.25. \tag{4}
$$

Table 1. XOR and a similar probability distribution

x_1	x_2	$P(\boldsymbol{x})$	x_1 XOR x_2
0	0	0.5	1
0	1	0	0
1	0	0	0
1	1	0.5	1

It is important to note that no two-dimensional distributions are arbitrarily far from being independent, and in general from Theorem 1 (3) and (2) there are no distributions (in the context of Section 2) that are arbitrarily bad in the sense that the maximal difference is equal to 1, although as Figure 3 shows for **high** dimensional distributions the worst case is pretty close to one.

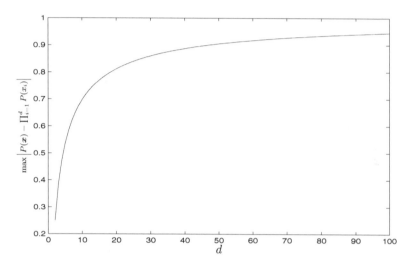

Fig. 1. The maximum prediction error as a function of d

The result in Theorem 1 also holds asymptotically in density estimation (learning a density given samples) in a sense given now given. Let

$$\mathbb{1}_a(b) = \begin{cases} 1 \ a = b \\ 0 \ a \neq b \end{cases} \tag{5}$$

and for n independent identically distributed r.v.'s $\left(\boldsymbol{X}^{(l)}\right)_{l=1}^n$ let $S_{n,\boldsymbol{x}}(\boldsymbol{X}) = \mathbb{1}_{\boldsymbol{x}}\left(\boldsymbol{X}^{(1)}\right) + \ldots + \mathbb{1}_{\boldsymbol{x}}\left(\boldsymbol{X}^{(n)}\right)$. Then $S_{n,\boldsymbol{x}}(\boldsymbol{X})/n \overset{a.s.}{\rightarrow} P(\boldsymbol{x})$ as $n \rightarrow \infty$ and for $S_{n,x_i}(X_i) = \mathbb{1}_{x_i}\left(X_i^{(1)}\right) + \ldots + \mathbb{1}_{x_i}\left(X_i^{(n)}\right)$

$$\prod_{i=1}^d [S_{n,x_i}(X_i)/n] \overset{a.s.}{\rightarrow} \prod_{i=1}^d P(x_i); \text{ as } n \rightarrow \infty \ . \tag{6}$$

For the small sample behavior, Figure 2 shows the average from 1000 realizations of the left hand of (6), when the generating distribution is given by Table 1 ($d = 2$).

Hence we cannot expect that computation by independence assumption works better than in the prediction case when learning the distribution (1) (recall that here $d^{\frac{-1}{d-1}} - d^{\frac{-d}{d-1}} = 0.25$).

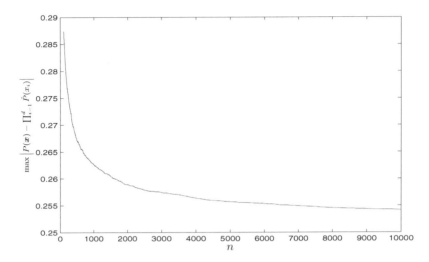

Fig. 2. Average of 1000 'learning' realizations of (6)

4 Pattern Recognition

In $0 - 1$ classification $(C, \boldsymbol{X})$ is a r.v. with $C \in \{0, 1\}$ such that the class is denoted by c when observed. A classifier is a function $\hat{c} : \mathcal{X} \to C$ such that given $\boldsymbol{x}$, $\hat{c}(\boldsymbol{x})$ is an estimate of c. One such function is the Bayes classifier, here defined as

$$\hat{c}_B(\boldsymbol{x}) = \begin{cases} 1; P(\boldsymbol{x}|C = 1)P(C = 1) > P(\boldsymbol{x}|C = 0)P(C = 0) \\ 0; \text{otherwise} \end{cases}.$$

Bayes classifier can be defined using the posteriour $P(C|\boldsymbol{X})$ directly, but this may lead to difficult computations [10]. It is well known (i.e. see [8]) that Bayes classifier is at least as good as **any** other $\hat{c}(\boldsymbol{x})$ in the sense that

$$P(\hat{c}(\boldsymbol{X}) = C) \leqslant P(\hat{c}_B(\boldsymbol{X}) = C) . \tag{7}$$

When approximating P by $\hat{P}$ it is standard to use the plug-in function

$$\hat{c}_{\hat{B}}(\boldsymbol{x}) = \begin{cases} 1; \hat{P}(\boldsymbol{x}|C = 1)\hat{P}(C = 1) > \hat{P}(\boldsymbol{x}|C = 0)\hat{P}(C = 0) \\ 0; \text{otherwise} \end{cases}. \tag{8}$$

The approximation in question is independence, which in the context of classifiers often is called the 'Naïve Bayes' classifier. This is due to the 'naïve' assumption that

$$\hat{P}(\boldsymbol{x}|c) = \prod_{i=1}^{d} P(x_i|c). \tag{9}$$

The Naïve Bayes assumption for specific data sets can actually perform better than a plug-in classifier incorporating some dependencies as shown in [11]. In

[12] Naïve Bayes has been reported as performing worse than taking dependence into account (but not on all data sets), and even then the difference was in many cases not large. In [9] it is found as suboptimal in most data sets. A more in-depth Meta study on the subject is [4].

From a theoretical standpoint there exist cases when it is possible to prove that independence works in pattern recognition. For example:

1. The distribution $P(\boldsymbol{x}|c)$ is very concentrated [13, 14]
2. The distribution $P(\boldsymbol{x}|c)$ is very non-concentrated [15]
3. The margin $|P(\boldsymbol{x}|C = 0) - P(\boldsymbol{x}|C = 1)|$ is large [13, 14, 16]
4. The very part in 2,3 is reduced in the case of partial independence [14, 17]

However, an important question that naturally arises is if class conditional independence can be proven to be good enough for classification **in general**. That this is not the case will be shown after the next definition.

Definition 2. *The Hamming distance between two states $\boldsymbol{x}_j$ and $\boldsymbol{x}_l$, here denoted by $d(\boldsymbol{x}_j, \boldsymbol{x}_l)$ is the number of elements where $\boldsymbol{x}_j$ and $\boldsymbol{x}_l$ are different.*

Example 1. Let $d = 3, r = 2$ and $\hat{P}(c) = P(c) = \frac{1}{2}$ for all $c \in \{0, 1\}$. Let $P(\boldsymbol{x}|C = 0)$ be the distribution in (1). Since

$$P(x_i|C = 0) = \begin{cases} 3^{-\frac{1}{2}} & x_i = 0 \\ 1 - 3^{-\frac{1}{2}} & x_i = 1 \end{cases} \tag{10}$$

the distribution imposed by the independence assumption is as follows

$$\hat{P}(\boldsymbol{x}|C = 0) = \prod_{i=1}^{3} \left(3^{-\frac{1}{2}}\right)^{1-x_i} \left(1 - 3^{-\frac{1}{2}}\right)^{x_i} \ .$$

For $C = 1$ let

$$P(\boldsymbol{x}|C = 1) = \begin{cases} 0 & d(\boldsymbol{x}, [0,0,0]) = 0 \\ 3^{-\frac{1}{2}} - \frac{1}{3} & d(\boldsymbol{x}, [0,0,0]) = 1 \\ \frac{2}{3} - 3^{-\frac{1}{2}} & d(\boldsymbol{x}, [0,0,0]) = 2 \\ 0 & d(\boldsymbol{x}, [0,0,0]) \geqslant 2 \end{cases} \ , \tag{11}$$

then $\sum_{\boldsymbol{x} \in \mathcal{X}} P(\boldsymbol{x}) = 3 \left(3^{-\frac{1}{2}} - \frac{1}{3}\right) + 3 \left(\frac{2}{3} - 3^{-\frac{1}{2}}\right) = 1$ and

$$P(X_i = 1|C = 1) = 3^{-\frac{1}{2}} - \frac{1}{3} + \binom{2}{1}\left(\frac{2}{3} - 3^{-\frac{1}{2}}\right) = \frac{2}{3} - \left(3^{-\frac{1}{2}} - \frac{1}{3}\right) = 1 - 3^{-\frac{1}{2}}$$

as well as

$$P(X_i = 0|C = 1) = \binom{2}{1}\left(3^{-\frac{1}{2}} - \frac{1}{3}\right) + \left(\frac{2}{3} - 3^{-\frac{1}{2}}\right) = 3^{-\frac{1}{2}} - \frac{2}{3} + \frac{2}{3} = 3^{-\frac{1}{2}} \ .$$

In summary $\hat{P}(\boldsymbol{x}|C = 1) = \hat{P}(\boldsymbol{x}|C = 0)$ yielding for $P(\hat{c}(\boldsymbol{X}) = C) = P(C = 0) = \frac{1}{2}$ for the plug-in Bayes classifier in (8). This is a bad case computation by independence in the sense that construction of an estimate based on features yields the same performance/classifier as not using the features at all ($\hat{c}_{\hat{B}}(\boldsymbol{x}) = 0$ for all $\boldsymbol{x} \in \mathcal{X}$). It also has the same performance as choosing a class randomly with equal probability, for example by using a fair coin.

Supervised classification poses different challenges for computation under the independence assumption. In this context learning means that a rule is learned from n independent identically distributed r.v.'s $(\boldsymbol{X}^{(l)}, C^{(l)})_{l=1}^{n}$. We will now show that a generating model with no independent variables translated to a rule $c_F^{(n)}$ need not be optimal. In this context an optimal rule is a rule $\hat{c}_*^{(n)}$ such that for all other $\hat{c}^{(n)} \in \mathcal{B}$

$$P\left[\hat{c}^{(n)}\left(X_n \middle| \left(C^{(l)}, X^{(l)}\right)_{l=1}^{n-1}\right) = C^{(n)}\right] \leqslant P\left[\hat{c}_*^{(n)}\left(X_n \middle| \left(C^{(l)}, X^{(l)}\right)_{l=1}^{n-1}\right) = C^{(n)}\right].$$

One rule that takes dependence into account is the fundamental rule, which is defined as

$$\hat{c}_F^{(n)}(\boldsymbol{x}|(\boldsymbol{x}^{(l)}, c^{(l)})_{l=1}^{n}) := \begin{cases} 1 & \sum_{l=1}^{n} \mathbb{1}_{(\boldsymbol{x},1)}\left(\boldsymbol{x}^{(l)}, c^{(l)}\right) > \sum_{l=1}^{n} \mathbb{1}_{(\boldsymbol{x},0)}\left(\boldsymbol{x}^{(l)}, c^{(l)}\right) \\ 0 & \text{otherwise} \end{cases}.$$

The supervised classifier that assumes corresponding (class conditional) independent rule is $\hat{c}_I^{(n)}(\boldsymbol{x}|(\boldsymbol{x}^{(l)}, c^{(l)})_{l=1}^{n}) :=$

$$\begin{cases} 1 & \hat{P}(C = 1)\hat{P}(x_1|C = 1)\hat{P}(x_2|C = 1) > \hat{P}(C = 0)\hat{P}(x_1|C = 0)\hat{P}(x_2|C = 0) \\ 0 & \text{otherwise} \end{cases},$$

where

$$\hat{P}(C = c) = \begin{cases} \frac{\sum_{l=1}^{n} \mathbb{1}_c\left(c^{(l)}\right)}{n}, & n > 1 \\ 0 & \text{otherwise} \end{cases}$$

and

$$\hat{P}(x_i|C = j) = \begin{cases} \frac{\sum_{l=1}^{n} \mathbb{1}_{x_i}\left(x_i^{(l)}, c^{(l)}\right)}{\sum_{l=1}^{n} \mathbb{1}_j\left(c^{(l)}\right)} & \sum_{l=1}^{n} \mathbb{1}_j\left(c^{(l)}\right) > 0 \\ 0 & \text{otherwise} \end{cases}. \tag{12}$$

Example 2. In this example $r = 2, d = 2$ and $P(c) = \frac{1}{2}$ for all $c \in \{0,1\}$, where the class conditional probabilities are tabulated in Table 2.

With the independence assumption one gets the class conditional probabilities in Table 3.

While the class conditional probabilities are not correct given the independence assumption it is true that

$$\arg\max_{c \in \mathcal{C}} P(c)P(\boldsymbol{x}|c) = \arg\max_{c \in \mathcal{C}} P(c)P(x_1|c)P(x_2|c)$$

for all $x_1 \in \mathcal{X}_1$ and $x_2 \in \mathcal{X}_2$. So here Bayes **classifier** is equivalent to the Naïve Bayes **classifier**. For $[0,0],[1,0]$ Bayes Classifier will choose class 1, otherwise class 2. Hence $P(\hat{c}_B(\boldsymbol{X}) = C)$

$$= P(C = 0)\left[P([0,0]|0) + P([1,0]|0)\right] + P(C = 1)\left[P([0,1]|1) + P([1,1]|1)\right] = 0.7.$$

Table 2. Class conditional probabilities

$C = 0$			$C = 1$		
$P(\boldsymbol{X}\vert C=0)$	$X_1 = 0$	$X_1 = 1$	$P(\boldsymbol{X}\vert C=1)$	$X_1 = 0$	$X_1 = 1$
$X_2 = 0$	0.4	0.3	$X_2 = 0$	0.1	0.2
$X_2 = 1$	0.2	0.1	$X_2 = 1$	0.3	0.4

Table 3. Class conditional probabilities with the Independence assumption

	$X_i = 0$	$X_i = 1$		$X_i = 0$	$X_i = 1$
$P(X_1\vert C=0)$	0.6	0.4	$P(X_1\vert C=1)$	0.4	0.6
$P(X_2\vert C=0)$	0.7	0.3	$P(X_2\vert C=1)$	0.3	0.7
$C = 0$			$C = 1$		
$P(X_1\vert C)P(X_2\vert C)$	$X_1 = 0$	$X_1 = 1$	$P(X_1\vert C)P(X_2\vert C)$	$X_1 = 0$	$X_1 = 1$
$X_2 = 0$	0.42	0.28	$X_2 = 0$	0.12	0.18
$X_2 = 1$	0.18	0.12	$X_2 = 1$	0.28	0.42

This is the absolute maximum that can be achieved by $\hat{c}^{(n)}$ and/or $\hat{c}_I$ in the sense of (7).

Figures 3 contains plots on the Monte Carlo simulated effect of the fundamental rule versus the independent rule. Note that the number of training data affects which rule is best (of the two rules simulated). In particular there exists an n where the independent rule has a higher probability of classifying correctly than the fundamental rule. Figure 4 shows the explicit difference with confidence interval, thus showing that the results are **very probable**. Hence taking the dependencies in the generating model into account is not optimal in general.

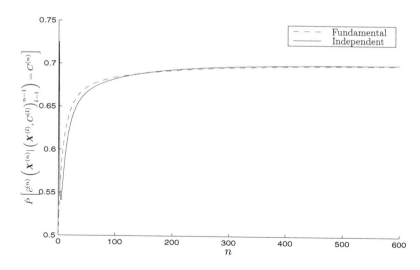

Fig. 3. Average performance of rules as function of n

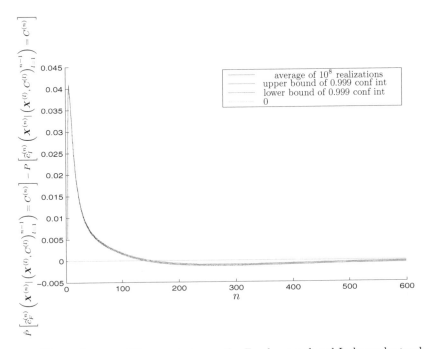

Fig. 4. Average difference between the Fundamental and Independent rule

5 Conclusions

The computational independence assumption is widely used in one form or another. When using independence or improving upon computational independence it is useful to know of its limitations. A class of distributions that maximizes the state wise probabilistic difference to its product of marginals has been derived. It has been shown that in low dimensions that in the context of Section 2 there are no distributions arbitrarily far from being independent, in particular for low dimensions. The use of MSDI distributions has been demonstrated in learning densities and classification. It has been shown that in supervised classification computational independence can outperform independence assumptions based on the generating models and that this performance may not only depend on the generating model but also on the number of samples.

References

[1] Russell, S., Norvig, P.: Artificial intelligence: a modern approach. Prentice-Hall, Englewood Cliffs (1995)
[2] Chow, C., Liu, C.: Approximating discrete probability distributions with dependency trees. IEEE Transactions on Information Theory 14(3), 462–467 (1968)
[3] Heckerman, D., Geiger, D., Chickering, D.: Learning Bayesian networks: The combination of knowledge and statistical data. Machine Learning Journal 20(3), 197–243 (1995)

[4] Hand, D., Yu, K.: Idiot's bayes–not so stupid after all? International Statistical Review 69(3), 385–398 (2001)

[5] Lewis, P.: Approximating probability distributions to reduce storage requirements. Information and Control 2, 214–225 (1959)

[6] Vapnik, V.: Statistical Learning Theory. Wiley, Chichester (1998)

[7] Catoni, O.: Statistical Learning Theory and Stochastic Optimization. Springer, Heidelberg (2004)

[8] Devroye, L., Györfi, L., Lugosi, G.: A Probabilistic Theory of Pattern Recognition. Springer, Heidelberg (1996)

[9] Huang, K., King, I., Lyu, M.: Finite mixture model of bounded semi-naive Bayesian network classifier. In: Kaynak, O., Alpaydın, E., Oja, E., Xu, L. (eds.) ICANN 2003 and ICONIP 2003. LNCS, vol. 2714, Springer, Heidelberg (2003)

[10] Ripley, B.: Pattern Recognition and Neural Networks. Cambridge University Press, Cambridge (1996)

[11] Titterington, D., Murray, G., Murray, L., Spiegelhalter, D., Skene, A., Habbema, J., Gelpke, G.: Comparison of discrimination techniques applied to a complex data set of head injured patients. Journal of the Royal Statistical Society. 144(2), 145–175 (1981)

[12] Chickering, D.: Learning equivalence classes of bayesian-network structures. The Journal of Machine Learning Research 2, 445–498 (2002)

[13] Rish, I., Hellerstein, J., Thathachar, J.: An analysis of data characteristics that affect naive bayes performance. Technical Report RC21993, IBM (2001)

[14] Ekdahl, M.: Approximations of Bayes Classifiers for Statistical Learning of Clusters. Licentiate thesis, Linköpings Universitet (2006)

[15] Ekdahl, M., Koski, T., Ohlson, M.: Concentrated or non-concentrated discrete distributions are almost independent. IEEE Transactions on Information Theory, submitted (2006)

[16] Domingos, P., Pazzani, M.: On the optimality of the simple bayesian classifier under zero-one loss. Machine Learning 29(2), 103–130 (1997)

[17] Ekdahl, M., Koski, T.: Bounds for the loss in probability of correct classification under model based approximation. Journal of Machine Learning Research 7, 2473–2504 (2006)

[18] Hagerup, T., Rub, C.: A guided tour of Chernoff bounds. Information Processing Letters 33, 305–308 (1989)

A   Technical Results Used in the Proof of Theorem 1

We will prove Theorem 3 (used in the proof of Theorem 1) through extending the theory in [15]. Let $p_j = P(\boldsymbol{x}_j)$, $p := \max_{\boldsymbol{x} \in \mathcal{X}} P(\boldsymbol{x})$ and

$$h = \arg \max \left\{ g | g \in \{0, 1, \ldots, d\}, \sum_{b=1}^{g} (r-1)^b \binom{d}{b} \leqslant \frac{1 - p_1}{p} \right\} . \qquad (13)$$

Definition 3. *Let IHB be defined by*

$$
\begin{cases}
\left(1 - \dfrac{1-p}{d-1}\right)^{d-1}(1-p)\,; \text{ for } p \geqslant \dfrac{1}{d} & (14) \\[3mm]
\left(1 - \dfrac{1}{d}\right)^{d}\,; \text{ for } \dfrac{1}{(r-1)d} \leqslant p \leqslant \dfrac{1}{d} & (15) \\[3mm]
\displaystyle\max_{p_1 \in \{\max(0,1-p(|\mathcal{X}|-1)),p\}} \left(\left[1 - p\sum_{b=1}^{h}(r-1)^b\binom{d-1}{b-1}\right. \right. \\[3mm]
\left. \left. - \left(1 - p\sum_{b=1}^{h}(r-1)^b\binom{d}{b}\right) - p_1\right) \dfrac{h+1}{d}\right]^{d} - p_1 \right)\,; \text{ for } p \leqslant \dfrac{1}{(r-1)d} & (16)
\end{cases}
$$

Theorem 2. *[15] For all* $\boldsymbol{x} \in \mathcal{X}, P \in \mathcal{P}$

$$
\left| P(\boldsymbol{x}) - \prod_{i=1}^{d} P(x_i) \right| \leqslant \begin{cases} IHB & \text{for } P(\boldsymbol{x}) \leqslant \prod_{i=1}^{d} P(x_i) & (17) \\ p - p^d & \text{otherwise.} & (18) \end{cases}
$$

A theory will be developed that will show that $\max\left(p - p^d, IHB\,[p]\right) = d^{\frac{-1}{d-1}} - d^{\frac{-d}{d-1}}$, then it only remains to find a distribution such that the evaluation $\max_{\boldsymbol{x} \in \mathcal{X}}\left|P(\boldsymbol{x}) - \prod_{i=1}^{d} P(x_i)\right| = d^{\frac{-1}{d-1}} - d^{\frac{-d}{d-1}}$ holds.

Lemma 1. $\max_{p \in [0,1]}\left(p - p^d\right) = d^{\frac{-1}{d-1}} - d^{\frac{-d}{d-1}}$

Proof. Define $g(p) := p - p^d$. The only extreme point is given by $\frac{d}{da}g(a) = 1 - d \cdot a^{d-1} = 0 \Leftrightarrow a = d^{\frac{-1}{d-1}}$. Finally $\frac{d^2}{da^2}g(a) = -d(d-1)a^{d-2} < 0$ when $a > 0$, so this point gives a maximum.

The maximization of $IHB(p)$ part of $\max\left(p - p^d, IHB(p)\right)$ is carried out through showing that

$$
\max_{p \in \mathcal{A}} IHB(p) = \left(1 - \frac{1}{d}\right)^{d}
$$

for both $\mathcal{A} = \left[\frac{1-p_1}{|\mathcal{X}|-1}, \frac{1}{d}\right]$ and $\mathcal{A} = \left[\frac{1}{d}, 1\right]$.

Lemma 2

$$
\max_{p \in \left[\frac{1}{d}, 1\right]} IHB(p) = \left(1 - \frac{1}{d}\right)^{d} \tag{19}
$$

Proof From Definition 3 (14) we have $IHB(p) = \left(1 - \frac{1-p}{d-1}\right)^{d-1}(1-p)$ when $p \in \left[\frac{1}{d}, 1\right]$. Furthermore

$$
\frac{d}{dp}IHB(p) = -\left(1 - \frac{1-p}{d-1}\right)^{d-1} + \left(1 - \frac{1-p}{d-1}\right)^{d-2}(1-p)
$$

$$= \left(1 - \frac{1-p}{d-1}\right)^{d-2} \left(-1 + \frac{(1-p)}{d-1} + (1-p)\right)$$

$$= \left(1 - \frac{1-p}{d-1}\right)^{d-2} \left(\frac{(1-p) - p(d-1)}{d-1}\right) = \left(1 - \frac{1-p}{d-1}\right)^{d-2} \left(\frac{1 - d \cdot p}{d-1}\right) \leqslant 0$$

when $p \in \left[\frac{1}{d}, 1\right]$. Now (14) for $\frac{1}{(r-1)d} \leqslant p \leqslant \frac{1}{d}$ yields (19)

$$\max_{p \in \left[\frac{1}{d}, 1\right]} IHB(p) = IHB\left(\frac{1}{d}\right) = \left(1 - \frac{1}{d}\right)^d.$$

We continue to maximize $IHB(p)$ for small values of p.

Definition 4. *Given p_1 let $\{p_i\} \subset \left[\frac{1}{|\mathcal{X}|}, \frac{1}{(r-1)d}\right]$ be the biggest set such that for all i*

$$h = \arg \max \left\{ g \middle| g \in \{0, 1, \ldots, d\}, \sum_{b=1}^{g} (r-1)^b \binom{d}{b} \leqslant \frac{1 - p_1}{p_i} \right\}$$

$$\Rightarrow \sum_{b=1}^{h} (r-1)^b \binom{d}{b} = \frac{1 - p_1}{p_i} \qquad (20)$$

Now all changes from $h = k$ to $h = k + 1$ can be modeled by $a \downarrow 0$ for $p_i + a$, since

$$\sum_{b=1}^{h} (r-1)^b \binom{d}{b} = \frac{1 - p_1}{p_i} \Rightarrow \sum_{b=1}^{h} (r-1)^b \binom{d}{b} > \frac{1 - p_1}{p_i + a}$$

for all $a > 0$.

Lemma 3. *Let $0 < c < \infty$, $g \in \mathbb{N}$ and let p_i be as in Definition 4. Let*

$$\lim_{a \downarrow 0} c \left[1 - (p_i + a) \sum_{b=1}^{k} (r-1)^b \binom{d-1}{b-1} \right.$$

$$- \left(1 - p_1 - (p_i + a) \sum_{b=1}^{k} (r-1)^b \binom{d}{b}\right) \frac{k+1}{d} \Bigg]^g$$

$$\rightarrow c \left[1 - p_i \sum_{b=1}^{k+1} (r-1)^b \binom{d-1}{b-1} \right]^g. \qquad (21)$$

Proof. $f(p_i + a) :=$

$$1 - (p_i + a) \sum_{b=1}^{k} (r-1)^b \binom{d-1}{b-1} - \left(1 - p_1 - (p_i + a) \sum_{b=1}^{k} (r-1)^b \binom{d}{b}\right) \frac{k+1}{d}$$

Here a $\delta(\varepsilon) > |a|$ is found such that

$$\delta(\varepsilon)\left|f(p_i + a) - \left[1 - p_i\sum_{b=1}^{k+1}(r-1)^b\binom{d-1}{b-1}\right]\right| < \varepsilon.$$

We have $\left|f(p_i + a) - \left[1 - p_i\sum_{b=1}^{k+1}(r-1)^b\binom{d-1}{b-1}\right]\right|$

$$= \left|-(p_i + a)\sum_{b=1}^{k}(r-1)^b\binom{d-1}{b-1} - \left(1 - p_1 - (p_i + a)\sum_{b=1}^{k}(r-1)^b\binom{d}{b}\right)\frac{k+1}{d}\right.$$

$$\left. + p_i\sum_{b=1}^{k+1}(r-1)^b\binom{d-1}{b-1}\right|.$$

That $1 - p_i = p_i\left(\frac{1-p_1}{p_i}\right)$ and collecting all the p_i and a terms and term yields

$$= \left|p_i\left(-\sum_{b=1}^{k}(r-1)^b\binom{d-1}{b-1}\right) + \left(-\frac{1-p_1}{p_i} + \sum_{b=1}^{k}(r-1)^b\binom{d}{b}\right)\frac{k+1}{d} + \right.$$

$$\left.\sum_{b=1}^{k+1}(r-1)^b\binom{d-1}{b-1}\right) + a\left(-\sum_{b=1}^{k}(r-1)^b\binom{d-1}{b-1} + \frac{k+1}{d}\sum_{b=1}^{k}(r-1)^b\binom{d}{b}\right)\right|.$$

Further simplification of the p_i factor yields

$$= \left|\frac{p_i(k+1)}{d}\left(-\frac{1-p_1}{p_i} + \sum_{b=1}^{k}(r-1)^b\binom{d}{b} + (r-1)^{k+1}\binom{d}{k+1}\right)\right.$$

$$\left. + a\left(-\sum_{b=1}^{k}(r-1)^b\binom{d-1}{b-1} + \frac{k+1}{d}\sum_{b=1}^{k}(r-1)^b\binom{d}{b}\right)\right|.$$

Then Definition 4 (20) reduced the expression to

$$\leq a\left|-\sum_{b=1}^{k}(r-1)^b\binom{d-1}{b-1} + \frac{k+1}{d}\sum_{b=1}^{k}(r-1)^b\binom{d}{b}\right|$$

With

$$\delta = \frac{\varepsilon}{\left|-\sum_{b=1}^{k}(r-1)^b\binom{d-1}{b-1} + \frac{k+1}{d}\sum_{b=1}^{k}(r-1)^b\binom{d}{b}\right|}$$

$\delta > |a|$ yields $\left|f(p_i + a) - \left[1 - p_i\cdot\sum_{b=1}^{k+1}(r-1)^b\binom{d-1}{b-1}\right]\right| < \varepsilon$, hence the $\lim_{a\downarrow 0} f(p_i + a) = f(p_i)$. Then the fundamental limit theorem gives (21).

Lemma 4

$$\max_{p\in\left[\frac{1}{|\mathcal{X}|},\frac{1}{(r-1)d}\right]} IHB(p) = \left(1-\frac{1}{d}\right)^d$$

Proof Let

$$f(p) = \left[1 - p\cdot\sum_{b=1}^{h}(r-1)^b\binom{d-1}{b-1} - \left(1 - p\sum_{b=1}^{h}(r-1)^b\binom{d}{b} - p_1\right)\frac{h+1}{d}\right]^d$$

$$= \left[1 - (1-p_1)\frac{h+1}{d} + p\sum_{b=1}^{h}(r-1)^b\left(\frac{h+1}{d}\binom{d}{b} - \binom{d-1}{b-1}\right)\right]^d$$

since $\frac{h+1}{d}\binom{d}{b} = \frac{(h+1)d!}{d\cdot b!(d-b)!} = \frac{(h+1)(d-1)!}{b\cdot(b-1)!(d-b)!} = \frac{h+1}{b}\binom{d-1}{b-1}$ and then compacting using constants c_1 and $c_2 \geqslant 0$

$$= \left[\frac{d + p_1(h+1)}{d} + p\cdot\sum_{b=1}^{h}(r-1)^b\binom{d-1}{b-1}\left(\frac{h+1}{b}-1\right)\right]^d = [c_1 + p\cdot c_2]^d .$$

Finally $c_1 + p\cdot c_2 \geqslant 0$ since it represents a probability (see the proof of Lemma 10 in [15]) so

$$\frac{d}{dp}f(p) = d\cdot c_2\cdot[c_1 + p\cdot c_2]^{d-1} \geqslant 0.$$

To maximize $f(p)$ p should chosen to be as big as possible for each fixed p_1 and h. Also when h changes so we use Lemma 3 to show that this is not a problem, i.e. $\lim_{a\downarrow0} f(p_i + a) \to f(p_i)$ as well as $\lim_{a\downarrow0}\frac{d}{dp}f(p_i + a) \to \frac{d}{dp}f(p_i)$. Hence

$$\max_{p\in\left[\frac{1}{|\mathcal{X}|},\frac{1}{(r-1)d}\right]} IHB(p) = \max_{p\in\left[\frac{1}{|\mathcal{X}|},\frac{1}{(r-1)d}\right],p_1\in[\ldots]} f(p) =$$

$$= \max_{p_1\in[\ldots]}\left[\max_{p\in\left[\frac{1}{|\mathcal{X}|},\frac{1}{(r-1)d}\right]} f(p)\right] = \max_{p_1\in[\ldots]}\left[IHB\left(\frac{1}{(r-1)d}\right)\right]$$

$$= \max_{p_1\in[\ldots]}\left[\left(1-\frac{1}{d}\right)^d\right] = \left(1-\frac{1}{d}\right)^d .$$

Now both $p - p^d$ (Lemma 1) and $IHB(p)$ (Lemmas 2 and 4) have been maximized.

Theorem 3

$$\max\left(\max_{p\in\left[\frac{1}{|\mathcal{X}|},1\right]} IHB(p), \max_{p\in\left[\frac{1}{|\mathcal{X}|},1\right]}(p - p^d)\right) = d^{\frac{-1}{d-1}} - d^{\frac{-d}{d-1}} . \tag{22}$$

Proof. By Lemma 1, 4 and Definition 3

$$\max_{p\in\left[\frac{1}{|\mathcal{X}|},1\right]} IHB(p) = \left(1 - \frac{1}{d}\right)^d$$

and by Lemma 1 $\max_{p\in\left[\frac{1}{|\mathcal{X}|},1\right]} (p - p^d) = d^{\frac{-1}{d-1}} - d^{\frac{-d}{d-1}}$. Let $f(a) = \left(1 - \frac{1}{a}\right)^a$ for $a \in \mathbb{R}$, then $\lim_{a\to\infty} f(a) \uparrow e^{-1} \approx 0.3679$ (see for example [18]). Hence $f(d) < 0.37$ for all $d \geqslant 2$. Let $f_2(d) = d^{\frac{-1}{d-1}} - d^{\frac{-d}{d-1}}$ then we have $f_2(2) = f(2)$ and $f_2(3) \approx 0.3849$. Using (4) we get for $a \geqslant 2$

$$\frac{d}{da}f_2(a) = \frac{\ln(a)\,a^{\frac{-1}{a-1}}}{(a-1)a} \geqslant 0 \ .$$

Hence $f_2(d) \geqslant \left(1 - \frac{1}{d}\right)^d$ and (22) follows.

Comparison of a Novel Combined ECOC Strategy with Different Multiclass Algorithms Together with Parameter Optimization Methods

Marco Hülsmann[1,2] and Christoph M. Friedrich[2]

[1] Universität zu Köln, Germany
[2] Fraunhofer-Institute for Algorithms and Scientific Computing (SCAI), Schloß Birlinghoven, 53754 Sankt Augustin, Germany

Abstract. In this paper we consider multiclass learning tasks based on Support Vector Machines (SVMs). In this regard, currently used methods are *One-Against-All* or *One-Against-One*, but there is much need for improvements in the field of multiclass learning. We developed a novel combination algorithm called *Comb-ECOC*, which is based on posterior class probabilities. It assigns, according to the Bayesian rule, the respective instance to the class with the highest posterior probability. A problem with the usage of a multiclass method is the proper choice of parameters. Many users only take the default parameters of the respective learning algorithms (e.g. the regularization parameter C and the kernel parameter γ). We tested different parameter optimization methods on different learning algorithms and confirmed the better performance of *One-Against-One* versus *One-Against-All*, which can be explained by the maximum margin approach of SVMs.

1 Introduction

All multiclass learning methods considered here are based on Support Vector Machines, which are presented for example by Schölkopf and Smola [22] and Vapnik [25]. Mostly, several binary classifications are resolved by an SVM, which are then combined to a multiclass solution. Our goal is to present improved methods in the open-research field of multiclass learning.

In section 2 we start with the presentation of different state-of-the-art multiclass algorithms. We consider the standard methods *One-Against-All (OAA)* and *One-Against-One (OAO)* using implementations of the *libsvm* [11] and *SVMlight* [15]. We continue with two direct approaches, which are not based on several binary optimization problems: the algorithm by Crammer and Singer [4] and *SVMmulticlass* based on the theory of [24]. Furthermore we use the exhaustive ECOC algorithm introduced by Dieterich and Bakiri [5] and present a novel combination approach of ECOC, OAA and probability predictions. This method ist called *Comb-ECOC* and has been developed in [14]. The probability predictions are based on Bradley-Terry models described in [13].

The next principal subject of this paper will be the parameter optimization. In many applications, Support Vector Machines are used with their default

P. Perner (Ed.): MLDM 2007, LNAI 4571, pp. 17–31, 2007.
© Springer-Verlag Berlin Heidelberg 2007

parameters. Optimizing the parameters improves the classification performance drastically, which will be shown in section 3. We consider different optimization methods, such as the common grid search and the *SVMpath* algorithm introduced in [10].

Finally, we give results obtained from different test runs with all considered multiclass algorithms and parameter optimization methods in section 3, together with the practical confirmation of the maximum margin explanation in the case of *One-Against-One* and *One-Against-All*. A final discussion is carried out in section 4.

2   Theoretical Background

In this section we consider the multiclass algorithms from a theoretical point of view. We shortly describe the principal ideas of both the learning methods and the parameter optimization. In order to avoid later misunderstandings, we here already enumerate the used algorithms in Table 1. Detailed experiments and comparisons can also be found in [9] and [12]. In the latter, especially *One-Against-One* is suggested for real-world applications, which will be confirmed by our analysis.

We start with a general description of the multiclass learning task, in order to appoint the notations used in this paper: We consider an input space $\mathcal{X} = \{x_1, ..., x_m\}$ and assign $k > 2$ classes to this set, so that each element in $\mathcal{X}$ belongs to exactly one class. The goal is to find a decision function $f : \mathcal{X} \to \{1, ..., k\}$ to get a pair $(x_i, f(x_i))$ for all $i = 1, ..., m$. The assigned class $f(x_i) = r \in \{1, ..., k\}$ is also called *label*. We furthermore distinguish between the *input space* and the *feature space*. The input space $\mathcal{X}$ can be anything. It is not necessary that it consists of vectors or numerical values. In contrast, the feature space $\mathcal{F}$ is a high dimensional vector space. In this paper, let its dimension be defined as n. A map $\Phi : \mathcal{X} \to \mathcal{F}$ is used to assign an element of the input space to a vector in the feature space. In order to avoid the computation of Φ, we simply use a kernel function $k(x, x') = \langle \Phi(x), \Phi(x') \rangle$ for all $x, x' \in \mathcal{X}$. Here we only give results obtained by using Gaussian kernels $k(x, x') = \exp(-\gamma ||x - x'||^2)$, $\gamma \in \mathbb{R}$, because pursuant to [14], it delivers the best results and the corresponding SVM algorithm works faster in comparison to the polynomial and the sigmoid kernel. Instead of predicting classes we also use methods which predict posterior class probabilities.

2.1   Standard Multiclass Algorithms

One-Against-All (OAA). A well-known simple approach for the assignment of instances to several classes is to separate each class from all the other classes. This method is called *One-Versus-the-Rest* or *One-Against-All* and bears on k binary classificators $f^1, ..., f^k$, where k is the number of classes. In order to classify a test point x, one computes the decision function

$$f(x) = \arg\max_{j=1,...,k} \sum_{i=1}^{m} y_i \alpha_i^j k(x, x_i) + b^j, \tag{1}$$

Table 1. Description of the different multiclass learning algorithms

Algorithm	Description	Section
OAA	One-Against-All classification by *libsvm* using binary class probabilities	2.1
OAO	One-Against-One classification by *libsvm*, voting	2.1
SVMpath	SVMpath algorithm described in [10] using the appropriate predict function	2.6
SVMlight-OAA	Interface between R and *SVMlight*, One-Against-All classification	2.1
SVMlight-OAO	Interface between R and *SVMlight*, One-Against-One classification	2.1
Crammer-Singer	Multiclass Algorithm by Crammer and Singer [4]	2.2
SVMmulticlass	Direct Algorithm by Thorsten Joachims [24]	2.2
ECOC	Standard Exhaustive ECOC Algorithm by Dieterich and Bakiri [5] using the *libsvm* for binary classification	2.3
ECOC-SVMpath	ECOC with SVMpath as binary predictor	2.3, 2.6
Comb-ECOC	Combined ECOC Algorithm, combination with OAA predicting posterior class probabilities	2.4

where the index j refers to the binary separation of class j from the rest. The coefficients α_i and b stem from the dual optimization problem, which is set up in the context of binary Support Vector Machine classification, see e.g. [3]. We use this method in two implementations: First, we consider the method named *OAA* (see Table 1), which uses binary class probabilities computed – pursuant to [19] – by:

$$P(j|x) = \frac{1}{1 + \exp(Af^j(x) + B)}, \qquad (2)$$

where f^j is the binary classificator that separates class j from the rest. A and B are parameters obtained by the minimization of a negative log-likelihood function. Then the class is computed by:

$$f(x) = \arg \max_{j=1,\dots,k} P(j|x). \qquad (3)$$

The second implementation called *SVMlight-OAA* is based on the interface between R [21] and *SVMlight* [15] (website: http://svmlight.joachims.org) provided by the R-library *klaR* (see [20]). It calls *SVMlight* k times for all k binary classifications and computes the class by equation (1).

One-Against-One (OAO). The idea of this method is to extract all pairs of classes and accomplish a binary classification between the two classes in each pair. Altogether, there are $\binom{k}{2} = \frac{k(k-1)}{2}$ binary classifications. The training set contains only elements of two classes. The other training instances are eliminated from the set. This results in a smaller complexity in comparison to the *One-Against-All* method, but the number of classes is $\mathcal{O}(k^2)$ instead of $\mathcal{O}(k)$. The assignment of a class to a test point occurs by voting. Pursuant to [16], an advantage of the pairwise classification is that in general, the margin is larger than in the OAA case, as we can see from Figure 1. Moreover, the difference of the margin sizes is bigger in the OAO case. Very large margins are possible to appear.

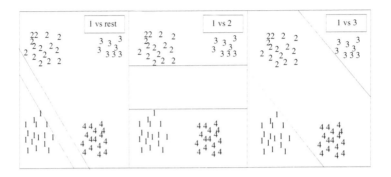

Fig. 1. Comparison of the margin size in the case of OAA and OAO (4 classes): In the left picture class 1 is separated from the rest. The margin is small. In the middle class 1 is separated from class 2 and in the right picture class 1 is separated from class 3. The margin is larger in general. The margin size differs more in the OAO case. The right picture shows a very large margin.

We use OAO from the *libsvm* (see [17]) as a function called *svm* from the *R*-package *e1071* (see [6]).

2.2   Direct Multiclass Algorithms

In our evaluation, we use two direct methods. Direct means that there is only one optimization problem to solve instead of multiple binary ones.

The Algorithm by Crammer and Singer. The direct approach consists in formulating one quadratic program with constraints for each class. The algorithm developed by Crammer and Singer [4] is based on the same idea as the OAA-approach, namely to look for a function $H_M : \mathcal{F} \to \{1, ..., k\}$ of the form

$$H_M(x) = \arg\max_{r=1}^{k}\{M_r \cdot x\}, \tag{4}$$

where M is a matrix with k rows and n columns and M_r the rth row of M. The aim is to determine M, so that the training error is minimized.

We take the implementation of this algorithm from the *R*-library *kernlab*. The used function is named *ksvm*.

The Algorithm by Thorsten Joachims. This method is based on [24], where the composition of the optimization problem and its solution are described in detail. The first reason for finding a more general direct multiclass algorithm is the fact that in real world examples, one does not only have labels in $\mathbb{N}$. The label set can be arbitrary. For instance, it can consist of structured output trees. The output set $\mathcal{Y}$ does not have to be a vector space. The second reason is allowing different loss functions. Most multiclass algorithms only minimize the zero-one-loss.

We use the software $SVM^{multiclass}$ implemented by Thorsten Joachims (see http://www.cs.cornell.edu/People/tj/svm_light/svm_multiclass.html).

2.3 Error Correcting Output Codes

A very simple but efficient method is based on **E**rror **C**orrecting **O**utput **C**odes (the so-called *ECOC* method) and has been developed by Dietterich and Bakiri [5]. ECOC is not a direct application but uses several binary classifications like OAA and OAO. The main difference between OAA, OAO and ECOC is that these are not predisposed. They can be chosen arbitrarily and are determined by a coding matrix. The advantage of the usage of error correcting output codes is that several binary classification errors can be handled, so that no error occurs in the multiclass problem. We use an *exhaustive* code matrix, as proposed in [5]. The *ECOC* algorithm is part of our own implementation.

2.4 Combined ECOC Method

The novel *Comb-ECOC* algorithm (see also [14]) is a mixture of OAA, ECOC and probability prediction. The results of three different partial algorithms are combined by a combination method. We use three code matrices, one defined by the user himself and two generated at random. The random code matrices are composed, so that the minimum Hamming distance is equal to $\lceil \frac{\ell}{2} \rceil$. The number of binary classificators ℓ is a random integer between $2k$ and 2^k, so the algorithm can be NP-complete, like the *Exhaustive-ECOC* algorithm of section 2.3. This disadvantage can be reduced by precalculating the code matrices for a given k. Furthermore we use the binary class probabilities computed by equation (2) for each of the ℓ binary classificators and apply Bradley-Terry methods to estimate the multiclass probabilities following [13]. The optimization problem results in a fixpoint algorithm. The complete procedure of *Comb-ECOC* is described in [14].

All three initial code matrices contain OAA columns, that means columns which define an OAA classification (one entry 1 and the rest 0). By a small modification of the convergence proof in [13], one can show that the fixpoint algorithm converges with this assumption. A test point is assigned to the class with the highest posterior probability.

The fact that we consider three different code matrices delivers three different multiclass models m_i, $i = 1, ..., 3$. These models can be combined in order to be able to compensate the weaknesses of particular models, as described in [8]. We illustrate the idea of combination by the following example:

Example 1 (Combination of multiclass models). *Consider k classes and μ multiclass models. Let $p_i(x_{j,r})$ be the posterior probability with which a test point*

x_j *belongs to a class* $r \in \{1, ..., k\}$ *predicted by the multiclass model* m_i, $i \in \{1, ..., \mu\}$. *Let furthermore* $p_{comb}(x_{j,r})$ *be the resulting combination classificator. Then* x_j *is assigned to a class as follows:*

$$class(x_j) = \arg \max_{r=1,...,k} p_{comb}(x_{j,r}). \qquad (5)$$

We consider different combination models for p_{comb}:

1. The maximum: $p_{comb}(x_{j,r}) = \max_{i=1,...,\mu} p_i(x_{j,r})$
2. The minimum: $p_{comb}(x_{j,r}) = \min_{i=1,...,\mu} p_i(x_{j,r})$
3. The average: $p_{comb}(x_{j,r}) = \frac{1}{\mu} \sum_{i=1}^{\mu} p_i(x_{j,r})$
4. The median:

$$p_{comb}(x_{j,r}) = \begin{cases} \frac{p_{\frac{\mu}{2}}(x_{j,r}) + p_{\frac{\mu}{2}+1}(x_{j,r})}{2} & : \quad \mu \text{ is even} \\ p_{\frac{\mu+1}{2}}(x_{j,r}) & : \quad \mu \text{ is odd} \end{cases} \qquad (6)$$

5. The entropy:

$$p_{comb}(x_{j,r}) = H_r(x_j) = \frac{1}{\mu} \sum_{i=1}^{\mu} \frac{p_i(x_{j,r})}{-\sum_{s=1}^{k} p_i(x_{j,s}) \log p_i(x_{j,s})} \qquad (7)$$

6. The product:

$$p_{comb}(x_{j,r}) = \prod_{i=1}^{\mu} p_i(x_{j,r}) \qquad (8)$$

2.5   Parameter Optimization Via Grid Search

In order to show that it is not sufficient to use the default parameters of the respective methods, we performed a parameter optimization via grid search. The parameters we consider are the SVM regularization parameter C and the Gaussian kernel parameter γ. We first defined a training set and a test set of the original dataset. To warrant the comparability of the results we used the same training set and the same test set for all algorithms and optimization methods. For the parameter optimization itself, we determined 10 bootstrap replications on the training set (for details concerning the bootstrap see [7]). Each of them is evaluated with different parameter pairs. The parameter pair with the smallest mean error over the 10 bootstrap replications is taken for predictions on the test set.

The parameters are defined on a grid, that means that we only allow a final number of parameters for the optimization. For computational reasons, we only use $C \in \mathcal{C} = \{2^{-2}, 2^{-1}, ..., 2^3, 2^4\}$ and $\gamma \in \mathcal{G} = \{2^{-3}, 2^{-2}, ..., 2^2, 2^3\}$. In total, have $7 \cdot 7 \cdot 10 = 490$ bootstrap replications in our optimization process. If we have two or more pairs (C, γ) with the same mean error rate over the 10 bootstrap replications, we take the pair with the largest C to facilitate the choice of the maximum margin classification. We differenciate between two methods: *global* and *local*. The *global* method tunes the whole multiclass algorithm globally, that means it takes one pair (C, γ) for the entire algorithm, i.e. the same pair for all contained binary classifications. However, the *local* method tunes all binary classificators, making a grid search for each binary classification. Therefore, in the global case we have a *for*-loop over C and over γ and evaluate the multiclass algorithm with C and γ. The result is one parameter pair (C, γ). In the local case we implemented a *for*-loop over all ℓ binary classificators, which in turn contains the C- and the γ-*for*-loops. Then the binary classification $i \in \{1, ..., \ell\}$ is made with C and γ. The result is a vector $(C_i, \gamma_i)_{i=1,...,\ell}$. The computational complexity of the global optimization is $|\mathcal{C}| \cdot |\mathcal{G}| \cdot \#$(op. in the multiclass algorithm). The complexity of the local optimization is $\ell \cdot |\mathcal{C}| \cdot |\mathcal{G}| \cdot \#$(op. in the binary algorithm). But especially in the case of the *ECOC* algorithms: $\#$(op. in the multiclass algorithm) $\geq \ell \cdot \#$(op. in the binary algorithm). Therefore the local optimization method is faster in general.

2.6   Parameter Optimization Via SVMpath

In [10], a different method to optimize the cost parameter C is suggested. They compute the entire regularization path for a binary Support Vector Machine algorithm. The advantage of the calculation of this path is the fact that the complexity is as large as the one of a usual SVM algorithm. Consider the decision function $f(x) = \beta_0 + g(x)$, where

$$\forall_{i=1,...,m} \ g(x_i) = \frac{1}{\lambda} \sum_{j=1}^{m} \alpha_j y_j k(x_i, x_j) \tag{9}$$

with α_j the Lagrangian multipliers of the SVM optimization problem and $\lambda = \frac{1}{C}$. Then the following definitions are made:

- $\mathcal{E} := \{i | y_i f(x_i) = 1, 0 \leq \alpha_i \leq 1\}$ is called *Elbow*
- $\mathcal{L} := \{i | y_i f(x_i) < 1, \alpha_i = 1\}$ is called *Left of the Elbow*
- $\mathcal{R} := \{i | y_i f(x_i) > 1, \alpha_i = 0\}$ is called *Right of the Elbow*.

For the calculation of the optimal λ a Linear Equation System has to be solved. The respective matrix can be singular, which is a problem of the *SVMpath* algorithm.

3   Results

Finally, we show some results from the evaluation of the algorithms indicated in Table 1. We took the *Glass* dataset for the evaluation. It consists of 241 instances, 9 features and 6 classes and stems from the *UCI* repository of machine learning databases [18]. The task of the *Glass* problem is to discriminate between different glass types to support criminological and forensic research. As mentioned in section 2.5, we divided the dataset into a training and a test set. Table 2 shows the classification error rates ($= 1 -$ test accuracy) for the application of the

Table 2. Error Rates, runtime in CPU seconds (on an Intel Pentium IV processor with 3.06 GHz and 1 GB RAM) and number of support vectors for *Glass* dataset with default and optimal parameters for each algorithm and each optimization method. The mean of the support vectors (over all binary classifications) with the standard deviations in parentheses are mostly indicated. Exceptions: *OAO-global* (total number of support vectors, specific output of *libsvm*), *Crammer-Singer/SVMmulticlass* (Note that these are direct algorithms. See [4] and [24] for more details.) Note that in the case of *Comb-ECOC* we have three different code matrices: the user defined matrix (U) and two randomly defined matrices (R1, R2).

Results with Default Parameters of each Multiclass Algorithm			
Algorithm	Error Rate	Runtime	Number of SVs
OAA	**0.27**	0.53	57.17 (38.85)
OAO	0.34	**0.09**	132.00
SVMlight-OAA	0.34	0.98	85.17 (29.53)
SVMlight-OAO	0.38	2.27	32.27 (17.14)
Crammer-Singer	0.28	0.46	480.00
SVMmulticlass	**0.27**	3.33	430.00
ECOC	0.45	1.81	84.42 (33.51)
Comb-ECOC	0.48	189.92	U: 96.07 (28.95)
			R1: 101.00 (24.92)
			R2: 101.00 (24.91)

Error Rates and Runtime with Parameter Optimization						
	Optimization Method					
Algorithm	Global		Local		SVMpath	
OAA	0.30	947.64	0.38	1695.26	0.30	23.88
OAO	0.31	282.63	0.30	1062.92	**0.28**	5.63
SVMpath	–		–		0.30	**4.47**
SVMlight-OAA	0.33	**95.5**	0.30	1253.93	0.34	19.72
SVMlight-OAO	0.33	217.62	0.33	2522.45	0.30	10.34
SVMmulticlass	**0.25**	817.79	–		–	
ECOC	0.47	996.54	0.44	**944.09**	0.52	81.29
ECOC-SVMpath	–		–		0.48	87.36
Comb-ECOC	0.28	57822.65	**0.23**	5279.15	0.31	458.38

Number of Support Vectors with Parameter Optimization						
	Optimization Method					
Algorithm	Global		Local		SVMpath	
OAA	70.00	(37.36)	64.00	(40.63)	48.00	(34.04)
OAO	135.00		28.47	(18.58)	21.13	(15.22)
SVMpath	–		–		14.8	(8.67)
SVMlight-OAA	82.00	(24.47)	72.17	(30.31)	34.00	(21.73)
SVMlight-OAO	32.6	(13.83)	24.27	(15.90)	14.8	(9.50)
SVMmulticlass	570.00		–		–	
ECOC	101.84	(22.31)	88.87	(28.34)	94.03	(15.81)
ECOC-SVMpath	–		–		61.81	(21.87)
Comb-ECOC	U: 20.35	(4.66)	U: 33.10	(2.45)	U: 42.47	(22.05)
	R1: 21.67	(3.62)	R1: 33.82	(2.55)	R1: 53.48	(22.65)
	R2: 21.65	(3.70)	R2: 33.64	(2.44)	R2: 53.48	(22.65)

model established by the *Glass* training set on the independent test set with the optimal parameters computed by the methods described in sections 2.5 and 2.6. The training set contains about 70% of the original dataset. It is guaranteed that all classes are contained both in the training set and in the test set.

The results are comparable with error rates obtained by authors of other reviews: Szedmak and Shawe-Taylor [23] got results in the range of 0.3-0.4 with standard OAA and OAO algorithms, García-Pedrajas and Ortiz-Boyer [9] a result of 0.28. Friedrich [8] used 50 bootstrap replications which evoked 50 error rates. Their mean was 0.39 with a standard deviation of 0.04, using a *k-Nearest-Neighbor* algorithm, 0.40 with a standard deviation of 0.07 using a *Decision Tree* based method and 0.41 with a standard deviation of 0.07 using a *Linear Discriminant Analysis*. The best result obtained by [8] was 0.23 with evolutionarily generated architectures of neural networks and combination models originated from bagging (see [2]). The best result of 0.23 in Table 2 is obtained by *Comb-ECOC* with a local parameter optimization. In general, the local optimization delivers the best results but also from *svmpath*, quite good results are achieved, especially with the simple multiclass algorithms that use the *libsvm*.

Table 2 also shows the results obtained with the default parameters of the methods. Note that *ECOC* and *Comb-ECOC* use the *libsvm* and *SVMlight*, respectively, as binary classificators. So we only took the binary default parameters. We also got a result from the method by Crammer and Singer described in section 2.2. Due to high runtimes during the parameter optimization, we did not indicate any results for it in the lower tabulars of Table 2.

If we compare these results with the error rates in the tabular below, we see that the results are much worse than with parameter optimization, except in the case of *OAA* and *ECOC* (global and svmpath optimization). Mainly in the case of the ECOC algorithms, a good parameter optimization is indispensable. To explain this, we differenciate between two kinds of classification errors:

1. Errors occuring because of overlapping classes
2. Errors occuring because of classes that are too far away from each other with other classes between them.

Especially errors of the second type can be serious. Depending on how the classes are distributed in the input space and which classes have to be merged by a binary classificator, the Gaussian kernel may not accomplish several classifications anymore. Therefore, the choice of better parameters is necessary.

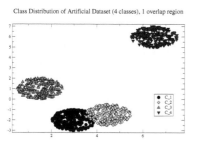 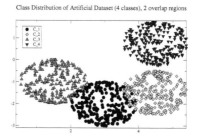

Fig. 2. Artificial Datasets with different margin sizes for OAA and OAO. Each class is defined by a unit circle. The first dataset has one, the second has two overlap regions.

It is not excluded that the default parameters are better than the optimized ones. This can have two reasons:

1. They are not taken from a grid but defined by another way, so their domain is different.
2. As we use bootstrapping for finding the optimal parameters, the parameters can differ from run to run, as the bootstrap replicates are defined at random.

Table 3. Error Rates, runtime and number of support vectors for the artificial dataset in Figure 2 with 1 overlap region with optimal parameters for each algorithm and each optimization method. The mean of the support vectors (over all binary classifications) with the standard deviations in parentheses are mostly indicated. Exception: *OAO-global* (total number of support vectors, specific output of *libsvm*).

Error Rates and Elapsed Runtime in CPU seconds						
	Optimization Method					
Algorithm	Global		Local		SVMpath	
OAA	0.0233	477.08	**0.0233**	**136.37**	0.0233	177.52
OAO	**0.0200**	**40.11**	0.0267	157.93	**0.0167**	**40.52**
SVMlight-OAA	0.0267	290.11	**0.0233**	2980.90	**0.0167**	381.90
SVMlight-OAO	0.0267	368.49	**0.0233**	3999.49	**0.0167**	108.60
Number of Support Vectors						
	Optimization Method					
Algorithm	Global		Local		SVMpath	
OAA	58.25	(41.92)	70.75	(40.05)	22.5	(23.04)
OAO	53.00		41.17	(38.94)	13.67	(19.56)
SVMlight-OAA	82.00	(61.16)	131.25	(76.68)	33.75	(35.53)
SVMlight-OAO	29.50	(18.43)	85.67	(14.25)	13.83	(13.17)

Table 4. Error Rates, runtime and number of support vectors for the artificial dataset in Figure 2 with 2 overlap regions with optimal parameters for each algorithm and each optimization method. The mean of the support vectors (over all binary classifications) with the standard deviations in parentheses are mostly indicated. Exception: *OAO-global* (total number of support vectors, specific output of *libsvm*). Note that in the OAA case, the matrix of the resulting Linear Equation System is singular.

Error Rates and Elapsed Runtime in CPU seconds						
	Optimization Method					
Algorithm	Global		Local		SVMpath	
OAA	0.0267	509.11	0.0233	166.41	singular system	
OAO	0.0200	**48.33**	0.0267	157.09	**0.0267**	**60.39**
SVMlight-OAA	0.0233	10748.38	**0.0200**	3081.58	singular system	
SVMlight-OAO	**0.0133**	450.48	**0.0200**	3848.86	0.0267	142.03
Error Rates and Elapsed Runtime in CPU seconds						
	Optimization Method					
Algorithm	Global		Local		SVMpath	
OAA	184.25	(5.85)	133.5	(34.85)	singular system	
OAO	113.00		58.83	(31.69)	23.83	(32.96)
SVMlight-OAA	111.00	(8.29)	133.25	(69.01)	singular system	
SVMlight-OAO	37.50	(19.69)	89.83	(24.84)	17.00	(23.68)

However, the higher runtime of a parameter optimization is worth getting better models for classification tasks.

Beside the test accuracies, one also has to account the number of support vectors, in order to exclude overfitting. They are also shown in Table 2. If we confront the OAO algorithms with the OAA algorithms, we see that in the OAO case, the number of support vectors is much lower than in the OAA case, as in general, the margin is much larger in the first case (see Figure 1). Another explanation might be the larger size of the training sets in the OAA case. Secondly, if we compare $ECOC$ and $Comb\text{-}ECOC$, we see that the bad test accuracies of $ECOC$ in Table 2 are related to the high number of support vectors, whereas the good results of $Comb\text{-}ECOC$ are connected with a small number of support vectors.

Note that the complexities of the standard algorithms OAA and OAO are polynomial in the number of classes. Also the algorithms by Crammer and Singer [4] and by Thorsten Joachims [24] can be run in polynomial time. The methods using error correcting output codes are NP-complete. As $Comb\text{-}ECOC$ consists of three runs defined by three different code matrices, each of which can have an exponential number of columns, its elapsed runtime is the largest by far, followed by $ECOC$ and $SVMmulticlass$. OAA is not always faster than OAO, even if its complexity is $\mathcal{O}(k)$ instead of $\mathcal{O}(k^2)$. Note that the number of data points is smaller in the case of pairwise classification.

Finally, we compared the OAA and OAO algorithms on two artificially composed datasets, which are plotted in Figure 2. It is a realization of the theoretical idea formulated by Figure 1. The margin sizes of the OAA algorithm are much smaller than the ones of the OAO algorithm. Furthermore, we have different margin sizes in the OAO case.

Each of the two datasets is two-dimensional and consists of 1000 instances. The 4 classes are equally distributed. They are divided into a training set with 700 and a test set with 300 instances. Table 3 shows the results for the first and Table 4 for the second dataset. In the case of one overlap region, we cannot see any difference between OAA and OAO yet, except the fact that in the OAO case, the number of support vectors is much smaller. The separations 'Class 1 vs rest', 'Class 2 vs rest' and 'Class 1 vs Class 2' will produce training errors. All the other cases are separable. Figure 3 shows the binary decision boundaries of the first two OAA separations and the multiclass decision boundaries of the OAO algorithm. In the OAO case, the margins are much larger than in the OAA case.

The results for the second dataset with two overlap regions are summarized in Table 4. Here, we see a difference between $SVMlight\text{-}OAA$ and $SVMlight\text{-}OAO$: As in Table 3, the number of support vectors is always smaller in the OAO case. Furthermore the error rate is smaller using $SVMlight\text{-}OAO$. Also the runtime is much higher in the OAA case. That is because the optimization problems are more difficult to solve. The individual binary separations are more complicated. The corresponding feature spaces will have very high dimensions. For the second dataset, the separations 'Class 1 versus the rest', 'Class 2 versus the rest', 'Class 3

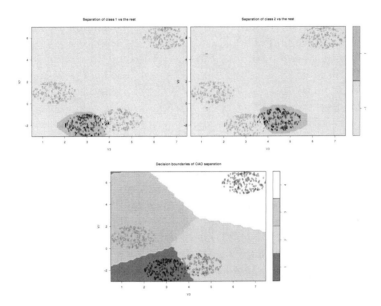

Fig. 3. Decision boundaries for the separations 'Class 1 vs rest', 'Class 2 vs rest' and OAO decision boundaries for the first dataset in Figure 2. We see that the OAA margins are much narrower than the OAO margins. The support vectors are highlighted by crosses.

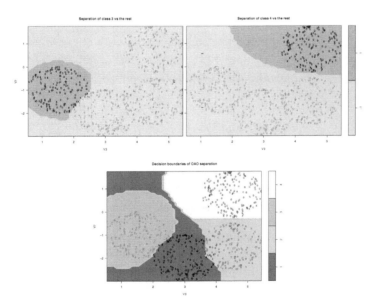

Fig. 4. Decision boundaries for the separations 'Class 3 vs rest', 'Class 4 vs rest' and OAO decision boundaries for the second dataset in Figure 2. We see that the OAA separations are much narrower than the OAO margins. The support vectors are highlighted by crosses.

versus the rest', 'Class 1 versus Class 2' and 'Class 1 versus Class 3' cause positive training errors. Figure 4 shows the binary decision boundaries of the last two OAA separations and the multiclass decision boundaries of the OAO algorithm. The margin is smaller in the OAA case, as expected.

In order to complete the experimental proof that the OAO algorithm performs better than the OAA algorithm in the aforesaid special geometric case, we executed pairwise t-tests on 30 bootstrap replications, using a confidence level $\alpha = 0.05$. Especially in the global optimization case the t-tests were significant (p-value: $p \ll \alpha$), in favor of the OAO algorithm.

4   Discussion and Conclusion

In this paper, we present a multiclass combination method named *Comb-ECOC* and compare it statistically with other existing algorithms enlisted in Table 1. Furthermore, we oppose different parameter optimization methods.

Comb-ECOC delivers good results. It performed best for the *Glass* dataset with a special kind of parameter optimization. Its advantages are

1. the usage of error correcting output codes,
2. the prediction of posterior class probabilities,
3. its robustness and
4. the combination of three multiclass runs defined by three different code matrices.

Of course, there are also some disadvantages. The result for the *Glass* dataset is the best, but from this fact, one cannot interprete that it always outperforms the other algorithms. Pursuant to [26], this also depends on the respective dataset. The main problem of *Comb-ECOC* lies in the precoding. In order to maintain the robustness against several binary classification errors, a large minimum Hamming distance between the rows of the code matrices must always be guaranteed. Some approaches can be found in [1], but it is indispensable to use algebraic methods from Galois theory. Especially if the number of k is large (e.g. $k = 15$), a large minimum Hamming distance is not procurable in an easy way, and therefore *Comb-ECOC* will fail. But with a good precoding strategy its performance will be enhanced.

The second disadvantage is its high runtime: For the *local* parameter optimization it takes 1.5 hours and for the *global* optimization even 16 hours. However, from Table 2 (first tabular) we can see that one modelling and classification process without parameter optimization before only takes about 3 minutes.

The second subject of this paper is the parameter optimization. We suggest three methods: *global, local* and *svmpath*. Intuitively, the *local* optimization should perform better than the *global* optimization, because each classificator gets its own optimal parameters. The danger of overfitting is excluded, because we do not minimize training errors but test errors occuring during several bootstrap replicates. As we can see from Table 2, the results are mostly better in the local case.

Another way is to take into account the contrast between a small training error and a large margin instead of only considering several test errors. *SVM-path* does so. Furthermore, it has the advantage of a complexity that is not higher than a usual Support Vector Machine. But as we see in this paper, it does not perform well on *foreign* classification algorithms. It should be used for algorithms implemented in the *libsvm*. The best solution would be to use its own prediction function. A disadvantage of *SVMpath* is the fact that it may compute negative regularization constants which are useless for Support Vector Machines. We found this problematic using other datasets.

Despite the advantages that SVMs show for binary classifications, other methods like *Neural Networks*, *Linear Discriminant Analysis* or *Decision Trees* should be considered in multiclass scenarios.

References

1. Bose, R.C., Ray-Chaudhuri, D.K.: On A Class of Error Correcting Binary Group Codes. Information and Control 3 (1960)
2. Breiman, L.: Bagging Predictors. In: Machine Learning 24, 123–140 (1996)
3. Christianini, N., Shawe-Taylor, J.: An Introduction to Support Vector Machines. Cambridge University Press, Cambridge (2000)
4. Crammer, K., Singer, Y.: On the Algorithmic Implementation of Multiclass Kernel-based Vector Machines. Journal of Machine Learning Reseach 2, 265–292 (2001)
5. Dieterich, T., Bakiri, G.: Solving Multiclass Learning Problems via Error-Correcting Output Codes. Journal of Artificial Intelligence Research 2, 263–286 (1995)
6. Dimitriadou, E., Hornik, K., Leisch, F., Meyer, D., Weingessel, A.: The e1071 package. Manual (2006)
7. Efron, B., Tibshirani, R.: An Introduction to the Bootstrap. Chapman & Hall/CRC (1993)
8. Friedrich, C.: Kombinationen evolutionär optimierter Klassifikatoren. PhD thesis, Universität Witten/Herdecke (2005)
9. García-Pedrajas, N., Ortiz-Boyer, D.: Improving Multiclass Pattern Recognition by the Combination of Two Strategies IEEE Transactions on Pattern Analysis and Machine Intelligence 28 (2006)
10. Hastie, T., Rosset, S., Tibshirani, R., Zhu, J.: The Entire Regularization Path for the Support Vector Machine. Technical Report, Statistics Department, Stanford University (2004)
11. Hsu, C.-W., Chang, C.-C., Lin, C.-J.: A Practical Guide to Support Vector Classification. Department of Computer Science and Information Engineering, National Taiwan University (2006)
12. Hsu, C.-W., Lin, C.-J.: A comparison of methods for multi-class Support Vector Machines. IEEE Transactions on Neural Networks 13, 415–425 (2002)
13. Huang, T.-J., Weng, R.C., Lin, C.-J.: Generalized Bradley-Terry Models and Multiclass Probability Estimates. Journal of Machine Learning Research 7, 85–115 (2006)
14. Hülsmann, M.: Vergleich verschiedener kernbasierter Methoden zur Realisierung eines effizienten Multiclass-Algorithmus des Maschinellen Lernens. Master's thesis, Universität zu Köln (2006)

15. Joachims, T.: Making large-Scale SVM learning practical. In: Advances in Kernel Methods – Support Vector Learning, pp. 41–56. MIT Press, Cambridge (1999)
16. Mencía, E.L.: Paarweises Lernen von Multilabel-Klassifikatoren mit dem Perzeptron-Algorithmus. Master's thesis, Technische Universität Darmstadt (2006)
17. Meyer, D.: Support Vector Machines, the Interface to libsvm in package e1071. Vignette (2006)
18. Newman, D.J., Hettich, S., Blake, C.L., Merz, C.J.: UCI Repository of machine learning databases (1998),
    http://www.ics.uci.edu/~mlearn/MLRespository.html
19. Platt, J.C.: Probabilistic Ouputs for Support Vector Machines and Comparisons to Regularized Likelihood Methods. In: Proceedings of Advances in Large-Margin Classifiers, pp. 61–74. MIT Press, Cambridge (1999)
20. Roever, C., Raabe, N., Luebke, K., Ligges, U., Szepanek, G., Zentgraf, M.: The klaR package. Manual (2006)
21. Ihaka, R., Gentleman, R.R.: A Language for Data Analysis and Graphics. Journal of Computational and Graphical Statistics 5, 299–314 (1996)
22. Schölkopf, B., Smola, A.: Learning with Kernels: Support Vector Machines, Regularization, Optimization and Beyond. MIT Press, Cambridge (2002)
23. Szedmak, S., Shawe-Taylor, J.: Multiclass Learning at One-Class Complexity. Information-Signals, Images, Systems (ISIS Group), Electronics and Computer Science. Technical Report (2005)
24. Tsochantaridis, I., Hofmann, T., Joachims, T., Altun, Y.: Support Vector Machine Learning for Interdependent and Structured Output Spaces. In: Proceedings of the 21th International Conference on Machine Learning. Banff, Canada (2004)
25. Vapnik, V.: The Nature of Statistical Learning Theory. Springer, Heidelberg (1995)
26. Wolpert, D.H.: No Free Lunch Theorems for Optimization. In: Proceedings of IEEE Transactions on Evolutionary Computation 1, pp. 67–82 (1997)

Multi-source Data Modelling: Integrating Related Data to Improve Model Performance

Paul R. Trundle[1], Daniel C. Neagu[1], and Qasim Chaudhry[2]

[1] University of Bradford, Richmond Road, Bradford, West Yorkshire, BD7 1DP, UK
{p.r.trundle, d.neagu}@bradford.ac.uk
[2] Central Science Laboratory, Sand Hutton, York, YO41 1LZ, UK
q.chaudhry@csl.gov.uk

Abstract. Traditional methods in Data Mining cannot be applied to all types of data with equal success. Innovative methods for model creation are needed to address the lack of model performance for data from which it is difficult to extract relationships. This paper proposes a set of algorithms that allow the integration of data from multiple datasets that are related, as well as results from the implementation of these techniques using data from the field of Predictive Toxicology. The results show significant improvements when related data is used to aid in the model creation process, both overall and in specific data ranges. The proposed algorithms have potential for use within any field where multiple datasets exist, particularly in fields combining computing, chemistry and biology.

Keywords: Data Integration, Data Mining, Machine Learning, Multi-Species Modelling.

1 Introduction

Current methods in Data Mining allow accurate and robust predictions to be made by models built for a wide variety of data. Unfortunately the process of building a reliable predictive model is not straightforward for certain types of data. In these situations novel and innovative techniques are needed to overcome problems with the quality or availability of data. Poor quality data can yield models that are overwhelmed by inputs which are unrelated to the prediction target, resulting in poor quality predictions [1]. A lack of available data also causes problems, as model building algorithms often fail to find underlying mathematical relationships between inputs and output(s) when insufficient amounts of training data are available. One field of research that often suffers from both of these problems is Predictive Toxicology; the study of chemical and biological data in order to make predictions about particular biological effects of (new) chemical compounds for a target species.

Datasets containing large amounts of high quality data, both in terms of inputs and target output(s) can be difficult to find for many reasons. Inputs are often values calculated by software packages, and their monetary expense means that these calculated values are not freely released by users. The accuracy of the available data

P. Perner (Ed.): MLDM 2007, LNAI 4571, pp. 32–46, 2007.
© Springer-Verlag Berlin Heidelberg 2007

is also subject to some debate [2], further complicating the choice of which source of data to use for any given task. Political, governmental and ethical factors obviously limit the amount of new experiments that are carried out, forcing researchers to rely on fairly limited amounts of data that may or may not be shared by the relevant owners. Further to this, the quality of these experiments must be taken into consideration. Documentation detailing the procedures used in the experiments can be difficult to find, despite attempts to make good experimental practices a priority [3], but examination of the methods used is necessary to determine the inherent reliability of the results. The reproducibility of such results is also highly relevant [4].

Our original approach attempts to address particular difficulties facing model builders for certain applications where the development of new and innovative algorithms and techniques are vital. With this in mind, we exemplify using chemical and biological datasets which have the potential to be exploited to improve the predictive power of models, particularly in situations where traditional model building techniques have failed. For any given chemical substance in a toxicology dataset, or indeed other collections of measurements of biological effects for chemical substances, there can be a toxicity value recorded for multiple different species or methods of administration. A particular species, with a chemical substance administered in a particular way (e.g. oral, intravenous, dietary etc.) over a given time-frame is known as an *endpoint*. These figures, when taken separately, can be used to make predictions for unknown values within the dataset. Traditional model creation techniques usually involve building models on data from a single endpoint, as the biological differences between two species, as well as changes in magnitude of measured effects caused by different methods of administration mean that predictions for one endpoint cannot be scientifically justified when directly applied to any other endpoint [5]. Whilst some success has been reported with so called inter-species extrapolation [6], [7], it remains difficult to intelligently use multi-endpoint data values in an accurate and justifiable way. The authors have attempted to overcome the problems associated with integrating data in this way, learning from existing techniques such as the work on meta-learning described in [8], but with a more data-driven approach. Real-world data is used rather than artificial or trivial datasets, with model accuracy rather than computational efficiency as the goal.

This paper proposes three algorithms that allow multiple endpoints to be used co-operatively for the development of models. In Section 2 the algorithms are explained theoretically, and in Section 3 results are presented from a real-world application using Predictive Toxicology data. Section 4 summarises the conclusions drawn from interpretation of the results presented.

2 Methods

The inherent difficulties in using data from similar but separate endpoints can make the development of techniques and algorithms to integrate them complicated, but not impossible. The three original algorithms proposed in this paper are explained in Sections 2.1, 2.2 and 2.3.

2.1 Toxicity Class Value Integration

The first technique we propose uses the toxicity data from different endpoints together to build models that make predictions for a single endpoint. The toxicity values, which are usually used solely as a target for prediction, are taken as an additional input for another endpoint. For example, suppose we have two datasets, containing numerical input/output values for two endpoints, known as A and B. Each dataset has a set of inputs, I_A and I_B respectively. Each dataset also has a toxicity value (output), used as a target for prediction for created models, defined as T_A and T_B respectively. The two datasets have input and output values for a number of chemicals, and there is a degree of overlap between the two, such that there are three distinct subsets of chemicals. Let the subset of chemicals for which output values exist only in A be known as S_A, the subset where they exist only in B be known as S_B and the subset of chemicals where they exist both in A and B be known as S_{AB}:

$$S_A = (A \cup B) - B \tag{1}$$

$$S_B = (A \cup B) - A \tag{2}$$

$$S_{AB} = (A \cap B) \tag{3}$$

Suppose that we wish to create a model that predicts values for the toxicity of dataset A, T_A. A traditional model might use the input data from A (I_A) in order to find some mathematical relationship between those values and the prediction target. However, using this method of model creation we lose the potential extra information stored in dataset B, the data from a different but related endpoint. Since the input values are likely to be the same in both A and B we must consider the usefulness of the toxicity values from B for building a model to predict the toxicity value from A. The subset of data that is of interest is the intersection of chemicals in both datasets; S_{AB}. Due to the aforementioned differences between two distinct endpoints, they cannot be directly applied as output values for another given endpoint. Despite this, it stands to reason that in two species which are somewhat biologically similar a correlation may be observed between the toxicity values recorded for both species. Any correlations are, of course, subject to variations in both the chemical structure of a given substance and any biological differences between the two target species, but the usefulness of these values should not be underestimated.

The algorithm works by including the toxicity class value from another endpoint as an additional input in the model building process: the toxicity class information for the secondary endpoint will be added to the instances about chemicals common to both endpoints to be used for obtaining a model predicting toxicity for the primary endpoint. With the inclusion of a secondary set of toxicity class values as an input, it is expected that they will form a strong basis for prediction for the output of created models. This applies when the values are taken in conjunction with traditional chemical descriptors; the resulting predictions could be stronger than those made using either the descriptors or the secondary toxicity class alone. An effective model creation algorithm should be capable of determining which descriptors have a significant effect on the correlation of the additional secondary endpoint data and the prediction target, thus using both information sources to their maximum potential. The main drawback of this technique is the restriction that only chemicals in the subset S_{AB} may be used, due to the need for toxicity values to exist for both endpoints.

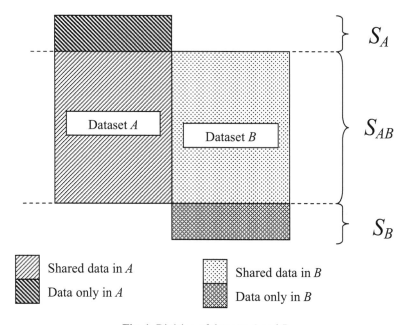

Fig. 1. Division of datasets A and B

2.2   Integration of Predicted Toxicity Class Values

Our second algorithm addresses a significant problem highlighted in the previous section. Whilst the potential for increases in overall classification accuracy of models exists, it is limited to those chemicals within the subset of instances for which toxicity values exist for both endpoints; namely subset S_{AB}. This could prove to be problematic in sparse or small datasets where subset S_{AB} may be too small for any significant learning to take place during the model building process. In order to counter this limitation, the authors propose the use of toxicity class values that were produced by a *model*, in addition to experimental values determined by the use of living organisms in laboratory tests.

Suppose we wish to build a model using dataset A, and we wish to include data from the target outputs of dataset B as additional inputs. Suppose also that we wish to include *all* the instances in A as inputs for our model, including those for which no data exists in B (subset S_A). In order to do this we must fill in the missing target output data from B with *artificial* prediction values. This involves the creation of a model using all the instances in B, and using it to make predictions about the output values for the chemicals in S_A. Once we have these predictions, we can use the artificial data for subset S_A along with the real values for subset S_{AB} and build models for A using the same algorithm described in section 2.1. The creation of artificial data is illustrated in Figure 2.

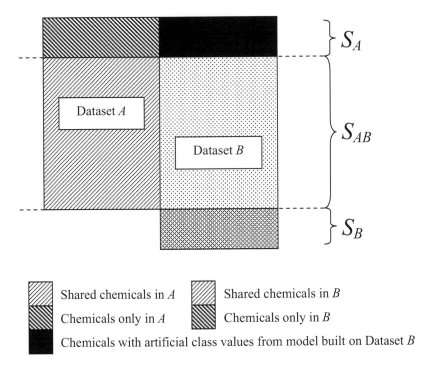

Fig. 2. Expansion of dataset *B* using artificial prediction data

2.3   Fusion of Connectionist Models

The third algorithm proposed herein further explores the potential of integrating data from multiple sources by replacing many multi-layer single-output neural networks [9] with a larger, multi-layer multi-output network. The process of constructing this new network involves fusing together the datasets used in all of the single output networks and constructing the larger multi-output network. The motivation behind this algorithm is that if many separate single-output neural networks can make predictions, and we know that integrating data from multiple endpoints can yield better results; it should be possible to create a single larger network that can make predictions for many outputs, utilising information from multiple sources.

Whilst the technique is applicable to any finite number of single output networks, for the sake of simplicity the authors exemplify using just two networks. If we have a neural network built to make predictions for a single source of data, and a second network built to make predictions for a second, but related, set of data, it is not straightforward to directly fuse these two networks together to provide a combined model. However at the architectural and learning level we propose to fuse them in a new connectionist multi-input multi-output model which we create. The aim is that the new network can make more accurate predictions for both of the outputs based on data from both sources.

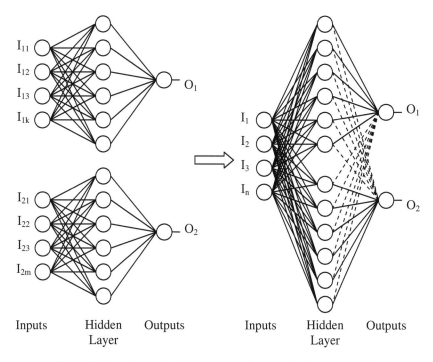

Fig. 3. Fusion of two related data sources into one dual-output model

The fusion network allows sharing of knowledge from both sources. Figure 3 demonstrates the fusion algorithm; initial inputs are defined as I_{ab}, where a is the network that uses the input, and b is the number of the given input for the specific model. Note that the inputs to each of the single output networks may be identical (as is assumed in Figure 3) so simply using the original inputs from one of the single networks is acceptable. In cases where the single-output networks use non-identical sets of inputs, the union of all input datasets is supplied as the training data for the larger multi-output model. Feature reduction techniques can be applied to this dataset to address any resulting redundancy of features or noise. The two outputs of the networks are defined as O_1 and O_2. The new connections formed when constructing the larger, combinatorial network are shown as dotted lines between neurons.

3 Experiments

In order to test the effectiveness of the techniques proposed in Section 2, a set of experiments were carried out using real-world toxicity datasets from the DEMETRA project [10]. The results of these experiments are detailed in sections 3.1, 3.2 and 3.3. All of the experiments used one or more of the following three endpoints, which were selected based on the amount of chemicals for which toxicity values existed, the amount of overlap of chemicals between the datasets and the biological similarities of the species used in the experiments: LC_{50} 96h Rainbow Trout acute toxicity (Trout), EC_{50} 48h Water Flea acute toxicity (Daphnia), and LD_{50} 14d Oral Bobwhite Quail (Oral_Quail). LC_{50}, EC_{50} and LD_{50} describe a Lethal Concentration, an Effect Concentration producing a measurable effect and a Lethal Dosage of a specific substance that affects 50% of test subjects respectively. For each endpoint a numerical value existed describing the magnitude of toxic effect. This value is used to categorise each chemical as belonging to one of four classes: Class_1 (very high toxicity), Class_2 (high toxicity), Class_3 (medium toxicity) or Class_4 (low toxicity).

In combination with the three sets of endpoints, nineteen separate sets of input values were used (named in Table 1), all of which were calculated from a variety of software packages using chemical structure information. The calculated inputs included both 2D and 3D attributes (known as descriptors) for each chemical. They describe information such as the molecular weight of a substance, numbers of various constituent atoms, solubility in water, and numerous other measurable attributes.

3.1 Toxicity Class Value Integration

The first experiment implements the algorithm defined in Section 2.1, and uses all three endpoints described in Section 3. The Trout endpoint was used as a basis for comparison, and traditional models using only chemical descriptors were built using six different well known Machine Learning algorithms: Bayesian Networks (Bayes) [11], Artificial Neural Networks (ANN) [9], Support Vector Machines (SVM) [12], k-Nearest Neighbour (KNN) [13], Decision Trees (DT) [14], and Rule Induction (RI) [15]. All six algorithms were implemented using the Weka software package [16]. The results from these initial tests are presented in Table 1. The classification accuracy is defined as the percentage of correctly classified instances using 10-fold cross validation [17].

The same tests were carried out using the same datasets, but this time the toxicity class values from the Daphnia endpoint were added as a new input for model creation. A third set of tests were carried out: Daphnia toxicity class values were replaced with those from the Oral_Quail endpoint. The results of these tests were compared to the results in Table 1, and any change in the classification accuracies noted. Table 2 presents results from the Daphnia tests, where each cell shows the increase/decrease for each combination of input set and model creation algorithm. Note that a decrease in accuracy is shown in italics, and that the cell in the lower-right corner shows the overall change in classification accuracy compared to the traditional Trout models built in the first set of tests. The results for the addition of Oral_Quail data are not shown, as no significant increases or decreases in accuracy were recorded.

The models have been generated with default parameters and no optimization procedure has been applied to increase model performances as reported below. Our objective was focused on general performance into the context of data integration.

Table 1. Percentage of correctly classified instances for Trout endpoint

Descriptors	Bayes	ANN	SVM	KNN	DT	RI	Average
2DABLegend	41.05	39.47	44.74	46.32	39.47	38.42	**41.65**
2DACD	46.84	40.53	43.68	42.11	45.26	45.26	**44.06**
2DCodessa	37.89	40.53	42.11	47.37	42.11	45.26	**41.58**
2DDragonA	40.53	43.16	45.79	39.47	46.84	45.26	**43.61**
2DDragonB	44.74	47.37	49.47	45.26	48.95	45.79	**46.17**
2DDragonC	36.84	35.26	44.21	46.84	43.68	44.74	**40.90**
2DDragonD	35.26	45.79	48.42	45.79	41.58	41.58	**43.76**
2DPallas	44.21	40.53	44.74	42.11	43.16	50.00	**43.68**
3DCache	38.77	47.34	51.26	46.84	38.60	43.45	**44.56**
3DCodessaA	48.36	47.37	51.78	49.01	45.73	48.54	**48.30**
3DCodessaB	41.90	51.64	45.82	44.59	50.09	49.97	**46.22**
3DCodessaC	39.27	44.62	38.63	42.98	41.35	41.35	**41.68**
3DCodessaD	41.46	43.68	43.51	49.53	44.12	40.76	**43.71**
3DDragonA	35.94	45.73	45.26	43.60	37.51	41.26	**41.98**
3DDragonB	44.09	42.40	52.78	41.96	45.18	44.65	**44.30**
3DDragonC	41.46	43.88	48.98	41.40	38.71	40.26	**42.31**
3DDragonD	36.08	40.85	42.98	44.53	35.35	39.80	**39.36**
3DMMP	41.46	40.41	42.46	42.40	40.67	46.81	**41.91**
3DRecon	42.13	41.81	52.19	50.12	39.88	39.21	**43.48**
Average	**40.96**	**43.28**	**46.25**	**44.85**	**42.54**	**43.81**	**43.33**

Table 2 shows that a significant improvement (7.71%) in overall classification accuracy across all input sets and algorithms was recorded when Daphnia toxicity class values were used to create models for the prediction of Trout endpoint toxicity. The addition of Oral_Quail toxicity class values showed a slight decrease (-0.15%) in overall accuracy across all datasets and algorithms, though this value is not highly significant and could be due to random variations in the model creation processes rather than a direct result of the addition of the Oral_Quail values. When these results are taken in conjunction with the correlation values of the toxicity class values between the three endpoints, the reasons for the differences in accuracy are more easily explained. Each chemical is placed into a class based on the numerical value of its toxicity; determined by the experiments for each endpoint. The numerical values were compared for Trout and Daphnia and a correlation coefficient of 0.49 was found (0 being no correlation and 1 being perfect correlation). The same coefficient was calculated for Trout and Oral_Quail values and was found to be 0.16; a significantly lower correlation than for Daphnia.

The differences in both correlation and accuracy increases/decreases are very likely based upon the biological similarities of the species in question. The Trout and Daphnia endpoints both deal with aquatic species, and certain chemical features are more important than others in determining the magnitude of any biological effects for

Table 2. Change in classification accuracy when Daphnia toxicity class values were added as an additional input to Trout models

Descriptors	Bayes	ANN	SVM	KNN	DT	RI	Average
2DABLegend	13.16	12.63	10.00	11.05	1.58	8.95	**9.56**
2DACD	1.05	16.84	13.68	9.47	13.68	1.05	**9.30**
2DCodessa	12.11	7.37	17.37	13.16	1.58	2.11	**8.95**
2DDragonA	14.21	9.47	11.58	7.37	14.21	-4.74	**8.68**
2DDragonB	10.00	8.42	7.89	8.42	10.00	-1.58	**7.19**
2DDragonC	2.11	15.79	5.26	16.84	2.63	0.00	**7.11**
2DDragonD	5.79	7.89	8.42	2.63	3.68	0.53	**4.82**
2DPallas	2.63	9.47	13.68	12.63	11.05	6.84	**9.39**
3DCache	9.85	7.05	8.04	13.01	3.10	7.63	**8.11**
3DCodessaA	0.53	5.47	3.22	6.11	2.25	5.00	**3.76**
3DCodessaB	6.64	5.06	11.20	5.56	8.77	-2.60	**5.77**
3DCodessaC	11.46	12.02	12.54	-0.03	17.37	5.94	**9.88**
3DCodessaD	14.59	10.12	9.74	9.80	5.44	4.85	**9.09**
3DDragonA	8.68	8.19	7.08	6.61	10.38	7.60	**8.09**
3DDragonB	4.91	8.65	-0.03	13.13	11.93	3.30	**6.98**
3DDragonC	5.94	3.46	6.02	8.04	5.85	4.85	**5.69**
3DDragonD	2.89	5.91	9.85	13.04	10.12	10.91	**8.79**
3DMMP	12.89	12.92	11.90	18.39	2.84	3.39	**10.39**
3DRecon	1.64	4.44	1.78	13.27	2.78	5.76	**4.95**
Average	**7.43**	**9.01**	**8.91**	**9.92**	**7.33**	**3.67**	**7.71**

aquatic toxicity. Oral_Quail experiments were carried out on a species of bird, and the chemical features affecting the relative toxicity of different substances on this species are likely to be very different than those for aquatic species. It would appear that the biological similarity of the candidate species for use with this technique, as well as the correlation between the toxicity values of the endpoints, is a critical factor in determining whether any improvement is likely to occur when new data is added.

Further to this, examining the improvements observed when Daphnia values were added shows that certain datasets and algorithms appeared to benefit more overall from the additional input. From Tables 1 and 2 we can see that descriptor sets with low average increases in accuracy, such as 3DCodessaA and 3DCodessaB (3.76% and 5.77% improvements respectively) had relatively high average accuracies for models built *without* Daphnia data. The 3DMMP dataset, which showed the largest average improvement (10.39%), had a relatively low average accuracy in Table 1 of just 41.91% when compared to the average of all datasets (43.33%). Logically we can assume that, in many cases, this technique could be used most effectively when traditional model creation techniques have proved inadequate for creating effective models. Also shown in Table 1 is the apparent difference in the success of the technique across the various model creation algorithms. The lowest overall improvement accuracy for a particular algorithm over all datasets is 3.67% for the

Rule Induction technique. Obviously this is significant when compared to the second lowest average improvement of 7.33% for the Decision Tree algorithm, and the overall average improvement across all data of 7.71%. This leads the authors to the conclusion that particular model building techniques may be more suitable for producing improvements in the prediction accuracy of models.

3.2 Integration of Predicted Toxicity Class Values

The experiments detailed in this section implemented the algorithm described in Section 2.2. Once again the Trout and Daphnia endpoints were used, as they had shown potential for continued development in the previous set of experiments (see section 2.1). The inclusion of the Oral_Quail endpoint was discounted due to the previous lack of improvement shown. The same six algorithms were applied, and the same descriptor sets (see Table 1) were used, with a single exception: 3DDragonA, -B, -C and -D were combined into three new descriptor sets: 3DDragonA, -B and -C. Using only the Daphnia toxicity data, all combinations of algorithms and descriptor sets were used to build a total of 108 candidate models for the creation of the artificial toxicity class values. The models were judged according to the overall classification accuracy for all data using 10-fold cross validation, and the model with the highest overall accuracy was chosen, as well as the model with the accuracy closest to the average accuracy of all 108 models. These were an Artificial Neural Network and a Bayesian Network respectively, both of which used the 3DCodessaB descriptor set as input values.

Both models were used to create predicted toxicity class values for chemicals in the Trout dataset that had no corresponding toxicity class value in the Daphnia dataset. These missing toxicity values correspond to chemicals that were used in the Trout experiments, but not used in the Daphnia experiments.

Once again a variety of models were built using Trout data, combining all eighteen descriptor sets and all six algorithms, but this time all toxicity values for chemicals in the entire Trout dataset were used. 10-fold cross validation was carried out and the accuracies of the created models were recorded, and are presented in Table 3.

With the results in Table 3 to be used as a benchmark, two further sets of tests were carried out using the new toxicity class values from the Daphnia dataset. The two sets of values, one expanded using the Artificial Neural Network predictions and the other expanded using the Bayesian Net predictions, were added as new inputs for use during model creation. Once again the entire range of descriptor sets and algorithms were used for both of these sets of tests.

In Table 3 the lower-right cell shows the overall classification accuracy across all datasets and algorithms, which is 50.41%. The addition of the Artificial Neural Network and Bayesian Network Daphnia prediction values resulted in an overall improvement in prediction accuracy, with >2.5% and >5% improvements respectively. Note however that the largest overall improvement comes from the Bayesian Network data, which was not considered the best model in terms of its predictive power when selected. This suggests that improvements in prediction power may be more related to the type of model used as apposed to an objective measure of the models performance without considering the way predictions are calculated.

Table 3. Average classification accuracy for entire Trout dataset

Descriptors	Bayes	ANN	SVM	KNN	DT	RI	Average
2DABLegend	47.78	46.78	47.78	35.89	42.78	45.67	**44.44**
2DACD	56.56	57.67	47.89	55.56	48.89	55.33	**53.65**
2DCodessa	47.78	40.22	48.89	50.22	45.89	47.00	**46.67**
2DDragonA	47.78	58.56	53.11	49.00	49.11	45.44	**50.50**
2DDragonB	46.67	51.11	55.22	55.22	57.89	59.89	**54.33**
2DDragonC	46.89	48.33	48.00	49.11	49.11	43.67	**47.52**
2DDragonD	45.56	53.44	54.22	48.89	40.00	45.44	**47.93**
2DPallas	52.22	50.33	45.78	51.33	53.44	52.11	**50.87**
3DCache	48.33	56.00	61.56	58.44	52.67	49.56	**54.43**
3DCodessaA	58.67	52.78	59.44	57.44	53.22	54.11	**55.94**
3DCodessaB	63.00	65.00	60.56	62.78	63.00	56.22	**61.76**
3DCodessaC	47.11	46.33	49.33	47.22	47.22	47.22	**47.41**
3DCodessaD	47.22	45.11	47.22	45.00	44.00	42.89	**45.24**
3DDragonA	45.00	52.89	47.22	48.44	40.78	51.56	**47.65**
3DDragonB	46.33	53.78	41.78	47.33	40.67	50.33	**46.70**
3DDragonC	47.33	46.22	56.11	57.22	50.67	51.67	**51.54**
3DMMP	43.89	44.22	55.89	49.44	39.67	57.11	**48.37**
3DRecon	38.33	54.00	59.56	57.33	48.33	57.44	**52.50**
Average	**48.69**	**51.27**	**52.20**	**51.44**	**48.19**	**50.70**	50.41

As in the previous experiment, there are differences in the relative increases/ decreases in performance across the various datasets and algorithms, signifying the dependence that these techniques have upon the particular algorithms and sets of inputs used. Rule Induction models again performed worse overall when the Daphnia data was used as an input, with an average accuracy across all datasets of 50.07% using the Artificial Neural Network data, and of 49.26% using the Bayesian Network predictions. When compared to the value of 50.70% from models built using just Trout data, this again suggests its possible unsuitability for these techniques.

3.3 Fusion of Connectionist Models

The experiments in this section implement the algorithm proposed in section 2.3, and involved creating a variety of single output artificial neural networks, each of which predicted numerical values (as apposed to class values as in the previous two experiments) for a single endpoint and set of descriptors. Seven of the previous descriptor sets were used in this experiment: 2DABLegend, 2DACD, 2DCodessa, 2DDragonA, 2DDragonB, 2DDragonC, and 2DDragonD. This resulted in fourteen neural networks, seven for the Trout endpoint and seven for the Daphnia endpoint. In addition to this, seven larger neural networks were created, each of which was built using the same inputs as the single-output networks, but used both sets of toxicity values as targets for prediction and thus had two outputs. Each of the larger networks used twice as many hidden neurons in the hidden layer as the smaller networks, and of course had two output neurons.

The fourteen Trout and Daphnia datasets in the experiment (combinations of the two sets of toxicity values and the seven descriptor sets) were divided into a training subset (75% of the chemicals) and a testing subset (25% of the chemicals). The training subset was used to train the networks, and the accuracy of the predictions was measured using the instances in the testing subset. By comparing the results from each of the single-output networks with those of the dual-output networks, it is possible to see if any significant improvements in accuracy can be made by using the larger networks. Table 4 shows the improvements in the Mean Absolute Error of the dual-output networks over the single-output network; so a positive value indicates that the dual-output network performed better than the single-output network, and a negative value indicates the opposite.

Table 4. Improvement in accuracy on chemicals in testing subset when using dual-output networks compared to using single-output networks

Descriptors	Trout Improvement	Daphnia Improvement
2DABLegend	-0.44	5.07
2DACD	0.38	-0.52
2DCodessa	0.75	0.23
2DDragonA	1.88	0.57
2DDragonB	0.62	0.37
2DDragonC	1.40	-0.39
2DDragonD	1.56	-1.58
Average	**0.88**	**0.54**

As the table shows, an overall improvement across all the descriptor sets for both endpoints can bee seen. There is however, a large variation in the accuracies reported. Once again this would seem to indicate a high dependence on the particular dataset used with this technique. A close look at some of the results shows us the potential for improvements with this method. The results for 2DDragonA are amongst the highest for both endpoints, and Figures 4 and 5 show us these results in more detail. The figures show scatter plots of the predictions made by both the single- and dual-output networks plotted against the actual toxicity values for each endpoint. The x-axis measures the actual value and the y-axis measures the predicted values made by the networks.

As we can see from the figures, the outputs of the two types of networks are somewhat different across the range of toxicity values. Figure 4 shows the values for the Trout endpoint, and the predictions from the single-output network (squares) are frequently found in the upper section of the graph, distant from the broken line denoting where the actual and predicted values would be equal (perfect prediction).

The predictions from the dual-output network (diamonds) are generally found much closer to this line, and none are found in the upper portion of the graph where the x-axis exceeds a value of 4. Note that any value that falls above the broken line

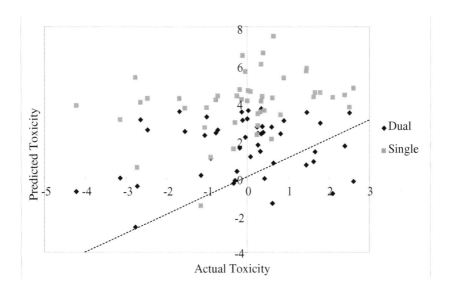

Fig. 4. Scatter plot of dual-output network and single-output network predictions against actual toxicity values for Trout endpoint

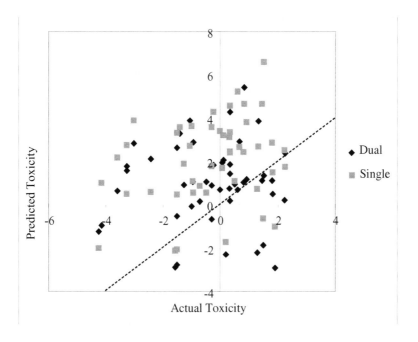

Fig. 5. Scatter plot of dual-output network and single-output network predictions against actual toxicity values for Daphnia endpoint

has underestimated the toxicity of a chemical, which is a highly undesirable characteristic of potential toxicity models. The single output network has points much further to the top and left than the dual-output network, particularly to the upper left of the graph where the y-axis records values of less than -3.

In Figure 5 we again can see an imbalance in the distribution of predicted values between the two networks. Close to the line of perfect prediction we can observe significantly more data points in the area contained within the values of -2 and 2 on both the x- and y-axis. There are also many more data points from the single-output network in the outlying areas towards the edges of the graph, far from the line of perfect prediction.

4   Conclusions

As the results show, this technique shows real potential for improvement of prediction accuracy both in general terms and for particular aspects of a given dataset, such as particular ranges of data that may be of interest. In all three experiments, it was found that the potential for improvement of models needs to be weighed against the sensitivity of the methods to the use different model creation algorithms and sets of descriptor values. There is also the drawback of the somewhat limited scope of applicability of these techniques, with a great deal of standard datasets having no related data to integrate. However, for applications where multiple related collections of data exist, the algorithms can be applied with only minor modification, as long as the underlying principles remain consistent, i.e. multiple sets of inputs that have some relationship to the target attribute of each collection of data, which are themselves related in some meaningful way. The authors propose the application of the techniques to a more diverse range of datasets as future work.

Using appropriate types of data, particularly in the fields combining computing with biology and chemistry, and with particular attention paid to finding an appropriate combination of descriptors and a machine learning algorithm, Multi-Source Data Modelling could be of real use in improving the quality of predictive models in the future.

Acknowledgements

The authors would like to thank the EPSRC for providing funding as part of the CASE project "Hybrid Intelligent Systems Applied to Predict Pesticides Toxicity", as well as the Central Science Laboratory, York, for supplying further funding and data. The authors also acknowledge the support of the EU project DEMETRA.

References

1. Blum, A., Langley, P.: Selection of Relevant Features and Examples in Machine Learning. Artifical Intelligence 97, 245–271 (1997)
2. Helma, C., Kramer, S., Pfahringer, B., Gottmann, E.: Data Quality in Predictive Toxicology: Identification of Chemical Structures and Calculation of Chemical Properties. Environmental Health Perspectives 108(11), 1029–1103 (2000)

3. Cooper-Hannan, R., Harbell, J.W., Coecke, S., Balls, M., Bowe, G., Cervinka, M., Clothier, R., Hermann, F., Klahm, L.K., de Lange, J., Liebsch, M., Vanparys, P.: The Principles of Good Laboratory Practice: Application to In Vitro Toxicology Studies. In: The Report and Recommendations of ECVAM Workshop 37, in ATLA vol. 27, pp. 539–577 (1999)

4. Gottmann, E., Kramer, S., Pfahringer, B., Helma, C.: Data Quality in Predictive Toxicology: Reproducibility of Rodent Carcinogenicity Experiments. Environmental Health Perspectives 109(5), 509–514 (2001)

5. Hengstler, J.G., Van Der Burg, B., Steinberg, P., Oesch, F.: Interspecies Differences in Cancer Susceptibility and Toxicity. Drug Metabolism Reviews 31(4), 917–970 (1999)

6. Gold, L.S., Bernstein, L., Magaw, R., Slone, T.H.: Interspecies Extrapolation in Carcinogenesis: Prediction Between Rats and Mice. Environmental Health Perspectives. 81, 211–219 (1989)

7. Mineau, P., Collins, B.T., Baril, A.: On the Use of Scaling Factors to Improve Interspecies Extrapolation of Acute Toxicity in Birds. Regulatory Toxicology and Pharmacology 24(1), 24–29 (1996)

8. Chan, P., Stolfo, S.: Learning Arbiter and Combiner Trees From Partitioned Data for Scaling Machine Learning. In: Proceedings of the 1st International Conference on Knowledge Discovery and Data Mining (KDD'95), pp. 39–44 (1995)

9. Haykin, S.: Neural Networks: A Comprehensive Foundation, 2nd edn. 842 pages. Prentice-Hall, Englewood Cliffs (1998)

10. http://www.demetra-tox.net/ (accessed February 2007)

11. Cooper, G.F., Herskovits, E.: A Bayesian Method for the Induction of Probabilistic Networks from Data. Machine Learning 9(4), 309–347 (1992)

12. Cristianini, N., Shawe-Taylor, J.: An Introduction to Support Vector Machines and Other Kernel-based Learning Methods, 204 pages. Cambridge University Press, Cambridge (2000)

13. Aha, D.W., Kibler, D., Albert, M.K.: Instance-based Learning Algorithms. Machine Learning 6(1), 37–66 (1991)

14. Quinlan, J.R.: Induction of Decision Trees. Machine Learning 1(1), 81–106 (1986)

15. Cohen, W.W.: Fast Effective Rule Induction. In: Proceeding of the 12th International Conference on Machine Learning (1995)

16. Witten, I.H., Frank, E.: Data Mining: Practical Machine Learning Tools and Techniques, 2nd edn. 416 pages. Morgan Kaufmann, Seattle (2005)

17. Kohavi, R.: A Study of Cross-validation and Bootstrap for Accuracy Estimation and Model Selection. In: Proceedings of the 14th International Joint Conference on Artificial Intelligence, pp. 1137–1143. Morgan Kaufmann, Seattle (1995)

An Empirical Comparison of Ideal and Empirical ROC-Based Reject Rules

Claudio Marrocco, Mario Molinara, and Francesco Tortorella

DAEIMI, Università degli Studi di Cassino
Via G. Di Biasio 43, 03043 Cassino (FR), Italia
{c.marrocco,m.molinara,tortorella}@unicas.it

Abstract. Two class classifiers are used in many complex problems in which the classification results could have serious consequences. In such situations the cost for a wrong classification can be so high that can be convenient to avoid a decision and reject the sample. This paper presents a comparison between two different reject rules (the Chow's and the ROC rule). In particular, the experiments show that the Chow's rule is inappropriate when the estimates of the a posteriori probabilities are not reliable.

Keywords: ROC curve, reject option, two-class classification, cost-sensitive classification, decision theory.

1   Introduction

Frequently, in two class classification problems, the cost for a wrong classification could be so high that it should be convenient to introduce a reject option [1]. This topic has been addressed by Chow in [2]. The rationale of the Chow's approach relies on the exact knowledge of the a posteriori probabilities for each sample to be recognized. Under this hypothesis, the Chow's rule is optimal because minimizes the error rate for a given reject rate (or vice versa). For the two-class classification cases in which the ideal setting assumed by the Chow's rule is not guaranteed and a real classifier must be used, an alternative method has been proposed in [3] where the information provided about the classifier performances by the empirical ROC curve is used to draw a reject rule which minimizes the expected cost for the application at hand. In [4] a review of the reject rule based on the empirical ROC and a comparison with the Chow's rule is presented; in the paper the authors claim to demonstrate the theoretical equivalence between the two rules and suggest that the Chow's reject rule should produce lower classification costs than those obtained by means of the reject rule based on the empirical ROC, even when real classifiers are employed.

A first comparison between the two approaches has been already proposed in [5] with reference to Fisher LDA. The experiments presented show that the empirical ROC reject rule works better than the Chow's rule in the majority of the cases considered.

P. Perner (Ed.): MLDM 2007, LNAI 4571, pp. 47–60, 2007.
© Springer-Verlag Berlin Heidelberg 2007

In this paper we aim to analyze more extensively the two rules and to compare, by means of thorough experiments, their behavior in order to demonstrate that the Chow's rule is inappropriate when the the distributions of the two classes are not perfectly known. In the next sections we resume the main features of the two reject rules while in the last section the experiments performed on both artificial and real data sets are reported.

2   Two Class Classification and the ROC Curve

2.1   The Ideal Case

In two class classification problems, the goal is to assign a pattern $\mathbf{x}$ coming from an instance space X to one of two mutually exclusive classes that can be generically called Positive (P) class and Negative (N) class; in other words, $X = P \cup N$ and $P \cap N = \emptyset$. Let us firstly consider a typical Decision Theory scenario and suppose to have a complete knowledge of the distributions of the samples within X, i.e. we know the a priori probabilities of the two classes (π_P, π_N) and the class conditional densities $f_P(\mathbf{x}) = p(\mathbf{x} | \mathbf{x} \in P)$ and $f_N(\mathbf{x}) = p(\mathbf{x} | \mathbf{x} \in N)$. If $\begin{bmatrix} \lambda_{NN} & \lambda_{NP} \\ \lambda_{PN} & \lambda_{PP} \end{bmatrix}$ is the cost matrix defined for the problem at hand (where λ_{AB} is the cost of assigning a pattern to the class B when it actually belongs to the class A), the conditional risk associated to the classification of a given sample $\mathbf{x}$ is minimized by a decision rule which assigns the sample $\mathbf{x}$ to the class P if

$$lr(\mathbf{x}) = \frac{f_P(\mathbf{x})}{f_N(\mathbf{x})} > \frac{(\lambda_{NP} - \lambda_{NN}) \pi_N}{(\lambda_{PN} - \lambda_{PP}) \pi_P}$$

where $lr(\mathbf{x})$ is the likelihood ratio evaluated for the sample $\mathbf{x}$. A way to assess the quality of such rule as the costs and the a priori probabilities vary, is to evaluate the performance obtained on each class by the rule $lr(\mathbf{x}) > t$ as the threshold t is varied. For a given threshold value t, two appropriate performance figures are given by the *True Positive Rate* $TPR(t)$, i.e. the fraction of actually-positive cases correctly classified and by the *False Positive Rate* $FPR(t)$, given by the fraction of actually-negative cases incorrectly classified as "positive". If we consider the class-conditional densities of the likelihood ratio $\varphi_P(\tau) = p(lr(\mathbf{x}) = \tau | \mathbf{x} \in P)$ and $\varphi_N(\tau) = p(lr(\mathbf{x}) = \tau | \mathbf{x} \in N)$, $TPR(t)$ and $FPR(t)$ are given by:

$$TPR(t) = \int_{t}^{+\infty} \varphi_P(\tau) \, d\tau \qquad FPR(t) = \int_{t}^{+\infty} \varphi_N(\tau) \, d\tau \qquad (1)$$

Taking into account the samples with likelihood ratio less than t, it is possible to evaluate the *True Negative Rate* $TNR(t)$ and the *False Negative Rate* $FNR(t)$, defined as:

$$TNR(t) = \int_{-\infty}^{t} \varphi_N(\tau) \, d\tau = 1 - FPR(t)$$

$$FNR(t) = \int_{-\infty}^{t} \varphi_P(\tau) \, d\tau = 1 - TPR(t)$$

$$(2)$$

As it is possible to note from eq.(2), the four indices are not independent and the pair $(FPR(t), TPR(t))$ is sufficient to completely characterize the performance of the decision rule for a given threshold t. Most importantly, they are independent of the a priori probability of the classes because they are separately evaluated on the different classes. The *Receiver Operating Characteristic* (*ROC*) curve plots $TPR(t)$ vs. $FPR(t)$ by sweeping the threshold t into the whole real axis, thus providing a description of the performance of the decision rule at different operating points. An important feature of the ROC curve is that the slope of the curve at any point $(FPR(t), TPR(t))$ is equal to the threshold required to achieve the FPR and TPR of that point [6]. Therefore, the corresponding operating point on the ROC curve is the one where the curve has gradient $\frac{(\lambda_{NP} - \lambda_{NN})\pi_N}{(\lambda_{PN} - \lambda_{PP})\pi_P}$; such point can be easily found moving down from above in the ROC plane a line with slope $\frac{(\lambda_{NP} - \lambda_{NN})\pi_N}{(\lambda_{PN} - \lambda_{PP})\pi_P}$ and selecting the point in which the line touches the ROC curve [1]. The ROC generated by the decision rule based on the likelihood ratio is the optimal ROC curve, i.e. the curve which, for each $FPR \in [0, 1]$, has the highest TPR among all possible decision criteria employed for the classification problem at hand. This can be proved if we recall the *Neyman Pearson* lemma which can be stated in this way: if we consider the decision rule $lr(x) > \beta$ with β chosen to give $FPR = \varepsilon$, there is no other decision rule providing a TPR higher than $TPR(\beta)$ with a $FPR \leqslant \varepsilon$. The demonstration of the lemma can be found in [8,9]. The shape of the optimal ROC curve depends on how the class-conditional densities are separated: two perfectly distinguished densities produce an ROC curve that passes through the upper left corner (where $TPR = 1.0$ and $FPR = 0.0$), while the ROC curve generated by two overlapped densities is represented by a 45° diagonal line from the lower left to the upper right corner. Qualitatively, the closer the curve to the upper left corner, the more discriminable the two classes.

2.2 The Empirical Approach

The ideal scenario of the Bayesian Decision Theory considered so far unfortunately cannot be applied to the most part of real cases where we rarely have this kind of complete knowledge about the probabilistic structure of the problem. As a consequence, in real problems the optimal ROC is unknown since the actual class conditional densities are not known. In this case, the decision is performed by means of a trained classifier. Without loss of generality, let us assume that the classifier provides, for each sample $\mathbf{x}$, a value $\omega(\mathbf{x})$ in the range $(-\infty, +\infty)$ which can be assumed as a confidence degree that the sample belongs to one of the two classes, e.g. the class P. The sample should be consequently assigned to the class N if $\omega(\mathbf{x}) \to -\infty$ and to the class P if $\omega(\mathbf{x}) \to +\infty$. Also in this case it is possible to plot an ROC curve by considering the outputs provided by the trained classifier on a validation set V containing n_+ positive samples and n_- negative samples $V = \{p_i \in P, i = 1 \ldots n_+\} \cup \{n_j \in N, j = 1 \ldots n_-\}$. In this way, we obtain an empirical estimator of the optimal ROC curve by evaluating, for each possible value of a threshold t in the range $(-\infty, +\infty)$, the empirical true and false positive rates as follows:

$$TPR(t) = \frac{1}{n_+} \sum_{i=1}^{n_+} S(\omega(p_i) > t) \qquad FPR(t) = \frac{1}{n_-} \sum_{j=1}^{n_-} S(\omega(n_j) > t)$$

where $S(.)$ is a predicate which is 1 when the argument is true and 0 otherwise. Let us call the obtained curve empirical ROC curve (see fig. 1) in order to distinguish it from the ideal ROC.

There are some differences to be highlighted between the empirical and the optimal ROC:

- once the two classes N and P have been specified through their conditional densities $f_P(\mathbf{x})$ and $f_N(\mathbf{x})$, the ideal ROC is unique, while different classifiers trained on the same problem have different empirical ROCs;
- for a continuous likelihood ratio, the ideal ROC is continuous and its slope in a particular point equals the value of the threshold required to achieve TPR and FPR of that point [1]; the empirical ROC is instead a discrete function and the relation between slope and threshold does not hold. However it is still possible to find the optimal operating point also on the empirical ROC by moving down from above in the ROC plane a line with slope and selecting the point in which the line touches the ROC curve. Provost and Fawcett [7] have shown that the point is one of the vertices of the convex hull which contains the empirical ROC curve (see fig. 2);
- the ideal ROC is the optimal ROC curve, i.e. the curve which, for each FPR [0,1], has the highest TPR among all possible decision criteria employed for the classification problem at hand. In other words, the ideal ROC curve dominates every empirical ROC and consequently has the highest *area under the ROC curve (AUC)* attainable.

3   Two-Class Classification with Reject

When dealing with cost sensitive applications which involve a reject option, the possible outcomes of the decision rule include the reject and thus the cost matrix changes accordingly (see tab. 1).

Table 1. Cost matrix for a two-class problem with reject

		Predicted Class		
		N	P	R
True	N	λ_{NN}	λ_{NP}	λ_R
Class	P	λ_{PN}	λ_{PP}	λ_R

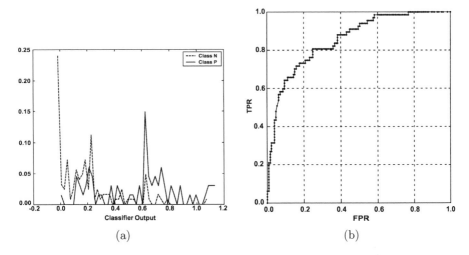

          (a)   (b)

Fig. 1. (a) The densities of the confidence degree obtained by the classifier output on real data and (b) the corresponding ROC curve

It is worth noting that λ_{NN} (cost for a True Negative) and λ_{PP} (cost for a True Positive) are negative costs since related to benefits, while λ_{NN} (cost for a False Negative) and λ_{NP} (cost for a False Positive) are positive costs. λ_R weights the burden of managing the reject (e.g. by calling another, more proficient classifier) and thus it is positive but smaller than the error costs.

3.1 The Ideal Case

Let us firstly suppose that we are working within a Bayesian scenario, i.e. we know the a priori probabilities of the two classes (π_P, π_N) and the class conditional densities $f_P(\mathbf{x})$ and $f_N(\mathbf{x})$. In this ideal setting, the classification cost is minimized by the Chow's rule [2] which can be expressed in terms of the likelihood ratio $lr(\mathbf{x})$ as follows:

$$\mathbf{x} \rightarrow N \quad \text{if} \;\; lr(\mathbf{x}) < \frac{\pi_N}{\pi_P} \frac{\lambda_R - \lambda_{NN}}{\lambda_{PN} - \lambda_R} = u_1$$

$$\mathbf{x} \rightarrow P \quad \text{if} \;\; lr(\mathbf{x}) > \frac{\pi_N}{\pi_P} \frac{\lambda_{NP} - \lambda_R}{\lambda_R - \lambda_{PP}} = u_2 \qquad (3)$$

$$reject \quad \text{if} \;\; u_1 \leq lr(\mathbf{x}) \leq u_2$$

The rule can be also defined in terms of the a *posteriori* probability $\Pr(P \,|\mathbf{x})$. If we recall that

$$\Pr(P \,|\mathbf{x}) = \frac{\pi_P f_P(\mathbf{x})}{\pi_N f_N(\mathbf{x}) + \pi_P f_P(\mathbf{x})} = \frac{\pi_P lr(\mathbf{x})}{\pi_N + \pi_P lr(\mathbf{x})}$$

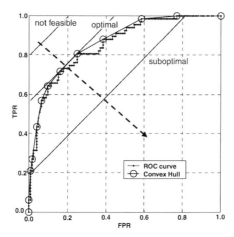

Fig. 2. The ROC curve shown in fig. 1 and its convex hull. Three level lines with the same slope are also shown: the line touching the ROC convex hull determines the optimal operating point since it involves the minimum risk. The line above the optimal line does not determine any feasible point, while the line below identifies only suboptimal points.

the Chow's rule can be written as:

$$
\begin{aligned}
\mathbf{x} \rightarrow N \quad \text{if} \ \Pr\left(P\,|\mathbf{x}\right) &< \frac{\lambda_R - \lambda_{NN}}{\lambda_{PN} - \lambda_{NN}} = t_1 = \frac{\pi_P u_1}{\pi_N + \pi_P u_1} \\
\mathbf{x} \rightarrow P \quad \text{if} \ \Pr\left(P\,|\mathbf{x}\right) &> \frac{\lambda_{NP} - \lambda_R}{\lambda_{NP} - \lambda_{PP}} = t_2 = \frac{\pi_P u_2}{\pi_N + \pi_P u_2} \\
reject \quad \text{if} \ t_1 &\leq \Pr\left(P\,|\mathbf{x}\right) \leq t_2
\end{aligned}
\tag{4}
$$

It is worth noting that in the ideal scenario, the slope of the ROC curve at any point is equal to the threshold on the likelihood ratio which has generated that point [6], and thus the points corresponding to the two thresholds u_1 and u_2 can be easily identified on the ideal ROC.

3.2   The Empirical Approach

In real problems, however, the class conditional densities are not available and thus the optimal decision rule in 3 or in 4 cannot be applied. In such cases, the typical approach is to train a classifier $\omega\left(\mathbf{x}\right)$ on a set of samples representative of the classes to be discriminated and to use it to estimate the class of new samples. Even though the Chow's rule cannot be directly applied, a reject option can be still defined on the empirical ROC, as it has been shown in [3]. The decision rule is still based on two thresholds τ_1 and τ_2 applied on the output of the classifier $\omega\left(\mathbf{x}\right)$:

$$\begin{aligned}
\mathbf{x} &\to N \quad \text{if } \omega(\mathbf{x}) < \tau_1 \\
\mathbf{x} &\to P \quad \text{if } \omega(\mathbf{x}) > \tau_2 \\
reject &\quad \text{if } \tau_1 \leq \omega(\mathbf{x}) \leq \tau_2
\end{aligned} \tag{5}$$

As a consequence, the values of TPR and FPR change as:

$$TPR(\tau_2) = \frac{1}{n_+} \sum_{i=1}^{n_+} S(\omega(p_i) > \tau_2) \qquad FPR(\tau_2) = \frac{1}{n_-} \sum_{j=1}^{n_-} S(\omega(n_j) > \tau_2) \tag{6}$$

It is worth noting that the condition described in eq.(2) is no more satisfied since now there are two thresholds. In fact, the values of TNR and FNR are given by:

$$\begin{aligned}
FNR(\tau_1) &= \frac{1}{n_+} \sum_{i=1}^{n_+} S(\omega(p_i) < \tau_1) \\
TNR(\tau_1) &= \frac{1}{n_-} \sum_{j=1}^{n_-} S(\omega(n_j) < \tau_1)
\end{aligned} \tag{7}$$

Moreover, we have now a portion of samples rejected given by:

$$\begin{aligned}
RP(\tau_1, \tau_2) &= \frac{1}{n_+} \sum_{i=1}^{n_+} S(\tau_1 \leqslant \omega(p_i) \leqslant \tau_2) = 1 - TPR(\tau_2) - FNR(\tau_1) \\
RN(\tau_1, \tau_2) &= \frac{1}{n_-} \sum_{j=1}^{n_-} S(\tau_1 \leqslant \omega(n_j) \leqslant \tau_2) = 1 - TNR(\tau_1) - FPR(\tau_2)
\end{aligned} \tag{8}$$

As a consequence, the classification cost obtained when imposing the threshold τ_1 and τ_2 is given by:

$$\begin{aligned}
C(\tau_1, \tau_2) = {}&\pi_P \cdot \lambda_{PN} \cdot FNR(\tau_1) + \pi_N \cdot \lambda_{NN} \cdot TNR(\tau_1) + \\
&\pi_P \cdot \lambda_{PP} \cdot TPR(\tau_2) + \pi_N \cdot \lambda_{NP} \cdot FPR(\tau_2) + \\
&\pi_P \cdot \lambda_R \cdot RP(\tau_1, \tau_2) + \pi_N \cdot \lambda_R \cdot RN(\tau_1, \tau_2)
\end{aligned} \tag{9}$$

The values of the thresholds should be chosen in order to minimize $C(\tau_1, \tau_2)$; to this aim the classification cost can be written as:

$$C(\tau_1, \tau_2) = k_2(\tau_2) - k_1(\tau_1) + \pi_P \cdot \lambda_{PN} + \pi_N \cdot \lambda_{NN} \tag{10}$$

where:

$$\begin{aligned}
k_1(\tau_1) &= \pi_P \cdot \lambda'_{PN} \cdot TPR(\tau_1) + \pi_N \cdot \lambda'_{NN} \cdot FPR(\tau_1) \\
k_2(\tau_2) &= \pi_P \cdot \lambda'_{PP} \cdot TPR(\tau_2) + \pi_N \cdot \lambda'_{NP} \cdot FPR(\tau_2)
\end{aligned}$$

and

$$\begin{aligned}
\lambda'_{PP} &= \lambda_{PP} - \lambda_R & \lambda'_{PN} &= \lambda_{PN} - \lambda_R \\
\lambda'_{NN} &= \lambda_{NN} - \lambda_R & \lambda'_{NP} &= \lambda_{NP} - \lambda_R
\end{aligned}$$

In this way, the optimization problem can be simplified and the optimal values for thresholds τ_{1opt} and τ_{2opt} can be separately obtained by maximizing $k_1(\tau_1)$ and minimizing $k_2(\tau_2)$:

$$\begin{aligned}
\tau_{1opt} &= \arg\max_\tau \left[\pi_P \cdot \lambda'_{PN} \cdot TPR(\tau_1) + \pi_N \cdot \lambda'_{NN} \cdot FPR(\tau_1) \right] \\
\tau_{2opt} &= \arg\min_\tau \left[\pi_P \cdot \lambda'_{PP} \cdot TPR(\tau_2) + \pi_N \cdot \lambda'_{NP} \cdot FPR(\tau_2) \right]
\end{aligned} \tag{11}$$

As described in [3], the optimal thresholds can be found by considering the empirical ROC evaluated on a validation set; in particular, they correspond to the points T_1 and T_2 of the empirical ROC touched by two lines with slopes:

$$m_1 = -\frac{\pi_N}{\pi_P}\frac{\lambda_{NN} - \lambda_R}{\lambda_{PN} - \lambda_R} \qquad\qquad m_2 = -\frac{\pi_N}{\pi_P}\frac{\lambda_{NP} - \lambda_R}{\lambda_{PP} - \lambda_R} \qquad (12)$$

and the thresholds are the values of the confidence degree which have generated those points (see fig. 3).

It is worth noting that τ_{1opt} must be less than τ_{2opt} to achieve the reject option, i.e. the slopes must be such that $m_1 < m_2$. If $m_1 \geqslant m_2$, the reject option is not practicable and thus the best choice is to work at 0 reject. Taking into account eq.(12), the condition for the reject option to be applicable is $\frac{\lambda_{NN}-\lambda_R}{\lambda_{PN}-\lambda_R} > \frac{\lambda_{NP}-\lambda_R}{\lambda_{PP}-\lambda_R}$. This condition depends only on the cost values; however, there could be situations in which the condition is verified but the geometry of the ROC curve of the classifier at hand is such that the level curves corresponding to m_1 and m_2 touch the same point [3]. In other words, in spite of the costs which could allow the reject, the characteristics of the classifier could make not applicable the reject option.

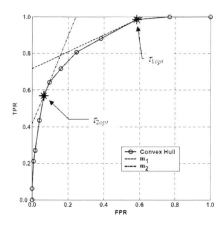

Fig. 3. The ROC curve, the level curves and the optimal thresholds for a given cost combination

3.3 Ideal and Empirical ROC Reject Rules. Are They Equivalent?

It is worth noting that, in the empirical ROC, the values of the thresholds are not an immediate function of the slopes (like in the ideal case) but the relation between the slopes (and the costs) and the threshold values is provided by the geometry of the ROC curve and after all by the output of the classifier. As a consequence, such values change when considering a different classifier.

The empirical rule reduces to the Chow's rule when dealing with the ideal ROC instead of an empirical ROC. In fact, in the ideal case, the lines with slopes in eq. (12) identify two points in which the likelihood ratio has the same value of the two slopes. In other words, $u_2 = m_2$, $u_1 = m_1$ and the reject rule in eq. (5) reduces to the reject rule in eq. (3). This means that the empirical rule is certainly suboptimal with respect to the Chow's rule, but this latter does not work when the ideal setting assumptions (i.e. that the distributions of the two classes are completely known) are not verified and a real dichotomizer should be used instead of the optimal decision rule based on the likelihood ratio or on the *aposteriori* probabilities.

However, on the basis of this observation, authors in [4] claim to demonstrate the theoretical equivalence between the two rules and suggest that the adoption of the ideal thresholds in eq.(3) (or thresholds derived from those in eq. (4) if post probabilities are adopted in the decision rule instead of the likelihood ratio) should produce lower classification costs than those obtained by means of the reject rule based on the empirical ROC. This would mean that the reject thresholds are independent of the classifier chosen or, in other words, that every real classifier can be considered an effective estimator of the true likelihood ratio or of the true post probabilities.

Such assumption does not hold at all for a large class of classification systems such as margin based classifiers which do not provide an estimate of the post probabilities. In these cases the Chow's rule does not work, while the reject rule based on the empirical ROC is still applicable (see, e.g., [10]). However, such assumption seems to be excessive even for classification systems which provide estimates of the likelihood ratio or of the post probabilities (e.g. Multi Layer Perceptrons), since there are many elements (e.g. the limited size of the training set, the learning parameters which are not univocal, etc.) which make the estimate not very accurate.

In particular, the differences between the two rules should be higher and higher in favor of the empirical ROC-based rule as the ideal setting assumption becomes less verified. To experimentally prove such hypothesis, we have designed a set of experiments using both synthetic data sets and real data sets. The synthetic data sets are built by adding noise to the post probabilities generated from some chosen distributions. The aim is to simulate in a controlled way a realistic situation, i.e. a classification problem in which the post probabilities cannot be exactly obtained since the distributions of the two classes are not completely known and must be estimated by means of some method which inevitably provides a certain amount of error. For the real data sets, we train some well known classifiers and we use the outputs of the classifiers as estimates of the post probabilities. For each sample to be classified, the output is compared with the thresholds provided by the two rules thus obtaining two decisions which can be compared.

In the next section we present the methodology adopted in the experiments and the results obtained.

4   Experiments

4.1   Synthetic Data Set

To create an artificial problem, a gaussian model for the distribution of the samples of the two classes has been adopted. In particular, we simulate the output of a classifier ω as $\omega(\mathbf{x}) = Pr(P|\mathbf{x}) + \varepsilon(\mathbf{x})$ where $Pr(P|\mathbf{x})$ is the a posteriori probability of the class P given the input vector $\mathbf{x}$ and $\varepsilon(\mathbf{x})$ is the error associated to that sample. In our framework the distribution of the two classes is supposed to be gaussian with known mean and covariance matrix $\Sigma = I$.

In particular, we generate the likelihood probabilities for the two classes P and N:

$$f_N(\mathbf{x}) = (2\pi)^{-K/2} \exp\left(-\frac{1}{2}(\mathbf{x} - \mu_N)^T (\mathbf{x} - \mu_N)\right)$$

$$f_P(\mathbf{x}) = (2\pi)^{-K/2} \exp\left(-\frac{1}{2}(\mathbf{x} - \mu_P)^T (\mathbf{x} - \mu_P)\right) \tag{13}$$

and from the Bayes theorem we find the a posteriori probability $Pr(P|\mathbf{x})$. Knowing the distribution of the samples it is possible to vary the vector of the mean μ, so as to create different data sets according to a value M that measures the distance between the means of the distributions of the two classes. In this paper, we considered three cases of interest: $M = 4.5$, i.e. the two classes are completely separated; $M = 3$, i.e. the two classes are partially overlapped and $M = 1.5$, i.e. the two classes are quite completely overlapped. Then, the term $\varepsilon(\mathbf{x})$ that simulates the error committed by a classifier is modeled according to two distributions: a gaussian distribution with zero mean and variance varying among 0 and 1 with step 0.1 and an uniform distribution varying among 0 and 1 with step 0.1. For each value of M 1000 samples have been generated and the distributions have been truncated so that each output ω is in the interval [0,1].

4.2   Real Data Set

Four data sets publicly available at the UCI Machine Learning Repository [11] have been used in the following experiments; all of them have two output classes and numerical input features. All the features were previously rescaled so as to have zero mean and unit standard deviation. More details for each data sets are reported in table 2. The employed classifiers are neural networks and Support Vector Machines (SVM). In particular, four Multi Layer Perceptron (MLP) with a variable number of units in the hidden layer between two and five have been trained for 10000 epochs using the back propagation algorithm with a learning rate of 0.01 and four Radial Basis Function (RBF) have been built with a variable number of units in the hidden layer between two and five. Then, four Support Vector Machine (SVM) with different kernels have been used;in particular, the kernels used were linear, polynomial of degree 2, RBF with $\sigma = 1$ and sigmoidal with $\sigma = 0.01$.

Table 2. Data sets used in the experiments

Data Sets	Features	Samples	% Major Class	Train Set	Valid. Set	Test Set
Pima	8	768	65.1	538	115	115
German	24	1000	70.0	700	150	150
CMC	9	1473	57.3	1031	221	221
Heart	13	303	54.1	213	45	45

4.3   Results

In the comparison of the two reject rules 12 runs of a multiple hold out procedure have been performed to reduce the bias in the data. In each run, each data set has been divided into three subsets: a training set used to train the classifiers, a validation set to evaluate the thresholds of the empirical reject rule and a test set to compare the two methods. In the experiments with artificial data we had only the validation and the test set containing respectively the 23% and the 77% of the whole data set. In the experiments with real data, the three subsets contain respectively the 70%, the 15% and the 15% of the samples of the whole data set. In this way, 12 different values of the required costs have been obtained.

Another possible cause of bias is given by the employed cost matrix. To achieve a result independent of the particular cost values, we have used a matrix (called cost model) in which each cell contains a distribution instead of a fixed value. In this way, 1000 different cost matrices have been randomly generated on the basis of the cost model adopted. In our experiments, an uniform distribution over the interval $[-10, 0]$ for λ_{PP} and λ_{NN}, over the interval $[0, 50]$ for λ_{NP} and λ_{PN} and over the interval $[0, 30]$ for the reject cost λ_R.

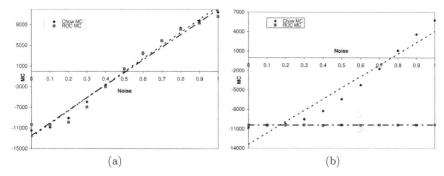

Fig. 4. Results obtained on artificial data sets: (a) $M = 4.5$ and additive Gaussian noise, (b) $M = 4.5$ and additive uniform noise

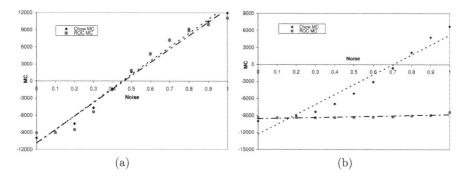

Fig. 5. Results obtained on artificial data sets: (a) $M = 3$ and additive Gaussian noise, (b) $M = 3$ and additive uniform noise

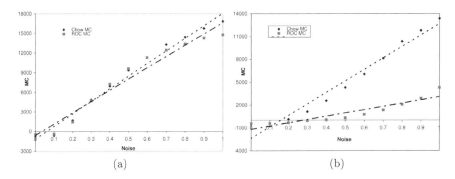

Fig. 6. Results obtained on artificial data sets: (a) $M = 1.5$ and additive Gaussian noise, (b) $M = 1.5$ and additive uniform noise

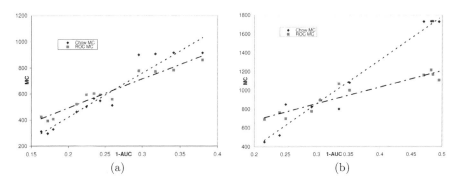

Fig. 7. Results obtained on real data sets: (a) Pima, (b) German

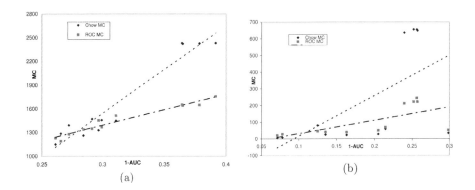

Fig. 8. Results obtained on real data sets: (a) CMC, (b) Heart

The obtained results are shown in figs. 4-6 for the synthetic data and in figs. 7-8 for the real data sets. In both cases we report the comparison in terms of mean cost (MC) intended as the average of the classification costs obtained on the 12 runs of the hold out procedure and on the 1000 cost matrices employed on the considered problem. To order the classifier performance in the artificial case we refer to the added noise since we have a complete knowledge of the post probabilities and in particular we refer to the variance for the Gaussian distribution and to the width of the interval in the uniform distribution. However, while in the synthetic model we know the effective distribution of the classes and so it is possible to relate the obtained results to the noise added to the true post probabilities, when dealing with the real data sets we cannot do the same unless we have a measure of the accuracy with which the real classifier estimates the true post probabilities. To this aim, the AUC can be seen as a reliable estimate of the discriminating quality of the classifier [12]. Moreover, since the ideal ROC curve represents the "upper bound" of any empirical ROC curves (i.e. it is dominant with respect to any empirical ROC curve), we can reasonably assume that the greater is the AUC, the closer is the empirical to the ideal ROC curve and the better is the estimate of the true post probabilities. For this reason, in the graphs the value 1-AUC is reported on x-axis to be consistent with the previous figures. In each graph, beyond the scatter plot of the mean costs values also the regression lines are reported to emphasize the trend of the mean costs obtained by the two analyzed rules.

If we look at the behavior of the two rules on the synthetic data sets it is possible to note that the Chow reject rule outperforms the ROC rule only when the added noise is low. On the contrary, when the noise becomes higher the empirical ROC rule becomes better since the estimate of the post probabilities becomes worse and worse. This behavior is more visible when the added noise follows an uniform distribution (figs. 4-(b), 5-(b), 6-(b)) while a similar behavior is shown if the added noise is gaussian (figs. 4-(a), 5-(a), 6-(a)) because it produces less bias in the data.

The same behavior obtained on the artificial data is shown on the real data sets (figs. 7, 8) where the improvement obtained with the ROC rule is very evident when AUC decreases, i.e. when the classifier is not able to estimate a reliable post probability for the two classes and the ideal conditions are less verified.

5 Conclusions and Future Work

In this paper we have experimentally compared the Chow's reject rule and the ROC based reject rule presented in [3]. Despite what claimed in [4] we have found that the Chow's rule is inappropriate when the estimates of the a posteriori probabilities are not sufficiently accurate, while the ROC based reject rule gives good results. One could argue that such result could be not surprising, but we believe that the strong assertion about the robustness of the Chow's rule made in [4] is worth a critical analysis based on the evidence of specific experiments besides theoretical arguments. Finally, the analysis begun in this paper has pointed out the need of a further investigation to characterize the type of situations when one rule has advantage over another.

References

1. Webb, A.R.: Statistical Pattern Recognition. John Wiley and Sons Ltd, West Sussex (2002)
2. Chow, C.K.: On optimum recognition error and reject tradeoff. IEEE Trans. Information Theory, IT10, 41–46 (1970)
3. Tortorella, F.: A ROC-based reject rule for dichotomizers. Pattern Recognition Letters 26, 167–180 (2005)
4. Santos-Pereira, C.M., Pires, A.M.: On optimal reject rules and ROC curves, Pattern Recognition Letters. Pattern Recognition Letters 26, 943–952 (2005)
5. Xie, J., Qiu, Z., Wu, J.: Bootstrap methods for reject rules of Fisher LDA. In: Proc. 18th Int. Conf. on Pattern Recognition, 3rd edn. pp. 425–428. IEEE Press, NJ (2006)
6. van Trees, H.L.: Detection, Estimation, and Modulation Theory. Wiley, New York (1968)
7. Provost, F., Fawcett, T.: Robust classification for imprecise environments. Machine Learning 42, 203–231 (2001)
8. Mukhopadhyay, N.: Probability and Statistical Inference. Marcel Dekker Inc. New York (2000)
9. Garthwaite, P.H., Jolliffe, I.T., Jones, B.: Statistical Inference, 2nd edn. Oxford University Press, Oxford (2002)
10. Tortorella, F.: Reducing the classification cost of support vector classifiers through an ROC-based reject rule. Pattern Analysis and Applications 7, 128–143 (2004)
11. Blake, C., Keogh, E., Merz, C.J.: UCI Repository of Machine Learning Databases, Irvine, University of California, Department of Information and Computer Science (1998), http://www.ics.uci.edu/mlearn/MLRepository.html
12. Huang, J., Ling, C.X.: Using AUC and accuracy in evaluating learning algorithms, IEEE Trans. Knowledgde and Data Engineering 17, 299–310 (2005)

Outlier Detection with Kernel Density Functions

Longin Jan Latecki[1], Aleksandar Lazarevic[2], and Dragoljub Pokrajac[3]

[1] CIS Dept. Temple University Philadelphia, PA 19122, USA
latecki@temple.edu
[2] United Technology Research Center 411 Silver Lane, MS 129-15 East Hartford,
CT 06108, USA
aleks@cs.umn.edu
[3] CIS Dept. CREOSA and AMRC, Delaware State University
Dover DE 19901, USA
dpokrajac@desu.edu

Abstract. Outlier detection has recently become an important problem in many industrial and financial applications. In this paper, a novel unsupervised algorithm for outlier detection with a solid statistical foundation is proposed. First we modify a nonparametric density estimate with a variable kernel to yield a robust local density estimation. Outliers are then detected by comparing the local density of each point to the local density of its neighbors. Our experiments performed on several simulated data sets have demonstrated that the proposed approach can outperform two widely used outlier detection algorithms (LOF and LOCI).

1 Introduction

Advances in data collection are producing data sets of massive size in commerce and a variety of scientific disciplines, thus creating extraordinary opportunities for monitoring, analyzing and predicting global economical, demographic, medical, political and other processes in the World. However, despite the enormous amount of data available, particular events of interests are still quite rare. These rare events, very often called outliers or anomalies, are defined as events that occur very infrequently (their frequency ranges from 5% to less than 0.01% depending on the application). Detection of outliers (rare events) has recently gained a lot of attention in many domains, ranging from video surveillance and intrusion detection to fraudulent transactions and direct marketing. For example, in video surveillance applications, video trajectories that represent suspicious and/or unlawful activities (e.g. identification of traffic violators on the road, detection of suspicious activities in the vicinity of objects) represent only a small portion of all video trajectories. Similarly, in the network intrusion detection domain, the number of cyber attacks on the network is typically a very small fraction of the total network traffic. Although outliers (rare events) are by definition infrequent, in each of these examples, their importance is quite high compared to other events, making their detection extremely important.

P. Perner (Ed.): MLDM 2007, LNAI 4571, pp. 61–75, 2007.

Data mining techniques that have been developed for this problem are based on both supervised and unsupervised learning. Supervised learning methods typically build a prediction model for rare events based on labeled data (the training set), and use it to classify each event [1,2]. The major drawbacks of supervised data mining techniques include: (1) necessity to have labeled data, which can be extremely time consuming for real life applications, and (2) inability to detect new types of rare events. On the other hand, unsupervised learning methods typically do not require labeled data and detect outliers (rare events) as data points that are very different from the normal (majority) data based on some pre-specified measure [3]. These methods are typically called outlier/anomaly detection techniques, and their success depends on the choice of similarity measures, feature selection and weighting, etc. Outlier/anomaly detection algorithms have the advantage that they can detect new types of rare events as deviations from normal behavior, but on the other hand suffer from a possible high rate of false positives, primarily because previously unseen (yet normal) data are also recognized as outliers/anomalies, and hence flagged as interesting.

Outlier detection techniques can be categorized into four groups: (1) statistical approaches; (2) distance based approaches; (3) profiling methods; and (4) model-based approaches. In statistical techniques [3,6,7], the data points are typically modeled using a stochastic distribution, and points are labeled as outliers depending on their relationship with the distributional model.

Distance based approaches [8,9,10] detect outliers by computing distances among points. Several recently proposed distance based outlier detection algorithms are founded on (1) computing the full dimensional distances among points using all the available features [10] or only feature projections [8]; and (2) on computing the densities of local neighborhoods [9,35]. Recently, LOF (Local Outlier Factor) [9] and LOCI (Local Correlation Integral) [35] algorithms have been successfully applied in many domains for outlier detection in a batch mode [4,5,35]. In addition, clustering-based techniques have also been used to detect outliers either as side products of the clustering algorithms (as points that do not belong to clusters) [11] or as clusters that are significantly smaller than others [12].

In profiling methods, profiles of normal behavior are built using different data mining techniques or heuristic-based approaches, and deviations from them are considered as outliers (e.g., network intrusions). Finally, model-based approaches usually first characterize the normal behavior using some predictive models (e.g. replicator neural networks [13] or unsupervised support vector machines [4,12]), and then detect outliers as deviations from the learned model.

In this paper, we propose an outlier detection approach that can be classified both into statistical and density based approaches, since it is based on local density estimation using kernel functions. Our experiments performed on several simulated data sets have demonstrated that the proposed approach outperforms two very popular density-based outlier detection algorithms, LOF [9] and LOCI [35].

2   Local Density Estimate

We define outlier as an observation that deviates so much from other observations to arouse suspicion that it was generated by a different mechanism [13]. Given a data set $D = \{\mathbf{x}_1, \mathbf{x}_2, ..., \mathbf{x}_n\}$, where n is the total number of data samples in Euclidean space of dimensionality dim, we propose the algorithm that can identify all outliers in the data set D. Our first step is to perform density estimate. Since we do not make any assumption about the type of the density, we use a nonparametric kernel estimate [39] to estimate the density of *majority* data points $q(\mathbf{x})$, also referred to as a ground truth density. Consequently, all data samples that appear not to be generated by the ground truth density $q(\mathbf{x})$ may be considered as potential outliers.

However, it is impossible to directly use density estimate to identify outliers if the estimated distribution is multimodal, which mostly is the case. Data points belonging to different model components may have different density without being outliers. Consequently, normal points in some model components may have lower density than outliers around points from different model components.

In order to detect outliers, we compare the estimated density at a given data points to the average density of its neighbors. This comparison forms the basis of most unsupervised outlier detection methods, in particular of LOF [9]. The key difference is that we compare densities, which have solid statistical foundation, while the other methods compare some local properties that are theoretically not well understood.

One of our main contributions is to provide proper evaluation function that makes outlier detection based on density estimate possible.

There is a large body of published literature on non-parametric density estimation [39]. One of the best-working non-parametric density estimation methods is the variable width kernel density estimator [39]. In this method, given n data samples of dimensionality dim, the distribution density can be estimated as:

$$\tilde{q}(\mathbf{x}) = \frac{1}{n} \sum_{i=1}^{n} \frac{1}{h(\mathbf{x}_i)^{dim}} K\left(\frac{\mathbf{x} - \mathbf{x}_i}{h(\mathbf{x}_i)}\right), \tag{1}$$

where K is a kernel function (satisfying non-negativity and normalization conditions) and $h(\mathbf{x}_i)$ are the bandwidths implemented at data points $\mathbf{x}_i$. One of the main advantages of this sample smoothing estimator is that $\tilde{q}(\mathbf{x})$ is automatically a probability density function [39] if K is a probability density function. In our case, K is a multivariate Gaussian function of dimensionality dim with zero mean and unit standard deviation:

$$K(\mathbf{x}) = \frac{1}{(2\pi)^{dim}} \exp\left(-\frac{||\mathbf{x}||^2}{2}\right), \tag{2}$$

where $||\mathbf{x}||$ denotes the norm of the vector. The simplest version of the bandwidth function $h(\mathbf{x}_i)$ is a constant function $h(\mathbf{x}_i) = h$, where h is a fixed bandwidth. However, for real data sets, local sample density may vary. Therefore, it is necessary to have a method that is adaptive to the local sample density. This may be achieved for $h(\mathbf{x}_i) = h d_k(\mathbf{x}_i)$, where $d_k(\cdot)$ denotes the distance to the kth nearest neighbor of point $\mathbf{x}_i$. The usage of the kth nearest neighbor in kernel density estimation was first proposed in [38] (see also [37]).

Since we are interested in detecting outlier data samples based on comparing them to their local neighborhood, the sum in Eq. 1 needs only to be taken over a sufficiently large neighborhood of a point x. Let $mNN(\mathbf{x})$ denotes the m nearest neighbors of a sample $\mathbf{x}$. Thus, from Eq. 1 and 2 we obtain the following formula for distribution density at data sample $\mathbf{x}_j$:

$$\tilde{q}(\mathbf{x}_j) \propto \frac{1}{m} \sum_{\mathbf{x}_i \in mNN(\mathbf{x}_j)} \frac{1}{h(\mathbf{x}_i)^{dim}} K\left(\frac{\mathbf{x}_j - \mathbf{x}_i}{h(\mathbf{x}_i)}\right)$$

$$= \frac{1}{m} \sum_{\mathbf{x}_i \in mNN(\mathbf{x}_j)} \frac{1}{(2\pi)^{\frac{dim}{2}} h(\mathbf{x}_i)^{dim}} \exp\left(-\frac{d(\mathbf{x}_j, \mathbf{x}_i)^2}{2h(\mathbf{x}_i)^2}\right). \tag{3}$$

Here,

$$d(\mathbf{x}_j, \mathbf{x}_i) = ||\mathbf{x}_j - \mathbf{x}_i||^2 \tag{4}$$

is the squared Euclidean distance between samples $\mathbf{x}_i$ and $\mathbf{x}_j$. Restricting the sum in Eq. 1 to a local neighborhood as in Eq. 3 has a computational advantage. While the computation of $\tilde{q}$ for all data points has $O(n^2)$ complexity, the average computation in Eq. 3 can be accomplished in $O(mn \log n)$ time, where n is the number of data samples in a data set D and $O(m \log n)$ refers to the cost of search for m nearest neighbors of a data sample if a hierarchical indexing structure like R-tree is used [46].

Observe that Euclidean distance from Eq. 4 may be very small if there is a neighbor $\mathbf{x}_i$ very close to sample $\mathbf{x}_j$. In such a case, it is possible to misleadingly obtain a large density estimate $\tilde{q}(\mathbf{x}_j)$. To prevent such issues and increase the robustness of the density estimation, following the LOF approach [9], we compute reachability distance for each sample $\mathbf{y}$ with respect to data point $\mathbf{x}$ as follows:

$$\mathbf{rd}_k(\mathbf{y}, \mathbf{x}) = \max(d(\mathbf{y}, \mathbf{x}), d_k(\mathbf{x})), \tag{5}$$

where $d_k(x)$ is the distance to kth nearest neighbor of point x. Eq. 5 prevents the distance from $\mathbf{y}$ to $\mathbf{x}$ to become too small with respect to the neighborhood of point $\mathbf{x}$.

We obtain our local density estimate (LDE) by replacing the Euclidean distance in Eq. 3 with the reachability distance:

$$LDE(\mathbf{x}_j) \propto \frac{1}{m} \sum_{\mathbf{x}_i \in mNN(\mathbf{x}_j)} \frac{1}{(2\pi)^{\frac{dim}{2}} h(\mathbf{x}_i)^{dim}} \exp\left(-\frac{\mathbf{rd}_k(\mathbf{x}_j, \mathbf{x}_i)^2}{2h(\mathbf{x}_i)^2}\right)$$

$$= \frac{1}{m} \sum_{\mathbf{x}_i \in mNN(\mathbf{x}_j)} \frac{1}{(2\pi)^{\frac{dim}{2}} (h \cdot d_k(\mathbf{x}_i))^{dim}} \exp\left(-\frac{\mathbf{rd}_k(\mathbf{x}_j, \mathbf{x}_i)^2}{2(h \cdot d_k(\mathbf{x}_i))^2}\right). \tag{6}$$

The name of local density estimate (LDE) is justified by the fact that we sum over a local neighborhood mNN compared to the sum over the whole data set commonly used to compute the kernel density estimate (KDE), as shown in Eq. 1.

LDE is not only computationally more efficient than the density estimate in Eq. 1 but yields more robust density estimates. LDE is based on the ratio of two kinds of distances: the distance from a point $\mathbf{x}_j$ to its neighbors $\mathbf{x}_i$ and distances of the neighboring points $\mathbf{x}_i$ to their k-th neighbors. Namely, the exponent term in Eq. 6 is a function of the ratio $\frac{\mathbf{rd}_k(\mathbf{x}_j,\mathbf{x}_i)}{d_k(\mathbf{x}_i)}$, which specifies how is the reachability distance from $\mathbf{x}_j$ to $\mathbf{x}_i$ related to the distance to the k-th nearest neighbor of $\mathbf{x}_i$. In fact, we use $d_k(\mathbf{x}_i)$ as a "measuring unit" to measure the Euclidean distance $d(\mathbf{x}_j,\mathbf{x}_i)$. If $d(\mathbf{x}_j,\mathbf{x}_i) \leq d_k(\mathbf{x}_i)$, then the ratio $\frac{\mathbf{rd}_k(\mathbf{x}_j,\mathbf{x}_i)}{d_k(\mathbf{x}_i)}$ is equal to one (since $\mathbf{rd}_k(\mathbf{x}_j,\mathbf{x}_i) = d_k(\mathbf{x}_i)$), which yields the maximal value of the exponential function $(\exp(-\frac{1}{2h^2}))$. Conversely, if $d(\mathbf{x}_j,\mathbf{x}_i) > d_k(\mathbf{x}_i)$, then the ratio is larger than one, which results in smaller values of the exponent part.

The bandwidth h specifies how much weight is given to $d_k(\mathbf{x}_i)$. The larger h, the more influential are the k nearest neighbors that are further away. The smaller h, the more we focus on k nearest neighbors.

Observe that we compare a given point $\mathbf{x}_j$ to its neighbors in $mNN(\mathbf{x}_j)$. It is important that the neighborhood $mNN(\mathbf{x}_j)$ is not too small (otherwise, the density estimation would not be correct). Overly large m does not influence the quality of the results, but it influences the computing time (to retrieve m nearest neighbors).

Having presented an improved local version of a nonparametric density estimate, we are ready to introduce our method to detect outliers based on this estimate. In order to be able to use LDE to detect outliers, the local density values $LDE(\mathbf{x}_j)$ need to be related to the LDE values of neighboring points. We define **Local Density Factor (LDF)** at a data point as the ratio of average LDE of its m nearest neighbors to the LDE at the point:

$$LDF(\mathbf{x}_j) \propto \frac{\sum_{\mathbf{x}_i \in mNN(\mathbf{x}_j)} \frac{LDE(\mathbf{x}_i)}{m}}{LDE(\mathbf{x}_j) + c \cdot \sum_{\mathbf{x}_i \in mNN(\mathbf{x}_j)} \frac{LDE(\mathbf{x}_i)}{m}}. \tag{7}$$

Here, c is a scaling constant (in all our experiments we used $c = 0.1$). The scaling of LDE values by c is needed, since $LDE(\mathbf{x}_j)$ may be very small or even equal to zero (for numerical reasons), which would result in very large or even infinity values of LDF if scaling is not performed, i.e., if $c = 0$ in Eq. 7. Observe that the LDF values are normalized on the scale from zero to $1/c$. Value zero means that $LDE(\mathbf{x}_j) \gg \sum_{\mathbf{x}_i \in mNN(\mathbf{x}_j)} \frac{LDE(\mathbf{x}_i)}{m}$ while value $1/c$ means that $LDE(\mathbf{x}_j) = 0$. The higher the LDF value at a given point (closer to $1/c$) the more likely the point is an outlier.

The normalization of LDE values makes possible to identify outliers with a threshold $LDF(\mathbf{x}_j) > T$ chosen independently for a particular data set.

Observe that it is possible to use the Eq. 6 with covariance matrix of the Gaussian that automatically adjusts to the shape of the whole neighborhood mNN. Let $\mathbf{\Sigma}_i$ be the covariance matrix estimated on the m data points in

$mNN(\mathbf{x}_i)$. If we use a general Gaussian kernel with covariance matrices $\mathbf{\Sigma}_i$, then Eq. 3 becomes:

$$\tilde{q}(\mathbf{x}_j) \propto \sum_{i=1}^{n} \frac{1}{h^{dim} |\mathbf{\Sigma}_i|^{\frac{1}{2}}} \exp(-\frac{d_{\Sigma}(\mathbf{x}_j, \mathbf{x}_i)^2}{2h^2}), \tag{8}$$

where $d_{\Sigma}(\mathbf{x}, \mathbf{y})^2 = (\mathbf{x} - \mathbf{y})^T \mathbf{\Sigma}_i^{-1}(\mathbf{x} - \mathbf{y})$ is the Mahalanobis distance of vectors $\mathbf{x}$ and $\mathbf{y}$. It can be shown that

$$d_{\Sigma}(\mathbf{x}, \mathbf{y})^2 = (\mathbf{x}^* - \mathbf{y}^*)^T \cdot (\mathbf{x}^* - \mathbf{y}^*). \tag{9}$$

Here,

$$\mathbf{x}^* \equiv (\mathbf{\Lambda}^T)^{-\frac{1}{2}} \mathbf{V}^T (\mathbf{x}^* - \mu), \tag{10}$$

where $\Lambda = diag(\lambda_1, \ldots, \lambda_k)$ is the diagonal matrix of eigenvalues and $\mathbf{V} = [\mathbf{v}_1, \ldots, \mathbf{v}_k]$ is the matrix od corresponding eigenvectors of $\mathbf{\Sigma}_i$ and μ is the mean of the vectors in the mNN neighborhood. Therefore, Eq. 8 can be, using Eq. 9 and Eq. 4 represented in the form:

$$\tilde{q}(\mathbf{x}_j) \propto \sum_{i=1}^{n} \frac{1}{h^{dim} |\mathbf{\Sigma}_i|^{\frac{1}{2}}} \exp(-\frac{d(\mathbf{x}^*_j, \mathbf{x}^*_i)^2}{2h^2}). \tag{11}$$

Now, analogous to Eq.6, we may generalize the LDE measure to:

$$LDE(\mathbf{x}_j) \propto \frac{1}{m} \sum_{\mathbf{x}_i \in NN(\mathbf{x}_j)} \frac{1}{(2\pi)^{\frac{dim}{2}} h^{dim} |\mathbf{\Sigma}_i|^{\frac{1}{2}}} \exp(-\frac{\mathbf{rd}_k(\mathbf{x}^*_j, \mathbf{x}^*_i)^2}{2h^2}) \tag{12}$$

Equation 12 can be replaced within Eq. 7 to obtain generalized measure of the local density factor.

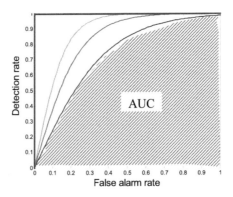

Fig. 1. The ROC curves for different detection algorithms

3   Performance Evaluation

Outlier detection algorithms are typically evaluated using the detection rate, the false alarm rate, and the ROC curves [44]. In order to define these metrics, let us look at a confusion matrix, shown in Table 1. In the outlier detection problem, assuming class "C" as the outlier or the rare class of the interest, and "NC" as a normal (majority) class, there are four possible outcomes when detecting outliers (class "C")-namely true positives (TP), false negatives (FN), false positives (FP) and true negatives (TN). From Table 1, detection rate and false alarm rate may be defined as follows:

$$DetectionRate = \frac{TP}{TP + FN}$$

$$FalseAlarmRate = \frac{FP}{FP + TN}.$$

Table 1. Confusion matrix defines four possible scenarios when classifying class "C"

	Predicted Outliers –Class C	Predicted Normal Class–NC
Actual Outliers –Class C	True Positives (TP)	False Negatives (FN)
Actual Normal –Class NC	False Positives (FP)	True Negatives (TN)

Detection rate gives information about the relative number of correctly identified outliers, while the false alarm rate reports the number of outliers misclassified as normal data records (class NC). The ROC curve represents the trade-off between the detection rate and the false alarm rate and is typically shown on a $2-D$ graph (Fig. 1), where false alarm rate is plotted on x-axis, and detection rate is plotted on y-axis. The ideal ROC curve has 0% false alarm rate, while having 100% detection rate (Fig. 1). However, the ideal ROC curve is hardly achieved in practice. The ROC curve can be plotted by estimating detection rate for different false alarm rates (Fig. 1). The quality of a specific outlier detection algorithm can be measured by computing the area under the curve (AUC) defined as the surface area under its ROC curve. The AUC for the ideal ROC curve is 1, while AUCs of "less than perfect" outlier detection algorithms are less than 1. In Figure 1, the shaded area corresponds to the AUC for the lowest ROC curve.

4   Experiments

In this section, we compare the performance of the proposed LDF outlier detection measures (Eq. 7) to two state of the art outlier detection algorithms

LOF [9] and LOCI [35] on several synthetic data sets. In all of our experiments, we have assumed that we have information about the normal behavior (normal class) and rare events (outliers) in the data set. However, we did not use this information in detecting outliers, i.e. we have used completely unsupervised approach.

Recall that LOF algorithm [9] has been designed to properly identify outliers as data samples with small local distribution density, situated in vicinity of dense clusters. To compare LOF to our proposed LDF algorithm, we created two data sets *Dataset*1 and *Dataset*2. *Dataset*1 shown in Fig. 2(a) has two clusters of non-uniform density and sizes (with 61 and 27 data samples) and two clear outliers A and B (marked with stars in Fig. 2(a)). Data sample C does not belong to the second cluster, but as argued in [9] is should not be regarded as an outlier, since its local density is similar to its neighbors' local densities. Although points A and C have equal distances to their closest clusters (*cluster*1 and *cluster*2 correspondingly), the difference in clusters density suggests that A is an outlier while C is a normal data point. Recall that one of the main motivations for LOF in [9] is based on a data set of this kind.

As shown in Fig. 3, both methods LOF and the proposed LDF correctly identify the outliers A and B in *Dataset*1 without classifying C as an outlier (as in all figures presented here, the larger the circle and the darker its color, the higher the outlier factor value). However, observe that LOF assigns a significantly smaller *LOF* value to point B than A. This is counter intuitive, since point B is definitely the strongest outlier, and may lead to incorrect outlier detection results.

We illustrate the main problem of LOF on the second data set with two clusters of different densities shown in Fig. 2(b). The data set contains 41 points in sparse *cluster*1, 104 points in the dense *cluster*2, and four outstanding outliers A, B, C and D (marked with stars). While samples C and D are clearly outliers, we regard samples A and B as outliers in analogy to sample A from *Dataset*1 (see Fig. 2(a)). Like sample A in *Dataset*1, their local density is lower then the local density of their neighbors from *cluster*1. In other words, samples A and B are too far from the closet cluster to be regarded as normal data points. However, the outlier values for points C and D should be significantly larger than for points A and B.

LOF was not able to detect points A and B as outliers for any value of its parameter k. We illustrate this fact in Fig. 4 for $k = 5$ and 20. Observe also that for larger k values, the *LOF* value of point C actually decreases. In contrast, as shown in Fig.5(a), LDF is able to clearly identify all four outliers. Fig. 5 also illustrates a multiscale behavior of LOF as a function of the bandwidth parameter h. For small values of h, more weight is given to close neighbors of a sample, while for larger values of h, the more distant neighbors also receive higher weight. In other words, with smaller h values, we have higher sensitivity to local situations, and therefore are able to detect all four outliers in Fig. 5(a) for $h = 0.5$. In contrast, with larger h, we smooth local variations.

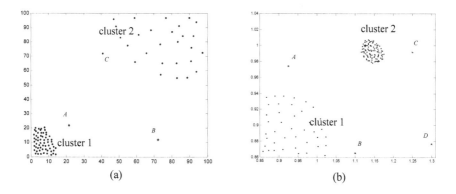

(a)   (b)

Fig. 2. Two simulated data sets with two clusters of different densities. (a) *Dataset1*: Two outliers A and B are marked with stars. (b) *Dataset2*: Four outliers are marked with A, B, C, and D.

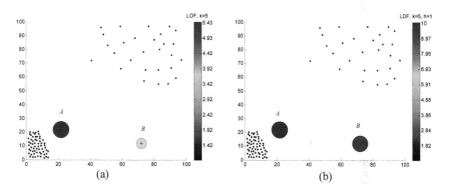

(a)   (b)

Fig. 3. Results on two cluster data set in Fig. 2(a) for $k = 5$: (a) LOF. (b) LDF

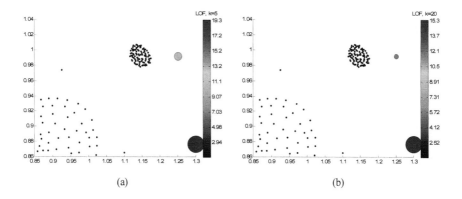

(a)   (b)

Fig. 4. LOF results on the data set in Fig. 2(b) for $k = 5$ and 20

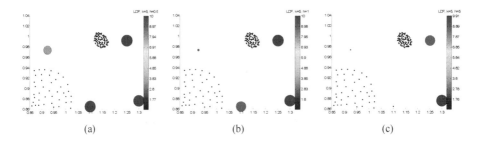

(a)   (b)   (c)

Fig. 5. LDF results on two cluster data set in 2(b) for $k = 5$ and bandwidth (a) $h = 0.5$, (b) $h = 1$, (c) $h = 5$

Consequently, for $h = 5$, LDF detects only two outliers, while for $h = 1$, LDF detects all four outliers, while assigning higher LDF values for the two clear outliers C and D.

To further compare the results of the proposed algorithm with existing algorithms [9,35], we generated synthetic data sets similar to those used in [35] (original data from this reference were not available to us). The data set $Dens$ contains two uniformly distributed rectangular clusters (coordinates $(12, 22; 15, 25)$ and $(80, 120; 30, 70)$ respectively) with 200 samples in each and one outlier at coordinates $(32, 20)$. The second data set $Multimix$ contains a Gaussian cluster, two uniform clusters, 3 outstanding outliers (with coordinates $(80, 110)$, $(85, 110)$ and $(20, 50)$ and three points linearly positioned on top of the uniform cluster. The Gaussian cluster has 250 samples with mean at $(20,110)$ and diagonal covariance matrix with both variances equal to 5. The first uniform cluster has 400 samples uniformly distributed in the rectangle $(130, 150; 95, 105)$. The second uniform cluster had 200 points uniformly distributed in the circle with center at $(80, 50)$ and radius 20.

In Fig. 6(a,b), we demonstrate the performance of the LDF algorithm with parameters $h = 1, k = 10, m = 30$ on these data sets. We compare results of the proposed algorithm with LOF [9]. Fig. 6(c,d) contains results of executing LOF algorithm for the same value of $k = 10$.

As we can see, the proposed LDF and the LOF algorithm performed similarly. LDF values for samples on the boundaries of the Gaussian cluster of $Multimix$ tend to be higher, but the computed high rank correlation [48] between LDF and LOF values (0.85) indicates similar order performance (since the outlier detection is performed by thresholding). We also compare the performance of the proposed algorithm with exact LOCI algorithm with parameters suggested in [35]. LOCI results are shown in Fig. 6(e,f) for $n_{min} = 20, \alpha = 2, k_\sigma = 2$. The visualization in Fig. 6(e,f) is different from (a-d), since LOCI outputs only a binary classification for each data point (outlier or not an outlier). As can be clearly seen, LOCI has trouble with data points on cluster boundaries. It tends to identify samples on boundaries of clusters as outliers.

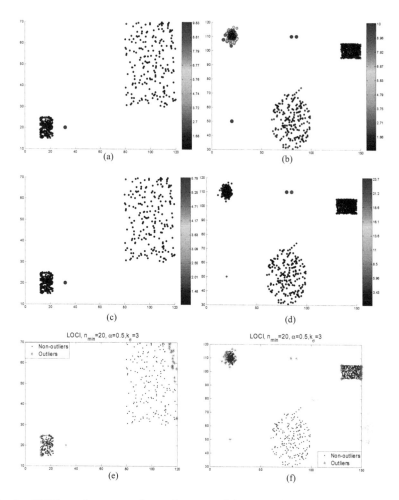

Fig. 6. LDF results on synthetic data sets (a) *Dens* (b) *Multimix*. Corresponding results for LOF are in (c) and (d), and for LOCI in (e) and (f)

5   Conclusions

A novel outlier detection framework is presented that is closely related to statistical nonparametric density estimation methods. Experimental results on several synthetic data sets indicate that the proposed outlier detection method can result in better detection performance than two state-of-the-art outlier detection algorithms (LOF and LOCI). Data sets used in our experiments contained different percentage of outliers, different sizes and different number of features, thus providing a diverse test bed and illustrating wide capabilities of the proposed framework. Although performed experiments have provided evidence that the proposed method can be very successful for the outlier detection task, future work is needed to fully characterize the method in real life data, especially in very large and high dimensional databases, where new methods for estimating data densities are worth considering. It would also be interesting to examine the influence of irrelevant features to detection performance of LDF method as well as to investigate possible algorithms for selecting relevant features for outlier detection task.

6   Reproducible Results Statement

All data sets used in this work are available by emailing the authors.

Acknowledgments

D. Pokrajac has been partially supported by NIH (grant $\sharp 2P20RR016472 - 04$), DoD/DoA (award 45395-MA-ISP) and NSF (awards $\sharp$ 0320991, $\sharp$0630388, $\sharp$ HRD-0310163). L. J. Latecki was supported in part by the NSF Grant IIS-0534929 and by the DOE Grant $DE - FG52 - 06NA27508$.

We would like to thank Guy Shechter for providing his KDTREE software in Matlab and for extending it by a KDNN function.

References

1. Joshi, M., Agarwal, R., Kumar, V., Nrule, P.: Mining Needles in a Haystack: Classifying Rare Classes via Two-Phase Rule Induction. In: Proceedings of the ACM SIGMOD Conference on Management of Data, Santa Barbara, CA (May 2001)
2. Chawla, N., Lazarevic, A., Hall, L., Bowyer, K.: SMOTEBoost: Improving the Prediction of Minority Class in Boosting. In: Lavrač, N., Gamberger, D., Todorovski, L., Blockeel, H. (eds.) PKDD 2003. LNCS(LNAI), vol. 2838, pp. 107–119. Springer, Heidelberg (2003)
3. Barnett, V., Lewis, T.: Outliers in Statistical Data. John Wiley and Sons, New York, NY (1994)
4. Lazarevic, A., Ertoz, L., Ozgur, A., Srivastava, J., Kumar, V.: A comparative study of anomaly detection schemes in network intrusion detection. In: Proceedings of the Third SIAM Int. Conf. on Data Mining, San Francisco, CA (May 2003)
5. Lazarevic, A., Kumar, V.: Feature Bagging for Outlier Detection. In: Proc. of the ACM SIGKDD Int. Conf. on Knowledge Discovery and Data Mining, Chicago, IL (August 2005)
6. Billor, N., Hadi, A., Velleman, P.: BACON: Blocked Adaptive Computationally-Efficient Outlier Nominators. Computational Statistics and Data Analysis 34, 279–298 (2000)
7. Eskin, E.: Anomaly Detection over Noisy Data using Learned Probability Distributions. In: Proceedings of the Int. Conf. on Machine Learning, Stanford University, CA (June 2000)

8. Aggarwal, C.C., Yu, P.: Outlier detection for high dimensional data. In: Proceedings of the ACM SIGMOD International Conference on Management of Data (2001)
9. Breunig, M.M., Kriegel, H.P., Ng, R.T., Sander, J.: Identifying, L.O.F.: Density Based Local Outliers. In: Proceedings of the ACM SIGMOD Conference, Dallas, TX (May 2000)
10. Knorr, E., Ng, R.: Algorithms for Mining Distance based Outliers in Large Data Sets. In: Proceedings of the Very Large Databases (VLDB) Conference, New York City, NY (August 1998)
11. Yu, D., Sheikholeslami, G., Zhang, A.: FindOut: Finding Outliers in Very Large Datasets. The Knowledge and Information Systems (KAIS) 4, 4 (2002)
12. Eskin, E., Arnold, A., Prerau, M., Portnoy, L., Stolfo, S.: A Geometric Framework for Unsupervised Anomaly Detection: Detecting Intrusions in Unlabeled Data. In: Jajodia, S., Barbara, D. (eds.) Applications of Data Mining in Computer Security, Advances In Information Security, Kluwer Academic Publishers, Boston (2002)
13. Hawkins, S., He, H., Williams, G., Baxter, R.: Outlier Detection Using Replicator Neural Networks. In: Kambayashi, Y., Winiwarter, W., Arikawa, M. (eds.) DaWaK 2002. LNCS, vol. 2454, pp. 170–180. Springer, Heidelberg (2002)
14. Medioni, G., Cohen, I., Hongeng, S., Bremond, F., Nevatia, R.: Event Detection and Analysis from Video Streams. IEEE Trans. on Pattern Analysis and Machine Intelligence 8(23), 873–889 (2001)
15. Chen, S.-C., Shyu, M.-L., Zhang, C., Strickrott, J.: Multimedia Data Mining for Traffic Video Sequences. MDM/KDD pp. 78–86 (2001)
16. Chen, S.-C., Shyu, M.-L., Zhang, C., Kashyap, R.L.: Video Scene Change Detection Method Using Unsupervised Segmentation And Object Tracking. ICME (2001)
17. Tao, Y., Papadias, D., Lian, X.: Reverse kNN search in arbitrary dimensionality. In: Proceedings of the 30th Int. Conf. on Very Large Data Bases, Toronto, Canada (September 2004)
18. Singh, A., Ferhatosmanoglu, H., Tosun, A.: High Dimensional Reverse Nearest Neighbor Queries. In: Proceedings of the ACM Int. Conf. on Information and Knowledge Management (CIKM'03), New Orleans, LA (November 2003)
19. Stanoi, I., Agrawal, D., Abbadi, A.E.: Reverse Nearest Neighbor Queries for Dynamic Databases. In: ACM SIGMOD Workshop on Research Issues in Data Mining and Knowledge Discovery, Dalas, TX (May 2000)
20. Anderson, J., Tjaden, B.: The inverse nearest neighbor problem with astrophysical applications. In: Proceedings of the 12th Symposium of Discrete Algorithms (SODA), Washington, DC (January 2001)
21. Pokrajac, D., Latecki, L.J., Lazarevic, A., et al.: Computational geometry issues of reverse-k nearest neighbors queries, Technical Report TR-CIS5001, Delaware State University (2006)
22. Conway, J., Sloane, N.H.: Sphere Packings, Lattices and Groups. Springer, Heidelberg (1998)
23. Preparata, F.P., Shamos, M.I.: Computational Geometry: an Introduction, 2nd Printing. Springer, Heidelberg (1988)
24. Roussopoulos, N., Kelley, S., Vincent, F.: Nearest neighbor queries. In: Proceedings of the ACM SIGMOD Conference, San Jose, CA, pp. 71–79 (1995)
25. Beckmann, N., Kriegel, H.-P., Schneider, R., Seeger, B.: The R*-tree: an efficient and robust access method for points and rectangles. SIGMOD Rec. 19(2), 322–331 (1990)

26. Berchtold, S., Keim, D.A., Kriegel, H.-P.: The X-tree: An index structure for high-dimensional data. In: Vijayaraman, T.M., Buchmann, A.P., Mohan, C., Sarda, N.L. (eds.) Proceedings of the 22nd International Conference on Very Large Databases, San Francisco, U.S.A, pp. 28–39. Morgan Kaufmann Publishers, Seattle (1996)
27. Weber, R., Schek, H.-J., Blott, S.: A quantitative analysis and performance study for similarity-search methods in high-dimensional spaces. In: VLDB '98: Proceedings of the 24rd International Conference on Very Large Data Bases, San Francisco, CA, USA, pp. 194–205. Morgan Kaufmann, Seattle, Washington (1998)
28. DeMenthon, D., Latecki, L.J., Rosenfeld, A., Stückelberg, M.V.: Relevance Ranking of Video Data using Hidden Markov Model Distances and Polygon Simplification. In: Laurini, R. (ed.) VISUAL 2000. LNCS, vol. 1929, pp. 49–61. Springer, Heidelberg (2000)
29. Latecki, L.J., Miezianko, R., Megalooikonomou, V., Pokrajac, D.: Using Spatiotemporal Blocks to Reduce the Uncertainty in Detecting and Tracking Moving Objects in Video. Int. Journal of Intelligent Systems Technologies and Applications 1(3/4), 376–392 (2006)
30. Jolliffe, I.T.: Principal Component Analysis, 2nd edn. Springer, Heidelberg (2002)
31. Lippmann, R.P., Fried, D.J., Graf, I.J., et al.: Evaluating Intrusion Detection Systems: The 1998 DARPA Off-line Intrusion Detection Evaluation. In: Proc. DARPA Information Survivability Conf. and Exposition (DISCEX) 2000, vol. 2, pp. 12–26. IEEE Computer Society Press, Los Alamitos (2000)
32. Tcptrace software tool, www.tcptrace.org
33. UCI KDD Archive, KDD Cup Data Set (1999), www.ics.uci.edu/~kdd/databases/kddcup99/kddcup99.html
34. Tang, J., Chen, Z., Fu, A., Cheung, D.: Enhancing Effectiveness of Outlier Detections for Low Density Patterns. In: Chen, M.-S., Yu, P.S., Liu, B. (eds.) PAKDD 2002. LNCS(LNAI), vol. 2336, pp. 535–548. Springer, Heidelberg (2002)
35. Papadimitriou, S., Kitagawa, H., Gibbons, P.B., Faloutsos, C.: LOCI: Fast Outlier Detection Using the Local Correlation Integral. In: Proc. of the 19th Int. Conf. on Data Engineering (ICDE'03), Bangalore, India (March 2003)
36. Bay, S.D., Schwabacher, M.: Mining distance-based outliers in near linear time with randomization and a simple pruning rule. In: Proceedings of the Ninth ACM SIGKDD Int. Conf. on Knowledge Discovery and Data Mining, New York, NY (2003)
37. Breiman, L., Meisel, W., Purcell, E.: Variable kernel estimates of multivariate densities. Technometrics 19(2), 135–144 (1977)
38. Loftsgaarden, D.O., Quesenberry, C.P.: A nonparametric estimate of a multivariate density function. Ann. Math. Statist. 36, 1049–1051 (1965)
39. Terrell, G.R., Scott, D.W.: Variable kernel density estimation. The Annals of Statistics 20(3), 1236–1265 (1992)
40. Maloof, M., Langley, P., Binford, T., Nevatia, R., Sage, S.: Improved Rooftop Detection in Aerial Images with Machine Learning. Machine Learning 53(1-2), 157–191 (2003)
41. Michalski, R., Mozetic, I., Hong, J., Lavrac, N.: The Multi-Purpose Incremental Learning System AQ15 and its Testing Applications to Three Medical Domains. In: Proceedings of the Fifth National Conference on Artificial Intelligence, Philadelphia, PA, pp. 1041–1045 (1986)
42. van der Putten, P., van Someren, M.: CoIL Challenge 2000: The Insurance Company Case, Sentient Machine Research, Amsterdam and Leiden Institute of Advanced Computer Science, Leiden LIACS Technical Report 2000-09 (June 2000)

43. Ertoz, L.: Similarity Measures, PhD dissertation, University of Minnesota (2005)
44. Provost, F., Fawcett, T.: Robust Classification for Imprecise Environments. Machine Learning 42(3), 203–231 (2001)
45. Blake, C., Merz, C.: UCI Repository of machine learning databases (1998),
    http://www.ics.uci.edu/~mlearn/MLRepository.html
46. Roussopoulos, N., Kelly, S., Vincent, F.: Nearest Neighbor Queries. In: Proc. ACM SIGMOD, pp. 71-79 (1995)
47. Devore, J.: Probability and Statistics for Engineering and the Sciences, 6th edn. (2003)
48. Conover, W.J.: Practical Nonparametric Statistics, 3rd edn. (1999)

Generic Probability Density Function Reconstruction for Randomization in Privacy-Preserving Data Mining

Vincent Yan Fu Tan[1,*] and See-Kiong Ng[2]

[1] Massachusetts Institute of Technology (MIT), Cambridge, MA 02139
vtan@mit.edu
[2] Institute for Infocomm Research (I²R), Singapore 119613
skng@i2r.a-star.edu.sg

Abstract. Data perturbation with random noise signals has been shown to be useful for data hiding in privacy-preserving data mining. Perturbation methods based on additive randomization allows accurate estimation of the Probability Density Function (PDF) via the Expectation-Maximization (EM) algorithm but it has been shown that noise-filtering techniques can be used to reconstruct the original data in many cases, leading to security breaches. In this paper, we propose a *generic* PDF reconstruction algorithm that can be used on non-additive (and additive) randomization techiques for the purpose of privacy-preserving data mining. This two-step reconstruction algorithm is based on Parzen-Window reconstruction and Quadratic Programming over a convex set – the probability simplex. Our algorithm eliminates the usual need for the iterative EM algorithm and it is generic for most randomization models. The simplicity of our two-step reconstruction algorithm, without iteration, also makes it attractive for use when dealing with large datasets.

Keywords: Randomization, Privacy-preserving data mining, Parzen-Windows, Quadratic Programming, Convex Set.

1 Introduction

Consider the following scenario: There are two hospitals which seek to predict new patients' susceptibility to illnesses based on existing data. It would be useful for the hospitals to pool their data, since data mining tasks can often benefit from a large training dataset. However, by law, the hospitals cannot release private patient data. Instead, some form of sanitized data has to be provided to a centralized server for further analysis. It is thus imperative to discover means to protect private information and be able to perform data mining tasks with a masked version of the raw data. Can privacy and accuracy co-exist? This is the fundamental question in privacy-preserving data mining [2,3].

---

* Vincent Tan is supported by the Agency for Science, Technology and Research (A*STAR), Singapore. He performed this work at I²R, A*STAR.

P. Perner (Ed.): MLDM 2007, LNAI 4571, pp. 76–90, 2007.

Randomization has been shown to be a useful technique for hiding data in privacy-preserving data mining. The basic concept is to sufficiently mask the actual values of the data by perturbing them with an appropriate level of noise that can still allow the underlying Probability Density Function (PDF) of the original dataset to be adequately estimated from the randomized data. A balance has to be achieved between two conflicting concerns in such approaches. On one hand, the *confidentiality* of the precise information has to be protected *i.e.* to minimize *privacy loss*. On the other hand, the *utility* of the aggregate data statistics has to be maintained *i.e.* to minimize *information loss*.

The use of randomization for preserving privacy was studied extensively in the framework of statistical databases [1]. It typically involves a trusted centralized database in which the data are already fully known before they are randomized and released for publication (e.g. census data). As such, privacy-preserving transformations such as sampling [24] and swapping [24] are more suitable for perturbing the data as they can incorporate knowledge about the aggregate characteristics of the dataset. In privacy preserving data mining (PPDM), we consider both (trusted) centralized database scenarios as well as distributed scenarios in which there is one (untrusted) central server that needs pieces of private information from multiple clients to build a aggregate model for the data, and the clients would each perturb the information before releasing them to the server to preserve privacy.

The early attempts by the pioneering authors in PPDM [2] applied additive white noise (e_i), generated from a pre-determined distribution, to the true data (x_i) and then transmitting the sum $(z_i = x_i + e_i)$ instead of the raw data. As it was shown that the distribution of the original data $f_X(x)$ can be reconstructed to a high accuracy *i.e.* low information loss, data mining can then be done satisfactorily using the sum (z_i) instead of the original data values (x_i). The reconstruction process hinges on the use of the *iterative* Expectation-Maximization (EM) algorithm taking the original values x_i as the latent variables.

However, Kargupta *et al.* [15] showed that such methods risk privacy breaches as the additive noise can be filtered off leaving a reasonably good estimation of the original data in many cases. Thus, other randomization models, such as using multiplicative noise, have been suggested [15].

Motivated by this, we develop a novel *non-iterative* PDF reconstruction scheme based on Parzen-Window reconstruction and Quadratic Programming (QP) optimization with only one equality constraint and one inequality constraint. These constraints define the probability simplex, which is a convex set. Convex programming/optimization [5,7] has been widely studied and efficient methods can be employed to estimate the PDF. As far as we know, currently only the mean and the variance in a multiplicative model can be estimated accurately [16]. To the best of our knowledge, for the first time, our method can allow the underlying PDF of the original dataset to be accurately reconstructed from randomized data set perturbed with multiplicative noise, additive noise or other noise models. From the estimated PDF, *all* the statistics can be inferred for distribution-based data mining purposes. Our approach therefore provides a complete description of the original data without compromising on the privacy.

Other randomization/reconstruction methods based on multiplicative noise have been proposed [22] but the implementation of the reconstruction method was very computationally demanding. As such, we have also made sure that our reconstruction method can be efficiently implemented, avoiding using the iterative Expectation-Maximization algorithm employed in many reconstruction approaches for perturbation-based privacy-preserving data mining [2]. This makes our method attractive for use with the increasingly large datasets that have become commonplace in recent years.

More recently, [18] proposed a data perturbation approach in which the data is multiplied by a randomly generated matrix, hence preserving privacy by effectively projecting the data into a lower dimension subspace. As the transformation is distance-preserving, the authors showed that it is possible to estimate from the perturbed data various distance-related statistical properties of the original data. We consider non-distance-preserving randomization models in this paper because the distance-preserving nature of the randomization scheme in [18] may result in security breaches if some private data is also revealed.

In short, our reconstruction algorithm has two main advantages:

1. Unlike EM, it is non-iterative and can handle large datasets.
2. More importantly, it can be applied to generic (non-additive) randomization models, including multiplicative noise models.

The rest of this paper is organized as follows: We define the generic perturbation model and state some assumptions in Section 2. We describe the Parzen-Window and Quadratic Programming reconstruction algorithm in Section 3. In Section 4 we describe the evaluation metrics. We then present extensive evaluation results on both simulated and real data sets to validate our technique in Section 5. Finally, we conclude in Section 6 and provide some discussions on future work.

2   Problem Definition

The current PPDM framework consists of two processes: a randomization process, followed by a reconstruction process. First, the source data is randomized at possibly multiple client sites. The randomized data are then transmitted to a centralized server which attempts to recover the PDF of the original data for aggregate analyses. In the next two sections, we will first formally define the randomization model for privacy-preserving preservation, followed by the basic assumptions that are necessary for the subsequent reconstruction process.

2.1   Randomization Model

The generic randomization problem can be stated, succintly and generally, using the following mathematical model. Consider a set of N original scalars representing a particular private attribute (e.g. income) $x_1, \ldots, x_N$, which are drawn from independent and identically distributed (IID) random variables $X_1, \ldots, X_N$. These random variables X_i follow a common PDF $f_X(x)$. To create the perturbation, we consider the generic two-variable randomization model:

$$z_i = \mathcal{Z}(e_i, x_i), \qquad \forall\, i \in \{1, \ldots, N\} \tag{1}$$

where the $e_1, \ldots, e_N$ are realizations of known IID random variables $E_1, \ldots, E_N$. $\mathcal{Z}(\cdot, \cdot)$ is a deterministic, possibly nonlinear, randomization operator. The e_i's are sampled from a specified uniform distribution. Therefore $E_i \sim \mathcal{U}(e; a_E, b_E)$, where $f_E(e) = \mathcal{U}(e; a_E, b_E)$ is the uniform distribution parameterized by lower and upper limits a_E and b_E respectively.

2.2   Reconstruction of PDF and Assumptions

Given the perturbed values $z_1, \ldots, z_N$ and the noise distribution, the reconstruction task is to obtain an estimate for the original PDF, which we denote $\hat{f}_X(x)$[1]. We make the following simple assumptions for recovering the PDF of X, $f_X(x)$:

$\mathcal{A}1$. The random variables X and E are statistically independent (SI) *i.e.* the joint distribution $f_{X,E}(x, e) = f_X(x)f_E(e)$ is equal to the product of the marginals.

$\mathcal{A}2$. The PDFs of X and E are finitely supported by $\mathcal{D}_X$ and $\mathcal{D}_E$ respectively. Outside these domains, $f_X(x) = f_E(e) = 0$.

Assumption $\mathcal{A}1$ is a common assumption in privacy-preserving data mining using randomization. It basically implies that the perturbing and original distributions are SI, which is a reasonable assumption. Assumption $\mathcal{A}2$ simplifies the computation for the reconstruction of the original PDF $\hat{f}_X(x)$ without loss of generality. This will be evident in Section 3, where the reconstruction algorithm is presented.

3   Randomization and Reconstruction Algorithms

Given the original data x_i, we will generate random numbers from a known uniform distribution to obtain the randomized data values z_i (c.f. Section 2.1). Because we are applying the noise e_i element-wise (as in Eq (1)), our randomization and reconstruction algorithm can be applied to both the centralized the distributed scenarios. It has been suggested [15] that the use of multiplicative noise is better than the additive model for minimizing risk of security breaches. In fact, our model goes beyond multiplicative noise. Any noise model of the form $z_i = \mathcal{Z}(e_i, x_i)$ can be used.

The key here, is whether we can effectively reconstruct the PDF of original data from the perturbed data. In this section, we will show how this can be done effectively and efficiently, without the need of the commonly-used iterative EM

---

[1] In this paper, estimates of functions, vectors and other variables are denoted with a overhead hat. For example, $\hat{a}$ is the estimate for a.

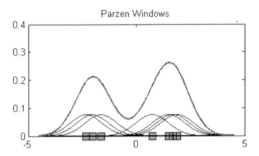

Fig. 1. Illustration of Parzen-Windows for estimation of the multimodal PDF. The *boxes* are the $N = 7$ independent realizations of the multimodal random variable. The individual Gaussian kernels are centered at the realizations. Their *sum*, as detailed in Eq (2) and indicated by the bold line, is the Parzen-Window approximation [20].

reconstruction algorithm. The general idea is as follows. Given the perturbed values z_i, we will first obtain an estimate of $f_Z(z)$ via Parzen-Windows [20]. Following the estimation of $f_Z(z)$, we will use Quadratic Programming (QP) to obtain an estimate of $f_X(x)$.

3.1   Estimate PDF of Perturbed Samples $f_Z(z)$ Via Parzen-Windows

The first step of the reconstruction algorithm is to estimate $\hat{f}_Z(z)$ using Parzen density estimation [20]. In this step, we are given N perturbed samples $z_1, \ldots, z_N$. They follow a common random variable Z, with true PDF $f_Z(z)$.

Parzen-Windows. The Parzen-Window approximation of the PDF of the perturbed samples is

$$\hat{f}_Z(z) = \frac{1}{N} \sum_{i=1}^{N} \frac{1}{\sigma_p \sqrt{2\pi}} \exp\left[-\frac{(z - z_i)^2}{2\sigma_p^2}\right], \tag{2}$$

where σ_p is the standard deviation or 'width' of the kernel. This estimator uses the Gaussian kernel function to smooth the raw sample set, placing more probability mass in regions with many samples, which is intuitively evident.

Example 1. An illustration of how the Parzen-Window method works for $N = 7$ is shown in Fig 1. We show the samples drawn from an arbitrary distribution. The Parzen approximation is the sum of the individual Gaussian kernels of equal standard deviations σ_p.

Remark 1. For Parzen-Window estimation, the quality of the estimate depends on the number of samples N as well as the standard deviation (SD) σ_p. If σ_p is too small, the Parzen approximation suffers from too much statistical variability and if σ_p is too large, the Parzen approximation is over-smoothed. Hence, we will now turn our attention to the selection of the optimal value of σ_p.

Cross-validation scheme for σ_p. In our experiments, we will use a cross-validation scheme that guarantees an optimal value of σ_p [4,21] in the l_2 sense. In this univariate optimization procedure, we seek to minimize the Integrated Squared Error (ISE) between the estimated PDF $\hat{f}_Z(z)$ and the actual PDF $f_Z(z)$:

$$\text{ISE} \triangleq \int_{\mathcal{D}_Z} \left(\hat{f}_Z(z) - f_Z(z) \right)^2 dz. \tag{3}$$

The ISE can be simplified to given the 'leave-one-out' (LOO) cross-validation criterion

$$\sigma_p^* = \underset{\sigma_p}{\text{argmin}} \; E_{LOO}(\sigma_p), \tag{4}$$

with $E_{LOO}(\sigma_p)$ defined as

$$E_{LOO}(\sigma_p) \triangleq \frac{1}{N^2} \sum_{i=1}^{N} \sum_{j=1}^{N} \mathcal{N}(z_i; z_j, \sqrt{2}\sigma_p) - \frac{2}{N(N-1)} \sum_{i=1}^{N} \sum_{\substack{j=1 \\ j \neq i}}^{N} \mathcal{N}(z_i; z_j, \sigma_p), \tag{5}$$

and $\mathcal{N}(x; \mu, c) = (c\sqrt{2\pi})^{-1} \exp\left[-(x-\mu)^2/2c^2\right]$ is the Gaussian kernel with mean μ and variance c^2. The optimization problem in Eq (4) is one-dimensional and efficient line search methods [17] will yield sufficiently accurate solutions.

3.2   Estimate Original PDF $f_X(x)$ Via Quadratic Programming (QP)

Equipped with an estimate of the perturbed PDF $\hat{f}_Z(z)$, we are ready to estimate the original PDF $f_X(x)$.

Theorem 1. *Let $Z = \mathcal{Z}(X, E)$ be the result of a function of two random variables that can also be expressed as $E = \mathcal{E}(X, Z)$ i.e. given $X = x$, the transformation is one-to-one. Then, if assumptions $\mathcal{A}1$ and $\mathcal{A}2$ (c.f. Section 2.2) are satisfied, the Probability Density Function (PDF) of Z, $\hat{f}_Z(z)$ can be written as*

$$\hat{f}_Z(z) = \int_{\mathcal{D}_X} f_X(x) f_E[\mathcal{E}(x, z)] \left| \frac{\partial \mathcal{E}(x, z)}{\partial x} \right| dx. \tag{6}$$

Proof. See Appendix A.   □

The assumption that the transformation from Z to E given $X = x$ is one-to-one is made *without any loss of generality*. This is because we can represent the set $\mathcal{A} = \{(x, e)\}$ of input variables as the union of a finite number, say K, of mutually disjoint subsets $\{\mathcal{A}_k\}_{k=1}^{K}$ such that the transformation is one-to-one in each of $\mathcal{A}_k$ onto $\mathcal{B} = \{(v, z)\}$. We focus on the one-to-one case for notational simplicity but note that it is straightforward to extend the argument to the case where the transformation is not one-to-one. For example, the randomization model $z_i = \mathcal{Z}(e_i, x_i) = x_i e_i + x_i^2 e_i^4$ is not one-to-one. Nonetheless, it is still possible to apply our reconstruction algorithm, with appropriate modifications to Eq (6). We refer the reader to the excellent treatment of functions of random variables by Hogg and Craig [14, Chapter 4].

QP Formulation. Using Theorem 1, we can formulate the Quadratic Program to estimate the optimal $f_X(x)$. Discretizing[2] the integral in Eq. (6) yields

$$\hat{f}_Z(z) \approx \sum_{n\Delta x \in \mathcal{D}_X} f_X(n\Delta x) f_E \left[\mathcal{E}(n\Delta x, z)\right] \left|\frac{\partial \mathcal{E}(x, z)}{\partial x}\right|_{x=n\Delta x} \Delta x, \qquad (7)$$

where $\Delta x > 0$ is the step size of $f_X(x)$ and $n\Delta x \in \mathcal{D}_X$ and $\mathcal{D}_X \stackrel{\triangle}{=} \{(n_0 + 1) \Delta x, \ldots, (n_0 + N_x)\Delta x\}$ is the set of discretized points contained in the finitely-supported[3] domain of X. Then for $z \in \{z_1, \ldots, z_{N_z}\}$, Eq. (7) can be written as

$$\hat{\mathbf{f}}_Z = \mathbf{G}_E \, \mathbf{f}_X, \qquad (8)$$

where the length N_z vector $\hat{\mathbf{f}}_Z$, length N_x vector $\mathbf{f}_X$ and the N_z by N_x matrix $\mathbf{G}_E$ are defined as

$$[\hat{\mathbf{f}}_Z]_j \stackrel{\triangle}{=} \hat{f}_Z(z_j), \qquad (9)$$

$$[\mathbf{f}_X]_i \stackrel{\triangle}{=} f_X((n_0 + i)\Delta x), \qquad (10)$$

$$[\mathbf{G}_E]_{ij} \stackrel{\triangle}{=} f_E \left[\mathcal{E}((n_0 + i)\Delta x, z_j)\right] \left|\frac{\partial \mathcal{E}(x, z_j)}{\partial x}\right|_{x=(n_0+i)\Delta x} \Delta x, \qquad (11)$$

and $[\mathbf{v}]_k$ is the kth element of the vector $\mathbf{v}$ and $[\mathbf{M}]_{ij}$ is the element in the ith row and j^{th} column of the matrix $\mathbf{M}$ and $i \in \{1, \ldots N_z\}$ and $j \in \{1, \ldots N_x\}$. Eq. (8) can be converted into the canonical cost function in a Quadratic Program as shown in Appendix B.

Example 2. If as in [3], we use an additive scheme *i.e.* $z_i = \mathcal{Z}_{add}(e_i, x_i) = x_i + e_i$, then Eq (11), together with the convolution formula [19], simplifies to give

$$[\mathbf{G}_E]_{ij} \stackrel{\triangle}{=} f_E \left[z_j - (n_0 + i)\Delta x\right]. \qquad (12)$$

Example 3. If instead we use a multiplicative scheme [16] *i.e.* $z_i = \mathcal{Z}_{mul}(e_i, x_i) = e_i \times x_i$, then Eq (11) together with the result in [13] simplifies to give

$$[\mathbf{G}_E]_{ij} \stackrel{\triangle}{=} \left|\frac{1}{n_0 + i}\right| f_E \left[\frac{z_j}{(n_0 + i)\Delta x}\right], \qquad n_0 \neq -i. \qquad (13)$$

Constraints. As $f_X(x)$ is a PDF, it has to satisfy the stochastic constraints $f_X(x) \geq 0, \forall x \in \mathcal{D}_X$ and $\int_{\mathcal{D}_X} f_X(x)\, dx = 1$. This places an inequality and an equality constraint on the vector $\mathbf{f}_X$, which can be easily incorporated into the QP as:

$$\mathbf{f}_X \geq \mathbf{0}_{N_X \times 1}, \qquad \sum_{n\Delta x \in \mathcal{D}_X} f_X(n\Delta x) = \frac{1}{\Delta x}. \qquad (14)$$

---

[2] This is done using the Rectangular rule. We can alternatively use the Trapezoidal, Simpson or Quadrature rules [8] to discretize the integral. Our experimental results, however, show that the performances of these discretization rules are very similar and hence for simplicity, we shall only present Rectangular rule.

[3] By assumption $\mathcal{A}2$.

Sufficient conditions for QP. We now derive sufficient conditions for the optimization problem. Because the constraint set is particularly simple, we can obtain the optimal solution without the use of iterative methods (such as gradient projection or modern interior points methods). Consider our quadratic program:

$$\min_{\mathbf{f}_X} \quad J(\mathbf{f}_X) = \frac{1}{2}\mathbf{f}_X^\mathsf{T}\mathbf{H}\mathbf{f}_X + \mathbf{h}^\mathsf{T}\mathbf{f}_X, \tag{15}$$

subject to

$$\mathbf{f}_X \in \mathcal{C}, \quad \text{with} \quad \mathcal{C} = \left\{ \mathbf{f}_X \;\middle|\; \mathbf{f}_X \geq \mathbf{0},\ \sum_{i=1}^{N_x}[\mathbf{f}_X]_i = 1 \right\}, \tag{16}$$

for appropriately chosen $\mathbf{H}$ and $\mathbf{h}$ (as shown in Appendix B). Then, the necessary condition for $\mathbf{f}_X^*$ to be a local minimum over a convex set [6, Section 2.1] is

$$\sum_{i=1}^{N_x} \frac{\partial J(\mathbf{f}_X^*)}{\partial[\mathbf{f}_X]_i}([\mathbf{f}_X]_i - [\mathbf{f}_X^*]_i) \geq 0, \quad \forall\, \mathbf{f}_X \in \mathcal{C}. \tag{17}$$

Subsequent simplification yields the condition

$$[\mathbf{f}_X^*]_i > 0 \Rightarrow \frac{\partial J(\mathbf{f}_X^*)}{\partial[\mathbf{f}_X]_i} < \frac{\partial J(\mathbf{f}_X^*)}{\partial[\mathbf{f}_X]_j} \Leftrightarrow \sum_{k=1}^{N_x}[\mathbf{H}]_{ik}[\mathbf{f}_X]_k + [\mathbf{h}]_i < \sum_{k=1}^{N_x}[\mathbf{H}]_{jk}[\mathbf{f}_X]_k + [\mathbf{h}]_j, \ \forall\, j. \tag{18}$$

Thus, all coordinates which are (strictly) positive at the optimum must have minimal (and equal) partial cost derivates [6]. Since $\mathbf{G}_E$ only contains real entries, the Hessian matrix $\mathbf{H} = \mathbf{G}_E^\mathsf{T}\mathbf{G}_E$ of the QP is positive semidefinite. Consequently, the cost function $J(\cdot)$ is convex [7] and any local optimum of Eq (15) is also a global optimum, which implies that the cost value is equal for all local optima. Moreover, the set of local optima is always convex.

We exploit the convexity of the cost function to conclude that Eq (18) is also a sufficient condition for global optimality of $\mathbf{f}_X^* = \hat{\mathbf{f}}_X$.

3.3   Discussion

We have completed the discussion of our non-iterative PDF reconstruction for generic randomization schemes for privacy-preserving data mining. There are two steps: Firstly, we build the Parzen-Window of the perturbed samples $\hat{f}_Z(z)$. Secondly, we perform a QP over the probability simplex to reconstruct an estimate of the original PDF $\hat{f}_X(x)$. Our algorithm is summarised in Fig 2. We conclude this section with two comments on our algorithm.

1. Discretizing the integral in Eq (6) is, in general, intractable if we are reconstructing PDFs of high dimensions as the problem suffers from the 'curse of dimensionality'. We can mitigate the effects of the curse by assuming the dimensions are independent, if possible. Using this naïve approach, we estimate the PDF in each dimension before taking their product to form the joint density. Alternatively, we can project the data onto a lower dimensional subspace and perform the same analysis in that subspace.

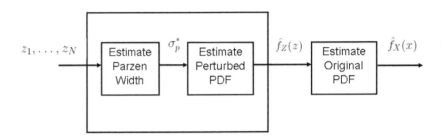

Fig. 2. The PDF reconstruction algorithm. There are two main steps. We reconstruct $\hat{f}_Z$ via Parzen-Windows. Then we estimate of $\hat{f}_X(x)$ using the QP.

2. The reconstruction algorithm can handle large datasets. One approach is to find a *random subset* of the samples from the dataset z_i to build the Parzen-Window and to perform the QP. This is known as the reduced Parzen-Window and is discussed in more detail in [12].

4   Performance Metrics

As mentioned earlier, there are two competing issues. Firstly, we hope to minimize the *privacy loss* so that individual information is not revealed. At the same time, we want to preserve the structure and the aggregate statistics of the underlying data. In other words, we also hope to minimize the *information loss*.

4.1   Privacy Loss

In this section, we will quantify privacy loss using mutual information. It was argued in [2] that the mutual information between two random variables X and Z measures the degree of independence between the random variables and hence, the privacy loss for X when Z is revealed.

The mutual information $I(X; Z)$ tells us how much information one random variable tells about another one. In other words, $I(X; Z)$ is the amount of uncertainty in X, which is removed by knowing Z. When X and Z are independent, $I(X; Z) = 0$. The lower the value of $I(X; Z)$, the better the privacy gain via the given perturbation scheme, the more the privacy is preserved. This leads us to the notion of the privacy loss $\mathcal{P}(X|Z)$ of X when Z is known. It is defined as:

$$\mathcal{P}(X|Z) \triangleq 1 - 2^{-I(X;Z)}. \tag{19}$$

By definition, $0 \leq \mathcal{P}(X|Z) \leq 1$. $\mathcal{P}(X|Z) = 0$ if and only if X and Z are SI.

Remark 2. Privacy breach [9], based on worst-case information loss, has also been suggested as an alternative privacy measure. However, in our work, we consider an *average* disclosure measure – mutual information. Also, the privacy breach [10] measure is typically used in the context of association-rule mining, which is not applicable in our context.

4.2   Information Loss

In this section, we will define information loss, which is a measure of the effectiveness and accuracy of the reconstruction algorithm. It is clear that given the perturbed values $z_1, \ldots, z_N$, it is, in general, not possible to reconstruct the original density $f_X(x)$ with arbitrary precision. The lack of precision in estimating $f_X(x)$ from the perturbed values is referred to as information loss. The closer our estimate $\hat{f}_X(x)$ is to the actual PDF $f_X(x)$, the lower the information loss. We use the following *universal* metric suggested in [2] to quantify the information loss in the reconstruction of $f_X(x)$.

$$\mathcal{I}(f_X, \hat{f}_X) \triangleq \frac{1}{2} \mathbf{E} \left[\int_{\mathcal{D}_X} \left| f_X(x) - \hat{f}_X(x) \right| \, dx \right], \qquad (20)$$

where $\hat{f}_X(x)$ is the estimate for the PDF of the random variable X. It is easy to see that $0 \leq \mathcal{I}(f_X, \hat{f}_X) \leq 1$. We will see that our algorithm produces an accurate original PDF that is amendable to various distribution-based data mining tasks.

5   Experiments

We conducted two main experiments to demonstrate the efficiency and accuracy of the PQP reconstruction algorithm.

1. Firstly, we examine the tradeoff between the privacy loss and information loss. In Section 5.1, we show empirically that our generic PDF reconstruction algorithm performs as well as the additive randomization-EM algorithm suggested in [2]. We emphasize that our PDF reconstruction algorithm is applicable to all randomization models that can be expressed in the form Eq (1).
2. Secondly, we applied our algorithm to a *real dataset* and demonstrate that privacy can be preserved and, at the same time, the aggregate statistics can be mined. The results are discussed in Section 5.2.

5.1   Privacy/Accuracy Tradeoff

As mentioned previously, data perturbation based approaches typically face a privacy/accuracy loss tradeoff. In this section, we shall examine this tradeoff and compare it to existing technologies. We used two different randomization models – multiplicative and additive and examine the efficacy of the PDF reconstruction algorithm ('PQP'). The results are summarized in Fig 3.

We observe that our reconstruction algorithm performs as well as EM with the added bonus that it is generic. It can be applied to multiplicative, additive and other randomization models. Besides, it is non-iterative.

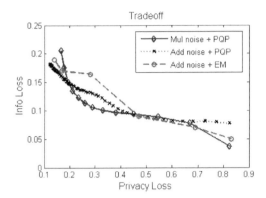

Fig. 3. Plot of the tradeoff between information loss $\mathcal{I}(f_X, \hat{f}_X)$ and privacy loss $\mathcal{P}(X|Z)$. Our PDF reconstruction algorithm ('PQP') performs just as well as EM but has the added bonus of being a generic reconstruction method.

Table 1. Information Losses resulting from the various perturbation/reconstruction methods. Privacy Loss is kept constant at $\mathcal{P}(X|Z) = 0.330$. We observe that the PQP reconstruction algorithm gives a superior (lower) information loss as compared to EM.

Method	Mul + PQP	Add + PQP	Add + EM [2]
$\mathcal{I}(f_X, \hat{f}_X)$	0.1174	**0.0957**	0.1208

5.2   Application to Real Data

We applied the Parzen-Window and QP reconstruction ('PQP') algorithm to real data obtained from The U.S. Department of Housing and Urban Development's (USDHUD's) Office of Policy Development and Research (PD&R) [23]. As with the previous experiment, we perturbed the data with multiplicative noise and additive noise. Other randomization techniques are also applicable.

The data in [23] provides us with the median income of all the counties in the 50 states in the U.S in 2005. The length of the dataset is $N = 3195$. This is plotted as a histogram with 75 bins in Figure 4(a). We multiplied each data value with samples drawn from a uniform distribution with domain $1 \le e \le 3$ giving a privacy loss value of $\mathcal{P}(X|Z) = 0.330$.

In addition to using the multiplicative randomization and PQP reconstruction algorithm, we also ran the PQP algorithm on the data corrupted by additive noise. The level of noise was adjusted such that the privacy loss is kept constant at $\mathcal{P}(X|Z) = 0.330$. Finally, we implemented the additive noise and EM reconstruction algorithm [2] on the data.

We averaged our results over 500 independent runs and the results are tabulated in Table 1. The results showed that our PDF reconstruction algorithm ('PQP') performed better than additive/EM [2] on the real data. The added

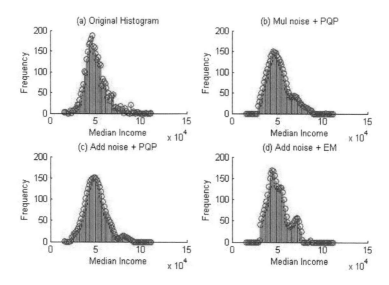

Fig. 4. (a) Original histogram of Median Income of Counties in the U.S. [23] (b) Reconstructed histogram after Multiplicative Randomization and our PDF reconstruction algorithm ('PQP'). (c) Reconstructed histogram after Additive Randomization and 'PQP'. (d) Reconstructed histogram after Additive Randomization and EM [2]. Note the accuracy of our PDF reconstruction algorithm.

advantage here is that our novel non-iterative PDF reconstruction algorithm can be applied to *all* randomization models of the form Eq (1).

6   Conclusions and Further Work

In this paper, we have devised a novel PDF reconstruction scheme for privacy-preserving data mining. This scheme is based on Parzen-Window reconstruction and Quadratic Programming (with a positive semidefinite Hessian) over a convex set. For the first time, the original PDF $f_X(x)$ can be approximated from the samples which have been perturbed by any type of noise (even multiplicative) that follows the generic randomization equation $z_i = \mathcal{Z}(e_i, x_i)$. We performed extensive numerical experiments demonstrating the efficacy of our algoritm. There are two distinct advantages over the existing PDF reconstruction algorithms which are based on the iterative EM algorithm.

1. Firstly, our proposed two-step reconstruction algorithm eliminated the common need for the *iterative* Expectation-Maximization (EM) algorithm. This is essential for problems which involve larger datasets, as it circumvents the need for iteration. It only involves two steps: Parzen-Window reconstruction and Quadratic Programming. The QP is particularly easy to solve because of the nature of the constraints – the (convex) probability simplex.

2. Secondly, our reconstruction method is also *generic*. Theorem 1 shows that the algorithm can be applied to many other randomization models as long as the perturbing random variable E and the underlying random variable X are SI, which is a common assumption for randomization methods in privacy-preserving data mining. We emphasize that although we examined the multiplicative and additive models only in Section 5, our reconstruction algorithm can be applied to all randomization models of the form Eq (1).

A natural extension to this work is to examine even more randomization models and reconstruction algorithms. For instance, we can parameterize Eq (1) as follows: $z_i = \mathcal{Z}(e_i(\psi), x_i; \psi)$ where ψ is an unknown but deterministic/non-random parameter. This adds an additional layer of privacy and the PDF can be estimated using a combination of our PQP reconstruction algorithm and maximum-likelihood methods. Finally, a question of paramount importance that researchers can try to decipher is: Does a fundamental relation between the privacy loss and information loss exist? We believe this needs to be answered precisely in order to unlock the promising future in privacy-preserving data mining.

Acknowledgments. The first author would like to thank Dr. Mafruzzaman Ashrafi of the Institute for Infocomm Research (I^2R) for his helpful comments. An anonymous reviewer made several valuable comments.

References

1. Adam, N.R., Worthmann, J.C.: Security-control methods for statistical databases: A comparative study. ACM Comput. Surv. 21, 515–556 (1989)
2. Agrawal, D., Aggarwal, C.C.: On the design and quantification of privacy preserving data mining algorithm. In: Symposium on Principles of Database Systems, pp. 247–255 (2001)
3. Agrawal, R., Srikant, R.: Privacy-preserving data mining. In: Proc. of the ACM SIGMOD Conference on Management of Data, pp. 439–450. ACM Press, New York (2000)
4. Assenza, A., Archambeau, C., Valle, M., Verleysen, M.: Assessment of probability density estimation methods: Parzen-Window and Finite Gaussian Mixture ISCAS. In: IEEE International Symposium on Circuits and Systems, Kos (Greece) (2006), pp. 3245–3248 (2006)
5. Bertsekas, D., Nedic, A., Ozdaglar, A.E.: Convex Analysis and Optimization Athena Scientific (2003)
6. Bertsekas, D.: Nonlinear Programming Athena Scientific (2004)
7. Boyd, S., Vandenberghe, L.: Convex Optimization. Cambridge University Press, Cambridge (2004)
8. Davis, P.J., Rabinowitz, P.: Methods of Numerical Integration. Academic Press, San Diego (1984)
9. Evfimievski, A., Gehrke, J., Srikant, R.: Limiting Privacy Breaches in Privacy Preserving Data Minin. In: Proc. of ACM SIGMOD/PODS Conference, pp. 211–222 (2003)

10. Evfimievski, A., Srikant, R., Agrawal, R., Gehrke, J.: Privacy Preserving Mining of Association Rule. In: Proceedings of the 8th ACM SIGKDD International Conference on Knowledge Discovery in Databases and Data Mining, Edmonton, Alberta, Canada, pp. 217–228 (2002)

11. Fessler, J.A.: On transformations of random vectors. In: Technical Report 314, Comm. and Sign. Proc. Lab. Dept. of EECS, Univ. of Michigan, Ann Arbor, MI, 48109-2122 (1998)

12. Fukunaga, K.: Statistical Pattern Recognition, 2nd edn. California Academic Press, San Diego (1990)

13. Glen, A., Leemis, L., Drew, J.: Computing the distribution of the product of two continuous random variables. Computational statistics & data analysis 44, 451–464 (2004)

14. Hogg, R.V., Craig, A.T.: Introduction to Mathematical Statistics, 5th edn. Prentice-Hall, Englewood Cliffs, NJ (1995)

15. Kargupta, H., Datta, S., Wang, Q., Sivakumar, K.: On the privacy preserving properties of random data perturbation technique. In: Proceedings of the 3rd IEEE International Conference on Data Mining, Washington, DC, USA, pp. 99–106 (2003)

16. Kim, J.J., Winkler, W.E.: Multiplicative noise for masking continuous data Technical Report Statistics #2003-01, Statistical Research Division, U.S. Bureau of the Census, Washington, DC, USA (2003)

17. Luenberger, D.G.: Linear and Nonlinear Programming. Addison-Wesley, London (1984)

18. Liu, K., Kargupta, H., Ryan, J.: Random Projection-Based Multiplicative Data Perturbation for Privacy Preserving Distributed Data Mining IEEE Transactions on Knowledge and Data Engineering (TKDE), 18, pp. 92–106 (2006)

19. Oppenheim, A.V., Willsky, A.S.: Signals and Systems. Prentice-Hall, Englewood Cliffs (1996)

20. Parzen, E.: On the estimation of a probability density function and mode. Annals of Mathematical Statistics 33, 1065–1076 (1962)

21. Silverman, B.W.: Density Estimation for Statistics and Data Analysis. Chapman & Hall, London (1986)

22. Makov, U.E., Trotini, M., Fienberg, S.E., Meyer, M.M.: Additive noise and multiplicative bias as disclosure limitation techniques for continuous microdata: A simulation study. Journal of Computational Methods in Science and Engineering 4, 5–16 (2004)

23. U.S. Department of Housing, Urban Developments (USDHUDs) Office of Policy Development and Research (PD&R). (2005), http://www.huduser.org/datasets/il/IL_99_05_REV.xls

24. Verykios, V.S., Bertino, E., Fovino, I.N., Provenza, L.P., Saygin, Y., Theodoridis, Y.: State-of-the-art in Privacy Preserving Data Mining. ACM SIGMOD Record 3, 50–57 (2004)

Appendix

A   Proof of Theorem 1

Proof. Our proof is adapted from [11] and [13]. Using the transformation technique [14], the transformation $V = X$ and $Z = \mathcal{Z}(X, E)$ constitutes a one-to-one mapping from from $\mathcal{A} = \{(x, e)\}$ to $\mathcal{B} = \{(v, z)\}$. Let u denote the transformation and w the inverse transformation. The transformation and its inverse can be written as:

$$v = u_1(x, e) = x, \quad z = u_2(x, e) = \mathcal{Z}(x, e), \tag{A.1}$$

$$x = w_1(v, z) = z, \quad e = w_2(v, z) = \mathcal{E}(v, z). \tag{A.2}$$

Consequently, the Jacobian determinant can be expressed in the form

$$J = \begin{vmatrix} \frac{\partial x}{\partial z} & \frac{\partial x}{\partial v} \\ \frac{\partial e}{\partial z} & \frac{\partial e}{\partial v} \end{vmatrix} = \left| \frac{\partial \mathcal{E}(v, z)}{\partial v} \right| = \left| \frac{\partial \mathcal{E}(x, z)}{\partial x} \right|. \tag{A.3}$$

The marginal density of Z, which can be obtained through Parzen reconstruction from the samples of z_i can be found by integrating the joint density of V and Z

$$\hat{f}_Z(z) = \int_{\mathcal{D}_V} f_{V,Z}(v, z) \, dv. \tag{A.4}$$

Application of the transformation from $\mathcal{B}$ to $\mathcal{A}$ yields

$$\hat{f}_Z(z) = \int_{\mathcal{D}_V} f_{X,E}(w_1(v, z), w_2(v, z)) \left| \frac{\partial \mathcal{E}(x, z)}{\partial x} \right| dv, \tag{A.5}$$

A further simplification and the use of the statistical independence of X and E (Assumption $\mathcal{A}1$) gives Eq. (6). $\qquad\qquad\square$

B   Detailed Formulation of the Quadratic Program

The canonical QP can be written as

$$\boldsymbol{\theta}^* = \operatorname*{argmin}_{\boldsymbol{\theta}} \left\{ \frac{1}{2} \boldsymbol{\theta}^{\mathrm{T}} \mathbf{H} \boldsymbol{\theta} + \mathbf{h}^{\mathrm{T}} \boldsymbol{\theta} \right\}, \tag{B.1}$$

subject to

$$\mathbf{A}\boldsymbol{\theta} \leq \mathbf{b}, \quad \mathbf{A}_{\mathbf{eq}}\boldsymbol{\theta} = \mathbf{b}_{\mathbf{eq}}, \tag{B.2}$$

where $\mathbf{H}$, $\mathbf{A}$ and $\mathbf{A}_{\mathbf{eq}}$ are matrices and $\mathbf{h}$, $\mathbf{b}$, $\mathbf{b}_{\mathbf{eq}}$ and $\boldsymbol{\theta}$ are vectors, all appropriately sized. To optimize for a solution to Eq (8), we can write it in terms of an cost function

$$J(\mathbf{f}_X) = \frac{1}{2} \left\| \hat{\mathbf{f}}_Z - \mathbf{G}_E \mathbf{f}_X \right\|_2^2, \tag{B.3}$$

where $\| \cdot \|_2$ is the l_2 norm. Eq (B.3) can be can be simplified to give

$$J(\mathbf{f}_X) = \frac{1}{2} \mathbf{f}_X^{\mathrm{T}} \mathbf{G}_E^{\mathrm{T}} \mathbf{G}_E \mathbf{f}_X - \hat{\mathbf{f}}_Z^{\mathrm{T}} \mathbf{G}_E \mathbf{f}_X + c, \tag{B.4}$$

where c is some constant independent of $\mathbf{f}_X$. Hence, by comparing Eq (B.1) and Eq (B.4), we observe that $\boldsymbol{\theta} = \mathbf{f}_X$ is the vector of control variables and

$$\mathbf{H} = \mathbf{G}_E^{\mathrm{T}} \mathbf{G}_E, \quad \mathbf{h} = -\mathbf{G}_E^{\mathrm{T}} \hat{\mathbf{f}}_Z, \tag{B.5}$$

are the matrix (Hessian) and vector that define the cost function. Also, comparing the constraints in Eq (14) to the constraints in the canonical QP, we obtain

$$\mathbf{A} = -\mathbf{I}_{N_X \times N_X}, \quad \mathbf{b} = \mathbf{0}_{N_X \times 1}, \quad \mathbf{A}_{\mathbf{eq}} = (\Delta x) \mathbf{1}_{1 \times N_X}, \quad \mathbf{b}_{\mathbf{eq}} = 1. \tag{B.6}$$

An Incremental Fuzzy Decision Tree Classification Method for Mining Data Streams*

Tao Wang[1], Zhoujun Li[2], Yuejin Yan[1], and Huowang Chen[1]

[1] Computer School, National University of Defense Technology, Changsha, 410073, China
[2] School of Computer Science & Engineering, Beihang University, Beijing, 100083, China
InsistStar@nudt.edu.cn

Abstract. One of most important algorithms for mining data streams is VFDT. It uses Hoeffding inequality to achieve a probabilistic bound on the accuracy of the tree constructed. Gama et al. have extended VFDT in two directions. Their system VFDTc can deal with continuous data and use more powerful classification techniques at tree leaves. In this paper, we revisit this problem and implemented a system fVFDT on top of VFDT and VFDTc. We make the following four contributions: 1) we present a threaded binary search trees (TBST) approach for efficiently handling continuous attributes. It builds a threaded binary search tree, and its processing time for values inserting is $O(nlogn)$, while VFDT's processing time is $O(n^2)$. When a new example arrives, VFDTc need update $O(logn)$ attribute tree nodes, but fVFDT just need update one necessary node. 2) we improve the method of getting the best split-test point of a given continuous attribute. Comparing to the method used in VFDTc, it improves from $O(nlogn)$ to $O(n)$ in processing time. 3) Comparing to VFDTc, fVFDT's candidate split-test number decrease from $O(n)$ to $O(logn)$. 4) Improve the soft discretization method to be used in data streams mining, it overcomes the problem of noise data and improve the classification accuracy.

Keywords: Data Streams, Incremental, Fuzzy, Continuous Attribute, Threaded Binary Search Tree.

1 Introduction

Decision trees are one of the most used classification techniques for data mining. Tree models have high degree of interpretability. Global and complex decisions can be approximated by a series of simpler and local decisions. Algorithms that construct decision trees from data usually use a divide and conquer strategy. A complex problem is divided into simpler problems and recursively the same strategy is applied to the sub-problems. The solutions of sub-problems are combined in the form of a tree to yield the solution of the complex problem [3, 20, 22].

---

* This work was supported by the National Science Foundation of China under Grants No. 60573057, 60473057 and 90604007.

P. Perner (Ed.): MLDM 2007, LNAI 4571, pp. 91–103, 2007.

More recently, the data mining community has focused on a new model of data processing, in which data arrives in the form of continuous streams [1, 3, 9, 11, 12, 16, 28, 29]. The key issue in mining on data streams is that only one pass is allowed over the entire data. Moreover, there is a real-time constraint, i.e. the processing time is limited by the rate of arrival of instances in the data stream, and the memory and disk available to store any summary information may be bounded. For most data mining problems, a one-pass algorithm cannot be very accurate. The existing algorithms typically achieve either a deterministic bound on the accuracy or a probabilistic bound [21, 23].

Domingos and Hulten [2, 6] have addressed the problem of decision tree construction on data streams. Their algorithm guarantees a probabilistic bound on the accuracy of the decision tree that is constructed. Gama et al. [5] have extended VFDT in two directions: the ability to deal with continuous data and the use of more powerful classification techniques at tree leaves.

Peng et al.[30]propose the soft discretization method in traditional data mining field,it solve the problem of noise data and improve the classification accuracy.

The rest of the paper is organized as follows. Section 2 describes the related works that is the basis for this paper. Section 3 presents the technical details of fVFDT. The system has been implemented and evaluated, and experimental evaluation is done in Section 4. Last section concludes the paper, resuming the main contributions of this work.

2 Related Work

In this section we analyze the related works that our fVFDT bases on.

Decision trees support continuous attributes by allowing internal nodes to contain tests of the form $A_i \leq T$ (the value of attribute i is less than threshold T). Traditional induction algorithms learn decision trees with such tests in the following manner. For each continuous attribute, they construct a set of candidate tests by sorting the values of that attribute in the training set and using a threshold midway between each adjacent pair of values that come from training examples with different class labels to get the best split-test point.

There are several reasons why this standard method is not appropriate when learning from data streams. The most serious of these is that it requires that the entire training set be available ahead of time so that split thresholds can be determined.

2.1 VFDT

VFDT(Very Fast Decision Tree) system[2], which is able to learn from abundant data within practical time and memory constraints. In VFDT a decision tree is learned by recursively replacing leaves with decision nodes. Each leaf stores the sufficient statistics about attribute-values. The sufficient statistics are those needed by a heuristic evaluation function that evaluates the merit of split-tests based on attribute-values. When an example is available, it traverses the tree from the root to a leaf, evaluating the appropriate attribute at each node, and following the branch corresponding to the attribute's value in the example. When the example reaches a leaf, the sufficient statistics are updated. Then, each possible condition based on attribute-values is evaluated. If there is enough statistical support in favor of one test over the others, the leaf is changed to a decision node. The new decision node will have as many descendant leaves as the number of possible values for the chosen attribute (therefore this tree is not necessarily binary). The decision nodes only maintain the information about the split-test installed in this node. The initial state of the tree consists of a single leaf: the root of the tree. The heuristic evaluation function is the Information Gain (denoted by G(.). The sufficient statistics for estimating the merit of a discrete attribute are the counts nijk, representing the number of examples of class k that reach the leaf, where the attribute j takes the value i. The Information Gain measures the amount of information that is necessary to classify an example that reach the node: G(Aj)=info(examples)-info(Aj). The information of the attribute j is given by:

$$\inf o(A_j) = \sum_i P_i (\sum_k -P_{ik} \log(P_{ik}))$$

where $P_{ik} = n_{ijk} / \sum_a n_{ajk}$, is the probability of observing the value of the attribute i

given class k and $P_i = \sum_a n_{ija} / \sum_a \sum_b n_{ajb}$ is the probabilities of observing the

value of attribute i.

As mentioned in Catlett and others [23], that it may be sufficient to use a small sample of the available examples when choosing the split attribute at any given node. To determine the number of examples needed for each decision, VFDT uses a statistical result known as Hoeffding bounds or additive Chernoff bounds. After n independent observations of a real-valued random variable r with range R, the Hoeffding_bound ensures that, with confidence 1-δ, the true mean of r is at least $r - \varepsilon$, where r is the

observed mean of samples and $\varepsilon = \sqrt{\dfrac{R^2 \ln(1/\delta)}{2n}}$. This is true irrespective of the

probability distribution that generated the observations.

Let G($\cdot$) be the evaluation function of an attribute. For the information gain, the range R, of G($\cdot$) is log_2 #classes. Let x_a be the attribute with the highest G($\cdot$), x_b the attribute with second-highest G($\cdot$) and $\_\Delta G = \overline{G}(x_a) - \overline{G}(x_b)$, the difference between the two better attributes. Then if $\Delta G > \varepsilon$ with n examples observed in the leaf, the Hoeffding bound states with probability 1-δ that x_a is really the attribute with highest value in the evaluation function. In this case the leaf must be transformed into a decision node that splits on x_a.

For continuous attribute, whenever VFDT starts a new leaf, it collects up to M distinct values for each continuous attribute from the first examples that arrive at it. These are maintained in sorted order as they arrive, and a candidate test threshold is maintained midway between adjacent values with different classes, as in the traditional method. Once VFDT has M values for an attribute, it stops adding new candidate thresholds and uses additional data only to evaluate the existing ones. Every leaf uses a different value of M, based on its level in the tree and the amount of RAM available when it is started. For example, M can be very large when choosing the split for the root of the tree, but must be very small once there is a large partially induced tree, and many leaves are competing for limited memory resources. Notice that even when M is very large (and especially when it is small) VFDT may miss the locally optimal split point. This is not a serious problem here for two reasons. First, if data is an independent, identically distributed sample, VFDT should end up with a value near (or an empirical gain close to) the correct one simply by chance. And second, VFDT will be learning very large trees from massive data streams and can correct early mistakes later in the learning process by adding additional splits to the tree.

Thinking of each continuous attribute, we will find that the processing time for the insertion of new examples is $O(n^2)$, where n represents the number of distinct values for the attribute seen so far.

2.2 VFDTc

VFDTc is implemented on top of VFDT, and it extends VFDT in two directions: the ability to deal with continuous attributes and the use of more powerful classification techniques at tree leaves. Here, we just focus on the handling of continuous attributes.

In VFDTc a decision node that contains a split-test based on a continuous attribute has two descendant branches. The split-test is a condition of the form $attrib_j \leq T$. The two descendant branches correspond to the values $TRUE$ and $FALSE$ for the split-test. The cut point is chosen from all the possible observed values for that attribute. In order to evaluate the goodness of a split, it needs to compute the class distribution of the examples at which the attribute-value is less than or greater than the cut point. The counts n_{ijk} are fundamental for computing all necessary statistics. They are kept with the use of the following data structure: In each leaf of the decision tree it maintains a vector of the classes' distribution of the examples that reach this leaf. For each continuous attribute j, the system maintains a binary attribute tree structure. A node in the binary tree is identified with a value i(that is the value of the attribute j seen in an example), and two vectors (of dimension k) used to count the values that go through that node. Two vectors, VE and VH contain the counts of values respectively $\leq i$ and $> i$ for the examples labeled with class k. When an example reaches leaf, all the binary trees are updated. In [5], an algorithm of inserting a value in the binary tree is presented. Insertion of a new value in this structure is $O(nlogn)$ where n represents the number of distinct values for the attribute seen so far.

To obtain the Information Gain of a given attribute, VFDTc uses an exhaustive method to evaluate the merit of all possible cut points. Here, any value observed in the examples seen so far can be used as cut point. For each possible cut point, the information of the two partitions is computed using equation 1.

$$\inf o(A_j(i)) = P(A_j \leq i) * iLow(A_j(i)) + P(A_j > i) * i\,High(A_j(i)) \quad (1)$$

Where i is the cut point, $iLow(A_j(i))$ the information of $A_j \leqslant i$ (equation 2) and $iHigh(A_j(i))$ the information of $A_j > i$ (equation 3).

$$iLow(A_j(i)) = -\sum_K P(K = k \mid A_j \leq i) * \log(P(K{=}k \mid A_j \leq i)) \qquad (2)$$

$$iHigh(A_j(i)) = -\sum_K P(K = k \mid A_j > i) * \log(P(K{=}k \mid A_j {>} i)) \qquad (3)$$

VFDTc only considers a possible *cut_point* if and only if the number of examples in each of subsets is higher than Pmin (a user defined constant) percentage of the total number of examples seen in the node. [5] Presents the algorithm to compute #(Aj≤ i) for a given attribute j and class k. The algorithm's processing time is O(logn), so the best split-test point calculating time is O(nlogn). Here, n represents the number of distinct values for the attribute seen so far at that leaf.

2.3 Soft Discretization

Soft discretization could be viewed as an extension of hard discretization, and the classical information measures defined in the probability domain have been extended to new definitions in the possibility domain based on fuzzy set theory [13]. A crisp set A_c is expressed with a sharp characterization function $A_c(a) : \Omega \to \{0,1\} : a \in \Omega$, alternatively a fuzzy set A is characterized with a membership function $A(a) : \Omega \to [0,1] : a \in \Omega$. The membership $A(a)$ is called the possibility of A to take a value $a \in \Omega$ [14]. The probability of fuzzy set A is defined, according to Zadeh [15], by $P_F(A) = \int_\Omega A(a)dP$, where dP is a probability measure on Ω, and the subscript F is used to denote the associated fuzzy terms. Specially, if A is defined on discrete domain $\Omega = \{a_1, ..., a_i, ..., a_m\}$, and the probability of $P(a_i) = p_i$ then its probability is $P_F(A) = \sum_{i=1}^{m} A(a_i) p_i$.

Let $Q = \{A_1, ..., A_k\}$ be a family of fuzzy sets on Ω. Q is called a fuzzy partition of Ω [16] when $\sum_{r=1}^{k} A_r(a) = 1, \forall a \in \Omega$.

A hard discretization is defined with a threshold T, which generates the boundary between two crisp sets. Alternatively, a soft discretization is defined by a fuzzy set pair which forms a fuzzy partition. In contrast to the classical method of non-overlapping partitioning, the soft discretization is overlapped. The soft discretization is defined with three parameters/functions, one is the cross point T, the other two are the membership functions of the fuzzy set pair A_1 and A_2: $A_1(a)+A_2(a)=1$. The cross point T, i.e. the localization of soft discretization, is determined based on whether it can maximize the information gain in classification, and the membership functions of the fuzzy set pair are determined according to the characteristics of attribute data, such as the uncertainty of the associated attribute.

3   Technique Details

Improving soft discretizaiont method, we implement a system named fVFDT on top of VFDT and VFDTc. It handles continuous attributes based on threaded binary search trees, and uses a more efficient best split-test point calculating method.

For discrete attributes, they are processed using the algorithm mentioned in VFDT [2]. Our fVFDT specially focus on continuous attribute handling.

3.1   Threaded Binary Search Tree Structure for Continuous Attributes

fVFDT maintains a threaded binary search tree for each continuous attribute. The threaded binary search tree data structure will benefit the procedure of inserting new example and calculating best split-test point.

For each continuous attribute i, the system maintains a threaded binary search tree structure. A node in the threaded binary search tree is identified with a value $keyValue$ (that is the value of the attribute i seen in the example), and a vector(of dimension k) used to count the values that go through that node. This vector $classTotals[k]$ contains the counts of examples which value is $keyValue$ and class labeled with k. A node manages $left$ and $right$ pointers for its left and right child, where its left child corresponds to $\leq keyValue$, while its right child corresponds to $>keyValue$. For the goodness of calculating the best split-test point, a node contains $prev$ and $next$ pointers for the previous and next node. At most, three nodes` $prev$ and $next$ pointers will be updated while new example arrives.

fVFDT maintains a $head$ pointer for each continuous attribute to traverse all the threaded binary trees.

3.2   Updates the Threaded Search Binary Tree While New Examples Arrives

One of the key problems in decision tree construction on streaming data is that the memory and computational cost of storing and processing the information required to obtain the best split-test point can be very high. For discrete attributes, the number of distinct values is typically small, and therefore, the class histogram does not require much memory. Similarly, searching for the best split predicate is not expensive if number of candidate split conditions is relatively small.

However, for continuous attributes with a large number of distinct values, both memory and computational costs can be very high. Many of the existing approaches are scalable, but they are multi-pass. Decision tree construction requires a preprocessing phase in which attribute value lists for continuous attributes are sorted [20]. Preprocessing of data, in comparison, is not an option with streaming data, and sorting during execution can be very expensive. Domingos and Hulten have described and evaluated their one-pass algorithm focusing only on discrete attributes [2], and in later version they uses sorted array to handle continuous attribute. This implies a very high memory and computational overhead for inserting new examples and determining the best split point for a continuous attribute.

In fVFDT a Hoeffding tree node manages a threaded binary search tree for each continuous attribute before it becomes a decision node.

Procedure InsertValueTBSTree(x, k, TBSTree)
  Begin
   while (TBSTree ->right != NULL || TBSTree ->left != NULL)
     if (TBSTree ->keyValue == x)   then   break;
     Elseif (TBSTree ->keyValue > x)   then TBSTree = TBSTree ->left;
     else TBSTree = TBSTree ->right;
   Creates a new node curr based on x and k;
   If (TBSTree.keyValue == x)   then   TBSTree.classTotals[k]++;
   Elesif (TBSTree.keyValue > x)   then   TBSTree.left = curr;
   else   TBSTree.right = curr;
   Threads the tree ;(The details of threading is in figure2)
  End

Fig. 1. Algorithm to insert value x of an example labeled with class k into a threaded binary search tree corresponding to the continuous attribute i

In the induction of decision trees from continuous-valued data, a suitable threshold T, which discretizes the continuous attribute i into two intervals: $atrr_i \leq T$ and $atrr_i > T$, is determined based on the classification information gain generated by the corresponding discretization. Given a threshold, the test $atrr_i \leq T$ is assigned to the left branch of the decision node while $atrr_i > T$ is assigned to the right branch. As a new example (x,k) arrives, the threaded binary search tree corresponding to the continuous attribute i is update as Figure 1.

In [5], when a new example arrives, $O(logn)$ binary search tree nodes need be updated, but fVFDT just need update a necessary node here. VFDT will cost $O(n^2)$, and our system fVFDT will just cost $O(nlogn)$ (as presented in Figure 1) in execution time for values inserting, where n represents the number of distinct values for the given attribute seen so far.

3.3 Threads the Binary Tree While New Example Arrives

fVFDT need thread the binary search trees while new example arrives. If the new example's value is equal to an existing node's value, the threaded binary tree doesn't need be threaded. Otherwise, the threaded binary tree need be threaded as Figure 2.

At most, three relevant nodes need be updated here. This threading procedure mentioned in Figure 2 can be embedded in the procedure presented in Figure 1, and the inserting procedure's processing time is still $O(nlogn)$.

```
                Procedure TBSTthreads()
                Begin
                if (new node curr is left child of ptr)
                        curr->next = ptr;
                        curr->nextValue = ptr->keyValue;
                        curr->prev = ptr->prev;
                        ptr->prev->next = curr;
                        prevPtr->nextValue = value;
                        ptr->prev = curr;
                        if (new node curr is right child of ptr)
                                curr->next = ptr->next;
                                curr->nextValue = ptr->nextValue;
                                curr->prev = ptr;
                                ptr->next->prev = curr;
                                ptr->nextValue = value;
                                ptr->next = curr;
        End
```

Fig. 2. Algorithm to thread the binary search tree while new example arrives

3.4 Soft Discretization of Continuous Attributes

Taking advantage of threaded binary search tree, we use a more efficient method to obtain the fuzzy information gain of a given attribute.

Assuming we are to select an attribute for a node having a set S of N examples arrived, these examples are managed by a threaded binary tree according to the values of the continuous attribute i ; and an ordered sequence of distinct values $a_1, a_2 \ldots a_n$ is formed. Every pair of adjacent data points suggests a potential threshold $T = (a_i + a_{i+1})/2$ to create a cut point and generate a corresponding partition of attribute i. In order to calculate the goodness of a split, we need to compute the class distribution of the examples at which the attribute value is less than or greater than threshold T. The counts *TBSTree.classTotals[k]* are fundamental for computing all necessary statistics.

To take the advantage of threaded binary search tree, we record the *head* pointer of each attribute's threaded binary search tree. As presented in Figure 3, traversing from the *head* pointer to the tail pointer, we can easily compute the fuzzy information of all the potential thresholds. fVFDT implies soft discretization by managing Max/Min value and example numbers.

```
    Procedure BSTInorderAttributeSplit(TBSTtreePtr ptr,int *belowPrev[])
    Begin
        if ( ptr->next == NULL) then    break;
        for ( k = 0 ; k < count ; k++)
                *belowPrev[k] += ptr->classTotals[k];
        Calculates the information gain using *belowPrev[];
            BSTInorderAttributeSplit( ptr->next,int *belowPrev[]);
        End
```

Fig. 3. Algorithm to compute the information gain of a continuous attribute

Here, VFDTc will cost $O(n \log n)$, and our system fVFDT will just cost $O(n)$ in processing time, where n represents the number of distinct values for the given continuous attribute seen so far.

3.5 Classify a New Example

The classification for a given unknown object is obtained from the matching degrees of the object to each node from root to leaf. The possibility of an object belonging to class C_i is calculated by a fuzzy product operation $\otimes$. In the same way, the possibility of the object belonging to each class can be calculated, $\{\Pi_i\}_{i=1...k}$. If more than one leaf are associated with a same class C_i, say, the value of $\Pi_i = \oplus(\Pi_j)$ will be considered as the possibility of the corresponding class, where the maximum operation is used as the fuzzy sum operation $\oplus$ In the end, if one possibility value, such as Π_k , is much higher than others, that is $\Pi_k \gg \Pi_{i...k}$, then the class will be assigned as the class of the object, otherwise the decision tree predicts a distribution over all the classes.

4 Evaluation

In this section we empirically evaluate fVFDT. The main goal of this section is to provide evidence that the use of threaded binary search tree decreases the processing time of VFDT, while keeps the same error rate and tree size. The algorithms` processing time is listed in Table 1.

Table 1. Algorithm's processing time

Algorithm Name	Inserting time	Best split-test point calculating time
VFDT	$O(n^2)$	$O(n)$
VFDTc	$O(n \log n)$	$O(n \log n)$
fVFDT	$O(n \log n)$	$O(n)$

We first describe the data streams used for our experiments. We use a tool named *treeData* mentioned in [2] to create synthetic data .It creates a synthetic data set by sampling from a randomly generated decision tree. They were created by randomly generating decision trees and then using these trees to assign classes to randomly generated examples. It produced the random decision trees as follows. Starting from a tree with a single leaf node (the root) it repeatedly replaced leaf nodes with nodes that tested a randomly selected attribute which had not yet been tested on the path from the root of the tree to that selected leaf. After the first three levels of the tree each selected leaf had a probability of f of being pre-pruned instead of replaced by a split (and thus of remaining a leaf in the final tree). Additionally, any branch that reached a depth of 18 was pruned at that depth. Whenever a leaf was pruned it was randomly (with uniform probability) assigned a class label. A tree was completed as soon as all of its leaves were pruned.

VFDTc`s goal is to show that using stronger classification strategies at tree leaves will improve classifier's performance. With respect to the processing time, the use of naïve Bayes classifier will introduce an overhead [5], VFDTc is slower than VFDT. In order to compare the VFDTc and fVFDT , we implement the continuous attributes solving part of VFDTc ourselves.

We ran our experiments on a Pentium IV/2GH machine with 512MB of RAM, which running Linux RedHat 9.0.

Table 2 shows the processing (excluding I/O) time of learners as a function of the number of training examples averaged over nine runs. VFDT and fVFDT run with

parameters $\delta = 10^{-7}, \tau = 5\%, n_{\min} = 300, example\ number = 100000K$, no

leaf reactivation, and no rescan. Averagely, comparing to VFDT, fVFDT`s average reduction of processing time is 16.66%, and comparing to VFDTc, fVFDT`s average reduction is 6.25%.

Table 2. The comparing result of processing time

time(seconds) / example numbers	VFDT	VFDTc	fVFDT
10000	4.66	4.21	3.75
20736	9.96	8.83	8.12
42996	22.88	20.59	18.57
89156	48.51	43.57	40.87
184872	103.61	93.25	87.12
383349	215.83	187.77	175.23
794911	522.69	475.65	441.61
1648326	1123.51	1022.39	939.35
3417968	2090.31	1839.45	1758.89
7087498	3392.94	3053.65	2882.23
14696636	5209.47	4688.53	4389.35
30474845	8203.05	7382.75	6850.12
43883922	13431.02	11953.61	11068.23
90997707	17593.46	15834.12	15020.46
100000000	18902.06	16822.86	15986.23

In this work, we measure the size of tree models as the number of decision nodes plus the number of leaves. As for dynamic data stream with 100 million examples, we notice that the two learners similarly have the same tree size. We have done another experiment using 1 million examples generated on disk, and the result shows that they have same tree size.

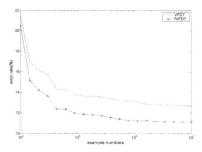

Fig. 4. Error rate as a function of the examples numbers

Figure 4 shows the error rate curves of VFDT and fVFDT. Both algorithms have 10% noise data, VFDT`s error rate trends to 12.5%, while the fVFDT`s error rate trends to 8%. Experiment results show that fVFDT get better accuracy by using soft discretization, and it overcomes the problem of noise.

5   Conclusions and Future Work

On top of VFDT and VFDTc, improve the soft discretization method, we propose a system fVFDT. Focusing on continuous attribute, we have developed and evaluated a new technique named TBST to insert new example and calculate best split-test point efficiently. It builds threaded binary search trees, and its processing time for values insertion is $O(nlogn)$. Comparing to the method used in VFDTc, it improves from $O(nlogn)$ to $O(n)$ in processing time for best split-test point calculating. As for noise data, we improve the soft discretization method in traditional data mining field, so the fVFDT can deal with noise data efficiently and improve the classification accuracy.

In the future, we would like to expand our work in some directions. First, we do not discuss the problem of time changing concept here, and we will apply our method to those strategies that take into account concept drift [4, 6, 10, 14, 15, 19, 24, 25]. Second, we want to apply other new fuzzy decision tree methods in data streams classification [8, 13, 17, 18, 26].

References

[1] Babcock, B., Babu, S., Datar, M., Motawani, R., Widom, J.: Models and Issues in Data Stream Systems. In: PODS (2002)
[2] Domingos, P., Hulten, G.: Mining High-Speed Data Streams. In: Proceedings of the Association for Computing Machinery Sixth International Conference on Knowledge Discovery and Data Mining, pp. 71–80 (2000)

[3] Mehta, M., Agrawal, A., Rissanen, J.: SLIQ: A Fast Scalable Classifier for Data Mining. In: Proceedings of The Fifth International Conference on Extending Database Technology, Avignon, France, pp. 18–32 (1996)

[4] Fan, W.: StreamMiner: A Classifier Ensemble-based Engine to Mine Concept Drifting Data Streams, VLDB'2004 (2004)

[5] Gama, J., Rocha, R., Medas, P.: Accurate Decision Trees for Mining High-Speed Data Streams. In: Domingos, P., Faloutsos, C. (eds.) Proceedings of the Ninth International Conference on Knowledge Discovery and Data Mining, ACM Press, New York (2003)

[6] Hulten, G., Spencer, L., Domingos, P.: Mining Time-Changing Data Streams, ACM SIGKDD (2001)

[7] Jin, R., Agrawal, G.: Efficient Decision Tree Construction on Streaming Data. In: proceedings of ACM SIGKDD (2003)

[8] Last, M.: Online Classification of Nonstationary Data Streams. Intelligent Data Analysis 6(2), 129–147 (2002)

[9] Muthukrishnan, S.: Data streams: Algorithms and Applications. In: Proceedings of the fourteenth annual ACM-SIAM symposium on discrete algorithms (2003)

[10] Wang, H., Fan, W., Yu, P., Han, J.: Mining Concept-Drifting Data Streams using Ensemble Classifiers. In: the 9th ACM International Conference on Knowledge Discovery and Data Mining, Washington DC, USA. SIGKDD (2003)

[11] Arasu, A., Babcock, B., Babu, S., Datar, M., Ito, K., Nishizawa, I., Rosenstein, J., Widom, J.: STREAM: The Stanford Stream Data Manager Demonstration Description –Short Overview of System Status and Plans. In: Proc. of the ACM Intl Conf. on Management of Data (SIGMOD 2003) (June 2003)

[12] Aggarwal, C., Han, J., Wang, J., Yu, P.S.: On Demand Classification of Data Streams. In: Proc. 2004 Int. Conf. on Knowledge Discovery and Data Mining (KDD'04), Seattle, WA (2004)

[13] Guetova, M., Holldobter, Storr, H.-P.: Incremental Fuzzy Decision Trees. In: 25th German conference on Artificial Intelligence (2002)

[14] Ben-David, S., Gehrke, J., Kifer, D.: Detecting Change in Data Streams. In: Proceedings of VLDB (2004)

[15] Aggarwal, C.: A Framework for Diagnosing Changes in Evolving Data Streams. In: Proceedings of the ACM SIGMOD Conference (2003)

[16] Gaber, M.M., Zaslavskey, A., Krishnaswamy, S.: Mining Data Streams: a Review, SIGMOD Record, vol. 34(2) (June 2005)

[17] Cezary, Janikow, Z.: Fuzzy Decision Trees: Issues and Methods. IEEE Transactions on Systems, Man, and Cybernetics 28(1), 1–14 (1998)

[18] Utgoff, P.E.: Incremental Induction of Decision Trees. Machine Learning 4(2), 161–186 (1989)

[19] Xie, Q.H.: An Efficient Approach for Mining Concept-Drifting Data Streams, Master Thesis

[20] Quinlan, J.R.: C4.5: Programs for Machine Learning. Morgan Kaufmann, San Mateo, CA (1993)

[21] Hoeffding, W.: Probability Inequalities for Sums of Bounded Random Variables. Journal of the American Statistical Association 58, 13–30 (1963)

[22] Breiman, L., Friedman, J.H., Olshen, R.A., Stone, C.J.: Classification and Regression Trees, Wadsworth, Belmont, CA (1984)

[23] Maron, O., Moore, A.: Hoeffding Races: Accelerating Model Selection Search for Classification and Function Approximation. In: Cowan, J.D., Tesauro, G., Alspector, J. (eds.) Advances in Neural Information Processing System (1994)

[24] Kelly, M.G., Hand, D.J., Adams, N.M.: The Impact of Changing Populations on Classifier Performance. In: Proc. of KDD-99, pp. 367–371 (1999)
[25] Black, M., Hickey, R.J.: Maintaining the Performance of a Learned Classifier under Concept Drift. Intelligent Data Analysis 3, 453–474 (1999)
[26] Maimon, O., Last, M.: Knowledge Discovery and Data Mining, the Info-Fuzzy Network(IFN) Methodology. Kluwer Academic Publishers, Boston (2000)
[27] Fayyad, U.M., Irani, K.B.: On the Handling of Continuous-valued Attributes in Decision Tree Generation. Machine Learning 8, 87–102 (1992)
[28] Wang, T., Li, Z., Yan, Y., Chen, H.: An Efficient Classification System Based on Binary Search Trees for Data Streams Mining, ICONS (2007)
[29] Wang, T., Li, Z., Hu, X., Yan, Y., Chen, H.: A New Decision Tree Classification Method for Mining High-Speed Data Streams Based on Threaded Binary Search Trees, PAKDD (2007) workshop (2007)
[30] Peng, Y.H., Flach, P.A.: Soft Discretization to Enhance the Continuous Decision Tree Induction. In: Proceedings of ECML/PKDD-2001 Workshop IDDM-2001, Freiburg, Germany (2001)

On the Combination of Locally Optimal Pairwise Classifiers

Gero Szepannek, Bernd Bischl, and Claus Weihs

Department of Statistics
University of Dortmund
44227 Dortmund
szepannek@statistik.uni-dortmund.de

Abstract. If their assumptions are not met, classifiers may fail. In this paper, the possibility of combining classifiers in multi-class problems is investigated. Multi-class classification problems are split into two class problems. For each of the latter problems an optimal classifier is determined. The results of applying the optimal classifiers on the two class problems can be combined using the *Pairwise Coupling* algorithm by Hastie and Tibshirani (1998).

In this paper exemplary situations are investigated where the respective assumptions of Naive Bayes or the classical Linear Discriminant Analysis (LDA, Fisher, 1936) fail. It is investigated at which degree of violations of the assumptions it may be advantageous to use single methods or a classifier combination by Pairwise Coupling.

1 Introduction

When talking about ensemble methods one usually has in mind very popular principles like *Bagging* (Breiman, 1996) or *Boosting* (Freund and Shapire, 1997). Both rely on combinations of different classification rules that are build on sampled or weighted instances of the original data. In this paper a somewhat different perspective to combining classifiers for multi-class problems is worked out: the basic observation is that classifiers only work well if their underlying assumptions hold, e.g. classwise independent features in the case of a *Naive Bayes* classifier. This may be the case for some but not necessarily all of the classes.

Pairwise Coupling (PWC) generates $K(K-1)/2$ subsamples of the data (K being the number of classes) each consisting only of objects of one specific *pair of classes*. For these two classes, an optimal classifier is determined, e.g. using cross-validation. According to the thoughts presented above, the optimal classifier may be a different one for different class pairs.

When classifying new data, of course in general no prior information is available to which pair of classes an object belongs. Thus, one can make use of the *Pairwise Coupling* algorithm of Hastie and Tibshirani (1998) and apply the prediction models for all class pairs and then construct posterior probabilities (and thus a multi-class classification rule) from the results. The Pairwise Coupling algorithm will be explained in Section 2.

P. Perner (Ed.): MLDM 2007, LNAI 4571, pp. 104–116, 2007.

In this paper, the principle of combining pairwise optimal classifiers is investigated for the case of two very common classification methods, namely *Naive Bayes* and *Linear Discriminant Analysis* (Fisher, 1936). Both methods are briefly described in Section 3. In a simulation study the degree of violation of the assumption of both methods is varied. The results give quite an interesting indication of the robustness of both methods as well as they produce a 'map' that shows when to use whether one of the single classifiers or a combination. It turns out that in some situations a combination of pairwise optimized classifiers can strongly improve the classification results if the assumptions of single methods do not hold for all classes.

The following pseudo-code summarizes the steps of the suggested proceeding:

Build classification model *(data, set of classification methods)*

1. For each pair of two classes do
2. (a) Remove temporarily all observations that do not belong to one of both classes from *data*: return *newdata*.
   (b) For each *classifier* in *set of classification methods*
      − Build *classifier* on *newdata*.
      − Validate *classifier* e.g. using cross-validation.
      − Store Results temporally in *classifier results*.
   (c) Choose best *classifier* according to *classifier results* return *classifier of class-pair*.
   (d) Train *classifier of class-pair* on *newdata*:
   (e) Return *model of class-pair*.
3. Return the whole model consisting of *model of class-pair* for all pairs of classes.

Predict class *(new object, models of class-pairs)*

1. For each pair of subclasses do
2. (a) Calculate the posterior probabilities for *new object* assuming the object being of one of the currently considered two classes according to *model of class-pair*.
   (b) Return the *class pair posterior probabilities*.
3. Use the Pairwise Coupling algorithm to calculate the posterior probabilities for all K classes from the set of all estimated pairs of conditional *class pair posterior probabilities*.
4. Return the predicted class k with maximal *class posterior probability*.

The following section describes a solution to the problem of gaining the vector of posterior probabilities form the pairwise classification models built with possibly different classifiers.

2   Pairwise Coupling

2.1   Definitions

We now tackle the problem of finding posterior probabilities of a K-(sub)class classification problem given the posterior probabilities for all $K(K-1)/2$ pairwise comparisons. Let us start with some definitions.

Let $p(x) = p = (p_1, \ldots, p_K)$ be the vector of (unknown) posterior probabilities. p depends on the specific realization x. For simplicity in notation we will omit x. Assume the "true" conditional probabilities of a pairwise classification problem to be given by

$$\mu_{ij} = Pr(i|i \cup j) = \frac{p_i}{p_i + p_j} \ . \tag{1}$$

Let r_{ij} denote the estimated posterior probabilities of the two-class problems. The aim is now to find the vector of probabilities p_i for a given set of values r_{ij}.

Example 2: Let $p = (0.7, 0.2, 0.1)$. The μ_{ij} can be calculated according to equation 1 and can be presented in a matrix:

$$(\mu_{ij})_{i,j} = \begin{pmatrix} . & 7/9 & 7/8 \\ 2/9 & . & 2/3 \\ 1/8 & 1/3 & . \end{pmatrix} . \tag{2}$$

Example 3: The inverse problem does not necessarily have a proper solution, since there are only $K - 1$ free parameters but $K(K-1)/2$ constraints. Consider

$$(r_{ij})_{i,j} = \begin{pmatrix} . & 0.9 & 0.4 \\ 0.1 & . & 0.7 \\ 0.6 & 0.3 & . \end{pmatrix} \tag{3}$$

where the row i contains the estimated conditional pairwise posterior probabilities r_{ij} for class i. It can be easily checked that the linear system resulting from applying equation 1 cannot be solved.

From Machine Learning, majority voting ("Which class wins most comparisons ?") is a well known approach to solve such problems. But here, it will not lead to a result since any class wins exactly one comparison. Intuitively, class 1 may be preferable since it dominates the comparisons the most clearly.

2.2   Algorithm

In this section we present the Pairwise Coupling algorithm of Hastie and Tibshirani (1998) to find p for a given set of r_{ij}. They transform the problem into an iterative optimization problem by introducing a criterion to measure the fit between the observed r_{ij} and the $\hat{\mu}_{ij}$, calculated from a possible solution $\hat{p}$. To measure the fit they define the weighted Kullback-Leibler distance:

$$l(\hat{p}) = \sum_{i<j} n_{ij} \left(r_{ij} \ln \left(\frac{r_{ij}}{\hat{\mu}_{ij}} \right) + (1 - r_{ij}) \ln \left(\frac{1 - r_{ij}}{1 - \hat{\mu}_{ij}} \right) \right) \ . \tag{4}$$

n_{ij} is the number of objects that fall into one of the classes i or j.

The best solution $\hat{p}$ of posterior probabilities is found as in Iterative Proportional Scaling (IPS) (for details on the IPS-method see e.g. Bishop, Fienberg and Holland, 1975). The algorithm consists of the following three steps:

1. Start with any $\hat{p}$ and calculate all $\hat{\mu}_{ij}$.
2. Repeat until convergence $i = (1, 2, \ldots, K, 1, \ldots)$:

$$\hat{p}_i \leftarrow \hat{p}_i \frac{\sum_{j \neq i} n_{ij} r_{ij}}{\sum_{j \neq i} n_{ij} \hat{\mu}_{ij}} , \qquad (5)$$

   renormalize $\hat{p}$ and calculate the new $\hat{\mu}_{ij}$.
3. Finally scale the solution to $\hat{p} \leftarrow \frac{\hat{p}}{\sum_i \hat{p}_i}$.

Motivation of the algorithm: Hastie and Tibshirani (1998) show that $l(p)$ increases at each step. For this reason, since it is bounded above by 0, $l(p)$ converges. providing $\hat{\mu}_{ij} = r_{ij} \, \forall \, i \neq j$, it will be found.

Even if the choice of $l(p)$ as optimization criterion is rather heuristic, it can be motivated in the following way: consider a random variable $n_{ij}r_{ij}$, being the number of observations of class i among the n_{ij} observations of class i and j. This random variable can be considered to be binomially distributed $n_{ij}r_{ij} \sim B(n_{ij}, \mu_{ij})$ with "true" (unknown) parameter μ_{ij}. Since the same (training) data is used for all pairwise estimates r_{ij}, the r_{ij} are not independent, but if they were, $l(p)$ of equation 4 would be equivalent to the log-likelihood of this model (see Bradley and Terry, 1952). Then, maximizing $l(p)$ would correspond to maximum-likelihood estimation for μ_{ij}.

Going back to example 3, we obtain $\hat{p} = (0.47, 0.25, 0.28)$, a result being consistent with the intuition that class 1 may be slightly preferable.

In Wu et al. (2004) several methods for multi-class probability estimation by Pairwise Coupling algorithms are presented and compared. For the simulations of this paper, the method of Hastie and Tibshirani (1998) is used.

3   Implemented Methods

3.1   Linear Discriminant Analysis

In its classical form *Linear Discriminant Analysis* was constructed by R. Fisher in 1936 for linear reduction of dimensionality to maximize the distance of class means w.r.t. the covariance structure of the data.

The method is shown to be optimal in the sense that it minimizes the *Bayes Risk* if the underlying class distributions follow normal law but have equal co-variance matrices for all classes (see e.g. Hastie and Tibshirani, 2001, p.95).

The classification for an object x is obtained by maximizing the decision rule $\hat{d}_k(x)$ over all classes k:

$$\hat{d}_k(x) = \bar{x}_k \hat{\Sigma}^{-1} x - \frac{1}{2} \bar{x}_k \hat{\Sigma}^{-1} \bar{x}_k + \ln(\pi(k)) \qquad (6)$$

with $\pi(k)$ being the class prior membership probabilities, $\bar{x}_k$ denoting the mean of class k and $\hat{\Sigma}$ being the *pooled covariance matrix*

$$\hat{\Sigma} = \frac{1}{N - K} \sum_{k=1}^{K} \sum_{n=1}^{N} I_{[k]}(k_n)(x_n - \bar{x}_k)(x_n - \bar{x}_k)'. \qquad (7)$$

Here $I_{[k]}(k_n)$ represents the indicator function that becomes 1 if object n of the training data is of class k and 0 if not. The term pooled covariance follows from the fact that equation 7 can be reformulated in terms of the classwise covariance estimations $\hat{\Sigma}_k$:

$$\hat{\Sigma} = \frac{1}{N-K} \sum_{k=1}^{K} n_k \hat{\Sigma}_k \tag{8}$$

where n_k denotes the size of class k in the training data.

The classification rule linearly partitions the feature space. This is shown in Figure 1 for the first two dimensions of the well known iris data from Fisher (1936).

Hastie and Tibshirani (2001, p.89) mention that Linear Discriminant Analysis often shows good results and is among top 3 classifiers for 7 of 22 real world data data sets of the Statlog project (Michie et al., 1994).

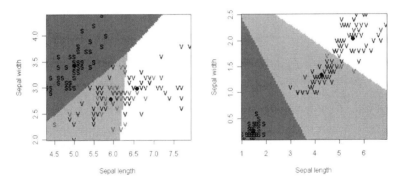

Fig. 1. Two-dimensional projections of the partition of the feature space using Linear Discriminant Analysis on Iris data

3.2 Naive Bayes

When using the *Naive Bayes* method features are assumed to be conditionally independent given the class. For each class k and variable d mean $\hat{\mu}_{d,k}$ and covariance $\hat{\sigma}_{d,k}$ are estimated.

For a new observation x the likelihood $P_d(x|k)$ of its realization in variable d given class k can be calculated then assuming normal distribution.

Finally, the predicted class is obtained my maximizing the decision rule

$$\hat{d}_k(x) = \pi(k) \prod_d P_d(x|k). \tag{9}$$

with $\pi(k)$ again denoting the prior probability of class k. Doing so implicitly assumes no correlations between the different variables d given the class: the covariance matrix of class k Σ_k is assumed to be 0 for all elements except for the main diagonal elements.

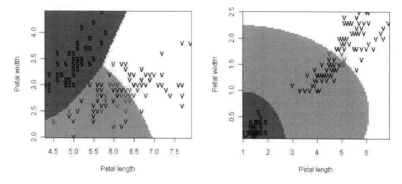

Fig. 2. Two-dimensional projections of the partition of the feature space using Naive Bayes on Iris data.

This dramatically decreases the number of free model parameters, especially if the number of features is large. Another advantage of the Naive Bayes method may be that equal variances are not assumed as it is done in LDA. Nevertheless, it may be disadvantageous if there are strong correlations among the predictor variables.

4   Simulation Study

4.1   An Introductory Example

To gain some insight into the merit of the method a synthetic example was constructed. This example consisted of four equally large classes in two-dimensional space, all normally distributed (see Fig. 3, the different classes are labelled with numbers from 1 to 4).

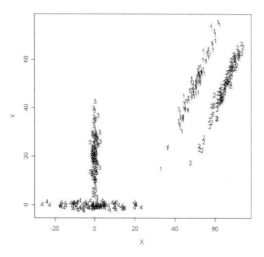

Fig. 3. First example of simulated data

Classes 1 and 2 have an equal covariance structure and can therefore be optimally separated by an LDA classifier, but not by the Naive Bayes method since the input variables are not independent given the class.

Likewise, as classes 3 and 4 have uncorrelated features given the class, they can therefore be optimally identified by the Naive Bayes method, but LDA will produce a higher error because the underlying normal distributions do not have an equal covariance matrix.

Table 1. Test error rates on synthetic example of Fig. 3 (400 samples per class, 2/3 training data and 1/3 test data)

Method	Test Error
LDA	0.07
Naive Bayes	0.14
PWC	**0.01**

It is now conjectured that by training a PWC classifier on the dataset, a LDA-classifier is chosen to separate the first pair of classes and a Naive Bayes classifier for the latter pair. This expected behaviour can be observed on the simulated data. The results show a strong increase in classification performance on separately simulated test data when combing both classifiers as opposed to use only the base methods (see table 3).

4.2 Experimental Setting

In oder to investigate when it is beneficial to use one of the base methods or their classification using Pairwise Coupling (and choosing the pairwise optimal classifier based on cross-validated error rates) a study is performed with simulated data as in Section 4.1 but with varying degree of violated assumptions for both methods:

Four normally distributed classes are generated with class expectations:

$$\mu_1 = (50, 50)'$$
$$\mu_2 = (65, 50)'$$
$$\mu_3 = (0, 20)'$$
$$\mu_4 = (0, 0)'.$$

The class covariance matrices are constructed as a convex combination of four extreme cases:

$$\Sigma_1^*(\rho) = \Sigma_2^*(\rho) = \begin{pmatrix} \sigma_1^2 & \rho\sigma_1\sigma_2 \\ \rho\sigma_1\sigma_2 & \sigma_2^2 \end{pmatrix}$$

and

$$\Sigma_3^*(\rho) = \Sigma_3^* = \begin{pmatrix} \sigma_3^2 & 0 \\ 0 & \sigma_4^2 \end{pmatrix}$$

$$\Sigma_4^*(\rho) = \Sigma_4^* = \begin{pmatrix} \sigma_4^2 & 0 \\ 0 & \sigma_3^2 \end{pmatrix}$$

with $\sigma_1 = 5$, $\sigma_2 = 10$, $\sigma_3 = 1$ and $\sigma_4 = 10$.

The covariance matrices of class 1 and 2 exactly hold the assumptions that underly Linear Discriminant Analysis, since covariances are greater 0 but the same for both classes. The covariance matrices of class 3 and 4 represent the 'Naive Bayes - case' since the variables are independent but have different variances for both classes.

The covariance Σ_i of class i is set to be

$$\Sigma_i(\alpha, \rho) = \alpha \Sigma_1^*(\rho) + (1 - \alpha)\Sigma_i^*(\rho) \tag{10}$$

The parameter $\alpha \in [0, 1]$ determines how equal the class covariance matrices look like, the larger α is the more equal they are. For $\alpha = 1$ all classes' covariances equal to $\Sigma_1^*(\rho)$. Then, the assumptions of LDA holds.

The free parameter $\rho \in [0, 1]$ determines the correlation in $\Sigma_1^*(\rho)$. $\rho = 0$ means independent variables for all classes as it is assumed for the Naive Bayes method. Four exemplary situations are shown in Figure 4: The upper left figure shows

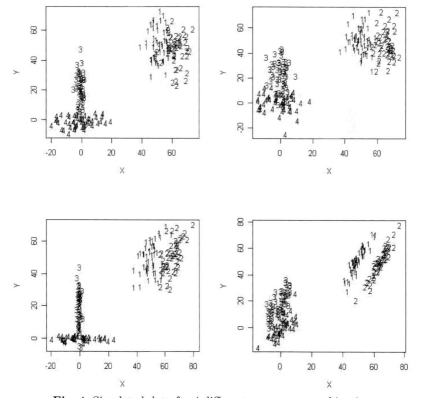

Fig. 4. Simulated data for 4 different parameter combinations

simulated data for $\alpha = 0.1, \rho = 0.1$: all classes possess quite specific covariance matrices with very small correlations among the variables. This should be a case where the Naive Bayes method can be assumed to produces good results. The upper right figure illustrates simulated data for $\alpha = 0.5, \rho = 0$: for all classes the variables are completely uncorrelated but the class-specific covariance structure is not as present as in the example before. The bottom left figure illustrates the data situation for $\alpha = 0, \rho = 0.5$: The covariance matrices of the classes are unique and the variables of class 3 and 4 are correlated. Both parameters are set to $\alpha = \rho = 0.9$ in the bottom right figure: The covariance shapes of the classes look very similar and contain strong correlations. In this situation, the assumptions of Linear Discriminant Analysis are quite well met.

For our simulation study both parameters α and ρ are varied in the interval $[0, 1]$. For each simulation 400 observations are generated for each class. The data are split into $2/3$ training data. The last third is used for testing. The locally optimal classifiers are chosen by 3-fold cross-validation.

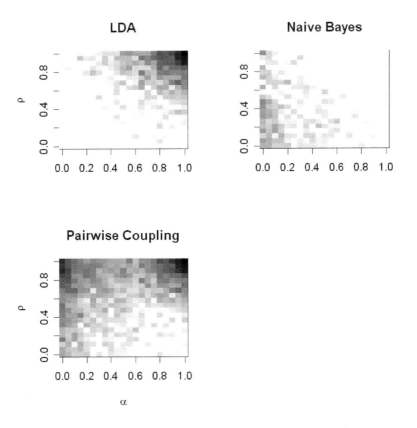

Fig. 5. Results of the simulation study for the different methods (scaled between 0 error (black) and the worst result (white)

4.3   Results

The results of the simulation study are shown in Figure 5. The first three plots show the results of the two base methods as well as their combination using PWC for our simulated data.

For each pair of simulation parameters (α, ρ) the results are scaled: black indicates a test error rate of 0 while white denotes the worst obtained result.

It can be easily recognized that LDA performs best for both high parameters of ρ and α, i.e. equal covariance matrices of all classes and strong correlations between the variables. Using Naive Bayes is advantageous for a low parameter α, i.e. strongly differing covariance matrices of the classes, especially if there are furthermore low correlations in the variables. Combining both classifiers is a good compromise in most situations except if there are equal covariance matrices with small correlations of all classes. A strong benefit can be obtained if the covariance matrices of the classes are not equal and there are also strongly correlated variables, i.e. if the assumptions of both base methods do not hold.

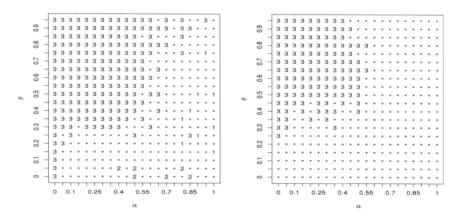

Fig. 6. Best method in dependence of the parameter if its test error is siginifanctly better than the error of the best competitor. * indicates that none of the methods significantly outperforms the other methods. Method 1: LDA; 2: Naive Bayes and 3: PWC. Left: simple significances, right: using Bonferroni-Holm correction for multiple testing.

To determine whether one method significantly outperforms the other two, the above mentioned simulation was conducted 30 times and for each pair of parameters a paired t-test between the test error rates winning and the second best method was applied (Dietterich, 1995). Figure 6 shows the results. In the left figure tests were performed with a simple significance level of 0.05, while in the right figure - in order to cope with the problem of multiple testing - results are given after adapting the significance levels by the Bonferroni-Holm method. One can observe that there are situations (strongly differing covariances between the classes combined with correlations that appear in the data) where the PWC approach leads to significant improvements of the misclassification rate. The base methods show a possible advantage in the areas where their underlying assumptions are met (low ρ for Naive Bayes and and high α for LDA implying equal covariance matrices in the classes). But these advantages proved not to be significant after adapted significance levels and might be caused by falsely rejecting the null hypothesis of equal mean error rates due to multiple testing.

5   Application to Real World Data

For not to restrict our analysis on the simulated data we also applied the methods to several real world multi-class data from the *UCI Machine Learning Repository*. An overview over some characteristics of the chosen data sets is given in Table 2. For an explicit description of the data sets see Michie et al. (1994) and Merz and Murphy (1998).

Table 2. Statistics of data sets

	Satellite	Vehicle	Nursery	Vowel
classes	6	4	5	11
features	36	18	8	10
examples	6435	846	12960	990

In each experiment the data were randomly split into a training and test set (2/3 and 1/3), except for the Satellite set, where the same 4435 examples as in Statlog (Michie et al., 1994)were used for training and the remaining 2000 examples for testing.

The results are given in Table 3 in terms of test error rates for both base methods as well as their combination. For the Satellite data, the error rates of the Naive Bayes method can be improved by a combined classifier but LDA performs overall best. For the Vehicle data set, Naive Bayes shows very bad results. The rates of LDA can even be slightly improved using a PWC classifier combination. For the Nursery data LDA shows very bad results. The error rates of Naive Bayes here can be improved by Pairwise Coupling. Finally, for the Vowel data set the recognition rates of both methods can be dramatically improved using a classifier combination. As a conclusion, the proposed local combination

Table 3. Test error rates MLBench

Method	Satellite	Vehicle	Nursery	Vowel
LDA	**0.15**	0.26	0.47	0.42
Naive Bayes	0.20	0.57	0.10	0.52
PWC	0.18	**0.23**	**0.08**	**0.17**

of classifiers sometimes yielded a large improvement of the results but never showed very bad performance compared to the winning base method method. This result is in harmony with the observations made in Section 4.

6 Summary

Classifier combination for multi-class classification problems is proposed in different way compared to the very common Bagging and Boosting approaches: for each pair of classes an optimal classifier is determined using cross-validation and class pairwise models are trained.

A new object is labelled by applying all classifiers for each class pair and then combining the results by Pairwise Coupling (Hastie and Tibshirani, 1998).

Such a proceeding may be advantageous in situations where the assumptions of the different base methods hold for different classes.

The benefit of such a classifier combination is investigated for two very common methods, namely Linear Discriminant Analysis and Naive Bayes. A simulation study is performed where the degree of violation of the specific assumptions for both methods is varied and finally a map is obtained that indicates when it is better to implement a single one of these methods or their combination.

Furthermore, the methods are applied to common real world problems from the UCI Machine Learning Repository. Recapitulating the results, it turned out that sometimes large improvements of the misclassification rate are achieved by using PWC while its results were never much worse than the winning base method.

It should also be mentioned that Moreira and Mayoraz (1998) proposed a different approach to build classifiers from class pairwise rules by calculating conditional probabilities for the membership of a new object to a class pair. A comparison to this approach was not the main interest of this study but may be a topic of further investigation as well as the investigation of the principle for other classifiers.

Finally – referring to the work of Dietterich and Bakiri (1995) – multiclass-classification problems can also be solved by transforming them into several binary classification problems using the method of *Error-Correcting Output Codes*. There basically, in every binary classification problem the K classes are grouped into two sets of classes which are then separated. The result is a sequence of binary classifiers. Each of the classes is coded by a vector of the binary group-labels. Prediction of an object is done by applying all classifiers and choosing the class with the most similar code vector.

PWC can be embedded in this context according to Allwein et al. (2000) and thus an extension of the suggested approach towards Error-Correcting Output Codes may also be topic of further investigation.

Acknowledgment. This work has been supported by the Collaborative Research Center 'Reduction of Complexity in Multivariate Data Structures' (SFB 475) of the German Research Foundation (DFG).

References

Allwein, E., Schapire, R., Singer, Y.: Reducing Multiclass to Binary: A Unifying Approach for Margin Classifiers. Journal of Machine Learning Research 1, 113–141 (2000)

Bishop, Y., Fienberg, S., Holland, P.: Discrete multivariate analysis. MIT Press, Cambridge (1975)

Bradley, R., Terry, M.: The rank analysis of incomplete block designs, i. the method of paired comparisons. Biometrics, pp. 324–345 (1952)

Breiman, L.: Bagging Predictors. Machine Learning 24(2), 123–140 (1996)

Dietterich, T., Bakiri, G.: Solving Multiclass Learning Problems via Error-Correcting Output Codes. Journal of Artificial Intelligence Research 2, 263–286 (1995)

Dietterich, T.: Statistical tests for comparing supervised classification learning algorithms. OR 97331, Department of Computer Science, Oregon State University 2 (1996)

Fisher, R.: The use of multiple measures in taxonomic problems. Annals of Eugenics 7, 179–188 (1936)

Freund, Y., Schapire, R.: A Decision-Theoretic Generalization of on-Line Learning and an Application to Boosting. Journal of Computer and System Sciences 55(1), 119–139 (1997)

Hastie, T., Tibshirani, R.: Classification by Pairwise Coupling. Annals of Statistics 26(1), 451–471 (1998)

Hastie, T., Tibshirani, R.: The elements of statistical learning - data mining, inference and prediction. Springer, NY (2001)

Merz, C., Murphy, P.: UCI Repository of Machine Learning Data Bases: Irvine, CA: University of California, Dept. of Information and Computer Science (1998), http://www.ics.uci.edu/~mlearn/mlrepository.html

Michie, D., Spiegelhalter, D., Taylor, C.: Machine Learning, Neural and Statistical Classification. Ellis Horwood Limited, Hertfordshire (1994)

Moreira, M., Mayoraz, E.: Improved Pairwise Coupling Classification with Correcting Classifiers. In: European Conference on Machine Learning, pp. 160–171 (1998)

Wu, T.-F., Lin, C.-J., Weng, R.: Probability Estimates for Multi-class Classification by Pairwise Coupling. Journal of Machine Learning Research 5, 975–1005 (2004)

An Agent-Based Approach to the Multiple-Objective Selection of Reference Vectors

Ireneusz Czarnowski and Piotr Jędrzejowicz

Department of Information Systems, Gdynia Maritime University
Morska 83, 81-225 Gdynia, Poland
{irek,pj}@am.gdynia.pl

Abstract. The paper proposes an agent-based approach to the multiple-objective selection of reference vectors from original datasets. Effective and dependable selection procedures are of vital importance to machine learning and data mining. The suggested approach is based on the multiple agent paradigm. The authors propose using JABAT middleware as a tool and the original instance reduction procedure as a method for selecting reference vectors under multiple objectives. The paper contains a brief introduction to the multiple objective optimization, followed by the formulation of the multiple-objective, agent-based, reference vectors selection optimization problem. Further sections of the paper provide details on the proposed algorithm generating a non-dominated (or Pareto-optimal) set of reference vector sets. To validate the approach the computational experiment has been planned and carried out. Presentation and discussion of experiment results conclude the paper.

1 Introduction

As it has been observed in [9], in supervised learning, a machine-learning algorithm is shown a training set, which is a collection of training examples called instances. After learning from the training set, the learning algorithm is presented with additional input vectors, and the algorithm must generalize, that is to decide what the output value should be.

It is well known that in order to avoid excessive storage and time complexity and to improve generalization accuracy by avoiding noise and overfitting, it is often advisable to reduce original training set by removing some instances before learning phase or to modify the instances using a new representation.

Instances reduction, often referred to as a selection of reference vectors, becomes especially important in case of large data sets, since overcoming storage and complexity constraints might become computationally very expensive. Although a variety of instance reduction methods has been so far proposed in the literature (see, for example the review [9]), no single approach can be considered as superior nor guaranteeing satisfactory results and a reduction of the learning error or increased efficiency of the supervised learning. Therefore, the problem of selecting the reference instances remains an interesting field of research.

P. Perner (Ed.): MLDM 2007, LNAI 4571, pp. 117–130, 2007.
© Springer-Verlag Berlin Heidelberg 2007

One of the most important application areas of the machine learning methods and tools is data mining understood as the extraction of implicit, previously unknown, and potentially useful information from data. Unfortunately, several useful machine learning tools and techniques as for example neural networks, support vector machines or statistical methods do not provide explanations on how they solve problems. In some application areas like medicine or safety assurance this may cause some doubts or even lower the trust of the users. In such cases users may prefer approaches where the process of knowledge extraction from data is easier comprehensible by human beings. An obvious approach would be using methods leading to the extraction of some logical rules representing the knowledge about phenomenon at hand. Extracting precise, reliable, useful and easy to comprehend rules from datasets is not a trivial task [10][13].

Most widely used techniques for the rules generation, such as, for example, algorithms C4.5 and CART [14][15], are based on decision trees. However in case of the large datasets the resulting decision tree might become very complex making it difficult to understand and evaluate by the human being. Possible way to overcome the problem is to select a set of reference vectors as an input to the decision tree generating algorithm producing than, so called, prototype-based model [10]. It is expected that instance reduction through selection of reference vectors may bring about several benefits including increased quality of generalization, easier to comprehend set of rules, decreased requirements for storage and computational resources and increased simplicity of the extracted knowledge.

Selecting reference vectors is inherently a multiple-objective problem. The resulting set should be evaluated not only in terms of generalization (classification) quality of the prototype model, but also in terms of the resulting number of rules, their complexity, data compression level, computational time required etc. Considering the above, in this paper the selection of reference vectors is seen as a multi-objective optimization problem which solution is a non-dominated (or Pareto-optimal) set of reference vector sets. To obtain solutions to such problems an agent-based approach is suggested.

The paper proposes the multiple-objective agent-based optimization of reference vectors selection algorithm, implemented using the JABAT environment. JABAT is a middleware supporting the construction of the dedicated A-Team architectures that can be used for solving a variety of computationally hard optimization problems [3].

The paper is organized as follows. Section 2 reviews briefly a general multiple-objective, optimization problem. Section 3 of the paper contains formulation of the multiple-objective, agent-based, reference vectors selection optimization problem. Section 4 provides details on the proposed algorithm generating a non-dominated (or Pareto-optimal) set of reference vector sets. To validate the approach the computational experiment has been planned and carried out. Its results are presented and discussed in Section 5. Finally, in the last section some conclusions are drawn and directions for future research are suggested.

2   Multiple-Objective Optimization

The general multiple-objective optimization problem is formulated following [11] as:

$$max\{z_1, \ldots, z_J\} = max\{f_1(x), \ldots, f_J(x)\} \tag{1}$$

or

$$min\{z_1, \ldots, z_J\} = min\{f_1(x), \ldots, f_J(x)\} \tag{2}$$

where $x \in D$ and solution $x = [x_1, \ldots, x_l]$ is a vector of decision variables, D is the set of feasible solutions and z is a vector of objective functions z_j, $j = 1, \ldots, J$. The type of the variables may describe different classes of problems. When the variables are discrete the multiple-objective optimization problem is called as multiple-objective combinational optimization problem.

The image of a solution x in the objective space is a point $z^* = [z_1^*, \ldots, z_J^*]$, where $z_j^* = f(x_j)$, $j = 1, \ldots, J$.

Point z dominates z', if, for the maximization case, $z_j \geq z_j'$ (for each j) and $z_j > z_j'$ for at least one j, and vice versa for the minimization problem.

A solution $x \in D$ is Pareto-optimal, if there is no $x' \in D$ that dominates x. A point being an image of Pareto-optimal solution is called non-dominated. The set of all Pareto-optimal solutions is called the Pareto-optimal set. The image of the Pareto-optimal set in objective space is called the non-dominated set.

An approximation of the non-dominated set is a set A of feasible points such that $\neg \exists z_1, z_2 \in A$ such that z_1 dominated z_2.

Weighted linear scalarizing functions are defined as:

$$s_l(z, \Lambda) = \sum_{j=1}^{J} \lambda_j z_j, \tag{3}$$

where $\Lambda = [l_1, \ldots, l_J]$ is a weight vector such that $\lambda_j \geq 0$ and $\sum_{j=1}^{J} \lambda_j = 1$.

Others scalarizing functions are based on calculation of distances between z_j and z_j^0, where z^0 is a references point. The weighted Tchebycheff scalarizing function may serve as an example of such a function. It is defined as follows:

$$s_\infty(z, z^0, \Lambda) = \max_j \{\lambda_j (z_j^0 - z_j)\}. \tag{4}$$

Further details in respect to the multiple-objective optimization can be found, for example, in [11].

3   Multiple-Objective Selection of Reference Instances

Instance reduction problem concerns removing a number of instances from the original training set T and thus producing the reduced training set S. Let N denote the number of instances in T and n-the number of attributes. Total length of each instance (i.e. training example) is equal to $n + 1$, where element numbered $n + 1$ contains the output value. Let also $X = \{x_{ij}\}$ $(i = 1, \ldots, N,$ $j = 1, \ldots, n+1)$ denote a matrix of $n+1$ columns and N rows containing values of all instances from T.

Usually, instance reduction algorithms are based on distance calculation between instances in the training set. In such a case selected instances, which are situated close to the center of clusters of similar instances, serve as the reference instances. The approach requires using some clustering algorithms. Other methods, known as similarity-based methods, remove k nearest neighbors from a given category based on an assumption that all instances from the neighbor will be, after all, correctly classified. The third group of methods eliminates training examples based on an evaluation using some removal criteria [9] [12].

In this paper instance reduction (or reference vector selection) is seen as a multiple-objective optimization problem. It can be solved by producing a set of Pareto-optimal solution instances each being a non-dominated set of reference vectors. The following criteria are used to evaluate reference vectors:

- Classification quality - f_1
- Data compression level - f_2
- Number of rules - f_3
- Length of rules -f_4

It is clear that the above set of criteria represents a situation with several conflicting goals. Selection of the preferred reference vector from the set of Pareto-optimal ones is left to the user. Hence, solving an instance reduction problem is seen as generating a set of non-dominated solutions each, in turn, representing a set of the selected reference vectors.

4 Agent-Based Algorithm for Generating Pareto-Optimal Sets of Reference Vectors

4.1 Instance Reduction Algorithm

It is proposed to base instance reduction on the idea of Instance Reduction Algorithm (IRA) proposed in the earlier paper of the authors [12]. The IRA was originally proposed as a tool for solving a single objective version of instance reduction problem. It was shown in [12] that the approach can result in reducing the number of instances and still preserving a quality of the data mining results. It has been also demonstrated that in some cases reducing the training set size can increase efficiency of the supervised learning. The proposed algorithm is based on calculating, for each instance from the original set, the value of its similarity coefficient, and then grouping instances into clusters consisting of instances with identical values of this coefficient, selecting the representation of instances for each cluster and removing the remaining instances, thus producing the reduced training set. The algorithm involves the following steps:

Stage 1. Transform X normalizing value of each x_{ij} into interval $[0, 1]$ and then rounding it to the nearest integer, that is 0 or 1.

Stage 2. Calculate for each instance from the original training set the value of its similarity coefficient I_i:

$$I_i = \sum_{j=1}^{n+1} x_{ij} s_j, i = 1, \ldots, N, \tag{5}$$

where:

$$s_j = \sum_{i=1}^{N} x_{ij}, j = 1, \ldots, n+1. \tag{6}$$

Stage 3. Map input vectors (i.e. rows from X) into t clusters denoted as Y_v, $v = 1, \ldots, t$. Each cluster contains input vectors with identical value of the similarity coefficient I and t is a number of different values of I.

Stage 4. Select input vectors to be retained in each cluster. Let $|Y_v|$ denote a number of input vectors in cluster v. Then the following rules for selecting input vectors are applied:

- If $|Y_v| = 1$ then $S = S \cup Y_v$.
- If $|Y_v| > 1$ then $S = S \cup \{x_j^v\}$, where x_j^v are reference instances from the cluster Y_v selected by applying the JABAT and where the number of selected instances corresponds to multi objective optimization problem.

4.2   Overview of the JABAT

The single objective instance reduction is a combinatorial and computationally difficult problem [12]. Its multiple-objective version can not be computationally easier. To deal with the multiple-objective instance reduction it is proposed to use the population-based approach with optimization procedures implemented as an asynchronous team of agents (A-Team), originally introduced by Talukdar [2]. An A-Team is a collection of software agents that cooperate to solve a problem by dynamically evolving a population of solutions. An A-Team usually uses combination of approaches inspired by natural phenomena including, for example, insect societies [4], evolutionary processes [5] or particle swarm optimization [7], as well as local search techniques like, for example, tabu search [6].

An A-Tam is a cyclic network of autonomous agents and shared, common memories. Each agent contains some problems solving skills and each memory contains a population of temporary solutions to the problem to be solved. All the agents can work asynchronously and parallel. During their works agents cooperate by selecting and modifying these solutions. In the reported approach the A-Team was designed and implemented using JADE-based A-Team (JABAT) environment.

JABAT is a middleware allowing to design and implement an A-Team architecture for solving combinatorial optimization problems. The main features of JABAT include:

- The system can in parallel solve instances of several different problems.

- A user, having a list of all algorithms implemented for given problem may choose how many and which of them should be used.
- The optimization process can be performed on many computers. The user can easily adjoin or delete a computer from the system. In both cases JABAT will adapt to the changes, commanding the agents working within the system to migrate.

The JABAT produces solutions to combinatorial optimization problems using a set of optimising agents, each representing an improvement algorithm. To escape getting trapped into a local optimum an initial population of solutions called individuals is generated or constructed. Individuals forming an initial population are, at the following computation stages, improved by independently acting agents, thus increasing chances for reaching a global optimum.

Main functionality of the proposed environment is searching for the optimum solution of a given problem instance through employing a variety of the solution improvement algorithms. The search involves a sequence of the following steps:

- Generation of an initial population of solutions.
- Application of solution improvement algorithms which draw individuals from the common memory and store them back after attempted improvement, using some user defined replacement strategy.
- Continuation of the reading-improving-replacing cycle until a stopping criterion is met.

The above functionality is realized by the two main types of classes. The first one includes *OptiAgents*, which are implementations of the improvement algorithms. The second are *SolutionManagers*, which are agents responsible for maintenance and updating of individuals in the common memory. All agents act in parallel. Each *OptiAgent* is representing a single improvement algorithm (for example simulated annealing, tabu search, genetic algorithm, local search heuristics etc.). An *OptiAgent* has two basic behaviors defined. The first is sending around messages on readiness for action including the required number of individuals (solutions). The second is activated upon receiving a message from some *SolutionManager* containing the problem instance description and the required number of individuals. This behaviour involves improving fitness of individuals and resending the improved ones to a sender. A *SolutionManager* is brought to life for each problem instance. Its behaviour involves sending individuals to *OptiAgents* and updating the common memory.

Main assumption behind the proposed approach is its independence from a problem definition and solution algorithms. Hence, main classes *Task* and *Solution* upon which agents act, have been defined at a rather general level. Interfaces of both classes include function *ontology()*, which returns JADE's ontology designed for classes *Task* and *Solution*, respectively. Ontology in JADE is a class enabling definition of the vocabulary and semantics for the content of message exchange between agents. More precisely, an ontology defines how the class is transformed into the text message exchanged between agents and how the text message is used to construct the class (here either *Task* or *Solution*).

4.3   Implementation of the Multiple-Objective Instance Reduction Algorithm

The JABAT environment has served as the tool for solving instances of the multiple-objective instance reduction problem. All the required classes have been defined in the package called MORIS (*multiple-objective reference instances selection*). The MORIS includes the following classes: *MORIS_Task* inheriting form the *Task* class, *MORIS_Solution* inheriting from the *Solution* class. The *MORIS_Task* identifies data set and creates the clusters of potential reference instances. *MORIS_Solution* contains representation of the solution. It consists of the list of the selected references instances from original data set and the values of the cost factors corresponding respectively to the classification accuracy, the percentage of compression of the training set and the number of rules. To obtain values of these factors the C 4.5 classification tool is used. For each decision tree produced by the C 4.5 the size of rules is additionally calculated and recorded.

To communication between optimization agents and the solution manager the *MORIS_TaskOntology* and *MORIS_SolutionOntology* classes have been also defined through over-ridding the *TaskOntology* and *SolutionOntology*, respectively. The *TaskOntology* is needed to enable sending between agents and the common memory task parameters and instance numbers belonging to respective clusters and representing potential reference instances. The *SolutionOntology* is needed to enable sending around potential solutions.

Each optimization agent operates on one individual (solution) provided and randomly selected form the population by the *SolutionManager*. Its role is to improve quality of the solution. After the stopping criterion has been met, each agent resends individuals to the *SolutionManager*, which, in turn, updates common memory by replacing randomly selected individual with the improved ones. Generally, the *SolutionManager* manages the population of solutions, which on initial phase is generated randomly. The generation of an initial population of solutions is designed to obtain a population consisting of solutions with different number of reference instances in each clusters. The *SolutionManager*, after adding to the population a solution received from the *OptiAgent*, overwrides and updates the set of potentially Pareto-optimal solutions.

To solve the discussed multiple objective problem two types of agents representing different improvement procedures have been implemented. In each case the agent's classes are inherited from the *OptiAgent* class. Both procedures aim at improving current solution through modification and exchange of the reference vectors in different clusters. After having received a solution to be improved an optimization agent generates random vector of weights Λ. It is used to obtain the normalized function $s(z, \Lambda)$, which, in turn, is used to evaluate potential solutions.

The first optimization agent - local search with tabu list (in short: RLS), modifies the current solution by removing the randomly selected reference vector from the randomly chosen cluster and replacing it with some other randomly chosen reference vector thus far not included within the improved solution. The modification takes place providing the vector to be replaced is not on the tabu list. After the modification the newly added reference vector is placed on the tabu list and remains there for a given number of iterations. This number depends on the cluster size and decreases for smaller clusters. The modified solution replaces the current one if it is evaluated as a better one using the current normalized function $s(z, \Lambda)$.

The second optimization agent - incremental/decremental local serach (in short: IDLS), modifies the current solution either by removing the randomly selected reference vector from the randomly chosen cluster or by adding some other randomly chosen reference vector thus far not included within the improved solution. Increasing or decreasing a number of reference vectors within clusters is a random move executed with equal probabilities. Pseudo-codes showing both types of the discussed optimization agents are shown in Example 1 and 2.

Example 1: Pseudo code of the RLS type optimization agent

```
public class RandomLocalSearch extends OptiAgent {
 public void improveSolution() {
  Initiate the list of tabu moves;
  Draw at random a weight vector L;
  MORIS_Solution x = (MORIS_Solution)solution.clone();
  /*where x is the solution that has been sent to optimize*/
  do{
     Select randomly cluster from x;
     Select randomly n, where n corresponds to instance number
     from selected cluster;
     If (n is not on the list of tabu active moves){
       Select randomly n', where n' corresponds to instance
       number which is not represented within x;
       Remove n from x and add n' to x producing x';
       Calculate fitness of the x' on s(z,L);
       if(x' is better on s(z,L) then x) x=x';
       Add n to the list of tabu moves and during next s
       iterations do not change this instance number;
     }
     Update the list of tabu moves;
   }while (!terminatingCondition);
   /*solution is ready to be sent back*/
   solution = x;}
}
```

Example 2: Pseudo code of the IDLS type optimization agent

```
public class IncDecLocalSearch extends OptiAgent {
 public void improveSolution() {
  Draw at random a weight vector L;
  Set s as a parameter determining decremental/incremental phase;
  MORIS_Solution x = (MORIS_Solution)solution.clone();
```

```
/*where x is the solution that has been sent to optimize*/
do{
  counter=0;
  Select randomly cluster from x;
  if(( counter % s ) == 0)
  {
    Generate a random binary digit;
    if(a random digit is 0)
    {
     Select randomly n, where n corresponds to instance number
     which is not represented within x;
     Add n to x;
    }
    else
    {
    Select randomly n, where n corresponds to instance
    number from selected cluster;
    Remove n from x;
    }
  }
  Select randomly n, where n corresponds to instance number
  from selected cluster;
  Select randomly n', where n' corresponds to instance
  number which is not represented within x;
  Remove n from x and add n' to x producing x';
  Calculate fitness of the x' on s(z,L);
  if (x' is better on s(z,L) then x) x=x';
  counter++;
  }while (!terminatingCondition);
  /*solution is ready to be sent back*/
  solution = x;}
}
```

5   Computational Experiment Results

To validate the proposed approach several benchmark instances have been solved.
The main aim of the experiment has been to evaluate usefulness and effectiveness
of the agent-based approach to solving the problem of multiple-objective selection
of reference vectors. This has been achieved through establishing experimentally
how different strategies of selecting and using optimization agents affect the computation results.

The proposed approach has been used to solve four well known classification
problems - Cleveland heart disease (303 instances, 13 attributes, 2 classes), credit
approval (690, 15, 2), Wisconsin breast cancer (699, 9, 2) and sonar problem
(208, 60, 2). The respective datasets have been obtained from [8].

Experiment plan has been based on the 10-cross-validation approach. Each thus obtained training set T has been then reduced to a subset S containing reference vectors. Each reference vectors set has been, in turn, used to produce a decision tree. This has been evaluated from the point of view of the four criteria discussed in Section 3. Each decision tree was created using only the instances in S and each C 4.5 classifier was trained without pruned leaves.

For each benchmarking problem the experiment has been repeated 50 times and the reported values of the quality measures have been averaged over all runs. All optimization agents have been allowed to continue iterating until 100 iterations have been performed. The common memory size in JABAT was set to 100 individuals. The number of iterations, the size of common memory and selection criteria have been set out experimentally at the fine-tuning phase. The search for solutions was satisfactory performed at reasonable computation time.

In order to evaluate the resulting Pareto-optimal sets approximations two quality measures have been used [11]. The first measure is the average of the best values of weighted Tchebycheff scalarizing function over a set of systematically generated normalized weight vectors. The set of such weight vectors is denoted and defined as $\Psi_s = \{\Lambda = [\lambda_1, \ldots, \lambda_J] \in \Psi | \lambda_j \in \{0, \frac{1}{k}, \frac{2}{k}, \ldots, \frac{k-1}{k}, 1\}\}$, where Ψ is the set of all normalized weight vectors and k is a sampling parameter.

Finally, the measure is calculated in the following way:

$$R(A) = 1 - \frac{\sum_{\Lambda \in \Psi_s} s_\infty^*(z^0, A, \Lambda)}{|\Psi_s|}, \tag{7}$$

where $s_\infty^*(z^0, A, \Lambda) = min_{z \in A}\{s_\infty(z, z^0, \Lambda)\}$ and is the best value achieving by function $s_\infty(z, z^0, \Lambda)$ on set A. Before calculating the value of this measure the reference point z^0 was set as an ideal point.

Table 1. Performance of different agent combinations measured using average values of C and R

Optimizing	C(RLS, IDLS)	C(IDLS, RLS)	C(RLS +IDLS, RLS)	C(RLS, RLS +IDLS)	C(RLS +IDLS, IDLS)	C(IDLS, RLS +IDLS)	RLS	IDLS	RLS+IDLS
	C measure and standard deviations						R measure and standard deviations		
f_1, f_2, f_3, f_4	0,464	0,618	0,862	0,208	0,760	0,328	0,858	0,857	0,859
	±0,12	±0,073	±0,149	±0,084	±0,254	±0,11	±0,003	±0,003	±0,002
f_1, f_2	0,361	0,798	0,735	0,430	0,867	0,422	0,732	0,732	0,734
	±0,172	±0,155	±0,087	±0,096	±0,153	±0,183	±0,004	±0,004	±0,003
f_1, f_3	0,523	0,728	0,827	0,390	0,824	0,435	0,959	0,961	0,959
	±0,103	±0,052	±0,158	±0,139	±0,084	±0,194	±0,003	±0,003	±0,007

The second measure is the coverage of the two approximations of the non-dominated set and is defined as:

$$C(A, B) = \frac{|\{z'' \in B\}| \exists z' \in A : z' \succ z''|}{|B|},\tag{8}$$

where the value $C(A, B) = 1$ means that all points in B are dominated by or are equal to some points in A. The value $C(A, B) = 0$ means that no point in B is covered by any point in A.

Experiment results for different combinations of optimization agents averaged over all benchmark datasets and instances are shown in Table 1. The cost factors (optimization criteria) include classification accuracy, percentage of compression of the training set, number of rules and size of the decision tree. Values of the R measure have been calculated with the sampling parameter k set to 100 and 5 for the bi-objective and four-objective cases, respectively.

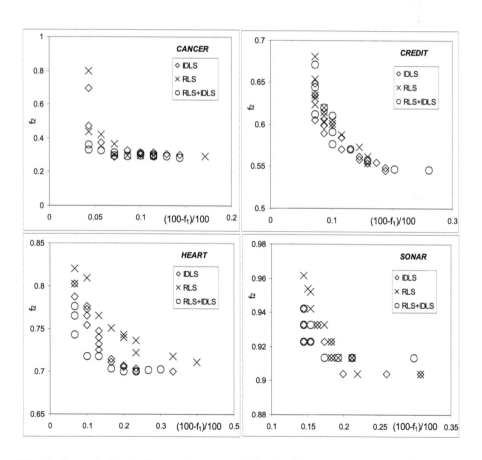

Fig. 1. Example Pareto fronts - instances of the bi-objective optimization (f_1 and f_2)

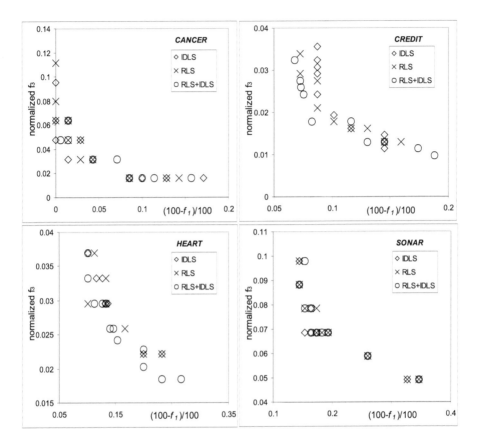

Fig. 2. Example Pareto fronts - instances of the bi-objective optimization (f_1 and f_3)

The results of Pareto-optimal set approximations using the R measure indicate that each combination of agents produces similar results. There are no statistically significant differences between average values of the R measure for all investigated combination of agents.

The results of Pareto-optimal set approximations using the C measure indicate that IDLS produces a better coverage then RLS and RLS+IDLS better coverage then either RLS or IDLS. This observation holds for all investigated cases i.e. multi-objective optimization with two and four objectives and is independent on dimensionality of problems. Thus RLS+IDLS generates best approximation of the Pareto-optimal (non-dominated set).

The values of C measure have been also used to carry a pair-wise comparison of average performance of different combinations of optimization agents. It has been observed that the following inequalities are statistically significant:

- $C(IDLS, RLS) > C(RLS, IDLS)$,
- $C(RLS + IDLS, RLS) > C(RLS, RLS + IDLS)$,
- $C(RLS + IDLS, IDLS) > C(IDLS, RLS + IDLS)$.

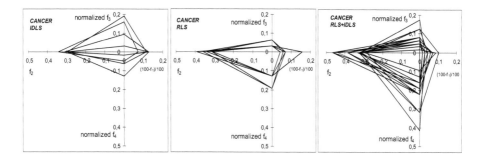

Fig. 3. Example approximations of Pareto-optimal sets - an instance of the four-objective optimization problem

In Fig. 1 and 2 example Pareto fronts obtained by solving a single instance of each of the considered problem types are shown. Each set of points has been obtained in a single run for the respective bi-objective optimization problem.

In Fig. 3 example approximations of Pareto-optimal sets produced by different combination of agents for an instance of the four-objective selection of reference vector problem are presented.

6   Conclusion

The paper proposes an agent-based multiple-objective approach to the selection of reference vectors from original datasets. Effective and dependable selection procedures are of vital importance to machine learning and data mining. The suggested approach is based on the multiple agent paradigm. Using a team of agents brings about several advantages including better use of computational resources, flexibility and ability to carry computations in the distributed environment. The focus of the paper is however not on efficiency of the agent based approach but rather on the methodology of dealing with the multiple-objective selection of reference vectors through employing a team of agents. It has been shown that there exist adequate methodology and suitable tools allowing to obtain good approximations of the Pareto-optimal solutions to problems of the discussed type. The proposed method and tools can be used to design customized machine learning and data mining systems corresponding better to the user requirements and needs. The approach allows also for discovery of interactions between composition of various vector selection optimization procedures and a quality of generalization measured using multiple criteria. Such knowledge can be used for evaluation and selection of optimization agents and procedures.

Future research should focus on refining the theoretical framework for agent-based, multiple-objective optimization of reference vector selection as well as on designing more user friendly tools for solving practical multiple objective reference vectors selection problems.

References

1. Bellifemine, F., Caire, G., Poggi, A., Rimassa, G.: JADE. A White Paper. Exp 3(3), 6–20 (2003)
2. Talukdar, S., Baerentzen, L., Gove, L., de Souza, P.: Asynchronous Teams: Cooperation Schemes for Autonomous, Computer-Based Agents, Technical Report EDRC 18-59-96, Carnegie Mellon University, Pittsburgh (1996)
3. Jędrzejowicz, P., Wierzbowska, I.: JADE-Based A-Team Environment, pp. 719–726. Springer, Berlin, Heidelberg (2006)
4. Oster, G.F., Wilson, E.O.: Caste and Ecology in the Social Insect, vol. 8. Princeton University Press, Princeton, NJ (1978)
5. Davis, L. (ed.): Handbook of Genetic Algorithms, Van Nostrand Reinhold (1991)
6. Glover, F.: Tabu Search. Part I and II, ORSA Journal of Computing. 1(3), Summer (1990) and 2(1) Winter (1990)
7. Kennedy, J., Eberhart, R.C.: Particle swarm optimisation. In: Proc. of IEEE International Conference on Neural Networks, Piscataway, N.J. pp. 1942-1948 (1995)
8. Merz, C.J., Murphy, P.M.: UCI Repository of Machine Learning Databases Irvine, CA: University of California, Department of Information and Computer Science (1998), http://www.ics.uci.edu/~mlearn/MLRepository.html
9. Wilson, D.R., Martinez, T.R.: Reduction techniques for instance-based learning algorithm. In: Machine Learning, vol. 33(3), pp. 33–33. Kluwer Academic Publishers, Boston (2000)
10. Duch, W., Blachnik, M., Wieczorek, T.: Probabilistic distance measure for prototype-based rules. In: Proc. of the 12 International Conference on Neural Information Processing, ICONIP, pp. 445–450 (2005)
11. Jaszkiewicz, A.: Multiple objective metaheuristic algorithms for combinational optimization. Habilitation thesis, 360, Pozna University of Technology, Poznań (2001)
12. Czarnowski, I., Jędrzejowicz, P.: An Approach to instance reduction in supervised learning. In: Coenen, F., Preece, A., Macintosh, A. (eds.) Research and Development in Intelligent Systems, vol. XX, pp. 267–282. Springer, London (2004)
13. Stefanowski, J.: Algorytmy indukcji regu decyzyjnych w odkrywaniu wiedzy, Habilitation thesis, 361, Pozna University of Technology, Poznań (in Polish) (2001)
14. Quilan, J.R.: C 4.5: programs for machine learning, San Matoe. Morgan Kaufman, Seattle (1993)
15. Breiman, L., Friedman, J.H., Oslhen, R.A., Stone, C.J.: Classification and Regression Trees. Belmont, CA: Wadsworth International Group (1984)

On Applying Dimension Reduction
for Multi-labeled Problems*

Moonhwi Lee and Cheong Hee Park

Dept. of Computer Science and Engineering
Chungnam National University
220 Gung-dong, Yuseong-gu
Daejeon, 305-763, Korea
{moone81,cheonghee}@cnu.ac.kr

Abstract. Traditional classification problem assumes that a data sample belongs to one class among the predefined classes. On the other hand, in a multi-labeled problem such as text categorization, data samples can belong to multiple classes and the task is to output a set of class labels associated with new unseen data sample. As common in text categorization problem, learning a classifier in a high dimensional space can be difficult, known as the curse of dimensionality. It has been shown that performing dimension reduction as a preprocessing step can improve classification performances greatly. Especially, Linear discriminant analysis (LDA) is one of the most popular dimension reduction methods, which is optimized for classification tasks. However, in applying LDA for a multi-labeled problem some ambiguities and difficulties can arise. In this paper, we study on applying LDA for a multi-labeled problem and analyze how an objective function of LDA can be interpreted in multi-labeled setting. We also propose a LDA algorithm which is effective in a multi-labeled problem. Experimental results demonstrate that by considering multi-labeled structures LDA can achieve computational efficiency and also improve classification performances greatly.

Keywords: Dimension Reduction, Linear Discriminant Analysis, Multi-labeled Problems.

1   Introduction

While traditional classification problem assumes that a data sample belongs to only one class among the predefined classes, a multi-labeled problem can arise in real situation where a data sample is associated with multiple class labels. For example, in text categorization documents can be classified to multiple categories of topics [1]. In bioinformatics, each gene is associated with a set of functional classes [2]. In a multi-labeled problem, the main task is to output a set of class labels associated with a new unseen data sample. One common way to deal with a multi-labeled problem is to transform it to several binary problems. In other words, for each class a binary problem is constructed where data samples belonging to the class compose the positive set and the

---

* This work was supported by the Korea Research Foundation Grant funded by the Korean Government(MOEHRD)(KRF-2006-331-D00510).

P. Perner (Ed.): MLDM 2007, LNAI 4571, pp. 131–143, 2007.

remaining data makes up the negative set. Single-label classification methods such as Support vector machines and k-nearest neighbor classifier can be applied independently for each binary problem [3,4,5]. A maximum entropy based method was also developed which explored correlations among classes [6]. Outputs from each binary classifier are combined to produce a set of class labels for new data.

When the data dimensionality is high as common in text categorization problem, learning a classifier in a high dimensional space can be difficult, known as the curse of dimensionality. It has been shown that performing dimension reduction as a preprocessing step can improve classification performances greatly [7,8,9,10]. By extracting a small number of most optimal features for an intended main task, original data is transformed to a low dimensional space where learning process can be performed more efficiently. Among several statistical dimension reduction methods, Linear discriminant analysis (LDA) performs dimension reduction to maximize class separability in the reduced dimensional space [11]. Due to this objective criterion, LDA can be most optimal for classification tasks.

While many generalized LDA algorithms have been shown to be efficient for high dimensional undersampled problems [12,13,14,15], LDA has not been applied for a multi-labeled problem. Since the objective function of LDA was originally developed for a single-labeled problem, LDA has been considered nonapplicable for a multi-labeled problem. In this paper, we first derive formulations for applying LDA in a multi-labeled problem. We also propose a computationally efficient LDA algorithm for a multi-labeled problem with a small sample size. The proposed method can save computational costs and memory requirements by utilizing QR-decomposition. And through the generation of semi-artificial data samples, it overcomes problems occurred due to a small sample size and multiple class labels. Experimental results demonstrate that by considering multi-labeled structures LDA can achieve computational efficiency and also improve classification performances greatly.

The paper is organized as follows. In Section 2, brief reviews for a multi-labeled problem and generalized LDA algorithms are given. In Section 3, we propose an efficient dimension reduction method which is more effective in multi-labeled data sets. Experimental results in Section 4 compare performances of dimension reduction methods under various conditions. Conclusions follow in Section 5.

2   Multi-labeled Classification and Linear Discriminant Analysis

2.1   Multi-labeled Classification

Let $X = \{\mathbf{x}_1, \mathbf{x}_2, \cdots, \mathbf{x}_k\}$ be a set of data samples, and a data sample is represented as a vector in a m-dimensional space such as $\mathbf{x}_i = [x_{i1}, \cdots, x_{im}]^T$. The notation T denotes the transpose of a vector or a matrix. We also assume that data samples can have one or more class labels assigned to them among the predefined r classes. Let $Y = \{\mathbf{y}_1, \mathbf{y}_2, \cdots, \mathbf{y}_k\}$ be the set of class label vectors corresponding to the data samples in X. Each $\mathbf{y}_i = [y_{i1}, \cdots, y_{ir}]$ is a vector denoting class labels associated with $\mathbf{x}_i$ such that $y_{ij} = 1$ if $\mathbf{x}_i$ belongs to the class j, and $y_{ij} = 0$ if $\mathbf{x}_i$ does not belong to the class j.

In a multi-labeled problem, the classification task is to output a set of class labels associated with a new unseen data sample. Instead of making a hard decision whether a data sample belongs to a class or not, multi-labeled classification can produce confidence level at which a data sample is assigned to each class so that r class labels can be ordered according to their confidence levels. By setting a threshold to accept the class label, ranking based classification can be transformed to hard-decision-making classification. Also note that a single-label problem is a special case of a multi-label problem, in which each data sample is to have only one positive class label.

Classification performance in a multi-labeled problem can be evaluated by several measures. Among them, we introduce f1-measure and one-error. We refer to the papers [16,17] for more details about evaluation measures. The f1-measure was originally used in information retrieval. For each class, the precision (p), recall (r) and f1-measure are defined such as

$$p = \frac{TP}{TP + FP}, \quad r = \frac{TP}{TP + FN}, \quad f1 = \frac{2pr}{p + r}, \tag{1}$$

where TP represents the number of positive samples which are predicted as positive, FP is the number of negative samples which are predicted as positive, and FN is the number of positive samples which are predicted as negative. The f1 value averaged over all the classes is called the macro-averaged f1 measure. On the other hand, when TP, FP, FN are first summed over all the classes respectively and then f1 is calculated based on them. It is called micro-averaged f1 measure.

One-error is used for multi-labeled classification which produces only one positive class label. For a ranking based classifier, the class label with the highest rank is only considered in one-error measurement. Let us assume that $\mathcal{T}$ is a collection of new unseen data samples. For each $\mathbf{x}$ in $\mathcal{T}$, $t_{\mathbf{x}}$ denotes the set of the true class labels of $\mathbf{x}$ and $p_{\mathbf{x}}$ is the predicted class label. One-error measures the probability that the predicted class label is not one of the true class labels as follows.

$$one - error = \frac{1}{|\mathcal{T}|} |\{\mathbf{x} \in \mathcal{T} | p_{\mathbf{x}} \notin t_{\mathbf{x}}\}|, \tag{2}$$

where $|\cdot|$ means the cardinality of the set. One-error is actually same as prediction error in a single-labeled problem. In our experiments, we used the micro-averaged f1 measure and one-error to evaluate performances.

2.2 On Applying LDA for Multi-labeled Data

LDA utilizes the between-class scatter and within-class scatter as a means to measure class separability. When the distance between classes is maximal and the scatterness within classes is minimal, it is an ideal clustering structure for classification. LDA finds a linear transformation to a low dimensional space that maximizes class separability.

We first review LDA in a single-labeled problem [11]. Let us represent the data set as

$$A = \{\mathbf{a}_1^1, \cdots, \mathbf{a}_{n_1}^1, \cdots\cdots, \mathbf{a}_1^r, \cdots, \mathbf{a}_{n_r}^r\}, \tag{3}$$

where $\{\mathbf{a}_1^i, \cdots, \mathbf{a}_{n_i}^i\}$ are the data samples belonging to the class i and the total number of data is $n = n_1 + \cdots + n_r$. The between-class scatter matrix S_b, the within-class scatter matrix S_w, and the total scatter matrix S_t are defined as

$$S_b = \sum_{i=1}^r n_i (\mathbf{c}_i - \mathbf{c})(\mathbf{c}_i - \mathbf{c})^T, \quad S_w = \sum_{i=1}^r \sum_{j=1}^{n_i} (\mathbf{a}_j^i - \mathbf{c}_i)(\mathbf{a}_j^i - \mathbf{c}_i)^T, \quad (4)$$

$$S_t = \sum_{i=1}^r \sum_{j=1}^{n_i} (\mathbf{a}_j^i - \mathbf{c})(\mathbf{a}_j^i - \mathbf{c})^T,$$

using the class centroids $\mathbf{c}_i = \frac{1}{n_i} \sum_{j=1}^{n_i} \mathbf{a}_j^i$ and the global centroid $\mathbf{c} = \frac{1}{n} \sum_{i=1}^r \sum_{j=1}^{n_i} \mathbf{a}_j^i$. The *trace* [18] which is defined as the sum of the diagonal components of a matrix gives measures for the between-class scatter and the within-class scatter such as

$$trace(S_b) = \sum_{i=1}^r n_i \|\mathbf{c}_i - \mathbf{c}\|^2, \quad trace(S_w) = \sum_{i=1}^r \sum_{j=1}^{n_i} \|\mathbf{a}_j^i - \mathbf{c}_i\|^2.$$

One of optimization criteria in LDA is to find a linear transformation G^T which maximizes

$$J = trace((G^T S_w G)^{-1}(G^T S_b G)), \quad (5)$$

where $G^T S_i G$ for $i = b, w$ is the scatter matrix in the transformed space by G^T. It is well known [11] that J in (5) is maximized when the columns of G are composed of the eigenvectors $\mathbf{g}$ corresponding to the $r - 1$ largest eigenvalues λ of

$$S_w^{-1} S_b \mathbf{g} = \lambda \mathbf{g}. \quad (6)$$

Now suppose data samples can have more than one class label. Let $X = \{\mathbf{x}_1, \mathbf{x}_2, \cdots, \mathbf{x}_k\}$ and $Y = \{\mathbf{y}_1, \mathbf{y}_2, \cdots, \mathbf{y}_k\}$ denote data samples and their class label vectors respectively as in section 2.1. One way to apply LDA for multi-labeled data would be to transform a multi-labeled problem to r binary problems and perform LDA for each binary problem. For each class a binary problem is constructed where data samples belonging to the class compose the positive set and the remaining data makes up the negative set. LDA reduces the data dimension by finding optimal projective directions and in the reduced dimensional space classification is performed. We call this approach *LDA-BIN* in order to distinguish it from the method suggested next.

In the approach of LDA-BIN, LDA is applied independently for each binary problem and it may not reflect correlations among classes. And also dimension reduction should be performed as many times as the number of classes, since a binary problem is constructed for each class. Now we propose to apply the objective function (5) directly to a multi-labeled data instead of constructing multiple binary problems. A data sample with multiple class labels should contribute to all classes for which it is a positive sample. We compose the data set A in (3) by taking all data samples positive to each class. Hence a data sample with multiple class labels can appear several times in A and the total number $n \equiv n_1 + \cdots + n_r$ can be greater than the number of the original data samples. LDA is applied to the composed set A as in a single-labeled problem. We call this approach *LDA-ALL*.

Dimension reduction is most effective in high dimensional data as in text categorization. Often high dimensional data is closely related with undersampled problems, where the number of data samples is smaller than the data dimension. In the classical LDA, S_w is assumed to be nonsingular and the problem is to compute eigenvectors of $S_w^{-1}S_b$ as in (6). However, in undersampled problems, all of the scatter matrices become singular and the classical LDA is difficult to apply. In order to make LDA applicable for undersampled problems, several methods have been proposed [19,12,20,21,13,14]. In the next section, we analyze generalized LDA algorithms and propose a LDA algorithm which is effective for a multi-labeled problem.

3   On Applying LDA-ALL for Undersampled Multi-labeled Problems

In undersampled problems, the minimization of the within-class scatter can be accomplished by using the null space of the within-class scatter matrix S_w, since the projection by the vectors in the null space of S_w makes the within-class scatter zero. It has been shown that a linear transformation based on the null space of S_w can improve classification performance greatly [20,21,13,14]. On the other hand, the maximization of trace($G^T S_b G$) suggests that the column vectors of G should come from the range space of S_b, since for any $\mathbf{g} \in null(S_b)$

$$trace(\mathbf{g}^T S_b \mathbf{g}) = trace(\sum_{i=1}^r n_i \mathbf{g}^T (\mathbf{c}_i - \mathbf{c})(\mathbf{c}_i - \mathbf{c})^T \mathbf{g}) = \sum_{i=1}^r n_i \|\mathbf{g}^T \mathbf{c}_i - \mathbf{g}^T \mathbf{c}\|^2 = 0, \quad (7)$$

and therefore all the class centroids become equal to the global centroid in the projected space.

Generalization of LDA for undersampled problems can be characterized by the two-step process of the minimization of the within-class scatter and the maximization of the between-class scatter where one of them is performed after the other. In the next sections, based on which one in two steps is first applied, we explore the applicability of generalized LDA algorithms for a multi-labeled problem, and we also propose an efficient algorithm.

3.1   Maximizing the Between-Class Distance in the First Stage

The method by Yu and Yang [12] first transforms the original space by using a basis of range(S_b). Then in the transformed space the minimization of the within-class scatter is performed by the eigenvectors corresponding to the smallest eigenvalues of the within-class scatter matrix. This method is called *Direct LDA* (or DLDA). The computation in DLDA can be efficient by taking advantage of the singular value decomposition (SVD) for the smaller matrix $H_b^T H_b$ instead of $S_b = H_b H_b^T$, where

$$H_b = [\sqrt{n_1}(\mathbf{c}_1 - \mathbf{c}), \cdots, \sqrt{n_r}(\mathbf{c}_r - \mathbf{c})]$$

for the class centroids $\mathbf{c}_i$ and the global centroid $\mathbf{c}$. Our experiments in Section 4 show that in LDA-ALL approach, DLDA obtains competitive performances while maintaining low computational complexities. Algorithm 1 summarizes DLDA for a multi-labeled problem.

Algorithm 1. DLDA-ALL for multi-labeled problems

For a data set $X = \{\mathbf{x}_1, \mathbf{x}_2, \cdots, \mathbf{x}_k\}$ and a set of class label vectors $Y = \{\mathbf{y}_1, \mathbf{y}_2, \cdots, \mathbf{y}_k\}$ as in Section 2.1, compose a data matrix $A = [\mathbf{a}_1^1, \cdots, \mathbf{a}_{n_1}^1, \cdots\cdots, \mathbf{a}_1^r, \cdots, \mathbf{a}_{n_r}^r]$ where $\mathbf{a}_1^i, \cdots, \mathbf{a}_{n_i}^i$ are the data samples belonging to the class i.

1. Construct the between-class scatter matrix S_b from A.
2. Compute the eigenvector u_i's of S_b corresponding to the largest nonzero eigenvalue λ_i's.
3. Let $U = [\lambda_1^{-1/2} u_1, \cdots, \lambda_{r-1}^{-1/2} u_{r-1}]$. Compose the within-class scatter matrix S_w^* in the data space $U^T A$.
4. Compute the eigenvector v_i's of S_w^* corresponding to the smallest nonzero eigenvalue μ_i's.
5. The linear transformation matrix G is composed as $G = UV$ where $V = [\mu_1^{-1/2} v_1, \cdots, \mu_s^{-1/2} v_s]$.

3.2 Minimizing the Within-Class Scatter in the First Stage

Unlike DLDA, the methods in [20,21,13,14] utilize the null space of S_w, more specifically, null$(S_w) \cap$ range(S_b). Most of generalized LDA algorithms mainly rely on the Singular value decomposition (SVD) in order to compute eigenvectors. But, the computational complexities and memory requirements for the SVD can be very demanding, especially for high dimensional data. Zheng et al.'s method called GSLDA [14] obtains computational efficiency by using QR-decomposition which is cheaper than the SVD. However GSLDA assumes that the given data samples are independent. Hence it can not work well in LDA-ALL approach. Even in LDA-BIN setting, it performs poorly when data samples are nearly dependent as will be demonstrated in our experiments.

Now we discuss some difficulties with using the null space of S_w in a multi-labeled problem. Since range(S_w) is the orthogonal complement of null(S_w),

$$\mathbb{R}^m = range(S_w) \oplus null(S_w), \tag{8}$$

where $\oplus$ denotes the direct sum of the vector spaces, range(S_w) and null(S_w) [22]. Since $null(S_t) \subset null(S_b)$ and using any vector in null(S_b) for the projection is undesirable as described in (7), considering only the space $range(S_t)$ does not make any effects. Hence from (8) we have

$$range(S_t) = (range(S_t) \cap range(S_w)) \oplus (range(S_t) \cap null(S_w)) \text{ and}$$
$$dim(range(S_t)) = dim(range(S_t) \cap range(S_w)) + dim(range(S_t) \cap null(S_w)). \tag{9}$$

As the ratio of data samples which are not independent or belong to multiple classes increases, the dimension of $range(S_t)$ gets reduced severely from the number of data samples. And also the dimension of $range(S_t) \cap null(S_w)$ becomes lower, possibly to zero. It is because the subtraction of class centroids from each data sample makes rank reduction in S_w less than in S_t. In order to visualize it, using a real data set we test effects of multi-labeled data on the space $range(S_t) \cap null(S_w)$. From the largest eight classes in the Reuter-21578 text dataset, independent 286 documents were chosen to construct a base set. While adding documents with multiple class labels to a base set, dim(range(S_t)) and $dim(range(S_t) \cap range(S_w))$ were computed. As shown in

Figure 1, as data samples with multiple class labels are added to the base set more and more, the dimension of $range(S_t) \cap range(S_w)$ gets to close to the dimension of $range(S_t)$, and therefore the dimension of $range(S_t) \cap null(S_w)$ goes down to near zero. It implies that GSLDA or any other algorithms utilizing the null space of S_w will suffer from the shrinking of $range(S_t) \cap null(S_w)$. In the next section, we propose a new method which can overcome such a problem.

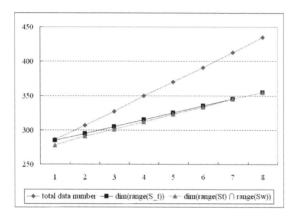

Fig. 1. The effects of multi-labeled data on $dim(range(S_t) \cap null(S_w))$

3.3   An Efficient LDA Algorithm for Multi-labeled Problems

Note that the scatter matrices in (4) can be computed as a product of the smaller matrices as follows:

$$S_t = H_t H_t^T, \quad S_w = H_w H_w^T, \quad S_b = H_b H_b^T,$$

where

$$H_t = [\mathbf{a}_1^1 - \mathbf{c}, \cdots, \mathbf{a}_{n_1}^1 - \mathbf{c}, \cdots \cdots, \mathbf{a}_1^r - \mathbf{c}, \cdots, \mathbf{a}_{n_r}^r - \mathbf{c}],$$
$$H_w = [\mathbf{a}_1^1 - \mathbf{c}_1, \cdots, \mathbf{a}_{n_1}^1 - \mathbf{c}_1, \cdots \cdots, \mathbf{a}_1^r - \mathbf{c}_r, \cdots, \mathbf{a}_{n_r}^r - \mathbf{c}_r],$$
$$H_b = [\sqrt{n_1}(\mathbf{c}_1 - \mathbf{c}), \cdots, \sqrt{n_r}(\mathbf{c}_r - \mathbf{c})]$$

and $n = n_1 + \cdots + n_r$. In order to obtain an orthonormal basis of $range(S_w) = range(H_w)$, we apply QR-decomposition with column pivoting [18] for $H_w \in \mathbb{R}^{m \times n}$ such as

$$H_w \Pi_1 = Q_1 R_1, \tag{10}$$

where Π_1 is a column permutation in H_w. When $t = rank(H_w)$, $Q_1 \in \mathbb{R}^{m \times t}$ has orthonormal columns and $R_1 \in \mathbb{R}^{t \times n}$ has zeros below the main diagonal. The columns of Q_1 make an orthonormal basis of $range(H_w)$, which is also an orthonormal basis of $range(S_w)$. Note that $null(S_w)$ is the orthogonal complement of $range(S_w)$. Hence any vector x is uniquely expressed as the sum of the orthogonal projection of x onto the range space of S_w and the orthogonal projection of x onto the null space of S_w [22]. For $1 \leq i \leq r$ and $1 \leq j \leq n_i$,

$$\mathbf{a}_j^i - \mathbf{c} = proj_{range(S_w)}(\mathbf{a}_j^i - \mathbf{c}) + proj_{null(S_w)}(\mathbf{a}_j^i - \mathbf{c}), \tag{11}$$

where $proj_W(\mathbf{x})$ denotes the orthogonal projection of $\mathbf{x}$ onto the space W. Since the columns of Q_1 are an orthonormal basis for $range(S_w)$, $Q_1 Q_1^T$ is the orthogonal projection onto $range(S_w)$ [18]. Hence from (11), we have

$$proj_{null(S_w)}(\mathbf{a}_j^i - \mathbf{c}) = (\mathbf{a}_j^i - \mathbf{c}) - proj_{range(S_w)}(\mathbf{a}_j^i - \mathbf{c})$$
$$= (\mathbf{a}_j^i - \mathbf{c}) - Q_1 Q_1^T(\mathbf{a}_j^i - \mathbf{c}) \in range(S_t) \cap null(S_w).$$

Let

$$X = H_t - Q_1 Q_1^T H_t. \tag{12}$$

Then QR-decomposition with column pivoting for $X \in \mathbb{R}^{m \times n}$ gives an orthonormal basis Q_2 of $range(S_t) \cap null(S_w)$ as

$$X \Pi_2 = Q_2 R_2. \tag{13}$$

In the above process, QR-decomposition can save computational complexities and memory requirement greatly compared with the SVD. However, there still exist some problems related with multi-labeled data. As shown in Fig 1, as the number of data samples with multiple class labels gets increased, the dimension of $range(S_t) \cap null(S_w)$ becomes zero. Therefore the above process cannot completely resolve the shrinking of $range(S_t) \cap null(S_w)$ caused by multi-labeled data. Now we propose a method to overcome the problem through the generation of semi-artificial data.

Let P be any column permutation matrix on H_t. Instead of $X = H_t - Q_1 Q_1^T H_t$ in (12), we shuffle the columns of H_t in the second term by using the permutation P:

$$\hat{X} = H_t - Q_1 Q_1^T H_t P. \tag{14}$$

It means that instead of

$$proj_{null(S_w)}(\mathbf{a}_j^i - \mathbf{c}) = (\mathbf{a}_j^i - \mathbf{c}) - Q_1 Q_1^T(\mathbf{a}_j^i - \mathbf{c}) \in range(S_t) \cap null(S_w),$$

we create artificial data samples

$$\hat{x} \equiv (\mathbf{a}_j^i - \mathbf{c}) - Q_1 Q_1^T(\mathbf{a}_s^l - \mathbf{c}) \in range(S_t)$$

for some $1 \leq l \leq r$ and $1 \leq s \leq n_l$. The generated data may deviate from the space $range(S_t) \cap null(S_w)$. However, this deviation from $range(S_t) \cap null(S_w)$ can prevent the shrinking of $range(S_t) \cap null(S_w)$ occurred by multi-labeled data. In this sense, we call $\hat{X}$ the *complement* to $range(S_t) \cap null(S_w)$. Now from QR-decomposition with column pivoting for $\hat{X}$,

$$\hat{X} \hat{\Pi}_2 = \hat{Q}_2 \hat{R}_2, \tag{15}$$

where $\hat{\Pi}_2$ is a column permutation of $\hat{X}$, we obtain the *complement* to a basis of $range(S_t) \cap null(S_w)$, $\hat{Q}_2$.

Algorithm 2. LDA for multi-labeled problems using QR-decomposition

---

For a given data set $A = \{\mathbf{a}_j^i | 1 \leq i \leq r, 1 \leq j \leq n_i\}$, this algorithm computes a transformation matrix G by which dimension reduction for multi-labeled data is performed.

1. Compute QR-decomposition with column pivoting for $H_w : H_w \Pi_1 = Q_1 R_1$.
   Π_1 is a column permutation and the columns of Q_1 are an orthonormal basis of range(H_w)=range(S_w).
2. Let P be any column permutation on H_t.
3. Compute QR-decomposition with column pivoting for $\hat{X} \equiv H_t - Q_1 Q_1^T H_t P : \hat{X} \hat{\Pi}_2 = \hat{Q}_2 \hat{R}_2$. $\hat{\Pi}_2$ is a permutation matrix and $\hat{Q}_2$ gives the complement to an orthonormal basis of range(S_t) $\cap$ null(S_w).
4. Compute the SVD of $\hat{Q}_2^T H_b : \hat{Q}_2^T H_b = U_2 \Sigma_2 V_2^T$.
   Let U_{21} be composed of the columns of U_2 corresponding to nonzero diagonal components of Σ_2.
5. $G^T = (\hat{Q}_2 U_{21})^T$ gives a transformation for dimension reduction.

In the projected space by $\hat{Q}_2$, the maximization of the between-class scatter is pursued. Let the SVD of $\hat{Q}_2^T H_b$ be $\hat{Q}_2^T H_b = U_2 \Sigma_2 V_2^T$. Then $\hat{Q}_2^T S_b \hat{Q}_2 = U_2 (\Sigma_2 \Sigma_2^T) U_2^T$ is the SVD of $\hat{Q}_2^T S_b \hat{Q}_2$. When U_{21} consists of the columns of U_2 corresponding to nonzero diagonal components of Σ_2, the projection by U_{21} realizes the maximization of the between-class scatter. Hence $G = \hat{Q}_2 U_{21}$ gives a transformation matrix for dimension reduction. Algorithm 2 summarizes the proposed method.

Supervised learning is based on the expectation that new unseen data would come from the same distribution as training data. The vector from range(S_t) $\cap$ null(S_w) is ideal projective direction for training data, but it is too optimistic to expect that new data would perfectly fit to the data model based on small training data. Overfitting to the training data leads to generalization errors. In (14), using a permutation and generating artificial data which slightly deviates from range(S_t) $\cap$ null(S_w) can reduce the generalization errors caused by small sample size.

4   Experimental Results

We performed experiments to compare performances of dimension reduction methods in text categorization. The first data set is from Reuters corpus volume I which is a text categorization test collection of about 800,000 stories from Reuters newswire. We used a publically available version of it, RCV1-v2 [1]. As in [23], choosing a small part of data and picking up topics with more than 50 documents and words occurred in more than 5 documents, we composed a data set of 3228 documents with 4928 terms over 41 topics. A document belongs to 2.88 topics on average. One third of the data was used as a training set and two thirds was for a test set, and this splitting was randomly repeated ten times.

Several LDA algorithms including LDA [12], GSLDA [14], PCA-LDA [21] and Algorithm 2 in Section 3.3 were applied. Dimension reduction is performed in two ways, LDA-ALL and LDA-BIN, as explained in Section 2.2. In both cases, a nearest neighbor classifier and microaveraged f1-measure were used for performance evaluation. Table 1

Table 1. Microaveraged f1-measures in the first data set from RCV1-v2

		LDA-ALL	LDA-BIN
1-NN in the original space		0.640 (0.005)	-
Using range(S_b)	DLDA	**0.737** (0.011)	0.682 (0.008)
Methods	GSLDA	0.449 (0.016)	0.694 (0.049)
using null(S_w)	PCA-LDA	0.657 (0.010)	0.314 (0.130)
	Algorithm 2	**0.747** (0.010)	0.685 (0.007)

Table 2. Microaveraged f1-measures in the second data set from Reuter-21578

		LDA-ALL	LDA-BIN
1-NN in the original space		0.782 (0.005)	-
Using range(S_b)	DLDA	**0.837** (0.006)	0.790 (0.006)
Methods	GSLDA	0.599 (0.017)	0.531 (0.034)
using null(S_w)	PCA-LDA	0.718 (0.015)	0.715 (0.035)
	Algorithm 2	**0.850** (0.006)	0.790 (0.006)

summarizes the experimental results, where mean and standard deviations of f1 values obtained from 10 times splitting are shown. For Algorithm 2, we tested the following three approaches in choosing a permutation P.

1. Use any random permutation.
2. Use random permutations within each class.
3. Use the permutation matrix Π_1 obtained by QR-decomposition of H_w in (10).

In our experiments, the three approaches did not make noticeable difference in performance and we only report the results from the third method of using Π_1 obtained by QR-decomposition of H_w in (10).

The second data set was composed from Reuter-21578 [1] which has been a very popular benchmark data set in text mining. Similar as in the first data set, taking topics with more than 50 documents and words occurred in more than 5 documents, we composed a data set with 6537 documents and 4347 terms over 23 topics. A document belongs to 1.63 topics on average. Experimental setting was done same as in the first data set. The result is shown in Table 2.

In both tables, DLDA and the proposed Algorithm 2 demonstrate the best performance in LDA-ALL, compared with other methods utilizing the null space of the within-class scatter matrix. In LDA-BIN, since the data set in Table 1 has 41 topic classes, dimension reduction process was repeated 41 times for the constructed binary problems. While it makes the time complexity of LDA-BIN worse than LDA-ALL, the performance of LDA-BIN was also not good as in LDA-ALL. Also note that in both cases the dimension reduction by DLDA and Algorithm 2 improved the performances compared with those by a nearest neighbor classification in the original space, as shown in the first rows of Table 1 and 2. By dimension reduction, training a classifier is performed in very low dimensional space instead of high dimensional original

---

[1] http://www.research.att.com/~lewis

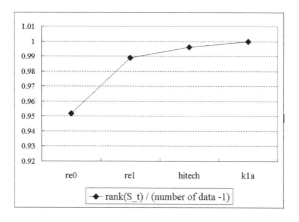

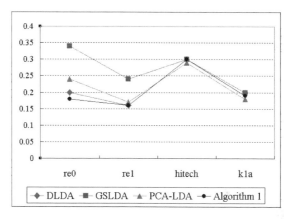

Fig. 2. Data independence (top figure) and one-errors (bottom figure)

space. Therefore it can save computational costs and also circumvent the curse of dimensionality.

We also tested how the proposed method works for independent or nearly dependent data in a single-labeled problems. In this experiment, we used four text data sets[2] which do not contain multi-labeled data. Each text data set was randomly split to the training and test set with the ratio of $1 : 1$ and it is repeated ten times. One-error in (2) was used as a performance measure. In order to measure data independency, for each data set

$$I = \frac{rank(S_t)}{total number of data - 1} \tag{16}$$

was computed. The range of I is $[0, 1]$, and if the data is independent, then I equals to 1. The top figure in Figure 2 compares the I values for each data set and the figure in the bottom plots one-errors. In the data sets re0 and re1 whose I values indicate

---

[2] The data sets were downloaded and preprocessed from http://www-users.cs.umn.edu/~karypis/cluto/download.html

data dependence, Algorithm 2 achieves comparable performances. It indicates that the proposed method can be effective for dependent data sets in a single-labeled problem as well.

5   Conclusion

In this paper, we explored the application of LDA for multi-labeled problems. We showed that instead of constructing multiple binary problems and performing dimension reduction to each binary problem, LDA can be applied directly and efficiently to a multi-labeled problem. In composing scatter matrices for LDA, a data sample with multiple class labels contributes equally to all classes for which it is a positive sample. In that way, LDA needs to be performed only once.

If a data sample belongs to multiple classes, what role does it play in separating the classes maximally and minimizing scatter within each class at the same time? In the dimension reduced space, data samples belonging to multiple classes are placed in the boundary areas of the classes, while the maximization of the between-class distance and the minimization of the within-class scatter are optimized. Experimental results demonstrate that the classification performance can be improved when multiple class labels are considered all together as in LDA-ALL rather than when it is transformed to binary problems.

References

1. Lewis, D., Yang, Y., Rose, T., Li, F.: Rcv1: a new benchmark collection for text categorization research. Journal of Machine learning research 5, 361–397 (2004)
2. Pavlidis, P., Weston, J., Cai, J., Grundy, W.: Combining microarray expression data and phylogenetic profiles to learn functional categories using support vector machines. In: Proceedings of the 5th Annual international conference on computational biology, Montreal, Canada (2001)
3. Elisseeff, A., Weston, J.: A kernel method for multi-labeled classification. Advances in neural information processing systems 14, 681–687 (2002)
4. Zhang, M., Zhou, Z.: A k-nearest neighbor based algorithm for multi-label classification. In: 2005 IEEE International Conference on Granular Computing (2005)
5. Godbole, S., Sarawagi, S.: Discriminative methods for multi-labeled classification. In: Dai, H., Srikant, R., Zhang, C. (eds.) PAKDD 2004. LNCS(LNAI), vol. 3056, pp. 22–30. Springer, Heidelberg (2004)
6. Zhu, S., Ji, X., Xu, W., Gong, Y.: Multi-labelled classification using maximum entropy method. In: SIGIR 05, Salvador, Brazil (2005)
7. Torkkola, K.: Linear discriminant analysis in document classification. In: IEEE ICDM-2001 Workshop on Text Mining (TextDM'2001), San Jose, CA (2001)
8. Belhumeur, P.N., Hespanha, J.P., Kriegman, D.J.: Eigenfaces v.s. fisherfaces: Recognition using class specific linear projection. IEEE transactions on pattern analysis and machine learning 19(7), 711–720 (1997)
9. Nguyen, D., Rocke, D.: Tumor classification by partial least squares using microarray gene expression data. Bioinformatics 18(1), 39–50 (2002)
10. Park, C.H., Park, H., Pardalos, P.: A comparative study of linear and nonlinear feature extraction methods. In: Fourth IEEE International Conference on Data Mining, Brighton, United Kingdom, pp. 495–498 (2004)

11. Fukunaga, K.: Introduction to Statistical Pattern Recognition, 2nd edn. Acadamic Press, San Diego (1990)
12. Yu, H., Yang, J.: A direct lda algorithm for high-dimensional data- with application to face recognition. pattern recognition 34, 2067–2070 (2001)
13. Howland, P., Park, H.: Generalizing discriminant analysis using the generalized singular value decomposition. IEEE transaction on pattern analysis and machine intelligence 26(8), 995–1006 (2004)
14. Zheng, W., Zou, C., Zhao, L.: Real-time face recognition using gram-schmidt orthogonalization for lda. In: the Proceedings of the 17th International Conference on Pattern Recognition (2004)
15. Ye, J., Janardan, R., Park, C.H., Park, H.: An optimization criterion for generalized discriminant analysis on undersampled problems. IEEE Transactions on Pattern Analysis and Machine Intelligence 26(8), 982–994 (2004)
16. Schapire, R., Singer, Y.: Boostexter: a boosting-based system for text categorization. Machine learning 39, 135–168 (2000)
17. Luo, X., Zincir-Heywood, N.: Evaluation of two systems on multi-class multi-label document classification. In: ISMIS05, New York, USA (2005)
18. Golub, G.H., Van Loan, C.F.: Matrix Computations. Johns Hopkins University Press, Baltimore (1996)
19. Friedman, J.H.: Regularized discriminant analysis. Journal of the American statistical association 84(405), 165–175 (1989)
20. Chen, L., Liao, H.M., Ko, M., Lin, J., Yu, G.: A new lda-based face recognition system which can solve the small sample size problem. pattern recognition 33, 1713–1726 (2000)
21. Yang, J., Yang, J.-Y.: Why can lda be performed in pca transformed space? Pattern Recognition 36, 563–566 (2003)
22. Kolman, B., Hill, D.: Introductory linear algebra, 8/e edn. Prentice-Hall, Englewood Cliffs (2005)
23. Yu, K., Yu, S., Tresp, V.: Multi-label informed latent semantic indexing. In: SIGIR'05, Salvador, Brazil (2005)

Nonlinear Feature Selection by Relevance Feature Vector Machine*

Haibin Cheng[1], Haifeng Chen[2], Guofei Jiang[2], and Kenji Yoshihira[2]

[1] CSE Department, Michigan State University
East Lansing, MI 48824
chenghai@msu.edu
[2] NEC Laboratories America, Inc.
4 Independence Way, Princeton, NJ 08540
{haifeng,gfj,kenji}@nec-labs.com

Abstract. Support vector machine (SVM) has received much attention in feature selection recently because of its ability to incorporate kernels to discover nonlinear dependencies between features. However it is known that the number of support vectors required in SVM typically grows linearly with the size of the training data set. Such a limitation of SVM becomes more critical when we need to select a small subset of relevant features from a very large number of candidates. To solve this issue, this paper proposes a novel algorithm, called the 'relevance feature vector machine'(RFVM), for nonlinear feature selection. The RFVM algorithm utilizes a highly sparse learning algorithm, the relevance vector machine (RVM), and incorporates kernels to extract important features with both linear and nonlinear relationships. As a result, our proposed approach can reduce many false alarms, e.g. including irrelevant features, while still maintain good selection performance. We compare the performances between RFVM and other state of the art nonlinear feature selection algorithms in our experiments. The results confirm our conclusions.

1 Introduction

Feature selection is to identify a small subset of features that are most relevant to the response variable. It plays an important role in many data mining applications where the number of features is huge such as text processing of web documents, gene expression array analysis, and so on. First of all, the selection of a small feature subset will significantly reduce the computation cost in model building, e.g. the redundant independent variables will be filtered by feature selection to obtain a simple regression model. Secondly, the selected features usually characterize the data better and hence help us to better understand the data. For instance, in the study of genome in bioinformatics, the best feature (gene) subset can reveal the mechanisms of different diseases[6]. Finally, by eliminating the irrelevant features, feature selection can avoid the problem of "curse

---

* The work was performed when the first author worked as a summer intern at NEC Laboratories America, Inc.

P. Perner (Ed.): MLDM 2007, LNAI 4571, pp. 144–159, 2007.

of dimensionality" in case when the number of data examples is small in the high-dimensional feature space [2].

The common approach to feature selection uses greedy local heuristic search, which incrementally adds and/or deletes features to obtain a subset of relevant features with respect to the response[21]. While those methods search in the combinatorial space of feature subsets, regularization or shrinkage methods [20][18] trim the feature space by constraining the magnitude of parameters. For example, Tibshirani [18] proposed the Lasso regression technique which relies on the polyhedral structure of L_1 norm regularization to force a subset of parameter values to be exactly zero at the optimum. However, both the combinatorial search based methods and regularization based methods assume the linear dependencies between features and the response, and can not handle their nonlinear relationships.

Due to the sparse property of support vector machine (SVM), recent work [3][9] reformulated the feature selection problem into SVM based framework by switching the roles of features and data examples. The support vectors after optimization are then regarded as the relevant features. By doing so, we can apply nonlinear kernels on feature vectors to capture the nonlinear relationships between the features and the response variable. In this paper we utilize such promising characteristic of SVM to accomplish nonlinear feature selection. However, we also notice that in the past few years the data generated in a variety of applications tend to have thousands of features. For instance, in the gene selection problem, the number of features, the gene expression coefficients corresponding to the abundance of mRNA, in the raw data ranges from 6000 to 60000 [19]. This large number of features presents a significant challenge to the SVM based feature selection because it has been shown [7] that the number of support vectors required in SVM typically grows linearly with the size of the training data set. When the number of features is large, the standard SVM based feature selection may produce many false alarms, e.g. include irrelevant features in the final results.

To effectively select relevant features from vast amount of attributes, this paper proposes to use the "Relevance Vector Machine"(RVM) for feature selection. Relevance vector machine is a Bayesian treatment of SVM with the same decision function [1]. It produces highly sparse solutions by introducing some prior probability distribution to constrain the model weights governed by a set of hyper-parameters. As a consequence, the selected features by RVM are much fewer than those learned by SVM while maintaining comparable selection performance. In this paper we incorporate a nonlinear feature kernel into the relevance vector machine to achieve nonlinear feature selection from large number of features. Experimental results show that our proposed algorithm, which we call the "Relevance Feature Vector Machine"(RFVM), can discover nonlinear relevant features with good detection rate but low rate of false alarms. Furthermore, compared with the SVM based feature selection methods [3][9], our proposed RFVM algorithm offers other compelling benefits. For instance, the parameters in RFVM are automatically learned by the maximum likelihood estimation rather than the time-consuming cross validation procedure as does in the SVM based methods.

The rest of the paper is organized as follows. In Section 2, we will summarize the related work of nonlinear feature selection using SVM. In Section 3, we extend the relevance vector machine for the task of nonlinear feature selection. The experimental results and conclusions are presented in Section 4 and Section 5 respectively.

2   Preliminaries

Given a data set $D = \left[X_{n\times m}, \boldsymbol{y}_{n\times 1} \right]$, where $X_{n\times m}$ represents the n input examples with m features and $\boldsymbol{y}_{n\times 1}$ represents the responses, we first describe definitions of *feature space* and *example space* with respect to the data. In the feature space, each dimension is related to one specific feature, the data set is regarded as a group of data examples $D = [(\boldsymbol{x}_1, y_1), (\boldsymbol{x}_2, y_2), \cdots, (\boldsymbol{x}_n, y_n)]^T$, where $\boldsymbol{x}_i$s are the rows of X, $X = [\boldsymbol{x}_1^T, \boldsymbol{x}_2^T, \cdots, \boldsymbol{x}_n^T]^T$. The sparse methods such as SVM in the feature space try to learn a sparse example weight vector $\overline{\boldsymbol{\alpha}} = [\alpha_1, \alpha_2, \cdots, \alpha_n]$ associated with the n data examples. The examples with nonzero values α_i are regarded as support vectors, which are illustrated as solid circles in Figure 1(a). Alternatively, each dimension in the example space is related to each data sample $\boldsymbol{x}_i$, and the data is denoted as a collection of features $X = [\boldsymbol{f}_1, \boldsymbol{f}_2, \cdots, \boldsymbol{f}_m]$ and response $\boldsymbol{y}$. The sparse solution in the example space is then related to a weight vector $\boldsymbol{w} = [w_1, w_2, \cdots, w_m]^T$ associated with m features. Only those features with nonzero elements in $\boldsymbol{w}$ are regarded as relevant ones or "support features". If we use SVM to obtain the sparse solution, those relevant features

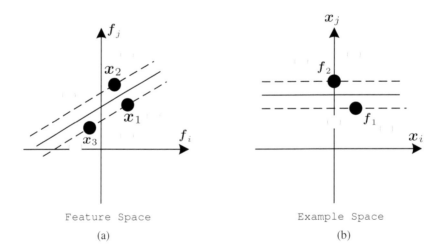

Feature Space

(a)

Example Space

(b)

Fig. 1. (a) The feature space where each dimension is related to one feature ($\boldsymbol{f}$) in the data. SVM learns the sparse solution (denoted as black points) of weight vector $\overline{\boldsymbol{\alpha}}$ associated with data examples $\boldsymbol{x}_i$. (b) The example space in which each dimension is a data example $\boldsymbol{x}_i$. The sparse solution (denoted as black points) of weight vector $\boldsymbol{w}$ is associated with related features ($\boldsymbol{f}$).

are derived from the support features as shown in Figure 1(b). In this section, we first describe feature selection in the SVM framework. Then we will present nonlinear feature selection solutions.

2.1 Feature Selection by SVM

Support Vector Machine [13] is a very popular machine learning technique for the task of classification and regression. The standard SVM-regression [14] aims to find a predictive function $f(\boldsymbol{x}) = \langle \boldsymbol{x}, \boldsymbol{w} \rangle + b$ that has at most ϵ deviation from the actual value y and is as flat as possible, where $\boldsymbol{w}$ is the feature weight vector as described before and b is the offset of function f. If the solution can be further relaxed by allowing certain degree of error, the optimization problem of SVM-regression can be formulated as

$$min\ \frac{1}{2}||\boldsymbol{w}||^2 + C\boldsymbol{1}^T(\boldsymbol{\xi}^+ + \boldsymbol{\xi}^-) \tag{1}$$

$$sub. \begin{cases} \boldsymbol{y} - \langle X, \boldsymbol{w} \rangle - b\boldsymbol{1} \le \epsilon\boldsymbol{1} + \boldsymbol{\xi}^+ \\ \langle X, \boldsymbol{w} \rangle + b\boldsymbol{1} - \boldsymbol{y} \le \epsilon\boldsymbol{1} + \boldsymbol{\xi}^- \\ \boldsymbol{\xi}^+, \boldsymbol{\xi}^- \ge 0 \end{cases}$$

where $\boldsymbol{\xi}^+$ and $\boldsymbol{\xi}^-$ represent the errors, C measures the trade-off between error relaxation and flatness of function, and $\boldsymbol{1}$ denotes the vector whose elements are all 1s. Instead of solving this optimization problem directly, it is usually much easier to solve its dual form [14] by SMO algorithm. The dual problem of the SVM-regression can be derived from Lagrange optimization with KKT conditions and Lagrange multipliers $\boldsymbol{\alpha}^+, \boldsymbol{\alpha}^-$:

$$min\ \frac{1}{2}(\boldsymbol{\alpha}^+ - \boldsymbol{\alpha}^-)^T \langle X, X^T \rangle (\boldsymbol{\alpha}^+ - \boldsymbol{\alpha}^-)$$
$$-\boldsymbol{y}^T(\boldsymbol{\alpha}^+ - \boldsymbol{\alpha}^-) + \epsilon\boldsymbol{1}^T(\boldsymbol{\alpha}^+ + \boldsymbol{\alpha}^-) \tag{2}$$
$$sub.\ \boldsymbol{1}^T(\boldsymbol{\alpha}^+ - \boldsymbol{\alpha}^-) = 0, 0 \le \boldsymbol{\alpha}^+ \le C\boldsymbol{1}, 0 \le \boldsymbol{\alpha}^- \le C\boldsymbol{1}$$

The dual form also provides an easy way to model nonlinear dependencies by incorporating nonlinear kernels. That is, a kernel function $K(\boldsymbol{x}_i, \boldsymbol{x}_j)$ defined over the examples $\boldsymbol{x}_i, \boldsymbol{x}_j$ is used to replace the dot product $\langle \boldsymbol{x}_i, \boldsymbol{x}_j \rangle$ in equation (2). The term $\epsilon\boldsymbol{1}^T(\boldsymbol{\alpha}^+ + \boldsymbol{\alpha}^-)$ in (2) works as the shrinkage factor and leads to the sparse solution of the example weight vector $\overline{\boldsymbol{\alpha}} = (\boldsymbol{\alpha}^+ - \boldsymbol{\alpha}^-)$, which is associated with data examples in the feature space.

While the SVM algorithm is frequently used in the *feature space* to achieve sparse solution $\overline{\boldsymbol{\alpha}}$ for classification and regression tasks, the paper [3] employed SVM in the *example space* to learn a sparse solution of feature weight vector $\boldsymbol{w}$ for the purpose of feature selection by switching the roles of features and data examples. After data normalization such that $X^T\boldsymbol{1} = 0$ and thus $X^Tb\boldsymbol{1} = 0$, the SVM based feature selection described in [3] can be formulated as the following optimization problem.

$$min\ \frac{1}{2}||X\boldsymbol{w}||^2 + C\boldsymbol{1}^T(\boldsymbol{\xi}^+ + \boldsymbol{\xi}^-) \tag{3}$$

$$sub. \begin{cases} \langle X^T, \boldsymbol{y} \rangle - \langle X^T, X \rangle \boldsymbol{w} \le \epsilon \mathbf{1} + \boldsymbol{\xi}^+ \\ \langle X^T, X \rangle \boldsymbol{w} - \langle X^T, \boldsymbol{y} \rangle \le \epsilon \mathbf{1} + \boldsymbol{\xi}^- \\ \boldsymbol{\xi}^+, \boldsymbol{\xi}^- \ge 0 \end{cases}$$

The above equation (3) makes it easy to model nonlinear dependencies between features and response, which has also been explored in the work [9]. Similarly, the dual problem of (3) can also be obtained with Lagrange multiplies $\boldsymbol{w}^+, \boldsymbol{w}^-$ and KKT conditions

$$min \ \frac{1}{2}(\boldsymbol{w}^+ - \boldsymbol{w}^-)^T \langle X^T, X \rangle (\boldsymbol{w}^+ - \boldsymbol{w}^-)$$
$$- \langle \boldsymbol{y}^T, X \rangle (\boldsymbol{w}^+ - \boldsymbol{w}^-) + \epsilon \mathbf{1}^T (\boldsymbol{w}^+ + \boldsymbol{w}^-) \qquad (4)$$
$$sub. \ 0 \le \boldsymbol{w}^+ \le C\mathbf{1}, 0 \le \boldsymbol{w}^- \le C\mathbf{1}$$

The intuition behind the dual optimization problem (4) is very obvious. It tries to minimize the mutual feature correlation noted as $\langle X^T, X \rangle$ and maximize the response feature correlation $\langle \boldsymbol{y}^T, X \rangle$. The parameter "C" in equation (4) controls the redundancy of the selected features. Small value of "C" reduces the importance of mutual feature correlation $\langle X^T, X \rangle$ and thus allow more redundancy. The term $\epsilon \mathbf{1}^T (\boldsymbol{w}^+ + \boldsymbol{w}^-)$ in the above dual form (4) achieves the sparseness of the feature weight vector $\boldsymbol{w} = (\boldsymbol{w}^+ - \boldsymbol{w}^-)$. After optimization, the nonzero elements in $\boldsymbol{w}$ are related to the relevant features in the example space. For the detailed explanation about the derivation of (3) and (4), please see [3].

2.2   Nonlinear Feature Selection

If we set $\epsilon = \frac{\lambda}{2}$ and ignore the error relaxation in the primal problem (3), the optimization form (3) can be rewritten in the example space using features $X = [\boldsymbol{f}_1, \boldsymbol{f}_2, \cdots, \boldsymbol{f}_m]$ and the response $\boldsymbol{y}$

$$min \ \frac{1}{2} \sum_{i=1}^{m} \sum_{j=1}^{m} w_i w_j \langle \boldsymbol{f}_i, \boldsymbol{f}_j \rangle \qquad (5)$$

$$sub. \ | \sum_{i=1}^{m} w_i \langle \boldsymbol{f}_j, \boldsymbol{f}_i \rangle - \langle \boldsymbol{f}_j, \boldsymbol{y} \rangle | \le \frac{\lambda}{2}, \quad \forall j$$

The optimization problem in (5) has been proved in [9] to be equivalent to the Lasso regression (6) [18] which has been widely used for linear feature selection

$$min \|X\boldsymbol{w} - \boldsymbol{y}\|^2 + \lambda \|\boldsymbol{w}\|_1 . \qquad (6)$$

While the Lasso regression (6) is performed in the *feature space* of data set to achieve feature selection, the optimization (5) formulates the feature selection problem in the *example space*. As a consequence, we can define nonlinear kernels over the feature vectors to model nonlinear interactions between features. For the feature vectors $\boldsymbol{f}_i$ and $\boldsymbol{f}_j$ with nonlinear dependency, we assume that they can be projected to a high dimensional space by a mapping function ϕ so that they interact linearly in the mapped space. Therefore the nonlinear dependency can be represented by introducing the feature kernel $K(\boldsymbol{f}_i, \boldsymbol{f}_j) = \phi(\boldsymbol{f}_i)^T \phi(\boldsymbol{f}_j)$. If we replace the dot product $\langle,\rangle$ in (5) with the feature kernel K, we can obtain its nonlinear version:

$$min \ \frac{1}{2} \sum_{i=1}^{m} \sum_{j=1}^{m} w_i w_j K(\boldsymbol{f}_i, \boldsymbol{f}_j) \tag{7}$$

$$sub. \ |\sum_{i=1}^{m} w_i K(\boldsymbol{f}_j, \boldsymbol{f}_i) - K(\boldsymbol{f}_j, \boldsymbol{y})| \le \frac{\lambda}{2}, \quad \forall j \ .$$

In the same way, we can incorporate nonlinear feature kernels into the general expression (4) and obtain

$$min \ \frac{1}{2} \sum_{i=1}^{m} \sum_{j=1}^{m} (w_i^+ - w_i^-) K(\boldsymbol{f}_i, \boldsymbol{f}_j)(w_j^+ - w_j^-)$$

$$- \sum_{i=1}^{m} K(\boldsymbol{y}, \boldsymbol{f}_i)(w_i^+ - w_i^-) + \epsilon \sum_{i=1}^{n} (w_i^+ + w_j^-) \tag{8}$$

$$sub. \ 0 \le w_i^+ \le C, 0 \le w_i^- \le C, \ \forall i$$

Both (7) and (8) can be used for nonlinear feature selection. However, they are both derived from the SVM framework and share the same weakness of standard SVM algorithm. For instance, the number of support features will grow linearly with the size of the feature set in the training data. As a result, the provided solution in the example space is not sparse enough. This will lead to a serious problem of high false alarm rate, e.g. including many irrelevant features, when the feature set is large. To solve this issue, this paper proposes a RVM based solution for nonlinear feature selection, which is called "Relevance Feature Vector Machine". RFVM achieves more sparse solution in the example space by introducing priors over the feature weights. As a result, RFVM is able to select the most relevant features as well as decrease the number of false alarms significantly. Furthermore, we will also show that RFVM can learn the hyper-parameters automatically and hence avoids the effort of cross validation to determine the trade-off parameter "C" in SVM optimization (8).

3   Relevance Feature Vector Machine

In this section, we will investigate the problem of using Relevance Vector Machine for nonlinear feature selection. We will first introduce the Bayesian framework of standard Relevance Vector Machine algorithm [1]. Then we present our Relevance Feature Vector Machine algorithm which utilizes RVM in the example space and exploits the mutual information kernel for nonlinear feature selection.

3.1   Relevance Vector Machine

The standard RVM [1] is to learn the vector $\tilde{\boldsymbol{\alpha}}_{(n+1)\times 1} = [\alpha_0, \overline{\boldsymbol{\alpha}}]$ with $\alpha_0 = b$ denoting the "offset" and $\overline{\boldsymbol{\alpha}} = [\alpha_1, \alpha_2, \cdots, \alpha_n]$ as the "relevance feature weight vector" associated with data examples in the feature space. It assumes that the response y_i is sampled from the model $f(\boldsymbol{x}_i)$ with noise ϵ, and the model function is expressed as

$$f(\boldsymbol{x}) = \sum_{j=1}^{n} \alpha_j \langle \boldsymbol{x}, \boldsymbol{x}_j \rangle + \alpha_0 + \epsilon \tag{9}$$

where ϵ is assumed to be sampled independently from a Gaussian distribution noise with mean zero and variance σ^2. If we use kernel to model the dependencies between the examples in the feature space, we can get the $n \times (n+1)$ 'design' matrix Φ:

$$\Phi = \begin{bmatrix} 1 & K(\boldsymbol{x}_1, \boldsymbol{x}_1) & K(\boldsymbol{x}_1, \boldsymbol{x}_2) & \cdots & K(\boldsymbol{x}_1, \boldsymbol{x}_n) \\ 1 & K(\boldsymbol{x}_2, \boldsymbol{x}_1) & K(\boldsymbol{x}_2, \boldsymbol{x}_2) & \cdots & K(\boldsymbol{x}_2, \boldsymbol{x}_n) \\ \vdots & & & \\ 1 & K(\boldsymbol{x}_n, \boldsymbol{x}_1) & K(\boldsymbol{x}_n, \boldsymbol{x}_2) & \cdots & K(\boldsymbol{x}_n, \boldsymbol{x}_n) \end{bmatrix}$$

In order to estimate the coefficients $\alpha_0, \cdots, \alpha_n$ in equation (9) from a set of training data, the likelihood of the given data set is written as

$$p(\boldsymbol{y}|\tilde{\boldsymbol{\alpha}}, \sigma^2) = (2\pi\sigma^2)^{-\frac{n}{2}} exp\left\{ -\frac{1}{\sigma^2} ||\boldsymbol{y} - \Phi\tilde{\boldsymbol{\alpha}}||^2 \right\} \tag{10}$$

In addition, RVM defines prior probability distributions on parameters $\tilde{\boldsymbol{\alpha}}$ in order to obtain sparse solutions. Such prior distribution is expressed with $n+1$ hyper-parameters $\tilde{\boldsymbol{\beta}}_{(n+1)\times 1} = [\beta_0, \beta_1, \cdots, \beta_n]$:

$$p(\tilde{\boldsymbol{\alpha}}|\tilde{\boldsymbol{\beta}}) = \prod_{i=0}^{n} N(\alpha_i|0, \beta_i^{-1}) \tag{11}$$

The unknowns $\tilde{\boldsymbol{\alpha}}, \tilde{\boldsymbol{\beta}}$ and σ^2 can be estimated by maximizing the posterior distribution $p(\tilde{\boldsymbol{\alpha}}, \tilde{\boldsymbol{\beta}}, \sigma^2|\boldsymbol{y})$, which can be decomposed as:

$$p(\tilde{\boldsymbol{\alpha}}, \tilde{\boldsymbol{\beta}}, \sigma^2|\boldsymbol{y}) = p(\tilde{\boldsymbol{\alpha}}|\boldsymbol{y}, \tilde{\boldsymbol{\beta}}, \sigma^2)p(\tilde{\boldsymbol{\beta}}, \sigma^2|\boldsymbol{y}) . \tag{12}$$

Such decomposition allows us to use two steps to find the solution $\tilde{\boldsymbol{\alpha}}$ together with hyper-parameters $\tilde{\boldsymbol{\beta}}$ and σ^2. For details of the optimization procedure, please see [1]. Compared with SVM, RVM produces a more sparse solution $\tilde{\boldsymbol{\alpha}}$ as well as determines the hyper-parameters simultaneously.

To the best of our knowledge, current RVM algorithm is always performed in the feature space in which the relevance weight vector $\overline{\boldsymbol{\alpha}}$ in RVM is associated with data examples. This paper is the first to utilize the promising characteristics of RVM for feature selection. In the next section, we reformulate the Relevance Vector Machine in the example space and incorporate nonlinear feature kernels to learn nonlinear "relevant features".

3.2 Nonlinear Feature Selection with Relevance Feature Vector Machine

This section presents the relevance feature vector machine (RFVM) algorithm, which utilizes RVM in the example space to select relevant features. We will also show how the kernel trick can be applied to accomplish nonlinear feature selection. Again, we assume the data $(X, \boldsymbol{y})$ is standardized. We start by rewriting the function (9) into an equivalent form by incorporating the feature weight vector $\boldsymbol{w}$

$$\boldsymbol{y} = \sum_{j=1}^{m} w_j \boldsymbol{f}_j + \boldsymbol{\epsilon} \tag{13}$$

The above formula assumes the linear dependency between features and the response. When such relationship is nonlinear, we project the features and responses into high dimensional space by a function ϕ so that the dependency in the mapped space becomes linear

$$\phi(\boldsymbol{y}) = \sum_{j=1}^{m} w_j \phi(\boldsymbol{f}_j) + \boldsymbol{\epsilon} \ . \tag{14}$$

Accordingly the likelihood function given the training data can be expressed as

$$p(\phi(\boldsymbol{y})|\boldsymbol{w}, \sigma^2) = (2\pi\sigma^2)^{-\frac{n}{2}} exp\left\{ -\frac{||\phi(\boldsymbol{y}) - \phi(X)\boldsymbol{w}||^2}{\sigma^2} \right\} \tag{15}$$

where $\phi(X) = [\phi(\boldsymbol{f}_1), \phi(\boldsymbol{f}_2), \cdots, \phi(\boldsymbol{f}_m)]$. We expand the squared error term in the above likelihood function and replace the dot product with certain feature kernel K to model the nonlinear interaction between the feature vectors and response, which results in

$$\begin{aligned}
&||\phi(\boldsymbol{y}) - \phi(X)\boldsymbol{w}||^2 \\
&= (\phi(\boldsymbol{y}) - \phi(X)\boldsymbol{w})^T (\phi(\boldsymbol{y}) - \phi(X)\boldsymbol{w}) \\
&= \phi(\boldsymbol{y})^T \phi(\boldsymbol{y}) - 2\boldsymbol{w}^T \phi(X)^T \phi(\boldsymbol{y}) + \boldsymbol{w}^T \phi(X)^T \phi(X)\boldsymbol{w} \\
&= K(\boldsymbol{y}^T, \boldsymbol{y}) - 2\boldsymbol{w}^T K(X^T, \boldsymbol{y}) + \boldsymbol{w}^T K(X^T, X)\boldsymbol{w}
\end{aligned}$$

where:

$$K(X^T, \boldsymbol{y}) = \begin{bmatrix} K(\boldsymbol{y}, \boldsymbol{f}_1) \\ K(\boldsymbol{y}, \boldsymbol{f}_2) \\ \vdots \\ K(\boldsymbol{y}, \boldsymbol{f}_m) \end{bmatrix}$$

and

$$K(X^T, X) = \begin{bmatrix} K(\boldsymbol{f}_1, \boldsymbol{f}_1) & K(\boldsymbol{f}_1, \boldsymbol{f}_2) & \cdots & K(\boldsymbol{f}_1, \boldsymbol{f}_m) \\ K(\boldsymbol{f}_2, \boldsymbol{f}_1) & K(\boldsymbol{f}_2, \boldsymbol{f}_2) & \cdots & K(\boldsymbol{f}_2, \boldsymbol{f}_m) \\ \vdots & & & \\ K(\boldsymbol{f}_m, \boldsymbol{f}_1) & K(\boldsymbol{f}_m, \boldsymbol{f}_2) & \cdots & K(\boldsymbol{f}_m, \boldsymbol{f}_m) \end{bmatrix}$$

After some manipulations, the likelihood function (15) can be reformulated as

$$\begin{aligned}
p(\phi(\boldsymbol{y})|\boldsymbol{w}, \sigma^2) = (2\pi\sigma^2)^{-\frac{n}{2}} exp\big\{ \big(-K(\boldsymbol{y}^T, \boldsymbol{y}) + \\
2\boldsymbol{w}^T K(X^T, \boldsymbol{y}) - \boldsymbol{w}^T K(X^T, X)\boldsymbol{w}\big)/\sigma^2 \big\}
\end{aligned} \tag{16}$$

Note that RFVM differs from traditional RVM in that the prior $\boldsymbol{\beta} = [\beta_1, \beta_2, \cdots, \beta_m]$ is defined over the relevance feature vector weight $\boldsymbol{w}$.

$$p(\boldsymbol{w}|\boldsymbol{\beta}) = \prod_{i=1}^{m} N(w_i|0, \beta_i^{-1}) \tag{17}$$

The sparse solution $\boldsymbol{w}$ corresponding to relevant features can be obtained by maximizing

$$p(\boldsymbol{w}, \boldsymbol{\beta}, \sigma^2|\phi(\boldsymbol{y})) = p(\boldsymbol{w}|\phi(\boldsymbol{y}), \boldsymbol{\beta}, \sigma^2)p(\boldsymbol{\beta}, \sigma^2|\phi(\boldsymbol{y})) \tag{18}$$

Similar to RVM, we use two steps to find the maximized solution. The first step is now to maximize

$$p(\boldsymbol{w}|\phi(\boldsymbol{y}), \boldsymbol{\beta}, \sigma^2) = \frac{p(\phi(\boldsymbol{y})|\boldsymbol{w}, \sigma^2)p(\boldsymbol{w}|\boldsymbol{\beta})}{p(\phi(\boldsymbol{y})|\boldsymbol{\beta}, \sigma^2)}$$

$$= (2\pi)^{-\frac{n+1}{2}}|\Sigma|^{-\frac{1}{2}}exp\left\{-\frac{1}{2}(\boldsymbol{w} - \boldsymbol{\mu})^T|\Sigma|^{-1}(\boldsymbol{w} - \boldsymbol{\mu})\right\} \tag{19}$$

Given the current estimation of $\boldsymbol{\beta}$ and σ^2, the covariance Σ and mean $\boldsymbol{\mu}$ of the feature weight vector $\boldsymbol{w}$ are

$$\Sigma = (\sigma^{-2}K(X^T, X) + B)^{-1} \tag{20}$$

$$\boldsymbol{\mu} = \sigma^{-2}\Sigma K(X^T, \boldsymbol{y}) \tag{21}$$

and $B = diag(\beta_1, \cdots, \beta_n)$.

Once we get the current estimation of $\boldsymbol{w}$, the second step is to learn the hyper-parameters $\boldsymbol{\beta}$ and σ^2 by maximizing $p(\boldsymbol{\beta}, \sigma^2|\phi(\boldsymbol{y})) \propto p(\phi(\boldsymbol{y})|\boldsymbol{\beta}, \sigma^2)p(\boldsymbol{\beta})p(\sigma^2)$. Since we assume the hyper-parameters are uniformly distributed, e.g. $p(\boldsymbol{\beta})$ and $p(\sigma^2)$ are constant, it is equivalent to maximize the marginal likelihood $p(\phi(\boldsymbol{y})|\boldsymbol{\beta}, \sigma^2)$, which is computed by:

$$p(\phi(\boldsymbol{y})|\boldsymbol{\beta}, \sigma^2) = \int p(\phi(\boldsymbol{y})|\boldsymbol{w}, \sigma^2)p(\boldsymbol{w}|\boldsymbol{\beta})d\boldsymbol{w}$$

$$= (2\pi)^{-\frac{n}{2}}|\sigma^2\boldsymbol{I} + \phi(X)B^{-1}\phi(X)^T|^{-\frac{1}{2}}$$

$$*exp\left\{-\frac{1}{2}\boldsymbol{y}^T(\sigma^2\boldsymbol{I} + \phi(X)B^{-1}\phi(X)^T)^{-1}\boldsymbol{y}\right\} \tag{22}$$

By differentiation of equation (22), we can update the hyper-parameters $\boldsymbol{\beta}$ and σ^2 by:

$$\beta_i^{new} = \frac{1 - \beta_i N_{ii}}{\mu_i{}^2} \tag{23}$$

$$\sigma^{2new} = \frac{||\phi(\boldsymbol{y}) - \phi(X)\boldsymbol{\mu}||^2}{n - \sum_i(1 - \beta_i N_{ii})} \tag{24}$$

where N_{ii} is i_{th} diagonal element of the covariance from equation (20) and $\boldsymbol{\mu}$ is computed from equation (21) with current $\boldsymbol{\beta}$ and σ^2 values. The final optimal set of $\boldsymbol{w}$, $\boldsymbol{\beta}$ and σ^2 are then learned by repeating the first step to update the covariance Σ (20) and mean $\boldsymbol{\mu}$ (21) of the feature weight vector $\boldsymbol{w}$ and the second step to update the hyper-parameters $\boldsymbol{\beta}$ (23) and σ^2 (24) iteratively.

RFVM learns a sparse feature weight vector $\boldsymbol{w}$ in which most of the elements are zeros. Those zero elements in $\boldsymbol{w}$ indicate that the corresponding features are irrelevant and should be filtered out. On the other hand, large values of elements in $\boldsymbol{w}$ indicate high importance of the related features. In this paper we use mutual information as the kernel function $K(\cdot, \cdot)$, which will be introduced in the following section. In that case, $K(X^T, \boldsymbol{y})$ actually measures the relevance between the response $\boldsymbol{y}$ and features in the data matrix X and $K(X^T, X)$ indicates the redundancy between features in the data matrix X. The likelihood maximization procedure of RFVM tends to maximize the relevance between the features and response and minimize the mutual redundancy within the features.

3.3 Mutual Information Feature Kernel

While kernels are usually defined over data examples in the feature space, the RFVM algorithm places the nonlinear kernel over the feature and response vectors for the purpose of feature selection. As we know, the mutual information [16] of two variables measures how much uncertainty can be reduced about one variable given the knowledge of the other variable. Such property can be used as the metric to measure the relevance between features. Given two discrete variables U and V with their observations denoted as u and v respectively, the mutual information I between them is formulated as

$$I(U,V) = \sum_{u \in U} \sum_{y \in V} p(u,v) log_2 \frac{p(u,v)}{p(u)p(v)} \tag{25}$$

where $p(u,v)$ is the joint probability density function of U and V, and $p(u)$ and $p(v)$ are the marginal probability density functions of U and V respectively.

Now given two feature vectors $\boldsymbol{f}_u$ and $\boldsymbol{f}_v$, we use the following way to calculate the value of their mutual information kernel $K(\boldsymbol{f}_u, \boldsymbol{f}_v)$. We regard all the elements in the vector $\boldsymbol{f}_u$ (or $\boldsymbol{f}_v$) as multiple observations of a variable $\boldsymbol{f}_u$ (or $\boldsymbol{f}_v$), and discretize those observations into bins for each variable. That is, we sort the values in the feature vectors $\boldsymbol{f}_u$ and $\boldsymbol{f}_v$ separately and discretize each vector into N bins, with the same interval for each bin. For example, if the the maximal value of $\boldsymbol{f}_u$ is u_{max} and the minimal value is u_{min}, the interval for each bin of feature vector $\boldsymbol{f}_u$ is $(u_{max} - u_{min})/N$. Now for each value u in feature vector $\boldsymbol{f}_u$ and v in feature vector $\boldsymbol{f}_v$, we assign $u = i$ and $v = j$ if u falls into the i_{th} bin and v falls into the j_{th} bin of their discretized regions respectively. The probability density functions $p(\boldsymbol{f}_u, \boldsymbol{f}_v)$, $p(\boldsymbol{f}_u)$ and $p(\boldsymbol{f}_v)$ are calculated as the ratio of the number of elements within corresponding bin to the length of vector n. As a result, we have

$$p(u = i) = counts(u = i)/n$$
$$p(v = j) = counts(v = j)/n$$
$$p(u = i, v = j) = counts(u = i \ and \ v = j)/n$$

and

$$K(\boldsymbol{f}_u, \boldsymbol{f}_v) = \sum_{i=1}^{N} \sum_{j=1}^{N} p(u=i, v=j) log_2 \frac{p(u=i, v=j)}{p(u=i)p(v=j)} \qquad (26)$$

The mutual information kernel is symmetric and non-negative with $K(\boldsymbol{f}_u, \boldsymbol{f}_v)$ $= K(\boldsymbol{f}_v, \boldsymbol{f}_u)$ and $K(\boldsymbol{f}_u, \boldsymbol{f}_v) \geq 0$. It also satisfies the mercer's condition [13], which guarantees the convergence of the proposed RFVM algorithm. In this paper we set the number of bins for discretization as $log_2(m)$, where m is the number of features.

4   Experimental Results

Experimental results are presented in this section to demonstrate the effectiveness of our proposed RFVM algorithm. We compare RFVM with other two state of the art nonlinear feature selection algorithms in [3]and [9]. To be convenient, we call the algorithm proposed in [3] as P-SVM and that in [9] as FVM algorithm. All the experiments are conducted on a Pentium 4 machine with 3GHZ CPU and 1GB of RAM.

4.1   Nonlinear Feature Selection by RFVM

In order to verify that the proposed RFVM is able to catch the nonlinear dependency between the response and feature vectors, we simulated 1000 data examples with 99 features and one response. The response y is generated by the summation of three base functions of f_1, f_2, f_3 respectively, together with the Gaussian noise ϵ distributed as $N(0, 0.005)$.

$$y = f(f_1, f_2, f_3, \cdots, f_{99})$$
$$= 9f_1 + 20(1 - f_2)^3 + 17\sin(80 * f_3 - 7) + \epsilon$$

The three base functions are shown in Figure 2(b)(c)(d), in which the first is a linear function and the other two are nonlinear. Figure 2(a) also plots the distribution of y with respect to the two nonlinear features f_2 and f_3. The values of features f_1, f_2, f_3 are generated by a uniform distribution in $[0, 1]$. The other 96 features $f_4, f_5, \cdots, f_{99}$ are generated uniformly in $[0, 20]$ and are independent with the response y.

We modified the MATLAB code provided by Mike Tipping [17] to implement RFVM for the task of feature selection. The RFVM is solved by updating the posterior covariance Σ in equation (20) and the mean $\boldsymbol{\mu}$ in equation (21) along with the hyper-parameters $\boldsymbol{\beta}$ in equation (23) and σ^2 in equation (24) iteratively using the two step procedure. The nonlinear dependencies between response and features by using mutual information kernel in RFVM. That is, we replace the dot product of the features and response, $< X^T, y >$ and $< X^T, X >$, by the precomputed mutual information kernel $K(X^T, y)$ and $K(X^T, X)$. The optimal

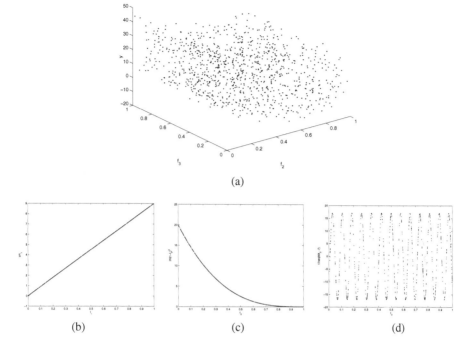

(a)

(b)   (c)   (d)

Fig. 2. (a) The distribution of response y with respect to two nonlinear features f_2 and f_3. The bottom three figures illustrate the three components of the simulated function f: (b) linear, (c) cubic and (d) sin.

set of feature vector weight w along with the hyper-parameters β and σ^2 in RFVM are automatically learned by a two step updating procedure. The initial values of the hyper-parameters are set as $\beta = 10$ and $\sigma^2 = std(y)/10$.

Figure 3(a) and (b) plot the values of feature weights computed by linear RVM and nonlinear RFVM over the simulated data. From Figure 3(a), we see that the linear RVM can detect the linear dependent feature f_1, as well as the feature f_2 which has cubical relationship. The reason that f_2 is also detected by linear RVM is that the cubical curve can be approximated by a linear line in certain degree, which is shown in Figure 2(b). However, RVM missed the feature f_3 completely, which is a highly nonlinear feature with periodical sin wave. On the other hand, the nonlinear RFVM detects all the three features successfully, which is shown in Figure 3(b). Furthermore, the detected feature set is pretty sparse compared with the results of linear RVM.

4.2   Performance Comparison

This section compares the performance of RFVM algorithm with other nonlinear feature selection algorithms such as FVM in [9] and P-SVM in [3]. To demonstrate that RFVM is able to select most relevant features with much lower false

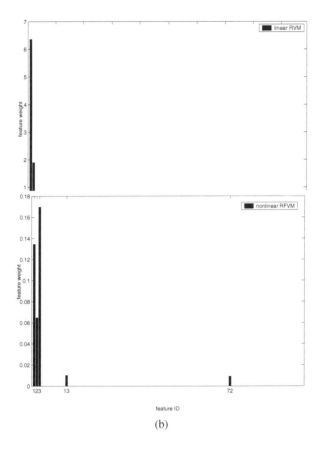

Fig. 3. (a) The histogram of feature weights from linear RVM. It detects f_1 and f_2 but misses the highly nonlinear relevant feature f_3. (b) The histogram of feature weights from nonlinear RVM. It detects all the three features.

alarm rate, we simulate another data set with 2000 data examples and 100 features. The first 20 features are simulated uniformly in $[-0.5, 0.5]$ and the rest are generated uniformly in $[0, 20]$ with Gaussian noise. The response y is the summation of functions $F_i(\cdot)$ on the first 20 features

$$y = \sum_{i=1}^{20} F_i(f_i) \ . \tag{27}$$

The basis function $F_i(\cdot)$ is randomly chosen from the pool of eight candidate functions

$$F_i(f_i) \in \{F_1(f_i), F_2(f_i), \cdots, F_8(f_i)\} \tag{28}$$

where the expressions of those candidate functions are described in Table 1. As you can see our simulation covers almost all kinds of common nonlinear relationships.

Table 1. The 8 basis function

$j =$	1	2	3	4
$F_j(f_i) =$	$40f_i$	$20(1 - f_i^2)$	$23f_i^3$	$20\sin(40f_i - 5)$
$j =$	5	6	7	8
$F_j(f_i) =$	$20e^{f_i}$	$-\log_2(|f_i|)$	$20\sqrt{1 - f_i}$	$20\cos(20f_i - 7)$

We divide the data into two parts, the first 1000 examples are used as training data to determine the parameter λ in FVM and ϵ, C in P-SVM by 10 fold cross validation. The rest 1000 data examples are used for test. The performances of those algorithms are compared in terms of detection rate and false alarm rate. We run 20 rounds of such simulations and present the results in Figure 4. Figure 4(a) plots the detection rate of RFVM together with those of FVM and P-SVM. It shows that RFVM maintains comparable detection rate as the other

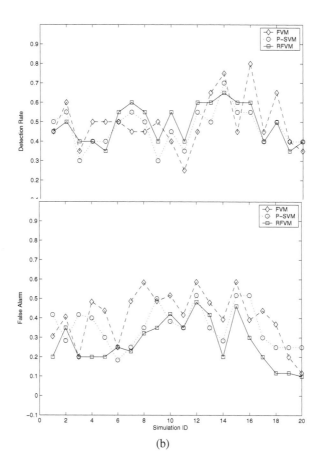

(b)

Fig. 4. (a) The detection rates of FVM, P-SVM and RFVM. (b) The false alarm rates of FVM, P-SVM and RFVM.

two algorithms. Note since the nonlinear relationship (27) generated in our simulation is very strong, the range of detection rates for those algorithms is reasonable. Figure 4(b) plots the false alarm rates of FVM, P-SVM and RFVM algorithms. It demonstrates that RFVM has lower false alarm rate generally, which is due to the sparseness of RFVM in the example space compared with FVM and P-SVM. Also note in the experiment we don't need to predetermine any parameters in RFVM since the parameters are automatically learned by two step maximum likelihood method, while FVM and P-SVM are both sensitive to parameters and need extra efforts of cross validation to determine those values.

5   Conclusions

This paper has proposed a new method, the "Relevance Feature Vector Machine"(RFVM), to detect features with nonlinear dependency. Compared with other state of the art nonlinear feature selection algorithms, RFVM has two unique advantages based on our theoretical analysis and experimental results. First, by utilizing the highly sparseness nature of RVM, the RFVM algorithm reduces the false alarms in feature selection significantly while still maintains desirable detection rate. Furthermore, unlike other SVM based nonlinear feature selection algorithms whose performances are sensitive to the selection of parameter values, RFVM learns the hyper-parameters automatically by maximizing the "marginal likelihood" in the second step of the two-step updating procedure. In the future, we will apply RFVM to some real applications to further demonstrate the advantages of our algorithm.

References

1. Tipping, M.E.: Sparse Bayesian learning and the relevance vector machine. Journal of Machine Learning Research 1, 211–244 (2001)
2. Bellman, R.E.: Adaptive Control Processes. Princeton University Press, Princeton, NJ (1961)
3. Hochreiter, S., Obermayer, K.: Nonlinear feature selection with the potential support vector machine. In: Guyon, I., Gunn, S., Nikravesh, M., Zadeh, L. (eds.) Feature extraction, foundations and applications, Springer, Berlin (2005)
4. Figueiredo, M., Jain, A.K.: Bayesian Learning of Sparse Classifiers. In: Proc. IEEE Computer Soc. Conf. Computer Vision and Pattern Recognition, vol. 1, 35–41 (2001)
5. Figueiredo, M.A.T.: Adaptive sparseness for supervised learning. IEEE Transactions on Pattern Analysis and Machine Intelligence 25(9), 1150–1159 (2003)
6. BΦ, T.H., Jonassen, I.: New feature subset selection procedures for classification of expression profiles, Genome Biology, 3 research 0017.1-0017.11 (2000)
7. Burges, C.: Simplified support vector decision rules. In: Proc. of the Thirteenth International Conf. on Machine Learning, pp. 71–77. Morgan Kaufmann, Seattle (1996)
8. Aizerman, M.E., Braverman, Rozonoer, L.: Theoretical foundations of the potential function method in pattern recognition learning. Automation and Remote Control 25, 821–837 (1964)

9. Li, F., Yang, Y., Xing, E.P.: From Lasso regression to Feature Vector Machine, Advances in Neural Information Processing Systems, 18 (2005)
10. Faul, A., Tipping, M.E.: Analysis of sparse bayesian learning. In: Dietterich, T., Becker, S., Ghahramani, Z. (eds.) Advances in Neural Information Processing Systems 14, pp. 383–389. MIT Press, Cambridge, MA (2002)
11. Faul, A., Tipping, M.: A variational approach to robust regression, in Artificial Neural Networks. In: Dorffner, G., Bischof, H., Hornik, K. (eds.), pp. 95–202 (2001)
12. Roth, V.: The Generalized LASSO, V. IEEE Transactions on Neural Networks, Dorffner, G. vol. 15(1). (2004)
13. Vapnik, V.N.: The Nature of Statistical Learning Theory. Springer, Heidelberg (1995)
14. Smola, A.J., Scholkopf, B.: A tutorial on support vector regression, NEUROCOLT Technical Report NC-TR-98-030, Royal Holloway College, London (1998)
15. Long, F., Ding, C.: Feature Selection Based on Mutual Information: Criteria of Max-Dependency, Max-Relevance, and Min-Redundancy. IEEE Transactions on Pattern Analysis and Machine Intelligence 27(8), 1226–1238 (2005)
16. Guiasu, Silviu.: Information Theory with Applications. McGraw-Hill, New York (1977)
17. Tipping, M.E.: Microsoft Corporation, http://research.microsoft.com/MLP/RVM/
18. Tibshirani, R.: Regression Shrinkage and Selection via the Lasso. Journal of the Royal Statistical Society. Series B (Methodological) 58(1), 267–288 (1999)
19. Guyon, I., Elisseeff, A.: An Introduction to Variable and Feature Selection. Journal of Machine Learning Research 3, 1157–1182 (2003)
20. Bishop, C.M.: Neural Networks for Pattern Recognition. Oxford University Press, Oxford (1995)
21. Reeves, S.J., Zhao, Z.: Sequential algorithms for observation selection. IEEE Transactions on Signal Processing 47, 123–132 (1999)

Affine Feature Extraction: A Generalization of the Fukunaga-Koontz Transformation

Wenbo Cao and Robert Haralick

Department of Computer Science,
Pattern Recognition Laboratory,
The Graduate Center, City University of New York
365 Fifth Avenue, New York, NY 10016, USA

Abstract. Dimension reduction methods are often applied in machine learning and data mining problems. Linear subspace methods are the commonly used ones, such as principal component analysis (PCA), Fisher's linear discriminant analysis (FDA), et al. In this paper, we describe a novel feature extraction method for binary classification problems. Instead of finding linear subspaces, our method finds lower-dimensional affine subspaces for data observations. Our method can be understood as a generalization of the Fukunaga-Koontz Transformation. We show that the proposed method has a closed-form solution and thus can be solved very efficiently. Also we investigate the information-theoretical properties of the new method and study the relationship of our method with other methods. The experimental results show that our method, as PCA and FDA, can be used as another preliminary data-exploring tool to help solve machine learning and data mining problems.

1   Introduction

Because of the curse of dimensionality and the concern of computational efficiency, dimensionality reduction methods are often used in machine learning and data mining problems. Examples are face recognition in computer vision [3, 20], electroencephalogram (EEG) signal classification in Brain-Computer Interface (BCI)[5, 16] and microarray data analysis [4]. Linear subspace methods have been widely used for the purpose of dimension reduction. We give a brief review of the most commonly used ones.

Principal component analysis (PCA) and independent component analysis (ICA) are unsupervised linear subspace methods for dimension reduction. PCA tries to find linear subspaces such that the variance of data are maximally preserved. ICA is a way of finding linear subspaces in which the second- and higher-order statistical dependencies of the data are minimized; that is the transformed variables are as statistically independent from each other as possible. Note that, as unsupervised methods, neither PCA nor ICA use label information, which is crucial for classification problems. Consequently, PCA and ICA are optimal for pattern description, but not optimal for pattern discrimination.

Fisher's discriminant analysis (FDA) finds linear subspaces in which the distance between the means of classes is maximized and the variance of each class

P. Perner (Ed.): MLDM 2007, LNAI 4571, pp. 160–173, 2007.

is minimized at the same time. An important drawback of FDA is that, for K-class classification problems, it can only find $K - 1$ dimensional subspaces. This becomes more serious when binary classification problems are considered, for which FDA can only extract one optimal feature. Canonical correlation analysis (CCA) is a method for finding linear subspaces to maximize the correlation of the observation vectors and their labels. It has been known for a long time that FDA and CCA indeed give identical subspaces for the dimension reduction purpose [2].

Recently there has been some interest in partial least squares (PLS) [18]. Only recently, it has been shown that PLS has a close connection with FDA [1]. PLS finds linear subspaces by iteratively maximizing the covariance of the deflated observation vectors and their labels. In one mode, PLS can be used to extract more than one feature for binary classification. The main concern in PLS is the efficiency issue, since in each iteration one has to subtract the observation matrix by its rank-one estimation found in the previous iteration, and generate deflated observation vectors.

Linear subspaces are specific instances of affine subspaces. In this study, we propose a novel affine feature extraction (AFE) method to find affine subspaces for classification. Our method can be seen as a generalization of the Funkunaga-Koontz transformation (FKT) [9]. We investigate the information-theoretical properties of our method and study the relationship of AFE and other similar feature extraction methods.

Our paper is organized as follows: in section 2, we present the main result of our work: the motivation of the study, the AFE method and its closed-form solutions. We investigate the information-theoretical properties of AFE and the relationship of AFE with other linear subspace dimension reduction methods in section 3. We present experimental results in section 4, and conclude the study with the summary of our work, and possible future directions in section 5.

2   Affine Feature Extraction

Consider a binary classification problem, which is also called *discriminant analysis* in statistics. Let $\{(\mathbf{x}_j, g_j) \in \mathbb{R}^m \times \{1, 2\} | j = 1, 2, \cdots, N\}$ be a training set. $\mathbf{x}_j$ and g_j are the *observation vector* and the corresponding *class label*. For simplicity, we assume the training set is permuted such that observations 1 to N_1 have label 1, and observations $N_1 + 1$ to $N_1 + N_2$ have label 2. Define a *data matrix* as

$$\mathbf{X} = (\mathbf{x}_1, \mathbf{x}_2, \cdots, \mathbf{x}_N) = (\mathbf{X}_1, \mathbf{X}_2),$$

where $\mathbf{X}_1 = (\mathbf{x}_1, \mathbf{x}_2, \cdots, \mathbf{x}_{N_1})$, and $\mathbf{X}_2 = (\mathbf{x}_{N_1+1}, \mathbf{x}_{N_1+2}, \cdots, \mathbf{x}_N)$. For the convenience of future discussion, we define *augmented observation vectors* as

$$\mathbf{y}_i = \begin{pmatrix} \mathbf{x}_i \\ 1 \end{pmatrix}. \tag{1}$$

We can similarly define an *augmented data matrix* $\mathbf{Y}_i$ for class i as $\mathbf{Y}_i^T = (\mathbf{X}_i^T, \mathbf{1})$. Throughout this paper, we use the following conventions: (1) vectors are column vectors; (2) $\mathbf{1}$ is a vector of all ones; (3) $\mathbf{I}$ is an identity matrix; (4) $\square^T$ is the transpose of a vector or matrix $\square$; and (5) tr($\square$) is the trace of a matrix $\square$.

2.1   Background

In this subsection, we give a brief introduction of dimension reduction for classical discriminant analysis. Due to the limitation of space, we cannot provide complete details for classical discriminant analysis. We refer to section 4.3 of [11] for a nice treatment on this topic. This subsection also serves as our motivation to carry on this study.

Before going on further, let us define the sample *mean*, *covariance* and *second-moment* for class i as follows:

$$\text{mean}\quad \hat{\mu}_i = \frac{1}{N_i}\mathbf{X}_i\mathbf{1}, \tag{2}$$

$$\text{covariance}\quad \hat{\mathbf{\Sigma}}_i = \frac{1}{N_i}\mathbf{X}_i(\mathbf{I} - \frac{1}{N_i}\mathbf{1}\mathbf{1}^T)^2\mathbf{X}_i^T, \tag{3}$$

$$\text{second-moment}\quad \hat{\mathbf{M}}_i = \frac{1}{N_i}\mathbf{X}_i\mathbf{X}_i^T. \tag{4}$$

One essential assumption of classical discriminant analysis is that the probability density for each class can be modeled as a multivariate normal distribution, i.e. $\mathcal{N}(\mu_i, \mathbf{\Sigma}_i)$ $(i = 1, 2)$. Equations 2 and 3 can be seen as the empirical estimations of classical density parameters μ_i and $\mathbf{\Sigma}_i$, respectively. Without losing generality, let us consider how to find a one-dimensional linear subspace for classical discriminant analysis; that is to find a linear transformation for observations:

$$z_i = \mathbf{w}^T\mathbf{x}_i,$$

where $\mathbf{w}^T$ is a m-dimensional vector.

When the two classes have a common covariance, i.e. $\mathbf{\Sigma}_1 = \mathbf{\Sigma}_2 = \mathbf{\Sigma}$, the problem is relatively easy. It is not hard to show that the optimal $\mathbf{w}^*$ is the eigenvector of $\mathbf{\Sigma}^{-1}(\mu_2 - \mu_1)(\mu_2 - \mu_1)^T$. FDA essentially capture this situation by solving the following problem:

$$\max \frac{\mathbf{w}^T(\hat{\mu}_2 - \hat{\mu}_1)(\hat{\mu}_2 - \hat{\mu}_1)^T\mathbf{w}}{\mathbf{w}^T\hat{\mathbf{\Sigma}}\mathbf{w}}, \tag{5}$$

where $N\hat{\mathbf{\Sigma}} = N_1\hat{\mathbf{\Sigma}}_1 + N_2\hat{\mathbf{\Sigma}}_2$.

When $\mathbf{\Sigma}_1 \neq \mathbf{\Sigma}_2$, to find an optimal linear subspace is hard. The only known closed-form solution is that $\mathbf{w}^*$ is the eigenvector of $\mathbf{\Sigma}_1^{-1}\mathbf{\Sigma}_2 + \mathbf{\Sigma}_2^{-1}\mathbf{\Sigma}_1$, which has the largest eigenvalue. It can be shown that, when $\mu_1 = \mu_2 = 0$, the solution optimizes the Kullback-Leibler KL divergence and the Bhattacharyya distance, (c.f. Section 10.2 of [8]). The KL distance and the Bhattacharyya distance are approximations of the Chernoff distance, which is the best asymptotic error exponent of a Bayesian approach. Therefore the optimizing of these distances serves as the theoretical support to use it as a dimension reduction method. Widely used in EEG classification problems, common spatial pattern (CSP) solves the following problem:

$$\max \frac{\mathbf{w}^T \hat{\boldsymbol{\Sigma}}_1 \mathbf{w}}{\mathbf{w}^T \hat{\boldsymbol{\Sigma}}_2 \mathbf{w}} \quad \text{or} \quad \max \frac{\mathbf{w}^T \hat{\boldsymbol{\Sigma}}_2 \mathbf{w}}{\mathbf{w}^T \hat{\boldsymbol{\Sigma}}_1 \mathbf{w}}. \tag{6}$$

Therefore CSP only works well when the difference between the class means is small, i.e. $|\mu_2 - \mu_1| \approx 0$. For many classification problems, this restriction is unrealistic. Furthermore, unlike FDA, CSP has no natural geometrical interpretation.

The FKT method can be seen as an extension of CSP by shrinking $\hat{\mu}_i$ to zero. It can be seen as a rough shrinkage estimation of the mean for high dimensional data. FKT solves the following problem:

$$\max \frac{\mathbf{w}^T \hat{\mathbf{M}}_1 \mathbf{w}}{\mathbf{w}^T \hat{\mathbf{M}}_2 \mathbf{w}} \quad \text{or} \quad \max \frac{\mathbf{w}^T \hat{\mathbf{M}}_2 \mathbf{w}}{\mathbf{w}^T \hat{\mathbf{M}}_1 \mathbf{w}} \tag{7}$$

Taking a closer look at the criterion of FKT, we note that the criterion max $\frac{\mathbf{w}^T \hat{\mathbf{M}}_1 \mathbf{w}}{\mathbf{w}^T \hat{\mathbf{M}}_2 \mathbf{w}}$ can be written as

$$\min \quad \mathbf{w}^T \hat{\mathbf{M}}_2 \mathbf{w}$$
$$\text{s.t. } \mathbf{w}^T \hat{\mathbf{M}}_1 \mathbf{w} = 1.$$

Note $\mathbf{w}^T \hat{\mathbf{M}}_i \mathbf{w} = \frac{1}{N_i} \sum_{j=k_i+1}^{k_i+N_i} z_j^2$, where $k_1 = 1$, $k_2 = N_1$ and $i = 1, 2$. That is: $\mathbf{w}^T \hat{\mathbf{M}}_i \mathbf{w}$ is the mean of square transformed observations, i.e. z_j^2, of class i. Therefore FKT can be interpreted as finding a linear subspace in which one can maximize the distance of the means of square transformed observations. However FKT may ignore important discriminant information for some cases, for example, the one proposed in [7].

2.2 Method

Let $z_i = v_0 + \mathbf{v}_1^T \mathbf{x}_i$ be an affine transformation for observations $\mathbf{x}_i$, where $\mathbf{v}_1$ is a m dimensional vector. Linear transformations are a special form of affine transformations, where $v_0 = 0$. Now denoting $\mathbf{w}^T = (\mathbf{v}_1^T, v_0)$, we have $z_i = \mathbf{w}^T \mathbf{y}_i$. Note that we have abused the notation of $\mathbf{w}$. From now on, we shall use $\mathbf{w}$ for affine transformations unless specified otherwise. Define a sample *augmented second moment* matrix as

$$\hat{\boldsymbol{\Xi}}_i = \frac{1}{N_i} \mathbf{Y}_i \mathbf{Y}_i^T. \tag{8}$$

The relation of augmented second moment matrix, covariance matrix and mean can be found in appendix A. Motivated by FKT, we use the following objective function to find the optimal one-dimensional affine subspace

$$\max \xi \frac{\mathbf{w}^T \hat{\boldsymbol{\Xi}}_1 \mathbf{w}}{\mathbf{w}^T \hat{\boldsymbol{\Xi}}_2 \mathbf{w}} + (1 - \xi) \frac{\mathbf{w}^T \hat{\boldsymbol{\Xi}}_2 \mathbf{w}}{\mathbf{w}^T \hat{\boldsymbol{\Xi}}_1 \mathbf{w}}, \tag{9}$$

where $0 \leq \xi \leq 1$. We use the sum of ratios to measure the importance of $\mathbf{w}$ instead of two separated optimization problems in FKT. And ξ can be used to balance the importance of different classes and thus is useful for asymmetric learning problems.

Now let us consider how to find higher dimensional affine subspaces. Let $\mathbf{W} = (\mathbf{w}_1, \mathbf{w}_2, \cdots, \mathbf{w}_d) \in \mathbb{R}^{(m+1) \times d}$ be a low-rank affine transformation matrix. Let $\mathbf{z}_i$ be the lower-dimensional representation of $\mathbf{x}_i$, i.e. $\mathbf{z}_i = \mathbf{W}^T \mathbf{y}_i$. We propose the following optimization problem to find $\mathbf{W}$:

$$\max \xi \sum_{i=1}^{d} \frac{\mathbf{w}_i^T \hat{\mathbf{\Xi}}_1 \mathbf{w}_i}{\mathbf{w}_i^T \hat{\mathbf{\Xi}}_2 \mathbf{w}_i} + (1 - \xi) \sum_{i=1}^{d} \frac{\mathbf{w}_i^T \hat{\mathbf{\Xi}}_2 \mathbf{w}_i}{\mathbf{w}_i^T \hat{\mathbf{\Xi}}_1 \mathbf{w}_i}$$

$$\text{s.t.} \qquad \mathbf{w}_i^T \hat{\mathbf{\Xi}}_t \mathbf{w}_j = \delta_{ij},$$

where $N\hat{\mathbf{\Xi}}_t = N_1 \hat{\mathbf{\Xi}}_1 + N_2 \hat{\mathbf{\Xi}}_2$, and δ_{ij} is 1 if $i = j$, and 0 otherwise. Let $\hat{\mathbf{\Pi}}_i = \mathbf{W}^T \hat{\mathbf{\Xi}}_i \mathbf{W}$. It is easy to recognize that $\hat{\mathbf{\Pi}}_i$'s are indeed the second moment matrices in the lower dimensional space. Now we can write the problem more compactly:

$$\max \xi \operatorname{tr}(\hat{\mathbf{\Pi}}_1^{-1} \hat{\mathbf{\Pi}}_2) + (1 - \xi) \operatorname{tr}(\hat{\mathbf{\Pi}}_2^{-1} \hat{\mathbf{\Pi}}_1)$$

$$\text{s.t} \qquad \mathbf{W}^T \hat{\mathbf{\Xi}}_t \mathbf{W} = \mathbf{I},$$

Generally speaking, we want to generate compact representations of the original observations. Therefore it is desirable to encourage finding lower dimensional affine subspaces. Motivated by the Akaike information criterion and Bayesian information criterion, we propose the following objective function that is to be maximized:

$$C(\mathbf{W}; \xi, d) = \xi \operatorname{tr}(\hat{\mathbf{\Pi}}_1^{-1} \hat{\mathbf{\Pi}}_2) + (1 - \xi) \operatorname{tr}(\hat{\mathbf{\Pi}}_2^{-1} \hat{\mathbf{\Pi}}_1) - d, \qquad (10)$$

where $0 \leq \xi \leq 1$, d $(1 \leq d \leq m)$ is the number of features we want to generate. We see that high dimensional solutions are penalized by the term $-d$. Hyperparameter ξ may be tuned via standard cross-validation methods [11]. In principal, the optimum d can also be determined by cross-validation procedures. However such a procedure is often computationally expensive. Therefore we propose the following alternative: define $C_0(\xi) = C(\mathbf{I}; \xi, m)$; we select the smallest d such that C is large enough, i.e. $d^* = \inf\{d | C(\mathbf{W}; \xi, d) \geq \beta C_0\}$, where β is a constant.

The constraint $\mathbf{W}^T \hat{\mathbf{\Xi}}_t \mathbf{W} = \mathbf{I}$ is necessary in our generalization from the one dimensional to the high dimensional formulation, but it does not generate mutually orthogonal discriminant vectors. Obtaining orthogonal discriminant vectors basis is geometrically desirable. Therefore we introduce another orthogonality constraint $\mathbf{W}^T \mathbf{W} = \mathbf{I}$. We refer to [6] for a geometrical view of the roles of the two constraints in optimization problems. To summarize, we are interested in two different kinds of constraints as follows:

1. $\hat{\mathbf{\Xi}}_t$-orthogonal constraint: $\mathbf{W}^T \hat{\mathbf{\Xi}}_t \mathbf{W} = \mathbf{I}$;
2. Orthogonal constraint: $\mathbf{W}^T \mathbf{W} = \mathbf{I}$.

2.3   Algorithms

In this subsection, we show how to solve the proposed optimization problems. Define function f as:

$$f(x;\xi) = \xi x + (1-\xi)\frac{1}{x}. \tag{11}$$

Let $0 < a \leq x \leq b$. Note f is a convex function, and thus achieves its maximum at the boundary of x, i.e. either a or b.

Define $\mathbf{\Lambda} = \mathrm{diag}(\lambda_1, \lambda_2, \cdots, \lambda_{m+1})$, and λ_i's are the eigenvalues of $(\hat{\mathbf{\Xi}}_1, \hat{\mathbf{\Xi}}_2)$ $(i = 1, 2, \cdots, m+1)$, i.e. $\hat{\mathbf{\Xi}}_1 \mathbf{u}_i = \lambda_i \hat{\mathbf{\Xi}}_2 \mathbf{u}_i$. Let $\lambda_i(\xi)$'s be the ordered eigenvalues of $(\hat{\mathbf{\Xi}}_1, \hat{\mathbf{\Xi}}_2)$ with respect to $f(\lambda; \xi)$. That is: define $f_i(\xi) = f(\lambda_i(\xi); \xi)$, then we have $f_1(\xi) \geq f_2(\xi) \geq \cdots \geq f_{m+1}(\xi)$. The following lemma for nonsingular symmetric $\hat{\mathbf{\Xi}}_1$ and $\hat{\mathbf{\Xi}}_2$ can be found in [10]:

Lemma 1. *There exists nonsingular matrix* $\mathbf{U} \in \mathbb{R}^{(m+1)\times(m+1)}$ *such that*

$$\mathbf{U}^T \hat{\mathbf{\Xi}}_2 \mathbf{U} = \mathbf{I}, \quad \mathbf{U}^T \hat{\mathbf{\Xi}}_1 \mathbf{U} = \mathbf{\Lambda}.$$

In Appendix C, we show that:

$$C(\mathbf{W};\xi,d) \leq \sum_{i=1}^{d} f_i(\xi) - d, \tag{12}$$

Note that: if $\mathbf{W}_1$ maximizes $C(\mathbf{W};\xi,d)$, then $\mathbf{W}_1\mathbf{R}$ also maximizes $C(\mathbf{W};\xi,d)$, where $\mathbf{R}$ is a nonsingular matrix. The proof is straight forward and therefore is omitted.

Proposition 1. *Let* $\mathbf{U}_\xi = (\mathbf{u}_1^\xi, \mathbf{u}_2^\xi, \cdots, \mathbf{u}_d^\xi)$, *where* $\mathbf{u}_i^\xi$ *is the eigenvector of* $(\hat{\mathbf{\Xi}}_1, \hat{\mathbf{\Xi}}_2)$ *and has eigenvalue* $\lambda_i(\xi)$. *Let* $\mathbf{R}$ *be a nonsingular matrix. Then* $\mathbf{W} = \mathbf{U}_\xi \mathbf{R}$ *maximize* $C(\mathbf{W};\xi,d)$.

Proof. It is enough to prove $\mathbf{U}_\xi$ maximizes $C(\mathbf{W};\xi,d)$. Note $\mathbf{U}_\xi^T \hat{\mathbf{\Xi}}_2 \mathbf{U}_\xi = \mathbf{I}$ and $\mathbf{U}_\xi^T \hat{\mathbf{\Xi}}_1 \mathbf{U}_\xi = \mathrm{diag}(\lambda_1(\xi), \lambda_2(\xi), \cdots, \lambda_d(\xi))$. Then it is easy to affirm the proposition.

Let $\mathbf{U}_\xi = \mathbf{Q}\mathbf{R}$, where $\mathbf{Q}$ and $\mathbf{R}$ are the thin QR factorization of $\mathbf{U}_\xi$; then $\mathbf{W}_1 = \mathbf{U}_\xi \mathbf{R}^{-1}$ maximizes $C(\mathbf{W};\xi,d)$ and satisfies the orthogonal constraint. Let $\mathbf{W}_2 = \mathbf{U}_\xi \mathbf{\Gamma}^{-\frac{1}{2}}$, where

$$\mathbf{\Gamma} = \frac{1}{N}\mathrm{diag}(N_1\lambda_1(\xi) + N_2, N_1\lambda_2(\xi) + N2, \cdots, N_1\lambda_d(\xi) + N_2). \tag{13}$$

It can be easily shown that $\mathbf{W}_2$ maximizes $C(\mathbf{W};\xi,d)$ and satisfies the $\hat{\mathbf{\Xi}}_t$-orthogonal constraint. In practice, we only need to check the largest d and the smallest d eigenvalues and eigenvectors of $(\hat{\mathbf{\Xi}}_1, \hat{\mathbf{\Xi}}_2)$ in order to generate d features. The pseudo-code of the algorithm is given in Table 1. Practically, we may need to let $\hat{\mathbf{\Xi}}_i \leftarrow \hat{\mathbf{\Xi}}_i + \alpha_i \mathbf{I}$ to guarantee the positive definiteness of $\hat{\mathbf{\Xi}}_i$, where α_i is a small positive constant.

Table 1. Pseudo-code for feature extraction

Algorithm for feature extraction

Input: Data sample $\mathbf{x}_1, \mathbf{x}_2, \cdots, \mathbf{x}_n$
Output: Transformation matrix $\mathbf{W}$

---

1. Calculate the augment second moment matrices $\hat{\boldsymbol{\Xi}}_1$, and $\hat{\boldsymbol{\Xi}}_2$;
2. Compute the largest d and the smallest d eigenvalues and eigenvectors of $(\hat{\boldsymbol{\Xi}}_1, \hat{\boldsymbol{\Xi}}_2)$;
3. Sort $2d$ eigenvalues and eigenvectors with respect to Eq. 11;
3. Selected the largest d eigenvectors to form $\mathbf{U}_\xi$;
4*. (For orthogonal constraint) apply the thin QR factorization on $\mathbf{U}_\xi$, i.e. $\mathbf{U}_\xi = \mathbf{QR}$;
5*. (For orthogonal constraint) Let $\mathbf{W} = \mathbf{Q}$;
6**. (For $\hat{\boldsymbol{\Xi}}_t$-orthogonal constraint), calculate $\boldsymbol{\Gamma}$ as Eq. 13;
7**. (For $\hat{\boldsymbol{\Xi}}_t$-orthogonal constraint), Let $\mathbf{W} = \mathbf{U}_\xi \boldsymbol{\Gamma}^{-\frac{1}{2}}$;
6. Return $\mathbf{W}$.

---

3   Discussion

In this section, we investigate the properties of our proposed method, and study the relationship of the new proposed method with other dimension reduction methods. For simplicity, we assume that $\hat{\boldsymbol{\Xi}}_i$'s are reliably estimated. Therefore we shall use $\boldsymbol{\Xi}_i$ in our discussion directly.

3.1   Information Theoretical Property of the Criterion

The KL divergence of two multivariate normal distributions p_i and p_j has a closed expression as:

$$L_{ij} = \frac{1}{2}\{\log(|\boldsymbol{\Sigma}_i^{-1}\boldsymbol{\Sigma}_j|) + \text{tr}(\boldsymbol{\Sigma}_i\boldsymbol{\Sigma}_j^{-1}) + (\mu_i - \mu_j)^T\boldsymbol{\Sigma}_j^{-1}(\mu_i - \mu_j) - m\}; \quad (14)$$

where $p_i = \mathcal{N}(\mu_i, \boldsymbol{\Sigma}_i)$. The symmetric KL divergence is defined as $J_{ij} = L_{ij} + L_{ji}$. It is easy to obtain

$$J_{12} = \frac{1}{2}\{\text{tr}(\boldsymbol{\Sigma}_2^{-1}\boldsymbol{\Sigma}_1) + \text{tr}(\boldsymbol{\Sigma}_1^{-1}\boldsymbol{\Sigma}_2) + \text{tr}[(\boldsymbol{\Sigma}_1^{-1} + \boldsymbol{\Sigma}_2^{-1})(\mu_2 - \mu_1)(\mu_2 - \mu_1)^T] - 2m\}. \quad (15)$$

Using formulas in Appendix A, one can easily get that

$$J_{12} = C_0(\frac{1}{2}) - 1; \quad (16)$$

That is, when ξ is $1/2$, C_0 is equivalent to the symmetric KL divergence (up to a constant) of two normal distributions. The solution of maximizing C can be seen as finding an affine subspace that maximally preserves C_0, i.e. an optimal truncated spectrum of J_{12}.

The KL divergence can be seen as a distance measure between two distributions, and therefore a measure of separability of classes. Traditional viewpoints aim at maximizing the KL divergence between classes in lower dimensional linear subspaces, see [8] for an introduction and [14] for the recent development. It is easy to show that maximizing the lower-dimensional KL divergence in [14] is equivalent to our proposed problem with an additional constraint

$$\mathbf{W}^T = (\mathbf{V}^T, \mathbf{e}) \tag{17}$$

where $\mathbf{V} \in \mathbb{R}^{m \times d}$, and $\mathbf{e}^T = (0, 0, \cdots, 1)$. With the additional constraint, a closed-form solution cannot be found. By relaxing $\mathbf{e} \in \mathbb{R}^{m \times 1}$, we can find closed-form solutions.

3.2 Connection to FDA and CSP

Without losing generality, let us consider the one dimensional case in this subsection. Let $\mathbf{w}^T = (\mathbf{v}_1^T, v_0)$. Then we have $Z = \mathbf{v}_1^T X + v_0$, where X and Z are random covariate in higher- and lower-dimensional spaces. Displacement v_0 is the same for both classes, and therefore plays no important role for final classifications. In other words, the effectiveness of the generated feature is solely determined by $\mathbf{v}_1$. Let $\mathbf{v}_1^*$ be an optimal solution.

Consider maximizing $C(\mathbf{W}; 1/2, d)$. We know that $\mathbf{w}^*$ is the eigenvector of $\boldsymbol{\Xi}_1^{-1}\boldsymbol{\Xi}_2 + \boldsymbol{\Xi}_2^{-1}\boldsymbol{\Xi}_1$ with the largest eigenvalue.

First, let us consider $\mu_1 = \mu_2 = \mu$. Using formulas in Appendix A, we can simplify $\boldsymbol{\Xi}_1^{-1}\boldsymbol{\Xi}_2 + \boldsymbol{\Xi}_2^{-1}\boldsymbol{\Xi}_1$ as

$$\boldsymbol{\Xi}_1^{-1}\boldsymbol{\Xi}_2 + \boldsymbol{\Xi}_2^{-1}\boldsymbol{\Xi}_1 = \begin{pmatrix} \boldsymbol{\Sigma}_1^{-1}\boldsymbol{\Sigma}_2 + \boldsymbol{\Sigma}_2^{-1}\boldsymbol{\Sigma}_1 & 0 \\ 2\mu^T - \mu^T(\boldsymbol{\Sigma}_1^{-1}\boldsymbol{\Sigma}_2 + \boldsymbol{\Sigma}_2^{-1}\boldsymbol{\Sigma}_1) & 1 \end{pmatrix}$$

Then by simple linear algebra, we can show that $\mathbf{v}_1^*$ is also the eigenvector of $\boldsymbol{\Sigma}_1^{-1}\boldsymbol{\Sigma}_2 + \boldsymbol{\Sigma}_2^{-1}\boldsymbol{\Sigma}_1$ with the largest eigenvalue.

Second, let us consider $\boldsymbol{\Sigma}_1 = \boldsymbol{\Sigma}_2 = \boldsymbol{\Sigma}$. In this case, it is easy to verify the following:

$$\boldsymbol{\Xi}_2^{-1}\boldsymbol{\Xi}_1 - \mathbf{I} = \begin{pmatrix} \boldsymbol{\Sigma}^{-1}(\mu_1 - \mu_2)\mu_1^T & \boldsymbol{\Sigma}^{-1}(\mu_1 - \mu_2) \\ \mu_1^T - \mu_2^T - \mu_2^T\boldsymbol{\Sigma}^{-1}(\mu_1 - \mu_2)\mu_1^T & -\mu_2^T\boldsymbol{\Sigma}^{-1}(\mu_1 - \mu_2) \end{pmatrix}$$

$$\boldsymbol{\Xi}_1^{-1}\boldsymbol{\Xi}_2 - \mathbf{I} = \begin{pmatrix} \boldsymbol{\Sigma}^{-1}(\mu_2 - \mu_1)\mu_2^T & \boldsymbol{\Sigma}^{-1}(\mu_2 - \mu_1) \\ \mu_2^T - \mu_1^T - \mu_1^T\boldsymbol{\Sigma}^{-1}(\mu_2 - \mu_1)\mu_2^T & -\mu_1^T\boldsymbol{\Sigma}^{-1}(\mu_2 - \mu_1) \end{pmatrix}$$

Then we have

$$\boldsymbol{\Xi}_1^{-1}\boldsymbol{\Xi}_2 + \boldsymbol{\Xi}_2^{-1}\boldsymbol{\Xi}_1 = \begin{pmatrix} \mathbf{A} & 0 \\ 0 & \mathbf{B} \end{pmatrix} + 2\mathbf{I},$$

where $\mathbf{A} = \boldsymbol{\Sigma}^{-1}(\mu_1 - \mu_2)(\mu_1 - \mu_2)^T$ and $\mathbf{B} = (\mu_1 - \mu_2)^T\boldsymbol{\Sigma}^{-1}(\mu_1 - \mu_2)^T$. It is then not hard to show that $\mathbf{v}_1^*$ is the eigenvector of $\mathbf{A}$ with the largest eigenvalue.

In summary, we show that FDA and CSP are special cases of our proposed AFE for normally distribute data. Therefore, theoretically speaking AFE is more flexible than FDA and CSP.

4   Experiments

In order to compare our method with PCA and FDA, a 7-dimensional toy data set has been generated. The toy data set contains 3-dimensional relevant components, while the others are merely random noise. The 3 relevant components form two concentric cylinders. The generated data are spread along the surfaces of the cylinders. Figure 1 illustrates the first two features found by PCA, FDA and our new approach AFE. As a result of preserving the variance of data, PCA projects data along the surfaces, and thus does not reflect the separation of the data (Figure 1(a)). Figure 1(b) shows that FDA fails to separate the two classes. On the other hand, Figure 1(c) shows that our method correctly captures the discriminant information in the data.

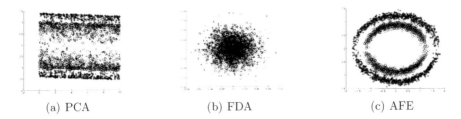

(a) PCA   (b) FDA   (c) AFE

Fig. 1. Comparision of features found by PCA, FDA, and Our method. Star and circle points belong to different classes.

We selected three benchmark data sets: German, Diabetes and Waveform. The dimensionality of these data sets are 20, 8, and 21 respectively. They can be freely downloaded from   http://ida.first.fraunhofer.de/projects/bench/benchmarks. htm. The data sets had been preprocessed and partitioned into 100 training and test sets (about 40% : 60%). They have been used to evaluate the performance of kernel FDA [15], kernel PLS [19] and soft-margin AdaBoost [17].

We compared our new approach with FDA, CSP, and FKT. For convenience, AFE1 and AFE2 are used for orthogonal and Ξ_t-orthogonal AFE algorithms. We used FDA, CSP, FKT, AFE1 and AFE2 to generate lower-dimensional features; the features are then used by linear support vector machines (SVM) to do classifications. To measure the discriminant information of the data set, we also classified the original data set via linear SVMs, which we denote FULL in the reported figures. Feature extraction and classification are trained on training sets, and test-set accuracy (TSA) are calculated with predictions on test sets. Statistical boxplots of TSAs are shown in Figures 2, 3 and 4 for the three chosen data sets. The poor performance of FDA, CSP and FKT affirms that first-order or second-order statistics alone cannot capture discriminant information contained in the data sets. By comparing AFE1 and AFE2 with FULL, we see that AFE1 and AFE2 are capable of extracting the discriminant information of the chosen data. AFE1 and AFE2 can be used to generate much compact discriminant features, for example, the average dimensionality of extracted features for German, Diabetes and Waveform are 8.16, 3.18 and 1.2, respectively.

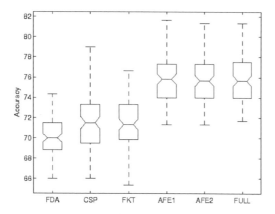

Fig. 2. Test set accuracy for German data sets. See text for notations and details.

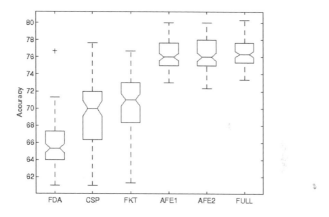

Fig. 3. Test set accuracy for Diabetes data set. See text for notations and details.

We conducted preliminary experiments with AFE1 and AFE2 on data sets Tübingen:1a and Berlin:IV from BCI competition 2003 [1]. We used AFE1 and AFE2 to generate low-dimensional representations and then apply logistic regression on the extracted features. For data set Tübingen:1a, we obtained TSA as 77.13% and 85.32% for AFE1+ and AFE2+logistic regression, respectively. The results are comparable with the ones of rank 11 and rank 4 of the competition, correspondingly. For data set Berlin:IV, we obtained TSA 71% for both AFE1+ and AFE2+logistic regression, which are comparable with rank 8 of the competition.

---

[1] see http://ida.first.fraunhofer.de/projects/bci/competition_ii/results/index.html

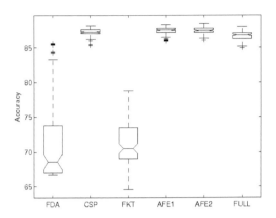

Fig. 4. Test set accuracy for Waveform data set. See text for notations and details.

5 Conclusions

In this study, we proposed a novel dimension reduction method for binary classification problems. Unlike traditional linear subspace methods, the new proposed method finds lower-dimensional affine subspaces for data observations. We presented the closed-form solutions of our new approach, and investigated its information-theoretical properties. We showed that our method has close connections with FDA, CSP and FKT methods in the literature. Numerical experiments show the competitiveness of our method as a preliminary data-exploring tool for data visualization and classification.

Though we focus on binary classification problems in this study, it is always desirable to handle multi-class problems. One can extend AFE to multi-class problems by following the work presented in [5]. Here we proposed another way to extend AFE to multi-class. Let J_{ij} be the symmetric KL distance of classes i and j, and assume class i, $(i = 1, 2, \cdots, K)$, can be modeled by multivariate normal distribution. Then we have

$$\sum_{i=1}^{K} \Xi_i^{-1} \Xi_t \propto \sum_{i,j=1}^{K} J_{ij},$$

where Ξ_i is the augmented second moment matrix for class i and $N\Xi_t = \sum_{i=1}^{K} N_i \Xi_i$. Therefore we may calculate the truncated spectrum of $\sum_{i=1}^{K} \Xi_i^{-1} \Xi_t$ for the lower-dimensional representations.

Another more important problem is to investigate the relationship of our new proposed method with quadratic discriminant analysis (*QDA*). It has long been known that FDA is an optimal dimension reduction method for linear discriminant analysis (*LDA*) [11]. But there is no well-accepted dimension reduction method for QDA in the literature. Recently, Hou et al. proposed that FKT might be seen as an optimal one for QDA under certain circumstance [13]. Our future work will be dedicated to finding the relationship of AFE and QDA.

Acknowledgement

The authors thank anonymous reviewers for constructive comments on improving the presentation of this work.

References

[1] Barker, M., Rayens, W.: Partial least squares for discrimination. Journal of Chemometrics 17(3), 166–173 (2003)

[2] Bartlett, M.S.: Further aspects of the theory of multiple regression. In: Proc. Camb. Phil. Soc. 34 33–40 (1938)

[3] Belhumeur, P.N., Hespanha, J., Kriegman, D.J.: Eigenfaces vs. Fisherfaces: Recognition using class specific linear projection. IEEE Transactions on Pattern Analysis and Machine Intelligence 19(7), 711–720 (1997)

[4] Dai, J.J, Lieu, L., Rocke, D.: Dimension reduction for classification with gene expression microarray data. Stat Appl Genet Mol Biol, 5 Article6 (2006)

[5] Dornhege, G., Blankertz, B., Curio, G., Müller, K.-R.: Increase information transfer rates in BCI by CSP extension to multi-class. In: Advances in Neural Information Processing Systems 16 (2004)

[6] Edelman, A., Arias, T.A., Smith, S.T.: The geometry of algorithms with orthogonality constraints. SIAM J. Matrix Anal. Appl. 20(2), 303–353 (1999)

[7] Foley, D.H., Sammon Jr, J.W.: An optimal set of discriminant vectors. Computers, IEEE Transactions on C-24, 281–289 (1975)

[8] Fukunaga, K.: Introduction to statistical pattern recognition, 2nd edn. Academic Press, London (1990)

[9] Fukunaga, K., Koontz, W.: Application of the Karhunen-Loéve expansion to feature selection and ordering. Computers, IEEE Transactions on C-19, 311–318 (1970)

[10] Golub, G.H., Van Loan, C.F.: Matrix Computations. The Johns Hopkins University Press, Baltimore (1996)

[11] Hastie, T., Tibshirani, R., Friedman, J.H.: The Elements of Statistical Learning. Springer, Heidelberg (2001)

[12] Horn, R.A., Johnson, C.R.: Matrix Analysis. Cambridge University Press, Cambridge (1990)

[13] Huo, X., Elad, M., Flesia, A.G., Muise, B., Stanfill, R., Friedman, J., Popescu, B., Chen, J., Mahalanobis, A., Donoho, D.L.: Optimal reduced-rank quadratic classifiers using the Fukunaga-Koontz transform, with applications to automated target recognition. In: SPIE's 7th International Symposium on Aerospace/Defense Sensing

[14] De la Torre, F., Kanade, T.: Multimodal oriented discriminant analysis. In: ICML '05: Proceedings of the 22nd international conference on Machine learning, New York, NY, USA, pp. 177–184. ACM Press, New York (2005)

[15] Mika, S., Ratsch, G., Weston, J., Scholkopf, B., Muller, K.: Fisher discriminant analysis with kernels. In: Proceedings of IEEE Neural Networks for Signal Processing Workshop 1999 (1999)

[16] Ramoser, H., Müller-Gerking, J., Pfurtscheller, G.: Optimal spatial filtering of single trial EEG during imagined hand movement. IEEE Trans. Rehab. Eng. 8, 441–446 (2000)
[17] Rätsch, G., Onoda, T., Müller, K.-R.: Soft margins for adaboost. Mach. Learn. 42(3), 287–320 (2001)
[18] Rosipal, R., Krämer, N.: Overview and recent advances in partial least squares. In: Saunders, C., Grobelnik, M., Gunn, S., Shawe-Taylor, J. (eds.) SLSFS 2005. LNCS, vol. 3940, pp. 34–51. Springer, Heidelberg (2005)
[19] Rosipal, R., Trejo, L.J., Matthews, B.: Kernel PLS-SVM for linear and nonlinear classification. In: Fawcett, T., Mishra, N. (eds.) ICML, pp. 640–647. AAAI Press, Stanford (2003)
[20] Yang, M.-H., Kriegman, D.J., Ahuja, N.: Detecting faces in images: A survey. IEEE Trans. Pattern Anal. Mach. Intell. 24(1), 34–58 (2002)

Appendix A

Let X be a random covariate which has probability distribution p. So we have

$$\mu = E_{X \sim p} X,$$
$$\mathbf{\Sigma} = E_{X \sim p}(X - \mu)(X - \mu)^T,$$
$$\mathbf{\Xi} = E_{X \sim p}\left\{ \begin{pmatrix} X \\ 1 \end{pmatrix} (X^T, 1) \right\},$$

where μ, $\mathbf{\Sigma}$ and $\mathbf{\Xi}$ are, respectively, the mean, covariance and augmented second moment of X. When μ and $\mathbf{\Sigma}$ are finite, i.e. $\|\mu\| < \infty$ and $\|\mathbf{\Sigma}\| < \infty$, we have

$$\mathbf{\Xi} = \begin{pmatrix} \mathbf{\Sigma} + \mu\mu^T & \mu \\ \mu^T & 1 \end{pmatrix}$$

Assuming $\mathbf{\Sigma}$ is positive definite, we have the inverse of $\mathbf{\Xi}$ as follows:

$$\mathbf{\Xi}^{-1} = \begin{pmatrix} \mathbf{\Sigma}^{-1} & -\mathbf{\Sigma}^{-1}\mu \\ -\mu^T\mathbf{\Sigma}^{-1} & 1 + \mu^T\mathbf{\Sigma}^{-1}\mu \end{pmatrix}.$$

Appendix B

Lemma 2. *Let $\mathbf{A}$ be an $r \times s$ matrix, $(r \geq s)$, and $\mathbf{A}^T\mathbf{A} = I$. Let $\mathbf{\Lambda}$ be a diagonal matrix. Then*

$$\xi \, tr(\mathbf{A}^T\mathbf{\Lambda}\mathbf{A}) + (1 - \xi) tr([\mathbf{A}^T\mathbf{\Lambda}\mathbf{A}]^{-1}) \leq \sum_{i=1}^{s} f_i(\xi);$$

Proof. By the Poincaré separation theorem (c.f. [12] P190), we know the eigenvalues of $\mathbf{A}^T\mathbf{\Lambda}\mathbf{A}$ interlaces with those of $\mathbf{\Lambda}$. That is, for each integer j, $(1 \leq j \leq s)$, we have

$$\lambda_j \leq \tau_j \leq \lambda_{j+r-s},$$

where τ_j is the eigenvalue of $\mathbf{A}^T \mathbf{\Lambda} \mathbf{A}$. Then it is obvious that

$$\xi\mathrm{tr}(\mathbf{A}^T\mathbf{\Lambda}\mathbf{A}) + (1-\xi)\mathrm{tr}([\mathbf{A}^T\mathbf{\Lambda}\mathbf{A}]^{-1})$$
$$= \sum_{i=1}^s [\xi\tau_i + (1-\xi)\tfrac{1}{\tau_i}]$$
$$\leq \sum_{i=1}^s f_i(\xi);$$

Appendix C

Proof. Let $\mathbf{U}$ be a nonsingular matrix such that $\mathbf{U}^T\hat{\mathbf{\Xi}}_2\mathbf{U} = \mathbf{I}$ and $\mathbf{U}^T\hat{\mathbf{\Xi}}_1\mathbf{U} = \mathbf{\Lambda}$. Then we have

$$\hat{\mathbf{\Pi}}_2 = \mathbf{W}^T(\mathbf{U}^{-1})^T\mathbf{U}^T\hat{\mathbf{\Xi}}_2\mathbf{U}\mathbf{U}^{-1}\mathbf{W} = \mathbf{V}^T\mathbf{V}$$
$$\hat{\mathbf{\Pi}}_1 = \mathbf{W}^T(\mathbf{U}^{-1})^T\mathbf{U}^T\hat{\mathbf{\Xi}}_1\mathbf{U}\mathbf{U}^{-1}\mathbf{W} = \mathbf{V}^T\mathbf{\Lambda}\mathbf{V},$$

where $\mathbf{V} = \mathbf{U}^{-1}\mathbf{W} \in \mathbb{R}^{(m+1)\times k}$. Then we can get

$$C(\mathbf{W};\xi,d) = \xi\mathrm{tr}[(\mathbf{V}^T\mathbf{V})^{-1}\mathbf{V}^T\mathbf{\Lambda}\mathbf{V}] + (1-\xi)\mathrm{tr}[(\mathbf{V}^T\mathbf{\Lambda}\mathbf{V})^{-1}\mathbf{V}^T\mathbf{V}].$$

Applying SVD on $\mathbf{V}$, we get $\mathbf{V} = \mathbf{A}\mathbf{D}\mathbf{B}^T$. Here $\mathbf{A}$ and $\mathbf{B}$ are $(m+1)\times d$ and $d \times d$ orthogonal matrices, i.e. $\mathbf{B}^T\mathbf{B} = \mathbf{I}$, $\mathbf{B}\mathbf{B}^T = \mathbf{I}$, and $\mathbf{A}^T\mathbf{A} = \mathbf{I}$. $\mathbf{D}$ is a $d \times d$ diagonal matrix. Therefore we have:

$$\mathrm{tr}[(\mathbf{V}^T\mathbf{V})^{-1}\mathbf{V}^T\mathbf{\Lambda}\mathbf{V}] = \mathrm{tr}[\mathbf{V}(\mathbf{V}^T\mathbf{V})^{-1}\mathbf{V}^T\mathbf{\Lambda}]$$
$$= \mathrm{tr}(\mathbf{A}\mathbf{A}^T\mathbf{\Lambda})$$
$$= \mathrm{tr}(\mathbf{A}^T\mathbf{\Lambda}\mathbf{A}).$$

$$\mathrm{tr}[(\mathbf{V}^T\mathbf{\Lambda}\mathbf{V})^{-1}\mathbf{V}^T\mathbf{V}] = \mathrm{tr}[\mathbf{V}(\mathbf{V}^T\mathbf{\Lambda}\mathbf{V})^{-1}\mathbf{V}^T]$$
$$= \mathrm{tr}[\mathbf{A}(\mathbf{A}^T\mathbf{\Lambda}\mathbf{A})^{-1}\mathbf{A}^T]$$
$$= \mathrm{tr}[(\mathbf{A}^T\mathbf{\Lambda}\mathbf{A})^{-1}].$$

Thus by Lemma 2, we know that

$$C(\mathbf{W};\xi,d) = \mathrm{tr}[\xi\mathbf{A}^T\mathbf{\Lambda}\mathbf{A} + (1-\xi)(\mathbf{A}^T\mathbf{\Lambda}\mathbf{A})^{-1}] - d$$
$$\leq \sum_{i=1}^d f_i(\xi) - d.$$

A Bounded Index for Cluster Validity

Sandro Saitta, Benny Raphael, and Ian F.C. Smith

Ecole Polytechnique Fédérale de Lausanne (EPFL)
Station 18, 1015 Lausanne, Switzerland
sandro.saitta@epfl.ch, bdgbr@nus.edu.sg, ian.smith@epfl.ch

Abstract. Clustering is one of the most well known types of unsupervised learning. Evaluating the quality of results and determining the number of clusters in data is an important issue. Most current validity indices only cover a subset of important aspects of clusters. Moreover, these indices are relevant only for data sets containing at least two clusters. In this paper, a new bounded index for cluster validity, called the score function (SF), is introduced. The score function is based on standard cluster properties. Several artificial and real-life data sets are used to evaluate the performance of the score function. The score function is tested against four existing validity indices. The index proposed in this paper is found to be always as good or better than these indices in the case of hyperspheroidal clusters. It is shown to work well on multidimensional data sets and is able to accommodate unique and sub-cluster cases.

Keywords: clustering, cluster validity, validity index, k-means.

1 Introduction

The goal of clustering [1,2] is to group data points that are similar according to a chosen similarity metric (Euclidean distance is commonly used). Clustering techniques have been applied in domains such as text mining [3], intrusion detection [4] and object recognition [5]. In these fields, as in many others, the number of clusters is usually not known in advance.

Several clustering techniques can be found in the literature. They usually belong to one of the following categories [6]: partitional clustering, hierarchical clustering, density-based clustering and grid-based clustering. An additional category is the mixture of Gaussian approach. Since its computational complexity is high, it is not likely to be used in practice. All these categories have drawbacks. For example, hierarchical clustering has a higher complexity. Density-based clustering algorithms often require tuning non-intuitive parameters. Finally, density-based clustering algorithms do not always reveal clusters of good quality. The K-means [1] algorithm, part of the partitional clustering, is the most widely used. Advantages of K-means include computational efficiency, fast implementation and easy mathematical background. However, K-means also has limitations. They include a random choice of centroid locations at the beginning of the procedure, treatment of categorical variables and an unknown number of clusters k.

P. Perner (Ed.): MLDM 2007, LNAI 4571, pp. 174–187, 2007.

Concerning the first limitation, multiple runs may be a solution. The paper by Huang [7] contains a possible solution to the second limitation through the use of a matching dissimilarity measure to handle categorical parameters. Finally, the third issue is related to the number of clusters and therefore cluster validity.

Clustering is by definition a subjective task and this is what makes it difficult [8]. Examples of challenges in clustering include i) the number of clusters present in the data and ii) the quality of clustering [9]. Elements of answers to these two issues can be found in the field of cluster validation. Other challenges such as initial conditions and high dimensional data sets are of importance in clustering. The aim of cluster validation techniques is to evaluate clustering results [6,8,10]. This evaluation can be used to determine the number of clusters within a data set. Current literature contains several examples of validity indices [9,11,12,13]. Recent work has also been done on evaluating them [14].

The Dunn index [11] combines dissimilarity between clusters and their diameters to estimate the most reliable number of clusters. As stated in [6], the Dunn index is computationally expensive and sensitive to noisy data. The concepts of dispersion of a cluster and dissimilarity between clusters are used to compute the Davies-Bouldin index [12]. The Davies-Bouldin index has been found to be among the best indices [14]. The Silhouette index [13] uses average dissimilarity between points to identify the structure of the data and highlights possible clusters. The Silhouette index is only suitable for estimating the first choice or the best partition [15]. Finally, the Maulik-Bandyopadhyay index [9] is related to the Dunn index and involves tuning of a parameter.

All of these indices require the specification of at least two clusters. As noted in [16], the one cluster case is important and is likely to happen in practice. As a prerequisite to the identification of a single cluster, a definition of what is a cluster is important. Among those that exist in the literature, a possible definition is given in [17]. Briefly, it states that a cluster is considered to be "real" if it is significantly compact and isolated. Concepts of compactness and isolation are based on two parameters that define internal properties of a cluster. While this definition is precise, it is often too restrictive since few data sets satisfy such criteria. More details of single cluster tests can be found in [16]. Other validity indices exist in the literature. Some are computationally expensive [6] while others are unable to discover the real number of clusters in all data sets [14]. This paper proposes a new validity index that helps overcome such limitations.

This article is organized as follows. Section 2 describes existing validity indices from the literature. Section 3 proposes a new validity index, named the score function. Performance of the score function is described in Section 4. The last Section provides conclusions and directions for future work.

2   Existing Indices

In this Section, four validity indices suitable for hard partitional clustering are described. These indices serve as a basis for evaluating results from the score function on benchmark data sets. Notation for these indices have been adapted to provide a coherent basis. The metric used on the normalized data is the standard Euclidean distance defined as $||x - y|| = \sqrt{\sum_{i=1}^{d}(x_i - y_i)^2}$ where x and y are data points and d is the number of dimensions.

Dunn index: One of the most cited indices is proposed by [11]. The Dunn index (DU) identifies clusters which are well separated and compact. The goal is therefore to maximize the inter-cluster distance while minimizing the intra-cluster distance. The Dunn index for k clusters is defined by Equation 1:

$$DU_k = \min_{i=1,\ldots,k} \left\{ \min_{j=1+1,\ldots,k} \left(\frac{diss(c_i, c_j)}{\max_{m=1,\ldots,k} diam(c_m)} \right) \right\} \tag{1}$$

where $diss(c_i, c_j) = \min_{x \in c_i, y \in c_j} ||x - y||$ is the dissimilarity between clusters c_i and c_j and $diam(c) = \max_{x,y \in c} ||x-y||$ is the intra-cluster function (or diameter) of the cluster. If Dunn index is large, it means that compact and well separated clusters exist. Therefore, the maximum is observed for k equal to the most probable number of clusters in the data set.

Davies-Bouldin index: Similar to the Dunn index, Davies-Bouldin index [12] identifies clusters which are far from each other and compact. Davies-Bouldin index (DB) is defined according to Equation 2:

$$DB_k = \frac{1}{k} \sum_{i=1}^{k} \max_{j=1,\ldots,k,i\neq j} \left\{ \frac{diam(c_i) + diam(c_j)}{||c_i - c_j||} \right\} \tag{2}$$

where, in this case, the diameter of a cluster is defined as:

$$diam(c_i) = \left(\frac{1}{n_i} \sum_{x \in c_i} ||x - z_i||^2 \right)^{1/2} \tag{3}$$

with n_i the number of points and z_i the centroid of cluster c_i. Since the objective is to obtain clusters with minimum intra-cluster distances, small values for DB are interesting. Therefore, this index is minimized when looking for the best number of clusters.

Silhouette index: The silhouette statistic [13] is another well known way of estimating the number of groups in a data set. The Silhouette index (SI) computes for each point a width depending on its membership in any cluster. This silhouette width is then an average over all observations. This leads to Equation 4:

$$SI_k = \frac{1}{n} \sum_{i=1}^{n} \frac{(b_i - a_i)}{\max(a_i, b_i)} \tag{4}$$

where n is the total number of points, a_i is the average distance between point i and all other points in its own cluster and b_i is the minimum of the average dissimilarities between i and points in other clusters. Finally, the partition with the highest SI is taken to be optimal.

Maulik-Bandyopadhyay index: A more recently developed index is named the I index [9]. For consistence with other indices it is renamed MB. This index, which is a combination of three terms, is given through Equation 5:

$$MB_k = \left(\frac{1}{k} \cdot \frac{E_1}{E_k} \cdot D_k \right)^p \tag{5}$$

where the intra-cluster distance is defined by $E_k = \sum_{i=1}^{k} \sum_{x \in c_i} ||x - z_i||$ and the inter-cluster distance by $D_k = \max_{i,j=1}^{k} ||z_i - z_j||$. As previously, z_i is the center of cluster c_i. The correct number of clusters is estimated by maximizing Equation 5. In this work, p is chosen to be two.

Discussion: Although all these indices are useful in certain situations, they are not of general-purpose. For example, Dunn index is computationally heavy and has difficulty to deal with noisy data. It is useful for identifying clean clusters in data sets containing no more than hundreds of points. Davies-Bouldin index gives good results for distinct groups. However, it is not designed to accommodate overlapping clusters. The Silhouette index is only able to identify the first choice and therefore should not be applied to data sets with sub-clusters. The Maulik-Bandyopadhyay index has the particularity of being dependent on a user specified parameter.

3 Score Function

In this paper, we propose a function to estimate the number of clusters in a data set. The proposed index, namely the score function (SF), is based on inter-cluster and intra-cluster distances. The score function is used for two purposes: i) to estimate the number of clusters and ii) to evaluate the quality of the clustering results. The score function is a function combining two terms: the distance between clusters and the distance inside a cluster. The first notion is defined as the "between class distance" (*bcd*) whereas the second is the "within class distance" (*wcd*) .

Three common approaches exist to measure the distance between two clusters: single linkage, complete linkage and comparison of centroids. This proposal is based on the third concept since the first two have computational costs that are too high [6]. In this work, the *bcd* is defined by Equation 6:

$$bcd = \frac{\sum_{i=1}^{k} ||z_i - z_{tot}|| \cdot n_i}{n \cdot k} \tag{6}$$

where k is the number of clusters, z_i the centroid of the current cluster and z_{tot} the centroid of all the clusters. The size of a cluster, n_i is given by the number of points inside it. The most important quantity in the *bcd* is the distance between z_i and z_{tot}. To limit the influence of outliers, each distance is weighted by the cluster size. This has the effect to reduce the sensitivity to noise. Through n, the *bcd* sensitivity to the total number of points is avoided. Finally, values for k are used to penalize the addition of a new cluster. This way, the limit of one point per cluster is avoided. The wcd is given in Equation 7:

$$wcd = \sum_{i=1}^{k} \left(\frac{1}{n_i} \sum_{x \in c_i} ||x - z_i|| \right) \tag{7}$$

Computing values for wcd involves determining the distance between each point to the centroid of its cluster. This is summed over the k clusters. Note that $||x - z_i||$ already takes into account the size of the corresponding cluster. As in bcd (Equation 6), the cluster size in the denominator avoids the sensibility to the total number of points. With Equations 6 and 7, bcd and wcd are independent of the number of data points.

For the score function to be effective, it should i) maximize the bcd, ii) minimize the wcd and iii) be bounded. Maximizing Equation 8 satisfies the above conditions:

$$SF = 1 - \frac{1}{e^{e^{bcd-wcd}}} \tag{8}$$

The higher the value of the SF, the more suitable the number of clusters. Therefore, with the proposed SF, it is now possible to estimate the number of clusters for a given set of models. Difficulties such as perfect clusters ($wcd = 0$) and unique cluster ($bcd = 0$) are overcome. Moreover, the proposed score function is bounded by $]0, 1[$. The upper bound allows the examination of how close the current data set is to the perfect cluster case. Thus we seek to maximize Equation 8 to obtain the most reliable number of clusters. As can be seen through Equations 6 and 7, computational complexity is linear. If n is the number of data points, then the proposed score function has a complexity of $O(n)$. In the next Section, the score function is tested on several benchmark problems and compared with existing indices.

4   Results

In this Section, the performance of validity indices are compared. For this purpose, the standard K-means algorithm is used. K-means is a procedure that iterates over k clusters in order to minimize their intra-cluster distances. The K-means procedure is as follows. First, k centroids are chosen randomly over all the points. The data set is then partitioned according to the minimum squared distance. New centroid positions are calculated according to the points inside clusters. The process of partitioning and updating is repeated until a stopping criterion is reached. This happens when either the cluster centers or the intra-cluster distances do not significantly change over two consecutive iterations.

To control the randomness of K-means, it is launched 10 times from k_{min} to k_{max} clusters. The optimum - minimum or maximum, depending on the index-is chosen as the most suitable number of clusters. The indices for comparison have been chosen according to their performance and usage reported in the literature (see Section 1). Selected indices are Dunn (DU), Davies-Bouldin (DB),

Silhouette (SI) and Maulik-Bandyopadhyay (MB). These are compared with the
Score Function (SF). Results according to the number of clusters identified for
the proposed benchmarks are shown next. Particularities of the score function
such as perfect and unique clusters as well as hierarchy of clusters are then
tested. Finally, examples of limitations concerning the score function are given.

4.1   Number of Clusters

The score function has been tested on benchmark data sets and results are
compared with other indices. k_{min} and k_{max} are taken to be respectively 2 and
10. Artificial data sets used in this Section are composed of 1000 points in two
dimensions.

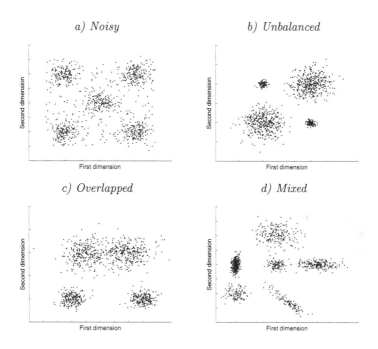

Fig. 1. Four artificial data sets, namely *Noisy, Unbalanced, Overlapped* and *Mixed*. All
of these data sets contains 1000 points in 2D space.

Example 1: In the first data set, *Noisy*, five clusters in a noisy environment
are present (see Figure 1a). It is improbable that a data set contains no noise.
Therefore, clusters are frequently surrounded by noise. Table 1 shows that, unlike
other indices, the Dunn index is not able to estimate correctly the number of
clusters (five). This result confirms the idea that the Dunn index is sensitive to
noise [6].

Example 2: The second data set, *Unbalanced*, consists of four clusters (see
Figure 1b). These clusters are of different sizes and densities. According to [18],

Table 1. Results of the five validity indices on the *Noisy* data set (example 1). The data set is shown in Figure 1a. Bold numbers show maximum values for all indices except DB, where minimum value is desired. This indication is used for Tables 1-6. The correct number of clusters is five.

k	2	3	4	5	6	7	8	9	10
DU	0.018	0.016	0.019	0.019	0.032	**0.035**	0.027	0.028	0.023
DB	1.060	0.636	0.532	**0.440**	0.564	0.645	0.665	0.713	0.729
SI	0.534	0.573	0.719	**0.821**	0.785	0.768	0.733	0.706	0.669
MB	1.314	2.509	3.353	**5.037**	4.167	3.323	2.898	2.515	2.261
SF	0.424	0.489	0.553	**0.592**	0.584	0.578	0.575	0.573	0.572

clusters of varying densities are of importance. Table 2 shows the results for this data set. Whereas DU underestimates the number of clusters, MB overestimates it. This is not the case for DB, SI and SF which correctly identify four clusters.

Table 2. Results of the five validity indices on the *Unbalanced* data set (example 2). The data set is shown in Figure 1b. The correct number of clusters is four.

k	2	3	4	5	6	7	8	9	10
DU	**0.154**	0.066	0.025	0.024	0.016	0.018	0.014	0.012	0.016
DB	0.739	0.522	**0.347**	0.552	0.633	0.712	0.713	0.722	0.733
SI	0.709	0.688	**0.803**	0.689	0.704	0.701	0.679	0.683	0.590
MB	3.900	3.686	4.795	4.751	**4.941**	4.844	4.540	3.575	3.794
SF	0.549	0.563	**0.601**	0.593	0.591	0.589	0.589	0.588	0.589

Example 3: This data set, named *Overlapped*, contains four clusters, two of them overlap. It can be seen in Figure 1c. Two clusters are likely to overlap in real-life data sets. Therefore, the ability to deal with overlapping cluster is one of the best ways to compare indices [19]. Table 3 contains the results for this data set. It can be seen that DU and DB underestimate the correct number of clusters. Only SI, MB and SF are able to identify four clusters.

Table 3. Results of the five validity indices on the *Overlapped* data set (example 3). The data set is shown in Figure 1c. The correct number of clusters is four.

k	2	3	4	5	6	7	8	9	10
DU	**0.030**	0.025	0.013	0.013	0.012	0.019	0.021	0.012	0.012
DB	0.925	**0.451**	0.482	0.556	0.701	0.753	0.743	0.774	0.761
SI	0.635	0.740	**0.818**	0.728	0.713	0.669	0.683	0.669	0.656
MB	1.909	3.322	**5.755**	5.068	4.217	3.730	3.527	3.150	3.009
SF	0.452	0.555	**0.610**	0.601	0.593	0.589	0.588	0.585	0.584

Example 4: The following data set, named *Mixed*, contains six clusters. They have different size, compactness and shape. The data set is shown in Figure 1d. Table 4 presents results. First, it can be seen that DU is maximum for two consecutive values (although not the correct ones). MB is the only index to overestimate the correct number of clusters. Finally, only DB, SI and SF are able to identify correctly six clusters.

Table 4. Results of the five validity indices on the *Mixed* data set (example 4). The data set is shown in Figure 1d. The correct number of clusters is six.

k	2	3	4	5	6	7	8	9	10
DU	0.015	**0.041**	**0.041**	0.027	0.018	0.020	0.014	0.018	0.017
DB	1.110	0.751	0.630	0.575	**0.504**	0.554	0.596	0.641	0.662
SI	0.578	0.616	0.696	0.705	**0.766**	0.744	0.758	0.730	0.687
MB	1.523	1.574	2.379	2.813	3.389	3.661	**3.857**	3.490	3.236
SF	0.442	0.492	0.540	0.559	**0.583**	0.579	0.577	0.576	0.579

Example 5: The data set used in this example, *Iris* is one of the most used real-life data sets in the machine learning and data mining communities [20]. It is composed of 150 points in four dimensions. *Iris* contains three clusters (two of them are not linearly separable). It is a good example of a case where the dimension is more than two and clusters overlap. Table 5 shows the index values for this data set. In this case, only SF is able to correctly identify the three clusters. The overlap is too strong for other tested indices to enumerate the clusters.

Table 5. Results of the five validity indices on the *Iris* data set (example 5). The data set is made by 150 points in a 4D space. The correct number of clusters is three (two of them overlap).

k	2	3	4	5	6	7	8	9	10
DU	**0.267**	0.053	0.070	0.087	0.095	0.090	0.111	0.091	0.119
DB	**0.687**	0.716	0.739	0.744	0.772	0.791	0.833	0.752	0.778
SI	**0.771**	0.673	0.597	0.588	0.569	0.561	0.570	0.535	0.580
MB	**8.605**	8.038	6.473	6.696	5.815	5.453	4.489	4.011	4.068
SF	0.517	**0.521**	0.506	0.507	0.503	0.503	0.497	0.510	0.513

Example 6: The next data set, named *Wine*, is also a real-life data set [20]. It contains 178 points in 13 dimensions. *Wine* data set contains three clusters. Results of the five indices are given in Table 6. Whereas DU overestimates the correct number of clusters, MB underestimates it. DB, SI and SF are able to discover the three clusters.

Table 6. Results of the five validity indices on the *Wine* data set (example 6). The data set is made of 178 points in a 13 dimension space. The correct number of clusters is three.

k	2	3	4	5	6	7	8	9	10
DU	0.160	0.232	0.210	0.201	0.202	0.208	**0.235**	0.206	0.214
DB	1.505	**1.257**	1.499	1.491	1.315	1.545	1.498	1.490	1.403
SI	0.426	**0.451**	0.416	0.394	0.387	0.347	0.324	0.340	0.288
MB	**5.689**	5.391	3.548	2.612	2.302	2.124	1.729	1.563	1.387
SF	0.131	**0.161**	0.151	0.146	0.143	0.145	0.147	0.149	0.150

Table 7 summarizes the results of the application of the five indices to four artificial and two real-life data sets. Among the five indices tested, SF has the best performance. SF correctly identified the number of clusters in all six data sets. The SF successfully processes the standard case with clusters and noise (*Noisy*), clusters of different size and compactness (*Unbalanced*), overlapped clusters (*Overlapped*), multiple kind of clusters (*Mixed*) and multidimensional data (*Iris* and *Wine*).

Table 7. Estimated number of clusters for six data sets and five cluster validity indices. Notation indicates when the correct number of clusters has been found (O) or not (X).

Data Sets	DU	DB	SI	MB	SF
Noisy	7(X)	5(O)	5(O)	5(O)	5(O)
Unbalanced	2(X)	4(O)	4(O)	6(X)	4(O)
Overlapped	2(X)	3(X)	4(O)	4(O)	4(O)
Mixed	3/4(X)	6(O)	6(O)	8(X)	6(O)
Iris	2(X)	2(X)	2(X)	2(X)	3(O)
Wine	8(X)	3(O)	3(O)	2(X)	3(O)

4.2   Perfect Clusters

Since the score function is bounded, its upper limit (1.0) can be used to estimate the closeness of data sets to perfect clusters. The next two data sets are used to test how the SF deals with perfect clusters. The data sets *Perfect3* and *Perfect5* are made of 1000 points in 2D and contain three and five clusters respectively which are near to perfect (i.e. with a very high compactness). Although the number of clusters is correctly identified, it is interesting to note that the maximum value for the SF is different in both cases. In the three cluster case, the maximum (0.795) is higher than in the second one (0.722). This is due to the dependence of the SF on the number of clusters k. This can be seen in the denominator of Equation 6. Nevertheless, the SF gives an idea of how good clusters are through the proximity of the value of the index to its upper bound of unity.

4.3   Unique Cluster

An objective of the SF is to accommodate the unique cluster case. This case is not usually treated by others. In this subsection, k_{min} and k_{max} are taken to be respectively, 1 and 8. When the SF is plotted against the number of clusters, two situations may occur. Either the number of clusters is clearly located with a local maximum (Figure 2, left) or the SF grows monotonically between k_{min} and k_{max} (Figure 2, right).

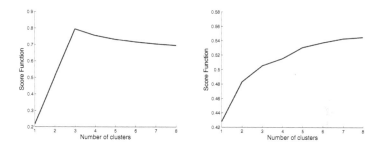

Fig. 2. Difference of the SF trend with a data set containing three clusters (left) and one cluster (right)

Since in the first situation, the number of clusters is identifiable, the challenge lies in the second situation. There are three possible cases. They are: i) no structure in the data, ii) data that forms one cluster and iii) the correct number of clusters is higher than k_{max}. The first situation is out of the scope of this article. More details of whether the data is structured or not, known as cluster tendency, can be found in [1]. In the last two situations, the SF grows monotonically with the number of clusters.

Two observations have been noticed. First, in the unique cluster cases, the value of the SF when $k = 2$, denoted as SF_2 is closer to the value for $k = 1$ (SF_1) than in other data sets. Second, the SF is dependent upon the dimensionality of the data set. Therefore, the slope between SF_2 and SF_1 weighted by the dimensionality of the data set is used as an indicator. To test the unique cluster case, two new data sets are introduced: *UniqueN* is a unique cluster with an added noise and *Unique30* is a unique cluster in a 30 dimensional space. Results of this indicator on all data sets are given in Table 8.

According to Table 8, it is empirically stated that the data set is likely to contain more than one cluster if Equation 9 is satisfied.

$$(SF_2 - SF_1) \cdot d > 0.2 \qquad (9)$$

where d is the dimensionality of the data, SF_2 and SF_1 are respectively the value for SF when $k = 2$ and $k = 1$. Only two data sets containing unique clusters do not satisfy the condition in Equation 9. Therefore, the index SF is the only one, among all tested indices, that is able to identify a unique cluster situation.

Table 8. Results of the indicator $(SF_2 - SF_1) \cdot d$ for eight benchmark data sets

Data sets	Indicator	Data sets	Indicator
Noisy	0.37	*UniqueN*	**0.11**
Unbalanced	0.65	*Unique30*	**0.10**
Overlapped	0.45	*Iris*	1.49
Mixed	0.41	*Wine*	1.31

4.4   Sub-clusters

Another interesting study concerns the sub-cluster case. This situation occurs when existing clusters can be seen as a cluster hierarchy. If this hierarchy can be captured by the validity index, more information about the structure of the data can be given to the user. Data set *Sub-cluster* in Figure 3 is an example of this situation. The index SF is compared with the previously mentioned indices on this topic. Figure 3 shows the evolution of each validity index with respect to the number of clusters.

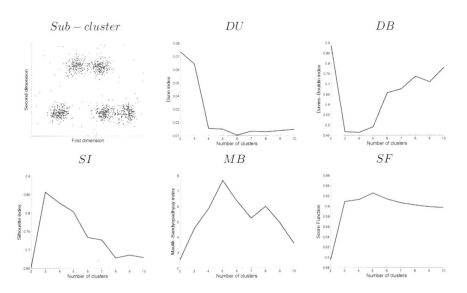

Fig. 3. Comparison of DU, DB, SI, MB and SF for the sub-cluster case. DB must be minimized

DU is not able to find the correct number of clusters (neither the sub-clusters, nor the overall clusters). Although MB finds the sub-clusters, no information about the hierarchy is visible. In the case of DB, even if it is not able to find the five clusters (it finds four), the sub-cluster hierarchy is visible because the value of the index drops rapidly at three clusters. The SI index is not able to recover the correct number of clusters (i.e. the sub-clusters) although it can find the three overall clusters. Finally, the only index which is capable of giving the correct five clusters as well as an indication for the three overall clusters is SF.

4.5   Limitations

In the above subsections, data sets used to test the different indices contain hyperspheroidal clusters. In this subsection, arbitrarily-shaped clusters are briefly studied using two new data sets. *Pattern* is a data set containing 258 points in 2D. It contains three clusters with a specific pattern and different shapes. *Rectangle* is made of 1000 points in 2D that represent three rectangular clusters. These data sets are shown in Figure 4.

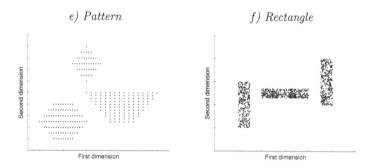

Fig. 4. Two new artificial data sets. *Pattern* and *Rectangle* contain respectively 258 and 1000 points in 2D.

Regarding the *Pattern* data set, all indices are able to find the correct number of clusters (3). The proposed shapes and the pattern do not reveal weaknesses in any index. Concerning the *Rectlangle* data set, results are different. The proposed score function is not able to discover the three clusters. All other tested indices fail as well. All indices overestimates the correct number of clusters: DU (9), DB (7), SI (8), MB (8) and SF (10). A likely explanation is that clusters are far from hyperspheroidal. Therefore, a limitation of the score function, as well as other tested indices, is their restriction to data sets containing hyperspheroidal clusters.

5   Conclusions

Although there are several proposals for validity indices in the literature, most of them succeed only in certain situations. A new index for hard clustering - the score function (SF) - is presented and studied in this paper. The proposed index is based on a combination of the within and between class distances. It can accommodate special cases such as the unique cluster and perfect cluster cases. The SF is able to estimate correctly the number of clusters in several artificial and real-life data sets. The SF has successfully estimated the number of clusters in data sets containing unbalanced, overlapped and noisy clusters. In addition, the SF has been tested successfully on multidimensional real-life data sets. No other index performed as well on all data sets. Finally, in the case of sub-cluster hierarchies, only the SF was able to estimate five clusters and overall, three groups. Therefore, the index SF outperforms four other validity indices (Dunn, Davies-Bouldin, Silhouette and Maulik-Bandyopadhyay) for the k-means algorithm on hyperspheroidal clusters. The proposed index can also accommodate perfect and unique cluster cases. In order to identify the one cluster case, an empirical condition has been formulated. Finally, determining values for the index is computationally efficient.

Several extensions to the present work are in progress. For example, a theoretical justification for the unique cluster condition (Equation 9) is under study. More extensive testing on arbitrarily shaped clusters is necessary. Finally, studies of other clustering algoritms are also under way.

Acknowledgments

This research is funded by the Swiss National Science Foundation (grant no 200020-109257). The authors recognize Dr. Fleuret for fruitful discussions and the two anonymous reviewers for their helpful comments.

References

1. Jain, A., Dubes, R.: Algorithms for Clustering Data. Prentice-Hall, Englewood Cliffs (1988)
2. Webb, A.: Statistical Pattern Recognition. Wiley, Chichester (2002)
3. SanJuan, E., Ibekwe-SanJuan, F.: Text mining without document context. Inf. Process. Manage. 42(6), 1532–1552 (2006)
4. Perdisci, R., Giacinto, G., Roli, F.: Alarm clustering for intrusion detection systems in computer networks. Engineering Applications of Artificial Intelligence 19(4), 429–438 (2006)
5. Jaenichen, S., Perner, P.: Acquisition of concept descriptions by conceptual clustering. In: Perner, P., Amiya, A. (eds.) MLDM 2005. LNCS (LNAI), vol. 3587, pp. 153–162. Springer, Berlin, Heidelberg (2005)
6. Halkidi, M., Batistakis, Y., Vazirgiannis, M.: On clustering validation techniques. Journal of Intelligent Information Systems 17(2-3), 107–145 (2001)
7. Huang, Z.: Extensions to the k-means algorithm for clustering large data sets with categorical values. Data Mining and Knowledge Discovery 2(3), 283–304 (1998)
8. Jain, A.K., Murty, M.N., Flynn, P.J.: Data clustering: a review. ACM Computing Surveys 31(3), 264–323 (1999)
9. Maulik, U., Bandyopadhyay, S.: Performance evaluation of some clustering algorithms and validity indices. IEEE Transactions Pattern Analysis Machine Intelligence 24(12), 1650–1654 (2002)
10. Bezdek, J., Pal, N.: Some new indexes of cluster validity. IEEE Transactions on Systems, Man and Cybernetics 28(3), 301–315 (1998)
11. Dunn, J.: Well separated clusters and optimal fuzzy partitions. Journal of Cybernetics 4, 95–104 (1974)

12. Davies, D., Bouldin, W.: A cluster separation measure. IEEE PAMI 1, 224–227 (1979)
13. Kaufman, L., Rousseeuw, P.: Finding Groups in Data: an Introduction to Cluster Analysis. John Wiley & Sons, West Sussex (1990)
14. Kim, M., Ramakrishna, R.: New indices for cluster validity assessment. Pattern Recognition Letters 26(15), 2353–2363 (2005)
15. Bolshakova, N., Azuaje, F.: Cluster validation techniques for genome expression data. Signal Processing 83(4), 825–833 (2003)
16. Gordon, A.: Cluster Validation. In: Hayashi, C., Yajima, K., Bock, H.H., Ohsumi, N., Tanaka, Y., Baba, Y. (eds.) Data science, classification and related methods, pp. 22–39. Springer, Heidelberg (1996)
17. Ling, R.: On the theory and construction of k-clusters. Computer Journal 15, 326–332 (1972)
18. Chou, C., Su, M., Lai, E.: A new cluster validity measure and its application to image compression. Pattern Analysis Applications 7(2), 205–220 (2004)
19. Bouguessa, M., Wang, S., Sun, H.: An objective approach to cluster validation. Pattern Recognition Letters 27(13), 1419–1430 (2006)
20. Merz, C., Murphy, P.: UCI machine learning repository (1996), http://www.ics.uci.edu/~mlearn/MLSummary.html

Varying Density Spatial Clustering Based On a Hierarchical Tree

Xuegang Hu[1], Dongbo Wang[1], and Xindong Wu[1,2]

[1] School of Computer Science and Information Engineering,
Hefei University of Technology, Anhui 230009, China
[2] Department of Computer Science, University of Vermont, Burlington, VT 50405, USA
jsjxhuxg@hfut.edu.cn, wangdb@msn.com, xwu@cs.uvm.edu

Abstract. The high efficiency and quality of clustering for dealing with high-dimensional data are strongly needed with the leap of data scale. Density-based clustering is an effective clustering approach, and its representative algorithm DBSCAN has advantages as clustering with arbitrary shapes and handling noise. However, it also has disadvantages in its high time expense, parameter tuning and inability to varying densities. In this paper, a new clustering algorithm called VDSCHT (Varying Density Spatial Clustering Based on a Hierarchical Tree) is presented that constructs a hierarchical tree to describe subcluster and tune local parameter dynamically. Density-based clustering is adopted to cluster by detecting adjacent spaces of the tree. Both theoretical analysis and experimental results indicate that VDSCHT not only has the advantages of density-based clustering, but can also tune the local parameter dynamically to deal with varying densities. In addition, only one scan of database makes it suitable for mining large-scaled ones.

Keywords: High-dimensional, Hierarchical Tree, Density-based Clustering, Varying Density.

1 Introduction

Clustering groups similar objects together. As an important research area in data mining, clustering is extensively used in many diversified applications such as pattern recognition, image processing, business intelligence etc. Clustering can also be used as a pretreatment for other data mining tasks. The exponential growth of data scale and the enrichment of data types have put forward the following requirements on clustering algorithms: scalability, noise handling, dealing with different kinds of attributes, dealing with multi-dimensional data, discovery of clusters with arbitrary shapes, minimum dependence of domain knowledge or the user to determine the input parameters, insensitivity to the input order, dealing with restrained clustering, and interpretability and usability of the clustering results.

Many clustering algorithms have recently been proposed in the literature, among which the density-based method shows obvious advantages in efficiency and effect. DBSCAN, proposed in [3], which is the representative algorithm for density-based clustering, can discover clusters of arbitrary shapes and handle noise. Meanwhile,

P. Perner (Ed.): MLDM 2007, LNAI 4571, pp. 188–202, 2007.

there are three main defects in DBSCAN: (1) a high time expense - especially when the data size is large, a significant amount of time is consumed in iterative detection; (2) the limitation of parameter tuning - the crucial parameter that directly influences the clustering results must be specified by the user; and (3) the inability to deal with varying density - a global parameter is used for clustering, but it does not work well under varying-density data environments.

With the above defects of DBSCAN, many improvements have been proposed in recent years. The OPTICS algorithm presented in [4] does not form clusters explicitly, but calculates a cluster order. The discovery of the clusters is much more convenient based on the order, which can also be used to solve the parameter tuning problem in DBSCAN. The equivalence in structure makes OPTICS and DBSCAN have a similar time complexity; however, because the former adopts a complicated processing method and requires extra I/O operations, its actual running speed is well below the latter. Another clustering algorithm presented in [10] enhances DBSCAN by first partitioning the dataset in order to reduce the search space of the neighborhoods. Instead of examining the whole dataset, the enhanced method searches only in the objects within each partition. A merging stage is needed to reach the final natural number of clusters. Other approaches do not aim at producing the exact hierarchical clustering structure, but an approximate one, like sampling-based and grid-based clustering. The former clustering applies an expensive procedure to a subset of the database by sampling, whose clustering quality depends much on the quantity and quality of sampling. Details can be found in [11]. The data space is partitioned into a number of grid cells in the latter clustering, which can be used as a filter step for range queries to accelerate query processing. One of the classic algorithms called CLIQUE is presented in [8].

The above algorithms mostly concentrate on improving the performance of DBSCAN by different techniques; however, the problems with parameter tuning with, high time expense and inability to deal with varying-density clustering have not been solved, especially for large-scaled, high-dimensional and varying-density databases.

In this paper, a Hierarchical Tree model, or H-Tree, is proposed to describe subclusters and the original dataset, and a new density-based clustering algorithm, based on the H-Tree, called VDSCHT, is presented. VDSCHT is of multiple phases: it first scans the dataset to get subcluster information and builds an H-Tree, and the relevant parameter (or threshold) is adaptively adjusted in the tree building process. A density-based clustering procedure is adopted subsequently, and the essential parameter of clustering is locally and dynamically confirmed by the distribution situations of data objects in subclusters and the available information of the H-Tree. DBSCAN detects clusters by scanning the original dataset over and again, but VDSCHT detects and gets the full and natural clusters from the adjacent leaf nodes of the H-Tree by the local density parameter. Both theoretical analysis and experimental results indicate that VDSCHT not only possesses the advantages of density-based clustering (e.g. DBSCAN) including discovering clusters with arbitrary shapes and handling noise, but can also cluster well in varying-density datasets. It has a near linear complexity on both time and space to the size and dimensionality of the dataset and does not depend much on the user or domain knowledge.

The rest of this paper is organized as follows. Section 2 details the idea and description of the VDSCHT algorithm. An algorithm analysis is provided in

Section 3, and experimental results are presented to demonstrate the effectiveness and efficiency of the VDSCHT algorithm in Section 4. Finally, Section 5 concludes this paper and outlines some issues for future research.

2   The VDSCHT Algorithm

In this section, we review and define related concepts at first, and then present the idea and details of the VDSCHT clustering algorithm. Some useful notations and their meanings are shown in Table 1.

Table 1. Notations and their meanings

notation	meaning
$DDR\ (q,\ p)$	Object p is directly density-reachable from object q
$N_\varepsilon(q)$	$N_\varepsilon(p) = \{q \in D \mid dist(p,q) \le \varepsilon\}$
$MinPts$	The minimum number of objects to form a cluster.
B	The maximum number of sub-trees at each non-leaf node in H-Tree.
radius of cluster: R	$R = \left(\dfrac{1}{N} \sum_{i=1}^{N} \left(\vec{X}_i - \vec{X}_0 \right)^2 \right)^{1/2}$

2.1   Related Concepts

Definition 1 (Directly Density-Reachable) [3, 4]. Object p is directly density-reachable from object q (written as $DDR(q, p)$ for short) wrt. ε and MinPts in a set of objects D if:

1) $p \in N_\varepsilon(q)$ ($N_\varepsilon(q)$ is a subset of D contained in the ε -neighborhood of q).

2) $Card(N_\varepsilon(q)) \ge MinPts$ ($Card(N)$ denotes the cardinality of set N).

The condition $Card(N_\varepsilon(q)) \ge MinPts$ is called the "core object condition". If this condition holds for an object q , then we call q a "core object". Only from core objects, other objects can be directly density-reachable.

Definition 2 (Core Distance) [4]. Let p be an object from a database D , ε be a distance value, $N_\varepsilon(p)$ be the ε -neighborhood of p , $MinPts$ be a natural number, and MinPts_Distance (p) be the distance from p to its MinPts' neighbor. Then, the core_distance of p is defined as:

$$core\_distance_{\varepsilon,MinPts}(p) = \begin{cases} UNDEFINED, if \quad Card(N_\varepsilon(p)) < MinPts \\ MinPts\_distance(p), otherwise \end{cases}$$

The core-distance of an object p is simply the smallest distance ε' between p and an object in its ε -*neighborhood* such that p would be a core object wrt. ε' if this neighbor is in $N_\varepsilon(p)$. Otherwise, the core-distance is *undefined*.

Definition 3 (*HSCF*, Hierarchical Subcluster Feature). For a given N d-dimensional data points $\{\vec{X}_i\}$ (where i = 1, 2... N) in a cluster, the *HSCF* vector of the cluster is defined as a triple $HSCF = (N, \overrightarrow{LS}, core\_dist)$, where N is the number of data points in the cluster, $\overrightarrow{LS}$ is the linear sum of the N data points, i.e. $\sum_{i=1}^{N} \vec{X}_i$, and *core_dist* saves the core distance of data objects in the corresponding subcluster. Commonly, *core_dist* is often shared by every data object in any subcluster.

Definition 4 (H-Tree, Hierarchical Tree). An H-Tree is a height-balanced tree with two parameters: a branching factor B defining the maximal number of sub-trees at each non-leaf node, and a threshold T defining the maximum diameter (or radius) of every subcluster. Each node of the tree contains B *HSCFs*, shown in Fig. 1.

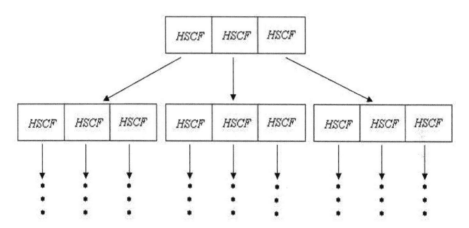

Fig. 1. H-Tree Structure with B = 3; a node is divided as B parts (subclusters or HSCFs)

Definition 5 (Candidate Cluster and Candidate Object). For a given dataset D, if the number of data objects in one subcluster C is not less than the specific threshold (*MinPts*), then C is called as a candidate cluster, and each data point in C is called candidate object.

Definition 6 (Cluster Subset and Non-Cluster Subset). For a given dataset D_1 which can form a full and natural cluster $C(D_1)$, dataset D_1' is a subset of the given dataset D_1, i.e., $D_1' \subseteq D_1$; if D_1' can form a candidate cluster, then D_1' is a cluster subset of $C(D_1)$; otherwise, D_1' is a non-cluster subset of $C(D_1)$.

2.2 Description of VDSCHT

The clustering process of VDSCHT is of three phases: clustering preprocessing, parameter confirming and clustering detection. A high-level description of the VDSCHT algorithm is described as follows.

```
VDSCHT Algorithm
{
  Clustering preprocessing with a hierarchical
  clustering method;
  Confirming parameter ε locally and dynamically based on
  the information of preprocessing;
  Clustering detection among the adjacent subclusters
  with a density-based method;
}.
```

The main steps of VDSCHT are discussed in the following subsections.

2.2.1 Clustering Preprocessing

Scan the dataset and build an H-Tree with a small threshold value T defined in Definition 4 as follows.

```
Building H-Tree Algorithm
{
  Step 1: Initialize the root of the H-Tree;
  Repeat scanning dataset; // Scan only one time.
  {
    Repeat inserting data
    {// Find an appropriate HSCF at a leaf node and
    // insert the new object into the H-Tree.
      Step 2: Identify the most appropriate leaf:
              starting from the root, recursively
              descends H-Tree by choosing the closest
              child node according to distance;
      Step 3: Modify the leaf: there are three kinds of
              situations to deal with according to the
              status of the leaf node from Step 2:
          a): The leaf absorbs the current data point:
              update the HSCF information of the leaf and
              the relevant parent nodes;
          b): Create a new leaf; and insert the current
              data object into the new position by B+
              tree rules;
          c): Split the leaf by choosing the farthest
              pair of entries (HSCF) as seeds, and
              redistribute the remaining entries based on
              the closest criterion;
    } Until needs rebuilding H-Tree;
    // Rebuild when specific situation is satisfied.
```

Step 4: Rebuild the H-Tree: tune the *threshold* and insert the entries of nodes into the new H-Tree based on the structure of the original tree;
      } Until all data objects have been scanned;
   } // Building H-Tree ends.

The computation of the mean distance between all data objects (including the current object) and the centroid of each subcluster at leaf nodes is inevitable to judge whether a certain subcluster can absorb the new data object O in Step 4. The mean distance, i.e. **radius**, is defined as $R = \left(\dfrac{1}{N} \sum_{i=1}^{N} \left(\vec{X}_i - \vec{X}_0 \right)^2 \right)^{1/2}$, where $\vec{X}_0$ is the centroid of the cluster. It is obvious that the computation of the radius of the cluster is expensive when the cluster includes a large number of data objects. Therefore, we provide Definition 7 and Theorem 1 as follows.

Definition 7. For a given dataset D, if the mean distance between all objects and the centroid of cluster C equals to a given parameter value δ, then we call this cluster C is **saturated**.

Theorem 1. For a given value δ, if cluster C is saturated, C can be abstracted as the centroid $\overrightarrow{Center} = \dfrac{1}{N} \sum_{i=1}^{N} \vec{X}_i$ (where N is the number of objects in C). The distance d between the new object O and $\overrightarrow{Center}$ can be used to judge whether C can absorb O: if $d \leq \sqrt{1 + \frac{1}{N}} \cdot \delta$, C can absorb O, i.e., the radius of the newly formed cluster is not larger than δ; otherwise, if $d > \sqrt{1 + \frac{1}{N}} \cdot \delta$, C cannot absorb O, i.e., the radius of the newly formed cluster is larger than δ.

Proof: The radius of cluster C is defined as $R = \left(\dfrac{1}{N} \sum_{i=1}^{N} \left(\vec{X}_i - \vec{X}_0 \right)^2 \right)^{1/2}$. If C can absorb the new object O, the centroid of the newly formed cluster C' is $\vec{X}_0' = \dfrac{1}{N+1} \sum_{i=1}^{N+1} \vec{X}_i$, and the radius of this new cluster C' is $R' = \left(\dfrac{1}{N+1} \sum_{i=1}^{N+1} \left(\vec{X}_i - \vec{X}_0' \right)^2 \right)^{1/2}$. Thus, the proof of whether C can absorb O can be translated as follows: if $R' - R \leq 0$, C can absorb O; otherwise, if $R' - R > 0$, C can not absorb data object O.

$$R'^2 - R^2 = \frac{\sum_{i=1}^{N+1}\left(\vec{X}_i - \vec{X}'_0\right)^2}{N+1} - \frac{\sum_{i=1}^{N}\left(\vec{X}_i - \vec{X}_0\right)^2}{N} = \frac{N\left(d^2 - R^2\right) - R^2}{(N+1)^2}$$. Here,

d is the Euclidean distance between O and the centroid of C defined before. If C is saturated, i.e. $R^2 = \delta^2$, we set $d^2 = (1+t)\cdot\delta^2\,(t \geq -1)$, then $R'^2 - R^2 = (N+1)^{-2}\cdot(N\cdot t - 1)\cdot\delta^2$, we can discuss this question with two conditions:

1). If $-1 \leq t \leq \dfrac{1}{N}$, then $d \leq \sqrt{1 + \dfrac{1}{N}}\cdot\delta$, $\because R' \geq 0, R \geq 0$

apparently $R'^2 - R^2 \leq 0 \Leftrightarrow R' - R \leq 0$, so O can be absorbed in cluster C.

2). If $t > \dfrac{1}{N}$, then $d > \sqrt{1 + \dfrac{1}{N}}\cdot\delta$, apparently $R'^2 - R^2 > 0 \Leftrightarrow R' - R > 0$,

so O can not be absorbed in cluster C.   **End**.

Accordingly, the computation of the distance between the new object O and the centroid of each cluster is much simpler than the computation of the radius of the cluster. In fact, to verify whether a certain cluster is saturated, we only need to check whether the radius approximately equals to the given value if there is no strict requirement on the quality of clustering.

Building the H-Tree is only a data preprocessing step to find all the subcluster information in our VDSCHT design. The calculation is much simplified compared with the Birch algorithm in [2], for the following reasons. First, we only calculate the $\overrightarrow{LS}$ in $HSCF$ of each subcluster in Step 3 and need not calculate the $SS = \sum_{i=1}^{N}\overrightarrow{Xi}^2$ as Birch does, which will save time and space when $HSCF$ includes much more data and we need to update the information of nodes. Second, the "Closest Criterion" is straightforward according to the distance between the new inserted data object and the centroid of $HSCF$ by Theorem 1 when identifying the most appropriate leaf node in the H-Tree. Therefore, the time expense of finding the most appropriate leaf is much reduced.

Dealing with three situations of modifying the leaf node in Step 3 is similar to B-Tree, and rebuilding the H-tree when needed is simple, based on the processes in the original H-Tree structure, in a similar way to the reconstruction of the CF-Tree in [2]. These two procedures are not detailed here due to space limitations. However, "*how to determine the timing of the tree rebuilding?*" is a significant issue.

Undoubtedly, "how to select the appropriate threshold value and determine the timing of tree rebuilding" is a very difficult problem. Clustering in the face of massive and complicated high-dimensional data without a sufficient understanding on the data is a real challenge. Paper [2] did not elaborate on the threshold selection and tree reconstruction.

In this paper, an abnormality-support factor (ab_limit) is defined as:

$$ab\_limit = \frac{ab\_number}{N},$$

which is used to evaluate the validity of the H-Tree, where ab_number figures the number of possible abnormality data objects included in the $HSCF$ of H-Tree leaf nodes, i.e.,

$$ab\_number = HSCF_1 + \sum_{\alpha=2}^{\lfloor MinPts/2 \rfloor} HSCF_\alpha,$$

where $HSCF_1$ figures the number of $HSCFs$ that each contain only one data object and $HSCF_\alpha$ figures the number of $HSCFs$ that contain α data objects and the mean distance between these α data objects is distinctly greater than the threshold value. Our experiments have shown that the constructed H-Tree can better reflect the overall distribution of data objects and suit the next phase of density clustering to discover higher-quality clustering results when $ab\_limit \approx 9\% \pm 3\%$.

Otherwise, $ab\_limit > 12\%$ indicates a smaller threshold resulting in some points of certain clusters may have been wrongly dealt as possibly abnormal data, and it therefore would be appropriate to increase the threshold value to reconstruct a new H-Tree. Also, $ab\_limit < 6\%$ shows that a large threshold leads to the expansion of the cluster and the wrong absorption of surrounding outliers, and at this time, the threshold should be appropriately reduced to reconstruct a new H-Tree. Naturally, a more accurate threshold value can be specified in practical applications when the customer has a deep understanding of the database.

Different from increasing the threshold driven by memory or stored pages to rebuild the tree in [2], we tune the threshold value based on the data distribution of the dataset and the initial clustering results. Also, the reconstruction process by increasing the threshold value can be based on the original tree structure and can be quickly completed; however, the reconstruction by decreasing the threshold value will consume more time. Therefore, it is preferable to specify a small threshold value to begin with in order to improve the efficiency of the follow-up rebuilding when needed in the tree building process.

The leaf nodes in the H-Tree including all the subcluster information can be divided into three kinds of forms: a cluster subset, a non-cluster subset of the natural clusters, and noisy data objects (outliers). Accordingly, the insertion of a data object always starts from the root of the H-Tree and finds a cluster according to the "closest criterion". In the tree building process, when a natural cluster is divided into several subclusters by hierarchical clustering, these subclusters correspond to leaf nodes and are always adjacent in the H-Tree. Therefore, our iterative clustering detection at a later stage of VDSCHT does not need to scan the original dataset over and again as DBSCAN does when discovering clusters, but only detects clusters among the adjacent leaf nodes of the H-Tree. This reduces the time expense of discovering clusters significantly.

2.2.2 Parameter Confirming

Density-based clustering (e.g. DBSCAN) needs two essentially important parameters: MinPts and ε. MinPts determines the minimum number of data objects included in the data space to form a cluster, and ε defines the size of the data space. MinPts is easy to confirm and [3, 4, 6] consider that setting MinPts as 5 is suitable even for large-scale databases. We have set MinPts as 5 in this paper, unless otherwise stated. The ε parameter is difficult to specify and it directly influences the final results of clustering. DBSCAN and its recent variations push away the task of establishing ε to the user. The results of clustering will be undoubtedly much more accurate if the user knows the distribution and density of the whole data very well (in which case, a natural question would be: *if so, why do they need computer clustering?*).

In addition, high-dimensional data is often sparse and has a varying density. The global ε confirmed by the user can reduce the adaptability and quality of clustering algorithms. Another method of establishing ε is to calculate the $k-th$ nearest distance of each data object in the whole dataset, draw a k-dist chart, and then get the user to assign the ε value according to the tendency of the chart (which mostly corresponds to the inflexion). However, after a large number of experiments, we have found that there are many similar "inflexions" in the k-dist chart in most cases, especially when the data set is high-dimensional and has a varying density. All of these observations show that the global ε is hard and infeasible to specify. Furthermore, the course of drawing a K-dist chart is expensive.

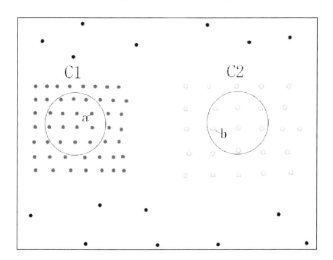

Fig. 2. Two clusters with varying density

For the dataset shown in Fig. 2, DBSCAN specifies the global ε parameter to cluster the dataset, then clustering result will be: (i) if the ε parameter is suitable for point a in cluster C1, a can be used to detect all the red points around it; however, the ε is awful for point b in cluster C2, as b can be used to detect nothing although other points in C2 around b can obviously form a cluster much like C1 except the density; (ii) otherwise, if the ε parameter is suitable for point b in cluster C2, C1 will inevitably absorb the surround noise data. Accordingly, a local ε is indeed indispensable for finding out accurate and natural clustering results, especially under varying density environments.

Then, *how to specify the local density parameter ε* ? First of all, a straightforward observation is that the distance between any two closest data objects in a certain cluster is approximately equal; otherwise, they are wrongly specified in one cluster when this distance is much different from others. Therefore, the local density parameter ε will be similar or even the same when detecting from data objects in one certain cluster; but are mostly specified differently when detecting from different clusters. Second, each data object used to detect a cluster must be a core object. Apparently, it is so natural and rational to set ε as the core distance that is shared with all data objects detected from a candidate cluster, i.e.,

$$\varepsilon = core\_dist,$$

where core_dist is saved in HSCF. Finally, the calculation of the core distance does not need scanning the whole dataset, but only through the adjacent *HSCFs* and/or nodes for the insertion of all data objects into the H-Tree based on the "closest criterion", which ensures that data objects close to each other in the dataset are distributed in adjacent *HSCFs* and leaf nodes. Thus, we can calculate the local ε parameter for every subcluster under varying density environments simply and efficiently.

Take the dataset in Fig. 2 as an example. We dynamically specify different local ε_a and ε_b values for detecting clusters from points a and b, and the objects in the same cluster share the same local ε value. In this way, both points a and b can detect its own cluster naturally and correctly by using a different local ε.

Preprocessing the data by hierarchical clustering and building the H-Tree at first, we can then specify ε as a local value based on the preprocessing and the core distance of the candidate point dynamically. This does not require the user to specify the ε parameter, avoids a possible serious distortion of the final clustering results, and improves the quality and adaptability of clustering at the same time.

2.2.3 Clustering Detection

Clustering detection of every data object in all leaf nodes uses the "far to near" criterion according to the distance between the current data object and the centroid with density-based clustering. It starts from a data object in the candidate cluster to improve the space and time efficiencies. For convenience, we select a random candidate point P from a candidate cluster to start the clustering detection process. If P belongs to a certain cluster, it does not change its attachment to this certain cluster; otherwise, if P does not belong to any cluster, this indicates that the candidate cluster in which P is located is a newly found cluster. When another data object Q is detected from P, clustering detection can be divided into the following three kinds of possible situations for different treatments.

(a) Q does not belong to any cluster using DDR (Q, P), and Q is not a candidate point. The cluster to which P attaches absorbs Q in this case, i.e., P and Q are included in the same cluster, as shown in left of Fig. 3 (define MinPts = 3, hereinafter);

(b) Q does not belong to a certain cluster, DDR (Q, P), and Q is a candidate point. The cluster to which P belongs absorbs the candidate cluster Q, i.e., a subcluster merger takes place, shown in middle of Fig. 3.

(c) Q belongs to a certain cluster (i.e. Q has been detected) and DDR (Q, P). The cluster to which Q belongs absorbs the cluster to which P is attached (or oppositely, the cluster where P is located absorbs the other cluster). I.e., the two clusters where P and Q are located are amalgamated into one cluster, as shown in right of Fig. 3.

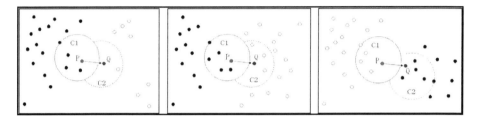

Fig. 3. Three kinds of possible situations of clustering detection

The treatments of situations (a) and (b) provide the VDSCHT algorithm with the ability to discover clusters of arbitrary shapes. The treatment of situation (c) shields the sensitivity to input order of the DBSCAN algorithm.

VDSCHT preprocesses data objects by hierarchical clustering to find the overall information of subclusters, and then locally and dynamically determines the crucial parameter ε which is indispensable to follow-up density-based clustering at a later stage. With the aim to improve the efficiency of clustering, detection only proceeds with candidate points among the adjoining areas. Non-candidate points are always far from the integrated and natural clusters; not to mention the outliers. However, if a non-candidate point is close to a cluster, it must be detected from a candidate point. Clustering detection always follows the "from far to near" criterion during the process, because the detection scope of points far from the centroid frequently covers the scope of points close to the centroid. Therefore, following this criterion, it can detect data points in other subclusters with more possibilities and extensions. The points close to the centroid can either be selected or never detected according to the actual conditions. Our experimental results indicate that detection by the criterion can improve the quality and efficiency of clustering significantly.

3   Complexity Analysis

The time complexity of the hierarchical clustering we have used to preprocess the data is $O\bigl(d \cdot N \cdot B(1 + \log_B S)\bigr)$ (see [2] for details), where S is the maximal size of the H-tree; B is the branching factor of the H-Tree, defined in Definition 2; d is the data dimensionality; and N is the number of data objects. The follow-up clustering detection checks L adjacent neighbors among K candidate clusters with a complexity of $O(K \cdot L)$. Therefore, the total time complexity is $O\bigl(d \cdot N \cdot B(1 + \log_B S) + K \cdot L\bigr)$ (commonly L<<K<<N).

VDSCHT uses an array list to store the original dataset, and the nodes on the H-Tree to record the positions of the corresponding data objects in the array list.

Therefore, the total space complexity is $O(P + d \cdot N)$, but P is always $O(N)$, so the total space complexity is $O(d \cdot N)$ (where P is the number of nodes, d is the dimensionality and N is the number of data objects in the original dataset; commonly $P \ll N$).

4 Performance Evaluation

In this section, we present a comprehensive evaluation of VDSCHT using several synthetic databases, the SEQUOIA 2000 database benchmark, and the Weka 3.4 data [13]. VDSCHT and its rival algorithm DBSCAN are implemented in JAVA. All experiments are run on a 2.0 GHz CPU and 256 MB RAM.

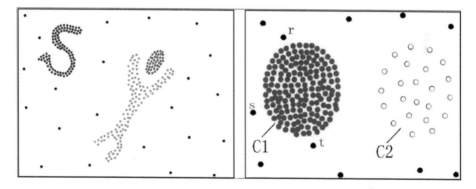

Fig. 4. Two databases used for comparing the clustering accuracy of VDSCHT and DBSCAN

To confirm the accuracy and integrality of the VDSCHT algorithm, we run it on two sample databases (Database 1 and Database 2 in Fig. 4). Database 1 includes three equal-density clusters with arbitrary shapes, and VDSCHT and DBSCAN can find these three clusters completely as shown in Fig. 4. Database 2 contains two varying-density clusters C1 and C2, also shown in Fig. 4, however, the experimental results is disparate.

Our experiments show that DBSCAN always finds one cluster, but there are differences in the cluster's content, which can be generally divided into three kinds of situations. (1) When ε is small, DBSCAN finds cluster C1 but other points (including all points in cluster C2 and three points of r, s, t) are considered as noise. (2) When the ε value increases, the cluster found by DBSCAN includes all points in C1 and the three points of r, s, t. The reason is that these three points can be directly density-reachable from some points in cluster C1, and are therefore included in the same cluster. (3) When ε is relatively large, the cluster found by DBSCAN contains all points in C1, some points in C2 and the three points of r, s, t. Here not only the three points, but also some points of C2 close to C1 are also density-reachable from some points in C1, and therefore they are all combined into one cluster.

As mentioned in Section 0, we dynamically specify the local ε by the subcluster information of the H-Tree. In other words, VDSCHT specifies a different value of ε

as the core distance of data objects in every cluster. Therefore, the three points of r, s, t cannot be reachable from any points in C1 because of the small ε when detecting from cluster C1, and therefore, they will not be wrongly absorbed in cluster C1. When detecting from cluster C2, the value of ε increases with the core distance of points in cluster C2, which ensures that all points in C2 can be detected completely. It is obvious that when clustering under varying density environments, VDSCHT has a better quality and a higher adaptability compared with DBSCAN.

We perform comparative studies using the three databases in Table 2. The Weka_2 database has the most number of dimensionality, which is used to verify the time performance with varying dimensions. The test data size is 600 and its dimensionality is from 5 to 40.

Table 2. Databases for experiments

Dataset	Number of Objects	Dimensionality
SEQUOIA 2000	20000	2
Weka_1	10000	19
Weka_2	5000	40

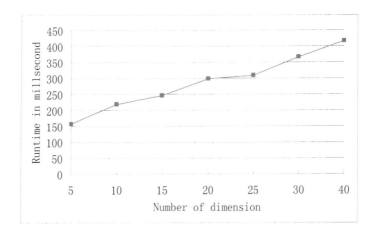

Fig. 5. Run time w.r.t. varying dimensions

Our experimental results show a linear increase in Fig. 5 and demonstrate that VDSCHT has the advantage of clustering high-dimensional data. The performance of VDSCHT is evaluated and compared with DBSCAN with three databases in Table 1. The former only detects clusters from candidate clusters in adjacent neighbors to discover the whole natural and integral clusters; however, the latter detects clusters from every core point and scans the database over and again. Therefore, VDSCHT is more efficient than DBSCAN and displayed in Fig. 6.

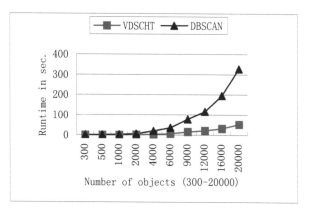

1). Sequoia 2000 database, 2 dim.

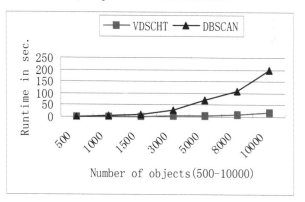

2). Weka_1 database, 19 dim.

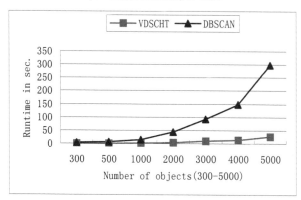

3). Weka_2 database, 40 dim.

Fig. 6. Run time w.r.t. different databases

5 Conclusions

This paper has analyzed density-based clustering and probed into the advantages and shortcomings of DBSCAN, and then proposed a varying density space clustering algorithm VDSCHT using a hierarchical tree. VDSCHT possesses the complementary advantages of density-based clustering and hierarchical clustering. Also, the crucial parameter ε is locally and dynamically determined on the basis of clustering preprocessing. In addition, the characteristic of only one scan of the database makes VDSCHT suitable for mining large-scale and high-dimensional data. The theoretical analysis and experimental results have both confirmed the above conclusions. Factual data are often complicated, incomplete and distributed; and how to discover the natural clusters efficiently is the main target of our future research.

References

1. Han, J.W., Kanber, M.: Data Mining: Concepts and Techniques. Morgan Kaufmann Publishers, Seattle (2001)
2. Zhang, T., Ramakrishnan, R., Livny, M.: BIRCH: An efficient data clustering method for very large databases [C]. In: Proc. 1996 ACM-SIGMOD Int. Conf. Management of Data (SIGMOD'96), pp. 103–114, Montreal, Canada (June 1996)
3. Ester, M., Kriegel, H.-P., Sander, J., Xu, X.: DBSCAN: A density-based algorithm for discovering clusters in large spatial databases with noise. In: Proc. 1996 Int. Conf. Knowledge Discovery and Data Mining (KDD'96), pp. 226–231, Portland, OR (August 1996)
4. Ankerst, M., Bruenig, M., Kreigel, H.-P., Sander, J.: OPTICS: Ordering points to identify the clustering structure. In: Proc. 1999 ACM-SIGMOD Int. Conf. Management of Data (SIGMOD'99), pp. 49–60, Philadelphia, PA (June 1999)
5. Dash, M., Liu, H., Xu, X.: '1+1>2': merging distance and density based clustering. In: Proc. 2001 Int. Conf. Database Systems for Advanced Applications, pp. 32–39, Hong Kong, China (April 2001)
6. Brecheisen, S., Kriegel, H.-P., Pfeifle, M.: Efficient density-based clustering of complex objects. In: Proc. 2004 Int. Conf. Data Mining (ICDM'04), pp. 43–50 (November 2004)
7. Brecheisen, S., Kriegel, H.-P., Pfeifle, M.: Multi-step density-based clustering. Knowledge and Information Systems 9(3), 284–308 (2006)
8. Agrawal, R., Gehrke, J., Gunopulos, D., Raghavan, P.: Automatic Subspace Clustering of High Dimensional Data. Data Mining and Knowledge Discovery 11, 5–33 (2005)
9. Guha, S., Rastogi, R., Shim, K.: CURE: An efficient clustering algorithm for large databases. In: Proc. 1998 ACM-SIGMOD Int. Conf. Management of Data (SIGMOD'98), pp. 73–84, Seattle, WA (June 1998)
10. Yasser, E.-S., Ismail, M.A., Farouk, M.: An Efficient Density Based Clustering Algorithm for Large Databases. In: Proc. 2004 16th IEEE Int. Conf. Tools with Artificial Intelligence (ICTAI 2004)
11. Borah, B., Bhattacharyya, D.K.: An improved sampling-based DBSCAN for large spatial databases. Intelligent Sensing and Information Processing (2004)
12. Stonebraker, M., Frew, J., Gardels, K., Meredith, J.: The SEQUOIA 2000 Storage Benchmark. In: Proc. ACM SIGMOD Int. Conf. on Management of Data, pp. 2–11, Washington, DC (1993)
13. http://www.cs.waikato.ac.nz/ml/weka/

Kernel MDL to Determine the Number of Clusters

Ivan O. Kyrgyzov[1], Olexiy O. Kyrgyzov[2], Henri Maître[1],
and Marine Campedel[1]

[1] Competence Centre for Information Extraction
and Image Understanding for Earth Observation,
GET/Télécom Paris - LTCI, UMR 5141, CNRS
46, rue Barrault, 75013, Paris, France
name@enst.fr
[2] Department of Computer Science and Electrical Engineering, OGI School of Science
and Engineering, Oregon Health and Science University, 20000 NW Walker Road,
Beaverton, OR, USA, 97006
name@csee.ogi.edu

Abstract. In this paper we propose a new criterion, based on Minimum
Description Length (MDL), to estimate an optimal number of clusters.
This criterion, called Kernel MDL (KMDL), is particularly adapted to
the use of kernel K-means clustering algorithm. Its formulation is based
on the definition of MDL derived for Gaussian Mixture Model (GMM).
We demonstrate the efficiency of our approach on both synthetic data
and real data such as SPOT5 satellite images.

1 Introduction

We are interested in knowledge extraction from a SPOT5 satellite image
database. One of our tasks is to find categories of images and to classify them
without prior knowledge on the type or number of these categories. Considering
the amount of available data we are concerned in using simple, fast and efficient
clustering algorithms. K-means is one of them but suffers from several drawbacks: i) it cannot adapt to any cluster shape ii) the knowledge of number of
clusters is necessary iii) the result strongly depends on the initialization process.

To answer the first problem, a classical solution is to use Kernel K-means algorithm [9] [14]. During the last decade kernel-based algorithms attracted lots of
researchers who applied them to various tasks such as machine learning, pattern
recognition, computer vision, *etc*. The success of these approaches is related
to the fact that using a kernel (see definition and properties of kernel in [13]
[14]) is equivalent to defining a feature space transform; the resulting feature
space is tuned to simplify the classification process and allows efficient classical
algorithms (like K-means) processing. This feature space depends on kernel pa-
rameter(s); several approaches are proposed in the literature to determine the
optimal parameter(s) [3]: in this work we use one kernel with fixed parameter.

P. Perner (Ed.): MLDM 2007, LNAI 4571, pp. 203–217, 2007.
© Springer-Verlag Berlin Heidelberg 2007

To answer the second and third problems we propose to use a standard approach such as selection of a clustering solution obtained using different number of clusters and initializations. This selection is based on the minimum of our KMDL criterion. It allows us to stabilize clustering results and to have a smoothed KMDL curve.

Our proposition about using MDL criteria to determine the number of clusters is based on several arguments. Firstly, MDL is able to give access to an optimal code or an optimal data representation for a certain model of data [10], e.g. for GMM in our case. Secondly, this criterion works well when lots of data are available [6]. This is our case because we have a huge storage of satellite images. Finally, in the literature we have not found previous works about applying MDL criteria to Kernel K-means to find the optimally associated number of clusters. It gives us the motivation to formulate MDL criteria for Kernel K-means clustering.

We revise the main definition of MDL for GMM and we show a simplification of MDL through the complete log-likelihood of GMM in Sect. 2. The objective function for Kernel K-means is presented in Sect. 3. Then we formulate KMDL in Sect. 4 using the simplified MDL for GMM. Results on synthetic data and real satellite images are presented in Sect. 5 and Sect. 6, respectively. Conclusions are in Sect. 7.

2   MDL for the Gaussian Mixture Model

2.1   Gaussian Mixture Model

The finite mixture model is widely used to represent data in statistical pattern recognition. Let $X = \{X_1, ..., X_I\}$ denote the data set of samples X_i, where each X_i is a vector $X_i = (X_{i1}, ..., X_{iD})$ of feature values X_{id}. The set X is modelled by a finite mixture model consisting of two parts [10]:

1. the prior probability $P(X_i \in j \mid \Theta_j) = \alpha_j$ that every sample X_i is a member of only one mixture component j, $(j = 1, ..., J)$, where $\alpha_j = n_j/I$, (n_j denoting the number of samples belonging to the mixture component j);
2. the conditional probability modelling each component j by the parameterized probability density function (pdf) $P_j(X_i \mid \Theta_j)$, where Θ_j denotes the parameter set.

Let $P_j(X_i \mid \Theta_j)$ denote the class-probability of observing the sample X_i conditional to X_i belonging to the component j. The finite mixture model expresses the probability of observing the sample X_i as a sum of pdf:

$$P(X_i \mid \Theta) = \sum_{j=1}^{J} \alpha_j P_j(X_i \mid \Theta_j). \tag{1}$$

An important sub-class of mixture models is the multivariate Gaussian distribution, based on a Gaussian class-distribution:

$$P_j(X_i \mid \Theta_j) = \mathcal{N}(X_i \mid \mu_j, \Sigma_j) = \frac{e^{-\frac{1}{2}\left((X_i - \mu_j)^T \Sigma_j^{-1}(X_i - \mu_j)\right)}}{(2\pi)^{D/2} \mid \Sigma_j \mid^{1/2}}, \tag{2}$$

where μ_j and Σ_j are the mean and the covariance matrix of the jth component, respectively. Estimates of the jth mean and covariance matrix are classically obtained as:

$$\mu_j = \frac{1}{n_j} \sum_{l=1}^{n_j} X_l \,, \tag{3}$$

$$\Sigma_j = \frac{1}{n_j} \sum_{l=1}^{n_j} (X_l - \mu_j)^T (X_l - \mu_j) \,, \tag{4}$$

where $X_l \subseteq j$.

With the assumption that the data instances X_i are independently distributed, the joint data probability (probability of observing data set X or likelihood function) is the product of the individual instance probabilities:

$$P(X \mid \Theta) = \prod_{i=1}^{I} \sum_{j=1}^{J} \alpha_j P_j(X_i \mid \Theta_j) \,. \tag{5}$$

The Expectation-Maximization (EM) algorithm [10] can be used to estimate the optimal parameters Θ_j of GMM. Without loss of generality we say that the jth component of GMM models the jth cluster.

The purpose of clustering data is to simplify their representation in the feature space by replacing each sample by a generic class which is likely to express all the properties of the samples. However, when substituting a sample by its model, an error is introduced. The more complex the model, the less the error. The "model complexity" is well expressed by the number of parameters needed to build the model. In the mixture of Gaussians case where every cluster is given by its mean (3) and its covariance matrix (4), the more clusters are used, the more complex the model is, and the less error between data and model. A method to choose the optimal number of clusters consists in selecting the number that most efficiently codes the data, i.e. that provides the shortest description when representing the samples using models and the errors to the model. This method, named Minimum Description Length (MDL), was proposed by Rissanen [2], [11], [12]. MDL is defined as [12]:

$$\min_{k,\Theta} -log(P(X|\Theta)) + \frac{1}{2}klog(I) \,, \tag{6}$$

where $log(P(X \mid \Theta))$ is the log-likelihood of the mixture model (5) and $\frac{1}{2}klog(I)$ is a penalty function with k parameters.

2.2   MDL for the Complete Log-Likelihood of GMM

Let see the log-likelihood for the mixture of Gaussian distributions in more details. To complete the likelihood $P(X|\Theta)$ (5) of the finite mixture expressed by (1), we should introduce the hidden variable z which attribute any sample to a class: $z = \{z_1, ..., z_i, ..., z_I\}$ [4] [5]. Label z_i is coded as a binary vector

$z_i = [z_{i1}, ..., z_{ij}, ..., z_{iJ}]$, where $z_{ij} = 1$ if sample i belongs to cluster j, or 0 if not. Using (5), the complete log-likelihood $log(P(X, z|\Theta))$ becomes [4] [5]:

$$log\left(P(X, z \mid \Theta)\right) = log\left(\prod_{i=1}^{I}\sum_{j=1}^{J} z_{ij}\alpha_j P_j(X_i \mid \Theta_j)\right) =$$

$$\sum_{i=1}^{I} z_{ij}log(\alpha_j P_j(X_i \mid \Theta_j)). \tag{7}$$

By substituting the multivariate Gaussian distribution $P_j(X_i \mid \Theta_j)$ (2) in the complete log-likelihood (7), we obtain:

$$\sum_{i=1}^{I} z_{ij}log(\alpha_j\mathcal{N}(X_i \mid \mu_j, \Sigma_j)) = \sum_{i=1}^{I} z_{ij}log\left(\alpha_j\frac{e^{-\frac{1}{2}\left((X_i-\mu_j)^T\Sigma_j^{-1}(X_i-\mu_j)\right)}}{(2\pi)^{D/2} \mid \Sigma_j \mid^{1/2}}\right) =$$

$$\sum_{i=1}^{I} z_{ij}\left(log\left(\frac{\alpha_j}{\mid \Sigma_j \mid^{1/2}}\right) - \frac{D}{2}log(2\pi) - \frac{1}{2}\left((X_i-\mu_j)^T\Sigma_j^{-1}(X_i-\mu_j)\right)\right) =$$

$$\frac{1}{2}\sum_{i=1}^{I} z_{ij}log\left(\frac{\alpha_j^2}{\mid \Sigma_j \mid}\right) - \frac{1}{2}\sum_{i=1}^{I} z_{ij}D\log(2\pi)$$

$$-\frac{1}{2}\sum_{i=1}^{I} z_{ij}\left((X_i-\mu_j)^T\Sigma_j^{-1}(X_i-\mu_j)\right). \tag{8}$$

In this equation, some terms are constant:

$$-\frac{1}{2}\sum_{i=1}^{I} z_{ij}D\log(2\pi) = -\frac{1}{2}\sum_{j=1}^{J} n_jD\log(2\pi) = -\frac{1}{2}ID\log(2\pi) = const_1. \tag{9}$$

Moreover, to calculate the matrix Σ_j (4) the only samples from the cluster j are needed, therefore:

$$-\frac{1}{2}\sum_{i=1}^{I} z_{ij}\left((X_i-\mu_j)^T\Sigma_j^{-1}(X_i-\mu_j)\right) = -\frac{1}{2}\sum_{j=1}^{J} n_jDI = -\frac{DI^2}{2} = const_2. \tag{10}$$

Then, the complete log-likelihood $log(P(X, z|\Theta))$ (7) may be written as:

$$\frac{1}{2}\sum_{i=1}^{I} z_{ij}log\left(\frac{\alpha_j^2}{\mid \Sigma_j \mid}\right) + const = \frac{1}{2}\sum_{j=1}^{J} n_jlog\left(\frac{\alpha_j^2}{\mid \Sigma_j \mid}\right) + const. \tag{11}$$

In the right part of the MDL definition (6), $\Bbbk$ is the model free parameters number. In case of Gaussian mixture model free parameters are:

- $J - 1$ parameters for J weights α_j (since $\sum \alpha_j = 1$);
- D parameters for each mean μ_j;
- $D(D + 1)/2$ parameters for each covariance matrix Σ_j.

Therefore, the number of free parameters is:

$$\Bbbk = J - 1 + J(D + D(D + 1)/2) = J(D^2 + 3D + 2)/2 - 1. \qquad (12)$$

Using the complete log-likelihood (11) and the free parameter number of (12), the description length (6) of Gaussian mixture model with J clusters is:

$$-\frac{1}{2} \sum_{j=1}^{J} n_j log \left(\frac{\alpha_j^2}{| \Sigma_j |} \right) + (J(D^2 + 3D + 2)/2 - 1)log(I)/2 + const. \qquad (13)$$

The *const* term having no influence on MDL for different cluster numbers and as $\alpha_j = n_j/I$, we may minimize:

$$\Lambda = - \sum_{j=1}^{J} n_j log \left(\frac{n_j^2}{| \Sigma_j |} \right) + J(D^2 + 3D + 2)log(I)/2. \qquad (14)$$

Equation (14) shows that a quality of clustering only depends on the weighted determinants of the covariance matrices which express the square errors between data and model. Estimating the covariance matrices Σ_j and the populations of each cluster n_j, we can draw the MDL curve Λ as a function of the cluster number J. The minimum on this curve indicates the optimal description of the data set X, i.e. the minimum error with the minimum model complexity.

The MDL criterion (14) may be applied to any clustering method: to EM, which, as said before, provides the best clustering, given a number of clusters, or to simpler algorithms - like K-means which may be seen as a simplified version of EM [10], or Kernel K-means, which is an extension of K-means. Based on this remark, we propose first to define an MDL optimization of Kernel K-means.

3 Kernel K-Means Algorithm

In the case where data have a complex structure (e.g. data are non linearly separable), a direct application of K-means is not suit because of its tendency to group data into globe-shaped clusters [10]. To solve this problem, data may be mapped by a transformation into a new feature space where samples are linearly separable [14]. The transformation is defined by a kernel $K(\cdot)$ as the inner product:

$$K(X_k, X_l) = \langle \phi(X_k), \phi(X_l) \rangle, \qquad (15)$$

where $\phi(\cdot)$ is a mapping of X to an inner product feature space [14] and k, l take values $[1, ..., I]$. The simplest kernel is a linear:

$$K(X_k, X_l) = X_k X_l, \qquad (16)$$

and one of the frequently used kernels is the Gaussian kernel:

$$K(X_k, X_l) = e^{-\dfrac{\|X_k - X_l\|^2}{2\sigma^2}}, \tag{17}$$

where σ is a kernel parameter. Kernel K-means minimizes an optimization function on the transformed data space [14]:

$$\min \sum_{j=1}^{J} \sum_{k \subseteq j} \| \phi(X_k) - \bar{\phi}(X_k) \|^2, \tag{18}$$

where $\bar{\phi}(X_k) = \frac{1}{n_j} \sum_{X_k \subseteq j} \phi(X_k)$ is the jth cluster mean. One of the advantages of using the kernel function is that we can solve (18) (e.g. for the Gaussian kernel (17) without the explicit representation of function $\phi(\cdot)$. The distance $\| \phi(X_k) - \bar{\phi}(X_k) \|^2$ may be calculated with the inner product $\langle \phi(\cdot)\phi(\cdot)\rangle$. With this objective, the standard steps of K-means algorithm are applied [14]. As can be seen Kernel K-means algorithm is equal to K-means when the linear kernel (16) is used.

4   Kernel MDL

Taking advantage of the formulation of (14), we propose to derive now a more general form for MDL.

From (14) it has been said that the simplified MDL is depending on the determinants of the $|\Sigma_j|$ matrices which describe the model to data error. This error may be determined in the original space X, as well as in the transform space after kernel transformation. Therefore, we propose to define a general MDL, similar to (14), as:

$$-\sum_{j=1}^{J} n_j log \left(\frac{n_j^2}{Dist(X_k, X_l | k, l \subseteq j)} \right) + P(J, D, I) \tag{19}$$

where $Dist(X_k, X_l | k, l \subseteq j)$ is the error function for sample X_k being represented by the jth cluster (for instance, the distance between X_k and the mean of cluster j) and $P(J, D, I)$ is a penalty function.

The simplest error function is the Euclidean distance which may be calculated using the kernel K (15). The sum-squares distances from patterns to their corresponding j^{th} cluster centroid was presented in [14] as the optimization function for Kernel K-means:

$$S_j = \frac{1}{n_j D} \sum_{k \subseteq j} \left(K(X_k, X_k) - \frac{1}{n_j} \sum_{l \subseteq j} K(X_k, X_l) \right). \tag{20}$$

In case where K is the linear kernel, S equals the variance in the original space X as expressed by (16). To obtain the complete MDL formulation of (14), supposing the variances of a cluster equal for each dimension, we may rewrite the determinant of covariance matrix Σ_j as:

$$| \Sigma_j | = S_j^D . \tag{21}$$

As the error S_j (20) may be derived for any kernel, *e.g.* Gaussian (17), we may substitute the determinant (21) in the MDL expression (14) to obtain the kernel MDL:

$$\text{KMDL} = - \sum_{j=1}^{J} n_j log \left(\frac{n_j^2}{S_j^D} \right) + J(D^2 + 3D + 2)log(I)/2 . \tag{22}$$

For the following experiments the same penalty function as in (14) have been used. The derivation of an alternative penalty is not addressed in this paper. One of the main advantages of this formulation lies in that the explicit mean of a cluster j is not needed. This point is important when this mean has no physical meaning, as it is often the case for non-convex clusters. To calculate MDL criterion for the mixture of Gaussians in the original space X the distance between samples and the nearest cluster centroid must be calculated. Problems may appear in case of data distributed on clusters with holes as in Fig. 1-d.

5   Experiments with Synthetic Data

We tested our approach on synthetic data before applying it to real data such as satellite images. The simplest and often used example of synthetic data are using Gaussian distributions where each distribution is a cluster. When working on satellite images, we expect to have a large number of clusters because of the great variety of possible scenes. Therefore we demonstrate the potential of the method with a rather large number of clusters, larger than in the usual literature [8]. We make use of 20 Gaussian distributions as presented in Fig. 1-a with 100 samples per cluster. EM algorithm run 20 times for each cluster number, with a different random initialization. Two curves are presented in Fig. 1-b, showing the results of clustering using either MDL (14) or KMDL (22) with Gaussian kernel and parameter $\sigma = 2$. For all curves of KMDL a constant is added to better visualise with MDL. As expected, both curves exhibit a well defined minimum, with an optimal number of clusters equals to 20.

The same experiments were done for another toy example having clusters with a complex structure. Points of this cluster are distributed on a circle. Here again, EM-algorithm and Kernel K-means with Gaussian kernel ($\sigma = 0.5$) have been used. Optimal results are presented Fig. 1-c and Fig. 1-d. From Fig. 1-e, it may be observed that EM with MDL detects more clusters than expected because of the difficulty to linearly separate a cluster with a complex structure (also seen in Fig. 1-c where the circle is split into 4 clusters). On the contrary Kernel K-means with the Gaussian kernel optimally separates the mixture in Fig. 1-d, and KMDL determinates the true number of clusters.

The last experiment concerns two real world data sets Iris and Thyroid taken from the UCI machine learning repository. Iris data contain 3 classes, 50 samples

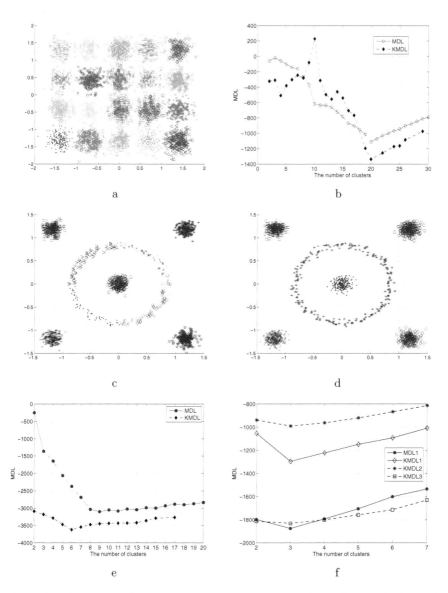

Fig. 1. Synthetic examples. In a: synthetic example 1 with 20 clusters. In b: results on clustering example 1. Detection of the optimal number of clusters by MDL (14) (solid line) and by KMDL (22) (dashed line). In c: example 2 with a circular cluster as clustered by EM. In d: the same as clustered by Kernel K-means. In e: curves drawn for example 2. In f: Optimal number of clusters for Thyroid and Iris data. MDL (14) (solid line with points) and KMDL1 (22) with $\sigma = 5$ (solid line with diamonds) propose 3 as an optimal number of clusters for Thyroid data set. KMDL2 (22) with (16) (dashed line with stars) and KMDL3 (22) with (17) $\sigma = 4$ (dashed line with squares) propose 3 as an optimal number of clusters for Iris data set.

per class and 4 features per sample. The minimum of KMDL (22) with the linear kernel (16) and the Gaussian kernel (17) determines the true number of clusters as three Fig. 1-f. Thyroid data have 3 classes: 150, 35 and 30 samples per class, respectively, and 5 features per sample. Both criteria KMDL (22) with the Gaussian kernel (17) and MDL (14) determine the true number of clusters as three Fig. 1-f.

From this set of experiments, several practical rules have been observed. At first, it seems that it is better to start from high values of cluster number to progressively reduce it in order to have a less chaotic behaviour of the curve. Then we observe that the MDL is often unequivocal, allowing to use speeding search techniques like dichotomy for instance.

6   Experiments with Real Data: Satellite Images

6.1   The Experiment

In the framework of the CNES-DLR Competence Centre we are interested in information extraction and image understanding for Earth observation with high resolution images [1]. In order to reduce the amount of information carried by an image, we propose to categorize satellite images. To avoid bias and omissions due to human expertise, we investigate unsupervised image category extraction. In this scope we consider each cluster as a category. The optimal number of clusters obtained from a given set of images is therefore an important clue which cannot be arbitrarily fixed. The previous approach (with simplified MDL (14) and KMDL (22)) will be our guideline to determine this number.

We are working with images from the SPOT 5 satellite, they are panchromatic images with a ground resolution of 5m per pixel. Each original image is very large (12000 × 12000 pixels) and quite complex; therefore we extract smaller images (1024 × 1024 pixels) with rather homogeneous content on urban areas. These (1024 × 1024) images will, from now on, be named "the images" since the original large images will no longer be used in the rest of this document. The images represent 6 cities: Copenhagen (Denmark), Istanbul (Turkey), Los Angeles (USA), La Paz (Mexico), Madrid (Spain), Paris (France). We assume that because of geography, culture and history each image has different surface textures. Sub-samples of images are presented in Fig. 2. From these images, we form a database of samples by cutting each image into 400 samples, each of size 64 × 64 pixels. Samples overlap by 13 pixels. The composed database contained 2400 samples, 6 cities and 400 samples per city. From each sample, 202 features have been extracted: statistics issued from Quadratic Mirror Filters filtering, statistics from Gabor filters, statistics from Haralick co-occurrence matrix descriptors and geometrical features. 15 features were automatically selected from the initial features using unsupervised feature extraction [9].

The data matrix of size 2400 × 15 is clustered with two algorithms: EM-algorithm [10] with GMM and Kernel K-means [14] with the Gaussian kernel (17) and parameter $\sigma = 15$. 50 random initializations were performed and the best clustering was chosen. In our experiments the data were normalised in a

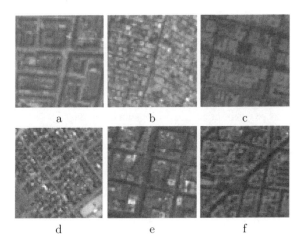

Fig. 2. Samples of SPOT5 images (64 × 64 pixels per sample) : a - Copenhagen (Denmark), b - Istanbul (Turkey), c - Los Angeles (USA), d - La Paz (Mexique), e - Madrid (Spain), f - Paris (France). ©Copyright CNES.

such a way that their mean equals 0 and the standard deviation of each column is 1, so that the weight of each feature be the same.

$$\mu_d = \frac{1}{I}\sum_{i=1}^{I} X_{id}, \tag{23}$$

$$\sigma_d = \sqrt{\frac{1}{I}\sum_{i=1}^{I}(X_{id} - \mu_d)^2}, \tag{24}$$

$$\tilde{X}_{id} = \frac{X_{id} - \mu_d}{\sigma_d} \tag{25}$$

Setting in (17) σ as the data dimension ($\sigma = D$), we obtain the curves shown in Fig. 3 for MDL and for KMDL (22). For EM-algorithm the optimal number of clusters is 9 whereas for Kernel K-means it is 11. We may present these optimal clusterings as distribution matrices (as in Tables 1 and 2 respectively), where each column corresponds to a city in the same order as in Fig. 2, and each line represents a cluster.

6.2 Discussion

In the ideal case, where all the cities would be perfectly different, we could consider that the clustering is good if each cluster consists of one city only. From the classification matrices Tables 1 and 2 we can see that the EM-algorithm and Kernel K-means give almost the same clusters. But EM-algorithm finds cluster 4 as a mixture of two cities (Los Angeles and Paris), although these cities exhibit

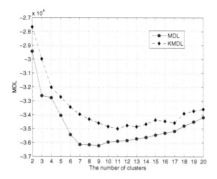

Fig. 3. Detection of the optimal number of clusters by MDL (solid line) and KMDL (dashed line) criteria for SPOT 5 image textures

Table 1. Clustering matrix for 6 cities with EM-algorithm

	Cities						
Clusters	Copenhagen	Istanbul	Los Angeles	La Paz	Madrid	Paris	$\sum$
1	2	3	2	4	155	6	172
2	117	14	0	0	0	0	131
3	86	131	1	0	5	6	229
4	6	3	253	20	24	251	557
5	131	221	0	0	0	0	352
6	0	0	5	256	7	32	300
7	28	11	7	20	32	48	146
8	30	17	132	4	177	56	416
9	0	0	0	96	0	1	97
	400	400	400	400	400	400	

Table 2. Clustering matrix for 6 cities with Kernel K-means algorithm

	Cities						
Clusters	Copenhagen	Istanbul	Los Angeles	La Paz	Madrid	Paris	$\sum$
1	0	0	0	94	0	1	95
2	28	10	6	22	31	49	146
3	0	0	19	24	9	259	311
4	67	123	1	0	4	6	201
5	112	27	0	0	1	0	140
6	0	0	4	252	5	28	289
7	20	16	72	4	172	34	318
8	13	2	296	0	35	19	365
9	2	2	2	4	142	4	156
10	114	208	0	0	1	0	323
11	44	12	0	0	0	0	56
	400	400	400	400	400	400	

Table 3. Texture examples of clusters, Kernel K-means

Clusters	Texture examples
1	
2	
3	
4	
5	
6	

rather different structures Fig. 2. The classification matrix of Kernel K-means (Table 2) shows that these two cities are separated (clusters 3 and 8). Even if we set the number of clusters to 12 for the EM-algorithm the confusion between these cities remains. This confusion disappears when the number of clusters is

Table 4. Texture examples of clusters, Kernel K-means

Clusters	Texture examples

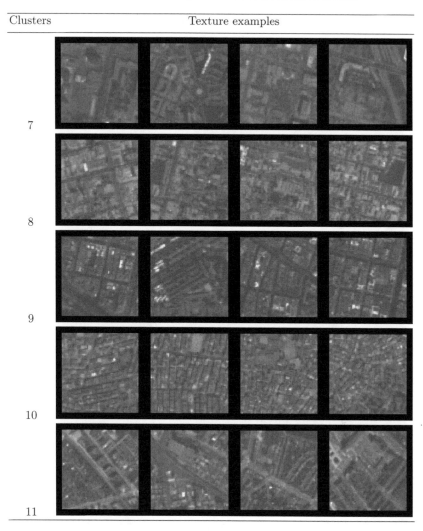

15, but it will not be an optimal clustering in terms of MDL. We consider that Kernel K-means better clusters data than EM-algorithm because clusters better correspond to cities. Some texture examples of clustered cities (4 textures per cluster) by Kernel K-means are presented in Tables 3 and 4. The samples closest from the centre of the corresponding clusters have been chosen. Each row of Table 3 has 4 texture examples for clusters from 1 to 6 and Table 4 for clusters from 7 to 11. We analyze visually this examples using classification matrix in Table 2. The first and sixth rows of Table 3 correspond to 4 textures of La Paz. These clusters show two different surfaces for this city. The second row has samples from every city and corresponds to large places which are likely to be similar almost everywhere around the world. The third column is a typical examples of Paris city blocks and we see from the classification matrix in Table 2 that cluster 3 collects nearly all samples of this city. Cluster 4 has mixed samples from Istanbul and Copenhagen with a domination of Istanbul (see cluster 4 in Table 2). These textures represent both urban and rural areas. Cluster 5 has also similar urban textures from these cities but with a domination of Copenhagen. Cluster 7 in Table 4 has mainly textures from Madrid but also from other cities. Los Angeles is represented by cluster 8 with its typical square streets. Half textures of Madrid are represented by cluster 9. Dense areas of Istanbul correspond to cluster 10. Cluster 11 has textures which contain wide roads. From this early interpretation of classification results, we are quite satisfied by the way the textures have been grouped and the homogeneity of the obtained classes. Results of clusterings in Tables 1 and 2 show that several clusters have redundant information. It means that for different clusterings there are clusters which have the same samples. It will be useful for data mining to combine samples that always belong to common clusters that may reduce redundant information and find some interesting particular clusters in data [7].

7   Conclusions

In this paper we proposed a new criterion called Kernel MDL (KMDL) to estimate the optimal number of clusters for the Kernel K-means algorithm. This criterion is derived from a simplified formulation of the classical MDL for the Gaussian Mixture Model. Both KMDL and the simplified MDL allow to determine the optimal number of clusters using simply the error function between the data and the model of clusters. To adapt the criterion to the Kernel K-means algorithm we defined this error function as the corresponding optimized criterion.

The error can be calculated on the kernel function with the Kernel K-means algorithm. The advantage of this approach is that Kernel K-means can linearly separate data which are non linearly separable in the original space. As we can see from experimental results the two criteria MDL and KMDL work well and give optimal numbers of clusters each for its own algorithm. Kernel K-means algorithm with KMDL shows superior results than EM with MDL for synthetic data as well as real data.

Acknowledgements. This study[1] was done with the financial support of Centre National d'Etudes Spatiales (CNES-France). The authors[1] would like to thank M. Datcu and O. Cappé for fruitful discussions.

References

1. http://www.coc.enst.fr/
2. Barron, A., Rissanen, J., Yu, B.: The minimum description length principle in coding and modeling. IEEE Trans. Inform. Theory 44(6), 2743–2760 (1998)

3. Chapelle, O., Vapnik, V., Bousquet, O., Mukherjee, S.: Choosing multiple parameters for support vector machines. Machine Learning 46, 131–159 (2002)
4. Figueiredo, A.K., Jain, M.A.F.: Unsupervised learning of finite mixture models. IEEE Transactions on Pattern Analysis and Machine Intelligence 24(3), 381–396 (2002)
5. Govaert, G.: Analyse des données. Lavoisier (2003)
6. Heas, P., Datcu, M.: Modelling trajectory of dynamic clusters in image time-series for spatio-temporal reasoning. IEEE Transactions on Geoscience and Remote Sensing 43(7), 1635–1647 (2005)
7. Maître, H., Kyrgyzov, I., Campedel, M.: Combining clustering results for the analysis of textures of spot5 images. In: ESA-EUSC: Image Information Mining (2005)
8. Jain, A., Dubes, R.C.: Algorithms for Clustering Data. Prentice-Hall, Englewood Cliffs, NJ (1988)
9. Maître, H., Campedel, M., Moulines, E., Datcu, M.: Feature selection for satellite image indexing. In: ESA-EUSC: Image Information Mining (2005)
10. MacKay, D.J.C.: Information Theory, Inference, and Learning Algorithms. Cambridge University Press, Cambridge (2003)
11. Rissanen, J.: Modeling by shortest data description. Automatica 14, 465–471 (1978)
12. Rissanen, J.: Universal coding, information, prediction, and estimation. IEEE Trans. Inform. Theory 30(4), 629–636 (1984)
13. Scholkopf, B., Smola, A.J.: Learning with Kernels: Support Vector Machines, Regularization, Optimization, and Beyond. MIT Press, Cambridge, MA, USA (2001)
14. Shawe-Taylor, J., Cristianini, N.: Kernel Methods for Pattern Analysis. Cambridge University Press, Cambridge (2004)

Critical Scale for Unsupervised Cluster Discovery

Tomoya Sakai[1], Atsushi Imiya[1], Takuto Komazaki[2], and Shiomu Hama[2]

[1] Institute of Media and Information Technology, Chiba University, Japan
{tsakai,imiya}@faculty.chiba-u.jp
[2] Graduate School of Science and Technology, Chiba University, Japan

Abstract. This paper addresses the scale-space clustering and a validation scheme. The scale-space clustering is an unsupervised method for grouping spatial data points based on the estimation of probability density function (PDF) using a Gaussian kernel with a variable scale parameter. It has been suggested that the detected cluster, represented as a mode of the PDF, can be validated by observing the lifetime of the mode in scale space. Statistical properties of the lifetime, however, are unclear. In this paper, we propose a concept of the 'critical scale' and explore perspectives on handling it for the cluster validation.

1 Introduction

Cluster discovery is an essential approach to data mining. Most of the clustering methods are based on either or both of a distance measure and estimation of the probability density function (PDF) for a dataset. While the geometric distance measure enables us to quantify internal cohesion and external isolation of clusters in the dataset, such cluster characteristics are statistically governed by the PDF.

The PDF-based methods treat the dataset as a set of instances of random points distributed in a feature space. In the case that a model of the PDF is not presumable, the PDF is estimated by a nonparametric approach [1]. The clusters are generated according to the structure of the PDF. Therefore, clustering is essentially a structural analysis of the estimated PDF. The details of the PDF structure, however, are controlled by the cardinality of the dataset. In other words, a finite number of data points provide the geometric structure of the PDF with some resolution or scale.

An unsupervised clustering method on the basis of the PDF estimation using a Gaussian kernel with a variable kernel width, i.e. scale, is known as the scale-space clustering [2,3,4,6]. The scale-space clustering can be interpreted as an extraction of a hierarchical structure of the PDF on the basis of hierarchical relationships among the data points in a scale space. We focus on an important property that the modes of the estimated PDF in scale space are deterministic above a certain *critical scale*, even though the positions of the data points are stochastic. By selecting the scale for clustering above such critical scale, we can obtain valid clusters without prior knowledge of the number of clusters or their locations.

P. Perner (Ed.): MLDM 2007, LNAI 4571, pp. 218–232, 2007.

In this paper, we first review the Gaussian scale-space theory and introduce the scale-space clustering. A hierarchical clustering is achieved by constructing a mode tree. Second, we define the critical scale and describe its concept. The cluster validity is examined from the viewpoint of the PDF structure. We experimentally show that the *lifetime* of the invalid cluster exhibits unimodal distributions, which are indispensable for statistical validations. Finally, we demonstrate the scale-space clustering and recursive validation using the mode tree.

2   Scale-Space Analysis

2.1   Gaussian Scale Space

The scale space is classically explained as Gaussian blurring [13,14,16]. The Gaussian convolution derives a one-parameter family of functions from a given positive function $f(\mathbf{x})$.

$$f(\mathbf{x}, \tau) = G(\mathbf{x}, \sqrt{2\tau}) * f(\mathbf{x}) \tag{1}$$

Here, $G(\mathbf{x}, \sqrt{2\tau})$ is the isotropic Gaussian function. In this paper, we call $f(\mathbf{x}, \tau)$ the generalised function in scale space $(\mathbf{x}, \tau)$. The parameter τ is called the "scale", which can be regarded as an inversion of the resolution if $f(\mathbf{x})$ is an image. The scale plays a role of the kernel bandwidth if $f(\mathbf{x}, \tau)$ represents a kernel density estimate, which we will treat in later sections.

The isotropic Gaussian convolution satisfies the following axioms [16,17].

- Non-negative intensity.
- Linearity.
- Closedness under affine transformations.
- Associative (or semigroup) property.

Equivalently, the scale space can also be defined as a space in which a spatial function is governed by a diffusion equation with respect to the scale τ.

$$\frac{\partial f}{\partial \tau} = \Delta f. \tag{2}$$

Any function described by the Gaussian convolution satisfies the linear diffusion equation.

2.2   Hierarchical Structure

A remarkable geometric feature of the generalised function is a set of critical points, where the spatial gradient of the function vanishes.

$$P(\tau) = \{\mathbf{x}(\tau) | \nabla f(\mathbf{x}, \tau) = \mathbf{0}\} \tag{3}$$

Local extrema are representative of regions with high and low values of $f(\mathbf{x}, \tau)$. Saddle points reside between the local extrema.

The trajectories of the critical points observed in the scale space are called the critical curves. The critical curves (a.k.a stationary curves) are solutions to the equation

$$\mathbf{H}\frac{d\mathbf{x}(\tau)}{d\tau} = -\nabla\Delta f(\mathbf{x}(\tau), \tau),\qquad(4)$$

where $\mathbf{H} = \nabla\nabla^\top f$ is the Hessian matrix of the generalised function [15,18].

The hierarchical structure of $f(\mathbf{x}, \tau)$ in scale space has been investigated by various authors [15,19,20,21,22,23,24]. Especially, the critical curves in scale space indicate the topological relationships among the critical points. The critical curves start at the critical points $P(0)$, and end at scales where the critical points are annihilated by blurring. Generically, a local extremum and a saddle meet at a singular point where $\det\mathbf{H} = 0$. Equation (4) indicates that the spatial velocity of the critical point with respect to scale becomes infinite at the annihilation point.

We have investigated the hierarchical structure on the basis of links of singular points, analysing the spatial gradient field of $f(\mathbf{x}, \tau)$ in the Gaussian scale space [23,24]. Each singular point generically has a considerable gradient field curve, which we call the antidirectional figure-flow curve. The antidirectional figure-flow curve defines the link of the singular point to another local extremum. Consequently, the hierarchical relationships among the critical points can be determined by the critical curves across scales and the antidirectional figure-flow curves at fixed scales.

3   Clustering Based on PDF Estimation in Scale Space

3.1   PDF Estimation with Gaussian Kernel

For a dataset in a feature space,

$$P = \{\mathbf{x}_n | \mathbf{x}_n \sim p(\mathbf{x}), n \le N, n \in \mathbb{N}, \mathbf{x} \subset \mathbb{R}^d\},$$

a kernel density estimate of the probability density function (PDF) [1] is defined as

$$\tilde{p}(\mathbf{x}; \sigma) = \frac{1}{N}\sum_{n=1}^{N} K(\mathbf{x} - \mathbf{x}_n; \sigma).\qquad(5)$$

Here, K is the kernel function and σ is its bandwidth. This PDF estimation is known as a nonparametric method, and $\tilde{p}(\mathbf{x}; \sigma)$ with a suitable kernel function converges to the true PDF $p(\mathbf{x})$ if $\sigma \to 0$ when the cardinality N of the dataset approaches to infinity.

The Gaussian function with the scale parameter σ is a popular kernel for the PDF estimation. We justify the use of an isotropic Gaussian kernel,

$$K(\mathbf{x}; \sigma) = G(\mathbf{x}, \sigma) = \frac{1}{\sqrt{2\pi\sigma^2}^d}e^{-\frac{|\mathbf{x}|^2}{2\sigma^2}},\qquad(6)$$

for the following reasons.

- $G(\mathbf{x}, \sigma)$ is radial and unimodal. The radial function can express dominance of a point at its center.
- Let $\sigma = \sqrt{2\tau}$ and define

$$f(\mathbf{x}, \tau) = \tilde{p}(\mathbf{x}; \sqrt{2\tau}) = \frac{1}{N\sqrt{4\pi\tau}^d} \sum_{n=1}^{N} e^{-\frac{|\mathbf{x}-\mathbf{x}_n|^2}{4\tau}}. \tag{7}$$

Then, $f(\mathbf{x}, \tau)$ satisfies the linear diffusion equation (2). This property implies that the data points disperse by Brownian motion. Initial positions at $\tau = 0$ are given by P, and a superposition of the Gaussian functions represents uncertainty of the location of the points after the time τ.
- $\tilde{p}(\mathbf{x}; \sqrt{2\tau})$ satisfies the scale-space axioms. The parameter τ can be regarded as the scale. Scale-space analyses are available for the estimated PDF.
- The number of modes of the homoscedastic Gaussian mixture seldom increases as the scale σ increases [7]. That is, mode creation is less expected if the Gaussian functions are unequally weighted. It is known that non-isotropic Gaussian mixtures with different covariances yield spurious modes outside the convex hull of P.

3.2 Scale Selection Problem

In practice, we need to estimate a finitesimal value of σ for P with a finite cardinality. If σ is too small relative to data intervals, each data point $\mathbf{x}_n$ approximately maximises $\tilde{p}(\mathbf{x}; \sigma)$ in its neighbourhood. $\tilde{p}(\mathbf{x}; \sigma)$ with such a small σ is not a feasible PDF estimate of $p(\mathbf{x})$, since geometric features of $\tilde{p}(\mathbf{x}; \sigma)$, such as configurations of the peaks of $\tilde{p}(\mathbf{x}; \sigma)$ and their topologies, are stochastic depending on randomness of P. In contrast, the geometric features of $\tilde{p}(\mathbf{x}; \sigma)$ with a large σ are expected to be deterministic although the PDF $\tilde{p}(\mathbf{x}; \sigma)$ does not provide the detail of distribution of the data points.

An essential approach to the scale selection problem in the kernel density estimation is to analyse the structure of PDF $\tilde{p}(\mathbf{x}; \sigma)$ at all bandwidths σ, simultaneously. Observation of $\tilde{p}(\mathbf{x}; \sigma)$ in $(\mathbf{x}, \sigma)$-space clarifies a hidden structure across scales. The Gaussian scale-space analysis discussed in section 2 is of great help if we equate the kernel density estimate with the generalised function in scale space.

3.3 Scale-Space Clustering

Several authors has elaborated the clustering of data points based on the PDF estimation in scale space [2,3,4]. Their brief concept is to trace the modes of $\tilde{p}(\mathbf{x}; \sigma)$ from $\sigma = 0$ in scale space. The modes are the local maxima of the PDF, which represent the regions where the data points are dense. That is, the data clusters are detected as the modes of the estimated PDF.

Hierachical Clustering. The number of traceable modes of $\tilde{p}(\mathbf{x}; \sigma)$ from the positions of data points at $\sigma = 0$ decreases with increasing bandwidth σ. The data points are hierarchically clustered according to the hierarchical relationships among modes, which are described as the local maximum curves in scale space. The mode hierarchy is symbolically described as a tree. The algorithm of the tree construction is as follows.

ALGORITHM I – Mode Tree

1. Set $\operatorname{card}(P)$ nodes with labels k $(k = 1, \ldots, \operatorname{card}(P))$ to be leaves of a tree T.
2. Let $\hat{P} = P$, and $\sigma = \sqrt{2\tau} = 0$.
3. Increase the scale τ by $\Delta\tau$, which is a small value so that $\sqrt{2\Delta\tau}$ is negligible compared to the space intervals of the points in $\hat{P}$.
4. For each point $\hat{\mathbf{p}}_i \in \hat{P}$, update $\hat{\mathbf{p}}_i$ by maximising $\tilde{p}(\mathbf{x}; \sigma)$ with $\hat{\mathbf{p}}_i$ as the initial position.
5. For each point $\hat{\mathbf{p}}_i \in \hat{P}$, if $\exists \hat{\mathbf{p}}_j \in \hat{P}$, $j \neq i$, $|\hat{\mathbf{p}}_i - \hat{\mathbf{p}}_j| < \varepsilon\sigma$, where ε is the tolerance of maximisation, then remove $\hat{\mathbf{p}}_i$ from $\hat{P}$, and add a new node with two branches attached to the nodes labelled i and j in T. The new node inherits the label j, and contains the values $\hat{\mathbf{p}}_i$ and $\tau_i = \tau$.
6. If $\operatorname{card}(\hat{P}) = 1$ then stop; otherwise go to Step 3.

The iteration to update the dataset $\hat{P}$ is equivalent to performing the so-called mean-shift clustering [9,10,11], and the mean-shift algorithms are available for the maximisation in Step 4 [8]. Depending on the step size $\Delta\tau$, Algorithm I functions as the nested and nonnested versions of hierarchical clustering [6].

Cluster Validation. The scale-space clustering finds the data clusters at any scale. The detected clusters have been validated by several properties of the clusters: the number of clusters vs. scale, compactness, isolation, lifetime and birthtime [2,3,4,6]. As suggested in [3,4,5], the decrease in the total number of clusters pauses over a scale interval where valid clusters survive. However, such interval is detectable only if the valid clusters are simultaneously stable and each cluster keeps a constant cardinality. This observation fails to detect valid clusters each of which is stable at different scales.

The scale at which the cluster number varies indicates the death of a cluster. The cluster lifetime [3,4,6], which refers to a range of the scale interval where the cluster survives, is more essential than the cluster number. A cluster with long lifetime is considered to be valid. Although the cluster validity is quantified by the lifetime, it is still unclear how to distinguish valid clusters from invalid ones. We need adequate criteria to identify the valid clusters. It is preferable to derive such criteria from the nature of scale space, rather than strategical, empirical or deliberate measures.

4   Critical Scale

4.1   Concept

The dataset P is an instance of a set of points which stochastically located in a space according to the true PDF $p(\mathbf{x})$. An important fact is that the cardinality of the dataset plays a role of resolution of the PDF. The detail of geometric features of the estimated PDF is dependent on the cardinality of the dataset. If the structure of the true PDF is so complicated that the dataset P cannot express the PDF in detail, the structure of the estimated PDF are stochastic at small scales.

It should be emphasised that the geometric features of PDF is provided from coarse to fine by data samples, and a finite number of samples represent the PDF, incompletely. The estimated PDF $\tilde{p}(\mathbf{x}; \sigma)$ reveals the structure of true PDF $p(\mathbf{x})$ from top to bottom with increasing cardinality. For the PDF described by the dataset with a finite cardinality, there exists a critical lower bound of scale, above which the structure is deterministic, and under which the structure is stochastic. Clustering should be employed above such a critical scale.

Definition 1. *The critical scale is a threshold of the spatial measure above which the dataset or its subset is informative and under which results of pattern analyses lose statistical significance.*

4.2   Non-structured PDF

The dataset P potentially contains valid clusters iff the true PDF $p(\mathbf{x})$ has meaningful features such as modes. Contrapositively, if $p(\mathbf{x})$ is featureless, or the uniform distribution, then no valid cluster exists in P. The dataset generated from such non-structured PDF should be classified into a single cluster with all data points. Any small cluster detected in the uniformly distributed data points is said to be invalid.

The arrangement of data points which achieves the uniform distribution is not unique. We present three examples of uniformly distributed random points in Fig. 1: the perfectly random arrangement (a.k.a the Poisson random arrangement), the quasi-random arrangement, and the regular triangular grid points. Each dataset consists of a thousand points. For the dataset in the perfectly random arrangement, the number of points within an area S obeys the Poisson distribution with mean $\lambda = \rho S$, where ρ is the point density. The dataset in the quasi-random arrangement is generated using the Sobol's sequence [25]. The quasi-random arrangement has a property of filling vacant spaces among the previously generated points, uniformly. The grid points in Fig. 1(c) is also considered to be the uniformly distributed random points if the direction θ_G and the origin O_G of the grid are uniformly random so that $0 \leq \theta_G < 2\pi$ and $O_G \in \mathbb{R}^2$. Note that none of the three datasets has valid clusters, statistically.

4.3   Lifetime Histogram

In this paper, we define the lifetime as follows.

Definition 2. *The lifetime of a data point $\mathbf{p}_i \in P$ is defined as $\sqrt{2\tau_i}$, where τ_i is the terminating scale of the critical curve of local maximum whose starting point is $(\mathbf{p}_i, 0)$ in the scale space.*

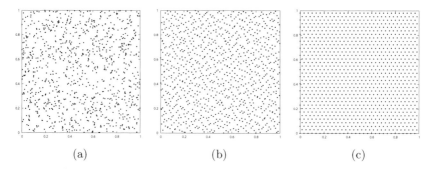

Fig. 1. Examples of uniformly distributed random points. (a) Perfectly random arrangement, (b) quasi-random arrangement, and (c) regular triangular grid points. The grid points in (c) can be regarded as random points if the direction and the origin of the grid are random.

We consider the distribution of the lifetimes. It has been suggested that the number of modes of $\tilde{p}(\mathbf{x}; \sigma)$ exponentially decays with increasing scale if the dataset does not contain clusters [3,6]. The scales at which the number of clusters varies indicate the lifetimes of modes. According to those antecedent works, the lifetime distribution is also expected to be in an exponential shape function.

The exponential decay, however, is not guaranteed depending on the arrangement of points. Figure 2 is the lifetime histograms for the perfectly random points and the quasi-random points. We have averaged the frequencies of the lifetimes over a hundred of datasets, each of which consists of a thousand data points distributed in a unit square $[0, 1] \times [0, 1]^1$. As shown in Fig. 2(a), the lifetime histogram for the perfectly random dataset shows an unimodal shape, which is similar to the Weibull distribution. The lifetime distribution of the quasi-random dataset, shown in Fig. 2(b), has a sharp peak at $\sqrt{2\tau} \approx 0.01$, which is dependent only on the point density. The assumption of the exponential decay in the antecedent works may be based on the skirt of these distributions.

4.4   Detection of Valid Clusters

Detection of the valid clusters is to distinguish the structured PDF of data points from the non-structured uniform distribution. The lifetime histogram is available to detect the meaningful structure. Multimodality and statistic outliers in the lifetime data indicate the non-uniformity of the PDF. The outlying lifetime values approximate the sizes of the valid clusters. The peak and decay of the lifetime distribution are found only in small scale relative to the outlying lifetimes. In other words, the critical scale lies after the decay of the lifetime distribution.

In practice, the critical scale can be roughly estimated using the lifetime histogram. If there exist valid clusters, one can find outstanding lifetimes above

---

[1] For each dataset, we count the lifetimes of the points within $[1/3, 2/3] \times [1/3, 2/3]$ to suppress the boundary effect, although it is negligible at small scales.

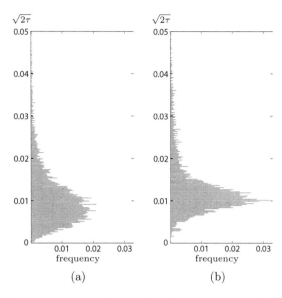

Fig. 2. Lifetime histograms for uniformly distributed points in (a) perfectly random arrangement, and (b) quasi-random arrangement

the critical scale. Statistical outlier detection is also possible if the underlying lifetime distribution of the invalid clusters is assumable. The lifetimes of valid clusters found in a critical region (a.k.a rejection region) are rejected by the statistical hypothesis testing. The critical scale acts as critical value in statistics.

The number of clusters is determined by selecting the scale τ so that $\sqrt{2\tau}$ is greater than the critical scale. All data points are classified into a universal cluster at the coarsest scale, which is represented by the root node of the mode tree. If the scale is sufficiently large, the position of one remaining local maximum converges to the barycentre of the cluster. As the scale decreases, new modes appear one after another. The appearance of the mode indicates that a cluster splits into subclusters, which are represented by the nodes of subtrees of the mode tree.

4.5 Recursive Validation

Since the hierarchical relationships among the clusters are explicitly described as the mode tree, we can recursively validate whether a cluster can be split into deterministic subclusters. Construct a histogram of the lifetime stored in a subtree corresponding to the cluster. If a critical scale is found in that histogram, then there exist valid subclusters with outstanding lifetime values above the critical scale.

5   Experimental Examples

5.1   Artificial Dataset: Cluster Discovery and Validity

We demonstrate the clustering for artificial datasets with different cardinalities shown in Fig. 4. These datasets are generated from a PDF in Fig. 3(a). The

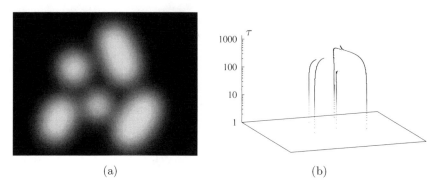

(a)   (b)

Fig. 3. PDF and its critical curves in scale space. (a) The PDF to be estimated. The brightness indicates the probability density of the data points. (b) Critical curves corresponding to local maxima of the five blobs.

PDF consists of five elliptic blobs, so the expected number of clusters is five. The critical curves of local maxima corresponding to the five blobs are found in the scale space as shown in Fig. 3(b).

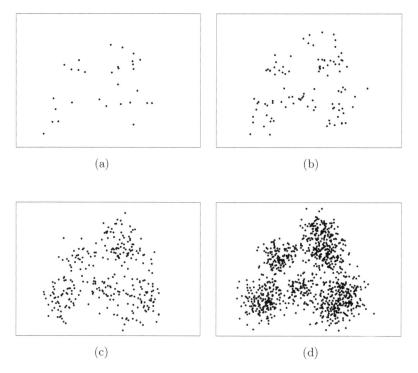

(a)   (b)

(c)   (d)

Fig. 4. Artifical dataset. (a) P_{30}, (b) P_{100}, (c) P_{300}, and (d) P_{1000}. card(P_{30}) = 30, card(P_{100}) = 100, card(P_{300}) = 300, and card(P_{1000}) = 1000.

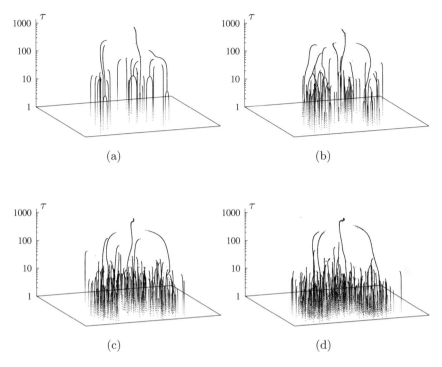

Fig. 5. Critical curves of local maxima for (a) $f(P_{30}; \mathbf{x}, \tau)$, (b) $f(P_{100}; \mathbf{x}, \tau)$, (c) $f(P_{300}; \mathbf{x}, \tau)$, and (d) $f(P_{1000}; \mathbf{x}, \tau)$.

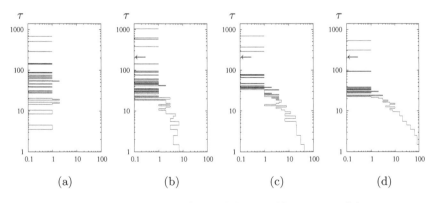

Fig. 6. Histograms of lifetimes for (a) P_{30}, (b) P_{100}, (c) P_{300}, and (d) P_{1000}. Roughly estimated critical scale is indicated by the arrow. Note that both axes have a logarithmic scale.

For each dataset, the critical curves of local maxima in scale space and the lifetime histogram are shown in Fig. 5 and 6, respectively. For the dataset P_{30}, five critical curves of local maxima seem to represent the true five clusters. Their

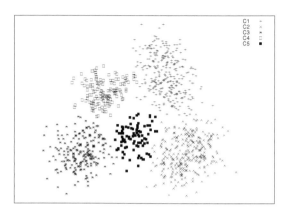

Fig. 7. Clustering result for P_{1000}

lifetimes, however, are not outstanding in the histogram in Fig. 6(a). Besides, the hierarchy indicated by these five critical curves is different from that of the five blobs of the true PDF in Fig. 3. Therefore, the dataset P_{30} is too poor to estimate the dominance of each cluster, correctly.

Each of datasets P_{100}, P_{300} and P_{1000} has a histogram with five outstanding lifetimes as shown in Fig. 6(b), 6(c) and 6(d). They are well-detached and so distinguishable from the others related to invalid small clusters. The increase in the cardinality does not affect the number of valid clusters but clarify the decay under the critical scale.

Nevertheless, any hierarchy of the valid clusters estimated by P_{100} and P_{300} disagree with that of the five blobs in the true PDF. We confirm that P_{1000} successfully estimates both of the number of clusters and the hierarchy. The clustering result for P_{1000} is shown in Fig. 7. A larger number of data points would be required for the estimation if the structure of the true PDF was more complicated.

5.2   Data Clustering with Recursive Validation

We apply the scale-space clustering to the breast-cancer-wisconsin dataset [26]. This dataset consists of 683 instances each of which has 9 attributes of breast cancer. The attribute values are integers ranging from 1 to 10. We treat all the instances as the spatial data points in 9-dimensional space.

Figure 8 shows a few levels of mode tree and lifetime histograms for the detected clusters. The lifetime histogram for all data points, which is plotted with broken line in Fig. 8(b), exhibits a bimodal distribution. This indicates the existence of valid clusters. Accordingly, the universal cluster splits into two major clusters C_B and C_M at $\tau = 4.86$ in the mode tree.

As shown in Fig. 8(b), the cluster C_B is mainly composed of data points with small lifetime values, and no outstanding lifetimes are found in its histogram. On

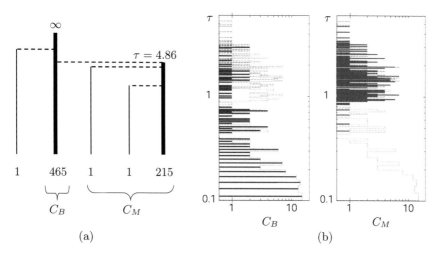

Fig. 8. Clustering of Wisconsin breast cancer dataset. (a) Mode tree. Most of data samples are classified into C_B and C_M. (b) Histograms of lifetimes for the samples in C_B and C_M. The broken line in both histograms indicates the lifetime histogram for all samples.

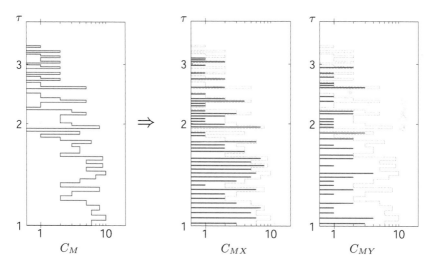

Fig. 9. Lifetime histograms for cluster C_M and its invalid subclusters C_{MX} and C_{MY}

the other hand, the histogram for C_M consists of large lifetime values compared to C_B. Therefore, C_M is larger and more sparse than C_B.

All instances in the breast-cancer dataset are labelled as either benign or malignant. We have confirmed that 94.0 belong to the benign and malignant classes, respectively. Figure 9 replots the lifetime histogram for C_M in a suitable range to validate its subclusters. No significant lifetime is found in the histogram. Although the cluster C_M splits into subclusters C_{MX}, C_{MY} and five outliers, there is no evidence to justify the validity of these subclusters. We note that the invalid cluster C_{MY} has only malignant instances. The subclusters could turn out to be valid if the dataset has a larger cardinality.

6   Concluding Remarks

Unsupervised clustering achieves data mining. We develop a clustering method based on the PDF estimation in a scale space. The modes of the estimated PDF is its geometric feature points, which correspond to clusters in the spatial dataset. The nonparametric PDF estimation using the Gaussian kernel satisfies the scale-space axioms. Reducing the spatial resolution for the dataset should not be merging the data points based on user-defined distance measures, but describing the PDF at coarser scales. In the same manner of the scale-space analysis, the PDF structure across scales can be clarified in the scale space. Since the Gaussian filtering gradually averages out the randomness of the dataset, the geometric features of the PDF are established from coarse to fine. This scale-space filtering reveals the hierarchical relationships among the modes, which provides us with the top-down approach to identifying valid clusters of data points.

The scale-space analysis of the PDF clarifies how the statistically deterministic features of the dataset appear in higher scales even though the positions of the data points are stochastic. In this paper, we proposed a concept of the critical scale, which discriminates between deterministic and stochastic features of the spatial dataset. The data clustering should be employed above the critical scale.

We showed that the uniformly distributed data points having no cluster yield a unimodal lifetime distribution. It has been believed by many authors that the lifetime distribution is exponential. In order to discover the valid clusters, it is essential to analyse the lifetime distribution and prove the statistical significance of individual clusters. Although we have not presented statistical algorithms, we should remark the importance of distinguishing the structured data from unstructured one in scale space.

The scale-space clustering has potential to discover and validate unknown clusters in the dataset without any prior information. We demonstrated the clustering for artifical datasets and a practical medical dataset. The experimental examples clarified our clustering concepts and produced results in substantial agreement.

References

1. Parzen, E.: On the estimation of a probability density function and mode. Annals of Mathematical Statistics 33, 1065–1076 (1962)
2. Chakravarthy, S.V., Ghosh, J.: Scale-Based Clustering Using the Radial Basis Function Network. IEEE Trans. on Neural Networks 7(5), 1250–1261 (1996)

3. Roberts, S.J.: Parametric and non-parametric unsupervised cluster analysis. Pattern Recognition 30(2), 261–272 (1997)
4. Nakamura, E., Kehtarnavaz, N.: Determining number of clusters and prototype locations via multi-scale clustering. Pattern Recognition Letters 19(14), 1265–1283 (1998)
5. Hinneburg, A., Keim, D.A.: An Efficient Approach to Clustering in Large Multimedia Databases with Noise. In: Proc. 4th International Conference on Knowledge Discovery and Data Mining, pp. 58–65 (1998)
6. Leung, Y., Zhang, J.-S., Xu, Z.-B.: Clustering by scale-space filtering. IEEE Trans. on Pattern Analysis and Machine Intelligence 22(12), 1396–1410 (2000)
7. Carreira-Perpiñán, M.Á., Williams, C.K.I.: On the number of modes of a Gaussian mixture. In: Griffin, L.D, Lillholm, M. (eds.) Scale Space Methods in Computer Vision. LNCS, vol. 2695, pp. 625–640. Springer, Heidelberg (2003)
8. Carreira-Perpiñán, M.Á.: Fast nonparametric clustering with Gaussian blurring mean-shift. In: ACM International Conference Proceeding Series, ICML2006, vol. 148, pp. 153–160 (2006)
9. Fukunaga, K., Hostetler, L.D.: The estimation of the gradient of a density function, with applications in pattern recognition. IEEE Trans. on Information Theory 21(1), 32–40 (1975)
10. Cheng, Y.: Mean shift, mode seeking, and clustering. IEEE Trans. on Pattern Analysis and Machine intelligence 17(8), 790–799 (1995)
11. Comaniciu, D., Meer, P.: Mean shift: A robust approach toward feature space. IEEE Trans. on Pattern Analysis and Machine Intelligence 24(5), 603–619 (2002)
12. Griffin, L.D., Lillholm, M.: Mode estimation using pessimistic scale space tracking. In: Griffin, L.D, Lillholm, M. (eds.) Scale Space Methods in Computer Vision. LNCS, vol. 2695, pp. 266–280. Springer, Heidelberg (2003)
13. Witkin, A.P.: Scale space filtering. In: Proc. of 8th IJCAI, pp. 1019–1022 (1983)
14. Koenderink, J.J.: The structure of images. Biological Cybernetics 50, 363–370 (1984)
15. Zhao, N.-Y., Iijima, T.: Theory on the method of determination of view-point and field of vision during observation and measurement of figure. IEICE Japan, Trans. D (in Japanese) J68-D, 508–514 (1985)
16. Weickert, J., Ishikawa, S., Imiya, A.: Linear scale-space has first been proposed in Japan. Journal of Mathematical Imaging and Vision 10, 237–252 (1999)
17. Sporring, J., Nielsen, M., Florack, L.M.J., Johansen, P. (eds.): Gaussian Scale-Space Theory. Computational Imaging and Vision Series. Kluwer, Dordrecht (1997)
18. Lindeberg, T.: Scale-Space Theory in Computer Vision. Kluwer, Boston (1994)
19. Johansen, P.: On the classification of toppoints in scale space. Journal of Mathematical Imaging and Vision 4(1), 57–67 (1994)
20. Griffin, L.D., Colchester, A.: Superficial and deep structure in linear diffusion scale space: Isophotes, critical points and separatrices. Image and Vision Computing 13(7), 543–557 (1995)
21. Florack, L.M.J., Kuijper, A.: The topological structure of scale-space images. Journal of Mathematical Imaging and Vision 12(1), 65–79 (2000)
22. Kuijper, A., Florack, L.M.J., Viergever, M.A.: Scale space hierarchy. Journal of Mathematical Imaging and Vision 18(2), 169–189 (2003)
23. Sakai, T., Imiya, A.: Figure field analysis of linear scale-space image. In: Kimmel, R., Sochen, N.A., Weickert, J. (eds.) Scale-Space 2005. LNCS, vol. 3459, pp. 374–385. Springer, Heidelberg (2005)

24. Sakai, T., Imiya, A.: Scale-space hierarchy of singularities. In: Olsen, O.F., Florack, L.M.J., Kuijper, A. (eds.) DSSCV 2005. LNCS, vol. 3753, pp. 181–192. Springer, Heidelberg (2005)
25. Bratley, P., Fox, B.: Algorithm 659: Implementing Sobol's Quasirandom Sequence Generator. ACM Trans. on Mathematical Software 14(1), 88–100 (1988)
26. Mangasarian, O.L., Setiono, R., Wolberg, W.H.: Pattern recognition via linear programming: Theory and application to medical diagnosis. In: Proc. of the Workshop on Large-Scale Numerical Optimization, pp. 22–31 (1989)

Minimum Information Loss Cluster Analysis for Categorical Data*

Jiří Grim[1] and Jan Hora[2]

[1] Institute of Information Theory and Automation
of the Czech Academy of Sciences,
P.O. BOX 18, 18208 Prague 8, Czech Republic
grim@utia.cas.cz
http://ro.utia.cas.cz/mem.html

[2] Faculty of Nuclear Science and Physical Engineering
Czech Technical University,
Trojanova 13, CZ-120 00 Prague 2, Czech Republic
hora@utia.cas.cz

Abstract. The EM algorithm has been used repeatedly to identify latent classes in categorical data by estimating finite distribution mixtures of product components. Unfortunately, the underlying mixtures are not uniquely identifiable and, moreover, the estimated mixture parameters are starting-point dependent. For this reason we use the latent class model only to define a set of "elementary" classes by estimating a mixture of a large number components. We propose a hierarchical "bottom up" cluster analysis based on unifying the elementary latent classes sequentially. The clustering procedure is controlled by minimum information loss criterion.

1 Introduction

concept of cluster analysis is closely related to the similarity of objects or distance of data vectors defined by a metric. The cluster analysis of categorical (nominal, qualitative) data is difficult because the standard arithmetical operations are undefined and also there is no generally acceptable definition of distance for multivariate categorical data. For these reasons the available methods of cluster analysis cannot be applied directly to categorical data.

At present the standard approach to cluster analysis of categorical data is to introduce some similarity measure or distance function in a heuristical manner. It appears that the only statistically justified method to analyze multivariate categorical data is the latent class model of Lazarsfeld [10]. Motivated by sociological research he proposed the fitting of multivariate Bernoulli mixtures

---

* This research was supported by the grant GACR 102/07/1594 of the Czech Grant Agency and by the projects of the Grant Agency of MŠMT 2C06019 ZIMOLEZ and 1M0572 DAR.

to binary data with the aim to identify possible latent classes of respondents. Serious drawback of the Lazarsfeld's idea has been the tedious and somewhat arbitrary methods used for fitting the models. The numerical problems have been removed by the computationally efficient EM algorithm [3]. In the last years the original idea of Lazarsfeld has been widely applied and frequently modified by different authors (cf. e.g. [8] and [14] for extensive references).

A serious problem of the latent class model is the validity of the resulting clusters. There are at least three different sources of uncertainty which may influence the result of latent class cluster analysis. First, in application the number of classes is usually unknown and there are no exact means to make a qualified choice. Further, the EM algorithm may converge to a local maximum of the maximized log-likelihood criterion and therefore the estimated parameters may be starting-point dependent. Finally, the discrete mixtures of product components are known to be non-identifiable and therefore the resulting latent class model is not defined uniquely. It appears that the problem of a proper number of components can be managed by additional computational constraints [7] but the missing identifiability is a serious theoretical drawback (cf. [8]).

In this paper we propose a hierarchical approach to cluster analysis of categorical data in the context of data mining. Applying the latent class model to large multivariate databases we assume a large number of classes ($M \approx 10^1 \div 10^2$) with the aim to approximate the unknown probability distribution. The EM algorithm yields different parameter estimates but the approximation accuracy of the estimated mixture is comparable. The initial parameters of the estimated mixture can be chosen randomly without affecting the quality of estimates essentially. Unlike the latent class analysis we use the estimated mixture components only to identify "elementary" latent classes with the posterior component weights playing the role of membership functions. The underlying decision problem can be characterized by the statistical decision information. We assume that the statistical properties of data can be described by the estimated mixture even if the "elementary" components are not defined uniquely. As the estimated mixture model is the only information source about the structural properties of data. We assume that potential clusters can be identified by the optimal decomposition of the estimated mixture into sub-mixtures. We propose a hierarchical clustering procedure based on sequential unifying of the elementary latent classes. The procedure is controlled by the minimum information loss criterion.

The paper is organized as follows. We first describe the idea of latent class analysis and the related problem of estimating discrete product mixtures by means of EM algorithm (Sec. 2). In Sec. 3 we discuss the problem of unique identification of latent classes and the specific features of multidimensional datamining problems. Section 4 introduces the statistical information criterion and Sec. 5 describes the method of hierarchical cluster analysis. The application of the method is illustrated by numerical examples in Sec. 6. Finally we discuss the main results in the Conclusion.

2   Latent Class Model

Let us suppose that some objects are described by a vector of discrete variables taking values from finite sets:

$$\boldsymbol{x} = (x_1, \ldots, x_N), \quad x_n \in \mathcal{X}_n, \quad |\mathcal{X}_n| < \infty, \quad \boldsymbol{x} \in \mathcal{X} = \mathcal{X}_1 \times \cdots \times \mathcal{X}_N. \quad (1)$$

We assume that the variables are categorical (i.e. non-numerical, nominal, qualitative) without any type of ordering. Considering the problem of cluster analysis we are given a set of data vectors

$$\mathcal{S} = \{\boldsymbol{x}^{(1)}, \ldots, \boldsymbol{x}^{(K)}\}, \quad \boldsymbol{x}^{(k)} \in \mathcal{X} \quad (2)$$

and the goal of cluster analysis is to partition the set $\mathcal{S}$ into "natural" well separated subsets of similar objects

$$\Re = \{\mathcal{S}_1, \mathcal{S}_2, \ldots, \mathcal{S}_M\}, \quad \mathcal{S} = \cup_{j=1}^{J} \mathcal{S}_j, \quad \mathcal{S}_i \cap \mathcal{S}_j = \emptyset, \quad \text{for } i \neq j. \quad (3)$$

In this sense the concept of cluster analysis is closely related to some similarity or dissimilarity measures. Unfortunately, in case of categorical variables the arithmetical operations are undefined and therefore we cannot compute means and variances nor there is any generally acceptable way to define distance for the categorical data vectors $\boldsymbol{x} \in \mathcal{X}$. Binary data, as a special case, may appear to be naturally ordered, however, the values 0 and 1 are often assigned quite arbitrarily. For these reasons the available algorithms of cluster analysis are not directly applicable to categorical data.

The standard way to avoid this difficulty is to introduce a similarity measure or distance function for categorical data in a heuristical manner. It may appear quite easy to define a distance table for a single categorical variable, especially in case of some well interpretable values. However, in a multidimensional space the problem of distance definition becomes difficult because of uneasy foreseen consequences of interference of different distance tables.

As it appears the only statistically justified approach to clustering categorical data can be traced back to the latent structure analysis of Lazarsfeld [10] who proposed to identify latent classes in binary data by estimating multivariate Bernoulli mixtures. The method is easily generalized to categorical variables and it is often applied in different modifications as "latent class analysis" [14]. The latent class model is defined as a finite mixture of a given number of product components

$$P(\boldsymbol{x}) = \sum_{m \in \mathcal{M}} w_m F(\boldsymbol{x}|m), \quad \boldsymbol{x} \in \mathcal{X}, \quad \mathcal{M} = \{1, \ldots, M\}. \quad (4)$$

Here w_m are non-negative probabilistic weights

$$\sum_{m \in \mathcal{M}} w_m = 1, \quad 0 < w_m < 1, \quad m \in \mathcal{M}, \quad (5)$$

$F(\boldsymbol{x}|m)$ are the mixture components defined as products of univariate conditional (component specific) discrete distributions $f_n(x_n|m)$

$$F(\boldsymbol{x}|m) = \prod_{n \in \mathcal{N}} f_n(x_n|m), \quad \mathcal{N} = \{1, \ldots, N\} \tag{6}$$

and $\mathcal{M}, \mathcal{N}$ are the index sets of components and variables respectively.

Let us recall that a set of multivariate categorical data has no internal structure in itself. Formula (4) is the only source of the structural properties of the data set $\mathcal{S}$ and therefore it should be justified by some theoretical arguments. From the probabilistic point of view Eq. (4) can be interpreted as a model of conditional independence with respect to the index variable m which is sometimes called the latent variable. In view of Eq. (4), the statistical relations among the variables x_n are fully explained by their dependence on the latent variable m. Given the value of the latent variable $m \in \mathcal{M}$, the variables x_n are statistically independent, i.e. their mutual dependence is removed. In this sense the latent variable m can identify some "hidden causes" which remove the statistical dependencies between the observed variables x_n. Once specified, the hidden cause $m \in \mathcal{M}$ would permit us to treat the visible variables x_n in a simple way as if they were mutually independent (cf. [11],[12]).

The latent class model (4) naturally defines a statistical decision problem. Having estimated the mixture parameters we can compute the conditional probabilities

$$q(m|\boldsymbol{x}) = \frac{w_m F(\boldsymbol{x}|m)}{\sum_{j \in \mathcal{M}} w_j F(\boldsymbol{x}|j)}, \quad \boldsymbol{x} \in \mathcal{X}, \quad m \in \mathcal{M} \tag{7}$$

which can be viewed as membership functions of the estimated latent classes. They are particularly useful if there is some interpretation of the mixture components, e.g. if the components can be shown to correspond to some real "latent classes" [10], "hidden causes" [11] or "clusters" having a specific meaning.

A unique classification of data vectors $\boldsymbol{x} \in \mathcal{X}$ can be obtained by means of Bayes decision function (with the ties arbitrarily decided)

$$d(\boldsymbol{x}) = \arg \max_{j \in \mathcal{M}} \{q(j|\boldsymbol{x})\}, \quad \boldsymbol{x} \in \mathcal{X}. \tag{8}$$

By using the Bayes decision function $d(\boldsymbol{x})$ we obtain the elementary "latent class" partition $\Re$ of the set $\mathcal{S}$ by classifying the points $\boldsymbol{x} \in \mathcal{S}$:

$$\Re = \{\mathcal{S}_1, \mathcal{S}_2, \ldots, \mathcal{S}_M\}, \quad \mathcal{S}_m = \{\boldsymbol{x} \in \mathcal{S} : d(\boldsymbol{x}) = m\}, \quad m \in \mathcal{M}. \tag{9}$$

In other words the partition $\Re$ is defined by the maximum posterior weights $q(m|\boldsymbol{x})$ and represents the result of latent class analysis in the original form as proposed by Lazarsfeld (cf. [10], [8]), [14]). The latent class model (4) seems to be one of the most widely applicable tools of cluster analysis of categorical data. The original idea of Lazarsfeld has been used by many authors to identify individual classes of bacteria (cf. e.g. [8]) and more recently Vermunt et al. [14] describe different modifications of the latent class analysis as applied in diverse fields.

3   Non-unique Identification of Latent Classes

The standard way of estimating mixtures is to use EM algorithm (cf. [3], [9]). In particular to compute maximum-likelihood estimates of mixture parameters we maximize the log-likelihood function

$$L = \frac{1}{|\mathcal{S}|} \sum_{x \in \mathcal{S}} \log P(x) = \frac{1}{|\mathcal{S}|} \sum_{x \in \mathcal{S}} \log \left[\sum_{m \in \mathcal{M}} w_m F(x|m) \right] \tag{10}$$

by means of the basic EM iteration equations:

$$q(m|x) = \frac{w_m F(x|m)}{\sum_{j \in \mathcal{M}} w_j F(x|j)}, \quad w_m' = \frac{1}{|\mathcal{S}|} \sum_{x \in \mathcal{S}} q(m|x), \quad m \in \mathcal{M}, \tag{11}$$

$$f_n'(\xi|m) = \frac{1}{\sum_{x \in \mathcal{S}} q(m|x)} \sum_{x \in \mathcal{S}} \delta(\xi, x_n) q(m|x), \quad \xi \in \mathcal{X}_n, \quad n \in \mathcal{N}. \tag{12}$$

Here w_m', $f_n'(\cdot|m)$ are the new parameter values and $\delta(\xi, x_n)$ denotes the usual delta-function, i.e. $\delta(\xi, x_n) = 1$ for $\xi = x_n$ and otherwise $\delta(\xi, x_n) = 0$. The number of mixture components M is a parameter which is assumed to be known or has to be specified in advance.

The sequence of log-likelihood values $\{L^{(t)}\}_0^\infty$ produced by EM algorithm is nondecreasing and converges to a local or global maximum in the parameter space. The final estimates of parameters are therefore starting-point dependent (cf. e.g. [9] for a more detailed discussion of convergence properties).

A serious disadvantage of the latent class analysis relates to the fact that the resulting clusters may be non-unique. It is obvious that, if the estimated mixture is not defined uniquely, then the corresponding interpretation of data in terms of latent classes may become questionable. Unfortunately, there are at least three sources of uncertainty which may influence the resulting mixture parameters. First, there is no exact method to choose the proper number of mixture components (cf. [9]). Another source of multiple solutions is the existence of local maxima of the log-likelihood function (10). For this reason we can expect different locally optimal solutions depending on the chosen initial parameters. However, even if we succeed to manage the computational aspects of mixture estimation, there is still the well known theoretical problem that the latent class model is not identifiable (cf. [1], [8], [13]). In particular it is easily verified that any non-degenerate mixture (4) can be expressed equivalently in infinitely many different ways [7].

In practice the non-identifiability of latent class models does not seem to have serious consequences since the classes can often be uniquely identified [2]. As it appears, in many practical problems well separated components can be identified spontaneously or by means of external knowledge or additional constraints (cf. [2], [7]). However, in the context of data-mining the problem of non-identifiable latent classes becomes more essential because of high dimensionality of data spaces and in view of a large number of components to be expected.

In case of large multidimensional real-life databases, which are typical for data-mining problems, the form of the estimated distribution is generally unknown. Obviously, in the present context, the primary condition of a successful cluster analysis is a high approximation accuracy of the latent class model (4). Multidimensional spaces are "spars" and therefore the estimated components will be nearly non-overlapping. For this reason, in order to achieve reasonable approximation accuracy, we have to assume relatively large number of components ($M \approx 10^1 \div 10^3$). Some of the resulting components usually have very low weights and may be omitted without observable consequences. In this sense the exact number of components M is less relevant. According to our practical experience (cf. [4], [5], [6]) there are usually numerous local maxima of the likelihood function having similar values. The corresponding mixtures may include different components but their approximation quality is comparable. From the point of view of approximation accuracy the influence of initial parameters is negligible and the EM algorithm can be initialized randomly.

We can conclude that in case of data-mining we have to consider multivariate latent class models with a large number of classes. There is usually large variability of the estimated parameters which, on the other hand, correspond to similar values of the log-likelihood function and provide comparable approximation accuracy of the estimated mixture. In this sense the latent classes themselves are not suitable to define directly the latent structure of large multivariate categorical data sets. The basic idea of the proposed method is to apply the latent class model only to define a sufficiently large set of elementary latent classes. Considering the global quality of the estimated mixture we assume that possible clusters can be constructed by unifying the elementary latent classes even if the underlying mixture components are not defined uniquely. We propose a hierarchical "bottom up" clustering procedure which consists in sequential pairwise unifying of the elementary latent classes. The process of hierarchical cluster analysis is controlled by minimizing the information loss criterion.

4   Minimum Information Loss Mixture Decomposition

In view of Sec. 2 the latent class model (4) is the only information source about the structural properties of the data set $\mathcal{S}$. For this reason we identify the clusters by means of the optimal decomposition of mixture (4) into sub-mixtures.

Recall that having estimated the mixture parameters we can define the elementary latent classes by classifying the data vectors $\boldsymbol{x} \in \mathcal{S}$ according to the maximum posterior weight $q(m|\boldsymbol{x})$ (cf. (8), (9)). The underlying decision problem can be characterized by the statistical decision information. By using the Shannon formula we can write

$$I(\mathcal{X}, \mathcal{M}) = H(\mathcal{M}) - H(\mathcal{M}|\mathcal{X}), \qquad H(\mathcal{M}) = \sum_{m \in \mathcal{M}} -w_m \log w_m, \qquad (13)$$

$$H(\mathcal{M}|\mathcal{X}) = \sum_{\boldsymbol{x} \in \mathcal{X}} P(\boldsymbol{x}) H_x(\mathcal{M}) = \sum_{\boldsymbol{x} \in \mathcal{X}} P(\boldsymbol{x}) \sum_{m \in \mathcal{M}} -q(m|\boldsymbol{x}) \log q(m|\boldsymbol{x}). \qquad (14)$$

Here $H(\mathcal{M})$ is the uncertainty connected with estimating the outcome $m \in \mathcal{M}$ of a random experiment with the probabilities $\{w_1, \ldots, w_M\}$ without any other knowledge. Given a vector $\boldsymbol{x} \in \mathcal{X}$ we can improve the estimation accuracy by computing the more specific conditional probabilities $q(m|\boldsymbol{x})$. The statistical decision information $I(\mathcal{X}, \mathcal{M})$ contained in the latent class model is defined as the difference between the a priori entropy $H(\mathcal{M})$ and the mean conditional entropy $H(\mathcal{M}|\mathcal{X})$ which corresponds to the knowledge of $\boldsymbol{x} \in \mathcal{X}$.

It can be seen that a partition $\mathcal{U}$ of the index set $\mathcal{M}$

$$\mathcal{U} = \{\mathcal{M}_1, \mathcal{M}_2, \ldots, \mathcal{M}_C\}, \qquad \bigcup_{c=1}^{C} \mathcal{M}_c = \mathcal{M}, \quad i \neq j \Rightarrow \mathcal{M}_i \cap \mathcal{M}_j = \emptyset. \quad (15)$$

actually defines a decomposition of the estimated mixture into sub-mixtures:

$$P(\boldsymbol{x}) = \sum_{c=1}^{C} \sum_{m \in \mathcal{M}_c} w_m F(\boldsymbol{x}|m) = \sum_{c=1}^{C} P(\boldsymbol{x}|\mathcal{M}_c)\, p(c), \qquad \boldsymbol{x} \in \mathcal{X}, \quad (16)$$

$$P(\boldsymbol{x}|\mathcal{M}_c) = \sum_{m \in \mathcal{M}_c} \frac{w_m}{p(c)} F(\boldsymbol{x}|m), \quad p(c) = \sum_{m \in \mathcal{M}_c} w_m, \quad c = 1, \ldots, C. \quad (17)$$

Again the sub-mixtures $P(\boldsymbol{x}|\mathcal{M}_c)$ can be used to define the partition of the data set $\mathcal{S}$ into corresponding clusters. We can write

$$p(c|\boldsymbol{x}) = \frac{p(c)P(\boldsymbol{x}|\mathcal{M}_c)}{P(\boldsymbol{x})} = \frac{\sum_{m \in \mathcal{M}_c} w_m F(\boldsymbol{x}|m)}{P(\boldsymbol{x})} = \sum_{m \in \mathcal{M}_c} q(m|\boldsymbol{x}), \quad (18)$$

$$d(\boldsymbol{x}|\mathcal{U}) = \arg\max_c \{p(c|\boldsymbol{x})\}, \quad \boldsymbol{x} \in \mathcal{X} \quad (19)$$

and by using the decision function $d(\boldsymbol{x}|\mathcal{U})$ we obtain the partition

$$\Re_{\mathcal{U}} = \{\mathcal{S}_c, c = 1, \ldots, C\}, \quad \mathcal{S}_c = \{\boldsymbol{x} \in \mathcal{S} : d(\boldsymbol{x}|\mathcal{U}) = c\}. \quad (20)$$

Here the clusters $\mathcal{S}_c \in \Re_{\mathcal{U}}$ correspond to the respective sub-mixtures $P(\boldsymbol{x}|\mathcal{M}_c)$.

By using the Shannon formula we can express the statistical decision information contained in the decomposed mixture. In analogy with (13), (14) we can write

$$I(\mathcal{X}, \mathcal{U}) = H(\mathcal{X}) - H(\mathcal{X}|\mathcal{U}), \qquad H(\mathcal{X}) = \sum_{\boldsymbol{x} \in \mathcal{X}} -P(\boldsymbol{x}) \log P(\boldsymbol{x}), \quad (21)$$

$$H(\mathcal{X}|\mathcal{U}) = \sum_{c \in \mathcal{U}} p(c) H_{\mathcal{M}_c}(\mathcal{X}), \qquad H_{\mathcal{M}_c}(\mathcal{X}) = \sum_{\boldsymbol{x} \in \mathcal{X}} -P(\boldsymbol{x}|\mathcal{M}_c) \log P(\boldsymbol{x}|\mathcal{M}_c). \quad (22)$$

Intuitively it is clear that by fusing sub-mixtures (or components) we loose some decision information. Indeed, we can easily verify that the decision information decreases if we join any two subset $\mathcal{M}_i, \mathcal{M}_j \in \mathcal{U}$ of a given partition $\mathcal{U}$.

In particular, let $\mathcal{U}'$ be a partition which derives from $\mathcal{U}$ by unifying two subset $\mathcal{M}_i, \mathcal{M}_j \in \mathcal{U}$., i.e. the partition $\mathcal{U}'$ contains only one subset $\mathcal{M}_i \cup \mathcal{M}_j$

instead on the two original subsets $\mathcal{M}_i$ and $\mathcal{M}_j$. We can show that the union of the two corresponding sub-mixtures $P(\boldsymbol{x}|\mathcal{M}_i)$, $P(\boldsymbol{x}|\mathcal{M}_j)$ is connected with some information loss. Considering the difference

$$\Delta I_{ij} = I(\mathcal{X},\mathcal{U}) - I(\mathcal{X},\mathcal{U}') = H(\mathcal{X}|\mathcal{U}') - H(\mathcal{X}|\mathcal{U}) \tag{23}$$

we can write (cf. (22))

$$\Delta I_{ij} = [p(i) + p(j)] \, H_{\mathcal{M}_i \cup \mathcal{M}_j}(\mathcal{X}) - p(i)H_{\mathcal{M}_i}(\mathcal{X}) - p(j)H_{\mathcal{M}_j}(\mathcal{X}) \tag{24}$$

and by using relation

$$\left[p(i) + p(j)\right] P(\boldsymbol{x}|\mathcal{M}_i \cup \mathcal{M}_j) = p(i)P(\boldsymbol{x}|\mathcal{M}_i) + p(j)P(\boldsymbol{x}|\mathcal{M}_j) \tag{25}$$

we can rewrite Eq. (24) in the form

$$\Delta I_{ij} = \sum_{c=i,j} p(c) \sum_{\boldsymbol{x} \in \mathcal{X}} P(\boldsymbol{x}|\mathcal{M}_c) \log \frac{P(\boldsymbol{x}|\mathcal{M}_c)}{P(\boldsymbol{x}|\mathcal{M}_i \cup \mathcal{M}_j)} \geq 0. \tag{26}$$

It can be seen that the last sum in the expression above represents the Kullback-Leibler information divergence which is non-negative for any two distributions $P(\boldsymbol{x}|\mathcal{M}_i)$, $P(\boldsymbol{x}|\mathcal{M}_i \cup \mathcal{M}_j)$ and therefore the information loss ΔI_{ij} accompanying the fusion of the two sub-mixtures is non-negative.

5   Minimum Information Loss Cluster Analysis

In view of the above equations any cluster analysis based on mixture decomposition is connected with some information loss from the point of view of the underlying decision problem. Naturally we are interested to minimize the information loss caused by clustering the data and therefore the elementary information loss (26) is a suitable criterion to control the process of sequential fusion of the components and/or sub-mixtures in the original latent class model.

Let us note that the criterion (26) includes summing over the whole data space $\mathcal{X}$ and therefore it is not suitable from the computational point of view. For this reason we express the corresponding information loss equivalently in the form (cf. (17), (18))

$$\Delta I_{ij} = - \sum_{c=i,j} p(c) \log \frac{p(c)}{p(i) + p(j)} + \sum_{\boldsymbol{x} \in \mathcal{X}} P(\boldsymbol{x}) \sum_{c=i,j} p(c|\boldsymbol{x}) \log \frac{p(c|\boldsymbol{x})}{p(i|\boldsymbol{x}) + p(j|\boldsymbol{x})} \tag{27}$$

By using a simple estimate in the last equation we can write the resulting criterion in the following more suitable form

$$Q_{ij} = - \sum_{c=i,j} p(c) \log \frac{p(c)}{p(i) + p(j)} + \frac{1}{|\mathcal{S}|} \sum_{\boldsymbol{x} \in S} \sum_{c=i,j} p(c|\boldsymbol{x}) \log \frac{p(c|\boldsymbol{x})}{p(i|\boldsymbol{x}) + p(j|\boldsymbol{x})}. \tag{28}$$

The described criterion Q_{ij} is an estimate of the information loss arising after union of the two subsets $\mathcal{M}_i, \mathcal{M}_j$ in the partition $\mathcal{U}$, i.e. by the fusion of the two sub-mixtures $P(\boldsymbol{x}|\mathcal{M}_i), P(\boldsymbol{x}|\mathcal{M}_j)$. In the following we use the estimated information loss Q_{ij} as a criterion for the optimal choice of the pair of subsets to be unified. In other words, in each step of the procedure we unify the two sets $\mathcal{M}_i, \mathcal{M}_j \in \mathcal{U}$ for which the resulting information loss Q_{ij} is minimized. Let us note (cf. (26)) that the criterion Q_{ij} tends to fuse similar sub-mixtures preferably with small weights $p(i), p(j)$.

In the considered decision-making framework a natural goal of cluster analysis is to preserve maximum decision information with a minimum number of clusters. Let us remark that the most general result of the above algorithm is the sequence of information loss values $\{Q_{ij}^{(k)}\}_{k=1}^{K}$ produced by the hierarchical clustering procedure. The form of the sequence suggests different possibilities of final clustering and simultaneously it can be seen how justified are the resulting clusters. For a given mixture (4) the sequence $\{Q_{ij}^{(k)}\}_{k=1}^{K}$ is defined uniquely and the form of the sequence should be similar for comparably good estimates of the underlying latent class model.

The proposed method of cluster analysis of categorical data can be summarized as follows:

Algorithm

1. Estimation of the latent class model (4) for the categorical data set $\mathcal{S}$ by means of EM algorithm for a sufficiently large M.
2. Definition of the basic latent class partition $\mathcal{U} = \{\{1\}, \{2\}, \ldots, \{M\}\}$.
3. Hierarchical cluster analysis by sequential unifying the most similar subsets $\mathcal{M}_i, \mathcal{M}_j \in \mathcal{U}$ for which the resulting information loss Q_{ij} (cf. (28)) is minimal.
4. Choice of the optimal partition $\mathcal{U}^*$ according to the point of the increasing information loss $Q_{ij}^{(k)}$.
5. Definition of the resulting clusters in $\mathcal{S}$ by means of the decision function $d(\boldsymbol{x}|\mathcal{U}^*)$.

6   Numerical Experiments

6.1   Artificial Data

In order to illustrate the properties of the proposed method of cluster analysis we have chosen a discrete data problem with the possibility of a visual verification of results. The data vectors describe images on a square raster and the discrete (binary) variables correspond to grey levels of the raster fields. However, the algorithm treats the variables as general discrete categorical data without any use of their physical meaning.

In particular, we have used an idea similar to that of the paper [2]. We have constructed a multivariate Bernoulli mixture of M=8 components with uniform weights $w_m = 0.125$ and with the dimension N=256

$$P(\boldsymbol{x}) = \sum_{m=1}^{8} w_m \prod_{n=1}^{256} \theta_{mn}^{x_n}(1 - \theta_{mn})^{1-x_n}, \qquad 0 \le \theta_{mn} \le 1. \qquad (29)$$

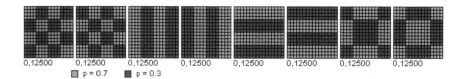

0,12500   0,12500   0,12500   0,12500   0,12500   0,12500   0,12500   0,12500

▨ p = 0.7   ▦ p = 0.3

Fig. 1. Parameters of the original mixture - dimension $N = 256$, number of components $M = 8$ and identical component weights $w_m = 0.125$. The component parameters are set equal either to $\theta_{mn} = 0.7$ or $\theta_{mn} = 0.3$ according to simple geometrical patterns.

The parameters θ_{mn} have a simple geometrical meaning. For each komponent the parameters θ_{mn} have only two possible values $\theta_{mn} = 0.7$ or $\theta_{mn} = 0.3$ and in the square raster arrangement they correspond to eight basic geometrical patterns as shown in Fig. 1.

By random sampling from the distribution (29) we have generated a sample of $|\mathcal{S}| = 100000$ multidimensional binary data vectors to be the subject of cluster analysis. Unlike usual benchmarking data the set $\mathcal{S}$ contains a large number of independent observations of a random vector which are identically distributed according to the Bernoulli mixture (29). In view of the well differentiated probabilities $\theta_{mn}, 1 - \theta_{mn}$ the randomly generated data vectors more or less correspond to the original patterns of Fig. 1 but simultaneously they are quite noisy as it can be seen in Fig. 2. In each row of Fig. 2 there is first the pattern of component parameters and then the examples of the corresponding randomly generated bi-

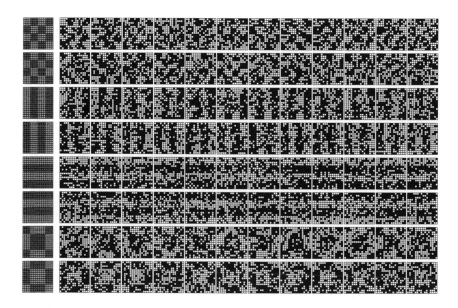

Fig. 2. A sample of randomly generated data set. In each row there is first the component pattern and then the examples of the corresponding randomly generated binary patterns.

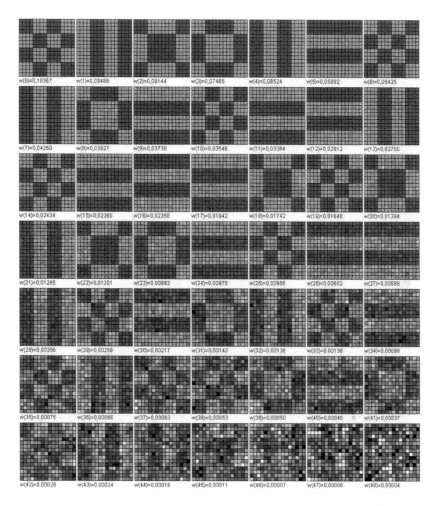

Fig. 3. The component parameters (in square arrangement) of the mixture of 49 components estimated from the randomly generated data

nary data vectors. According to the chosen source mixture the data set S should contain eight well defined latent classes which correspond to the respective mixture components. Note that the two possible values of θ_{mn} affect the distance between components and also the amount of noise in the generated vectors. Applying the clustering algorithm we have estimated first the latent class model (29). The number of components has been set to $M = 49$. We have achieved satisfactory convergence after 40 iterations. The resulting component parameters in square raster arrangement are shown in Fig. 3. The estimated latent class model has been subject of the hierarchical clustering procedure. By unifying the mixture components sequentially we have obtained a sequence of 48 possible decompositions of the original mixture. From the corresponding sequence of information loss values $Q_{ij}^{(k)}$ (cf. (28)) we have found that the information loss essentially increases when the number of clusters is less than eight. The resulting sub-mixtures can be characterized by the respective cluster means as shown in the first column of Fig. 4. The error matrix shows how the resulting clusters coincide with the randomly generated classes. In total there are 363 erroneously classified data vectors (0.36 %) which correspond mainly to the last column of the matrix. Probably the related sub-mixture does not describe the underlying class with sufficient accuracy.

class 0	12562	0	0	0	0	0	0	5
class 1	0	12275	0	0	0	0	0	161
class 2	0	0	12305	0	0	0	0	23
class 3	0	0	0	12553	0	0	0	25
class 4	0	1	0	0	12376	0	0	138
class 5	0	0	0	0	0	12568	0	1
class 6	0	0	0	0	0	0	12482	9
class 7	0	0	0	0	0	0	0	12516

Fig. 4. Resulting cluster means for the final eight clusters. The error matrix illustrates the coincidence of the resulting clusters with the original randomly generated classes. There are 363 erroneously classified data vectors (0.36 %) mainly concentrated in the last column.

6.2 Handwritten Non-stylized Numerals

In the second example the proposed minimum information loss cluster analysis has been applied to classification of handwritten non-stylized numerals on a binary rastr. We have used 400 000 numerals from the NIST database uniformly representing the classes 0,1,...,9. Each of the numerals in the data base has been normalized to a square 16×16 binary raster, i.e. it has been represented by a 256-dimensional binary vector.

Normally the NIST numerals are used as a benchmark problem for a supervised pattern recognition. The supervised classifier is trained for each class separately with the resulting relatively low classification error. Obviously, the non-supervised solution of the problem cannot be expected to achieve comparable accuracy, however, from the point of view of cluster analysis, we have again the possibility of a visual inspection of results.

Fig. 5 illustrates the properties of the NIST database. In the rows there are examples of numerals from the database. Again we have estimated the latent class model in the form of a 256-dimensional Bernoulli mixture (29). We have

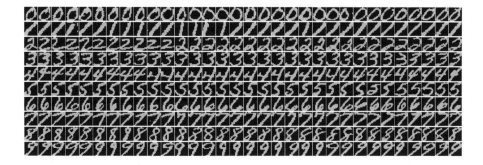

Fig. 5. Examples of numerals from the NIST database normalized to the 16x16 binary raster

chosen a model of $M = 60$ components. The EM algorithm has been initialized randomly with the uniform component weights. The EM algorithm has been stopped after 30 iterations and the estimated parameters θ_{mn} (in the raster arrangement) are shown in Fig. 6. The estimated elementary latent classes as characterized by the components in Fig. 6 have been unified sequentially by using the algorithm of Sec. 5. The hierarchical procedure based on pairwise unifying the most similar sub-mixtures has been stopped at the level of 12 clusters which precedes a local increase of the information loss Q_{ij}.

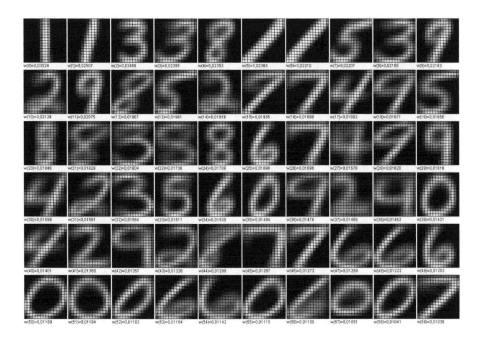

Fig. 6. The component parameters (in square arrangement) of the mixture of 60 components estimated from the NIST database

Fig. 7 describes the properties of the resulting clusters. The number of clusters is higher than 10 because for some numerals there are different variants which are too dissimilar in the high-dimensional description. Every column shows the distribution of data vectors in the resulting clusters with respect to the true classes. It can be seen that digits with specific shape (e.g. one) are separated well, whereas the others form clusters according to similarities in their shapes (e.g. zero, two and six in the eight column). From the point of view of unsupervised clustering this can be treated as an acceptable result.

class 0	26112	9	38	300	887	20	58	12331	425	43	133	7
class 1	91	40474	554	68	1035	71	60	154	29	306	1771	91
class 2	273	6	26457	635	713	19	281	11072	21	213	358	24
class 3	147	4	355	10937	12354	68	15828	351	68	158	679	163
class 4	180	15	106	5348	248	16732	2	513	99	12043	463	3405
class 5	358	1	15	5317	3492	65	17760	507	7338	167	1556	30
class 6	4516	48	50	161	736	4	190	17001	17098	12	120	1
class 7	5167	53	180	893	86	1930	2	102	0	23747	700	9033
class 8	61	136	112	3104	10361	100	974	439	137	194	23513	448
class 9	131	53	58	556	550	8544	109	98	7	10433	793	18201

Fig. 7. Resulting cluster means for the final 12 clusters. The matrix illustrates the coincidence of the resulting clusters with the original classes.

7   Conclusion

The latent class models have been used repeatedly as a tool of cluster analysis of multivariate categorical data since the standard approaches are usually not directly applicable. Unfortunately, the underlying discrete distribution mixtures with product components are not uniquely identifiable. In order to avoid the problem of identifiability the latent class model is applied only to identify elementary latent classes. We assume that the potential clusters can be constructed by unifying the elementary classes even if they are not defined uniquely. A hierarchical procedure is proposed to define the optimal decomposition of the underlying mixture. The hierarchical cluster analysis is controlled by minimum information loss criterion.

References

1. Blischke, W.R.: Estimating the parameters of mixtures of binomial distributions. Journal Amer. Statist. Assoc. 59, 510–528 (1964)
2. Carreira-Perpignan, M.A., Renals, S.: Practical identifiability of finite mixtures of multivariate Bernoulli distributions. Neural Computation 12, 141–152 (2000)

3. Dempster, A.P., Laird, N.M., Rubin, D.B.: Maximum likelihood from incomplete data via the EM algorithm. J. Roy. Statist. Soc. B 39, 38 (1977)
4. Grim, J., Haindl, M.: Texture Modelling by Discrete Distribution Mixtures. Computational Statistics and Data Analysis 41(3-4), 603–615 (2003)
5. Grim, J., Kittler, J., Pudil, P., Somol, P.: Multiple classifier fusion in probabilistic neural networks. Pattern Analysis & Applications 5(7), 221–233 (2002)
6. Grim, J., Somol, P., Haindl, M., Pudil, P.: A statistical approach to local evaluation of a single texture image. In: Proceedings of the Sixteenth Annual Symposium of the Pattern Recognition Association of South Africa, PRASA 2005. Nicolls, F. (ed.). University of Cape Town, Cape Town 2005, pp. 171–176 (2005)
7. Grim, J.: EM cluster analysis for categorical data. In: Yeung, D.-Y., Kwok, J.T., Fred, A., Roli, F., de Ridder, D. (eds.) Structural, Syntactic, and Statistical Pattern Recognition. LNCS, vol. 4109, pp. 640–648. Springer, Heidelberg (2006)
8. Gyllenberg, M., Koski, T., Reilink, E., Verlaan, M.: Non-uniqueness in probabilistic numerical identification of bacteria. Journal of Applied Prob. 31, 542–548 (1994)
9. McLachlan, G.J., Peel, D.: Finite Mixture Models. John Wiley & Sons, New York, Toronto (2000)
10. Lazarsfeld, P.F., Henry, N.W.: Latent structure analysis. Houghton Miflin, Boston (1968)
11. Pearl, J.: Probabilistic reasoning in intelligence systems: networks of plausible inference. Morgan-Kaufman, San Mateo, CA (1988)
12. Suppes, P.A.: Probabilistic theory of causality. North-Holland, Amsterdam (1970)
13. Teicher, H.: Identifiability of mixtures of product measures. Ann. Math. Statist. 39, 1300–1302 (1968)
14. Vermunt, J.K., Magidson, J.: Latent Class Cluster Analysis. In: Hagenaars, J.A., McCutcheon, A.L. (eds.) Advances in Latent Class Analysis, Cambridge University Press, Cambridge (2002)

A Clustering Algorithm Based on Generalized Stars

Airel Pérez Suárez and José E. Medina Pagola

Advanced Technologies Application Centre (CENATAV)
7a #21812 e/ 218 y 222, Rpto. Siboney, Playa. C.P. 12200, C. Habana, Cuba
{asuarez, jmedina}@cenatav.co.cu

Abstract. In this paper we present a new algorithm for document clustering called Generalized Star (GStar). This algorithm is a generalization of the Star algorithm proposed by Aslam *et al.*, and recently improved by them and other researchers. In this method we introduced a new concept of star allowing a different star-shaped form with better overlapping clusters. The evaluation experiments on standard document collections show that the proposed algorithm outperforms previously defined methods and obtains a smaller number of clusters. Since the GStar algorithm is relatively simple to implement and is also efficient, we advocate its use for tasks that require clustering, such as information organization, browsing, topic tracking, and new topic detection.

Keywords: Clustering, Data mining, Document processing and recognition.

1 Introduction

Clustering is the process of grouping the data into classes or clusters so that objects within a cluster have high similarity in comparison to one another, but are very dissimilar to objects in other clusters. Dissimilarities are assessed based on the attribute values describing the objects. Often, distance measures are used. Clustering has its roots in many areas, including data mining, statistics, biology, and machine learning. Cluster analysis has been widely used in numerous applications, including pattern recognition, data analysis, image processing, and market research. By clustering, one can identify crowded and sparse regions and, therefore, discover overall distribution patterns and interesting correlations among data attributes.

Initially, document clustering was evaluated for improving the results in information retrieval systems [11]. Clustering has been proposed as an efficient way of finding automatically related topics or new ones; in filtering tasks [2] and grouping the retrieved documents into a list of meaningful categories, facilitating query processing by searching only clusters closest to the query [12].

Several algorithms have been proposed for document clustering. One of these algorithms is Star, presented and evaluated by Aslam *et al.* [1]. They shown that the Star algorithm outperforms other methods such as Single Link, Average Link [7] in different tasks [1]; however, this algorithm depends on data order and produces illogical clusters. Another method that improves the Star algorithm is the Extended Star method proposed by Gil *et al.* [8]. The Extended Star method outperforms the original Star algorithm, reducing considerably the number of clusters; nevertheless this algorithm can leave uncovered objects and in some cases produce unnecessary clusters. Another version of the Extended Star method was proposed by Gil *et al.* to construct a parallel algorithm [9]. However, this version also has some drawbacks; first of all, it can produce illogical clusters and in some cases unnecessary clusters.

In this paper we propose a clustering method, called Generalized Star or GStar, which solves the drawbacks above mentioned. In GStar, we introduced a definition of star allowing a different star-shaped sub-graph form with better overlapping clusters.

The experimentation - comparing our proposal against the original Star and the Extended algorithms - shows that our method outperforms those algorithms. Since the GStar algorithm is relatively simple to implement and is also efficient, we advocate its use for task that require clustering, such as information organization, browsing, topic tracking, and new topic detection.

The basic outline of this paper is as follows. Section 2 is dedicated to related work. Section 3 contains the description of the GStar method. The experimental results are discussed in section 4. The conclusions of the research and some ideas about future directions are exposed in section 5.

2 Related Work

In this section we analyze the Star algorithm and two proposed versions of the Extended Star method for document clustering, and we show their drawbacks.

2.1 Star Algorithm

The Star algorithm was proposed by Aslam *et al.* in 1998 [1], with several extensions and applications in filtering and information organization tasks [2], [3]. They formalized the problem representing the document collection by its similarity graph, finding overlaps with dense sub-graphs; it is done so because the clique cover of the similarity graph is an NP-complete problem, and it does not admit polynomial time approximation algorithms. With this cover approximation by dense sub-graphs, in spite of loosing intra-cluster similarity guarantees, we can gain in computational efficiency.

We call similarity graph to an undirected and weighted graph $G = (V, E, w)$, where vertices in the graph correspond to documents and each weighted edge corresponds to a similarity measure between two documents.

G is a complete graph with edges of varying weight. Nevertheless, we can consider a minimum similarity measure to obtain a thresholded graph. Let $V = \{d_1, ..., d_n\}$ be the universe of documents in study. Besides, let $Sim(d_i, d_j)$ be a similarity (symmetric) function between documents d_i and d_j, and σ a similarity threshold defined by the user. The thresholded graph G_σ is an undirected graph obtained from G by eliminating all the edges whose weights are lower than σ.

In the Star algorithm, a clique cover is approximated by covering the associated thresholded similarity graph with star-shaped sub-graphs. A star-shaped sub-graph on $m+1$ vertices consists of a single *star center* and m *satellite* vertices, where the *star center* is the vertex that has a degree greater or equal than any other vertex in the

sub-graph and there are edges between the star center and each of the satellite vertices
[3]. It would appear at first glance that finding star-shaped sub-graphs provides
similarity guarantees between the star center and each of its satellite vertices, but not
such similarity guarantees between satellite vertices. However, Aslam et al. showed
that the pairwise similarity between satellite vertices in a star-shaped sub-graph is
high and a cover with these sub-graphs is an accurate method for clustering a set of
documents [1], [3].

A thresholded similarity graph may have several different star covers because when
there are several vertices of the same highest degree, this algorithm arbitrarily chooses
one of them as a star center, whichever shows up first in the sorted list of vertices.

This algorithm encodes the vertices by assigning the types "center" and "satellite"
(which is the same as "not center"). It generates a correct star cover assigning the
types "center" and "satellite" in such a way that (i) a star center is not adjacent to any
other star center and (ii) every satellite vertex is adjacent to at least one center vertex
of equal or higher degree. Using this star cover, the algorithm creates a set of clusters,
where each cluster is made of a center vertex and its adjacent vertices.

The Star algorithm is different to others, for example, Scatter/Gather [6],
Charikar's et al. [5] and classical K-means algorithms, because it does not impose a
fixed number of clusters a priori as a constraint on the solution. The clusters created
by this algorithm can be overlapped, that is why it was recommended in document
processing systems, since documents can have multiple topics. Nevertheless, this
algorithm has some drawbacks.

The problems of the Star algorithm are illustrated in Fig. 1, where the dark vertices
are the star centers. First of all, the obtained clusters depend on the data order. On the
left graph (A), the Star algorithm takes first the vertex of the center (vertex 5).
However, if vertex 2 (or 7) is processed before vertex 5, the algorithm obtains the
clusters shown on graph (B). As we can see, the obtained clusters are different.

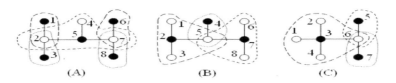

Fig. 1. Drawbacks of Star algorithm

It is important to note that this example shows not only how different the resulting
clusters can be depending on the data order, but how this dependence can affect the
quality of the clusters.

The second main drawback is that it can produce "illogical" clusters, regardless of
data order, since two star centers are never adjacent; see graph (C) in Fig. 1. As we
can see, vertex 6 should be a star center and its neighbours with less degree should be
its satellites.

2.2 Extended Star Algorithm

The Extended Star algorithm was proposed by Gil et al. to solve the aforementioned
drawbacks [8]. They represent also the document collection by its thresholded

similarity graph, defining a new notion of star center obtaining, as a consequence, different star-shaped clusters that are independent of data order.

Previously, they defined the concept of Complement Degree. The *Complement Degree* (CD) of a vertex is the quantity of neighbours (adjacent vertices) not included yet in any cluster. Notice that the complement degree of a vertex can decrease during the clustering process as more vertices are included in clusters.

Then, the extended star definition is presented in the following way: An object o is considered a star (center) if it has at least a neighbour o' (with less or equal degree than o) that satisfies one of the following conditions:

- Object o' does not have a neighbour marked as star.
- The neighboring star center of o' with the highest degree has a degree not greater than the degree of o.

The logic of the algorithm is to generate cluster centers in an iterative process – while there are non-clustered documents – where vertices are selected, from the set of candidates L with maximum complement degree and degree, if they meet Extended Star definition without cluster redundancy.

Unlike the Star algorithm, the obtained clusters are independent of data order. Nevertheless, the Extended Star algorithm has also some drawbacks. First of all, it can leave uncovered vertices, producing an infinite loop. This situation is illustrated in Fig. 2 (A).

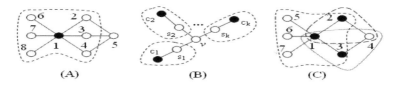

Fig. 2. Drawbacks of Extended algorithm

In the first iteration, vertex 1 of graph (A) is selected as star center. In the second iteration, vertices 2, 3 and 4 have the highest complement degree (CD = 1) and also the highest degree, but none of them satisfies the center condition, remaining vertices 5, 6, 7 and 8 as the candidates for the next iteration. In the third iteration, all have CD = 0, but 5 has the highest degree; nevertheless, it does not satisfy the center condition and it is removed from L. In the fourth iteration, the remaining vertices do not satisfy the condition and all are removed from L, reaching L the empty set. Henceforth, the loop will not finish because a non-clustered object exists (vertex 5) but it will not be covered by any cluster. This infinite loop would be avoided if the algorithm also checks that L is not empty, but even so vertex 5 remains as uncovered.

This situation can be generalized. Indeed, any time that a vertex v exists, such as the illustrated in graph (B) of Fig. 2, the algorithm produces an infinite loop, leaving the vertex v uncovered. In this graph, each s_i represents the corresponding neighbours (adjacent vertices) of v, and c_i, is the adjacent center of s_i with highest degree. If vertex v satisfies the conditions described in (1), then it has a set of neighbours with less degree than v, but not satisfying the Extended Star condition; therefore, neither v nor its neighbours can be selected as center, remaining v as an uncovered vertex.

$$\forall s_i, 1 \le i \le k, |v.Adj| > |s_i.Adj| \wedge \forall c_i, 1 \le i \le k, |c_i.Adj| > |v.Adj| \qquad (1)$$

In this and in the following expressions, $x.Adj$ represents the set of adjacent vertices of the vertex x.

The second drawback of this algorithm is that it can produce unnecessary clusters, since more than one center can be selected at the same time. As can be noticed in graph (C) of Fig. 2, vertex 2 and vertex 3 should not be centers at the same time because we only need one of them to cover vertex 4.

A different version of the Extended Star algorithm was proposed by Gil *et al.* to construct a parallel approach [9]. This version also uses the complement degree of a vertex, but it does not apply the Extended Star definition for the selection of the star centers; as a consequence, the obtained clusters are different from those obtained by the original version.

This new version is also independent of data order, and solves the first drawback of the former Extended Star algorithm, but it maintains some drawbacks too. As the first Extended Star version, it can produce unnecessary clusters, since more than one center can be selected at the same time (see graph (C) of Fig. 2).

The second drawback of this version is that it can produce, like the original Star method, illogical (less dense) clusters. This situation is illustrated in Fig. 3.

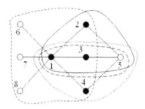

Fig. 3. Drawback of the new version of the Extended Star algorithm

In the first iteration, vertex 1 is selected as star center. In the second iteration, vertices 2, 3 and 4 have the highest complement degree (CD = 1) and also the highest degree, so they are selected as centers. After this iteration, all vertices are covered and the algorithm stops.

This behavior, in our opinion, is caused because it was not considered the Extended Star condition to obtain the clusters, observing behaviors similar to the original Star method.

3   A Generalized Star Method

In this section we introduce a new concept of star allowing a different star-shaped form and a new method, called Generalized Star or GStar, with better performance than the original Star and the Extended Star methods. As the aforementioned algorithms, we represent the document collection by its thresholded similarity graph G_σ. In order to define this new star concept and to describe the method, we define the following sets: Weak Satellites, Potential Satellites, and Potential Centers.

The set of *Weak Satellites* (WeakSats) of o is the set of all adjacent vertices with (standard) degrees not greater than the degree of o (2).

$$o.\text{WeakSats} = \left\{ s \in o.\text{Adj} / |o.\text{Adj}| \geq |s.\text{Adj}| \right\} \qquad (2)$$

The *WeakSats degree* of a vertex o is defined as the quantity of vertices included in its set of Weak Satellites, i.e. the length of its WeakSats set.

The set of *Potential Satellites* (PotSats) of o is the set of all adjacent vertices with WeakSats degrees not greater than the WeakSats degree of o (3).

$$o.\text{PotSats} = \left\{ s \in o.\text{Adj} / |o.\text{WeakSats}| \geq |s.\text{WeakSats}| \right\} \qquad (3)$$

Considering the aforementioned sets, we can define the Generalized Star-shaped sub-graph concept as follows: A *Generalized Star-shaped sub-graph* of $m+1$ vertices consists of a single star center c and m adjacent vertices, verifying the property described in (4).

$$\forall s \in c.\text{PotSats}, |c.\text{PotSats}| \geq |s.\text{PotSats}| \qquad (4)$$

The set of *Potential Centers* (PotCenters) of o is the set of all adjacent vertices that potentially can be centers of some star-shaped sub-graphs that include o, i.e. the set of adjacent vertices which has at least one vertex in its PotSats set (5).

$$o.\text{PotCenters} = \left\{ c \in o.\text{Adj} / |c.\text{PotSats}| \neq 0 \right\} \qquad (5)$$

Starting from this definition and guaranteeing a full cover, this method should satisfy the following post-conditions:

$$\forall x \in V, (x \in C) \vee (x.\text{PotCenters} \cap C \neq \phi), \qquad (6)$$

$$\forall c \in C, \forall s \in c.\text{PotSats}, |c.\text{PotSats}| \geq |s.\text{PotSats}|, \qquad (7)$$

where V is the set of vertices of the thresholded similarity graph G_σ and C is the set of centers of each cluster obtaining by the GStar algorithm.

The first condition (6) guarantees that each object of the collection belongs at least to one group, as a center or as a satellite. Besides, the condition (7) indicates that all the centers satisfy the generalized star-shaped sub-graph definition (4).

3.1 Initial Considerations

It is very important, before the definition of the GStar algorithm, to analyze which drawbacks will be solved and, more important, how they will be solved.

First of all, the GStar method intends to solve the drawbacks that directly affect the quality of the obtained clusters; these are:

- Illogical clusters (Star and Extended).
- Uncovered vertices (Extended).
- Unnecessary clusters (Extended).

The generalized star-shaped sub-graph concept and the post-conditions (6) and (7) above mentioned, will have the GStar algorithm solves these three drawbacks.

The dependence on data order is a property that the Extended Star method certainly solves. Nevertheless, as we had previously indicated, it is necessary only when that dependence affects the quality of the resulting clusters.

If we consider the graph of Fig. 4, and assume that we want to avoid unnecessary clusters, several "acceptable" solutions could be obtained.

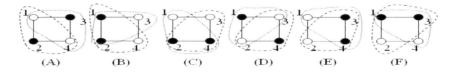

Fig. 4. Symmetric or similar solutions

The answer to this situation is that these solutions are rather symmetric or similar, and any of them is a good result. This situation is solved by Extended Star method selecting all the potential centers (in this case all the vertices) as effective ones. We have considered that this solution is not a correct alternative.

3.2 Generalized Star Algorithm

In order to define the GStar algorithm, we introduce the concept of Necessary Satellite.

The set of *Necessary Satellites* (NecSats) of o is the set of its adjacent vertices that could depend on o to be covered. This concept is necessary only during the cluster generation. Initially, NecSats takes the value of PotSats; but, it can decrease during the clustering process as more documents are covered by stars.

The Generalized Star algorithm is summarized in Fig. 5.

Algorithm: GStar

Input: $V = \{d_1, ..., d_n\}$ – Set of vertices
σ - Similarity threshold

Output: SC – Set of cluster

```
 1)    forall vertex dᵢ ∈ V do
 2)        dᵢ.Adj = {dⱼ ∈ V / Sim(dᵢ, dⱼ) ≥ σ};
 3)    forall vertex d ∈ V do
 4)        d.WeakSats = {s ∈ d.Adj / |d.Adj| ≥ |s.Adj|};
 5)    forall vertex d ∈ V do begin
 6)        d.PotSats = {s ∈ d.Adj/|d.WeakSats|≥|s.WeakSats|};
 7)        d.NecSats = d.PotSats;
 8)    end
 9)    forall vertex d ∈ V do
10)        d.PotCenters = {c ∈ d.Adj / |c.PotSats| ≠ 0};
11)    L = V;
12)    C = ∅;
```

```
13)  while L ≠ ∅ do begin
14)      d = Arg max|dᵢ.PotSats| ;
              dᵢ∈L
15)      Update(d, C, L);
16)  end
17)  "Sort C in ascending order by PotSats";
18)  forall center c ∈ C do
19)      if c.PotCenters ∩ C ≠ ∅ and
20)          ∀ s ∈ c.PotSats, (s ∈ C or
                              s.PotCenters ∩ C \ {c} ≠ ∅)
21)      then C = C \ {c};
22)  forall center c ∈ C do begin
23)      create a cluster cl = {c} ∪ c.adj
24)      SC = SC ∪ cl;
25)  end
```

Fig. 5. Pseudo code of Generalized Star algorithm

The procedure Update (see Fig. 6) is applied to mark a vertex as center, deleting it from L, and for updating the set NecSats on each of its necessary satellites.

Procedure: Update
Input: d – vertex to be a cluster center
Input/Output: C – Set of cluster centers
                  L – Set of unprocessed vertices

```
1)   C = C ∪ {d};
2)   L = L \ {d};
3)   forall s ∈ d.NecSats do begin
4)       s.NecSats = s.NecSats \ {d};
5)       if s.NecSats = ∅ then L = L \ {s};
6)       forall c ∈ s.PotCenters \ {d} do begin
7)           c.NecSats = c.NecSats \ {s};
8)           if c.NecSats = ∅ and
                  c.PotCenters ∩ C ≠ ∅
9)           then L = L \ {c};
10)      end
11)  end
```

Fig. 6. Pseudo code of Update Procedure

The algorithms works as follows: using a list L, containing the vertices that can be selected as center (initialized at step 11 in the way that it contains all the vertices), the algorithm iterates over the list L until it be empty; this means that all the vertices belong to at least one cluster, as a center or satellite. In each iteration, the algorithm will select, amongst all the contained vertices in L, a vertex d which has the higher number of elements in its PotSats set, selecting that vertex as center. In this way we guarantee that every selected center satisfies (7).

After that selection, it is necessary to update the set NecSats of some vertices, according to the concept of Necessary Satellites; this process is made by the Update procedure considering the following situations:

- As vertex d (the new selected center) has been clustered, all its necessary satellites should eliminate d from their NecSats sets. After that operation, if any necessary satellite w of d gets an empty NecSats set, then w can be removed from L since no vertex needs for it to be clustered and, besides, it has been already clustered by d.
- As each necessary satellite w of d has been clustered, all the potential centers (excepting d) of w should eliminate w from their NecSats sets. In the same way, if any of these potential centers z gets an empty NecSats set and also had been already clustered by any other center, then z can be removed from L.

It is important to notice that in each iteration the list L decreases its length at least in one unit, because of the deletion of the selected center. Besides, as we explained above, the length of list L is also decreased every time that we detect that some vertex is not necessary to cluster either at least one of the vertices in its NecSats sets or itself. The speed of the decrease of L depends on the number of vertices that were eliminated in the updating process of the NecSats set of the neighboring vertices of the selected center and this number will be increased along the iterations.

The importance of the decrease of the length of L is that in this way we guarantee that the process of clustering (or the process of centers selection) will stop when L be empty, and that means that all the vertices either are center or belong to at least one center, i.e. all vertices satisfies (6). Besides we avoid the apriority selection of centers that will make redundant clusters.

At the end, we sort the set of centers in ascending order according to the length of the PotSats set of each vertex, and then we check to eliminate the redundant centers that would emerge at the end of the selection process.

3.3 General Considerations of GStar Method

The GStar method - as the original Star algorithm and the two versions of the Extended algorithm - generates clusters which can be overlapped and guarantees also that the pairwise similarity between satellites vertices in a generalized star-shaped sub-graph be high.

Unlike the Star algorithm, and as the two versions of the Extended Star method, the GStar algorithm allows centers as adjacent vertices. Besides, the generalized star-shaped sub-graph definition (4) guarantees a better cluster generation than the Extended Star method; particularly, it solves the drawbacks of the first version of the Extended Star algorithm.

Fig. 7 shows the solutions to uncovered vertices (A) and to unnecessary clusters (B), presented in graphs (A) and (C) of Fig. 2.

The GStar method can not produce illogical clusters, because all the centers must satisfy the generalized star-shaped sub-graph definition (4). Note that in the process of selection of the candidate center, the strategy is to select the vertex with the higher length of its PotSats set; so, in this way we guarantee that the cluster is not illogical and also the correctness of the resultant set of clusters.

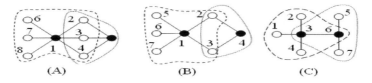

Fig. 7. Solutions to uncovered vertices (A), unnecessary clusters (B) and illogical clusters (C)

Unlike the Star algorithm and the second version of the Extended Star method, GStar solves this drawback. Graphs (A) and (C) of Fig. 7 show the solution to illogical clusters presented in the graph (C) of Fig. 1 and in the graph of Fig. 3.

4 Experimental Results

In this section we present the experimental evaluation of our method and document representation, comparing its results against the Extended Star method and the original Star algorithms. The produced clustering results are evaluated by the same method and criterion to ensure a fair comparison across all algorithms.

Two data sets widely used in document clustering research were used: TREC-5 and Reuters-21578. These are heterogeneous with respect to document size, cluster size, number of classes, and document distribution. The data set TREC-5 contains news in Spanish published by AFP during 1994 (http://trec.nist.gov); Reuters-21578 was obtained from http://kdd.ics.uci.edu. We excluded from data sets the empty documents and also those documents do not have an associated topic.

In our experiments, the documents are represented using the traditional vector space model. The index terms of documents represent the lemmas of the words appearing in the texts. Stops words, such as articles, prepositions and adverbs are removed from document vectors. Terms are statistically weighted using the term frequency. We use the traditional cosine measure to compare the documents.

The main characteristics of these collections are summarized in Table 1. We also included in this table the number of overlapping documents for each collection.

Table 1. Characteristics of document collections

Collect.	Doc.	Overlap. doc.	Topics	Lang.
AFP	695	16	25	Span.
Reuters	10377	1722	119	Engl.

The literature abounds in measures defined by multiple authors to compare two partitions on the same set. The most used are: Jaccard index, and F-measure.

Jaccard index. This index (noted j) takes into account the objects simultaneously joined [10]. It is defined as follows:

$$j(A,B) = \frac{n_{11}}{\frac{N*(N-1)}{2} - n_{00}} \tag{8}$$

In this index, n_{11} denotes the number of pairs of objects which are both in the same cluster in A and are also both in the same cluster in B. Similarly, n_{00} is the number of pairs of objects which are in different clusters in A and are also in different clusters in B.

The performances of the algorithms in the document collections considering Jaccard index are shown in Fig. 8.

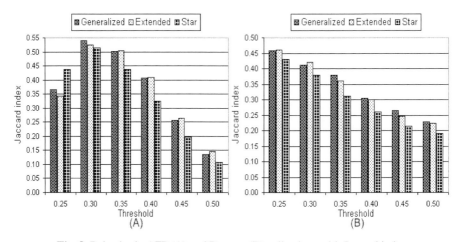

Fig. 8. Behavior in AFP (A) and Reuters (B) collections with Jaccard index

F-measure. The aforementioned index and others are usually applied to partitions. In order to make a better evaluation of overlapping clustering, we have considered F-measure calculated over pairs of points, as defined in [4].

For each pair of points that share at least one cluster in the overlapping clustering results, these measures try to estimate whether the prediction of this pair as being in the same cluster was correct with respect to the underlying true categories in the data. Precision is calculated as the fraction of pairs correctly put in the same cluster, recall is the fraction of actual pairs that were identified, and F-measure (noted *Fmeasure*) is the harmonic mean of precision and recall (9).

$$Fmeasure = \frac{2 * Precision * Recall}{Precision + Recall} \tag{9}$$

$$Precision = \frac{n_{11}}{Number\ of\ Identified\ Pairs} \qquad Recall = \frac{n_{11}}{Number\ of\ True\ Pairs}$$

The performances of the algorithms in the document collections considering F-measure are shown in Fig. 9.

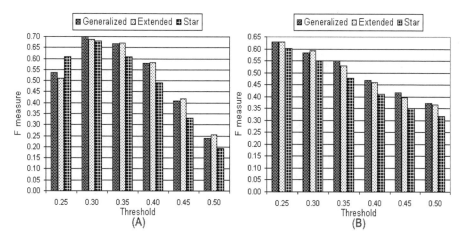

Fig. 9. Behavior in AFP (A) and Reuters (B) collections with F-measure

As can be noticed, the accuracy obtained using the GStar method is in most cases (for all the indexes) comparable with that obtained from the other methods investigated; moreover, our proposal can outperform those methods for all the indexes. But, this behavior is not homogeneous for all σ similarity thresholds; for each collection, there is a minimum value for which GStar outperforms previous Star methods. Starting from this minimum value, the accuracy of GStar is in general as good as, or even in many cases higher than, the others. Besides, this minimum value is generally lesser than the intra-similarity average in all the manually labeled topics.

Furthermore, GStar in all cases obtains lesser clusters than the other algorithms (see Fig. 10), and in most cases obtains denser clusters (see Table 2). This behavior could be of great importance for obtaining a minimum quantity of clusters without loosing the precision.

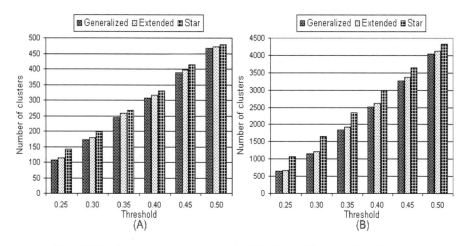

Fig. 10. Number of generated clusters in AFP (A) and Reuters (B) collections

Table 2. Average of elements per clusters in AFP and Reuters collections

Collect.	Algorithm	Threshold					
		0.25	0.30	0.35	0.40	0.45	0.50
AFP	Generalized	18,73	11,54	7,10	5,01	3,43	2,12
	Extended	18,61	11,43	6,85	4,92	3,42	2,20
	Star	9,32	5,43	3,57	2,83	2,13	1,73
Reuters	Generalized	227,36	90,29	51,00	24,57	19,72	12,89
	Extended	176,27	72,74	40,40	23,34	16,61	11,81
	Star	47,21	24,97	15,77	10,93	7,74	6,49

Table 3. Number of redundant clusters generated by the Extended algorithm in both collections

Collect.	Threshold					
	0.25	0.30	0.35	0.40	0.45	0.50
AFP	7	5	13	10	8	3
Reuters	25	65	82	94	104	74

Besides, we include also in Table 3 the number of redundant clusters generated by the Extended algorithm in both collections.

It is important to notice that the Extended algorithm could cover all the vertices, but only in these experiments. Nevertheless, as it was explained, theoretically the Extended algorithm may fail with other repositories.

Despite the experiments carried out by Aslam et al. in [1], and in order to ensure the effectiveness of our proposed algorithm, we made a new experimentation to compare the performance of GStar algorithm against the Single Link and Average Link [7] algorithms, which uses different cost functions. For a fair comparison across all algorithms, we used the same thresholds of the previous experiments, stopping the execution of the Single Link and Average Link algorithms when the two selected clusters to be joined do not satisfy the current threshold, meaning that the evaluation of the cost function for all pair of clusters in the current algorithm return a value greater than the selected threshold. After that, we evaluated each algorithm considering the Jaccard index and F-measure, and we selected the average value of each algorithm for the selected measures for all thresholds.

The performances of the algorithms in the document collections considering Jaccard index, and F-measure are shown in Fig. 11 and Fig. 12.

As we can see, our proposal also outperforms the Single Link and Average Link algorithms in both collections. Thus, the GStar algorithm represents a 68.2% improvement in performance compared to average link and a 42.3% improvement compared to single link in AFP collection considering the Jaccard index; if we consider F-measure, then the GStar algorithm represents a 57.6% improvement in performance compared to average link and a 33.3% improvement compared to single link in the same collection. In the case of the Reuters collection the improvements are higher and even in some cases it doubles the result.

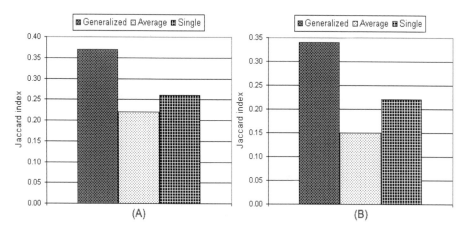

Fig. 11. Behavior in AFP (A) and Reuters (B) collections with Jaccard index

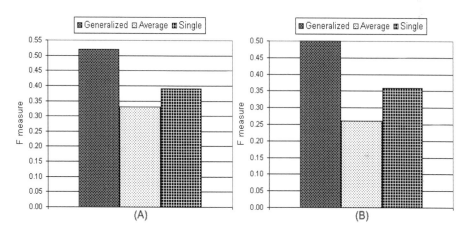

Fig. 12. Behavior in AFP (A) and Reuters (B) collections with F-measure

5 Conclusion

In this paper we presented a new clustering algorithm called Generalized Star (GStar). As a consequence, we obtained different star-shaped clusters. This algorithm solves several drawbacks observed in Star and Extended Star methods; particularly, the dependence on data order (for non symmetric or similar solutions), the generation of illogical and redundant clusters. Besides, our proposal does not leave uncovered vertices.

We compared the GStar algorithm with the original Star and the Extended Star methods in two standard document collections considering two different measures. The experimentation shows that our proposal outperforms previous methods for all the measures. Besides, we also evaluated GStar with the previous methods considering the number and density of the generated clusters. In these experiments,

GStar algorithm generally outperforms previous methods producing lesser and denser clusters. These performances prove the validity of our algorithm for clustering tasks.

This algorithm can be used for organizing information systems in several ways; for example, as a pre-processing step in a static information system or as a post-processing step on the specific documents retrieved by a query. The algorithm can also be used for browsing, topic tracking and new topic detection.

Although we employ our algorithm to cluster documents collections, its use is not restricted to this area, since it can be applied to any problem of pattern recognition where clustering is needed.

As a future work, we will try to develop an incremental version of this algorithm and maybe try to use it as a clustering routine in a dynamic hierarchical clustering algorithm. We also want to do some other experiments considering other representations of the documents and other similarity measures. These experiments could help us to decide how to choose a priory the threshold value in order to obtain the best performance of our algorithm.

References

1. Aslam, J., Pelekhov, K., Rus, D.: Static and Dynamic Information Organization with Star Clusters. In: Proceedings of the 1998 Conference on Information Knowledge Management, Baltimore (1998)
2. Aslam, J., Pelekhov, K., Rus, D.: Using Star Clusters for Filtering. In: Proceedings of the Ninth International Conference on Information and Knowledge Management, USA (2000)
3. Aslam, J., Pelekhov, K., Rus, D.: The Star Clustering Algorithm for Static and Dynamic Information Organization. Journal of Graph Algorithms and Applications 8(1), 95–129 (2004)
4. Banerjee, A., Krumpelman, C., Basu, S., Mooney, R., Ghosh, J.: Model Based Overlapping Clustering. In: Proceedings of International Conference on Knowledge Discovery and Data Mining (KDD) (2005)
5. Charikar, M., Chekuri, C., Feder, T., Motwani, R.: Incremental Clustering and Dynamic Information Retrieval. In: Proceedings of the 29th Symposium on Theory of Computing (1997)
6. Cutting, D., Karger, D., Pedersen, J.: Constant Interaction-time Scatter/Gather Browsing of Very Large Document Collections. In: Proceedings of the 16th SIGIR (1993)
7. Duda, R., Hart, P., Stork, D.: Pattern Classification. John Wiley & Sons Inc. West Sussex (2001)
8. Gil-García, R.J., Badía-Contelles, J.M., Pons-Porrata, A.: Extended Star Clustering Algorithm. In: Sanfeliu, A., Ruiz-Shulcloper, J. (eds.) CIARP 2003. LNCS, vol. 2905, pp. 480–487. Springer, Heidelberg (2003)
9. Gil-García, R.J., Badía-Contelles, J.M., Pons-Porrata, A.: Parallel Algorithm for Extended Star Clustering. In: Sanfeliu, A., Martínez Trinidad, J.F., Carrasco Ochoa, J.A. (eds.) CIARP 2004. LNCS, vol. 3287, pp. 402–409. Springer, Heidelberg (2004)
10. Kuncheva, L., Hadjitodorov, S.: Using Diversity in Cluster Ensembles. In: Proceedings of IEEE SMC 2004, The Netherlands (2004)
11. van Rijsbergen, C.J.: Information Retrieval, 2nd edn. Buttersworth, London (1979)
12. Zhong, S., Ghosh, J.: A Comparative Study of Generative Models for Document Clustering. In: Proceedings of SDM Workshop on Clustering High Dimensional Data and Its Applications (2003)

Evolving Committees of Support Vector Machines*

D. Valincius[1,3], A. Verikas[1,2], M. Bacauskiene[1], and A. Gelzinis[1]

[1] Department of Applied Electronics, Kaunas University of Technology,
Studentu 50, LT-51368, Kaunas, Lithuania
[2] Intelligent Systems Laboratory, Halmstad University,
Box 823, S-30118 Halmstad, Sweden
[3] UAB "Elinta", Pramones pr. 16E, LT-51187 Kaunas
donatas.valincius@elinta.lt, antanas.verikas@ide.hh.se, mabaca@ktu.lt,
adas.gelzinis@ktu.lt

Abstract. The main emphasis of the technique developed in this work for evolving committees of support vector machines (SVM) is on a two phase procedure to select salient features. In the first phase, clearly redundant features are eliminated based on the paired t-test comparing the SVM output sensitivity-based saliency of the candidate and the noise feature. In the second phase, the genetic search integrating the steps of training, aggregation of committee members, and hyper-parameter as well as feature selection into the same learning process is employed. A small number of genetic iterations needed to find a solution is the characteristic feature of the genetic search procedure developed. The experimental tests performed on five real world problems have shown that significant improvements in correct classification rate can be obtained in a small number of iterations if compared to the case of using all the features available.

1 Introduction

Aggregating outputs of multiple predictors into a committee output is one of the most important techniques for improving prediction accuracy [1,2,3]. An efficient committee should consist of predictors that are not only very accurate, but also diverse in the sense that the predictor errors occur in different regions of the input space [4,5]. Manipulating training data set, using different architectures, and employing different subsets of variables are the most popular approaches used to achieve the diversity. To promote diversity of neural networks aggregated into a committee, Liu and Yao [6,7] proposed the so-called *Negative correlation learning* approach, according to which, all individual networks in the committee are trained simultaneously, using an error function augmented with a correlation penalty term. In [8], aiming to find a trade-off between the accuracy and diversity

---

* We gratefully acknowledge the support we have received from the agency for international science and technology development programmes in Lithuania (EUREKA Project E!3681).

P. Perner (Ed.): MLDM 2007, LNAI 4571, pp. 263–275, 2007.

of committee networks, the approach was extended by integrating into the same learning process also the feature selection step. However, to assess and control diversity of predictors and to find the trade-off between the accuracy and diversity is not a trivial task [9,10]. For instance, feature selection may influence the quality of a committee in several ways, namely by reducing model complexity, promoting diversity of committee members, and affecting the trade-off between the accuracy and diversity of committee members. Therefore it seems promising to integrate the steps of training, hyper-parameter and feature selection, and aggregation of members into a committee into the same learning process and to use the prediction accuracy to assess the quality of the committee.

This paper is concerned with such an approach to evolving committees of support vector machines for classification. The main emphasis of the paper is on feature selection for classification committees. A large variety of feature selection techniques have been proposed for a single predictor [11,12], ranging from the sequential forward selection or backward elimination [13,14], sequential forward floating selection [15] to the genetic [16] or tabu search [17]. However, works on feature selection for classification or regression committees are very scarce [5]. It has been demonstrated that even simple random selection of feature subsets may be an effective technique for increasing the accuracy of classification committees [18,19].

One needs to assess the feature saliency when selecting features. The Predictor output sensitivity [20,21,22,23] is the most popular measure used to assess the saliency. Eq. 1 exemplifies such a measure [20,21]

$$ \Upsilon_i = \frac{1}{QP} \sum_{j=1}^{Q} \sum_{p=1}^{P} \left| \frac{\partial y_{jp}}{\partial x_{ip}} \right| \qquad (1) $$

where y is the predictor output, Q is the number of outputs, P is the number of training samples, and x_{ip} is the ith component of the pth input vector $\mathbf{x}_p$. However, a saliency measure alone does not indicate how many of the candidate features should be used. Therefore, some of feature selection procedures are based on making comparisons between the saliency of the candidate and the noise feature [20,21]. Nonetheless the usefulness of such comparisons, the measure does not have direct relation to the prediction error.

The procedure developed in this work for evolving classification committees consists of two phases. In the first phase, clearly redundant features are eliminated based on the paired t-test comparing the saliency of the candidate feature and the noise feature in a single classifier. Then, in the second phase, the genetic search integrating the steps of training, aggregation of committee members into a committee, search for the optimal hyper-parameter values, and selection amongst the remaining features into the same learning process is employed. The committee prediction accuracy is the measure used to assess the committee quality in the genetic search. A small number of genetic iterations needed to find a solution is the characteristic feature of the genetic search procedure developed. The rationale of using the first phase of the procedure is to reduce the computation time needed for the genetic search. If the computation time is not a problem, the first phase of the procedure can be skipped. We use an SVM as a committee member in our tests. However, other types of classifiers can also be utilized.

2   Procedure

The procedure for evolving classification committees is summarized in the following steps.

1. Augment the input vectors with one additional noise feature.
2. Train the model.
3. Calculate the saliency score Γ_i,

$$\Gamma_i = \frac{\Upsilon_i}{\max_{l=1,...,N} \Upsilon_l}, \quad i = 1, ..., N \tag{2}$$

   where Υ_i is given by (1) and N is the number of features.
4. Repeat Steps 2 to 3 K times using different random data partitioning into training, validation and test sets.
5. Eliminate features the saliency of which, do not exceed the saliency of the noise feature. Use the paired t-test to compare the saliency values.
6. Choose the number of committee members L. Construct a chromosome characterizing feature inclusion/noninclusion, regularization and kernel parameters of all the committee members. More details on the chromosome definition are given in Section 2.3.
7. Perform the genetic search.
8. The committee is given by the parameters encoded in the "best" chromosome found during the genetic search.

2.1   The Paired t-Test

To assess the equality of the mean saliency of ith feature μ_{Γ_i} and the noise μ_{Γ_n} the paired t-test is defined as suggested in [21]: Null Hypothesis $\mu_{D_i} = 0$, Alternative Hypothesis $\mu_{D_i} > 0$, where $\mu_{D_i} = \mu_{\Gamma_i} - \mu_{\Gamma_n}$. To test the null hypothesis, a t^* statistic

$$t^* = \frac{\overline{D}_i}{S_{\overline{D}_i}} \tag{3}$$

is evaluated, where $\overline{D}_i = K^{-1} \sum_{j=1}^{K} D_{ij}$, $D_{ij} = \Gamma_{ij} - \Gamma_{nj}$, Γ_{ij} and Γ_{nj} are the saliency scores computed using (2) for the ith and the noise feature, respectively, in the jth loop, and

$$S_{\overline{D}_i} = \sqrt{\frac{\sum_{j=1}^{K}(D_{ij} - \overline{D}_i)^2}{K(K-1)}} \tag{4}$$

Under the null hypothesis, the t^* statistic is t distributed. If $t^* > t_{crit}$, the hypothesis that the difference in the means is zero is rejected, where t_{crit} is the critical value of the t distribution with $\nu = K - 1$ degrees of freedom for a significance level of α: $t_{crit} = t_{1-\alpha,\nu}$.

2.2   The SVM Output Sensitivity, an Example

The output of a support vector machine $y(\mathbf{x})$ is given by:

$$y(\mathbf{x}) = \sum_{j=1}^{N_s} \alpha_j^* d_j \kappa(\mathbf{x}_j, \mathbf{x}) + b \tag{5}$$

where N_s is the number of support vectors, $\kappa(\mathbf{x}_j, \mathbf{x})$ is a kernel, d_j is a target value ($d_j = \pm 1$), and the threshold b and the parameter α_j^* values are found as a solution to the optimization problem defined by the type of SVM used. In this work, we used the 1-norm soft margin SVM [24]. The parameters α_j satisfy the following constrains:

$$\sum_{j=1}^{N_s} \alpha_j y_j = 0, \quad \sum_{j=1}^{N_s} \alpha_j = 1, \quad 0 \le \alpha_j \le C, \quad j = 1, ..., N_s \tag{6}$$

with C being the regularization constant.

For a Gaussian kernel $\kappa(\mathbf{x}_j, \mathbf{x}_k) = \exp\{-||\mathbf{x}_j - \mathbf{x}_k||^2/\sigma\}$, where σ is the standard deviation of the Gaussian, having the jth input vector $\mathbf{x}_j$ presented to the input, the derivative of the output with respect to the ith feature is given by:

$$\frac{\partial y(\mathbf{x}_j)}{\partial x_{ij}} = -\frac{2}{\sigma} \sum_{k=1}^{N_s} \alpha_k^* d_k (x_{ij} - x_{ik}) \exp\left\{ -\sum_{n=1}^{N} \frac{(x_{nj} - x_{nk})^2}{\sigma} \right\} \tag{7}$$

2.3   Genetic Search

Chromosome design, initial population generation, evaluation, selection, crossover, mutation, and reproduction are the issues to consider when designing a genetic search algorithm. We divide **the chromosome** into sections and each section into parts. The number of sections is equal to the number of committee members L. There are three parts in each section. One part encodes the regularization constant C, one the kernel width σ, and the third one encodes the inclusion/noninclusion of features. The binary encoding scheme has been adopted in this work. Fig. 1 illustrates the chromosome structure, where NC and $N\sigma$ stand for the number of bits used to encode the regularization constant C and the kernel width σ, respectively and N is the number of features.

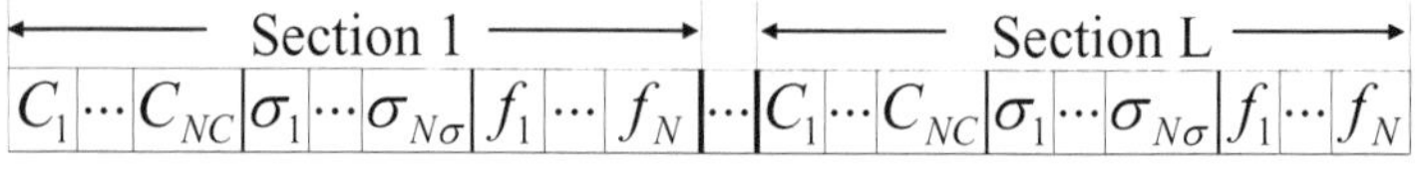

←	Section 1	→	←	Section L	→
$C_1 \cdots C_{NC}$	$\sigma_1 \cdots \sigma_{N\sigma}$	$f_1 \cdots f_N$	$\cdots C_1 \cdots C_{NC}$	$\sigma_1 \cdots \sigma_{N\sigma}$	$f_1 \cdots f_N$

Fig. 1. The structure of the chromosome consisting of L sections

To generate the **initial population**, information obtained from the first feature selection phase, namely, the values of C and σ, and the maximum number of features, is exploited. The maximum number of features allowed for one committee member is equal to the number of features determined in the first phase. In the initial population, the features are masked randomly and values of the parameters C and σ are chosen randomly from the interval $[C_0 - \Delta C, C_0 + \Delta C]$ and $[\sigma_0 - \Delta\sigma, \sigma_0 + \Delta\sigma]$, respectively, where C_0 and Δ_0 are the parameter values obtained from the first phase.

The **fitness function** used to evaluate the chromosomes is given by the correct classification rate of the validation set data. In this study, the committee output was obtained by averaging the outputs of committee members. To distinguish between more than two classes, the one vs one pairwise-classification scheme has been used.

The **selection process** of a new population is governed by the fitness values. A chromosome exhibiting a higher fitness value has a higher chance to be included in the new population. The selection probability of the ith chromosome p_i is given by

$$p_i = \frac{r_i}{\sum_{j=1}^{M} r_j} \tag{8}$$

where r_i is the correct classification rate obtained from the model based on the ith chromosome and M is the population size.

The **crossover operation** for two selected chromosomes is executed with the probability of crossover p_c. If a generated random number from the interval $[0,1]$ is larger than the crossover probability p_c, the crossover operation is executed. Crossover is performed separately in each section of a chromosome. In the "feature mask" and two parameter parts of each section, the crossover point is randomly selected and the corresponding parts of two chromosomes selected for the crossover operation are exchanged at the selected point.

The **mutation operation** adopted is such that each gene is selected for mutation with the probability p_m. The mutation operation is executed independently in each part of each chromosome section. If the gene selected for mutation is in the feature part of the chromosome, the value of the bit representing the feature in the feature mask is reversed. To execute mutation in the parameter part of the chromosome, to choices are possible: i. to reverse the value of the bit in the parameter representation determined by the selected gene; ii. to mutate the value of the offspring parameter determined by the selected gene by $\pm\Delta\gamma$, where γ stands for C or σ, as the case may be. The mutation sign is determined by the fitness values of the two chromosomes, namely the sign resulting into a higher fitness value is chosen. The way of determining the mutation amplitude $\Delta\gamma$ is somewhat similar to that used in [25] and is given by

$$\Delta\gamma = w\beta(\max(|\gamma - \gamma_{p1}|, |\gamma - \gamma_{p2}|)) \tag{9}$$

where γ is the actual parameter value of the offspring, $p1$ and $p2$ stand for parents, $\beta \in [0, 1]$ is a random number, and w is the weight decaying with the iteration number:

$$w = k(1 - t/T) \tag{10}$$

where t is the iteration number, k is a constant, and T is the total number of iterations.

In the **reproduction process**, the newly generated offspring replaces the chromosome with the smallest fitness value in the current population, if a generated random number from the interval [0,1] is larger than the reproduction probability p_r or if the fitness value of the offspring is larger than that of the chromosome with the smallest fitness value.

3   Experimental Investigations

In all the tests, we run an experiment 30 times with different random partitioning of the data set into `<Learning>`, D_l, `<Validation>`, D_v, and `<Test>`, D_t data sets. The mean values and standard deviations of the correct classification rate presented in this paper were calculated from these 30 trials. The parameter values used in the genetic search have been found experimentally. The following values worked well in all the tests: $p_c = 0.05$, $p_m = 0.02$, and $p_r = 0.05$.

3.1   Data Used

To test the approach we used five real-world problems. Data characterizing four of the problems: *US congressional voting records problem, The diabetes diagnosis problem, Wisconsin breast cancer problem,* and *Wisconsin diagnostic breast cancer problem* are available at: `www.ics.uci.edu/~mlearn/`. The fifth problem concerns classification of laryngeal images [26].

Laryngeal images. The task is to automatically categorize colour laryngeal images (images of vocal folds) into the *healthy, nodular,* and *diffuse* decision classes [26]. Fig. 2 presents characteristic examples from the three decision classes considered.

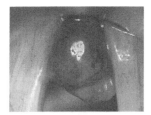 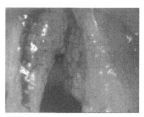 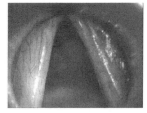

Fig. 2. Images from the *nodular* (left), *diffuse* (middle), and *healthy* (right) classes

Due to a large variety of appearance of vocal folds, the categorization task is sometimes difficult even for a trained physician. Fig. 3 provides an example of such a task. The image placed on the right-hand side of the figure comes from the *nodular* class, while the other two are taken from the *healthy* vocal folds. In this case, the only discriminative feature is the slightly convex vocal fold edges in the upper part of the image coming from the *nodular* class.

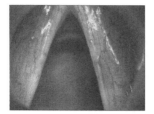 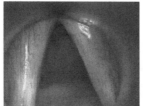 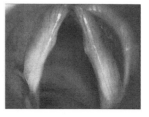

Fig. 3. Three examples of laryngeal images

Aiming to obtain a comprehensive description of laryngeal images, multiple feature sets exploiting information on image colour, texture, geometry, image intensity gradient direction, and frequency content are extracted [27]. Image colour distribution, distribution of the image intensity gradient direction, parameters characterizing the geometry of edges of vocal folds, distribution of the spectrum of the Fourier transform of the colour image complex representation (two types of the frequency content based features), and parameters calculated from multiple co-occurrence matrices are the feature types used to describe laryngeal images [27]. A separate SVM is used to categorize features of each type into the decision classes. The final image categorization is then obtained based on the decisions provided by a committee of support vector machines. In this work, there were 49 images from the *healthy* class, 406 from the *nodular* class, and 330 from the *diffuse* class. Out of the 785 images available, 650 images were assigned to the set D_l.

3.2   Results

First, the average test data set correct classification rate obtained from a single SVM without any involvement of the designing procedure proposed was estimated. The optimal values of the regularization constant C and the kernel width σ have been selected experimentally. Table 1 presents the average test data set correct classification rate obtained for the first four data sets from a single SVM when using all the original features in the classification process. The number of classes and the number of features available are also given in the table. In the parentheses, the standard deviation of the correct classification rate is provided. The average test data set correct classification rate obtained when using a separate SVM for each type of features extracted from the laryngeal images is shown in Table 2.

In the next experiment, we studied the effectiveness of the feature selection procedure applied to single SVMs. Table 3 summarizes the results of the test concerning the first four problems. Apart from the average test data set correct classification rate obtained using the selected features, the table also provides the number of selected features and the number of genetic iterations required to achieve the solution. The number of features eliminated in the first selection phase has been equal to 1, 1, 6, and 12 for the *Diabetes*, *WBCD*, *Voting*,

Table 1. The average test data set correct classification rate obtained for the different data sets from a single SVM when using all the original features

Data set	Number of Classes	Number of features	Classification rate
Diabetes	2	8	76.87 (1.60)
WBCD	2	9	96.86 (0.79)
Voting	2	16	95.49 (1.03)
WDBC	2	30	97.23 (1.01)

Table 2. The average test data set correct classification rate obtained when using a separate SVM for each type of features extracted from the laryngeal images

Feature type	Number of classes	Number of features	Classification rate
Gradient	3	1000	52.30 (5.80)
Co-occurrence	3	42	83.63 (3.17)
Frequency (F1)	3	180	83.38 (3.43)
Frequency (F2)	3	40	78.02 (3.04)
Geometrical	3	18	69.19 (3.48)
Colour	3	50	91.80 (2.69)

and *WDBC* databases, respectively. Observe that the first two problems are characterized by 8 and 9 features, respectively. Thus, there are very few clearly redundant features. The larger number of features eliminated in the first phase for the other two problems significantly speeds up the genetic search executed in the second phase.

Table 3. The average test data set correct classification rate obtained for the different data sets from a single SVM when using the selected features

Data set	Average number of selected features	Average number of iterations	Classification rate
Diabetes	4	8	77.64 (1.50)
WBCD	6	7	97.20 (0.75)
Voting	3	12	96.30 (0.96)
WDBC	17	20	98.06 (0.73)

As it can be seen from Table 1 and Table 3, for all the databases, the average correct classification rate obtained from the single SVMs trained on the selected feature sets is higher than that achieved using all the features available. The number of genetic iterations needed to achieve the solutions is very small. The number of attempts made to make the crossover operation during one genetic iteration is equal to the population size, which was set 50 in all the tests. Fig. 4 provides two graphs plotting the correct classification rate as a function of the number of genetic iterations for the *WDBC* and *Voting* databases. For each genetic iteration, the performance of the best (*max*), the average (*mean*) and the worst (*min*) population member is shown in Fig. 4. The performance achieved by the best member at the end of the search procedure is also shown.

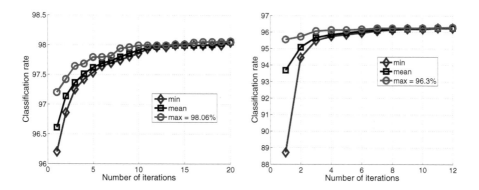

Fig. 4. The test data set correct classification rate obtained from a single SVM as a function of the number of genetic iterations for the Wisconsin diagnostic breast cancer (*left*) and the US congressional voting records (*right*) data sets

Table 4. The average test data set correct classification rate obtained for the different types of features extracted from laryngeal images when using a separate SVM for each type of selected features

Feature type	Average number of selected features	Average number of iterations	Classification rate
Gradient	362	17	83.65 (4.40)
Co-occurrence	28	13	85.48 (3.63)
Frequency (F1)	78	37	89.68 (2.36)
Frequency (F2)	29	13	79.56 (3.47)
Geometrical	10	13	72.12 (3.53)
Colour	42	13	92.74 (2.58)

The results obtained for the different feature sets characterizing the laryngeal images are summarized in Table 4. The number of features eliminated in the first feature selection phase ranged from 5 to over 400. As it can be seen from Table 2 and Table 4, a considerable improvement in classification accuracy has been obtained using the proposed SVM designing approach. The number of features chosen is considerably lower than that presented in Table 2, especially for the *Gradient* and *Frequency* (*F1*) feature types. On average, a very small number of genetic iterations was required to find the solutions. Fig. 5 provides two graphs plotting the correct classification rate as a function of the number of genetic iterations for the two types of frequency features. For each genetic iteration, the performance of the best (*max*), the average (*mean*) and the worst (*min*) population member is shown.

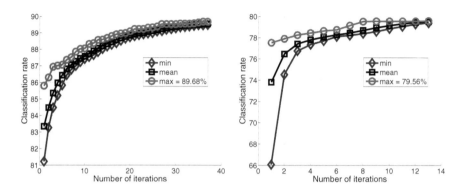

Fig. 5. The test data set correct classification rate obtained from a single SVM as a function of the number of genetic iterations for the two types of frequency features extracted from the laryngeal images

In the last experiment, the effectiveness of the feature selection procedure applied to SVM committees has been studied. Table 5 summarizes the results of the experiment.

Table 5. The average test data set correct classification rate obtained for the different data sets from a committee when using the selected features

Data set	Average number of selected features	Average number of iterations	Classification rate
Diabetes	5	8	77.66 (1.50)
WBCD	5	14	97.27 (0.59)
Voting	6	37	96.62 (0.79)
WDBC	9	20	98.31 (0.46)
Laryngeal	95	8	95.04 (1.88)

All the committees were made of six members. All six members of the committees built for solving the first four problems used the same initial feature set. Each member of the committee built for solving the Laryngeal problem utilized a different feature set—one of the six available types. The average test data set correct classification rate, the average number of features used by one committee member, and the number of iterations needed to obtain the solution are given in Table 5. As it can be seen from Table 5, the technique developed is capable of evolving accurate classification committees in a small number of genetic iterations. The relatively large average number of features used by the "laryngeal" committee is due to the large number of "gradient" features selected. Fig. 6 provides two graphs plotting the test data set correct classification rate obtained from the committees as a function of the number of genetic iterations for the Laryngeal (*left*) and the Wisconsin diagnostic breast cancer (*right*) problems.

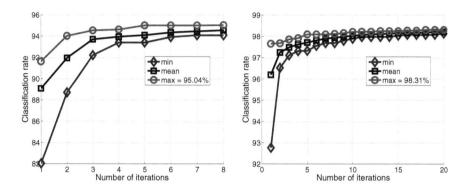

Fig. 6. The test data set correct classification rate obtained from the committee as a function of the number of genetic iterations for the Laryngeal (*left*) and the Wisconsin diagnostic breast cancer (*right*) data sets

4   Conclusions

A technique for evolving committees of support vector machines has been presented in this work. The main emphasis of the technique is on selection of salient features. Elimination of clearly redundant features in the first phase of the procedure developed speeds up the genetic search executed in the second phase of the designing process. The genetic search integrating the steps of training, aggregation of committee members, and hyper-parameter as well as feature selection into the same learning process allows creating effective models in a small number of genetic iterations. The experimental tests performed on five real world problems have shown that considerable improvements in classification accuracy can be obtained using the proposed SVM designing approach.

References

1. Gader, P.D., Mohamed, M.A., Keller, J.M.: Fusion of handwritten word classifiers. Pattern Recognition Letters 17, 577–584 (1996)
2. Liu, C.L.: Classifier combination based on confidence transformation. Pattern Recognition 38, 11–28 (2005)

3. Verikas, A., Lipnickas, A., Malmqvist, K., Bacauskiene, M., Gelzinis, A.: Soft combination of neural classifiers: A comparative study. Pattern Recognition Letters 20, 429–444 (1999)
4. Krogh, A., Vedelsby, J.: Neural network ensembles, cross validation, and active learning. In: Tesauro, G., Touretzky, D.S., Leen, T.K. (eds.) Advances in Neural Information Processing Systems, vol. 7, pp. 231–238. MIT Press, Cambridge (1995)
5. Bacauskiene, M., Verikas, A.: Selecting salient features for classification based on neural network committees. Pattern Recognition Letters 25, 1879–1891 (2004)
6. Liu, Y., Yao, X.: Ensemble learning via negatine correlation. Neural Networks 12, 1399–1404 (1999)
7. Liu, Y., Yao, X., Higuchi, T.: Evolutionary ensembles with negative correlation learning. IEEE Trans on Evolutionary Computation 4, 380–387 (2000)
8. Bacauskiene, M., Cibulskis, V., Verikas, A.: Selecting variables for neural network committees. In: Wang, J., Yi, Z., Zurada, J.M., Lu, B.-L., Yin, H. (eds.) ISNN 2006. LNCS, vol. 3971, pp. 837–842. Springer, Heidelberg (2006)
9. Kuncheva, L.I., Whitaker, C.J.: Measures of diversity in classifier ensembles and their relationship with the ensemble accuracy. Machine Learning 51, 181–207 (2003)
10. Brown, G., Wyatt, J., Harris, R., Yao, X.: Diversity creation methods: a survey and categorisation. Information Fusion 6, 5–20 (2005)
11. Kudo, M., Sklansky, J.: Comparison of algorithms that select features for pattern classifiers. Pattern Recognition 33, 25–41 (2000)
12. Verikas, A., Bacauskiene, M.: Feature selection with neural networks. Pattern Recognition Letters 23, 1323–1335 (2002)
13. Mao, K.Z.: Orthogonal forward selection and backward elimination algorithms for feature subset selection. IEEE Trans Systems, Man, & Cybernetics—Part B: Cybernetics 34, 629–634 (2004)
14. Guyon, I., Weston, J., Barnhill, S., Vapnik, V.: Gene selection for cancer classification using support vector machines. Machine Learning 46, 389–422 (2002)
15. Pudil, P., Novovicova, J., Kittler, J.: Floating search methods in feature selection. Pattern Recognition Letters 15, 1119–1125 (1994)
16. Yu, S., Backer, S.G., Scheunders, P.: Genetic feature selection combined with composite fuzzy nearest neighbor classifiers for hyperspectral satellite imagery. Pattern Recognition Letters 23, 183–190 (2002)
17. Zhang, H., Sun, G.: Feature selection using tabu search method. Pattern Recognition 35, 701–711 (2002)
18. Ho, T.K.: The random subspace method for constructing decision forests. IEEE Trans Pattern Analysis and Machine Intelligence 20, 832–844 (1998)
19. Tsymbal, A., Puuronen, S., Patterson, D.W.: Ensemble feature selection with simple Bayesian classification. Information Fusion 4, 87–100 (2003)
20. Priddy, K.L., Rogers, S.K., Ruck, D.W., Tarr, G.L., Kabrisky, M.: Bayesian selection of important features for feedforward neural networks. Neurocomputing 5, 91–103 (1993)
21. Steppe, J.M., Bauer, K.W.: Improved feature screening in feedforward neural networks. Neurocomputing 13, 47–58 (1996)
22. Acir, N., Guzelis, C.: Automatic recognition of sleep spindles in EEG via radial basis support vector machine based on a modified feature selection algorithm. Neural Computing & Applications 14, 56–65 (2005)
23. Evgeniou, T., Pontil, M., Papageorgiou, C., Poggio, T.: Image representations and feature selection for multimedia database search. IEEE Trans Knowledge and Data Engineering 15, 911–920 (2003)

24. Shawe-Taylor, J., Cristianini, N.: Kernel Methods for Pattern Analysis. Cambridge University Press, Cambridge, UK (2004)

25. Leung, K.F., Leung, F.H.F., Lam, H.K., Ling, S.H.: Application of a modified neural fuzzy network and an improved genetic algorithm to speech recognition. Neural Computing & Applications 16 (2007)

26. Verikas, A., Gelzinis, A., Bacauskiene, M., Uloza, V.: Integrating global and local analysis of colour, texture and geometrical information for categorizing laryngeal images. International Journal of Pattern Recognition and Artificial Intelligence 20, 1187–1205 (2006)

27. Verikas, A., Gelzinis, A., Valincius, D., Bacauskiene, M., Uloza, V.: Multiple feature sets based categorization of laryngeal images. Computer Methods and Programs in Biomedicine 85, 257–266 (2007)

Choosing the Kernel Parameters for the Directed Acyclic Graph Support Vector Machines

Kuo-Ping Wu and Sheng-De Wang

Department of Electrical Engineering
National Taiwan University
Taipei, Taiwan
sdwang@ntu.edu.tw

Abstract. The directed acyclic graph support vector machines (DAGSVMs) have been shown to be able to provide classification accuracy comparable to the standard multiclass SVM extensions such as Max Wins methods. The algorithm arranges binary SVM classifiers as the internal nodes of a directed acyclic graph (DAG). Each node represents a classifier trained for the data of a pair of classes with the specific kernel. The most popular method to decide the kernel parameters is the grid search method. In the training process, classifiers are trained with different kernel parameters, and only one of the classifiers is required for the testing process. This makes the training process time-consuming. In this paper we propose using separation indexes to estimate the generalization ability of the classifiers. These indexes are derived from the inter-cluster distances in the feature spaces. Calculating such indexes costs much less computation time than training the corresponding SVM classifiers; thus the proper kernel parameters can be chosen much faster. Experiment results show that the testing accuracy of the resulted DAGSVMs is competitive to the standard ones, and the training time can be significantly shortened.

1 Introduction

The support vector machines (SVMs) are originally designed for binary classification problems [1]. To solve the multiclass classification problems, the SVMs should be extended. The most often used extensions of the SVMs are the one-against-one [2] and one-against-all strategies [3]. A decision-tree-based modification of the one-against-one strategy is the directed acyclic graph SVM (DAGSVM) [4]. These algorithms have been shown to perform well in real world applications [5]. Among these extensions, the performance of the DAGSVM is as good as that of the one-against-one strategy, while the testing time of DAGSVM is reduced. The DAGSVM constructs the directed tree structure with the pairwise classifiers as its nodes. Therefore, designing a DAGSVM for a multiclass problem is to train the standard SVM classifiers for each node.

P. Perner (Ed.): MLDM 2007, LNAI 4571, pp. 276–285, 2007.

The training algorithms of SVMs look for the optimal separating hyperplane which has a maximized margin between the hyperplane and the data and thus minimizing the upper bound of the generalization error. The separating hyperplane is represented by a small number of training data, called support vectors (SVs). Since the real data are often linearly inseparable, the data are mapped into a higher dimensional space, the feature space, in which the data are possibly more separable. In practice, a kernel function is incorporated to simplify the computation of the inner product value of the transformed data.

Although the performance of an SVM depends largely on the kernel, there is no theoretical method for determining a kernel and its parameters. Many existing approaches to determining the kernel parameters are mentioned in section 2. Most of them look for good parameter combinations by training SVMs with all parameter combinations in selected intervals, resulting in a very time-consuming total training process. In this paper we propose using some indexes to predict a good choice of the kernel parameters. The indexes derived from the inter-cluster distances are proposed to estimate the performance of the classifier generalization ability. If the pairwise classes data are considered as two labeled clusters, a kernel parameter combination that leads to a good separation of data in the resulted feature space would be a good choice. Meanwhile, the classifier being found in that feature space is likely to have larger margin and thus will have a higher generalization ability. The indexes we proposed are with $O(l^2)$ computational complexity for the sample size l. Although the time complexity is about the same as training an SVM, the experimental results being listed in section 4 show that the actual computation time for the indexes are much less than that for SVM training. According to the indexes, the kernel parameters we choose from the possible candidates result in DAGSVMs which perform as well as the ones being generated from the widely used grid search method.

2   SVM, DAGSVM and Kernel Selection

The SVM is designed for binary-classification problems, assuming the data are linearly separable. Given the training data $(x_i, y_i), i = 1, \ldots, l, x_i \in R^n, y_i \in \{+1, -1\}$, where the R^n is the input space, x_i is the sample vector and y_i is the class label of x_i, the separating hyperplane (w, b) is a linear discriminating function that solves the optimization problem:

$$\min_{w,b} \langle w, w \rangle,$$
$$\text{subject to } y_i \left(\langle w, x_i \rangle + b \right) \geq 1, i = 1, \ldots, l \ . \tag{1}$$

$\langle .,. \rangle$ indicates the inner product operation. The minimal distance between the samples and the separating hyperplane, i.e. the margin, is $1/\|w\|$.

In order to relax the margin constraints for the non-linearly separable data, the slack variables are introduced into the optimization problem:

$$\min_{\xi,w,b} \langle w, w \rangle + C \sum_{i=1}^{l} \xi_i,$$

$$subject\ to\ y_i\left(\langle\omega,x_i\rangle+b\right)\geq 1-\xi_i, i=1,\ldots,l, \xi_i\geq 0 \tag{2}$$

which is the soft margin SVM being generally discussed and applied. The resulted classifier is called the 1-norm soft margin SVM, and C is the penalty parameter of error. The decision function of the classifier is

$$sign\left(\sum_{x_i:SV} y_i\alpha_i\langle x_i,x\rangle+b\right). \tag{3}$$

In practice, since the real data are often not linearly separable in the input space, the data can be mapped into a high dimensional feature space, in which the data are sparse and possibly more separable. The mapping is often not explicitly given. Instead, a kernel function is incorporated to simplify the computation of the inner product value of the transformed data in the feature space.

When using a function $\phi : X \to F$ to map the data into a high dimensional feature space, the decision function of the classifier becomes

$$sign\left(\sum_{x_i:SV} y_i\alpha_i\langle \phi\left(x_i\right),\phi\left(x\right)\rangle+b\right). \tag{4}$$

The mapping ϕ is not given explicitly in most cases. Instead, a kernel function $K(x,x')=\langle\phi\left(x\right),\phi\left(x'\right)\rangle$ gives the inner product value of x and x' in the feature space. Choosing a kernel function is therefore choosing a feature space and the decision function becomes

$$sign\left(\sum_{x_i:SV} y_i\alpha_i K\left(x_i,x\right)+b\right). \tag{5}$$

The most often used kernel functions are:

- linear: $K\left(x,z\right)=\langle x,z\rangle$
- polynomial: $K\left(x,z\right)=\left(\gamma\langle x,z\rangle+r\right)^d, \gamma>0$
- radial basis function (RBF): $K\left(x,z\right)=e^{\left(-\frac{\|x-z\|^d}{2\sigma^2}\right)}, \sigma>0$
- sigmoid: $K\left(x,z\right)=\tanh\left(\gamma\langle x,z\rangle+r\right)$

For certain parameters, the linear kernel is a special case of RBF kernels [6], and the sigmoid kernel behaves like the RBF kernel [7]. When the data are linearly inseparable, a non-linear kernel such as RBF that maps the data into the feature space non-linearly can handle the data better than the linear kernels.

When using the DAGSVM for the multiclass data, an n-class problem is considered as $n(n-1)/2$ 2-class problems. The data from each pair of the classes are used to train a binary classifier, and all the binary classifiers form the internal nodes of a rooted binary directed acyclic graph. Classifying a sample is going through a path from the root to a leaf. At each node, the sample is classified by the classifier of that node. The classification output decides moving to the left or right child node, as illustrated in Fig. 1.

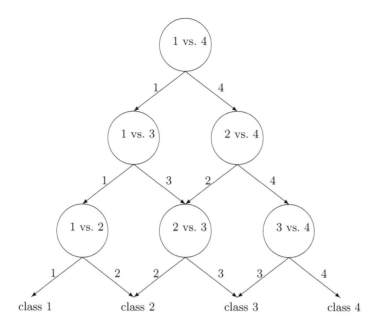

Fig. 1. A DAG with a predefined node order

The most often used kernel function in an SVM is the RBF kernel. When using the RBF kernel, the parameters $\langle d, \sigma \rangle$ should be set properly. Generally d is set to be 2, thus the kernel value is related to the Euclidean distance between the two samples. σ is related to the kernel width. The most straightforward way to set the parameters is the grid search method [5]. This method trains SVMs with all preferred parameter combinations. The classifier with the best accuracy is applied in the testing process. The genetic algorithms (GAs) can be applied to the SVM parameter search [8] [9] [10]. These parameter selection methods are time-consuming because SVMs are trained for many combinations. The training time complexity of the SVMs is $O(l^4)$ in worst case [11] or is experimentally shown to be $O(l)$–$O(l^{2.2})$ [12] [13], depends on the data. Debnath and Takahashi use the principle components of data in the feature space to create an index for choosing proper kernel parameters [14]. Bi et al propose a method for selecting a suitable σ for RBF kernels according to the relationship of boundary data in the feature space [15]. These methods calculate kinds of indexes as the heuristics of choosing kernel parameters. The actual computation time is relatively short in comparison to training an SVM. Based on the data geometry and distribution in the feature space, we propose using the indexes derived from the inter-cluster distances in the feature space as the heuristics to choose the kernel parameters.

3   Inter-cluster Distance and the Proposed Indexes

Generalization is the ability that a trained model gives the target value of an input sample which is not in the training set. Many indexes can be used to assess the generalization ability. The most direct one is the validation accuracy. For example, the training process of the grid search method uses the validation accuracy to indicate the generalization ability of the classifier to the validation data. With this method, a classifier is generated for each parameter combination and the classifier with the best validation accuracy is chosen for generalization usage. In addition to the training and testing samples, the validation samples are reserved for validating the trained model to see whether the model overfits the training samples. Therefore, it reflects the generalization ability of the model to unknown data.

There are some other SVM-related indexes that can estimate the generalization ability. Takahashi [16] used the ratio of the numbers of support vectors to the training samples as an index. This suggests a useful index as $\frac{M_{ij}}{SV_{ij}}$ for the classifier of classes i and j, where M_{ij} is the number of training data and SV_{ij} is the number of support vectors. Phetkaew [17] proposed using the SVM margin to identify a classifier that causes wrong classifications. As the margin decrease, the distance between the nearest two samples from different classes decreases and the more confusion between the two classes will be. These indexes require the information of the trained classifiers, so they are as time-consuming as the grid search is. Similar to the margin stands for, our previous work [18] proposed a separation index which indicates the separation of two classes in the feature space. We extend the previous proposed index to the following ones. Bezdek and Pal [19] mentioned several inter-cluster distances. Considering the classes data as the labeled clusters data, these distances can be modified and be used for estimating the separation degree of two classes. The following two are robust to noise:

$$\delta_4\left(X_+, X_-\right) = d\left(\bar{x_+}, \bar{x_-}\right)$$

$$= d\left(\frac{\sum\limits_{x_+ \in X_+} x_+}{l_+}, \frac{\sum\limits_{x_- \in X_-} x_-}{l_-}\right), \tag{6}$$

$$\delta_5\left(X_+, X_-\right) = \frac{1}{l_+ + l_-}\left(\sum\limits_{x_+ \in X_+} d\left(x_+, \bar{x_-}\right) + \sum\limits_{x_- \in X_-} d\left(x_-, \bar{x_+}\right)\right), \tag{7}$$

where X_+ and X_- are positive and negative classes, l_+ and l_- are sample sizes of X_+ and X_-, and $\bar{x_+}$ and $\bar{x_-}$ are the class means of X_+ and X_-, respectively. $d(x, z)$ is the preferred function which calculates the distance between two vectors. A class mean is the arithmetic mean of the data in the same class. It can be thought as the center of a class and thus stands for the class. The distance index δ_4 is the distance between the two means of the classes. The distance index δ_5 is the average distance between each sample to its opposite class mean. Both the indexes reflect the distance between the two classes in some ways, thus they are estimators of the separation degree of two classes.

In equation (6) and (7), $d(x, z)$ is calculated with the data in the input space in general. It represents the spatial relationship of two vectors in the input space.

Since the data are mapped into the feature space first for the SVM training, the indexes which are calculated in the input space may not represent the separation degree of the two classes in the feature space. Referring to our previous work [18], the distance indexes can be calculated in the feature space using the L^2-norm with the kernel function incorporated. The inter-cluster distance in the feature space can be evaluated by applying the L^2-norm on the mapped data. The kernel function gives the inner product values of vectors in the feature space. We define an index, δ_{4FSQ}, which corresponds to the distance measure δ_4 in the feature space using the L^2-norm.

$$
\begin{aligned}
&\delta_{4FSQ}\left(X_+, X_-\right) \\
&= d\left(\hat{x_+}, \hat{x_-}\right) \\
&= \sqrt{\left|\hat{x_+} - \hat{x_-}\right|_2^2} \\
&= \sqrt{\left|\frac{\sum\limits_{x_+ \in X_+} \phi(x_+)}{l_+} - \frac{\sum\limits_{x_- \in X_-} \phi(x_-)}{l_-}\right|_2^2} \\
&= \sqrt{\frac{\sum\limits_{\substack{x_{+i} \in X_+ \\ x_{+j} \in X_+}} K(x_{+i}, x_{+j})}{l_+^2} + \frac{\sum\limits_{\substack{x_{-p} \in X_- \\ x_{-q} \in X_-}} K(x_{-p}, x_{-q})}{l_-^2} - \frac{2 \sum\limits_{\substack{x_{+m} \in X_+ \\ x_{-n} \in X_-}} K(x_{+m}, x_{-n})}{l_+ l_-}},
\end{aligned}
$$
(8)

where $\hat{x_+}$ and $\hat{x_-}$ are the class means of the mapped X_+ and X_- data, respectively. Similarly, δ_{5FSQ} corresponding to δ_5 can be

$$
\begin{aligned}
&\delta_{5FSQ}\left(X_+, X_-\right) \\
&= \frac{1}{l_+ + l_-}\left(\sum\limits_{x_+ \in X_+} \sqrt{\left|\phi(x_+) - \hat{x_-}\right|_2^2} + \sum\limits_{x_- \in X_-} \sqrt{\left|\phi(x_-) - \hat{x_+}\right|_2^2}\right),
\end{aligned}
$$
(9)

where

$$
\begin{aligned}
&\left|\phi(x_+) - \hat{x_-}\right|_2^2 \\
&= \left|\phi(x_+) - \frac{\sum\limits_{x_- \in X_-} \phi(x_-)}{l_-}\right|_2^2 \\
&= K(x_+, x_+) + \frac{\sum\limits_{\substack{x_{-p} \in X_- \\ x_{-q} \in X_-}} K(x_{-p}, x_{-q})}{l_-^2} - \frac{2 \sum\limits_{x_{-n} \in X_-} K(x_+, x_{-n})}{l_-}.
\end{aligned}
$$
(10)

and $\left|\phi(x_-) - \hat{x_+}\right|_2^2$ can be expressed in the similar way. δ_{4FSQ} and δ_{5FSQ} use the L^2-norm to calculate $d(x, z)$ for two mapped vectors. Computing the L^2-norm values involves many inner products of mapped vectors, which can be in

Table 1. Problem statistics

data set	# training data	# testing data	# classes	# attributes
dna	2000	1186	3	180
letter	15000	5000	26	16
satimage	4435	2000	6	36

Table 2. Training Process 1: Ordinary DAGSVM training

for each $< C, \sigma >$ combination
    for each pair of classes
        train a classifier with the pairwise training data
    construct the DAG with all the pairwise classifiers
    classify the whole validation data by the DAG
the DAG with highest validation accuracy is for testing process usage

turn expressed by the kernel function. Therefore, the separation index values for data in the feature space can be obtained. Calculating these indexes are with $O(l^2)$ computational complexity. Since these indexes are related to inter-cluster distances, they can represent the separation degree of the classes and thus can estimate the classifier generalization ability.

4   Experiments and Results

In our experiments, we use the data sets dna, satimage and letter from the Statlog collection [20]. The database statistics are listed in Table 1. The original training data are scaled to be in $[-1, 1]$ and are partitioned into training and validation data sets [5]. The partitioned data are available on [21]. All the pairwise classifiers are trained with the 1-norm soft margin SVM. We use the RBF kernel with $d = 2$, $\sigma \in [2^{-4}, 2^{-3}, \ldots, 2^5]$ and C is set with the values $[2^{-4}, 2^{-3}, \ldots, 2^{10}]$.

We first find the best $< C, \sigma >$ combination for the whole model. The corresponding training process (TP1) is listed in Table 2. For each parameter combination, all the pairwise classifiers are trained to construct a DAG corresponding to the parameter combination. The validation data are classified by each DAG to obtain the validation rate of the whole DAG. As there is no such limitation that all the pairwise classifiers must use the same parameter combination, we also test the performance of the DAGSVM with each classifiers using different parameter combinations. The relaxed training process (TP2) is listed in Table 3. For a pair of classes, the grid search method is applied to the training process to choose the best classifier for the two classes. Only the validation data from the two classes are classified by the classifier. All the best classifiers for different pair of classes together construct the DAG for testing. In practice, TP1 and TP2 can share the same training and validation results with different arrangements, and therefore they can have the same training time. However, since TP1 validates each DAG with all validation data while TP2 validates each pairwise

Table 3. Training Process 2: DAGSVM with each pairwise classifier choosing parameters separately

for each pair of classes
for each $< C, \sigma >$ combination
train a classifier with the pairwise training data
classify the validation data corresponding to the two classes
choose the classifier with highest validation accuracy
construct the DAG by the chosen classifiers for testing process usage

Table 4. Training Process 3: DAGSVM using proposed indexes for choosing kernel parameters

for each pair of classes
for each σ
calculate the separation index
choose the σ with the highest index value
with the chosen σ, for each C
train a classifier with the pairwise training data
classify the validation data corresponding to the two classes
choose the classifier with highest validation accuracy
construct the DAG by the chosen classifiers for testing process usage

classifier with limited validation data, the resulted DAGs can be different in chosen classifiers and testing accuracy.

Since the validation accuracy is an estimation of classifier generalization ability, it can be substituted by using the proposed indexes as the heuristics to choose the kernel parameters. Because the kernel functions are involved in calculating the index values, the kernel parameters can be chosen according to the index values. There left only C to be chosen by the validation process, and the time needed for the training process for different kernel parameters can be saved. The training process (TP3) is presented in Table 4. We apply δ_{4FSQ} to TP3 to show that the heuristic works. The resulted DAG is validated with all the validation data.

Table 5 lists the testing accuracy and training time of the trained DAGSVMs. As TP1 and TP2 use the identical training and validation results, these two training process are identical in the time consumption. Although TP1 uses the global information while TP2 and TP3 don't, the testing accuracy of TP2 and TP3 is competitive to TP1 because the best pairwise classifiers are used. Since it is not necessary to train models for the kernel parameters other than the ones chosen by the heuristics, the training time of TP3 is significantly shorter than TP1 and TP2. As we use ten candidates of σ and TP3 trains one DAGs only in our experiment, the training process is about ten-time sped up. If a finer grid for searching the parameters is used in TP1 and TP2, more candidates of the kernel parameters are required to be validated and the proposed method would speed up the training process more. We also observed that the kernel parameters

Table 5. Testing rates (%) and training time (in seconds) for the DAGSVMs using TP1, TP2 and TP3

data set	TP1 testing rate	TP1 training time	TP2 testing rate	TP2 training time	TP3 testing rate	TP3 training time	speed up factor
dna	94.35	3277.2	94.69	3277.2	94.60	162.1	20.2
letter	97.02	62004.3	95.42	62004.3	95.64	5827.8	10.6
satimage	90.45	7066.0	90.80	7066.0	89.85	874.9	8.1

chosen by the proposed heuristics are often the same ones or close to the ones chosen by the validation accuracy. According to this, the proposed indexes can also be used to suggest the starting point or searching center for the grid search process.

5   Conclusion

We propose using the inter-cluster distance based index to choose the kernel parameters for each classifier in DAGSVMs. Choosing the kernel parameters for classifiers individually can possibly lead to a DAGSVM which performs even better than a DAGSVM with the same parameter for all classifiers. Meanwhile, it is not necessary to choose the kernel parameters for the pairwise classifiers by the time-consuming grid search method. Using the proposed index, the proper kernel parameters can be chosen. Since the index can be calculated much faster than the grid search method, the training process for the proposed method can be significantly faster than the generally used grid search method. Currently the penalty parameter C is not incorporated into the proposed strategy; thus it is our future work to also predict the penalty parameter.

Acknowledgments. The work was partially supported by National Science Council, Taiwan under the grant NSC 95-2221-E-002-061.

References

1. Vapnik, V.: Statistical Learning Theory. Wiley, New York (1998)
2. Kreß el, U.: Pairwise classification and support vector machines. In: Schölkopf, B., Burges, C.J.C., Smola, A.J. (eds.) Advances in Kernel Methods – Support Vector Learning, MIT Press, Cambridge, MA (1999)
3. Bottu, L., Cortes, C., Denker, J., Drucker, H., Guyon, I., Jackel, L., LeCun, Y., Muller, U., Sackinger, E., Simard, P., Vapnik, V.: Comparison of classifier methods: A case study in handwritten digit recognition. In: Proc. Int. Conf. Pattern Recognition, pp. 77–87 (1994)
4. Platt, J.C., Cristianini, N., Show-Tayler, J.: Large margin DAG's for multiclass classification. In: Advances in Neural Information Processing Systems, vol. 12, pp. 547–553. MIT Press, Cambridge, MA (2000)

5. Hsu, C.-W., Lin, C.-J.: A comparison of methods for multiclass support vector machines. IEEE trans. Neural Networks 13(2), 415–425 (2002)
6. Keerthi, S.S., Lin, C.-J.: Asymptotic behaviors of support vector machines with Gaussian kernel. Neural Computation 15(7), 1667–1689 (2003)
7. Lin, H.-T., Lin, C.-J.: A study on sigmoid kernels for SVM and the training of non-PSD kernels by SMO-type methods, Technical report, Department of Computer Science and Information Engineering, National Taiwan University
8. Rojas, S.A., Fernandez-Reyes, D.: Adapting Multiple Kernel Parameters for Support Vector Machines using Genetic Algorithms. The 2005 IEEE Congress on Evolutionary Computation 1, 626–631 (2005)
9. Liang, X., Liu, F.: Choosing multiple parameters for SVM based on genetic algorithm. In: 6th International Conference on Signal Processing, vol. 1, pp. 117–119 (August 26-30, 2002)
10. Liu, H.-J., Wang, Y.-N., Lu, X.-F.: A Method to Choose Kernel Function and its Parameters for Support Vector Machines. In: Proceedings of 2005 International Conference on Machine Learning and Cybernetics, vol. 7, pp. 4277–4280 (August 18–21, 2005)
11. Hush, D., Scovel, C.: Polynomial-time decomposition algorithms for support vector machines. Machine Learning 51, 51–71 (2003)
12. Platt, J.C.: Sequential minimal optimization: A fast algorithm for training support vector machines, Microsoft Research, Technical Report MST-TR-98-14 (1998)
13. Maruyama, K.-I., Maruyama, M., Miyao, H., Nakano, Y.: A method to make multiple hypotheses with high cumulative recognition rate using SVMs. Pattern Recognition 37, 241–251 (2004)
14. Debnath, R., Takahashi, H.: An efficient method for tuning kernel parameter of the support vector machine. In: IEEE International Symposium on Communications and Information Technology, vol. 2, pp. 1023–1028 (October 26–29, 2004)
15. Bi, L.-P., Huang, H., Zheng, Z.-Y., Song, H.-T.: New Heuristic for Determination Gaussian Kernel's Parameter. In: Proceedings of 2005 International Conference on Machine Learning and Cybernetics, vol. 7, pp. 4299–4304 (August 18–21, 2005)
16. Takahashi, F., Abe, S.: Optimizing directed acyclic graph support vector machines. In: Proc. Artificial Neural Networks in Pattern Recognition (ANNPR 2003), pp. 166–170 (September 2003)
17. Phetkaew, T., Kijsirikul, B., Rivepiboon, W.: Reordering adaptive directed acyclic graphs: An improved algorithm for multiclass support vector machines. In: Proc. Internat. Joint Conf. on Neural Networks (IJCNN 2003), vol. 2, pp. 1605–1610 (2003)
18. Wu, K.-P., Wang, S.-D.: Choosing the Kernel parameters of Support Vector Machines According to the Inter-cluster Distance. In: Proc. Internat. Joint Conf. on Neural Networks (IJCNN 2006) (2006)
19. Bezdek, C., Pal, N.R.: Some New Indexes of Cluster Validity. IEEE Transaction on Systems, Man, And Cybernetics-Part B: Cybernetics 28(3), 301–315 (1998)
20. Michie, D., Spiegelhalter, D.J., Taylor, C.C.: Machine learning, neural and statistical classification (1994), [Online available]:
http://www.amsta.leeds.ac.uk/~charles/statlog/
21. Chang, C.-C., Lin, C.-J.: LIBSVM: a library for support vector machines, [Online] available from World Wide Web:
http://www.csie.ntu.edu.tw/~cjlin/libsvmtools/datasets/

Data Selection Using SASH Trees
for Support Vector Machines

Chaofan Sun and Ricardo Vilalta

Department of Computer Science, University of Houston
4800 Calhoun Rd., Houston TX, 77204-3010
{cfsun, vilalta}@cs.uh.edu
Center for Research and Advanced Studies (CINVESTAV)
Av. Científica 1145, Guadalajara, México, 45010
rvilalta@gdl.cinvestav.mx

Abstract. This paper presents a data preprocessing procedure to select support vector (SV) candidates. We select decision boundary region
vectors (BRVs) as SV candidates. Without the need to use the decision
boundary, BRVs can be selected based on a vector's nearest neighbor of
opposite class (NNO). To speed up the process, two spatial approximation sample hierarchical (SASH) trees are used for estimating the BRVs.
Empirical results show that our data selection procedure can reduce a full
dataset to the number of SVs or only slightly higher. Training with the
selected subset gives performance comparable to that of the full dataset.
For large datasets, overall time spent in selecting and training on the
smaller dataset is significantly lower than the time used in training on
the full dataset.

Keywords: sampling methods, support vector machines.

1 Introduction

Support vector machines (SVMs) [1] are a popular approach to machine learning
because of their solid analytical foundation and frequent high generalization
power. However, standard maximal margin SVMs face the difficulty of solving
a quadratic programming (QP) problem with time and space complexities of
$O(n^3)$ and $O(n^2)$ respectively (where n is the size of the input dataset). When
the dataset is relatively large, training SVMs becomes intolerable slow and often
results in memory shortage. Multiple efforts have been undertaken to overcome
these difficulties. Previous work has focused on how to simplify or modify the
QP problem (e.g., the sequential minimum optimization (SMO) algorithm [2]).
Another line of research has focused on reducing the training dataset by selecting
only a set of potential support vectors (SVs) to define the decision boundary.

In terms of data selection, existing methodologies can be roughly divided
into three categories: (1) data sampling, (2) neighborhood-based selection, and
(3) boundary-based selection. In the first category, statistical techniques such
as random sampling and stratified sampling [3,4,5] are employed to do data

P. Perner (Ed.): MLDM 2007, LNAI 4571, pp. 286–295, 2007.
© Springer-Verlag Berlin Heidelberg 2007

selection. These techniques are simple but tend to miss SVs and select non-SVs. In the second category, data selection is done by focusing on regions populated by k-mean clusters ([6,7]), or on vector neighborhoods (such as k-NNs [8,9], or Gabriel neighbors [10]). Using clustering to do data selection relies heavily on the existence of regions with high sample density; sparse data makes it difficult to identify clusters even when some of these data should be kept as SVs candidates. Hence, using clusters to select SVs candidates tends to be overly conservative. In addition, using k-NNs to select SVs candidates is based on the assumption that data close to the decision boundary should have a mixed number of class labels among their neighbors; a common research problem is to measure the degree of k-NN purity, but a major problem is the time used to find all k-NNs which is at least $O(n^2)$ without using an index. In the third category, data selection techniques find the decision boundary, such as [11,12,13]. The general procedure follows two steps: 1) find a tentative decision boundary; and 2) use the boundary to find potential SVs. The selected vectors are used again during the first step to update the decision boundary. The two steps are repeated until no more updating is necessary. Even though this approach will eventually generate a decision boundary, it may need a long search and do multiple scans over the dataset before it converges.

In this paper we propose the following two steps: (1) select the decision boundary region vectors (BRVs) as the SVs candidates without using iterative decision boundary, and (2) use the spatial approximation sample hierarchy (SASH [14]) tree structure to speed up this process. In section 2, we provide a detailed account of our approach followed by an empirical evaluation in Section 3. Section 4 gives conclusions and discusses future work.

2   Data Selection Based on Boundary Region Vectors

Most vectors far from the decision boundary are not SVs and can be safely removed. Without knowing the exact position of the decision boundary, however, we can always find a vector close to the boundary followed by a search for its nearest neighbor of opposite class (NNO); we can then consider them both as SV candidates. The concept of NNO is used for data condensing in instance-based learning (IBL). Data condensing aims at selecting the minimal subset of data that preserves the same accuracy to that obtained when invoking a 1-NN algorithm on the full dataset. Methodologies developed in data condensing mainly focus on the quality (minimal or not) and accuracy (consistent or not) of the condensed set. For example, Dasarathy's MCS (minimum consistent set [15]) condenses a dataset based on k-NNs of the same class and its nearest opposite-class neighbors.

Different from data condensing, our purpose using NNOs is to select SV candidates similar to those identified by SVMs. A positive vector $\mathbf{x}_i^+$'s NNO can be obtained by comparing the distances from this vector to all negative vectors $\mathbf{x}_j^-$:

$$NNO_i = \arg\min_{\mathbf{x}_j^-} ||\mathbf{x}_i^+ - \mathbf{x}_j^-||.$$

Algorithm 1. Finding BRVs using a distance matrix $D \in \mathbb{R}^{n_p \times n_n}$. In each step, the set of marked vectors M is reduced to size k_n or k_p, respectively.

Given: $k_p \geq 1$, $k_n \geq 1$
Output: B, the set of BRVs
$B = \{\}$
for $1 \leq i \leq n_p$:
  $d_i = k_n$th smallest distance in row i
  $M = \{\mathbf{x}_j \mid D_{i,j} \leq d_i\}$
  $B = B \cup M$
for $1 \leq j \leq n_n$:
  $d_j = k_p$th smallest distance in column j
  $M = \{\mathbf{x}_i \mid D_{i,j} \leq d_j\}$, where
  $B = B \cup M$

The same can be done to find a negative vector's NNO. We define the union of NNOs for both classes as boundary region vectors (BRVs); these are assumed to be within or close to the region around the decision boundary delineated by the margin.

2.1   Finding BRVs Using a Pairwise Matrix

Our approach begins by splitting the full dataset into positive and negative subsets, with size n_p and n_n respectively. A 2-d distance matrix D with n_p lines and n_n columns is then created. Along the ith row of the matrix, the jth element is the distance from positive vector $\mathbf{x}_i^+$ to negative vector $\mathbf{x}_j^-$. The $\mathbf{x}_i^+$'s NNO can be found by looking for the shortest distance in the i row. Suppose this shortest distance is found on the vth column; vector $\mathbf{x}_v^-$ is then the NNO. The procedure iterates until all NNOs in both classes are found and selected as BRVs (see Algorithm 1). In this algorithm, more than one NNO is allowed to be selected for each vector, i.e. $k_p \geq 1$ and $k_n \geq 1$. We will discuss this in the next section.

Algorithm 1 is an effective approach to finding nearest neighbors across the decision boundary. The time and space complexities of this algorithm are both of $O(n^2)$ because we need to compute every pairwise distance and store that in memory. Large datasets demand an index to speed up the search for BRVs.

2.2   Approximating BRVs Using SASH Trees

There are many indexing structures that can be used to reduce the complexity of NN search, such as R-tree. In this study, we use SASH trees because they require tuning fewer parameters than other methods. Since SV data selection is a data preprocessing step, computing minimal and consistent subsets is unnecessary.

To guide the NN search, a SASH tree assumes that transitivity holds for the NN relation. Specifically, if vector $\mathbf{x}$ is a NN vector of $\mathbf{y}$, and $\mathbf{y}$ is a NN vector

Algorithm 2. Finding the approximate BRVs given positive and negative SASH trees T^+ and T^-. The dataset X is split into classes X^+ and X^-. The function k-CLOSEST(k, u, X) returns the k closest elements in the set X to the node u.

Given: number of NNOs in each class, $k_+ \geq 1, k_- \geq 1$
Output: approximate BRVs, S
$S = \{\}$
for $c \in \{+, -\}$:
  $\bar{c} = \{+, -\} - c$
  for $u \in X^{(c)}$:
    $P_1(u) = $ root of $T^{(\bar{c})}$
    for $2 \leq i \leq $ HEIGHT$(T^{(\bar{c})})$:
      $C_i(u) = \{v \in P_{i-1}(u)\}$
      $P_i(u) = k$-CLOSEST$(p, u, C_i(u))$
    $P(u) = \bigcup_i P_i(u)$
    $s(u) = k$-CLOSEST$(k_{\bar{c}}, u, P(u))$
    $S = S \cup s(u)$

---

of $\mathbf{z}$, then $\mathbf{x}$ is likely to be a NN vector of $\mathbf{z}$. As a result, only approximate NNs in adjacent levels are connected to each other. Although a SASH tree cannot guarantee finding all exact NNs, it limits the NNs search in each level of the tree to a small number of candidates, thus significantly reducing search time. Similar work [16] uses a clustering feature hierarchy as index but guides the data selection process through a tentative decision boundary.

Construction of SASH trees. A SASH tree consists of n nodes, where n is the number of data vectors. Each vector is randomly assigned to a node, such that starting from the bottom, there are $\frac{n}{2}, \frac{n}{4}, \ldots, 1$ nodes in each level. Tree levels are numbered starting from 1 for the top level until h for the bottom level, where $h \approx \log_2 n$ is the height of the tree. Each node is allowed to have at most p parents and c children. The leaf nodes have no children. The root node has no parent and is fully connected to all nodes in level 2.

Links for the interior nodes are created through the following iterative process: (1) Interior nodes in levels $3 \leq i \leq h$ are only connected to p parent nodes in level $i-1$. Parent nodes are selected from a pool of pc candidates. The candidates are obtained by selecting at most pc nearest nodes in each level, starting from the root. (2) For each parent node, all but the c nearest child nodes are removed.

Because pc is much smaller than the actual number of nodes in most levels, finding the p nearest parents substantially reduces search time. The time complexity for connecting one node is $O(pch)$, or $O(pc \log_2 n)$. Tree construction has time complexity $O(n \log_2 n)$ and space complexity $O(n)$.

Approximating BRVs using SASH tress. To speed up our BRVs search, we need to construct positive and negative SASH trees. After each SASH tree is constructed, we find BRVs by querying the opposite tree for each node. Querying is a top-down search similar to finding the parents for a child node. For each node in the ith level of the positive SASH tree, searching starts from the root node of the negative SASH tree. In each level, at most pc distances are compared and only the p closest parents are collected. The final k_n NNOs of the positive node are selected from the p closest nodes obtained from the search. Through this process, BRVs are selected for all nodes, as described in Algorithm 2. The time complexity for BRVs querying is $O(n \log_2 n)$.

3   Experiments and Discussions

3.1   Experiment Setup

In this study, we consider only binary-class datasets. Data are taken from UCI repository with each variable scaled to $[-1, 1]$. For each dataset, we do the following: (1) select BRVs, (2) train on the selected subset of data and save the resulting model, (3) assess the model on the testing dataset and obtain accuracy results, and (4) stop if there is no performance improvement in two consecutive runs; otherwise go back to step (1) and select more data after increasing the value of k (parameter of nearest neighbor).

For our implementation, we used the publicly available LIBSVM library, C++ version 2.83, to train the SVMs [17]. SASH trees are also implemented in C++. Experiments were run on a PC with a 2.2 GHz Pentium CPU and 1GB of RAM. The classifier used is C-SVM with L_2 penalty for noisy data and radial basis functions (RBF) as kernel. With this kernel, there are two user-defined parameters: (C, γ). We invoked a grid search to find the best-performing (C, γ) values for each dataset.

3.2   Support Vector Recovery

We detail the BRV selection process on the breast-cancer dataset. To test our SV recovery ratio, we trained on the full dataset. The dataset consists of 683 vectors and 10 features each. Using (C, γ) values of $(1.0, 0.022)$ we found 87 SVs. Of these, 44 are positive and 43 are negative, which accounts for only 12.7% of all vectors.

We use Algorithm 1 to select BRVs and compare them to the actual 87 SVs. We first select only one NNO for each vector. In this case, BRVs can only recover 34.5% of the set of SVs (see Table 1). The corresponding SVM's accuracy is very low. As we increase the number of NNOs for each vector, k (here $k_p = k_n = k$), the number of selected BRVs increases and the SV recovery ratio grows, while accuracy improves. When $k = 8$, the selected BRVs (74 positive and 69 negative vectors) recover 90.8% of SVs, and accuracy reaches its highest value. We find that selecting more data does not necessarily yield better accuracy even when the number of recovered SVs increases. When this occurs, the ratio of BRVs (selected to reach the same accuracy as SVM on the full dataset) is termed the *critical ratio*. In this case, the critical ratio is 20.9%. The remaining 79.1% of

Table 1. SV recovery from breast cancer dataset (683 vectors with 10 features)

$k : k_p(k_n)$	1(1)	3(3)	5(5)	7(7)	8(8)	10(10)	15(15)	-
BRVs: p(n)	17(18)	40(35)	54(52)	66(64)	74(69)	83(80)	95(96)	239(444)
selection ratio	5.1	11.0	15.5	19.0	20.9	23.1	22.1	100
SV recovered (%)	34.5	64.4	79.3	86.2	90.8	92.0	94.3	100
SVM test acc(%)	65.0	95.9	96.2	96.5	97.4	97.4	97.4	97.4

Table 2. *Critical ratios* and SVM performance for different datasets

datasets	full data set			selected BRVs		
$(n \times d)$	#SVs	SV ratio (%)	SVM acc. (%)	#BRVs	*critical ratio* (%)	SVM acc. (%)
breast cancer(683x10)	87	12.3	97.4	143	20.9	97.4
diabetes(768x8)	467	60.8	78.1	497	64.7	77.3
heart(270x13)	103	38.1	85.2	134	49.6	85.2
ionosphere (351x34)	194	55.3	100	225	64.1	98.9
mushrooms(8124x112)	313	3.8	100	1093	13.5	100

data are redundant and can be removed. This critical ratio is 8% higher than the SV ratio.

The critical ratio is slightly higher than the actual SV ratio for two main reasons. (1) some BRVs are not SVs but lie close to the decision boundary; (2) noisy data causes a vector to be selected as an NNO. Noise has two effects in our data selection process: *(i)* some NNOs of a noisy vector may be non-noisy. These are selected in our approach and cause the critical ratio to increase; *(ii)* a noisy vector may mislead the search of real NNOs if it is closer to a vector than the actual NNO. In this case we have to increase the k value to select more data; this helps to recover the real SVs, but increases the critical ratio.

Additional experimental results are shown in Table 2. Results show that critical ratios are 4-10% higher than SV ratios. For large datasets having low SVs ratio, the reduction is substantial. For example, the mushroom dataset can be reduced from 8124 vectors to 1093 vectors, where about 86% of the data are eliminated without performance degradation.

Our investigation also shows that data selected according to the critical ratio does not recover 100% of all SVs (again see Table 1). As we look for the best (C, γ), the highest performance contour lies in a flat region within which all SVMs have almost the same performance. This means that the best (C, γ) can change within small ranges without performance degradation. However, the SVs set is very sensitive to (C, γ) values both in terms of size and vectors contained. Within different sets of SVs, many SVs are shared; others are mutual NNs of common SVs. Shared SVs are essential to defining the decision boundary. Other SVs can be replaced by their neighbors without changing overall performance. This explains why a slight different BRVs set, even if it does not cover 100% SVs, does not affect SVMs consistency.

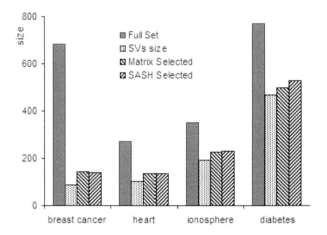

Fig. 1. Comparison of different sizes (SASHs with $p = 2$ and $c = 8$) for different datasets

3.3 Accuracy of SASH Approximation

As mentioned earlier, SASH only finds approximate nearest neighbors. There are two ways to obtain more accurate approximations. One way is to increase the values of p and c. This method will increase the time of SASH tree construction and the time of k-NNs querying. Alternatively, one can keep p and c fixed and select more SV candidates with a larger value of k-NNOs. This produces a larger critical ratio and will increase SVM's training time. Using either method can make the selected subset of data generate models comparable to those obtained with the full dataset. In practice, $p = 2$ and $c = 8$ results in a good approximation. Figure 1 compares the size of the full dataset to the subset selected by the matrix approach and the subset selected by SASHs for four datasets. Experiments show that subsets obtained from SASHs are slightly larger than those obtained from the matrix approach. However the difference is small, which leads us to conclude that SASH approximation is very similar to the matrix approach.

3.4 Overall Time Savings

To check the overall improvement in computational time, we compare the time it takes to train SVMs on a full dataset to the time it takes to do SASH data selection and training on the selected subsets. Our experiments test the Adult dataset with varying sizes (from 1600 to 32561 vectors, available from the LIB-SVM website [17]). These data have 123 features (scaled to $[-1, 1]$). Experimental results are shown in Figure 2. This figure shows that when the size of the dataset is small, the overhead for data selection dominates. As the size of data increases, the time for data selection is shorter than the time needed to train on the full dataset. Training time on the reduced dataset is even shorter. Overall time savings are significant as the dataset becomes larger than 30,000 vectors. On

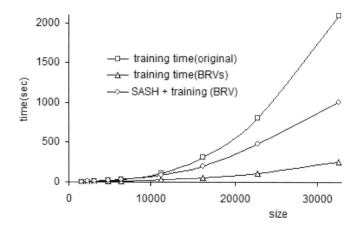

Fig. 2. Comparison of different CPU times

Table 3. Overall time comparison

dataset	full dataset			selected BRVs				overall time
$(n \times d)$	#SVs	t-time (sec)	SVM acc. (%)	#BRVs	s-time (sec)	t-time (sec)	SVM acc. (%)	saving (%)
mushrooms (8,124x112)	313	30	100	1,093	23	2	100	16.6
adult-9 (32,561x123)	1,1505	2,098	85.1	11,733	742	263	84.5	52.1
SensIT (30,000x100)	9,561	1,810	87.8	11,257	561	172	86.9	59.5
SensIT (40,000x100)	12,329	3,465	88.0	14,667	964	417	87.2	60.1
SensIT (50,000x100)	15,415	5,658	88.0	18,431	1360	823	87.2	61.4

(t-time: training time on 10-run average; s-time: selecting time on 10-run average)

additional datasets (see Table 3), we can observe that time savings are significant, particularly for over-sized datasets.

Part of our proposed strategy is based on computing and comparing the distance between two vectors in input space. When it comes to finding NNs, using Radial Basis Functions RBF as kernels does not change the NNOs nor BRVs. For two vectors $\mathbf{x}$ and $\mathbf{y}$, the distance over the input space is $d^2 = ||\mathbf{x} - \mathbf{y}||^2$. When the vectors are mapped to $\phi(\mathbf{x})$ and $\phi(\mathbf{y})$, the distance in kernel space can be written as $d_k^2 = ||\phi(\mathbf{x}) - \phi(\mathbf{y})||^2 = K(\mathbf{x}, \mathbf{x}) - 2K(\mathbf{x}, \mathbf{y}) + K(\mathbf{y}, \mathbf{y})$. Using RBF, $K(\mathbf{x}, \mathbf{y}) = e^{-\gamma ||\mathbf{x} - \mathbf{y}||^2}$, we have $K(\mathbf{x}, \mathbf{x}) = K(\mathbf{y}, \mathbf{y}) = 1$. In that case the distance is $d_k^2 = 2 - 2e^{-\gamma ||\mathbf{x} - \mathbf{y}||^2}$. This shows how distance order does not change in NN search. If we select k-NNOs, we will find the same vectors in input space as those found in the kernel space. Thus we do not need to compute a kernel function to select data. This can save a lot of time as compared to the case where data selection is carried out in the kernel space.

As final remarks, we add that data selection is used to select potential SVs but will not produce gainings in time savings if a given dataset has a high SV ratio. Data with high noise ratio suffers of the same problem. For particular distributions such as equally spaced vectors, our approach is not an appropriate solution because we select data based on differences in distance.

4   Conclusions and Future Work

Data selection based on BRVs can recover most SVs when we select slightly more data than the number of actual SVs. Training on selected data subsets significantly reduces the amount of training time without degrading performance. Since the decision boundary can be defined with different set of SVs, we do not need to recover 100% SVs. Instead, selecting BRVs as SVs candidates produces almost-identical results. For large datasets, the SASH tree can be used to approximate all BRVs. Time saving are significant compared to training an SVM on the full dataset.

Future work will focus on studying performance on extremely large datasets, and in using fixed-sized SASH trees to approximate BRVs. As noisy data causes more data to be selected, we will consider different techniques to detect and remove noise vectors. For the case of highly unbalanced datasets, selecting the same k-NNOs for both classes can have a strong impact on the example distribution. Future work will study how to address the class imbalance problem using our proposed techniques for data selection.

References

1. Vapnik, V.N.: Statistical Learning Theory. Addison-Wiley, New York, NY (1998)
2. Platt, J.: Sequential minimal optimization: A fast algorithm for training support vector machines. In: Microsoft Research Technical Report MSR-TR-98-14 (1998)
3. Bazzani, A., Bevilacqua, A., Bollini, D., Brancaccio, R., Campanini, R., Lanconelli, N., Riccardi, A., Romani, D.: An SVM classifier to separate false signals from microcalcifications in digital mammograms. vol 46, pp. 1651–1663 (2001)
4. Schohn, G., Cohn, D.: Less is more: Active learning with support vector machines. In: Proceedings of the 17th International Conference on Machine Learning (ICML), pp. 839–846 (2000)
5. Milenova, Boriana, L., Yarmus, J.S., Campos, M.M.: SVM in oracle database 10g: Removing the barriers to widespread adoption of support vector machines. In: Proceedings of the 31st VLDB Conference, Trondheim, Norway (2005)
6. Almeida, M., Braga, A.P., Braga, J.P.: SVM-KM: Speeding SVMs learning with a priori cluster selection and k-means. In: Proceedings of the 6th Brazilian Symposium on Neural Networks, pp. 162–167 (2000)
7. Koggalage, R., Halgamuge, S.: Reducing the number of training samples for fast support vector machine classification. Neural Information Processing - Letters and Reviews 2, 57–65 (2004)

8. Shin, H., Cho, S.: Fast pattern selection for support vector classifiers. In: Whang, K.-Y., Jeon, J., Shim, K., Srivastava, J. (eds.) PAKDD 2003. LNCS(LNAI), vol. 2637, pp. 376–387. Springer, Heidelberg (2003)

9. Wang, J., Neskovic, P., Cooper, L.N.: Training data selection for support vector machines. In: Wang, L., Chen, K., Ong, Y.S. (eds.) ICNC 2005. LNCS, vol. 3610, pp. 554–564. Springer, Heidelberg (2005)

10. Zhang, W., King, I.: A study of the relationship between support vector machine and gabriel graph. In: IEEE, pp. 239–245 (2002)

11. Roobaert, D.: DirectSVM: A fast and simple support vector machine perceptron. In: Proceedings. IEEE Int. Workshop Neural Networks for Signal Processing, Sydney, Australia, pp. 356–365 (2000)

12. Vishwanathan, S.V.N., Murty, N.M.: A simple SVM algorithm. In: Proceedings 2002 International Joint Conference on Neural Networks. IJCNN '02. vol. 3, Honolulu, Hawaii, pp. 2393–2398 (2002)

13. Raicharoen, T., Lursinsap, C.: Critical support vector machine without kernel function. In: Proc. of 9th International Conference on Neural Information. vol. 5, pp. 2532–2536 (2002)

14. Houle, M.E.: SASH: A spatial approximation sample hierarchy for similarity search. Technical Report pages 16, IBM Tokyo Research Laboratory Report RT-0517 (March 5, 2003)

15. Dasarathy, B.V., Sanchez, J.S., Townsend, S.: Nearest neighbour editing and condensing tools-synergy exploitation. In: Pattern Analysis & Applications, vol. 3, pp. 19–30. Springer, London (2000)

16. Yu, H., Yang, J., Han, J.: Classifying large data sets using SVM with hierarchical clusters. In: SIGKDD '03 Washington, DC, USA (2003)

17. Chang, C.C., Lin, C.J.: LIBSVM: A library for support vector machines (2001)

Dynamic Distance-Based Active Learning with SVM

Jun Jiang[1] and Horace H.S. Ip[1,2]

[1] Image Computing Group, Department of Computer Science
[2] Center for Innovative Applications of Internet and Multimedia Technologies
(AIMtech Centre), City University of Hong Kong, Hong Kong
jamesj@student.cityu.edu.hk, cship@cityu.edu.hk

Abstract. In this paper, we present a novel active learning strategy, named dynamic active learning with SVM to improve the effectiveness of learning sample selection in active learning. The algorithm is divided into two steps. The first step is similar to the standard distance-based active learning with SVM [1] in which the sample nearest to the decision boundary is chosen to induce a hyperplane that can halve the current version space. In order to improve upon the learning efficiency and convergent rates, we propose in the second step, a dynamic sample selection strategy that operates within the neighborhood of the "standard" sample. Theoretical analysis is given to show that our algorithm will converge faster than the standard distance-based technique and using less number of samples while maintaining the same classification precision rate. We also demonstrate the feasibility of the dynamic selection strategy approach through conducting experiments on several benchmark datasets.

Keywords: Active Learning, Dynamic Active Learning, Distance-based Active Learning, Information Retrieval, Support Vector Machine.

1 Introduction

Active learning is a learning model in which the training samples will be selected by the learner, labeled by an oracle and then added into the current training set to form the enlarged training one on which the classifier will be improved. Instead of passive learning, Active Learning is similiar to the human learning mode and has become an efficient tool for solving semi-supervised learning problems, especially when the initial training set only includes a few labeled samples. Since the last decade, Support Vector Machine (SVM) has been increasingly adopted as an universal classification tool for its excellent performance and its strong theoretical basis. More recently, some work has been done to combine active learning strategy with SVM and some successes have been reported for application domains such as information retrieval (including image and text retrievals), data mining, etc [1, 2, 3, 4, 5].

Because of its' good performance and low additional computational requirements, distance-based active learning with SVM [1, 13] in particular is one of the more well-known active learning methods. In terms of the Version Space Theory [1], the current version space can be approximately halved by the hyperplane induced by the sample

P. Perner (Ed.): MLDM 2007, LNAI 4571, pp. 296–309, 2007.

nearest to the decision boundary after making several weak assumptions. In distance-based strategy, this sample will be queried and added into the training set after labeling.

In this paper, we propose to improve upon the standard distance-based active learning in terms of learning efficiency and convergent rates by incorporating a dynamic sample selection strategy to the standard distance-based active learning method. Particularly, our approach selects, during each training iteration, the sample in the neighborhood of the sample nearest to the current decision boundary that induces the hyperplane which maximally reduces the surface area of the version space. Thus the approach can lead to faster convergence as well as improved learning efficiency. In the following section, we will present the theoretic foundation of the algorithm followed by experiments conducted on benchmark datasets to demonstrate the feasibility of the proposed method.

2 Sample Selection in Version Space for SVM Training

Version space [6] is the set of all hypotheses which satisfy all the samples in the training set. We suppose that the target function can be well studied, i.e., the target function can be perfectly expressed by one hypothesis in the version space. Thus when the area of the version space becomes smaller and smaller, the resulting classifier trained on the training set will approximate the target function more and more accurately, formally speaking, the generalization error will decrease with the area of the version space [7]. However, the version space for a typical classification problem is so complex that directly computing its' surface area is in general impossible. It is fortunate that, in practice, we only need to calculate the rate at which the surface area of the version space decreases, instead of directly and explicitly computing its' surface area. Since the total surface area of the version space is limited by the samples, this rule can be exploited to design active learning methods. Version space-based active learning is a strategy that the sample which can reduce the area of the version space is queried and added into the current training set after being labeled by an oracle. While for general classifiers, such as Artificial Neural Network, it is still an open-problem to find the relation between the version space and the samples, but the problem is tractable for SVM. In the following, we first give a description of the existing distance-based active learning with SVM, and then describe our improved algorithm.

2.1 Support Vector Machine and Its Representation in Version Space

Because of their strong mathematics foundation and excellent performances in practice, SVMs have received much attention in the computer vision community since the 90s. A tutorial on SVM can be found in [8], and other introductions of SVM can also be found in [9, 10]. A SVM classifier is a linear classifier where the separating hyperplane is chosen to minimize the expected classification error of unseen test patterns. For popular C-Support Vector Classification (C-SVC) [11], it can be obtained by solving the following optimization problem:

$$\min_{w,b,\xi} \frac{1}{2}\|w\|^2 + C(\sum_{i=1}^{l}\xi_i)^k$$

$$\text{s.t.}\quad y_i(w \bullet \Phi(x_i)+b) \geq 1-\xi_i,$$

$$\xi_i \geq 0,\quad i=1,...,l$$

(1)

where w denotes the coefficient of the classifier $c(x)=\text{sgn}(w \bullet x + b)$, $\frac{1}{2}\|w\|^2$

denotes the structure risk, $(\sum_{i=1}^{l}\xi_i)^k$ is the training error, l is the number of training

samples, and C is a trade-off parameter between these two terms.

For separable cases, the model can be simplified to:

$$\min_{w,b,\xi} \frac{1}{2}\|w\|^2$$

$$\text{s.t.}\quad y_i(w \bullet \Phi(x_i)+b) \geq 1,\quad i=1,...,l$$

(2)

As we know, $\Phi(x)$ can be seen as an implicit feature map. In the mapped feature

space, the classifier becomes linear and $\dfrac{|w \bullet \Phi(x_i)+b|}{\|w\|}$ denotes the distance from

the point $\Phi(x_i)$ to the decision boundary $w \bullet \Phi(x)+b=0$. When $\|w\|$ is

minimized, the largest margin can be obtained. Thus if we suppose $\|w\|=1$, model

(2) can be transformed [1] as follows:

$$\max_{w \in W} \ \text{margin} = \min_i \frac{\{|y_i(w \bullet \Phi(x_i)+b)|\}}{\|w\|}$$

$$= \min_i \{y_i(w \bullet \Phi(x_i)+b)\}$$

(3)

$$\text{s.t.}\quad \|w\|=1$$

$$y_i(w \bullet \Phi(x_i)+b) > 0,\quad i=1,...,l$$

where W is the parameter space which equals to the mapped space $F = \Phi(x)$.

For one classifier corresponds to one w in the parameter space, according to its' definition and model (3), its version space can be expressed as follow:

$$V = \{w \in W \mid \|w\|=1,\ y_i(w \bullet \Phi(x_i)+b) > 0,\ i=1,...,l\}$$

(4)

By assuming $\|\Phi(x_i)\|=1$ and $y_i = \pm 1$, we also have

$$\text{margin} = \min_i \frac{\{|w \bullet (y_i \Phi(x_i)) + y_i b|\}}{\|y_i \Phi(x_i)\|} = \min_i \{|w \bullet (y_i \Phi(x_i)) + y_i b|\}$$

$$= \min_i \{y_i (w \bullet \Phi(x_i) + b)\}$$

(5)

which means that the margin of the SVM also equals to the minimum distance from one point, w , to the hyperplane $w \bullet (y_i \Phi(x_i)) + b = 0$ (normal vector $\equiv y_i \Phi(x_i)$) in the parameter space W .

From the above analysis, we can see that the goal of SVM is to find the center of the largest radius hypersphere whose center can be placed in the version space and whose surface does not intersect with the hyperplanes induced by the labeled instances. It follows that the hyperplanes touched by the maximal radius hypersphere correspond to the support vectors and that the radius of the hypersphere is the margin of the SVM.

For un-separable cases, as noted in [15], it is possible to modify any kernel so that the data in the new induced feature space is linearly separable.

2.2 Distance-Based Active Learning with SVM

In the mapped feature space F , SVM becomes a linear classifier. So there exists a duality between the feature space F and the parameter space W : points in F correspond to hyperplanes in W and vice versa. (These rules were already embodied in the above section.) Based on the duality property, any samples near the decision boundary can approximately halve the current version space if the later is symmetric. The situation where the version space is asymmetric is outside the scope of this paper and will be discussed in a separate work.

Tong [12] proposed a lemma:

Lemma 1. *Suppose we have an input space X , finite dimensional feature space F (induced by a kernel k), and parameter space W . Suppose active learner l^* always queries instances whose corresponding hyperplanes in W halves the area of the current version space. Let l be any other active learner. Denote the version spaces of l^* and l after i pool-queries as V_i^* and V_i respectively. Let P denote the set of all conditional distributions of y given x. Then*

$$\forall i \in N^+ \quad \sup_{P \in P} E_P[Area(V_i^*)] \le \sup_{P \in P} E_P[Area(V_i)],$$

with strict inequality whenever there exist a query j by l that does not halve version space V_{j-1} .

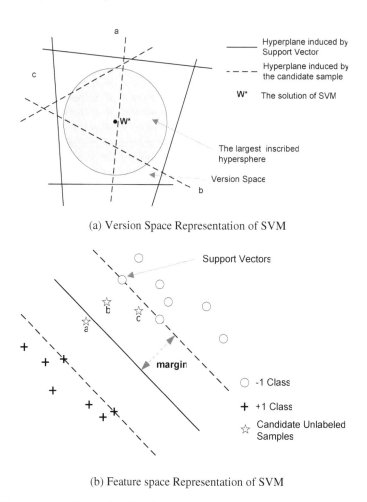

(a) Version Space Representation of SVM

(b) Feature space Representation of SVM

Fig. 1. Illustration of distance-based active learning with SVM (sample 'a' will be selected as query instance)

Based on an assumption that the version space is symmetric and Lemma 1, the sample nearest to the decision boundary will be queried, and be added into the current training set after being labeled by the oracle (figure 1). This strategy is called distance-based active learning with SVM.

2.3 Theoretic Derivation and Analysis of Dynamic Sample Selection Strategy

Since training a SVM is usually expensive, one of the goals of active learning is using the fewest number of samples to train a classifier which can approximate the target function as much as possible. As to the version-space-based active learning, the target function can be well-defined by one point in the hypothesis space. Thus the goal is equivalent to finding the fewest number of samples which can maximally reduce the surface area of the version space. Based on this idea, we propose a dynamic sample selection strategy to choose the sample which can reduce a larger area of the version space than the sample nearest to decision boundary could achieve.

We can see that distance-based active learning strategy combines the following two steps into one atom operation:

Step 1: query the sample nearest to the current decision boundary, and get its' label from the oracle;
Step 2: add the labeled sample into the current training set to get enlarged training set, and the classifier will be updated on the enlarged training set;

Our approach splits the above atom operation to two separate steps. In the conventional distance-based strategy, it should be noted that, after Step 1, the label of the sample nearest to the decision boundary becomes known, i.e., the normal vector of the hyperplane induced by this sample is known, and adding it immediately into the current training set as defined in Step 2 may be not the best choice yet.

We will first derive the theoretical conditions for the choice of sample in dynamic sample selection and then state the associated algorithm. (Here we only consider the problem of classification, and, without loss of generality, assume the label of positive and negative category is "1" and "-1")

Definition 1. Version space is the set of all hypotheses which satisfy all the samples in the training set. For separable SVM, it can be expressed by the following formulation (rewriting equation (4)):

$$V = \left\{ w \in W \middle|\ \|w\| = 1,\ \ y_i(w \bullet \Phi(x_i) + b) > 0,\ \ i = 1, \ldots, l \right\} \tag{6}$$

where l is the number of the current training samples.

Definition 2. Area of the version space $Area(V)$ is the surface area that the version space occupies on the hypersphere $\|w\| = 1$.

Definition 3. Distance from the point x to the current decision boundary $w * \bullet \Phi(x) + b = 0$ is $D(x) = w * \bullet \Phi(x) + b$ ($w *$ is the solution of the SVM).

Proposition 1

For separable SVM, given l training samples which form the symmetric version space in the form of equation (6), V_1^{new} and V_2^{new} are the new updated version after adding sample (x^1, y^1) and (x^2, y^2) respectively, then

1) if $y^1, y^2 = 1$, and $D(x^1) > D(x^2)$, then $Area(V_1^{new}) > Area(V_2^{new})$;
2) if $y^1, y^2 = -1$, and $D(x^1) > D(x^2)$, then $Area(V_1^{new}) < Area(V_2^{new})$;

Proof

After adding a new sample (x, y) in the margin band of the current SVM, the current version space is divided, by the hyperplane $y(w \bullet \Phi(x) + b) = 0$ induced by the sample (x, y), into two portions: one is the new updated version space expressed by equation (7) and another is the removed version space expressed by equation (8).

$$V^{new} = \left\{ w \in W \left| \begin{array}{l} \|w\| = 1, \ \ (y_i (w \bullet \Phi(x_i) + b) > 0, \ \ i = 1,...,l) \\ and \ \ \ y(w \bullet \Phi(x) + b) > 0 \end{array} \right. \right\} \tag{7}$$

$$V^{remove} = \left\{ w \in W \left| \begin{array}{l} \|w\| = 1, \ \ (y_i (w \bullet \Phi(x_i) + b) > 0, \ \ i = 1,...,l) \\ and \ \ \ y(w \bullet \Phi(x) + b) < 0 \end{array} \right. \right\} \tag{8}$$

So after adding (x^1, y^1) and (x^2, y^2) respectively ($y^1 = y^2$ here), we get their new updated version space as follows:

$$V_1^{new} = \left\{ w \in W \left| \begin{array}{l} \|w\| = 1, \ \ (y_i (w \bullet \Phi(x_i) + b) > 0, \ \ i = 1,...,l) \\ and \ \ \ y^1 (w \bullet \Phi(x^1) + b) > 0 \end{array} \right. \right\} \tag{9}$$

$$V_2^{new} = \left\{ w \in W \left| \begin{array}{l} \|w\| = 1, \ \ (y_i (w \bullet \Phi(x_i) + b) > 0, \ \ i = 1,...,l) \\ and \ \ \ y^2 (w \bullet \Phi(x^2) + b) > 0 \end{array} \right. \right\} \tag{10}$$

It is reasonable to conceive that there exist a hyperplane in the parameter space which has the same distance to the current solution of the SVM $w*$ as the hyperplane induced by the sample (x^1, y^1), and the same normal vector as the hyperplane induced by the sample (x^2, y^2) (figure 2). Thus this conceivable hyperplane can be expressed with $y^2 (w \bullet \Phi(x^2) + b_{temp}) = 0$ (b_{temp} is offset constant). If this hyperplane corresponds to the sample (x_{temp}, y_{temp}) ($y_{temp} = y^1 = y^2$), we get that

$$D(x^1) = D(x_{temp}) = w* \bullet \Phi(x^2) + b_{temp} \tag{11}$$

And the newly formed version space after adding the sample (x_{temp}, y_{temp}) into the current training set is listed as follows:

$$V_{temp}^{new} = \left\{ w \in W \left| \begin{array}{l} \|w\| = 1, \ \ (y_i (w \bullet \Phi(x_i) + b) > 0, \ \ i = 1,...,l) \\ and \ \ \ y^2 (w \bullet \Phi(x^2) + b_{temp}) > 0 \end{array} \right. \right\} \tag{12}$$

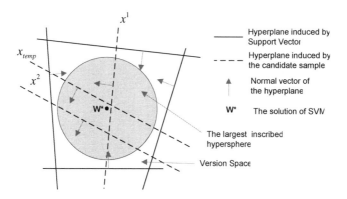

Fig. 2. Illustration of the new updated version space after adding the sample (x^1, y^1), (x^2, y^2) respectively (here $y^1 = y^2 = 1$ and $D(x^1) > D(x^2)$, the sample (x^1, y^1) is selected into the current training set by standard distance-based active learning, while dynamic distance-based active learning selects the sample (x^2, y^2) into the current training set)

Considering the conditions that the current version space is symmetric, and $w*$ is almost the center of the current version space (model (3)), it can be easily followed that

$$Area(V_{temp}^{new}) = Area(V_1^{new}) \tag{13}$$

1) Substituting equation (11) into $D(x^1) > D(x^2)$, we can infer that:

$$w* \bullet \Phi(x^2) + b_{temp} > w* \bullet \Phi(x^2) + b \Rightarrow b_{temp} > b \tag{14}$$

By combining the condition $y^1 = y^2 = 1$, we can get

$$y^2(w \bullet \Phi(x^2) + b_{temp}) > y^2(w \bullet \Phi(x^2) + b)$$

$$\Downarrow \tag{15}$$

$$y^2(w \bullet \Phi(x^2) + b) > 0 \Rightarrow y^2(w \bullet \Phi(x^2) + b_{temp}) > 0$$

Considering their definitions of the new updated version space in the equation (10) and (12), we can conclude $V_2^{new} \subset V_{temp}^{new}$, thus $Area(V_2^{new}) < Area(V_{temp}^{new})$. Combined with equation (13), it is evidently that $Area(V_2^{new}) < Area(V_1^{new})$.

2) The proof processes are similar to item (1). After substituting equation (11) into $D(x^1) > D(x^2)$, we can infer that:

$$w * \bullet \Phi(x^2) + b_{temp} > w * \bullet \Phi(x^2) + b \Rightarrow b_{temp} > b \qquad (16)$$

By combining the condition $y^1 = y^2 = -1$, we can get

$$y^2(w \bullet \Phi(x^2) + b) > y^2(w \bullet \Phi(x^2) + b_{temp})$$

$$\Downarrow \qquad\qquad (17)$$

$$y^2(w \bullet \Phi(x^2) + b_{temp}) > 0 \Rightarrow y^2(w \bullet \Phi(x^2) + b) > 0$$

Considering their definitions of the new updated version space in the equation (10) and (12), we can conclude $V_2^{new} \supset V_{temp}^{new}$, thus $Area(V_2^{new}) > Area(V_{temp}^{new})$. Combined with equation (13), it is evident that $Area(V_1^{new}) < Area(V_2^{new})$. ■

From proposition 1, we know that if the sample (x^1, y^1) and (x^2, y^2) are both belong to the positive category, i.e., $y^1 = y^2 = 1$, and $D(x^1) > D(x^2)$, then $Area(V_1^{new}) > Area(V_2^{new})$. It means that the sample (x^2, y^2) can reduce a larger area of the version space than the sample (x^1, y^1) can do. However, as the sample (x^1, y^1) is nearest to the current decision boundary and will hence be added into the current training set using the standard distance-based active learning with SVM (illustrated in figure 2).To overcome this problem, we propose one dynamic distance-based selecting strategy whose goal is to choose the sample which can reduce the version space as much as possible. Our dynamic distance-based active learning algorithm can be detailed in the following two steps:

Step 1: query the sample nearest to the current decision boundary (without losing generality, we call it sample "a"), and get its' label from the oracle; then create an short ascended list based on the samples' distance to the current decision boundary, which is only consisted of the neighbors of sample "a";

Step 2: if the label of sample "a" is "+1" (i.e., sample "a" belongs to positive category), the sample in the former position of the sample "a" in the list will be queried, if its' label is still "+1", repeat querying its' former sample until reaching one sample whose label is "-1" or the end of the list, then the last positive sample will be added into the current training set; if the label of sample "a" is "-1", similar processes can be implemented; (illustrated in figure 3).

As we will show later, the list is normally short, and it can be shown to the oracle in one go. So in practice, the oracle only needs to label the last sample which has the same label with the sample nearest to the current decision boundary (such as sample "d" in figure 3).

After several additional queries, we can find the sample in the neighborhood of the nearest sample 'a' that can maximally reduce the current version space (such as sample "b" in figure 4). Since every added sample selected by this process can reduce the area of the version space more than the sample selected by the standard distance-based active learning, the convergent rate of the version space will also be faster than that of the standard approach. It means that the classification accuracy performance of the resulting classifier trained by this dynamic strategy will be better than the one trained by the standard strategy using the same number of training samples.

the label of sample "a" is "1"			the label of sample "a" is "-1"		
(sample "a" is the one nearest to the current decision boundary)					
ascend list	label	description	ascend list	label	description
f	?	unlabeled	...	?	unlabeled
e	-1	Sample "d"	a	-1	Sample "d"
d	1	selected by	b	-1	selected by
...	1	dynamic	c	-1	dynamic
c	1		...	-1	
b	1	(D(f)<D(e)	d	-1	(D(f)>D(e)
a	1	<...<D(a))	e	+1	>...>D(a))
...	?	unlabeled	f	?	unlabeled

Fig. 3. List-based illustration of dynamic distance-based active learning with SVM (usually the list is short, and the letters "a", "b", ... denote different samples)

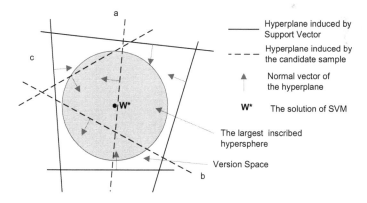

Fig. 4. Graphical-based illustration of dynamic distance-based active learning with SVM (sample 'b' will be selected as query instance, because it can reduce more area of the version space than sample 'a', although the later is nearest to the decision boundary. Here, we assume 'a' and 'b' have '+1' label +1, and sample 'c' has '-1' label)

From the experiment shown in figure 5, it can be easily seen that the sample selected by our dynamic distance-based strategy is nearer to the true decision boundary than the sample selected by the standard distance-based strategy. It also shows that the dynamic strategy can make the version space converge to the target function quicker compared with the standard distance-based active learning.

In the dynamic approach, the only additional computation is creating an ascending list. Since the initial sample of the list, nearest to the current decision boundary, is usually near to the true decision boundary, it is enough to set up a short list. Moreover, the members of the list must be in the margin band of the current classifier. Thus the running time on this operation is negligible compare to the total amount of computations required of training the SVM. It means that our dynamic method can achieve better performance than the standard distance-based active learning with almost the same spending.

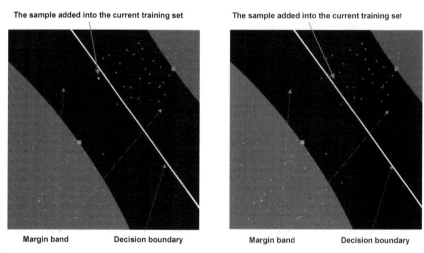

Figure 3a. Dynamic distance-based active learning

Figure 1b. Standard distance-based active learning

Fig. 5. Comparison of the dynamic distance-based active learning and the standard distance-based active learning methods

3 Experiments and Conclusions

Several experiments were conducted using four datasets: three of them have been downloaded from the benchmark websites and one is generated in this work. All datasets are first randomly divided into training and test sets. The training set is further divided into three portions: P-Set, N-Set and Unlabeled-Set. Table 1 describes the datasets. In the "Triangle" dataset, there are two classes of data. One category of the data is distributed in the input space, and is shaped likes two triangles, thus the dataset is named 'triangle' dataset (figure 6). It is particularly challenging due to the fact that the same category of data is distributed in two disjoint clusters in the input space and is therefore not separable using a single decision boundary.

The procedures of the experiment are listed as follows:

Step 1: Initial step: (1) prepare initial training set: one positive sample from P-Set, and other negative samples from N-Set, the size is ten for triangle dataset and twenty for the other datasets. (2) train an initial classifier on the initial training set and test it on the corresponding test dataset.

Step 2: Based on different strategies, one sample will be selected, labeled by the oracle and added in the current training set (the standard and dynamic distance-based active learning strategy are utilized respectively here).

Step 3: Re-train the classifier on the enlarged training set and test it on the corresponding test dataset.

Step 4: Repeat step 2 and step 3 until there is no samples in the margin band of the SVM classifier.

In our experiments, C-SVC is utilized, kernel function deployed is the Radial Basis Function ($\exp(-\gamma * \|x - y\|^2)$), and the trade-off parameter C equals to 100. The results are listed in figure 7. Since the P-Set only contains ten samples, the final value of the precision rate is averaged on ten tests which are initialized by ten different initial training sets. Figure 7a, 7b and 7c show clearly that dynamic distance-based active learning outperforms the standard technique, i.e., dynamic strategy uses less number of iterations than the standard strategy to achieve the same precision ratio. In figure 7d, it is interesting to note that the standard strategy outperforms the dynamic approach in several initial iterations. The reason may be that the version space is un-symmetric in the initial period. In this case, the sample selected by the dynamic strategy cannot guarantee to reduce more area of the version space than the sample selected by the standard method. It is interesting to

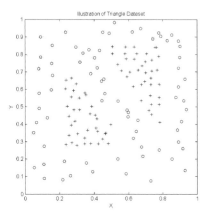

Fig. 6. Illustration of Triangle dataset (one category data is distributed in the input space, and is shaped likes two triangles)

Table 1. Description of the datasets

Name	P-Set	N-Set	Unlabeled-Set	Test	Source
Australian	10	40	295	345	Statlog
Fourclass	10	40	381	431	TKH96a
Breast Cancer	10	40	291	342	UCI
Triangle	10	16	145	579	Self-create

(Test denotes test dataset, unlabeled-Set is used as selection pool).

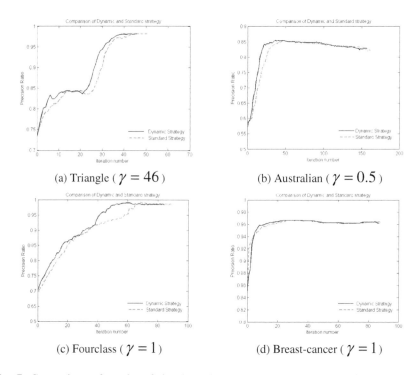

(a) Triangle ($\gamma = 46$)   (b) Australian ($\gamma = 0.5$)

(c) Fourclass ($\gamma = 1$)   (d) Breast-cancer ($\gamma = 1$)

Fig. 7. Comparison of results of the dynamic distance-based active learning and standard distance-based active learning on different datasets (The dimension of Triangle, Australian, Fourclass and Breast-cancer is two, fourteen, two and ten respectively)

note that, despite of this, the situation of dynamic strategy is still better than the standard one.

In this paper, we present a novel dynamic distance-based active learning with SVM. Both the theoretical analysis and experiments show that our dynamic strategy outperforms the standard distance-based approach. The algorithm can be applied in information retrieval, data mining, etc., and is an efficient method to deal with the large scale problem of SVM [14]. When the version space is non-symmetric, the merit of our dynamic will become weaker. Future work includes extending this approach for non-symmetric version space.

Acknowledgments. The authors would like thank the SVM groups at National Taiwan University which provided the code for the implementation of the SVMs for public use on the World Wide Web at the URL address: http://www.csie.ntu.edu. tw/~cjlin/. This work is fully supported by CityU Strategic Research Grant: 7001840.

References

1. Tong, S., Chang, E.: Support Vector Machine Active Learning for Image Retrieval. In: Proceedings of the Ninth ACM International Conference on Mulitimedia, pp. 107–118 (2001)
2. Campbell, C., Cristianini, N., Smola, A.: Query Learning with Large Margin Classifiers. In: Proceedings of ICML-00 17th International Conference on Machine Learning, pp. 111–118. Morgan Kaufmann, Seattle (2000)
3. Jiang, W., Er, G., Dai, Q.: Boost SVM Active Learning for Content-Based Image Retrieval, Conference Record of the Asilomar Conference on Signals, Systems and Computers, v 2, Conference Record of the Thirty-Seventh Asilomar Conference on Signals, Systems and Computers, pp. 1585–1589 (2003)
4. Cheng, J., Wang, K.: Active Learning for Image Retrieval with Co-SVM. Pattern Recognition 40(1), 330–334 (2007)
5. Brinker, K.: Incorporating Diversity in Active Learning with Support Vector Machines. In: 12th International Conference on Machine Learning, pp. 59–66 (2003)
6. Mitchell, T.: Generalization as Search. Artificial Intelligence 28, 203–226 (1982)
7. Cohn, D., Atlas, L., Ladner, R.: Improving Generalization with Active Learning. Machine Learning 15, 201–221 (1994)
8. Christopher, J.C., Burges, A.: Tutorial on Support Vector Machines for Pattern Recognition. Data Mining and Knowledge Discovery 2, 121–167 (1998)
9. Vapnik, V.: Statistical Learning Theory. John Wiley & Sons, New York (1998)
10. Scholkopf, B., Smola, A.J.: Learning with Kernels. MIT Press, Cambridge (2002)
11. Cortes, C., Vapnik, V.: Support Vector Network. Machine Learning 20, 273–297 (1995)
12. Tong, S., Koller, D.: Support Vector Machine Active Learning with Application to Text Classification. In: Proceedings of the 17th International Conference on Machine Learning, pp. 401–412 (2000)
13. Chang, E.Y., Tong, S., Goh, K., Chang, C.-W.: Support Vector Machine Concept-Dependent Active Learning for Image Retrieval (accepted 2005)
14. Bordes, A., Ertekin, S., Weston, J., Bottou, L.: Fast Kernel Classifiers with Online and Active Learning. Journal of Machine Learning Research 6, 1579–1619 (2005)
15. Shave-Taylor, J., Cristianini, N.: Further Results on the Margin Distribution. In: Proceedings of the 12th Annual Conference on Computational Learning Theory, pp. 278–285 (1999)

Off-Line Learning with Transductive Confidence Machines: An Empirical Evaluation

Stijn Vanderlooy[1], Laurens van der Maaten[1], and Ida Sprinkhuizen-Kuyper[2]

[1] MICC-IKAT, Universiteit Maastricht, P.O. Box 616,
6200 MD Maastricht, The Netherlands
{ s.vanderlooy,l.vandermaaten } @micc.unimaas.nl
[2] NICI, Radboud University Nijmegen, P.O. Box 9104,
6500 HE Nijmegen, The Netherlands
i.kuyper@nici.ru.nl

Abstract. The recently introduced transductive confidence machines (TCMs) framework allows to extend classifiers such that they satisfy the calibration property. This means that the error rate can be set by the user prior to classification. An analytical proof of the calibration property was given for TCMs applied in the on-line learning setting. However, the nature of this learning setting restricts the applicability of TCMs. In this paper we provide strong empirical evidence that the calibration property also holds in the off-line learning setting. Our results extend the range of applications in which TCMs can be applied. We may conclude that TCMs are appropriate in virtually any application domain.

1 Introduction

Machine-learning classifiers are common in many real-life applications. Many of these applications are characterized by high error costs, indicating that incorrect classifications can have serious consequences. It is therefore desired to have classifiers that output reliable classifications. One way to achieve this is to complement each classification with a confidence value. Classifications with a low confidence value are not reliable and should be handled with caution.

For some classifiers (such as the naive Bayes classifier) a measure of confidence is readily available, but for many other classifiers this is not the case. The recently introduced transductive confidence machines (TCMs) framework allows for an efficient way to provide confidence values produced by virtually any classifier [8,17]. The essential property of TCMs is that their error rate is controlled by the user prior to classification. For example, if the user specifies an error rate of 0.05, then at most 5% of the classifications made by a TCM are incorrect. This property is called the calibration property and has been proven to hold in the on-line learning setting. However, this learning setting restricts the applicability of TCMs. In the paper we investigate to what extent the calibration property holds in the off-line learning setting. We investigate this by means of a systematic empirical evaluation of TCMs using six different classifiers on various real-world datasets.

P. Perner (Ed.): MLDM 2007, LNAI 4571, pp. 310–323, 2007.

The remainder of the paper is organized as follows. Section 2 defines the learning setting that we consider. Section 3 explains TCMs and the calibration property. It also provides implementations of six classifiers in the TCM framework. Section 4 investigates to what extent the calibration property holds in the off-line learning setting. Section 5 provides a final discussion on TCMs. Section 6 concludes that TCMs satisfy the calibration property in the off-line learning setting.

2   Learning Setting

We consider the supervised machine-learning setting. The instance space is denoted by $\mathcal{X}$ and corresponding label space by $\mathcal{Y}$. An example is of the form $z = (x, y)$ with $x \in \mathcal{X}$ and $y \in \mathcal{Y}$. The symbol $\mathcal{Z}$ will be used as a compact notation for $\mathcal{X} \times \mathcal{Y}$. Training data are considered as a sequence of examples:

$$S = (x_1, y_1), \ldots, (x_n, y_n) = z_1, \ldots, z_n, \tag{1}$$

where each example is generated by the same unknown probability distribution P over $\mathcal{Z}$. We assume that this distribution satisfies the *exchangeability assumption*. This assumption states that the joint probability of a sequence of random variables is invariant under any permutation of the indices of these variables. In other words, the information that the z_i's provide is independent of the order in which they are collected. Formally, we write:

$$P(z_1, \ldots, z_n) = P(z_{\pi(1)}, \ldots, z_{\pi(n)}), \tag{2}$$

for all permutations π on the set $\{1 \ldots, n\}$.[1]

We apply a classifier in the *off-line learning setting* (batch setting): the classifier is learned on training data and subsequently used to classify instances one-by-one. The true labels of instances are not returned. This is in contrast to the *on-line learning setting* where the true label of each instance is provided after prediction. The classifier is then retrained after each prediction since new information is available. Clearly, the on-line learning setting restricts the applicability of classifiers since any form of feedback can be very expensive.

3   Transductive Confidence Machines

Traditionally, classifiers assign a single label to an instance. In contrast, transductive confidence machines (TCMs) are allowed to assign a set of labels to each instance. Such a *prediction set* contains multiple labels if there is uncertainty in the true label of the instance [7,8,17]. Subsection 3.1 explains the construction of prediction sets. Subsection 3.2 discusses the calibration property. Subsection 3.3 outlines six practical implementations of TCMs.

---

[1] Note that exchangeable random variables are identically distributed and not necessarily independent from each other. Therefore, identically and independently distributed (iid) random variables are also exchangeable. The exchangeability assumption is thus weaker (i.e., more general) than the iid assumption.

3.1   Construction of Prediction Sets

To construct a prediction set for an unlabeled instance x_{n+1}, TCMs operate in a transductive manner. Each possible label $y \in \mathcal{Y}$ is tried as a label for instance x_{n+1}. In each try we form the example $z_{n+1} = (x_{n+1}, y)$ and add it to S. Then we measure how likely it is that the resulting sequence is generated by the underlying distribution P. To this end, each example in the *extended sequence*:

$$(x_1, y_1), \ldots, (x_n, y_n), (x_{n+1}, y) = z_1, \ldots, z_{n+1}, \tag{3}$$

is assigned a nonconformity score by means of a nonconformity measure. This measure defines how nonconforming an example is with respect to other available examples. We require that it is irrelevant in which order the nonconformity scores of the examples are calculated (due to the exchangeability assumption).

Definition 1. *A nonconformity measure is a measurable mapping:*

$$A : \mathcal{Z}^{(*)} \times \mathcal{Z} \to \mathbb{R} \cup \{\infty\}, \tag{4}$$

with output indicating how nonconforming an example is with respect to all other examples. The symbol $\mathcal{Z}^{(*)}$ denotes the set of all bags of elements of $\mathcal{Z}$. A bag is denoted by $\lfloor \cdot \rfloor$.

Definition 2. *Given a sequence of examples $z_1, \ldots, z_{n+1}$ with $n \geq 1$, the nonconformity score of example z_i $(i = 1, \ldots, n)$ is defined as:*

$$\alpha_i = A(\lfloor z_1, \ldots, z_{i-1}, z_{i+1}, \ldots, z_{n+1} \rfloor, z_i), \tag{5}$$

and the nonconformity score of example z_{n+1} is defined as:

$$\alpha_{n+1} = A(\lfloor z_1, \ldots, z_n \rfloor, z_{n+1}). \tag{6}$$

Nonconformity scores can be scaled arbitrarily by multiplying with a fixed nonzero number. Therefore, to know how nonconforming the created example z_{n+1} is in the extended sequence, the nonconformity score α_{n+1} is compared to all other α_i $(i = 1, \ldots, n)$.

Definition 3. *Given a sequence of nonconformity scores $\alpha_1, \ldots, \alpha_{n+1}$ with $n \geq 1$, the p-value of label y assigned to unlabeled instance x_{n+1} is defined as:*

$$p_y = \frac{|\{i = 1, \ldots, n+1 : \alpha_i \geq \alpha_{n+1}\}|}{n+1}. \tag{7}$$

If the p-value is close to its lower bound $1/(n+1)$, then example z_{n+1} is very nonconforming. The closer the p-value is to its upper bound 1, the more conforming example z_{n+1} is. Hence, the p-value indicates how likely it is that the tried label for an unlabeled instance is in fact the true label. A TCM outputs the set of labels with p-values above a predefined significance level ϵ.

Definition 4. *A transductive confidence machine determined by some noncon-formity measure is a function that maps each sequence of examples $z_1, \ldots, z_n$ with $n \geq 1$, unlabeled instance $\dot{x}_{n+1}$, and significance level $\epsilon \in [0, 1]$ to the pre-diction set:*

$$\Gamma^\epsilon(z_1, \ldots, z_n, x_{n+1}) = \{y \in \mathcal{Y} \mid p_y > \epsilon\}. \tag{8}$$

There may be situations in which many training examples have nonconformity score equal to the score of example z_{n+1}. The p-value is then large, but caution is needed since many examples are equally nonconforming, making it impossible to discriminate between them. To alleviate this problem, a randomized version of the p-value has been proposed [17, p. 27].

Definition 5. *Given a sequence of nonconformity scores $\alpha_1, \ldots, \alpha_{n+1}$ with $n \geq 1$, the randomized p-value of label y assigned to unlabeled instance x_{n+1} is defined as:*

$$p_y^\tau = \frac{|\{i = 1, \ldots, n+1 : \alpha_i > \alpha_{n+1}\}| + \tau |\{i = 1, \ldots, n+1 : \alpha_i = \alpha_{n+1}\}|}{n+1},$$
$$\tag{9}$$

with τ a random number uniformly sampled from $[0, 1]$ for instance x_{n+1}.

Definition 6. *A randomized transductive confidence machine determined by some nonconformity measure is a function that maps each sequence of examples $z_1, \ldots, z_n$ with $n \geq 1$, unlabeled instance x_{n+1}, uniformly distributed random number $\tau \in [0, 1]$, and significance level $\epsilon \in [0, 1]$ to the prediction set:*

$$\Gamma^{\epsilon, \tau}(z_1, \ldots, z_n, x_{n+1}) = \{y \in \mathcal{Y} \mid p_y^\tau > \epsilon\}. \tag{10}$$

A randomized TCM treats the borderline cases $\alpha_i = \alpha_{n+1}$ more carefully. Instead of increasing the p-value with $1/(n+1)$, the p-value is increased with a random amount between 0 and $1/(n+1)$. In the following, we employ randomized TCMs, although for brevity we simply call them TCMs.

3.2   Calibration Property

In the on-line learning setting, TCMs have been proven to satisfy the *calibration property* [17, p. 20-22 & p. 193]. This property states that the long run error rate of a TCM with significance level ϵ equals ϵ:

$$\limsup_{n \to \infty} \frac{Err_n^\epsilon}{n} = \epsilon, \tag{11}$$

with Err_n^ϵ the number of prediction sets that do not contain the true label, given the first n prediction sets.[2] The idea of the proof is to show that the sequence of prediction outcomes (i.e., whether the prediction set contains the true label or not) is a sequence of independent Bernoulli random variables with parameter ϵ. From (11) follows that the significance level has a frequentist interpretation as

---

[2] In case of non-randomized TCMs, the equality sign in (11) is replaced by the $\leq$ sign.

the limiting frequency of errors. It allows to control the number of errors prior to classification. The calibration property holds regardless of which nonconformity measure is used.

In the off-line learning setting there theoretically exists a small probability that TCMs are not well-calibrated (the training data is kept fixed, and therefore the prediction outcomes are not independent) [17, p. 111]. Section 4 investigates empirically whether TCMs are well-calibrated in the off-line learning setting.

3.3   Implementations

This subsection shows that virtually any classifier can be plugged into the TCM framework. Nonconformity measures are formulated for the following six classifiers: (1) k-nearest neighbour, (2) nearest centroid, (3) linear discriminant, (4) naive Bayes, (5) kernel perceptron, and (6) support vector machine. Although the nonconformity measures are based on specific classifier characteristics, they can readily be applied to similar classifiers. In addition, they provide clear insights.in how to define new nonconformity measures.

The implementation of TCMs based on linear discriminant, kernel perceptron, and support vector machine considers binary classification tasks. This is due to the nature of these classifiers. We denote the binary label space as $\mathcal{Y} = \{-1, +1\}$. Extensions to multilabel learning are well-known and therefore not discussed in the paper. We implemented TCMs that can incrementally learn and decrementally unlearn a single instance, hereby keeping time complexity low. Pseudo codes of these efficient implementations are found in a technical report [16].

k-Nearest Neighbour. The k-nearest neighbour classifier (k-NN) classifies an instance by means of majority vote among the labels of the k nearest neighbours ($k \geq 1$) [4]. An example is nonconforming when it is far from nearest neighbours with identical label and close to nearest neighbours with different label.

A nonconformity measure can model this as follows. Given example $z_i = (x_i, y_i)$, define an ascending ordered sequence $D_i^{y_i}$ with distances from instance x_i to its k nearest neighbours with label y_i. Similarly, let $D_i^{-y_i}$ contain ordered distances from instance x_i to its k nearest neighbours with label different from y_i. The nonconformity score is then defined as:

$$\alpha_i = \frac{\sum_{j=1}^{k} D_{ij}^{y_i}}{\sum_{j=1}^{k} D_{ij}^{-y_i}}, \tag{12}$$

with subscript j representing the j-th element in a sequence [12]. Clearly, the nonconformity score is monotonically increasing when distances to the k nearest neighbours with identical label increase and/or distances to the k nearest neighbours with different label decrease.

Nearest Centroid. The nearest centroid classifier (NC) learns a Voronoi partition on the training data. It assumes that examples cluster around a class centroid. An example is nonconforming when it is far from the class centroid of its label and close to the class centroids of other labels. Therefore, the nonconformity score of example $z_i = (x_i, y_i)$ can be defined as the distance from x_i to the class centroid of y_i relative to the minimum distance from x_i to all other class centroids [2]. Formally, we write:

$$\alpha_i = \frac{d(\mu_{y_i}, x_i)}{\min_{y \neq y_i} d(\mu_y, x_i)},\tag{13}$$

with μ_y the class centroid of label y which is defined as:

$$\mu_y = \frac{1}{|C_y|} \sum_{i \in C_y} x_i,\tag{14}$$

with C_y the set of indices of instances with label y.

Linear Discriminant. The linear discriminant classifier (LDC) learns a separating hyperplane by maximizing the between scatter of instances with different labels while minimizing the within scatter of instances with identical labels [6]. Instances close to the hyperplane are classified with low confidence since a small change in the hyperplane can result in a different classification of nearby instances. Therefore, a natural nonconformity score of example $z_i = (x_i, y_i)$ is the signed perpendicular distance from x_i to the hyperplane:

$$\alpha_i = -y_i \left(\langle w, x_i \rangle + b \right),\tag{15}$$

with w and b the normal vector and intercept of the hyperplane, and $\langle \cdot, \cdot \rangle$ the inner product. If a classification is correct, then the nonconformity score is negative. A larger distance to the hyperplane represents more confidence in a correct classification, and consequently a lower nonconformity score is obtained. If a classification is incorrect, then the nonconformity score is positive and monotonically increasing with larger distances to the hyperplane.

Naive Bayes. The naive Bayes classifier (NB) is a probabilistic classifier that applies Bayes theorem with independence assumptions [5]. A valid nonconformity score is large if the label of an instance is strange under the Bayesian model [17, p. 102]. We use the following as nonconformity score of example $z_i = (x_i, y_i)$:

$$\alpha_i = 1 - \mathbb{P}(y_i),\tag{16}$$

with $\mathbb{P}(y_i)$ the conditional probability of label y_i that is estimated from the training data and instance x_i, i.e., $\mathbb{P}(\cdot)$ is the posterior label distribution computed by the naive Bayes classifier.[3]

---

[3] It is tempting to believe that the probabilities $\mathbb{P}(\cdot)$ are confidence values. However, it has been verified that these probabilities are overestimated in case of an incorrect prior, e.g., classifying with a probability of 0.7 does not mean that the true label is predicted 70% of the time [10].

Kernel Perceptron. The kernel perceptron (KP) learns a separating hyperplane by updating a weight vector in a high-dimensional space during training [9]. The weight vector represents the normal vector and intercept of the hyperplane. The expansion of the weight vector in dual form is:

$$w = \sum_{i=1}^{n+1} \lambda_i y_i \Phi(x_i), \qquad (17)$$

with λ_i the dual variable for instance x_i and Φ the mapping to the high-dimensional space. It is easily verified that λ_i encodes the number of times that instance x_i is incorrectly classified during training [15, p. 241-242]. Hence, the nonconformity score of example $z_i = (x_i, y_i)$ can be defined as $\alpha_i = \lambda_i$ [10]. However, such a nonconformity score is not valid in the sense that the KP solution depends on the ordering of the training examples. Different KP runs result in different nonconformity scores. In our experiments we show that this violation of the exchangeability assumption does not have any effect in practice.

Support Vector Machine. The support vector machine (SVM) finds a separating hyperplane with maximum margin using pairwise inner products of instances mapped to a high-dimensional space. The inner products are efficiently computed using a kernel function. The maximum margin hyperplane is found by solving a quadratic programming problem in dual form [15, Ch. 7].

In this optimization problem, the Lagrange multipliers $\lambda_1, \lambda_2, \ldots, \lambda_{n+1}$ associated with examples $z_1, \ldots, z_{n+1}$ take values in the domain $[0, C]$ with C the SVM error penalty. Examples with $\lambda_i = 0$ lie outside the margin and at the correct side of the hyperplane. Examples with $0 < \lambda_i < C$ also lie at the correct hyperplane side, but on the margin. Examples with $\lambda_i = C$ can lie inside the margin and at the correct side of the hyperplane, or they can lie at the incorrect side of the hyperplane. Clearly, larger Lagrange multipliers represent more nonconformity and therefore they are valid nonconformity scores, i.e., we define $\alpha_i = \lambda_i$ as the nonconformity score of example $z_i = (x_i, y_i)$ [13,14].

4   Experiments

The previous section discussed technical properties and practical implementations of TCMs. This section empirically investigates whether the calibration property holds when TCMs are applied in the off-line learning setting. We performed experiments with TCMs on a number of benchmark datasets. Subsection 4.1 briefly describes the datasets that we used. Subsection 4.2 outlines the experimental setup. Subsection 4.3 presents the results of the experiments.

4.1   Benchmark Datasets

In the following, we denote the aforementioned TCM implementations by the classifier name and the prefix TCM, e.g., TCM-kNN is the TCM based on the k-NN nonconformity measure.

We tested the six TCMs on ten well-known binary datasets from the UCI benchmark repository [11]. The datasets are: `heart statlog`, `house votes`, `ionosphere`, `liver`, `monks1`, `monks2`, `monks3`, `pima`, `sonar`, and `spect`. Some datasets such as `liver` and `sonar` are known to be highly non-linear. For these non-linear datasets, it is especially challenging to verify if TCM-LDC satisfies the calibration property. The `monks` datasets are datasets for which distance-based classifiers can have difficulties [3].

As a preprocessing step, all instances with missing feature values are removed as well as duplicate instances. Features are standardized to have zero mean and unit variance to remove possible effects caused by features with different orders of magnitude.

4.2   Experimental Setup

The classifiers TCM-kNN, TCM-KP, and TCM-SVM require the selection of one or more parameters. We performed model selection by applying a ten-fold cross validation process that was repeated for five times. The chosen parameter values are those for which the number of prediction sets with multiple labels is minimized for significance levels in the domain $[0, 0.2]$.[4] The number of nearest neighbours for TCM-kNN is restricted to $k = 1, 2, \ldots, 10$. For TCM-SVM and TCM-KP we tested polynomial and Gaussian kernels with exponent values $e = 1, 2, \ldots, 10$ and bandwidth values $\sigma = 0.001, 0.01, 0.03, 0.06, 1, 1.6$ respectively. The SVM error penalty C is kept fixed at value 10.

Once the parameter values are chosen, TCMs are applied in the off-line learning setting with ten-fold cross validation. To ensure that results are independent of the order of examples in the training folds, the experiments were repeated five times with random permutations of the data. We report the average performance of all experiments and test folds.

The performance of TCMs is measured by two key statistics. First, the percentage of prediction sets that do not contain the true label is measured. This is the *error rate* measured as a percentage. Second, we measure *efficiency* to indicate how useful the prediction sets are. Efficiency is given by the percentages of three types of prediction sets. The first type are prediction sets with one label. These prediction sets are called certain predictions. Second, uncertain predictions correspond to prediction sets with two labels and indicate that both labels are likely to be correct. Third, prediction sets can also be empty. Clearly, certain predictions are preferred.

4.3   Results

In this section we report our empirical results of off-line TCMs on the ten benchmark datasets. To visualize performance of a TCM, we follow the convention as defined in [17]. Results are shown as graphs indicating four values for each significance level: (1) percentage of incorrect predictions, (2) percentage of uncertain

---

[4] The conclusions based on our experiments do not depend on the chosen parameter values. Other values simply result in more prediction sets with multiple labels.

predictions, (3) percentage of empty predictions, and (4) percentage of incorrect predictions that are allowed at the significance level. The first value represents the error rate as a percentage, while the second and third values represent efficiency.[5] The line connecting the percentage of incorrect predictions allowed at each significance level is called the *error calibration line*. As an example, Fig. 1 shows the result of applying TCM-kNN and TCM-NC on the `ionosphere` dataset. Graphs of all TCMs and datasets are found in a technical report [16]. In the following we first focus our attention to the calibration property, then we give some remarks about efficiency.

TCMs satisfy the calibration property if the percentage of incorrect predictions at each significance level lies on the error calibration line. From Fig. 1 follows that the corresponding TCMs are well-calibrated up to neglectable statistical fluctuations (the empirical error line can hardly be distinguished from the error calibration line). For example, at $\epsilon = 0.05$ approximately 5% of the prediction sets do not contain the true label. Table 1 verifies the calibration property for all TCMs and datasets by reporting the average deviation between empirical errors and the the error calibration line for $\epsilon = 0, 0.01, \ldots, 0.50$. We do not consider significance levels above 0.5 since these result in classifiers for which more than 50% of the prediction sets do not contain the true label. Deviations are given in percentages and are almost zero, indicating that TCMs satisfy the calibration property when they are applied in the off-line learning setting. Note that we included datasets for which some classifiers have difficulties to achieve a low error rate (Subsection 4.1). Even for these datasets and classifiers, Table 1 reports deviations that are almost zero.

To measure efficiency we note that the percentage of uncertain predictions is 100% when $\epsilon = 0$ since the computed prediction sets contain all labels. We allow for more incorrect predictions when the significance level is set to a higher value. Therefore, the percentage of uncertain predictions monotonically decreases with higher significance levels. How fast this decline goes depends on the performance of the classifier plugged into the TCM framework. This means that k-NN performs significantly better than NC on the `ionosphere` dataset, as illustrated by Fig. 1. The percentage of empty predictions starts to occur at approximately the significance level for which there are no more uncertain predictions. The percentage of empty predictions monotonically increases after this significance level, moving closer to the error calibration line to eventually lie on this line. To summarize efficiency for the `ionosphere` dataset, we consider four significance levels that we believe to be of interest in many practical situations: $\epsilon = 0.20, 0.15, 0.10, 0.05$. For these significance levels, Table 2 reports means and standard deviations for the percentage of incorrect, certain, and empty predictions of all six TCMs. Of course, Table 2 again verifies that the calibration property holds. The reported standard deviations may not seem that small. However, the number of instances in a single test fold is small for the `ionosphere`

---

[5] The percentage of certain predictions is trivially derived from the reported percentages of the other types of prediction sets. Note that the percentage of empty predictions is at most the percentage of incorrect predictions.

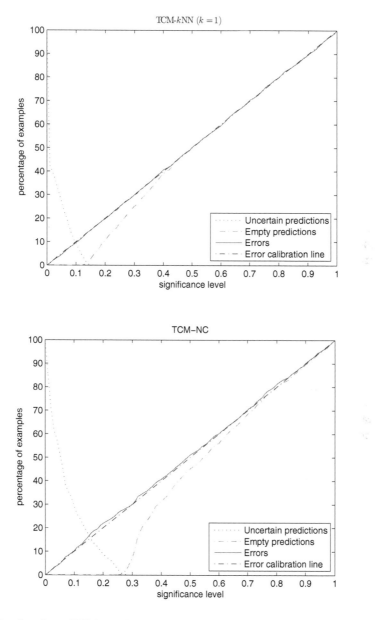

Fig. 1. Results of two TCMs applied on the `ionosphere` dataset in the off-line learning setting: (a) TCM-kNN and (b) TCM-NC

dataset (35 test instances). All values correspond to our discussion of efficiency. Efficiency results for the other datasets are similar and presented in a technical report [16].

Table 1. The deviations between empirical errors and the error calibration line. Values are reported as percentages.

	TCM-kNN	TCM-NC	TCM-LDC	TCM-NB	TCM-KP	TCM-SVM
heart statlog	0.34	0.59	0.35	0.20	0.25	0.31
house votes	0.33	0.27	0.38	0.29	0.53	0.28
ionosphere	0.21	0.81	0.31	0.28	0.33	0.38
liver	0.62	1.35	0.35	0.43	0.47	0.23
monks1	0.98	1.02	0.40	0.60	0.26	0.40
monks2	0.49	1.29	0.46	0.29	0.27	0.36
monks3	0.32	0.51	0.22	0.52	0.21	0.45
pima	0.21	0.28	0.13	0.16	0.16	0.16
sonar	0.59	1.09	0.38	0.32	0.46	0.67
spect	0.35	1.06	0.36	0.58	0.51	0.61

5   Discussion

This section elaborates more on the difference between randomized and non-randomized TCMs, and on the meaning of empty prediction sets.

In our experiments with non-randomized TCMs, we found that the line connecting the empirical errors of a non-randomized TCM-SVM is a step function that tends to stay below the error calibration line (results not shown, see [16] for an example). The reason for this observation is as follows. There are two possible scenarios when a new example is added to the training examples. First, the new example may be a support vector. The difference between the randomized p-value and the non-randomized p-value is then small since the number of support vectors with equal nonconformity score is only a small fraction of the available examples. Second, the new example may be a non-support vector. The randomized p-value is then significantly smaller than the non-randomized p-value since all non-support vectors have equal nonconformity score. This implies that the non-randomized TCM-SVM will compute less empty prediction sets than the randomized TCM-SVM. Therefore, the empirical error line becomes a step function since empty prediction sets are counted as errors. A similar reasoning holds for the difference between a non-randomized TCM-KP and a randomized TCM-KP. For the remaining TCM implementations, a non-randomized version did not led to significantly different results than a randomized version. Indeed, when the nonconformity scores take values in a large domain, then the difference between non-randomized and randomized TCMs is neglectable.

Empty prediction sets indicate that the classification task has become too easy: we can afford the luxury of refusing to make a prediction. Thus, empty prediction sets are a tool to satisfy the calibration property for high significance levels. In fact, the significance level for which empty prediction sets start to arise is approximately equal to the error rate of the classifier when it is not plugged into the TCM framework. To avoid empty predictions, TCMs can be modified to include the label with highest p-value into the prediction set, even though this p-value can be smaller than or equal to the significance level. In this situation,

Table 2. Results of the six TCMs applied on the `ionosphere` dataset in the off-line learning setting

classifier	% error		% certain		% empty	
ϵ	mean	std	mean	std	mean	std
TCM-kNN						
0.20	19.71	7.66	89.43	5.75	10.57	5.75
0.15	14.86	7.05	97.26	3.96	2.63	3.99
0.10	9.66	5.59	90.69	6.34	0.00	0.00
0.05	4.46	4.25	72.97	8.62	0.00	0.00
TCM-NC						
0.20	21.94	7.86	91.14	4.87	0.00	0.00
0.15	15.40	6.80	82.80	5.87	0.00	0.00
0.10	10.23	6.00	70.86	6.88	0.00	0.00
0.05	4.69	4.27	48.00	8.79	0.00	0.00
TCM-LDC						
0.20	19.71	6.86	87.60	5.96	12.34	5.96
0.15	14.69	6.38	93.71	3.74	5.43	3.75
0.10	10.00	5.24	95.31	3.86	0.11	0.57
0.05	5.14	4.32	81.71	6.76	0.00	0.00
TCM-NB						
0.20	19.88	8.20	95.42	4.20	4.57	4.20
0.15	14.74	7.12	93.82	4.91	0.00	0.00
0.10	9.71	5.88	83.82	7.67	0.00	0.00
0.05	4.80	4.18	71.82	8.62	0.00	0.00
TCM-KP						
0.20	20.11	6.88	88.86	5.51	11.09	5.59
0.15	14.74	5.90	96.11	3.20	2.06	2.77
0.10	8.80	5.22	89.83	5.39	0.00	0.00
0.05	5.37	4.57	70.40	10.01	0.00	0.00
TCM-SVM						
0.20	20.06	8.67	81.14	8.26	18.86	8.26
0.15	15.31	7.48	86.06	6.97	13.31	7.11
0.10	10.29	6.85	77.03	5.91	7.20	5.42
0.05	5.31	4.55	52.34	9.51	2.69	3.48

the percentage of empirical errors will also become a step function below the error calibration line since an empty prediction set was previously counted as an error. The significance level now gives an upper bound on the error rate, although we do not know how tight this bound is. The resulting TCMs are called *forced TCMs* and they are said to be conservatively well-calibrated [1].

6   Conclusions

In this paper we focused on the applicability and validity of transductive confidence machines (TCMs) applied in the off-line learning setting. TCMs allow to make predictions such that the error rate is controlled a priori by the user. This property is called the calibration property. An analytical proof of the calibration property exists when TCMs are applied in the on-line learning setting. However, this learning setting restricts the applicability of TCMs.

We provided an extensive empirical evaluation of TCMs applied in the off-line learning setting. Six TCM implementations with different nonconformity measures were applied on ten well-known benchmark datasets. From the results of our experiments we may conclude that TCMs satisfy the calibration property in the off-line learning setting, hereby strongly extending the range of tasks in which they can be applied. TCMs have a significant benefit over conventional classifiers for which the error rate cannot be controlled by the user prior to classification, especially in tasks where reliable instance classifications are desired.

Since TCMs have now been shown to be widely applicable and well-calibrated in virtually any application domain, our future work focuses on efficiency. We noticed that the chosen nonconformity measure affects efficiency while it does not violate the upper bound on the error rate. Our next goal is to minimize the size of the computed prediction sets, especially in case of multilabel learning. We believe that this can be achieved with a new nonconformity measure. Our interest is a measure that is independent of the specific TCM implementation and that is designed to provide a confidence value on nonconformity scores too.

Acknowledgments

The first author is supported by the Dutch Organization for Scientific Research (NWO), ToKeN programme, grant nr: 634.000.435. The second author is supported by NWO, CATCH programme, grant nr: 640.002.401.

References

1. Bellotti, T.: Confidence Machines for Microarray Classification and Feature Selection. PhD thesis, Royal Holloway University of London, London, UK (February 2006)
2. Bellotti, T., Luo, Z., Gammerman, A., Van Delft, F., Saha, V.: Qualified predictions for microarray and proteomics pattern diagnostics with confidence machines. International Journal of Neural Systems 15(4), 247–258 (2005)
3. Blanzieri, E., Ricci, F.: Probability based metrics for nearest neighbor classification and case-based reasoning. In: Althoff, K.-D., Bergmann, R., Branting, L.K. (eds.) ICCBR 1999. LNCS (LNAI), vol. 1650, pp. 14–28. Springer, Heidelberg (1999)
4. Cover, T., Hart, P.: Nearest neighbor pattern classification. IEEE Transactions on Information Theory 13(1), 21–27 (1967)
5. Domingos, P., Pazzani, M.: On the optimality of the simple Bayesian classifier under zero-one loss. Machine Learning 29(2), 103–130 (1997)
6. Fisher, R.: The use of multiple measurements in taxonomics problems. Annals of Eugenics 7, 178–188 (1936)
7. Gammerman, A., Vovk, V.: Prediction algorithms and confidence measures based on algorithmic randomness theory. Theoretical Computer Science 287(1), 209–217 (2002)

8. Gammerman, A., Vovk, V., Vapnik, V.: Learning by transduction. In: Cooper, G., Moral, S. (eds.) UAI 1998. 14th Conference on Uncertainty in Artificial Intelligence, Madison, WI, July 24-26 1998, pp. 148–155. Morgan Kaufmann, San Francisco (1998)

9. Khardon, R., Roth, D., Servedio, R.: Efficiency versus convergence of boolean kernels for on-line learning algorithms. Journal of Artificial Intelligence Research 24, 341–356 (2005)

10. Melluish, T., Saunders, C., Nouretdinov, I., Vovk, V.: Comparing the Bayes and typicalness frameworks. In: Flach, P.A., De Raedt, L. (eds.) ECML 2001. LNCS (LNAI), vol. 2167, pp. 360–371. Springer, Heidelberg (2001)

11. Newman, D., Hettich, S., Blake, C., Merz, C.: UCI repository of machine learning databases (1998)

12. Proedrou, K., Nouretdinov, I., Vovk, V., Gammerman, A.: Transductive confidence machines for pattern recognition. Technical Report 01-02, Royal Holloway University of London, London, UK (2001)

13. Saunders, C., Gammerman, A., Vovk, V.: Transduction with confidence and credibility. In: Dean, T. (ed.) IJCAI 1999. 16th International Joint Conference on Artificial Intelligence, Stockholm, Sweden, July 31 - August 6, 1999, pp. 722–726. Morgan Kaufmann, San Francisco (1999)

14. Saunders, C., Gammerman, A., Vovk, V.: Computationally efficient transductive machines. In: Okamoto, T., Hartley, R., Kinshuk, Klus, J. (eds.) ICALT 2000. 11th International Conference on Algorithmic Learning Theory, Madison, WI, August 6-8, 2000, pp. 325–333. IEEE Computer Society Press, Los Alamitos (2000)

15. Shawe-Taylor, J., Cristianini, N.: Kernel Methods for Pattern Analysis. Cambridge University Press, Cambridge (2004)

16. Vanderlooy, S., van der Maaten, L., Sprinkhuizen-Kuyper, I.: Off-line learning with transductive confidence machines: an empirical evaluation. Technical Report MICC-IKAT 07-03, Universiteit Maastricht, Maastricht, The Netherlands (2007)

17. Vovk, V., Gammerman, A., Shafer, G.: Algorithmic Learning in a Random World. LNCS. Springer, Heidelberg (2005)

Transductive Learning from Relational Data

Michelangelo Ceci, Annalisa Appice, Nicola Barile, and Donato Malerba

Dipartimento di Informatica, Università degli Studi di Bari
via Orabona, 4 - 70126 Bari - Italy
{ceci,appice,malerba}@di.uniba.it, n.barile@gmail.com

Abstract. Transduction is an inference mechanism "from particular to particular". Its application to classification tasks implies the use of both labeled (training) data and unlabeled (working) data to build a classifier whose main goal is that of classifying (only) unlabeled data as accurately as possible. Unlike the classical inductive setting, no general rule valid for all possible instances is generated. Transductive learning is most suited for those applications where the examples for which a prediction is needed are already known when training the classifier. Several approaches have been proposed in the literature on building transductive classifiers from data stored in a single table of a relational database. Nonetheless, no attention has been paid to the application of the transduction principle in a (multi-)relational setting, where data are stored in multiple tables of a relational database. In this paper we propose a new transductive classifier, named TRANSC, which is based on a probabilistic approach to making transductive inferences from relational data. This new method works in a transductive setting and employs a principled probabilistic classification in multi-relational data mining to face the challenges posed by some spatial data mining problems. Probabilistic inference allows us to compute the class probability and return, in addition to result of transductive classification, the confidence in the classification. The predictive accuracy of TRANSC has been compared to that of its inductive counterpart in an empirical study involving both a benchmark relational dataset and two spatial datasets. The results obtained are generally in favor of TRANSC, although improvements are small by a narrow margin.

1 Introduction

In the usual inductive classification setting, data is supposed to have been generated independently and identically distributed (i.i.d.) from an unknown probability distribution P on some domain X and are labeled according to an unknown function g. The domain of g is spanned by m independent (predictor) random variables X_i (either numerical or categorical), that is, $X = X_1, X_2, \ldots, X_m$. The range of g is a finite set $Y = \{C_1, C_2, \ldots, C_L\}$, where each C_i is a distinct class label. After being inputted a training sample $S = \{(x, y) \in X \times Y | y = g(x)\}$, an inductive learning algorithm returns a function f that is hopefully close to g on the domain X. However, there are many cases in which the goal is to estimate the value of the unknown function g at a given set of points of a working

P. Perner (Ed.): MLDM 2007, LNAI 4571, pp. 324–338, 2007.
© Springer-Verlag Berlin Heidelberg 2007

sample $W \subseteq X$ based on the training sample S. The usual way of estimating these values consists in first finding an approximation g' to the desired function g and then using this approximation to get the required estimates. This approach is not always the best when the cardinality of the training sample S is much smaller than that of the working sample W, which is often the case in many real-world situations. It characterizes the traditional inductive learning setting, which uses only labeled examples to generate a classifier and discards a large amount of information potentially conveyed by the unlabeled instances to be classified. Conversely, the idea of *transductive inference* (or *transduction*) [20] is to analyze both the labeled (training) data S and the unlabeled (working) data W to build a classifier whose main goal is that of classifying (only) the unlabeled data W as accurately as possible.

Several transductive learning methods have been proposed in the literature for support vector machines [1] [10] [13] [6], for k-NN classifiers [14] and even for general classifiers [15]. However, despite the growing interest of the scientific community for transductive inference, all of those transductive learning algorithms are based on the *single-table assumption* [22], according to which the training/test data are represented in a single table (or database relation) whose rows (or tuples) represent independent units of the sample population, while columns correspond to properties of these units. This classic tabular representation of data, also known as *propositional* or *feature-vector* representation, turns out to be too restrictive for some complex applications. For instance, in spatial data mining, different spatial objects may have distinctive properties, which can be properly modeled by as many data tables as the number of object types. Moreover, attributes of the neighbors of spatial objects may affect each other (spatial autocorrelation), hence the need for representing object interactions by additional data tables. Although several methods have been proposed to transform a *(multi-)relational* (or *structural*) representation of training data into a single table, this approach (known as *propositionalization*) is fraught with many difficulties in practice [7,11].

In this paper, we propose a novel transductive classification algorithm, named TRANSC (TRANsductive Structural Classifier), that exploits the expressive power of Multi-Relational Data Mining (MRDM) to deal with relational data in their original form. This means that knowledge on the relational data model (e.g., foreign key constraints) is obtained free of charge from the database schema and used to guide the search process. The method works in a transductive setting and employs a probabilistic approach to classification. Information on the potential uncertainty of classification conveyed by probabilistic inference is useful when small changes in the attribute values of a test case may result in sudden changes of the classification. It is also useful when missing (or imprecise) information may prevent a new object from being classified at all [5].

The rest of the paper is organized as follows. In the next section, the background of this research and some related works are introduced, while the (multi-)relational transductive learning problem solved by TRANSC is formally defined in Section 3. In Section 4 experimental results are reported for both a benchmark dataset typically used in MRDM and for two spatial datasets. Finally, Section 5 concludes and discusses ideas for further work.

2   Background and Related Work

The combination of relational representation with principled probabilistic and statistical approaches to inference and learning has been deeply investigated. In particular, relational naïve Bayesian classifiers have been designed to perform probabilistic classification tasks.

Given a feature-vector representation of a test data x, a classical naïve Bayesian classifier assigns x to the class C_i that maximizes the *posterior probability* $P(C_i|x)$. By applying the Bayes theorem, $P(C_i|x)$ is expressed as follows:

$$P(C_i|x) = \frac{P(C_i)P(x|C_i)}{P(x)}. \tag{1}$$

Under the conditional independence (or *naïve*) assumption of object attributes, the likelihood $P(x|C_i)$ can be factorized as follows:

$$P(x|C_i) = P(x_1, \ldots, x_m|C_i) = P(x_i|C_i) \times \ldots \times P(x_m|C_i) \tag{2}$$

where $x_1, \ldots, x_m$ represent the attribute values different from the class label used to describe the object x. Surprisingly, naïve Bayesian classifiers have been proved accurate even when the conditional independence assumption is grossly violated. This is due to the fact that when the assumption is violated, although the estimates of posterior probabilities may be poor, the correct class still has the highest estimate. This leads to correct classifications [8].

The above formalization of a naïve Bayesian classifier is clearly limited to propositional representations. In the case of relational representations, some extensions are necessary. The basic idea is that of using a set of relational patterns to describe an object to be classified, and then to define a suitable decomposition of the likelihood $P(x|C_i)$ *à la* naïve Bayes to simplify the resolution of the probability estimation problem.

An example of relational pattern considered in this work is the following:
$molecule\_Atom(A, B) \wedge molecule\_Type(B, [22, 27])$
$$\Rightarrow molecule\_Attribute(A, active).$$

This is a relational classification rule generated for the Mutagenesis dataset considered in Section 4.1. The literal $molecule\_Attribute(A, active)$ in the consequent of the rule represents the class label (i.e. "active") associated to the molecule A. The literal $molecule\_Atom(A, B)$ in the antecedent of the rule is a *structural characteristic* representing the foreign-key constraint between the tables *Molecule* and *Atom*, while the literal $molecule\_Type(B, [22, 27])$ is a *property* stating that the value of the attribute *Type* of the atom B (composing the molecule A) is a number in the interval [22,27].

Each $P(x|C_i)$ is computed on the basis of a set $\Re = \{A_j \Rightarrow y(X, C_i)\}$ of relational classification rules, where $C_i \in Y$, $y(\_, \_)$ is a binary predicate representing the class label for an example X and the antecedent A_j is a conjunction of literals describing both relations and properties of objects. More precisely, if $\Re(x) \subseteq \Re$ is the set of rules whose antecedent covers the reference object x, then:

$$P(x|C_i) = P(\bigwedge_{R_k \in \Re(x)} antecendent(R_k)|C_i). \qquad (3)$$

This extension of the naïve Bayesian classifier to the case of multi-relational data was originally proposed by Pompe and Kononenko [18] and was recently reworked by Flach and Lachiche [9]. In both works, the conditional independence assumption is straightforwardly applied to all literals in $\bigwedge_{R_k \in \Re(x)} antecedent(R_k)$. However, this may lead to underestimate $P(x|C_i)$ when several similar rules in $\Re$ are considered for the class C_i. Therefore, in this study, we employ a less biased procedure for the computation of the probabilities 3, namely that adopted in the multi-relational naïve Bayesian classifier Mr-SBC [5].

All above mentioned works on relational naïve Bayesian classifiers ignore unlabeled data when mining the classifier. In *semi-supervised learning* approaches, both labeled and unlabeled data are used for training, but the inferential principle is still inductive, that is, a general rule hopefully valid for the whole instance space is generated. An example of semi-supervised learning algorithm has been proposed by Nigam et al. [16], who combine the the naïve Bayesian classifier with the Expectation-Maximization (EM) algorithm. The former is trained on labeled data and provides an initial classification of unlabeled data, while the latter is used to perform hill-climbing in data likelihood space, finding the classifier parameters that locally maximize the likelihood of all the data, both the labeled and the unlabeled.

Vapnik [20] has introduced the transductive Support Vector Machines (SVMs), which take into account a particular test set and try to minimize the misclassification rate of just those particular examples. A different approach has been proposed by Blum and Chawla [2], who uses a similarity measure to construct a graph and then partitions the graph in such a way that it minimizes (roughly) the number of similar pairs of examples that are given different labels. An evolution of this work is the transductive version of k-NN, which has been designed to avoid the myopia of the greedy search strategy adopted in graph partitioning by efficiently and globally solving an optimization problem via spectral methods [14].

Finally, some studies on transductive inference have investigated the opportunity of applying transduction to evaluate the predictive reliability of a real-valued regression model. The basic idea in [3] is to construct transductive predictors and to establish a connection between initial and transductive predictions. An initial predictor is obtained as the model that best fits the training set. It is used to assign a label to a single unlabeled example to be included in the training set and the new training set is used to obtain the final transductive predictor in an iterative process.

3   Probabilistic Transduction in TRANSC

Let $D = \{(x, y) \in X \times Y | y = g(x)\}$ be a dataset labeled according to an unknown function g whose range is a finite set $Y = \{C_1, C_2, \ldots, C_L\}$. Our transductive classification problem is formalized as follows:

Input
- a training set $S \subset D$ and
- the projection of the working set $W = D - S$ on X;

Output: a prediction of the class value (y) of each example in the working set W which is as accurate as possible.

The learner receives full information (including labels) on the examples in S and partial information (only that concerning the independent variables X_i) on the examples in W and is required to predict the class values only of the examples that W consists of. The original formulation of the problem of function estimation in a transductive (*distribution-free*) setting requires that S be sampled from D without replacement. This means that, unlike the standard inductive setting, the examples in the training (and working) set are supposed to be mutually dependent. Vapnik also introduced a second (*distributional*) transduction setting in which the learner receives training and working sets, which are assumed to be drawn i.i.d. from some unknown distribution. As shown in [20] (Theorem 8.1), error bounds for learning algorithms in the distribution-free setting also apply to the more popular distributional transductive setting. Therefore, in this work we focus our attention to the first setting.

In the case of relational data, the problem of transductive classification we aim at solving can be formulated as follows:
Given:

- a database schema S which consists of a set of h relational tables $\{T_0, \ldots, T_{h-1}\}$, a set PK of primary key constraints on the tables in S, and a set FK of foreign key constraints on the tables in S
- a target relation $T \in S$
- a target discrete attribute y in T, different from the primary key of T, whose domain is the finite set $\{C_1, C_2, \ldots, C_L\}$
- the projection T' of T on all attributes of T except y
- a training (working) set that is an instance TS (WS) of the database schema S with known values for y

Find: the most accurate prediction of the values of y for examples in WS represented as a tuple of $t \in WS.T'$ and all tuples related to t in WS according to FK.

This problem is solved by TRANSC by accessing, as in the propositional case, both the full representation of instances in the training set (including that of y) and the partial representation of instances in the working set (represented by T' and its joined tables).

In keeping with the main idea adopted in [13], we iteratively refine the classification by changing the classification of training and working examples in the "borderline" of the class that would be more likely subject to errors. In particular, we propose an algorithm (see Algorithm 1) which starts with a given classification and, at each iteration, alternates a step during which examples are reclassified and a step during which the class of "borderline" examples is changed.

Algorithm 1. Top level transductive algorithm description

1: **transductiveClassifier**(initialClassification, TS, WS)
2: classification1 ← initialClassification;
3: changedExamples ← ϕ;
4: i ← 0;
5: **repeat**
6:    prevClassification ← classification1;
7:    prevChangedExamples ← changedExamples;
8:    classification2 ← reclassifyExamplesKNN(classification1, TS, WS);
9:    (classification1, changedExamples) ← changeClassification(classification2);
10: **until** ((++i ≥ MAX_ITERS) OR
     (computeOverlap(prevChangedExamples,changedExamples) ≥ MAXOVERLAP))
11: **return** prevClassification

The initial classification of an example $E \in WS \cup TS$ is obtained according to the following classification function:

$$preclass(E) = \begin{cases} class(E) & \text{if } E \in TS \\ BayesianClassification(E) & \text{if } E \in WS \end{cases}$$

where $BayesianClassification(E)$ is the classification function corresponding to the initial inductive classifier built from the training set TS. Such an initial classifier is obtained by means of an improved version of the relational probabilistic learning algorithm Mr-SBC [5] whose search strategy is enhanced by considering cyclic paths in the set of foreign keys FK.

The examples are then reclassified by means of a version of the k-NN algorithm tailored for transductive inference in MRDM. The idea is to classify each example $E \in WS \cup TS$ on the basis of a k-sized neighborhood $N_k(E) = \{E_1, \ldots, E_k\}$ consisting of the k examples included in $WS \cup TS$ closest to E with respect to a dissimilarity measure d. This step aims at identifying the value y' of the L-dimensional class probability vector associated to the example E, that is $y' = (y_1(E), \ldots, y_L(E))$, where each $y_i(E) = P(class(E) = C_i)$ is estimated based on $N_k(E)$.

Each probability $P(class(E) = C_i)$ is estimated as follows:

$$P(class(E) = C_i) = \frac{|\{E_j \in N_k(E) | C_{E_j} = C_i\}|}{k} \qquad (4)$$

such that:

- $P(class(E) = C_i) \geq 0$ for each $i = 1, \ldots, L$,
- $\sum_{i=1,\ldots,L} P(class(E) = C_i) = 1$.

In Equation (4), C_{E_j} is the generic class value associated to the example E_j at the previous step; at the first step, C_{E_j} is the class label returned by $preclass(E_j)$. It should be noted that $P(class(E) = C_i)$ is estimated according to the transductive inference principle, as both training and working set are taken into account in the process.

The $changeClassification$ procedure is in charge of changing the classification of the examples on the borderline of a class. Unlike what proposed in [13], where support vectors are used to identify examples on the border, in our case we consider examples for which the entropy of the decision taken by the classifier is maximum. The entropy for each example E is computed from the probabilities associated with each class C_i:

$$Entropy(E) = - \sum_{i=1,\ldots,L} P(class(E) = C_i) \times log(P(class(E) = C_i)) \quad (5)$$

The examples are ordered according to the entropy function and the class label of at most the first k examples having $Entropy(E) > MINENTROPY$ is changed. The class to which each selected example E is assigned is the most likely class C_i for E among those remaining after the the old class of E has been excluded. The threshold k is necessary in order to avoid changing the class of several examples that would lead to erroneously change class of entire "clusters".

In Algorithm 1, two distinct stopping criteria are used. The first criterion stops the execution of the algorithm when the maximum number of iterations (MAX_ITERS) is reached. This guarantees the termination of the algorithm. Indeed, our experiments showed that this criterion is rarely attained when the parameter MAX_ITERS is as small as 10.

The second criterion aims at stopping execution when a cycle insists on the same examples of the previous one. For this purpose, the overlap between two sets of examples is determined. The $computeOverlap$ function returns the ratio between the cardinality of the intersection between the sets of examples and the cardinality of their union.

The classifier returned by Mr-SBC starting from the training set TS is not just employed to pre-classify the working examples in WS. Indeed, the initial Mr-SBC classifier includes a set of first-order classification rules used to represent the examples to be classified. TRANSC reuses such rules to derive a boolean feature-vector representation of each example in WS on which the similarity function subsequently determined is based.

More formally, let $\Re = \{A_j \Rightarrow y(X, C_i)\}$ be the set of classification rules extracted by Mr-SBC, where $C_i \in Y$, $y(\_,\_)$ is a binary predicate representing the class label for an example X and the antecedent A_j is the conjunction of at most MAX_LEN_PATH literals describing both relations and properties of objects. Then each example $E \in WS$ is described by a boolean feature-vector

V_E composed by $|\Re|$ elements, that is, $A_1, \ldots, A_{|\Re|}$. If the antecedent of a rule $(A_j \Rightarrow y(X, C_i)) \in \Re$ covers E, that is, a substitution θ exists such that $A_j \theta \subseteq E$, then the j-th element of V_E is set to *true*; otherwise, it is set to *false*.

The similarity between two examples E_1 and E_2 is determined by matching the *true* values of the corresponding vectors V_{E_1} and V_{E_2}. More precisely, by computing Jaccard's similarity coefficient, which is defined as follows:

$$s(E_1, E_2) = \frac{cardinality(V_{E_1} \; AND \; V_{E_2})}{cardinality(V_{E_1} \; OR \; V_{E_2})} \tag{6}$$

where $cardinality(\bullet)$ returns the number of *true* values included in a boolean vector. Coefficient 6 takes values in the unit interval: $s(E_1, E_2) = 1$ if the two vectors match perfectly, while $s(E_1, E_2) = 0$ if the two vectors are orthogonal or in the degenerate case of no *true* value occurring in both vectors. The dissimilarity between two examples is then defined as follows:

$$d(E_1, E_2) = 1 - s(E_1, E_2) \tag{7}$$

4 Experiments

An empirical evaluation of our algorithm was carried out on both the Mutagenesis dataset, which have been used extensively in testing MRDM algorithms, and on two real-world spatial data collections concerning North West England Census data and Munich Census data, respectively.

We compared the performance of TRANSC to that of Mr-SBC in order to identify the advantages of employing a transductive reformulation of the problem of relational probabilistic classification in real-world applications where few labeled examples are available and manual annotation is fairly expensive.

The two algorithms are compared on the basis of the average misclassification error on the same K-fold cross validation of each dataset. For each dataset, the target table is first divided into K blocks of nearly-equal size and then a subset of tuples related to the tuples of the target table block by means of foreign key constraints are extracted. This way, K database instances are created. For each trial, both TRANSC and Mr-SBC are trained on a single database and tested on the hold-out $K - 1$ database instances forming the working set. It should be noted that the error rates reported in this work are significantly higher than those reported in other literature [5] [4] because of this peculiar experimental design. Indeed, unlike the standard cross-validation approach, here one fold at a time is set aside to be used as the *training set* (and not as the *test set*). Small training set sizes allows us to validate the transductive approach but result in high error rates as well.

A non-parametric Wilcoxon two-sample paired signed rank test [17] is employed to perform a pairwise comparison of the two algorithms. In this test, the summations on both positive (W+) and negative (W-) ranks determine the winner.

It should be noted that in our experiments the size of the working set is one order of magnitude greater than the size of the training set; this is something rather different from what usually happens when testing algorithms developed according to the inductive paradigm. Since the performance of the transductive classifier TRANSC may vary significantly depending on the size (k) of the neighborhood used to predict the class value of each working example, experiments for different k are performed in order to set the optimal value. In theory, we should experiment with each value of k ranging in the interval $[1, |D|]$ where D is the labeled data set. However, as observed in [21] it is not necessary to consider all possible values of k during cross-validation to obtain the best performance. The best performances are obtained by means of cross-validation on no more than approximately ten values of k. A similar consideration has also been reported in [12], where it is shown that the search for the optimal k can be substantially reduced from $[1, |D|]$ to $[1, \sqrt{|D|}]$, without loosing too much accuracy of the approximation. Hence, we have decided to consider in our experiments only $k = \eta i$ such that i value ranges on the sample $[1, \sqrt{|D|}/h]$ and η is the step value.

Classifiers mined in all experiments in this study are obtained by setting $MAX\_LENGTH\_PATH = 3$, $MAX\_ITERS = 10$, $MINENTROPY = 0.65$ and $MAXOVERLAP = 0.5$. The step η is different for each dataset.

4.1   Benchmark Relational Data Application

The Mutagenesis dataset concerns the problem of identifying some mutagenic compounds. We have considered, similarly to most experiments on data mining algorithms reported in literature, the "regression friendly" dataset consisting of 188 molecules. A study on this dataset [19] has identified five levels of background knowledge. Each subset is constructed by augmenting a previous subset and provides richer descriptions of the examples. Table 1 shows the first three sets of background knowledge, the ones we have used in our experiments, where $BK_i \subset BK_{i+1}$ for $i = 0, 1$. The larger the background knowledge set, the more complex the learning problem. All experiments consist in a 10-fold cross validation ($K = 10$).

Table 1. Background knowledge for Mutagenesis data

Background	Description
BK_0	Data obtained with the molecular modeling package QUANTA. For each compound it obtains the atoms, bonds, bond types, atom types, and partial charges on atoms.
BK_1	Definitions in BK_0 plus indicators *ind1* and *inda* in molecule table.
BK_2	Variables (attributes) *logp* and *lumo* are added to definitions in BK_1.

The predictive accuracy of TRANSC was measured by considering the values $k \in \{2, 4, 6, 8, 10, 12\}$. For each setting BK_i ($i = 0, 1, 2$), the average misclassification error of both TRANSC and Mr-SBC is reported in Figure 1. Results show that with BK_0, TRANSC performs better than Mr-SBC, although the improvement is not statistically significant (see Table 2). The results in the BK_1 and

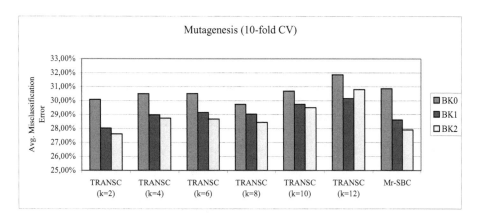

Fig. 1. TRANSC vs. Mr-SBC: average misclassification error on the working sets of Mutagenesis 10-CV data

BK_2 settings suggest different conclusions. As also shown in [5], the predictive accuracy of Mr-SBC increases so significantly when background knowledge is increased (BK_1 and BK_2 setting), that the consideration of unlabeled examples in a neighborhood can even lead to a deterioration in predictive accuracy. In this case, we obtain the best results when k is the lowest.

Table 2. Mutagenesis dataset: results of the Wilcoxon test (p-value) on average accuracy of TRANSC vs. Mr-SBC. The statistically significant p-values (< 0.05) are in italics. The sign + (-) indicates that TRANSC outperforms Mr-SBC (or vice-versa).

BK/k	2	4	6	8	10	12
BK_0	0.23 (+)	0.65 (+)	0.73 (+)	0.19 (+)	0.84 (+)	0.25 (-)
BK_1	0.42 (+)	0.65 (-)	0.76 (-)	0.55 (-)	0.35 (-)	0.2 (-)
BK_2	1.0 (+)	0.13 (-)	0.38 (-)	0.64 (-)	*0.02* (-)	*0.001* (-)

4.2   Spatial Data Application

We have also tested our transductive algorithm on two different spatial data collections, that is, the North-West England Census Data and the Munich Census Data.

The North-West England Census data are obtained from both census and digital maps data provided by the European project SPIN! (http://www.ais. fraunhofer.de/KD/SPIN/project.html). These data concern Greater Manchester, one of the five counties of North West England (NWE). Greater Manchester is divided into ten metropolitan districts, each of which is in turn decomposed into censual sections (wards), for a total of two hundreds and fourteen wards. Census data are available at ward level and provide socio-economic statistics

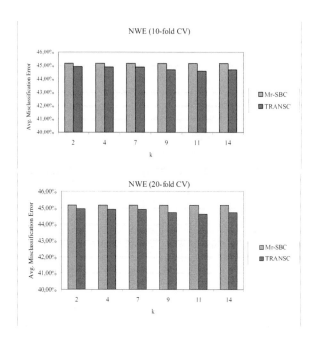

Fig. 2. TRANSC vs. Mr-SBC on NWE census data: average misclassification error on the working sets for 10-fold and 20-fold cross-validation

(e.g. mortality rate – the percentage rate of deaths with respect to the number of inhabitants) as well as some measures of the deprivation of each ward according to information provided by Census combined into single index scores. We have employed Jarman Underprivileged Area Score (which is designed to estimate the need for primary care), the indices developed by Townsend and Carstairs (used to perform health-related analyses), and the Department of the Environment's (DoE) index (which is used in targeting urban regeneration funds). The higher the index value the more deprived the ward. The mortality percentage rate takes values in the finite set $\{low = [0.001, 0.01], high =]0.01, 0, 18]\}$.

The goal of the classification task is to predict the value of the mortality rate by exploiting both deprivation factors and geographical factors represented in some linked topographic maps. Spatial analysis is possible thanks to the availability of vectorized boundaries of the 1998 census wards as well as of other Ordnance Survey digital maps of NWE, where several interesting layers such as urban area (115 lines), green area (9 lines), road net (1687 lines), rail net (805 lines) and water net (716 lines) can be found. The objects on each layer have been stored as tuples of relational tables including also information on the object type (TYPE). For instance, an urban area may be either a "large urban area" or a "small urban area". Topological relationships between wards and objects in all these layers are materialized as relational tables (WARDS_URBAN_AREAS, WARDS_GREEN_AREAS, WARDS_ROADS, WARDS_RAILS and WARDS_WATERS) expressing non-disjointing relations.

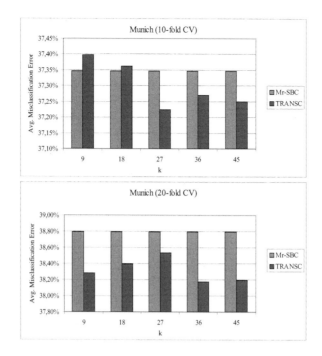

Fig. 3. TRANSC vs. Mr-SBC on Munich census data: average misclassification error on the working sets for 10-fold and 20-fold cross-validation

The number of materialized "non disjoint" relationships is 5313 (381 wards-urban areas, 13 wards-green areas, 2798 wards-roads, 1054 wards-rails and 1067 wards-waters).

The Munich Census Data concern the level of monthly rent per square meter for flats in Munich expressed in German Marks (http://www.di.uniba.it/∼ceci/ mic Files/munich_db.tar.gz). The data have been collected in 1998 by Infratest Sozialforschung to develop the 1999 Munich rental guide. This dataset contains 2180 geo-referenced flats situated in the 446 subquarters of Munich obtained by first dividing the Munich metropolitan area up into three areal zones and then by decomposing each of these zones into 64 districts. The vectorized boundaries of subquarters, districts and zones as well as the map of public transport stops consisting of public train stops (56 subway (U-Bahn) stops, 15 rapid train (S-Bahn) stops and 1 railway station) within Munich are available for this study. The objects included in these layers are stored in different relational tables (SUB-QUARTERS, TRANSPORT_STOPS and FLATS). Information on the "area" of subquarters is stored in the corresponding table. Transport stops are described by means of their type (U-Bahn, S-Bahn or Railway station), while flats are described by means of their "monthly rent per square meter", "floor space in square meters" and "year of construction".

The target attribute was represented by the "monthly rent per square meter", whose values have been discretized into the two values $low = [2.0, 14.0]$

or $high =]14.0, 35.0]$. The spatial arrangement of data is defined by both the "close_to" relation between Munich metropolitan subquarters areas and the "inside" relation between public train stops and metropolitan subquarters. Both of these topological relations are materialized as relational tables (CLOSE_TO and INSIDE).

The average misclassification error of TRANSC and Mr-SBC on both NWE Census Data and Munich Census Data is reported in Figure 2 and Figure 3, respectively. The reported results refer to both a 10-fold cross validation (CV) of the data and 20-fold cross validation of the same data. When experimenting on the NWE Census Data, we set $k \in \{2, 4, 7, 9, 1, 14\}$, while when experimenting on the Munich Census Data we set $k \in \{9, 18, 27, 36, 45\}$.

The results of Wilcoxon test are reported in Table 3 for the NWE Census Data and in Table 4 for the Munich Census Data. The results showed a slight improvement in the predictive accuracy of the transductive classifier over its inductive counterpart. Considering that both datasets are characterized by a strongly relevant structural component, these results confirm what observed with the Mutagenesis dataset, that is, the transductive approach we are proposing is beneficial when structural information is strongly relevant for the task at hand.

Table 3. TRANSC vs. Mr-SBC on NWE census data: results of the Wilcoxon test. Statistically significant p-values (< 0.05) are in italics. The sign + (-) indicates that TRANSC outperforms Mr-SBC (or vice-versa).

Experiment/k	2	4	6	8	10	12
10-fold CV	0.43 (+)	0.84 (+)	0.31 (+)	0.29 (+)	0.21 (+)	0.37 (+)
20-fold CV	0.12 (+)	0.17 (+)	0.36 (+)	0.12 (+)	0.09 (+)	0.16 (+)

Table 4. TRANSC vs. Mr-SBC on Munich census data: results of the Wilcoxon test. Statistically significant p-values (< 0.05) are in italics. The sign + (-) indicates that TRANSC outperforms Mr-SBC (or vice-versa).

Experiment/k	9	18	27	36	45
10-fold CV	0.42 (-)	0.74 (-)	*0.04* (+)	0.25 (+)	0.20 (+)
20-fold CV	*0.0019* (+)	*0.03* (+)	0.1 (+)	*0.00012* (+)	*0.00006* (+)

5   Conclusions

In this work we have investigated the combination of transductive inference with principled probabilistic MRDM classification in order to face the challenges posed by real-world applications characterized by both complex and heterogeneous data, which are naturally modeled as several tables of a relational database, and the availability of a small (large) set of labeled (unlabeled) data. Our proposed algorithm builds on an initial inductive classifier, namely a multi-relational naïve Bayesian classifier (Mr-SBC), learned from the training (i.e., labeled) examples and used to perform a preliminary labeling of the working (i.e., unlabeled) data. The initial classification of the examples comprising the working set is then refined iteratively over a finite number of steps, each of which consists in a k-NN classification of all unlabeled examples and a subsequent reclassification of some "borderline" unlabeled examples. Neighbors are determined by computing a distance measure on a propositionalized representation of working examples. Propositionalization is based on the set of multi-relational rules mined by Mr-SBC.

The proposed transductive multi-relational classifier (TRANSC) has been compared to its inductive counterpart (Mr-SBC) in an empirical study involving both a benchmark relational dataset and two spatial datasets. The results of the experiments conducted on the benchmark dataset are in favor of TRANSC only when no background knowledge is considered (setting BK_0). Experimental results on spatial data are generally in favor of TRANSC and statistically significant in the case of the largest disproportion between training and working set (Munich census data with 20-fold cross validation). However, the improvements over the inductive counterpart are small. This findings confirm for the relational framework what already established for the propositional case [14], where similar small improvements have been observed when comparing SVMs in the inductive and transductive setting (SVMs vs TSVMs). Nonetheless, we intend to perfect our work in order to corroborate our intuition that transductive inference has benefits over inductive inference when applied to situations, like text mining, where the unlabeled examples heavily outnumber the labeled ones.

Acknowledgment

This work partially fulfills the research objective of ATENEO-2006 project titled "Metodi di scoperta di conoscenza per ubiquitous computing".

References

1. Bennett, K.P.: Combining support vector and mathematical programming methods for classification. pp. 307–326 (1999)
2. Blum, A., Chawla, S.: Learning from labeled and unlabeled data using graph mincuts. In: Proceedings of 18th International Conf. on Machine Learning, pp. 19–26. Morgan Kaufmann, San Francisco (2001)
3. Bosnic, Z., Kononenko, I., Robnic-Sikonja, M., Kukar, M.: Evaluation of prediction reliability in regression using the transduction principle. In: The IEEE Region 8 EUROCON 2003, pp. 99–103. IEEE Computer Society Press, Los Alamitos (2003)
4. Ceci, M., Appice, A.: Spatial associative classification: propositional vs structural approach. Journal of Intelligent Information Systems 27(3), 191–213 (2006)
5. Ceci, M., Appice, A., Malerba, D.: Mr-SBC: a multi-relational naive bayes classifier. In: Lavrač, N., Gamberger, D., Todorovski, L., Blockeel, H. (eds.) PKDD 2003. LNCS (LNAI), vol. 2838, pp. 95–106. Springer, Heidelberg (2003)
6. Chen, Y., Wang, G., Dong, S.: Learning with progressive transductive support vector machines. Pattern Recognition Letters 24, 1845–1855 (2003)

7. De Raedt, L.: Attribute-value learning versus inductive logic programming: the missing links. In: Page, D. (ed.) Inductive Logic Programming. LNCS, vol. 1446, pp. 1–8. Springer, Heidelberg (1998)
8. Domingos, P., Pazzani, M.: On the optimality of the simple bayesian classifier under zeo-ones loss. Machine Learning 28(2-3), 103–130 (1997)
9. Flach, P.A., Lachiche, N.: Naive bayesian classification of structured data. Machine Learning 57(3), 233–269 (2004)
10. Gammerman, A., Azoury, K., Vapnik, V.: Learning by transduction. In: UAI 1998. Proc. of the 14th Annual Conference on Uncertainty in Artificial Intelligence, pp. 148–155. Morgan Kaufmann, San Francisco (1998)
11. Getoor, L.: Multi-relational data mining using probabilistic relational models: research summary. In: Knobbe, A., Van der Wallen, D.M.G. (eds.) Proc.of the 1st Workshop in Multi-relational Data Mining, Freiburg, Germany (2001)
12. Gora, G., Wojna, A.: RIONA: A classifier combining rule induction and k-nn method with automated selection of optimal neighbourhood. In: Elomaa, T., Mannila, H., Toivonen, H. (eds.) ECML 2002. LNCS (LNAI), vol. 2430, pp. 111–123. Springer, Heidelberg (2002)
13. Joachims, T.: Transductive inference for text classification using support vector machines. In: ICML 1999. Proc. of the 16th International Conference on Machine Learning, pp. 200–209. Morgan Kaufmann, San Francisco (1999)
14. Joachims, T.: Transductive learning via spectral graph partitioning. In: ICML 2003. Proc. of the 20th International Conference on Machine Learning, Morgan Kaufmann, San Francisco (2003)
15. Kukar, M., Kononenko, I.: Reliable classifications with machine learning. In: Elomaa, T., Mannila, H., Toivonen, H. (eds.) ECML 2002. LNCS (LNAI), vol. 2430, pp. 219–231. Springer, Heidelberg (2002)
16. Nigam, K., McCallum, A.K., Thrun, S., Mitchell, T.M.: Text classification from labeled and unlabeled documents using EM. Machine Learning 39(2/3), 103–134 (2000)
17. Orkin, M., Drogin, R.: Vital Statistics. McGraw Hill, New York (1990)
18. Pompe, U., Kononenko, I.: Naive bayesian classifier within *ilpr*. In Raedt, L.D. (ed) Proc. of the 5th Int. Workshop on Inductive Logic Programming, Dept. of Computer Science, Katholieke Universiteit Leuven, pp. 417–436 (1995)
19. Srinivasan, A., King, R.D., Muggleton, S.: The role of background knowledge: using a problem from chemistry to examine the performance of an ILP program. In Technical Report PRG-TR-08-99, Oxford University Computing Laboratory, Oxford (1999)
20. Vapnik, V.: Statistical Learning Theory. Wiley, Chichester (1998)
21. Wettschereck, D.: A study of Distance-Based Machine Learning Algorithms. PhD thesis, PhD thesis, Department of Computer Science, Oregon State University, Corvalis, OR (1994)
22. Wrobel, S.: Relational Data Mining. In: chapter Inductive logic programming for knowledge discovery in databases. LNCS (LNAI), pp. 74–101. Springer, Heidelberg (2001)

A Novel Rule Ordering Approach in Classification Association Rule Mining

Yanbo J. Wang[1], Qin Xin[2], and Frans Coenen[1]

[1] Department of Computer Science, The University of Liverpool,
Ashton Building, Ashton Street, Liverpool, L69 3BX, UK
{jwang,frans}@csc.liv.ac.uk
[2] Department of Informatics, University of Bergen,
P.B.7800, N-5020 Bergen, Norway
Xin@ii.uib.no

Abstract. A Classification Association Rule (CAR), a common type of mined knowledge in Data Mining, describes an implicative co-occurring relationship between a set of binary-valued data-attributes (items) and a pre-defined class, expressed in the form of an "antecedent $\Rightarrow$ consequent-class" rule. Classification Association Rule Mining (CARM) is a recent Classification Rule Mining (CRM) approach that builds an Association Rule Mining (ARM) based classifier using CARs. Regardless of which particular methodology is used to build it, a classifier is usually presented as an ordered CAR list, based on an applied rule ordering strategy. Five existing rule ordering mechanisms can be identified: (1) Confidence-Support-size_of_Antecedent (CSA), (2) size_of_Antecedent-Confidence-Support (ACS), (3) Weighted Relative Accuracy (WRA), (4) Laplace Accuracy, and (5) χ^2 Testing. In this paper, we divide the above mechanisms into two groups: (i) pure "support-confidence" framework like, and (ii) additive score assigning like. We consequently propose a hybrid rule ordering approach by combining one approach taken from (i) and another approach taken from (ii). The experimental results show that the proposed rule ordering approach performs well with respect to the accuracy of classification.

Keywords: Classification Association Rules, Classification Association Rule Mining, Data Mining, Rule Ordering.

1 Introduction

Classification Rule Mining (CRM) [15] is a well-known Data Mining technique for the extraction of hidden Classification Rules (CRs) from a given database that is coupled with a set of pre-defined classes, the objective being to build a classifier to classify "unseen" data records. One recent approach to CRM is to employ Association Rule Mining (ARM) [1] techniques to identify the desired CRs, i.e. Classification Association Rule Mining (CARM). In [9], Coenen *et al.* suggest that results presented in [13] and [14] show that CARM seems to offer greater accuracy of classification, in many cases, than other CRM methods such as C4.5 [15]. CARM mines a set of Classification Association Rules (CARs) from a class transaction database (the

P. Perner (Ed.): MLDM 2007, LNAI 4571, pp. 339–348, 2007.

well-established transaction database in a class fashion), where a CAR describes an implicative co-occurring relationship between a set of binary-valued data attributes (items in a transaction database) and a pre-defined class, expressed in the form of an "antecedent ⇒ consequent-class" rule. Regardless of which particular methodology is used to generate CARs, a classifier is usually presented as an ordered CAR list, based on an applied rule ordering mechanism. In [7] Coenen and Leng evaluated a number of alternative case satisfaction and rule ordering strategies. They indicate that (1) three common case satisfaction approaches are best first rule, best *K* rule, and all rules; and (2) five existing rule ordering mechanisms are Confidence-Support-size_of_Antecedent (CSA), size_of_Antecedent-Confidence-Support (ACS), Weighted Relative Accuracy (WRA), Laplace Accuracy, and χ^2 Testing. In this paper, we further divide (2) into two groups: (i) pure "support-confidence" framework like, and (ii) additive score assigning like. We consequently propose a hybrid rule ordering approach by combining one mechanism taken from (i) and another mechanism taken from (ii). The experimental results show good performance regarding the accuracy of classification when using the proposed rule ordering approach with the best first rule case satisfaction.

2   Related Work

2.1   An Overview of CARM Algorithms

The idea of CARM was first presented in [3]. Subsequently a number of alternative approaches have been described. Broadly CARM algorithms can be categorized into two groups according to the way that the CARs are generated:

- **Two stage algorithms** where a set of CARs are produced first (stage 1), which are then pruned and placed into a classifier (stage 2). Examples of this approach include CBA [14] and CMAR [13]. CBA (Classification Based Associations), developed by Liu *et al.* in 1998, is an Apriori [2] based CARM algorithm, which (1) applies its CBA-GR procedure for CAR generation; and (2) applies its CBA-CB procedure to build a classifier based on the generated CARs. CMAR (Classification based on Multiple Association Rules), introduced by Han and Jan in 2001, is similar to CBA but generates CARs through a FP-tree [11] based approach.
- **Integrated algorithms** where the classifier is produced in a single processing step. Examples of this approach include TFPC[1] [7] [9], and induction systems such as FOIL [16], PRM and CPAR [17]. TFPC (Total From Partial Classification), proposed by Coenen *et al.* in 2004, is a Apriori-TFP [8] based CARM algorithm, which generates CARs through efficiently constructing both P-tree and T-tree set enumeration tree structures. FOIL (First Order Inductive Learner) is an inductive learning algorithm for generating CARs developed by Quinlan and Cameron-Jones in 1993. This algorithm was later developed by Yin and Han to produce the PRM (Predictive Rule Mining) CAR generation

---

[1] TFPC may be obtained from http://www.csc.liv.ac.uk/~frans/ KDD/Software.

algorithm. PRM was then further developed, by Yin and Han in 2003 to produce CPAR (Classification based on Predictive Association Rules).

2.2 Case Satisfaction Approaches

In [7] Coenen and Leng summarized three case satisfaction approaches that have been employed in different CARM algorithms for utilizing the resulting classifier to classify "unseen" data. These three case satisfaction approaches are itemized as follows (given a particular case):

- **Best First Rule:** Select the first best rule that satisfies the given case according to some ordering imposed on the CAR list. The ordering can be defined according to many different ordering mechanisms, including: (1) CSA – combinations of confidence, support and size of antecedent, with confidence being the most significant factor (used in CBA, TFPC and the early stages of processing of CMAR); (2) ACS – an alternative to CSA that considers the size of the rule antecedent as the most significant factor; (3) WRA – which reflects a number of rule "interestingness" measures as proposed in [12]; (3) Laplace Accuracy – as used in PRM and CPAR; (5) χ^2 Testing – χ^2 values as used, in part, in CMAR; etc.
- **Best K Rules:** Select the first best K rules that satisfy the given case and then select a rule according to some averaging process as used for example, in CPAR. The term "best" in this case is defined according to an imposed ordering of the form described in Best First Rule.
- **All Rules:** Collect all rules in the classifier that satisfy the given case and then evaluate this collection to identify a class. One well-known evaluation method in this category is WCS (Weighted χ^2) testing as used in CMAR.

3 Rule Ordering Approaches

As noted in the previous section five existing rule ordering mechanisms are identified to support the best first rule case satisfaction strategy. Each can be further separated into two stages: (1) a rule weighting stage where each CAR is labeled with a weighting score that represents the significance of this CAR indicates a predefined class; and (2) a rule re-ordering stage, which sorts the original CAR list in a descending manner, based on the score assigned in stage (1), of each CAR. Based on (1) we divide these existing rule ordering mechanisms into two groups: (i) pure "support-confidence" framework like, and (ii) additive score assigning like. With regards to both stages of rule weighting and rule re-ordering, each rule ordering mechanism can be described in more detail as follows:

(i) Pure "support-confidence" framework like

- **CSA:** The CSA rule ordering mechanism is based on the well-established "support-confidence" framework. It does not assign an additive weighting score to any CAR in its rule weighting stage, but simply gathers the value of confidence and support, and the size of the rule antecedent to "express" a weighting score for each CAR. In the rule re-ordering stage, CSA generally sort the original CAR list in a descending order based on the value of confidence of each CAR. For these CARs that share a common value of confidence, CSA sorts them in a descending order based on their support value. Furthermore for these CARs that share common values for both confidence and support, CSA sorts them in an ascending order based on the size of the rule antecedent.

- **ACS:** The ACS rule ordering mechanism is a variation of CSA. It takes the size of the rule antecedent as its major factor (using a descending order) followed by the rule confidence and support values respectively. This rule ordering mechanism ensures that "specific rules have a higher precedence than more general rules" [7].

(ii) Additive score assigning like

- **WRA:** The use of WRA can be found in [12], where this technique is used to determine an expected accuracy for each CAR. In its rule weighing stage, WRA assigns an additive weighting score to each CAR. The calculation of the value of a CAR r, confirmed in [7], is: $wra(r)$ = support (r.antecedent) * (confidence (r) – support (r.consequent)). In the rule re-ordering stage the original CAR list is simply sorted in a descending order based on the assigned wra value of each CAR.

- **Laplace Accuracy:** The use of the *Laplace expected error estimate* [5] can be found in [17]. The principle of applying this rule ordering mechanism is similar to WRA. The calculation of the *Laplace* value of a CAR r is: $Laplace(r)$ = (support (r.antecedent $\cup$ r.consequent) + 1) / (support (r.antecedent) + c), where c represents the number of pre-defined classes.

- **χ^2 Testing:** χ^2 Testing is a well known technique in statistics, which can be used to determine whether two variables are independent of one another. In χ^2 Testing a set of observed values (O) is compared against a set of expected values (E) – values that would be estimated if there were no associative relationship between the variables. The value of χ^2 is calculated as: $\sum_{[i = 1...n]} (O_i - E_i)^2 / E_i$, where n is the number of observed/expected values, which is always 4 in CARM. If the χ^2 value between two variables (the antecedent and consequent-class of a CAR) above a given threshold value (for CMAR the chosen threshold is 3.8415), thus it can be concluded that there is a relation between the rule antecedent and consequent-class, otherwise there is not a relation. After assigning an additive χ^2 value to each CAR, it can be used to re-order the CAR list in a descending basis.

4 The Hybrid Rule Ordering Strategy

In [17] Yin and Han believe that there are only a limited number, say at most K in each class, of CARs that are required to distinguish between classes and should be thus used to make up a classifier. They suggest a value of 5 as an appropriate value for K, and employ the Laplace accuracy (a method in the additive score assigning like) to estimate the accuracy of CARs. With respect to the above suggestions, we propose a hybrid rule ordering strategy by combining one rule ordering mechanism taken from (i) the pure "support-confidence" framework like, and another rule ordering mechanism taken from (ii) the additive score assigning like. We sketch the process of the proposed rule ordering approach in Figure 1 as follows.

```
Procedure HYBRID RULE ORDERING;
Input:  a list of CARs ℜ (in a CSA or ACS
            ordering manner);
Output: a re-ordered list of CARs ℜ⁺;

(1) begin
(2)     ℜ⁺ := {∅};
(3)     ℜ⁰ := {∅};
(4)     ℜ⁰ ← catch the best K rules in ℜ using
                a method in (ii);
(5)     ℜ⁰ ← re-order ℜ⁰ based on a method in (i);
(6)     ℜ⁺ ← ℜ⁰ + ℜ;
(7)     return (ℜ⁺);
(8) end
```

Fig. 1. The HYBRID RULE ORDERING Procedure

Six different schemes can be identified in this hybrid approach:

- **Hybrid CSA/WRA:** Selects the best K rules in a WRA manner, and re-orders both the best K CAR list and the original CAR list in a CSA fashion. (Note: we assume that both CAR lists use the same ordering fashion as either CSA or ACS);
- **Hybrid CSA/Laplace:** Selects the best K rules in a Laplace manner, and re-orders both the best K CAR list and the original CAR list in a CSA fashion;
- **Hybrid CSA/χ^2:** Selects the best K rules in a χ^2 manner, and re-orders both the best K CAR list and the original CAR list in a CSA fashion;
- **Hybrid ACS/WRA:** Selects the best K rules in a WRA manner, and re-orders both the best K CAR list and the original CAR list in an ACS fashion.
- **Hybrid ACS/Laplace:** Selects the best K rules in a Laplace manner, and re-orders both the best K CAR list and the original CAR list in an ACS fashion; and
- **Hybrid ACS/χ^2:** Selects the best K rules in a χ^2 manner, and re-orders both the best K CAR list and the original CAR list in an ACS fashion.

5 Experimental Results

In this section, we aim to evaluate the proposed hybrid rule ordering approach with respect to the accuracy of classification. All evaluations were obtained using the TFPC algorithm coupled with the best first rule case satisfaction strategy, although any other CARM classifier generator, founded on the best first rule strategy, could equally well be used. Experiments were run on a 1.20 GHz Intel Celeron CPU with 256 Mbyte of RAM running under Windows Command Processor.

The experiments were conducted using a range of datasets taken from the LUCS-KDD discretised/normalized ARM and CARM Data Library [6]. The chosen datasets are originally taken from the UCI Machine Learning Repository [4]. These datasets have been discretised and normalized using the LUCS-KDD DN software[2], so that data are then presented in a binary format suitable for use with CARM applications. It should be noted that the datasets were rearranged so that occurrences of classes were distributed evenly throughout the datasets. This then allowed the datasets to be divided in half with the first half used as the training set and the second half as the test set. Although a "better" accuracy figure might have been obtained using Ten-Cross Validation [10], it is the relative accuracy that is of interest here and not the absolute accuracy.

The first set of evaluations undertaken used a confidence threshold value of 50% and a support threshold value 1% (as used in the published evaluations of CMAR [11], CPAR [17], TFPC [7] [9]). The results are presented in Table 1 where 120 classification accuracy values are listed based on 24 chosen datasets. The row labels describe the key characteristics of each dataset: for example, the label

Table 1. Classification accuracy – five existing rule ordering approaches

DATASETS	CSA	ACS	WRA	Laplace	χ^2
adult.D97.N48842.C2	80.83	73.99	81.66	76.07	76.07
anneal.D73.N898.C6	91.09	75.50	87.75	77.51	77.51
auto.D137.N205.C7	61.76	53.92	50.00	47.06	50.00
breast.D20.N699.C2	89.11	89.11	87.68	65.62	65.62
connect4.D129.N67557.C3	65.83	64.83	67.93	65.83	65.83
cylBands.D124.N540.C2	65.93	42.59	64.07	57.78	57.78
flare.D39.N1389.C9	84.44	83.86	84.15	84.44	84.44
glass.D48.N214.C7	58.88	43.93	50.47	52.34	50.47
heart.D52.N303.C5	58.28	28.48	55.63	54.97	54.97
hepatitis.D56.N155.C2	68.83	48.05	71.43	79.22	79.22
horseColic.D85.N368.C2	72.83	40.76	79.89	79.89	63.04
ionosphere.D157.N351.C2	85.14	61.14	86.86	64.57	64.57
iris.D19.N150.C3	97.33	97.33	97.33	97.33	97.33
led7.D24.N3200.C10	68.38	61.38	63.94	63.88	65.56
letRecog.D106.N20000.C26	31.13	26.21	26.33	26.33	28.52
mushroom.D90.N8124.C2	99.21	65.76	98.45	98.45	49.43
nursery.D32.N12960.C5	80.35	55.88	70.17	70.17	70.17
pageBlocks.D46.N5473.C5	90.97	90.97	90.20	89.80	89.80
pima.D38.N768.C2	73.18	71.88	72.92	65.10	65.10
soybean-large.D118.N683.C19	86.22	79.77	36.36	36.07	77.42
ticTacToe.D29.N958.C2	71.61	36.12	68.06	65.34	65.34
waveform.D101.N5000.C3	61.56	47.96	56.24	57.84	57.28
wine.D68.N178.C3	56.18	37.08	80.90	73.03	70.79
zoo.D42.N101.C7	80.00	42.00	56.00	42.00	42.00
Average	**74.13**	**59.10**	**70.18**	**66.28**	**65.34**

adult.D97.N48842.C2 denotes the "adult" dataset, which includes 48,842 records in 2 pre-defined classes, with attributes that for the experiments described here have been discretised and normalized into 97 binary categories.

From Table 1 it can be seen that with a 50% confidence threshold and a 1% support threshold the CSA rule ordering mechanism worked better than other alternative approaches. When applying the CSA rule ordering mechanism, the average accuracy of classification throughout the 24 datasets is 74.13%, whereas using ACS is 59.10%, WRA is 70.18%, Laplace is 66.28%, and χ^2 is 65.34%.

The second set of evaluations undertaken used a confidence threshold value of 50%, a support threshold value of 1%, and a value of 5 as an appropriate value for K when selecting the best K rules (as suggested by Yin and Han in [17]). The results are presented in Table 2 where 144 classification accuracy values are listed based on 24 chosen datasets.

Table 2. Classification accuracy – six hybrid rule ordering schemes

DATASETS	CSA/ WRA	CSA/ Laplace	CSA/ χ^2	ACS/ WRA	ACS/ Laplace	ACS/ χ^2
adult.D97.N48842.C2	83.33	79.95	79.95	78.56	83.76	80.14
anneal.D73.N898.C6	91.09	91.54	91.54	80.40	80.62	88.20
auto.D137.N205.C7	59.80	58.82	53.92	55.88	54.90	52.94
breast.D20.N699.C2	89.11	88.54	89.11	89.11	88.54	89.11
connect4.D129.N67557.C3	67.67	65.83	65.83	64.88	64.88	64.88
cylBands.D124.N540.C2	67.04	69.26	57.78	61.11	70.00	53.33
flare.D39.N1389.C9	84.29	84.44	84.44	83.86	83.86	83.86
glass.D48.N214.C7	66.36	66.36	66.36	65.42	65.42	68.22
heart.D52.N303.C5	55.63	56.95	58.94	52.32	50.33	50.33
hepatitis.D56.N155.C2	84.42	84.42	84.42	63.64	71.43	68.83
horseColic.D85.N368.C2	83.15	83.15	79.89	75.00	83.15	71.20
ionosphere.D157.N351.C2	90.29	89.71	88.00	90.29	89.71	88.00
iris.D19.N150.C3	97.33	97.33	97.33	97.33	97.33	97.33
led7.D24.N3200.C10	68.19	68.19	68.38	62.06	62.06	62.31
letRecog.D106.N20000.C26	31.49	31.49	31.56	27.39	27.39	28.41
mushroom.D90.N8124.C2	98.45	98.82	98.45	98.45	98.82	98.45
nursery.D32.N12960.C5	78.86	78.86	78.86	66.73	66.73	66.73
pageBlocks.D46.N5473.C5	90.97	90.97	90.97	90.97	90.97	90.97
pima.D38.N768.C2	73.18	73.18	72.66	73.18	73.18	72.66
soybean- large.D118.N683.C19	80.94	80.94	82.11	75.66	75.66	78.01
ticTacToe.D29.N958.C2	74.95	74.74	72.65	60.75	70.35	67.22
waveform.D101.N5000.C3	57.96	57.96	60.60	59.20	59.20	60.60
wine.D68.N178.C3	77.53	77.53	77.53	77.53	77.53	77.53
zoo.D42.N101.C7	84.00	90.00	72.00	80.00	80.00	80.00
Average	**76.50**	**76.62**	**75.14**	**72.07**	**73.58**	**72.47**

From Table 2 it can be seen that with a 50% confidence threshold, a 1% support threshold, and 5 as the value of K, the approach hybrid CSA/Laplace preformed better than other alternative hybrid schemes. When applying the hybrid CSA/Laplace, the average accuracy of classification throughout the 24 datasets is 76.62%. Let CSA and Laplace be the "parents" of the hybrid CSA/Laplace, we indicate that the classification accuracy obtained using the hybrid CSA/Laplace is significantly higher than using its "parents", where CSA is 74.13% and Laplace is 66.28%. Furthermore we identify:

- The classification accuracy of the hybrid CSA/WRA is significantly higher than the accuracies of its "parents", where the average accuracy of the hybrid CSA/WRA is 76.50% whereas CSA is 74.13% and WRA is 70.18%.
- The classification accuracy of the hybrid CSA/χ^2 is significantly higher than the accuracies of its "parents", where the accuracy of the hybrid CSA/χ^2 is 75.14% whereas CSA is 74.13% and χ^2 is 65.34%;
- The accuracy of the hybrid ACS/WRA is significant higher than the accuracies of its "parents", where the hybrid ACS/WRA is 72.07% whereas ACS is 59.10% and WRA is 70.18%;
- The accuracy of the hybrid ACS/Laplace is significantly higher than its "parents", where the hybrid ACS/Laplace is 73.58% whereas ACS is 59.10% and Laplace is 66.28%; and
- The accuracy of the hybrid ACS/χ^2 is significantly higher than its "parents", where the hybrid ACS/χ^2 is 72.47% whereas ACS is 59.10% and χ^2 is 65.34%.

6   Conclusion

This paper is concerned with an investigation of CARM. An overview of alternative CARM algorithms was provided in Section 2.1, and three current case satisfaction strategies were reviewed in Section 2.2. In Section 3 with regards to both stages of rule weighting and rule re-ordering, we described the existing rule ordering mechanisms in groups (the "support-confidence" framework like vs. the additive score assigning like). A hybrid rule ordering approach was proposed in Section 4, which combines an approach taken from the "support-confidence" framework like, and another approach taken from the additive score assigning like. Subsequently six hybrid rule ordering schemes were introduced. From the experimental results (see Section 5), all six hybrid schemes presented good classification accuracy – the accuracy is significantly higher than the accuracies obtained by their "parent" rule ordering approaches. Further research is suggested to identify the improved rule ordering approach to give a better performance.

Acknowledgments

The authors would like to thank Prof. Paul Leng and Dr. Robert Sanderson of the Department of Computer Science at the University of Liverpool for their support with respect to the work described here.

References

1. Agrawal, R., Imielinski, T., Swami, A.: Mining Association Rules between Sets of Items in Large Databases. In: Buneman, P., Jajodia, S. (eds.) SIGMOD-93. Proceedings of the 1993 ACM SIGMOD International Conference on Management of Data, Washington, DC, May 1993, pp. 207–216. ACM Press, New York (1993)
2. Agrawal, R., Srikant, R.: Fast Algorithm for Mining Association Rules. In: Bocca, J.B., Jarke, M., Zaniolo, C. (eds.) VLDB-94. Proceedings of the 20th International Conference on Very Large Data Bases, Santiago de Chile, Chile, September 1994, pp. 487–499. Morgan Kaufmann Publishers, San Francisco (1994)
3. Ali, K., Manganaris, S., Srikant, R.: Partial Classification using Association Rules. In: Heckerman, D., Mannila, H., Pregibon, D., Uthurusamy, R. (eds.) KDD-97. Proceedings of the Third International conference on Knowledge Discovery and Data Mining, Newport Beach, California, August 1997, pp. 115–118. AAAI Press, Menlo Park (1997)
4. Blake, C.L., Merz, C.J.: UCI Repository of Machine Learning Databases. Department of Information and Computer Science, University of California, Irvine, CA, United States (1998), http://www.ics.uci.edu/ mlearn/MLRepository.html
5. Clark, P., Boswell, R.: Rule Induction with CN2: Some Recent Improvement. In: Kodratoff, Y. (ed.) Machine Learning - EWSL-91. LNCS, vol. 482, pp. 111–116. Springer, Heidelberg (1991)
6. Coenen, F.: The LUCS-KDD Discretised/Normalised ARM and CARM Data Library. Department of Computer Science, The University of Liverpool, UK (2003), http://www.csc.liv.ac.uk/~frans/KDD/Software/LUCS-KDD-DN
7. Coenen, F., Leng, P.: An Evaluation of Approaches to Classification Rule Selection. In: ICDM-04. Proceedings of the 4th IEEE International Conference on Data Mining, Brighton, November 2004, pp. 359–362. IEEE Computer Society Press, Los Alamitos (2004)
8. Coenen, F., Leng, P., Ahmed, S.: Data Structure for Association Rule Mining: T-trees and P-trees. IEEE Transactions on Knowledge and Data Engineering 16(6), 774–778 (2004)
9. Coenen, F., Leng, P., Zhang, L.: Threshold Tuning for Improved Classification Association Rule Mining. In: Ho, T.-B., Cheung, D., Liu, H. (eds.) PAKDD 2005. LNCS (LNAI), vol. 3518, pp. 216–225. Springer, Heidelberg (2005)
10. Freitas, A.A.: Data Mining and Knowledge Discovery with Evolutionary Algorithms. Springer, Heidelberg (2002)
11. Han, J., Pei, J., Yin, Y.: Mining Frequent Patterns without Candidate Generation. In: Chen, W., Naughton, J.F., Bernstein, P.A. (eds.) SIGMOD-00. Proceedings of the 2000 ACM SIGMOD International Conference on Management of Data, Dallas, TX, May 2000, pp. 1–12. ACM Press, New York (2000)
12. Lavrac, N., Flach, P., Zupan, B.: Rule Evaluation Measures: A Unifying View. In: Džeroski, S., Flach, P.A. (eds.) Inductive Logic Programming. LNCS (LNAI), vol. 1634, pp. 174–185. Springer, Heidelberg (1999)
13. Li, W., Han, J., Pei, J.: CMAR: Accurate and Efficient Classification based on Multiple Class-association Rules. In: Cercone, N., Lin, T.Y., Wu, X. (eds.) ICDM-01. Proceedings of the 2001 IEEE International Conference on Data Mining, San Jose, November 29 – December 2, 2001, pp. 369–376. IEEE Computer Society Press, Los Alamitos (2001)
14. Liu, B., Hsu, W., Ma, Y.: Integrating Classification and Association Rule Mining. In: Agrawal, R., Stolorz, P.E., Piatetsky-Shapiro, G. (eds.) KDD-98. Proceedings of the Fourth International Conference on Knowledge Discovery and Data Mining, New York City, August 1998, pp. 80–86. AAAI Press, Menlo Park (1998)

15. Quinlan, J.R.: C4.5: Programs for Machine Learning. Morgan Kaufmann Publishers, San Francisco (1993)
16. Quinlan, J.R., Cameron-Jones, R.M.: FOIL: A Midterm Report. In: Brazdil, P.B. (ed.) Machine Learning: ECML-93. LNCS, vol. 667, pp. 3–20. Springer, Heidelberg (1993)
17. Yin, X., Han, J.: CPAR: Classification based on Predictive Association Rules. In: Barbara, D., Kamath, C. (eds.) SDM-03. Proceedings of the Third SIAM International Conference on Data Mining, San Francisco, May 2003, SIAM, Philadelphia pp. 331–335 (2003)

Distributed and Shared Memory Algorithm for Parallel Mining of Association Rules

J. Hernández Palancar, O. Fraxedas Tormo, J. Festón Cárdenas,
and R. Hernández León

Advanced Technologies Application Center (CENATAV), 7a ♯ 21812 e/ 218 y 222,
Rpto. Siboney, Playa, C.P. 12200, La Habana, Cuba
{jpalancar,ofraxedas,jfeston,rhernandez}@cenatav.co.cu

Abstract. The search for frequent patterns in transactional databases is considered one of the most important data mining problems. Several parallel and sequential algorithms have been proposed in the literature to solve this problem. Almost all of these algorithms make repeated passes over the dataset to determine the set of frequent itemsets, thus implying high I/O overhead. In the parallel case, most algorithms perform a sum-reduction at the end of each pass to construct the global counts, also implying high synchronization cost. We present a novel algorithm that exploits efficiently the trade-offs between computation, communication, memory usage and synchronization. The algorithm was implemented over a cluster of SMP nodes combining distributed and shared memory paradigms. This paper presents the results of our algorithm on different data sizes experimented on different numbers of processors, and studies the effect of these variations on the overall performance.

1 Introduction

The discovery of Association Rules is one of the most productive fields in the development of sequential algorithms as well as parallel algorithms for Data Mining. Simultaneously, with the evolution of these algorithms the possible applications of Association Rule Mining (ARM) has also been extended together with a corresponding increase in the volume of the databases to be mined. As a consequence of the latter, even using the most efficient sequential ARM algorithms, it is not possible to reduce the support threshold to the desired level without causing a combinatorial explosion in the number of identified frequent itemsets coupled with a corresponding computational overhead.

The situation described above confirms the relevance of the application of Parallel Computing for Association Rule Mining, which is a very active global research area. The main challenges of Parallel Computing are: load balancing, minimization of the inter-process communication overhead, the reduction of synchronization requirements and effective use of the memory available to each processor.

These issues must be taken into account in the development of efficient parallel algorithms for Association Rule Mining; basic references to consider are [8,9,10].

P. Perner (Ed.): MLDM 2007, LNAI 4571, pp. 349–363, 2007.

The prototypical ARM application is the analysis of sales or *basket data* [1]. The task can be broken into two steps. The first step consists of finding the set of all frequent sets of items that can be the transaction database. The second step consists of forming implication rules among the sets of items found; the latter can be done in a straightforward manner so we will focus on the first step.

In previous papers [11,12], we proposed a new algorithm called CBMine (Compressed Binary Mine) for mining association rules and frequent itemsets. Its efficiency is based on a compressed vertical binary representation of the database. CBMine has been compared with other efficient ARM algorithms to obtain frequent itemsets, including: Fp-growth (implementation of Bodon), MAFIA and Patricia Trie . The experimental results obtained showed that CBMine gives the best performance in most cases, especially on big and sparse databases.

In this paper we propose a new parallel algorithm based in CBMine named ParCBMine(Parallel Compressed Binary Mine). ParCBMine exploits efficiently the trade-offs between computation, communication, memory usage and synchronization. The algorithm was implemented over a cluster of SMP nodes combining distributed and shared memory paradigms. Section 5 of this paper shows the experimental results of our algorithm on different data sizes, evaluated on different numbers of processors, and studies about the effect of these variations on the overall performance.

The paper is organized as follows: the next section is dedicated to related work; in section 3 we give a formal definition of association rules; section 4 contains a description of ParCBMine algorithm; experimental results are discussed in section 5; and some conclusions are presented in section 6.

2   Related Work

Until now the great majority of the parallel algorithms for Association Rule Mining are based on the sequential Apriori algorithm. An excellent survey made by Zaki in 1999 [18] classifies different algorithms up to that date, according to the load balance strategy, the architecture and the type of parallelism used in the algorithm. Other important references are [2,15,16,19,20,22].

Apriori algorithm has been the most significant of all sequential algorithms proposed in the literature. Yet, directly adapting an Apriori-like algorithm will not significantly improve performance over frequent itemsets generation. To perform better than Apriori-like algorithms, we must focus on the disadvantages associated with this approach. The main challenges include synchronization, communication minimization, work-load balancing, finding good data layout and data decomposition, and disk I/O minimization.

Recently interesting parallel ARM has increased as a result of this early work. We can identify a number of early ARM algorithms: *Count Distribution*, *Data Distribution* and *Candidate Distribution*. These algorithms were first presented in [2] and offer a fairly simple parallelization of Apriori using different paradigms of parallelization; namely data-parallelism and control-parallelism, or a combination of both. In Count Distribution the dataset is partitioned equally among the nodes of the parallel system. Each of these nodes computes the local support for every candidate k-itemset in the iteration k. At the end of each iteration by exchanging the local supports the global support is generated and the frequent itemsets determined. The nodes must be synchronized to receive the candidate itemsets and the coordinator node must wait for all local counts to generate the global support. The former factors affects communication cost and load-balancing; however the Count Distribution algorithm represents a good first step and can be the core of subsequent implementations that address these issues. In Data Distribution the set of candidate itemset is partitioned into disjoint sets and these are sent to different nodes. The problem in this parallel version of Apriori is the magnitude of the huge communications required at the end of each iteration. In Candidate Distribution load-balancing is thus the main target, selectively replicating the dataset so that each processor proceeds independently. The algorithm requires redistribution of the dataset at level l, this is identified using a heuristic approach.

There are other parallel versions of well-known sequential algorithms like PDM (parallelizing DHP) [3]. But this was not a successful attempt due to its poor performance with respect to the above algorithms. Other algorithms that address the size of candidacy and better pruning techniques are DMA and FDM presented in [4,5]. In [7] the Optimized Distributed Association Mining (ODAM) algorithm is proposed based on Count Distribution which reduce both the size of the average transaction and the number of message exchanges among nodes in order to achieve better performance.

The Eclat(Equivalence CLass Transformation) algorithm [17] uses an itemset grouping scheme based on equivalence classes and partitions them into disjoint subsets among the processors. At the same time Eclat makes use of a kind of vertical representation of the dataset and then selectively replicates it so that each processor has the portion of the dataset it needs for calculations. After the initial phase the algorithm eliminates the need for later communication or synchronization. The algorithm scans the local partition of the dataset three times, therefore diminishing the I/O overhead. Unlike other earlier algorithms, Eclat uses simple intersection operations to compute frequent itemsets and does not use complex hash tables structures. The main deficiency of this algorithm lies in the need for a proper heuristic to achieve a suitable load balance among the processors as of the L_2 partitioning, because the equivalence classes do not have the same cardinality.

In [15] a collection of algorithms with different partitioning and candidate itemsets count schemes are described. Like Eclat, all of them assume a vertical representation of the dataset (tidlits per item), which facilitates the intersection operation of tids of items that make up an itemset. The dataset is duplicated in a selective fashion to reduce synchronization. Two of these algorithms (Par-Eclat and Par-MaxEclat) are based on the classes of equivalence formed by the candidates first item, whereas the other two algorithms (Par-Clique and Par-MaxClique) use the maximum closed hypergraph to partition the candidates.

In [21], a parallel algorithm is proposed for Association Rule Mining that uses a classification hierarchy named HPGM (Hierarchical Hash Partitioned Generalized Association Rule Mining). In this algorithm, the available memory space is completely used identifying the frequent occurrence of candidates itemsets and replicating them to all processors, considering that frequent itemsets can be locally processed without communication. This way the load asymmetry among processors can be effectively reduced.

3   Problem Definition

In this section we define some necessary terminology to facilitate understanding of the following sections. In this context it should be noted that we are only focused on the problem of identifying frequent itemsets on large databases.

A *dataset* is a set of *transactions* and each of these is composed by a transaction identifier (TID) and a set of *items*. The items in a transaction may represent a shopping list in a supermarket by a customer (known as basket data) or words in a document or stocks movements. A set of items, called *itemset* is *frequent* if it is contained in a number of transactions above a user-specified threshold (minimum support-*minsup*).

An itemset with k items will be referred to as k-*itemset* and its support will be denoted as $X.sup$, where X is the k-itemset in question; support is represented as a percentage rather than an absolute number of transactions.

More formally: $I = \{i_1, i_2, ..., i_n\}$ be a set of n distinct items. Each transaction T in the dataset D contains a set of items, such that $T \subseteq I$. An itemset is said to have a support s if $s\%$ of the transaction in D contains the itemset.

4   ParCBMine

In this section we describe the parallel version of the CBMine algorithm which we have named ParCBMine (Parallel CBMine).

ParCBMine takes advantage of the vertical representation of the dataset as in CBMine and combines suitably the parallel programming models of shared and distributed memory using the libraries *pthreads* (for multithreads programming) and MPI (for message passing programming) respectively.

The mixture of multithreads programming and message passing in ParCBMine was done based on the fact that the algorithm was implemented over a cluster with SMP (Symmetric Multi-Processing) nodes for the parallel processing managed by a GNU/Linux operating system; each node is composed by a dual processor. All processors are Intel Xeon with *hyperthreading* technology, which provides up to four threads on each node.

Although the algorithm is not tied to the number of real threads (processors) that could be deployed on each node, this is an important element on the scalability of ParCBMine, because it allows the use of global information in the shared memory of each node in a better way, i.e., in candidate generation and support counting. Many authors refer to this as "intra-node parallelism", and in certain way we have incorporated some aspects of the Candidate Distribution algorithm, in this case by means of multithread programming using the Pthreads library.

4.1   CBMine Algorithm

CBMine is a breadth-first search algorithm with a VTV organization, that uses compressed integer-lists for itemset representation.

Let T be the binary representation of a database, with n filtered items and m transactions. Taking from T the columns associated with frequent items, each item j can be represented as a list I_j of integers (integer-list) of word size w, as follows:

$$I_j = \{W_{1,j}, \ldots, W_{q.j}\}, q = \lceil m/w \rceil, \tag{1}$$

where each integer of the list can be defined as:

$$W_{s,j} = \sum_{r=1}^{min(w,m-(s-1)*w)} 2^{(w-r)} * t_{((s-1)*w+r),j}. \tag{2}$$

The upper expression $min(w, m - (s - 1) * w)$ is included to consider the case in which the transaction number $(s - 1) * w + r$ does not exist due to the fact that it is greater than m. The value $t_{i,j}$ is the bit value of term j in the transaction i.

This algorithm iteratively generates a prefix list PL_k. The elements of this list have the format: $\langle Prefix_{k-1}, CA_{Prefix_{k-1}}, Suffixes_{Prefix_{k-1}} \rangle$, where $Prefix_{k-1}$ is a $(k-1)$-itemset, $CA_{Prefix_{k-1}}$ is the corresponding compressed integer-list, and $Suffixes_{Prefix_{k-1}}$ is the set of all suffix items j of k-itemsets extended with the same $Prefix_{k-1}$, where j is lexicographically greater than every item in the prefix and the extended k-itemsets are frequent. This representation not only reduces the required memory space to store the integer-lists but also eliminates the Join step described in Apriori algorithm.

The Prune step of Apriori algorithm is optimized by generating PL_k as a sorted list according to the prefix field and, for each element, by the suffix field.

In order to determine the support of an itemset with a compressed integer-list CA, the following expression is considered:

$$Support(CA) = \sum_{\langle s,B_s \rangle \in CA} BitCount(B_s), \tag{3}$$

where $BitCount(B_s)$ represents a function that calculates the Hamming Weight of each B_s.

Although this algorithm uses compressed integer-lists of non null integers (CA) for itemset representation, in order to improve the efficiency, we maintain the initial integer-lists (including the null integers) $I_j = \{W_{1,j}, \ldots, W_{q,j}\}$ associated with each large 1-itemset j. This consideration allows direct accessing for any I_j the integer position defined in CA.

The above allows us to define the following formula (notice that this function represents a significant difference and improvement with respect to other methods):

$$CompAnd(CA, I_j) = \{\langle s, B'_s \rangle | \langle s, B_s \rangle \in CA, B'_s = (B_s andW_{s,j}), B'_s \neq 0\}. \quad (4)$$

Note that the cardinality of CA is reduced as the size of the itemsets increases due to the downward closure property; thus the application of identities 3 and 4 becomes more efficient.

The complete CBMine algorithm is presented in Table 1.

Table 1. CBMine algorithm

Algorithm 1: CBMine

```
1  L₁ = {large 1-itemsets} ;                        // Scanning the database
2  PL₂ = {⟨Prefix₁, CA_Prefix₁, Suffixes_Prefix₁⟩};
3  for k = 3;PL_{k-1} ≠ ∅; k + + do
4      forall ⟨Prefix, CA, Suffixes⟩ ∈ PL_{k-1} do
5          forall item j ∈ Suffixes do
6              Prefix' = Prefix ∪ {j};
7              CA' = CompactAnd(CA,I_j);
8              forall (j' ∈ Suffixes) and (j' > j) do
9                  if Prune(Prefix' ∪ {j'}, PL_{k-1}) and Support(CompactAnd(CA',
                     I_{j'})) ≥ minsup then
10                     | Suffixes' = Suffixes' ∪ {j'};
11                 end
12                 if Suffixes' ≠ ∅ then
13                     | PL_k = PL_k ∪ {⟨Prefix', CA', Suffixes'⟩};
14                 end
15             end
16         end
17     end
18 end
19 Answer=⋃_k L_k ;                                   // L_k is obtained from PL_k
```

Note that this algorithm only scans the dataset once in the first step.

4.2   Intelligent Block Partitioning

Given PL_{k-1} we need to partition it among the threads in the most efficient manner. In the literature we can identified several partitioning techniques, such as Bitonic Partitioning from Zaki [19]. Nevertheless, given the features of the sequential algorithm, we achieve the best results making a block partitioning, so the information to be processed by each thread was not fragmented.

For this purpose we develop Intelligent Block Partitioning (IBP), dynamically recomputing the load balance for each thread in a straightforward manner. We use equation 5 to compute the work load generated by a Prefix based on the size of its Suffixes. The pseudo-code of IBP is given in Table 2.

$$G(x) = \frac{x(x-1)}{2} \tag{5}$$

Table 2. Intelligent Block Partitioning algorithm

Algorithm 2: IBP
---

1 Total $= \sum_{j=0}^{|PL_{k-1}|} G(|Suffixes_{Prefix_j}|);$ /* Total Work Load */

2 Ideal $= \frac{Total}{MaxThreads};$ /* Ideal Work Load for each thread */

3 $i = 1;$

4 $load = starts[0] = 0;$

5 **forall** $\langle Prefix, CA, Suffixes \rangle \in PL_{k-1}$ **do**

6 $\quad load = load + G(|Suffixes|);$

7 $\quad$ **if** $load > $ Ideal **then**

$\quad\quad$ /* Set block boundaries   */

8 $\quad\quad start[i] = stop[i-1] = \langle Prefix, CA, Suffixes \rangle;$

$\quad\quad$ /* Dynamically recompute Total and Ideal Work Load   */

9 $\quad\quad$ Total $=$ Total$- load + G(|Suffixes|);$

10 $\quad\quad$ Ideal $= \frac{Total}{MaxThreads-i};$

11 $\quad\quad load = G(|Suffixes|);$

12 $\quad\quad i++;$

13 $\quad$ **end**

14 **end**

15 $stop[i] = \langle Prefix_{|PL_{k-1}|}, CA_{Prefix_{|PL_{k-1}|}}, Suffixes_{Prefix_{|PL_{k-1}|}} \rangle;$

The aforementioned partition strategy is one of the improvements ParCB-Mine introduces over its sequential counterpart, and this can be verified in the experimental results.

4.3 ParCBMine Algorithm

Considering a master-slave framework, typical of parallel clusters, in the first pass, the master node or coordinator determines the global L_1 and partitions the dataset D in N equitable segments and sends each one of them to the corresponding node that makes up the cluster, of this way ParCBMine like Count Distribution, adopts a horizontal partitioning of the dataset thus using "inter-node parallelism", in this case the communication among the nodes is made by means of message passing using the MPI library.

The first pass is special. For all other passes $k > 1$, the algorithm works as follows:

1. Each master-thread process $P_j(j = 1, N)$ generates all the set C_k, using all the frequent itemsets L_{k-1} created at the end of pass $k-1$. Notice

that every process has the same L_{k-1}, so they will generate identical C_k. Threads $P_i(i = 1, N \times MaxThreads)$ running in the same node, share the same memory structure for L_{k-1}, C_k and $D_j(j = 1, N)$.

2. The master-thread process P_j creates $MaxThreads-1$ new threads and each one of these makes a pass over D_j data partition and develops local support counts for a portion of the candidates in C_k which was previously partitioned using the IBP strategy. With this, the local C_k at node j, is partitioned equitably and each thread of the process develops the support count of its candidates without making any synchronization to access the memory, since the support count is developed on a reserved memory structure for each candidate, taking advantage of the vertical representation of the dataset.

3. The master-thread process P_j sends the local counts of C_k to the master node or coordinator, in order to make an *all-reduce* operation to generate the global counts of C_k. Master-thread processes are forced to synchronize in this step.

4. The master node or coordinator computes L_k from C_k. If L_k is not empty the coordinator sends it to the master-thread process P_j and continues on to the next pass.

Notice that unlike Count Distribution we have replaced the word processor by process, since given the characteristics of the hardware of our cluster the amount of processes is greater than the amount of processors, and can be expressed by $N \times MaxThreads$, where: N is the number of nodes and $MaxThreads$ is the maximum number of threads per node, in our particular case $MaxThreads = 4$ considering the use of the Hyperthreading technology.

Unlike PAR-DCI algorithm [13], in which the local dataset is partitioned yet again into as many portions as threads that were possible to deploy, in step 2 of ParCBMine a more efficient solution was adopted. We distribute the candidate support count among the threads by partitioning the candidate set into disjoint parts of approximately the same size, without the need for semaphoric operations to control memory access.

4.4   Complexity Analysis

In this section we will evaluate the complexity of our algorithm in three different contexts, first assuming the use of shared memory model, second employing the distributed memory model, and lastly the solution proposed by us of fusing the models of shared memory and distributed memory.

Given that our algorithm is intended for a parallel framework based on an SMP cluster, it is important to indicate that if we had used only a distributed memory model based on message passing, like other authors have done, the performance of the algorithm would have suffered considerably.

It is well known that in any algorithm based on Count Distribution the scalability degrades as the number of dataset partitions increases, due to the amount of information that each MPI process receives in each pass when synchronizing the processes in order to develop global support counts of itemsets in the candidate set. The amount of information is $(N - 1) \times |C_k|$, where: N is the number of MPI processes with a dataset portion assigned to it and $|C_k|$ is the cardinality of the candidate set generated in each pass and for which each MPI process computes a local support count. Bear in mind that whatever the value of N is, the cardinality of C_k does not change. This is the reason why the reduction of local set C_k is an issue that has been and continues to be a research objective. From the literature one suggested approach involves the use of probabilistic estimations of local support, see [6,14].

Given that we are using SMP nodes the advantages offered by the data locality would be wasted since all the MPI processes running in each node will try to equitably distribute the total physical memory of the node, reserving equal amounts of memory for data structures to store L_{k-1} and C_k, as well as for the dataset partition assigned to the node. For the analysis lets assume that the problem size remains constant, so the amount of candidates will be the same in each case and will be denoted as $|C_k|$.

In the development of a parallel algorithm the most common notation for the execution times are (if we consider a problem of size m running in p processors): Sequential computation denoted by $\sigma(m)$, Parallel execution time (computation that can be performed in parallel) denoted by $\varphi(m)$ and Parallel overhead (communication and synchronization, etc) denoted by $\kappa(m)$. For the experiments performed, the sequential plus the parallel execution time was considered as the parallel execution time because of the characteristic of the CBMine, the part that can not be parallelized is less than 1% of total execution time. For that reason, the two times measured were: **Parallel execution time** and **Parallel overhead**.

Shared Memory: L_{k-1} is partitioned in disjoint sets using IBP, support is develop from the common data base.

Algorithm 3: Shared Memory
$while(L_{k-1} \neq \emptyset)$;   /* Level Iterator */
$C_k^t = IBP(L_{k-1}, t)$;   /* $C_k = \bigcup C_k^t$ t=1,..,MaxThreads */
$foreach(X \in C_k^t)$;   /* Count each X in the DB */
$if(sup(X, DB) \geq minsup)$ $L_k = L_k \bigcup \{X\}$

where: $\varphi(m) = |C_k| * |DB|$, $\kappa(m) = \emptyset$.

$$ShTime = \sum_k \frac{|C_k| * |DB|}{p} = \ldots = \sum_k max|C_k^t| * |DB| \tag{6}$$

Distributed Memory: The DB is partitioned among the quantity process denoted by P, each one has a copy of L_{k-1}, count the local support and exchange it (i.e. Message Passing). For example: if the nodes are single processor $P = N$; in case that the nodes are SMP then $P = N \times p$, where: N is the quantity nodes, and p is the number of processors in each node.

Algorithm 4: Distributed Memory

$DB_{id} = Partition(DB, id)$;   /* Horizontal partitioning */
$while(L_{k-1} \neq \emptyset)$;   /* Level Iterator */
   $C_k = GenerateCandidate(L_{k-1})$;   /* C_k is the same for each
process */
   $foreach(X \in C_k)$;   /* Count each X in the DB_{id} */
     $local[x] = sup(X, DB_{id})$;   /* Local support */
   $global = InterchangeAndSum(local)$;   /* All to All */
   $foreach(X \in C_k)$;   /* global support */
     $if(global[X] \geq minsup)$ $L_k = L_k \bigcup \{X\}$

---

where: $\varphi(m) = |C_k| * |DB|$, $\kappa(m) = \sum_k InterchangeAndSum = \sum_k |C_k| * 2 * P = \sum_k |C_k| * 2 * N * p$.

$$DsTime = \sum_k \frac{|C_k| * |DB|}{N * p} + \kappa(m) = \sum_k |C_k| * max|DB_{id}| + \kappa(m) \quad (7)$$

Share + Distributed Memory Solution (Hybrid memory): In the previous cases the P processes were sharing the memory or completely distributed, in this case there will be N MPI-processes in correspondence with the quantity of nodes and in each node p processes sharing memory, for that reason $P = N$, because for the communication among nodes is not considering the quantity of processes in each node. In this case a process master for each node is in charge of the communication with the remaining nodes and of distributing tasks to the other processes that are in its node.

Algorithm 5: Hybrid Memory

$DB_N = Partition(DB, N)$;   /* Horizontal partitioning */
$while(L_{k-1} \neq \emptyset)$;   /* Level Iterator */
   $C_k^t = IBP(L_{k-1}, t)$;   /* $C_k = \bigcup C_k^t$ t=1,..,MaxThreads */
   $foreach(X \in C_k^t)$;   /* Count each X in the DB_N */
     $local[x] = sup(X, DB_N)$;   /* Local support */
   $if(master(t))$ then $global = InterchangeAndSum(local)$
   $foreach(X \in C_k^t)$;   /* global support */
     $if(global[X] \geq minsup)$ $L_k = L_k \bigcup \{X\}$

---

where: $\varphi(m) = |C_k| * |DB|$, $\kappa(m) = \sum_k InterchangeAndSum = \sum_k |C_k| * 2 * N$.

$$HyTime = \sum_k \frac{|C_k| * |DB|}{N * p} + \kappa(m) = \sum_k max|C_k^t| * max|DB_N| + \kappa(m) \quad (8)$$

If we are using the same processes quantity for each memory model it is very simple to observe that:

$$ShTime < HyTime < DsTime \tag{9}$$

5   Results

All the experiments described in this section were performed on a SMP cluster of which we used 6 nodes: the master node and five working nodes. Each working node is equipped with two Intel Xeon processors at 2.4 GHz based on hyperthreading technology, 512 MB of RAM, 40 GB of disk space and 1 Gb/s Fast Ethernet card. The working nodes are connected to a master node by a network switch Gigabit Ethernet. The master node is equipped with two Intel Xeon processors at 3.06 GHz based on hyperthreading technology too, 2 GB of RAM, and a disk array of five disks, 36.4 GB of disk space each (total 145.6 GB).

We ran two versions of the parallel algorithm: one using the distributed memory model implemented with MPI, so there were two processes for each node, i.e. one process by physical CPU; and the other combining the distributed memory model (MPI again) and the shared memory model implemented using Pthreads, in this case there were 4 threads per node sharing the same memory, considering the use of hyperthreading technology.

The experiments were made with one synthetic (T40I10D600K composed by 600000 transactions and 999 items) and one textual dataset (Kosarak composed by 990007 transactions and 41935 items) (available from FIMI repository-http://fimi.cs.helsinki.fi). The Kosarak Dataset was provided by Ferenc Bodon and contains (anonymized) click-stream data of a Hungarian on-line news portal. The T40I10D600K was created using an IBM generator(www.almaden.ibm.com/cs/quest/syndata.html).

In the first experiment we compared the execution times between ParCBMine using MPI plus Threads and ParCBMine using MPI only. The Figure 1 (a) and Figure 1 (b) show that the parallel execution time is reduced to half when the number of processors is doubled, for both implementations. The communication time overhead is stable in the first case (with the use of MPI + Tthreads) but increases linearly in the second (MPI only).

The second experiment was performed to analyze the SpeedUp (Figure 2) and Efficiency (Figure 3) of both implementations of ParCBMine algorithm. Figure 2 shows that the implementation of ParCBMine algorithm using MPI plus Threads scales better when the number of processors is increased in spite of the communication time overhead. Likewise, notice that in Figure 3 the degradation of the efficiency for ParCBMine implementation using MPI plus Threads is much slower with the increase of the communication time overhead.

In the third experiment we analyzed the algorithm scalability, thus we considered the case where both datasets were so big that they could not fit in the main memory of any node, increasing databases size in proportion with the number of nodes (N), these datasets were named T40I10D600KxN and KosarakxN. The

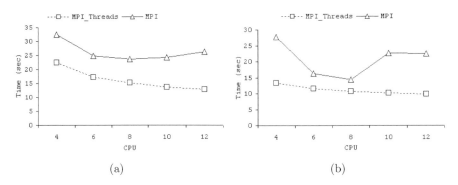

(a)   (b)

Fig. 1. Execution time comparison: (a) T40I10D600K, $minsup = 0.01$, (b) Kosarak, $minsup = 0.002$

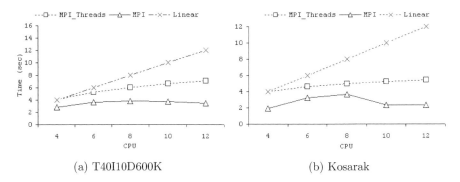

(a) T40I10D600K   (b) Kosarak

Fig. 2. SpeedUp comparison

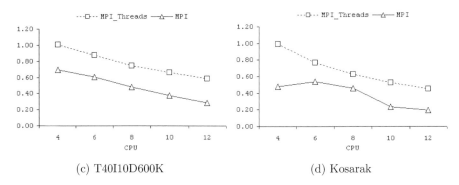

(c) T40I10D600K   (d) Kosarak

Fig. 3. Efficiency comparison

minimum support values used in each case were the smallest that the sequential version could process (for T40I10D600KxN the minimum support was set to 0.005 and for KosarakxN the minimum support was set to 0.003).

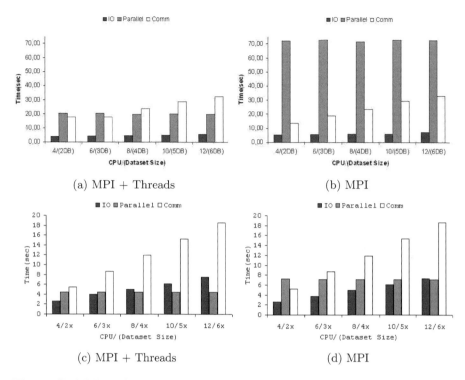

Fig. 4. Scalability Analysis of ParCBMine in T40I10D600KxN (a y b) and KosarakxN(c y d)datasets

Observe in Figure 4 that the parallel execution time remains constant, thus the scalability of the algorithm does not depend on the database used.

As a conclusion of these experiments we can affirm that the shared-distributed memory combination proved to be an effective way to avoid high traffic of data and drastic reduction of the efficiency of the parallel algorithm.

6   Conclusions

The algorithms proposed by Rakesh Agrawal and John Shafer in [2] are recognized as benchmarks for the development of parallel algorithms for Association Rules Mining.

Making a general assessment of these algorithms we can say that the Count Distribution reduces the communication overhead at the expense of ignoring the system physical memory. In a cluster of workstations environment, with monoprocessor nodes, this is probably the best approach; nevertheless it may not be the best solution in the case where nodes are SMP, because it would not take advantage of the combination of shared and distributed memory models. In order to reach efficient implementations based on Count Distribution, determining new heuristics that allow a reduction of the cardinality of the local C_k obtained by each processor continues to be a latent problem.

The Data Distribution algorithm can help us to explore this feature by fully exploiting the physical memory with the risk of increasing communication overhead. The ability to count in a single pass T times as many candidates as Count Distribution makes this algorithm a strong contender.

If we include detailed background knowledge of the problem in the Candidate Distribution, the joint benefits of Count Distribution and Data Distribution [2] can be obtained. Yet, there are still some challenges for researchers in parallel algorithms for association rule mining: to find a heuristic that allows (from a step $k = l$) candidate itemsets partitioning so that synchronization among processors is not needed, and to obtain a suitable load balance among the processors.

In conclusion we suggest that the purposed parallel algorithm described here ParCBMine based on the sequential algorithm CBMine, and sustained on the principles of Count Distribution in which some features of Candidate Distribution are also introduced, suitably combines the parallel programming based on the message passing model with multithread programming. ParCBMine continues to be developed and in the future we expect to present new results oriented to the reduction of the computational effort at synchronization level and to reach a better load balance among processors.

References

1. Agrawal, R., Mannila, H., Srikant, R., Toivonen, H., Verkamo, A.I.: Fast discovery of association rules. In: Fayyad, U., et al. (eds.) Advances in Knowledge Discovery and Data Mining, MIT Press, Cambridge (1996)
2. Agrawal, R., Shafer, J.: Parallel mining of association rules. In IEEE Trans. on Knowledge and Data Engg. 8(6), 962–969 (1996)
3. Park, J.S., Chen, M., Yu, P.S.: Efficient parallel data mining for association rules. In: ACM Intl. Conf. Information and Knowledge Management (November 1995)
4. Cheung, D., Ng, V., Fu, A., Fu, Y.: Efficient mining of association rules in distributed databases. In IEEE Trans. on Knowledge and Data Engg. 8(6), 911–922 (1996)
5. Cheung, D., Han, J., Ng, V., Fu, A., Fu, Y.: A fast distributed algorithm for mining association rules. In: 4th Intl. Conf. Parallel and Distributed Info. Systems (December 1996)
6. Cheung, D.W., Xiao, Y.: Effect of Data Skewness in Parallel Mining of Association Rules. In: Proceedings of the 2nd Pacific-Asia Conference on Knowledge Discovery and Data Mining, pp. 48–60, Melbourne, Australia (April 1998)
7. Ashrafi, M.Z., Taniar, D., Smith, K.A.: ODAM: An Optimized Distributed Association Rule Mining Algorithm. IEEE Distributed Systems Online (5) (2004)
8. Freitas, A.A., Lavington, S.H.: Mining Very Large Databases with Parallel Processing. Kluwer Academic Publishers, Boston (1998)
9. Freitas, A.A.: A Survey of Parallel Data Mining. In: Proc. 2nd Int. Conf. on the Practical Applications of Knowledge Discovery and Data Mining (1998)
10. Skillicorn, D.: Parallel Data Mining, Department of Computing and Information Science Queen's University, Kingston (1999)

11. Palancar, J.H., León, R.H., Pagola, J.M., Díaz, A.H.: Mining Frequent Patterns Using Compressed Vertical Binary Representations In: Lin, T.Y., Xie, Y. (eds.) Proceedings of a Workshop Foundation of Semantic Oriented Data and Web Mining, held in Conjunction with the Fifth IEEE International Conference on Data Mining, Houston, Texas, USA, pp. 29–33 (November 27-30, 2005) ISBN 0-9738918-7-4

12. Palancar, J.H., León, R.H., Pagola, J.M., Díaz, A.H.: A Compressed Vertical Binary Algorithm for Mining Frequent Patterns. In: the book Data Mining: Foundations and Practice Lin, T.Y., Wasilewska, A., Petry, F., Xie, Y.(eds.) Springer-Verlag, Accepted for publication (to appear)

13. Orlando, S., Palmerini, P., Perego, R., Silvestri, F.: A Scalable Multi-Strategy Algorithm for Counting Frequent Sets Washington, USA, pp. 19–30. In: Proceedings of the 5th Workshop on High Performance Data Mining, in conjunction with Second International SIAM Conference on Data Mining (April 2002)

14. Schuster, A., Wolff, R.: Communication-Efficient Distributed Mining of Association Rules. Data Mining and Knowledge Discovery, 8(2) (March 2004)

15. Zaki, M.J., Parthasarathy, S., Ogihara, M., Li, W.: New algorithms for fast discovery of association rules. In: Heckerman, D., Mannila, H., Pregibon, D., Uthurusamy, R. (eds.) Proceedings of the Third International Conference on Knowledge Discovery and Data Mining (KDD-97), p. 283. AAAI Press, Stanford (1997)

16. Zaki, M.J., Parthasarathy, S., Ogihara, M., Li, W.: New Parallel Algorithms for Fast Discovery of Association Rules. Data Mining and Knowledge Discovery 1(4), 343–373 (1997)

17. Zaki, M.J., Parthasarathy, S., Li, W.: A Localized Algorithm for Parallel Association Mining. In: Proceedings of the 9th ACM Symposium on Parallel Algorithms and Architectures (1997)

18. Zaki, M.J.: Parallel and Distributed Association Mining: A Survey. IEEE Concurrency. (October- December 1999)

19. Zaki, M., Parthasarhaty, S., Ogihara, M., Li, W.: Parallel Data Mining for Association Rules on Shared Memory Systems. (February 28, 2001)

20. Zaiane, O.R., El-Hajj, M., Lu, P.: Fast Parallel Association Rule Mining without candidacy generation. Techical Report TR01-12. Department of Computing Sciences, University of Alberta, Canada (2001)

21. Shintani, T., Kitsuregawa, M.: Parallel Mining Algorithms for Generalized Association Rules with Classification Hierarchy. In: Proceedings ACM SIGMOD International Conference on Management of Data, SIGMOD 1998, Seattle, Washington, USA. (June 2-4, 1998)

22. Sam, E.-H., Karypis, H.G., Kumar, V.: Scalable Parallel Data Mining for Association Rules. Department of Computer Science. University of Minnesota (1997)

Analyzing the Performance of Spam Filtering Methods When Dimensionality of Input Vector Changes

J.R. Méndez[1], B. Corzo[2], D. Glez-Peña[1], F. Fdez-Riverola[1], and F. Díaz[3]

[1] Computer Science Dept., University of Vigo, Escuela Superior de Ingeniería Informática,
Edificio Politécnico, Campus Universitario As Lagoas s/n, 32004, Ourense, Spain
{moncho.mendez,dgpena,riverola}@uvigo.es
[2] Dept. Advertising Graphics, Arts College of Oviedo,
C/ Julián Clavería, 12, 33006, Oviedo, Spain
beatrizcom@educastur.princast.es
[3] Computer Science Dept., University of Valladolid, Escuela Universitaria de Informática,
Plaza Santa Eulalia, 9-11, 40005, Segovia, Spain
fdiaz@infor.uva.es

Abstract. Spam is a complex problem that makes difficult the exploitation of Internet resources. In this sense, several authorities have alerted about the dimension of this problem and aim everybody to fight against it. In this paper we present an extensive analysis showing how the effect of changing the dimensionality of message representation influences the accuracy of some well-known classical spam filtering techniques. The conclusions drawn from the experiments carried out will be useful for building a comparison of the dimensionality reorganization effects between classical filtering techniques and a successful spam filter model called SPAMHUNTING.

1 Introduction and Motivation

A lot of techniques used to distribute information in a massive form by using the newest technologies (like SMS or Internet) are considered spam behaviours. These approaches are used by malicious users to advertise illegal products and other little real value stuff. Spam can be easily found in blogs, posts, newsgroups, search engines, mobile messages and mainly in e-mails.

The use of Internet as an advertising platform is related to the way in which it is financed. Internet is a large set of interconnected networks, where each user takes on the cost of the connection between his network and other nodes. Every time an e-mail is delivered, it is routed through several sub-networks using the infrastructure financed by all of its users. Keeping in mind these ideas, spam is considered an irritating problem for all the Internet community.

The most common form of spam is the distribution of illegitimate e-mail messages. Nowadays, the majority of the messages delivered through Internet are spam. As a consequence final users are not able to take advantage of the new forms of communication through Internet and newest technologies because spam is limiting their function.

P. Perner (Ed.): MLDM 2007, LNAI 4571, pp. 364–378, 2007.
© Springer-Verlag Berlin Heidelberg 2007

In the last years, some strategic plans have been introduced by local Administrations and Government in order to promote Internet and the newest technologies as a way for developing an economy based on the knowledge. These plans have been adopted by different authorities including the Minister Council of Spain through the implantation of the AVANZA program [1], and the European Commission through eEurope 2002 plan [2], approved during the year 2001. Since these plans have been properly executed and successful results have been reported, some similar programs were newly introduced: Ingenio-2010 [3] in Spain, eEurope 2005 [4] and i2010 [5] in Europe.

Recently (17th November 2006), the European Commission made public an alarming study showing a huge increment of the amount of spam messages received through Internet. The 85 percent of the messages received by Internet European users are spam. Vivian Reding, commissioner for the Information Society of the European Union, addressed the member states advising the development and deployment of anti-spam strategies. She congratulates the initiatives for spam filtering of Finland and Netherlands, particularly the OPTA Dutch anti-spam unit. Also the commissioner aims to the rest of member states to develop similar filtering systems and to use all available tools within the framework of the law.

Spam e-mail is the principal cause of several associated drawbacks: (*i*) privacy problems (when the spammer obtains an e-mail address he will not stop sending spam messages to its owner), (*ii*) an increment in the costs supported by Internet providers and final users and (*iii*) an extensive waste of time. Moreover, spam activity is an important way of achieving illegal sales of fraudulent products and drugs and it is frequently used to attract people for carry out illicit activities. Finally, spam represents an important obstacle for the diffusion of new technologies based on Internet and mobile phone platforms.

In this work we show an extensive study of how changes in message representation dimensionality influence the accuracy of existing spam filtering models. For this purpose, we have carried out several experiments using a representative selection of well-known techniques for spam e-mail classification. In our test-bed we have considered: (*i*) Naïve Bayes [6] and (*ii*) Flexible Bayes [7] approaches, (*iii*) SVM (*Support Vector Machines*) [8], (*iv*) AdaBoost [9] and (*v*) SPAMHUNTING, our previous successful spam filtering model [10, 11].

In order to properly select appropriate public available corpus and input data dimensionality, we have taken into account some relevant conclusions from previous successful research works such as [12, 13]. The major findings and conclusions of this work will be useful to compare and optimize several dimensionality issues of our previous spam filtering model. The SPAMHUNTING system has been developed using a CBR (*Case-Based Reasoning*) approach [14] and is founded on the use of a disjoint knowledge representation mechanism and an indexation structure known as EIRN (*Enhanced Instance Retrieval Network*).

The rest of the paper is structured as follows: Section 2 presents a brief outline of our previous related work in the spam filtering domain. Section 3 summarizes the state of art in spam filtering models, whereas Section 4 introduces some relevant information about publicly available experimental data. Section 5 describes the experiments carried out and presents the results obtained discussing the major findings. Section 6 presents how dimensionality affects to our SPAMHUNTING system. Finally, Section 7 summarizes the main conclusions and details the further work.

2 Previous Work

This section summarizes previous results in spam filtering domain that are relevant to this work. Taking into account the purpose of this proposal, we need to look back on issues from past works used to evaluate the impact of the pre-processing steps [12] and the performance of the distinct feature selection methods [13].

In [12] we performed a sound analysis of different tokenizing schemes and introduced the tokenizing by using only blank characters as separators. We also checked the advisability of using stemming [15, 16] and stopword [17] techniques in spam filtering domain. For this purpose, we have chosen a representative set of spam filtering models (including Naïve Bayes, Adaboost, SVM, and three CBR systems) in order to measure their accuracy when the context of tokenising, stopword removal and stemming changes. Results of this previous work have shown the convenience of using only blanks as token separator allowing the preservation of noise data. The noise data is helpful for spam detection, in fact, it is distinctive between spam and legitimate classes. Moreover, the use of stemming techniques is not advisable whereas a stopword removal process can significantly improve the performance of the vast majority of techniques.

In a previous work [13] we analysed the strengths and weaknesses of different feature selection strategies used in text categorization when they are applied to the spam filtering domain. In this contribution, we showed the results obtained by using different anti-spam content-based techniques when changing the feature selection of available training corpus. The selected feature selection approaches were: (*i*) IG (*Information Gain*), (*ii*) DF (*Document Frequency*), (*iii*) MI (*Mutual Information*) and (*iv*) χ^2 (*Chi Square*) [18]. For comparison purposes, we have used Naïve Bayes, Boosting Trees, SVM and a CBR system named ECUE [13]. This work confirms that the usage of IG and χ^2 approaches are the most reliable methods for feature selection purposes with no statistical significant differences between them. We also found that IG achieves better precision (security) while χ^2 is slightly superior in effectiveness. Therefore, as security is a critical feature in spam filtering, we highlighted the results achieved by using IG. Moreover, we showed a clear disadvantage of the utilization of MI method in spam filtering. The bad results were motivated by the tricks used by spammers for term obfuscation. Regarding the DF feature selection method and despite its simplicity, the results were in general, good. Nevertheless, the quality of these results does not reach the performance achieved by using other feature selection methods such as IG or χ^2.

As a result of the previously commented conclusions, we have introduced a new feature selection technique in [19] able to adequately manage data affected by concept drift. In this work we argue that one of the most relevant issues in domains affected by the concept drift problem is the passage of time.

Expression 1 defines a measure capturing the amount of information achieved when a term w is selected for representing a message e. In this expression, K represents the knowledge that has been acquired until the message e arrived, *length* (w) is the length of the word w, P $(w \mid e)$ represents the frequency of the term w in the message e, and finally, P (w | S, K) and P (w | L, K) stands for the frequency of finding the term w in legitimate and spam messages, respectively.

$$AI\left(w,e|\boldsymbol{K}\right)=P\left(w|e\right)\cdot\left[1-\frac{1}{length(w)}\right]\cdot\left[\frac{\left|P\left(w|S,\boldsymbol{K}\right)-P\left(w|L,\boldsymbol{K}\right)\right|}{P\left(w|S,\boldsymbol{K}\right)+P\left(w|L,\boldsymbol{K}\right)}\right] \tag{1}$$

In order to compute the achieved information (AI) of a term t when representing a message e, we suggest combining the relevance of the term w in the message e (estimated by the frequency), the length of the term t (in order to discard shorter terms) and the capacity of discriminating between spam and legitimate messages (the last part of Equation 1).

In order to represent each e-mail, we use the set of most relevant terms having an amount of AI greater than a certain percentage of the information achieved by all the terms belonging to the given message. The results achieved by using this approach as feature selection strategy are very promising and will be employed in the experimentation of the current work.

3 Successful Spam Filtering Models

In this section we present a brief recompilation of previous successful works on spam filtering. We highlight some differences that can be found in these approaches such as the stage when the feature selection is carried out, the technique type, the capabilities of handling concept drift and the learning strategy they applied. Table 1 summarizes the above mentioned aspects.

In first place we have introduced previous successful collaborative approaches. We have also included the white list approach, which is based on trusting the e-mail addresses belonging to the senders. The main disadvantage of these models is its limited generalization ability. They are based on mechanically generating a classification for a given message whereas content-based approaches include machine learning capabilities. The results achieved by using collaborative-based approaches present a higher safety level (small amount of false positive errors) and a reduced effectiveness (most of the spam messages are not detected). In the collaborative approach feature vectors are not used and therefore, the dimensionality analysis proposed in this work makes no sense.

In our study, we have also included some classical techniques such as Naïve and Flexible Bayes, SVM or AdaBoost. Due to its generalization capabilities and its simplicity, the first two approaches have been widely used and included into well-known e-mail clients (such as Mozilla Thunderbird). Owning to the difficulties of SVM and AdaBoost algorithms, their relevance has been primarily limited to the scientific field. Content-based approaches have a higher degree of effectiveness but the safety level is more reduced than the achieved by using collaborative approaches. Content-based filtering techniques use feature vectors and a previous feature selection stage as their main strategy for learn from data. Therefore, we will use them for testing purposes.

Table 1 also includes a novel technique called Chung-Kwey, which is able to address the problem of spam classification. It is based on the study of a lot of spam messages in order to detect patterns that appear in illegitimate e-mails. Despite of the great results achieved by using this technique, a large spam e-mail corpus is needed during the learning stage. Due to the fact that this approach is not based on the use of feature vectors, it has not been considered for our experimental setup.

Table 1. Successful spam filtering techniques

Model	Authors	Feature Selection	Knowledge Representation	Type	Concept Drift	Attributes	Learning Strategy
SBL	[20]	-	IP address	Black list / Collaborative	-	Server IP addresses	None
MAPS	[21]	-	IP address	Black list / Collaborative	-	Server IP addresses	None
White lists	[22]	-	E-mail address	White list	-	E-mail addresses	None
Razor	[23]	-	E-mail hashes	Black list / Collaborative	-	Body and subject	None
Pyzor	[24]	-	E-mail hashes	Black list / Collaborative	-	Body and subject	None
DCC	[25]	-	E-mail hashes	Black list / Collaborative	-	Body and subject	None
Naïve Bayes	[6]	Before learning	Probabilities	Content-based	-	Body and subject	Eager learning (training)
Flexible Bayes	[7]	Before learning	Probabilities	Content-based	No	Body and subject	Eager learning (training)
Adaboost	[9]	Before learning	Combination of weak classifiers	Content-based	No	Body and subject	Eager learning (training)
SVM	[8]	Before learning	Straight line dividing the data and non-linear transformation over the input space	Content-based	No	Body and subject	Eager learning (training)
Chung-Kwey	[26]	-	Spam patterns	Black list / Content based	-	Body and subject	Eager learning (training)
ECUE_prev	[27]	Before learning	Instances	Content-based	Yes	Body and subject	Lazy learning
ECUE	[28]	Before learning	Instances	Content-based	Yes	Body and subject	Lazy learning
Similarity Cases	[29]	Before learning	Classes of spam and legitimate messages	Content-based	No	Body and subject	Lazy learning
SPAMHUNTING	[11]	Every time a message is received	Instances (EIRN Model)	Content-based and potentially collaborative	Yes	Body and subject	Lazy learning (through the system operation)

The most recent and successful approaches for spam labelling and filtering are based on the use of the CBR methodology [14]. Despite of the high accuracy and effectiveness level achieved by them, they are not yet included in commercial software due to their recently appearance. The comparison between classical and CBR alternatives in the field of spam filtering is basically an analysis of the convenience of use lazy or eager learning. After the successful application of CBR techniques (especially, SPAMHUNTING [11] and ECUE [28] models), we highlight the benefits of using lazy learning strategies. Finally, from the experiments and results achieved we should mention SPAMHUNTING as a great alternative for spam filtering due to its capability of unify both content-based and collaborative approaches. For the work presented in this article, we have discarded the ECUE_prev. version, ECUE and Similarity Cases techniques because they get poor results in comparison with SPAMHUNTING system as showed in previous works [19, 29].

4   Available Corpus and E-mail Representation

Despite privacy issues, a large number of corpuses like SpamAssassin[1], Ling-Spam[2], DivMod[3], SpamBase[4] or JunkEmail[5] can be downloaded from Internet. In [13] a detailed description of the above mentioned corpus can be found. In this section, we outline the corpus used to carry out the experiments in our current work.

Keeping in mind the nature of this work, we have considered using the SpamAssassin corpora, merging e-mails from 2002 and 2003 versions. Table 2 shows the structure of the SpamAssassin corpus focussing in the spam and legitimate ratio and the distribution form.

Table 2. Message distribution of the SpamAssassin corpus

Downloaded filename	Year	Message type	Number of e-mails
20021010_easy_ham.tar.bz2	2002	legitimate	2552
20021010_hard_ham.tar.bz2	2002	legitimate	251
20030228_easy_ham.tar.bz2	2003	legitimate	2500
20030228_easy_ham_2.tar.bz2	2003	legitimate	1400
20030228_hard_ham.tar.bz2	2003	legitimate	250
20021010_spam.tar.bz2	2002	spam	502
20030228_spam.tar.bz2	2003	spam	500
20030228_spam_2.tar.bz2	2003	spam	1397

The merged SpamAssassin corpus contains 9332 different messages from January 2002 up to and including December 2003. This corpus has not been pre-processed by the author and the messages are distributed in RFC-822 format [30]. This fact represents and advantage in comparison with other corpus that are pre-processed and distributed as feature vectors. Therefore, SpamAssassin corpus becomes appropriate to carry out an analysis of the impact of changing input vector dimensionality.

5   Experimental Setup and Results

In this section we describe the experimental setup for analyzing the impact of changing feature vector dimensionality in the performance of different spam classifiers. Then, we introduce some specific issues about model configuration and finally, we present and discuss in detail the experimental results.

As we commented earlier, we have chosen the following techniques: (*i*) Naïve Bayes, (*ii*) Flexible Bayes, (*iii*) AdaBoost and (*iv*) SVM. For each experiment we have tested all the algorithms working with 20 different feature vector sizes. We started the experiments using vectors with a hundred of features and we incremented the size in 100 features each time until reach the top of 2000 features. In order to configure the AdaBoost classifier we have used Decision Stumps [31] as weak learner with 150 boost iterations whilst the SVM algorithm has been tested by using a polynomial kernel.

---

[1] Available at http://www.spamassassin.org/publiccorpus/
[2] Available at http://www.iit.demokritos.gr/
[3] Available at http://www.divmod.org/cvs/corpus/spam/
[4] Available at http://www.ics.uci.edu/~mlearn/MLRepository.html
[5] Available at http://clg.wlv.ac.uk/projects/junk-e-mail/

For each experiment we have used a 10 stratified fold cross-validation in order to increase the confidence level of the outcomes obtained [32]. The results showed in this work represent the mean value of the 10 tests computed in each fold-cross validation.

Table 3 shows the percentage of correct classifications and error rate specifying the proportion of false positives and false negatives. From results, we can conclude that SVM is able to achieve a greater amount of correct classifications using small feature vector dimensionalities. Moreover, Flexible Bayes is always the best approach if the main goal is the reduction of false positive errors.

Table 3. Percentages of correct classifications, false positives and false negatives

Features	Naïve Bayes			Flexible Bayes			AdaBoost			SVM		
	%OK	%FP	%FN	%OK	%FP	%FN	%OK	%FP	%FN	%OK	%FP	%FN
100	89.62	6.42	3.96	88.85	0.26	10.89	91.86	2.39	5.75	94.16	1.74	4.10
200	89.64	6.52	3.85	91.25	0.27	8.49	92.77	2.12	5.11	96.16	1.13	2.71
300	89.76	6.43	3.81	93.12	0.23	6.65	93.52	1.81	4.67	97.02	0.81	2.16
400	89.96	6.30	3.74	93.92	0.16	5.91	93.50	1.80	4.70	97.46	0.70	1.84
500	89.88	6.41	3.71	94.65	0.13	5.22	93.53	1.84	4.63	97.65	0.70	1.65
600	89.72	6.55	3.73	95.18	0.17	4.65	93.56	1.77	4.67	97.81	0.63	1.55
700	89.81	6.55	3.64	95.33	0.10	4.58	93.85	1.60	4.55	98.05	0.48	1.47
800	89.83	6.57	3.60	95.77	0.11	4.13	94.15	1.49	4.36	98.25	0.57	1.18
900	89.98	6.52	3.50	96.24	0.15	3.61	93.99	1.41	4.60	98.25	0.57	1.18
1000	90.02	6.50	3.47	96.53	0.19	3.28	93.89	1.46	4.65	98.28	0.55	1.17
1100	90.11	6.43	3.46	96.71	0.19	3.10	93.95	1.48	4.58	98.27	0.60	1.13
1200	90.10	6.46	3.44	96.82	0.18	3.00	93.97	1.49	4.54	98.37	0.54	1.09
1300	90.17	6.40	3.43	96.89	0.19	2.91	93.97	1.51	4.52	98.36	0.55	1.09
1400	90.51	5.98	3.51	97.01	0.21	2.78	93.91	1.55	4.53	98.38	0.55	1.07
1500	90.37	6.25	3.39	97.18	0.20	2.61	93.92	1.58	4.50	98.42	0.64	0.93
1600	90.29	6.30	3.41	97.30	0.19	2.51	93.99	1.47	4.53	98.46	0.54	1.007
1700	90.38	6.20	3.42	97.43	0.21	2.36	93.98	1.44	4.59	98.46	0.51	1.03
1800	90.43	6.15	3.42	97.67	0.19	2.13	93.92	1.43	4.65	98.47	0.46	1.07
1900	90.42	6.15	3.43	97.59	0.17	2.24	93.85	1.51	4.64	98.50	0.51	0.99
2000	90.45	6.14	3.41	97.72	0.16	2.12	93.83	1.49	4.68	98.53	0.53	0.93

From another interesting point of view, Table 3 shows that incrementing the dimensionality while using Naïve Bayes does not result in a relevant improvement on the accuracy. Nevertheless, increasing the dimensionality when using SVM and AdaBoost techniques can slightly improve the results achieved. In the case of Flexible Bayes, we can realize that it is especially sensitive to the increment of the vector dimensionality.

Despite current version of our SPAMHUNTING system achieves greater results than Flexible Bayes technique [33], we are really impressed by the small ratio of false positives achieved when using Flexible Bayes model and its capacity of improving the results by increasing the feature vector dimensionality.

From another perspective, Table 4 shows the *recall* and *precision* scores [16] achieved by using the selected techniques with the different configurations. In order to understand the obtained results we should keep in mind the underground idea of these measures. *Recall* stands for the effectiveness (amount of spam messages successfully detected) while *precision* is indicative of security (avoid false positive errors).

Table 4. *Recall* and *precision* scores obtained from the different analyzed scenarios

Features	Naïve Bayes		Flexible Bayes		AdaBoost		SVM	
	Recall	Precision	Recall	Precision	Recall	Precision	Recall	Precision
100	0.845	0.771	0.573	0.983	0.774	0.893	0.839	0.925
200	0.849	0.769	0.667	0.985	0.799	0.906	0.894	0.953
300	0.850	0.772	0.739	0.988	0.817	0.920	0.915	0.966
400	0.853	0.776	0.768	0.992	0.816	0.921	0.928	0.971
500	0.855	0.773	0.795	0.993	0.818	0.919	0.935	0.972
600	0.854	0.769	0.818	0.992	0.817	0.922	0.939	0.974
700	0.857	0.770	0.821	0.996	0.822	0.930	0.942	0.980
800	0.859	0.770	0.838	0.995	0.829	0.935	0.954	0.977
900	0.863	0.772	0.858	0.993	0.820	0.937	0.954	0.977
1000	0.864	0.773	0.871	0.991	0.817	0.935	0.954	0.978
1100	0.864	0.775	0.879	0.991	0.821	0.934	0.956	0.976
1200	0.855	0.774	0.882	0.992	0.822	0.934	0.957	0.979
1300	0.866	0.776	0.886	0.992	0.823	0.933	0.957	0.978
1400	0.862	0.789	0.891	0.991	0.822	0.931	0.958	0.978
1500	0.867	0.780	0.897	0.991	0.824	0.930	0.963	0.975
1600	0.866	0.779	0.901	0.992	0.822	0.935	0.961	0.979
1700	0.866	0.781	0.908	0.991	0.820	0.936	0.959	0.979
1800	0.866	0.783	0.916	0.992	0.818	0.936	0.958	0.981
1900	0.866	0.783	0.912	0.993	0.818	0.933	0.961	0.980
2000	0.866	0.783	0.917	0.993	0.816	0.934	0.963	0.979

From the *recall* measure showed in Table 4 we confirm the findings achieved from the analysis of Table 3. Using higher dimensionality of input vector, Naïve Bayes can not improve effectiveness whilst AdaBoost and SVM get a small improvement. As in the previous case, Flexible Bayes is able to achieve the best effectiveness increment.

Analysing *precision* scores from Table 4 we can see that, in general, the use of higher dimensionality vectors is not useful for improving the security achieved by the different approaches. We highlight the security achieved by Flexible Bayes. As we can see from results, any *precision* value achieved by Flexible Bayes goes beyond the one obtained by the use of any other configuration.

We have also included a study of the *balanced f-score* measure [34] with the following three different values of β: 1 (equivalent to *f-score* [16]), 1.5 and 2. *f-score* measure is used to combine *precision* and *recall* into a single piece of information. Moreover, the β parameter is used to weight the importance of effectiveness and security. When $\beta=1$, security and effectiveness have the same weight while a greater value stands for a higher importance of precision. In the case of β being smaller than 1, then *recall* is assumed to be more important than *precision*.

Expression 2 defines the *balanced f-score* measure used to combine *precision* and *recall*. Moreover, Figure 1 shows a detailed comparative of *balanced f-score* measures in all the analyzed scenarios.

$$f - score_{\beta} = \frac{\left(\beta^2 + 1\right) \cdot precision \cdot recall}{\beta^2 \cdot precision + recall} \tag{2}$$

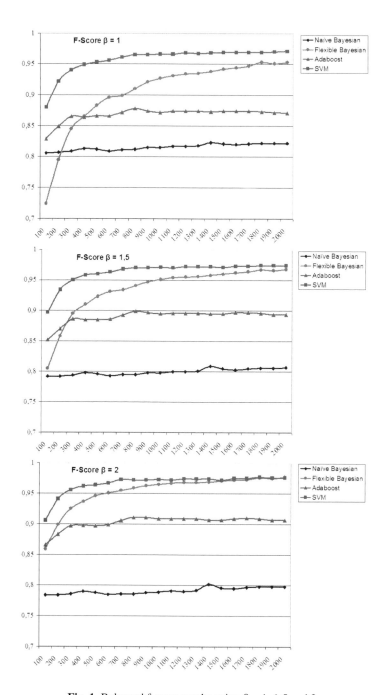

Fig. 1. Balanced f-score results using β = 1, 1.5 and 2

Figure 1 shows the *balanced f-score* measure computed by using the selected values for the β parameter and considering the dimensionality variable. When effectiveness and security are considered with the same relevance ($\beta =1$), we can see that Naïve Bayes does not achieve a substantial improvement even though the dimensionality is incremented. Moreover, SVM and AdaBoost algorithms present a relevant improvement until a dimensionality of 400 is reached. When the dimensionality is over this value, the benefits of using a higher dimensionality are very poor. Finally, Flexible Bayes presents a significant performance increment when we increase the dimensionality.

In the case of giving priority to security against effectiveness, AdaBoost, Naïve Bayes and SVM models show a similar behaviour. Nevertheless, the results presented by Flexible Bayes are comparable to those achieved by SVM. Despite the highest security of Flexible Bayes model, SVM technique is clearly the best approach being able to make correct classifications with a reduced number of features.

From another perspective, TCR (*Total Cost Ratio*) score [6] shows the suitability of a technique keeping in mind the cost of the different kind of errors. For use it, we need to define the value of a λ parameter indicating the proportion between the costs assigned to a false positive and a false negative. If the value of λ is 9, we assume that a false positive cause the same damage as nine false negatives. The most common values for the λ parameter are 1, 9 and 999. Table 5 shows the TCR scores for the different analyzed scenarios using the above values for the λ parameter.

Table 5. TCR results obtained from the different analyzed scenarios

Features	Naïve Bayes			Flexible Bayes			AdaBoost			SVM		
	$\lambda=1$	$\lambda=9$	$\lambda=999$	$\lambda=1$	$\lambda=9$	$\lambda=999$	$\lambda=1$	$\lambda=9$	$\lambda=999$	$\lambda=1$	$\lambda=9$	$\lambda=999$
100	2.465	0.416	0.004	2.304	1.957	0.338	3.144	0.960	0.011	4.420	1.321	0.015
200	2.467	0.411	0.004	2.942	2.388	0.124	3.549	1.084	0.013	6.797	2.300	0.032
300	2.499	0.417	0.004	3.787	3.085	0.652	3.970	1.254	0.015	8.864	2.882	0.035
400	2.552	0.426	0.004	4.321	3.644	1.513	3.974	1.292	0.016	10.368	3.326	0.040
500	2.533	0.419	0.004	4.877	4.142	1.912	3.986	1.245	0.015	11.401	3.396	0.039
600	2.491	0.410	0.004	5.382	4.274	1.171	4.011	1.309	0.016	12.088	3.748	0.044
700	2.511	0.410	0.004	5.566	4.817	2.668	4.177	1.385	0.017	13.589	4.772	0.061
800	2.215	0.409	0.004	6.148	5.176	2.453	4.411	1.503	0.019	15.998	4.668	0.056
900	2.554	0.414	0.004	6.909	5.532	1.430	4.298	1.357	0.019	15.997	4.668	0.056
1000	0.565	0.414	0.004	7.564	5.231	0.764	4.217	1.483	0.019	16.019	4.666	0.056
1100	2.589	0.419	0.004	8.020	5.497	0.782	4.270	1.499	0.018	16.300	4.562	0.053
1200	0.583	0.417	0.004	8.203	5.823	0.982	4.288	1.492	0.019	16.653	4.862	0.060
1300	2.606	0.422	0.004	8.474	5.773	1.655	4.289	1.482	0.018	16.859	4.734	0.056
1400	2.737	0.484	0.005	8.756	5.607	0.140	4.253	1.446	0.018	17.896	5.163	0.068
1500	2.660	0.432	0.004	9.252	5.897	0.144	4.264	1.437	0.018	18.167	4.945	0.063
1600	2.640	0.428	0.004	9.679	6.162	0.156	4.308	1.509	0.019	20.204	5.760	0.069
1700	2.662	0.434	0.004	10.200	6.166	0.133	4.272	1.522	0.019	19.666	4.988	0.055
1800	2.676	0.438	0.004	11.184	6.939	0.157	4.239	1.524	0.019	20.944	5.529	0.064
1900	2.675	0.438	0.004	10.993	7.213	1.129	4.194	1.463	0.018	19.692	5.460	0.065
2000	0.684	0.439	0.004	11.572	7.600	1.183	4.176	1.477	0.019	20.600	5.246	0.061

From Table 5, we can deduce that SVM and Flexible Bayes models are widely affected by dimensionality when the cost is weighted. Moreover, the results achieved by AdaBoost and Naïve Bayes can not be improved by using high dimensionality vectors.

From another point of view, results from Table 5 show that Flexible Bayes technique can achieve the most safety results reducing the global cost of errors. Moreover, AdaBoost and specially Naïve Bayes achieve the worst results while SVM and Flexible Bayes demonstrate their suitability as reliable alternatives for spam filtering.

6 SPAMHUNTING and Dimensionality

In the previous section we have presented a wide analysis of the effect of changing the dimensionality of input vector in four well-known techniques usually applied for spam filtering. Keeping in mind those results, this section carries out a particular study of the same matter in our most recent version of SPAMHUNTING system [19].

In order to study the behaviour of our system in relation to the amount of AI (available information), we have tested the performance achieved by using different percentages of AI starting from 25% and up to 70% using steps of five units (see Section 2 for details).

Table 6 shows the results achieved by using SPAMHUNTING with the selected configurations. As in previous experiments, we have used the following measures: (*i*) percentages of correct classifications, false positives and false negatives, (*ii*) recall and precision, (*iii*) balanced f-score with β values assigned to 1, 1.5 and 2, and (*iv*) TCR scores using λ values 1, 9 and 999.

Table 6. SPAMHUNTING results obtained from the different analyzed scenarios

Features	Percentages			Effectiveness and Security		Balanced f-score			TCR		
	%OK	%FP	%FN	Recall	Precision	$\beta = 1$	$\beta = 1.5$	$\beta = 2$	$\lambda = 1$	$\lambda = 9$	$\lambda = 999$
25 %	96.61	0.20	3.18	0.88	0.99	0.93	0.95	0.97	7.70	5.80	3.31
30 %	96.85	0.18	2.87	0.88	0.99	0.93	0.96	0.97	8.36	6.25	2.80
35 %	97.05	0.13	2.82	0.89	0.99	0.94	0.96	0.97	8.83	6.85	3.62
40 %	97.40	0.14	2.46	0.90	0.99	0.95	0.96	0.97	10.07	7.98	4.49
45 %	97.54	0.10	2.37	0.91	0.99	0.95	0.97	0.98	10.74	9.08	6.83
50 %	97.69	0.10	2.22	0.91	0.99	0.95	0.97	0.98	11.45	9.76	7.47
55 %	97.78	0.10	2.12	0.92	0.99	0.95	0.97	0.98	11.84	9.51	6.10
60 %	97.76	0.13	2.11	0.92	0.99	0.95	0.97	0.98	11.83	9.33	6.48
65 %	97.70	0.11	2.20	0.91	0.99	0.95	0.97	0.98	11.35	8.87	5.75
70 %	97.84	0.10	2.07	0.92	0.99	0.96	0.97	0.98	12.12	9.67	6.14
75 %	97.97	0.11	1.92	0.92	0.99	0.96	0.97	0.98	13.11	9.68	4.93

As we can realize from Table 6, the percentage results achieved by SPAMHUNTING are very similar in all the tested configurations. The FP error ratio is always very small and outperforms the results achieved by other well-known analyzed techniques (see Table 3). From another point of view, only SVM achieves a better percentage of correct classifications.

From recall and precision scores, we can see that SPAMHUNTING presents the highest security level (see Tables 4 and 6 for details). Despite the change of dimensionality, precision scores have small variability and outperform the ones computed by using another technique. Moreover, as we can realize from Table 6, recall gets slightly higher values with the increment of the dimensionality.

Analyzing balanced f-score obtained values from Table 6 we can see that an increment of the dimensionality can slightly improve the obtained results for the considered β values. Moreover, variability remains small due to the quality of these results. Finally, balanced f-score achieved values are better than the ones computed from other techniques (see Figure 1 for details).

Finally, from TCR scores showed on Table 6, we can see that SPAMHUNTING is able to effectively reduce the cost of the different kind of errors. This finding can be anticipated due to the small ratio of FP errors and the great precision level. From another point of view, the increment of the dimensionality allows the achievement of higher TCR scores.

7 Conclusions and Further Work

In this work we have compared and analyzed the effects of changing the dimensionality of input vector over four well-known spam filtering techniques. For this issue, we have summarized our previous findings in the spam filtering domain and included a brief description of the current available techniques for spam identification and classification. We have also introduced the publicly available corpus used for experimental purposes and finally, we have summarized details about the setup of our experimentation.

In order to carry out the experiments, we have considered several scenarios related to the dimensionality (number of selected features) of the terms representing a whole corpus. For this purpose, we used input vectors starting from 100 features and ending in 2000. We carried out a 10 stratified fold cross-validation for studying the evolution of several well-known metrics, including percentages, *precision*, *recall*, *balanced f-score* and *TCR* scores.

Results show that Naïve Bayes approach does not improve the security neither the effectiveness ratio. The difference between using feature vectors with 100 and 2000 features for representing the SpamAssassin corpus is very small (see Table 4 for details). These findings backup the ideas exposed in [7], where it is pointed out that assuming normality in the data and modelling each conditional distribution with a single Gaussian makes up a hard simplification of the problem that can significantly deteriorate the obtained results.

Flexible Bayes classifier always presents the highest precision rate. Nevertheless, using this technique with input vectors having small dimensionality is not a good idea for effective filter of spam e-mails. In these situations, despite of the small number of false positive errors, the amount of detected spam messages is limited (see Figure 1 for details). In the other hand, working with high dimensionality feature vectors, Flexible Bayes classifier gets the most remarkable values for error cost (see Table 5 for a thoroughly analysis).

The main difference between Naïve and Flexible Bayes classifiers is the amount of messages classified as spam. The Naïve Bayes approach needs less evidence for classifying a message as spam than Flexible Bayes. This issue is due to the mechanism used to estimate the probability of a message being spam. From this fact we conclude that if we use vectors having small dimensionality a Naïve Bayes approach is more adequate for achieving good results. Nevertheless, if high dimensionality is used, the Flexible Bayes approach is the most suitable.

Summarizing the results of the AdaBoost classifier, it can be observed that it presents the smallest effectiveness ratio within all the tested scenarios (see recall measures from Table 4). Despite this circumstance, the precision score can get better by using a higher dimensionality. From another point of view, the security level of this technique is lower than the achieved when Flexible Bayes or SVM are applied. Keeping in mind these findings, this approach should not be used for spam filtering, especially when the dimensionality is small. The mixture of successful classifiers is less suitable for spam filtering than other techniques such as the combination of probabilities (Flexible Bayes) and/or the use of a hyperplane for dividing the areas where spam and legitimate messages are situated.

From experimental results we have observed that dimensionality is a very important issue for using an SVM classifier. Both the recall and precision scores get better with the increment in the dimensionality of input vector (see recall and precision from Table 4). The security level obtained by this classifier is lower than the achieved by using Flexible Bayes. Nevertheless, this technique has obtained the best effectiveness ratio in the experiments carried out. The underground idea behind SVM classifier is that it is very effective in order to achieve higher effectiveness and security rates.

In order to choose the best choice for spam filtering from classical models, we should keep in mind our study about dimensionality issues. If a lower dimensionality is used, SVM is recommended. Nevertheless, if a high dimensionality is used we should decide between maximize security or the effectiveness ratio. In the first case Flexible Bayes classifier should be used, whereas in the second case the SVM approach is more appropriate.

Analyzing the results achieved by our SPAMHUNTING system, it can be seen that this technique obtains the maximum confidence level. This issue occurs because SPAMHUNTING is able to internally detect when there is no enough available knowledge for correctly classifying a message (for instance, when no similar instances are retrieved from their memory given target e-mail). Moreover, SPAMHUNTING classifies messages using instances representing entire e-mails as the smallest piece of information, whereas classical approaches use individual terms for this goal. This issue allows more confidence level on decisions while slightly reduces the effectiveness of the filter. The exhibit measures show SPAMHUNTING as the most reliable technique for spam filtering having a high effectiveness level (see Table 6 for details). Moreover, using SPAMHUNTING a high dimensionality representation is not required for getting a high confidence level but the effectiveness can be improved by using it.

The introduction of SPAMHUNTING system has evidenced the importance of lazy learning and continuous updating capabilities in spam filtering. Moreover, it also shows the importance of using messages as the minimum information units as a way to improve the security. Finally, SPAMHUNTING has introduced a new way of doing feature selection based on the disjoint representation of the messages able to successfully address the concept drift problem.

Talking about future work, we think that new improvements should be introduced in the pre-processing stage, especially for successfully address the noise present in spam messages. We are convinced that the most important difficulty that should be overcome is relative to the successful addressing of noise both in lexical and semantic form. Therefore, we think that newer tokenizing and information extraction techniques should be introduced for improving the results of the current state-of-the-art spam filtering techniques.

Acknowledgments

This work has been supported by the University of Vigo project SAEICS: *sistema adaptativo con etiquetado inteligente para correo spam / Adaptive system with intelligent labeling for spam e-mails.*

References

1. AVANZA Plan: Spanish Council of Industry, Tourism and Trade (2006), http://www.planavanza.es
2. eEurope 2002 Plan: European Commission (2002), http://europa.eu.int/information_society/eeurope/2002/index_en.htm
3. Ingenio-2010 Plan: Spanish Council of Presidency (2007), http://www.ingenio2010.es
4. eEurope 2005 Plan: European Commission (2005), http://europa.eu.int/information_society/eeurope/2005/index_en.htm
5. i2010 Plan. European Commission (2007), http://europa.eu.int/information_society/eeurope/i2010/index_en.htm
6. Androutsopoulos, I., Koustias, J., Chandrinos, K.V., Paliouras, G., Spyropoulos, C.: An Evaluation of Naïve Bayesian Anti-Spam Filtering. In: Proc. of the Workshop on Machine Learning in the New Information Age at 11th European Conference on Machine Learning, pp. 9–17 (2000)
7. John, G., Langley, P.: Estimating Continuous Distributions in Bayesian Classifiers. In: Proc. of the 11th Conference on Uncertainty in Artificial Intelligence, pp. 338–345 (1995)
8. Freund, Y., Schapire, R.E.: A Decision-Theoretic Generalization on On-Line Learning and Application to Boosting. Journal of Computer and System Sciences 55(1), 119–139 (1997)
9. Vapnick, V.: The nature of Statistical Learning Theory. In: Statistic for Engineering and Information Science, 2nd edn. Springer, Heidelberg (1999)
10. Fdez-Riverola, F., Iglesias, E.L., Díaz, F., Méndez, J.R., Corchado, J.M.: Applying Lazy Learning Algorithms to Trackle Concept Drift in Spam Filtering. Expert Systems With Applications 33(1), 36–48 (2007)
11. Iglesias, E.L., Fdez-Riverola, F., Díaz, F., Méndez, J.R., Corchado, J.M.: Spam Hunting: An Instance-Based Reasoning System for Spam Labelling and Filtering. Decision Support Systems. (2007) (In press)
12. Méndez, J.R., Iglesias, E.L., Fdez-Riverola, F., Díaz, F., Corchado, J.M.: Tokenising, Stemming and Stopword Removal on the Spam Filtering Domain. In: Proc. of the 11th Conference of the Spanish Association for Artificial Intelligence, pp. 449–458 (2005)
13. Méndez, J.R., Fdez-Riverola, F., Iglesias, E.L., Díaz, F., Corchado, J.M.: A Comparative Performance Study of Feature Selection Methods for the Anti-Spam Filtering Domain. In: Proc. of the 6th Industrial Conference on Data Mining, pp. 106–120 (2006
14. Watson, I.: Case-based reasoning is a methodology not a technology. Knowledge-Based Systems 12(5-6), 303–308 (1999)
15. Porter, M.: An Algorithm for Suffix Stripping. Program 14(3), 130–137 (1980)
16. Rijsbergen, C.J.: Information Retrieval. Ed. Butterworth (1979)

17. Baeza-Yates, R., Ribeiro-Neto, B.: Modern Information Retrieval. Addison-Wesley, London (1999)
18. Yang, Y., Pedersen, J.O.: A Comparative Study of Feature Selection in Text Categorization. In: Proc. of the 14th International Conference on Machine Learning, pp. 412–420 (1997
19. Méndez, J.R., Fdez-Riverola, F., Díaz, F., Iglesias, E.L., Corchado, J.M.: Tracking Concept Drift at Feature Selection Stage in SpamHunting: an Anti-Spam Instance-Based Reasoning System. In: Proc. of the 8th European Conference on Case-Based Reasoning, pp. 504–518 (2006)
20. SpamHaus: SBL, SpamHaus Block List (2005), http://www.spamhaus.org/sbl
21. Trend Micro Incorporated: MAPS (2005), http://www.mail-abuse.com
22. SpammerX: Inside the Spam Cartel: Trade Secrets from the Dark Side. Syngress Publishing (2004)
23. Prakash, V.V.: Vipul's Razor (2005), http://razor.sf.net
24. Pyzor: Pizor Page (2005), http://pyzor.sf.net
25. Rhyolite Software: DCC: Distributed Checksum ClearingHouse (2000), http://www.rhyolite.com/anti-spam/dcc/
26. Rigoutsos, I., Huynh, T.: Cheng-Kwey: a Pattern-Discovery-Based System for the Automatic Identification of Unsolicited E-Mail Messages (Spam). In: Proc. of the 1st Conference on Email and Anti-Spam (CEAS-2004) (2004), http://www.ceas.cc/papers-2004/index.html
27. Cunningham, P., Nowlan, N., Delany, S.J., Haahr, M.: A Case-Based Approach to Spam Filtering that Can Track Concept Drift. In: Proc. of the 5th International Conference on Case-Based Reasoning, pp. 115–123 (2003)
28. Delany, S.J., Cunningham, P., Coile, L.: An Assesment of Case-Based Reasoning for Spam Filtering. In: Proc. of the 15th Irish Conference on Case-Based Reasoning, pp. 9–18 (2004)
29. Kinley, A.: Acquiring Similarity Cases for Classification Problems. In: Proc. of the 6th International Conference on Case-Based Reasoning, pp. 327–338 (2005
30. Crocker, D.: Standard for the Format of ARPA Internet Text Messages. STD 11, RFC 822 (1982), http://www.faqs.org/rfcs/rfc822.html
31. Oliver, J.J., Hand, D.: Averaging over decision stumps. In: Proc. of the 7th European Conference on Machine Learning, pp. 231–241 (1994)
32. Kohavi, R.: A study of cross-validation and bootstrap for accuracy estimation and model selection. In: Proc. of the 14th International Joint Conference on Artificial Intelligence, pp. 1137-1143 (1995)
33. Méndez, J.R.: Adaptative system with intelligent labelling for spam e-mail classification. Ph Dissertation. Computer Science Department. University of Vigo (2006)
34. Shaw, W.M., Burgin, R., Howell, P.: Performance standards and evaluations in IR test collections: Cluster-based retrieval models. Information Processing and Management 33(1), 1–14 (1997)

Blog Mining for the Fortune 500

James Geller, Sapankumar Parikh, and Sriram Krishnan

College of Computing Sciences,
Department of Computer Sciences,
New Jersey Institute of Technology
Newark, NJ 07102

Abstract. In recent years there has been a tremendous increase in the number of users maintaining online blogs on the Internet. Companies, in particular, have become aware of this medium of communication and have taken a keen interest in what is being said about them through such personal blogs. This has given rise to a new field of research directed towards mining useful information from a large amount of unformatted data present in online blogs and online forums. We discuss an implementation of such a blog mining application. The application is broadly divided into two parts, the indexing process and the search module. Blogs pertaining to different organizations are fetched from a particular blog domain on the Internet. After analyzing the textual content of these blogs they are assigned a sentiment rating. Specific data from such blogs along with their sentiment ratings are then indexed on the physical hard drive. The search module searches through these indexes at run time for the input organization name and produces a list of blogs conveying both positive and negative sentiments about the organization.

1   Introduction

A blog, which is a collection of web pages on some website or portal, serves as an online diary maintained by an individual to share his thoughts and ideas and express his feelings. Blogs are powerful in the sense that they allow individuals over the globe to bring forth their ideas and garner feedback from other Internet users. Blogs appeared in the late 1990s but have since seen an unprecedented increase. Given the astounding blog-posting frequency and the amount of information communicated through online blogs, they are being viewed as potentially valuable resources of research. Furthermore, many people seem to get their news and form their opinions from authoritative blogs instead of standard media outlets, like broadcast news and newspapers [3].

Blog Mining techniques serve as an effective tool in social network analysis, economic research and network theory and form the basis for a myriad of services offered by popular blog analysis engines. For example, as a step towards social community mining, blog mining techniques can be used to find a community of bloggers who share a similar topic distribution in their blogs. This concept

P. Perner (Ed.): MLDM 2007, LNAI 4571, pp. 379–391, 2007.

is referred to as latent friend mining, wherein a latent friend is one who shares similar interests [2]. Determining communities of web pages based on named entity terms is another area where blog mining can prove to be useful. Named entity terms are names of persons, organizations, locations, etc. that occur in web documents with some relationships between them. Named entity terms are of high interest in web and blog search. While query strings can vary from a product model to a scientific concept, named entity terms are among the most frequently searched terms on the web [5].

"Reputation Management," is another technique under development by several companies [1]. Companies have now become aware of the potential of blogs to hurt them and have taken a two-pronged approach to dealing with the problem. On the one hand, many companies have created their own corporate blogs which have multiple purposes of keeping their customers and users informed, involved, and in some cases to garner feedback about products before releasing them to a wider audience. On the other hand, new blog mining startups have been created, which are hired by major companies to keep an eye on the blog space [8].

This paper discusses the latter kind of sentiment mining of the blog text. The architecture of the system is detailed in Figure 1.

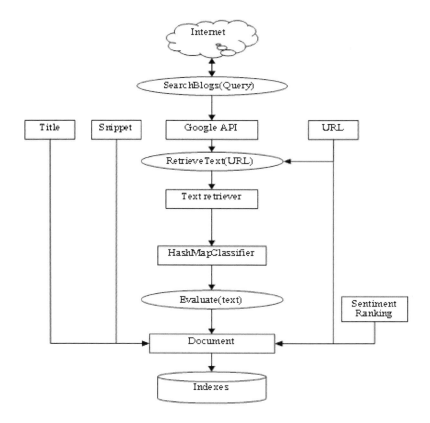

Fig. 1. System Architecture

The strength of positive or negative feelings is expressed by the sentiment ranking number that our program assigns to each blog. This sentiment ranking is a logarithmic number of positive or negative word counts from the blog text. The whole procedure of classifying a blog and giving it a ranking is done at index time, i.e. blogs are ranked and indexed prior to search time. This technique reduces the time required for searching results as they are already stored on the machine.

2 Blog Search

To search for blogs of a specific organization on the Internet we use APIs provided by Google. Google API is a lightweight framework developed for searching, made available by Google [6]. For implementation purposes, the Internet blog domain used is blogspot.com. The format of the query supplied to Google APIs is "company-name site:blogspot.com". The company names are obtained from a database table storing the list of Fortune 500 companies. The results returned are sorted according to Google's proprietary Page Rank method.

Each result returned consists of several parts as shown in Figure 2. These parts are:

summary - If the search result has a listing in the Open Directory Project directory, the ODP summary appears here as a text string. The Open Directory Project is a web directory of Internet resources. A web directory is something akin to a huge reference library. This directory is hierarchically arranged by subject - from broad to specific. The ODP is maintained by community editors who evaluate sites for inclusion in the directory. In Figure 2, the result does not have a summary from ODP.

URL - The URL of the search result, returned as text, with an absolute URL path. In Figure 2, "planetmath.org/encyclopedia/TimeComplexity.html" is a URL to reach this page.

snippet - A text excerpt from the results page that shows the query in context as it appears on the matching results page. This is formatted HTML and query terms are highlighted in bold in the results, and line breaks are included for proper text wrapping. If Google searched for stemmed variants of the query terms using its proprietary technology, those terms are also highlighted in bold in the snippet. Note that the query term does not always appear in the snippet. In Figure 2, "Time complexity refers to a function describing how much time it will take an algorithm ... The exact expression for time complexity of a particular sorting..." is the snippet. Snippet and summary are used interchangeably.

title - The title of the search result, returned as HTML. "PlanetMath: time complexity" is the title of the result shown in Figure 2.

cachedSize - Text (Integer + "k"). Indicates that a cached version of the URL is available; size is indicated in kilobytes. In the figure 22k is the size of the cached page.

relatedInformationPresent - Boolean indicating that the "related:" query term is supported for this URL. This is not a visible component of the results returned by Google.

hostName - When filtering occurs, a maximum of two results from any given host are returned. When this occurs, the second resultElement that comes from that host contains the host name in this parameter.

directoryTitle - If the URL for this resultElement is contained in the ODP directory, the title that appears in the directory appears here as a text string. Note that the directoryTitle may be different from the URL. [6]

PlanetMath: **time complexity**
Time complexity refers to a function describing how much **time** it will take an algorithm ... The exact expression   for   **time**   **complexity**   of   a   particular   sorting   ...
planetmath.org/encyclopedia/TimeComplexity.html - 22k - <u>Cached</u> - <u>Similar pages</u>

Fig. 2. Format of a result returned by Google API

3   Relevant Text Extraction

The URL of each result is used, in our program, to extract the text content of the corresponding web page. For this purpose we have used JTidy. JTidy is a Java port of HTML Tidy, an HTML syntax checker and pretty printer. Like its non-Java cousin, JTidy can be used as a tool for cleaning up malformed and faulty HTML. In addition, JTidy provides a Document Object Model interface (DOM) to the document that is being processed, which effectively makes you able to use JTidy as a DOM parser for real-world HTML [11]. The text retriever module uses JTidy to get a DOM representation of the web page and then extract text from it.

As seen on many blog portals, it is quite common to have several blogs listed on a single web page. These blogs are mostly written on different topics and are not related to each other. Of these, only a single topic or a few more may be relevant to the organization we are interested in. Doing a classification on all words of the web page yields highly erroneous results.

Hence, after fetching the text of the entire blog page, it is further processed to fetch pieces of text relevant to the corresponding organization. A record of the association between the query i.e. organization name and links corresponding to it, maintained during the Blog Search procedure described in Section 2, is utilized to extract such relevant text blocks. The complete text is split into various blocks with the company name as the delimiter. If the company name consists of more than one word then each individual block obtained in the first iteration is split further using any of the individual words in the company name as a delimiter.

For example, consider the text of a web page fetched for "Morgan Chase." The process of extracting relevant text is clarified in Figure 3.

After dividing the text into different blocks, the process of retrieving relevant text starts. Since there is a low probability that a "large" piece of text located before the first occurrence of the organization name would actually describe the organization, only 30 words are considered from the first block. Thereafter, each block is examined to check if its length is greater than 150 words. If no, the entire block is considered relevant. If yes, only the first 75 words from both the start and end of the block are considered relevant. Finally, for the last block, the first 75 words from the beginning of the block are taken. If the size of the last block is less than 75 words the entire block is used. To check for organization name matches within the text of the web page, regular expressions have been utilized which further improves the accuracy of the text extraction process.

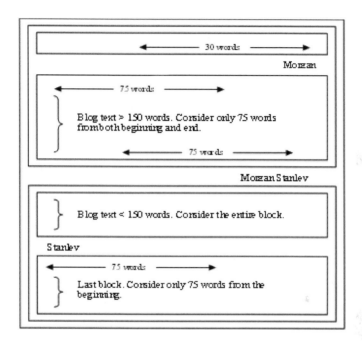

Fig. 3. Algorithm for extracting the relevant pieces of text

A sample run of the RelevantTextExtractor module on the content of a particular piece of blog text extracted from the web is shown in Figures 4 and 5. As shown in Figure 4, the blog is titled "J.P. Morgan posts 68% rise in quarterly profit" and the relevant text being extracted pertains to the company JP Morgan Chase. The RelevatTextExtractor module uses the procedure described in Figure 3 to search for occurrences of the phrase "Morgan Chase" or each individual word "Morgan" or "Chase" within the text, divide the text into blocks based on these occurrences and extract pieces of relevant text from the blocks. The output of the RelevantTextExtractor module is shown in Figure 5.

4   HashMap Classifier

4.1   Implementation

In order to determine the sentiment ranking of a piece of blog text, a table of words conveying either positive or negative sentiments was implemented. The

J.P. (Morgan) posts 68% rise in quarterly profit

J.P. (Morgan Chase) & Co. said its quarterly earnings soared 68% on strong investment banking growth and a gain from
the sale of the bank's corporate trust business. But credit quality weakened somewhat, as it has at other major banks,
suggesting that both commercial and individual customers are having more trouble keeping up with their bills.
The Manhattan-based bank, the nation's third largest, earned $4.53 billion, or $1.26 a share, in the fourth quarter,
up from $2.7 billion, or 76 cents a share, a year earlier. Excluding the $622 million after-tax gain on the sale of
its trust business, net income was $3.9 billion, or $1.09 a share. Revenue rose 19% to $16.05 billion.

Analysts had projected a profit of 95 cents a share. Some analysts believe the weakening housing market and slowing
economy are affecting credit quality and worry it could worsen in coming months. Mr. Dimon has been warning for some
time that the stellar credit conditions -- which reflected a growing economy and reform of national bankruptcy laws in
the fall of 2005 -- could not continue. In fact, weaker credit quality has begun showing in several divisions.

J.P. (Morgan)'s strongest fourth-quarter performance came from the investment bank, where net income grew 51% to $1
billion in the latest quarter. Still, the provision for credit losses in this division rose to $63 million from $7
million in the third quarter.

Fig. 4. Sample blog pertaining to the company J. P. Morgan Chase

Block 0--->   J.P.

Block 1--->   posts 68% rise in quarterly profit J.P.

Block 2--->   & Co. said its quarterly earnings soared 68% on strong investment banking growth and a gain
                from the sale of the bank's corporate trust business. But credit quality weakened somewhat, as it has
                at other major banks, suggesting that both commercial and individual customers are having more trouble
                keeping up with their bills.The Manhattan-based bank, the nation's third largest, earned $4.53
                billion, or $1.26 a share, in the fourth quarter, up from $2.7 billion,

Block 3--->   a profit of 95 cents a share. Some analysts believe the weakening housing market and slowing
                economy are affecting credit quality and worry it could worsen in coming months. Mr. Dimon has been
                warning for some time that the stellar credit conditions -- which reflected a growing economy and
                reform of national bankruptcy laws in the fall of 2005 -- could not continue. In fact, weaker credit
                quality has begun showing in several divisions. J.P.

Block 4--->   's strongest fourth-quarter performance came from the investment bank, where net income grew
                51% to $1 billion in the latest quarter. Still, the provision for credit losses in this division rose
                to $63 million from $7 million in the third quarter.

Fig. 5. Set of relevant text blocks extracted from Figure 4

table contains the actual word, its category i.e. positive or negative and its weight
which is a numerical value ranging between -1.5 and +1.5 in increments of 0.25,
excluding 0. Weights less than 0 are assigned to words that convey negative
sentiments and those greater than 0 are given to words that convey positive
sentiments. The value of the weight indicates the extent to which a word is
positive or negative. The table, at present, stores over 400 positive and negative
words. We initially settled on 6 positive and 6 negative steps. To avoid "infinite"
decimal numbers we chose the limit of 1.5 instead of 1, which will always provide
exact decimal numbers.

When starting the sentiment mining process on a given blog, a hash table is
created in memory, which contains the words conveying sentiments as keys and
weight numbers between -1.5 and 1.5 as values. The information loaded into this
hash table is extracted from the table mentioned above, which is implemented in
Oracle (see Figure 6). Every word in the relevant text blocks of a blog, extracted
as explained in Section 3, is then checked to see if it matches with the keys in the
hash map. Whenever a blog word matches a hash table key, the weight for this
key is returned. The weight of the entire set of relevant text blocks is obtained by
calculating the sum of weights of individual words that have a match in the hash
map. A negative word weight lowers the blog weight and a positive word weight
adds to it. Finally, the logarithm of the absolute value of the resultant sum is

used as the sentiment rating for the blog text. The complete blog is classified as positive or negative depending on whether the resultant logarithmic value is greater or less than zero, respectively. If the resultant value is 0 then the blog is neutral, in which case, it is skipped. However, this is a rare condition and the probability of occurrence of such blogs is minimal.

To get a closer match between the words in the text and those in the hash map, the base words and their variations have been included in the database. An example of this is also shown in the figure 6.

As shown in Figure 6, both "loves" and "loved" give a match with "love" in the hash map. If a particular word in the text is not found in the map, a check is performed to see if the word ends with any of the common suffixes. If a common suffix is found, it is stripped and the remaining part of the word is again checked for a match in the hash map. Common suffixes of 1, 2, 3 and 4 character lengths have been used for this purpose. These are listed in Figure 7. This helps limit the size of the wordlist table by reducing the number of variations to be entered for each word in the database. It also helps achieve better classification due to the probability of getting more matches.

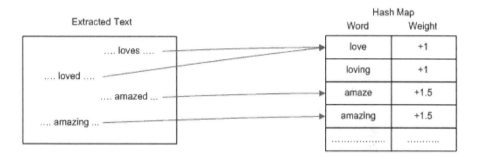

Fig. 6. Text word match with hash map keywords

4.2   Comparison of Classifiers

We have compared our HashMap classifier with two other classifiers, Dynamic Language Model Classifier and Binary Language Model Classifier, which are included in the LingPipe framework. LingPipe is a suite of natural language processing tools that performs tokenization, sentence detection, named entity detection, co-reference resolution, classification, clustering, part-of-speech tagging, general chunking, and fuzzy dictionary matching [12].

In order to compare our results with existing research, we needed a set of test documents which are publicly accessible and have already been classified as positive or negative. For that purpose, we have used LingPipe's language classification framework to perform the classification task on a set of test data, which is actually a collection of positive and negative movie reviews. Lee and

Suffixes having a single character length	s, d, y
Suffixes having a length of 2 characters	ed, er, ly
Suffixes having a length of 3 characters	ing, ent, ism, ive, ful, ous, ity
Suffixes having a length of 4 characters	ness, able, less, ment

Fig. 7. Table of common suffixes used

Pang have provided such movie review data for testing purposes. This data is already divided into two segments (slices), positive and negative. These reviews can be obtained from the link http://www.cs.cornell.edu/people/pabo/movie-review-data/review_polarity.tar.gz. This test data set contains a total of 2000 movie reviews, 1000 with positive sentiments and 1000 with negative sentiments. Following standard machine learning methods, for the purpose of training, 1800 movie reviews, 900 positive and 900 negative, are used. For testing, the remaining 200 movie reviews are used.

For each classifier and each training case, the number of training characters, the time to train, the correctness of the tests, and the total time for testing are given in Figure 8. The five cases of the experiment were done with different training and test data sets. For each case, 1800 training files were randomly picked and both the classifiers were trained with them. The remaining 200 files were used to test those trained classifiers. The highest accuracy achieved with the Dynamic-LM Classifier was about 83%, while the lowest was 77%. On the other hand, the accuracy of the BinaryLMClassifier was always 100%. The Binary-LM Classifiers also takes less time for training because it gets trained on only one category. This means that if the DynamicLMClassifier was trained on 1800 files then the Binary-LM Classifier was trained only on 900 files.

The Dynamic Language Model Classifier (DLM) is slow in both cases of training and evaluation. But the advantage of the Dynamic classifier against the Binary LM Classifier (BLM) is that it can support n classifications. That means that DLM can be trained to classify sentiments into several categories such as positive, extremely positive, neutral, negative, etc.

On the other hand the BLM is very fast. As the name suggests, it can classify only two categories, positive and negative. BLM only gets trained on one category and it rejects the other category. When a file is fed to it, it returns "true" or "false," true if the file belongs to the category it was trained on. For every other category it returns false. So, if there are only two categories this model works perfectly fine with a high level of precision.

We used our HashMap classifier for the movie review data set, without adapting the key words to the movie domain. We only used generic sentiment terms as we would use for any kind of blog mining. Looking at the results in Figure 8, it becomes clear that the HashMap classifier does not require any training time, as opposed to the Dynamic-LM and the Binary-LM classifier. Its classification performance of 66% is below the results for both the classifiers. However, HashMap classifier runs faster than both classifiers even during testing.

	Training Cases	Training Chars	Training Time (ms)	Evaluation Cases	Correct Evaluations	% Correct	Time to evaluate (ms)
Dynamic LM-Classifier							
Case 1	1800	7037315	138282	200	154	77	29375
Case 2	1800	7039594	138609	200	156	78	26610
Case 3	1800	7038941	143578	200	157	78.5	28891
Case 4	1800	7045936	163297	200	165	82.5	29344
Case 5	1800	7000543	140828	200	158	79.5	29047
Binary LM-Classifier							
Case 1	1800	7037315	13563	200	200	100	2625
Case 2	1800	7039594	4125	200	200	100	1109
Case 3	1800	7038941	4469	200	200	100	1218
Case 4	1800	7045936	4109	200	200	100	1250
Case 5	1800	7000543	4094	200	200	100	1297
HashMap Classifier	0	0	0	2000	1320	66	7140

Fig. 8. Statistics for different classifiers

Note that in Figure 8 the time for our HashMap classifier is given for 2000 movie reviews, while the times for the two other classifiers are given for 200 only. Thus, if we want to compare equal work loads, we need to divide the run time of 7140 ms by 10, giving 714 ms, which is faster than the fastest time of the Binary-LM Classifier.

5   Lucene Indexing

5.1   Indexing Records

The next step consists of indexing the Google API results with the sentiment rank. In order to do the indexing we have used Lucene [10]. Lucene is a free, open source information retrieval API originally implemented in Java by Doug Cutting. It is supported by the Apache Software Foundation and is released under the Apache Software License. While suitable for any application which requires full text indexing and searching capability, Lucene has been widely recognized for its utility in the implementation of Internet search engines and local, single-site searching.

At the core of Lucene is an index. Although Lucene is used for text indexing it does not index files, it indexes document objects. A document is a collection of fields which are nothing but name value pairs. An index in turn contains a set of documents.

The following piece of code demonstrates how document objects have been created in our module.

```
public Document indexThis(IndexWriter writer, String query,
GoogleSearchResultElement r, HashMap wordWeightMap)
throws Exception
{
    Document doc = new Document();
    Double reverseBoost;

    RelevantTextExtractor rte = new RelevantTextExtractor();

    String url = r.getURL().replaceAll("<b>", "");
    String snippet = r.getSnippet().replaceAll("<b>", "");
    String title = r.getTitle().replaceAll("</b>", "");

    String text = rte.parseThis(query, url).toString();
    Double boost =
    new Double(HashMapClassifier.evaluate(text, wordWeightMap));

    if(boost.doubleValue() >= 0)
        reverseBoost = new Double(100.0 - boost.doubleValue());
    else
        reverseBoost = new Double(100.0 + boost.doubleValue());

    doc.add(Field.Text("url", url));
    doc.add(Field.Text("snippet", snippet));
    doc.add(Field.Text("title", title));
    doc.add(Field.Text("rating", boost.toString()));
    doc.add(Field.Text("reverse_rating",
                        reverseBoost.toString()));

    return doc;
}
```

As shown in the code segment, a new instance of document object is created. The URL, snippet and title are extracted from each search result returned by the Google API. The URL is then passed to the RelevantTextExtractor module to retrieve the entire text of the blog web page and extract relevant text content from it. This relevant text is then passed to a static "evaluate" function of the HashMapClassifier class which returns a sentiment rating of the blog text. A reverse sentiment rating is also calculated, which is useful to sort blogs in descending order of their sentiment ratings. A document object is then created using the values of URL, snippet, title, rating and reverse rating. The document is then added to a new index or appended to one if the index already exists.

5.2   Searching Indexes

The search module provides a simple JSP page allowing a user to enter an organization name and search for it in the records already indexed on the hard drive. The snippet fields of all documents are checked for a match with the organization name. Once such records are found, positive and negative blogs

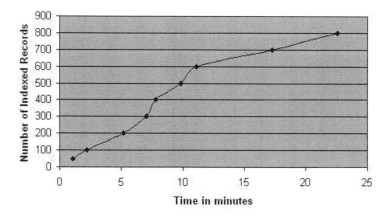

Fig. 9. Indexing Speed

are sorted and displayed separately on the JSP page. The displayed records are again in the form of small documents with a title, sentiment rating, snippet and URL.

6   Performance Statistics

Figure 9 shows the time required to complete the entire indexing process. This includes fetching blog URLs using Google APIs, extracting text from the blog web pages, determining sentiment rating to classify them as well as indexing blog data. To index blogs from about 800 web pages takes just over 22 minutes, which is fast considering the amount of processing that is done before indexing. Moreover, this does not affect run time performance since indexing is done offline, before hand.

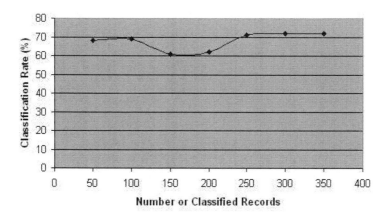

Fig. 10. Classification Rate

Figure 10 gives a plot of the number of blogs classified versus the percentage of correct classifications. As seen from the figure, this varies between 60-73%. Also, as seen from the comparison table in Figure 8, the average classification rate for 2000 movie reviews is 66%.

7 Conclusions and Future Work

For companies trying to determine where they stand in the market with respect to their customers, suppliers and many other stakeholders, blogs are becoming a prime medium. With the help of blog mining software, employees can easily go through the existing blogs to gain an insight into the feelings and opinions of users about the organization. This helps them make important decisions on improving product quality, increasing profit, market standing, and customer satisfaction.

Our implementation performs this kind of sentiment mining on blogs. The URLs of blogs specific to an organization are first fetched using APIs provided by Google. The entire text of the blog page corresponding to each URL is then extracted. The size of text is further reduced by determining pieces of text relevant to the organization. Using a table of words conveying sentiments the sentiment rating of the blog is determined and the blog is categorized as either positive or negative. The data for each blog is then indexed using Lucene APIs and stored on the hard drive. Since this entire classification takes place offline, the run time search procedure only needs to search through the indexed records already created, thereby keeping access time to a minimum. In our implementation, we have currently indexed 100 blogs for each of the Fortune 500 companies from the domain blogspot.com.

Using a hash map data structure for classification of blogs makes the indexing process very fast. However, since the classification relies only upon a list of positive and negative words the accuracy achieved is about 60-75%. To improve accuracy the size of the word list in the database needs to be increased by adding new positive and negative words as and when found. Note that due to the efficient retrieval of data from hash tables, adding new terms will, on an average, have a small effect on evaluation time. Thus, an expected increase in classification accuracy will maintain the positive timing characteristics of our approach. As an additional enhancement, a table of the most commonly occurring phrases conveying sentiments along with their weights can be created. An initial text weight can be determined first by checking for these common phrases in the extracted relevant text. These phrases can then be eliminated from the text and remaining text can be classified using our Hash Map classifier. This would further refine the classification task. Moreover, in addition to searching for only the company name, its products and/or services can also be used to fetch relevant text. This would result in a larger text on which classification would be more accurate.

References

1. Aschenbrenner, A., Miksch, S.: Blog Mining in a Corporate Environment, Technical Report ASGAARD-TR-2005-11, Technical University Vienna (September 2005) (accessed February 1, 2007), http://ieg.ifs.tuwien.ac.at/techreports/Asgaard-TR-2005-11.pdf
2. Shen, D., Sun, J.-T., Yang, Q., Chen, Z.: Latent Friend Mining from Blog Data. In: International Conference on Data Mining, pp. 552–561 (2006)
3. Tirapat, T., Espiritu, C., Stroulia, E.: Taking the community's pulse: one blog at a time. In: International Conference on Web Engineering, pp. 169–176 (2006)
4. Mishne, G.: Experiments with Mood Classification in Blog Posts. In: Style2005 the 1st Workshop on Stylistic Analysis of Text for Information Access, at SIGIR 2005 (August 2005)
5. Li, X., Liu, B., Yu, P.S.: Mining Community Structure of Named Entities from Web Pages and Blogs. In: AAAI Spring Symposium, Computational Approaches to Analyzing Weblogs, pp. 108–114 (2006)
6. Google Web APIs (March 10, 2006), http://code.google.com/apis
7. Fischer, I., Torres, E.: A Distributed Blog Search Platform (2006)
8. The Blog in the Corporate Machine, The Economist, (February 11, 2006)
9. Fortune 500 Full List, CNNMoney (April 17, 2006), http://money.cnn.com/magazines/fortune/fortune500/full_list
10. Hatcher, E., Gospodnetic, O.: Lucene in Action (2006)
11. JTidy - HTML Parser and Pretty-Printer in Java (March 10, 2006), http://jtidy.sourceforge.net
12. LingPipe (2007), http://www.alias-i.com/lingpipe

A Link-Based Rank of Postings in Newsgroup

Hongbo Liu[1], Jiahai Yang[1], Jiaxin Wang[2], and Yu Zhang[2]

[1] The Network Research Center
[2] Department of Computer Science and Technology
Tsinghua University, Beijing, China, 100084
liuhb1@gmail.com

Abstract. Discussion systems such as Usenet, BBS, Forum are important resources for information sharing, view exchanging, problem solving and product feedback, etc. on Internet. The postings in newsgroups on Usenet represents the judgments and choices of participators. The structure of postings could provide helpful information for the users. In this paper, we present a method called PostRank to rank the postings based on the structure of newsgroup. Its results correspond to the eigenvectors of the transition probability matrix and the stationary vectors of the Markov chains. It could provide useful global information for the newsgroup and it can be used to help the users access information in it more effectively and efficiently. This method can be also applied on other discussion systems. Some experimental results and discussions on real data sets collected by us are also provided.

Keywords: link analysis, rank, newsgroup, discussion systems.

1 Introduction

Usenet is a world-wide distributed discussion system, and it is one of the representative information resources on Internet. Usenet provides a convenient way for the communication and organization of discussions, which is much different from the World Wide Web (WWW) whose main purpose is information publishing. It consists of a set of newsgroups with names classified hierarchically by subject. In each group, postings are posted to the NNTP server and broadcast to other servers. With these servers, people all over the world can subscribe newsgroups they are interested in and can participate in the discussions.

Comparing with web pages, the content of postings on Usenet is generally more informal, brief and personalized. It contains rich information and ideas contributed by the participators. Due to the huge size of Usenet, people can only subscribe a few groups and generally read a small fraction of the postings. It may take quite much time for a newbie to familiar with a group and use it sensibly. It is difficult to access the required information efficiently from all the postings in a group because of its huge size and loose organization.

Information Retrieval (IR) techniques have been used on the web to make information more accessible. IR techniques have widely used the "bag-of-words model" for tasks such as document matching, ranking, and clustering [1]. On

P. Perner (Ed.): MLDM 2007, LNAI 4571, pp. 392–403, 2007.

WWW, because of the intrinsic hyperlink property of web pages, link analysis based on ideas of social networks has also been used in the ranking system of some search engines [2,3]. The simplicity, robustness and effectiveness of link-based ranking method have been witnessed with the success of Google, whose basis of ranking system is PageRank. Social networks have also been applied in other domains, such as marketing [4], email relationship [5], chat [6] and so on [7].

As most postings on the Usenet do not contain hypertexts, they can not be benefited from these link-based algorithms of WWW directly. Some IR techniques were used to improve the services of Usenet [8]. The briefness and casualty of newsgroup postings make it difficult for conventional text mining techniques. Some investigations based on social networks have also been done to extract useful information from the Usenet [9,10].

On Usenet, it is not easy to choose the postings before read when the users browse the newsgroup. Normally it is needed to read the postings throughout thread to get related information. It is time consuming and many postings are not very valuable. Some hints of postings may be greatly helpful to improve the efficiency of the Usenet users. For the Usenet search, the order of results is very important, which may be improve with a good ranking system. Therefore, good posting rank with intrinsic properties of a newsgroup can make the information on Usenet more accessible.

In this paper, according to the characteristics of Usenet, a link-based method to calculate the rank of postings on Usenet is proposed. Some mathematical analysis of this method is discussed and experimental results on real data sets are also given.

2   The Calculation of PostRank

2.1   Usenet Newsgroup and Its Representation

Unlike web pages, the postings on a newsgroup of Usenet are organized by threads. Each thread is invoked by one seed posting and followed by several response postings. The quantity and content of postings are determined by collaborative work of the participators along with the evolution of discussions.

Considering the posting v_i as node and the respondent relationship $e = \langle v_{i1}, v_{i2} \rangle$ as link, each thread can be abstracted as a rooted tree whose root is the seed posting and its descendants are the response postings of this thread. In the rooted tree, the leaf nodes are the postings without response. Since there are many threads in a newsgroup, its structure can be represented with forest $G(V, E)$. In this way, a newsgroup contains m postings with s seed postings and t leaf postings could be represented as a forest with m rooted trees and $m - s$ links.

Supposing posting v_i have c_i neighbors, due to the tree structure of thread there are two classes of neighbors for v_i according their relationships with v_i, i.e., the parent set $\mathcal{P}(v_i)$ containing a_i postings, and the offspring set $\mathcal{O}(v_i)$

containing b_i postings. a_i may be 0 or 1 depending on whether v_i is a seed posting. From the above description, we can get that

$$a_i + b_i = c_i, \sum_{v_i \in G} c_i = 2(m - s), \sum_{v_i \in G} |a_i - b_i| = 2t. \tag{1}$$

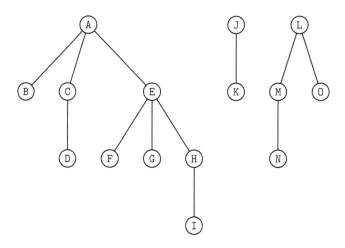

Fig. 1. graph representing a small newsgroup containing 3 threads with 15 postings

A small newsgroup containing 15 postings, including 3 seed postings and 8 leaf postings was shown in Fig. 1. It can be represented by the following adjacent posting matrix

$$
\mathbf{M} =
\begin{array}{c c}
 & \begin{array}{c c c c c c c c c c c c c c c} A & B & C & D & E & F & G & H & I & J & K & L & M & N & O \end{array} \\
\begin{array}{c} A \\ B \\ C \\ D \\ E \\ F \\ G \\ H \\ I \\ J \\ K \\ L \\ M \\ N \\ O \end{array} &
\left(\begin{array}{c c c c c c c c c c c c c c c}
0 & 0 & 0 & 0 & 0 & 0 & 0 & 0 & 0 & 0 & 0 & 0 & 0 & 0 & 0 \\
1 & 0 & 0 & 0 & 0 & 0 & 0 & 0 & 0 & 0 & 0 & 0 & 0 & 0 & 0 \\
1 & 0 & 0 & 0 & 0 & 0 & 0 & 0 & 0 & 0 & 0 & 0 & 0 & 0 & 0 \\
0 & 0 & 1 & 0 & 0 & 0 & 0 & 0 & 0 & 0 & 0 & 0 & 0 & 0 & 0 \\
1 & 0 & 0 & 0 & 0 & 0 & 0 & 0 & 0 & 0 & 0 & 0 & 0 & 0 & 0 \\
0 & 0 & 0 & 0 & 1 & 0 & 0 & 0 & 0 & 0 & 0 & 0 & 0 & 0 & 0 \\
0 & 0 & 0 & 0 & 1 & 0 & 0 & 0 & 0 & 0 & 0 & 0 & 0 & 0 & 0 \\
0 & 0 & 0 & 0 & 1 & 0 & 0 & 0 & 0 & 0 & 0 & 0 & 0 & 0 & 0 \\
0 & 0 & 0 & 0 & 0 & 0 & 0 & 1 & 0 & 0 & 0 & 0 & 0 & 0 & 0 \\
0 & 0 & 0 & 0 & 0 & 0 & 0 & 0 & 0 & 0 & 0 & 0 & 0 & 0 & 0 \\
0 & 0 & 0 & 0 & 0 & 0 & 0 & 0 & 0 & 1 & 0 & 0 & 0 & 0 & 0 \\
0 & 0 & 0 & 0 & 0 & 0 & 0 & 0 & 0 & 0 & 0 & 0 & 0 & 0 & 0 \\
0 & 0 & 0 & 0 & 0 & 0 & 0 & 0 & 0 & 0 & 0 & 1 & 0 & 0 & 0 \\
0 & 0 & 0 & 0 & 0 & 0 & 0 & 0 & 0 & 0 & 0 & 0 & 1 & 0 & 0 \\
0 & 0 & 0 & 0 & 0 & 0 & 0 & 0 & 0 & 0 & 0 & 1 & 0 & 0 & 0
\end{array}\right)
\end{array}
$$

According to the above descriptions, the following properties can be inferred directly.

Property 1. *(1) The diagonal elements of* $\mathbf{M}$ *are zero. (2)* $\mathbf{M}_{ij}$ *satisfies*

$$\sum_{j=1}^{m} \mathbf{M}_{ij} = a_i, \sum_{i=1}^{m} \mathbf{M}_{ij} = b_j, \sum_{i=1}^{m}\sum_{j=1}^{m} \mathbf{M}_{ij} = m - s$$

.

In this manuscript, the rooted trees are denoted with $Tk, k = 1, 2, \cdots, s$, and we use the symbols with subscript Tk to denote parameters of tree Tk. For tree Tk, posting is represented as $v_{i|Tk}$. The number of postings in Tk is m_{Tk}, and the number of leaf postings in Tk is t_{Tk}, and the level of Tk is l_{Tk}.

2.2 PostRank Calculation and Analysis

One intuitive idea to rank the postings based on G is that if a posting was responded by more postings, its rank is higher. Thus, b_i could be a candidate of the rank of v_i. This seems to be simple and feasible. However, this calculation works with the assumption that all postings have equal contribution to the rank, which is not reasonable enough. All postings are not created equal. A posting responded by a high rank posting might be more valuable than posting with a low rank response posting. It is better to retrieve the rank from the Usenet structure recursively.

Similar with some link-based rank methods on the web, the link from v_i to v_j in the forest G can be viewed as vote from v_i to v_j. If the rank of a posting is high, the postings it responding to and responded by provide some information intimately related with it, which might also be valuable for the users. When the users browse the Usenet, it is natural to read the parent or the offspring of his interested posting to get more detailed information. Thus, in our calculation, posting $v_{i|Tk}$ votes other postings according to their relationships with $v_{i|Tk}$. In the reverse direction, the rank of $v_{i|Tk}$ is determined by the ranks and relationships with other postings based on the structure of Usenet. Since the seed postings and the leaf postings are special in the Usenet, we add self-loop to them to give them additional bonuses. The rank of posting $v_{i|Tk}$ may contribute to its parent, its offspring, other postings in the same thread, and any posting in the newsgroup differently. In PostRank, we use $\alpha, \beta, \lambda, \eta$ to describe the difference of these relationships. Therefore, the PostRank $r_{i|Tk}$ of $v_{i|Tk}$ can be calculated as following:

$$r_{i|Tk} = \alpha \sum_{v_j \in \mathcal{P}(v_{i|Tk})} r_j/b_j + \beta \sum_{v_j \in \mathcal{O}(v_{i|Tk})} r_j + \lambda \sum_{v_j \in Tk} r_j/m_{Tk} + \eta, \qquad (2)$$

where $0 \le \alpha, \beta, \lambda, \eta < 1$ and $\alpha + \beta + \lambda + \eta = 1$.

As the PostRank vector $\mathbf{r}^T$ is a m dimensional row vector, it is convenient to use the matrix form of Eq. (2) when calculating. To illustrate this easily, two transformations were introduced. In the following discussions, by $\mathcal{M}, \mathcal{V}$ we denote matrix space and vector space respectively.

Transformation $\mathcal{D} : \mathscr{M} \to \mathscr{M}$ is defined as follows to distill the self-loops of seeding postings or leaf postings according to the posting matrix $\mathbf{M}$. For $\mathbf{A} \in \mathscr{M}$,

$$\mathcal{D}_{ij}(\mathbf{A}) = \begin{cases} 0 & \text{if } i \neq j \\ 0 & \text{if } i = j \text{ and } \sum_{j=1}^{m} \mathbf{A}_{ij} \neq 0 \\ 1 & \text{if } i = j \text{ and } \sum_{j=1}^{m} \mathbf{A}_{ij} = 0. \end{cases}$$

According to the definitions and Eq. (1), $\mathcal{D}(\mathbf{M})$ is the adjacent matrix of root self-loops and the $\mathcal{D}(\mathbf{M}^T)$ the adjacent matrix of leaf self-loops. Let square matrix

$$\mathbf{T} = \mathbf{M} + \mathcal{D}(\mathbf{M}). \tag{3}$$

$\mathbf{T}$ represents the new graph of G plus self-loops of root nodes. Since there is only one parent or self-loop for each node, each row of $\mathbf{T}$ contains and only contains one nonzero element.

Let l_{max} be the max level in G, i.e., $l_{max} = \max(l_{Tk})$. The whole thread Tk was included in the columns of seed postings of square matrix $\mathbf{T}^{l_{max}}$, and $\mathbf{T}^{l_{max}}$ could be used to indicate the correspondent relationships of postings and their seed postings.

Transformation $\mathcal{N} : \mathscr{M} \to \mathscr{M}$ is defined as the normalization of the matrix row vectors based on their l_1 norms. That is, for $\mathbf{A} \in \mathscr{M}$,

$$\mathcal{N}_{ij}(\mathbf{A}) = \mathbf{A}_{ij} / \sum_{j=1}^{n} \mathbf{A}_{ij}.$$

Therefore, the matrix form of PostRank can be represented with

$$\boldsymbol{r}^T = \boldsymbol{r}^T(\alpha \mathbf{T} + \beta \mathcal{N}(\mathbf{M}^T + \mathcal{D}(\mathbf{M}^T)) + \lambda \mathcal{N}(\mathbf{T}^{l_{max}})) + \eta \boldsymbol{w}^T, \tag{4}$$

where $\boldsymbol{w}^T$ is a m dimensional personalized row vector with $\boldsymbol{w} > 0$. $\boldsymbol{w}$ could be used to customize the PostRank vector for special demand. In Eq. (2), $\boldsymbol{w}^T = \boldsymbol{e}^T$, where $\boldsymbol{e}$ be a m dimensional column vector with all ones.

For the implement of PostRank calculation, Eq. (4) can be written in the form of iteration as

$$\boldsymbol{r}^T(k) = \boldsymbol{r}^T(k-1)(\alpha \mathbf{T} + \beta \mathcal{N}(\mathbf{M}^T + \mathcal{D}(\mathbf{M}^T)) + \lambda \mathcal{N}(\mathbf{T}^{l_{max}})) + \eta \boldsymbol{w}^T. \tag{5}$$

For the implementation of Eq. (5), its convergent property should be considered.

Property 2. *Let* $\mathbf{P} = \alpha \mathbf{T} + \beta \mathcal{N}(\mathbf{M}^T + \mathcal{D}(\mathbf{M}^T)) + \lambda \mathcal{N}(\mathbf{T}^{l_{max}})$. $\alpha + \beta + \lambda$ *is the spectrum radius of* $\mathbf{P}$.

Proof. Based on the above descriptions and Property 1,

$$\sum_{j=1}^{m} \mathbf{P}_{ij} = \alpha \sum_{j=1}^{m} \mathbf{T}_{ij} + \beta \sum_{j=1}^{m} \mathcal{N}(\mathbf{M}_{ij}^T + \mathcal{D}(\mathbf{M}_{ij}^T)) + \lambda \sum_{j=1}^{m} \mathcal{N}(\mathbf{T}_{ij}^{l_{max}})$$

$$= \alpha + \beta + \lambda, \tag{6}$$

so we have $\mathbf{P}e = (\alpha + \beta + \lambda)e$. Therefore $\alpha + \beta + \lambda$ is the eigenvalue of $\mathbf{P}$ and e is the corresponding right eigenvector.

According to the matrix property, the spectrum radius of $\mathbf{M}$

$$\rho(\mathbf{P}) \leq \|\mathbf{P}\|_\infty = \max_{i,j} \mathbf{P}_{ij}$$
$$\leq \alpha \max_{i,j}(\mathbf{T}_{ij}) + \beta \max_{i,j} \mathcal{N}(\mathbf{M}_{ij}^T + \mathcal{D}(\mathbf{M}_{ij}^T)) + \lambda \max_{i,j} \mathcal{N}(\mathbf{T}_{ij}^{l_{max}}) \quad (7)$$
$$= \alpha + \beta + \lambda.$$

Considering Eq. (6) and Eq. (7), $\rho(\mathbf{P}) = \alpha + \beta + \lambda$.

Starting from non-zero initial vectors $r^T(0)$, $r^T(k)$ can be calculated based on $r^T(k-1)$ using Eq. (5). Because $\rho(\mathbf{P}) = \alpha + \beta + \lambda < 1$ from Property 2, PostRank calculation can converge to their stable vector r^T, which is the solution satisfying Eq. (4).

The meaning of PostRank can also be understood with the discrete Markov model. Defining square matrix

$$\mathbf{Q} = \mathbf{P} + \eta e w^T, \quad (8)$$

when $\|r^T(0)\|_1 = 1$, Eq. (5) can be written as

$$r^T(k) = r^T(k-1)\mathbf{Q}. \quad (9)$$

When $\|w^T\|_1 = 1$, according to Property 2 we have

$$\mathbf{Q}e = \mathbf{P}e + \eta e w^T e = (\alpha + \beta + \lambda)e + \eta e = e. \quad (10)$$

Thus, $\mathbf{Q}$ is a stochastic matrix and the PostRank calculation build a Markov chain with transition probability matrix $\mathbf{Q}$. Since $\eta > 0$, from Eq. (8) $\mathbf{Q}$ is primitive. Hence the Markov chain can converge to its stationary vector, that is, the PostRank vector r^T. The Markov chain indicates random walk model that as the Usenet user read one posting, he may jump to the posting it responded to, the posting it responded by, any posting in the same thread or any posting in the newsgroup with different probability on the next. When $\|w^T\|_1 = 1$, with this model PostRank vector is the stationary probability distribution of all postings. From Eq. (10), PostRank vector is also the eigenvector corresponding to eigenvalue 1 of $\mathbf{Q}$ which was constructed on the newsgroup structure using PostRank equation. Therefore, PostRank vector r^T could reflect the nature features of G, and it is the intrinsic property and good measure of postings in a newsgroup.

In $m \times m$ matrix $\mathbf{M}$, there are only $m - s$ nonzero elements, which makes $\mathbf{M}$ very sparse. $\mathbf{P}$ can be obtained based on $\mathbf{M}$ before iteration using the definition of Property 2 and $nnz(\mathbf{P}) \leq 3m$, where $nnz(\mathbf{P})$ is the number of non-zeros in $\mathbf{P}$. The process of Eq. (5) iteration is matrix-free and only $nnz(\mathbf{P})$ multiplications are needed for each step. Only the storage of one vector $r^T(k)$ is required at each iteration. Thus, this algorithm is suitable for the large size and sparsity of the posting matrix of Usenet newsgroup. Some experiments were performed to acquire the PostRank vector on realistic datasets. The experimental results achieved will be discussed in the next section.

```
From: diffuser78@gmail.com
Newsgroups: comp.lang.python
Subject: Re: OS specific command in Python
Date: 21 Jun 2006 06:34:42 -0700
Organization: http://groups.google.com
Lines: 40
Message-ID: <1150896882.746722.95200@u72g2000cwu.googlegroups.com>
References: <1150781429.090359.148560@c74g2000cwc.googlegroups.com>
    <1150783324.258644.65770@u72g2000cwu.googlegroups.com>
    <4498dcd5$0$25503$626a54ce@news.free.fr>
NNTP-Posting-Host: 66.255.187.74
Mime-Version: 1.0
Content-Type: text/plain; charset="iso-8859-1"
X-Trace: posting.google.com 1150896887 8314 127.0.0.1 (21 Jun 2006 13:34:47 GMT)
X-Complaints-To: groups-abuse@google.com
NNTP-Posting-Date: Wed, 21 Jun 2006 13:34:47 +0000 (UTC)
In-Reply-To: <4498dcd5$0$25503$626a54ce@news.free.fr>
User-Agent: G2/0.2
Xref: news.edisontel.com comp.lang.python:41439
```

Fig. 2. The header of a typical posting on Usenet

3   Experiments and Their Results

3.1   Datasets Preparation

We wrote a bot program in Perl to download the postings from the NNTP server. The bot program communicates with NNTP server using socket connection following RFC 977 specification [11] and save the postings in text file with Mailbox format. Since only the headers are needed in our calculation, the headers were separated from the postings from the Mailbox file, and they are stored using CSV format after some text treatment. The header of a typical posting is shown in Fig. 2. The contents in CSV file were ordered and imported to the database. SQL statements were performed on the database by a Java program through JDBC interface to construct and extract the structure of newsgroup based on the header information. Postings in uncompleted thread were removed from the data sets during the structure extraction. The process of data sets collection and newsgroup structure extraction was shown in the diagram of Fig. 3.

Experiments were preformed on two data sets collected from comp.lang.perl. misc and comp.lang.python, which are two active newsgroups about computer languages on Usenet. The datasets are called DS1 and DS2 in the following.

DS1 contains 10532 postings including 1286 participators and 1774 threads of comp.lang.perl.misc from Mar 5, 2006 to Jun 27, 2006. DS2 contains 18821 postings including 2463 participators and 3408 threads of comp.lang.python from Mar 5, 2006 to Jun 27, 2006.

The probability distributions of response posting number b are shown in a log-log plot of Fig. 4. In DS1 and DS2, a few postings got many response postings and a lot of postings were only responded by few response posting or not responded. From the figure, the distributions exhibit power-law feature of $P(b) = b^{-\gamma}$ with $\gamma \simeq 4.1$ for both DS1 and DS2. Power-law distribution has also discovered on Internet and other systems[12], and it means the heterogeneity of network structure which is helpful for our PostRank.

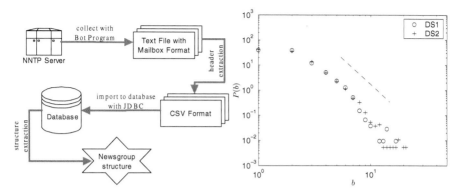

Fig. 3. The process of data collection and structure extraction of Usenet newsgroup

Fig. 4. The probability distribution of response posting number of postings in DS1 and DS2, the slope of dashed line is -4.1

3.2 Experimental Results

In our experiments, PostRank vector r^T was calculated with $\alpha = 0.25, \beta = 0.45, \lambda = 0.15, \eta = 0.15$. These parameters are determined from our experiments, and they can be adjusted to change the impacts of different kinds of postings. The personalized vector w^T was assigned e^T, so according to Eq. (4) the l_1 norm of PostRank vector $\|r^T\|_1 = \|w^T\|_1 = \|e^t\|_1 = m$.

We measure the rates of convergence using the l_1 norm of the residual vector, i. e.,

$$\Delta^{(k)} = \|r^T(k) - r^T(k-1)\|_1.$$

The convergence rates of in our experiments of DS1 and DS2 were plotted on a semi-log graph shown in Fig. 5. It could converge rapidly, which follows $O((\alpha + \beta + \lambda)^k)$.

Since most PostRank scores are small, the logarithms of PostRank of DS1 and DS2 are shown in the histograms of Fig. 6. In these histograms, there are few postings with high PostRank, and many PostRank scores are around the average value 1. Comparing with Fig. 4, we can see that they are very unalike. Effected by the number and PostRank scores of different kinds of related postings simultaneously, r_i is quite different with b_i for posting v_i. The relationship r_i and b_i is shown in Fig. 7, where each symbol represents a posting and the cycle symbols and plus symbols denote postings from DS1 and DS2 respectively. In this Figure, we could see that postings with same b_i may be very different in r_i, and vice versa. In DS1, the highest PostRank 37.145 is obtained by a seed posting titled "What is Expressiveness in a Computer Language", which was responded by only 4 postings with high PostRank scores. The thread it invoked contains 467 response postings, but this seeding posting is not very significant barely considering b_i. According to this figure, we get results alike for DS2.

In our example above, some seed postings with few b_i rank high mainly because they have a lot of descendants. The number of direct and indirect

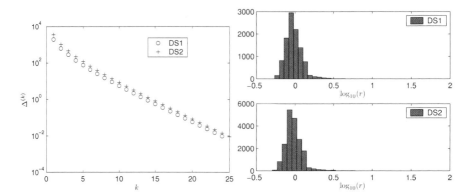

Fig. 5. The convergence rates of PostRank calculation of DS1 and DS2

Fig. 6. The PostRank histograms of postings in DS1 and DS2

descendants can be used as the candidate for the rank, which could also makes the seed postings rank high. However, this will lead to unreasonable results that parents always rank higher than their descendants. In PostRank, there is the opportunity for the descendants rank higher than their parents, such as when the offspring of descendants have high ranks. The influence of parents, descendants and other postings were considered simultaneously according to their distances and relationships with the posting being ranked, so PostRank is found to be a good ranking method for the postings in newsgroup.

As we discussed, PostRank can provide useful clues based on the newsgroup structure for the users to help them access the information more effectively. It can also be used in other applications of Usenet data mining. For example, we can obtain some helpful properties of the participators based on PostRank. On the Usenet, participators are judged only by his postings, irrespective of his social status or appearance. They are the soul of a newsgroup. The participators behave very differently owing to their character and knowledge background. Acquaintance and evaluation of participators in a newsgroup are very important for the users to use the newsgroup effectively. However, it may take quite much time, so some hints of participators may be of great help for this.

In the newsgroup, suppose there are n participators represented as $p_u, u = 1, 2, ..., n$. All postings posted by p_u is $\mathcal{T}(p_u)$, and the number of postings in $\mathcal{T}(p_u)$ is d_u. Define f_u^{sum} as the sum of PostRank of postings in $\mathcal{T}(p_u)$ and f_u^{ave} as the average PostRank of $\mathcal{T}(p_u)$, i.e.,

$$f_u^{sum} = \sum_{v_i \in \mathcal{T}(p_u)} r_i, f_u^{ave} = f_u^{sum}/d_u.$$

The relationship of f_u^{ave} and f_u^{sum} was shown in the log-log plot of Fig. 8 where each symbol represents a participator. The results from DS1 and DS2 are plotted in subgraphs respectively. In this figure, for the participators with high f_u^{sum} the average values of f_u^{ave} are about 1. Many participators with high f_u^{ave}

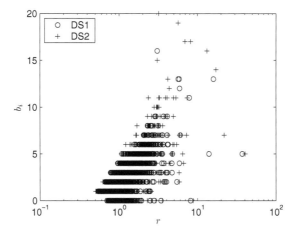

Fig. 7. Comparison of number of response posting and PostRank for postings in DS1 and DS2

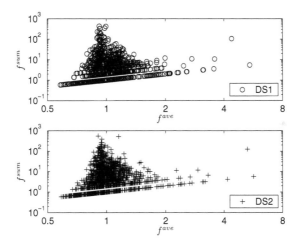

Fig. 8. The relationship of f^{sum} and f^{ave} of participators in DS1 and DS2

have small f_u^{sum}, which means they posted only a few postings. On the upper right of DS1 plot in Fig. 8, there is one participator whose f_u^{sum} and f_u^{ave} are both high, which make him relatively special. He is xah@xahlee.org, who own a homepage http://xahlee.org on computer and scientific art which was created in 1996 and visited by about 7 to 10 thousand unique visitors per day. Among 24 postings he posted in DS1, 4 postings get PostRank more than 10. Barely considering d_u, he is easy to be neglected. From Fig. 8, similar results can also be obtained for DS2.

4 Conclusions

In this paper, we proposed a method to calculate the PostRank vector of postings in a newsgroup based on the newsgroup structure. From the analysis, we can see that our method can converge rapidly. Its results correspond to the eigenvectors of the transition probability matrix and the stationary vectors of the Markov chains.

The calculation of PostRank is link-based and content independent, and it can be computed offline using only the posting headers. Therefore, it can be implemented on the servers of newsgroup services or on the newsgroup client softwares. It could provide useful intrinsic attribution for the postings and can be used in many applications including helping the organizing the search results, aiding the users in navigating newsgroup, mining the features of participators, investigating hot topics and their evolutions for some period and so on.

We provided a simple example in the experiments, and other applications of PostRank could be explored and developed. Our method provides an essential and simple way to determine the PostRank with link analysis on Usenet. Some improvements can be done to revise it or adjust the parameters according to the requirements.

On the WWW, hyperlinks indicate the choice of web page creators. It has been confirmed that link carries less noisy information than text, and the effectiveness of link analysis has been testified by some web search engines. Similar with web structure, the structure of newsgroup forms gradually along with the evolution of newsgroup. It represents the judgments and choices of participators and reflects the swarm intelligence of the newsgroup. Therefore, it could provide rich helpful information for the task of data mining on Usenet. Together with the IR methods based on text contents, link analysis can be used in the clustering, topic discovery, etc., to make full use of the rich resources on Usenet and other discussion systems.

References

1. Baeza-Yates, R.A., Ribeiro-Neto, B.A.: Modern Information Retrieval. ACM Press/ Addison-Wesley (1999)
2. Brin, S., Page, L., Motwanl, R., Winogard, T.: The pagerank citation ranking: Bring order to the web. Technical report, Stanford University, (1999), Available at http://dbpubs.stanford.edu:8090/pub/1999-66
3. Kleinberg, J.: Authoritative sources in a hyperlinked environment. Journal of the ACM 46(5), 604–632 (1999)
4. Domingos, P., Richardson, M.: Mining the network value of customers. In: Proc. of The Seventh ACM SIGKDD International Conference on Knowledge Discovery and Data Mining, New York, ACM Press, New York (2001)
5. Schwartz, M.F., Wood, D.C.M.: Discovering shared interests using graph analysis. Communications of the ACM 36(8), 78–89 (1993)
6. Tuulos, V.H., Tirri, H.: Combining topic models and social networks for chat data mining. In: Proc. of 2004 IEEE/WIC/ACM International Conference on Web Intelligence, pp. 206–213 (2004)

7. Wasserman, S., Faust, K.: Social Network Analysis: Methods and Applications. Cambridge University Press, Cambridge (1994)

8. Xi, W., Lind, J., Brill, E.: Learning effective ranking functions for newsgroup search. In: Proc. of the 27st Annual International ACM SIGIR Conference on Research and Development in Information Retrieval, New York, ACM Press, New York (2004)

9. Borgs, C., Chayes, J.T., Mahdian, M., Saberi, A.: Exploring the community structure of newsgroups. In: Proc. of the tenth ACM SIGKDD international conference on Knowledge discovery and data mining, pp. 783–787 (2004)

10. Agrawal, S.D., Rajagopaian, S., Srikant, R., Xu, Y.: Mining newsgroups using networks arising from social behavior. In: Proc. of the Twelfth International World Wide Web Conference, New York, ACM Press, New York (2003)

11. w3.org: Network news transfer protocol (Internet(WWW)), `http://www.w3.org/Protocols/rfc977/rfc977`

12. Albert, R., Barabási, A.L.: Statistical mechanics of complex networks. Reviews of Modern Physics 74, 47–97 (2002)

13. cpan.org: Comprehensive perl archive network (Internet(WWW)), `http://www.cpan.org`

14. Langville, A.N., Meyer, C.D.: Deeper inside pagerank. Internet Mathmatics 1, 335–380 (2003)

15. Tsaparas, P.: Link Analysis Ranking. PhD thesis, University of Toronto (2004)

16. Ng, A.Y., Zheng, A.X., Jordan, M.I.: Link analysis, eigenvectors, and stability. In: Proc. of International Joint Conference on Artificial Intelligence (2001)

A Comparative Study of Unsupervised Machine Learning and Data Mining Techniques for Intrusion Detection

Reza Sadoddin and Ali A. Ghorbani

Network Security Laboratory
University of New Brunswick
Fredericton, New Brunswick, Canada
{reza.sadoddin, ghorbani}@unb.ca

Abstract. During the past number of years, machine learning and data mining techniques have received considerable attention among the intrusion detection researchers to address the weaknesses of knowledgebase detection techniques. This has led to the application of various supervised and unsupervised techniques for the purpose of intrusion detection. In this paper, we conduct a set of experiments to analyze the performance of unsupervised techniques considering their main design choices. These include the heuristics proposed for distinguishing abnormal data from normal data and the distribution of dataset used for training. We evaluate the performance of the techniques with various distributions of training and test datasets, which are constructed from KDD99 dataset, a widely accepted resource for IDS evaluations. This comparative study is not only a blind comparison between unsupervised techniques, but also gives some guidelines to researchers and practitioners on applying these techniques to the area of intrusion detection.

1 Introduction

The significant increase of our everyday life dependency to Internet-based services has intensified the survivability of networks. On the other hand, the number of attacks on networks has dramatically increased during the recent years. Consequently, interest in network intrusion detection systems has increased among the researchers. At the core of an intrusion detection system relies the technique that is used for detecting intrusions. Two main approaches to intrusion detection have been proposed during the last decades. When an intrusion detection system learns about the normal behavior of the system or the network it monitors, it is categorized as an anomaly-based IDS. An anomaly is reported when the monitored behavior deviates significantly from the normal profile. A misuse detection approach, on the other hand, uses information about the known attacks and detects intrusions based on matches with existing signatures.

Traditional approaches to intrusion detection are based on expert knowledge for specification of what is normal or what is attack. This imposes a serious

P. Perner (Ed.): MLDM 2007, LNAI 4571, pp. 404–418, 2007.

limitations on intrusion detection systems considering the amount of knowledge that should be hard-coded in the system. The problem is exaggerated when the current knowledge of the system gets out of date. While the current attack signature should be updated in response to rapid development of new attacks, the specification of normal behavior of the system should cope with rapid changes of a dynamic environment. In recent years, considerable attention has been given to data mining and machine learning techniques to mitigate the problems of traditional intrusion techniques. Theses techniques have the capability of learning and discovering as apposed to hard-coding the specification of malicious or normal behaviors.

The machine learning and data mining techniques can be used in either supervised or unsupervised modes. However, providing the system with the labeled data is both time consuming and labor intensive. Unsupervised techniques are preferred over supervised techniques in this respect, but it doesn't come without cost. The algorithm should build the model without labels. In addition, the algorithm should be flexible to the distribution of the target dataset (in terms of relative population of normal and abnormal data) as not much is known about the target dataset in advance.

In this paper, we analyze the unsupervised techniques in terms of the main design choices one should make for using these techniques. We conduct experiments to compare the performance of different techniques with different labeling heuristics and also in direct and indirect application to test dataset. Furthermore, we analyze the sensitivity of the techniques to the distributions of training and test datasets.

The main goal of this study is to compare the performance of the unsupervised techniques based on common practices proposed for the applications of these technique to intrusion detection. To that end, we compare the performance of clustering techniques with two basic labeling heuristics and evaluate their performance in direct and indirect application to target datasets. Furthermore, the performance of unsupervised techniques in detecting different attack categories are compared with each other. We use different distributions of training and test datasets to evaluate the sensitivity of the techniques to these design choices. To the best of our knowledge, this work is the most comprehensive study in terms of both techniques that are considered and different types of experiments that are conducted.

The rest of this paper is organized as follows. We review the related works in Section 2. In Section 3, a brief overview of the techniques considered in this study is provided. In Section 4, we describe the steps that are taken to do the experiments and provide the results. The lessons learned in this study, concluding remarks and the future work are given in Section 5.

2 Related Work

Traditional works on anomaly detection [9, 25] are based on statistical methods for specifying the normal profile of the system. However, more and more machine learning and data mining techniques are proposed in the literature as the traditional methods cannot cope with the complexity of the problem. *Clustering* and *outlier detection* techniques are the main approaches to this problem.

A clustering method is used in direct or indirect mode for detecting intrusions. The indirect mode of clustering is a two-step process. In the training step, a set of clusters is created using the training dataset and clusters are labeled as normal or abnormal with a labeling heuristic. In the test step, each record of the test dataset is compared against the centroids of all clusters and will be assigned to the nearest one. In the direct mode, the clustering algorithm is applied directly to the target dataset and data are labeled using a labeling heuristic.

Portnoy et al. [19] presents a fix-width clustering technique in which sparse clusters are considered as anomalous based on a given threshold on the density of clusters. An extended version of the fix-width clustering is proposed in [6] by Chan et al. with the ability to estimate the width of clusters. The algorithm is applied directly to the target dataset and a labeling heuristic is proposed taking into account both density and average distance of a cluster from other clusters.

Guan et al. [12] propose Y-means, a new clustering technique based on K-means, which addresses the dynamic selection of the number of clusters. Ramadas et al. [20] use SOM (Self-organizing Map), a competitive learning technique, to detect anomalies using a two-step clustering technique. A new competitive learning technique is proposed in [17] by Lei et al., which mitigates the sensitivity of SOM to initialization of cluster centers.

Outlier detection techniques rely on the assumption that abnormal data can be detected based on their deviation from some common characteristic of normal data (e.g, belonging to one distribution, closeness, etc.). Three main approaches to outlier detection have been proposed in the literature. In *distribution-based* outlier detection techniques, outliers are detected by their significant deviation from the standard distribution presumably resembled by the majority of points in the target dataset. In *distance-based* outlier detection methods, outliers are detected based on their distances from their nearest neighbors. Finally, the main idea behind density-based techniques [4, 14] is to detect outliers with respect to local density of their neighborhood.

The most related works to ours are reported in [11], [16] and [26]. Eskin et al. [11] compare three outlier detection schemes including a fix-width clustering technique, K-nearest neighbor and unsupervised SVM on KDD99 dataset. Lazarevic et al. [16] report the comparative results of several distance-based (NN, KNN and Mahalanobis distance) and density-based (Local Outlier Function) outlier detection schemes as well as unsupervised SVM. In this work, the authors construct a dataset from DARPA98 dataset for training and test. which contains their own proposed features. Zhong et al. use KDD dataset to compare the performance of some clustering techniques (including k-means, Mixture-of-Spherical Gaussian, Self-organizing Map and Neural-Gas) using their proposed labeling heuristic.

3   Unsupervised Anomaly Detection Techniques

3.1   Clustering Techniques

The clustering techniques we have studied in our experiments include K-means, C-means, EM, Self-organizing Map (SOM), Y-means and Improved Competitive Learning Network (ICLN). K-means [18] is a very popular technique due to its simplicity and relatively fast convergence. Given a certain number of clusters (k), the algorithm starts with selecting k points from the input data as the initial centroids of the clusters. The main body of the algorithm is a convergence loop, in which the algorithm goes through two consecutive steps; *Assignment* updates the membership of each point based on its distance to the nearest cluster and *Relocation* deals with updating the centroid of the cluster so as to render the mean of belonged points.

Fuzzy C-means is a method of clustering, which allows one piece of data to belong to two or more clusters. Developed by Dunn [10] and improved later by Bezdek [3], it is used in applications for which the hard classification of data is not meaningful or difficult to achieve (e.g, pattern recognition). C-means algorithm is similar to K-Means except that membership of each point is defined based on a fuzzy function and all the points contribute to relocation of a cluster centroid based on their fuzzy membership to that cluster.

EM is another soft clustering method based on Expectation-Maximization meta algorithm [8]. Expectation-Maximization is an algorithm for finding maximum likelihood estimates of parameters in probabilistic models. EM clustering algorithm alternates between performing expectation (E) step, by computing an estimation of likelihood using current model parameters (as if they are known), and a maximization (M) step, by computing the maximum likelihood estimates of model parameters. The new estimations of model parameters contribute to expectation step of the next iteration.

SOM, proposed by Kohonen in [15], is a competitive learning technique that maps the input data from a possibly high-dimensional space to a low dimensional one (2 or 3-D) while preserving the topological properties of the input space. A SOM consists of *neurons* organized on a regular low-dimensional grid. Each neuron i is represented by a d-dimensional weight vector $W_i = [w_{i1}, ..., w_{id}]$. The neurons are connected to adjacent neurons by a neighborhood relation, which dictates the topology of the map.

SOM is trained iteratively. In each step, one sample from input vector is compared against all the neurons. The neuron whose distance from the input vector is minimum is selected as the *winner neuron* or the *Best Matching Unit* (BMU). After finding the BMU, the weight vector of the winner neuron and all of its neighbors are updated so as to resemble the current input vector. The SOM update rule for the weight vector of unit v is:

$$w_v(t+1) = w_v(t) + \eta(t)h_{cv}(t)(x_i(t) - w_v(t)),$$

where t denotes time, $x(t)$ is an input vector randomly selected from the dataset at time t, $h_{cv}(t)$ is the neighborhood kernel around the winner neuron c and $\alpha(t)$ is the learning rate at time t.

ICLN [17] is based on the Standard Competitive Learning Network (SCLN). SCLN is basically a single-layer neural network in which each output neuron is fully connected to the input node. In the SCLN, the output neurons compete to become active, very similar to the way the neurons of the map compete in SOM. However, the performance of SCLN is heavily dependent on the number of the output neurons and their initial weight vectors. A critical shortage of SCLN (and that of SOM) is that it may split one cluster into many small clusters (if several neurons are initialized in one cluster). ICLN improves the mentioned shortage by defining two different update rules for the winner neuron and loser ones. The winning neuron, updates its weight vector using the following formula:

$$w_j(t+1) = w_j(t) + \eta_1(t)(x - \omega_j(t)),$$

At the same time, other neurons update their weight vectors as follows:

$$w_j(t+1) = w_j(t) - \eta_2 K(d(x_i, w_j))(x_i(t) - w_j(t)),$$

where η_1 and η_2 are learning rates for winner and loser neurons, respectively, and $K(d(x, j))$ is a kernel function in which $d(x_i, w_j)$ is the distance between neuron j and input vector x. The new update rule for the losing neurons moves their weight vectors away from the input pattern. At the end, all the neurons with no associated input vectors are removed.

Proposed by Guan et al. in [12], Y-means is a dynamic clustering algorithm, which improves the K-means clustering algorithm in three aspects: 1) dependency on the initial number of clusters; 2) dependency on centroids' initialization and 3) degeneracy (i.e; ending up with some empty clusters). The algorithm starts by running *StabilizeWithKMeans* (an improved version of K-means which deals with empty clusters) on an arbitrary number of initial clusters. Then clusters are refined through two further steps. The *split* phase actually alternates between finding outliers of the clusters and running *StabilizeWithKMeans* on the new set of cluster centroids. To detect outliers, a *confident area* is defined around the centroid of each cluster whose radius is 2.32σ, where σ is the standard deviation of the points in that cluster. This criteria has its root from the "Cumulative Standardized Normal Distribution" table [5], meaning that 99% of the objects will lie within 2.32 standard deviation of the mean. At each step, the farthest outlier is excluded from the cluster and is considered as the centroid of a new cluster and possibly absorbs some points from the neighbor clusters in the following *StabilizeWithKMeans* step.

In the *merge* phase, two adjacent clusters with significant overlap are merged. The merge is performed to mitigate the problem of over-splitting which might have happened due to split. Two clusters x and y are merged into one cluster if the distance between their centroids is less than $1.44(\sigma_x + \sigma_y)$.

3.2   One-Class SVM

Unlike supervised SVM, the unsupervised SVM [22] is trained with an unlabeled dataset that contains both positive and negative data. However, the assumption is that the majority of training data are normal. Like supervised SVM, a kernel function is used for mapping the input data to a feature space. Using a constant kernel function (i.e, kernel functions that only depend on the distance between input vectors), all the input examples will be mapped onto the surface of a hypersphere centered around the origin in the feature space. The objective is to maximize the margin of separation from the origin or equivalently, find the smallest sphere enclosing the data. There is a tradeoff between maximizing the margin of separation and enclosing as much data points as possible. This trade-off is achieved by solving the following optimization problem that penalizes any points not separated from the origin while simultaneously trying to maximize the distance of the hyperplane from the origin:

$$min_{\omega,b,\xi,\rho} \frac{1}{2} < \omega, \omega > -\rho + \frac{1}{vl} \sum_{i=1}^{l} \xi_i,$$

subject to

$$< \omega, \phi(x) > \geq \rho - \xi_i, \quad \xi_i \geq 0,$$

where $< .,. >$ stands for a dot product kernel, ω is the normal vector to the separating hyperplane, l is the number of the training samples, ρ is the offset and $v \in (0,1)$ is a upper bound on the fraction of outliers. ξ_is are slack variables associated with each data example. The farther a data example is from the separating hyperplane on the wrong side, the greater is its slack variable. The slack variables penalize the objective function, but allow some of the training examples to be on the wrong side of the hyperplane. After solving the above optimization problem, the label for each point x would be determined using the following decision function:

$$f(x) = sgn((w.\phi(x)) - \rho)$$

3.3   K-Nearest-Neighbor

KNN is another method for estimating the degree of outlier of a point. *KNN-based indexes* assign an outlier degree to each point based on a function that takes into account the distance of the point with respect to its K nearest neighbors. We have selected *Kappa* and *Gamma* among the outlier indexes used in [13].

Kappa (κ) returns the distance of a point from its *k*th nearest neighbor,

$$\kappa(x) = \|x - z_k(x)\|,$$

where $z_i(x)$ is the *i*th nearest neighbor of point x. *Gamma* (γ) computes outlier degree of a point taking into account the distances of all of its k nearest neighbors as follows:

$$\gamma(x) = \frac{1}{k} \sum_{j=1}^{k} \|x - z_j(x)\|$$

4   Experiments

In this section, the steps that are taken for preparing the datasets and the results of different experiments are provided.

4.1   Experiment Setup

MIT Lincoln Laboratory conducted a project for evaluating network intrusion detection systems in 1998 and 1999. In this project, a simulation was performed on a military network in which victim machines were target of various attacks and normal traffic. Four categories of attacks were used in this simulation: 1) *Probe* in which a network of computers is scanned for information gathering; 2) *Denial of Service (DOS)* in which network services become unavailable due to excessive consumption of resources by the attacker; 3) *User to Root (U2R)* attacks, which result in super user privilege from a normal user privilege; and, 4) *Remote to User (R2L)* attacks, which result in a local account on a remote host.

The KDD CUP'99 data set was built based on the data captured in DARPA98 IDS evaluation program. Data packets that form a complete session were encoded in a single connection vector which contains 41 features. Further details on KDD99 features along with their descriptions can be found in [1]. Among the features of KDD, *protocol type*, *service type* and *status flag* are categorical features and the rest 38 features are continuous features. The categorical features are converted to continuous feature by using the frequency of different values appeared for these features (as proposed in [6]). Following that, all the features were normalized in the scale of 0 and 1 based on the formula proposed in [6].

To analyze the sensitivity of unsupervised techniques on distribution of the training and test dataset, we prepared datasets with different relative population of normal and attack records. The training and test data are selected randomly from the original 10% training and test datasets that are publicly available from UCI repository [1] (we have made sure that all 4 attack categories appear in the prepared datasets). We prepared five different datasets (*Train_8020*, *Train_8416, Train_8812, Train_9208, Train_9604*) for training and two datasets (*Test_8020* and *Test_9604*) for test (e.g. *Train_8020* is a dataset in which the percentage of normal and attack records is 80% and 20%, respectively).

We used Fuzzy Clustering and Data Analysis Toolbox [2], SOM Toolbox [23] and LIBSVM library [7] to carry out the experiments with C-Means, SOM and One-Class SVM, respectively. WEKA implementations of K-means and EM were used in our experiments [24] and KNN-indices were developed in a simple program in C. Source codes of Y-means and ICLN were provided by their respective authors.

All the clustering techniques were initialized with 50 clusters. For C-means, the weighting exponent was set to 2.0. We used a Hexagonal grid of size 5*10 with a *Gaussian* neighborhood function for SOM and set its initial learning rate to 0.5. For ICLN, the values of *reward* rate, *punish* rate and *minimum update* parameters were set to 0.1, 0.01 and 0.01, respectively. The *stability* threshold of Y-means was set to 0.01. For Unsupervised SVM, we used *RBF* as the kernel function and initialized its *gamma* parameter to 0.025 ($1/number of features$).

4.2   Experimental Results

The performance of intrusion detection techniques is evaluated based on two well-known criteria: *detection rate* and *false positive rate*. The detection rate

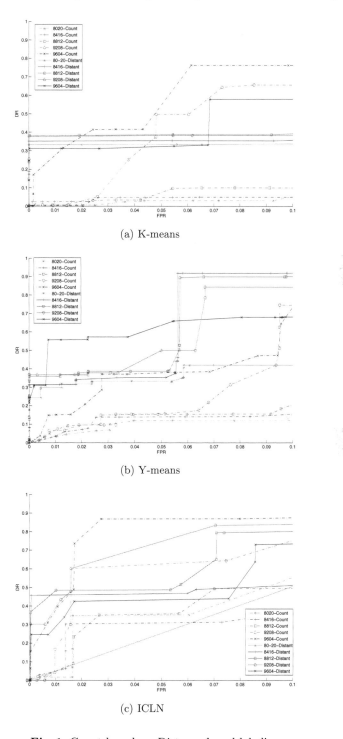

(a) K-means

(b) Y-means

(c) ICLN

Fig. 1. Count-based vs. Distance-based labeling

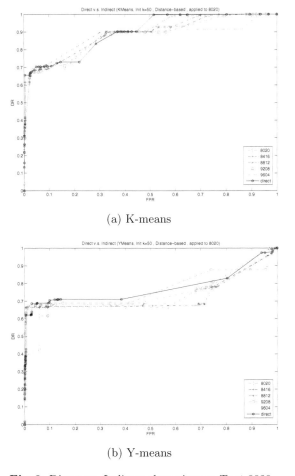

(a) K-means

(b) Y-means

Fig. 2. Direct vs. Indirect clustering on $Test\_8020$

represents the percentage of correctly detected attacks whereas, the false positive rate is the percentage of normal records detected incorrectly as attack.

In the first set of experiments, we evaluate the performance of each clustering technique with two labeling heuristics. The *count-based* heuristic labels clusters as normal or anomalous based on their populations, while in distance-based heuristic, clusters are labeled based on their distance from the center of other clusters. We use the *Inter cluster distance* (*ICD*) measurement proposed in [6]:

$$ICD_i = \frac{1}{|C|-1} \sum_{j, i \neq j} dist(c_i, c_j)$$

The results of the experiments are shown as ROC curves by changing the count or the distance thresholds in count-based and distance-based heuristics, respectively (The varying parameter is 'nu' for USVM). Figure 1 shows the results of comparison between count-based and distance-based heuristics for K-means, Y-means and ICLN when applied directly to training dataset. The results show

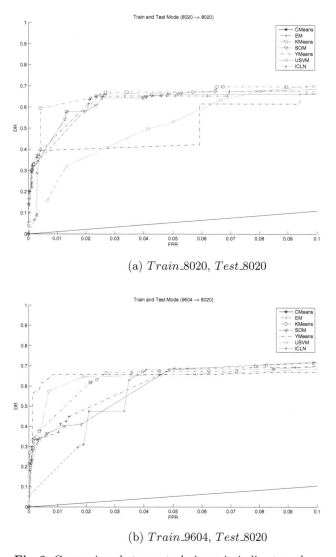

(a) $Train\_8020$, $Test\_8020$

(b) $Train\_9604$, $Test\_8020$

Fig. 3. Comparison between techniques in indirect mode

that distance-based heuristic delivers a better average performance. For most of the techniques (including EM, SOM and C-means), the distance-based heuristic is a clear dominant over count-based heuristic except possibly on datasets whose attack population is negligible (4%).

In the second set of experiments, the performance of each clustering technique is evaluated in direct versus indirect mode. Figure 2 shows the comparative results for K-means and Y-means. In indirect mode, the clusters were trained using different prepared training datasets. The initial number of clusters was set to 50 for all clustering techniques. The results show that direct application of clustering techniques perform on the average as good as trained clusters. Due to

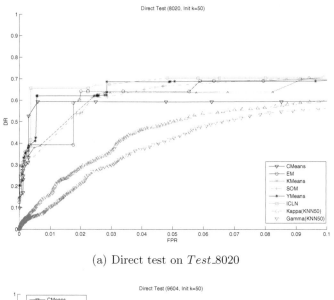

(a) Direct test on $Test\_8020$

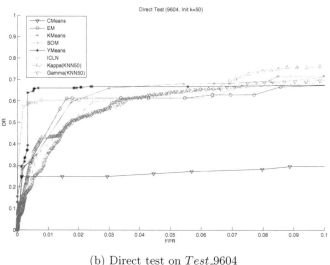

(b) Direct test on $Test\_9604$

Fig. 4. Comparison between techniques in direct mode

space limitations, only the results of two clustering techniques on $Test8020$ are shown here, however, similar results were seen for other clustering techniques.

The comparative results of unsupervised techniques in indirect mode are shown in Figure 3. The experiments include unsupervised SVM and other clustering techniques. The models are trained using two instances of training datasets and tested against two instances of test datasets. Either of the distance-based or count-based heuristic that is superior for a technique are used in the comparison. By comparing the curves, it is seen that all the techniques perform better when trained with $Train\_8020$ dataset except USVM and Y-means, which

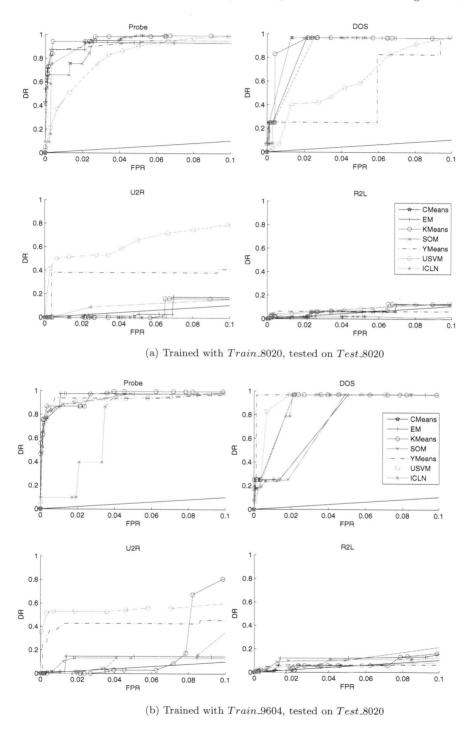

(a) Trained with $Train\_8020$, tested on $Test\_8020$

(b) Trained with $Train\_9604$, tested on $Test\_8020$

Fig. 5. Comparison on attack category detection

perform significantly better when trained with $Train\_9604$ dataset. For USVM, this behavior is expected since USVM is naturally an outlier detection technique. Obtaining similar results with Y-means can be attributed to the fact that in Y-means clusters are refined based on outlier detection schemes. As a result, it should deliver its best performance when abnormal data are rare in the training dataset. All in all, K-means delivers the best performance in this experiment. The results of experiments on $Test\_9604$ is quite similar to those of $Test\_8020$ though only the latter are shown here.

The clustering techniques as well as the KNN-based outlier detection techniques are compared with each other by direct application on test datasets. The results of these experiments are shown in Figure 4. By comparing these two figures, it is easy to see that performance of KNN-based outlier detection techniques (i.e, κ and γ) decreases as the population of attacks increases in the target dataset while the clustering techniques tend to be more robust in this respect. ICLN and Y-means delivers the best performance on $Test\_8020$ dataset while Y-means and K-means are superior on $Test\_9604$ dataset. Moreover, κ performs slightly better than γ.

Figure 5 compares the performance of the clustering techniques as well as USVM in detecting different attack categories. The models are trained with $Train\_8020$ and $Train\_9604$, respectively and tested against $Test\_8020$ dataset. Again, the superior labeling heuristic for each technique is selected in the comparison. The results obviously show that all the techniques perform poorly in detecting R2L attacks while most of them are good at detecting probe attacks. When trained with $Train\_8020$ dataset, K-means and SOM are dominant over other techniques in detecting DoS attacks, while Y-means and USVM deliver best performance when $Train\_9604$ dataset is used for training. Moreover, USVM and Y-means are clearly dominant in detecting U2R attacks in all experiments.

5   Concluding Remarks and Future Work

In this study, we conducted a set of blind experiments of unsupervised techniques on KDD99 dataset. Our main goal was to analyze common practices used in the literature in the application of unsupervised techniques for intrusion detection. These include the heuristics used for labeling the clusters, direct or indirect application of the techniques to target dataset, and using different distributions of datasets for training.

Our experiments show that distance-based heuristic is on the average dominant over count-based heuristic in almost all of the clustering techniques. Count-based heuristic does not deliver satisfactory results unless (possibly) for datasets in which attack population is negligible, while distance-based labeling is more robust to population of attacks in the target dataset.

Although the clustering techniques are used for intrusion detection in both direct and indirect modes, our experiments show that direct application of clustering techniques provides a comparable performance to that of trained clusters. Accordingly, it might be more efficient to apply the clustering techniques directly to the target dataset when normal data are dominant (the tolerance for attack population is at least 20% as our experiments show). Applied in indirect mode, the clustering techniques (except Y-means) perform better when trained with $Train\_8020$, while USVM and Y-means perform better when trained with $Train\_9604$. In direct mode, the performance of KNN-based outlier detection schemes decreases as the population of attack data increases in the target dataset. Again, Y-means dominates other techniques in $Test\_9604$ which supports the outlier detection nature of this technique.

Two observations can be highlighted in terms of detecting different attack categories. First, all techniques perform poorly in detecting $R2L$ attacks. Secondly, USVM and Y-means are clearly superior over other techniques in detecting $U2R$ attacks. We need further analysis to explain this behavior.

C-means delivers the worst results in almost all the experiments. It seems that fuzzy clustering is not suitable for distinguishing normal and abnormal data in intrusion detection.

The results of these experiments not only provide reliable guidelines for researchers and practitioners in applying unsupervised techniques to the area, but also reveals some new avenues of research. Further research is required on exploring more intelligent labeling heuristics for labeling the clusters. It seems that neither the count, nor the distance-based heuristic are sufficient and reliable criteria for labeling the clusters. This fact was supported in our experiments in that the overall detection rate of the clustering techniques hardly goes beyond 70%, even though the purity of clusters were satisfactory in some techniques. It is worthwhile to take into account other properties of clusters (e.g, density) and/or come up with more intelligent heuristics by combining the simple criteria.

Furthermore, the overall performance of unsupervised techniques might be enhanced by combining individual techniques. These *hybrid detectors* should be supported with intelligent voting policies that takes into account the strengths and weaknesses of each technique.

The poor performance of machine learning-based techniques in detecting $R2L$ and $U2R$ attacks has been reported in similar works [21]. This can be attributed to the fact that existing features in KDD99 dataset provide no or few information about these categories of attacks. Investigating more informative features for $R2L$ and $U2R$ attacks would be beneficial to this area.

References

[1] Available at http://kdd.ics.uci.edu//databases/kddcup99/kddcup99.html
[2] Balasko, B., Abonyi, J., Feil, B.: Fuzzy clustering and data analysis toolbox, Available at http://www.fmt.vein.hu/softcomp/fclusttoolbox
[3] Bezdek, J.C.: Pattern recognition with fuzzy objective function algorithms. Kluwer Academic Publishers, Norwell, MA, USA (1981)
[4] Breunig, M.M., Kriegel, H.P., Ng, R.T., Sander, J.: LOF: identifying density-based local outliers. SIGMOD Rec. 29(2), 93–104 (2000)
[5] Brownlee, K.: Statistical theory and methodology. John Wiley and Sons, New York (1967)
[6] Chan, P., Mahoney, M., Arshad, M.: Learning rules and clusters for anomaly detection in network traffic. Managing Cyber Threats: Issues, Approaches and Challenges, 81–99 (2003)

[7] Chang, C.-C., Lin, C.-J.: LIBSVM: a library for support vector machines (2001) Software available at http://www.csie.ntu.edu.tw/~cjlin/libsvm

[8] Dempster, A.P., Laird, N.M., Rubin, D.B.: Maximum likelihood from incomplete data via the em algorithm. J. Royal Stat. Soc. 39, 1–38 (1977)

[9] Denning, D.E.: An intrusion-detection model. IEEE Trans. Softw. Eng. 13(2), 222–232 (1987)

[10] Dunn, J.C.: A fuzzy relative of the isodata process and its use in detecting compact well-separated clusters. Journal of Cybernatics 3, 32–57 (1974)

[11] Eskin, E., Arnold, A., Prerau, M., Portnoy, L., Stolfo, S.: A geometric framework for unsupervised anomaly detection: Detecting intrusions in unlabeled data. Data Mining for Security Applications (2002)

[12] Guan, Y., Ghorbani, A., Belacel, N.: Y-Means: A clustering method for intrusion detection. In: Canadian Conference on Electrical and Computer Engineering, Montreal, Quebec, Canada (2003)

[13] Harmeling, S., Dornhege, G., Tax, D., Meinecke, F., Muller, K.: From outliers to prototypes: Ordering data. Neurocomputing 69(13-15), 1608–1618 (2006)

[14] Jin, W., Tung, A.K.H., Han, J.: Mining top-n local outliers in large databases. In: KDD '01. Proceedings of the seventh ACM SIGKDD international conference on Knowledge discovery and data mining, pp. 293–298. ACM Press, New York (2001)

[15] Kohonen, T.: Self-organizing map. Springer, Heidelberg (1997)

[16] Lazarevic, A., Ertoz, L., Kumar, V., Ozgur, A., Srivastava, J.: A comparative study of anomaly detection schemes in network intrusion detection. In: Proceedings of the Third SIAM International Conference on Data Mining (2003)

[17] Lei, J.Z., Ghorbani, A.: Network intrusion detection using an improved competitive learning neural network. In: CNSR, pp. 190–197 (2004)

[18] MacQueen, J.: Some methods for classification and analysis of multivariate observations. In: 5th Berkley Symposium on Math and Probability, pp. 281–297 (1967)

[19] Portnoy, L., Eskin, E., Stolfo, S.: Intrusion detection with unlabeled data using clustering. In: ACM Workshop on Data Mining Applied to Security (DMSA), ACM Press, New York (2001)

[20] Ramadas, M., Ostermann, S., Tjaden, B.: Detecting anomalous network traffic with self-organizing maps. In: Vigna, G., Krügel, C., Jonsson, E. (eds.) RAID 2003. LNCS, vol. 2820, Springer, Heidelberg (2003)

[21] Sabhnani, M., Serpen, G.: Application of machine learning algorithms to kdd intrusion detection dataset within misuse detection context. In: Proceedings of the International Conference on Machine Learning, Models, Technologies and Applications (MLMTA 2003), vol. 1, pp. 209–215 (2003)

[22] Scholkopf, B., Platt, J., Shawe-Taylor, J., Smola, A.J., Williamson, R.C.: Estimating the support of a high-dimensional distribution. Neural Computation 13(7), 1443–1472 (2001)

[23] Vesanto, J., Himberg, J., Alhoniemi, E., Parhankangas, J.: Som toolbox for matlab 5, Helsinki Univ. Technology (2000), Available at http://www.cis.hut.fi/projects/somtoolbox

[24] Witten, I.H., Frank, E.: Data mining: Practical machine learning tools and techniques, 2nd edn. Morgan Kaufmann, San Francisco (2005)

[25] Ye, N., Emran, S.M., Chen, Q., Vilbert, S.: Multivariate statistical analysis of audit trails for host-based intrusion detection. IEEE Trans. Comput. 51(7), 810–820 (2002)

[26] Zhong, S., Khoshgoftaar, T.M., Seliya, N.: Clustering-based network intrusion detection (2005)

Long Tail Attributes of Knowledge Worker Intranet Interactions

Peter Géczy, Noriaki Izumi, Shotaro Akaho, and Kôiti Hasida

National Institute of Advanced Industrial Science and Technology (AIST)
Tsukuba and Tokyo, Japan

Abstract. Elucidation of human browsing behavior in electronic spaces has been attracting substantial attention in academic and commercial spheres. We present a novel formal approach to human behavior analysis in web based environments. The framework has been applied to analyzing knowledge workers' browsing behavior on a large corporate Intranet. Analysis indicates that users form elemental and complex browsing patterns and achieve their browsing objectives via few subgoals. Knowledge workers know their targets and exhibit diminutive exploratory behavior. Significant long tail attributes have been observed in all analyzed features. A novel distribution that accurately models it has been introduced.

1 Introduction

"Nobody has really looked at productivity in white collar work in a scientific way." (Peter Drucker) [1]. Insufficient scientific evidence regarding knowledge worker productivity, efficiency, and their adequate measurement methods has been rising concerns in managerial circles [2]. Human dynamics [3] and behavior in electronic spaces [4], [5] have been rapidly gaining importance in a corporate sector. Corporations are eagerly exploiting ways to acquire and analyze large volumes of customer behavior data—primarily for commercial benefits [6].

Human behavior in electronic environments can be analyzed from observer studies or machine collected human–web interaction data. Observer studies are generally time consuming and resource demanding [7]. Only limited attempts have been made toward their automation [8]. Machine collected human–web interaction data has generally two forms: server-side data (web logs) and client-side data from script agents. The data is mined for user click-streams [9] and further analyzed. Various analysis methods have been applied to this end. Empirical analysis [10] provides generally rule-based results. For predictive purposes, statistical approaches have been favoring Markov models [11]. However, higher-order Markov models become exceedingly complex and computationally expensive. Less computationally intensive cluster analysis methods [12] and adaptive learning strategies [13] have scalability drawbacks. Mining only frequent patterns reduces the computational complexity and improves the speed, however, at the expense of substantial data loss [14]. A novel analysis framework and methods that effectively capture dimensions of human interactions in electronic environments are required.

P. Perner (Ed.): MLDM 2007, LNAI 4571, pp. 419–433, 2007.

2   Approach Formulation

We introduce the basic line of inquiry together with corresponding terminology. Definitions are accompanied by intuitive explanations that help us better understand the concept at higher formal level.

Click-stream sequences [15] of user page transitions are divided into sessions, and sessions are further divided into subsequences. Division of sequences into subparts is done with respect to the user activity and inactivity. Consider the conventional time-stamp click-stream sequence of the following form: $\{(p_i, t_i)\}_i$, where p_i denotes the visited page URL_i at the time t_i. For the purpose of analysis this sequence is converted into the form: $\{(p_i, d_i)\}_i$ where d_i denotes a delay between the consecutive views $p_i \rightarrow p_{i+1}$. User browsing activity $\{(p_i, d_i)\}_i$ is divided into subelements according to the periods of inactivity d_i.

Definition 1 *(Browsing Session, Subsequence, Train)*
Let $\{(p_i, d_i)\}_i$ be a sequence of pages p_i with delays d_i between consecutive transitions $p_i \rightarrow p_{i+1}$.

Browsing session is a sequence $B = \{(p_i, d_i)\}_i$ where each $d_i \leq T_B$. Length of the browsing session is $|B|$. Browsing session is often referred to simply as a **session**.

Subsequence of an individual browsing session B is a sequence $S=\{(p_i, dp_i)\}_i$ where each delay $dp_i \leq T_S$, and $\{(p_i, dp_i)\}_i \subset B$. The length of subsequence is $|S|$.

A browsing session $B = \{(S_i, ds_i)\}_i$ thus consists of a **train** of subsequences S_i separated by inactivity delays ds_i.

Sessions delineate tasks of various complexities users undertake in electronic environments. Subsequences correspond to session subgoals; e.g. subsequence S_1 is login, S_2 – document download, S_3 – search for internal resource, etc.

Important issue is determining the appropriate values of T_B and T_S that segment the user activity into sessions and subsequences. The former research [16] indicated that student browsing sessions last on average 25.5 minutes. However, we adopt the average maximum attention span of 1 hour as a value for T_B. If the user's browsing activity was followed by a period of inactivity greater than 1 hour, it is considered a single session, and the following activity comprises the next session.

Value of T_S is determined dynamically and computed as an average delay in a browsing session: $T_S = \frac{1}{N} \sum_{i=1}^{N} d_i$. If the delays between page views are short, it is useful to bound the value of T_S from below. This is preferable in environments with frame-based and/or script generated pages where numerous logs are recorded in a rapid transition. Since our situation contained both cases, we adjusted the value of T_S by bounding it from below by 30 seconds:

$$T_S = max\left(30, \frac{1}{N} \sum_{i=1}^{N} d_i\right). \tag{1}$$

Using these primitives we define navigation space and subspace as follows.

Definition 2 *(Navigation Space and Subspace)*
Navigation space is a triplet $\mathcal{G} = (\mathcal{P}, \mathcal{B}, \mathcal{S})$ where $\mathcal{P}$ is a set of points (e.g. URLs), $\mathcal{B}$ is a set of browsing sessions, and $\mathcal{S}$ is a set of subsequences.
Navigation subspace of $\mathcal{G}$ is a space A = (D,H,K) where $D \subseteq \mathcal{P}$, $H \subseteq \mathcal{B}$, and $K \subseteq \mathcal{S}$; denoted as $A \subseteq \mathcal{G}$.

Navigation space can be divided into subspaces based on the nature of detected or defined sequences. For example, human navigation space consists of human generated sequences, and machine navigation space may contain only the machine generated sequences. Different spaces may have distinctly different characteristics.

Another important aspect is to observe where the user actions are initiated and terminated. That is, to identify the starting and ending points of subsequences, as well as single user actions.

Definition 3 *(Starter, Attractor, Singleton)*
Let $\mathcal{G} = (\mathcal{P}, \mathcal{B}, \mathcal{S})$ be a navigation space and $B = \{(S_i, ds_i)\}_i^M$, $B \in \mathcal{B}$, be a browsing session, and $S = \{(p_k, dp_k)\}_k^N$, $S \in \mathcal{S}$, be a subsequence.

Starter is the first point of an element of subsequence or session with length greater that 1, that is, $p_1 \in \mathcal{P}$ such that there exist $B \in \mathcal{B}$ or $S \in \mathcal{S}$ where $|B| > 1$ or $|S| > 1$ and $(p_1, d_1) \in B$ or $(p_1, dp_1) \in S$.

Attractor is the last point of an element of subsequence or session with length greater that 1, that is, $p_N \in \mathcal{P}$ or $p_M \in \mathcal{P}$ such that there exist $B \in \mathcal{B}$ or $S \in \mathcal{S}$ where $|B| > 1$ or $|S| > 1$ and $(p_M, d_M) \in B$ or $(p_N, dp_N) \in S$.

Singleton is a point $p \in \mathcal{P}$ such that there exist $B \in \mathcal{B}$ or $S \in \mathcal{S}$ where $|B| = 1$ or $|S| = 1$ and $(p, d) \in B$ or $(p, dp) \in S$.

Starters refer to the starting navigation points of user actions, whereas attractors denote the users' targets. Singletons relate to the single user actions such as use of hotlists (e.g. history or bookmarks) [10].

We can formulate behavioral abstractions simply as pairs of starters and attractors. Then it is equally important to observe the connecting elements of transitions from one task (or sub-task) to the other.

Definition 4 *(SE Elements, Connectors)*
Let $B = \{(S_i, ds_i)\}_i$ be a browsing session with consecutive subsequences S_i and S_{i+1}, where $S_i = \{(p_{ik}, dp_{ik})\}_k^N$ and $S_{i+1} = \{(p_{i+1l}, dp_{i+1l})\}_l^M$.
SE element (start-end element) of a subsequence S_i is a pair $SE_i = (p_{i1}, p_{iN})$.
Connector of subsequences S_i and S_{i+1} is a pair of points $C_i = (p_{iN}, p_{i+1,1})$.

SE elements outline higher order abstractions of user subgoals. Knowing the starting point, users can follow various navigational pathways to reach the target. Focusing on the starting and ending points of user actions eliminates the variance of navigational choices. Connectors indicate the links between elemental browsing patterns. This enables us to observe formation of more complex behavioral patterns as interconnected sequences of elemental patterns.

3   Intranet and Data

Data used in this work was a one year period Intranet web log data of The National Institute of Advanced Industrial Science and Technology (Table 1–left). The majority of users are skilled knowledge workers. Intranet web portal had load balancing architecture comprising of 6 servers providing extensive range of web services and documents vital to the organization. Intranet services support managerial, administration and accounting processes, research cooperation with industry and other institutes, databases of research achievements, resource localization and search, attendance verification, and also numerous bulletin boards and document downloads. The institution has a number of branches at various locations throughout the country, thus certain services are decentralized. The size of visible web space was approximately 1 GB. Invisible web size was considerably larger, but difficult to estimate due to the distributed architecture and constantly changing back-end data.

Table 1. Basic information about raw and preprocessed data used in the study

Data Volume	~60 GB	Log Records	315 005 952
Average Daily Volume	~54 MB	Clean Log Records	126 483 295
Number of Servers	6	Unique IP Addresses	22 077
Number of Log Files	6814	Unique URLs	3 015 848
Average File Size	~9 MB	Scripts	2 855 549
Time Period	3/2005 — 4/2006	HTML Documents	35 532
		PDF Documents	33 305
		DOC Documents	4 385
		Others	87 077

Daily traffic was substantial and so was the data volume. It is important to note that the data was incomplete. Although some days were completely represented, every month there were missing logs from specific servers. Server side logs also suffered data loss due to caching and proxing. However, because of the large data volume, missing data only marginally affected the analysis. Web servers run open source Apache server software and the web log data was in the combined log format without referrer.

4   Data Preprocessing and Cleaning

Starting with the setup description we present the data preprocessing and initial cleaning. Row data contained large number of task irrelevant logs. Extracted clean data was structured, databased, and linked.

Setup. Extraction and analysis of knowledge worker navigation primitives from Intranet web logs was performed on Linux setup with MySQL database as a data storage engine for preprocessed and processed data. Analytic and processing routines were implemented in various programming languages and optimized for high performance. Processing of large data volumes was computationally and time demanding.

Preprocessing and Cleaning. Data fusion of web logs from 6 servers of a load balanced Intranet architecture was performed at the preprocessing level. Data was largely contaminated by logs from automatic monitoring software and required filtering. During the initial filtering phase logs from software monitors, invalid requests, web graphics, style sheets, and client-side scripts were eliminated. Access logs from scripts, downloadable and syndicated resources, and documents in various formats were preserved. The information was structured according to the originating IP address, complete URL, base URL, script parameters, date-time stamp, source identification, and basic statistics. Clean raw data was logged into database and appropriately linked.

Approximately 40.15% of the original log records remined after initial filtering (see Table 1–right). Major access to Intranet resources was via scripts (94.68%). Only relatively minor portions of accessible resources were HTML documents (1.18%), PDF documents (1.1%), DOC documents (0.15%), and others (2.89%), such as downloadable software, updates, spreadsheets, syndicated resources, etc. Detected IP address space (22077 unique IPs) consisted of both statically and dynamically assigned IP addresses. Smaller portion of IP addresses were static, and relatively uniquely associable with users.

5   Navigation Space Extraction

Complete navigation space extraction requires finding sessions and subsequences. Sessions and subsequences were extracted from clean log records. We observed that data contained machine generated logs. Separation of machine subspace from the human navigation subspace was carried out during the subsequence extraction. Detected machine generated traffic was eliminated from further analysis.

Session Extraction. Preprocessed and databased Apache web logs (in combined log format) did not contain referrer information. Click-stream sequences were reconstructed by ordering logs originating from unique IP addresses according to time-stamp information. Ordered log sequences from the specific IP addresses were divided into the browsing sessions as described in Definition 1. Divisor between sessions was the user inactivity period ds_i greater than $T_{BS} = 1\ hour$.

It is noticeable that user sessions on the corporate Intranet are on average longer (appx. 48.5 minutes) than those of students (appx. 25.5 minutes) reported in [16]. Average number of 156 sessions per IP address, and large variation in maximum and minimum number of sequences from distinct IP addresses, indicate that association of particular users with distinct IP addresses is relevant only for registered static IP addresses. Large number (3492) of single sessions only originated from distinct IP addresses due to wide DHCP use. It is possible to employ clustering techniques to identify reasonably diverse groups of users.

Table 2. Observed basic session data statistics

Number of Sessions	3 454 243
Number of Unique Sessions	2 704 067
Average Number of Sessions per Day	9 464
Average Session Length	36 [URL transitions]
Average Session Duration	2 912.23 [s] (48 min 32 sec)
Average Page Transition Delay per Session	81.55 [s] (1 min 22 sec)
Average Number of Sessions per IP Address	156
Maximum	1 553
Minimum	1

Subsequence Extraction. Each detected session was analyzed for subsequences as defined in Definition 1. Segmenting element dividing sessions into subsequences was the delay between page transitions $dp_i > T_S$, where T_S was determined according to (1). Lower bound of 30 seconds for the separating inactivity period dp_i was proper.

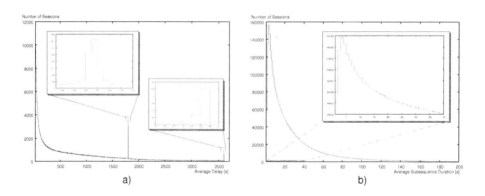

Fig. 1. Histograms: **a)** average delay between subsequences in sessions, **b)** average subsequence duration. There are noticeable spikes in chart **a)** around 1800 seconds (30 minutes) and 3600 seconds (1 hour). The detailed view is displayed in subcharts. Temporal variation of spikes corresponds to the peak average subsequence duration in chart **b)**. The spikes with relatively accurate delays between subsequences are due to machine generated traffic.

It has been observed that sessions contained machine generated subsequences. Periodic machine traffic with inactivity time less than the session separator delay could result in long session sequences. As seen in the histogram of average delays between subsequences (Figure 1-a), there was a disproportionally large number of sessions with average delays between subsequences around 30 minutes and 1 hour. This is indicated by spikes in the main chart of Figure 1-a. Detailed view (subcharts of Figure 1-a) revealed that the variation in the average delay between subsequences is approximately ± 3 seconds. It well corresponds to the peak in the histogram of average subsequence duration (Figure 1-b). It is highly unlikely that human generated traffic would produce this precision (although certain subsequences were legitimate).

Machine generated traffic contaminates the data and should be filtered, since we target primarily human behavior on the Intranet. We filtered two main groups of machine generated subsequences: login subsequences and subsesequences with delay periodicity around 30 minutes and 1 hour.

Every user is required to login into Intranet in order to access the services and resources. Login procedure involves validation and generates several log records with 0 delays. Records vary depending on whether the login was successful or unsuccessful. In both cases the log records and login related subsequences can be clearly identified and filtered.

The second group of machine generated traffic are subsequences with periodicity of 30 minutes and 1 hour. Direct way of identifying these subsequences is to search for sessions with only two subsequences having less than 1 second (or 0 second) duration (machines can generate requests fast and local Intranet servers are capable of responding within milliseconds) and delay ds_i between subsequences within the intervals: 1800 and 3600 ± 3 seconds. It has been discovered that substantial number of such sessions contained relatively small number (170) of unique subsequences. Furthermore, these subsequences contained only 120 unique URLs. Identified subsequences and URLs were considered to be machine generated and filtered from further analysis. Moreover, the subsequences with SE elements containing identified URLs were also filtered.

Table 3. Observed basic subsequence data statistics

Number of Subsequences	7 335 577
Number of Valid Subsequences	3 156 310
Number of Filtered Subsequences	4 179 267
Number of Unique Subsequences	3 547 170
Number of Unique Valid Subsequences	1 644 848
Average Number of Subsequences per Session	3
Average Subsequence Length	4.52 [URL transitions]
Average Subsequence Duration	30.68 [s]
Average Delay between Subsequences	388.46 [s] (6 min 28 sec)

Filtering of detected machine generated subsequences and their URLs significantly reduced the total number of subsequences - by 56.97% (from 7335577 to 3156310), as well as the number of unique subsequences - by 46.37% (from 3547170 to 1644848). Since the login sequences were also filtered, the number of subsequences per session decreased at least by 1. Reduction also occurred in the session lengths due to filtering of identified invalid URLs. Filtering did not significantly affect the duration of subsequences because the logs of machine generated subsequences occurred in rapid transitions with almost 0 durations and delays. It is noticeable that the average subsequence duration (30.68

seconds) is approximately equal to the chosen lower bound for ds_i (30 seconds). This empirically justifies the right choice of lower bound for T_S.

6 Knowledge Worker Browsing Behavior Analysis

By analyzing the navigation point characteristics (starters, attractors, and singletons) together with behavioral abstractions (SE elements and connectors) we infer several relevant observations. Exploratory analysis demonstrates usefulness of the approach in elucidating human browsing behavior in electronic spaces.

6.1 Starter, Attractor, and Singleton Analysis

Navigation point characteristics highlight initial (starters) and terminal targets (attractors) of knowledge worker activities, and also single-action behaviors (singletons). Starters, attractors, and singletons were extracted from subsequences.

Knowledge workers utilized small spectrum of starting navigation points and targeted relatively small number of resources during their browsing. The set of starters, i.e. the initial navigation points of knowledge workers' (sub-)goals, was approximately 3.84% of total navigation points. Although the set of unique attractors, i.e. (sub-)goal targets, was approximately three times higher than the set of initial navigation points, it is still relatively minor portion (appx. 9.55% of unique URLs). Knowledge workers aimed at relatively few resources.

Table 4. Statistics for starters, attractors, and singletons

	Starters	Attractors	Singletons
Total	7 335 577	7 335 577	1 326 954
Valid	2 392 541	2 392 541	763 769
Filtered	4 943 936	4 943 936	563 185
Unique	187 452	1 540 093	58 036
Unique Valid	115 770	288 075	57 894

Few resources were perceived of value to be bookmarked. Number of unique single user actions was minuscule. Single actions, such as use of hotlists [10], followed by delays greater than 1 hour are represented by singletons. Unique singletons accounted for only 1.92% of navigation points. If only small number of starters and/or attractors was perceived useful, there is a possibility that they were bookmarked and accessed directly in the following browsing experiences.

Knowledge workers had focused interests and diminutive exploratory behavior. Narrow spectrum of starters, attractors, and singletons was frequently used. Histograms and quantile characteristics of starters, attractors, and singletons (see Figure 2) indicate that higher frequency of occurrences is concentrated to relatively small number of elements. Approximately ten starters and singletons,

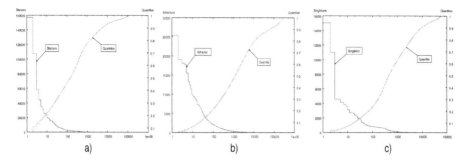

Fig. 2. Histograms and quantiles: **a)** starters, **b)** attractors, and **c)** singletons. Right y-axis contains a quantile scale. X-axis is in a logarithmic scale.

and fifty attractors were very frequent. About one hundred starters and singletons, and one thousand attractors were relatively frequent. Quantile analysis (Figure 2) reveals that ten starters (appx. 0.0086% of unique valid starters) and singletons (appx. 0.017% of unique valid singletons), and fifty frequent attractors (appx. 0.017% of unique valid attractors) accounted for about 20% of total occurrences. One hundred starters (appx. 0.086% of unique valid starters) and one thousand attractors (appx. 0.35% of unique valid attractors) constituted about 45% and 48% of total occurrences, respectively. Analogously, one hundred twenty singletons (appx. 0.21% of unique valid singletons) compounded to about 37% of total occurences.

6.2 SE Element and Connector Analysis

These components serve as higher order abstractions of knowledge worker behavior. SE elements represent starting and ending points of subsequences, or corresponding elemental patterns. Connectors delineate transitions between pattern primitives, and thus formation of more complex patterns.

Extraction of SE elements of subsequences and connectors between subsequences is relatively straightforward. SE elements and connectors also undergone filtering. If invalid URLs were present in at least one element of a pair, the respective SE element and/or connector was marked as invalid.

There is a noticeable reduction of SE elements and connectors due to the filtering. Number of SE elements decreased by 56.97% (from 7335577 to 3156310) and connectors by 40.63% (from 3952429 to 2346438). Similarly, reduction is evident in the number of unique SE elements (30.37%: from 1540093 to 1072340) and connectors (21.34%: from 1142700 to 898896).

Frequent users knew their targets and navigational paths to reach them. Duration of subsequences in sessions was short - with peak in the interval of two to five seconds (see histogram in Figure 1-b). During such short period users were able to navigate through four to five pages on average (see Table 3) in order to reach the target. Since there was approximately one second per page transition, there was virtually no time to thoroughly scan the page. Therefore

Table 5. Statistics for SE Elements and connectors

	SE Elements	Connectors
Total	7 335 577	3 952 429
Valid	2 392 541	2 346 438
Filtered	4 943 936	1 605 991
Unique	1 540 093	1 142 700
Unique Valid	1 072 340	898 896

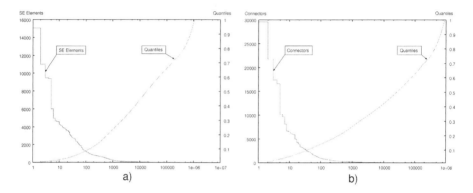

Fig. 3. Histograms and quantiles: **a)** SE elements, and **b)** connectors. Right y-axis contains a quantile scale. X-axis is in a logarithmic scale.

it is reasonable to assume knowledge workers knew where the next navigational point was located on the given page and proceed directly there. There was little exploratory behavior.

Session objective was accomplished via few subgoals. Average session (after filtering) contained three subsequences (see Table 3) where each subsequence can be considered a separate action and/or subgoal. Average knowledge worker spent about 30 seconds to reach the subgoal/resource, and additional 6.5 minutes before taking another action. Considering the number of unique valid subsequences (about 1.6 million) the complete population of users had relatively wide spectrum of browsing patterns. However, the narrow explored Intranet space of a single user suggests large diversification.

Small number of SE elements and connectors was frequently repetitive. Histogram and quantile charts in Figure 3 depict re-occurrence of SE elements and connectors. Approximately thirty SE elements and twenty connectors were very frequent (refer to left histogram curves of Figure 3). These thirty SE elements (appx. 0.0028% of unique valid SE elements) and twenty connectors (appx. 0.0022% of unique valid connectors) accounted for about 20% of total observations (see right quantile curves of Figure 3).

Knowledge workers formed frequent elemental and complex browsing patterns. Strong repetition of SE elements indicates that knowledge workers often initiated their browsing actions from the same navigation point and targeted the same resource. This underlines the elemental pattern formation. Relatively small number of elemental browsing patterns was frequently repeated. Re-occurrence of connectors suggests that after completing a browsing sub-task, by reaching the desired target, they proceeded to the frequent starting point of following sub-task(s). Frequently repeating elemental patterns interlinked with frequent transitions to other elemental sub-task highlights formation of more complex browsing patterns. Although the number of highly repetitive SE elements and connectors was small, knowledge workers exposed a spectrum of behavioral diversity in elemental as well as more complex behavioral patterns.

Formation of behavioral browsing patterns positively correlates with short peak average duration of subsequences (3 seconds). Knowledge workers with formed browsing patterns exhibited relatively fast page transitions. They also displayed shorter delays between subsequences.

7   Long Tails of Knowledge Worker Browsing Behavior

The term *long tail* colloquially refers to a feature of statistical distributions where the *head* contains a small number of high frequency elements that gradually progresses to the *long tail* of low frequency elements. The mass of a long tail can substantially outweigh the mass of a head. Numerous aspects of human dynamics have been observed to display such characteristics [3].

The former analysis indicates that the long tail characteristics are evident in knowledge worker browsing behavior. All histograms of starters, attractors, and singletons show long tails. The elemental behavioral abstractions, that is SE elements, and their connectors, throughout which users form more complex behavioral patterns, equally display long tails. Furthermore, even the complete sessions have this attribute. (Note that the histogram charts have x-axis in a logarithmic scale. It allows us to observe the details of heads of distributions.)

If the long tails are the common denominator of human browsing behavior in electronic spaces, what is the underlying functional law that accurately captures it? Conventionally, the heavy tails in human dynamics are modeled by Pareto distribution [17]. However, results of our analysis suggest better and more accurate novel distribution.

The novel distribution that efficiently captures the long tail features of a human browsing behavior in web environments is derived from analysis of log-log plots. Figure 4-a shows a log-log plot of attractor histogram. It is evident that the curve has a quadratic shape. Plots of other histograms have the same quadratic appearance. Nonlinearity is the reason why Pareto distribution (and other well known long tail distributions) is unsuitable since it only captures linear dependency. Models employing conventional distributions may display systematic deviations.

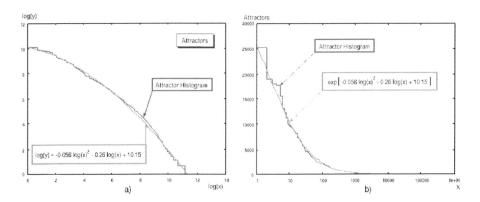

Fig. 4. Long tail analysis in attractor histogram: a) log-log plot, b) normal plot with x-axis in a logarithmic scale. Log-log plot clearly shows inverted quadratic characteristics. The distribution is well approximated by the LPE p.d.f. function $f(x) = \exp\left[-0.056 \log(x)^2 - 0.26 \log(x) + 10.15\right]$.

Expressing the quadratic characteristics of a log-log plot in an analytic form leads to the formula:

$$\log(y) = \sum_{i=0}^{2} \theta_i \log(x)^i.$$

Eliminating the logarithm on the left-hand-side of the equation, and presenting the generalized polynomial form results in the following expression:

$$f(x; \theta) = \exp\left[\sum_{i=0}^{n} \theta_i \log(x)^i\right]. \tag{2}$$

Naturally, even more generalized form can be obtained by not limiting i to non-negative integers, but considering it to be a real, $i \in R$.

The derived log-polynomial-exponential (LPE) function (2) appropriately represents the observed long tail dynamics of user browsing behavior. Although the general n-th order polynomial can be considered, the second order form was sufficient for modeling our observations (see Figure 4-b). When using the second order polynomial form, the common concave shape depicted in Figure 4-a suggests that the quadratic term will always be negative, $\theta_2 \in R^-$, and the offset at the origin always positive, $\theta_0 \in R^+$. One can also notice that LPE p.d.f. (2) is base independent. The estimation of parameters θ can be done by applying various statistical inference techniques.

Long tail characteristics of human browsing behavior in electronic spaces present both advantages and challenges. When the high (relative) frequency characteristics are important, the heads of long tail distributions may be considered beneficial, since they contain relatively small number of high frequency elements. The conventional clustering and classification methods that are essentially based on segmenting the observation domains with respect to the high frequency elements may be well applicable. Consider for example the application areas such as human-computer interfaces, and recommender systems. The conventional methods may be well suited for *one-fit-all* system designs, where the developers are often faced with lack of computing power, unavailability of more personalized information, and other limitations.

When the domain coverage is of importance, the challenging aspects of the long tail characteristics are the long tails themselves. *The effective domain coverage by features extractable from a head may be substantially smaller than the domain coverage by a spectrum of features extractable from a long tail.* Simply put, the cumulative power of a wide range of infrequent features may be higher than that of the very frequent ones. Observe for instance the quantile characteristics of SE elements and connectors. The heads cover approximately <30% of the domain, whereas the tails cover remaining >70%. For clustering and/or classification methods to reach significant populations in long tails may demand fine grained segmentation with respect to large number of attributes. This challenge calls for novel approaches, methods, and algorithms.

Significant finding was that *the underlying long tail characteristics generally hold for large user populations having mixed behavioral attributes.* They may no longer hold for behaviorally similar micro-groups of users. In other words, the massive user populations may exhibit typical long tail features, but the particular user group may have completely different characteristics. This has also been observed in the session dynamics.

To illustrate this finding we selected a frequent singleton that clearly corresponded to a single action associable with specific category of knowledge workers. This identified a distinct class of knowledge workers; denoted as user class A. When projecting user class A into IP address space, it has been detected that 382 unique IP addresses displayed the particular action. Additional browsing actions were shared among these users.

It can be seen from Figure 5-a that the session–IP dynamics for the complete user population have typical long tail characteristics. The user class A exhibited

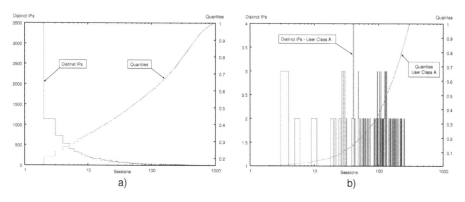

Fig. 5. Histogram and quantile analysis depicting the number of sessions with respect to distinct IPs: **a)** complete knowledge worker population, **b)** user class A. Right y-axis contains a quantile scale. X-axis is in a logarithmic scale.

noticeably different sessions–IP dynamics (see Figure 5-b). Rather than the long tail distribution, the user class A has a roughly Poisson shaped distribution (account for logarithmic x-axis). Analogously, quantile characteristics are significantly different.

Models suitable for large user populations may be unfit for focused user groups. Presented LPE distribution (2) may accurately model the browsing attributes of masses, however, may no longer be proper for behaviorally narrow user clusters. Implications of this important finding extend to numerous domains of web research, engineering, metrics, and design of applications and services. Researchers and practitioners often utilize the observations and models derived from massive user populations for design and development of personalized services, applications, and interfaces.

Personalization requires different approach. Globally revealed attributes, although valid, are reasonably applicable only to one-fit-all schemes. As the granularity of user populations increases one may experience a gradual divergence from generally observed distribution characteristics. The characteristics of narrow user groups may significantly deviate from large scale observations. It is therefore advisable to re-analyze micro-groups.

8   Conclusions and Future Work

A novel formal approach to analyzing human browsing behavior in electronic spaces based on navigation space construct has been introduced. The framework is not only applicable to analyzing and modeling human browsing behavior but also to engineering behaviorally centered algorithms.

The presented framework has been applied to elucidation of behavioral characteristics of knowledge workers on a large corporate Intranet. Exploratory analysis revealed several important behavioral aspects. Knowledge workers had generally well defined browsing targets and knew how to reach them. General browsing strategy of knowledge workers was remembering the starting point and recalling the navigational path to the target. The browsing objectives were accomplished via few subgoals. Knowledge workers had a significant tendency to form elemental and complex browsing patterns that were often reiterated. They had focused interests and effectively explored only diminutive range of resources.

All analyzed elements of knowledge worker browsing behavior exposed evident long tail characteristics. A novel distribution that accurately models it has been derived. Long tail aspects of human browsing behavior present new challenges and opportunities for development of novel behaviorally centered approaches and algorithms.

Focus of the future investigation is on further elucidation of the long tail characteristics in connection with content based analysis. Efficient utilization of long tails opens potential for design and development of the next generation personalization systems and tools.

Acknowledgment

The authors would like to thank Tsukuba Advanced Computing Center (TACC) for providing raw web log data.

References

1. Schlender, B.: Peter Drucker sets us straight. Fortune (December 29, 2003) http://www.fortune.com
2. Davenport, T.H.: Thinking for a Living - How to Get Better Performance and Results from Knowledge Workers. Harvard Business School Press, Boston (2005)
3. Barabasi, A.-L.: The origin of bursts and heavy tails in human dynamics. Nature 435, 207–211 (2005)
4. Park, Y.-H., Fader, P.S.: Modeling browsing behavior at multiple websites. Marketing Science 23, 280–303 (2004)
5. Géczy, P., Akaho, S., Izumi, N., Hasida, K.: Navigation space formalism and exploration of knowledge worker behavior. In: Kotsis, G., Taniar, D., Pardede, E., Ibrahim, I.K. (eds.) Information Integration and Web-based Applications and Services, pp. 163–172. OCG, Vienna (2006)
6. Moe, W.W.: Buying, searching, or browsing: Differentiating between online shoppers using in-store navigational clickstream. Journal of Consumer Psychology 13, 29–39 (2003)
7. Benbunan-Fich, R.: Using protocol analysis to evaluate the usability of a commercial web site. Information and Management 39, 151–163 (2001)
8. Norman, K.L., Panizzi, E.: Levels of automation and user participation in usability testing. Interacting with Computers 18, 246–264 (2006)
9. Bucklin, R.E., Sismeiro, C.: A model of web site browsing behavior estimated on clickstream data. Journal of Marketing Research 40, 249–267 (2003)
10. Thakor, M.V., Borsuk, W., Kalamas, M.: Hotlists and web browsing behavior–an empirical investigation. Journal of Business Research 57, 776–786 (2004)
11. Deshpande, M., Karypis, G.: Selective markov models for predicting web page accesses. ACM Transactions on Internet Technology 4, 163–184 (2004)
12. Wu, H., Gordon, M., DeMaagd, K., Fan, W.: Mining web navigaitons for intelligence. Decision Support Systems 41, 574–591 (2006)
13. Zukerman, I., Albrecht, D.W.: Predictive statistical models for user modeling. User Modeling and User-Adapted Interaction 11, 5–18 (2001)
14. Jozefowska, J., Lawrynowicz, A., Lukaszewski, T.: Faster frequent pattern mining from the semantic web. Intelligent Information Processing and Web Mining, Advances in Soft Computing, pp. 121–130 (2006)
15. Géczy, P., Akaho, S., Izumi, N., Hasida, K.: Extraction and analysis of knowledge worker activities on intranet. In: Reimer, U., Karagiannis, D. (eds.) Practical Aspects of Knowledge Management, pp. 73–85. Springer, Heidelberg (2006)
16. Catledge, L., Pitkow, J.: Characterizing browsing strategies in the world wide web. Computer Networks and ISDN Systems 27, 1065–1073 (1995)
17. Vazquez, A., Oliveira, J.G., Dezso, Z., Goh, K.-I., Kondor, I., Barabasi, A.-L.: Modeling bursts and heavy tails in human dynamics. Physical Review E73(19), 36127 (2006)

A Case-Based Approach to Anomaly Intrusion Detection

Alessandro Micarelli and Giuseppe Sansonetti

Department of Computer Science and Automation
Artificial Intelligence Laboratory
Roma Tre University
Via della Vasca Navale, 79, 00146 Rome, Italy
{micarel,gsansone}@dia.uniroma3.it

Abstract. The architecture herein advanced finds its rationale in the visual interpretation of data obtained from monitoring computers and computer networks with the objective of detecting security violations. This new outlook on the problem may offer new and unprecedented techniques for intrusion detection which take advantage of algorithmic tools drawn from the realm of image processing and computer vision. In the system we propose, the normal interaction between users and network configuration is represented in the form of snapshots that refer to a limited number of attack-free instances of different applications. Based on the representations generated in this way, a library is built which is managed according to a case-based approach. The comparison between the query snapshot and those recorded in the system database is performed by computing the Earth Mover's Distance between the corresponding feature distributions obtained through cluster analysis.

1 Introduction

Intrusion Detection Systems (IDSs) have the objective of detecting attacks launched against computers or computer networks. Their classification is usually based on the *audit source location* and on the *general detection strategy*. With respect to the first criterion, IDSs are divided into *host-based* techniques if the input information they analyze consists of audit trails and/or system logs and *network-based* techniques if it consists of network packets. According to the second criterion, IDSs are classified as *misuse-based* or *anomaly-based* techniques. The former use attack descriptions (*signatures*) in order to analyze the sequence of events obtained from monitoring a given network and single computers connected to it. If a known attack pattern is detected, an alarm is triggered. These systems are usually efficient and generate a limited number of false detections, called *false positives*. The main drawback of these systems lies in their inability to detect unknown attacks, *i.e.*, attacks for which there exists no prior information in the system database.

The anomaly-based IDSs follow an approach which is complementary to the previous one. They are based on models of the *normal* behavior (*profiles*) of users

P. Perner (Ed.): MLDM 2007, LNAI 4571, pp. 434–448, 2007.

and applications in order to detect anomalous activities which might provide an indication of an internal intrusion, launched by users attempting to abuse of their privileges, or of an external intrusion. The main advantage of this approach is the fact that it is capable of identifying unknown attacks. This advantage is however obtained at the price of a large number of false positives. Axelsson refers to it as "the limiting factor for the performance of an anomaly-based IDS" [2]. In addition to this, recent work [33,37] has shown that these systems are vulnerable to *mimicry attacks*, *i.e.*, attacks which aim at imitating normal activity, thereby avoiding identification by the system.

Nonetheless, we believe that the benefits offered by anomaly-based systems are such that a thorough critical analysis of the limits of the approaches advanced so far is needed in order to come up with adequate solutions. In particular, excluding some notable exceptions, most anomaly-based systems share these common characteristics:

1. They are based on a single feature, *i.e.*, they usually consider a single characteristic, based on which they assess the normality of a generic user-application interaction;
2. They have only one input since they examine only one data typology, relative either to the network or to a generic host and they do not propose the analysis of combined data;
3. The classification procedure, *i.e.*, the procedure whereby a generic event is considered part of an ongoing attack or not, once the relative anomaly score is known is trivial.

Concerning the first characteristic, most techniques used to date do not make appropriate tools available to take into account more than one element during the evaluation phase. It is therefore worthwhile to explore new techniques, inspired by different principles.

The rest of the paper is organized as follows. Section 2 outlines related work. Section 3 presents our intrusion detection system, in particular the case representation and the dissimilarity metric. Section 4 describes the experiments that were performed to evaluate the accuracy of the case-based classifier. Section 5 contains the conclusions and Section 6 discusses future directions of our research.

2   Related Work

In the field of Intrusion Detection, in addition to the traditional techniques used to date, various alternative solutions have recently been advanced which use, among others, *haptic technologies* [13], capable of generating tactile sensations, and *sonification techniques* [34], which make use of non-speech audio to convey information. *Visualization techniques* have also been proposed which operate the conversion from textual datasets to digital images [7,23,20]. There exist many advantages associated with this conversion. In particular, it has been observed that, from a physiological viewpoint, the interpretation of graphical images is a parallel process and, as such, it is much more efficient than reading textual information, which is an intrinsically serial process [22].

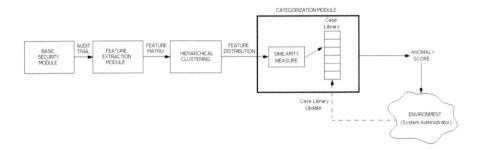

Fig. 1. The anomaly-detection system

Another major advantage is that a single image can convey several pieces of information simultaneously in a more structured and compact form than text [10]. There have been several contributions in the field of network-based visualization techniques, mainly aimed at representing relative performances and bandwidth usage in a graphical form [5,21].

Less attention has instead been devoted to Intrusion Detection. Among the early contributions, tools have been advanced to estimate the level of attack which a system being monitored undergoes [31,19]. Despite their usefulness, these tools only allow one to detect attacks which are already in progress, but they do not provide any proactive measure.

More recently, visual user interfaces have been devised which assist in the interpretation of data streams produced by IDSs [9,20,27,32]. If, on the one hand, these systems provide an important contribution, on the other hand, they can make the human interpretation of data easier but they do not replace it altogether.

To the best of our knowledge, there has been so far no contribution in automatic intrusion detection based on Image Processing and Computer Vision techniques. However, we believe that these fields have made an outstanding progress in providing useful tools in non-traditional application areas for these disciplines such as Intrusion Detection [12,30,17].

Much headway has been made since February 1992 when the National Science Foundation organized a workshop on *Visual Information Management Systems* in Redwood, California. The objective of this event was the identification of topic areas where to focus research aimed at designing and testing effective visual information management systems [24]. Such an interest was captured by the possibility to access large image databases where traditional query methods such as keywords and annotations cannot be used [6]. *Content-Based Image Retrieval* (CBIR) systems are nowadays largely in use. They exploit color, texture, shape information and spatial relations to represent and retrieve information [35]. Their large number and the excellent performances they can guarantee have inspired us to explore the use of such techniques in the arena of Intrusion Detection.

Concerning the use of Machine Learning techniques, several IDSs have resorted to them to improve their performance [18,8,3]. These systems can be grouped in two families: *rule-based* and *model-based* techniques. Even though these systems have been proven to be useful, they however suffer from the typical drawbacks of this kind of expert systems, *i.e.*, difficulties in the acquisition and representation of new knowledge.

Instead, *Case-Based Reasoning* (CBR) is a problem-solving paradigm which, rather than relying exclusively on general knowledge of the domain of interest or building associations through generalized relationships among problem descriptors and conclusions, it is capable of exploiting specific knowledge derived from situations (*cases*) already experienced and solved in the past [1].

In [14,11] a case-based reasoner (AUTOGUARD) for intrusion detection is presented. In AUTOGUARD, a translator model converts the low-level audit trail into high-level class representation of events. This information is recorded in the system as a collection of cases. In order to evaluate the similarity between the new case and every old case archived in the system library, the authors propose a fuzzy logic based approach. However, it is not clear if the design has been implemented altogether.

3   System Design

A block diagram representation of the system we have designed and implemented is shown in Figure 1. The input parameters are represented by the data obtained from monitoring computers connected to the network whereas the output parameter is the relative *anomaly score*. This value is given by the smallest value of dissimilarity obtained by comparing the input case with those stored in the database. This database is managed, queried, and updated according to modalities typical of the CBR approach.

It should be noted that the phase of relevance feedback is fundamental to keep the case record updated. In order for an input representation to be useful and, therefore, stored to optimize the system performance in case similar situations are encountered again, two requirements are necessary:

1. The environment (a term which also refers to human supervision, *e.g.*, the system administrator) has to confirm the system indications;
2. The input representation has to convey meaningful information, *i.e.*, in the database, there is no case capable of representing effectively the class the input snapshot belongs to.

Concerning the second objective, it is achieved using a second similarity threshold: in addition to an upper threshold (called *reliability threshold*) beyond which we can infer that the behavior being monitored is symptomatic of an attack underway, we have considered a lower threshold (called *identity threshold*) below which the input case is not kept. In other words, the input case is added to the knowledge base of the system, thereby assuming the characteristic of a profile, when its dissimilarity value with respect to all other cases archived in the library and relative to the same application is comprised between the two thresholds. This ensures that the cases, which are progressively added to the library, effectively reproduce a behavior not yet represented in the database. Therefore, they have to be recorded with the goal of optimizing the system performance in case a similar situation is encountered.

The need for carefully choosing the cases to keep stems from the need for optimizing the system resources, *i.e.*, memory support and processing time. Not only do these problems affect the system architecture herein proposed, but they also concern any case-based system. For this reason, they have been the object of research in the Artificial Intelligence (AI) community. There are several contributions suggesting memory models alternative to the simple flat memory. The interested reader is referred, for instance, to [25,38].

It should be noted that the domain expert possibility to intervene in the decision task is possible not only in the initial training phase of the system, but also during the verification phase for the classification response. The system is actually capable of acquiring knowledge also during its normal operation. The ease and quickness of the learning phase represent in fact some of the strong features of our case-based system.

In the following sections, we will analyze the key components of a typical case-based expert system, *i.e.*, the different case representations and the associated (dis)similarity metric.

3.1   Case Representation

The fundamental assumption of the proposed architecture is the following: in order for a program to effectively damage the system being monitored, it has to interact with the operating system through system calls.

Various host-based approaches to anomaly detection have been proposed which build profiles from the sequences of system calls [16,36]. Specifically, these systems are based on models of the system call sequences generated by the applications during the normal operation of the system. In the detection phase, every sequence being monitored which is not compliant with the profiles previously recorded is deemed a part of an attack. Later work has, however, shown that it is possible for the intruder to avoid this kind of detection [33,37].

An effective solution thus advocates the exploitation of additional information drawn from the audit files. In [26], the authors observe that the output parameters and the arguments of the system calls can play an important role in the intrusion detection process. Based on these considerations, we have decided to consider this information in our representation. Concerning the output parameters, namely the return value and the error status, their use is straightforward since they are already available in a numeric format.

The issue is more complicated with the system call arguments. These arguments can be divided into four categories: *file name, execution parameter, user ID*, and *flag* [26]. The first two are of `string` type, the other of `integer` type.

Fig. 2. Application audit trail

In this preliminary version of our system, we have considered only the `string` type, for which it is possible to consider three models, namely, the length, the character distribution and the grammar inference. The length and character distribution models can be applied straightforwardly, since, with reference to the second, we are only interested in the profile generated by the frequency of occurrence of the characters independent of their type.

Concerning the grammar inference, *i.e.*, the inference of the argument grammar, two processing steps are necessary. In the first, each character is replaced by the token corresponding to its class; in the second, the possible repetitions of elements belonging to the same class are merged [26].

Regarding the classes, we have considered three main groups of characters, namely `lowercase` letters, `uppercase` letters, and `digits`. Characters which do not belong to any of these classes are considered to belong to new classes. A different numeric identifier is associated with each class. For instance, assuming the following class-identifier association:

$$N_1: \texttt{lowercase letter}$$
$$N_2: \texttt{uppercase letter}$$
$$N_3: \texttt{digit}$$
$$N_4: \texttt{slash}$$
$$\ldots: \ldots$$

the string `/etc/usr/bin` is represented in terms of the these ten features:

$$N_4, N_1, N_4, N_1, N_4, N_1, 0, 0, 0, 0.$$

The input to the detection process is an ordered stream $X = \{x_1, x_2, \cdots\}$ of system call invocations representing the generic instance of an application. In

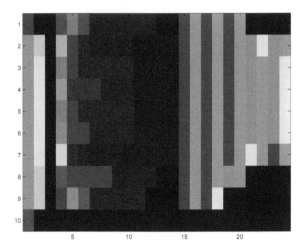

Fig. 3. Application snapshot

our system, based on the previous considerations, every system call invocation $x \in X$ is represented by means of the following features

$$< f_1^x, f_2^x, f_3^x, f_4^x, f_5^x, \cdots, f_{14}^x, f_{15}^x, \cdots, f_{24}^x >$$

where

$$
\begin{aligned}
f_1^x: &\quad \text{system call class} \\
f_2^x: &\quad \text{return value} \\
f_3^x: &\quad \text{error status} \\
f_4^x: &\quad \text{argument length} \\
f_5^x, \cdots, f_{14}^x: &\quad \text{argument character distribution} \\
f_{15}^x, \cdots, f_{24}^x: &\quad \text{argument grammar inference}
\end{aligned}
$$

In particular, we have monitored the following six system calls: `execve()`, `chmod()`, `chown()`, `exit()`, `open()`, `setuid()`, since these are the only ones deemed potentially dangerous. In [4], Axelsson points out that "this logging method consumes as little system resources as comparable methods, while still being more effective."

In order to spot the sequence of system calls within an audit trail of a generic application, it is sufficient to find the audit record representing the `execve()` system call in which the path name of the application of interest appears and to record the *process ID* assigned to the process by the operating system. The system calls to represent are all those which appear one after the other up to the record relative to the `exit()` command terminating the process having the ID under consideration. From the whole sequence of system calls we have represented only the six described above. For these audit events we have converted in numeric features only the pieces of information relative to the output parameters (second and third columns) and to the arguments (remaining columns).

Based on these considerations, for instance, the instance of the **ps** application comprised of the 43 system calls shown in Fig. 2, is associated with an $m \times n$ matrix of features where m is the number of system calls of the following types `execve()`, `chmod()`, `chown()`, `exit()`, `open()`, `setuid()` among the overall 43, 10 in this case, whereas n is fixed and equal to 24, *i.e.*, to the number of attributes which we have decided to consider and whose corresponding values constitute the matrix entries.

Fig. 3 shows the representation obtained with Matlab by interpreting each matrix entry as an index in the RGB color space. We have thus obtained a snapshot representing the temporal behavior of the **ps** application to monitor; this can then be compared against the profiles relative to the **ps** application stored in the system.

Cluster Analysis. In order to compare system call sequences which may be rather different in terms of their structure and of their number, a cluster analysis is needed. In particular, we have used *Hierarchical Clustering* with the *Jaccard Distance* in order to calculate the distance between every pair of objects. This distance is defined as one minus the *Jaccard coefficient*, that is the percentage of nonzero coordinates that differ from each other. Given an $m \times n$ feature matrix X representing the generic instance of an application and made up of m $1 \times n$ row vectors x_1, x_2, ..., x_m, representing the relative system calls, the Jaccard distance between the row vectors x_r e x_s has the following expression:

$$d_{rs} = \frac{\#\left[(x_{rj} \neq x_{sj}) \wedge ((x_{rj} \neq 0) \vee (x_{sj} \neq 0))\right]}{\#\left[(x_{rj} \neq 0) \vee (x_{sj} \neq 0)\right]} \tag{1}$$

where $\#$ is the cardinality.

Then we have set an *inconsistency coefficient* threshold to divide the objects in the hierarchical tree into clusters. This coefficient compares the height of a link in a cluster hierarchy with the average height of neighboring links. It is thus possible to identify the natural divisions in the dataset, but this involves a variable number of clusters for every instance of an application. Every instance of an application is represented by a set of a different number of clusters where each cluster is represented by the coordinates of its centroid and by a weight that, in the preliminary version of our system, is equal to the fraction of the distribution that belongs to that cluster (the procedure for assigning appropriate weights to the clusters will be one of the objectives of our future work). The information obtained in this way represents a case which is structured as a record comprising the following three fields:

- The first field is of **string** type and contains the name of the application to which it refers; it is obtained from the path of the relative **execve** system call;

- The second field is represented by an array of N records (where N is the number of clusters, a function of the threshold value for the inconsistency coefficient) having 24 fields of type `double`, which contain the values of the attributes represented and constitute the coordinates of the relative centroid;
- The third field is represented by an array of N values of type `double`, each expressing the weight of the corresponding cluster.

3.2   Dissimilarity Metric

Once the representation of an application instance has been generated according to the modalities discussed above, an appropriate dissimilarity metric has to be determined to compare the input case with those contained in the database.

Recently, the *Earth Mover's Distance* (EMD) [28] has been proposed to evaluate distribution dissimilarities. The EMD is based on the minimum cost associated with the transformation of one distribution into the other. In the case of Content-Based Image Retrieval, it has been proven to be more robust than the histogram-based techniques, since it is able to handle also representations with variable length. When used to compare distributions with the same overall mass, it can be readily shown that it is a real metric [29], which allows the use of more efficient data structures and query algorithms.

The EMD enables us to evaluate the dissimilarity between two multi-dimensional distributions. In our architecture, the two distributions are represented by two sets of weighted clusters that capture them. The clusters of any distribution can be in any number and the sum of their weights can be different than the sum of weights of the other distribution. This is the reason why a smaller sum appears at the denominator of the expression of the EMD. In order to calculate the EMD in some feature space, a distance measure (called *ground distance*) between single features must be defined.

The computation of the EMD value can be performed by solving the following linear programming problem: let X denote the distribution of the input instance of an application with m clusters,

$$X = \{(x_1, w_{x_1}), (x_2, w_{x_2}), \cdots, (x_m, w_{x_m})\} \tag{2}$$

where x_i represents the generic cluster and w_{x_i} the relative weight, and let Y denote the distribution of the generic instance of the same application in the archive of cases with n clusters

$$Y = \{(y_1, w_{y_1}), (y_2, w_{y_2}), \cdots, (y_n, w_{y_n})\} \tag{3}$$

Let $D = [d_{ij}]$ denote the ground distance matrix, d_{ij} being the ground distance between clusters x_i and y_j. The objective is to calculate the value of the flow $F = [f_{ij}]$ that minimizes the overall cost

$$WORK(X, Y, F) = \sum_{i=1}^{m} \sum_{j=1}^{n} f_{ij} d_{ij} \tag{4}$$

subject to the following constraints:

$$f_{ij} \geq 0 \qquad 1 \leq i \leq m, 1 \leq j \leq n \tag{5}$$

$$\sum_{j=1}^{n} f_{ij} \leq w_{x_i} \qquad 1 \leq i \leq m \tag{6}$$

$$\sum_{i=1}^{m} f_{ij} \leq w_{y_j} \qquad 1 \leq j \leq n \tag{7}$$

$$\sum_{i=1}^{m} \sum_{j=1}^{n} f_{ij} = min \left(\sum_{i=1}^{m} w_{x_i}, \sum_{j=1}^{n} w_{y_j} \right) \tag{8}$$

Once we have calculated the value of the flow that solves the above equations, the EMD has the following expression

$$EMD(X, Y) = \frac{\sum_{i=1}^{m} \sum_{j=1}^{n} f_{ij} d_{ij}}{\sum_{i=1}^{m} \sum_{j=1}^{n} f_{ij}} \tag{9}$$

4   Empirical Evaluation

In order to evaluate the accuracy of our case-based IDS, we performed experimental runs divided in a first training phase and in a second testing phase. During the training phase, a database of instances of every application was built, which represented normal behavior. The testing phase then ensued.

For the experiments we used the 1999 MIT Lincoln Lab Intrusion Detection Evaluation Data [15]. In particular, we employed data of two attack-free weeks (First Week and Third Week) to train the system and data of two other weeks (Fourth Week and Fifth Week) to test the ability of the proposed architecture to correctly classify applications with attacks and applications associated with the users' normal behavior. For some of the attacks in the evaluation data there is no evidence in the Solaris Basic Security Module (BSM) log, so we were not interested in them. Among the visible attacks in the BSM audit trail some are *policy violations*, in which the intruder tried to exploit possible system configuration made by the administrator. We did not try to detect this class of attacks with our system but we plan on performing this test in the future work. In particular, we were interested in detecting attacks based on *buffer overflow vulnerabilities*.

In our simulations, a value of 0.9 was chosen for the threshold of the inconsistency coefficient. As a distance for clustering, we have used the Jaccard distance whereas for the computation of the EMD we have chosen the Euclidean distance as the ground distance.

In particular, we have carried out two different experimental runs. In the first experiment we stored in the library case all the 117 instances of eject, fdformat, ffbconfig and ps applications encountered in the training phase.

Table 1. Experimental Results

	Total	With attack	Identified	False Alarms
eject	9	3	3	0
fdformat	9	6	6	0
ffbconfig	2	2	2	0
ps	315	14	14	0
	335	25	25	0

We have considered these four applications since these are the only ones subject to attack in the Lincoln Laboratory database. We have then tested the system by using a value of 5 for the reliabilty threshold: the input application has been compared with all the instances archived in the library and relative to the same application. If the minimum value obtained was lower than the threshold, the application was labeled attack-free, otherwise it was classified as containing an attack.

In the second experiment we started with an initially empty database. Every training input application was analyzed through hierarchical clustering and compared to all existing entries in the case memory. If a distribution was found in the database that was similar enough, *i.e.*, below the identity threshold set to 0.5, according the EMD similarity metric, this new case was discarded, because it was already adequately represented in the database. Otherwise, the distribution (clusters with their weights) that corresponded to the new input was included into the database. After the training phase, the library contained only 19 cases. A testing phase was then carried out by choosing the same parameters as those of the previous session.

The results obtained after the two experimental sessions are collected in Table 1. The fact that we have obtained the same values after the two testing runs confirms that recording only one case for each typology of situation encountered in the training phase, with the objective of improving the computational efficiency of the system, does not have any effect altogether on the system performance in terms of classification.

Concerning the experimental results, we did not obtain any false positives by testing the system with 335 instances of input applications and all 25 applications containing attacks have been correctly identified.

5   Conclusions

In this contribution, we have presented a case-based anomaly detection system which was inspired by the interpretation in the form of snapshots of system call sequences obtained from the log of the C2 BSM of a Solaris workstation and relative to different instances of applications. This allowed us to resort to Image Processing and Computer Vision techniques, in particular to methodologies drawn from Content-Based Image Retrieval for the implementation of our system. These techniques, together with a CBR approach in the management of the knowledge base and with a representation of the cases based on the information relative to output parameters and arguments of the system calls, enabled us to obtain no false positives, even with a limited number of cases in the library. In particular, it was possible to distinguish the 25 instances of applications affected by attacks from the 310 relative to the normal behavior of the system with very high accuracy. This was confirmed by the appreciable differences among the EMD values relative to the corresponding feature distributions.

Obtaining a null number of false positives represents a very important result, in consideration of the fact that achieving a small number of false positives constitutes one of the most difficult objectives of any anomaly detection system.

Furthermore, the possibility to intervene on different parameters of the classification procedure (inconsistency coefficient threshold, reliability threshold, identity threshold, etc.) allows one to conveniently change the sensitivity of the system, thereby increasing the probability to identify also the so-called *mimicry attacks*.

The procedure we have advanced has therefore allowed us to fully exploit the salient features of the user-network configuration interaction, enabling the accurate distinction between attacks and events associated with the normal behavior of the system.

6   Future Work

There are several research thrusts that we intend to pursue in the near future. First of all, we will focus our efforts on the clustering procedure, particularly on the weight assignment procedure. Even though the experimental results we have obtained are satisfactory, we intend to take into account other factors, such as the semantic difference between the various features and the presence of outliers obtained from monitoring the host.

We will continue our experimental evaluation of the system performance, using new benchmarks, in order to check its capability of recognizing also new classes of attacks in addition to buffer overflows already identified. In particular, we will tackle the so-called *policy violations*, which to not allow the intruders to directly upgrade their privileges, but have the objective of gaining classified information in order to exploit possible erroneous configurations of the system administrator. This class of attacks thus contain intrusions which do not exploit actual system flaws and turn out to be not easily detectable, since the intruders have access to classified information through the normal, although unintentional, behavior of the system. In order to achieve this goal, besides working on clustering, it is necessary to further develop the modalities for representing the cases, taking into account new models based on the information contained in the audit trails, such as, for instance, *execution parameter*, *user ID* and *flag*.

Another objective of our future research will be the integration of profiles with signatures relative to known attacks. Last but not least, we will work on the realization of a network-based version of our intrusion detection system, in order to realize a combined analysis of the data obtained from monitoring the whole network configuration.

Acknowledgements

We would like to thank Dick Kemmerer, Giovanni Vigna, and Luca Lucchese for the opportunity to work at the Reliable Software Laboratory, Computer Science Department, University of California, Santa Barbara (CA), USA, and at the Projects Laboratory, School of Electrical and Computer Science, Corvallis (OR), USA.

References

1. Aamodt, A., Plaza, E.: Case-based Reasoning: Foundational Issues, Methodological Variations and System Approaches. AICOM **7**(1), 39–59 (1994)
2. Axelsson, S.: Intrusion Detection Systems: A Survey and Taxonomy. In: Proceedings of the 6th ACM Conference on Computer and Communications Security, Singapore, November 1999, pp. 1–7. ACM Press, New York (1999)
3. Axelsson, S.: Intrusion Detection Systems: A Survey and Taxonomy. Technical Report 99-15, Department of Computer Engineering, Chalmers University (March 2000)
4. Axelsson, S., Lindqvist, U., Gustafson, U., Jonsson, E.: An Approach to UNIX Security Logging. In: Proceedings of the 21st NIST-NCSC National Information Systems Security Conference, Crystal City, VA, October 1998, pp. 62–75 (1998)
5. Becker, R., Eick, S.G., Wilks, A.: Visualizing Network Data. IEEE Transactions on Visualization and Computer Graphics **1**(1), 16–28 (1995)
6. Del Bimbo, A.: Visual Information Retrieval. Morgan Kaufmann Publishers, Inc. San Francisco, CA (1999)
7. Couch, A.: Visualizing Huge Tracefiles with Xscal. In: 10th Systems Administration Conference (LISA '96), Chicago, IL, October 1996, pp. 51–58 (1996)
8. Debar, H., Dacier, M., Wespi, A.: Towards a Taxonomy of Intrusion Detection Systems. Computer Networks 31(8), 805–822 (1999)
9. Erbacher, R.: Visual Traffic Monitoring and Evaluation. In: Proceedings of the Second Conference on Internet Performance and Control of Network Systems, Denver, CO, August 2001, pp. 153–160 (2001)
10. Erbacher, R., Frincke, D.: Visualization in Detection of Intrusions and Misuse in Large Scale Networks. In: Proceedings of the International Conference on Information Visualization '00, London, UK, July 2000, pp. 294–299 (2000)
11. Esmaili, M., Safavi-Naini, R., Balachandran, B.M.: AUTOGUARD: A Continuous Case-Based Intrusion Detection System. In: Proceedings of the 20th Australasian Computer Science Conference (1997)
12. Smeulders, A.W., et al.: Content-Based Image Retrieval at the End of the Early Years. IEEE Transactions on Pattern Analysis and Machine Intelligence 22(12), 1349–1380 (2000)
13. Nyarko, K., et al.: Network Intrusion Visualization with NIVA, an Intrusion Detection Visual Analyzer with Haptic Integration. In: Proceedings of the 10th Symposium on Haptic Interfaces for Virtual Environment and Teleoperator Systems, Orlando, FL (2002)
14. Esmaili, M., et al.: Case-Based Reasoning for Intrusion Detection. In: Proceedings of the 12th Annual Computer Security Applications Conference, San Diego, CA (1996)

15. Lippmann, R.P., et al.: Analysis and Results of the 1999 DARPA Off-Line Intrusion Detection Evaluation. In: Proceedings of Recent Advances in Intrusion Detection, Toulouse, France, pp. 162–182 (2000)
16. Forrest, S.: A Sense of Self for UNIX Processes. In: Proceedings of the IEEE Symposium on Security and Privacy, Oakland, CA, pp. 120–198. IEEE Computer Society Press, Los Alamitos (1996)
17. Forsyth, D., Ponce, J.: Computer Vision: A Modern Approach. Prentice-Hall, Inc., Upper Saddle River, NJ (2003)
18. Frank, J.: Artificial Intelligence and Intrusion Detection: Current and Future Directions. In: Proceedings of the 17th National Computer Security Conference, Washington, D.C., pp. 22–33 (1994)
19. Frincke, D., Tobin, D., McConnell, J., Marconi, J., Polla, D.: A Framework for Cooperative Intrusion Detection. In: Proceedings of the 21th National Information Systems Security Conference, Crystal City, VA, October 1998, pp. 361–373 (1998)
20. Girardin, L., Brodbeck, D.: A Visual Approach for Monitoring Logs. In: Proceedings of the Second Systems Administration Conference (LISA XII), Boston, MA, October 1998, pp. 299–308 (1998)
21. He, T., Eick, S.G.: Constructing Interactive Visual Network Interfaces. Bells Labs Technical Journal 3(2), 47–57 (1998)
22. Hendee, W., Wells, P.: The Perception of Visual Information. Springer, Heidelberg (1994)
23. Hughes, D.: Using Visualization in System and Network Administration. In: Proceedings of the 10th Systems Administration Conference (LISA '96), Chicago, IL, October 1996, pp. 59–66 (1996)
24. Jain, R.: Proceedings of US NSF Workshop Visual Information Management Systems (1992)
25. Kolodner, J.: Case-Based Reasoning. Morgan Kaufmann Publishers, Inc., San Mateo, CA (1993)
26. Kruegel, C., Mutz, D., Valeur, F., Vigna, G.: On the Detection of Anomalous System Call Arguments. In: Snekkenes, E., Gollmann, D. (eds.) ESORICS 2003. LNCS, vol. 2808, pp. 326–343. Springer, Heidelberg (2003)
27. Mizoguchi, F.: Anomaly Detection Using Visualization and Machine Learning. In: Proceedings of the 9th International Workshop on Enabling Technologies: Infrastructure for Collaborative Enterprises (WET ICE'00), Gaithersburg, MD, March 2000, pp. 165–170 (2000)
28. Rubner, Y., Tomasi, C., Guibas, L.J.: A Metric for Distributions with Applications to Image Databases. In: Proceedings of the IEEE International Conference on Computer Vision, Bombay, India, January 1998, pp. 59–66. IEEE Computer Society Press, Los Alamitos (1998)
29. Rubner, Y., Tomasi, C., Guibas, L.J.: The Earth Mover's Distance as a Metric for Image Retrieval. International Journal of Computer Vision 28(40), 99–121 (2000)
30. Shapiro, L.G., Stockman, G.C.: Computer Vision. Prentice-Hall, Inc., Upper Saddle River, NJ (2001)
31. Snapp, S.: DIDS (Distributed Intrusion Detection System): Motivation, Architecture and An Early Prototype. In: Proceedings of the National Information Systems Security Conference, Washington, D.C., October 1991, pp. 167–176 (1991)
32. Takada, T., Koike, H.: Tudumi: Information Visualization System for Monitoring and Auditing Computer Logs. In: Proceedings of the 6th International Conference on Information Visualization (IV'02), London, England, July 2002, pp. 570–576 (2002)

33. Tan, K., Killourhy, K., Maxion, R.: Undermining an Anomaly-Based Intrusion Detection System Using Common Exploits. In: Wespi, A., Vigna, G., Deri, L. (eds.) RAID 2002. LNCS, vol. 2516, Springer, Heidelberg (2002)

34. Varner, P.E., Knight, J.C.: Security Monitoring, Visualization, and System Survivability. In: 4th Information Survivability Workshop (ISW-2001/2002) Vancouver, Canada (March 2002) (2002)

35. Veltkamp, R.C., Tanase, M.: Content-Based Image Retrieval Systems: A Survey. Technical Report 2000-34, UU-CS, Utrecht, Holland (October 2000)

36. Wagner, D., Dean, D.: Intrusion Detection via Static Analysis. In: Proceedings of the IEEE Symposium on Security and Privacy, Oakland, CA, pp. 40–47. IEEE Computer Society Press, Los Alamitos (2001)

37. Wagner, D., Soto, P.: Mimicry Attacks on Host-Based Intrusion Detection Systems. In: Proceedings of the 9th ACM Conference on Computer and Communications Security, Washington, D.C., pp. 255–264. ACM Press, New York (2002)

38. Watson, I.: Case-Based Reasoning: Techniques for Enterprise Systems. Morgan Kaufmann Publishers, Inc., San Francisco (1997)

Sensing Attacks in Computers Networks with Hidden Markov Models

Davide Ariu, Giorgio Giacinto, and Roberto Perdisci

Department of Electrical and Electronic Engineering, University of Cagliari,
Piazza d'Armi, 09123 Cagliari, Italy
{davide.ariu, giacinto, roberto.perdisci}@diee.unica.it

Abstract. In this work, we propose an Intrusion Detection model for computer newtorks based on Hidden Markov Models. While stateful techniques are widely used to detect intrusion at the operating system level, by tracing the sequences of system calls, this issue has been rarely researched for the analysis of network traffic. The proposed model aims at detecting intrusions by analysing the sequences of commands that flow between hosts in a network for a particular service (e.g., an ftp session). First the system must be trained in order to learn the typical sequences of commands related to innocuous connections. Then, intrusion detection is performed by indentifying anomalous sequences. To harden the proposed system, we propose some techniques to combine HMM. Reported results attained on the traffic acquired from a European ISP shows the effectiveness of the proposed approach.

1 Introduction

The widespread diffusion of information systems in an increasing number of businesses, as well as for social and government services, requires incresing level of security. Very often, information resources are the core business of an organisation, or at least consitute one of the principal assets. The internal flow of information, and the external flow to customers and providers need to be deployed as a lightweigth service in order to be effective. As a consequence, information resources need to be easily reacheble and accessible, and the risk of misuse is increasing [18]. The adoption of best practices in the configuration and management of all the devices in the network is the first step to protect the information. Very often reported incidents in computer networks are related to the misconfiguration of:

- operating systems and the applications running on the hosts inside the network;
- routers and switches, which are the devices connecting the hosts in a local network, and the local network to the Internet;
- firewalls, which are the first line of defence used to protect a network from attempts of intrusions.

P. Perner (Ed.): MLDM 2007, LNAI 4571, pp. 449–464, 2007.

However, no matter how cleverly the network has been configured and managed, an intruder may find his path through the inevitable bugs and errors that are always present in software or may exploit legitimate services as security requires setting a trade-off bewteen protection of resources and their usability. As a consequence network analysis tools are needed to detect anomalous or intrusive traffic. These tools used to protect the network and its resources are called *Intrusion Detection Systems* (IDS). An IDS includes a set of tools that can be used to detect and stop attempts of intrusion. We can distinguish Intrusion Detection Systems between anomaly-based and misuse-based Systems. The anomaly-based approach has been the first to be developed as in principle this approach is able to detect intrusions never seen before [6]. These kind of IDS are based on a description of the normal behaviour. Starting from this description, the system classifies as anomalous all the behaviors that are different from the normal ones. Anomalous behaviors are typically related to intrusions, but they may also be related to normal activities as the definition of a good model of normal activities is far from being perfect. As a consequence, anomaly based systems may generate a very high percentage of false alarms. For this reason, the most widely used IDS model are based on misuse detection. Misuse-based systems perform a pattern matching between a set of rules (called signatures), which describes well known attacks, and currently observed patterns. If this process detects a matching between the observed behaviors and those encoded in the signatures, the system labels the observed patterns as an attempt of intrusion. It is easy to see that misuse based IDS can precisely detect known intrusions, but if the traits of attacks are only slightly modified the matching process is likely to fail. This is the case of so-called "polimorphic" attacks, where the code of the attack is changed in order to evade misuse-based IDS while retaining their malicious effect. The increasing number of these kind of attacks in recent years motivates a renewed interest on anomaly based IDS [19].

In this work, we propose an anomaly based IDS that analyzes sequences of commands exchanged between two hosts through a certain protocol (e.g., FTP, SMTP, HTTP, etc.), and produces an output score that is used to assess if the analyzed sequences are normal or anomalous. To model normal network traffic, we use Hidden Markov Models (HMM). After a sequence of commands is analysied, the HMM assigns a probability value that can be interpreted as the likelihood that the sequence is normal. By setting a threshold on this probability value, it is possible to flag anomalous traffic. However, the performances of HMM depend on the choice of the learning parameters as well as on the number of hiddden states. Thus, it may be difficult to design a model that meets the requirement of high detection rate and low false alarm rates. To solve this problem, we propose to use an ensemble of HMM created by using multiple training sets and multiple learning parameters. Experimental results show the effectiveness of the ensemble approach with respect to the use of a single model.

The paper is organised as follows. A review of the related works on stateful approaches to intrusion detection is reported in section 2. Section 3 summarises the basic concepts of Hidden Markov Models. The proposed IDS model is described in section 4, where the techniques used to design the ensemble of HMM are also reported. Experimental results related to the analysis of the FTP traffic of a European Internet Service Provider are reported in Section 5. Conclusions are drawn in Section 6.

2   Related Works

HMM have been successfully used in a numer of pattern recognition applications in the past years (e.g. Speech recognition, Motion recognition, etc.) . HMM have also been used for Intrusion Detection thanks to their ability to model time-series using a stateful approach where the role and meaning of the internal states are "hidden". In an Intrusions Detection problem these series may be sequences of events, commands or function running on a single host, or sequences of packets in a Network. The vast majority of studies that proposed HMM to implement IDS are related to host-based systems, i.e., IDS that analyzes the actions performed on a single host to detect attempts of intrusion[4][10][12][24]. The simplest way to detect an attempt of intrusion in a single host, is to analyze the log files that contain the traces of the system calls. In fact, when the goal of an intruder is to gain control of the operating system, typically it can be detected by analysing the sequence of system calls and comparing them to typical sequences observed during normal system usage [15].

The user's behavior can be described using different mechanisms of auditing. At the lower level the behavior of users is represented by the sequences of input characters, while, at higher levels, the behavior can be characterised by the sequence of input commands or by the characteristics of different work sessions (with a work session being usually defined as the set of simple operations that a user performs to carry out a more complex operation [5]).

When a sequence is evaluated by HMM, a value can be associated to the sequence, which denotes the probability that the sequence is produced by the process modeled by the HMM [21]. Also, the most likely sequence of states that generate the observed sequence of symbols can be compute. In this latter case, a database of normal sequences is needed to perform intrusion detection by the direct comparison of the sequence of states output by the HMM and the normal ones that are stored into the database [25].

To the best of our knowledge, only few works have proposed the use of HMM to analyse Network traffic [14] [11]. In addition, these works represent the traffic at the packet level using features as the source and destination ports, the values of flags, and the content of the message. Thus, according to these works, a probability value is assigned to each packet. On the other hand, in this work we propose a state model at the application level, where the traffic is characterised by the commands exchanged between hosts in the Internet.

3   Hidden Markov Models

Hidden Markov Models represent a very useful tool to model time-series, and to capture the underling structure of a set of strings of symbols. HMM is a stateful model, where the states are not observable (hidden). A probability density function is associated to each hidden state that provides the probability that a given symbol is emitted from that state. A Hidden Markov Model $\lambda = (S, V, A, B)$ is defined as (see figure 1):

- **S** $= \{S_1, ..., S_N\}$, the set of N hidden states in the model.
- **V** $= \{V_1, ..., V_M\}$, the set of M distinct observation symbols emitted from each state.
- **A** $= \{a_{i,j}\}$, a NxN matrix of transition probabilities between states, where $\{a_{i,j}\}$ is the probability of being in the state j at time $t + 1$ given that we were in state i at time t

  $a_{i,j} = P(q_{t+1} = S_j \mid q_t = S_i), 1 \leq i,j \leq N$, where q_t is the state at time t.

- The probability density function that describes the probability to emit symbols from each state of the HMM.

$$b_j(k) = P(V_k \mid q_t = S_j)$$
$$1 \leq j \leq N, 1 \leq k \leq M$$

- The probability of being in the state i at the beginning of the process (i.e., t=1) $\pi = \{\pi_i\}$.

$$\pi_i = P(q_1 = S_i) , 1 \leq i \leq N$$

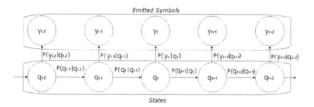

Fig. 1. The basic structure of HMM

HMM are based on the Markov's property whereby the probability of being in a state q_{t+1} at time t+1 depends only on the state q_t at time t. Accordingly, the joint probability of observable (emitted symbols y_i) and unobservable (hidden state q_i) variables can be expressed as:

$$P(y_1^T, q_1^T) = P(q_1) \prod_{t=1}^{T-1} P(q_{t+1} \mid q_t) \prod_{t=1}^{T} P(y_t \mid q_t)$$

The joint probability distribution is thus fully specified by: i) The initial state probability $P(q_1)$; ii) the transition probabilities $P(q_{t+1} \mid q_t)$; iii) the emission probabilities $P(y_t \mid q_t)$.

3.1   Basic HMM Problems

Three basic problems can be solved by Hidden Markov Models: the *Decoding* problem, the *Training* problem, and the *Evaluation* problem [22]. The *Decoding* problem is formulated as follows: given a sequence **O**, and a model λ, find the most likely sequence of states of λ that generated **O**. As this problem is not addressed in this paper, we will not provide details about it. On the other hand, we provide details of the *Training* and *Evaluation* procedures in the following subsections. Let us first describe the so-called Forward-Backward procedure, because forward and backward variables are used in the training and evaluation problems.

Forward-Backward Procedure. Let us consider the variable $\alpha_t(i)$ defined as

$$\alpha_t(i) = P(O_1, O_2, ..., O_t, q_t = S_i | \lambda),$$

This variable represents the probability of observing the sequence $\{O_1, O_2, ..., O_t\}$, given the model λ, and that the state variables at time t is $q_t = S_i$. The procedure of estimation of $P(O|\lambda)$ is made up of three steps:

1. *Initialization.* $\alpha_1(i) = \pi_i b_i(O_1)$, $1 \leq i \leq N$. This step initializes the forward probability α as the joint probability of the state S_i and the initial observation O_1.

2. *Induction.* $\alpha_{t+1}(j) = [\sum_{i=1}^{N} \alpha_t(i) a_{ij}] \cdot b_j(O_{t+1})$
   $1 \leq t \leq T - 1, \quad 1 \leq j \leq N$.

3. *Conclusion.* $P(O|\lambda) = \sum_{i=1}^{N} \alpha_T(i)$

The backward probability is computed in a similar way. The backward probability is defined as the probability that the last symbol of a sequence O_T is preceded by the sequence of symbols O_{T-1}, O_{T-2}, until the symbol O_{t+1}. The backward variable is defined as

$$\beta_t(i) = P(O_{t+1}, O_{t+2}, ..., O_T, q_t = S_i | \lambda)$$

and describes the probability of a subsequence of symbols within time t+1 and time T.

$\beta_t(i)$ can be calcultated by induction:

1. *Initialization* $\beta_T(i) = 1$, $1 \leq i \leq N$.

2. *Induction* $\beta_t(i) = \sum_{j=1}^{N} a_{ij} \cdot b_j(O_{t+1}) \beta_{t+1}(j)$,
   $t = \{T\text{-}1, T\text{-}2, ..., 1\}$, $1 \leq i \leq N$.

3.2   Evaluation

Given a model λ, and a sequence of symbols $\mathbf{O} = \{O_1, ..., O_T\}$, we want to compute the probability $P(O_| \lambda)$ that the sequence of symbols is emitted by the model. This probability provides a "matching value" between the model, and the sequence. This problem can be solved using the forward variables, because $P(O|\lambda)$ can be expressed as the sum of the terminal forward variables $\alpha_T(i)$:

$$P(O|\lambda) = \sum_{i=1}^{N} \alpha_T(i)$$

3.3   Training

Given a set of sequences $\{O_|\}$, we need to calculate the model λ which maximises $P(O_||\lambda)$. In this case, the problem is to find the set of parameters (A, B, π) that maximise the Emission Probabilities $P(\{O_t\}_||\lambda)$ of a given set of sequences $O_|$. The solution of the problem can be find through an iterative procedure aimed at finding a local maximization of $P(O|\lambda)$. One of the most widely used training procedure for HMM is the *Baum-Welch* algorithm [2], which is an *Expectation-Maximization* algorithm that computes the parameters of the model by maximizing the log-likelihood $\lambda = arg\,max\,log(P(\{O_t\}_||\lambda))$. At each iteration, a new estimation of the parameters is performed using the probability density functions estimated at the preceding iteration. Typically the initial values of the parameters are randomly chosen. More details on the Baum-Welch algorithm can be found in [2].

4   The Proposed IDS Model

The proposed IDS aims at analysing sequences of commands exchanged between pairs of hosts, in order to assess if the sequences represent attempts of intrusion or not. To perform this analysis, three problems must be addressed:

- the length of the sequences is not known in advance.
- the correlations between the elements in the sequences are not known in advance, so that we cannot use a window of fixed length to capture correlated elements.
- the internal state of the machine responding to the commands is unknown.

The first problem does not allow designing the IDS as a deterministic finite state machine, as for these state machines we must fix the initial and final states as well as the transitions between states. On the other hand, HMM is suitable for this purpose.

The basic idea of the proposed IDS is represented in figure 2.

The sequence of events we are interested in is the sequence of commands (**USER, PASS, PWD**, etc.) and numeric codes (**220, 231, 257**, etc.) exchanged between hosts. In particular, we are not interested in any argument asssociated to the command (e.g., the command "STORE xxx" is considered as "STORE"). In order to explain the characteristics of the proposed system, we will refer to the scheme reported in Figure 3 [25].

Fig. 2. Commands exchanged between hosts

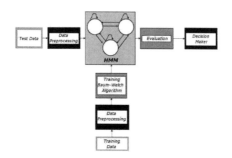

Fig. 3. The basic scheme of the proposed IDS

The scheme can be explained as follows:

1. The first component, called *Data-Preprocessing*, is a module that performs a number of preliminary operations on the sequence in order to make it suitable for the HMM. During the training phase, this module extracts the dictionary of symbols to be used by the HMM. Once the dictionary is created, all the test sequences are preprocessed in order to contain only the symbols that are in the dictionary. We will go into the details of the creation of dictionaries later on.

2. HMM are built using the Baum-Welch procedure, using a set of training sequences.

3. Once the model is built, its performances are assessed by a set of test sequences, using the *Evaluation Procedure*.

4. For each test sequence, the HMM outputs a probability value stating how likely the sequence is anomalous. By setting a decision threshold, the sequence can be labeled as normal or anomalous (i.e., potentially intrusive).

4.1   Creation of the Dictionaries of Symbols

To train and test HMM, we need to create the dictionaries of symbols. Such symbols are related to the commands exchanged by hosts for a given Internet service. Typically the number of commands defined by the RFC (i.e., the rules that define the protocols associated to services) is very large, compared to the number of commands that are actually used by applications. As a consequence, the use of a dictionary comprising all the possible commands would be ineffective as a large number of emission probabilities would be zero, and the computational load would be high. In addition, if a new command is added to the protocol, the HMM must be re-trained to take into account the new symbol. On the other hand, if we build the dictionary by using only the set of symbols of the sequences in the training set, some action must be perfomed for those test sequences that contain symbols that are not in the dictionary.

In this work, we propose to build the dictionaries by using only the symbols in the training sequences. We present two alternatives solutions for processing the test sequences that we call *Large Dictionary* and *Small Dictionary*.

Large Dictionary. The dictionary D=$\{S_1, S_2, ..., S_n, "NaS"\}$ contains all the symbols $S_1...S_n$ the are present in the training sequences, plus a special symbol "NaS" (Not a Symbol). Unknown symbols in the test set are managed by replacing them with "NaS". Of course this symbol is not a command defined by the RFC, but is simply used to replace all the symbols in the test sequences that don't belong to the dictionary of commands learned from the training sequences. As an example, if the HMM is trained using the dictionary of symbols "a","b","c","d", the sequence *[a-b-d-g-c-a]* cannot be analysed because there is the unknown symbol "g". If the dictionary is enlarged with the "NaS" symbol, so that the HMM is trained using the dictionary {"a","b","c","d","NaS"} , the symbol "g" into the test sequence can be replaced with "NaS": and the resulting sequence *[a-b-d-NaS-c-a]* can be analysed by the HMM.

Small Dictionary. In this second solution, we discard from the test sequences all the symbols that don't belong to the dictionary. Thus, the test sequence of the previous example *[a-b-d-g-c-a]* becomes *[a-b-d-c-a]*.

If we compare the two solutions we can observe that in the case of Large Dictionaries, all the test sequences that contain an unknown symbol are anomalous, as the symbol NaS is never encountered in the training set, and its associated probability of emission is 0. On the other hand, if the solution using a Small Dictionary is used, intrusions that contain unknown symbols cannot be detected if the sequence resulting after discarding the unknown symbols are similar to the normal ones. We can conclude that the fewer the number of erased symbols compared to the length of the sequence, the smaller the impact of the Small Dictionary solution. The use of Large Dictionaries on the other hand, allows producing an alert for each new symbol encountered. If the training set is highly representative, then the presence of an unknown symbol in a test sequence can be certainly related to some kind of anomaly.

4.2   Combination of HMM

As the performances of HMM are sensitive to the training set, and to the initial values of the parameters, in this work we explored the performances attained by combining an ensemble of HMM in order to attain low false alarms and high detection rates. To this end, we used three techniques for combining the outputs of HMM, namely:

- Arithmetic Mean
- Geometric Mean
- Decision Templates

The first two techniques simply combine the outputs by computing the average of the outputs:

$$P_{arithm}(O|\lambda) = 1/L \cdot \sum_{i=1}^{L} P(O|\lambda_i)$$

or the product of the outputs:

$$P_{geom}(O|\lambda) = \sqrt[L]{\prod_{i=1}^{L} P(O|\lambda_i)}$$

where $P(O|\lambda_i)$ is the probability that the sequence O has been emitted by the i-th HMM, and L represents the number of combined HMM.

The combination by *Decision Templates* is a more complex technique that has been first proposed in [16], and that has been used to combine HMM outputs [3] [7].

The *Decision Templates* method is based on a similarity measure between two vectors, called Decision Profile and Decision Template. The Decision Template is a vector whose elements represent the mean support given by each classifier to the N training sequences of each class. So, as in this case we are interested in modeling only one class, i.e. the normal class, the decision template represents the average of the emission probabilities of training sequences for each HMM. Let us define $dt_i(Z)$ as the average emission probability of the *i-th* HMM for the N sequences in the training set Z:

$$dt_i(Z) = 1/N \cdot \sum_{j \in Z} P(O_j/\lambda_i)$$

The Decision Template is thus defined as follows:

$$\mathbf{DT}(Z) = [dt_1(Z)...dt_k(Z)...dt_L(Z)]$$

Analougously the Decision Profile for a test sequence O_{test} is defined as follows:

$$\mathbf{DP}(O_{test}) = [P(O_{test}|\lambda_1)...P(O_{test}|\lambda_k)...P(O_{test}|\lambda_L)]$$

A soft label is then assigned to the test sequence O_{test} by means of a similarity measure between $DP(O_{test})$ and $DT(Z)$. We compute this similarity by the *Squared Euclidean Distance*:

$$Sim(DT(Z), DP(O_{test})) = 1 - \frac{1}{L} \sum_{i=1}^{L} (dt_i(Z) - P(O_{test}|\lambda_i))^2. \qquad (1)$$

5   Experimental Results

5.1   Dataset

The dataset used to test the proposed solution is made up of a set of sequences of FTP commands exchanged betwen a server and many clients. These sequence are extracted from the FTP traffic that is generated by the users that upload and download their resources on their own Web Space. The data were extracted from the network of the European ISP *Tiscali SpA*. The sequences of commands have been extracted by the live traffic using SNORT, a very popular open source IDS [23]. In order to filter out potentially intrusive sequences, we discarded all the sequences for which SNORT raised an alarm. The resulting dataset is made up of 40,000 sequences that have been used to train and test the HMM. First, we randomly extracted a training set made up of 80% of the traffic, the remaining 20% being used for testing. To avoid a bias in the evaluation, we repeated this subdivision five times. Thus, we created 5 different training sets, each one made up of 32,000 sequences, and 5 different test sets, each one made up of 8,000 sequences. Each of the 5 training set is further subdivided into 10 subsets (without replacement) of 3,200 sequences. Each of this sequences has been used to train distinct HMM. As a result, 50 different training set are availble for training HMMs. The main drawback of Hidden Markov Models is the computational cost of the training process, the larger the training set, the longer the training time. On the other hand, the training sets used to build Anomaly Based IDS are typically very large, so that normal activities overwhelm those anomalous events that can be present in the training traffic. In addition, the set of parameters used of a HMM trained on a large training set may not capture the structure of data. For this reason, it can be more effective to split the training set into a number of smaller subsets, and to use each subset to train different HMM. The outputs of these HMM can be then combined using the techniques outlined in the previous section, thus exploiting the information in the training set.

In order to create attack sequences, we used the simulator *IDS-Informer* [26] and added 22 attack sequences to the test set. It is worth nothing that the generation of attack sequences is not an easy task, because typically for each service a very small number of vulnerabilities can be actually exploited. This can be explained by the fact that software vendors and developers update frequently their products to correct known vulnerabilities. In addition the traffic sniffed in a network tipically contains a very small percentage of attacks. Thus, this experimental setup allows simulating a real network scenario.

5.2   Dictionaries of Symbols

In order to generate the Dictionaries of Symbols, we implemented the Large and Small Dictionaries described in the previous section. In particular, in the case of Large Dictionaries, we extracted the symbols from each of the 5 training sets made up of 32,000. Thus, in this case all the ten HMM extracted from the same training set, use the same dictionary. On the other hand, Small Dictionaries have been extracted from each of the 3,200 sequences used to train each HMM. As a result, the ten HMM extracted from the same training set, use ten different dictionaries.

5.3 Experimental Setup

For each of the five Training Set, the following simulations have been performed for each of the 10 HMMs: i) HMM have been created using both the Large and the Small Dictionary; ii) three values for the number of states of HMM have been considered, namely, 10, 20, and 30; iii) two different random initializations of the initial values of the emission and transition matrixes have been performed. Thus, for a given number of states of the HMM, and for a given dictionary of symbols, 100 HMM have been created. Finally, the number of iterations for the training algorithm has been set to 100.

5.4 Performances Evaluation

In order to evaluate the performances of individual HMM we decided to report the mean value and the standard deviation computed over all the 100 HMM with the same number of states and the same kind of dictionary. Combination techniques have been used to combine the 20 HMM generated for each training set. Results of combination are reported in terms of average and standard deviation computed over the five training set.

In order to evaluate the performance of the proposed IDS, we selected three measures:

- The Area under the ROC curve, where the ROC curve represents the performance of the HMM at different values of the decision threshold. In particular, the ROC curve represents the *False Acceptance Rate*, i.e., the rate of attacks classified as normal traffic Vs. the *True Positive Rate*, i.e. the rate of normal sequences classified as attacks. It is easy to see that the larger the AUC, the better the performance.
- The percentage of *real false alarms* measured on the test dataset when the *Detection Rate* is equal to 100%.
- The *Detection Rate*, when the percentage of *false alarms* measured on the training set is equal to 1%. The threshold has been calculated on the *Training Set*, so we evaluated the corresponding percentage of *false alarms* on the *Test Set*.

The second performance measure is used to assess the performance of the system in term of the number of false alarms that are produced if we wish to attain a 100% detection, while the third measure aims at assessing the performances when the false alarm rate is limited to 1%. This value typically represent an upper bound for the tolerable false positives for an IDS.

5.5 Nomenclature

Let us define some acronyms that are used in the tables where results are reported: i) **LD** and **SD** are used to denote respectively the use of a *Large* or

Small Dictionary; ii) *10s, 20s, and 30s* are used to specify the number of states of the HMM; iii) DR is used for the *Detection Rate*; iv) FA is used for the *False Alarms* rate.

5.6   Experimental Results

Experimental results pointed out that the use of Small Dictionaries provides significantly lower performances with respect to the use of Large Dictionaries. Thus, for the experiments relates to the use of Small Dictionaries, we decided to report only the best results attained by varying the number of states of the HMM. This result has been attained by setting the number of states of HMM to 30. Table 1 shows the performances of this configuration. Reported results clearly show that high vaues of AUC can be attained by combining the HMM using the Decision Template technique. Thus, as far as the AUC is concerned, combining an ensemble of HMM allows improving the performances. However, if we analyse the False Alarm rate produced when the decision threshold is set to have a 100% Detection Rate, we can easily see that these values cannot be accepted in a real working scenario, as more than 90% of normal sequences have been classified as intrusives. In addition, the combination of HMM provides less reliable results than those provided on average by individual HMM. On the hand, if we set the decision threshold (on the training set) so that the False Alarm rate is equal to 1%, we see that the performances of combination techniques are higher than those of individual HMM, the best performance being attained by the Geometric Mean. In the following we will see that the use of Large Dictionaries allows attaining higher performances. On the other hand, as far as the training time is concerned, the use of Small Dictionaries require a shorter training time than that needed when using Large Dictionaries.

Table 1. Simulations Small Dictionary 30 States

30 States SD		DR 100%		FA1%	
	AUC	FA(real%)	DR %	FA(real%)	
	mean(σ)	mean(σ)	mean(σ)	mean(σ)	
Mean 100 HMM	0.873 (0.006)	89.77 (4.14)	58.72 (2.52)	0.72 (0.23)	
Arithmetic Mean	0.874 (0.002)	95.27 (0.61)	63.63 (4.54)	0.31 (0.15)	
Geometric Mean	0.876 (0.002)	96.19 (1.50)	76.36 (2.03)	0.71 (0.20)	
Decision Templates	0.933 (0.004)	93.84 (8.40)	65.54 (3.80)	0.35 (0.16)	

If we analyse the performances of HMM using Large Dictionaries reported in Tables 2, 3, and 4, it is easy to see that performances are quite superior to those attained using Small Dictionaries. In particular, performances improved significantly by increasing the number of states from 10 to 20. On the other hand a further increase in the number of states from 20 to 30 does not provide significant improvements in performance, except for the standard deviation which is

Table 2. Simulations Large Dictionary 10 States

10 States LD		DR 100%	FA1%	
	AUC	FA(real%)	DR%	FA(real%)
	mean(σ)	mean(σ)	mean(σ)	mean(σ)
Mean 100 HMM	0.953 (0.006)	65.15 (2.09)	85.44 (1.73)	0.80 (0.23)
Arithmetic Mean	0.958 (0.002)	76.41 (1.22)	82.72 (2.03)	2.35 (1.24)
Geometric Mean	0.961 (0.001)	74.54 (1.07)	92.72 (4.06)	2.83 (1.22)
Decision Templates	0.958 (0.002)	74.89 (2.55)	83.62 (4.06)	2.42 (1.22)

Table 3. Simulations Large Dictionary 20 States

20 States LD		DR 100%	FA1%	
	AUC	FA(real%)	DR%	FA(real%)
	mean(σ)	mean(σ)	mean(σ)	mean(σ)
Mean 100 HMM	0.967 (0.004)	59.60 (4.46)	90.94 (1.27)	0.74 (0.22)
Arithmetic Mean	0.974 (0.002)	79.23 (3.02)	92.72 (6.1)	0.33 (0.16)
Geometric Mean	0.972 (0.001)	52.27 (3.06)	95.45 (0)	0.89 (0.09)
Decision Templates	0.965 (0.002)	52.34 (9.69)	95.45 (0)	0.41 (0.18)

Table 4. Simulations Large Dictionary 30 States

30 States LD		DR 100%	FA1%	
	AUC	FA(real%)	DR%	FA(real%)
	mean(σ)	mean(σ)	mean(σ)	mean(σ)
Mean 100 HMM	0.969 (0.003)	57.85 (3.57)	92.97 (0.65)	0.74 (0.16)
Arithmetic Mean	0.974 (0.0004)	55.92 (0.62)	95.45 (0)	0.53 (0.13)
Geometric Mean	0.971 (0.0008)	55.00 (1.20)	95.45 (0)	1.01 (0.10)
Decision Templates	0.962 (0.004)	86.02 (8.12)	95.45 (0)	0.62 (0.17)

smaller than that of HMM with 20 states. The values of AUC in the three cases are larger than 0.95, the combination by the arithmetic and geometric means providing the highest performances.

If we analyse the performances attained when the decision threshold is set so that the Detection Rate is equal to 100%, we see that the values of False Alarm rate are quite smaller than those attained using the Small Dictionaries, but still these values are not suited for a real operating environment. It is worth noting, however, that hardly any IDS is able to produce an acetable False Alarm rate when it is tuned to detect the 100% of attacks [1]. Thus the evaluation of IDS at 100% detection rate is just used to see the performances in the limit. From an operational point of view, it is more intersting to evaluate the detection rate when the false alarm rate is fixed at 1%. If we compare the values attained using 10, 20, and 30 states we can see that the detection rate increases as the number of states is increased. Again, the highest values are attained by combining HMM, reaching the value of 95.45%. If we compare this result with the false alarm rate attained at 100% detection rate, it is easy to see that a small increase in the detection rate is accompained by a very large increase in the false alarm rate. Finally, the tables also report the false alarm rate attained on the test set (FA(real%)) when the decision threshold is set to the value that produces the 1% false alarm rate on the training set. It can be seen that, apart from the case of 10 states, the false alarm rate on the test set is always smaller that 1%. Thus, the threshold estimated on the training set produces similar results on the test set.

6   Conclusions

This work proposed a novel technique to detect intrusions in computer networks, based on the analysis of sequences of commands exchanged between pairs of hosts. In particular we modelled sequences using HMM. For each command sequence, a probability value is assigned and a decision is taken according to some predefined decision threshold. We investigated different HMM models in terms of the dictionary of symbols, number of hidden states, and different training sets. We found that good performances can be attained by using dictionary of symbols made up of all symbols in the training set, and adding a NaS (not-a-symbol) symbol in account of symbols in the test set that are not represented in the training set. Performances can be further improved by combining different HMM. As the size of training sets in an intrusion detection application is typically large, we proposed to split the training set in a number of parts, training different HMM and then combining the output probabilities by three well known combination techniques. Reported results on a real dataset extracted from the live traffic of an ISP show the effectiveness of the proposed approach.

Future works should include the fusion of information from the proposed module, which analyses sequence of commands, with information from other modules devoted, for example, to the analysis of the arguments of commands (e.g., the name of files exchanged, subject of e-mails, etc.). In fact attacks can be reliably detected when multiple analysis are performed on the network traffic, and the partial results combined. We suspect the resulting IDS will not only produce a lower false alarm rate, but also be more robust to evasion activities, as the attacker should evade the detection capabilities of multiple modules working on different traffic characteristics.

References

1. Axelsson, S.: The Base-Rate Fallacy and its Implications for the Difficulty of Intrusion Detection. In: Proc. of RAID (May 1999)
2. Baum, L.E., Petrie, T., Soules, G., Weiss, N.: A maximization technique occurring in the statistical analysis of probabilistic functions of Markov chains. Ann. Math. Statist. 41(1), 164–171 (1970)
3. Bicego, M., Murino, V., Figueiredo, M.: Similarity-Based Clustering of Sequences Using Hidden Markov Models. In: Perner, P., Rosenfeld, A. (eds.) MLDM 2003. LNCS, vol. 2734, pp. 86–95. Springer, Heidelberg (2003)
4. Cho, S., Han, S.: Two sophisticated techniques to improve HMM-based intrusion detection systems. In: Vigna, G., Krügel, C., Jonsson, E. (eds.) RAID 2003. LNCS, vol. 2820, pp. 207–219. Springer, Heidelberg (2003)

5. Debar, H., Becker, M., Siboni, D.: A Neural Network Component for an Intrusion Detection System. In: Proc. of the IEEE Computer Society, Symposium on Research in Security and Privacy (1992)
6. Denning, D.E.: An Intrusion Detection Model. IEEE Trans. Software Eng. SE-13(2), 222–232 (1987)
7. Dietrich, C., Schwenker, F., Palm, G.: Classification of Time Series Utilizing Temporal and Decision Fusion. In: Kittler, J., Roli, F. (eds.) MCS 2001. LNCS, vol. 2096, pp. 378–387. Springer, Heidelberg (2001)
8. Dietterich, T.: Ensemble Methods in Machine Learning. In: Kittler, J., Roli, F. (eds.) MCS 2000. LNCS, vol. 1857, pp. 1–15. Springer, Heidelberg (2000)
9. Duda, R.O., Hart, P.E., Stork, D.G.: Pattern Classification. Wiley-Interscience, Chichester (2000)
10. Feng, H.H., Kolesnikov, O.M., Fogla, P., Lee, W., Gong, W.: Anomaly detection using call stack information. In: Proceedings of the 2003 IEEE Symposium on Security and Privacy, IEEE Computer Society Press, Los Alamitos (2003)
11. Gao, F., Sun, J., Wei, Z.: The prediction role of Hidden Markov Model in Intrusion Detection. In: Proc. of IEEE CCECE 2003, vol. 2, pp. 893–896 (May 2003)
12. Gao, D., Reiter, M., Song, D.: Behavioral Distance Measurement Using Hidden Markov Models. In: Zamboni, D., Kruegel, C. (eds.) RAID 2006. LNCS, vol. 4219, pp. 19–40. Springer, Heidelberg (2006)
13. Giacinto, G., Roli, F., Didaci, L.: Fusion of multiple classifiers for intrusion in computer networks. Pattern Recognition Letters 24(12), 1795–1803 (2003)
14. Hashem, M.: Network Based Hidden Markov Models Intrusion Detection Systems. IJICIS, 6(1) (2006)
15. Hoang, X.D., Hu, J.: An Efficient Hidden Markov Model Training Scheme for Anomaly Intrusion Detection of Server Applications Based on System Calls. In: Proc. of 12th IEEE Conference on Networks, 2004, vol. 2, pp. 470–474 (2004)
16. Kuncheva, L., Bezdek, J.C., Duin, R.P.W.: Decision Templates for Multiple Classifier Fusion. Pattern Recognition 34(2), 299–314 (2001)
17. Kuncheva, L.: Combining Pattern Classifiers: Methods and Algorithms. Wiley, Chichester (2004)
18. Mc Hugh, J., Christie, A., Allen, J.: Defending yourself: The role of Intrusion Detection Systems. IEEE Software 42–51 (September/October 2000)
19. Fogla, P., Sharif, M., Perdisci, R., Kolesnikov, O.M., Lee, W.: Polymorphic Blending Attack. In: USENIX Security Symposium (2006)
20. Proctor, P.E.: Pratical Intrusion Detection Handbook. Prentice-Hall, Englewood Cliffs (2001)
21. Qiao, Y., Xin, X.W., Bin, Y., Ge, S.: Anomaly Intrusion Detection Method Based on HMM. Electronic Letters 38(13) (June 2002)
22. Rabiner, L.R.: A tutorial on Hidden Markov Models and selected applications in speech recognition. In: Proc. of IEEE vol. 77(2), pp. 257–286 (February 1989)
23. Roesch, M.: Snort - Lightweight Intrusion Detection for Networks. In: Proc. of the 13th USENIX conference on System Administration, LISA '99
24. Warrender, C., Forrest, S., Pearlmutter, B.: Detecting intrusions using system calls: alternative data models. In: Proc. of the IEEE Symposium on Security and Privacy, IEEE Computer Society Press, Los Alamitos (1999)
25. Zhang, X., Fan, P., Zhu, Z.: A New Anomaly Detection Method Based on Hierarchical HMM. In: Proceedings of the 4th PDCAT conference (2003)
26. IDS-Informer, www.blade-software.com

FIDS: Monitoring Frequent Items over Distributed Data Streams

Robert Fuller and Mehmed Kantardzic

Computer Engineering and Computer Science Department
University of Louisville, Louisville, KY 40292
{rhfull01, mmkant01}@louisville.edu

Abstract. Many applications require the discovery of items which have occur frequently within multiple distributed data streams. Past solutions for this problem either require a high degree of error tolerance or can only provide results periodically. In this paper we introduce a new algorithm designed for continuously tracking frequent items over distributed data streams providing either exact or approximate answers. We tested the efficiency of our method using two real-world data sets. The results indicated significant reduction in communication cost when compared to naïve approaches and an existing efficient algorithm called Top-K Monitoring. Since our method does not rely upon approximations to reduce communication overhead and is explicitly designed for tracking frequent items, our method also shows increased quality in its tracking results.

1 Introduction

Many applications require the analysis of data streams. Data streams are sequences of data that arrive continuously over time. The properties of data streams impose many computational challenges. Some of these properties include:

1. Data may arrive at a very fast rate. Sometimes as fast as several gigabytes a second [10].
2. The final length of the stream is often times not known in advance. Therefore, they are treated as never-ending streams [17].
3. Analysis of the stream must be done in a single pass since the data is too vast to be stored [8].

A common data stream analysis task is to find items in the data which have occurred frequently. An item is defined to be frequent if it accounts for a high percentage of the total number of occurrences seen so far. Important applications of frequent item analysis include:

1. Web Advertising: Revenue may be increased by recognizing users who frequently click advertisements and displaying Pay-Per-Click advertisements when they visit your site [13].

P. Perner (Ed.): MLDM 2007, LNAI 4571, pp. 464–478, 2007.
© Springer-Verlag Berlin Heidelberg 2007

2. Network Flow Management: Generally only a few flows will account for a large portion of bandwidth in a network. Knowing these flows can be used to allocate bandwidth more fairly [16].
3. Detecting Network Anomalies: Some network attacks exhibit frequent characteristics. For example, worms can be detected by determining frequently occurring substring patterns in traffic flows [9].

This paper considers the problem of monitoring frequent items over distributed data streams. In this scenario frequently occurring items must be determined from multiple data streams originating from dispersed sources. The term monitoring, means that the up-to-date list of frequent items are displayed to the user continuously in real-time. This problem is difficult since it inherits the challenges of any data stream analysis task. Since the data is occurring very rapidly and possibly indefinitely, memory must be managed carefully. The high rate of data streams require very fast response time. The method must process the new data and determine if the frequency status of each item has changed very quickly to prevent getting overwhelmed. Finally, the problem is made more difficult by introducing multiple distributed streams. Information must be gathered from each source to determine which items are frequent. Thus communication must be limited to observe any imposed network constraints.

Due to the difficulties described above, it is not surprising that few solutions for this problem have been proposed [4,11]. Most available solutions for frequent item monitoring focus on computational constraints, and were not designed to operate in a distributed environment [1,6,7,8,10,12,13,16]. The solutions designed for this setting, are either expensive computationally or only report the frequent items periodically [4,11]. It is the goal of this work to build off prior solutions in order to provide a more comprehensive approach to the frequent item monitoring problem [3].

The remainder of this paper will be organized as follows. Section 2 gives a formal definition of our problem and the distributed architecture used. Section 3 discusses prior work in the defined problem domain for which we build off of. In Sect. 4 we describe our approach for monitoring frequent items over distributed data streams. We evaluate our method based on a series of criteria in Sect. 5. Finally, closing remarks are given in Sect. 6.

2   Formal Problem Statement

2.1   System Architecture

The distributed monitoring environment used in our method has been defined as a single-level hierarchical architecture [3]. It consists of $m + 1$ nodes and m distributed data streams. Of the nodes, $N_1, N_2, \ldots, N_m$ are used for summarizing the m data streams and are called monitoring nodes. Node N_0 is a specialized coordinator node. The coordinator node is responsible for displaying the set of frequent items over the union of the m distributed data streams. As in previous work [3,4,5], communication is conducted amongst the monitoring nodes and the coordinator. There is no direct communication between any two monitoring nodes. A schematic of this architecture can be seen in Fig. 1. Each of the distributed data streams $S_1, S_2, \ldots, S_m$, is used as input to corresponding monitoring nodes $N_1, N_2, \ldots, N_m$. The data streams consist of a sequence of tuples ordered by time of occurrence. Each tuple is of the form $\langle o_j, t_j \rangle$, where o_j is the unique identifier of a specific item of interest pulled from a finite (but possibly large) set of allowable identifiers U, and t_j is the timestamp of the tuple. Identifiers may be repeated any number of times in a data stream. An example of an input stream, corresponding to monitoring node N_1, may be $S_1 = \{\langle 2, 0.024 \rangle, \langle 2, 0.029 \rangle, \langle 1, 0.050 \rangle, \langle 0, 0.056 \rangle\}$ where $U = \{0, 1, 2, 3\}$.

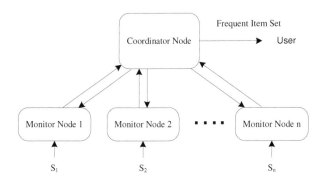

Fig. 1. Communication structure

As stated, each monitoring node maintains a summary of its corresponding data stream. This summary is made by managing a set of frequency counts $C_i = \{c_{1,i}, c_{2,i}, \ldots, c_{n,i}\}$, where each $c_{j,i} \in C_i$ corresponds to an item identifier from the set U. Initially each frequency count is equal to zero, and for each input tuple $\langle o_j, t_j \rangle$ to N_i, $c_{j,i}$ is incremented by one. Therefore, each frequency count in the set C_i maintains the number of occurrences of an item in the data stream S_i on monitoring node N_i. To extend the previous example, $C_1 = \{1, 1, 2, 0\}$.

2.2   Frequent Item Problem Definition

The purpose of the monitoring structure discussed above is to monitor frequent items over the union of the distributed data streams. Given an item o_j and corresponding counters $\{c_{j,1}, c_{j,2}, \ldots, c_{j,m}\}$, we call o_j frequent if $\sum_{1 \leq i \leq m} c_{j,i} \geq s \cdot N$, where $s \in (0, 1)$ is a user defined support parameter and N is the accumulative sum of all frequency counts across all monitoring nodes seen since the monitoring process was initiated. The set of all frequent items F, therefore, consists of all items which account for at least $s\%$ of the total number of item occurrences across the union of the m data streams.

To allow approximate frequency counts an extension has been proposed called the ϵ-deficient frequent items problem. The definition for this problem we used comes from the work of Manku and Motwani [12]. To extend the previous definition, the ϵ-deficient frequent items problem allows a degree of error on the frequency counts which is bounded by a user defined error tolerance parameter $\epsilon \ll s$. The membership of an item in the set F is modified with the following requirements:

1. Those whose true frequency exceeds $s \cdot N$ are in the frequent item set.
2. No item whose true frequency is less than $(s - \epsilon) \cdot N$ is in the frequent item set
3. Frequency counts are under counted or over counted by at most $\epsilon \cdot N$.

The resulting membership test derived from these three points is determined by whether the frequency counts are over estimated or under estimated. If the items are under estimated, an item is called frequent if $\sum_{1 \leq i \leq m} c_{j,i} \geq (s - \epsilon) \cdot N$, where each $c_{j,i}$ is an approximate frequency count underestimating the true frequency of o_j by at most $\epsilon \cdot |S_i|$. The advantage of allowing approximate frequency counts is the reduction of memory requirements across the monitoring structured. Although we will mainly focus on our previous definition throughout this paper, in Sect. 4.3 we will propose an extension to accommodate the ϵ-deficient frequent items problem.

2.3 Monitoring System Goals

As stated in the problem definition, the goal of the monitoring system is to continuously report the set of frequent items. However, in most scenarios the monitoring process is not the only responsibility of the nodes. They will usually be conducting other tasks (such as retrieving files requested by a user in the case of a file server). Therefore, the processing and memory requirements should be kept to the minimum. Finally, communication between the coordinator node and the monitoring nodes must be limited, to reduce overhead on the underlying network.

3 Prior Work

Prior work on monitoring frequent items over data streams focus on limiting space requirements [1,6,7,8,10,12,13,16]. Since the number of unique items to monitor may be very large, storing the frequencies of all these items may be unreasonable. To reduce the memory requirements, a relaxation of the original problem was created. The most common of which is the problem of finding the ϵ-deficient frequent items. This problem was addressed previously in Sect. 2.2.

Several algorithms have been proposed to solve the ϵ-deficient frequent items problem on a centralized monitoring environment. Manku and Motwani proposed two algorithms called Sticky Sampling and Lossy Counting to address this problem. The more popular of the two, Lossy Counting, requires $O(\frac{1}{\epsilon} \cdot \log(\epsilon N))$ space [12]. Demaine et al. proposed an algorithm known as Frequent which requires only $O(\frac{1}{\epsilon})$ space and $O(1)$ time [7]. Although this is an improvement in the worst case space requirement, Lossy Counting requires less space in practice on skewed data [1]. More recent work by Metwally et al. resulted in an algorithm called Space-Saving, in which they compared against several known ϵ-deficient frequent items algorithms. The final results showed that their method required more space and time than Frequent but provided better precision [13].

Although prior work has been completed on limiting space requirements in a centralized data stream environment, very little has been done on finding frequent items over distributed data streams. One paper to address this problem was proposed in [11]. In this paper frequent items were propagated up a hierarchical communication structure and displayed at a root node at the end of every T time units (ie. every 5 or 15 minutes). To reduce communication and space requirements, frequency precision at each level of the hierarchy was addressed. An obvious draw back of this method is that results are only given periodically. There may be cases when the set of frequent items have not changed between two or more consecutive time units, and thus the process results in wasted communication. Another problem is when the set of frequent items has changed between two consecutive time units. Some of these changes may not be detected by this approach. Another method proposed by Cormode and Garofalakis in [4] can be used to continuously monitor frequent items. Their method does this by maintaining at each monitoring node a summary of the input stream and a prediction sketch. If the summary varies from the prediction sketch by more than a user defined tolerance amount, the summary and (possibly) a new prediction sketch is sent to a coordinator node. The coordinator can use the information gathered from each monitoring node to continuously report frequent items. This method is very robust, being able to solve a range of monitoring tasks other than frequent item monitoring. One draw back of this approach is that the sketches maintained by each monitoring node require $O(\frac{1}{\epsilon^2} \log(\frac{1}{\delta}))$ space and $O(\log(\frac{1}{\delta}))$ time per update, where δ is a probabilistic confidence. Another drawback is that the total error tolerance must be high, otherwise small deviations of the stream summaries from the prediction sketches will result in communication with the coordinator.

Earlier work by Olston and Babcock in [3] addressed a similar problem. Their problem was to find the top-k items in a distributed stream environment. We believe this method, which we will call Top-K Monitoring, can also be used to monitor frequent items as defined in Sect. 2.2. One draw back of this method, however, is that the frequency count of every item encountered by a monitor is maintained in memory. Thus frequency counts can accumulate over time, placing a load on available memory.

It is the objective of this work to build off past experiences to develop a system to monitor frequent items. More specifically we examine the following three points.

1. The performance of Top-K Monitoring for tracking frequent items will be examined.
2. Top-K Monitoring will be modified to explicitly monitor frequent items, with the goal of reducing communication cost. We called this method FIDS (Monitoring Frequent Items over Distributed Data Streams).
3. Reduce memory requirements by proposing an extension to accommodate ϵ-deficient frequent items.

4 Frequent Item Monitoring

FIDS begins with an initialization phase. There are two ways to accomplish this task. One option is to issue an efficient one-time frequent item query. This method will reduce initial communication overhead since a number of update tuples will be summarized in a more condense fashion. The drawback of this approach is that the monitoring process will not begin until the initialization time period has passed. The second option is to forward all update tuples to the coordinator node. This will require more communication overhead but allow the monitoring process to begin immediately. Depending on the needs of the user any of these two methods can be used, although it is highly recommended that an initialization phase is used (the reasons will be clear later).

Once the initialization phase is completed the coordinator node sends to each monitor the current frequent item set F. Along with this, new values for the adjustment factors are assigned. The notation of adjustment factors are borrowed from Top-K Monitoring and are discussed below.

Each adjustment factor $\delta_{j,i}$ corresponds to an item o_j and node N_i, and are used to shift item occurrences amongst the nodes in the system to facilitate local constraint checking. Requirements for adjustment factors are that:

1. For each item o_j, its corresponding adjustment factors sum to zero across all nodes: $\sum_{0 \leq i \leq m} \delta_{j,i} = 0$.
2. For each item $o_f \in F$, its corresponding adjustment factor stored at the coordinator node is greater than or equal to zero: $\delta_{f,0} \geq 0$
3. For each item $o_{nf} \notin F$, its corresponding adjustment factor stored at the coordinator node is less than or equal to zero: $\delta_{nf,0} \leq 0$

After receiving the new adjustment factors and the current frequent item set, each monitor installs parameterized constraints which are used to determine if the validity of F has changed over time.

A key component to the parameterized constraints is a local threshold value T_i, kept by each corresponding monitoring node N_i. For each input tuple to N_i, T_i is incremented by the user defined support parameter s. By incrementing the threshold value in this fashion, it is clear that $T_i = s \cdot |S_i|$. Adding the threshold values for each monitoring node yields, $T = \sum_{1 \leq i \leq m} T_i = \sum_{1 \leq i \leq m} s \cdot |S_i| = s \cdot N$, by the definition of N. Thus, the local threshold value T_i represents the contribution N_i makes to the global threshold value, which defines the frequent item set.

With the definition of the local threshold values described, the parameterized constraints installed at each monitoring node can now be defined. For each item monitored at a node, the following constraints are installed:

1. If $o_j \in F$ then the installed constraint is defined, $c_{j,i} + \delta_{j,i} \geq T_i$, where $c_{j,i}$ is the frequency count of o_j and $\delta_{j,i}$ is the adjustment factor corresponding to o_j.
2. If $o_j \notin F$ then the installed constraint is defined, $c_{j,i} + \delta_{j,i} < T_i$.

If all the parameterized constraints hold for each node, then for every $o_j \in F$, $\sum_{1 \leq i \leq m} c_{j,i} + \sum_{0 \leq i \leq m} \delta_{j,i} \geq \sum_{1 \leq i \leq m} T_i$ or $\sum_{1 \leq i \leq m} c_{j,i} \geq T$. Likewise for every $o_j \notin F$, $\sum_{1 \leq i \leq m} c_{j,i} + \sum_{0 \leq i \leq m} \delta_{j,i} < \sum_{1 \leq i \leq m} T_i$ or $\sum_{1 \leq i \leq m} c_{j,i} < T$. Thus, as long as all constraints hold, the set of frequent items is guaranteed to be valid. In the event any one constraint is violated, the coordinator is notified that the current set may no longer be valid. At this point the coordinator begins a process called resolution to determine the new frequent item set.

4.1   Resolution

Whenever a local constraint is broken on any monitor node a three phase process called resolution is initiated. The purpose of this process is to determine if either the frequent item set has changed or it hasn't, and to install new parameterized constraints in such a fashion so that all constraints hold. This process is modified from Top-K Monitoring changing validation tests and message content. The changes made to the three phases are described below.

To begin the resolution process, in Phase 1 the monitor containing an invalid constraint N_I sends a message to the coordinator. This message contains a set of frequency counts, adjustment factors, and item identifiers which are involved in violated constraints. Also included in the message sent to the coordinator, is the local threshold value of the monitor. This value is used later when determining the value of the new adjustment factors.

It is important to note that the entire frequent item set does not need to be sent to the coordinator. The membership of an item in F is independent of any other item. As we will see later this is very important in reducing communication overhead, when comparing FIDS to Top-K Monitoring.

In Phase 2 the coordinator node determines if the frequent item set is still valid using information gathered from N_I and its own stored adjustment factors. For each violated constraint, the coordinator performs the following tests:

1. If $o_j \in F$ then the test performed is $c_{j,I} + \delta_{j,I} + \delta_{j,0} \geq T_I$.
2. If $o_j \notin F$ then the test performed is $c_{j,I} + \delta_{j,I} + \delta_{j,0} < T_I$.

In the event that all violated constraints passed their respective tests, a process called reallocation is initiated and resolution terminates. If any one test fails, however, Phase 3 is initiated instead. In Phase 3 of resolution, the coordinator contacts each monitoring node $N_i : i \neq I$ and collects the frequency counts, adjustment factors, and item identifiers corresponding to those involved in violated constraints on N_I. Also collected, are the local threshold values for each monitor contacted. Once all the values are collected the new frequent item set is determined, reallocation is initiated, and resolution terminates. Phase 3 of resolution can also be called a synchronization phase, as all monitors in the network are contacted to determine the new set F.

4.2   Reallocation

Once the new frequent item set is determined, adjustment factors are assigned to each node involved in resolution $\mathcal{N}$. The assignment is made so that all constraints defined become valid for the newly define set F. The process responsible for this is called reallocation. Like resolution, this process is a modification from the one used in Top-K Monitoring. The changes are described in the following paragraphs.

The first step of reallocation is to determine the accumulative threshold and the accumulative weighted frequency (frequency count plus adjustment factor) for each item o_j involved in an invalid constraint across each node in $\mathcal{N}$. Next the distance of each frequency with the threshold is determined $\triangle_j$. Whenever $\mathcal{N}$ contains all the monitoring nodes, $\triangle_j$ represents the amount an item is over or under the global threshold.

The third step of reallocation assigns new adjustment factors for each o_j involved in an invalid constraint. The assignment is made so that each local weighted frequency count is equal to the local threshold value. By doing this the constraints for each item in F are satisfied.

Finally, a portion of $\triangle_j$ is added to the new adjustment factor assigned in Step 3. The amount added is based on an allocation parameter $0 \leq F_i < 1$ corresponding to node N_i. Allocation parameters are set in a fashion to control the amount of $\triangle_j$ given to node N_i and it is required that $\sum_{0 \leq i \leq m} F_i = 1$. This notation is similar to that of Top-K Monitoring with exception that $F_0 \neq 1$. Assigning $F_0 = 1$ prevents any monitoring node from receiving a portion of $\triangle_j$. As a result, the constraints of items not in F may not be satisfied after the reallocation process terminates.

Given the description above, the reallocation procedure can be expressed formally with two expressions.

1. $\triangle_j = \sum_{i \in \mathcal{N}} c_{j,i} + \sum_{i \in \mathcal{N}} \delta_{j,i} - \sum_{i \in \mathcal{N}} T_i$.
2. $\delta_{j,i} = T_i - c_{j,i} + F_i \cdot \triangle_j$.

The first expression represents Step 1, while the second expression represents Steps 3 and 4. For each item o_j involved in a violated constraint and node in $\mathcal{N}$, both expressions are evaluated to determine the new adjustment factor $\delta_{j,i}$ where $i \in \mathcal{N}$ represents node N_i. Comparing these two equations to those used in Top-K Monitoring, will show that the reallocation method original designed can be re-used. Assigning the parameters used in Top-K Monitoring appropriately will result in the definitions given above.

4.3   Frequency Count Reduction

Thus far in this paper we have only examined the case when $\epsilon = 0$. Setting the error parameter in this fashion implies that every unique item observed has an associated frequency count. If the number of unique items is very large, this will result in impractical memory requirements.

To reduce memory requirements, the counting techniques discussed in Sect. 3 can be integrated into the system. If the counting method selected over counts, each frequency count will be over counted by at most $\epsilon \cdot |S_i|$. As a result, the accumulative frequency of an item will be at most over counted by $\sum_{1 \leq i \leq m} \epsilon \cdot |S_i| = \epsilon \cdot N$. Although memory on each monitoring node can be reduced in this fashion, the presence of adjustment factors introduces complications. First, if an item that is removed from memory and is globally frequent, removing it would invalidate our monitoring method. Secondly, if the item being removed contains a corresponding adjustment factor, deleting it would invalidate adjustment factor invariants. To alleviate these problems we will introduce a generic solution which can utilize any of the counting techniques discussed in Sect. 3 [7,12,13].

To prevent items that are globally frequent from being removed from memory, we require that these items always remain. Since there can be at most $\frac{1}{s}$ globally frequent items, this is the additional space is required. In the event that an item is being removed from memory is not frequent and contains an adjustment factor, a message is sent to the coordinator containing the identifier and its adjustment factor. The coordinator will then determine if this adjustment factor can be cancelled out using its own stored values. If it can, no response is given, otherwise a message is sent to the monitor in question. This response signifies to the monitor that it must poll the coordinator each time a new item is encountered to determine if a past adjustment factor resides on the coordinator. This process will continue until there are no adjustment factors remaining at the coordinator on behave of the monitoring node. Since only infrequent items are removed from memory and each local threshold increases over time, it can not be the case that an item that is removed from memory will ever exceed its local threshold.

Determining the memory bounds required for this solution depends upon the counting technique selected. If the counting technique used requires $\frac{1}{\epsilon}$ counters, each monitoring node will require $\frac{1}{\epsilon} + \frac{1}{s}$ counters or $O(\frac{1}{\epsilon})$ space. There can be at most $\frac{1}{\epsilon}$ different counters on each monitoring node, and thus at most $\frac{m^2}{\epsilon}$ adjustment factor assignments. Since the coordinator may store adjustment factors on behalf of a monitoring node, the coordinator will require at most $O(\frac{m^2}{\epsilon})$ space, where m is the number of monitoring nodes.

5   Experimental Evaluation

5.1   Data Sets

Two data sets were used to evaluate the performance of FIDS. The first data set consists of wide-area network traffic between Lawrence Berkeley Laboratory and the rest of the world [14]. The data set contains 1.8 million TCP packets with 1,622 unique user IDs. Records were evenly assigned to four monitoring nodes and frequent users of the network were tracked.

The second data set consists of 1998 World Cup web requests on 9th June [2]. The dataset contains approximately 20 million requests with 9,198 unique requested item IDs. On the particular day used 26 servers were active. Thus, frequently requested item IDs were tracked using 26 distributed monitoring nodes.

5.2   Performance Measures

To evaluate our algorithm a series of performance criteria was defined. The most important of these criteria is communication cost. In our studies $\epsilon = 0$, thus communication is only conducted during resolution. Communication cost is therefore defined as the ratio of the total number of elements sent to the coordinator during resolution over the number of update tuples in bits. The number of elements sent per resolution (EPR) in bits can be formally expressed with the following equation:

$$\text{EPR} = |\mathcal{F}| \cdot |\mathcal{N}'| \cdot (32 + 32 + 64) + |\mathcal{N}'| \cdot 64 + |\mathcal{F}| \cdot |\mathcal{N}'| \cdot (32 + 64) \ . \qquad (1)$$

In the equation $|\mathcal{F}|$ is defined as the total number of broken constraints and $|\mathcal{N}'|$ is the total number of monitoring nodes involved in resolution, where $\mathcal{N}' = \mathcal{N} - \{N_0\}$. In our equation we assumed adjustment factors and local thresholds require 64 bits and all other elements (including update tuples) require 32 bits.

Finally, two measures were used to compare the quality of the output between FIDS and a comparable method, Top-K Monitoring. The two measures used were precision and recall. Precision is defined as the percentage of correct items contained in the entire output. Similarly, recall is defined as the percentage of correct elements contained in the output to the number of total possible correct items [6]. It is sometimes helpful to combine these two measurements into a single value. This value gives the overall quality of the output and can be expressed in the following equation:

$$\text{F-Measure} = \frac{2PR}{(P + R)} \ . \qquad (2)$$

The equation used is derived from [15], and weighs precision and recall equally. In evaluation of FIDS we did not allow approximation in frequency counts, and since our method is explicitly designed for monitoring frequent items F-Measure = 1.

5.3   Experimental Results

The first experiments focused on the communication cost of our algorithm under varying parameter settings. The two parameters varied were the support value and coordinator allocation parameter F_0. In Fig. 2 and Fig. 3 we see the results using the two data sets described in Sect. 5.1. The results show that the effects of F_0 differ between the two data sets. We see for the Berkeley TCP data set that the allocation parameter increases communication cost as its value is increased. The opposite occurs with the World Cup data set. The communication cost is reduced with increased value.

As was seen in the analysis of Top-K Monitoring by Olston and Babcock [3], when $F_0 > 0$ reallocation can prevent reaching the expensive Phase 3 as often but constraints are broken more frequently. This same scenario occurs with FIDS. Since only four monitors were used with the Berkeley TCP data set, however, Phase 3 required little communication and the weaker constraints could not offset this cost. From these results we therefore recommend that F_0 be assigned a small value (< 0.3) when there are few nodes and a large value when there are many.

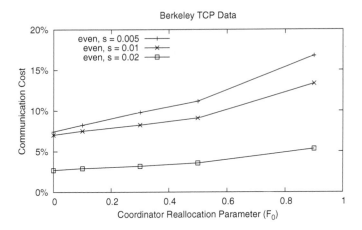

Fig. 2. Communication cost for Berkeley data set

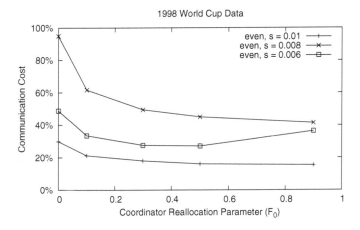

Fig. 3. Communication cost for '98 World Cup data set

Both results show that by raising the support parameter, communication cost decreases. This is not surprising as the average size of the frequent item set decreases with increased support. An anomaly did occur, however, in the World Cup data set when $s = 0.008$. In this scenario it is assumed that the frequent item set becomes more dynamic. This demonstrates the need for the data to maintain a degree of stability in order for the purposed method to significantly reduce communication cost. Also examined were effects of two reallocation heuristics on communication cost. These two methods are proportional allocation and even allocation, both defined in [3]. The experiments indicated no significant differences between the two methods. Therefore, even allocation was selected for our experiments.

Our second experiment focused on how communication cost accumulated over time to reach it final value. The World Cup data set was used for this experiment but execution was terminated after 500,000 update tuples. The coordinator re-allocation parameter was set to $F_0 = 0$, with the support value varying. To determine how communication accumulates we fixed the number of update tuples to the total number in the data set (about 20 million) in our communication cost formula. The results of this experiment are shown in Fig. 4. We see that a sudden spike in of communication cost occurs during the first 100,000 tuples, afterwards, only steadily rising to reach its final value. Extending the initialization phase to account for these tuples will reduce the communication cost of our results significantly.

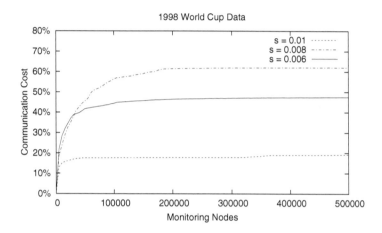

Fig. 4. Communication cost over time for '98 World Cup data set

5.4 Comparison

We compared FIDS to Top-K Monitoring observing differences in communication cost and in the output quality. Although Top-K Monitoring was not originally designed for finding frequent items, it can be used for this purpose. Assigning $K = \frac{1}{s}$ will guarantee that all frequent items are found. Setting K to a smaller value may introduce false negatives, but will likely increase the overall quality of the output and reduce communication cost. The effects of various settings of K were therefore examined to optimize the performance of Top-K Monitoring for our comparisons. Our experiments used the Berkeley TCP data set, measuring the recall and precision every 100,000 updates and then averaging the results. These two measures where then combined with equal weight in an F-measure. The results from Fig. 5 show that a good degree of quality can be obtained by assigning K appropriately, and will serve as a guide for selecting its value in our comparisons.

With the observation described above, we then began comparing Top-K Monitoring with our modifications. In all our experiments the Berkeley TCP data set was used. The results of our comparison are summarized in Table 1. Both methods were tested with three support values and the F_0 allocation setting

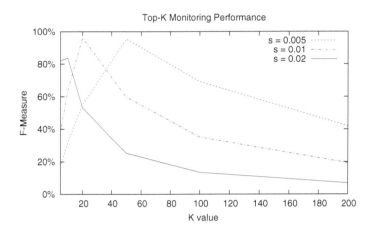

Fig. 5. Output quality of Top-K Monitoring with varying settings of K

which yielded lowest communication cost was selected. The K which yielded optimal output quality was selected for Top-K Monitoring. This value is indicated in parenthesis under the support column.

We see that in all scenarios communication cost is lower with our method. The differences between the two are greatest when a low support value is used. There was, however, one test scenario where Top-K Monitoring out-performed. When $K = 5$ communication cost is reduced by 0.1% when compared to FIDS using $s = 0.02$, but the quality of the output is less than optimal.

Although K was selected to yield optimal output quality, the results show it still yielded lower quality when compared to our method. This is not surprising though, since Top-K Monitoring was not specifically designed for finding frequent items (as defined in Sect. 2.2). It must be noted that when introducing error in frequency counts to reduce memory, the quality of the output of our modification will also decrease. These results, however, depend on the counting method selected.

For a final comparison, the scalability of each method was tested using the Berkeley TCP data set with varying number of monitoring nodes. The support value selected for these tests was $s = 0.01$ or $K = 20$. The results indicated that both methods grow linearly with the number of nodes, but the Top-k Monitoring method grew roughly four times faster.

We have seen in the previous paragraphs that FIDS out performs Top-K Monitoring in both communication cost and quality of results. FIDS is also comparable with methods [4] and [11] introduced in Sect. 3. The method in [4] can only provide the approximate frequent item set and relies on a generous error tolerance to significantly reduce communication cost. FIDS, however, can provide the exact frequent item set while yielding comparable communication cost. Finally, the method in [11] can only provide the frequent item set periodically. It is likely that the set will change before it is recomputed. FIDS can detect these changes when the periodic nature of [11] can not allow.

Table 1. Comparison of two approaches

Method	Support	Communication Cost	F-Measure
Top-K Monitoring	0.005 $(K = 50)$	143.34%	95.37%
	0.01 $(K = 20)$	46.66%	96.07%
	0.02 $(K = 10)$	12.20%	83.62%
FIDS	0.005	7.43%	100.00%
	0.01	7.03%	100.00%
	. 0.02	2.70%	100.00%

6   Conclusions

In this paper we studied the problem of continuously reporting frequent items over the union of distributed data streams. In our study we determined that Top-K Monitoring introduced in [3] can be used to solve this problem effectively. However, modifications to this algorithm can be made which can both reduce communication cost and improve the overall quality of the results.

Finally in an attempt to reduce memory requirements for both methods, we introduced a generic solution to prevent adjustment factor invalidation. This solution allows memory constraints on the nodes to be bounded but may introduce additional communication overhead. Future work consists of implementing this protocol and measuring the additional cost obtained. Additionally, a requirement for many monitoring tasks is to weigh newer occurrences more than older ones or to expire old item occurrences. Recency in frequency counts should be examined to determine the communication overhead required. This will also require new methods to reduce memory requirements while not invalidating adjustment factor invariants.

References

1. Arasu, A., Manku, G.: Approximate Counts and Quantiles over Sliding Windows. In: Proc. of the 23rd ACM Symposium on Principles of Database System (PDDS), pp. 286–296. ACM Press, New York (2004)
2. Arlitt, M., Jin, T.: 1998 World Cup Web Site Access Logs (1998), http://www.acm.org/sigcomm/ITA/
3. Babcock, B., Olston, C.: Distributed Top-k Monitoring. In: Proc. of ACM SIGMOD Intl. Conf. on Management of Data, pp. 28–39. ACM Press, New York (2003)
4. Cormode, G., Garofalakis, M.: Sketching Streams Through the Net: Distributed Approximate Query Tracking. In: Proc. of 31st Intl. Conf. on Very Large Data Bases, pp. 13–24 (2005)
5. Cormode, G., Garofalakis, M.: Efficient Strategies for Continuous Distributed Tracking Tasks. IEE Data Engineering Bulletin 28, 33–39 (2005)
6. Cormode, G., Muthukrishnan, S.: Whats Hot and Whats Not: Tracking Most Frequent Items Dynamically. In: Proc. of the 22nd ACM Symposium on Principles of Database Systems (PODS), pp. 296–306. ACM Press, New York (2003)

7. Demaine, E., Lopez-Ortiz, A., Munro, J.: Frequency estimation of internet packet streams with limited space. In: Proc. of the 10th Annual European Symposium on Algorithms, pp. 348–360 (2002)
8. Golab, L., DeHann, D., Demaine, E., Lopez-Ortiz, A., Munro, J.: Identifying Frequent Items in Sliding Windows over On-Line Packet Streams. In: Proc. of ACM Internet Measurements Conference (IMC), pp. 173–178. ACM Press, New York (2003)
9. Kim, H., Karp, B.: Autograph: Toward Automated Distributed Worm Signature Detection. In: Proc. of the 13th USENIX Security Symposium, pp. 271–286 (2004)
10. Lee, L.K., Ting, H.F.: A Simpler More Efficient Deterministic Scheme for Finding Frequent Items over Sliding Windows. In: Proc. of the 25th ACM Symposium on Principles of Database Systems (PODS), pp. 290–297. ACM Press, New York (2006)
11. Manjhi, A., Shkapenyuk, V., Dhamdhere, K., Olston, C.: Finding (Recently) Frequent Items in Distributed Data Streams. In: Proc. of Intl. Conf. on Data Engineering (ICDE), pp. 767–778 (2005)
12. Manku, G., Motwani, R.: Approximate Frequency Counts over Data Streams. In: Proceedings of 28th Intl. Conf. on Very Large Data Bases, pp. 364–357 (2002)
13. Metwally, A., Agrawal, D., Abbadi, A.: Computation of Frequent and Top-k Elements in Data Streams. In: Proceedings of the 10th ICDT. Intl. Conf. on Database Theory, pp. 398–412 (2005)
14. Paxson, V., Floyd, S.: Wide-Area Traffic: The Failure of Poisson Modeling. IEEE/ACM Trasactions on Networking 226–244 (1995)
15. van Rijsbergen, C.J.: Information Retrieval. Butterworths, London (1979)
16. Stanojevic, R.: Scalable Heavy-Hitter Identification
    http://www.hamilton.ie/person/rade/ScalableHH.pdf
17. Zhu, Y., Shasha, D.: StatStream: Statistical Monitoring of Thousands of Data Streams in Real Time. In: Proc. of the 28th Intl. Conf. on Very Large Databases, pp. 358–369 (2002)

Mining Maximal Frequent Itemsets in
Data Streams Based on FP-Tree

Fujiang Ao[1], Yuejin Yan[2], Jian Huang[1], and Kedi Huang[1]

[1] School of Mechanical Engineering and Automation, National University
of Defense Technology, Changsha, 410073, China
[2] School of Computer Science, National University of Defense
Technology, Changsha, 410073, China
fjao@nudt.edu.cn

Abstract. Mining maximal frequent itemsets in data streams is more difficult than mining them in static databases for the huge, high-speed and continuous characteristics of data streams. In this paper, we propose a novel one-pass algorithm called FpMFI-DS, which mines all maximal frequent itemsets in Landmark windows or Sliding windows in data streams based on FP-Tree. A new structure of FP-Tree is designed for storing all transactions in Landmark windows or Sliding windows in data streams. To improve the efficiency of the algorithm, a new pruning technique, extension support equivalency pruning (ESEquivPS), is imported to it. The experiments show that our algorithm is efficient and scalable. It is suitable for mining MFIs both in static database and in data streams.

Keywords: maximal frequent itemsets, data streams, FP-Tree, pruning technique.

1 Introduction

In recent years, data streams have been researched widely. The technologies about data streams are used in many applications. Examples of such applications include financial applications, network monitoring, security, telecommunications data management, web applications, manufacturing, sensor networks, and others [1]. In a word, a data stream is a real-time, continuous, ordered (implicitly by arrival time or explicitly by timestamp) sequence of items. The algorithm for mining data streams must be single-pass algorithm for the characters of data streams.

The time and space efficiency of data mining in data streams is more significant than that in static databases. The number of maximal frequent itemsets and closed frequent itemsets is much less than that of frequent itemsets. So, mining MFIs or CFIs can get better time and space efficiency than mining frequent itemsets. Mining maximal frequent itemsets [2][3][4] and mining closed frequent itemsets [5][6] in data streams is to be a tendency.

Many good algorithms have been developed for mining maximal frequent itemsets in static database, for example MaxMiner [7], DepthProject [8], GenMax [9], AFOPT

P. Perner (Ed.): MLDM 2007, LNAI 4571, pp. 479–489, 2007.

[10], FPMax* [11], FpMFI [12]. All these algorithms need to scan database more than one pass. They are not suitable for mining maximal frequent itemsets in data streams. In all these algorithms, FpMFI is almost fastest for all tested database [12]. The algorithm needs to scan database two passes. We reconstruct the algorithm to a single-pass one, called FpMFI-DS. To mining maximal frequent itemsets in Landmark windows or Sliding windows in data streams, we must store all transactions in the window. For Sliding windows, when transaction is out of window, it should be deleted from window. To satisfy with these requires, we designed a new structure of FP-Tree, which can store all transactions in Landmark windows or Sliding windows, and when transaction is out of Sliding windows, it can be deleted. To reduce search space of FpMFI-DS, a new pruning technique, extension support equivalency pruning, is added in the algorithm. The efficiency of FpMFI-DS is close to FPMax* and a little lower than that of FpMFI.

2 Preliminaries and Related Work

This section will formally describe the MFIs mining problem in data streams and the set enumeration tree that represents search space. Also the related works will be introduced in this section.

2.1 Problem Revisit

Let $I = \{i_1, i_2, ..., i_m\}$ be a set of m distinct elements, called *items*. A subset $X \subseteq I$ is called an *itemset*. An itemset with k items is called a k-itemset. Each transaction t is a set of items in I. A data stream, $DS = [t_1, t_2, ...t_N)$, is an infinite sequence of transaction. For all transactions in a given window W over data stream, the support of an itemset X, denoted as $sup(X) = D_x / |W|$, where D_x is the number of transactions in which X occurs as a subset and $|W|$ is the width of the window. For a given threshold *min_sup* in the range of [0,1], itemset X is frequent if $sup(X) \geq min\_sup$. If $sup(X) \geq min\_sup$ and for any $Y \supseteq X$, we have $sup(Y) < min\_sup$, then X is called maximal frequent itemset in window W.

From the definitions above, we can see that the selection of window W is important for an itemset X be a frequent one. In paper [13], three windows models are introduced, including landmark windows, sliding windows, damped windows. In this paper, we focus on mining the set of all maximal frequent itemsets in landmark windows or in sliding windows over data streams.

To get all maximal frequent itemsets, one method is to enumerate all itemsets that maybe be maximal frequent itemsets, count the support of these itemsets and decide whether they are maximal frequent itemsets. In paper [14], Rymon presents the concept of generic set enumeration tree search framework. The enumeration tree is a virtual tree. It is just used to illustrate how sets of items are to be completely enumerated in a search problem. The tree could be traversed depth-first, breadth-first, or even best-first as directed by some heuristic. In the domain of data mining, the set enumeration tree is also named after search space tree.

But, when the number of different items is big, the algorithm that searches all search space may suffer from the problem of combinatorial explosion. So the key to an efficient set-enumeration search is the pruning techniques that are applied to remove entire branches from consideration [7]. The two most often used pruning techniques, subset infrequency pruning and superset frequency pruning, are based on following two lemmas:

Lemma 1. A restricted subset of any frequent itemset is not a maximal frequent itemset.

Lemma 2. A subset of any frequent itemset is a frequent itemset, and a superset of any infrequent itemset is not a frequent itemset.

For example, for the dataset in the left, Fig. 1 shows the corresponding search space tree. In Fig. 1, we suppose $I = \{a,b,c,d,e\}$ is sorted in firm lexicographic order. The pruning techniques used in the tree includes *subset infrequency pruning* (SIP) and *superset frequency pruning* (SFP). The root of the tree represents the empty itemset, and the nodes at level k contain the k-itemsets. The itemset associated with each node, n, will be referred as the node's *head(n)*. The possible extensions of the itemset is denoted as *con_tail(n)*, which is the set of items after the last item of *head(n)*. The frequent extensions denoted as *fre_tail(n)* is the set of items that can be appended to *head(n)* to build the longer frequent itemsets. In depth-first traversal of the tree, *fre_tail(n)* contains only the frequent extensions of n. The itemset associated with each children node of node n is build by appended one of *fre_tail(n)* to *head (n)*. As example in Fig. 1, suppose node n is associated with $\{b\}$, then $head(n) = \{b\}$ and $con\_tail(n) = \{c,d,e\}$. For $\{e\}$ is not frequent, $fre\_tail(n) = \{c,d\}$. The children node of n, $\{b,c\}$, is build by appending c from *fre_tail(n)* to $\{b\}$.

The problem of MFI mining can be thought as to find a border of the tree, all the elements above the border are frequent itemsets, and others are not. All MFIs are near the border. As our examples in Fig. 1, itemsets in solid rectangle are MFIs.

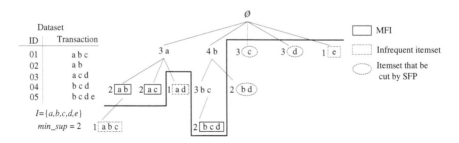

Fig. 1. The set enumeration tree built based on the dataset in the left

2.2 Related Work

Some one-pass algorithms for mining maximal frequent itemsets in data streams have been developed, for example, DSM-MFI [3], estDec+ [4] and INSTANT [2]. They are all approximate algorithm.

DSM-MFI mines the set of all maximal frequent itemsets in landmark windows over data streams. The algorithm is composed of four steps. First, it reads a window of transactions from the buffer in main memory, and sorts the items of transactions in a lexicographical order. Second, it constructs and maintains the in-memory summary data structure, *SFI-forest*. Third, it prunes the infrequent information from the summary data structure. Fourth, it searches the maximal frequent itemsets from the current summary data structure. Steps 1 and 2 are performed in sequence for a new incoming basic window. Steps 3 and 4 are usually performed periodically or when it is needed [3]. The experiment results in paper [3] show that DSM-MFI is efficient on both sparse and dense datasets, and scalable to very long data streams.

estDec+ use a structure, CP-tree (Compressed-prefix tree), to keep the supports of all the significant itemsets in main memory. It also consists of four phases: parameter updating, node restructuring, itemset insertion, and frequent itemset selection. When a new transaction T_k in a data stream D_{k-1} is generated, these phases except the frequent itemset selection phase are performed in sequence. The frequent itemset selection phase is performed only when the up-to-date result set of frequent or maximal frequent itemsets is requested. The main advantage of the algorithm is that it adopts an adaptive memory utilization scheme to maximize the mining accuracy for confined memory space at all times [4].

INSTANT mines maximal frequent itemsequences from data streams based on a new mining theory provided by paper [2]. Where an itemsequence is an ordered list of items. The main advantage of the algorithm is that it is an online algorithm, which can directly display current maximal frequent itemsequences while they are generated. But the time efficiency of the algorithm is affected.

Paper [12] proposed a MFIs mining algorithm, FpMFI. It is an improvement over FPMax* and outperforms FPMax* by 40% averagely. They all need to scan dataset two passes. In this paper, we propose an algorithm, FpMFI-DS, based on FpMFI. FpMFI-DS only need to scan dataset one pass. It is a one-pass and exact algorithm.

3   FpMFI-DS

In this section, FpMFI-DS algorithm is introduced in details.

3.1   The Construction of FP-Tree in FpMFI-DS

To construct FP-Tree, it usually needs to scan database two passes. The first scan of database derives a list of frequent items. Then it sorts the items by frequency descending order. The list of items in header table and each path of prefix-tree will follow this order. The second scan of database gets every transaction and inserts all frequent items in transaction into FP-Tree. During the process of mining, to construct the FP-Tree of node n, it needs to scan the *head(n)*'s conditional pattern base that comes from FP-tree of its parent node two passes[15]. Paper [11] improves this approach by adopting an array-based technique. It only needs to scan *head(n)*'s conditional pattern base one pass.

In FpMFI-DS, to implement one-pass algorithm, we must complete the construction of FP-Tree by only scanning dataset one-pass.

To mine maximal frequent itemsets in sliding windows, the FP-Tree of root should contain all the transactions in the Sliding windows. When a transaction comes to window, all items of the transaction are inserted to the FP-Tree of root, whether they are frequent or infrequent. And when a transaction is out of window, it should be deleted from the FP-Tree of root. So except for header table and prefix-tree, the FP-Tree of root in FpMFI-DS also contains a *tidlist*, a list of IDs of the transactions in window. Every item in the *tidlist* is composed of an ID of transaction (an integer) and a pointer to the last node of the transaction in the FP-Tree of root. For an one-pass algorithm, when adding the transaction to the root FP-Tree, we can't get the frequencies of items in all transactions. So the order of the items in the FP-Tree of root can't be frequency descending order. In FpMFI-DS, the order of the items in the FP-Tree of root is based on the lexicographical order of the items. When a transaction comes to window, all items of the transaction are inserted to FP-Tree by lexicographical order. When a transaction is out of window, the last item of the transaction in the FP-Tree of root can be found through the transaction's ID and pointer in the *tidlist*, then it is deleted from root FP-Tree. To mine maximal frequent itemsets in landmark windows, we only need to fixup beginning side of the window.

The subsequent FP-Tree during the process of mining is similar to that in FPMFI. To improve the effectiveness of superset frequency pruning, the order of the items in the subsequent FP-Tree also adopts frequency descending order.

For example, for data streams and window width in the left, Fig. 2 shows the FP-tree of root built based on transactions in first window. The FP-tree includes five transactions.

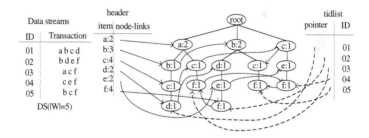

Fig. 2. The FP-Tree of root built based on transactions in first window

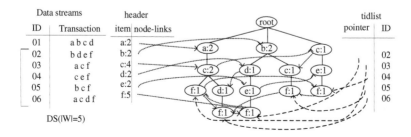

Fig. 3. The FP-Tree of root built based on transactions in second window

Fig. 3 shows the FP-tree of root built based on transactions in second window. In the FP-Tree, the first transaction is deleted from it and the sixth transaction is inserted into it.

When *min_sup* is 2, Fig. 4 shows the FP-tree of itemset {*f*} during the process of mining for data in Figure 3. The items order in Fig. 2 and Fig. 3 is based on the lexicographical order of the items, while that in Fig. 4 is based on frequency descending order.

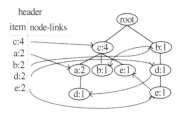

Fig. 4. The FP-Tree of itemset {*f*} during the process of mining for data in Figure 2

3.2 Pruning Techniques

FpMFI uses three pruning techniques, including subset infrequency pruning, superset frequency pruning, parent equivalence pruning. The efficiency of these pruning techniques is high for item ordering policy used by it. In FpMFI, since the item order in the FP-Tree of root is based on the lexicographical order, the item order in first level of search space tree has to accord with it. If only use these pruning techniques, the efficiency is lower than that in FpMFI, especially for dense dataset. For example, for dataset *MUSHROOM*, search space of FpMFI-DS is about as twice as that of FpMFI. So, Excepting for these pruning techniques, another pruning technique, ESEquivPS, is adopted by FpMFI-DS. ESEquivPS is firstly present in paper [16]. The pruning technique is described as following:

Supposed p and n are nodes in search space tree, and n is a children node of p. Let item $x \in fre\_tail(p)$ and $x \in fre\_tail(n)$. If $sup(head(p) \cup \{x\}) = sup(head(n) \cup \{x\})$, then any offspring node of p that contains item x and is in the right of node n can be pruned.

Proof. Let j be an item associated with node n, X and Y are itemsets associated with $head(p) \cup \{x\}$ and $head(n) \cup \{x\}$, respectively. Since $sup(X) = sup(Y)$, then any transaction T containing X must contains item j. Thus, the maximal frequent itemset containing X must containing j.In the p-subtree, the itemsets, which associated with the nodes that contain item x and are in the right of node n, must not contain item j for the character of search space tree. So, they can't be maximal frequent itemsets.

From experiments, we found that if the nodes in every level of search space tree is in the order of frequency descending, then the pruning technique is invalidity. Fortunately, the first level of search space tree in FpMFI-DS is in the lexicographical order of the items. Then we can use it for the first level of search space tree. The experiments show that for the dense datasets, *MUSHROOM*, the size of search space can be trimmed off by about 30%.

3.3 Algorithm FpMFI-DS

Fig. 5 shows algorithm FpMFI-DS. Though the items order in the first level of search space tree of FpMFI-DS is different from that of FpMFI, the mining procedure of the two algorithms is similar. The difference is that algorithm FpMFI-DS adopts the new pruning technique, ESEquivPS (line 4 to line 6).

PROCEDURE: FpMFI-DS Algorithm
INPUT:
    n: a node in search space tree that associated with a head itemset h,
      a FP-tree, a MFI-tree, and an $array$
    M-trees: MFI-trees of all ancestor nodes of n
1 For each item x from end to beginning in $header$ of $n.FP$-tree
2    $h'=h \cup \{x\}$ //h' identifies n'
3   if $(sup(h')<min\_sup)$ continue
4   if $(ESEquivPS\_cheching(x))$ continue
5   if $(Thirdlevel()$ and $sup(\{x\})==sup(h'))$
6   insert $true$ into respective position of a bool array for $ESEquivPS$
7   if x is not the end item of the header
8   if$(superset\_checking(con\_tail(n'),n.MFI$-$tree)$ return
9   if$(superset\_checking(con\_tail(n'),n.FP$-$tree)$
10   insert $h' \cup con\_tail(n')$ into M-trees return
11   if $n.array$ is not null
12      $fre\_tail(n') = \{$frequent items for x in $n.array\}$
13   else
14      $fre\_tail(n') = \{$frequent items in conditional pattern base of $h'\}$
15    $PeIs = \{$items whose count equal to the support of $h'\}$
16   if$(superset\_checking(fre\_tail(n'), n.MFI$-$tree)$
17   if the number of items before x in the header is $\mid fre\_tail(n')\mid$
18   return
19   else continue
20   if$(superset\_checking(fre\_tail(n'), n.FP$-$tree)$
21   insert $h' \cup fre\_tail(n')$ into M-trees
22   if the number of items before x in the header is $\mid fre\_tail(n')\mid$
23   return
24   insert $fre\_tail(n')$ into $n.MFI$-$tree$ continue
25    $h' = h' \cup PeIs$, $fre\_tail(n') = fre\_tail(n') - PeIs$
26   sort the items in $fre\_tail(n')$
27   construct the FP-tree of n'
28   if$(superset\_checking(fre\_tail(n'), n'.FP$-$tree)$
29   insert $h' \cup fre\_tail(n')$ into M-trees
30   if the number of items before x in theheader is $\mid fre\_tail(n')\mid$
31   return
32   insert $fre\_tail(n')$ into $n.MFI$-$tree$ continue
33   construct the MFI-tree of n'
34   M-$trees = M$-$trees \cup \{n.MFI$-$tree\}$
35   call FpMFI-DS$(n' ,M$-$trees)$

Fig. 5. Algorithm FpMFI-DS

To implement ESEquivPS, we use an integer array store the support of items in first level of search space tree and use a bool array denote if the respective items satisfy with the condition of ESEquivPS. When exploring the third level of search space tree, we check if the support of respective item equals to that of the corresponding item in the first level of search space tree. If they are, the corresponding position in the bool array is set *true* (line 5 to line 6). Before exploring any node in the search space tree, we first chech if the corresponding position in the bool array is set to *true*. If it is, the node should be cut off (line 4).

4   Experimental Evaluations

All the experiments were conducted with a 2.4 GHZ Pentium IV with 512 MB of DDR memory running a Redhat Linux 9.0 operation system. We implemented the code of FpMFI-DS by c++ and compiled it with the g++ 2.96 compiler.

4.1 Performance Comparisons

To evaluate the performance of FpMFI-DS, we have compared its performance with a representative algorithm, INSTANT [2]. The advantage of INSTANT is that it can directly display current maximal frequent itemsequence (not itemsets) while they are generated. For sparse datasets, the efficiency of the algorithm is high. But for dense datasets, the efficiency is not very good. The code of INSTANT was provided by its authors, Guojun Mao, etc. It is also written in c++ and compiled by g++ 2.96 compiler.

The dataset in the experiment is *T20I5D10K*, a dataset generated by IBM data generator [17]. The synthetic dataset *T20I5D10K* has average transaction size T of 20 items and the average size of frequent itemset I of 5 items and the number of transactions D of 10K. It is a sparse dataset.

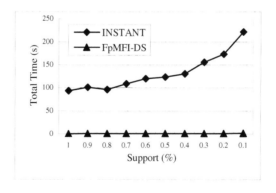

Fig. 6. Performance comparisons with INSTANT

Fig. 6 shows the result of performance comparisons for dataset *T20I5D10K*. The efficiency of FpMFI-DS is much higher than that of INSTANT. For the dataset, maximal total time of FpMFI-DS is lower than 2 seconds.

We also compared its performance with some multi-pass algorithms. Fig. 7 and Fig. 8 show the result of performance comparisons with algorithm FPMax*. The dataset in Fig. 7 is *T20I5D100K*, a sparse dataset and the dataset in Fig. 7 is

MUSHROOM, a dense dataset. The source code of algorithm FPMax* and dataset *MUSHROOM* were downloaded from [18]. Algorithm FPMax* is written in c++ and compiled with the g++ 2.96 compiler, too.

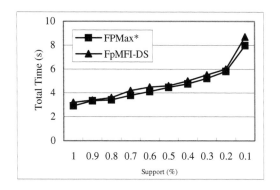

Fig. 7. Performance comparisons with FPMax* for dataset *T20I5D100K*

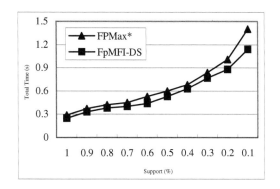

Fig. 8. Performance comparisons with FPMax* for dataset *MUSHROOM*

From Fig. 7 and Fig. 8, we can see that for dense dataset, the efficiency of FpMFI-DS is a little higher than that of FPMax*, and for sparse dataset, the efficiency of FpMFI-DS is a little lower than that of FPMax*. We can draw a conclusion that the efficiency of two algorithms is close. The result in paper [11] shows that the efficiency of FPMax* is very high. So the efficiency of FpMFI-DS is good, too.

4.2 Scalability of FpMFI-DS

To evaluate the scalability of FpMFI-DS, we use four huge datasets, *T10I5D1000K*, *T10I5D2000K*, *T10I5D3000K*, *T10I5D4000K*. The minimum support is 0.1%. From Fig. 9, we can see that the execution time grows smoothly as the dataset size increases from 1,000K to 4,000K. The algorithm has scalability.

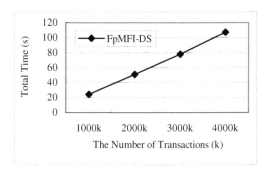

Fig. 9. The scalability of FpMFI-DS

5 Conclusions

In this paper, we proposed a novel one-pass algorithm, FpMFI-DS, which mines all set of the maximal frequent itemsets in data streams. To mine MFIs both in landmark windows and in sliding windows, we adopt a new structure of FP-Tree. To reduce the search space of the algorithm, a new pruning technique, ESEquivPS, is adopted by the algorithm. The experiments show that the algorithm is efficient on both sparse and dense datasets, and has good scalability.

Acknowledgements

We would like to thank Guojun Mao for providing the code of algorithm INSTANT. The work of this paper was supported by the Natural Science Foundation of China under the Grant No. 60573057.

References

1. Babcock, B., Babu, S., Datar, M., Motwani, R., Widom, J.: Models and issues in data stream systems. In: Proc. of the twenty-first ACM SIGMOD-SIGACTSIGART Symposium on Principles of Database Systems 2002, pp. 1–16 (2002)
2. Mao, G., Wu, X., Liu, C.: Online Mining of Maximal Frequent Itemsequences from Data Streams. University of Vermont, Computer Science Technical Report, CS-05-07 (2005)
3. Li, H., Lee, S., Shan, M.: Online mining (recently) maximal frequent itemsets over data streams. In: Proc. of the fifteenth International Workshops on Research Issues in Data Engineering: Stream Data Mining and Applications, Tokyo, Japan, pp. 11–18. IEEE Press, NJ (2005)
4. Lee, D., Lee, W.: Finding maximal frequent itemsets over online data streams adaptively. In: Proc. of the Fifth IEEE International Conference on Data Mining.Houston, USA, pp. 266–273. IEEE Press, NJ (2005)
5. Chi, Y., Wang, H., Yu, P S, Muntz, R.: Moment: maintaining closed frequent itemsets over a stream sliding window. In: Proc. of the fourth IEEE International Conference on Data Mining, UK, pp. 59–66. IEEE Press, NJ (2004)

6. Jiang, N., Gruenwald, L.: CFI-Stream: mining closed frequent itemsets in data streams. In: Proc. of the 12th ACM SIGKDD international conference on Knowledge discovery and data mining, Philadelphia, PA, USA, 2006, pp. 592–597 (2006)
7. Bayardo, R.: Efficiently mining long patterns from databases. In: ACM SIGMOD Conference (1998)
8. Agarwal, R., Aggarwal, C., Prasad, V.: A tree projection algorithm for generation of frequent itemsets. Journal of Parallel and Distributed Computing (2001)
9. Gouda, K., Zaki, M.J.: Efficiently Mining Maximal Frequent Itemsets. In: Proc. of the IEEE Int. Conference on Data Mining, San Jose (2001)
10. Rigoutsos, L., Floratos, A.: Combinatorial pattern discovery in biological sequences: The Teiresias algorithm. Bioinformatics 14(1), 55–67 (1998)
11. Grahne, G., Zhu, J.: Efficiently Using Prefix-trees in Mining Frequent Itemsets. In: Proc. of the IEEE ICDM Workshop on Frequent Itemset Mining Implementations, November 19, 2003, Melbourne, Florida, USA (2003)
12. Yan, Y., Li, Z., Chen, H.: Fast Mining Maximal Frequent ItemSets Based on FP-Tree. In: Webb, G.I., Yu, X. (eds.) AI 2004. LNCS (LNAI), vol. 3339, pp. 475–487. Springer, Heidelberg (2004)
13. Zhu, Y., Shasha, D.: StatStream: Statistical monitoring of thousands of data streams in real time. In: Bernstein, P., Ioannidis, Y., Ramakrishnan, R. (eds.) Proc. of the 28th Int'l Conf. on Very Large Data Bases, Hong Kong, pp. 358–369. Morgan Kaufmann, Seattle (2002)
14. Rymon, R.: Search through Systematic Set Enumeration. In: Proc. of Third Int'l Conf. on Principles of Knowledge Representation and Reasoning, pp. 539–550 (1992)
15. Han, J., Pei, J., Yin, Y.: Mining Frequent Patterns without Candidate Generation. In: Proc. 2000 ACM-SIGMOD Int. Conf. on Management of Data (SIGMOD'00), May 2000, Dallas, TX (2000)
16. Ma, Z., Chen, X., Wang, X.: Pruning strategy for mining maximal frequent itemsets. Journal of Tsinghua Univ 45(S1), 1748–1752 (2005)
17. Agrawal, R., Srikant, R.: Fast algorithms for mining association rules. In: Proc. Of the 20th Intl. Conf. on Very Large Databases (VLDB'94), Santiago, Chile, Sept, 1994, pp. 487–499 (1994)
18. Codes and datasets available at http://fimi.cs.helsinki.fi/

CCIC: Consistent Common Itemsets Classifier

Yohji Shidara, Atsuyoshi Nakamura, and Mineichi Kudo

Graduate School of Information Science and Technology,
Hokkaido University,
Kita 14, Nishi 9, Kita-ku, Sapporo 060-0814, Hokkaido, Japan
{shidara, atsu, mine}@main.ist.hokudai.ac.jp

Abstract. We propose a novel approach which extracts consistent (100% confident) rules and builds a classifier with them. Recently, associative classifiers which utilize association rules have been widely studied. Indeed, the associative classifiers often outperform the traditional classifiers. In this case, it is important to collect high quality (association) rules. Many algorithms find only high support rules, because decreasing the minimum support to be satisfied is computationally demanding. However, it may be effective to collect low support but high confidence rules. Therefore, we propose an algorithm that produces a wide variety of 100% confident rules including low support rules. To achieve this goal, we adopt a specific-to-general rule searching strategy, in contrast to the previous many approaches. Our experimental results show that the proposed method achieves higher accuracies in several datasets taken from UCI machine learning repository.

1 Introduction

As a new classification approach, *associative classifiers* which integrate association mining and classification have been widely studied [1,2,3,4,5,6]. According to several reports [1,2,3], higher classification accuracies are achieved by them compared with traditional classifiers such as C4.5 [7] and RIPPER [8]. The associative classifiers assume that the input dataset is a set of itemsets, that is, a transaction database (table-form databases are converted into transaction databases beforehand). Here, a transaction is a set of items. The pioneer of associative classification, CBA [1], builds a classifier from a set of association rules obtained by an association rule mining technique such as Apriori [9]. Here, only high quality rules are selected to construct a ruleset (a classifier).

In many mining techniques, the rules are found in a general-to-specific manner. In this case, starting from null item, the number of items used in a temporal rule increases during the rule narrowing process. Note that a rule with null item in the condition part explains every instance. In such an approach, we collect items such that a rule maintains a minimum support. Such a minimum-support requirement brings the efficiency of the mining process. One of challenging issues is to extract more rules having lower supports (below a required minimum support) but sufficiently high confidences to improve association classifiers. So far,

P. Perner (Ed.): MLDM 2007, LNAI 4571, pp. 490–498, 2007.
© Springer-Verlag Berlin Heidelberg 2007

this was not carried out due to the impractical increase of computational cost that is largely consumed to examine many combinations of items while keeping a small value of support. We resolve this difficulty by taking the reverse strategy, that is, a specific-to-general approach. In this approach, confidence is given more priority than support. As a result, regardless their values of support, more rules with a high confidence can be found efficiently.

The individual instance in a database can be regarded as a 100% confident rule itself, while the support value of the corresponding rule is quite low (maybe such a rule explains only one instance). They are most specific rules. We merge some instances by taking intersection (the set of common items among them), to make them more general. This merging is made only when the consistency (explaining instances of one class only) is kept. As a consequence, we obtain the most general rules, in the set inclusion relation, keeping consistency. In addition, such rules are expected to be highly interpretable because they represent some unique patterns for a class.

In the proposed method, we consider a combination of instances instead of a combination of items. The latter is a popular approach in the previous mining techniques. This exchange enables us to have an algorithm that runs in a linear order with respect to the number of items. Although the naive version of the proposed algorithm requires a combinatorial number of examinations of instances, we can reduce the computational cost by its randomized version.

Our contribution in this paper is summarized as follows: 1) a novel rule extraction method is proposed as an extension of the subclass method [10,11,12], by which a transaction database consisting of itemsets is now be able to be dealt with, 2) a classification rule combining obtained consistent (association) rules is proposed, and 3) the proposed classifier achieves higher classification accuracies in several datasets compared with the competitors, such as C4.5 [7], RIPPER [8], CBA [1], CMAR [2] and CPAR [3].

2   Methodology

Our method is built on the basis of subclass method [10,11,12] which is a classifier only applicable to numerical attributes. We expand the framework of the subclass method so as to deal with transaction databases.

2.1   Problem Setting

Let D be a dataset, $\mathcal{I} = \{I_1, I_2, \ldots, I_n\}$ be the set of all items in D, and $C = \{c_1, c_2, \ldots, c_k\}$ be the set of the class labels. We refer to a pair of an itemset $X \subseteq \mathcal{I}$ and a class label $c \in C$ as an *instance*, that is, $(X, c) \in D$ denotes an *instance*. For simplicity, we also call X *instance*. A rule is written as $r : A \to c$, where $A \subseteq \mathcal{I}$ and $c \in C$. If $A \subseteq X$ is held, we say that "itemset A covers (explains) instance X" or "rule $r : A \to c$ covers (explains) instance X". Let D_c denotes the family of the itemsets of the instances whose class label is c (*positive instances*), and let $D_{\bar{c}}$ be the itemsets of the rest of the instances (*negative instances*). If itemset A does not cover any instance of $D_{\bar{c}}$, then we say that itemset A is *consistent (to $D_{\bar{c}}$)*. For an itemset family S, let $R(S)$ denote the set of the common items seen in every instance of S, that is, $R(S) = \bigcap_{X \in S} X$.

For class c, *subclass S* is defined as a subset of D_c that satisfies:

1. $R(S)$ is consistent to $D_{\bar{c}}$.
2. S is maximal, that is, no instance in D_c can be added to S while keeping consistency of $R(S)$.

Here, $R(S)$ is called *Consistent Common Itemset (CCI)* in class c ($\in C$) if S is a subclass of C. The family of the subclasses for class c is denoted by Ω_c and the set of all subclasses for all $c \in C$ is denoted by Ω. Then, what we want to obtain is the family of subclasses Ω, as well as CCIs ($R(S)$'s for $S \in \Omega$) for all the classes.

Table 1. An illustrative dataset

ID	X	c
1	{sunny, hot, humid, windless}	not-play
2	{sunny, hot, humid, windy}	not-play
3	{rainy, cool, normal-humidity, windy}	not-play
4	{sunny, mild-temperature, humid, windless}	not-play
5	{rainy, mild-temperature, humid, windy}	not-play
6	{overcast, hot, humid, windless}	play
7	{rainy, mild-temperature, humid, windless}	play
8	{rainy, cool, normal-humidity, windless}	play
9	{overcast, cool, normal-humidity, windy}	play
10	{sunny, cool, normal-humidity, windless}	play
11	{rainy, mild-temperature, normal-humidity, windless}	play
12	{sunny, mild-temperature, normal-humidity, windy}	play
13	{overcast, mild-temperature, humid, windy}	play
14	{overcast, hot, normal-humidity, windless}	play

An example: For an illustrative dataset (Table 1), there are two CCIs for **not-play** class and five CCIs for **play** class as follows:

1. {sunny, humid} $\rightarrow$ not-play
2. {rainy, windy} $\rightarrow$ not-play
3. {windless, normal-humidity} $\rightarrow$ play
4. {overcast} $\rightarrow$ play
5. {rainy, windless} $\rightarrow$ play
6. {mild-temperature, normal-humidity} $\rightarrow$ play
7. {sunny, normal-humidity} $\rightarrow$ play

Here, each rule corresponds to one subclass (e.g., #1 rule corresponds to $S = \{1, 2, 4\}$) and each itemset in the condition part shows one CCI (e.g., #1 rule shows $R(S) = \{$sunny, humid$\}$). The procedure of finding these rules will be described in the following section.

2.2   Rule Extraction Procedure

We employ a randomized algorithm [10] to obtain a subset $\Omega' \subseteq \Omega$ to econo-mize the computational cost of enumerating all members of Ω. According to a theoretical analysis [10], the suboptimal subclass family Ω' (a subset of CCIs) obtained by this randomized algorithm with a fixed iteration number, has the following properties:

1. Larger subclasses (CCIs with larger coverage rates) are more probable to be found than smaller subclasses in the earlier iterations (the concrete procedure is given later).
2. Characteristic subclasses are also found in a higher probability. Here, a "char-acteristic subclass" is one that includes instances covered by only a few subclasses.

Now let us explain the algorithm briefly. The algorithm executes multiple scans for all the positive instances. The scanning is repeated t times for a given t. Each scan is made according to a random order, that is, a permutation $\sigma = (\sigma_1, \sigma_2, \ldots, \sigma_{|D_c|})$ randomly chosen. According to order σ, we merge the positive instances into S (initialized by the empty set), as long as the addition does not break the consistency of $R(S)$. Otherwise we skip the positive instance. Because of the fact that the merging process does not make $R(S)$ larger than before in the set inclusion relation and the fact that every positive instances are necessarily scanned, it is guaranteed that one subclass is necessarily found by one scan. Here, the dataset is assumed to be consistent in the weakest sense, that is, all the positive instances themselves are assumed to be consistent to the negative instances. We may obtain the same subclass for different σ's. Thus, the duplicated subclasses are removed in the last stage.

If we test all possible $|D_c|!$ permutations, we can obtain the complete family of the subclasses, Ω. However, even for not so large $|D_c|$, this number becomes infeasible. So we terminate the iteration by a given iteration number t. As de-scribed already, we can expect that almost all important subclasses are found even for a moderate iteration number, say $t = 1,000$, for each target class. For the constant t, the randomized algorithm runs in $O(|D_c||D_{\bar{c}}||\mathcal{I}|)$ for each target class, where $|D_c|$ is the number of instances in a target class, $|D_{\bar{c}}|$ is the number of instances in a non-target class and $|\mathcal{I}|$ is the number of the items.

An example: Let us show how to obtain a rule "{sunny, humid} → not-play" from Table 1.

Assume that the permutation is decided as $\sigma = (1, 2, 3, 4, 5)$.

1. We put instance #1 into S. Here, $S = \{1\}$ and $R(S) = \{$sunny, hot, humid, windless$\}$. This $R(S)$ is consistent, because no negative instance (Nos. 6–14) has all the items at once.
2. Putting instance #2 into S, $R(S)$ becomes {sunny, hot, humid}. $R(S)$ is still consistent.

3. If we put instance #3 into S, $R(S)$ becomes $\emptyset$, and the consistency is broken (because $\emptyset$ is included in any itemset). So we skip instance #3 for merging.
4. When we put instance #4 into S, $R(S)$ becomes {sunny, humid}. The consistency of the itemset is still kept.
5. If we put instance #5 into S, $R(S)$ becomes {humid}, then the consistency is again broken because the negative instances #6, #7 and #13 includes {humid}. So we skip this instance too.

Through this scan we obtain a CCI: $R(S) = $ {sunny, humid} with the corresponding of $S = \{1, 2, 4\}$.

Note that another scan with a different σ produces another rule such as "{rainy, windy} $\rightarrow$ not-play" for $\sigma = (3, 5, 1, 2, 4)$. Repeating scanning with all $120 (= 5!)$ permutations, we have all the subclasses for class **not-play**. In this example, there are only two subclasses for class **not-play** and five subclasses for class **play**.

2.3   Classification

Once CCIs have been obtained for each class, we can proceed to build a classifier. We design a classifier relying on the following belief (note that all the rules are 100% confident to the training set):

1. CCIs with larger coverage rates are more reliable.
2. Class assignment by a larger number of rules is more reliable.

In order to satisfy both of them, we introduce a score to a rule and sum up the scores of rules that explain a given instance. Here, the score of a rule is measured by its coverage rate of positive instances. According to the highest score, we assign a class to the instance.

Let us assume that a (class unknown) instance is given with an itemset A. Next, let $\mathcal{S}_{A,c}$ be the set of subclasses S ($\in \Omega_c$) whose $R(S)$ is included in A, that is,

$$\mathcal{S}_{A,c} = \{S \in \Omega_c \mid R(S) \subseteq A\}.$$

Then our classification rule is written as

$$\hat{c} = \arg\max_{c \in C} \sum_{S \in \mathcal{S}_{A,c}} \frac{|S|}{|D_c|}.$$

Here, $|S|/|D_c|$ ($0 < |S|/|D_c| \le 1$) is the score of subclass S (or equivalently CCI $R(S)$). Note that the score can be obtained without additional calculation during the rule generalization process by counting the number of the instances put into the subclass.

A tie-break is resolved by assigning it to the class with the largest population. If none of the rules is matched, the largest class is also chosen. We call this combining way the "Consistent Common Itemsets Classifier (shortly, CCIC)" approach.

3   Experimental Results

We conducted an experiment to evaluate the performance of CCIC approach. According to the literature [1,2,3], we used 26 datasets from UCI Machine Learning Repository [13]. The summary of datasets is shown in Table 2.

Every instance in the dataset is converted into an itemset. Numerical attributes are discretized into 5-bins, respectively. Here, the intervals of specifying the bins are taken so as to make the populations of the attribute values be equal in each attribute. The number of the iterations is set to $t = 1,000$. The negative instances that break the consistency of any positive instance are removed in order to keep the consistency of the dataset in a weakest sense.

We also present the accuracies of C4.5 [7], RIPPER [8], CBA [1], CMAR [2] and CPAR [3] as competitors. All of their results are copied from reference [3]. This is allowed because the experimental conditions are almost the same.

Table 2. Summary of the datasets. Three missing rates are: 1) the rate of attributes including missing values, 2) the rate of instances including missing values, and 3) the rate of missing values to the all values

dataset	#attr.	#attr. (cat.)	#attr. (num.)	#inst.	#classes	major class (%)	missing (attr.)	missing (inst.)	missing (val.)
anneal	38	32	6	898	6	0.76	0.763	1.000	0.650
austra	14	8	6	690	2	0.56	-	-	-
auto	25	10	15	205	7	0.33	0.280	0.224	0.012
breast	10	-	10	699	2	0.66	0.100	0.023	0.002
cleve	13	7	6	303	2	0.54	0.154	0.023	0.002
crx	15	9	6	690	2	0.56	0.467	0.054	0.006
diabetes	8	-	8	768	2	0.65	-	-	-
german	20	13	7	1000	2	0.70	-	-	-
glass	9	-	9	214	7	0.36	-	-	-
heart	13	-	13	270	2	0.56	-	-	-
hepati	19	13	6	155	2	0.79	0.789	0.484	0.057
horse	22	15	7	368	2	0.63	0.955	0.981	0.238
hypo	25	18	7	3163	2	0.95	0.320	0.999	0.067
iono	34	-	34	351	2	0.64	-	-	-
iris	4	-	4	150	3	0.33	-	-	-
labor	16	8	8	57	2	0.65	1.000	0.982	0.357
led7	7	7	-	3200	10	0.11	-	-	-
lymph	18	15	3	148	4	0.55	-	-	-
pima	8	-	8	768	2	0.65	-	-	-
sick	29	22	7	2800	2	0.94	0.276	1.000	0.056
sonar	60	-	60	208	2	0.53	-	-	-
tic-tac	9	9	-	958	2	0.65	-	-	-
vehicle	18	-	18	846	4	0.26	-	-	-
waveform	21	-	21	5000	3	0.34	-	-	-
wine	13	-	13	178	3	0.40	-	-	-
zoo	16	16	-	101	7	0.41	-	-	-

Table 3. Accuracy comparison. The best score is indicated in boldface. The column #CCIs is the average number of CCIs found by the proposed method. The column %drop is the avarage ratio of test instances that are not matched with any rule. The bottom row #bests shows the number of the datasets to which the method recorded the best accuracy.

dataset	C4.5	RIPPER	CBA	CMAR	CPAR	CCIC	#CCIs	%drop
anneal	0.948	0.958	0.979	0.973	**0.984**	0.966	128.0	0.02
austra	0.847	0.873	0.849	0.861	0.862	**0.877**	714.6	0.00
auto	0.801	0.728	0.783	0.781	**0.820**	0.787	334.2	0.07
breast	0.950	0.951	0.963	**0.964**	0.960	**0.964**	266.5	0.00
cleve	0.782	0.822	**0.828**	0.822	0.815	**0.828**	584.9	0.00
crx	0.849	0.849	0.847	0.849	0.857	**0.875**	716.8	0.00
diabetes	0.742	0.747	0.745	**0.758**	0.751	0.723	833.6	0.01
german	0.723	0.698	0.734	**0.749**	0.734	0.748	1635.6	0.00
glass	0.687	0.691	0.739	0.701	**0.744**	0.705	193.0	0.09
heart	0.808	0.807	0.819	0.822	0.826	**0.837**	548.0	0.00
hepati	0.806	0.767	0.818	0.805	0.794	**0.827**	270.3	0.01
horse	0.826	**0.848**	0.821	0.826	0.842	0.845	601.3	0.02
hypo	**0.992**	0.989	0.989	0.984	0.981	0.972	183.4	0.01
iono	0.900	0.912	0.923	0.915	**0.926**	0.923	999.7	0.00
iris	**0.953**	0.940	0.947	0.940	0.947	0.933	35.0	0.02
labor	0.793	0.840	0.863	**0.897**	0.847	0.833	77.0	0.04
led7	0.735	0.697	0.719	0.725	**0.736**	0.729	153.1	0.00
lymph	0.735	0.790	0.778	**0.831**	0.823	0.810	260.6	0.05
pima	**0.755**	0.731	0.729	0.751	0.738	0.732	829.2	0.01
sick	**0.985**	0.977	0.970	0.975	0.968	0.941	438.1	0.01
sonar	0.702	0.784	0.775	0.794	0.793	**0.836**	1655.2	0.00
tic-tac	0.994	0.980	**0.996**	0.992	0.986	0.989	268.9	0.00
vehicle	**0.726**	0.627	0.687	0.688	0.695	0.703	1715.2	0.00
waveform	0.781	0.760	0.800	**0.832**	0.809	0.802	2944.7	0.02
wine	0.927	0.916	0.950	0.950	0.955	**0.961**	407.8	0.00
zoo	0.922	0.881	0.968	**0.971**	0.951	0.891	8.8	0.11
average	0.8334	0.8293	0.8469	**0.8522**	0.8517	0.8476		
#bests	5	1	2	7	5	**8**		

Table 3 shows the summary of the results. The average accuracy is obtained by 10-fold cross validation. The average accuracy of CCIC was the third, while CCIC was best in the number of wins (8/26). As can be seen from those results, the performance of CCIC approach depends on the problems. So, let us examine the reasons why such a dependency happens.

Since CCIC uses only the consistent rules, the accuracy goes down if a sufficient number of consistent rules are not found. Even if many consistent rules are found, their coverage rates might be low, that is, the consistent rules may explain only a part of positive instances. In this case, an instance that is not matched by any rule is assigned to the largest population class. This is one possible reason of the performance degradation. Indeed, in some datasets such as auto, glass and zoo, more than 5% of the test instances were not matched with any rule (see column %drop in Table 3). This is a limitation of the proposed method relying only on the consistent rules. On the other hand, as seen in Table 3, CCIC shows better performance for datasets for which many consistent rules are found and almost all instances are covered by them. Adopting consistency-relaxed rules might improve the accuracy.

Another possible reason of the degradation of performance is that some CCIs fit too tightly to the training instances. In this case, a CCI may contain redundant items not contributable for classification. Such a situation occurs when the CCI is obtained from only a small number of positive instances. This problem may be resolved by removing such redundant items from the obtained CCIs.

Discretization of numerical values also affects the performance [14]. The optimal setting of bins is one of difficult problems. If we adopt too many bins, the common items among instances decreases. On the other hand, with too less bins, some instances would not be distinguished from the others. With a better selection of bins, the performance may be improved for datasets that include many numerical attributes such as diabetes, iris, and waveform.

All the experiments were performed on a 2GHz Intel Core Duo PC (running Mac OS X 10.4.8) with 1GB main memory. The implementation was not multithreaded. The most time-consuming dataset was waveform, its execution time of the 10-fold cross validation (including both training and test) was about 395 seconds. However, the algorithm is easily parallelized to reduce the running time with less overheads because each iteration is completely independent.

4   Conclusion

A novel classifier called CCIC has been proposed. The CCIC combines many consistent itemsets for classification. So, it is an associative classifier. The experimental results showed that CCIC outperformed the others in several datasets.

In the future works, we will consider more different combining ways of the consistent rules and adoption of consistency-relaxed rules to improve the performance of the classifier. In addition, rule selection should be considered in order to reduce redundancy of the ruleset.

References

1. Liu, B., Hsu, W., Ma, Y.: Integrating classification and association rule mining. In: Knowledge Discovery and Data Mining, pp. 80–86 (1998)
2. Li, W., Han, J., Pei, J.: CMAR: accurate and efficient classification based on multipleclass-association rules. In: Proceedings on IEEE International Conference on Data Mining (ICDM2001), pp. 369–376. IEEE Computer Society Press, Los Alamitos (2001)
3. Yin, X., Han, J.: CPAR: Classification based on predictive association rules. In: 3rd SIAM International Conference on Data Mining (SDM'03) (2003)

4. Zaiane, O.R., Antonie, M.-L.: Classifying text documents by associating terms with text categories. In: Proceedings of the 13th Australasian database conference, pp. 215–222 (2002)
5. Wang, Y., Wong, A.K.C.: From association to classification: Inference using weight of evidence. IEEE Transactions on Knowledge and Data Engineering 15(3), 764–767 (2003)
6. Dong, G., Zhang, X., Wong, L., Li, J.: CAEP: Classification by aggregating emerging patterns. Discovery Science, 30–42 (1999)
7. Quinlan, J.R.: C4.5: programs for machine learning. Morgan Kaufmann Publishers Inc, San Francisco (1993)
8. Cohen, W.W.: Fast effective rule induction. In: Proceedings of the 12th International Conference on Machine Learning, pp. 115–123 (1995)
9. Agrawal, R., Srikant, R.: Fast algorithms for mining association rules. In: Proceedings of the 20th International Conference on Very Large Data Bases (VLDB), pp. 487–499 (1994)
10. Kudo, M., Yanagi, S., Shimbo, M.: Construction of class regions by a randomized algorithm: A randomized subclass method. Pattern Recognition 29(4), 581–588 (1996)
11. Kudo, M., Shimbo, M.: Feature selection based on the structual indices of categories. Pattern Recognition 26(6), 891–901 (1993)
12. Kudo, M., Shimbo, M.: Analysis of the structure of classes and its applications – subclass approach. Current Topics in Pattern Recognition Research 1, 69–81 (1994)
13. Murphy, P.H., Aha, D.W.: UCI repository of machine learning databases http://www.ics.uci.edu/mlearn/MLRepository.html
14. Srikant, R., Agrawal, R.: Mining quantitative association rules in large relational tables. In: Jagadish, H.V., Mumick, I.S. (eds.) Proceedings of the 1996 ACM SIGMOD International Conference on Management of Data, Montreal, Quebec, Canada, 4–6 1996, pp. 1–12. ACM Press, New York (1996)

Development of an Agreement Metric Based Upon the RAND Index for the Evaluation of Dimensionality Reduction Techniques, with Applications to Mapping Customer Data

Stephen France and Douglas Carroll

Rutgers University, Graduate School of Management, Newark, New Jersey, 07102-3027
sfrance@andromeda.rutgers.edu, dcarroll@rci.rutgers.edu

Abstract. We develop a metric ψ, based upon the RAND index, for the comparison and evaluation of dimensionality reduction techniques. This metric is designed to test the preservation of neighborhood structure in derived lower dimensional configurations. We use a customer information data set to show how ψ can be used to compare dimensionality reduction methods, tune method parameters, and choose solutions when methods have a local optimum problem. We show that ψ is highly negatively correlated with an alienation coefficient K that is designed to test the recovery of relative distances. In general a method with a good value of ψ also has a good value of K. However the monotonic regression used by Nonmetric MDS produces solutions with good values of ψ, but poor values of K.

1 Introduction

Dimensionality reduction techniques have great applicability within marketing. Uses of these techniques include product placement, perceptual and cognitive mapping, and brand switching. These techniques have been developed in distinct traditions; those of psychometrics, statistics, and computer science. Most marketing applications of dimensionality reduction can be placed within the psychometric tradition though computer science based data-mining techniques have been applied to the large data sets typically found in customer systems.

There has been little work on the evaluation and comparison of dimensionality reduction techniques. We start by describing some Multidimensional Scaling based techniques referenced in the marketing literature; we then review some techniques from the data mining literature. We describe methods for evaluating the recovery of lower dimensional solutions and then develop and test metrics based upon the Agreement Rate described in [1,2,3].

2 Overview of Multidimensional Scaling

Multidimensional Scaling (MDS) can be described as a set of techniques for interpreting similarity or dissimilarity data. Typically MDS is used to take data of

P. Perner (Ed.): MLDM 2007, LNAI 4571, pp. 499–517, 2007.

high dimensionality and reduce the data to a more interpretable form, often but not always in one, two, or three dimensions. MDS has been applied to a large number of problems throughout the social and behavioral sciences, as well as to some extent in the biological and physical sciences. The technique has its basis in the mathematical psychology literature; the initial breakthrough paper is [40]. [37] introduces classical metric scaling, which is based on a singular value decomposition of a derived "scalar products" matrix. [33] shows that lower dimensional solutions can be extracted from ordinal scale, or "nonmetric" data, where the underlying latent dimensionality is lower than the observed dimensionality of the data. [21,22] introduce a computational algorithm inspired by the ideas of [33], which was specifically designed to handle such nonmetric proximity data.

3   Applications of Dimensionality Reduction in Marketing

Most marketing research involving dimensionality reduction has used some variant of MDS in order to produce perceptual maps. [17] gives a survey of perceptual and preference mapping uses of MDS, including product and brand mapping. [13] describes unfolding methodology for fitting choice data, [15] provides an analysis of brand switching data, [4] combines latent class choice models and latent class MDS models for empirical analysis of scanner data, and [13] develops spatial MDS models that account for the effects of brand size and buying power in consumer brand attraction. [26] combines econometric and MDS methodologies and uses panel data to create a time series of joint space maps, and then uses these maps in marketing response models. Other approaches for creating perceptual maps include corres-pondence analysis [9] and attribute elicitation mapping [35].

More recently, data warehousing has created huge repositories of customer data. Data mining researchers, typically with roots in computer science, have developed techniques for interpreting these data. Among these techniques are those of dimen-sionality reduction, which are typically used to create visual maps of customers, products, and services. [16] provides an overview of common dimensionality reduction techniques, including those based upon the psychometric and statistical literature, such as Principal Components Analysis (PCA), factor analysis, and MDS, and those from the data mining literature, such as kernel/nonlinear PCA, and self-organizing maps [20], which are often extensions of the original psychometric/statistical techniques. Other nonlinear dimensionality reduction techniques include PARAMAP [1,2,34], Isomap [36], and Local Linear Embedding (LLE) [31].

PARAMAP optimizes an index of continuity rather than a traditional MDS distance function. Isomap uses a shortest path algorithm to build geodesic distances and then performs classical MDS, as per the method usually associated with [37,38]. LLE produces a global lower dimensional embedding of higher dimensional data, while preserving local neighborhoods. Recent dimensionality reduction techniques include an incremental version of Isomap [24], diffusion maps and coarse-graining [23], and semi-definite programming techniques [39]. [18] shows that many of these techniques, including Isomap and LLE, can be described as Kernel PCA algorithms using Gram matrices.

4 Agreement Rate Metric (AR)

In order to test how well a lower dimensional configuration is related to a higher dimensional configuration, we need some sort of metric to compare the two solutions. This is not easy as most dimensionality reduction techniques maximize some measure of congruence, which is itself just such a metric, thus we need to develop a metric that in some sense is independent of the one on which the solution is based. The performance of dimensionality reduction techniques will depend on the structure of the data being transformed, the size of the data, and the number of target dimensions. We need some appropriate metrics to test the various dimensionality reduction techniques appropriate for a range of different problems.   When describing these comparison metrics, we will consider a higher dimensional solution A, and a derived lower dimensional solution B. Though we concentrate on comparing a higher dimensional solution with a derived lower dimensional solution, we could also compare two lower dimensional derived solutions.

Correlation based metrics have been used to compare configurations. [6] shows that the standard Pearson product moment correlation of data and obtained distances is inappropriate given that distances plus an additive constant are not invariant under a linear transformation; the distances will not correspond to distances among the points in the same configuration as the original untransformed distances. Also, the distances may not continue to satisfy the triangle inequality under such a transformation.

The paper goes on to describe a congruence coefficient. Take solution configure-tions A and B, with n points in each configuration. Calculate each of the n(n-1)/2 symmetric distances between solution points for both A and B; the resulting congruence coefficient is given in (1).

$$c = \sum_{i=1}^{n(n-1)/2} d_{Ai} d_{Bi} \left/ \left(\sum_{i=1}^{n(n-1)/2} d_{Ai} \sum_{i=1}^{n(n-1)/2} d_{Bi} \right)^{0.5} \right. \tag{1}$$

The coefficient is transformed into the alienation coefficient (2) in order to give a greater spread of values.

$$K = \left(1 - c^2\right)^{0.5} \tag{2}$$

[5] uses the alienation coefficient to test the recovery of perceptual maps, and also used a series of Monte Carlo tests on data sets of known dimensionality to test the recovery of true dimensionality for different variants of MDS. It should be noted that when testing with real life data sets, the intrinsic dimensionality of the data is unknown, through some techniques have been developed for estimating the intrinsic dimensionality of Manifolds [25,29].

[2] performs a comparison of two nonlinear dimensionality reduction techniques, PARAMAP and Isomap. To compare the solutions, Variance Accounted For (VAF) and the Agreement Rate (AR) were used. VAF can be used when testing different error perturbed versions of a base configuration. AR is more general, and can be used to compare any two solutions that have the same number of points. It is upon this idea of agreement rate that we will base our metrics for testing the recovery of lower dimensional solutions.

4.1 Description of the Agreement Rate Metric (AR)

The Agreement Rate metric is based upon the RAND index for comparing Clustering configurations [19,30] and was developed for comparing embeddings of sets of objects in [1,2]. Take solution configurations A and B. For each configuration, calculate the k nearest neighbors for each solution point i. For both A and B, we calculate the distances between each pair of items. This gives us derived $n(n-1)/2$ item distance matrices dA and dB. The k nearest neighbors for item i are the items with one of the k lowest values of dA_{ij} (for solution A) and the k lowest values of dB_{ij} (for solution B), where $i{\neq}j$. Denote the neighborhoods for item i as A_i and B_i. Let a_i represent the number of points in both A_i and B_i for point i. The agreement rate AR is equal to (3), where k is the size of neighborhood used and n is the number of points.

$$AR = \frac{1}{kn} \sum_{i=1}^{n} a_i \qquad (3)$$

4.2 Extension of the Agreement Rate Metric

Previous comparison work using AR such as [1,2,3] has tended to choose an arbitrary value of the neighborhood size k, based upon getting a good spread of results. Given n points in the embedding, we can calculate AR for neighborhood sizes from 1 to n-1. A neighborhood of size n-1 will include every other solution point, giving a value of $AR=1$. By results derived by [19] we can test the deviation from randomness using a chi-squared statistic. We assume however that our lower dimensional solution is non-randomly related to the higher dimensional one, so we want some relative measure of how good the solution is.

Theorem 1. AR is not monotonic with respect to k.

Proof. If k=n-1 then AR=1. Unless there is perfect agreement then for some k<n-1 then AR<1 so k does not decrease monotonically. Assume that AR increases monotonically with respect to k; we will give a simple counter example. Consider a higher dimensional solution:

$d(1,2)=1, d(1,3)=5, d(1,4)=6, d(2,3)=4.2\ d(3,4) = 1.5$

Consider also, the following derived one dimensional solution:

$d(1,2)=0.9, d(1,3)=5.7, d(1,4)=5.3, d(3,4) = 0.4$

Denoting AR(k) as the agreement rate for a neighborhood of size k we have:

$AR(1)=1,\ `AR(2)=0.75\ ,AR(3)=1$

A problem with using a specific value of k is it may favor one technique over another. Of course we can calculate the mean and standard deviation across all values of k. Taking the mean value of AR across all values of k gives us a measure of the performance of a technique over different neighborhood sizes, but for large values of neighborhood size AR has little discrimination value. We propose a statistic that takes account of the expected value of AR if the lower dimensional configuration is completely random. We need to find the proportion of the possible items in the neighborhood that agree. There are n-1 items in the potential neighborhoods of a single point in a lower dimensional solution. Given a neighborhood of size k and a randomly generated lower dimensional solution, each item in the higher dimensional neighborhood has a probability of being in the lower dimensional neighborhood of $p(A)=k/(n-1)$. Taking the expected value of AR over all n item configurations this gives the expected agreement rate for neighborhood size k in (4).

$$E\big[R(k)\big] = \frac{1}{kn}\sum_{i=1}^{n} a_i = \frac{1}{kn}\sum_{i=1}^{n} k \cdot p(A) = \frac{1}{kn}\sum_{i=1}^{n} k\left(\frac{k}{n-1}\right) = \frac{kn}{kn}\left(\frac{k}{n-1}\right) = \frac{k}{n-1} \tag{4}$$

Where $E[R(K)]$ is the expected agreement rate if neighborhood agreement is random, k is the size of the neighborhood, and n is the number of items in the solutions. We can define the improvement in agreement rate over that what would be expected from a random solution as $AR(k)-E[R(K)]$. If a solution is a perfect lower dimensional reconstruction then $AR(k)=1$ for all values of k. Thus the maximum possible increase in agreement rate from the $E[R(K)]$ is $1- E[R(K)]$. We define the statistic ψ as the sum of $AR(k)-E[R(K)]$ over k, divided by the sum of the $1- E[R(K)]$. over k, where k ranges from $k=1$ to $k=n-1$. The statistic is defined for $n>2$, as $n=1$ gives a neighborhood size of 0 and a denominator in (5) of 0. Negative values of ψ may occur if the solution is a worse representation than what would be expected randomly. A similar statistic, called the adjusted agreement rate, was developed in [1,2,3]. This statistic was defined for a single value of k and used an experimental rather than theoretical value for $E[R(K)]$.

$$\psi = \frac{\sum_{k=1}^{n-1}\big(AR(k) - E\big[R(k)\big]\big)}{\sum_{k=1}^{n-1}\big(1 - E\big[R(k)\big]\big)} \tag{5}$$

We can also calculate ψ for subsets of neighborhood size. We may wish to concentrate on a certain subset of k values as in (6). For example we may wish to choose the lowest 10% of k values, in order to check the preservation of local neighborhood structure.

$$\psi_S = \frac{\sum_{k \in S}\big(AR(k) - E\big[R(k)\big]\big)}{\sum_{k \in S}\big(1 - E\big[R(k)\big]\big)} \quad S \subseteq \{1,\cdots,n-1\} \tag{6}$$

We could also use a function to proportionally weight the agreement for different values of k as in (7).

$$\psi_{f(k)} = \frac{\sum_{k=1}^{n-1}\big(f(k)\big(AR(k) - E\big[R(k)\big]\big)\big)}{\sum_{k=1}^{n-1}\big(f(k)\big(1 - E\big[R(k)\big]\big)\big)} \tag{7}$$

Theorem 2. $\sup\{\psi_{f(k)}\} = 1$ and is independent of f(k).

Proof. For each k, $0 \le AR(k) \le 1$. As AR(k) is linear and in the numerator, $\sup\{\psi_{f(k)}\}$ occurs when $AR(k) = 1$ for each k.

$$\sup\{\psi_{f(k)}\} = \frac{\sum_{k=1}^{n-1}\left(f(k)\left(1 - E\left[R(k)\right]\right)\right)}{\sum_{k=1}^{n-1}\left(f(k)\left(1 - E\left[R(k)\right]\right)\right)} = 1 \tag{8}$$

Theorem 3. If f(k)=c for some constant c then $\inf\{\psi_{f(k)}\} = \dfrac{\Upsilon - n}{(n-2)}$ where

$$\Upsilon = 2\sum_{i=1}^{\lfloor n/2 \rfloor}\left(\frac{\left\lfloor \dfrac{n-1}{2} \right\rfloor + i}{n - \left(2\cdot\left\lceil (n-1)/2 \right\rceil\right) + 2\cdot i}\right) \Bigg/ \left(\frac{n}{\left\lfloor \dfrac{n-1}{2} \right\rfloor + i}\right)$$

Proof. As $AR(k)$ is linear and in the numerator, $\inf\{\psi_{f(k)=c}\}$ occurs when $AR(k)$ is minimized for each value of k. If $1 \le k \le \lfloor n/2 \rfloor$ then $\min\{AR(k)\} = 0$. If $\lfloor n/2 \rfloor > k$ then $\min\{AR(k)\}$ depends on the size of the overlap between the two solutions; the overlaps are of the form 1,3,5,.....,n-1 when n is even, and 2,4,.....,n-1 when n is odd.

$$\inf\{\psi_{f(k)=c}\} = \frac{\sum_{k=1}^{n-1}c\left(\left(\min\left[R(k)\right] - E\left[R(k)\right]\right)\right)}{\sum_{k=1}^{n-1}c\left(\left(1 - E\left[R(k)\right]\right)\right)} = \frac{\sum_{k=1}^{n-1}\min\left[R(k)\right]}{\sum_{k=1}^{n-1}\left(\left(1 - E\left[R(k)\right]\right)\right)} - \frac{\sum_{k=1}^{n-1}\left(E\left[R(k)\right]\right)}{\sum_{k=1}^{n-1}\left(\left(1 - E\left[R(k)\right]\right)\right)}$$

$$= \sum_{i=1}^{\lfloor n/2 \rfloor}\left(\frac{\left\lfloor \dfrac{n-1}{2} \right\rfloor + i}{n - \left(2\cdot\left\lceil (n-1)/2 \right\rceil\right) + 2\cdot i}\right) \Bigg/ \left(\frac{n-1}{\left\lfloor \dfrac{n-1}{2} \right\rfloor + i}\right) \Bigg/ \frac{(n-2)}{2} - \frac{n(n-1)/2}{(n-1)(n-2)/2} = \frac{\Upsilon - n}{(n-2)} \tag{9}$$

An example of a linear function, concentrating on preservation of local neighborhoods, is given in (10).

$$Let\ f(k) = \frac{(n-k)}{n} \tag{10}$$

When comparing any two techniques it may also be useful to check the sums of differences between the agreement rates. We split the differences into upper and lower sums.

$$U(1,2) = L(2,1) = \sum_{k=1}^{n-1}\max\{AR_1(k) - AR_2(k), 0\}$$

$$U(2,1) = L(1,2) = \sum_{k=1}^{n-1}\max\{AR_2(k) - AR_1(k), 0\} \tag{11}$$

If the lower sum is 0 then $AR_1(k) \geq AR_2(k)$ for all values of k and we can say a solution completely dominates another solution.

In Figure 1 we give an example of dimensionality reduction of 500 points, with two techniques, Metric and Nonmetric MDS. ψ is a discrete analogue of the area between the "Random" line and the technique line divided by the total area above the "Random" line. The upper and lower sums are the analogues of the areas between the two technique lines. For this example, $\psi(Metric)=0.545$, $\psi(Nonmetric)=0.582$, $U(Metric,Nonmetric)=0.623$, and $U(Nonmetric,Metric)=9.67$. We can see from this figure how ψ is a discrete analogue to the integral of the area between the random line and the solution line divided by the total area above the random line.

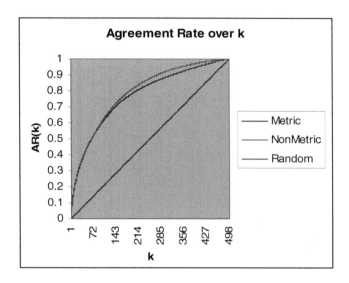

Fig. 1. Agreement Rate over k

5   Experimentation

In Section 1 we gave an overview of a selection of dimensionality reduction techniques. Many of the papers referenced test the lower dimensional embedding of certain standard data sets, such as the Swiss roll, sphere, 4-dimensional torus, and face data. The MDS marketing based papers tend to use psychometric data, where the variables are of uniform measurement type, and there are no missing observations.

In our experiments we aim to show how the metrics described in the previous section can be used, and how some common dimensionality reduction techniques perform on a set of realistic data of the type typically found in corporate customer information systems. The main thrust of this paper is to develop ideas on the comparison of dimensionality reduction techniques; we do not intend to provide a comparison of all of the techniques previously mentioned.

We tested our data with Principal Components Analysis (PCA), Metric MDS, Nonmetric MDS, and Isomap. PCA, and Isomap are guaranteed to give globally optimal solutions. The MDS procedures may have local optima, so multiple runs may be required to get a globally optimal solution. Isomap requires some parameter tuning for the size of the neighborhoods. We can therefore demonstrate the use of our metrics in both parameter tuning and in selecting solutions for the techniques with non-global optima.

PCA is based upon a simple Singular Value Decomposition (SVD). Let us define **Y** as the higher dimensional data matrix with m rows (items) and n columns (dimensions). We find the scalar product **A=YY'** and then decompose **Y** as **Y=UΣV'**, where **U** and **V** are orthogonal matrices of eigenvectors and Σ is a matrix of singular values. The derived lower dimensional solution is given as **X=UΣ$^{1/2}$**. PCA is equivalent to performing Classical Torgerson MDS (CMDS) on a proximity matrix calculated from the data using Euclidean distances. Isomap uses the same SVD, but the standard matrix **Y** is replaced by **Y***, where **Y*** contains geodesic distances (approximated by shortest path distances between points). A dynamic programming approach, such as Distikra's algorithm, is used to calculate the shortest paths.

The Metric and Nonmetric MDS techniques used both optimize the STRESS function given in (12).

$$ STRESS = \sqrt{\sum_i \sum_j \left(d_{ij} - \hat{d}_{ij}\right)^2 \Big/ \sum_i \sum_j d_{ij}^2} \tag{12} $$

Where $\hat{d}_{ij} = F\left(\delta_{ij}\right)$ is the best least squares approximation (attainable by transforming δ_{ij} by a function in the class of functions F) to the distance in the low dimensional space (d_{ij}). For Nonmetric MDS the transformation is monotone, and F is the best least squares monotonic function (non-decreasing if δ_{ij} is a dissimilarity measure). For Metric MDS F is a simple linear regression function.

5.1 Description of Data

In our experiments we mapped customer information from a specific data set. This was the "Churn Modeling Tournament" set supplied by the Duke University TerraData center. The dataset consists of 250,000 customer records. Each customer record has approximately 168 fields. These fields detail information such as demographics, customer purchase records, and behavioral information such as attitudes towards products. The data are of different measurement levels, for example there are purchase values (ratio), satisfaction ratings (interval or ordinal), and demographic groups (nominal).

The data are of variable quality, with some data values missing. The data set is introduced in [28]; we ignored the value of churn and concentrate on mapping customers based upon the other customer variables.

5.2 Experimentation and Results

We designed an experiment to test the recovery of lower dimensional solutions using the methods and data set previously described. As we present a marketing application, and dimensionality reduction in marketing is used to create a parsimonious visual representation of data, we restricted our lower dimensional derived solutions to 1, 2, and 3 dimensions. Out of the approximately 250,000 items, we randomly selected records for mapping. We selected 20 files (sets of records), each with 500 items. Most computational algorithms for dimensionality reduction techniques require ratio or interval scale data with no missing values. We ran the EM algorithm [12] to estimate parameters. We then took a single imputation of the data (if we were looking to hypothesize about the data then multiple imputations would be more appropriate, but we are only looking for a higher dimensional representation to transform).

We then used correspondence analysis [27] to transform the nominal and ordinal dimensions into a smaller number of ratio scale dimensions (taking all dimensions accounting for more than 1% of the variance). The resulting data set had 132 ratio and interval scale dimensions and no missing data. As the techniques tested utilize the Euclidean distance between items, some sort of transformation was required, so we standardized the variables to mean zero and variance one.

For each combination of technique and data set tested we produced lower dimensional representations in 1, 2, and 3 dimensions. We calculated the metric ψ across all values of k from 1 to n-1 (499). We also calculated the metric for each of the quartiles of the numbers 1-499 and calculated the alienation rate given in (1) and (2). We prescribed a randomized design (where File is the blocking random factor) and used MANOVA to analyze the experiments. The model specification is given in (13).

$$X_{ijk} = METHOD_i + DIM_j + FILE_k + METHOD_i \times DIM_j + \varepsilon_{ijk} \qquad (13)$$

Though the main purpose of the paper is not to compare dimensionality reduction algorithms, we tried to make our comparisons as fair as possible. We are testing a globally optimal technique (PCA), a globally optimal technique with parameter tuning (Isomap), and techniques utilizing gradient based optimization (Metric MDS and Nonmetric MDS). We did some initial experimentation to approximate the average time per run for each of the techniques. All techniques were tested using MATLAB implementations on a PC with a 2.8MHz Pentium 4 Zeon processor and 1GB of memory. As the value of k for Isomap could possibly affect the running time of the algorithm we averaged across values of k taken at intervals of 20, from 20 to 480. The average run times for each technique are given in Table 1.

Table 1.

Technique	Solution Time (seconds)
PCA	1.1
Isomap	13.0
Metric MDS	379.1
Nonmetric MDS	2029.1

We can see that one run of either Metric MDS or Nonmetric MDS takes many times the running time of the Isomap algorithm. To try and keep the comparison reasonable, when testing we took one run of the iterative MDS procedures (using the CMDS solution as a starting solution), and for Isomap we found the best solution from testing 30 different values of k (equidistantly spaced between 1 and n-1). This equates to running Isomap for approximately the same time as for Metric MDS. We can see from Table 1 that the Monotonic Regression used for Nonmetric MDS gives running times that are a factor of 7 to 8 greater than the running times for the standard linear regression function fitted by Metric MDS.

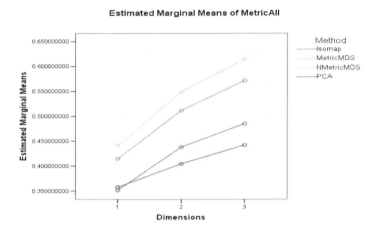

Fig. 2. Estimated Marginal Means of MetricAll

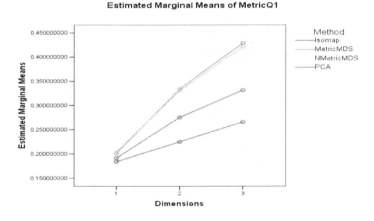

Fig. 3. Estimated Marginal Means of MetricQ1

We neglect to show the MANOVA table for the sake of brevity, but using Wilks's Lambda and an F-test based upon this statistic, all factors except one were significant with p<0.001. The one exception was the effect of the interaction term between method and dimension on the alienation dependent variable. Confidence intervals for the dependent variables are given in Table 2 and graphs of the estimated marginal means of the dependent variables are given in Figures 2-5. Figures are given for alienation; ψ across all k, ψ for the first quartile values of k, and ψ for the fourth quartile values of k.

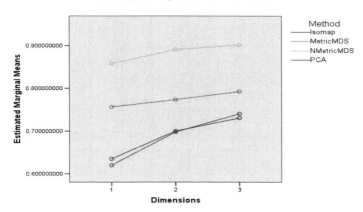

Fig. 4. Estimated Marginal Means of MetricQ4

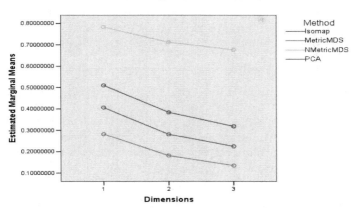

Fig. 5. Estimated Marginal Means of Alienation

We summarize the results of the Bonferroni post-hoc test for ψ across all values of k (MetricAll in table 2) in (14) and for the alienation metric in (15). We denote a>b to mean that technique "a" achieves a better result than "b", so the order of the alienation metric is reversed.

$$PCA<Isomap<Metric\ MDS<Nonmetric\ MDS \tag{14}$$

$$\text{Nonmetric MDS} < \text{PCA} < \text{Isomap} < \text{Metric MDS} \tag{15}$$

We can glean additional information from the quartile results and the graphs of marginal means given in Table 2. We can see that the performance of Metric MDS is better than Nonmetric MDS for quartile 1, but as the value of k increases Nonmetric MDS shows increasing advantage over Metric MDS. From Figure 2, we can see that this pattern is repeated across dimensions. Metric MDS is therefore strong at preserving local neighborhood structure, but weaker at preserving order for larger distances. Isomap does not perform as well as Metric MDS on either the agreement rate metric or the alienation metric, but the use of geodesic distances gives a bump to performance over PCA when reducing to 2 and 3 dimensional solutions.

Table 2. Confidence Intervals

Metric	Method	Mean	95% LB	95% UB
AMAll	Isomap	0.425	0.419	0.431
	MetricMDS	0.499	0.493	0.506
	NMetricMDS	0.535	0.529	0.541
	PCA	0.402	0.395	0.408
AMQ1	Isomap	0.266	0.261	0.272
	MetricMDS	0.322	0.316	0.327
	NMetricMDS	0.319	0.314	0.324
	PCA	0.225	0.220	0.230
AMQ2	Isomap	0.490	0.483	0.498
	MetricMDS	0.572	0.564	0.580
	NMetricMDS	0.614	0.607	0.622
	PCA	0.463	0.455	0.471
AMQ3	Isomap	0.686	0.678	0.694
	MetricMDS	0.774	0.766	0.783
	NMetricMDS	0.884	0.875	0.892
	PCA	0.688	0.680	0.697
Alien.	Isomap	0.304	0.283	0.325
	MetricMDS	0.199	0.178	0.220
	NMetricMDS	0.724	0.703	0.745
	PCA	0.404	0.383	0.425

The results throw up a paradox; why does Nonmetric MDS perform so well in terms of preserving neighborhoods but badly on the distance based alienation metric? Figures 8-9 show the solutions for the first random file for both 1 and 2 dimensions. We can see that the solutions for Nonmetric MDS have many points very closely clumped together, with small distances between points. These very small distances, which almost lead to a degenerate solution, when multiplied lead to a small numerator in (1) and thus a small congruence coefficient and large alienation coefficient. This lack of distance preservation is due to the monotonic regression only enforcing monotonicity in the distances, but not preserving relative distances. This illustrates a problem of using a nonmetric technique with essentially metric data. The resulting maps preserve distance order well, but the clumping of points makes the solutions almost impossible to interpret visually. [7] found similar degeneracy problems when visualizing Nonmetric MDS solutions. Comparing the other three techniques, all have more of a spread of points than Nonmetric MDS, giving more interpretable visualization. We can see that for PCA and Isomap, most of the variation is in the first dimension; this is to be expected given the nature of the SVD, finding maximal variance for each dimension in turn. For the Metric MDS solution the variance is much more evenly divided between dimensions, despite the fact that there may be some bias from the initial CMDS solution.

We explore the relationship between the agreement metrics by calculating the pairwise Pearson product moment correlation coefficient between the results for each pair of metrics. The correlation coefficients and significance values are summarized in Table 3. There is significant positive correlation between all the quartiles of the agreement metric, with p=0.000 for all pairwise comparisons. This suggests that for the techniques tested, good recovery of local neighborhoods implies good recovery of more distant neighborhoods.

Table 3. Correlations: All Methods

	AMAll	AMQ1	AMQ2	AMQ3	AMQ4
AMQ1	0.946				
	0.000				
AMQ2	0.995	0.920			
	0.000	0.000			
AMQ3	0.915	0.738	0.934		
	0.000	0.000	0.000		
AMQ4	0.835	0.638	0.848	0.954	
	0.000	0.000	0.000	0.000	
Alien.	0.026	-0.152	0.047	0.252	0.321
	0.652	0.008	0.418	0.000	0.000

```
AM=Agreement Metric
Alien.=Alienation Metric
Cell Contents: Pearson correlation
              P-Value
```

Table 3 shows no significant correlation exists between the agreement rate metric ψ and the alienation coefficient K. However, the correlations are skewed by the results for Nonmetric MDS. In Table 4 we calculate correlations between metric values for results and exclude Nonmetric MDS. All correlations are significant to $p<0.001$.

Table 4. Correlations: All Methods EXCEPT Nonmetric MDS

	AMAll	AMQ1	AMQ2	AMQ3	AMQ4
AMQ1	0.959				
	0.000				
AMQ2	0.993	0.929			
	0.000	0.000			
AMQ3	0.915	0.769	0.937		
	0.000	0.000	0.000		
AMQ4	0.848	0.703	0.856	0.929	
	0.000	0.000	0.000	0.000	
Alien.	-0.667	-0.667	-0.658	-0.569	-0.584
	0.000	0.000	0.000	0.000	0.000

We also calculate the correlations between metrics solely for the Nonmetric MDS runs. These correlations are given in Table 5. We find that the metrics are still significantly correlated, but this relationship is weaker (p=0.025) than for the other techniques.

Table 5. Correlations: Nonmetric MDS

```
        AMA11    AMQ1    AMQ2    AMQ3    AMQ4
AMQ1    0.995
        0.000
AMQ2    0.997   0.984
        0.000   0.000
AMQ3    0.964   0.934   0.974
        0.000   0.000   0.000
AMQ4    0.924   0.902   0.925   0.935
        0.000   0.000   0.000   0.000
Alien. -0.289  -0.276  -0.299  -0.301  -0.279
        0.025   0.033   0.020   0.019   0.031
```

Overall we have found that there is negative correlation between the neighborhood based metric ψ and the distance based alienation coefficient K. Generally, when comparing different techniques the higher the value of ψ for a technique, then the lower the value of K. This relationship holds for PCA, Isomap, and Metric MDS. The relationship does not hold for Nonmetric MDS; this technique produces a very high value of the agreement metric, but a poor value of alienation. This result shows that good values of ψ do not necessarily guarantee a mapping which preserves distances and is easy to interpret. However using a combination of ψ and K we can ensure solutions that preserve both neighborhood orderings and distances.

Given the success of Metric MDS in producing visually appealing solutions with good values of ψ and K for the CRM data tested, we decided to explore this technique further. As previously discussed the solution space produced by the STRESS function is non-convex, so any solution produced by the gradient based optimization procedure is not guaranteed to lead to a global optimum. The termination criterion set for the algorithm is STRESS<0.00001, so it is not possible to choose from random solutions based upon the value of STRESS.

For one of our test problems, we ran the Metric MDS algorithm 100 times. Table 6 summarizes how well the run with the CMDS starting point compares to the random solutions. This run is not guaranteed to be optimal, but for all metrics gives a solution within the top decile of the random solutions.

Table 6. Ranking of solution from CMDS start when combined with 100 random starts

Evaluation Techniques	Dimensions		
	1	2	3
Metric All	1	3	1
Metric Q1	1	3	9
Metric Q2	1	3	3
Metric Q3	1	4	7
Metric Q4	1	3	9
Alienation	1	2	1

Figures 6 and 7 show the normal curves for ψ and K respectively, separate curves are given for each dimension. We can see that the values of ψ have a much larger standard deviation than the values of K, giving a larger spread of values. A combination of the normalized values from ψ metrics and K can be used to select a solution based upon our desired solution characteristics.

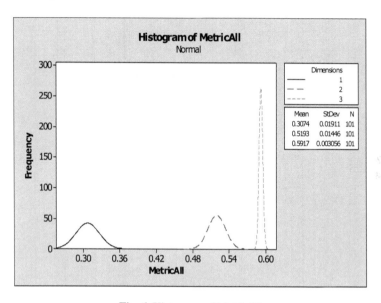

Fig. 6. Histogram of MetricAll

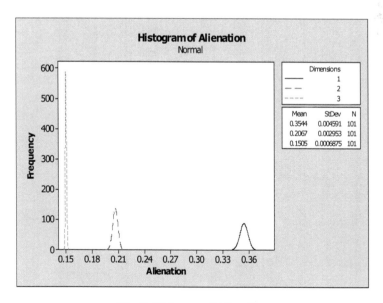

Fig. 7. Histogram of Alienation

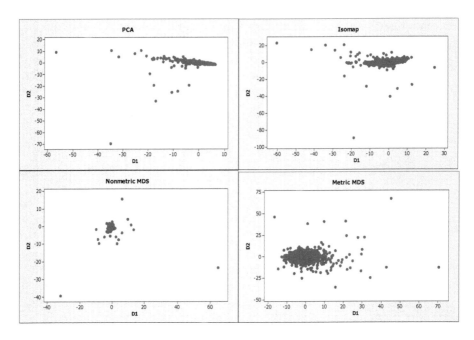

Fig. 8. 2D Solutions: Top Row (PCA,Isomap), Second Row (Nonmetric MDS, Metric MDS)

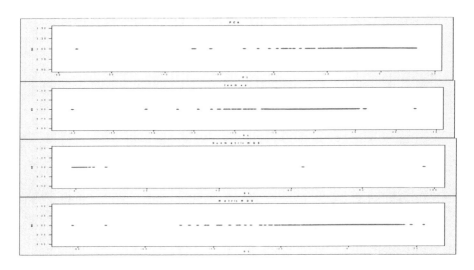

Fig. 9. 1D Solutions: From Top to Bottom: PCA→Isomap→Nonmetric MDS→Metric MDS

6 Conclusions and Future Work

We have introduced a metric based upon neighborhood agreement rate (ψ), and have shown how ψ can be used to help evaluate the performance of dimensionality reduction techniques. From our experimentation we propose three possible uses for the metric; comparing dimensionality reduction techniques, tuning parameters, and for techniques with local optima, selecting solutions.

We showed how ψ can be used to evaluate the reconstruction of local neighborhoods in lower dimensional solutions. We gave a typical marketing application, taking a set of customer data, and mapping consumers in 1, 2, and 3 dimensions. For our application, we found that performance for ψ correlates strongly with performance for the alienation coefficient K for metric measurement level based dimensionality reduction techniques, but that Nonmetric MDS performs strongly on the ψ measures and poorly on K. Thus ψ is a useful metric in evaluating lower dimensional solutions but must be used with care, particularly with techniques designed to measure rank order. To get an overall pattern of performance, ψ metrics may be combined with K. Future work on the metric could include developing a stronger theoretical basis for the statistic, possibly developing chi-squared tests based similar to those for the RAND index.

The calculation of ψ for large data sets could prove problematic. The algorithm developed to calculate ψ is of order $O(n^3)$. It may be worthwhile taking many samples of a small number of points, and then using a technique such as the bootstrap in order to calculate confidence intervals for the statistic.

In testing different dimensionality reduction methods on the customer data, Nonmetric MDS gave the best results for ψ, but produced solutions with points clumped together, giving poor visualization. Metric MDS gave strong all round results for both ψ and K. Fitting of Metric MDS has been somewhat ignored in the literature; this may be because most MDS literature is in the realm of psychometrics, and most psychological data are regarded as ordinal scale data, for which only the rank orders are meaningful, thus leading to a concentration on Nonmetric MDS techniques. Future Metric MDS research could involve improving the efficiency of the optimization algorithm, and fitting different regression functions (possibly using local kernel techniques or splines) to the data.

Isomap's main advantage is the globally optimal solution and the quick running time. Isomap produces solutions with a significant improvement in the values of ψ and K from PCA. Isomap would prove very useful for larger scale data sets where the optimization of the Stress function would be computationally expensive. A major advantage of Isomap for customer information data sets is that the information will change rapidly, necessitating rapid updates of the lower dimensional solutions, and thus an efficient updating technique. Once the Isomap neighborhood parameter is tuned for a specific type of data then solutions can be rapidly created. [24] gives an algorithm for the rapid updating of Isomap solutions, given changes in the higher dimensional data.

For all the methods tested, in order to visualize really large problems, memory storage for a distance matrix with n(n-1)/2 entries becomes a problem. Both Metric and Nonmetric MDS have local optimum problems. We showed how ψ and K can be combined and used to select solutions based on certain criteria, e.g. preservation of local neighborhoods, or reconstruction of distances. Given that different solutions may be invariant under Procrustes rotation, calculation of ψ may provide a quick way of finding similar solutions.

References

1. Akkucuk, U.: Nonlinear mapping: Approaches based on optimizing an index of continuity and applying classical metric MDS on revised distances. PhD dissertation: Rutgers University (2004)
2. Akkucuk, U., Carroll, J.D.: PARAMAP vs. ISOMAP: A Comparison of Two Nonlinear Mapping Algorithms Forthcoming, Journal of Classification (2007)
3. Akkucuk, U., Carroll, J.D.: Parametric Mapping (PARAMAP): An Approach to Nonlinear Mapping, 2006. In: Proceedings of the American Statistical Association, Section on Statistical Computing [CD-ROM], Alexandria, VA: American Statistical Association, pp. 1980–1986 (2006)
4. Andrews, R.L., Manrai, A.K.: MDS Maps for Product Attributes and Market Response: An Application to Scanner Panel Data. Marketing Science 18(4), 584–604 (1999)
5. Bijmolt, T.H.A., Wedel, M.: A comparison of Multidimensional Scaling Methods for Perceptual Mapping. Journal of Marketing Research. 36(2), 277–285 (1999)
6. Borg, I., Leutner, D.: Measuring the Similarity Between MDS Configurations. Multivariate Behavioral Research 20, 325–334 (1985)
7. Buja, A., Swayne, D.F.: Visualization Methodology for Multidimensional Scaling. Journal of Classification 19, 7–43 (2004)
8. Carroll, J.D., Arabie, P.: Multidimensional scaling. In: Birnbaum, M.H. (ed.) Handbook of Perception and Cognition. Measurement, Judgment and Decision Making, vol. 3, pp. 179–250. Academic Press, San Diego, CA (1998)
9. Carroll, J.D., Green, P.E., Schaffer, C.M.: Interpoint Distance Comparisons in Correspondence Analysis. Journal of Marketing Research 23(3), 271–280 (1986)
10. Carroll, J.D., Green, P.E.: Psychometric Methods in Marketing Research: Part II. Multidimensional Scaling, Journal of Marketing Research 34(2), 193–204 (1997)
11. Chen, L., Buja, A.: Local Multidimensional Scaling for Nonlinear Dimension Reduction, Graph Layout, and Proximity Analysis, Working Paper, University of Pennsylvania (2006)
12. Dempster, A.P, Laird, N.M, Rubin, D.B: Maximum likelihood from incomplete data via the EM algorithm. Journal of the Royal Statistical Society B, 39, 1–38 (1977)
13. DeSarbo, W.S., Kim, J., Choi, S.C., Spaulding, M.: A Gravity-Based Multidimensional Scaling Model for Deriving Spatial Structures Underlying Consumer Preference/Choice Judgements. Journal of Consumer Reseach 29(1), 91–100 (2002)
14. DeSarbo, W.S., Hoffman, D.L.: Constructing MDS Joint Spaces from Binary Choice Data: A Multidimensional Unfolding Threshold Model for Marketing Research. Journal of Marketing Research 26(1), 40–54 (1987)
15. DeSarbo, W.S., Manrai, A.K.: A new Multidimensional Scaling Methodology for the Analysis of Asymmetric Proximity Data in Marketing Research 11(1), 1–20 (1992)
16. Fodor, I.K.: A Survey of Dimension Reduction Techniques. LLNL technical report (2002)
17. Green, P.E.: Marketing Applications of MDS: Assessment and Outlook. Journal of Marketing 39, 24–31 (1975)
18. Ham, J., Lee, D.D., Mika, S., Schölkopf, B.: A kernel view of the dimensionality reduction of manifolds. In: Greiner, R., Schuurmans, D. (eds.) Proceedings of the Twenty-First International Conference on Machine Learning, pp. 369–376 (2006)
19. Hubert, L., Arabie, P.: Comparing Partitions. Journal of Classification 2, 193–218 (1985)
20. Kohonen, T.: Self-Organizing Map. Springer, New York (2001)
21. Kruskal, J.B.: Multidimensional scaling for optimizing a goodness of fit metric to a nonmetric hypothesis. Psychometrika 29, 1–27 (1964a)

22. Kruskal, J.B.: Nonmetric Multidimensional scaling: A numerical method. Psychometrika 29, 115–129 (1964b)
23. Lafon, S., Lee, A.B.: Diffusion Maps and Coarse-Graining: A Unified Framework for Dimensionality Reduction. Graph Partitioning, and Data Set Paramaterization 28(9), 1393–1403 (2006)
24. Law, M.H.C., Jain, A.K.: Incremental Nonlinear Dimensionality Reduction by Manifold Learning. IEEE Transactions on Pattern Analysis and Machine Intelligence 28(3), 377–391 (2006)
25. Levina, E.M, Bickel, P.J.: Maximum likelihood estimation of intrinsic dimension. In: Advances in Neural Information Processing Systems 17, MIT Press, Boston (2005)
26. Moore, W.L., Winer, R.S.: A Panel-Data Based Method for Merging Joint Space and Market Response Function Estimation. Marketing Science. 6(1), 25–42 (1987)
27. Murtagh, F,: Correspondence Analysis and Data Coding with R and Java, Chapman and Hall/CRC Press, London (2005)
28. Neslin, S.A., Gupta, S., Kamakura, W., Lu, J., Mason, C.H.: Defection Detection: Measuring and Understanding the Predictive Accuracy of Customer Churn Models. Journal of Marketing Research 43(2), 204–211 (2006)
29. Pettis, K., Bailey, T., Jain, A.K., Dubes, R.: An intrinsic dimensionality estimator from near-neighbor information. IEEE Transactions on Pattern Analysis and Machine Intelligence 1(1), 25–36 (1979)
30. Rand, W.M.: Objective Criteria for the Evaluation of Clustering methods. Journal of the American Statistical Association 66, 846–850 (1971)
31. Roweis, S.T., Saul, L.K.: Nonlinear Dimensionality Reduction by Locally Linear Embedding. Science 290(5500), 2323–2326 (2000)
32. Sha, F., Saul, L.K.: Analysis and Extension of Spectral Methods for Nonlinear Dimensionality Reduction. In: Proceedings of the 22nd International Conference on Machine Learning, Bonn, Germany (2005)
33. Shepard, R.N.: The analysis of proximities: Multidimensional scaling with an unknown distance function, Parts I and II. Psychometrika. 27, pp. 125–140, pp. 219–246 (1962)
34. Shepard, R.N., Carroll, J.D.: Parametric representation of nonlinear data structures. In: Krishnaiah, P.R. (ed.) Multivariate Analysis, pp. 561–592. Academic Press, New York (1966)
35. Steen, J.E.M., Trijup, H.C.M.V, Ten Berge, J.M.F.: Perceptual Mapping Based on Idiosyncratic Sets of Attributes. Journal of Marketing Research 31(1), 15–27 (1994)
36. Tenenbaum, J.B., de Silva, V., Langford, J.C.: A Global Geometric Framework for Nonlinear Dimensionality Reduction. Science 290(5500), 2319–2323 (2000)
37. Torgerson, W.S.: Multidimensional Scaling, I: theory and method. Psychometrika 17, 401–419 (1952)
38. Torgerson, W.S.: Theory and Methods of Scaling, vol. 32. Wiley, New York (1958)
39. Weinberger, K.Q., Saul, L.K.: Unsupervised Learning of Image Manifolds by Semidefinite Programming. International Journal of Computer Vision 70(1), 77–90 (2006)
40. Young, G., Householder, A.A.: Discussion of a Set of Points in Terms of their Mutual Distances. Psychometrika 3, 19–22 (1938)

A Sequential Hybrid Forecasting System for Demand Prediction

Luis Aburto[1] and Richard Weber[2]

[1] Penta Analytics, Santiago, Chile
[2] Department of Industrial Engineering, University of Chile
luaburto@analytics.cl, rweber@dii.uchile.cl

Abstract. Demand prediction plays a crucial role in advanced systems for supply chain management. Having a reliable estimation for a product's future demand is the basis for the respective systems. Various forecasting techniques have been developed, each one with its particular advantages and disadvantages compared to other approaches. This motivated the development of hybrid systems combining different techniques and their respective advantages. Based on a comparison of ARIMA models and neural networks we propose to combine these approaches to a sequential hybrid forecasting system. In our system the output from an ARIMA-type model is used as input for a neural network which tries to reproduce the original time series. The applications on time series representing daily product sales in a supermarket underline the excellent performance of the proposed system.

Keywords: Neural Networks, ARIMA, Demand Forecasting, Hybrid Forecasts.

1 Introduction

In time series forecasting we have seen the development of many different techniques. Especially those belonging to the ARIMA family have been applied successfully in various applications. The recent development of data mining has led to methods that differ conceptually from the ARIMA-type, such as e.g. regression trees, support vector regression, and neural networks [Han, Kamber 2001]. Each one of these techniques has its advantages and limitations compared to the others. Therefore, the development of hybrid forecasting systems combining two different methods was a natural consequence we could witness during recent years; see [Abraham et al. 2004].

In this paper we propose a sequential hybrid forecasting system (SHFS) and show its application to predict daily sales data from a Chilean supermarket. The message from our work is twofold: First, the proposed SHFS gives excellent results among all considered approaches for the application we worked on. Second – and maybe even more important - the generic idea of combining an ARIMA model sequentially with a neural network offers huge potential in other areas as well and should be considered in future forecasting applications.

Section 2 provides an overview on the state-of-the-art in hybrid forecasting models. Section 3 compares two techniques for time series prediction (ARIMA and neural

P. Perner (Ed.): MLDM 2007, LNAI 4571, pp. 518–532, 2007.

networks) and analyzes the respective strengths and weaknesses. Based on this analysis we develop a sequential hybrid forecasting system in section 4. Its advantages compared to traditional forecasting approaches as well as to already established hybrid models are presented in section 5 by way of an application for demand forecast. Section 6 concludes this work and points at future developments.

2   Overview on Hybrid Forecasting Systems

Hybrid forecasting systems combine different methods in order to improve the forecasting quality compared to the respective single techniques. Assigning weights to each model's forecast is one way to consolidate competing models into a single forecast. Several surveys on literature related to forecast-combination show approaches to solve the practical problem of determining such weights; see e.g. [Clemen 1989].

The forecasting literature contains also hybrid intelligent systems (HIS) where at least one so-called intelligent technique (e.g. neural network, fuzzy logic, genetic algorithm) is combined with another method. Such a HIS combines a neural network with fuzzy logic in order to model electricity demand [Abraham, Nath 2001]. This system outperformed pure neural networks as well as ARIMA models. Other forecasting applications of neural networks have been proposed for inventory control resulting in huge cost savings [Bansal et al. 1998] [Reyes-Aldasoro et al. 1999].

Additive hybrid forecasting systems (AHFS) combining ARIMA and neural networks have been used e.g. in [Zhang 2003] and [Aburto, Weber 2007] considering the original time series X(t) as a composition of two elements: the result from a linear model Y(t) and the non-linear part e(t) that cannot be modeled linearly as shown in the following equation: X(t) = Y(t) + e(t); see also [Donaldson, Kamstra 1996].

This AHFS has been applied to the following three well-known data sets: the Wolf's sunspot data, the Canadian lynx data, and the British pound/US dollar exchange rate data; see [Zhang 2003]. It has been shown that combining dissimilar models gives better results since error variance is reduced. Fitting the ARIMA model first to the data decreases also the overfitting effect neural networks can produce. On the other hand, the capabilities of neural networks are not exploited in the best possible way since only the error is presented to the network. This critique led to the sequential hybrid forecasting system proposed in this paper (see section 4) where both models, ARIMA as well as neural networks "see" the entire original time series.

3   Analysis and Comparison of ARIMA Models and Neural Networks

We describe briefly ARIMA models and neural networks for time series forecasting in order to have a basis for the proposed sequential hybrid model. We also provide a comparative analysis of these two techniques which motivated their combination.

3.1 ARIMA Models

Problems to predict time series have been solved mainly applying ARIMA models (Autoregressive Integrated Moving Average) proposed by [Box, Jenkins 1976]. Let X_t be the observation of a time series at time t with a probability distribution $f(X_t)$.

A is a time series of n white noise observations with average zero and variance σ^2_A
B is the delay operator, i.e. $BX_t = X_{t-1}$ and $BA_t = A_{t-1}$
$\nabla = 1-B$ is the differentiating operator, i.e. $\nabla X_t = (1-B)X_t = X_t - X_{t-1}$

An ARIMA process (p,d,q) is based on a series that has been differentiated d times, with p autoregressive terms and q mobile average terms. The respective equation is:

$$\phi_p(B)(\nabla^d X_t - \mu) = \theta_q(B)A_t \tag{1}$$

The result of these models is the real value μ and the parameter vectors θ_q (moving average) and ϕ_p (autoregressive) that best fit the data [Box, Jenkins 1976].

The process can be generalized even more when incorporating seasonal elements. The seasonal differentiating operator is defined as: $\nabla_s = 1 - B^s$, where s is the seasonal factor. Besides, the time series X_t can be explained by external variables or predictors (also called regressors). Based on these definitions, the most general model is expressed as SARIMAX (p,d,q) (sp,sd,sq) Y, where Y are the regressors. Finally, the general equation of the model is:

$$\phi_p(B)\Phi_{sp}(B)\left[\nabla^d \nabla_s^{sd}(X_t - \sum_{i=1}^{r} c_i Y_i) - \mu\right] = \theta_q(B)\Theta_{sq}(B)A_t \tag{2}$$

where $\Phi_{sp}(B)$ is the sp seasonal autoregressive polynomial, $\Theta_{sq}(B)$ is the sq seasonal mobile average polynomial and c_i are the regressors' coefficients.

3.2 Neural Networks

Neural networks are mathematical models that "learn" pattern from data and have proved to be very effective in order to solve classification and regression problems by handling non-linearity between input and output variables, being able to approximate any function under certain conditions [Hornik 1991]. Neural networks have performed very well e.g. in forecasting stock exchange indexes [Kuo, Reitsch 1995] and corporative bonds [Moody 1994]. One of the most popular models among neural networks is the Multi Layer Perceptron (MLP), which is trained usually with the back-propagation learning rule [Han, Kamber 2001].

A forecasting system using an MLP requires two parameters if no external variables are used. The first one is k, which indicates the size of the time window to be used as input. The second parameter is s, which is the number of intervals for which the time series is supposed to forecast in the future. The following figure presents a typical MLP-architecture used for time series forecasting with these parameters.

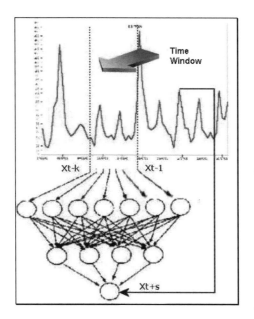

Fig. 1. MLP network for time series forecasting using parameters k and s

3.3 Comparing ARIMA and Neural Network Models

Many publications compare ARIMA and neural network models, both theoretically [Dorffner 1996][Wan 1993] and empirically [Faraway, Chatfield 1998][Kuo, Reitsch 1995]. According to [Dorffner 1996], the main limitation of ARIMA models is the linear relationship they assume between the independent and dependent variables. As [Wan 1993] states, MLP networks allow to model NARX processes, i.e. they are able to model autoregressive non-linear processes with exogenous variables.

On the other hand, the ARIMA family of models has also many advantages over neural networks, such as the information provided by the model. Analyzing the respective regressors' coefficients reveals the effect each independent variable has in relation to the dependent variable. Another disadvantage of neural networks is the high degree of freedom in their architecture. This implies several problems, such as e.g.:

- In order to obtain reliable results, a large number of training examples is needed.
- Having many weights can easily lead to overfitting ending up in local minima.

The following table summarizes the most important aspects of this comparison.

Table 1. Comparison between ARIMA and MLP models

ARIMA	Neural Networks (MLP)
Linear model: assumes certain behavior of the time series.	Nonlinear model: more degrees of freedom for the model.
The series has to be stationary.	Any time series can be analyzed.
Requires a lot of interaction with the user.	Requires less interaction with the user.
The model provides insight and information through its parameters.	Model difficult to interpret (black box).
No overfitting.	Overfitting is possible.

4   A Sequential Hybrid Forecasting System (SHFS)

Motivated by the above comparison we developed a sequential hybrid forecasting system (SHFS). The following observations led our development.

- SARIMAX processes are powerful tools to detect linear structures in time series.
- Neural networks, in particular a multilayer perceptron, have the capabilities of modeling non-linear relations between input and output vectors.
- If, however, we restrict a neural network only to the error from a SARIMAX process (as in additive hybrid forecasting systems), we loose a lot of their predictive power, because they do not "see" the entire time series, they "see" just the error.
- Neural networks do not perform well, if they have to work on raw time series. Instead they need "assistance" in modeling the time series properly, see e.g. [Hill et al. 1996]. Here is where a SARIMAX process can help.

As a conclusion of the above observations we developed the following system.

1. First, the original time series X(t) is modeled by a SARIMAX process, which provides as forecast the time series Y(t).
2. Then we train a neural network with the output from the SARIMAX process (Y(t)) as input and the original time series X(t) as desired output, i.e. we try to reproduce the original time series with the result from the SARIMAX process. With other words: We use the SARIMAX process as "filter" for the neural network which now receives preprocessed data.
3. In the final forecasting step we apply first SARIMAX as developed in Step 1 to the original time series and then the neural network as developed in Step 2 to the output from the SARIMAX process. The output from the neural network is the sequential hybrid forecast for the original time series.

The following figure illustrates the proposed system.

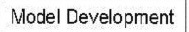

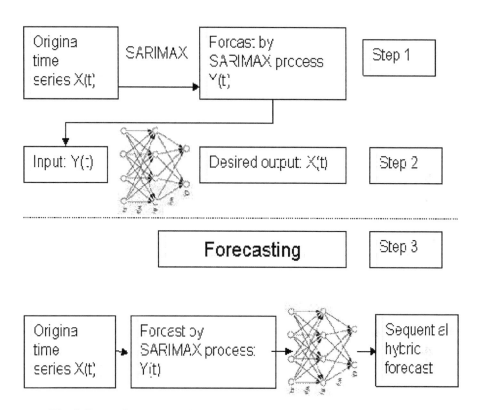

Fig. 2. Forecasting scheme using the sequential hybrid forecasting system (SHFS)

The proposed SHFS has several advantages since it combines "the best of both worlds". Applying a SARIMAX process to the original time series we have the advantages such a model offers, e.g. interpretability of the model parameters. The neural network, on the other hand, does not only forecast an error function (as e.g. in the additive hybrid system) but works on the entire time series. Instead of taking the raw time series as input, the network makes use of the already preprocessed data.

5 Application of the Sequential Hybrid Forecasting System and Comparison with Traditional Approaches

We show the advantages of our sequential hybrid forecasting system (SHFS) by comparing it to alternative approaches using data from the Chilean supermarket *Economax*.

Economax, as well as any other retail company, provides a broad range of products (about 5,000 different SKUs: stock keeping units) purchased from a large number of manufacturers and distributors. In order to offer such a variety of products to its customers at competitive prices, the supermarket and its providers have to manage efficiently the respective supply chain. Based on the data flow generated by the consumers the supermarket has to decide what, how much, and how often to buy.

In order to solve this problem satisfactorily, a reliable forecast of future demand is necessary. Determining future demand, however, is difficult since it depends on many factors, such as: past sales, prices, advertising campaigns, seasonality, holidays, weather, sales of similar products, competitors' promotions, among others. We applied the proposed hybrid sequential forecasting system to data provided by *Economax* and compared its results with those from the following techniques:

- Naïve forecasting
- Seasonal naïve forecasting
- Unconditional average
- Pure SARIMAX process
- Pure Multilayer Perceptron
- Additive Hybrid Forecasting Model (AHFS; see [Aburto, Weber 2007])

Section 5.1 presents the data set used; the respective preprocessing is described in 5.2. Section 5.3 exhibits the SARIMAX process we developed, while section 5.4 describes the pure neural network for demand forecasting. Section 5.5 presents the additive hybrid system and the sequential hybrid system. Section 5.6 provides the results obtained applying the mentioned techniques to various products' sales data.

5.1 Description of Analyzed Data Sets

We analyzed sales data for the 4 best-selling stock keeping units (SKU) in *Economax*. Below, we present demand forecasts for SKU 100595 (vegetal oil, 1 liter) exemplarily in more detail. The same procedure has been applied to the other 3 products. We had the following data for the 13 months between July 1, 2000 and July 31, 2001:

- Daily sales data for each SKU.
- Prices of the considered SKUs from *Economax* and of the respective products from the competitors in the micro market.

Additionally, we generated the following external binary variables:

- Payment: characterizing days at the end of each month when people receive monthly payment.
- Intermediate payment: characterizing days at the mid of each month when people receive 2-weekly payment.
- Before holiday: characterizing days before a holiday.
- Holiday: characterizing holidays.
- Independence: characterizing Chilean independence days (September 18).
- Santa: characterizing days of the week before eastern.

- Vacation: characterizing days that belong to the period of summer vacation.
- Summer: characterizing summer days (October 1 to March 31).
- New Year: characterizing the only day when supermarkets are closed (January 1).

The following figure shows daily sales data for SKU 100595.

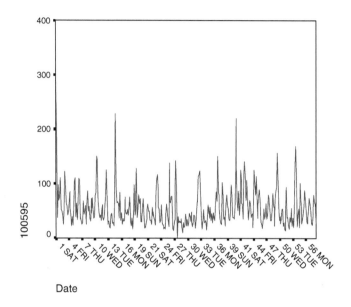

Fig. 3. Daily sales data for SKU 100595

5.2 Preprocessing of Available Data

Since we are interested in demand forecast we first had to correct the respective time series where "cero sales" appeared. This occurs when the respective product is out-of-stock and does not mean that there was no demand. Strictly speaking we do not want to forecast **sales**; we want to know future **demand** in order to manage the respective supply chain. These two variables take the same value if the respective product is available. In this case we assume the customers to find the product.

If sales for a certain day had value 0 we replaced it by the sales value of the same product for the same weekday one week before. The resulting time series have been normalized, i.e. sales data for each SKU has been transformed to the interval [0,1]. It has been shown that this kind of preprocessing improves significantly the forecasting results obtained with neural networks [Crone et al. 2006].

Using each SKU's prices from *Economax* as well as from its competitors in the respective micro market we calculated the following "derived price variables":

- PriceA = Original price of the SKU in *Economax*
- PriceB = Price A / Max(Price in micro market)
- PriceC = Price A / Min(Price in micro market)

This way the models receive implicitly the information if a certain product is the most expensive or cheapest one of all competing products. Additionally, the effect of price changes (e.g. promotions) is represented implicitly.

5.3 Application of SARIMAX Process

Since time series in the present application show a strong seasonal effect and external variables are important for forecasting we developed a SARIMAX model for SKU 100595 using the following 4 steps.

Step 1: Preliminary analysis
We determined the time series' autocorrelation as shown in the following figure.

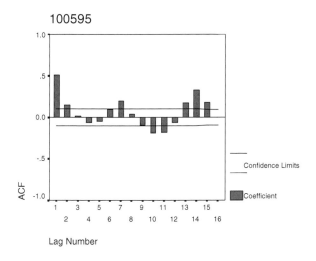

Fig. 4. Autocorrelation analysis for SKU 100595

Analyzing the time series with the preprocessed sales data and the autocorrelation function as shown above we can conclude:

- The series is stationary; i.e. we do not have to apply the differentiation operator.
- The time series shows a strong seasonality with time lags of 1, 7, and 14 days.

Step 2: Determine the order of the SARIMAX model
Given the above described autocorrelation analyses we developed a SARIMAX (1,0,0)(2,0,0) model with the external variables described in 5.1.

Step 3: Parameter estimation
We determined the parameters of our SARIMAX model using the software tool SPSS. The following table shows the final result of this parameter estimation.

Table 2. Variables of SARIMAX model

	B	**SEB**	**T-Ratio**	**Appr. Prob.**
AR1	.417463	.051085	8.1719215	.00000000
SAR1	.130165	.055043	2.3647876	.01866427
SAR2	.295255	.055631	5.3073729	.00000021
Payment	51.980466	5.876780	8.8450596	.00000000
Before Holiday	27.948165	6.911996	4.0434291	.00006672
Holiday	25.081074	6.947247	3.6102176	.00035735
SANTA	34.127746	17.129981	1.9922816	.04722842
Price B	14.634096	8.306789	1.7617031	.07911801
Price C	16.487375	8.888587	1.8548928	.06457326
Constant	35.859236	5.626348	6.3734482	.00000000

The second column contains the coefficient B for each selected variable. SEB is the error in estimating this coefficient, T-ratio the value of the t-test for B and SEB, and "Appr. Prob." is the probability that this coefficient has value 0.

The variables shown in table 2 have been selected applying the following steps. We started using all variables in the SARIMAX (1,0,0)(2,0,0) model and performed for each variable a t-test determining the probability that the respective coefficient of this variable has value 0 (last column of Table 2). Then we eliminated the variable with highest probability. We repeated these steps revising simultaneously the Akaike Information Criterion [Akaike, Kitagawa 1999; Akaike 1974], which helps to determine models with a good tradeoff between model fit and model complexity (number of parameters). Finally we obtained the variables shown in Table 2.

Step 4: Model validation

Given the residuals and autocorrelation shown in the following two figures we accept the SARIMAX (1,0,0)(2,0,0) model.

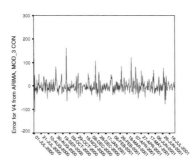

Fig. 5. Residuals for SARIMAX (1,0,0)(2,0,0) model

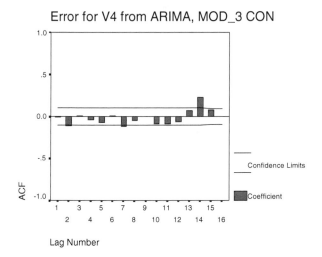

Fig. 6. Autocorrelation of residuals of SARIMAX(1,0,0)(2,0,0) model

5.4 Application of Neural Networks

In this section we describe the "pure" Multilayer Perceptron we applied for forecasting. Its architecture has also been used for the hybrid approaches presented below.

We used the following 32 input variables:

- Sales data from the past k=14 days, i.e. from a time window with size k=14.
- Nine binary variables characterizing special days as described in 5.1.
- Six binary variables characterizing the day of the week.
- Three price variables (Price A, Price B, Price C) as described in 5.2

In the hidden layer we had 16 neurons since (number of input variables+1)/2 hidden neurons are sufficient in order to approximate any non-linear function with a MLP [Hornik 1991]. We used one neuron in the output layer, i.e. we made a one-day forecast based on the information available the day before.

Regarding the neural network learning parameters we tested several settings. Particularly we analyzed learning rate, momentum, and pruning strategy. The latter is described by its relevance threshold (rt) and time constant (tc), which have the following interpretation: "If during tc epochs the absolute value of a weight is below rt the associated connection will be pruned."

The following parameter setting gave best results for the neural network models:

- Learning rate = 0.3
- Momentum = 0.8
- Using pruning with relevance threshold 0.001 and time constant 10.

In order to avoid overfitting of the network we tested its performance on a separate test set during training minimizing root-mean squared error (RMS) between calculated and desired output. We implemented all neural network models using the software tool DataEngine 4.0.

5.5 Application of Hybrid Forecasting Systems

We applied the two hybrid systems AHFS and SHFS as described next.

Additive Hybrid Forecasting System (AHFS: "Sarimax + MLP")

We first applied the SARIMAX model as presented in section 5.3 to forecast demand. The associated error has been considered as a time series by its own, which we forecast with an MLP. This neural network has the following characteristics:

- Input neurons:
  - error values in periods t-k (k=1, …, 14)
  - 15 binary variables characterizing the respective day t (as in 5.4)
  - 3 product prices (as in 5.4)
- Output neuron: error value in period t: e(t).

The network's architecture and parameter setting are as described in 5.4

Sequential Hybrid Forecasting System (SHFS: "First Sarimax then MLP")

We first applied the SARIMAX model as presented in section 5.3 to forecast demand. Then we took the respective forecast as new time series providing input values for the neural network. The network's desired output values, however, are the original sales values. This neural network has the following characteristics:

- Input variables:
  - demand forecast provided by SARIMAX for periods t-k (k=1, …, 14)
  - 15 binary variables characterizing the respective day t (as in 5.4)
  - 3 product prices (as in 5.4)
- Output neuron: original sales value in period t.

The network's architecture and parameter setting are as described in 5.4

5.6 Results

We applied the techniques the supermarket currently uses (naïve forecast, seasonal naïve, and unconditional average). Naïve forecast assumes that sales tomorrow will be the same as today whereas seasonal naïve presumes that sales tomorrow will be the same as the same weekday one week before. Unconditional average estimates tomorrow's sales as the average of previous sales data.

The performance of each system has been evaluated using as error function Mean Absolute Percentage Error (MAPE) and Normalized Mean Square Error (NMSE).

$$MAPE = \frac{1}{n}\sum_{k=1}^{n}\left|\frac{(X_k - \hat{X}_k)}{X_k}\right|$$

$$NMSE = \frac{\sum_k (X_k - \hat{X}_k)^2}{\sum_k (X_k - \overline{X})^2} = \frac{1}{\sigma^2 n}\sum_k (X_k - \hat{X}_k)^2$$

The following table shows these errors for the mentioned systems. The training set corresponds to the period July 1, 2000 to June 30, 2001; test set is the period between July 1, 2001 and July 31, 2001.

Table 3. Results from different forecasting systems for product 100595

		Training set		Test set	
	SKU 100595	MAPE (%)	NMSE	MAPE (%)	NMSE
M1	Naïve	44.28	0.6972	56.83	1.2481
M2	Seasonal Naïve	64.67	1.2212	45.75	1.9217
M3	Unconditional average	59.98	0.7759	48.54	0.9689
M4	SARIMAX (1,0,0)(2,0,0)	36.21	0.3301	40.49	0.6090
M5	MLP tw 14	31.15	0.3115	34.64	0.5703
M6	Additive hybrid (AHFS) ("M4 + M5")	30.13	0.2916	35.63	0.5177
M7	Sequential hybrid (SHFS) ("first M4 then M5")	23.97	0.1594	30.59	0.4262

As can be seen SARIMAX (model M4) and the "pure" neural network (model M5) outperform traditional techniques (models M1, M2, M3). The additive hybrid system (M6) gives better results than each of its components whereas the sequential hybrid forecasting system (M7) proposed in this paper gives best results among all considered approaches. We applied the systems M1, ..., M7 to three more products and obtained the results shown in the following table.

Table 4. MAPE test results (in %) for four SKUs

SKU	M4: SARIMAX	M5: MLP tw 14	M6: additive hybrid	M7: sequential hybrid
SKU1: 100595	40.49	34.64	35.63	<u>30.59</u>
SKU2: 108464	41.39	34.49	40.24	<u>33.60</u>
SKU3: 257842	41.34	39.81	<u>38.15</u>	38.29
SKU4: 262900	42.04	<u>34.25</u>	39.80	35.79
Average	41.32	35.80	38.46	<u>34.57</u>
Max	42.04	39.81	40.24	<u>38.29</u>
Min	40.49	34.25	35.63	<u>30.59</u>

The underlined values indicate the best result for the each row. From the previous table we can conclude:

- For 2 out of 4 cases SHFS (M7) gave best results.
- No other system gave better results for the remaining 2 SKUs.
- SHFS (M7) improved the SARIMAX process (M4) for all time series
- On average SHFS (M7) gave best results for the 4 SKUs.

6 Conclusions and Future Work

We proposed a sequential hybrid forecasting system (SHFS), which takes the output from a SARIMAX process as input for a neural network in order to reproduce the original time series. This system has been applied to forecast a supermarket's sales data. We compared the results with those from the two components of our hybrid system (i.e. SARIMAX and MLP), an additive hybrid system, and traditional techniques used by the supermarket so far. Neural networks outperformed the SARIMAX process and the proposed SHFS gave best results among all considered approaches.

The contribution of this paper is twofold. On one hand, the proposed SHFS provides very good results for the supermarket data we had. On the other hand - and maybe even more important - we should emphasize the generic idea of combining ARIMA-type approaches with neural networks in a sequential way. Whereas models from the first family have advantages in deriving interpretable linear models those from the latter family are stronger in modeling non-linear functions. One possible interpretation is that the result from a SARIMAX process provides richer input information for the neural network than original data itself. We think, this result is worth further research and application to other time series.

The proposed approach might show limitations in cases where the respective ARIMA-models do not provide reasonable results, i.e. the presented sequential hybrid forecasting system is strong in improving acceptable results using ARIMA-type forecasts which are the basis within our system. The issue of understanding better for which type of forecasting problems the presented system offers promising results should be studied further.

Acknowledgement. This project was supported by the Millennium Science Institute on Complex Engineering Systems (www.sistemasdeingenieria.cl), the Chilean Fondecyt project 1040926, and the Chilean Fondef project "Development of Management Tools for Improving Supply Chain Productivity in the Supermarket Industry"; project number D03I1057.

References

Abraham, A., Nath, B.: A neuro-fuzzy approach for modeling electricity demand in Victoria. Applied Soft Computing 1(2), 127–138 (2001)

Abraham, A., Philip, S., Mahanti, P.: Soft Computing Models for Weather Forecasting. International Journal of Applied Science and Computations 11(3), 106–117 (2004)

Aburto, L., Weber, R.: Improved Supply Chain Management based on Hybrid Demand Forecasts. Applied Soft Computing 7, 136–144 (2007)

Akaike, H., Kitagawa, G. (eds.): The Practice of Time Series Analysis. Springer, Berlin (1999)

Akaike, H.: A new look at the statistical model identification. IEEE Trans. Automatic Control AC-19, 716–723 (1974)

Bansal, K., Vadhavkar, S., Gupta, A.: Neural Networks Based Forecasting Techniques for Inventory Control Applications. Data Mining and Knowledge Discovery 2, 97–102 (1998)

Box, G.E.P., Jenkins, G.M.: Time Series Analysis: Forecasting and Control. Holden-Day, San Francisco (1976)

Chen, Y., Yang, B., Dong, J., Abraham, A.: Time-series forecasting using flexible neural tree model. Information Sciences 174(3-4), 219–235 (2005)

Clemen, R.T.: Combining forecasts: A review and annotated bibliography. International Journal of Forecasting 5, 559–581 (1989)

Crone, S., Guajardo, J., Weber, R.: A study on the ability of Support Vector Regression and Neural Networks to Forecast Basic Time Series Patterns. In: Bramer, M. (ed.) IFIP International Federation for Information Processing. Artificial Intelligence in Theory and Practice, vol. 217, pp. 149–158. Springer, Boston (2006) doi: http://dx.doi.org/10.1007/978-0-387-34747-9_16

Donaldson, R.G., Kamstra, M.: Forecast combining with neural networks. Journal of Forecasting 15, 49–61 (1996)

Dorffner, G.: Neural Networks for Time Series Processing. Neural Network World. 6(4), 447–468 (1996)

Faraway, J., Chatfield, C.: Time series forecasting with neural networks: a comparative study using the airline data. Applied Statistics 47(2), 231–250 (1998)

Han, J., Kamber, M.: Data Mining: Concepts and Techniques. Morgan Kaufmann Publishers, San Francisco (2001)

Hill, T., O'Connor, M., Remus, W.: Neural Networks for Time Series Forecasts. Management Science 42(7), 1082–1092 (1996)

Hornik, K.: Approximation capabilities of multilayer feedforward networks. Neural Networks 4, 251–257 (1991)

Kuo, C., Reitsch, A.: Neural networks vs. conventional methods of forecasting. The Journal of Business Forecasting Methods and Systems 14(4), 17–22 (1995)

Moody, J.: Prediction Risk and Architecture Selection for Neural Networks. In: Cherkassky, V., Friedman, J.H., Wechsler, H. (eds.) From Statistics to Neural Networks: Theory and Pattern Recognition Applications. NATO ASI Series F, Springer, Berlin (1994)

Reyes-Aldasoro, C.C., Ganguly, A., Lemus, G., Gupta, A.: A Hybrid Model Based on Dynamic Programming, Neural Networks and Surrogate Value for Inventory Optimisation Applications. Journal of Operational Research Society 50(1), 85–94 (1999)

Wan, E.: Finite impulse response neural networks for autoregressive time series prediction. In: Weigend, A., Gershenfeld, N. (eds.) Predicting the Future and Understanding the Past, Addison-Wesley, MA (1993)

Zhang, G.P.: Time series forecasting using a hybrid ARIMA and neural network model. Neurocomputing 50, 159–175 (2003)

A Unified View of Objective Interestingness Measures

Céline Hébert and Bruno Crémilleux

GREYC, CNRS - UMR 6072, Université de Caen
Campus Côte de Nacre
F-14032 Caen Cédex France
`Forename.Surname@info.unicaen.fr`

Abstract. Association rule mining often results in an overwhelming number of rules. In practice, it is difficult for the final user to select the most relevant rules. In order to tackle this problem, various interestingness measures were proposed. Nevertheless, the choice of an appropriate measure remains a hard task and the use of several measures may lead to conflicting information. In this paper, we give a unified view of objective interestingness measures. We define a new framework embedding a large set of measures called SBMs and we prove that the SBMs have a similar behavior. Furthermore, we identify the whole collection of the rules simultaneously optimizing all the SBMs. We provide an algorithm to efficiently mine a reduced set of rules among the rules optimizing all the SBMs. Experiments on real datasets highlight the characteristics of such rules.

1 Introduction

Exploring and analyzing correlations between features is on the core of KDD processes. Agrawal et al. [1] define association rules as the implications $X \rightarrow Y$ where X and Y represent one or several conjunctions of features (or attributes). However, among the overwhelming number of rules resulting from practical applications, it is difficult to determine the most relevant rules [10]. An essential task is to assist the user in selecting interesting rules.

Measuring the interestingness of discovered rules is an active and important area of data mining research. Interestingness measures are numerous and they are usually divided into two groups: subjective and objective measures. Whereas subjective measures take into account both the data and the user's expectations, objective measures are only based on raw data. In this paper, we focus on objective measures. Support and Confidence are probably the most famous ones [2], but there are more specific measures (e.g., Lift [6], Sebag and Schoenauer [18]). In practice, choosing a suitable measure and determining an appropriate threshold for its use is a challenge for the end user. Combining results coming from several measures is even much more difficult. Thus an important issue is to compare existing interestingness measures in order to highlight their similarities and differences and better understand their behaviors [17, 3]. The lack of generic

P. Perner (Ed.): MLDM 2007, LNAI 4571, pp. 533–547, 2007.
© Springer-Verlag Berlin Heidelberg 2007

results about the characteristics captured by interestingness measures was the starting point of this work.

Contributions. This paper deals with the behavior of objective interestingness measures when applied to association rules. Our main objective is to make clear the choice of such a measure. For this purpose, we design an original framework which gives a unified view of a large set of measures, the Simultaneously lower Bounded Measures (SBM). We demonstrate that SBMs have similar behaviors so that choosing an appropriate measure among them becomes a secondary issue. This framework shows that three parameters (the minimal threshold for the antecedent frequency γ, the maximal consequent frequency η and the maximal number of exceptions δ) are on the core of many measures. This formalization provides lower bounds for the SBMs according to these parameters and thus guarantees a minimal quality for the rules. Moreover, we provide an efficient method to mine a reduced set of rules simultaneously optimizing all the SBMs, which ensures to produce the best rules according to these measures.

In a previous work [12], we addressed the specific case of the so-called classification rules (i.e., rules concluding on a class label). In this context, we showed that most of the usual interestingness measures only depend on the rule antecedent frequency and the rule number of exceptions and that they have a similar behavior. This paper is a generalization of [12] to any association rule. This generalization is not straightforward because one key point in [12] is the fact that the rule consequent is a class label and thus its frequency is known. This is obviously no longer true when considering any association rule and the major difficulty is the lack of information about the consequent frequency. We overcome it by bounding the consequent frequency. The fact that any attribute may appear in a rule consequent also requires to design a new algorithm to mine the rules simultaneously optimizing all the measures of the framework.

Organization. The rest of the paper is organized as follows. Section 2 discusses related work on rule selection and gives preliminary definitions. Section 3 introduces our framework and the SBMs. Section 4 shows how the SBMs can be simultaneously lower bounded and studies their behavior. Section 5 presents our algorithm to mine a reduced set among the most significant rules from a database. Section 6 gives experimental results about the quality of the discovered rules.

2   Preliminaries

2.1   Related Work

Lossless cover. It is well known that the whole set of association rules contains a lot of redundant rules [1]. So several approaches (see [13] for a survey) propose to restrict the mining to a rule cover [22] like the *informative rules* [15] or the *informative generic base* [9]. These rules have minimal antecedents and maximal consequents. They are lossless and informative since they enable to regenerate the whole set of valid association rules and their exact support and confidence values. Our work is linked to this approach because we define informative SBM rules that have minimal antecedents and that simultaneously optimize the SBMs (see Section 5).

Selecting the most interesting rules with objective measures. As already told, researchers have proposed a lot of interestingness measures for various kinds of patterns. There is no widespread agreement on a formal definition of interestingness and several works attempt to define properties characterizing "good" interestingness measures [10, 16]. Piatetsky-Shapiro [16] proposes a framework with three properties and we set our work with respect to it. Other works compare interestingness measures to determine their differences and similarities, either in an experimental manner [20] or in a theoretical one [19, 8]. In [4], a visualization method is proposed to help the user in the rule exploration. There are also attempts to combine several measures to benefit from their joint qualities [7]. However choosing and using a measure remains a hard task. Our approach differs from these works : we argue that choosing the appropriate measure is a secondary issue because they all behave the same. We aim at analyzing the behavior of existing measures and showing their common features. We exhibit the minimal properties that a measure must satisfy to get a unified view of a lot of objective interestingness measures, the SBMs. Second, by simultaneously optimizing all the SBMs, our work combines the information brought by these measures.

2.2 Definitions

Basic definitions. A database $\mathcal{D}$ is a relation $\mathcal{R}$ between a set $\mathcal{A}$ of *attributes* and a set $\mathcal{O}$ of *objects*: for $a \in \mathcal{A}, o \in \mathcal{O}$, $a \mathcal{R} o$ if and only if the object o contains the attribute a. A *pattern* is a subset of $\mathcal{A}$. The frequency of a pattern X is the number of objects in $\mathcal{D}$ containing X; it is denoted by $\mathcal{F}(X)$. Table 1 shows an example of a database containing 8 attributes and 9 objects.

Table 1. An example of a database $\mathcal{D}$

$\mathcal{D}$	Attributes							
Objects	A	B	C	D	E	F	G	H
o_1	1	0	1	0	1	0	0	1
o_2	0	1	1	0	1	0	1	1
o_3	1	0	1	0	1	0	0	1
o_4	1	0	1	0	1	0	0	1
o_5	0	1	1	0	1	1	0	0
o_6	1	0	0	1	0	1	0	1
o_7	0	1	1	0	1	1	0	1
o_8	1	0	1	0	0	1	0	1
o_9	0	1	0	1	0	1	1	0

Association rules. An *association rule* $r : X \to Y$ is an implication where X and Y are patterns of $\mathcal{D}$. X is the *antecedent* of r and Y its *consequent*. $\mathcal{F}(XY)$

is the rule frequency, $\mathcal{F}(X)$ the antecedent frequency and $\mathcal{F}(Y)$ the consequent frequency. In Table 1, $r_1 : CG \rightarrow BEH$ and $r_2 : BCF \rightarrow E$ are association rules. The frequency of r_1 (resp. r_2) is equal to 1 (resp. 2), the frequency of its antecedent is 1 (resp. 2) and the frequency of its consequent is 2 (resp. 6).

Evaluating objective measures. An interestingness measure is a function which assigns a numerical value to an association rule according to its quality. A lot of interestingness measures are based on the rule, the antecedent and the consequent frequencies. We recall here the well-known Piatetsky-Shapiro's properties [16] which aim at specifying what a "good" measure is. In the next section, we will use properties P2 and P3 to define the SBMs.

Definition 1 (Piatetsky-Shapiro's properties). *Let* $r : X \rightarrow Y$ *be an association rule and* M *an interestingness measure.*

- *P1:* $M(r) = 0$ *if* X *and* Y *are statistically independent i.e. if* $|\mathcal{D}| \times \mathcal{F}(XY) = \mathcal{F}(X) \times \mathcal{F}(Y)$;
- *P2: When* $\mathcal{F}(X)$ *and* $\mathcal{F}(Y)$ *remain unchanged,* $M(r)$ *monotonically increases with* $\mathcal{F}(XY)$;
- *P3: When* $\mathcal{F}(XY)$ *and* $\mathcal{F}(X)$ *(resp.* $\mathcal{F}(Y)$*) remain unchanged,* $M(r)$ *monotonically decreases with* $\mathcal{F}(Y)$ *(resp.* $\mathcal{F}(X)$*).*

P2 ensures the increase of M according to the rule frequency and P3 the decrease of M according to the antecedent and the consequent frequencies. Most of usual measures satisfy P2 (e.g., support, confidence, interest, conviction). However, there are a few exceptions (e.g., J-measure, Goodman-Kruskal, Gini index). In [16], Piatetsky-Shapiro defines a measure called the Rule-Interest which satisfies the three properties P1, P2 and P3.

3   A Formal Framework for Objective Measures: The Set of Simultaneously Bounded Measures

This section presents our framework which gives a unified view of a large set of measures, the SBMs. The key idea is to express a measure according to variables which depend on frequencies in order to capture their joint effect. We will see in Section 4 that this rewriting provides lower bounds for the SBMs and highlights their behavior.

3.1   Measures as Functions

We rewrite any interestingness measure as a function according to the frequencies of a rule.

Definition 2 (Associated function). *Let* M *be an interestingness measure and* $r : X \rightarrow Y$ *an association rule.* $\Psi_M(x, y, z)$ *is the continuous function associated to* M *where* $x = \mathcal{F}(X)$ *and* $y = \mathcal{F}(Y)$ *and* $z = \mathcal{F}(XY)$.

For instance, the function associated to the Lift measure is: $\Psi_{Lift}(x, y, z) = \frac{z \times |\mathcal{D}|}{x \times y}$.

Let δ be the maximal authorized number of exceptions for a rule. Variables x, y and z are frequencies in the dataset and we only have to consider the case where they are greater than or equal to zero. Moreover, since the rules have less than δ exceptions, $z \geq 0$ implies $x \geq \delta$. Definition 3 underlines the influence of the rule number of exceptions.

Definition 3 (δ-dependent function). *Let M be an interestingness measure and $r : X \to Y$ an association rule. The δ-dependent function associated to M called $\Psi_{M,\delta}(x,y)$ is the two-variable function obtained by the change of variable $z = x - \delta$ in Ψ_M, i.e. $\Psi_{M,\delta}(x,y) = \Psi_M(x,y,x-\delta)$.*

Pursuing the Lift example, we obtain: $\Psi_{Lift,\delta}(x,y) = \frac{(x-\delta) \times |\mathcal{D}|}{x \times y}$.

3.2 Identifying Properties Shared by Measures

By using the previous definitions, we give now properties expressing basic characteristics of interestingness measures. These properties are on the core of our framework.

Property 1 (P2': weak P2). *Let M be an interestingness measure. Ψ_M increases with z.*

We call Property 1 *weak P2* since it is closely related to Piatetsky-Shapiro's property P2. The slight difference being it is not necessary that the measure monotonically increases.

Property 2 (P3': weak P3). *Let M be an interestingness measure. Ψ_M decreases with y.*

Property 2 is called *weak P3* since it corresponds to the first part of P3 (as well as P2, the definition does not require the *monotonical* decrease). Contrary to the Shapiro's set of properties, we do not make assumptions on the measure's behaviour according to the antecedent frequency. P3' only considers the consequent frequency and, unlike P3, does not require the symmetry between the antecedent and the consequent. As $\Psi_{Lift}(x,y,z)$ increases with z and decreases with y, it is immediate that the Lift satisfies P2' and P3'.

The link between the frequencies expressed by Definition 3 captures an important feature of an interestingness measure: its behavior with respect to the joint development of the antecedent and the consequent frequencies and the maximal rule number of exceptions. This characteristic is translated by Property 3 and we will use it in our framework.

Property 3 (P4: property of δ-dependent growth). *Let M be an interestingness measure. $\Psi_{M,\delta}$ increases with x.*

3.3 SBMs

Property 4 defines the SBMs. It establishes a powerful framework for analyzing the behavior of interestingness measures. Table 2 provides a sample of SBMs. The Rule-Interest measure (RI), which is a good measure according to Definition 1, belongs to this framework.

Property 4 (SBM). *An interestingness measure M is a simultaneously lower bounded measure (or SBM) if M satisfies P2', P3' and P4.*

Theorem 1 states that a linear combination of SBMs with positive coefficients is still a SBM. It also shows that the set of SBMs is infinite.

Table 2. A sample of SBMs

SBM	Definition
Support	$\dfrac{\mathcal{F}(XY)}{\lvert\mathcal{D}\rvert}$
Confidence	$\dfrac{\mathcal{F}(XY)}{\mathcal{F}(X)}$
Sensitivity	$\dfrac{\mathcal{F}(XY)}{\mathcal{F}(Y)}$
Specificity	$1 - \dfrac{\mathcal{F}(X) - \mathcal{F}(XY)}{\lvert\mathcal{D}\rvert - \mathcal{F}(Y)}$
Success Rate	$\dfrac{\lvert\mathcal{D}\rvert - \mathcal{F}(Y) - \mathcal{F}(X) + 2\mathcal{F}(XY)}{\lvert\mathcal{D}\rvert}$
Lift	$\dfrac{\lvert\mathcal{D}\rvert \times \mathcal{F}(XY)}{\mathcal{F}(Y) \times \mathcal{F}(X)}$
Rule-Interest [16]	$\mathcal{F}(XY) - \dfrac{\mathcal{F}(Y) \times \mathcal{F}(X)}{\lvert\mathcal{D}\rvert}$
Laplace (k=2)	$\dfrac{\mathcal{F}(XY) + 1}{\mathcal{F}(X) + 2}$
Odds ratio	$\dfrac{\mathcal{F}(XY)}{\mathcal{F}(X) - \mathcal{F}(XY)} \times \dfrac{\lvert\mathcal{D}\rvert - \mathcal{F}(Y) - \mathcal{F}(X) + \mathcal{F}(XY)}{\mathcal{F}(Y) - \mathcal{F}(XY)}$
Growth rate	$\dfrac{\mathcal{F}(XY)}{\mathcal{F}(X) - \mathcal{F}(XY)} \times \dfrac{\lvert\mathcal{D}\rvert - \mathcal{F}(Y)}{\mathcal{F}(Y)}$
Sebag & Schoenauer	$\dfrac{\mathcal{F}(XY)}{\mathcal{F}(X) - \mathcal{F}(XY)}$
Jaccard	$\dfrac{\mathcal{F}(XY)}{\mathcal{F}(Y) + \mathcal{F}(X) - \mathcal{F}(XY)}$
Conviction	$\dfrac{\lvert\mathcal{D}\rvert - \mathcal{F}(Y)}{\lvert\mathcal{D}\rvert} \times \dfrac{\mathcal{F}(X)}{\mathcal{F}(X) - \mathcal{F}(XY)}$
ϕ-coefficient	$\dfrac{\lvert\mathcal{D}\rvert \times \mathcal{F}(XY) - \mathcal{F}(Y) \times \mathcal{F}(X)}{\sqrt{\mathcal{F}(X) \times \mathcal{F}(Y) \times (\lvert\mathcal{D}\rvert - \mathcal{F}(X)) \times (\lvert\mathcal{D}\rvert - \mathcal{F}(Y))}}$
Added Value	$\dfrac{\mathcal{F}(XY)}{\mathcal{F}(X)} - \dfrac{\mathcal{F}(Y)}{\lvert\mathcal{D}\rvert}$
Certainty Factor	$\dfrac{\mathcal{F}(XY) \times \lvert\mathcal{D}\rvert - \mathcal{F}(X) \times \mathcal{F}(Y)}{\mathcal{F}(X) \times (\lvert\mathcal{D}\rvert - \mathcal{F}(Y))}$
Information Gain	$\log\left(\dfrac{\mathcal{F}(XY)}{\mathcal{F}(X)} \times \dfrac{\lvert\mathcal{D}\rvert}{\mathcal{F}(Y)}\right)$

Theorem 1. *Let $M_1, \ldots, M_n$ be SBMs and $\alpha_1, \ldots, \alpha_n$ be n positive real numbers. $\alpha_1 \times M_1 + \cdots + \alpha_n \times M_n$ is a SBM.*

The key idea of the proof relies on the fact that when multiplying a SBM M by a positive real number α, the associated function $\Psi_{\alpha M}$ and the δ-dependent function $\Psi_{\alpha M, \delta}$ behave like Ψ_M and $\Psi_{M, \delta}$.

Proof. We denote $\alpha_1 \times M_1 + \cdots + \alpha_n \times M_n$ by M. Let us show that M satisfies P2', P3' and P4. The following equalities hold: $\Psi_M = \alpha_1 \Psi_{M_1} + \cdots + \alpha_n \Psi_{M_n}$ and $\Psi_{M, \delta} = \alpha_1 \Psi_{M_1, \delta} + \cdots + \alpha_n \Psi_{M_n, \delta}$. Since $M_1, \ldots, M_n$ are SBMs, they satisfy P2', P3' and P4. $\Psi_{M_1}(x, y, z), \ldots, \Psi_{M_n}(x, y, z)$ increase with z and decrease with y, e.g., the partial derivatives of $\Psi_{M_1}, \ldots, \Psi_{M_n}$ w.r.t. z are positive and their partial derivatives w.r.t. y are negative. Thus the partial derivative of Ψ_M w.r.t. z remains positive and the partial derivative of Ψ_M w.r.t. y remains negative. We conclude that Ψ_M also increases with z and decreases with y. Thus M satisfies P2' and P3'. By the same reasoning, we prove that M satisfies P4 and we conclude that M is a SBM.   □

Theorem 1 can be used to define new SBMs or to check if a candidate interestingness measure is a SBM. For instance, the Novelty [14] (defined by $Nov(r) = \dfrac{\mathcal{F}(XY) \times |\mathcal{D}| - \mathcal{F}(X) \times \mathcal{F}(Y)}{|\mathcal{D}|^2}$) can be expressed according to the Rule-Interest since $Nov = \alpha \times RI$ with $\alpha = \frac{1}{|\mathcal{D}|}$. As the Rule-Interest is a SBM and α is a positive real number, Theorem 1 ensures that Novelty is a SBM as well.

4 SBMs' Bounds and Behavior

This section provides lower bounds for the SBMs. We show that all the SBMs can be simultaneously lower bounded and behave in a similar way. Let γ be the minimal antecedent frequency and η the maximal consequent frequency. Except for Property 5, the values of the parameters γ, η and δ are fixed.

4.1 Lower Bounds

Theorem 2 provides for each SBM its lower bound according to γ, η and δ. Such a bound expresses the minimal quality of a rule according to γ, η and δ. Table 3 gives the lower bounds for SBMs quoted in Table 2. With $\gamma = 3$, $\delta = 1$ and $\eta = 5$ in Table 1, the Lift lower bound is 4.6 and the Rule-Interest lower bound is $\frac{1}{27}$. Note that Theorem 1 enables to calculate $\Psi_{Nov, \delta}(\gamma, \eta)$ with $\alpha \times \Psi_{RI, \delta}(\gamma, \eta)$ where $\alpha = \frac{1}{|\mathcal{D}|}$.

Theorem 2 (Lower bounds). *Let M be a SBM. If $r : X \to Y$ is an association rule such that $\mathcal{F}(X) \geq \gamma$, $\mathcal{F}(Y) \leq \eta$ and r admits less than δ exceptions, then $M(r)$ is greater than or equal to $\Psi_{M, \delta}(\gamma, \eta)$.*

Proof. According to P2', $\Psi_M(x, y, z)$ increases with the variable z. Since $X \to Y$ admits less than δ exceptions, $\mathcal{F}(XY) \geq \mathcal{F}(X) - \delta$ and consequently $\Psi_M(x, y, z) \geq \Psi_M(x, y, x - \delta) = \Psi_{M,\delta}(x, y)$. A lower bound for x is γ and a upper bound for y is η thus, since $\Psi_{M,\delta}$ increases with x and decreases with y (consequence of weak P3), a lower bound for $\Psi_{M,\delta}(x, y)$ is $\Psi_{M,\delta}(\gamma, \eta)$.   □

Table 3. Lower bounds for SBMs defined in Table 2

SBM	Lower bound
Support	$\dfrac{\gamma - \delta}{|\mathcal{D}|}$
Confidence	$1 - \dfrac{\delta}{\gamma}$
Sensitivity	$\dfrac{\gamma - \delta}{\eta}$
Specificity	$1 - \dfrac{\delta}{|\mathcal{D}| - \eta}$
Success Rate	$1 + \dfrac{\gamma - 2\delta - \eta}{|\mathcal{D}|}$
Lift	$(1 - \dfrac{\delta}{\gamma}) \times \dfrac{|\mathcal{D}|}{\eta}$
Rule-Interest	$\gamma - \delta - \dfrac{\gamma \eta}{|\mathcal{D}|}$
Laplace (k=2)	$\dfrac{\gamma - \delta + 1}{\gamma + 2}$
Odds ratio	$[\dfrac{\gamma - \delta}{\eta - \gamma + \delta}] \times [\dfrac{|\mathcal{D}| - \eta - \delta}{\delta}]$
Growth rate	$\dfrac{\gamma - \delta}{\delta} \times \dfrac{|\mathcal{D}| - \eta}{\eta}$
Sebag & Schoenauer	$\dfrac{\gamma - \delta}{\delta}$
Jaccard	$\dfrac{\gamma - \delta}{\eta + \delta}$
Conviction	$\dfrac{|\mathcal{D}| - \eta}{|\mathcal{D}|} \times \dfrac{\gamma}{\delta}$
ϕ-coefficient	$\dfrac{\gamma \times (|\mathcal{D}| - \eta) - \delta \times |\mathcal{D}|}{\sqrt{\gamma \times (|\mathcal{D}| - \gamma) \times \eta \times (|\mathcal{D}| - \eta)}}$
Added Value	$\dfrac{\gamma - \delta}{\gamma} - \dfrac{\eta}{|\mathcal{D}|}$
Certainty Factor	$\dfrac{\gamma \times (|\mathcal{D}| - \eta) - \delta \times |\mathcal{D}|}{\gamma \times (|\mathcal{D}| - \eta)}$
Information Gain	$\log\left(\dfrac{\gamma - \delta}{\gamma} \times \dfrac{|\mathcal{D}|}{\eta}\right)$

As Theorem 2 is true for all the SBMs, we deduce that all the SBMs are simultaneously lower bounded. It means that the set of rules such that $\mathcal{F}(X) \geq \gamma$, $\mathcal{F}(Y) \leq \eta$ and admitting less than δ exceptions simultaneously satisfy minimal values according to all the SBMs. Thus, Theorem 2 enables to identify a set of "good" rules according to the SBMs because all the SBMs have high values, at least greater than or equal to their lower bounds. In the following, we are interested in *all* the rules r such that $M(r) \geq \Psi_{M,\delta}(\gamma, \eta)$ for all the SBMs:

Definition 4 (SBM rule). *The set of rules satisfying $M(r) \geq \Psi_{M,\delta}(\gamma, \eta)$ for all the SBMs is denoted by $\mathcal{R}_{SBM}$. A SBM rule is a rule belonging to $\mathcal{R}_{SBM}$.*

4.2 SBMs' Behavior

Property 5 specifies the behavior of the lower bounds according to γ, η and δ.

Property 5. $\Psi_{M,\delta}(\gamma, \eta)$ *increases with γ and decreases with η and δ.*

Proof. As M is a SBM, it is obvious that $\Psi_{M,\delta}(\gamma, \eta)$ increases with γ (P4) and decreases with η (weak P3). From weak P2, it follows that $\Psi_M(x, y, z)$ increases with z. Hence assuming $\delta_1 \geq \delta_2$ we have $\Psi_M(x, y, x - \delta_2) \geq \Psi_M(x, y, x - \delta_1)$ and $\Psi_{M,\delta_2}(\gamma, \eta) \geq \Psi_{M,\delta_1}(\gamma, \eta)$. Consequently, $\Psi_{M,\delta}(\gamma, \eta)$ decreases with δ. $\square$

Property 5 states that all the lower bounds behave in a similar way according to the parameters γ, η and δ. Consequently, it is possible to increase the rule quality according to the SBMs by increasing γ and decreasing η and δ.

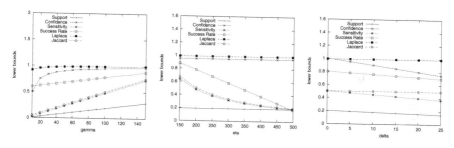

Fig. 1. Lower bounds according to γ, δ and η

For some usual measures, Figure 1 depicts the lowers bounds according to γ (with $\eta = 200$ and $\delta = 5$), η (with $\gamma = 100$ and $\delta = 5$) and δ (with $\eta = 200$ and $\gamma = 100$). These figures show the similar behavior of SBMs and that these measures can be simultaneously optimized.

5   Rule Mining

In this section, we start by characterizing $\mathcal{R}_{SBM}$. This characterization enables to infer an efficient rule mining algorithm.

5.1   Characterizing $\mathcal{R}_{SBM}$

Theorem 3 provides properties on the frequencies of a SBM rule.

Theorem 3. *If $r : X \to Y$ is a SBM rule then r satisfies the following conditions: $\mathcal{F}(X) \geq \gamma$, $\mathcal{F}(Y) \leq \eta$ and r admits less than δ exceptions.*

Proof. We define $M_1(r) = \mathcal{F}(X)$, $M_2(r) = \dfrac{1}{\mathcal{F}(Y)}$ and $M_3(r) = \dfrac{1}{\mathcal{F}(X) - \mathcal{F}(XY)}$. It is trivial to check that $M1$, $M2$ and $M3$ are SBMs. The inequalities $M_1(r) \geq \Psi_{M_1,\delta}(\gamma, \eta) = \gamma$, $M_2(r) \geq \Psi_{M_2,\delta}(\gamma, \eta) = \dfrac{1}{\eta}$ and $M_3(r) \geq \Psi_{M_3,\delta}(\gamma, \eta) = \dfrac{1}{\delta}$ immediately prove the result. $\square$

Theorem 3 is the converse of Theorem 2. These two theorems prove that $\mathcal{R}_{SBM}$ is equal to the set of rules having a γ-frequent antecedent, an η-infrequent consequent and less than δ exceptions. Thus, even if the set of SBMs is infinite, this characterization of $\mathcal{R}_{SBM}$ makes feasible the mining of the rules optimizing all SBMs and ensures the completeness of the mining. The next section shows that we can only mine a reduced set of rules having minimal antecedents among $\mathcal{R}_{SBM}$.

5.2   Informative Rules of $\mathcal{R}_{SBM}$

Section 2.1 has introduced the rule cover based on informative rules [15]. Informative rules are build with minimal patterns (also called *free* [5] or key patterns [15]) as antecedents and one part of their closures (see [21] for a definition) as consequents. By analogy in Definition 5, we call an informative SBM rule a rule having a minimal pattern (i.e., free pattern) as antecedent and one part of its closure as consequent.

Definition 5 (Informative SBM rules). *An informative SBM rule $r : X \to Y$ is a SBM rule such that X is a free pattern and XY is a closed pattern. Thus r satisfies:*

- *X is γ-frequent and free*
- *Y is η-infrequent*
- *$X \cap Y = \emptyset$*
- *XY is closed*
- *r has less than δ exceptions*

The set of informative SBM rules is denoted by $\mathcal{I}nf(\mathcal{R}_{SBM})$.

This definition is precious in practice to mine the informative SBM rules because there are efficient algorithms to extract the free or key patterns and their closures [5]. The next section provides an algorithm which mines the whole set of informative SBM rules.

5.3   Algorithm Mining $\mathcal{I}nf(\mathcal{R}_{SBM})$

This section gives the main features of our algorithm for mining $\mathcal{I}nf(\mathcal{R}_{SBM})$. The basic principle is to associate the free and the closed patterns given in input to build the informative SBM rules. Definition 4 states that the SBM rules satisfy the following constraints: $\mathcal{F}(X) \geq \gamma$, $\mathcal{F}(Y) \leq \eta$ and $\mathcal{F}(X) - \mathcal{F}(XY) \leq \delta$. These constraints lead to Property 6 which provides pruning conditions:

Property 6. *The SBM rules satisfy:*

1. $\gamma \leq \mathcal{F}(X) \leq \eta + \delta$
2. $\gamma - \delta \leq \mathcal{F}(XY) \leq \mathcal{F}(Y) \leq \eta$

Proof.

1. Since $Y \subset XY$, $\mathcal{F}(XY) < \mathcal{F}(Y) \leq \eta$. $\mathcal{F}(XY) \leq \eta$ is obvious. Thus $\mathcal{F}(X) \leq \mathcal{F}(XY) + \delta$ and we have $\mathcal{F}(X) \leq \eta + \delta$.
2. Since $Y \subset XY$, $\mathcal{F}(XY) < \mathcal{F}(Y)$. We have $\gamma - \delta \leq \mathcal{F}(X) - \delta \leq \mathcal{F}(XY)$. Thus $\gamma - \delta \leq \mathcal{F}(Y)$.

Algorithm 1 considers each pattern X in $\mathcal{F}ree_{(\gamma, \eta + \delta)}$, i.e., each free pattern having a frequency between γ and $\eta + \delta$. Then the closed patterns containing X and having a frequency between $\gamma - \delta$ and η are determined. $\mathcal{I}$ is the set of discovered informative SBM rules. Note that the antecedent and the rule satisfy the frequency constraints of Property 6 by construction. Then, the number of exceptions and the consequent frequency are checked. This latter is obviously greater than $\gamma - \delta$ but not necessarily less than η. The consequent frequency is computed by finding the smallest closed pattern containing the rule consequent. When discovered, a valid rule is added to $\mathcal{I}$. The algorithm stops when all the patterns in $\mathcal{F}ree_{(\gamma, \eta + \delta)}$ have been considered.

Data: $\mathcal{F}ree$ the set of free patterns, $\mathcal{C}losed$ the set of closed patterns
Result: the informative SBM rules $\mathcal{I}nf(\mathcal{R}_{SBM})$
1 **foreach** $X \in \mathcal{F}ree_{(\gamma, \eta + \delta)}$ such that $\mathcal{F}(X) - \mathcal{F}(Z) \leq \delta$ **do**
2   **foreach** $Z = XY \in \mathcal{C}losed_{(\gamma - \delta, \eta)}$ such that $\mathcal{F}(X) - \mathcal{F}(Z) \leq \delta$ **do**
3   **if** $\mathcal{F}(Y) \leq \eta$ **then**
4             $\mathcal{I} = \mathcal{I} \cup \{X \rightarrow Y\}$
5   **end**
6   **end**
7 **end**
8 **return** $\mathcal{I}$

Algorithm 1. Mining $\mathcal{I}nf(\mathcal{R}_{SBM})$

6 Experiments

The aim of the experiments is twofold: first, we quantify the size of $\mathcal{I}nf(\mathcal{R}_{SBM})$ according to the parameters of our framework, and second we observe the quality of the informative SBM rules mined in practice. Experiments are performed on a real data set, the hepatitis data collected at the Chiba University Hospital (Japan). These data are used in discovery challenges [11]. They contain the examinations of 499 patients which are described with 168 attributes.

Number of informative SBM rules. Figures on the top of Figure 2 plot on a logarithmic scale the size of $\mathcal{I}nf(\mathcal{R}_{SBM})$ i.e. the number of informative SBM rules according to the minimal antecedent frequency threshold γ (on the left) and the maximal number of exceptions δ (on the right) with $\eta = 200$. The figure on the bottom of Figure 2 plots the size of $\mathcal{I}nf(\mathcal{R}_{SBM})$ according to the maximal consequent frequency threshold η with $\gamma = 60$. As expected (cf. Property 5), the number of rules clearly decreases according to γ and increases both with η and δ. Nevertheless, these curves specify how these numbers vary.

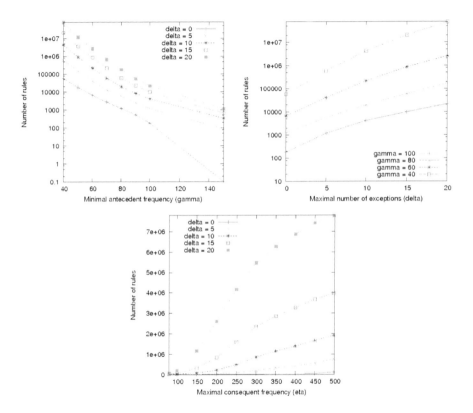

Fig. 2. Size of $\mathcal{I}nf(\mathcal{R}_{SBM})$ according to γ, δ and η

Figure 3 plots the number of rules with $\eta = 200$ and without maximal consequent frequency i.e., $\eta = 500$ (since the hepatitis data only contain 499 objects). It shows the reduction of the rule number due to η. For instance, with $\gamma = 80$, there are 5605 rules with $\eta = 200$ versus almost 200.000 rules with $\eta = 500$. Clearly, bounding the consequent frequency enables to drastically reduce the size of the output. This result is interesting because we know (thanks to Property 5) that the discarded rules have the worst values according to the set of SBMs.

Quality of the mined rules. We now focus on the quality of the informative SBM rules. With $\gamma = 60$, $\eta = 200$ and $\delta = 5$, $\mathcal{I}nf(\mathcal{R}_{SBM})$ includes 40697 rules. Table 4 indicates the minimal value, the lower bound (calculated with the expressions given in Table 3), the average value for the rules in $\mathcal{I}nf(\mathcal{R}_{SBM})$

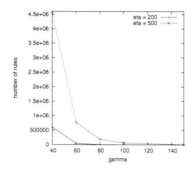

Fig. 3. Number of rules with and without a maximal consequent frequency

and the maximum value of a few SBMs. These results show for each SBM the minimal value guaranteed by our framework. Obviously, the average values are higher than the lower bounds and the difference between the average value and the lower bound depends on the measures. For instance, the Sensitivity ranges from 0 to 1. Its lower bound equals 0.275 and its average value is 0.411. For the Sebag & Schoenauer's measure (ranging from 0 to infinity), the lower bound is 11 while the average value is about 21.

Table 4. Minimum, lower bound, average value and maximum of a few SBMs

Measure	Support	Confidence	Sensitivity	Rule-Interest	Odds Ratio
Minimum	0	0	0	-0.25	0
Lower bound	0.1102	0.917	0.275	0.06203	22.303
Average value	0.134	0.962	0.411	0.089	74.038
Maximum	1	1	1	0.25	∞

Measure	GR	Sebag & Schoenauer	Jaccard	ϕ-Coefficient	Added Value
Minimum	0	0	0	-1	-0.5
Lower bound	16.45	11	0.300	0.452	0.599
Average value	43.616	21.274	0.406	0.540	0.626
Maximum	∞	∞	1	1	1

7 Conclusion and Future Work

Further work addresses the multi-criteria optimization of the SBMs. Theorem 1 shows that it is possible to combine several SBMs without loosing the properties of our framework. An approach is to get a lower bound for a weighted combination of SBMs in order to ensure a global quality for all SBMs. Another way is to automatically determine the parameters involved in the mining of the SBM rules in order to take into account the various semantics conveyed by the measures during the mining process.

Acknowledgements. The authors thank Nicolas Durand for preparing the hepatitis data. This work has been partially funded by the ACI "masse de données" (French Ministry of research), Bingo project (MD 46, 2004-2007).

References

[1] Agrawal, R., Imielinski, T., Swami, A.N.: Mining association rules between sets of items in large databases. In: Buneman, P., Jajodia, S. (eds.) SIGMOD'93 Conference, pp. 207–216. ACM Press, New York (1993)

[2] Agrawal, R., Srikant, R.: Fast algorithms for mining association rules in large databases. In: Bocca, J.B., Jarke, M., Zaniolo, C. (eds.) VLDB'94, Proceedings of 20th International Conference on Very Large Data Bases, Santiago de Chile, Chile, September 12-15, 1994, pp. 487–499. Morgan Kaufmann, San Francisco (1994)

[3] Bayardo, J.R.J., Agrawal, R.: Mining the most interesting rules. In: KDD'99, pp. 145–154 (1999)

[4] Blanchard, J., Guillet, F., Briand, H.: A user-driven and quality-oriented visualization for mining association rules. In: the 3rd IEEE International Conference on Data Mining (ICDM 2003), pp. 493–496. IEEE Computer Society Press, Los Alamitos (2003)

[5] Boulicaut, J.-F., Bykowski, A., Rigotti, C.: Approximation of frequency queries by means of free-sets. In: Zighed, D.A., Komorowski, H.J., Zytkow, J.M. (eds.) PKDD 2000. LNCS (LNAI), vol. 1910, pp. 75–85. Springer, Heidelberg (2000)

[6] Brin, S., Motwani, R., Silverstein, C.: Beyond market baskets: Generalizing association rules to correlations. In: Peckham, J. (ed.) SIGMOD 1997, Proceedings ACM SIGMOD International Conference on Management of Data, Tucson, Arizona, USA, May 13-15, 1997, pp. 265–276. ACM Press, New York (1997)

[7] Francisci, D., Collard, M.: Multi-criteria evaluation of interesting dependencies according to a data mining approach. In: Congress on Evolutionary Computation, Canberra, Australia, 12, pp. 1568–1574. IEEE Computer Society Press, Los Alamitos (2003)

[8] Fürnkranz, J., Flach, P.A.: Roc 'n' rule learning-towards a better understanding of covering algorithms. Machine Learning 58(1), 39–77 (2005)

[9] Gasmi, G., Yahia, S.B., Nguifo, E.M., Slimani, Y.: Igb: A new informative generic base of association rules. In: Ho, T.-B., Cheung, D., Liu, H. (eds.) PAKDD 2005. LNCS (LNAI), vol. 3518, pp. 81–90. Springer, Heidelberg (2005)

[10] Hilderman, R.J., Hamilton, H.J.: Measuring the interestingness of discovered knowledge: A principled approach. Intell. Data Anal. 7(4), 347–382 (2003)

[11] Hirano, S., Tsumoto, S.: Guide to the hepatitis data. In: ECML/PKDD'05 Discovery Challenge on hepatitis data co-located with the 9th European Conference on Principles and Practice of Knowledge Discovery in Databases (PKDD'05), Porto, Portugal, October 2005, pp. 120–124 (2005)

[12] Hébert, C., Crémilleux, B.: Optimized rule mining through a unified framework for interestingness measures. In: Tjoa, A.M., Trujillo, J. (eds.) DaWaK 2006. LNCS, vol. 4081, pp. 238–247. Springer, Heidelberg (2006)

[13] Kryszkiewicz, M.: Concise representations of association rules. In: Hand, D.J., Adams, N.M., Bolton, R.J. (eds.) Pattern Detection and Discovery. LNCS (LNAI), vol. 2447, pp. 92–109. Springer, Heidelberg (2002)

[14] Lavrac, N., Flach, P., Zupan, B.: Rule evaluation measures: a unifying view. In: Džeroski, S., Flach, P.A. (eds.) Inductive Logic Programming. LNCS (LNAI), vol. 1634, pp. 174–185. Springer, Heidelberg (1999)

[15] Pasquier, N., Bastide, Y., Taouil, R., Lakhal, L.: Discovering frequent closed itemsets for association rules. In: Beeri, C., Bruneman, P. (eds.) ICDT 1999. LNCS, vol. 1540, pp. 299–312. Springer, Heidelberg (1998)

[16] Piatetsky-Shapiro, G.: Discovery, analysis, and presentation of strong rules. In: Knowledge Discovery in Databases, pp. 229–248. AAAI/MIT Press, Cambridge (1991)

[17] Plasse, M., Niang, N., Saporta, G., Leblond, L.: Une comparaison de certains indices de pertinence des règles d'association. In: Ritschard, G., Djeraba, C. (eds.) EGC. Revue des Nouvelles Technologies de l'Information, vol. RNTI-E-6, pp. 561–568. Cépaduès-Éditions (2006)

[18] Sebag, M., Schoenauer, M.: Generation of rules with certainty and confidence factors from incomplete and incoherent learning bases. In: Boose, M. L. J., Gaines, B. (eds.)European Knowledge Acquisistion Workshop, EKAW'88, pages 28-1–28-20 (1988)

[19] Tan, P.-N., Kumar, V., Srivastava, J.: Selecting the right interestingness measure for association patterns. In: KDD, pp. 32–41. ACM, New York (2002)

[20] Vaillant, B., Lenca, P., Lallich, S.: A clustering of interestingness measures. In: The 7th International Conference on Discovery Science, 10, 2004, pp. 290–297 (2004)

[21] Wille, R.: chapter Restructuring lattice theory: an approach based on hierachies of concepts. In: Ordered sets, pp. 445–470. Reidel, Dordrecht (1982)

[22] Zaki, M.J.: Generating non-redundant association rules. In: KDD'00, pp. 34–43 (2000)

Comparing State-of-the-Art Collaborative Filtering Systems

Laurent Candillier, Frank Meyer, and Marc Boullé

France Telecom R&D Lannion, France
lcandillier@hotmail.com

Abstract. *Collaborative filtering* aims at helping *users* find *items* they should appreciate from huge catalogues. In that field, we can distinguish *user-based*, *item-based* and *model-based* approaches. For each of them, many options play a crucial role for their performances, and in particular the similarity function defined between users or items, the number of neighbors considered for user- or item-based approaches, the number of clusters for model-based approaches using clustering, and the prediction function used.

In this paper, we review the main collaborative filtering methods proposed in the litterature and compare them on the same widely used real dataset called *MovieLens*, and using the same widely used performance measure called *Mean Absolute Error* (MAE). This study thus allows us to highlight the advantages and drawbacks of each approach, and to propose some default options that we think should be used when using a given approach or designing a new one.

1 Introduction

Recommender systems [1] have known a growing interest in the last two decades, since the appearance of the first papers in the mid-1990s [2]. The aim of such systems is to help *users* find *items* they should appreciate from huge catalogues. To do this, three types of approaches are commonly used:

1. *collaborative filtering*,
2. *content-based filtering*,
3. and *hybrid filtering*.

In the first case, the input of the system is a set of ratings of users on sets of items, and the approach used to predict the rating of a given user on a given item is based on the ratings of a set of users who have already rated the given item and whose tastes are similar to the ones of the given user.

In the second case, the item descriptions are used to construct *user thematic profiles* (such as *"like comedy and dislike war"* when items are movies), and the prediction of interest of a user on a given item is based on the similarity between the item description and the user profile. In the third case of hybrid filtering, both information, collaborative and content-based, are used.

P. Perner (Ed.): MLDM 2007, LNAI 4571, pp. 548–562, 2007.

In this paper, we focus on the first type of techniques, because it is the most widely considered in the field of recommender systems, and yet many different collaborative filtering approaches are worth to be compared. We present here many different options of the three general approaches for collaborative filtering:

1. *user-based* approaches, that associate to each user its set of nearest neighbors, and then predict a user's rating on an item using the ratings, on that item, of its nearest neighbors,
2. *model-based* approaches, and more specifically those based on *clustering*, that construct a set of users groups, and then predict a user's rating on an item using the ratings, on that item, of the members of its group,
3. and *item-based* approaches, that associate to each item its set of nearest neighbors, and then predict a user's rating on an item using the ratings of the user on the nearest neighbors of the item considered.

We chose to focus on collaborative filtering instead of content-based filtering also because in many cases, well-structured item descriptions are hard to get, whereas collecting user ratings on items is easier, yet some real rating datasets are available for tests. We chose the most widely used one, called *MovieLens*, for our study. This dataset contains 1,000,209 ratings collected from 6,040 users on 3,706 items that represent movies. Then two ways for evaluating the performances of a collaborative filtering method can be used [3]:

1. evaluate its error rate in *cross-validation*,
2. or evaluate *user satisfaction* in the system.

We focus in this paper on the first approach that is the most widely used one, and less subjective. Many measures can then be used to compare the results of different collaborative filtering methods. The most widely used ones are:

1. *Mean Absolute Error* (MAE),
2. *Root Mean Squared Error* (RMSE),
3. and *Precision* and *Recall*.

The two first measures evaluate the capability of a method to predict if a user will like or dislike an item, whereas the third measure evaluates its capability of providing an ordered list of items that a user should like. So these measures carry different meanings [4]: in the first two cases, the method needs to be able to predict dislike, but there is no need for ordering items, whereas in the third case, the method only focuses on items users will like, but the order in which these items are ranked is important. In this paper, we focus on the MAE, that is the most widely used measure.

The rest of the paper is organized as follows: section 2 presents an overview of the principal approaches for collaborative filtering; we then report in section 3 the results of extensive experiments conducted using various collaborative filtering methods and various alternatives of each, on the *MovieLens* dataset using cross-validation and the MAE measure for comparison; finally, section 4 concludes the paper and proposes some default options that we think should be used when using a given collaborative filtering method, or designing a new one.

2   Collaborative Filtering Approaches

Let U be a set of N users and I a set of M items. v_{ui} denotes the rating of user $u \in U$ on item $i \in I$, and $S_u \subseteq I$ stands for the set of items that user u has rated. In the *MovieLens* dataset for example, ratings are integers ranging from 1 to 5.

2.1   User-Based Approaches

For user-based approaches [2], the prediction of rating p_{ai} of user a (*active*) on item i is computed using the sum of the user mean rating and the weighted sum of deviations from their mean rating of users that have rated item i. More formally, p_{ai} is computed as follows:

$$p_{ai} = \overline{v_a} + \frac{\sum_{\{u \in U \mid i \in S_u\}} w(a, u) \times (v_{ui} - \overline{v_u})}{\sum_{\{u \in U \mid i \in S_u\}} |w(a, u)|} \tag{1}$$

$\overline{v_u}$ represents the mean rating of user u:

$$\overline{v_u} = \frac{\sum_{i \in S_u} v_{ui}}{|S_u|} \tag{2}$$

And $w(a, u)$ stands for the similarity between users a and u, computed using *pearson* correlation in [2], that corresponds to the cosine of the users deviation from their mean:

$$w(a, u) = \frac{\sum_{i \in S_a \cap S_u} (v_{ai} - \overline{v_a})(v_{ui} - \overline{v_u})}{\sqrt{\sum_{i \in S_a \cap S_u} (v_{ai} - \overline{v_a})^2 \sum_{i \in S_a \cap S_u} (v_{ui} - \overline{v_u})^2}} \tag{3}$$

The influence of this similarity measure in the performances of this approach is very important. So many other measures have been considered in the litterature [5,6]. Let us introduce two of them:

– simple *cosine*:

$$w(a, u) = \frac{\sum_{i \in S_a \cap S_u} v_{ai} \times v_{ui}}{\sqrt{\sum_{i \in S_a \cap S_u} v_{ai}^2 \sum_{i \in S_a \cap S_u} v_{ui}^2}} \tag{4}$$

– and *constraint* pearson correlation, that corresponds to the cosine of users deviation from the mean rating, denoted by $\overline{v}$ (equal to 3 for a rating scale ranging from 1 to 5):

$$w(a, u) = \frac{\sum_{i \in S_a \cap S_u} (v_{ai} - \overline{v})(v_{ui} - \overline{v})}{\sqrt{\sum_{i \in S_a \cap S_u} (v_{ai} - \overline{v})^2 \sum_{i \in S_a \cap S_u} (v_{ui} - \overline{v})^2}} \tag{5}$$

Finally, a neighborhood for each user can be considered. In such a case, the neighborhood size K is then a system parameter that needs to be defined, and only the neighbors of the active user are considered for predictions.

The time complexity of user-based approaches is $O(N^2 \times M \times K)$ for the model construction, $O(K)$ for one rating prediction, and the space complexity is $O(N \times K)$.

2.2   Model-Based Approaches

Since predicting the rating of a given user on a given item requires the computation of the similarity between the given user and all its neighbors that have already rated the given item, its execution time may be long for huge datasets. In order to reduce such execution time, model-based approaches have been proposed [7]. The general idea is to derive off-line a model of the data in order to predict on-line ratings as fast as possible.

The first types of models that have been proposed consist in grouping the users using clustering and then predicting the rating of a given user on a given item using only the ratings of the users that belong to the same cluster. Then probabilistic clustering algorithms have been used in order to allow users to belong, at some level, to different groups of users [8,9]. Hierarchies of clusters have also been proposed, so that if a given cluster of users does not have opinion on a given item, its *super-cluster* can be considered [10].

In such approaches, the choice of the distance measure used to compare users is important. Let us present two widely used of them:

1. normalized *manhattan* distance:

$$dist(a,u) = \frac{\sum_{\{i \in S_a \cap S_u\}} |v_{ai} - v_{ui}|}{|\{i \in S_a \cap S_u\}|} \qquad (6)$$

2. and normalized *euclidian* distance:

$$dist(a,u) = \sqrt{\frac{\sum_{\{i \in S_a \cap S_u\}} (v_{ai} - v_{ui})^2}{|\{i \in S_a \cap S_u\}|}} \qquad (7)$$

The number of clusters considered is also of key importance. In many cases, different numbers of clusters are tested, and the one that led to the lowest error rate in cross-validation is kept. Clusters C_k are then generally represented by their centroid $\overrightarrow{\mu_k}$:

$$\mu_{ki} = \frac{\sum_{\{u \in C_k | i \in S_u\}} v_{ui}}{|\{u \in C_k | i \in S_u\}|} \qquad (8)$$

Then the predicted rating of a user to an item can be directly derived from the rating of its nearest centroid, or it can be computed using a sum on the ratings of all centroids, weighted by the distance between the given user and the centroids.

For this study, we implemented four clustering algorithms:

- *K-means*, the well-known full-space clustering algorithm based on the evolution of K centroids that represent the K clusters to be found,
- *Bisecting* K-means [11], based on the recursive use of (K=2)-means, by selecting at each step for next split the cluster that maximizes its inertia,
- *LAC* [12], that is based on K-means and adds a weight to each attribute, depending on the deviation of the cluster members from its mean,
- and *SSC* [13], that is a probabilistic clustering algorithm, based on a mixture of gaussians and the *EM* algorithm.

All these methods need to be run many times with random initial solutions in order to avoid local minimum solutions. We set the default number of runs to 10 in these experiments. The time complexity of cluster-based approaches is $O(K \times N \times M)$ for the model construction, $O(1)$ for one rating prediction, and the space complexity is $O(K \times M + N)$.

Finally, models based on item associations have also been considered. Bayesian models have been proposed to model dependencies between items [7]. The clustering of items have been studied in [14,15]. And models based on association rules have been studied in [16,17].

2.3   Item-Based Approaches

Then item-based approaches have known a growing interest [18]. Given a similarity measure between items (like cosine or pearson correlation presented earlier for user-based approaches), item-based approaches predict the rating of a given user on a given item using the ratings of the user on the items considered as similar to the target item. In [18], a weighted sum is used to predict the rating of active user a on item i, given $sim(i, j)$ a similarity measure between items:

$$p_{ai} = \frac{\sum_{\{j \in S_a | j \neq i\}} sim(i, j) \times v_{aj}}{\sum_{\{j \in S_a | j \neq i\}} |sim(i, j)|} \tag{9}$$

Two specific similarity measures have been proposed in [19,20] for item-based collaborative filtering methods:

- *adjusted* cosine, that corresponds to the cosine of items deviation from the user mean rating:

$$sim(i, j) = \frac{\sum_{\{u \in U | i \in S_u \& j \in S_u\}} (v_{ui} - \overline{v_u})(v_{uj} - \overline{v_u})}{\sqrt{\sum_{\{u \in U | i \in S_u \& j \in S_u\}} (v_{ui} - \overline{v_u})^2 \sum_{\{u \in U | i \in S_u \& j \in S_u\}} (v_{uj} - \overline{v_u})^2}} \tag{10}$$

- and a *probabilistic* similarity measure, that corresponds to the co-occurrence frequence of both items i and j, normalized by user frequences in order to enhance the contribution of users who have rated fewer items, and then normalized by the product of the frequences of both concerned items:

$$sim(i, j) = \frac{\sum_{\{u \in U | i \in S_u \& j \in S_u\}} v_{uj}/|S_u|}{|\{u \in U | i \in S_u\}| \times |\{u \in U | j \in S_u\}|} \tag{11}$$

Finally, as for user-based approaches, a neighborhood for each item can be considered. In such a case, the neighborhood size K is then a system parameter that needs to be defined, and only the neighbors of the target item are considered for predictions.

The time complexity of item-based approaches is $O(M^2 \times N \times K)$ for the model construction, $O(K)$ for one rating prediction, and the space complexity is $O(M \times K)$.

2.4   Complementary Approaches

Different default prediction techniques can also been considered, in particular when a method is not able to predict any rating, if a user has no rating, if it has no neighbor, if there is no rating on an item or if an item has no neighbor:

- *mean item* rating,
- *mean user* rating,
- *majority item* rating,
- and *majority user* rating.

We also propose an alternative approach where we consider the recommendation problem as a standard classification problem with two input variables, user u and item i, and one output variable, rating r. We apply the standard *Naive Bayes* approach, assuming that users and items are independent conditionally to the ratings. This approach is based on the following *Bayes* rule used to compute the probability of rating r for a given user u on a given item i:

$$P(r|u,i) = \frac{P(r|u) \times P(r|i)}{P(r)} \times \frac{P(u) \times P(i)}{P(u,i)} \tag{12}$$

$P(r|u)$ stands for the probability of rating r for user u, $P(r|i)$ the probability of rating r on item i, and $P(r)$ the global probability of rating r. The last three probabilities $P(u)$, $P(i)$ and $P(u,i)$ can be ignored since they are the same for all users and items. From these probabilities, we then propose three prediction schemes:

- predict the most probable rating, which corresponds to the *Maximum A Posteriori* (MAP) approach:

$$p_{ai} = Argmax_{r=1}^{5} P(r|a,i) \tag{13}$$

- compute the weighted sum of ratings, that corresponds to minimizing the expectation of *Mean Squared Error* (MSE):

$$p_{ai} = \sum_{r=1}^{5} r \times P(r|a,i) \tag{14}$$

– or select the rating that minimizes the expectation of *Mean Absolute Error* (MAE):

$$p_{ai} = Argmin_{r=1}^{5} \sum_{n=1}^{5} P(n|a, i) \times |r - n| \qquad (15)$$

The time complexity of bayes-based approaches is $O(R)$ for the model construction, with R the number of ratings in the dataset. The time complexity for one rating prediction is $O(1)$, and the space complexity is $O(N + M)$.

Model-based approaches can be combined with different default approaches, or with any user- or item-based approach. This is done by constructing local models from the different sub-datasets created using clustering.

Finally, since in many real datasets ratings are integer values, we can choose to round the predicted ratings instead of using their real values. Such a process improves the results when MAE is used, but not when RMSE is used.

3   Experiments

3.1   Parameters

Considering only the principal collaborative filtering approaches already leads us to a lot of choices and parameters. When implementing a user- or item-based approach, one may choose:

– a similarity measure: pearson (equation 3), cosine (4), constraint pearson (5), adjusted cosine (10), or probabilistic (11),
– a neighborhood size,
– and how to compute predictions: using a weighted sum of rating values (9), or using a weighted sum of deviations from the mean (1).

For model-based approaches, the following parameters need to be defined:

– the distance measure used: manhattan (6) or euclidian (7),
– the number of clusters,
– how to compute predictions in one cluster: using the mean rating of the cluster members on an item, using another default approach, or using a local user- or item-based approach,
– and how to compute predictions for one user: returning the prediction of its nearest cluster, or the weighted sum of predictions of each cluster.

Finally, in all cases, we can choose to round the results or not. As a default prediction scheme, if no prediction can be done for a given approach, the global mean item rating is returned, and if the item is not known by the system, then the mean user rating is returned.

3.2   Protocol

We conduct these experiments using the *MovieLens* dataset. We divided it into 10 parts in order to perform 10-fold cross-validations, training the chosen model using 9 parts and testing it on the last part. In all experiments, the division into 10 parts of the dataset is always the same, so that all approaches are evaluated under exactly the same conditions. Since the dataset size is important, the variance of the results over the 10 cross-validations is low.

Given $T = \{(u, i, r)\}$ the set of (user,item,rating) triplets used for test, the Mean Absolute Error Rate (MAE) and Root Mean Squared Error (RMSE) are used to evaluate the performances of the algorithms:

$$MAE = \frac{1}{|T|} \sum_{(u,i,r) \in T} |p_{ui} - r| \tag{16}$$

$$RMSE = \sqrt{\frac{1}{|T|} \sum_{(u,i,r) \in T} (p_{ui} - r)^2} \tag{17}$$

We also report the time spent for the model construction and for predictions.

3.3   Results

Let us start with the results of the default approaches, presented in table 1.

Table 1. Default approaches results measured using MAE and RMSE, when rounding (2) or not (1) the predicted ratings

	MeanItem	MeanUser	MajoItem	MajoUser	BayesMAP	BayesMSE	BayesMAE
MAE(1)	0.7821	0.8286	0.7702	0.8363	0.7159	0.7279	**0.6829**
MAE(2)	0.7501	0.7939	0.7702	0.8363	0.7159	0.6899	**0.6829**
RMSE(1)	0.9791	1.0350	1.0924	1.1991	1.0658	**0.9247**	0.9894
RMSE(2)	1.0182	1.0741	1.0924	1.1991	1.0658	**0.9684**	0.9894

We can thus already observe that the results are better when default ratings are based on item information than when they are based on user information, and that using the mean rating is better than using the majority rating. But default ratings using Bayes models lead to much better results. For such approaches, the MAE is minimized with the BayesMAE scheme (equation 15), but the RMSE is minimized with the BayesMSE sheeme (14). Rounding the predicted ratings improves the results when the MAE is used, but not when the RMSE is used. These two observations confirm the theory. In the following, we only report results using MAE and after the predicted ratings have been rounded. Figures 1 and 2 report such results using different user-based approaches.

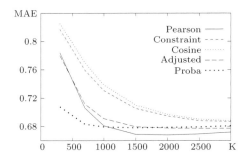

Fig. 1. Comparing similarity measures for user-based approaches using the *deviation* prediction scheme, and different neighborhood sizes (K)

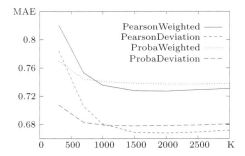

Fig. 2. Comparing prediction schemes for user-based approaches using *pearson* and *probabilistic* similarity measures, and different neighborhood sizes (K)

We can thus observe that the results are improved when many neighbors are considered. But of course the execution time is higher when more neighbors are used. The similarity measure that leads to the best results is pearson, according to figure 1. Predicting using weighted sum of deviations from the mean leads to better results than predicting using simple weighted sum according to figure 2. Rounding the predicted ratings improved the MAE from 2.5 to 5.5 percent. Figures 3 and 4 then report the results using item-based approaches.

We observe again from figure 4 that predicting using weighted sum of deviations from the mean leads to a lower MAE than predicting using simple weighted sum, no matter which similarity measure is used. But in that case, considering too much neighbors degrades the results, and the probabilistic similarity leads to the lowest MAE, according to figure 3. Rounding the predicted ratings improved the MAE from 2.3 to 6 percent. Finally, figures 5 and 6 report the results obtained using Kmeans-based approaches.

According to figure 5, we see that there is not a high difference in using manhattan or euclidian distance, although euclidian distance leads to slightly better results. In both cases, the optimal number of clusters is 6. On the contrary, predicting using BayesMAE (equation 15) leads to better results than when

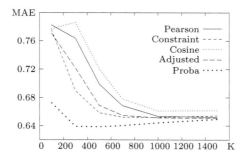

Fig. 3. Comparing similarity measures for item-based approaches using the *deviation* prediction scheme, and different neighborhood sizes (K)

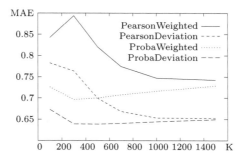

Fig. 4. Comparing prediction schemes for item-based approaches using *pearson* and *probabilistic* similarity measures, and different neighborhood sizes (K)

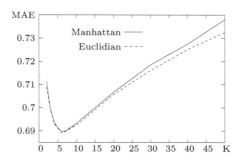

Fig. 5. Comparing distance measures for model-based approaches using the *mean item rating* prediction scheme, and different numbers of clusters (K)

MeanItem rating is used, according to figure 6, and in that case, the optimal number of clusters is 4. Those reported results concern predictions based on the nearest cluster rather than based on a weighted sum of predictions of each cluster because that first scheme led to better results. Figure 7 presents results using

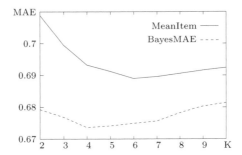

Fig. 6. Comparing prediction schemes for model-based approaches using the *euclidian* distance, and different numbers of clusters (K)

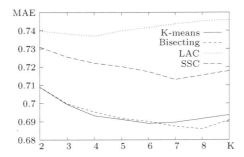

Fig. 7. Comparing different clustering algorithms for model-based approaches using the *euclidian* distance, the *mean item rating* prediction scheme, and different numbers of clusters (K)

other clustering algorithms than K-means, and shows that K-means outperforms LAC and SSC, but that Bisecting K-means can lead to better results when more clusters are considered.

Finally, table 2 summarizes the results of the best of each approach, including execution time and space complexity.

The *BestDefault* is BayesMAE: the default approach based on Bayes rule minimizing MAE (equation 15). The *BestUserBased* is the user-based approach based on pearson similarity (3) and 1500 neighbors. The *BestItemBased* is the item-based approach based on probabilistic similarity (11) and 400 neighbors. Both use predictions using weighted sum of deviations from the mean (1). Finally, the *BestModelBased* is the model-based approach using K-means with euclidian distance (7), 4 clusters and prediction scheme based on the nearest cluster and Bayes model minimizing MAE.

The best overall results are reached by the best item-based approach. It needs 170 seconds to construct the model and 3 seconds to predict 100,021 ratings. Then the best user-based approach has slightly lower MAE than model-based or default approaches, but for a model construction time of 730 seconds and

Table 2. Summary of the best approaches

	BestDefault	BestUserBased	BestItemBased	BestModelBased
model construction time (in sec.)	1	730	170	254
prediction time (in sec.)	1	31	3	1
MAE	0.6829	0.6688	**0.6382**	0.6736

prediction time of 31 seconds. On the other side, the best default approach only needs 2 seconds for both model construction and predictions, for a difference in MAE of only 0.0141.

In the litterature, [6] reported an MAE of 0.88 on the MovieLens dataset, [10] reported 0.73, and [19] 0.72. This last score is the best reported in the litterature, and corresponds to what we obtain with an item-based model using their proposed adjusted cosine similarity measure given by equation 10, their proposed weighted prediction scheme given by equation 9, 700 neighbors, and without rounding results.

Finally, we also tested the use of local item-based models constructed on the different user groups identified by clustering, but such an approach degrades the results of a global item-based approach.

4   Conclusion

Many approaches to collaborative filtering have been proposed in the last two decades. However, many different experimental protocols have been used to evaluate them, so that it is hard to compare them. In this paper, we chose to focus on the most widely used dataset and relevance measure in order to test the main methods for collaborative filtering and their main options.

According to our first results on default approaches, it seems that using Bayes model for default predictions is relevant since it has reasonable error rate for very low execution time. Besides, another important advantage of such a technique is that it is easily updatable since it is incremental, whereas the other approaches need to relearn their entire model in order to take into account new data.

For all experiments, rounding the predicted ratings led to an improvement ranging from 2 to 6 percent of the MAE. Besides, rounding ratings is natural in practice, since real users generally prefer rating scales based on natural numbers than on real numbers.

Computing predictions using weighted sum of deviations from the mean also led to better results than using simple weighted sum for both user- and item-based approaches. As far as we know, using such prediction scheme for item-based approaches is new, and that is what led us to the best results. The lowest error rates were reached using pearson similarity for user-based approaches and probabilistic similarity for item-based approaches.

Using Bayes default approach in order to predict ratings inside a given cluster leads to better results than when the mean item rating of the cluster members is used. Considering the prediction of the nearest cluster is better than computing a weighted sum of the predictions of each cluster. Finally, K-means led to better results than more sophisticated algorithms like LAC or SSC. We think it is not relevant, in the field of collaborative filtering, to assume that the ratings of users of a same cluster follow a normal distribution. In particular, we are faced with the problem that users generally use differently the rating scale: for example one user may rate 5 a movie he likes and 3 a movie he dislikes whereas another user with the same tastes will rate 4 a movie he likes and 1 a movie he dislikes.

More generally, item-based approaches seem the bests in our experiments. But these results need to be taken with precaution. Indeed, although in many cases the number of users is much more important than the number of items, in cases where there are more items than users, user-based approaches could lead to better results. On the same way, if there are some demographic information on users, results of user-based approaches can be improved [21]. On the other side, if some content information on items are available, results of item-based approaches can also be improved [22].

Besides their very good results, item-based approaches have other advantages: they seem to need fewer neighbors than user-based approaches, and such models are also appropriate for the navigation in item catalogues even when no information about the current user is available, since it can also present to a user the nearest neighbors of any item he is currently interested in.

For future work, it seems now interesting to study how these methods can be adapted to scale well when faced with huge datasets. The dataset provided by Netflix [23], a popular online movie rental service, can be used for such tests since it contains 100,480,507 movie ratings from 480,189 users on 17,770 movies.

In that field, [24] proposed a user selection scheme for user-based approaches, [11] proposed to create *super-users* by running a user-based approach considering as users the centroids found using a bisecting K-means clustering algorithm, [25] proposed to use *Principal Components Analysis* (PCA) and [26] proposed to use *Singular Value Decomposition* (SVD) in order to reduce the initial rating matrix size.

Unfortunately, such dimensionality reduction techniques then prevent us from presenting understandable results to the users because of the rating matrix transformation. So instead, we think it is interesting to study how bagging [27] could be used in collaborative filtering, and if using local item-based approaches in each cluster found using K-means still fails with huge datasets such as Netflix's one.

References

1. Adomavicius, G., Tuzhilin, A.: Toward the next generation of recommender systems: A survey of the state-of-the-art and possible extensions. IEEE Transactions on Knowledge and Data Engineering 17, 734–749 (2005)
2. Resnick, P., Iacovou, N., Suchak, M., Bergstrom, P., Riedl, J.: Grouplens: An open architecture for collaborative filtering of netnews. In: Conference on Computer Supported Cooperative Work, pp. 175–186. ACM Press, New York (1994)
3. Herlocker, J., Konstan, J., Terveen, L., Riedl, J.: Evaluating collaborative filtering recommender systems. ACM Transactions on Information Systems 22, 5–53 (2004)
4. McNee, S., Riedl, J., Konstan, J.: Being accurate is not enough: How accuracy metrics have hurt recommender systems. In: Extended Abstracts of the 2006 ACM Conference on Human Factors in Computing Systems, ACM Press, New York (2006)
5. Shardanand, U., Maes, P.: Social information filtering: Algorithms for automating word of mouth. In: ACM Conference on Human Factors in Computing Systems, vol. 1, pp. 210–217 (1995)
6. Weng, J., Miao, C., Goh, A., Shen, Z., Gay, R.: Trust-based agent community for collaborative recommendation. In: 5th International Joint Conference on Autonomous Agents and Multiagent Systems (2006)
7. Breese, J., Heckerman, D., Kadie, C.: Empirical analysis of predictive algorithms for collaborative filtering. In: 14th Conference on Uncertainty in Artificial Intelligence, pp. 43–52. Morgan Kaufman, San Francisco (1998)
8. Pennock, D., Horvitz, E., Lawrence, S., Giles, C.L.: Collaborative filtering by personality diagnosis: A hybrid memory- and model-based approach. In: 16th Conference on Uncertainty in Artificial Intelligence, pp. 473–480 (2000)
9. Kleinberg, J., Sandler, M.: Using mixture models for collaborative filtering. In: 36th ACM Symposium on Theory Of Computing, pp. 569–578. ACM Press, New York (2004)
10. Kelleher, J., Bridge, D.: Rectree centroid: An accurate, scalable collaborative recommender. In: Cunningham, P., Fernando, T., Vogel, C. (eds.) 14th Irish Conference on Artificial Intelligence and Cognitive Science, pp. 89–94 (2003)
11. Rashid, A.M., Lam, S.K., Karypis, G., Riedl, J.: ClustKNN: A highly scalable hybrid model- & memory-based CF algorithm. In: KDD Workshop on Web Mining and Web Usage Analysis (2006)
12. Domeniconi, C., Papadopoulos, D., Gunopulos, D., Ma, S.: Subspace clustering of high dimensional data. In: SIAM International Conference on Data Mining (2004)
13. Candillier, L., Tellier, I., Torre, F., Bousquet, O.: SSC: Statistical Subspace Clustering. In: Perner, P., Imiya, A. (eds.) MLDM 2005. LNCS (LNAI), vol. 3587, pp. 100–109. Springer, Heidelberg (2005)
14. Ungar, L., Foster, D.: Clustering methods for collaborative filtering. In: Workshop on Recommendation Systems, AAAI Press, Stanford, California (1998)
15. O'Conner, M., Herlocker, J.: Clustering items for collaborative filtering. In: ACM SIGIR Workshop on Recommender Systems, ACM Press, New York (1999)
16. Sarwar, B.M., Karypis, G., Konstan, J.A., Riedl, J.: Analysis of recommendation algorithms for e-commerce. In: ACM Conference on Electronic Commerce, pp. 158–167. ACM Press, New York (2000)
17. Lin, W., Alvarez, S., Ruiz, C.: Efficient adaptive-support association rule mining for recommender systems. In: Data Mining and Knowledge Discovery, vol. 6, pp. 83–105 (2002)

18. Karypis, G.: Evaluation of item-based top-N recommendation algorithms. In: 10th International Conference on Information and Knowledge Management, pp. 247–254 (2001)
19. Sarwar, B.M., Karypis, G., Konstan, J., Riedl, J.: Item-based collaborative filtering recommendation algorithms. In: 10th International World Wide Web Conference (2001)
20. Deshpande, M., Karypis, G.: Item-based top-N recommendation algorithms. ACM Transactions on Information Systems 22, 143–177 (2004)
21. Pazzani, M.J.: A framework for collaborative, content-based and demographic filtering. Artificial Intelligence Review 13, 393–408 (1999)
22. Vozalis, M., Margaritis, K.G.: Enhancing collaborative filtering with demographic data: The case of item-based filtering. In: 4th International Conference on Intelligent Systems Design and Applications, pp. 361–366 (2004)
23. NetflixPrize: (2006), http://www.netflixprize.com/
24. Yu, K., Xu, X., Tao, J., Ester, M., Kriegel, H.: Instance selection techniques for memory-based collaborative filtering. In: SIAM Data Mining (2002)
25. Goldberg, K., Roeder, T., Gupta, D., Perkins, C.: Eigentaste: A constant time collaborative filtering algorithm. Information Retrieval 4, 133–151 (2001)
26. Vozalis, M., Margaritis, K.: Applying SVD on item-based filtering. In: 5th International Conference on Intelligent Systems Design and Applications, pp. 464–469 (2005)
27. Breiman, L.: Bagging predictors. Machine Learning 24, 123–140 (1996)

Reducing the Dimensionality of Vector Space Embeddings of Graphs

Kaspar Riesen, Vivian Kilchherr, and Horst Bunke

Institute of Computer Science and Applied Mathematics, University of Bern,
Neubrückstrasse 10, CH-3012 Bern, Switzerland
{riesen,kilchher,bunke}@iam.unibe.ch

Abstract. Graphs are a convenient representation formalism for structured objects, but they suffer from the fact that only a few algorithms for graph classification and clustering exist. In this paper we propose a new approach to graph classification by embedding graphs in real vector spaces. This approach allows us to apply advanced classification tools while retaining the high representational power of graphs. The basic idea of our approach is to regard the edit distances of a given graph g to a set of training graphs as a vectorial description of g. Once a graph has been transformed into a vector, different dimensionality reduction algorithms are applied such that redundancies are eliminated. To this reduced vectorial data representation, pattern classification algorithms can be applied. Through various experimental results we show that the proposed vector space embedding and subsequent classification with the reduced vectors outperform the classification algorithms in the original graph domain.

1 Introduction

After many years of research, the fields of machine learning and data mining have reached a high level of maturity [1,2,3]. Powerful methods for clustering, classification, and other tasks have become available. However, the vast majority of these approaches rely on object representations given in terms of feature vectors. Such object representations have a number of useful properties. For example, object similarity, or distance, can be easily computed by means of the Euclidean distance or similar measures in the n-dimensional real space. Recently, however, a growing interest in graph-based object representation can be observed [4]. As a matter of fact, graph based object representations have a number of advantages over feature vectors. For example, graphs are able to represent not only the values of object properties, i.e. features, but can be used to explicitly model structural relations that may exist between different parts of an object. Furthermore, a graph may include an arbitrary number of nodes and edges. This is definitely more desirable than the case of feature vectors where the number of features, i.e. the dimensionality of the vector space, is fixed beforehand and one is confined to always using the same number of features regardless of the size or the complexity of the objects under consideration. There are many applications where the use of graphs as representation formalism is preferable over feature

P. Perner (Ed.): MLDM 2007, LNAI 4571, pp. 563–573, 2007.

vectors [5]. There are other applications, for example in computational chemistry or bioinformatics, where the use of a graph representation is indispensable [6,7].

Recently, various efforts have been made to bridge the gap between the domain of feature based and graph based object representations. In [8] an approach to graph embedding in vector spaces has been introduced. This method is based on algebraic graph theory and utilizes spectral matrix decomposition. Another approach for graph embedding has been proposed in [9]. It makes use of the relationship between the Laplace-Beltrami operator and the graph Laplacian to embed a graph onto a Riemannian manifold. In [10] an approach to graph embedding in an n-dimensional vector space by means of prototype selection and edit distance computation is described. The key idea of this approach is to use the distances of an input graph to a number of training graphs, termed prototype graphs, as vectorial description of the graph. An advantage of this method is the explicit use of graph edit distance, which allows us to deal with various kinds of graphs and utilize domains specific knowledge in defining the dissimilarity of nodes and edges through edit costs. However, an appropriate choice of the prototype graphs is one of the critical issues in this approach. A good selection seems to be crucial to succeed with the classification algorithm in the feature vector space. Different algorithms for solving this problem have been described [10,11,12]. However, it has turned out that there is no universally best method. The suitability of an algorithm for prototype selection depends on the underlying data set.

In this paper we propose a more principled way of embedding graphs in vector spaces. We also make use of graph edit distance to map graphs to vector spaces. In contrast to the above mentioned method, however, we avoid the difficult task of prototype selection by using the whole set of training graphs as prototypes. Of course, it can be expected that by using the whole training set as prototypes, we end up with feature vectors of very high dimensionality, which in turn may lead to redundancy and perhaps lower the performance as well as the efficiency of our algorithms in the features space. However, these problems may be overcome by feature selection. Popular methods for reducing the dimensionality of a feature space are Principal Component Analysis (PCA) and Fisher's Linear Discriminant Analysis (LDA) [13,14]. Both methods first combine the feature values and then project them onto a space of lower dimensionality. In this paper we make use of both approaches after we have mapped the considered graphs into a high dimensional feature space defined by the whole training set of graphs.

Our approach can be interpreted as a specified kind of graph kernel [15]. Graph kernels provide us with an embedding of the space of graphs into an inner product space. The main result of kernel theory is that it is not necessary to provide an explicit mapping from graphs to vectors as many algorithms can be formulated entirely in terms of inner products. Hence, knowing the value of the kernel function is sufficient. This procedure is termed *kernel trick* and has proved to be very powerful for classification tasks. In our case, the mapping from the graph domain to the feature space is explicitly given. Therefore, performing this mapping and subsequently computing scalar products represents a valid kernel function. Note, however, that the method proposed in this paper is even more flexible than graph kernels because we can also deal with algorithms that can't be kernelized, i.e. algorithms that need the image of a graph in the vector space in explicit form, or need functions in the feature space other than scalar products.

The rest of this paper is organized as follows. In Section 2, we describe our novel procedure of graph embedding and dimensionality reduction. Then a number of experimental results are presented, demonstrating the superior performance of the proposed method. Finally, we draw conclusions and discuss possible future work.

2 Graph Embedding and Subsequent Dimensionality Reduction

To embed an input graph in a vector space we assume a labeled set of training graphs, $T = \{g_1, \ldots, g_n\}$, and a dissimilarity measure $d(g_i, g_j)$ are given. In this paper we make use of *graph edit distance* as dissimilarity measure d [16,17]. The key idea of graph edit distance is to define the dissimilarity, or distance, of graphs by the amount of distortion that is needed to transform one graph into another. The distortions considered in this paper are insertions, deletions, and substitutions of nodes and edges. A sequence of edit operations that transforms a graph g_1 into another graph g_2 is called an *edit path* between g_1 and g_2. Costs are assigned to each individual edit operation and the cost of an edit path is the sum of the costs of its individual edit operations. Thus, the cost of an edit path represents the strength of the distortions of the corresponding edit sequence. Finally, the *edit distance* of two graphs is defined as the minimum cost, taken over all edit paths between two graphs under consideration.

Recently, it has been proposed to embed graphs in vector spaces [10]. The idea underlying this method was first developed for the embedding of real vectors in a dissimilarity space [11,12]. In our method, after having selected a set $P = \{p_1, \ldots, p_m\}$ of $m < n$ prototypes from T, we compute the dissimilarity of a graph $g \in T$ to each prototype $p \in P$. This leads to m dissimilarities, $d_1 = d(g, p_1), \ldots, d_m = d(g, p_m)$, which can be interpreted as an m-dimensional vector $(d_1, \ldots, d_m)$. In this way we can transform any graph from the training set, as well as any other graph from a validation or testing set, into a vector of real numbers. Note that whenever a graph from the training set, which has been choosen as a prototype before, is transformed into a vector $\mathbf{x} = (x_1, \ldots, x_m)$ one of the vector components is zero. Formally, if $T = \{g_1, \ldots, g_n\}$ is a training set of graphs and $P = \{p_1, \ldots, p_m\} \subseteq T$ is a set of prototypes, the mapping $t_m^P : T \to \mathbb{R}^m$ is defined as a function $t_m^P(g) \mapsto (d(g, p_1), \ldots, d(g, p_m))$ where $d(g, p_i)$ is the graph edit distance between the graph g and the i-th prototype.

One crucial question in this approach is how to find a subset P of prototypes that lead to a good performance of the classifier in the feature space. As a matter of fact, both the individual prototypes selected from T and their number have a critical impact on the classifier's performance. In [10,11,12] different prototype selection algorithms are discussed. It turns out that none of them is globally best, i.e. the quality of a prototype selector depends on the underlying data set. In this paper we propose a new approach where we use all available elements from the training set of prototypes, i.e. $P = T$ and subsequently apply dimensionality reduction methods. This process is much more principled than the previous approaches and allows us to completely avoid the dificult problem of heuristic prototype selection. For dimensionality reduction, we make use of the well known Principal Component Analysis (PCA) and Fisher's Linear Discriminant Analysis (LDA) [13,14].

2.1   Principal Component Analysis (PCA)

The Principal Component Analysis (PCA) [13,14] is a linear transformation. It seeks the projection which best represents the data. PCA is an unsupervised method which does not take any class label information into consideration. We first normalize the data by shifting the mean to the origin of the coordinate system and making the variance of each feature equal to one. Then we calculate the covariance matrix of the normalized data and determine the eigenvectors $\mathbf{e}_i$ and the eigenvalues λ_i of the covariance matrix. There exists one eigenvalue for each eigenvector. The eigenvectors are ordered according to decreasing magnitude of the corresponding eigenvalues, i.e. $\lambda_1 \geq \lambda_2 \geq \ldots \geq \lambda_n \geq 0$. The data is then represented in a new coordinate system defined by the eigenvectors. The eigenvectors are also called principal components. The first principal component points in the direction of the highest variance and, therefore, includes the most information about the data. The second principal component is perpendicular to the first principal component and points in the direction of the second highest variance and so on. For reducing the dimensionality of the transformed data we retain only the $m < n$ eigenvectors with the highest m eigenvalues. The transformation of a point in the original coordinate system into the new coordinate system is given in Eq. 1 where $\mathbf{x} = (x_{i1}, \ldots, x_{in})'$ denotes the untransformed vector of dimension n, and $\mathbf{y} = (y_{i1}, \ldots, y_{im})'$ the transformed vector which has dimension m $(m \leq n)$. The first row of the transformation matrix contains the values of the eigenvector $\mathbf{e}_1$, the second row the values of the eigenvector $\mathbf{e}_2$, and so on.

$$\begin{pmatrix} y_{i1} \\ y_{i2} \\ \vdots \\ y_{im} \end{pmatrix} = \begin{pmatrix} e_{11} & e_{12} & \cdots & e_{1n} \\ e_{21} & e_{22} & \cdots & e_{2n} \\ & & \vdots & \\ e_{m1} & e_{m2} & \cdots & e_{mn} \end{pmatrix} \begin{pmatrix} x_{i1} \\ x_{i2} \\ \vdots \\ x_{in} \end{pmatrix} \tag{1}$$

2.2   Fisher's Linear Discriminant Analysis (LDA)

Fisher's Linear Discriminant Analysis (LDA) [13,14] is a linear transformation as well. In contrast with PCA, LDA takes class label information into account. In its original form, LDA can be applied to two-class problems only. However, we make use of a generalization, called Multiple Discriminant Analysis (MDA), which can cope with more than two classes. In MDA, we are seeking the projection of the data which best separates the classes from each other. For this purpose, the expression given in Eq. 2 is maximized.

$$J(\mathbf{W}) = \frac{|\mathbf{W}^T \mathbf{S}_B \mathbf{W}|}{|\mathbf{W}^T \mathbf{S}_W \mathbf{W}|} \tag{2}$$

In this equation, $\mathbf{S}_B$ represents the between-class convariance matrix, $\mathbf{S}_W$ the within-class covariance matrix, and $\mathbf{W}$ represents the transpose of the transformation matrix. We first normalize the data by its mean and its variance. Then we calculate the matrices $\mathbf{S}_W$ and $\mathbf{S}_B$ and determine the eigenvectors and the eigenvalues of the matrix $\mathbf{S}_W^{-1}\mathbf{S}_B$. As matrix $\mathbf{S}_B$ has a maximal rank of $c - 1$, where c represents the number of classes, we have at most $c - 1$ different and nonzero eigenvalues and therefore, the transformed data points have a maximal dimensionality of $c - 1$. By means of the obtained eigenvectors and eigenvalues we get the transformed data points as described for the PCA in Subsection 2.1.

3   Experimental Results

The purpose of the experiments described in this section is to compare the classification accuracy of the proposed method with a reference system in the original graph domain. The classifier used in the vector space is a SVM with radial basis function kernel (RBF-kernel) [18,19]. This type of classifier has proven very powerful in various applications and has become one of the most popular classifiers in machine learning, pattern recognition, and related areas recently. As reference system, the k-nearest neighbor classifier in the graph domain is used. Note that as of today – up to very few exceptions, e.g. [20] – there exist no other classifiers for general graphs that can be directly applied in the graph domain.

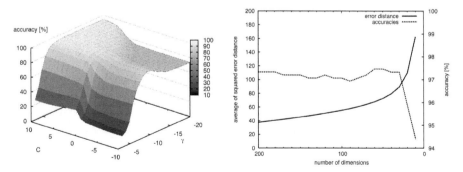

(a) Optimizing C and γ for a specific dimensionality.

(b) Optimizing the number of eigenvectors, i.e. the resulting dimensionality.

Fig. 1. SVM validation on the Letter data

In each of our experiments we make use of three disjoint graph sets, the *training set*, the *validation set* and the *test set*. The validation set is used to determine optimal parameter values for dimensionality reduction and for the classification task. Clearly, for the k-nearest neighbor classifier (reference system) the meta parameter k, i.e. the number of neighbors considered for classification, is optimized on the validation set. If we apply PCA based reduction, we have to determine the optimal dimensionality of the target space, i.e. the number of eigenvectors which have to be retained. For each dimensionality an individual SVM is trained. The RBF-kernel SVM used in this paper has parameters C and γ. C corresponds to the weighting factor for misclassification penalty and γ is used in our kernel function $K(\mathbf{u}, \mathbf{v}) = exp(-\gamma \cdot ||\mathbf{u} - \mathbf{v}||^2)$ [18,19]. These parameters are optimized, i.e. the parameter values that result in the lowest classification error on the validation set are applied to the independent test set. The procedure for the LDA transformed data differs from the validation on PCA data. First of all, the dimensionality is not validated for LDA reduction. We use the maximum possible dimension on each dataset, which is given by the number of classes minus one. Furthermore, for LDA it is more important to provide a large training set for transformation than optimizing the SVM parameter values. Hence, for LDA transformation we merge the validation and training set to one large set. Transformation is applied on this new set subsequently. Consequently, no validation set is available anymore and the standard parameter values for SVM classification are used.

3.1   Letter Database

The first database used in the experiments consists of graphs representing distorted letter drawings. In this experiment we consider the 15 capital letters of the Roman alphabet that consist of straight lines only (*A, E, F, ...*). For each class, a prototype line drawing is manually constructed. To obtain large sample sets of drawings with arbitrarily strong distortions, distortion operators are applied to the prototype line drawings. This results in randomly translated, removed, and added lines. These drawings are then converted into graphs in a simple manner by representing lines by edges and ending points of lines by nodes. Each node is labeled with a two-dimensional attribute giving its position. In Fig. 2 a graph representing the letter *A* is given under five distortion levels. The graph database used in our experiments consists of a training set, a validation set, and a test set, each of size 750 for each of a total of five different distortion levels. As mentioned above, there is no need for a validation set if one applies LDA transformation in our experiments. In this case, only two sets are necessary, a training set (1500 letters) and a test set (750 letters).

In Fig. 1 the PCA validation for all three parameter values is illustrated. An SVM for each dimensionality and each possible value of C and γ is trained. In Fig. 1 (a) we show the classification accuracy for various (C, γ) pairs on the validation set for one fixed value of the dimensionality. In Fig. 1 (b) (dashed line) the best classification results achieved on the validation set for various dimensionality values and the globally best (C, γ) pair are displayed. It turns out

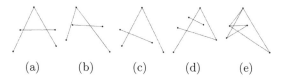

(a)   (b)   (c)   (d)   (e)

Fig. 2. Letter Database – Letter A under five distortion levels

(a) City   (b) Country   (c) People   (d) Streets   (e) Snowy

Fig. 3. Image Database – A sample image of the five classes

that only few dimensions are necessary to obtain good results. For instance, the accuracy is not drastically decreasing before the dimensionality is reduced to 20. Together with the accuracy a projection error curve is shown in the same figure (solid line). This curve displays the average of squared error distance of PCA reduced vectors as a function of the dimensionality m. This error distance for a PCA transformed vector $\mathbf{y}$ and subsequently dimensionality reduced vector $\hat{\mathbf{y}} = (y_1, \ldots, y_m, 0, \ldots, 0)$ is given by $|\mathbf{y} - \hat{\mathbf{y}}|$. As one expects, the error monotonically increases with a smaller number of dimensions. However, there is no clear cut-off point in the error curve. Hence, the optimal number of dimensions has to be found by means of a validation set.

Table 1. Letter Data: Classification accuracy in the graph and vector space

	Ref. System	Classifiers applied to reduced data			
Distortion	k-NN (graph)	PCA-SVM	(dim)	LDA-SVM	(dim)
0.1	98.27	98.53	(20)	99.07	(14)
0.3	97.60	98.40 ○	(30)	98.80 ○	(14)
0.5	94.00	97.20 ○	(30)	96.53 ○	(14)
0.7	94.27	95.20	(30)	94.80	(14)
0.9	90.13	93.73 ○	(100)	94.00 ○	(14)

○ Statistically significantly better than the reference system ($\alpha = 0.05$)

The parameter values that lead to the best recognition accuracy on the validation set are applied on the independent test set. The results of the experiments on the letter database are given in Table 1. In addition to the classification accuracy, the number of dimensions is indicated in bracktes for PCA-SVM and LDA-SVM. In Table 1, we observe that the new approach with PCA and LDA

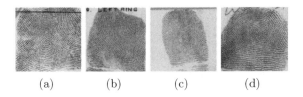

(a)   (b)   (c)   (d)

Fig. 4. NIST-4 fingerprint Database – A sample image of the classes *arch, left, right* and *whorl*

outperforms the reference system on all distortion levels. Note that six out of ten improvements are statistically significant. It is very interesting to note that in three out of five cases the results achieved by LDA are better than the results of PCA. But LDA uses only 14 dimensions on all distortion levels, while PCA reduction requires up to 100 dimensions.

3.2   Real World Data

For a more thorough evaluation of the proposed methods we additonally use three real world data sets. First we apply the proposed method to the problem of image classification. Images are converted into graphs by segmenting them into regions, eliminating regions that are irrelevant for classification, and representing the remaining regions by nodes and the adjacency of regions by edges [21]. The image database consists of five classes (*city, countryside, people, snowy, streets*) and is split into a training set, a validation set and a test set of size 54 each. In Fig. 3 a sample image of each class is given. The classification accuracies obtained by the different methods are given in the first row of Table 2. We note that both SVMs applied to the reduced data improve the accuracy compared to the reference system. But only the LDA reduced classifier achieves an improvement with statistical significance. Note that LDA uses only 4 dimensions while PCA makes use of 35 dimensions.

The second real world dataset is given by the NIST-4 fingerprint database [22]. We construct graphs from fingerprint images by extracting characteristic regions in fingerprints and converting the results into attributed graphs [23]. We use a validation set of size 300 and a test and training set of size 500 each. In this experiment we address the 4-class problem (*arch, left-loop, right-loop, whorl*). In Fig. 4 a fingerprint image of each class is given. The results achieved on this database are shown in the second row of Table 2. The SVM with the PCA reduced data achieves better results than the reference system, though not statistically significant. The accuracy achieved with LDA based SVM is lower than the reference system but uses only three dimensions. Compared to the PCA based SVM, where 150 dimensions are used, this is a vector space of very low dimensionality.

Finally, we apply the proposed method of graph embedding and subsequent SVM classification to the problem of molecule classification. To this end, we construct graphs from the AIDS Antiviral Screen Database of Active Compounds

[24]. Our molecule database consists of two classes (*active, inactive*), which represent molecules with activity against HIV or not. We use a validation set of size 250, a test set of size 1500 and training set of size 250. Thus, there are 2000 elements totally (1600 inactive elements and 400 active elements). The molecules are converted into graphs in a straightforward manner by representing atoms as nodes and the covalent bonds as edges. Nodes are labeled with the number of the corresponding chemical symbol and edges by the valence of the linkage. The results achieved on this database are shown in the third row of Table 2. The accuracy of the reference system in the graph domain is quite high, but it can be further improved by graph embedding and SVM classification. The PCA based SVM outperforms the reference system with statistical significance. The accuracy of the LDA based SVM is lower than the reference systems. Note, however, that only one dimension is used here. Taking this fact into account, an accuracy of 95.3% seems quite remarkable. For comparison, the PCA SVM uses 100 dimensions.

Table 2. Real World Data: Classification accuracy in the graph and vector space

	Ref. System	Classifiers applied to reduced data	
Database	k-NN (graph)	PCA-SVM (dim)	LDA-SVM (dim)
Image	57.4	61.1 (35)	68.5 ∘ (4)
Fingerprint	82.6	84.6 (150)	66.6 • (3)
Molecules	97.1	98.2 ∘ (100)	95.3 • (1)

∘ Statistically significantly better than the reference system ($\alpha = 0.05$)
• Statistically significantly worse than the reference systems ($\alpha = 0.05$)

4 Conclusions and Future Work

The main contribution of this paper is a general methodology for bridging the gap between statistical and structural pattern recognition. This is achieved through computing the graph edit distance to a training set of size n. As a result we obtain n real numbers which can serve as a high dimensional vectorial description of any given graph. In previous work about graph embedding, a selection of certain prototypes has to be done before graph embedding. However, the way of selecting these prototypes is critical. In the method proposed in this paper we avoid the difficult task of prototype selection by taking all available graphs from the training set as prototypes and reduce the dimensionality by applying the mathematically well founded dimensionality reduction algorithms PCA and LDA. This step can be interpreted as a sort of delayed prototype selection. With several experimental results we show that the performance of a k-nearest neighbor classifier in the graph domain, used as a reference system, can be outperformed with statistical significance. In case of classification problems with many classes, the LDA based system is preferable, while for a small number of classes the PCA based system is the method of choice. In either case, the new method allows us to combine the high representational power of graph representations with the good performance of SVM classifiers in vector spaces.

In future work we will investigate whether one can make further improvements in classification accuracy with kernelized versions of the dimensionality reduction algorithms. We will also study the problem of graph clustering using the proposed embedding method.

Acknowledgements

This work has been supported by the Swiss National Science Foundation (Project 200021-113198/1). Furthermore, we would like to thank R. Duin and E. Pekalska for valuable discussions and hints regarding our embedding methods. Finally, we thank B. Le Saux for making the image database available to us.

References

1. Perner, P. (ed.): ICDM 2006. LNCS (LNAI), vol. 4065. Springer, Heidelberg (2006)
2. Perner, P., Rosenfeld, A. (eds.): MLDM 2003. LNCS, vol. 2734. Springer, Heidelberg (2003)
3. Perner, P., Imiya, A. (eds.): MLDM 2005. LNCS (LNAI), vol. 3587. Springer, Heidelberg (2005)
4. Cook, D., Holder, L. (eds.): Mining Graph Data. Wiley, Chichester (2007)
5. Conte, D., Foggia, P., Sansone, C., Vento, M.: Thirty years of graph matching in pattern recognition. Int. Journal of Pattern Recognition and Artificial Intelligence 18(3), 265–298 (2004)
6. Borgwardt, K., Ong, C., Schönauer, S., Vishwanathan, S., Smola, A., Kriegel, H.-P.: Protein function prediction via graph kernels. Bioinformatics 21(1), 47–56 (2005)
7. Ralaivola, L., Swamidass, S.J., Saigo, H., Baldi, P.: Graph kernels for chemical informatics. Neural Networks 18(8), 1093–1110 (2005)
8. Wilson, R.C., Hancock, E.R., Luo, B.: Pattern vectors from algebraic graph theory. IEEE Trans. on Pattern Analysis ans Machine Intelligence 27(7), 1112–1124 (2005)
9. Robles-Kelly, A., Hancock, E.R.: A riemannian approach to graph embedding. Pattern Recognition 40, 1024–1056 (2007)
10. Riesen, K., Neuhaus, M., Bunke, H.: Graph embedding in vector spaces by means of prototype selection. Accepted for the 6th Int. Workshop on Graph-Based Representations in Pattern Recognition
11. Pekalska, E., Duin, R., Paclik, P.: Prototype selection for dissimilarity-based classifiers. Pattern Recognition 39(2), 189–208 (2006)
12. Duin, R., Pekalska, E.: The Dissimilarity Representations for Pattern Recognition: Foundations and Applications. World Scientific, Singapore (2005)
13. Duda, R., Hart, P., Stork, D.: Pattern Classification, 2nd edn. Wiley-Interscience, Chichester (2000)
14. Bishop, C.: Neural Networks for Pattern Recognition. Oxford University Press, Oxford (1996)
15. Gärtner, T.: A survey of kernels for structured data. SIGKDD Explorations 5(1), 49–58 (2003)

16. Bunke, H., Allermann, G.: Inexact graph matching for structural pattern recognition. Pattern Recognition Letters 1, 245–253 (1983)
17. Sanfeliu, A., Fu, K.S.: A distance measure between attributed relational graphs for pattern recognition. IEEE Transactions on Systems, Man, and Cybernetics (Part B) 13(3), 353–363 (1983)
18. Shawe-Taylor, J., Cristianini, N.: Kernel Methods for Pattern Analysis. Cambridge University Press, Cambridge (2004)
19. Schölkopf, B., Smola, A.: Learning with Kernels. MIT Press, Cambridge (2002)
20. Bianchini, M., Gori, M., Sarti, L., Scarselli, F.: Recursive processing of cyclic graphs. IEEE Transactions on Neural Networks 17(1), 10–18 (2006)
21. Le Saux, B., Bunke, H.: Feature selection for graph-based image classifiers. In: Marques, J., Pérez de la Blanca, N., Pina, P. (eds.) IbPRIA 2005. LNCS, vol. 3523(Part II), pp. 147–154. Springer, Heidelberg (2005)
22. Watson, C.I., Wilson, C.L.: NIST special database 4, fingerprint database. National Institute of Standards and Technology (March 1992)
23. Neuhaus, M., Bunke, H.: A graph matching based approach to fingerprint classification using directional variance. In: Kanade, T., Jain, A., Ratha, N.K. (eds.) AVBPA 2005. LNCS, vol. 3546, pp. 191–200. Springer, Heidelberg (2005)
24. Development Therapeutics Program DTP. Aids antiviral screen (2004)
    http://dtp.nci.nih.gov/docs/aids/aids_data.html

PE-PUC: A Graph Based PU-Learning Approach for Text Classification

Shuang Yu and Chunping Li

School of Software, Tsinghua University, Beijing 100084, China
yushuang@mails.tsinghua.edu.cn,
cli@tsinghua.edu.cn

Abstract. This paper presents a novel solution for the problem of building text classifier using positive documents (P) and unlabeled documents (U). Here, the unlabeled documents are mixed with positive and negative documents. This problem is also called PU-Learning. The key feature of PU-Learning is that there is no negative document for training. Recently, several approaches have been proposed for solving this problem. Most of them are based on the same idea, which builds a classifier in two steps. Each existing technique uses a different method for each step. Generally speaking, these existing approaches do not perform well when the size of P is small. In this paper, we propose a new approach aiming at improving the system when the size of P is small. This approach combines the graph-based semi-supervised learning method with the two-step method. Experiments indicate that our proposed method performs well especially when the size of P is small.

1 Introduction

Text classification is the technique of automatically assigning categories or classes to unlabeled documents. With the ever-increasing volume of text documents from various online sources, an automatic text classifier can save considerable time and human labor. Recently, a new direction of text classification problem becomes recognized, which is called PU-Learning [6] [7] [5] [8]. P represents the given labeled positive set; U represents the given unlabeled set, which is mixed with positive and negative documents. Usually, the positive set contains documents from a special topic and the negative set contains documents from diverse topics. PU-Learning is a special problem of text classification, where classifiers are built using labeled positive documents and unlabeled documents.

PU-Learning is of great use in the task of accurately labeling documents as positive and negative with respect to a special class. It is particularly useful when the user wants to find positive documents from many text collections or sources. For example, a student is interested in the field of text classification and has collected some papers of this field from ICML, now he wants to find papers of text classification from ICDM. At this time, PU-Learning is helpful. The papers collected from ICML are positive documents (P), all the papers in the ICDM are unlabeled documents (U).

P. Perner (Ed.): MLDM 2007, LNAI 4571, pp. 574–584, 2007.

With the help of a PU-Learning system, the user can get the papers he wants from ICDM automatically.

Recently several approaches have been proposed for solving the PU-Learning problem, such as typically the two-step methods such as PNB and PNCT [8] [3]. However, current existing methods can not perform well in some cases. Experiments show that especially when the labeled positive set P for training is relative small, the classification result is not satisfactory. The main reason is due to the uniqueness of PU-Learning: 1) a large portion of training documents is unlabeled and no labeled negative documents are given; 2) the positive class contains documents from a special topic while the negative class contains documents from diverse topics. When the size of P is small, it can hardly reflect the true feature distribution of the positive class. Small P and the high diversity of the negative class will make building a good classifier extremely difficult [8].

Graph based semi-supervised learning [11] is usually effective for the classification task in the case of small size of labeled training. This kind of methods assumes label smoothness over the graph. In other words, they are smooth with respect to the intrinsic structure revealed by the given labeled and unlabeled data. Current graph-based methods mainly include spectral methods [4], random walks [9], graph mincuts [1], Gaussian random field and harmonic functions [12], etc. The characteristics of the graph based semi-supervised motivate us think of using this kind of methods to solve the PU-Learning problem with small positive dataset P. In this paper, we propose a novel method aiming at solving the PU-Learning problem when the given positive dataset P is lacking. To overcome the difficulty caused by small size of positive dataset, we combine the graph-based method with classical two-step methods of PU-Learning in an effective way, and further present our approach called PE-PUC for constructing the *positive document enlarging PU classifier*.

The organization of the paper is as follows. In Section 2 we introduce the background of the two-step method for PU-Learning and the graph-based semi supervised learning. In Section 3, we present our PE-PUC approach by combining the graph based method to solve the PU-Learning problem with respect to text classification. In Section 4, we give the evaluation of our PE-PUC approach with experimental results. In Section 5, we have the concluding remarks and the future work.

2 PU-Learning and Graph Based Semi-supervised Learning

2.1 Two-Step Method of PU-Learning

The given training data for PU-Learning is the labeled positive dataset P and the unlabeled dataset U. The key feature of PU-Learning is that there is no labeled negative data for training, which makes the task of building classifier challenging. One class of algorithm for solving this problem is based on a two-step strategy.

Step 1: Identify a set of reliable negative documents RN, from the given unlabeled dataset U. In this step, several techniques can be used, such as naïve Bayesian approach, spy technique, 1-DNF and Rocchio algorithm, etc.

Step 2: Build a set of classifiers by applying a classification algorithm iteratively using the given labeled positive documents P, the extracted negative documents RN and the remaining unlabeled documents U-RN; at last, select a good classifier from the set. In this step, Expectation Maximization (EM) algorithm and Support Vector Machine (SVM) usually are used.

2.2 Graph-Based Semi-supervised Learning

Graph-based semi-supervised learning [11] [12] considers the problem of learning with labeled and unlabeled data. The problem can be described as follows.

Given a point set $\chi = \{x_1,...,x_l,x_{l+1},...,x_{l+n}\} \subset \mathbb{R}^m$ and a label set $C = \{1,...,c\}$, the first l points of χ are labeled as $y_i \in C$, here, each class of C at least has one point. The remaining n points are unlabeled. The task is to predict the label of unlabeled points. The graph-based method using the concept and characteristic of graph, compute the similarities between nodes and propagate according to a given rule until reach a global stable state. The points with high similarity are considered to have the same label.

3 PE-PUC Approach: Positive Document Enlarging PU Classifier

As indicated in Section 1, current two-step methods cannot work well when P is small. In order to solve the PU-Learning problem, the two-step methods first extract a set of reliable negative documents from U in Step 1. The key requirement for this step is that the identified negative documents from the unlabeled set must be reliable and relative pure, that is to say, with no or very few positive documents in RN. If not so, too many noisy documents will damage the performance of classifier, which is built in Step 2. When the size of P is small, P is too small to reflect the true feature distribution of the positive class. In the two-step methods, whatever technique we use in Step 1, it is difficult to get reliable RN. In other words, after Step 1, many positive documents may be extracted from U as negative ones and put into RN. In Step 2, the noisy RN and the small P will make it impossible to build good classifiers.

Our PE-PUC method proposes a solution of the PU-Learning problem with small P. Intuitively, if we can extract some positive documents from U to enlarge P, we will possibly extract RN with high precision in Step 1. However, it is difficult to extract positive documents from U because: 1) U is large in size and high in diversity; 2) only a small portion of U is positive. It is difficult to avoid importing some negative documents into P when enlarging P. Those noisy documents can not improve the system but make it even poorer. To enlarge P with high precision, we present the PE-PUC algorithm using the graph-based semi-supervised techniques with main steps in Figure 1.

PE-PUC (P,U)
Input: the given labeled positive documents, P, the given unlabeled documents, U;
Output: PU Classifier
1.   Based on P, extract a set of negative documents, RN, from U;
2.   Enlarge P: Extract a set of reliable positive documents, RP, from U-RN;
3.   P'= P∪RP, U'=U-RP, extract a set of negative documents, RN', from U';
4.   Build the final classifier using P', RN' and U'-RN'.

Fig. 1. PE-PUC algorithm

3.1 Extracting Negative Documents from U

We use the naïve Bayesian method to extract negative documents RN from U and get the remaining unlabeled dataset U-RN. The detail of the procedure is shown in Figure 2.

The reason for labeling each document in U with the class label "-1" is that the proportion of positive documents in U is usually very small. In order to build the naïve Bayesian classifier, we firstly assume U is negative. Since naïve Bayesian method can tolerate some noise, this assumption is feasible.

Extract RN (P, U)
Input: the given labeled positive documents, P, the given unlabeled documents, U;
Output: a set of reliable negative documents, RN, a set of remaining unlabeled documents, U-RN.
1.   Label each document in P with the class label 1;
2.   Label each document in U with the class label -1;
3.   Build a naïve Bayesian classifier, NB-C, using P and U;
4.   Classify U using NB-C;
5.   RN ← documents which are classified as negative; U-RN ← documents which are classified as positive;

Fig. 2. Algorithm for extracting reliable negative documents

3.2 Enlarge P: Extracting RP from U-RN

In order to solve the PU-Learning problem with small P, we try to enlarge P by extracting some reliable positive documents from U-RN. Now we give the detail of this procedure.

Given a point set $\chi = \{x_1,...,x_l,x_{l+1},...,x_{l+n}\} \subset \mathbb{R}^m$, the first l points of χ are labeled positive documents; the remaining points of χ are unlabeled documents which are to be ranked according to their relevance to the labeled positive documents. Let $d: \chi \times \chi \to \mathbb{R}$ denotes a matrix on χ, this matrix assigns each pair x_i, x_j the distance $d(x_i, x_j)$, and $f : \chi \to \mathbb{R}$ denotes a ranking function which assigns each data point of χ a ranking score. Finally, we define a vector $y = [y_1, \ldots y_{l+n}]^T$, in which $y_1, \ldots y_l = 1$, referring to the labeled positive documents, and $y_{l+1}, \ldots y_{l+n} = 0$, referring to the unlabeled documents.

Graph based semi-supervised learning is an effective approach to deal with small size of labeled training for the purpose of classification. But for PU-Learning based classification, the problem is that we don't have any negative documents for propagation. Thus, an improved graph-based algorithm for extracting RP from U-RN is proposed as shown in Figure 3. An intuitive description of the algorithm is to randomly select a set of positive documents from P and put them into PL, which is used as the seeds for propagation, and then a weighted graph is formed which takes each point in PL∪ (U-RN) as a vertex. A positive ranking score to each point in PL is further assigned while zero to the remaining ones, and all the points then spread their scores to the nearby points via the weighted graph. This spread process is repeated until a global stable state is reached, and all the points except the seed points will have their own scores according to which they will be ranked. The resultant ranking score of an unlabeled document in U-RN is in proportion to the probability that it is relevant to the positive class, with large ranking score indicating high probability. So, at last, we can choose a number of the top ranked documents as reliable positive documents and use them to enlarge P.

Enlarge P (P, U-RN, λ)

Input: a set of positive documents, P, a set of unlabeled documents, U-RN, the percentage of U-RN which will be extracted as positive documents, λ , $\lambda \in (0,1)$;

Output: a set of positive documents, RP;

1. RP $\leftarrow \varnothing$, n $\leftarrow$ the number of documents in U-RN;

2. Randomly select l documents from P and put them in PL;

3. Form the affinity matrix W , $W_{ij} = \exp\left[- \| x_i - x_j \|^2 / 2\sigma^2 \right]$ if $i \neq j$, $W_{ii} = 0$;

4. Symmetrically normalize W by $S = D^{1/2} W D^{1/2}$. D is the diagonal matrix with (i,i) -element equal to the sum of the ith row of W ;

5. $f^* = (1-\alpha)(1-\alpha S)^{-1} y$, $\alpha \in (0,1)$. Rank each document $x_i, i \in [l+1, l+n]$ according to the ranking score in f^* (largest ranked first);

6. RP $\leftarrow$ the top ranked documents in U-RN ($|RP| = \lceil \lambda \times |U - RN| \rceil$)

Fig. 3. Algorithm for enlarging P

But when P is extremely small, only several labeled positive documents are known for training. In this case, just extracting RP from U-RN to enlarge P may not improve the performance distinctly. Here we propose a repeated extraction approach to take place of the second step in PE-PUC, namely, enlarging P repeatedly. The procedure of repeated extraction is shown in Figure 4.

Repeated Extraction
Input: the given labeled positive documents, P, the given unlabeled documents, U, the number of iteration, m, $\Lambda = \{\lambda_1,...,\lambda_i,...,\lambda_m\}$;
Output: a set of reliable positive documents, RP;
1: **for** $i = 1: m$ **do**
2: get U-RN from Extract RN (U, P);
3: get RP from Enlarge P (P, U-RN, λ_i);
4: P←P∪RP, U←U-RP, $i = i + 1$;
5: **end for**
6: **return** RP

Fig. 4. Algorithm for repeated extraction

In PU-Learning problem, the negative class consists of diverse topics. It is the diversity that makes it difficult to extract RP from U. Thus, the main issue is to find a way to deal with the problem of diversity. The key to semi-supervised learning problem is the prior assumption of consistency: 1) nearby points are likely to have the same label; 2) points on the same structure (such as a cluster) are likely to have the same label. The classifying function, which is constructed by the graph-based method, is sufficiently smooth with respect to the intrinsic structure revealed by the given labeled and unlabeled data. Using this method to extract RP, the propagation of ranking score reflects the relationship of all the data points (each document in PL and U-RN now is looking as a point in the graph), since in the feature space, distant points will not have similar ranking scores unless they belong to the same cluster consisting of many points that help to link the distant points, and nearby points will have similar ranking scores unless they belong to different clusters. As we use positive documents as seeds for propagation, so after convergence, the documents with higher-ranking scores are more likely positive documents.

Another reason that we adopt the graph-based method is that it needs few labeled documents for propagation. This characteristic accords with our situation when the size of P is small. No matter how small |P| is, this method is relative feasible.

In this step, we extract positive documents from U-RN but not from U. This is reasonable. The given unlabeled set U is mixed with positive and negative documents. Usually the proportion of positive documents in U is quite small, and the negative documents are of high diversity, so it is difficult to extract positive documents from U with high precision. Moreover, the number of documents in U is quite large, which will make the computation complicated and time consuming. According to our experiments, when P is small, most of the negative documents in U are extracted into RN, and a lot of positive documents in U are also selected into RN. In other words, RN is of high recall but low precision. Under this circumstance, the number of documents in U-RN is much smaller than the number of documents in U, so the computation of the graph-based method is easy. In addition, the proportion of positive documents in U-RN is much larger than the proportion of positive documents in U, which makes the extraction with high precision possible.

3.3 Build the Final Classifier

In the process of enlarging P, a reliable positive set RP is extracted from U-RN and added to P. Then, we use P' and U' as the new input, and get the newly extracted negative documents, which is defined as set RN'. The final classifier is built based on P', RN' and U'-RN'. In our work, we use two techniques to build the final classifier, one is based on the naïve Bayesian method and the other is based on the Expectation-Maximization (EM) algorithm.

For the naïve Bayesian method, we directly build the final classifier with P' and RN'. For the EM algorithm, we build a set of classifiers using P', RN' and U'-RN'. EM iteratively runs naïve Bayesian algorithm to revise the probabilistic label of each document in set U'-RN'. The iteration of EM at each time generates a naïve Bayesian classifier. After convergence, we can get several classifiers. Since it is not easy to catch the best classifier, we choose the better one between the first classifier and the classifier at convergence as the final result.

4   Experiment and Evaluation

4.1   Experiment Setup

In the experiment, we use the 20 Newsgroups[1] dataset. For the 20newsgroup, there are totally 20 different classes, where each class contains about 1,000 documents. We use each newsgroup as the positive set and the rest of the 19 groups as the negative set, which creates 20 datasets. For each dataset, 30% of the documents are randomly selected as test documents, the rest (70%) are used as training documents. The training datasets are selected as follows. γ% of the documents from the positive class is first selected as the positive set P. The rest of the positive documents and negative documents are mixed to form the unlabeled set U. Our work focuses on the situation when |P| is small, so γ is ranged from 1% to 10% for evaluating our method.

In our experiment, we use NB-NB and NB-EM as the baseline systems which are adopted in [8]. In the process of enlarging P, 10 documents are randomly selected from P as the seeds for propagation. We compute the affinity matrix W with $\sigma = 1.0$ and iteration with $\alpha = 0.99$. The number of positive documents selected from U-RN is set according to |U-RN| and the parameter λ. We test different λ settings to get a better result. In the last step, naïve Bayesian and EM algorithms are used to build the final classifier, which are represented as PE-PUC-NB and PE-PUC-EM, respectively. We use the popular F-score on the positive class as the evaluation measure.

4.2   Result Evaluation

The PE-PUC Method Can Give Better Results When P Is Small
Table 1 is the average of F-scores of the 20 datasets for each γ setting. Columns 2 and 3 show the results of the baseline systems. Columns 4 and 5 show the results of our PE-PUC approach. The comparative result of the experiment is shown in Figure 5.

---

[1] http://www.cs.cmu.edu/afs/cs/project/theo-11/www/naive-bayes/20_newsgroups.tar.gz

Table 1. The results of Newsgroup20

$\gamma\%$	NB-NB	NB-EM	PE-PUC	
			NB	EM
1	0.063	0.429	0.272	0.451
2	0.155	0.499	0.474	0.512
3	0.192	0.511	0.538	0.538
4	0.253	0.524	0.597	0.597
5	0.321	0.530	0.648	0.648
6	0.370	0.531	0.625	0.625
7	0.421	0.568	0.611	0.627
8	0.464	0.590	0.630	0.666
9	0.497	0.599	0.642	0.679
10	0.530	0.625	0.657	0.690

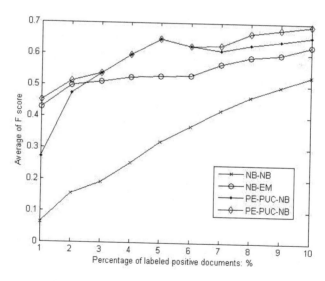

Fig. 5. Experiment results for the 20Newsgroup

The results indicate that our PE-PUC method performs better than the baseline systems significantly when P is small. For some cases, using the EM algorithm to build the final classifier can boost the systems. However, for the other cases, EM gives the same result as NB. As we use a classifier selection mechanism with the EM algorithm, which is able to select the first classifier if it is better than the one at convergence, so we can see, for some instances, the iteration of EM algorithm cannot boost the systems but degraded them.

Analysis of the Enlarging P Procedure

In our work, the graph-based method is used to extract positive documents from U-RN. Now we further analyze the function of this procedure according to the experimental results.

(1)   The number of positive documents extracted from U-RN affects the performance of the system.

We set the size of RP according to the number of documents in U-RN and the parameter λ, where $|RP| = \lceil \lambda \times |U - RN| \rceil$. Table 2 gives the results of different λ settings for $\gamma = 9\%$ in term of F-score. Due to the space limitation, here we just list out the results of the four classes. The results of the other classes behave in the similar way. From the results, we can observe that different λ produces different results.

Table 2. $\gamma = 9\%$, Results of different λ settings

| P \ λ | PE-PUC | | | | | |
| | NB-NB | | | NB-EM | | |
	0.6	0.7	0.8	0.6	0.7	0.8
Crypt	0.921	0.918	0.909	0.921	0.918	0.909
Electronics	0.437	0.449	0.461	0.576	0.578	0.580
Med	0.396	0.417	0.388	0.396	0.417	0.388
Space	0.601	0.641	0.673	0.781	0.785	0.790

(2) Enlarging P can help to extract negative documents with higher precision.

As indicated in Section 3, the key requirement for the extraction of RN is high precision, which is a main problem when P is small. Table 3 gives the results of the precision of RN, which is the average of 20 datasets for $\gamma = 10\%$. As we can see from Table 3, the enlarging P procedure can help to extract negative documents with higher precision.

Table 3. $\gamma = 10\%$, Precision of RN

| Method | PE-PUC | | | Two-step |
λ	$\lambda = 0.6$	$\lambda = 0.7$	$\lambda = 0.8$	
PrecisionOfRN	0.9771	0.9786	0.9756	0.9703

(3) Effectiveness of the repeated extraction approach

Another phenomenon shown in our experiment is that when the number of positive documents is extremely small, e.g. $\gamma \leq 5\%$, the number of documents in U-RN will be very small. The reason is that when P is extremely small, P is too small to represent the distribution of the positive class, so most of the documents in U will be extracted into RN as negative ones. In this case, the number of positive documents can be extracted from U-RN is small, which limits the performance of our PE-PUC approach. To solve this problem, we conceive the repeated extraction approach to gradually enlarge P. From Table 4 shows that our approach is effective when P is extremely small. The value in the form is the average of F-scores of the 20 datasets for each γ setting.

Table 4. The results of PE-PUC with Repeated Extraction Approach

$\gamma\%$	$m = 1$		$m = 2$		m=3	
	NB	EM	NB	EM	NB	EM
1	0.187	0.437	0.213	0.425	0.272	0.451
2	0.321	0.510	0.389	0.501	0.474	0.512
3	0.392	0.518	0.522	0.522	0.538	0.538
4	0.474	0.533	0.597	0.597	0.535	0.535
5	0.563	0.563	0.648	0.648	0.600	0.600

5 Conclusion and Future Work

In this paper, we present a novel approach called PE-PUC to solve the PU-Learning problem when the positive dataset P is small. PU-Learning refers to the problem of learning a classifier from positive and unlabeled data. A typical kind of method for solving this problem is a so called two-step method. However, the two-step method cannot perform well when the positive dataset P is small. In our PE-PUC approach, the graph-based method is combined with the two-step method, which is used to extract some reliable positive documents from the unlabeled dataset to enlarge P. A comprehensive evaluation shows that our PE-PUC approach outperforms current existing PU-Learning algorithms especially when positive dataset is small.

In the future work, our research will further focus on the parameter selection, namely, to effectively determine the most suitable λ and m settings in a pure mechanical way with respect to different datasets.

Acknowledgments. This work was finished with the supports of China 973 Research Project under Grant No. 2002CB312006.

References

[1] Blum, A., Chawla, S.: Learning from labeled and unlabeled data using graph minicuts. In: Proceedings of the 18th International Conference on Machine Learning, pp. 19–26 (2001)

[2] Denis, F., Gilleron, R., Tommasi, M.: Text classification and co-training from positive and unlabeled examples. In: Proceedings of the ICML-03 Workshop on Continuum from Labeled to Unlabeled Data, pp. 80–87 (2003)

[3] Denis, F., et al.: Learning from positive and unlabeled examples. Journal of Theoretical Computer Science 1(248), 70–83 (2005)

[4] Joachims, T.: Transductive learning via spectral graph partitioning. In: Proceedings of the 20th International Conference on Machine Learning, pp. 290–297 (2003)

[5] Lee, W.S., Liu, B.: Learning with positive and unlabeled examples using weighted logistic regression. In: Proceedings of the Twentieth International Conference on Machine Learning, 448–455 (2003)

[6] Li, X., Liu, B.: Learning to classify text using positive and unlabeled data. In: Proceedings of the 18th International Joint Conference on Artificial Intelligence, pp. 587–594 (2003)

[7] Liu, B., Lee, W.S., Yu, P., Li, X.: Partially supervised classification of text documents. In: Proceedings of the 19th International Conference on Machine Learning, pp. 387–394 (2002)

[8] Liu, B., et al.: Building text classifiers using positive and unlabeled examples. In: Proceedings of the Third IEEE International Conference on Data Mining, pp. 179–188 (2003)

[9] Szummer, M., Jaakkola, T.: Partially labeled classification with Markov random walks. Advances in Neural Information Processing Systems, 945–952 (2002)

[10] Yu, H., Han, J., Chang, K.: PEBL: Positive example based learning for Web page classification using SVM. In: Proc. ACM SIGKDD Int'l Conf. Knowledge Discovery in Databases, pp. 239–248 (2002)

[11] Zhou, D., et al.: Learning with local and global consistency. Advances in Neural Information Processing Systems, 321–328 (2003)

[12] Zhu, X., Ghahramani, Z., Lafferty, J.: Semi-supervised learning using Gaussian fields and harmonic functions. In: Proceedings of the 20th International Conference on Machine Learning, pp. 912–919 (2003)

Efficient Subsequence Matching Using the Longest Common Subsequence with a Dual Match Index

Tae Sik Han[1], Seung-Kyu Ko[1,*], and Jaewoo Kang[2,**]

[1] Dept. of Computer Science, North Carolina State University
Raleigh, NC 27569, USA
[2] Dept. of Computer Science and Engineering, Korea University, Seoul 136-705, Korea
kangj@korea.ac.kr

Abstract. The purpose of subsequence matching is to find a query sequence from a long data sequence. Due to the abundance of applications, many solutions have been proposed. Virtually all previous solutions use the Euclidean measure as the basis for measuring distance between sequences. Recent studies, however, suggest that the Euclidean distance often fails to produce proper results due to the irregularity in the data, which is not so uncommon in our problem domain. Addressing this problem, some non-Euclidean measures, such as *Dynamic Time Warping (DTW)* and *Longest Common Subsequence (LCS)*, have been proposed. However, most of the previous work in this direction focused on the whole sequence matching problem where query and data sequences are the same length. In this paper, we propose a novel subsequence matching framework using a non-Euclidean measure, in particular, *LCS*, and a new index query scheme. The proposed framework is based on the Dual Match framework where data sequences are divided into a series of disjoint equi-length subsequences and then indexed in an R-tree. We introduced similarity bound for index matching with *LCS*. The proposed query matching scheme reduces significant numbers of false positives in the match result. Furthermore, we developed an algorithm to skip expensive *LCS* computations through observing the warping paths. We validated our framework through extensive experiments using 48 different time series datasets. The results of the experiments suggest that our approach significantly improves the subsequence matching performance in various metrics.

Keywords: Subsequence matching, Longest Common Subsequence, Dual Match.

* He was supported by the IT Scholarship Program supervised by Institute for Information Technology Advancement and Ministry of Information and Communication in Republic of Korea.
** Corresponding author. His work was partially supported by the Microsoft Bioinformatics Award and the Korea University Research Grant.

P. Perner (Ed.): MLDM 2007, LNAI 4571, pp. 585–600, 2007.
© Springer-Verlag Berlin Heidelberg 2007

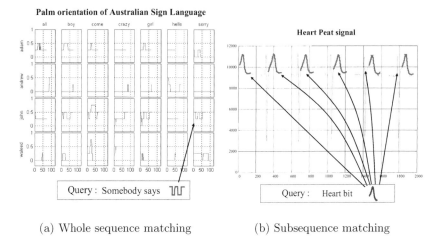

(a) Whole sequence matching   (b) Subsequence matching

Fig. 1. Whole sequence matching and Subsequence matching

1   Introduction

One of the basic problems in handling time series data is locating a pattern of interest from the long sequence of input data [1,2,7]. The sequence matching problem is largely classified into two categories: whole sequence matching and subsequence matching. Whole sequence matching involves finding, from the dataset, all sequence entries whose lengths are equal to the query and that fall within the similarity threshold specified by the user. For example, Figure 1(a) illustrates the whole sequence matching using the sign language palm orientation example. It shows the palm orientation readings from four different people (rows) using Australian Sign Language saying seven different words (columns)[4]. Each word from different signers has the same length and is searched for a given query.

Subsequence matching finds all subsequences from a longer data sequence that matches to the query. Figure 1(b) shows an example. It shows a short query sequence, one heart beat signal, and all matching regions from the longer data sequence. Subsequence matching is a more general problem than the whole sequence matching problem. However, most of the previous work has focused on the whole sequence matching problem [1,5,11]. While applying whole sequence matching techniques to the subsequence matching can be possible through GEMINI [2] framework, the application is not straightforward when non-Euclidean distance measures are used. Euclidean measure is sensitive to noise and due to the irregular nature of the data in sequence applications (e.g., moving object trajectories, query-by-humming, etc.), non-Euclidean measures are often desirable. The non-Euclidean distance measures such as DTW (*Dynamic Time Warping*) and LCS (*Longest Common Subsequence*) address some of the problems that Euclidean measure has [5,10].

In this work, we propose an efficient index searching framework for subsequence matching using LCS. We choose LCS because it is known to be more robust to the noise in the data than DTW [3,9] and yet to the best of our knowledge no previous work has considered it in the context of subsequence matching. We made the following contributions:

- We proposed a subsequence matching framework that employs a non-Euclidean distance measure LCS. It is for a more intuitive matching performance.
- We formally introduced the criteria for pruning the search space when using time series index with LCS similarity function.
- We introduced a new index query scheme, *multiple window sliding*, where several adjacent windows are queried and aggregated in order to improve the query performance.
- We proposed a new index search scheme that enables us to skip unnecessary similarity computations for the consecutive matching subsequences.

2 Background and Related Work

2.1 Notational Convenience

In order to state the problem and concepts clearly, we define some notations and terminologies in Table 1. In our work, we assume that a time series is a totally ordered set of real numbers and each real number element is collected from a single channel sensor device. A subsequence is a subset of a time series in contiguous time stamps.

Table 1. The basic notation

B	A time series data sequence, $< b_1, b_2, \ldots >$, each b_i is a real number at the ith time stamp.
$|B|$	Length of the sequence B
B_i	The ith subsequence of B when B is divided into disjoint subsequences of an equal length
Q	A query sequence, usually $|Q| \ll |B|$
$B[i:j]$	A subsequence of B from time stamp i to j

2.2 Subsequence Matching Framework (DualMatch vs. FRM)

There are at least two subsequence matching frameworks, FRM [2][1] and Dual Match [7]. Both of the matching processes are illustrated in Figure 2. Let n be the number of data points and w be the size of an index window. In FRM, the data sequence is divided into $n - w + 1$ sliding windows. Figure 2(a) shows the FRM indexing step. Every window is overlapped with the next window except the first data point. Whereas, query Q is divided into disjoint windows (Figure 2(b)), and each window is to be matched against the sliding windows of

---

[1] It is named after its authors.

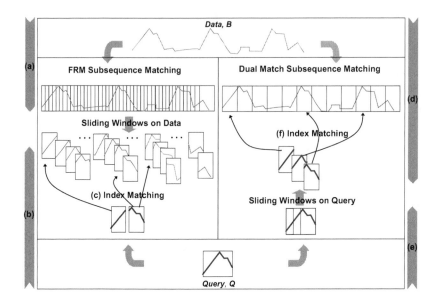

Fig. 2. Two Subsequence Matching Frameworks

the data sequence (Figure 2(c)). On the other hand, in Dual Match framework, data sequence is divided into disjoint windows (Figure 2(d)), and part of the query in its sliding window is matched to the data indices (Figure 2(e) and 2(f)). Since the Dual Match does not allow any overlap of the index windows, it needs less space for the index and, in consequence, index searching is faster than FRM. Through the index matching, we get a set of candidate matches and the actual similarity or distance is computed for them. Since the length of the data is usually very long, Dual Match framework reduces the indexing efforts. We employ the Dual Match as our indexing scheme.

2.3 Dual Match Subsequence Matching with Euclidean Distance

Dual Match consists of three steps. First, in the indexing step, data is decomposed into disjoint windows and each window is represented by a multi-dimensional vector. They are stored in a spatial index structure like R-tree. Second, query sequence is decomposed into a set of sliding windows and each window is transformed into the same dimensional vector representation as the index window. The size of the sliding window is the same as that of the index window. It is proven that if the length of the query is longer than twice of the index length, one of the sliding windows in the query is guaranteed to match to a data index that belongs to a subsequence that matches to the query [7]. The index matching always returns a super set of the true matching intervals since the similarity of the index and query sliding window is always larger than the similarity of the true match. Lastly, based on the positions of the matching sliding windows, whole matching intervals are decided and actual similarities are computed.

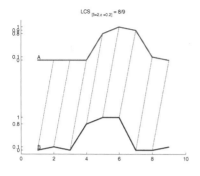

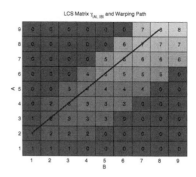

(a) Sequence A, B and warping path

(b) Sakoe-Chiba band in *LCS* warping path martix

Fig. 3. An example of LCS computation

2.4 A Non-euclidean Distance *LCS*

Non-Euclidean similarity measures such as *LCS* and *DTW* are useful to match two time series data when the data has irregularity. The *LCS* is known to be robust to the noise since it does not count the outliers in the sequence that fall out of the range (ϵ). Both use the same dynamic programming procedure to compute the optimal warping path within the time interval (δ). We chose *LCS* as our distance function and its definition is given below.

Definition 1. *[10] Let $Q=< q_1, q_2, ..., q_n >$ be a query and $B=< b_1, b_2, ..., b_n >$ be a data subsequence of time series. Given an integer δ and a real number $0 < \epsilon < 1$, we define the cumulative similarity $\gamma_{i,j}(Q, B)$ or $\gamma_{i,j}$ as*

$$\gamma_{i,j} = \begin{cases} 0, & \text{if } i, j = 0 \\ 1 + \gamma_{i-1,j-1} & \text{if } |q_i - b_j| \leq \epsilon \\ & \text{and } |i - j| \leq \delta \\ max(\gamma_{i,j-1}, \gamma_{i-1,j}) & \text{otherwise} \end{cases}$$

and using that, LCS similarity with δ and ϵ as

$$LCS_{\delta, \epsilon}(Q, B) = \gamma_{|Q|, |B|}$$

$LCS(Q, B)$ returns an integer between 0 and $min(|Q|, |B|)$. δ is the allowable matching interval in the time dimension and ϵ is the allowable error bound in the data value dimension. Here is an example of *LCS* match for the two sequences A and B of the same length where $A = <0, 0, 0, 0, 0.8, 1, 0.9, 0.1, 0>$ and $B = <0, 0.1, 0, 0.8, 1, 1, 0, 0, 0.1>$. Figure 3(a) shows the *LCS* warping path. Figure 3(b) shows the *LCS* computation process in the *LCS* warping path matrix. It is constructed by dynamic programming of the cumulative similarity $\gamma_{|A|, |B|}$. The non-zero boxes in light color in the *LCS* warping path matrix of Figure 3(b) is called a Sakoe-Chiba band [8].

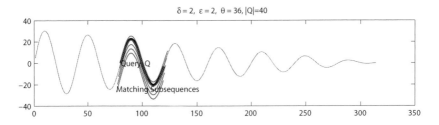

Fig. 4. Matching subsequences in subsequence matching

3   Problem Statement

The purpose of the subsequence matching is to find subsequences similar to the given query sequence. Subsequence matching framework with Euclidean distance has been already developed as we stated in the previous section. However, to the best of our knowledge, many things have not yet been considered when we apply non-Euclidean function to the subsequence matching. We need to improve the index search performance and we need to provide an index matching criteria that avoids expensive computation caused by non-Euclidean measures.

In order to describe what should be the output of the subsequence matching, we define matching subsequences for a query sequence Q in terms of $LCS_{\delta,\epsilon}$.

Definition 2. *Let $Q=< q_1, q_2, ...q_m >$ be a query and $B=< b_1, b_2, ...b_n >$ be a data subsequence of time series. Given an integer δ, a real number $0 < \epsilon < 1$ and user defined similarity threshold θ, we define the **matching subsequences**, $M = \{B[i:j] \mid LCS_{\delta,\epsilon}(Q, B[i:j]) \geq \theta\}$*

There may be many overlapping subsequences in the same region that exceed the similarity threshold θ. We restrict the scope of our work to find only the longest possible matching subsequences of the length $|Q| + 2\delta$. We do not return all matching subsequences that are properly contained in the longest possible one returned. It could be prohibitively expensive to find all matches of all lengths using a non-Euclidean measure. It makes sense to return only the longest matching subsequences since it contains all matching subsequences shorter than $|Q|+2\delta$ in the region. It is possible to search shorter matching subsequences, if needed, after the search process for the longest ones completes. In Figure 4, all the matching subsequences of size $|Q| + 2\delta$ are visualized in grey dotted lines.

Formally, our problem is defined as follows:Find all matching subsequences $B[i:j]$ of length $|Q| + 2\delta$ for data sequence B and query Q such that the similarity $LCS_{\delta,\epsilon}(Q, B[i:j])$ is no less than $s\%$ of the $|Q|$, $\frac{s}{100}|Q|$.

4   Subsequence Matching with *LCS*

4.1   Linear Search and Skipping *LCS* Computation

A straighforward approach to the subsequence matching is comparing the query subsequence Q to all of the candidate subsequences of the data sequence B in

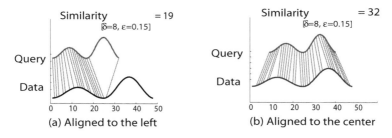

Fig. 5. Alignment with LCS when $|Query| = 32$ and $|Data| = 48$

a sequential manner. All the candidates can be chosen by sliding a fixed size window along the data sequence.

Alignment in LCS. When we compare query Q to a candidate data subsequence of length $|Q| + 2\delta$, we align the query in the middle of each candidate as illustrated in Figure 5(b). In the case of the whole sequence matching, alignment is not a problem since the query and data have the same length. However, in our subsequence matching, we need to locate the query in the candidate subsequence. If we align the query to the left side of a candidate, we may find a correct subsequence. In Figure 5(a), shorter query is not matched well to the longer data when aligned to the left. The right side of the query cannot be compared with the data since the δ is not big enough to cover all the matching points in the data. Larger δ increases the computational complexity of the matching process. Figure 5(a) shows that the query is correctly matched with the same δ when properly aligned.

Skipping LCS Computation. We can avoid expensive similarity computations of the adjacent subsequences by exploiting the LCS warping path and the local constraint such as the Sakoe-Chiba band. In the subsequence matching, we can think of the computation matrix as a moving window along the data sequence as shown in Figure 6.

Let us take a look at an example. Assume that $|Q| = 4$ and the user wants to find all the subsequences whose similarity is larger than or equal to 3. Figure 6(a) shows the LCS warping path which is represented as a set of arrows. In this case, $LCS(Q, B[1 : 6]) = 4$. Darker cells represent the Sakoe-Chiba band.

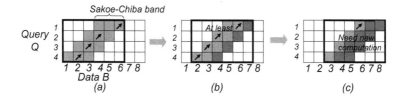

Fig. 6. An example of skipping LCS computation when $|Q| = 4$ and $\delta = 1$

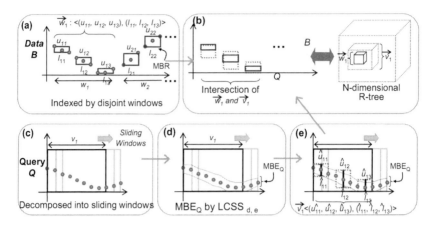

Fig. 7. Indexing and Index Matching where $w=9$ and $N=3$

In Figure 6(b), we move a sliding window by a time stamp. The Sakoe-Chiba band still includes the warping path. In this case, we don't have to compute the $LCS(Q, B[2:7])$ since the dynamic programming finds a maximum warping path in the Sakoe-Chiba band and $LCS(Q, B[2:7])$ must be larger than or equal to 4. In Figure 6(c),we need to compute $LCS(Q, B[3:8])$ since the first three warping steps now became invalid.

We can skip the computation of a sliding window by tracing the warping path. If we find that the Sakoe-Chiba band of the current LCS matrix includes the previous warping path more than or equal to the user defined threshold, then we can skip the LCS computation. The skipping goes until a Sakoe-Chiba band includes warping path less than the user defined threshold. It is a useful property to reduce the expensive similarity computation in the subsequence matching where the adjacent window usually has a similar similarity value.

4.2 Index Match

Indexing enables us to avoid unecessary similarity computations for true-negative candidates for subsequence matching. In order to do that, we compute the pruning criteria to choose candidate matching subsequences with LCS. We also introduce in this section a new framework to efficiently search the index.

Indexing. Data is divided into equi-length disjoint windows for indexing. Each window is then represented as a multi-dimensional vector. That is, data sequence B is divided into equi-length disjoint windows $< w_i >$. Let N be the dimensionality of the space we want to have indexed. An MBR, $Minimum Bounding Rectangle$, represents a dimension. N $MBRs$ for a w_i, are transformed into $\overrightarrow{w_i}$ $=< (u_{i1}, \ldots, u_{iN}), (l_{i1}, \ldots, l_{iN}) >$,where u_{ij} and l_{ij} represent the maximum and minimum values in the jth interval of w_i. $\overrightarrow{w_i}$ is stored in an N dimensional R-tree. An example is illustrated in Figure 7(a). In the figure, the data in the first window, $w_1 =< b_1, ..., b_9 >$ is transformed into $\overrightarrow{w_1} =< (u_{11}, u_{12}, u_{13}), (l_{11}, l_{12}, l_{13}) >$. It is stored in an R-tree as showin in Figure 7(b).

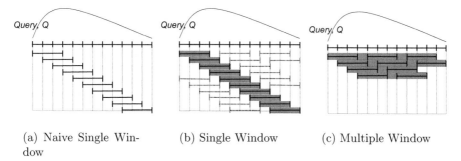

(a) Naive Single Window   (b) Single Window   (c) Multiple Window

Fig. 8. Window sliding schemes when $|v|=4$

Index Matching with *LCS*. Query Q is compared first to the index. Q is transformed into an *MBE, Minimum Bounding Envelope*, with $LCS_{\delta,\epsilon}$ function as illustrated in Figure 7(d). Let MBE_Q be an *MBE* for Q. Let the ith sliding window of Q be v_i. It is transformed into $\overrightarrow{v_i} =< (\hat{u}_{i1},\ldots,\hat{u}_{iN}),(\hat{l}_{i1},\ldots,\hat{l}_{iN}) >$, where $\hat{u}_{ij}$ and $\hat{l}_{ij}$ are the maximum and minimum values respectively in MBE_Q of the jth *MBR* of the v_i. This is illustrated in Figure 7(e). Since MBE_Q covers the whole possible matching area, any point that lies outside the MBE_Q is not counted for the similarity. The number of intersecting points between B and MBE_Q provides the upperbound for $LCS_{\delta,\epsilon}(B,Q)$ [10]. The number of intersections is counted through the R-tree operation as shown in Figure 7(b), which is the intersection of Figure 7(a) and Figure 7(e).

4.3   Window Sliding Schemes in Index Matching

There are three ways to slide query windows and choose the candidate matching subsequences: Naive Single Window Sliding, Single Window Sliding and Multiple Window Sliding. We explain each window sliding scheme and show how the the bounding similarity is computed.

Naive Single Window Sliding. In this scheme, as illustrated in Figure 8(a), we compare a sliding window of a query to index, which is first introduced in the Dual Match [6]. This overestimation method cannot be applied to the *LCS* based subsequence matching since it is based on the Euclidean distance. We should consider δ on both ends of the query sliding window. In Figure 9 (a), a sliding window v of a query Q is matched to a window w of the data sequence B. In actual index matching, near the ends of the point of the Q cannot be matched to the points of w as in Figure 9 (b). The data is just indexed by *MBR* that does not consider δ time shift.

We compute the similarity threshold for the naive single window sliding method.

Let v be a sliding window of Q. The minimum similarity, θ is

$$\theta = |v| - (|Q| - \frac{s}{100}|Q|) - 2\delta \tag{1}$$

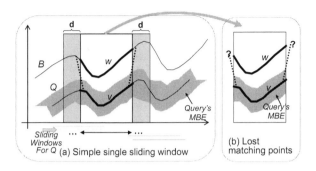

Fig. 9. Matching points not captured in the index matching using LCS

The term, $(|Q| - \frac{s}{100}|Q|)$ for the Equation (1) is subtracted from $|v|$ when all the mismatches can be found in the current window v. The last term 2δ is the maximum possible number of the lost matching points.

Single Window Sliding. When the query length is long enough to contain more than one sliding window, we can use the consecutive matching information as in Figure 8(b). Assume query Q and matching data subsequence B has M consecutive disjoint windows, B_i's and Q_i's. If some Q_i and B_i pairs are not similar, then the other Q_j and B_j pairs should be similar and we can recognize the B and Q pair is a candidate through B_j and Q_j. When all B_i and Q_i pairs have the same similarities, we should have the minimum value to decide the candidate for comparison. The *multiPiece* search [2] is proposed to choose candidates through this process. The same applies for the Euclidean distance measure. In the *multiPiece*, the two subsequences, B and Q, of the same length are given and each can be divided into p subsequences each of which has length l. $d(B, Q) < \epsilon \Rightarrow d(B_i, Q_i) < \frac{\epsilon}{\sqrt{p}}$ for some $1 \leq i \leq p$ where B_i, Q_i are i^{th} subsequence of the length l and $\epsilon > 0$. In the case of the Dual Match using Euclidean distance, we can count a candidate if the distance is less than or equal to $\frac{\epsilon}{\sqrt{p}}$.

Similarly, in the case of LCS, $LCS_{\delta,\epsilon}(B, Q) > \frac{s}{100}|Q| \Rightarrow LCS_{\delta,\epsilon}(v, Q[i : j]) > \frac{M|v|-(|Q|-\frac{s}{100}|Q|)-2\delta}{M}$ for some $j - i + 1 = |v|$. So the similarity threshold for single window sliding, θ_s is

$$\theta_s = |v| - \frac{(|Q| - \frac{s}{100}|Q|) + 2\delta}{M} \qquad (2)$$

As illustrated in Figure 8(b), M consecutive sliding windows are thought to be one big sliding window that might lose warping path at both ends. The threshold for the M sliding windows is $M|v|-(|Q|-\frac{s}{100}|Q|)-2\delta$ and it is divided by M for one sliding window. If one of the sliding windows among consecutive M sliding windows in Q is larger than or equal to θ_s, we can get a candidate and we don't have to do index matching for the remaining consecutive sliding windows at the same candidate location.

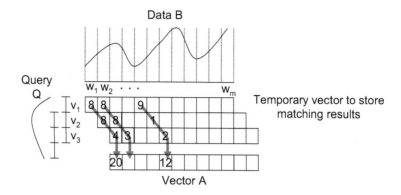

Fig. 10. Index matching result

Multiple Window Sliding. In this new window sliding scheme, as illustrated in Figure 8(c), the matching results of consecutive sliding windows in a query are aggregated. If we sum up the index matching result from M consecutive sliding windows, we can further reduce false positives. Let M be the number of consecutive windows fitted in a query Q. We vary M to contain the maximum number of sliding windows depending on the left most window.

The index matching results of each sliding window for all disjoint data windows are added up to get M consecutive sliding windows. In Figure 10, the aggregation is done by accumulating the results in a vector A of the size $\frac{|B|}{w}$. B is the data sequence and w is the length of an index window. Assume that $< v_1, \ldots, v_M >$ is a series of consecutive windows in the query Q. The index matching results of a query window v_j is placed in a temporary row vector in Figure 10. It is added to A and A is shifted to the right. The next matching result for v_{j+1} is placed in the temporary row vector. It is added to A and A is shifted right. In Figure 10, we get A such that

$$A[1] = LCS_{\delta,\epsilon}(\overrightarrow{v_1}, \overrightarrow{w_1}) + LCS_{\delta,\epsilon}(\overrightarrow{v}_2, \overrightarrow{w}_2) + LCS_{\delta,\epsilon}(\overrightarrow{v}_3, \overrightarrow{w}_3),$$
$$A[2] = LCS_{\delta,\epsilon}(\overrightarrow{v_1}, \overrightarrow{w_2}) + LCS_{\delta,\epsilon}(\overrightarrow{v}_2, \overrightarrow{w}_3) + LCS_{\delta,\epsilon}(\overrightarrow{v}_3, \overrightarrow{w}_4), \ldots$$
$$A[m] = LCS_{\delta,\epsilon}(\overrightarrow{v_1}, \overrightarrow{w_{m-2}}) + LCS_{\delta,\epsilon}(\overrightarrow{v}_2, \overrightarrow{w}_{m-1}) + LCS_{\delta,\epsilon}(\overrightarrow{v}_3, \overrightarrow{w}_m).$$

The shift operations aggregate the consecutive index matching results.

The similarity threshold for multiple sliding windows, θ_m, is computed as if the consecutive M windows move together like one big window.

$$\theta_m = M|v| - (|Q| - \frac{s}{100}|Q|) - 2\delta \tag{3}$$

θ_m is for an aggregate comparison of M consecutive sliding windows while θ_s is for one sliding window.

Through the aggregation of the consecutive index matching information, we can enhance the pruning power of the index. That is, we have less false alarms than the single window sliding scheme. In Figure 10, the diagonal sum illustrates the aggregatation of the consecutive index matching results. If $\theta_s = 8$, the first,

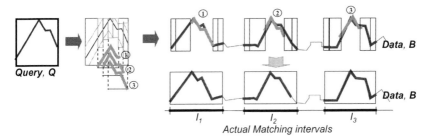

Fig. 11. Postprocessing to find whole length of the candidate matching subsequences

second and the fifth diagonals are selected as the candidates since one of the matches is greater than or equal to 8. However, in case of the multiple window sliding, if the $\theta_m = 20$, the fifth diagonal is not a candidate since the sum 12 is less than 20, so it has less false alarms than the single window sliding scheme.

Post-Processing. Post-processing is the final procedure to decide the whole length of the matching subsequence depending on the position of the matching sliding window and matching index. The actual similarity computation is done for the whole interval of the subsequence against the query. Figure 11 demonstrates the postprocessing. We intensionally omit the adjacent matching subsequences and show only one matching. Through the index matching process, matching indexes for each sliding window ①, ②, ③ are to be found and then whole length of the candidate subsequence is computed including 2δ area. In Figure 11, one candidate subsequence has an index matching area and a four box area.

Skipping _LCS_ computation. After deciding the whole length of the candidate subsequences, skipping _LCS_ computation is applied to reduce the computational load. Subsequence matching cannot avoid many adjacent matching subsequences where one subsequence is found. By tracing the warping path of the matching subsequences in its _LCS_ warping path matrix, we can reduce the _LCS_ computation.

5   Experiment

Experiments were conducted on a machine with 2.8 GHz pentium 4 processor and 2GB Memory using Matlab 2006a and Java. Here are the parameters to run the tests.

- **Dataset.** We used 48 different time series datasets[2] for evaluation. Each dataset has a different length of data and a different number of channels.

---

[2] http://www.cs.ucr.edu/ eamonn/TSDMA/UCR, The UCR Time Series Data Mining Archive.

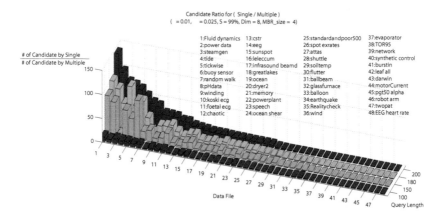

Fig. 12. Candidates generated by single window sliding and multiple window sliding

We set the length of each to 100,000 by attaching the beginning to the end so that all the datasets have the same length.

- **Index.** We set the dimension to 8 and *MBR* size to 4. Regarding the parameters to index dataset such as dimension, *MBR* and R-tree size need domain knowledge.
- **Query.** We choose 4 fixed length of queries, 100, 150, 180 and 200 so that each length includes 3,4,5 and 6 windows. 10 queries for each length are randomly selected from the data sequence.
- **Similarity.** ϵ is set to 1 % of the data range, δ is 2.5 % of the $|Q|$. Similarity threshold s is set to 99%.

5.1   Different Sliding Schemes and Candidates

We compare the performance of the two different index sliding schemes : single window sliding and multiple window sliding scheme. Figure 12 shows that the ratios, $\frac{\text{\# of candidates by single windows sliding}}{\text{\# of candidates by multiple windows sliding}}$ for different lengths of queries of each dataset. Ratios greater than one means that the multiple window sliding scheme generates less candidates than those of the single window sliding scheme. The multiple window sliding scheme has less false alarms than the single window sliding scheme in the tests. The ratio varies from 1 to 140. Multiple sliding window generates only $\frac{1}{140}$ of the single window sliding scheme in the Fluid dynamics dataset. Figure 13 shows the median values from the Figure 12 for each length of the queries. Figure 13 summarizes how much the performance is improved as the length of query gets longer in all of the datasets. It demonstrates that as the length of a query gets longer to include more index windows, we have less false alarms in the multiple window sliding than in the single window sliding.

However in the datasets such as EEG heart rate, two pat or robot arm, there is not much difference between the two methods. We can explain it in terms of the index. For these datasets, all of the disjoint data windows are very similar to each other. Figure 14 shows the first 500 points index of the best and the worst

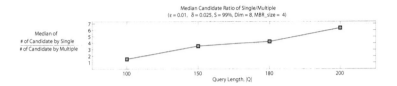

Fig. 13. Summary of Candidate generated in Figure 12

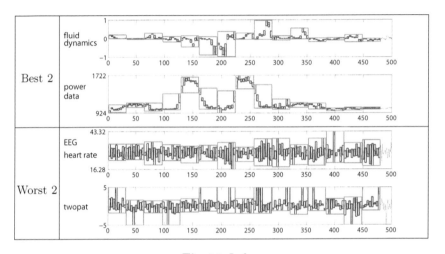

Fig. 14. Index

three datasets regarding the candidate generation. Comparing the index of the top three datasets to the bottom three, we cannot easily distinguish one window from another. It makes hard to search the index quickly even though multiple index information is used.

5.2 Goodness and Tightness

Goodness and tightness are metrics that shows how well the index works [5].

$$Goodness = \frac{\#\text{ of all true matches}}{\#\text{ of all candidates}} , Tightness = \frac{\text{Sum of all true similarity}}{\text{Sum of all estimated similarity}} \tag{4}$$

Goodness shows how much the index reduces the expensive computations. Tightness shows how the estimated values are close to the actual values in indexing [5]. If the tightness is 1.0 then it means estimation is perfect. In Figure 15, the multiple sliding window scheme shows higher goodness and tightness than that of the single window sliding scheme.

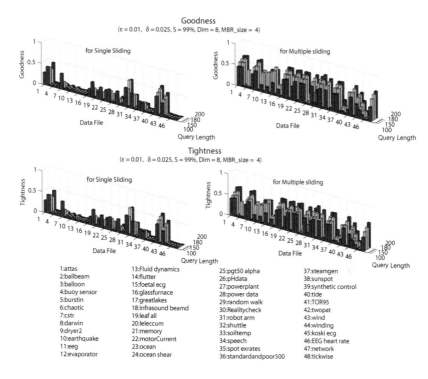

Fig. 15. Goodness and Tightness

5.3 Improving Performance by Skipping Similarity Computations

Figure 16 shows how the skipping of the similarity computation is effective. The chart shows that we can avoid many similarity computations as the length of the query gets longer.

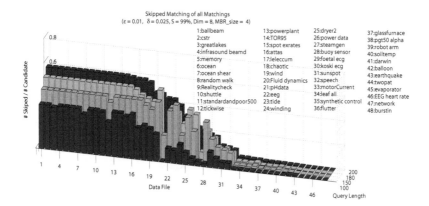

Fig. 16. Skipping Similarity Computations

However it also shows that the skipping mechanism does not work well for the datasets that cannot be properly indexed, since the index parameter captures all of the windows in the data as well as the ones similar to the LCS matrix.

6   Conclusion

We proposed a novel subsequence matching framework that employs a non-Euclidean distance, a multiple window sliding scheme and a similarity skipping idea. As validated through experiments with various datasets, proposed methods enable us to have more intuitive and efficient subsequence matching algorithms. The multiple window sliding scheme was more efficient than the single window sliding scheme for the longer query in candidate generation, goodness and tightness. In addition, skipping the LCS computation greatly reduces expensive similarity computations.

References

1. Agrawal, R., Faloutsos, C., Swami, A.N.: Efficient similarity search in sequence databases. In: Lomet, D.B. (ed.) FODO 1993. LNCS, vol. 730, pp. 69–84. Springer, Heidelberg (1993)
2. Faloutsos, C., Ranganathan, M., Manolopoulos, Y.: Fast subsequence matching in time-series databases. In: Proceedings 1994 ACM SIGMOD Conference, Mineapolis, MN, ACM Press, New York (1994)
3. Gunopoulos, D.: Discovering similar multidimensional trajectories. In: ICDE '02. Proceedings of the 18th International Conference on Data Engineering, p. 673. IEEE Computer Society Press, Los Alamitos (2002)
4. Kadous, M.: Grasp: Recognition of australian sign language using instrumented gloves (1995)
5. Keogh, E.J.: Exact indexing of dynamic time warping. In: VLDB, pp. 406–417 (2002)
6. Moon, Y.-S., Whang, K.-Y., Loh, W.-K.: Duality-based subsequence matching in time-series databases. In: Proceedings of the 17th ICDE, Washington, DC, pp. 263–272. IEEE Computer Society Press, Los Alamitos (2001)
7. Moon, Y.-S., Whang, K.-Y., Loh, W.-K.: Efficient time-series subsequence matching using duality in constructing window. Information Systems 26(4), 279–293 (2001)
8. Sakoe, H., Chiba, S.: Dynamic programming algorithm optimization for spoken word recognition, pp. 159–165 (1990)
9. Sankoff, D., Kruskal, J.: Time warps, string edits, and macromolecules: the theory and practice of sequence comparison. Addison-Wesley, Reading (1983)
10. Vlachos, M., Hadjieleftheriou, M., Gunopulos, D., Keogh, E.: Indexing multidimensional time-series with support for multiple distance measures. In: KDD '03, pp. 216–225. ACM Press, New York (2003)
11. Zhu, Y., Shasha, D.: Warping indexes with envelope transforms for query by humming. In: SIGMOD '03, pp. 181–192. ACM Press, New York (2003)

A Direct Measure for
the Efficacy of Bayesian Network Structures
Learned from Data*

Gary F. Holness

Quantum Leap Innovations
3 Innovation Way
Newark, DE. 19711
gfh@quantumleap.us

Abstract. Current metrics for evaluating the performance of Bayesian network structure learning includes order statistics of the data likelihood of learned structures, the average data likelihood, and average convergence time. In this work, we define a new metric that directly measures a structure learning algorithm's ability to correctly model causal associations among variables in a data set. By treating membership in a Markov Blanket as a retrieval problem, we use ROC analysis to compute a structure learning algorithm's efficacy in capturing causal associations at varying strengths. Because our metric moves beyond error rate and data-likelihood with a measurement of stability, this is a better characterization of structure learning performance. Because the structure learning problem is NP-hard, practical algorithms are either heuristic or approximate. For this reason, an understanding of a structure learning algorithm's stability and boundary value conditions is necessary. We contribute to state of the art in the data-mining community with a new tool for understanding the behavior of structure learning techniques.

1 Introduction

Bayesian networks are graphical models that compactly define a joint probability over domain variables using information about conditional independencies between variables. Key to the validity of a Bayesian Network is the Markov Condition [11]. That is, a network that is faithful to a given distribution properly encodes its independence axioms. Inducing Bayesian networks from data requires a scoring function and search over the space of network structures [10]. As a consequence of the Markov Condition, structure learning means identifying a network that leaves behind few unmodeled influences among variables in the modeled joint distribution.

* We acknowledge and thank the funding agent. This work was funded by the Office of Naval Research (ONR) Contract number N00014-05-C-0541. The opinions expressed in this document are those of the authors and do not necessarily reflect the opinion of the Office of Naval Research or the government of the United States of America.

P. Perner (Ed.): MLDM 2007, LNAI 4571, pp. 601–615, 2007.
© Springer-Verlag Berlin Heidelberg 2007

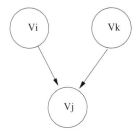

Fig. 1. Graphical representation for Bayesian Network

A Bayesian network takes form as a directed acyclic graph $G = (V, E)$ where the nodes $V_i \in V$ represent the variables in a data-set and the directed edges $(V_i, V_j) \in E$ encode the causal relationship between V_i and V_j. Dependencies among variables are modeled by a directed edge. As such, if edge $(V_i, V_j) \in E$, then V_j depends causally upon V_i or, similarly, V_i is the parent of V_j and V_j is the child of V_i. The graphical model for a simple Bayesian network appears in Figure 1. In this example, we have three nodes V_i, V_j, and V_k where V_j is causally dependent upon V_i and V_k. Causality is implied in edge directedness. Through its network structure, a Bayesian network model encodes the independence axioms of a joint distribution. Given a Bayesian network $G = (V, E)$ we compute the full joint distribution for variables $V_i \in V$ using the chain rule:

$$p(V_1, \ldots, V_n) = \prod_{i=1}^{n} p(V_i | pa_i)$$

where pa_i are the set of variables that are the parents of V_i. By expressing the joint distribution in terms of its conditionally independent factors, marginalization and inference are made more tractable.

Inference in Bayesian networks is well known to be an NP-hard problem both in the exact and approximate cases [2, 4]. Construction of Bayesian networks structures from data is also an NP-hard problem. The major classes of techniques for learning Baysian networks falls into two major categories. The first considers network construction as a constraint satisfaction problem [11, 14]. These methods compute independence statistics such as χ^2 test, KL-divergence, or entropy over variables and build networks that represent computed associations. The second considers network construction as an optimization problem. These methods search among candidate network structures for the optimum [3, 5, 15].

The search problem over Bayesian network structures is also an NP-hard problem. Heuristic approaches such as the $K2$ algorithm impose simplifying assumptions on the network in order to make learning and inference tractable [3]. In $K2$, nodes are are assumed to have a causal ordering. That is, a node appears later in an ordering than the nodes on which it depends. Additionally, the $K2$ algorithm also bounds the number of parent dependencies a node may have. In the recent $K2GA$ approach, the author employs a genetic algorithm to perform stochastic search simultaneously over the space of node orderings and network structures for an extension of the $K2$ algorithm. $K2GA$ has been found to perform competitively with respect to ground truth networks on benchmark data-sets [7]. Additional search techniques include greedy hill-climbing, simulated annealing and Markov Chain Monte Carlo (MCMC) [6, 13]. Other approaches make the problem more tractable by pruning the search space. For example, the sparse candidate algorithm uses mutual information between variables to prune the search space so that only a reduced set of potential parents are considered for each variable [9]. Another approach that has enjoyed success performs greedy search over equivalence classes of DAG patterns instead of the full DAG space representation [1].

The experiments described in this paper grew out of the need to characterize the performance of an implementation of $K2GA$. This work goes beyond measures of model fit and convergence time as typical in the Bayesian network literature to include measurements of stability. While we use $K2GA$ as the target system for evaluation, our techniques are generally applicable to any Bayesian network structure learning algorithm. In recent related work Shaughnessy and Livingston introduce a method for evaluating the causal explanatory value of structure learning algorithms [12]. Their approach begins with randomly generated ground truth networks involving three-valued discrete variables. Next they sample from them to produce small synthetic data-sets that are input to a structure learning algorithm. Finally, precision-recall measures are made from edge level statistics, such as *false positive edge count*, comparing the learned and ground truth networks. While this method evaluates different types of causal dependencies it cannot vary the strength of such dependencies and requires a sufficient number of samples. Because $K2GA$ is a stochastic algorithm, we set out to test if initial conditions and noise in the data affect the structure learner's ability to correctly capture variable dependencies.

In the sections that follow, we begin with a high level description of the stochastic algorithm K2GA. Then, we outline a method for testing how well the Bayesian network has modeled dependencies among variables. In doing so, we treat variable dependence as a retrieval problem and apply an ROC technique for measuring performance stability. Lastly, we describe our experiments and discuss results.

2   Structure Learning Using K2GA

$K2GA$ makes use of an alternate Bayesian network representation that encodes a DAG in terms of its undirected skeleton and the causal ordering of the nodes. Let $X = \{X_1, \ldots, X_N\}$ be a set of variables, $\Theta = \{\Theta_1, \ldots, \Theta_N\}$ be the ordering of nodes (where $\Theta_i \in [0,1]$), and $\mathcal{B}$ be the adjacency matrix for the undirected skeleton such that $\mathcal{B}_{ij} = 1$ if and only if X_i is related to X_j. Skeleton, $\mathcal{B}$, describes the dependency between two variables while Θ defines the edge directedness. For example, in the situation where X_j is causally dependent on X_i, we have $\mathcal{B}_{ij} = 1$ and $\Theta_i < \Theta_j$.

Given the exponential space of DAGS, a number of simplifying assumptions have been made to reduce the complexity of the search space to polynomial in the number of nodes. These include causal ordering of variables that participate in the model along with bounded in-degree between a node an its parents. The topological ordering, $\prec$, of graph nodes $\{X_1, \ldots, X_N\}$ is such that

$$\bigvee_{i,j} X_j \prec X_i \rightarrow X_j \in Ancestors(X_i)$$

Structure learning algorithms that assume the $K2$ heuristic search within a family of DAGS possible from fixed causal orderings. Given topological ordering, $\prec$, the set of all possible skeletons $\mathcal{S} = \{\mathcal{B}_1, \ldots, \mathcal{B}_L\}$ is defined by the number of unique skeletons that can be defined from the upper triangle of $\mathcal{B}$. Given N-variables, $|S| = 2^{\frac{N(N-1)}{2}}$. Since there are $N!$ orderings, this results in substantial reduction from a total of $N! \left(2^{\frac{N(N-1)}{2}}\right)$ possible DAG patterns. While a factorial reduction in search space is significant, the issue of which ordering to search remains. The $K2GA$ algorithm performs simultaneous search of the space of topological orderings and connectivity matrices. For more detailed descriptions of $K2GA$, We direct the reader to the original work [7].

3   Markov Blanket Retrieval: An Efficacy Measure

By extending the definition of a document in information retrieval, verification of a Bayesian network is treated as a retrieval problem where the information need is the set of causal dependencies for a given variable. This corresponds to the Markov blanket that most closely resembles the ground truth blanket for a given node. In using a vector space approach and ranking, we allow for partial similarity. This is particularly important for variables with weak dependence relationship.

This approach differs from traditional methods for verification of Bayesian networks in that we do not rely on samples from a hand constructed *gold standard network* for verification. Because such techniques rely on samples from the specified network, a sufficient sample size is required. Moreover, for nontrivial real-world problems, apriori knowledge of variable dependencies is difficult. Consider a real world complex data-set such as manufacturing or supply chain modeling scenario involving 100's or 1,000's of variables. It might be the case that the Bayesian network structure learned from data is correct, but its performance is discounted by a faulty hand constructed gold standard network. By exercising precise control over variable dependence and measuring resulting performance, we provide characterization of a Bayesian network learner's modeling stability using ROC analysis.

3.1   ROC Curves

More than just its raw performance numbers, an algorithm's quality is also measured in terms of sensitivity and specificity. A predictor's sensitivity measures the proportion of the cases picked out from a data set relative to the total number of cases that satisfy some test. Sensitivity is also called the true positive rate. A predictor's specificity measures its ability to pick out cases that do not satisfy some test. Specificity is also called the true negative rate. A receiver operating characteristic (ROC) curve is related to likelihood ratio tests in statistics and expresses how the relationship between sensitivity and specificity changes with system parameters [8].

In comparing Bayesian networks, we would like a single measure of predictive quality. Area under the curve (AUC) is a non-parametric approach for measuring predictive quality. AUC is simply the area under the ROC curve. This gives us a standard means of comparing performance. AUC varies in the closed interval $(0, 1)$ on the real number line and is interpreted as the rate of correct prediction.

As one could imagine, a good predictor is one that can correctly identify cases in the data that actually have the phenomenon under test. This corresponds to an AUC that is closer to 100%. AUC results are typically compared to the random performance. In an example where true positive and false negative are assumed equally likely, the ROC curve is a straight line with slope 45-degrees and AUC of 50%. Any method that cannot outperform random performance is not worth deployment.

For Bayesian network structures, in order to convert performance measures into likelihood ratio tests for the purpose of ROC analysis, we must compare structures learned from data with some notion of ground truth. This allows us to define what it means to have a true positive or a true negative.

3.2   Markov Blanket

A Markov blanket of a node, A, is defined as A's parents, children and spouses (the parents of A's children). The Markov blanket is the minimal set of nodes that give A conditional independence.

$$P(A|MB(A), B) = P(A|MB(A))$$

That is, A is conditionally independent of any node $B \notin MB(A)$ given $MB(A)$. The Markov blanket gives complete description of the variables upon which A depends. As depicted in Figure 2, these are the nodes that partition A from the rest of the nodes in the network.

The Markov blanket is related to d-separation in that given the set $Z = \{Z_i \in MB(A)\}$ and $C = \{X - A - MB(A)\}$ where X is the set of variables, it is the case that A is d-separated from C given $MB(A)$. Thus, the Markov blanket gives us the dependence relationship between a node and all other nodes in the Bayesian network. We use this to test $K2GA$'s efficacy in correctly modeling causal dependencies.

3.3   Ground Truth Causal Dependence

Controlling variable dependencies is accomplished by augmenting a data-set with synthetic variables. We treat synthetic variables X_{new} as queries. Because

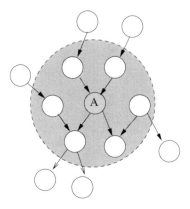

Fig. 2. A node and its Markov blanket

variables $X_1, \ldots, X_k$ are used to compute the synthetic variable, we know ground truth that $X_1, \ldots, X_k$ are in X_{new}'s Markov blanket. A synthetic variable also depends on a noise process ϵ used to control the strength of dependence on the ground truth parent variables. For strong dependence, the contribution to X_{new} by the noise process is dominated by $X_1, \ldots, X_k$. For weak dependence, the contribution to X_{new} by the ground truth parent variables is dominated by the noise process. Using random variable $A \sim Bernoulli(\alpha)$ taking on values $a \in \{0, 1\}$ we select

$$X_{new} = \begin{cases} f_{\boldsymbol{w}}(X_1, \ldots, X_k) & \text{if } a = 1 \\ \epsilon & o.w. \end{cases}$$

where parameter α is defined in the closed interval $(0, 1)$ on the real number line. Thus, α regulates the strength of causal dependence. Regardless of the strength of causal dependence, we know ground truth that $\{X_1, \ldots, X_k\} \subset MB_{\mathcal{B}}(X_{new})$.

The amount by which the synthetic variable depends on each of its ground truth parents is determined by a vector of weights. Given k ground truth parent variables, we have weight vector $\boldsymbol{w} = < w_1, \ldots, w_k >$ computed by uniform sampling from the unit simplex in k-dimensions. That is the series of weights from the set $\{< w_1, \ldots, w_k > | w_1 + \ldots + w_k = 1, 0 <= w_i <= 1, i = 1, \ldots, k\}$. Given Q-samples, this gives us representative coverage across the range of associations a dependent variable can have on k-parent variables. The dependent variable takes on values drawn from the union of the domains of its parents. In Figure 3 we list the values of the domain for three parents in rectangles along the top row and domain values of the dependent variable in rectangles along the bottom row. In this example, we have three parent variables whose domain sets have values $\{v_1, v_2\}$, $\{v_3, v_4, v_5\}$, and $\{v_6, v_7, v_8, v_9\}$ respectively. The dependent variable draws its values from the set $\{v_1, \ldots, v_9\}$ (Figure 3). This allows us to interpret the weight vector as the relative proportion of cases for which the value of the dependent variable is dictated by a given parent. An example of this appears in Figure 4. We list values for four cases by repeating the pair of rows

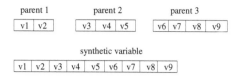

Fig. 3. The domain of a synthetic variable

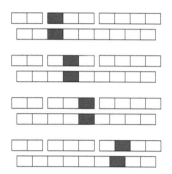

Fig. 4. Illustration of synthetic variable causally dependent on parents

from Figure 3 once for each case. The values taken by the three parents and the dependent variable are illustrated by shading in the appropriate positions in each row.

As can be seen in Figure 4, for the synthetic dependent variable, the first, second, and third cases are causally dependent on the second parent while the fourth is causally dependent on the third parent. These four example cases would correspond to a weight vector of $< 0.0, 0.75, 0.25 >$ with $\alpha = 1.0$. For $\alpha < 1$, we incorporate a noise process ϵ by selecting the dependent variable's value from its domain by sampling uniform at random for $(1 - \alpha)$ percent of the cases. We treat X_{new}'s Markov blanket computed from Bayesian network $\mathcal{B}$ as a document. The causal dependency set for each of the X_i is also treated as a document. This results in a collection of documents, one for X_{new} and each X_i.

3.4 The Retrieval Problem

A Markov blanket describes the complete set of dependencies for a given variable. By definition, the Markov blanket is a subset of the variables over the modeling domain. Let each variable, $X_1, \ldots, X_d$ (including X_{new}) in a data-set be an indexing term. A Markov blanket then becomes a simple document containing a subset of indexing terms. Define weight w_{ij} as the number of occurrences of term-i in document j. For a Markov blanket, because a variable occurs at most once, we have that $w_{ij} \in \{0, 1\}$. Given d-variables, the Markov blanket $MB(X_j)$ for variable X_j is compactly described by weight vector

$$MB(X_j) =< w_{1j}, \ldots, w_{dj} >$$

A ranking function $R(q_i, d_j)$ outputs a value along the real number line that defines an ordering of documents in terms of their relevance to the given information need. Define ranking function R:

$$R(d_j, q_i) = \frac{\sum_{k=1}^{d} w_{kj} q_{ki}}{|X|}$$

That is the proportion of the variables in the Markov blanket that satisfy the query. With this definition, we rank the d Markov blankets in a Bayesian network and select the Markov blanket *document* associated with the highest rank. The top ranked Markov blanket corresponds to the variable dependencies that are most relevant to the query. We then measure quality in modeling ground truth variable dependences using ROC curves.

We expand a query for X_{new} into known ground truth causal dependencies in vector form and search for the most relevant document in the collection. In our procedure, we create X_{new} randomly. Given Bayesian network $\mathcal{B}$ learned from a data set augmented with the synthetic variable, compute documents $d_i = MB_{\mathcal{B}}(X_i)$. Define the f-blanket for X_{new}, $MB_f(X_{new}) = \{X_1, \ldots, X_k\}$. Given a query expansion, q_i, the most relevant document, d_r, is returned:

$$d_r = argmax_j R(d_j, q_i)$$

That is the document with the highest rank. This corresponds to the Markov blanket in the learned network that most closely resembles the ground truth f-blanket. In using a vector space approach and ranking, we allow for partial similarity with a given query. This is particularly important for synthetic variables that are weakly dependent on their parents. By adding a set of synthetic variables whose dependence on $X_1, \ldots, X_d$ varies in the number parent nodes and strength of dependence, we can use the true positive and false positive rate for retrieval to measure the Bayesian network's ability to accurately model true causal dependencies.

We call our approach Markov blanket retrieval (MBR). The algorithm for MBR appears in Figure 5. Input parameters to MBR are a data-set $\mathbf{X}$ dependence strength α, and parent set size k. We begin by computing the number of cases and variables in steps 1 and 2. Measurements are made for a fixed number of Q queries (step 3). Each query consists of a synthetic variable whose k ground truth parents are selected randomly (step 4). For each selected parent set, we choose their dependence strengths by sampling from the unit simplex (step 5). Before constructing the synthetic variable, we first create its domain set by taking the union of the domains of its k-parent set (step 6). Then, looping over each of the N cases (step 7) we compute the value, $x_{i,new}$ of the synthetic variable using the mixture weights and the dependence strength α (step 8, 9, 10). This gives us a new column of data corresponding to the synthetic variable V_{new}. The augmented data-set $\mathbf{X}'$ is then constructed by including the column of values, X_{new}, for the synthetic variable among the columns $\{X_1, \ldots, X_L\}$ of the original data set (step 11). We run structure learning on the augmented data set and obtain a Bayesian network $\mathcal{B}$ (step 12). For each variable in the augmented

MARKOV-BLANKET-RETRIEVAL($k, \alpha, \mathbf{X}$)
```
 1   N ← |X|
 2   L ← num-variables(X)
 3   for q ← 1 to Q
 4   do sample {V₁, ..., Vₖ} ∈ X
 5       sample < w₁, ..., wₖ > from simplex
 6       domain(Vₙₑw) = ⋃ᵏⱼ₌₁ domain(Vⱼ)
 7       for i ← 1 to N
 8       do sample A ~ Bernoulli(α)
 9           if a = 1 then xᵢ,ₙₑw ← f(xᵢ,₁, ..., xᵢ,ₖ) //using mixture weights
10           else xᵢ,ₙₑw ← ε
11           X' = {X₁, ..., Xₗ, Xₙₑw} //augment data-set
12       B ← learn-structure(X')
13       for i ← 1 to L + 1
14       do
15           dᵢ = compute-document(MB_B(Xᵢ))
16           q = compute-document(MB_{ground−truth}(Xₙₑw))
17           dᵣ = argmaxₗ R(dₗ, q)
18           record ROC data
```

Fig. 5. algorithm for Markov blanket retrieval analysis of structure learner

data-set, we obtain the Markov blanked computed by the structure learner and compute a document (steps 13, 14, and 15). Given the ground truth Markov blanket for the synthetic variable, we expand it into a query (step 16). We then rank Markov blanket *documents* from step 14 and return the highest ranking document (step 17). We then record whether or not our result is a true positive, true negative, false positive or false negative and continue to the next query iteration (step 18).

4   Experiments

Our experimental goal was to uncover how $K2GA$'s ability to model causal dependence changed as we varied the genetic algorithm's population size and number of generations across data-sets of different complexities. We ran experiments using three data-sets from the UCI machine learning repository. We selected one nominal (zoo), one mixed nominal-integer (lymphoma), and one real valued (sonar) data set for experiments in order to have representation across different types of data-sets. We rank data-sets by their complexity defined in terms of the number of variables and the number of instances (Table 1).

In our ranking, we include the class label in our variable counts. Since optimization based approaches such as $K2GA$ bound the maximum in-degree of nodes in the Bayesian network, it is important demonstrate how in-degree for causal dependence affects performance. This means measuring performance as more parents nodes are recruited. We tested variable dependence by running experiments for f-blankets of size 1,2, and 3.

Table 1. Data set complexities

rank	data-set	number of variables	number of instances
1	zoo	17	101
2	lymphoma	19	148
3	sonar	61	208

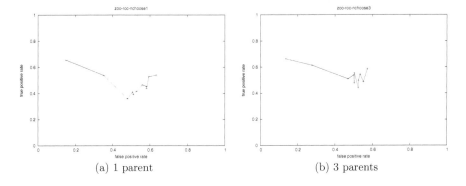

(a) 1 parent   (b) 3 parents

Fig. 6. ROC curve for zoo data-set with various parental causal dependencies for $K2GA$ at 50 generations and population size 10

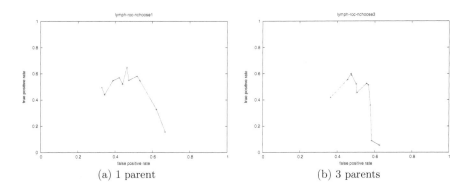

(a) 1 parent   (b) 3 parents

Fig. 7. ROC curve for lymphoma data-set with various parental causal dependencies for $K2GA$ at 50 generations and population size 10

A positive test instance is a synthetic variable for which a true causal dependency exists and a negative is a synthetic variable for which a dependence does not exist. We generated synthetic variables with 50% priors over positive instances. For the remaining 50%, we set thresholds for strength of causal dependence in regular increments for $\alpha = 0.0, 0.1, \ldots, 1.0$. Across all settings of α the expected generation rate for positive instances is $0.5 + \sum_{\alpha=0.0}^{1.0} 0.5\alpha = 0.75$.

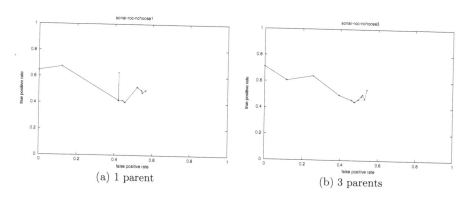

Fig. 8. ROC curve for sonar data-set with various parental causal dependencies for $K2GA$ at 50 generations and population size 10

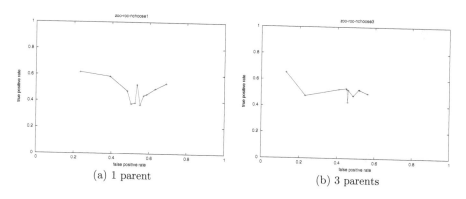

Fig. 9. ROC curve for zoo data-set with various parental causal dependencies for $K2GA$ at 100 generations and population size 20

An augmented data-set has $d + 1$ variables where d variables are from the original data-set, and the $d + 1$-th is the synthetic variable. Since a random approach must guess uniform at random which of the $d + 1$ Markov blanket *documents* matches the query, the probability of picking out the true positive is $\frac{1}{d+1}(0.5 + \sum_{\alpha=0.0}^{1.0} 0.5\alpha)$. Using the trapezoidal rule, we compute AUC for random performance as 0.5000.

K2GA performs optimization by stochastic search. K2GA is a genetic algorithm in which Bayesian network structure candidates are members of a population. Thus, the population size for $K2GA$ controls the number of frontiers along which stochastic search in the space of network structures is performed. The number of generations controls the number of optimization rounds for which search proceeds. We ran two versions of K2GA differing in population size and number of generations, one at 50 generations and population size of 10 and another at 100 generations and population size of 20. We refer to these as

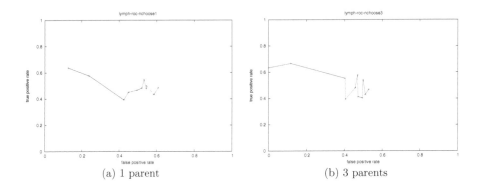

(a) 1 parent   (b) 3 parents

Fig. 10. ROC curve for lymphoma data-set with various parental causal dependencies for $K2GA$ at 100 generations and population size 20

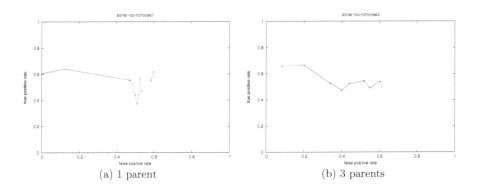

(a) 1 parent   (b) 3 parents

Fig. 11. ROC curve for sonar data-set with various parental causal dependencies for $K2GA$ at 100 generations and population size 20

K2GA-small and *K2GA-large*. We ran experiments for 100 queries for each setting of causal dependence strength using 5 fold cross validation on 10 random initializations of K2GA. Because validation is done directly on the resulting structure and not on the test set, we did not use the test set from each fold. We did this in order to train similarly to approaches that validate by partitioning data into training and testing sets.

This resulted in 5000 queries for each setting of α representing a total of $55,000$ total queries per experiment. ROC curves for f-blanket sizes 1 and 3 appear in Figure 6, 7, 8, 9, 10, 11. In each of our results, there was a dramatic decrease in the true positive rate once the false positive rate reached between 0.4 and 0.5. We compare K2GA-small with K2GA-large by their AUC scores (Table 2). We group the results of K2GA-small and K2GA-large and indicate the better performer in bold typeface.

Table 2. AUC scores for Markov-Blanket Retrieval

data-set	$K2GA$ setting	f-blanket size	AUC for MBR	AUC for random
zoo	50-gen 10-pop	1	**0.5802**	0.5000
zoo	100-gen 20-pop	1	0.5482	0.5000
zoo	50-gen 10-pop	2	**0.6338**	0.5000
zoo	100-gen 20-pop	2	0.6023	0.5000
zoo	50-gen 10-pop	3	**0.6359**	0.5000
zoo	100-gen 20-pop	3	0.5927	0.5000
lymph	50-gen 10-pop	1	0.5653	0.5000
lymph	100-gen 20-pop	1	**0.5737**	0.5000
lymph	50-gen 10-pop	2	**0.6305**	0.5000
lymph	100-gen 20-pop	2	0.6174	0.5000
lymph	50-gen 10-pop	3	0.6205	0.5000
lymph	100-gen 20-pop	3	**0.6529**	0.5000
sonar	50-gen 10-pop	1	0.6356	0.5000
sonar	100-gen 20-pop	1	**0.6709**	0.5000
sonar	50-gen 10-pop	2	0.6280	0.5000
sonar	100-gen 20-pop	2	**0.6640**	0.5000
sonar	50-gen 10-pop	3	**0.6652**	0.5000
sonar	100-gen 20-pop	3	0.6186	0.5000

We found that if the data-set contained a smaller number of variables as is the case with the zoo data-set (complexity rank 1), as we increase the number of parents upon which a variable can causally depend, K2GA-small consistently had higher AUC. A Bayesian network with larger node in-degree is a more complex model. Building more complex models require a larger number of training examples. The zoo data-set contains relatively few instances. Since K2GA-large searches twice as many frontiers for twice as many optimization rounds, it tends to over-fit the data. Therefore, its performance is worse on our simplest data-set as the f-blanket size increases. On the lymphoma data-set, we see a modest increase in the number of variables and 40% increase in number of instances. K2GA-large turns in its largest favorable difference in performance over K2GA-small when the f-blanket is 3. This coincides with K2GA-large's ability to search more complex models.

For the sonar data-set (rank 3), we find K2GA-large turns in a higher AUC for f-blanket sizes 1 and 2. The sonar data-set has $3x$ more variables. By searching twice as many frontiers for twice as many optimization rounds, K2GA-large is more able to consistently and stably (higher AUC) model causal linkages in complex data. When the f-blanket increases to 3, the number of instances becomes insufficient. Consider 2 Markov blankets each containing a child node with 3 parent nodes. Building network involves evaluating conditional probability tables. If each variable assumes only 2 states, we find the conditional probability table (CPT) for the child node has 2^4 entries. Across 2 Markov blankets, we have $(2^4)^2 = 256$ unique configurations. Estimating the CPTs for this example requires more than 208 instances. Because K2GA-small involves fewer frontiers and optimization rounds, it effectively builds lower complexity models. This gives us advantage when there are too few examples because it helps against over-fitting by early stopping. As we can see K2GA-small turns in a higher AUC than K2GA-large when the f-blanket is 3.

5   Discussion

We have presented a new tool for measuring the efficacy of structure learning algorithms in finding causal dependencies that exist within data. By treating membership in a Markov blanket as a retrieval problem and controlling for ground truth causal dependencies, we are able to borrow sound principles of ROC analysis to evaluate the structure learner's performance. Our measurements go beyond error by measuring stability across a range of dependence strengths using AUC. Our method measures structure learning efficacy directly from the learned structures themselves without use of a gold standard network. We have found from our experiments that Markov Blanket Retrieval (MBR) lends insight into parameter tuning and stability of a structure learning algorithm and feel it is a valuable tool for the data-mining community.

The goal for reported experiments was the development of a tool for comparing the performance of different parameterizations of a structure learner under varying dependence strengths. In complex real-world data-sets, some of the variables are correlated. Future investigation will include measurements for the effect of correlation between parent variables on modeling efficacy. We are encouraged by results for our measure on K2GA. A logical next step is to investigate MBR's utility in making fair comparison between different structure learning techniques. We represented Markov blankets using vector space and ranked documents based on a normalized inner product. This approach allowed us to observe the proportion of variables in the augmented data-set that matched the ground truth f-blanket. In future experiments we will extend our ranking approach to include measurement of graph properties as well as other distance measures.

References

[1] Chickering, D.: Learning equivalence classes of bayesian-network structures. Journal of Machine Learning Research 2, 445–498 (2002)
[2] Cooper, G.: Probabilistic inference using belief networks is np-hard. Artificial Intelligence 42(1), 393–405 (1990)
[3] Cooper, G., Herskovits, E.: A bayesian method for the induction of probabilistic networks from data. Mahcine Learning 9(4), 309–347 (1992)
[4] Dagum, P., Luby, M.: Approximating probabilistic inference in bayesian belief networks is np-hard. Artificial Intelligence 60(1), 141–153 (1993)
[5] Heckerman, D., Geiger, D., Chickering, D.: Learning bayesian networsk: The combination of knowledge and statistical data. Machine Learning 20, 197–243 (1995)
[6] Eaton, D., Murphy, K.: Bayesian structure learning using dynamic programming and mcmc. In: NIPS Workshop on Causality and Feature Selection (2006)

[7] Faulkner, E.: K2ga: Heuristically guided evolution of bayesian network structures from data. In: IEEE Symposium on Computational Intelligence and Data Mining, 3 Innovation Way, Newark, DE. 19702, April 2007, IEEE Computer Society Press, Los Alamitos (2007)

[8] Fawcett, T.: Roc graphs: Notes and practical considerations for data mining researchers. Technical Report HPL-2003-04, Hewlett Packard Research Labs (2003)

[9] Friedman, N., Nachman, I., Peér, D.: Learning bayesian network structure from massive datasets: The sparse candidate algorithm. In: Proceedings of UAI, pp. 206–215 (1999)

[10] Heckerman, D.: A tutorial on learning with bayesian networks (1995)

[11] Pearl, J., Verma, T.S.: A theory of inferred causation. In: Allen, J.F., Fikes, R., Sandewall, E. (eds.) KR'91: Principles of Knowledge Representation and Reasoning, San Mateo, California, pp. 441–452. Morgan Kaufmann, San Francisco (1991)

[12] Shaughnessy, P., Livingston, G.: Evaluating the causal explanatory value of bayesian network structure learning algorithms. In: AAAI Workshop on Evaluation Methods for Machine Learning (2006)

[13] Singh, M., Valtorta, M.: Construction of bayesian network structures from data: A brief survey and an efficient algorithm. International Journal of Approximate Reasoning 12(2), 111–131 (1995)

[14] Spirtes, P., Glymour, C., Scheines, R.: Causation, prediction, and search. Springer, Heidelberg (1993)

[15] Lam, W., Bacchus, F.: Learning bayesian belief networks: An approach based on the mdl principle. Comp. Int. 10, 269–293 (1994)

A New Combined Fractal Scale Descriptor for Gait Sequence

Li Cui[1] and Hua Li[2]

[1] School of Mathematics Sciences, Beijing Normal University, Beijing 100875
licui@bnu.edu.cn
[2] Institute of Computing Technology, Chinese Academy of Sciences, Beijing 100080
lihua@ict.ac.cn

Abstract. In this paper, we present a new combined fractal scale descriptor based on wavelet moments in gait recognition. This method is likely useful to general 2d objects pattern recognition. By introducing the Mallat algorithm of wavelet, it reduces the computational complexity compared with wavelet moments. Moreover, fractal scale has advantage on the self-similarity description of signals. And because it is based on wavelet moments, it is still translation, scale and rotation invariant, and have strongly anti-noise and occlusion handling performance. For completely decomposed signals, we get the new descriptor by combining the global and local fractal scale in each level. Experiments on a middle size database of gait sequences show that the new combined fractal scale method has simple computation and is an effective descriptor for 2-d objects.

1   Introduction

The classical moment invariants theory developed from Hu moment includes seven moment invariants for 2d image recognition [10]. The study of moment has been put more attention on because its effectiveness in pattern recognition. Many moments with different properties have been derived, such as Li moment, complex moment, Teague orthogonal moment, Zernike moments and Fourier-Mellin moments [6, 9, 15, 22]. And many descriptors based on moment for 2d patten recognition have been set up [21]. Moreover combining the descriptors of Zernike and Fourier moments, Zhang presented mixed descriptors, which work well in pattern recognition [26]. While sometimes these moments should compute some high-order moments. However the complexity and instability in computation of these moments increase when the order is high. In addition, the computation of high-order moments is still complex, though a few fast algorithms appear [20]. So it is hard to identify the similar objects with noise only using low order moments. Wavelet moments are new moments features, which combine the wavelet characteristic and moment trait [23]. Because wavelet transform is capable to provide both time and frequency localization [5], this characteristic is particularly suited to extract local discriminative features. But wavelet transform is not invariant to translation variations, some small translation can drastically change

P. Perner (Ed.): MLDM 2007, LNAI 4571, pp. 616–627, 2007.

wavelet feature. So simplex wavelet analysis is not widely used in the field of pattern recognition. Considering the respective characteristic of moments feature and wavelet analysis, wavelet moments resulted from the combination of moments feature and wavelet analysis, have many advantages: translation, scale and rotation invariance, strongly antinoise performance, and multi-resolution analysis. It overcomes the complex computation of high order moments and has the advantage on identification of the similar objects.

Recently gait recognition, as one branch of biometrics, is used to signify the identification of individuals in image sequences 'by the way they walk' [2]. From a surveillance perspective, gait recognition is an attractive modality because it may be performed at a distance, surreptitiously. Now, there are many methods contributing to the gait analysis, for example model-based [7, 14], appearance-based [1, 3, 8, 11, 13, 25]. In the image sequences, translation and scaling variations of walking people often exist, and moments is an efficient tool to deal with them. The application of classical moments to two dimensional images was first presented in the early sixties by Hu [10]. Little used moment based features to characterize optical flow for automatic gait recognition [16], thus linking adjacent images but not the complete sequence. Lee and Grimson computed a set of images features that are based on moments [14]. Liu et al. used the first and second moments of two binary silhouettes to determine an affine transformation that coarsely aligns them [17]. Shutler et al. developed new Zernike velocity moments to describe the motion throughout an image sequence to help recognize gait [24]. Zhao, Cui et al. combined wavelet velocity moment and reflective symmetry to analysis gait feature for human identification [27]. Moreover they proposed the wavelet reflective symmetry moments for gait recognition [28].

On the computation of Wavelet moments [23], only the wavelets with apparent formula are concerned. However the Mallat algorithm of wavelet [18] is not in use for computating the feature, which omits the filter character of wavelet. By introducing Mallat algorithm of wavelet and fractal scale, we have used fractal scale and wavelet moments in gait recognition [29]. In this paper, we develop a new combined fractal scale descriptor based on wavelet moments, which combine the global and local fractal scale in each level for completely decomposed gait sequence signals.

2 Combined Fractal Scale Descriptor

In mathematics, fractal scale is similar to Lipschitz exponent (Hölder exponent), which is used to discuss the global and local regularity of function. It is powerful to characterize the singularity of signals. While wavelet basis in $L^2(R)$ is a group of similar functions, which is formed by translation and scale of one function. So there is inner relation between wavelet and fractal scale. In [19], Mallat gave a method to compute the fractal scale by wavelet transform. And wavelet analysis with its fast Mallat algorithm provides multi-resolution, anti-noise and easy computation properties. Then solving the fractal scale of signals in this way has many advantages. Meanwhile, to recognize a signal, we use not only global fractal scale, but also local fractal scale. By completely decomposing signal with Mallat algorithm, we combine all fractal scale features in each level to represent a signal.

Because the researchful object of fractal scale is one dimension signals, we should get the adaptable signals from gait images first, then completely decompose the signals using Mallat algorithm. With the method in [19], we get the global and local fractal scale in each level. Then we combine all fractal scale to get the new descriptor for gait image. In gait recognition, we use the averaged features by normalizing with respect to the number of images in one period gait sequence. At last we analyze the time complexity of new combined features compared with wavelet moments and original fractal scale features.

2.1   Standardization and Sampling

To eliminate the influence of translation and scaling variations of walking person in image, we should standard the image first. The region of interest, i.e. the region of walking people, is mapped to the unit disc using the polar coordinates, where the centroid of walking region is the origin of the unit disc. Those pixels falling outside the unit disc are not used in the calculation. The coordinates are then described by the polar coordinate (r, θ). Translation invariance is achieved by moving the polar origin to the centroid of walking people by computing Hu's first moments. And scale invariance is implemented by scaling the walking region to the unit disc.

Rotation invariance is achieved by descriptor's definition. Shen [23] presented the wavelet moments in polar coordinate, which is easier to get rotate invariants. Our new descriptor is based on wavelet moments. It gets rotate invariants in the same way.

Definition 1. *Suppose $\psi(r)$ is a wavelet function. Then for each object $f(r, \theta)$ expressed in polar coordinates and defined on field Ω, its wavelet moments are*

$$F_{mnq} = \int \int_{\Omega} f(r, \theta)\psi_{mn}(r) \exp(-iq\theta)rdrd\theta \ ,$$

where $\psi_{mn}(r) = 2^{m/2}\psi(2^m r - n)$, $m, n, q \in Z^+$.

Here, q represents the frequency rank of image object. Descriptors $\|F_{mnq}\|_{mnq}$ ($\|F_{mnq}\|^2 := F_{mnq} * \overline{F_{mnq}}$) of f are invariant under the 2d rotations.

In the computation of our new descriptor, the first step is to compute the angular integral using FFT transform. After that we get a radial function, which is our sampling function. Signals are sampled along the radius direction.

Suppose (x, y) denotes pixel, $f(x, y)$ denotes 0 or 1 in black and white image, $\theta(x, y)$ is the polar angle in pixel (x, y), $r(x, y)$ is its polar radius and r_i is the average sampling point in radial direction. If the number of samples is 2^N, we divide the maximal radius into 2^N segments. In the kth segment, $MaxR \times (k/2^N) \leq R_k \leq MaxR \times ((k + 1)/2^N)$, after scale to an unit disc, $(k/2^N) \leq$

$r_k \leq ((k+1)/2^N)$, $MaxR$ is the maximum radius of current frame. $r = R/MaxR$, range of r is from 0 to 1, and R is from 0 to $MaxR$.

The sampling method is:

$$
\begin{aligned}
F_q(k) &= \int_{r_{k-1} \leq r \leq r_k} \int_{0 \leq \theta \leq 2\pi} f(r, \theta) \exp(-iq\theta) r \, dr \, d\theta \\
&= \int_{R_{k-1} \leq R \leq R_k} \int_{0 \leq \theta \leq 2\pi} f(x, y) \exp(-iq\theta(x,y))/MaxR^2 \, dx \, dy \\
&= \sum_x \sum_y f(x, y) \exp(-iq\theta(x,y)), \quad r_{k-1} \leq r \leq r_k \;,
\end{aligned}
$$

where $q = 1, 2, \cdots$.

In this way, we only need scan $f(x, y)$ by line in the processing of sampling but not compute the integral on θ in reality.

2.2 Completely Decomposition with Mallat Algorithm

Suppose $\varphi(x)$ is the scaling function, $\psi(x)$ is the wavelet function with compactly support, $\varphi_{j,k} = 2^{j/2}\varphi(2^j x - k)$, $\psi_{j,k} = 2^{j/2}\psi(2^j x - k)$, $c_k^j = \langle f, \varphi_{j,k} \rangle$, and $d_k^j = \langle f, \psi_{j,k} \rangle$, $k \in Z$.

For the sample signal of gait image, it is easy to suppose the number of sample is 2^N. In fact, $\{c_l^N\}_{l=0}^{2^N-1}$ is always known as the original sample signal. And the finite filters is concerned in practice. Completely decompose the signal into N levels with Mallat algorithm:

$$
c_k^{j-1} = \sum_{l \in Z} h_{l-2k} c_l^j
$$

$$
d_k^{j-1} = \sum_{l \in Z} g_{l-2k} c_l^j, \quad j = N, N-1, \cdots, 1, \quad k \in Z \;,
$$

where $\{h_l\}_l$ and $\{g_l\}_l$ are finite low-pass and high-pass filters. And if they are orthogonal, there is $g_l = (-1)^l h_{1-l}$. While in the bio-orthogonal condition there are four filters $\{h_l\}_l, \{g_l\}_l$(decomposition filters),$\{\tilde{h}_l\}_l, \{\tilde{g}_l\}_l$(reconstruction filters), and there are $g_l = (-1)^l \tilde{h}_{1-l}$, and $\tilde{g}_l = (-1)^l h_{1-l}$.

2.3 Computation of Global Fractal Scale

In [19], fractal scale is expressed in wavelet transform as follows.

If the signal $f \in L^2(R)$ is bounded and phase continuous and there exists certain α to make the wavelet transformation of f satisfy

$$
|\langle f, \psi_{j,k} \rangle| \leq c 2^{-j(\alpha+\frac{1}{2})}, \quad j = N-1, N-2, \cdots, N-M, \quad k \in Z \;,
$$

where $c > 0$ is constant, M is the level number of decomposition, N is the initial level number. Then the fractal scale of f is α.

Considering the high frequency part $\{d_k^j\}_{k,j}$, to get the global fractal scale is to solve the maximum α and minimum c, which satisfy the below inequations.

$$|d_k^j| \leq c2^{-j\alpha}, \quad j = N-1, N-2, \cdots, N-M, \quad k \in Z \ .$$

For the discrete initial signal $\{c_l^N, l = 0, \cdots, 2^N - 1\}$, we need to get the maximal high frequency signals in each level :

$$d_j^* = \max_k |d_k^j| \ ,$$

where $d_j^* > 0$. The problem becomes to solve c and α to satisfy

$$d_j^* \leq c2^{-j\alpha}, \quad j = N-1, N-2, \cdots, N-M.$$

Suppose $b_j^* = \log_2 d_j^*, b = \log_2 c$, and $\beta_j = b - j\alpha - b_j^*$, then using the least square estimation we can get α and b to minimize $\sum_j \beta_j^2$. To have enough data for least square estimation and for the stability of algorithm, only $M > 3$ is utilized in our implementation.

The fractal scale in the Mth level is $\alpha - \frac{1}{2}$. The global fractal scale in the Mth level is expressed as $\alpha_{M,global}$.

2.4   Computation of Local Fractal Scale

After original signals are decomposed completely, we get high frequency signals of former M levels and compute the local fractal scale in the Mth level ($M > 2$).

The high frequency signals of former M levels are

$$\begin{cases} d_k^{N-1}, & k = 0, 1, \cdots, 2^{N-1} - 1, \\ d_k^{N-2}, & k = 0, 1, \cdots, 2^{N-2} - 1, \\ \cdots, & \cdots, \\ d_k^j, & k = 0, 1, \cdots, 2^j - 1, \\ \cdots, & \cdots, \\ d_k^{N-M}, & k = 0, 1, \cdots, 2^{N-M} - 1 \ . \end{cases}$$

In the Mth level, the number of the local fractal scale is 2^{N-M}. And the No. l local fractal scale use the following wavelet coefficients, see Fig. 1.

$$\begin{cases} d_k^{N-M}, & k = l, \\ d_k^{N-M+1}, & k = 2l, 2l + 1, \\ \cdots, & \cdots \\ d_k^j, & k = 2^{j-N+M}l, 2^{j-N+M}l + 1, \cdots, 2^{j-N+M}l + 2^{j-N+M} - 1, \\ \cdots, & \cdots \\ d_k^{N-1}, & k = 2^{M-1}l, 2^{M-1}l + 1, \cdots, 2^{M-1}l + 2^{M-1} - 1 \ . \end{cases}$$

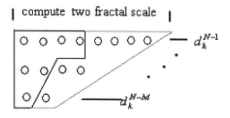

Fig. 1. Wavelet coefficients (The bottom row is d_k^{N-M}, the top row is d_k^{N-1}, from this partition, we can get two groups of high frequency signals corresponding to two blocks and so the two local fractal scale features)

To solve the lth local fractal scale in the Mth level is to solve the optimal α and c, which satisfy

$$|d_k^j| \leq c2^{-j\alpha}, \begin{array}{l} j = N - 1, N - 2, \cdots, N - M, \\ k = 2^{j-N+M}l, 2^{j-N+M}l + 1, \cdots, 2^{j-N+M}l + 2^{j-N+M} - 1 \end{array}.$$

The method is similar to the global one. The difference is the field of k becomes smaller. The No. l local fractal scale in M^{th} is expressed as $\alpha_{M,local}^l$ $(M > 3)$.

2.5 Combined Descriptor for Gait Sequence

In this way, with different frequency phase q in wavelet moments, we can compute the global and local fractal scale in each level larger than 3. Then combining all fractal features we get a new descriptor for 2d object, especially a new combined fractal scale descriptor for gait sequence.

Definition 2. *For each object $f(r, \theta)$ expressed in polar coordinates and defined on field Ω, compute*

$$s_q(r) = \int_0^{2\pi} f(r, \theta) \exp(-iq\theta) r d\theta, \quad q \in Z^+ .$$

Suppose the sample number of one dimensional signal $s_q(r)$ is 2^N, and completely decompose the sampling signal with Mallat algorithm. Solve the global and local fractal scale in each level. We get

$$\alpha_{M,global}^q, \alpha_{M,local}^{l,q}, \quad l = 0, 1, \cdots, 2^{N-M} - 1, \quad q = 1, 2, \cdots, n, \quad M = N - 1, \cdots, 4 ,$$

where $\alpha_{M,global}^q$ denotes $\alpha_{M,global}$ in phase q, $\alpha_{M,local}^{l,q}$ denotes $\alpha_{M,local}^l$ in the phase q. Then we call the vector $(\alpha_{M,global}^q, \alpha_{M,local}^{l,q})_{M,l,q}$ combined fractal scale descriptor.

In gait recognition, gait sequences are repetitive and exhibit nearly periodic behavior. So all features are averaged by normalizing with respect to the number of images in one period gait sequence. The new descriptor of gait is:

$$\begin{cases} \alpha^q_{M,global} = \sum_t \alpha^{t,q}_{M,global}/T \\ \alpha^{l,q}_{M,local} = \sum_t \alpha^{t,l,q}_{M,local}/T \end{cases}$$

where $\alpha^{t,q}_{M,global}$ and $\alpha^{t,l,q}_{M,local}$ respectively correspond to a certain image in one period gait sequence.

2.6   Time Complexity

The time complexity of our new descriptor will be discussed by comparing with the wavelet moments descriptor and the original fractal scale descriptors. Because the real difference focuses on the radial integral, we only consider the time complexity of the one dimension signal $s_q(r)$. That is to compare them in a certain phase q. Here $\psi(r)$ in fractal features can be any wavelet function. In fact we can use different filters to get fractal features. While in wavelet moments [23], only the cubic B-spline wavelet function is used, whose Gauss approximation form is

$$\psi(r) = \frac{4a^{d+1}}{\sqrt{2\pi(d+1)}} \cdot \sigma_\omega \cdot \cos(2\pi f_0(2r-1)) \cdot \exp(-\frac{(2r-1)^2}{2\sigma_\omega^2(d+1)}) \ ,$$

where $a = 0.697066, f_0 = 0.409177, \sigma_\omega^2 = 0.561145, d = 3, m = 0,1,2,3\cdots, n = 0,1,2,\cdots,2^{m+1}$.

According to this background, we discuss the time complexity of wavelet moments, original fractal scale descriptor and the combined fractal scale descriptors. Suppose the sampling number is 2^N, M is the level number in decomposition of wavelet moments and the length of low-pass and high-pass filers is n.

1. The computation of the wavelet moments descriptors (wavelet function is cubic B-spline): $(2^{M+2} + M - 1)2^N \cos(\cdot)\exp(\cdot)$, where we must compute two complex functions. The dimension of the descriptors is $2^{M+2} + M - 1$.
2. The computation of the original fractal scale descriptor in the Mth level includes two parts:
   (a) The complexity of Mallat algorithm is $2^N(1 - 2^{-M})2n$.
   (b) Solving the fractal scale mainly includes four $\log(\cdot)$ and four multiplication in least square estimation.
   So the computational complexity is $2^N(1-2^{-M})2n+(4+4\log(\cdot))(2^{N-M}+1)$. The dimension is $2^{N-M}+1$.
3. The computation of the new descriptor is combining all fractal scale in each level larger than 3. So the computational complexity of new descriptor is $(2^N - 1)2n + (4 + 4\log(\cdot))(2^{N-3} + N - 4)$. The dimension is $2^{N-3} + N - 4$.

From above analysis, the new combined fractal scale descriptor still has less computational complexity than the wavelet moments, though it is distinctly more complex than the original fractal descriptor in [29]. For higher recognition rate in gait recognition, the time cost of new combined feature is worthy. It is very important to process the huge data base.

3   Experiments

Our experiments is based on the background of gait recognition. To evaluate our method, we use CMU Motion of Body (MoBo) database (http://hid.ri. cmu. edu/ Hid/databases.html). CMU MoBo database includes 25 subjects who walk in treadmill. They are captured from 6 views. We extract 7 sequences for each subject in lateral and oblique view. Example image can be seen in Fig.2. The original image size is 486*640. To reduce the computation time, we resize the image to 180*240.

Rank order statistic [4] is used as performance measure. Due to a small number of examples, we hope to compute an unbiased estimate of the true recognition rate using a leave-one-out cross-validation method. That is, we first leave one example out, train the rest, and then classify or verify the omitted element according to its difference with respect to the rest examples.

Fig. 2. Image from CMU MoBo database (left is from oblique view and right is lateral view)

In [29], we have tested different filters with different sampling methods in a small UCSD (University of California, San Diego) database. The bi-orthogonal cubic B-spline filter with our sampling method works better. So we go on using bi-orthogonal cubic B-spline filters in this experiments. And the filters are:

$$h_{-1} = 0.25; h_0 = 0.75; h_1 = 0.75; h_2 = 0.25;$$
$$g_0 = 1; g_1 = -1.$$

To evaluate the robustness of our algorithm to occlusion occasions, we assume that in CMU database. We set occlusion in the central area of each frame for all sequences and test each occlusion subjects in other sequences database. This experiment to some extent simulates the segmentation errors which usually happen when people take something whose color or texture is similar to that of the background. Here we select the occlusion width 5 pixels, 11pixels and 15pixel, as shown in Fig.3. And to investigate the effects of noise, we added synthetic noise to each image. Fig.4 is the example images with added increasing noise level in CMU database.

Just like in [29], we use the same test method to get the local optimal parameters. So the radial sample number is 2^N, $N = 6$. In wavelet moments, the

Fig. 3. Occlusion images in CMU database(from left to right, 5 pixels, 11pixels and 15 pixels respectively)

Fig. 4. Noisy Data in CMU database(Noise level from left to right is from 0, to 10%, 20%, 30% and 50% respectively)

number of the decomposed level is 3. The angular phase number and the translative factor number are the same. While in the original fractal scale descriptors (with bi-orthogonal cubic B-spline filter) the angular phase number is $q_n = 8$, the number of the decomposed level is $M = 5$. With these parameters in new combined descriptor we get the features all together in $M = 4, 5, 6$ levels.

In this test, both the wavelet moments descriptors and the fractal scale descriptors adopts the following Hausdorff distance to measure the similarity. The Hausdorff distance for two sets of sequences $A = \{a_1, a_2 \cdots, a_n\}$ and $B = \{b_1, b_2 \cdots, b_m\}$. is defined as

$$\|A - B\|_H = max(max_{a \in A} D(a, B), max_{b \in B} D(b, A)) \ .$$

Where each sample is represented by fractal scale features, $D(p, Q)$ is the shortest Euclidean distance between sample p sequence and any sample in set Q. The Hausdorff distance has been proven to be an effective shape matching measure for object recognition [12].

In many aspects, we test our new combined fractal scale descriptor (CFS) in CMU database compared with B-spline wavelet moments descriptor (BWMs) [23] and original fractal scale descriptor (OFS)(using dual B-spline filters with our sampling method) [29], shown in Table 1. It can be seen that with the combined features, better performance in EER (Equal Error Rate), occlusion, speeding handling and anti-noise, has been achieved than in the case of only one level features. And in Table 2, we give the average time cost of every gait sequence

Table 1. Performance comparison in CMU database

Train	Test	BWMs	OFS	CFS
lateral view	lateral view	99.43%	98.86%	98.86%
EER(lateral view)		6.33%	6.14%	4.9%
lateral view	Occlusion 5p	100%	92%	100%
lateral view	Occlusion 11p	40%	60%	72%
lateral view	Occlusion 15p	20%	32%	60%
lateral view	Noise10%	100%	100%	92%
lateral view	Noise20%	92%	68%	76%
lateral view	Noise30%	72%	48%	48%
lateral view	Noise50%	36%	20%	28%
oblique view	oblique view	–	96%	98.29%
slow walk	fast walk	–	60%	60%
fast walk	slow walk	–	56%	72%

Table 2. Time comparison in CMU database (Unit: Second)

	BWMs	OFS	CFS
Time	26.038	7.5470	20.324

in one period using the three methods (on a 2.4GHz CPU, RAM 256M). Just like the analysis in section 2.6, the computation complexity of our new descriptor is acceptable in gait recognition. In fact, it is powerful and effective in our test.

4   Conclusions and Future Work

In this paper, we propose a new gait recognition method based on fractal scale and wavelet analysis. The new descriptors combine the fractal scale features from different decomposing levels together. The proposed method is still translation, scale and rotation invariants, and has strongly anti-noise and occlusion handling performance. And we analysis the time complexity in theory. The performance of new descriptors are evaluated in CMU databases.

Selection of coefficients, filters and sampling methods will influence recognition. In the future, it will be considered and discussed. Maybe we can try to find fit filters of fractal scale in gait. More conditions which can affect gait will be taken into account, experiment on bigger database and more changes will be done to test our method. And we would like to explore more general method and study more universal feature for recognition.

Acknowledgements

The authors would sincerely thank Dr. Guoying Zhao, who presents all experiment results. And we would like to thank Dr. R. Gross from the Carnegie

Mellon University for their help with the databases needed in the research. This work was supported by National High-Tech Research and Development Plan (grant No:2001AA231031), National Key Basic Research Plan(grant No: 2004CB318000) and National Special R&D Plan for Olympic Games(grant No: 2001BA904B08).

References

[1] Abdelkader, C.B., Cutler, R., Nanda, H., Davis, L.: EigenGait: Motion-based Recognition of People using Image Self-Similarity. In: Bigun, J., Smeraldi, F. (eds.) AVBPA 2001. LNCS, vol. 2091, pp. 284–294. Springer, Heidelberg (2001)

[2] BenAbdelkader, C., Cutler, R., Davis, L.: Motion-Based Recognition of People in EigenGait Space. In: Proc. of the fifth International Conference on Automatic Face and Gesture Recognition, pp. 254–259 (2002)

[3] Collins, R.T., Gross, R., Shi, J.: Silhouette-based Human Identification from Body Shape and Gait. In: Proc. of the fifth International Conference on Automatic Face and Gesture Recognition, pp. 366–371 (2002)

[4] Cutting, J.T., Proffitt, D.R., Kozlowski, L.T.: A biomechanical invariant for gait perception. J. Exp.Psych.:Human Perception and Performance, pp.357–372 (1978)

[5] Daubechies, I.: Ten Lectures on Wavelet, 2nd edn. Capital City press, Capital (1992)

[6] Derrode, S., Ghorbel, F.: Robust and efficient Fourier-Mellin transform approximations for invariant grey-level image description and reconstruction. Computer Vision and Image Understanding 83(1), 57–78 (2001)

[7] Dockstader, S.L., Murat Tekalp, A.: A Kinematic Model for Human Motion and Gait Analysis. In: Proc. of the Workshop on Statistical Methods in Video Processing (ECCV), Copenhagen, Denmark, June 2002, pp. 49–54 (2002)

[8] Hayfron-Acquah, J.B., Nixon, M.S., Carter, J.N.: Automatic Gait Recognition by Symmetry Analysis. In: Proceedings of the 2nd International conference on Audio and Video Based Person Authentication, pp. 272–277 (2001)

[9] Hew, P.C., Alder, M.D.: Recognition of Printed Digits using Zernike or Orthogonal Fourier-Mellon Moments, Perth, Western Australia, pp. 1–11 (1997)

[10] Hu, M.: Visual Pattern Recognition by Moment Invariant. IRE Trans. On Information Theory IT-8, 179–187 (1962)

[11] Huang, P.S.: Automatic Gait Recognition via Statistical Approaches for Extended Template Features. IEEE Transactions on Systems, Man, and Cybernetics-Part B: Cybernetics 31(5), 818–824 (2001)

[12] Huttenlocher, D.P., Klanderman, G.A., William, J.R.: Comparing images using the Hausdorff distance. IEEE Transactions on Pattern Analysis and Machine Intelligence 15(9), 850–863 (1993)

[13] Kale, A., Rajagopalan, A.N., Cuntoor, N., Kruger, V., Chellappa, R.: Identification of Humans Using Gait. IEEE Transactions on Image Processing, http://www.cfar.umd.edu/~kale/myproposal.pdf

[14] Lee, L., Grimson, W.E.L.: Gait Appearance for Recognition. Biometric Authentication, pp. 143–154 (2002)

[15] Li, Y.: Reforming the theory of invariant moments for pattern recognition. Pattern Recognition 25, 723–730 (1992)

[16] Little, J.J., Boyd, J.E.: Recognising people by their gait: The shape of motion. Videre 1(2), 2–32 (1998)

[17] Liu, Y., Collins, R., Tsin, Y.: Gait Sequence Analysis using Frieze Patterns. In: Heyden, A., Sparr, G., Nielsen, M., Johansen, P. (eds.) ECCV 2002. LNCS, vol. 2351, pp. 657–671. Springer, Heidelberg (2002)

[18] Mallat, S.: Multiresolution approximations and wavelet orthonormal bases of $L^2(R)$. A Trans. Amer. Math. Soc. 315, 69–87 (1989)

[19] Mallat, S., Huang, W.L.: Singularity Detection and Processing with Wavelets. IEEE Transaction on Information Theory 38(2), 617–643 (1992)

[20] Mohammed, A., Yang, J., Zhang, F.C.: A PC-Based Real-Time Computation of Moment Invariants. Journal of Software 13(9), 1165–1172 (2002)

[21] Prokop, R.J., Reeves, A.P.: A survey of moment-based techniques for unoccluded object representation and recognition. CVGIP: Graphical Models Image Process 54(5), 438–460 (1992)

[22] Teague, M.: Image analysis via the general theory of moments. J. Opt. Soc. Am. 70, 920–930 (1980)

[23] Shen, D.G., Ip, H.H.S.: Discriminative wavelet shape descriptors for recognition of 2D patterns. Pattern Recognition 32, 151–165 (1997)

[24] Shutler, J.D., Nixon, M.S.: Zernike Velocity Moments for Description and Recognition of Moving Shapes. In: BMVC, pp. 705–714 (2001)

[25] Wang, L., Ning, H., Hu, W., Tan, T.: Gait Recognition Based on Procrustes Shape Analysis. In: Proceedings of the 9th IEEE International Conference on Image processing, Rochester, New York, pp. 433–436 (2002)

[26] Zhang, D., Lu, G.: An Integrated Approach to Shape Based Image Retrieval. In: ACCV2002. The 5th Asian Conference on Computer Vision, Melbourne, Australia, pp. 652–657 (2002)

[27] Zhao, G., Cui, L., Li, H.: Combining Wavelet Velocity Moments and Reflective Symmetry for Gait Recognition. In: Li, S.Z., Sun, Z., Tan, T., Pankanti, S., Chollet, G., Zhang, D. (eds.) IWBRS 2005. LNCS, vol. 3781, pp. 205–212. Springer, Heidelberg (2005)

[28] Zhao, G., Cui, L., Li, H.: Automatic Gait Recognition Based on Wavelet Reflective Symmetry Moments. Journal of Information and Computational Science 2(2), 421–428 (2005)

[29] Zhao, G., Cui, L., Li, H.: Gait Recognition Using Fractal Scale and Wavelet Moments. The 2006 International Conference of Pattern Recognition (ICPR2006), pp. 693–705 (2006)

Palmprint Recognition by Applying Wavelet Subband Representation and Kernel PCA

Murat Ekinci and Murat Aykut

Computer Vision Lab.
Department of Computer Engineering,
Karadeniz Technical University, Trabzon, Turkey
ekinci@ktu.edu.tr

Abstract. This paper presents a novel Daubechies-based kernel Principal Component Analysis (PCA) method by integrating the Daubechies wavelet representation of palm images and the kernel PCA method for palmprint recognition. The palmprint is first transformed into the wavelet domain to decompose palm images and the lowest resolution subband coefficients are chosen for palm representation. The kernel PCA method is then applied to extract non-linear features from the subband coefficients. Finally, weighted Euclidean linear distance based NN classifier and support vector machine (SVM) are comparatively performed for similarity measurement. Experimental results on PolyU Palmprint Databases demonstrate that the proposed approach achieves highly competitive performance with respect to the published palmprint recognition approaches.

1 Introduction

Biometric approaches utilize the identity of a person with certain physiological or behavioral characteristics [1]. Palmprint is a relatively new biometric feature, and is regarded as one of the most unique, reliable, and stable personal characteristics [1]. Compared with other biometrics, the palmprints has several advantages: low-resolution imaging can be employed; low-cost capture devices can be used; it is difficult to fake a palmprint; the line features of the palmprints are stable, etc. [1]-[11]. It is for these reasons that palmprint recognition has recently attracted an increasing amount of attention from researchers.

There are many approaches for palmprint recognition based on line-based [6][4][5], texture-based [11][5], and appearance-based methods [3][10][9][8] in various literature. In the line-based approach, the features used such as principal lines, wrinkles, delta points, minutiae [6], feature points [4] and interesting points [5], are sometimes difficult to extract directly from a given palmprint image with low resolution. The recognition rates and computational efficiency are not strong enough for palmprint recognition. In the texture-based approach, the texture features [5][1] are not sufficient and the extracted features are greatly affected by the lighting conditions. From that disadvantages, researches have developed the appearance-based approaches.

P. Perner (Ed.): MLDM 2007, LNAI 4571, pp. 628–642, 2007.

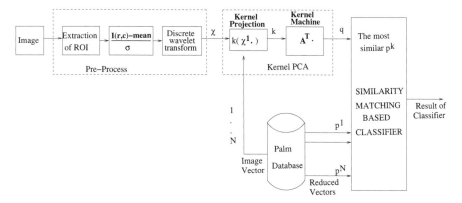

Fig. 1. Main steps in the proposed algorithm

The appearance-based approaches only use a small quantity of samples in each palmprint class randomly selected as training samples to extract the appearance features (commonly called algebraic features) of palmprints and form feature vector. Eigenpalms method [10], fisherpalms method [3], and eigen-and-fisher palms [9] are presented as the appearance-based approaches for palmprint recognition in literature. Basically, their representations only encode second-order statistics, namely, the variance and the covariance. As these second order statistics provide only partial information on the statistics both natural images and palm images, it might become necessary to incorporate higher order statistics as well. In other words, they are not sensitive to higher order statistics of features. A kernel fisherpalm [8] is presented as another work to resolve that problem. In addition, for palmprint recognition, the pixel wise covariance among the pixels may not be sufficient for recognition. The appearance of a palm image is also severely affected by illumination conditions that hinder the automatic palmprint recognition process.

Converging evidence in neurophysiology and psychology is consistent with the notion that the visual system analyses input at several spatial resolution scales [26]. Thus, spatial frequency preprocessing of palms is justified by what is known about early visual processing. By spatial frequency analysis, an image is represented as a weighted combination of basis functions, in which high frequencies carry finely detailed information and low frequencies carry coarse, shape-based information. Recently, there have been renewed interests in applying discrete transform techniques to solve some problems in face recognition [18][19][24], in palmprint recognition [2][24][25] and many real world problems. An appropriate wavelet transform can result in robust representations with regard to lighting changes and be capable of capturing substantial palm features while keeping computational complexity low.

From the considerations briefly explained above, we propose to use discrete wavelet transform (DWT) to decompose palm images and choose the lowest resolution subband coefficients for palm representation. We then apply kernel PCA as a nonlinear method to project palmprints from the high-dimensional palmprint space to a significantly lower-dimensional feature space, in which the palmprints from the different palms can be discriminated much more efficiently.

The block diagram of the main steps involved in developing the proposed palmprint algorithm is illustrated in Figure 1. The palm images are read from the digital system and used to extract a gray-level region of interest (ROI) depicting palmprint texture. The palm images are resized to 128 by 128 pixels, converted into column vector form and made zero mean and unit variance. In the process of feature extraction, the palmprint images are then decomposed into multi resolution representation by discrete wavelet transform (DWT). Then, the decomposed images in the lowest resolution subband coefficients are selected and are fed into a nonlinear method, Kernel PCA, computation. Therefore, we get the feature matrix of all training palmprint samples. The main contributions and novelties of the current paper are summarized as follows:

- To reliably extract palmprint representation, we adopt a template matching approach where the feature vector of a palm image is obtained through a multilevel two-dimensional discrete wavelet transform (DWT). The dimensionality of a palm image is greatly reduced to produce the *waveletpalm*.
- A nonlinear machine learning method, kernel PCA, is applied to extract palmprint features from the waveletpalm.
- The proposed algorithm is tested on two public palmprint databases, we called as PolyU-I and PolyU-II databases. We provide some quantitative comparative experiments to examine the performance of the proposed algorithm and different combinations of the proposed algorithm. Comparison between the proposed algorithm and other recent approaches is also given.

2   Discrete Wavelet Transform

The DWT was applied for different applications given in the literature e.g. texture classification [16], image compression [17], face recognition [18][19], because of its powerful capability for multi resolution decomposition analysis. The wavelet transform breaks an image down into four sub-sampled, or decimated, images. They are subsampled by keeping every other pixel. The results consist of one image that has been high pass filtered in both the horizontal and vertical directions, one that has been high pass filtered in the vertical and low pass filtered in the horizontal, one that has been lowpassed in the vertical and highpassed in the horizontal, and one that has been low pass filtered in both directions.

So, the wavelet transform is created by passing the image through a series of 2D filter bank stages. One stage is shown in Fig. 2, in which an image is first filtered in the horizontal direction. The filtered outputs are then down sampled by a factor of 2 in the horizontal direction. These signals are then each filtered by an identical filter pair in the vertical direction. Decomposed image into 4 subbands is also shown in Fig. 2. Here, H and L represent the high pass and low pass filters, respectively, and $\boxed{\downarrow 2}$ denotes the subsampling by 2. Second-level decomposition can then be conducted on the LL subband. Second-level structure

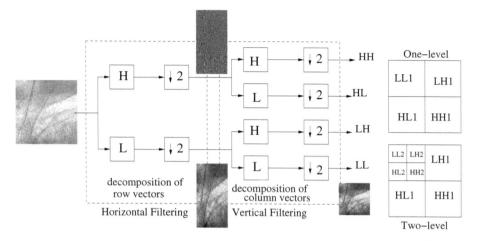

Fig. 2. One-level 2-D filter bank for wavelet decomposition and multi-resolution structure of wavelet decomposition of an image

of wavelet decomposition of an image is also shown in Fig. 2. This decomposition can be repeated for n-levels.

The proposed work based on the DWT addresses the four-level and six-level decomposition of images in Database I and Database II, respectively. Daubechies-4 and -8 low pass and high pass filters are also implemented [14]. Additionally, four- and six-levels of decompositions are produced, then 32 x 32 and 16 x 16 sub-images of 128 x 128 images in the wavelet is processed as useful features in the palmprint images. Reducing of the image resolution helps to decrease the computation load of the feature extraction process.

2.1   FFT and DCT

$F(u, v)$ and $C(u, v)$ are 2-D FFT and DCT coefficients of an W x H image $I(x, y)$, respectively. The feature sequence for each one is independently generated using the 2D-FFT and 2D-DCT techniques. The palmprint image (128 x 128) in the spatial domain is not divided into any blocks. The FFT and DCT coefficients for the palmprint image are first computed. In the FFT, the coefficients correspond to the lower frequencies than 3 x 3, and correspond to the higher frequencies than 16 x 16 in the FFT, are discarded by filtering. In other words, 247 coefficients correspond to the 6% coefficients in the frequency domain, are only implemented. These data are empirically determined to achieve best performance. Therefore, the palmprint image in the spatial domain is represented with a few coefficients, which is corresponding to 1.5% of the original size of image (128 x 128), by using filtered FFT based image representation. In DCT, low frequencies correspond to the 12.5% coefficients are also selected as useful features. Finally, $N = \mu$ x ν features form a vector $\chi \in \Re^N$, $\chi = (F_{0,0}, F_{0,1}, ...F_{\mu,\nu})$ for FFT, and form a vector $\chi = (C_{0,0}, C_{0,1}, ...C_{\mu,\nu})$ for DCT.

3   Kernel PCA

The kernel PCA (KPCA) is a technique for nonlinear dimension reduction of data with an underlying nonlinear spatial structure. A key insight behind KPCA is to transform the input data into a higher-dimensional **feature space** [12]. The feature space is constructed such that a nonlinear operation can be applied in the input space by applying a linear operation in the feature space. Consequently, standard PCA can be applied in feature space to perform nonlinear PCA in the input space.

Let $\chi_1, \chi_2, ..., \chi_M \in \Re^N$ be the data in the input space (the input space is 2D-DWT coefficients in this work), and let Φ be a nonlinear mapping between the input space and the feature space i.e. using a map $\Phi : \Re^N \rightarrow F$, and then performing a linear PCA in F. Note that, for kernel PCA, the nonlinear mapping, Φ, usually defines a kernel function [12]. The most often used kernel functions are polynomial kernels, Gaussian kernels, and sigmoid kernels [12]:

$$k(\chi_i, \chi_j) = \langle \chi_i, \chi_j \rangle^d, \tag{1}$$

$$k(\chi_i, \chi_j) = exp\left(-\frac{\|\chi_i - \chi_j\|^2}{2\sigma^2}\right), \tag{2}$$

$$k(\chi_i, \chi_j) = tanh(\kappa \langle \chi_i, \chi_j \rangle + \vartheta), \tag{3}$$

where d is a number in the set of natural numbers, e.g. $\{1,2,...\}$, $\sigma > 0$, $\kappa > 0$, and $\vartheta < 0$.

The mapped data is centered, i.e. $\sum_{i=1}^{M} \Phi(\chi_i) = 0$ (for details see [12]), and let D represents the data matrix in the feature space: $D = [\Phi(\chi_1)\Phi(\chi_2)\cdots\Phi(\chi_M)]$. Let $K \in \Re^{MxM}$ define a kernel matrix by means of dot product in the feature space:

$$K_{ij} = (\Phi(\chi_i) \cdot \Phi(\chi_j)). \tag{4}$$

The work in [12] shows that the eigenvalues, $\lambda_1, \lambda_2, \ldots, \lambda_M$, and the eigenvectors, $V_1, V_2, \ldots, V_M$, of kernel PCA can be derived by solving the following eigenvalue equation:

$$KA = MA\Lambda \tag{5}$$

with $A = [\alpha_1, \alpha_2, \ldots, \alpha_M]$ and $\Lambda = diag\{\lambda_1, \lambda_2, \ldots, \lambda_M\}$. A is MXM orthogonal eigenvector matrix, Λ is a diagonal eigenvalue matrix with diagonal elements in decreasing order ($\lambda_1 \geq \lambda_2 \geq \cdots \geq \lambda_M$), and M is a constant corresponds to the number of training samples. Since the eigenvalue equation is solved for α's instead of eigenvectors, $V = [V_1, V_2 \ldots V_M]$, of kernel PCA, first, A should be normalized to ensure that eigenvalues of kernel PCA have unit norm in the feature space, therefore $\lambda_i\|\alpha_i\|^2 = 1, i = 1, 2, \ldots, M$. After normalization the eigenvector matrix, V, of kernel PCA is then computed as follows:

$$V = DA \tag{6}$$

Now let χ be a test sample whose map in the higher dimensional feature space is $\Phi(\chi)$. The kernel PCA features of χ are derived as follows:

$$F = V^T \Phi(\chi) = A^T B \qquad (7)$$

where $B = [\Phi(\chi_1) \cdot \Phi(\chi) \Phi(\chi_2) \cdot \Phi(\chi) \cdots \Phi(\chi_M) \cdot \Phi(\chi)]^T$.

4   Similarity Measurement

When a palm image is presented to the wavelet-based kernel PCA classifier, the wavelet feature of the image is first calculated as detailed in Section 2, and the low-dimensional wavelet-based kernel PCA features, F, are derived using the equation 7. Let $M_k^0, k = 1, 2, .., L$, be the mean of the training samples for class w_k. The classifier applies, then, the nearest neighbor rule for classification using some similarity (distance) measure δ:

$$\delta(F, M_k^0) = min_j \delta(F, M_j^0) \longrightarrow F \in w_k, \qquad (8)$$

The wavelet-based kernel PCA feature vector, F, is classified as belong to the class of the closest mean, M_k^0, using the similarity measure δ.

Popular similarity measures include the Weighted Euclidean Distance (WED) [13] and Linear Euclidean Distance (LED) which are defined as follows:

$$WED : d_k = \sum_{i=1}^{N} \frac{(f(i) - f_k(i))^2}{(s_k)^2} \qquad (9)$$

where f is the feature vector of the unknown palmprint, f_k and s_k denote the kth feature vector and its standard deviation, and N is the feature length.

$$LED : d_{ij}(\mathbf{x}) = d_i(\mathbf{x}) - d_j(\mathbf{x}) = 0 \qquad (10)$$

where $d_{i,j}$ is the decision boundary separating class w_i from w_j. Thus $d_{ij} > 0$ for pattern of class w_i and $d_{ij} < 0$ for patterns of class w_j.

$$d_j(\mathbf{x}) = \mathbf{x}^T \mathbf{m}_j - \frac{1}{2}\mathbf{m}_j^T \mathbf{m}_j, \qquad j = 1, 2, ...M \qquad (11)$$

$$\mathbf{m}_j = \frac{1}{N_j} \sum_{\mathbf{x} \in w_j} \mathbf{x}, \qquad j = 1, 2, ..., M \qquad (12)$$

where M is the number of pattern classes, N_j is the number of pattern vectors from class w_j and the summation is taken over these vectors.

Support Vector Machines (SVMs) have recently been known to be successful in a wide variety of applications [12][20]. SVM-based and WED-based classifier are also compared in this work. In SVM, we first have a training data set, like, $D = \{(x_i, y_i)|x_i \in X, y_i \in Y, i = 1, ..., m\}$. Where X is a vector space of dimension d and $Y = \{+1, -1\}$. The basic idea of SVM consists in first mapping

x into a high dimension space via a function, then maximizing the margin around the separating hyper lane between two classes, which can be formulated as the following convex quadratic programming problem:

$$maximize \quad W(\alpha) = \sum_{i=1}^{m} \alpha_i - \frac{1}{2} \sum_{i,j=1}^{m} \alpha_i \alpha_j y_i y_j (K(x_i, x_j) + \frac{1}{C}\delta_{i,j}) \tag{13}$$

$$subject \quad to \quad 0 \le \alpha_i \le C, \forall_i, \tag{14}$$

$$and \quad \sum_{i}^{m} y_i \alpha_i = 0 \tag{15}$$

where $\alpha_i(\ge 0)$ are Lagrange multipliers. C is a parameter that assigns penalty cost to misclassification of samples. $\delta_{i,j}$ is the Kronecker symbol and $K(x_i, x_j) = \langle \phi(x_i) \cdot \phi(x_j) \rangle$ is the Gram matrix of the training examples. The form of decision function can be described as

$$f(x) = \langle w, \Phi(x) \rangle + b \tag{16}$$

where, $w = \sum_{i=1}^{m} \alpha_j^* y_i \Phi(x_i)$, and b is a bias term.

5   Experiments

The performance of the proposed method is evaluated with PolyU-I database [15] and PolyU-II database [11].

5.1   Database I

The samples in PolyU-I palmprint database were captured by a CCD based palmprint capture device [15]. The PolyU-I database contains 600 gray scale images of 100 different palms with six samples for each palm. Six samples from each of these palms were collected in two sessions, where the first three samples were captured in the first session, and the other three in the second session. The average interval between the first and the second session was two months. In our experiments, sub-image of each original palmprints was firstly cropped to the size of 128 x 128 by finger gaps using an algorithm similar to [21]. Figure 3 shows typical samples in the database in which the last two samples were captured from the same palm at different sessions. When palmprints are collected in different sessions, direction and amount of stretching of a palm may vary so that even palmprints from the same palm may have a little rotation and translation. Furthermore, palms differ in size and the lighting, translation, and orientation conditions in both sessions are very different. Hence they will effect the accuracy, if palmprint images are oriented and normalized before feature extractions and matching. But no any preprocessing step was done. It is directly processed to achieve recognition performances given in this paper. For instance, the palms

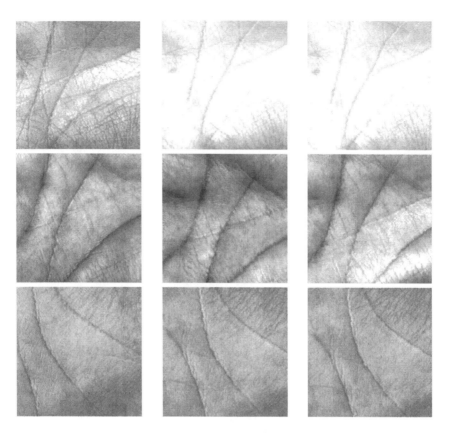

Fig. 3. Some typical restrictions in the the PolyU-I database. Left samples were cropped at first session. Last two samples were cropped from the same palm at second session. The restrictions are reported in this work as the different lighting (Top), orientation (Middle), and translation (Bottom) conditions.

shown in the first column in Figure 3 are used as training set, and the corresponding to the last two samples are also employed as testing set. In order to reduce the computation complexity, we independently adopted three different 2D discrete transforms (FFT, DCT, WT) to decompose the palm print image into lower resolution.

Two different experiments on this PolyU-I database were done to show the recognition performance of the proposed algorithm. The first experiment is the most challenging experiment which is in the case of that the palm images captured in the first session are chosen as training set, the other palms captured in the second session are selected as testing set. This is more realistic experiment and there are also more various problems such as lighting, orientation, and translation conditions because the training and test images were not obtained in the same session, which is always the case in a real world applications. During the experiments, the features are extracted by using the proposed method with

Table 1. Database I: Comparative performance evaluation for the different matching schemes with different feature lengths

Method	Feature length				
	25	75	125	200	300
PCA	212 (70.667)	231 (77.0)	229 (76.333)	227 (75.66)	231 (77.0)
KPCA	149 (49.67)	186 (62.0)	224 (74.667)	221 (73.667)	233 (77.667)
DCT+KPCA	218 (72.667)	231 (77.0)	240 (80.0)	240 (80.0)	242 (80.767)
FFT+KPCA	194 (64.667)	226 (75.333)	240 (80.0)	244 (81.333)	242 (80.667)
DWT+KPCA	215 (71.667)	234 (78.0)	237 (79.0)	242 (80.667)	244 (81.333)

length 25, 75, 125, 200, and 300. The WED is first used to cluster those features. The matching is separately conducted and the results are listed in Table 1. The numbers given in Table 1 show the correct recognition from 300 test samples. The entries in the brackets indicate the corresponding recognition accuracies. Kernel PCA gives higher performance than PCA when long feature lengths is used. A high recognition rate (81.333%) was achieved for the DCT+KPCA and DWT+KPCA, with feature lengths of 200 and 300, respectively.

Figure 4 shows the performance variation for WED and SVM classifiers with the increase in number of features produced by wavelet-based kernel PCA. The parameters of SVM employed in the experiments were empirically selected. The SVM using the radial basis function was implemented. The SVM training was achieved with C-SVM, a commonly used SVM classification algorithm [22]. The training parameter γ, ϵ and C were empirically fixed at 0.5, 0.1, and 100, respectively. When the number of features is less than about 60, the SVM-based classifier gives higher recognition rate. But while the number of features is higher than about 60, WED-based classifier gives higher accurate results.

The recognition accuracy (81.333%) in the first experiment may not be very encouraging. When the database is carefully investigated, there are translation,

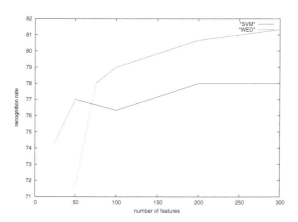

Fig. 4. Performance analysis of classifier with the number of features: DWT+ KPCA method using the SVM- and WED-based classifiers

Table 2. Database I: Recognition rate of different number of training samples(%)

Train	PCA		KPCA		FFT+KPCA		DCT+KPCA		DWT+KPCA	
Samples	LED	WED	LED	WED	LED	WED	LED	WED	LED	WED
1	66.0	80.2	73.6	80.6	82.0	82.8	80.0	82.4	82.8	82.0
2	74.25	93.5	83.75	94.0	92.25	98.0	89.5	95.75	89.5	95.75
3	77.3	95.0	84.0	96.67	90.6	98.67	90.0	97.33	90.3	97.67
4	70.0	97.5	82.0	98.0	92.0	**99.5**	90.5	98.0	91.0	98.0

rotation, or illumination changes in the input images at least 42 samples correspond to 14 persons (the database includes 100 persons). This is one of the main problem to obtain lower recognition rate than expected. We did not do more works in the pre-processing and at the palm image alignment for PolyU-I database, because we focused to another palmprint database which is developed by the PolyU [11] and includes more samples and persons, and we called PolyU-II to this database. However, we designed a second experiment for PolyU-I database to clarify the efficiency of the proposed algorithm. The experiments on the PolyU-II database will also be given in the next section.

The performance of the second experiment on the PolyU-I database is summarized in Table 2. Table 2 shows the different recognition rate with different number of training samples. Four kind of experiment schemes were designed: one (two, three, or four) sample(s) of each person was randomly selected for training, and other samples were used for authentication, respectively. Kernel PCA has given higher recognition rate than PCA. Discrete transform-based kernel PCA has increased the recognition rate. High recognition rates (99.5%) and (98.0%) were achieved by FFT+KPCA and DWT+KPCA, respectively, when the four samples were used as training. WED-based classifier has also given higher matching results than LED-based classifier.

5.2   Database II

The PolyU-II palmprint database [11] was also obtained by collecting palmprint images from 193 individuals using a palmprint capture device. People was asked to provide about 10 images, each of the left and right palm. Therefore, each person provided around 40 images, so that this PolyU database contained a total of 7,752 gray scale images from 386 different palms. The samples were collected in two sessions, where the first ten samples were captured in the first session and other ten in the second session. The average interval between the first and second collection was 69 days. The resolution of all original palmprint images is 384 x 284 pixels at 75 dpi. In addition, they changed the light source and adjusted the focus of the CCD camera so that the images collected on the first and second occasions could be regarded as being captured by two different palmprint devices. Typical samples captured under different lighting conditions on the second sessions of image capture could not be shown in this paper because of paper limitation, but they can be seen from [11]. Although the lighting conditions in the second

collection of palm images are quite different from the first collection, the proposed
method can still easily recognize the same palm.

At the experiments for PolyU-II database, we use the preprocessing technique
described in [11] to align the palmprints. In this technique, the tangent of the two
holes (they are between the forefinger and the middle finger, and between the
ring finger and the little finger) are computed and used to align the palmprint.
The central part of the image, which is 128 x 128, is then cropped to represent
the whole palmprint. Such preprocessing greatly reduces the translation and
rotation of the palmprints captured from the same palms. An example of the
palmprint and its cropped image is shown in Figure 5.

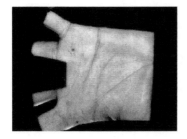

Fig. 5. Original palmprint and it's cropped image

Two different experiments were done on the PolyU-II database. In the first
experiment, the first session was used as training set, second session includes
3850 samples of 386 different palms was also used as testing set. In this experiment, the features are extracted by using the proposed kernel based eigenspace
method with length 50, 100, 200, 300, and 380. WED- and LED-based matching
were independently used to cluster those features. The matching is separately
conducted and the results are listed in Table 3. The number given in Table 3 represents the correct recognition samples in all test samples (3850). The entries in
brackets also represent corresponding the recognition rate. High recognition rates
93.454% and **93.168%** were achieved for the FFT+KPCA and DWT+KPCA,
with feature length of 300, respectively. A nearest-neighbor classifier based on

Table 3. Database II: Comparative performance evaluation for the different matching
schemes with different feature lengths. Train is first session, test is second session.

Method	Feature length				
	50	75	100	200	300
PCA	3411 (88.597)	3477 (90.311)	3498 (90.857)	3513 (91.246)	3513 (91.246)
DWT+PCA	3444 (89.454)	3513 (91.246)	3546 (92.103)	3570 (92.727)	3568 (92.675)
KPCA	3411 (88.597)	3481 (90.415)	3498 (90.857)	3508 (91.116)	3510 (91.168)
DCT+KPCA	3455 (89.74)	3528 (91.636)	3554 (92.311)	3595 (93.376)	3598 (93.454)
FFT+KPCA	2746 (71.324)	2933 (76.181)	3034 (78.805)	3174 (82.441)	3253 (84.493)
DWT+KPCA	3457 (89.792)	3531 (91.714)	3558 (92.415)	3584 (93.09)	3587 (93.168)

the weighted Euclidean distance (WED) is employed. It is evident that feature length can play an important role in the matching process. Long feature lengths lead to a high recognition rate.

The another interesting point, DCT+KPCA based method achieved highest recognition rate (93.454%) with feature length of 300, while it gave lowest accuracy for the first database as explained in previous section (see to the Table 1). Although FFT+KPCA based method achieved highest recognition rate for the first database, but it has given lowest recognition rate (84.493%) for the second database, with feature length of 300. DWT+KPCA based method has also achieved very close recognition rate to the highest recognition rates for both databases. For instance, at the experiments given in Table 3, although DWT+KPCA achieved the better performance than others for the feature lengths less than 300, but DCT+KPCA achieved higher recognition rate than DWT+KPCA for feature length of 300. Consequently, we propose DWT+KPCA based method for palmprint recognition because it has given stable experimental results on both databases.

The performance variation for WED-based nearest-neighbor (NN) and SVM classifiers with the increase in number of features are shown in Figure 6. The SVM using radial basis function was employed in the experiments and the parameters of SVM were empirically selected. The training parameter γ, ϵ and C were empirically fixed at 0.55, 0.001, and 100, respectively. As shown in Figure 6, the SVM classifier achieved higher recognition when 50 features were only implemented. For the feature lengths longer than 50, the WED-based NN classifier has achieved better performance.

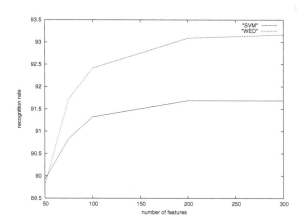

Fig. 6. Performance analysis of classifier with the number of features: DWT+ KPCA method using the SVM- and WED-based classifiers

In the literature by today, PolyU-I and PolyU-II databases are only published and public palmprint databases which include palm samples captured from the different sessions. The experimental results given in Table 3 are first candidate

Table 4. Testing results of the eight matching schemes with different feature lengths

Method		\multicolumn{5}{c}{Feature length}				
		50	100	200	300	380
PCA	LED	60.664 %	71.804 %	74.568 %	74.395 %	74.136 % (1717)
	WED	98.747 %	99.179 %	99.093 %	99.05 %	98.963 % (2292)
DWT+PCA	LED	59.542 %	71.459 %	87.305 %	87.737 %	87.737 % (2032)
	WED	98.834 %	99.309 %	99.352 %	99.352 %	99.395 % (2302)
KPCA	LED	63.557 %	73.661 %	75.82 %	74.697 %	73.92 % (1712)
	WED	98.877 %	99.222 %	99.05 %	99.006 %	98.92 % (2291)
DWT+KPCA	LED	83.462 %	86.01 %	86.01 %	87.435 %	88.039 % (2039)
	WED	98.747 %	99.309 %	99.568 %	**99.654 %**	**99.654** % (2308)

experimental results to be published in the literature, as we have followed the published papers in the literature. The published papers by [1][3][7] only worked on the palm samples collected from the one of the session. They used four samples as training set, and used remainder six samples as testing set. To compare the performance of the proposed algorithm with the published algorithms, a second experiment was designed in this section. In the second experiment which is same scenario to the experiments published in the literature, the palm images collected from the first session were only used to test the proposed algorithm. We use the first four palmprint images of each person as training samples and the remaining six palmprint images as the test samples. So, the numbers of training and test samples are 1544 and 2316. We also test the 8 approaches against conventional PCA method using different test strategies. Based on these schemes, the matching is separately conducted and the results are listed in Table 4. The meaning of LED and WED in Table 4 is linear Euclidean discriminant and the weighted Euclidean distance based nearest neighbor classifier, respectively. The entries in the brackets (in the last column) given in Table 4 indicate the number of the correct recognition samples in all 2316 palms used as test samples. A high recognition rate **(99.654 %)** was achieved for kernel PCA with 2D-DWT (abbreviated as DWT+KPCA) and WED-based classifier approach, with feature length of 300. One of the important conclusion from Table 4 is that, long feature lengths still lead to a high recognition rate. However, this principle only holds to a certain point, as the experimental results summarized in Table 4 show that the recognition rate remain unchanged, or even become worse, when the feature length is extended further.

A comparison has been finally conducted among our method and other methods published in the literature, and is illustrated in Table 5. The databases given in the Table 5 are defined as the numbers of the different palms and whole samples tested. The data represent the recognition rates and given in Table 5 is taken from experimental results in the cited papers. In biometric systems, the recognition accuracy will decrease dramatically when the number of image classes increase [1]. Although the proposed method is tested on the public database includes highest number of different palms and samples, the recognition rate of our method is more efficient, as illustrated in Table 5.

Table 5. Comparison of different palmprint recognition methods

		Method									
		Proposed	In [4]	In [5]	In [3]	In [10]	In [8]	In [9]	In [23]	In [24]	In [25]
Data palms		386	3	100	300	382	160	100	100	190	50
base samples		3860	30	200	3000	3056	1600	600	1000	3040	200
Recog. Rate(%)		**99.654**	95	91	99.2	99.149	97.25	97.5	95.8	98.13	98

6   Conclusion

This paper presents a new appearance-based non-linear feature extraction (kernel PCA) approach to palmprint identification that uses low-resolution images. We first transform the palmprints into wavelet domain to decompose the original palm images. The kernel PCA method is then used to project the palmprint image from the very high-dimensional space to a significantly lower-dimensional feature space, in which the palmprints from the different palms can be discriminated much more efficiently. WED based NN classifier is finally used for matching. The feasibility of the wavelet-based kernel PCA method has been successfully tested on two data sets from the PolyU-I and PolyU-II databases, respectively. The first data set contains 600 images of 100 subjects, while the second data set consists of 7752 images of 386 subjects. Experimental results demonstrate the effectiveness of the proposed algorithm for the automatic palmprint recognition.

Acknowledgments. This research is partially supported by The Research Foundation of Karadeniz Technical University (Grant No: KTU-2004.112.009.001). The authors would like to thank to Dr. David Zhang from the Hong Kong Polytechnic University, Hung Hom, Hong Kong, for providing us with the PolyU palmprint databases.

References

1. Zhang, D., Jing, X., Yang, J.: Biometric Image Discrimination Technologies. Computational Intelligence and Its Application Series. Idea Group Publishing (2006)
2. Li, W., Zhang, D., Xu, Z.: Palmprint Identification Using Fourier Transform. Int. Journal of Pattern Recognition and Artificial Intelligence 16(4), 417–432 (2002)
3. Wu, X., Zhang, D., Wang, K.: Fisherpalms Based Palmprint Recognition. Pattern Recognition Letters 24(15), 2829–2838 (2003)
4. Duta, N., Jain, A.K., Mardia, K.V.: Matching of palmprint. Pattern Recognition Letters 23(4), 477–485 (2002)
5. You, J., Li, W., Zhang, D.: Hierarchical palmprint identification via multiple feature extraction. Pattern Recognition 35(4), 847–859 (2002)
6. Zhang, D., Shu, W.: Two novel characteristics in palmprint verification:Datum point invariance and line feature matching. Pattern Recognition 32, 691–702 (1999)

7. Lu, G., Wang, K., Zhang, D.: Wavelet Based Independent Component Analysis for Palmprint Identification. In: IEEE Proc. of the 3rd. Int. Conf. on Machine Learning and Cybernetics, vol. 6, pp. 3547–3550. IEEE, Los Alamitos (2004)
8. Wang, Y., Ruan, Q.: Kernel Fisher Discriminant Analysis for Palmprint Recognition. In: ICPR'06. The 18th Int. Conference on Pattern Recognition, pp. 457–460 (2006)
9. Jiang, W., Tao, J., Wang, L.: A Novel Palmprint Recognition Algorithm Based on PCA and FLD. IEEE, Int. Conference. on Digital Telecommunications. IEEE Computer Society Press, Los Alamitos (2006)
10. Lu, G., Zhang, D., Wang, K.: Palmprint Recognition Using Eigenpalms Features. Pattern Recognition Letters 24(9-10), 1463–1467 (2003)
11. Zhang, D., Kongi, W., You, J., Wong, M.: Online Palmprint Identification. IEEE Trans. on Pattern Analysis and Machine Intelligence 25(9), 1041–1049 (2003)
12. Scholkopf, B., Somala, A.: Learning with Kernel: Support Vector Machine, Regularization, Optimization and Beyond. MIT Press, Cambridge (2002)
13. Zhu, Y., Tan, T.: Biometric Personal Identification Based on Handwriting. Pattern Recognition (2), 797–800 (2000)
14. Daubechies, I.: Ten Lectures on Wavelets. Philadelphia. SIAM, PA (1992)
15. PolyU Palmprint Database available: (2004),
    http://www.comp.ployu.edu.hk/biometrics/
16. Chang, T., Kuo, C.J.: Texture Analysis and Classification with Tree-Structured Wavelet Transform. IEEE Transactions on Image Processing 2(4), 429–441 (1993)
17. Averbuch, A., Lazar, D., Israeli, M.: Image Compression Using Wavelet Transform and Multiresolution Decomposition. IEEE Trans. Image Processing 5(1), 4–15 (1996)
18. Zhang, B., Zhang, H., Sam, S.: Face Recognition by Applying Wavelet Subband Representation and Kernel Associative Memory. IEEE Trans. on Neural Networks 15(1), 166–177 (2004)
19. Chien, J., Wu, C.C.: Discriminant Waveletfaces and Nearest Feature Classifier for Face Recognition. IEEE Trans. on Pattern Analysis and Machine Intelligence 24(12), 1644–1649 (2002)
20. Li, W., Gong, W., Yang, L., Chen, W., Gu, X.: Facial Feature Selection Based on SVMs by Regularized Risk Minimization. In: The 18th Conf. on Pattern Recognition (ICPR'06), vol. 3, pp. 540–543. IEEE, Los Alamitos (2006)
21. King-Kong, W.: Palmprint Texture Analysis Based on Low-Resolution Images for Personal Identification. In: IEEE 16th Int. Conf. on Pattern Recognition, pp. 807–810. IEEE Computer Society Press, Los Alamitos (2002)
22. Cristianini, N., Taylor, J.S.: An Introduction to Support Vector Machines. Cambridge University Press, Cambridge (2001)
23. Kumar, A., Zhang, D.: Personal Recognition Using Hand Shape and Texture. IEEE Transactions on Image Processing 5(8), 2454–2460 (2006)
24. Jing, X.Y., Zhang, D.: A Face and Palmprint Recognition Approach Based on Discriminant DCT Feature Extraction. IEEE Trans. on Systems, Man, and Cybernetics-Part B:Cybernetics 34(6), 2405–2415 (2004)
25. Zhang, L., Zhang, D.: Characterization of Palmprints by Wavelet Signatures via Directional Context Modeling. IEEE Trans. on Systems, Man, and Cybernetics-Part B: Cybernetics 34(3), 1335–1347 (2004)
26. Valentin, T.: Face-space models of face recognition. In: Computational, Geometric, and Process Perspectives on Facial Cognition: Context and Challenges, Lawrence Erbaum, Hillsdale, NJ (1999)

A Filter-Refinement Scheme for 3D Model Retrieval Based on Sorted Extended Gaussian Image Histogram

Zhiwen Yu, Shaohong Zhang, Hau-San Wong, and Jiqi Zhang

Department of Computer Science,
City University of Hong Kong
{cshswong, yuzhiwen}@cs.cityu.edu.hk

Abstract. In this paper, we propose a filter-refinement scheme based on a new approach called Sorted Extended Gaussian Image histogram approach (SEGI) to address the problems of traditional EGI. Specifically, SEGI first constructs a 2D histogram based on the EGI histogram and the shell histogram. Then, SEGI extracts two kinds of descriptors from each 3D model: (*i*) the descriptor from the sorted histogram bins is used to perform approximate 3D model retrieval in the filter step, and (*ii*) the descriptor which records the relations between the histogram bins is used to refine the approximate results and obtain the final query results. The experiments show that SEGI outperforms most of state-of-art approaches (e.g., EGI, shell histogram) on the public Princeton Shape Benchmark.

Keywords: Filter-refinement, Extended Gaussian Image.

1 Introduction

In the past few years, there exist a lot of 3D model collections, such as the Public Princeton Shape Benchmark (PSB) database, the Protein Data Bank, the Digital Michelangelo Project archive, and so on. In order to perform indexing and retrieval of the 3D models in 3D model collections, an efficient and effective search engine needs to be designed. One of the most important factors to determine the performance of the search engine is the 3D shape descriptor. Although there exist a lot of approaches to extract the descriptor from the model shape [1]-[12], such as D2 [1][2], EGI (Extended Gaussian Image) [4], VEGI (Volumetric Extended Gaussian Image)[3], Geometrical Moments [7], Shell histogram[6], Sector histogram[6], and so on, finding a suitable descriptor to capture the global or local properties of the shape is still a challenging problem.

In this paper, we focus on the EGI approach. Traditional EGI approach has three limitations: (*i*) it is sensitive to the rotation transformation; (*ii*) it requires the pose alignment process to align the 3D models before extracting the descriptor, which takes a lot of computational effort; and (*iii*) it cannot distinguish between the convex object and the concave object. In order to solve the limitation of EGI, we propose a new approach called Sorted Extended Gaussian Image histogram approach (SEGI). Specifically, SEGI first constructs a 2D

P. Perner (Ed.): MLDM 2007, LNAI 4571, pp. 643–652, 2007.
© Springer-Verlag Berlin Heidelberg 2007

histogram based on the EGI histogram (as the rows) and the shell histogram (as the columns). Then, it sorts the bins in each row of the 2D histogram in descending order. Next, SEGI extracts the value of the histogram bins in the first l-th columns as the descriptors for the filtering step. Finally, SEGI records the relations of these selected histogram bins in the original 2D histogram as the descriptor for the refinement step. In addition, a filter-refinement scheme is designed for 3D model retrieval based on the new descriptors. The experiments on the public Princeton Shape Benchmark (PSB) database shows that SEGI works well.

The remainder of the paper is organized as follows. Section 2 describes the SEGI approach. Section 3 presents the filter-refinement scheme. Section 4 evaluates the effectiveness of our approach through the experiments. Section 5 concludes the paper and describes possible future works.

2   The SEGI Approach

2.1   Preprocess

SEGI first triangulates the 3D mesh models by the approach in [13]. Then, SEGI calculates the EGI histogram and the shell histogram by considering the triangles on the surface of the 3D model one by one.

EGI Histogram: SEGI subdivides the sphere into $n_\alpha \times n_\beta$ cells, which are denoted as $(\alpha_i, \beta_j)(i \in [0, n_\alpha - 1],\ j \in [0, n_\beta - 1])$ as shown in Figure 1 (a).

$$\alpha_i = \frac{\pi}{n_\alpha}(i + \frac{1}{2}), \quad i \in [0, n_\alpha - 1] \tag{1}$$

$$\beta_j = \frac{\pi}{n_\beta}(2j + 1), \quad j \in [0, n_\beta - 1] \tag{2}$$

where α_i is the latitudinal angle, and β_j is the longitudinal angle. The feature of each triangle on the surface of the 3D model will be mapped to the corresponding cells in the EGI histogram according to the outward normal of the triangle (here the feature is the normalized surface area of the triangle, which is the ratio between the surface area of the triangle and the total area of all the triangles in the 3D model).

Shell Histogram: The 3D model is decomposed by n_s concentric shells (n_s is the number of concentric shells) around the centroid of the 3D model as shown in Figure 1 (b). The feature of each triangle on the surface of 3D model will be mapped to the corresponding shell in the shell histogram according to the normalized distance d_{normal} between the centroid of the 3D model and the geometric center of the triangle.

$$d_{normal} = \frac{d - d_{min}}{d_{max} - d_{min}} \tag{3}$$

where d_{max} and d_{min} are the maximum distance and the minimum distance among all the triangles respectively, and d is the distance without normalization.

$$h = \begin{cases} 1 & \text{if } d_{normal} = 0 \\ \lceil d_{normal} \times n_s \rceil & \text{Otherwise} \end{cases} \tag{4}$$

where h denotes the h-th shell that the triangle mapped to.

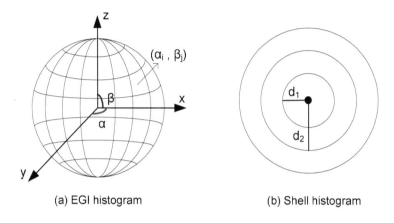

(a) EGI histogram   (b) Shell histogram

Fig. 1. EGI Histogram and Shell histogram

Next, SEGI constructs a 2D histogram whose row consists of the bins in the EGI histogram and whose column consists of the bins in the shell histogram. The feature value in the 2D histogram bin is the normalized surface area.

Traditional EGI approach cannot distinguish between the convex object and the concave object, since it only considers the normals of the surface. With the help of the shell histogram, SEGI not only considers the normals of the surface, but also the distance from the surface to the centroid of the 3D model, which make it capable of identifying the convex object and the concave object. This is the first motivation of the paper.

2.2   Rotation Invariance

In the next step, SEGI sorts the values of the bins in each row of the 2D histogram. The original EGI requires pose alignment, since it is sensitive to rotation. Unfortunately, the process of pose alignment (i) takes a lot of computational time and (ii) cannot accurately align all the 3D models due to the limitation of the pose alignment approach. SEGI performs the sorting operation on each row of the original 2D histogram. The 2D histogram after sorting satisfies rotation invariance, which is the second motivation of the paper. As a result, the descriptor obtained by SEGI from the 2D histogram after sorting satisfies translation invariance, rotation invariance and scaling invariance.

2.3   Dimensionality Reduction

A feature vector of the 3D model can be obtained by concatenating the values of the bins in the 2D histogram, but the dimension of the feature vector is very high, which is equal to $n_\alpha \times n_\beta \times n_s$. As a result, we need to reduce the dimension of the feature vector.

In this paper, we only focus on reducing $n_\alpha \times n_\beta$, which is the number of the bins in the EGI histogram. The optimal $n_\alpha \times n_\beta$ is 64×64 in [4], which forms a histogram consisting of 64×64 bins as shown in Figure 2(a). The histogram bins in Figure 2(a) are sorted in descending order according to the values (the normalized surface areas) of the bins. Only the values in a very small number of the histogram bins on the right hand side of Figure 2(a) is useful for identifying different models, while the values of other bins are close to zero. Figure 2(b) enlarges the values in the first 100 histogram bins. Table 1 illustrates the effect of the number of bins with respect to the sum of the normalized surface area $(A(n_b))$:

$$A(n_b) = \frac{\sum_{i=1}^{n_b}}{A} \times 100\% \tag{5}$$

where n_b denotes the first n_b histogram bins after sorting in the descending order and A is the total surface area of the 3D model. For example, when the $n_b = 32$, 69% of the surface areas in the 3D model maps to the first 32 histogram bins, while 31% of the surface areas in the 3D model falls into the remaining 4064 histogram bins.

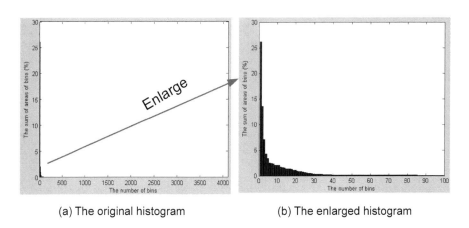

(a) The original histogram   (b) The enlarged histogram

Fig. 2. The motivation of dimensionality reduction

As a result, SEGI extracts a $l \times n_s$ dimensional feature vector from each 3D model (where $l = 32$ in this paper) in the fifth step, which is the descriptor (called the area descriptor) in the filtering step during the retrieval process.

Table 1. The effect of the number of bins

NO. of bins (n_b)	8	16	32	48	64
Sum of the normalized area	37%	53%	69%	76%	82%

2.4   The Relation Between Histogram Bins

SEGI considers the relation between the bin with the maximum value and the remaining $l - 1$ selected bins in each row of the histogram. We assume that (i) the selected l bins in the h-th row can be represented by the tuple (α_i, β_j, r_h) $(i \in \prod_\alpha, j \in \prod_\beta, \prod_\alpha$ and $\prod_\beta$ denotes the set of α values and β values of the selected l bins respectively), which corresponds to the cell (α_i, β_j) in the EGI histogram and the h-th shell with the radius r_h in the shell histogram; and (ii) the bin with the maximum value in the hth row ($h \in [1, n_s]$) is represented by $(\alpha_{i^*}, \beta_{j^*}, r_h)$. The normal of the bin with the maximum value $(\alpha_{i^*}, \beta_{j^*}, r_h)$ and other selected bins (α_i, β_j, r_h) $(i, i^* \in \prod_\alpha, j, j^* \in \prod_\beta, i \neq i^*, j \neq j^*)$ can be expressed as follows:

$$\overrightarrow{n}_{(\alpha_{i^*}, \beta_{j^*}, r_h)} = [r_h sin\beta_{j^*}, r_h cos\beta_{j^*} sin\alpha_{i^*}, r_h cos\beta_{j^*} cos\alpha_{i^*}] \qquad (6)$$

$$\overrightarrow{n}_{(\alpha_i, \beta_j, r_h)} = [r_h sin\beta_j, r_h cos\beta_j sin\alpha_i, r_h cos\beta_j cos\alpha_i] \qquad (7)$$

The angle θ between two normals $\overrightarrow{n}_{(\alpha_{i^*}, \beta_{j^*}, r_h)}$ and $\overrightarrow{n}_{(\alpha_i, \beta_j, r_h)}$ can be calculated by the following equations:

$$\theta = arc\ cos\ \theta \qquad (8)$$

$$cos\theta = \frac{1}{2}(sin\beta_{j^*} \times sin\beta_j + cos\beta_{j^*} sin\alpha_{i^*} \times cos\beta_j sin\alpha_i$$
$$+ cos\beta_{j^*} cos\alpha_{i^*} \times cos\beta_j cos\alpha_i) \qquad (9)$$

SEGI obtains a $(l - 1) \times n_s$ dimensional feature vector consisting of the angle values from each row in the 2D histogram, which is the descriptor (called the angle descriptor) in the refinement step in the process of 3D model retrieval.

3   A Filter-Refinement Scheme

Figure 3 provides an overview of the filter-refinement scheme. Given a new 3D query model, the 3D search engine retrieves k relevant models from the 3D model database, which are the most similar to the query model. In the filter-refinement scheme, the 3D search engine first extracts the area descriptor and the angle descriptor from the query model by SEGI. Then, it uses the area descriptor of the query model to retrieve $c \cdot k$ (c is a constant and $c > 1$, we set $c = 5$ in the experiment) 3D models as the approximate query results in the filtering step. In the refinement step, it uses the angle descriptor to select k

models from the approximate query results as the final query result. We adopt the Minkowski distance (L_p norm) as the similarity measure between the descriptors. The Minkowski distance between two descriptors $\overrightarrow{f_i}$ and $\overrightarrow{f_j}$ is defined as follows:

$$L_p = (\sum_{h=1}^{m} |\overrightarrow{f_{ih}} - \overrightarrow{f_{jh}}|^p)^{\frac{1}{p}} \qquad (10)$$

where m denotes the number of dimensions of the descriptor, and h denotes the h-th dimension. In the experiment, we set $p = 1$ which is the Manhattan distance (L_1 norm).

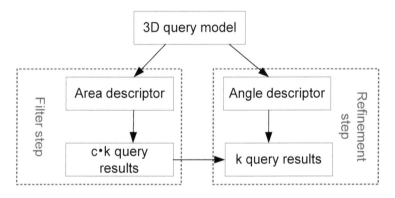

Fig. 3. The filter-refinement scheme

4 Experiments

4.1 Data Set and Experimental Setting

Our 3D model database comes from the test dataset of the public Princeton Shape Benchmark (PSB) database [2]. There are 907 3D models in our 3D model database, which can be categorized into 131 classes, such as computer, airplane, car, plane, animal, architecture, and so on. The vertices contained in the 3D models range from 108 to 160940, with a median of 1005 vertices per model. Figure 4 illustrates examples of the front views of the 3D models in some of the classes.

We compare SEGI with D2 [1][2], EGI (Extended Gaussian Image) [3], shell histogram (SH), geometric moment (GM), and the random approach in the

Fig. 4. Examples of 3D models

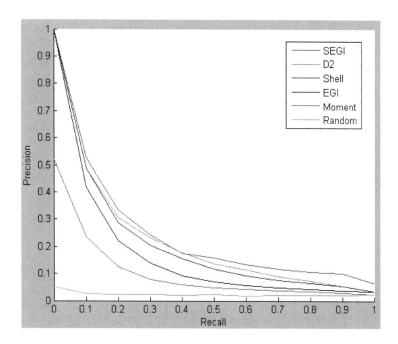

Fig. 5. Comparison of the precision-recall curves among different approaches

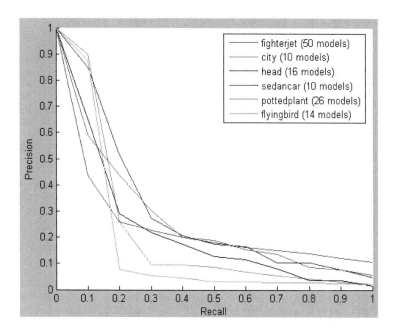

Fig. 6. Comparison of the precision-recall curves among different classes

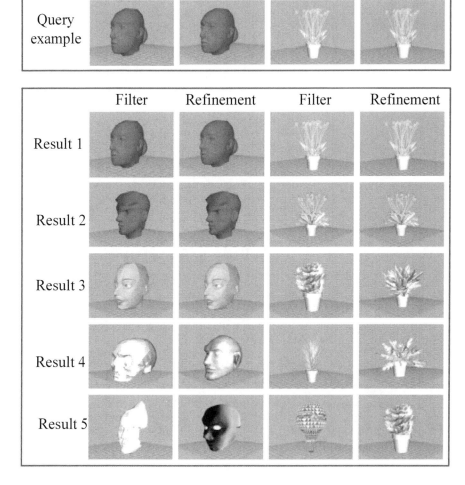

Fig. 7. Comparison of the results between the filter step and the refinement step

experiment. The parameters of SEGI are 32 for l (l is the selected number of bins in each row of the 2D histogram) and 16 for n_s (n_s is the number of concentric spheres).

The performance of the approaches based on different descriptors are measured by the precision-recall curve. The precision p and the recall r is defined as follows:

$$p = \frac{\rho}{\varrho}, \quad r = \frac{\rho}{\sigma} \tag{11}$$

where ρ denotes the number of the retrieved 3D models which is relevant to the query model, ϱ is the total number of the retrieved 3D models, and σ denotes the number of relevant models in the 3D model database.

4.2   Experimental Results

Figure 5 compares the performance of different approaches according to the precision-recall curves. The SEGI approach which combines the EGI histogram and the shell histogram outperforms the EGI approach which only uses the EGI histogram and the SH approach which only adopts the shell histogram, due to its ability to combine the advantages of the EGI histogram and the shell histogram. The SEGI approach is also better than the D2 approach and the GM approach.

Figure 6 shows the precision-recall curves among six classes by the SEGI approach, while Figure 7 compares the query results of two query examples between the filtering step and the refinement step. The result in both examples in the refinement step are correct, while the fifth result in both examples in the filtering step are not correct.

5   Conclusion and Future Work

In this paper, we investigate the problem of content based 3D model retrieval. Although there exist a lot of approaches for 3D model retrieval, few of them can guarantee translation invariance, scaling invariance and rotation invariance. Our major contribution consists of proposing: (i) a new approach called Sorted Extended Gaussian Image Histogram (SEGI) to extract the descriptor from 3D models and (ii) a filter-refinement scheme for 3D model retrieval based on the new descriptors. In the future, we will explore how to assign different weights to the descriptors extracted from different rows.

Acknowledgments. The work described in this paper was fully supported by grants from the Research Grants Council of Hong Kong Special Administrative Region, China [Project No. CityU 1197/03E and CityU 121005].

References

1. Osada, R., Funkhouser, T., Chazelle, B., Dobkin, D.: Shape distributions. ACM Transactions on Graphics 21(4), 807–832 (2002)
2. Funkhouser, T., Min, P., Kazhdan, M., Chen, J., Halderman, A., Dobkin, D., Jacobs, D.: A search engine for 3D models. ACM Transactions on Graphics 22(1), 83C105 (2003)
3. Zhang, J., Wong, H.-S., Yu, Z.: 3D model retrieval based on volumetric extended gaussian image and hierarchical self organizing map, ACM Multimedia 2006, Santa Barbara, Califonia, pp. 121–124 (2006)
4. Wong, H.-S., Cheung, K.K.T., Ip, H.H.-S.: 3D head model classification by evolutionary optimization of the Extended Gaussian Image representation. Pattern Recognition 37(12), 2307–2322 (2004)
5. Kazhdan, M., Funkhouser, T.: Harmonic 3D Shape Matching. SIGGRAPH 2002. Technical Sketch (2002)
6. Kriegel, H.-P., Brecheisen, S., Kroger, P., Pfeile, M., Schubert, M.: Using sets of feature vectors for similarity search on voxelized CAD objects. In: SIGMOD 2003 (2003)

7. Elad, M., Tal, A., Ar, S.: Directed search in a 3d objects database using svm. Technical report, HP Laboratories, Israel (2000)
8. Bespalov, D., Shokoufandeh, A., Regli, W.C.: Reeb graph based shape retrieval for CAD. In: DETC-03 (2003)
9. De Alarcon, Pascual Montano, P. A., Carazo, J.M.: Spin images and neural networks for efficient content-based retrieval in 3d object databses. In: Lew, M.S., Sebe, N., Eakins, J.P. (eds.) CIVR 2002. LNCS, vol. 2383, Springer, Heidelberg (2002)
10. Chen, D.Y., Tian, X.P., Shen, Y.T., Ouhyoung, M.: On visual similarity based 3d model retrieval. In: Computer Graphics Forum (EUROGRAPHICS 2003), September 2003, vol. 22, pp. 223–232 (2003)
11. Hilaga, M., Shinagawa, Y., Kohmura, T.: Topology matching for fully automatic similarity estimation of 3D shapes. In: SIGGRAPH 2001, pp. 203–212 (2001)
12. Cyr, C.M., Kimia, B.: 3D object recognition using shape similiarity-based aspect graph. In: ICCV01, vol. I, pp. 254–261 (2001)
13. Fournier, A., Montuno, D.Y.: Triangulating Simple Polygons and Equivalent Problems. ACM Trans. Graphics 3, 153–174 (1984)

Fast-Maneuvering Target Seeking Based on Double-Action Q-Learning

Daniel C.K. Ngai and Nelson H.C. Yung

Department of Electrical & Electronic Engineering
The University of Hong Kong, Pokfulam Road, Hong Kong
ckngai@eee.hku.hk, nyung@eee.hku.hk

Abstract. In this paper, a reinforcement learning method called DAQL is proposed to solve the problem of seeking and homing onto a fast maneuvering target, within the context of mobile robots. This Q-learning based method considers both target and obstacle actions when determining its own action decisions, which enables the agent to learn more effectively in a dynamically changing environment. It particularly suits fast-maneuvering target cases, in which maneuvers of the target are unknown a priori. Simulation result depicts that the proposed method is able to choose a less convoluted path to reach the target when compared to the ideal proportional navigation (IPN) method in handling fast maneuvering and randomly moving target. Furthermore, it can learn to adapt to the physical limitation of the system and do not require specific initial conditions to be satisfied for successful navigation towards the moving target.

Keywords: Moving object navigation, reinforcement learning, Q-learning.

1 Introduction

Seeking and homing onto a moving target is an interesting and essential issue for a mobile robot. The navigation problem is no longer a global path planning matter, but the agent (robot) now requires to constantly re-plan its navigational tactic with respect to the motion feature of the target. Common applications include soccer robotics, autonomous vehicle navigation, and missile guidance. Although navigation methods such as potential field [1] and cell decomposition method [2] are sophisticated and well studied, most of these researches concentrate on stationary target only. When a moving target is concerned, most of these methods designed for stationary targets do not work. Therefore, a new set of strategy has to be developed, from which new methods are to be evaluated for this complex scenario.

Briefly, moving targets may be classified into two categories. The first one is slow-maneuvering target, which allows a more accurate prediction of the target's motion. The second one is fast-maneuvering target, which moves quickly and randomly such that accurate prediction on the target's motion is not always possible. For slow-maneuvering targets, a common approach is to equip the navigation system with a prediction capability to derive a more accurate and optimal path to intercept the

P. Perner (Ed.): MLDM 2007, LNAI 4571, pp. 653–666, 2007.
© Springer-Verlag Berlin Heidelberg 2007

target. Examples of navigation system using this approach include the fuzzy logic control method [3], [4] and the artificial vision method [5].

When fast-maneuvering targets are concerned and prediction is not available, approaches such as the potential field method [6], [7] and the line of sight navigation method [8] have been adopted. Because of their computational efficiency, real time application is possible. In [7], the robot navigates towards the moving goal by considering the potential field with the velocity of both the robot and the target taken into account. However, the potential field method suffers from the local minima problem as in stationary target. The line of sight navigation method in [8] is based on the line of sight guidance law which aims at minimizing the distance between the robot and target by making the expected intercept time as small as possible. However, it requires the notion of an observer as a control station or reference point which may not always exist in real application.

The proportional navigation (PN) [9] guidance law for missile guidance is another approach used for fast-maneuvering targets and is well studied and developed. In PN, the interceptor is guided by applying a commanded acceleration proportional to the rate of rotation of the line of sight and acting perpendicular along a direction defined by the specific variant of PN. There are two generic classes of PN, 1) the true proportional navigation (TPN) [10] method which had the commanded acceleration applied in the direction normal to the line of sight between the interceptor and the target and, 2) the pure proportional navigation (PPN) [10] method which had the commanded acceleration applied in the direction normal to the interceptor velocity. Although these PN methods have the advantage of implementation simplicity in practice, they are sensitive to the initial condition and may fail to reach the target if requirements for the initial condition are not satisfied [10]. For missile guidance, the problem on initial condition may not be critical as the operators can make fine adjustments on the initial velocity and direction for the missile before it is launched. However, for a general purpose autonomous navigation application, it is reasonable to assume that the agent starts from rest and not necessarily at a heading angle pointing towards the target. Therefore, TPN and PPN may not be completely suitable for autonomous navigation.

Another variation of PN called the ideal proportional navigation (IPN) method, which has the commanded acceleration applied in the direction normal to the relative velocity between the interceptor and the target, has been presented in [11]. IPN is less dependent on the initial condition of the interceptor and thus more applicable in robotics applications. In [12], [13], Mehrandezh et al. presented an IPN based guidance system for moving object interception. Practical issues such as limitations to the velocity and acceleration of the robot were considered to enable IPN to be used in a robotics system.

Motivated by the fact that there are still rooms for improvement for current approaches on the issue of moving target navigation, we regard it as an intelligent control problem and apply reinforcement learning to deal with the problem. Reinforcement Learning (RL) [14] aims to find an appropriate mapping from situations to actions in which a certain reward is maximized. It can be defined as a class of problem solving approaches in which the learner (agent) learns through a series of trial-and-error searches and delayed rewards [14]-[17]. The purpose is to maximize not only the immediate reward, but also the cumulative reward in the long run, such that the agent can learn to approximate an optimal behavioral strategy by continuously interacting with the environment. This allows the agent to work in a previously unknown environment by learning to adapt to it gradually. The advantage of using RL is that we do not have to make any assumptions regarding the underlining kinematics model used by the agent and it should work on any autonomous vehicles given the state information from the environment. The agent can therefore learn to adapt to the physical limitations of the system and not require any specified initial condition to be satisfied for successful navigation towards the moving target.

In this paper, we propose a solution for the problem of fast-maneuvering target seeking using a RL approach called Double Action Q-Learning (DAQL) [18], [19]. The proposed method uses DAQL to consider the response or action of the target when deriving the agent's next action. In the case of seeking the fast-maneuvering target, the agent naturally selects a more appropriate action when compare with other RL or non-RL methods that do not consider target actions. DAQL also works with or without prediction. In this paper, we present an approach without using the prediction technique for convenience sake. In addition, the target is assumed to be moving without known maneuvers. By using a reinforcement learning approach, the proposed method is suitable for application in general autonomous navigation, with the ability to learn to adapt to the limitations and underlining kinematics of the system. When compared to the PN methods, the proposed method is insensitive to initial conditions and therefore more suitable for the application of autonomous navigation.

This paper is organized as follows: an overview of the proposed system is given in Section 2. Following that, section 3 discusses the geometrical relations between agent and target. Section 4 introduces our approach by showing how the problem of moving target navigation can be accomplished with DAQL. Section 5 presents the simulation environment and the result in different simulation cases. Finally, the conclusion is given in Section 6.

2   System Overview

Fig. 1. depicts an overview of the proposed RL approach. The DAQL method is designed to handle fast-maneuvering target, or obstacles. As DAQL requires a discrete state input, a state matrix is used to quantize the continuous input into discrete states. The states information together with the reward received is used to update the Q-values by using the DAQL update rule. Finally, an exploration policy is employed to select an action and carry out proper exploration. Detailed explanations on each part will be discussed in the following section.

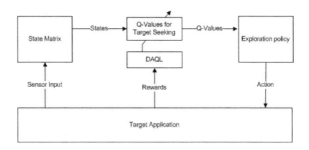

Fig. 1. Overview of the reinforcement learning approach

3   Geometrical Relations Between Agent and Target

The control variables of the agent and the target at time t are depicted in Fig.2. It is assumed that the location information of the target can be detected by sensory system (e.g. GPS). The agent and target are assumed to be circles with radius r_a and r_T respectively. We assume that the agent is $d_T \in \Re^+$ away from the target and is at an angle $\phi \in \Theta$ where $\Theta = [0, 2\pi] \subset \Re$. The two parameters: d_T, and ϕ are quantized into states through the state matrix, where the actual number of states is determined by how DAQL are realized. The state set for the relative location of the destination is $l_T \in L_T$ where $L_T = \left\{ \left(\tilde{d}_T, \tilde{\phi} \right) \mid \tilde{d}_T \in D_T \text{ and } \tilde{\phi} \in \Theta_q \right\}$, $D_T = \{ i \mid i = 0, 1, \ldots, 5 \}$ and $\Theta_q = \{ j \mid j = 0, 1, \ldots, 15 \}$. Quantization is achieved as follows:

$$\tilde{d}_T = \begin{cases} \lfloor d_T / 20 \rfloor & \text{for } d_T < 100 \\ 5 & \text{for } d_T \geq 100 \end{cases}, \tag{1}$$

$$\tilde{\phi} = \begin{cases} \left\lfloor \left| \dfrac{\phi + \pi/16}{\pi/8} \right| \right\rfloor & \text{for } 0 \leq \phi < 31\pi/16 \\ 0 & \text{for } 31\pi/16 \leq \phi < 2\pi \end{cases}. \tag{2}$$

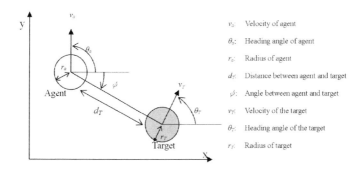

Fig. 2. Control variables of agent and the target

There are altogether 96 states for L_T. The output actions are given by $a^1 \in A$ where $A = \left\{ \left(|v_a|, \theta_a \right) \mid |v_a| \in V_a \text{ and } \theta_a \in \Theta_a \right\}$, $V_a = \{ m \times v_{a,max} / 5 \mid m = 0, 1, \ldots, 5 \}$, $\Theta_a = \{ n\pi/8 \mid n = 0, 1, \ldots, 15 \}$, and $v_{a,max}$ is the maximum velocity of the agent. For $|\vec{v}_a| = 0$, the agent is at rest despite of θ_a, resulting in only 81 actions. For DAQL, we assume that the target have velocity $v_T \in \Re^+$ and heading angle $\theta_T \in \Theta$. They are quantized to $a^2 \in A_T$ where $A_T = \left\{ \left(\tilde{v}_T, \tilde{\theta}_T \right) \mid \tilde{v}_T \in V_q \text{ and } \tilde{\theta}_T \in \Theta_q \right\}$ and $V_q = \{ l \mid l = 0, 1, \ldots, 10 \}$. Quantization is achieved as follows:

$$\tilde{v}_T = \begin{cases} \lfloor v_T + 5 \rfloor & \text{for } 0 \le v_T < 105 \\ 10 & \text{for } v_T \ge 105 \end{cases}, \tag{3}$$

$$\tilde{\theta}_T = \left\lfloor \frac{\theta_T}{\pi/8} \right\rfloor, \tag{4}$$

where there are altogether 161 actions for the target as observed by the agent.

4 Target Seeking Using DAQL

In order to navigate towards the moving target, the DAQL method is adopted such that the agent learns to select the most appropriate action according to the current state. DAQL is a modified version of QL [20] that works in a dynamic environment by essentially having the actions performed by both the agent and the target taken in its formulation. The DAQL update rule shown in Eqt. (5) is used to update the Q-values ($Q(l_{T,t}, a_t^1, a_t^2)$) which represent the action values in different states.

$$Q(l_{T,t}, a_t^1, a_t^2) \leftarrow$$
$$Q(l_{T,t}, a_t^1, a_t^2) + \alpha \left[r_{t+1} + \gamma \max_{a_{t+1}^1} Q(l_{T,t+1}, a_{t+1}^1, a_{t+1}^2) - Q(l_{T,t}, a_t^1, a_t^2) \right] \tag{5}$$

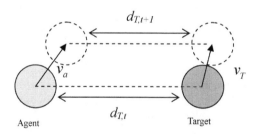

Fig. 3. Change in d_T form t-1 to t

where $l_{T,t}$, $l_{T,t+1}$ are the input states, a_t^1, a_{t+1}^1 are the action of the agent , a_t^2, a_{t+1}^2 are the action of the target in t and $t+1$ respectively. α and γ are the weighting parameter and discount rate respectively, which both range from 0 to 1. r_{t+1} describes the reward received by the agent generated from the reward function at time $t+1$. Reward is given to the agent to encourage it to travel towards the target using the shortest path with the maximum velocity. With reference to Fig. 3, let us define $\Delta d_T = d_{T,t} - d_{T,t+1}$, where $d_{T,t}$ is the distance between the agent and destination at t, $d_{T,t+1}$ is the distance at $t+1$; and the agent travels at v_a from t to $t+1$. Therefore, Δd_T will be larger if the agent maximizes the distance it can travel in one time step with higher velocity. The normalized reward function of the agent is thus defined as:

$$r_{t+1} = (\Delta d_T) / (v_{a,\max} T) \tag{6}$$

When r_{t+1} is available, the agent uses the DAQL update rule to learn the navigation task, as depicted in Fig. 4. Given obstacles' actions in two time steps (t & $t+1$), the agent updates its Q-values ($Q(l_{T,t}, a_t^1, a_t^2)$) at $t+2$ and the values are stored in the Q-value table.

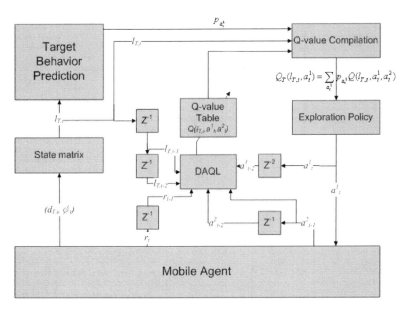

Fig. 4. DAQL for moving target navigation. Q-values updated at t.

Apart from learning, the agent needs to determine its own action in the current time step. Given the state information of the current time step, the agent can use it together with the Q-values in the Q-value table to determine an action that is most appropriate for the navigation task. That is, the agent needs to determine an action a_t^1 given $Q(l_{T,t}, a_t^1, a_t^2)$. However, since a_t^2 is not known at t, it has to be predicted, which can be treated independently from RL, i.e. the agent predicts from the environment's historical information, or it can be based on concepts (rules learn from examples) and instances (pools of examples). In this paper, we assume that a_t^2 has equal probability in taking any of the 161 actions. To incorporate the predicted a_t^2, the corresponding Q-value can be acquired as follows:

$$Q_T(l_{T,t}, a_t^1) = \sum_{a_t^2} p_{a_t^2} Q(l_{T,t}, a_t^1, a_t^2) \tag{7}$$

where $p_{a_t^2}$ is the probability that the target performs the action a_t^2. The expected value of the overall Q-value is obtained by summing the product of the Q-value of the target when it takes action $a_{t,t}^2$ with its probability of occurrence. Assuming an evenly distributed probability, we have $p_{a_t^2} = 1/N_A$, where N_A is the number of actions that the target can perform ($N_A = 161$), and $Q_T(a_t^1)$ can now be expressed as follows:

$$Q_T(l_{T,t}, a_t^1) = \frac{1}{N_A} \sum_{a_t^2} Q(l_{T,t}, a_t^1, a_t^2) \tag{8}$$

The final decision of the agent is made by using the ε-greedy policy as shown in Eqt. (9). Exploration policy allows the agent to learn the different response from the environment so that the agent can discover better approaches to move toward the target.

$$a_t^1 = \begin{cases} \arg\max_{a_t^1} Q_T(l_{T,t}, a_t^1) & \text{with probability } 1 - \varepsilon \\ \text{random} & \text{with probability } \varepsilon \end{cases} \tag{9}$$

5   Simulation and Result

In this simulation, length has unit of cm and time has unit of second. The agent is assumed to be circular with radius r_a of 50cm while the target is also assumed to be circular with radius r_t of 50cm. Physical limitations have been applied on the agent so that it only has a maximum acceleration of 20cm/s^2 and maximum velocity of 50cm/s. The agent is required to start from rest, and hit the target, i.e, the distance d_T between them is reduced to zero. To acquire environmental information, a sensor simulator has been implemented to measure distances between the agent and target. The sensor simulator can produce distance measurements, at T interval (typically 1 s) to simulate practical sensor limitations. The other parameters for simulations are α=0.6, γ=0.1, ε=0.5, and T=1s.

We considered a number of scenarios here including when the target moves in a straight line, circular path, and randomly. A comparison with the IPN method is conducted to illustrate the difference in navigation behavior of the RL method and the IPN method. To distinguish between the two methods, we called the agent which uses the proposed approach the RL agent while the agent that uses the IPN method the IPN agent.

Before evaluating the performance of the RL agent, the agent is trained in an environment that contains randomly moving targets with the origin of both parties randomly placed. The agent is trained for 10000 episodes in which each episode is defined by the agent being able to hit the target. The learning and exploration process of the agent is then stopped and different cases of simulations are performed and discussed in the next section.

For the IPN method, as it only defines the magnitude and direction of the acceleration that should be applied to the agent but without the indication of any physical limitations to the agent itself. For fairness of comparison, the same limitations on maximum velocity and acceleration used by the RL agent are applied to the IPN agent. The method used for limiting the acceleration command is the same as described in [12], where the magnitude of the acceleration is scaled down to the maximum value if the acceleration limit is violated. The velocity limit is applied to the agent similar to that of the acceleration limit. Besides, as the IPN agent is sensitive to its initial condition, i.e., when it starts from rest, it needs quite a long time to accelerate to its maximum velocity, we therefore used a relatively large value of λ =50 to alleviate this problem.

5.1   Case 1 - Target Moving in a Straight Line

In this case, the target moves in a straight line, from left to right, with a constant velocity of v_t=30cm/s. The aim in this example is to demonstrate how the agent seeks the target and navigates toward a non-accelerating moving target. The paths of the RL

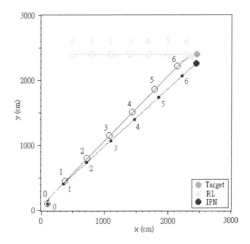

Fig. 5. Agents' path for a target that moves with constant velocity, observed at t=66s. The initial position of the agents and target are (100,100) and (500,2400) respectively.

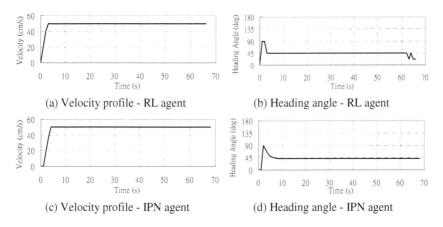

(a) Velocity profile - RL agent   (b) Heading angle - RL agent

(c) Velocity profile - IPN agent   (d) Heading angle - IPN agent

Fig. 6. Velocity profile and heading direction for RL and IPN agent when target moves in a straight line

agent, IPN agent and the target are depicted in Fig. 5. The numbers in the figure represent the location of the agents and targets in every 10s. The velocity and heading angle profile of the RL agent and IPN agent are shown in Fig. 6. It can be observed that although both agents share a similar path in the first 10 seconds, their path deviate thereafter. The agents are able to reach the moving target at $t=66$s and $t=68$s for the RL and IPN respectively. The minor difference is probably due to that the IPN agent required slightly more time to turn and steer towards the target. The RL agent uses a straight path to approach the target with a nearly constant heading angle of 45° while the IPN agent also travel straightly to the target but with a slightly smaller angle of 41.55°. This is understandable as the IPN method applies a commanded acceleration in the direction normal to the relative velocity between the agent and the target, which makes the agent travel further towards the direction which the target is approaching.

Comparing Fig. 6(a) and 6(c), it can be observed that the velocity of the IPN agent is still zero in the first second after the simulation started. This is because both the agent and the target are assumed to be at rest initially and thus the relative velocity between the agent and the target is zero which makes the IPN agent slow to apply a commanded acceleration right after the simulation begins. The initial condition problem is a common nature to the PN methods and the IPN agent can be initialized by using other methods to solve the problem [12]. Comparing Fig. 6(b) and 6(c), the two heading direction functions are very similar, except that the IPN function is continuous, while the RL function is discrete, which is its inherent feature.

5.2 Case 2 - Target Moving in a Circle

In this case, the target moves in a circle with radius of 200cm. Velocity of the target is $v_T=30$cm/s. Two scenarios are illustrated in Fig. 7(a) and 7(b), where the target moves clockwise and counterclockwise respectively. The velocity and heading angle profiles of the RL agent and IPN agent are depicted in Fig. 8. The agents are able to reach the moving target at $t=26$s and $t=33$s for the RL and IPN agent respectively in the

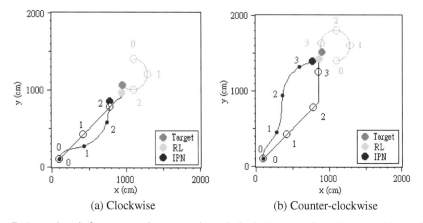

(a) Clockwise   (b) Counter-clockwise

Fig. 7. Agents' path for a target that moves in a circle. In (a), snapshot taken at $t=26$s. In (b), snapshot taken at $t=34$s. The initial position of the agents and target are (100,100) and (1100,1400) respectively.

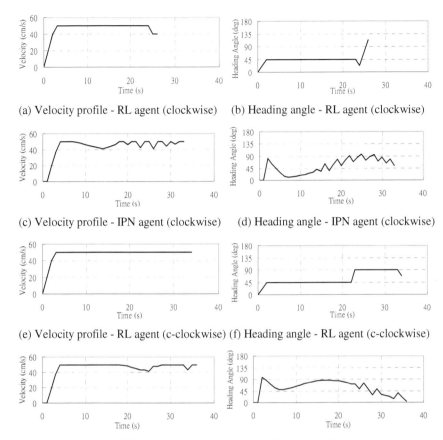

(a) Velocity profile - RL agent (clockwise)   (b) Heading angle - RL agent (clockwise)

(c) Velocity profile - IPN agent (clockwise)   (d) Heading angle - IPN agent (clockwise)

(e) Velocity profile - RL agent (c-clockwise) (f) Heading angle - RL agent (c-clockwise)

(g) Velocity profile - IPN agent (c-clockwise) (h) Heading angle - IPN agent (c-clockwise)

Fig. 8. Velocity profile and heading direction for RL and IPN agent when target moves in a circle clockwisely

clockwise case and at $t=34s$ and $t=36s$ for the counterclockwise case. It can be observed that the RL agent is able to reach the target using more constant heading angle and velocity along the path when compared to the IPN agent.

The IPN agent, on the contrary, adopted a more convoluted path with frequent change in velocity and heading angle. This is due to the nature of IPN method which applies commanded acceleration in a direction normal to the relative velocity between the agent and the target. Since this relative velocity changes continuously as the target moves in a circular path, the path of the IPN agent thus changes accordingly, resulting in the using of non-optimal speed, and a relatively large change in heading angle along the path.

5.3 Case 3 – Target Moving Randomly

In this case, the target moves randomly, so that it has an equal probability in performing one of the actions (a^2) in A_T in each time step. We tested the performance of the two agents for 1000 episodes of simulations. One episode defines the success of reaching the target for the agents. The initial positions of the agents and the target are fixed in each episode and a sample of an episode is depicted in Fig. 9. In this example, the agents are able to reach the moving target at $t=42s$ and $t=44s$ for the RL and IPN method respectively. From the velocity and heading angle profiles as depicted in Fig. 10 for both agents, we observed that although both agents maintains almost maximum speed through the journey, they have quite a different behavior in selecting the heading angle. It is shown in Fig 10(b) and 10(d) that the RL agent has a constant heading angle most of the time and has a large variation only if it is close enough to the target. On the contrary, the IPN agent changes its heading angle frequently and thus applied a more convoluted path to reach the target. This depicts that the IPN agent is more reactive to the short term change in velocity of the target and requires longer time to reach the target.

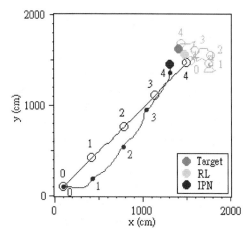

Fig. 9. Agents' path for a target moving randomly. Observed at $t=42s$. The initial position of the agents and target are (100,100) and (1500,1500) respectively.

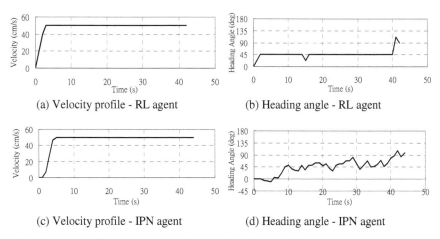

(a) Velocity profile - RL agent   (b) Heading angle - RL agent

(c) Velocity profile - IPN agent   (d) Heading angle - IPN agent

Fig. 10. Velocity profile and heading direction for RL and IPN agent when target moves randomly

Table 1. Average path time and the corresponding path time for the two agents in navigating towards a randomly moving target

Agent	Average path time (s)	Standard deviation path time
RL	44.18	6.08
IPN	48.62	7.16

Table 2. Summary on the simulation on the 1000 episodes of simulation with a randomly moving target

RL reaches target faster	IPN reaches target faster	Both reach target with the same time
926	24	50

A summary of the results from the 1000 episodes of simulations are shown in Tables 1 and 2. Table 1 shows the average number of time required for the agents to reach the target and the corresponding standard deviation. It illustrates that the RL agent is able to reach the target in an average time of 44.18s while the IPN agent needs an average time of 48.62s. Table 2 depicts that in 926 out of the 1000 episodes, the RL agent is able to reach the target faster while the IPN agent can reach the target faster in 24 of the episodes. This suggested that the RL method is less affected by the sudden changes of a randomly moving and unpredictable target.

5.4 In the Presence of Obstacles

In environment with the presence of obstacles, the proposed method for moving target navigation is also able to find a collision-free path to reach the target. This can

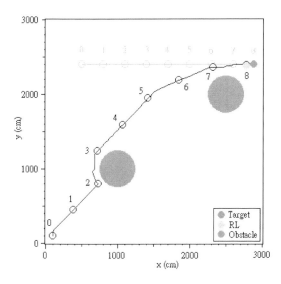

Fig. 11. Agents' path for a target moving in a straight line, with the presence of static obstacles. Observed at t=80s.

be done by combining the proposed moving target navigation method with the obstacle avoidance method proposed in [19]. Fig. 9 depicts a situation same as case 1 but with the presence of two static obstacles. Since there are obstacles in the environment, the agent applies a more convoluted path to avoid colliding with the obstacles while reaching the moving target at the same time. The agent is able to reach the target successfully at t=80s.

6 Conclusion

In this paper, we presented a method for solving the problem of fast-maneuvering target seeking in general robotics applications. The target's maneuvers are assumed unknown to the agent and no predictions on target's maneuvers are required. The proposed method applies the DAQL reinforcement learning approach with implicit consideration on the actions performed by the target. Through reinforcement learning, the agent learns to seek the moving target by receiving rewards when the environment reacts upon the agent's actions. Thus, no kinematics models on the agent and target are required. Simulation results prove that the RL agent is able to reach the moving target successfully with its path less affected to the sudden change in motion of the target. Initial condition is not critical for the RL agent to find a path towards the target as it learns and reacts with the most suitable action in each time instant. The RL agent can also learn to adapt to the physical limitations of the system during the navigation task. For future research, some kind of prediction should be incorporated with the agent to enhance its ability further.

References

1. Ge, S.S., Cui, Y.J.: New potential functions for mobile robot path planning. IEEE Transactions on Robotics and Automation 16, 615–620 (2000)
2. Conte, G., Zulli, R.: Hierarchical Path Planning in a Multirobot Environment with a Simple Navigation Function. IEEE Transactions on Systems Man and Cybernetics 25, 651–654 (1995)
3. Li, T.H.S., Chang, S.J., Tong, W.: Fuzzy target tracking control of autonomous mobile robots by using infrared sensors. IEEE Transactions on Fuzzy Systems 12, 491–501 (2004)
4. Luo, R.C., Chen, T.M., Su, K.L.: Target tracking using hierarchical grey-fuzzy motion decision-making method. IEEE Transactions on Systems Man and Cybernetics Part a-Systems and Humans 31, 179–186 (2001)
5. Dias, J., Paredes, C., Fonseca, I., Araujo, H., Batista, J., Almeida, A.T.: Simulating pursuit with machine experiments with robots and artificial vision. IEEE Transactions on Robotics and Automation 14, 1–18 (1998)
6. Adams, M.D.: High speed target pursuit and asymptotic stability in mobile robotics. IEEE Transactions on Robotics and Automation 15, 230–237 (1999)
7. Ge, S.S., Cui, Y.J.: Dynamic motion planning for mobile robots using potential field method. Autonomous Robots 13, 207–222 (2002)
8. Belkhouche, F., Belkhouche, B., Rastgoufard, P.: Line of sight robot navigation toward a moving goal. IEEE Transactions on System Man and Cybernetic Part b-Cybernetic 36, 255–267 (2006)

9. Yang, C.D., Yang, C.C.: A unified approach to proportional navigation. IEEE Transactions on Aerospace and Electronic Systems 33, 557–567 (1997)
10. Shukla, U.S., Mahapatra, P.R.: The Proportional Navigation Dilemma - Pure or True. IEEE Transactions on Aerospace and Electronic Systems 26, 382–392 (1990)
11. Yuan, P.J., Chern, J.S.: Ideal Proportional Navigation. Journal of Guidance Control and Dynamics 15, 1161–1165 (1992)
12. Borg, J.M., Mehrandezh, M., Fenton, R.G., Benhabib, B.: An Ideal Proportional Navigation Guidance system for moving object interception-robotic experiments. In: Systems, Man, and Cybernetics, 2000 IEEE International Conference, vol. 5, pp. 3247–3252 (2000)
13. Mehrandezh, M., Sela, N.M., Fenton, R.G., Benhabib, B.: Robotic interception of moving objects using an augmented ideal proportional navigation guidance technique. In: IEEE Transactions on Systems Man and Cybernetics Part a-Systems and Humans, vol. 30, pp. 238–250 (2000)
14. Sutton, R.S., Barto, A.G.: Reinforcement learning: an introduction. MIT Press, Cambridge (1998)
15. Kaelbling, L.P.: Learning in embedded systems. MIT Press, Cambridge (1993)
16. Kaelbling, L.P., Littman, M.L., Moore, A.W.: Reinforcement learning: A survey. Journal of Artificial Intelligence Research 4, 237–285 (1996)
17. Sutton, R.S.: Reinforcement Learning. The International Series in Engineering and Computer Science, vol. 173. Kluwer Academic Publishers, Dordrecht (1992)
18. Ngai, D.C.K., Yung, N.H.C.: Double action Q-learning for obstacle avoidance in a dynamically changing environment. In: Proceedings of the 2005 IEEE Intelligent Vehicles Symposium, Las Vegas, pp. 211–216 (2005)
19. Ngai, D.C.K., Yung, N.H.C.: Performance Evaluation of Double Action Q-Learning in Moving Obstacle Avoidance Problem. In: Proceedings of the 2005 IEEE International Conference on Systems, Man, and Cybernetics, Hawaii, October 2005, pp. 865–870 (2005)
20. Watkins, C.J.C.H., Dayan, P.: Technical Note: Q-learning. Machine Learning 8, 279–292 (1992)

Mining Frequent Trajectories of Moving Objects for Location Prediction

Mikołaj Morzy

Institute of Computing Science
Poznań University of Technology
Piotrowo 2, 60-965 Poznań, Poland
Mikolaj.Morzy@put.poznan.pl

Abstract. Advances in wireless and mobile technology flood us with amounts of moving object data that preclude all means of manual data processing. The volume of data gathered from position sensors of mobile phones, PDAs, or vehicles, defies human ability to analyze the stream of input data. On the other hand, vast amounts of gathered data hide interesting and valuable knowledge patterns describing the behavior of moving objects. Thus, new algorithms for mining moving object data are required to unearth this knowledge. An important function of the mobile objects management system is the prediction of the unknown location of an object. In this paper we introduce a data mining approach to the problem of predicting the location of a moving object. We mine the database of moving object locations to discover frequent trajectories and movement rules. Then, we match the trajectory of a moving object with the database of movement rules to build a probabilistic model of object location. Experimental evaluation of the proposal reveals prediction accuracy close to 80%. Our original contribution includes the elaboration on the location prediction model, the design of an efficient mining algorithm, introduction of movement rule matching strategies, and a thorough experimental evaluation of the proposed model.

1 Introduction

Moving objects are ubiquitous. Portable devices, personal digital assistants, mobile phones, laptop computers are quickly becoming affordable, aggressively entering the market. This trend is parallel to the widespread adoption of wireless communication standards, such as GPRS, Bluetooth, or Wi-Fi networks. Recent advances in positioning technology compel manufacturers to equip their devices with positioning sensors that utilize Global Positioning System (GPS) to provide accurate location of a device. Accurate positioning of mobile devices paves the way for the deployment of location-based services and applications. Examples of location-based services include location-aware information retrieval, emergency services, location-based billing, or tracking of moving objects. It is important to note that location-based services are not limited to mobile devices, such as mobile phones, PDAs or laptops; these services can be successfully deployed for other types of moving objects, e.g., vehicles or even humans. In order to fully exploit the possibilities offered by location-aware services, it is crucial to determine the current position of a moving object at any given point in time.

P. Perner (Ed.): MLDM 2007, LNAI 4571, pp. 667–680, 2007.

Typically, a moving object is equipped with a transmitting device that periodically signals its position to the serving wireless carrier. Between position disclosures the exact location of a moving object remains unknown and can be determined only approximately. Unfortunately, the periodicity of position acknowledgments can be interrupted by several factors. For instance, the failure can be caused by power supply shortage of a moving object. Positioning systems have known limitations that can result in communication breakdown. Signal congestions, signal losses due to natural phenomena, or the existence of urban canyons lead to temporal unavailability of a moving object positioning information. Whenever the location of a moving object is unknown, a robust method of possible location prediction of a moving object is required.

Predicting the location of a moving object can be a difficult task. Firstly, the sheer amount of data to be processed precludes using traditional prediction methods known from machine learning domain. The stream of data generated by positioning sensors of thousands of moving objects requires new, robust and reliable data mining processing methods. The location prediction mechanism must allow for fast scoring of possible moving object location. The method must work online and should not require expensive computations. Furthermore, the performance of the prediction method should not degrade significantly with the increase of the number of moving objects. We also require that, given the prediction accuracy is satisfactory and does not drop below a given threshold, the prediction method should favor prediction speed over prediction accuracy. We believe that this feature is crucial for the development of successful location-based services. The success of a location-based service depends on whether the service is delivered to a particular object at a particular location and on particular time. If objects move quickly and change their location often, then the speed of computation must allow to deliver the service while the object still occupies a relevant location. For instance, complex models of movement area topology and movement interactions between objects may produce accurate results, but their computational complexity is unfeasible in mobile environment. Similarly, prediction methods based on simulation strongly depend on numerous input parameters that affect the quality of the resulting movement model. The cost of computing the model can be prohibitively high and the model itself may not scale well with the number of moving objects.

Another important drawback of currently used prediction methods is the fact that most of these methods do not utilize historical data. The raw data collected from moving objects hide useful knowledge patterns that describe typical behavior of moving objects. In particular, trajectories frequently followed by moving objects can be mined to discover movement patterns. Movement patterns, represented in the form of human-readable movement rules can be used to describe and predict the movement of objects.

Data mining techniques have been long considered inappropriate and unsuitable for online location prediction due to long processing times and computational expensiveness of these techniques. In this paper we prove that this assumption is entirely incorrect and that data mining techniques can be successfully used for location prediction. We build a probabilistic model of an unknown position of a moving object based on historical data collected from other objects moving on the same area. We mine logs of historical position acknowledgments to discover frequent trajectories of objects representing popular movement routes, and then we transform frequent trajectories into movement rules. In order to predict the location of a moving object, for which only a part of its movement history is known, we score the movement history of the object against the database of movement rules to find possible locations of the object. For each possible location we compute the probability of prediction correctness based on the support and confidence of discovered movement rules. Our method is fast and reliable. Frequent trajectories and movement rules are discovered periodically in an offline manner. The scoring process is performed online. Our experiments show that the scoring process can be performed within milliseconds. The presented method is independent of the movement area topology and scales well with the number of moving objects. The idea of using movement rules for location prediction was first presented in [15]. The work presented in this paper continues and extends our previous initial findings in a number of ways. The original contribution of this paper includes:

- refinement of the frequent trajectory model,
- design of an efficient *Traj-PrefixSpan* algorithm for mining frequent trajectories,
- modification of the *FP-Tree* index structure for fast lookup of trajectories,
- experimental evaluation of the proposal.

The paper is organized as follows. Section 2 presents the related work on the subject. In section 3 we introduce notions and definitions used throughout the paper. The *Traj-PrefixSpan* algorithm and frequent trajectory matching methods are presented in section 4. Section 5 contains the results of the experimental evaluation of our proposal. The paper concludes in section 6 with a brief summary.

2   Related Work

Both spatial data mining and mobile computing domains attract significant research efforts. The first proposal for spatial data mining has been formulated in [11]. Since then, many algorithms for spatial data mining have been proposed [5]. Authors in [6] introduce a spatial index for mining spatial trends using relations of topology, distance, and direction. A comprehensive overview of current issues and problems in spatial and spatio-temporal databases can be found in [7], and recent advances in spatio-temporal indexing are presented in [14]. However, the problem of mining trajectories of moving objects in spatial databases remained almost unchallenged until recently. Examples of advances in this field include the idea of similar trajectory clustering [12] and the proposal to use periodic trajectory patterns for location prediction [13]. The aforementioned works extend basic frameworks of periodic sequential patterns [8] and frequent sequential patterns [1].

An interesting area of research proposed recently focuses on moving object databases [2]. In [17] authors consider the effect of data indeterminacy and fuzziness on moving objects analysis. According to the authors, an inherent uncertainty of moving objects data influences attribute values, relations, timestamps, and time intervals. Advances in mobile object databases can be best illustrated by the development of the Path-Finder system, a prototype moving object database capable of mining moving object data. The idea of using floating car data of an individual moving object to describe movement patterns of a set of objects is presented in [3].

Several proposals come from mobile computing domain. Most notably, tracking of moving objects resulted in many interesting methods for location prediction. Authors in [10] present a probabilistic model of possible moving object trajectories based on road network topology. The solution presented in [21] advocates to use time-series analysis along with simulation of traveling speed of moving objects to determine possible trajectory of an object. A modification of this approach consisting in using non-linear functions for movement modeling is presented in [19]. A movement model that employs recursive motion functions mimicking the behavior of objects with unknown motion patterns is introduced in [18]. Another complex model with accuracy guarantees is presented in [20]. Recently, [22] consider predicting location in presence of uncertain position signals from moving objects. The authors present a *min-max* property that forms the basis for their *TrajPattern* algorithm for mining movement sequences of moving objects.

3 Definitions

Given a database of moving object locations, where the movement of objects is constrained to a specified area A. Let $O = \{o_1, \ldots, o_i\}$ be the set of moving objects. Let p denote the position of a moving object w.r.t. a system of coordinates W, $p \in W$. The *path* $P = (p_1, \ldots, p_n)$ is an ordered n-tuple of consecutive positions of a moving object. Unfortunately, the domain of position coordinates is continuous and the granularity level of raw data is very low. Therefore, any pattern discovered from raw data cannot be generalized. To overcome this problem we choose to transform original paths of moving objects into trajectories expressed on a coarser level. The *net* divides the two-dimensional movement area A into a set of rectangular regions of fixed size. We refer to a single rectangular region as a *cell*. Each cell has four *edges*. Cells form a two-dimensional matrix covering the entire area A, so each cell is uniquely identified by discrete coordinates $\langle i, j \rangle$ describing the position of the cell in the matrix. A moving object always occupies a single cell at any given point in time. When moving, an object crosses edges between neighboring cells. Each edge can be traversed in two directions, vertical edges can be traversed eastwards and westwards, whereas horizontal edges can be traversed northwards and southwards.

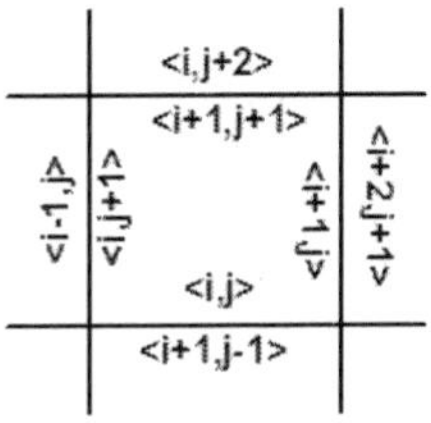

Fig. 1. Edge enumeration

Figure 1 presents the enumeration scheme of edges of the cell $\langle i, j \rangle$ used by our algorithm. Intuitively, the enumeration scheme preserves the locality of edges and allows for a fast lookup of all edges adjacent to a given edge. By an adjacent edge we mean an edge that can be traversed next after traversing a given edge. Each edge receives two sets of coordinates that are relative to its cell coordinates. The two sets of coordinates represent two possible directions of edge traversal. For instance, consider an object occupying the cell $\langle 2, 4 \rangle$. When the object moves northwards, it traverses an edge labeled $\langle 3, 5 \rangle$. The same edge, when traversed southwards, is identified as the edge $\langle 2, 6 \rangle$. The reason for this enumeration scheme is straightforward. An edge cannot have a single coordinate, because the set of possible adjacent edges depends on the direction of traversal. We have also considered other enumeration schemes, such as Hilbert curve or z-ordering. The main advantage of the presented edge enumeration scheme is the fact that any two neighboring edges differ by at most 2 on a dimension. In addition, any two edges within a single cell differ by at most 1 on any dimension.

Each path P_i of a moving object o_i can be unambiguously represented as a sequence of traversed edges. A *trajectory* of an object o_i is defined as an ordered tuple $R_i = (E_1, E_2, \ldots, E_n)$ of edges traversed by the path P_i. The *length* of a trajectory R_i, denoted $length(R_i)$, is the number of edges constituting the trajectory R_i. We refer to a trajectory of the length n as n-trajectory. We say that the trajectory $X = (X_1, X_2, \ldots, X_m)$ is contained in the trajectory $Y = (Y_1, Y_2, \ldots, Y_n)$, denoted $X \subseteq Y$, if there exist $i_1 < i_2 < \ldots < i_m$ such that $X_1 = Y_{i_1}$, $X_2 = Y_{i_2}$, $\ldots$, $X_m = Y_{i_m}$. A trajectory X is *maximal* if it is not contained in any other trajectory. We say that a trajectory Y *supports* a trajectory X if $X \subseteq Y$. The *concatenation* Z of trajectories $X = (X_1, X_2, \ldots, X_m)$ and $Y = (Y_1, Y_2, \ldots, Y_n)$, denoted $Z = X \otimes Y$, is the trajectory $Z = (X_1, X_2, \ldots, X_m, Y_1, Y_2, \ldots, Y_n)$. Given a database of trajectories $D_T = \{R_1, \ldots, R_q\}$. The *support* of a trajectory R_i is the percentage of trajectories in D_T that support the trajectory R_i.

$$support(R_i) = \frac{|\{R_j \in D_T : R_i \subseteq R_j\}|}{|D_T|}$$

A trajectory R_i is *frequent* if its support exceeds user-defined threshold of minimum support, denoted *minsup*. Given a trajectory $R_i = (E_1, E_2, \ldots, E_n)$. The *tail* of the trajectory R_i, denoted $tail(R_i, m)$, is the trajectory $T_i = (E_1, E_2, \ldots, E_m)$. The *head* of the trajectory R_i, denoted $head(R_i, m)$, is the trajectory $H_i = (E_{m+1}, E_{m+2}, \ldots, E_n)$. Concatenation of the tail and head yields the original trajectory, i.e., $tail(R_i, m) \otimes head(R_i, m) = R_i$.

Frequent trajectories are transformed into movement rules. A movement rule is an expression of the form $T_i \Rightarrow H_i$ where T_i and H_i are frequent adjacent trajectories and $T_i \otimes H_i$ is a frequent trajectory. The trajectory T_i is called the tail of the rule, the trajectory H_i is called the head of the rule. Contrary to the popular formulation from association rule mining, we do not require the tail and the head of a rule to be disjunct. For instance, an object may traverse edges E_i, E_j, E_k and then make a U-turn to go back through edges E_k, E_j. Thus, the same edge may appear both in the tail and the head of a rule. However, this difference does not affect the definition of statistical measures applied to movement rules, namely, support and confidence.

The *support* of the movement rule $T_i \Rightarrow H_i$ is defined as the support of $T_i \otimes H_i$,

$$support \, (T_i \Rightarrow H_i) = \frac{|T_j \in D_T : T_j \supseteq (T_i \otimes H_i)|}{|D_T|}$$

The *confidence* of the movement rule $T_i \Rightarrow H_i$ is the conditional probability of H_i given T_i,

$$confidence\,(T_i \Rightarrow H_i) = P\,(H_i|T_i) = \frac{support\,(T_i \otimes H_i)}{support\,(T_i)}$$

4 Proposed Solution

Formally, the location prediction problem can be decomposed into two subproblems:

- discover movement rules with support and confidence greater than user-defined thresholds of *minsup* and *minconf*, respectively,
- match movement rules against the trajectory of a moving object for which the current location is to be determined.

In section 4.1 we present the *Traj-PrefixSpan* algorithm that aims at efficient discovery of frequent trajectories and movement rules. Section 4.2 describes the modified *FP-Tree* index structure. In section 4.3 we introduce three matching strategies for movement rules.

4.1 Traj-PrefixSpan Algorithm

The algorithm presented in this section is a modification of a well-known *PrefixSpan* algorithm [16]. The difference consists in the fact that, contrary to the original formulation, we do not allow multiple edges as elements of the sequence (each element of a sequence is always a single edge). In addition, each sequence is grown only using adjacent edges, and not arbitrary sequence elements. The following description presents an overview of the *PrefixSpan* algorithm, already augmented to handle trajectories.

Given a trajectory $X = (X_1, X_2, \ldots, X_n)$, the *prefix* of the trajectory X is a trajectory $Y = (Y_1, Y_2, \ldots, Y_m)$, $m \leq n$, such that $Y_i = X_i$ for $i = 1, 2, \ldots, m - 1$. The *projection* of the trajectory X over prefix Y is a sub-trajectory X' of the trajectory X, such that Y is the prefix of X' and no trajectory X'' exists such that Y is the prefix of X'', X'' is the sub-trajectory of X, and $X'' \neq X'$.

Let $X' = (X_1, \ldots, X_n)$ be a projection of X over $Y = (Y_1, \ldots, Y_{m-1}, X_m)$. The trajectory $Z = (X_{m+1}, \ldots, X_n)$ is a postfix of X over the prefix Y, denoted $Z = X/Y$. In other words, for a given prefix Y and a given postfix Z, $X = Y \otimes Z$.

Let Y be a frequent trajectory in the database of trajectories D_T. An *Y-projected trajectory database*, denoted by $D_{T/Y}$, is the set of all postfixes of trajectories in D_T over the prefix Y. Let X be a trajectory with the prefix Y. The *support count* of X in Y-projected trajectory database, denoted by $support_{D_{T/Y}}(X)$, is the number of trajectories Z in $D_{T/Y}$, such that X is a sub-trajectory of $Y \otimes Z$.

Traj-PrefixSpan algorithm consists of three phases. In the first phase the algorithm performs a full scan of the trajectory database D_T to discover all frequent 1-trajectories. In the second phase each frequent 1-trajectory Y is used to create an Y-projected trajectory database. Every pattern contained in an Y-projected trajectory database must have the prefix Y. The third phase of the algorithm consists in recursive generation of

procedure $TrajPrefixSpan(Y, l, D_{T/Y})$

  1: scan $D_{T/Y}$ to find edges e such, that
       if $(l > 0)$ **then** e is adjacent to the last edge of Y
       else Y can be extended by e to form a frequent trajectory
  2: **foreach** edge e create $Y' = Y \otimes e$
  3: **foreach** Y' build $D_{T/Y'}$
  4: **run** $TrajPrefixSpan(Y', l + 1, D_{T/Y'})$
end procedure

Fig. 2. Traj-PrefixSpan algorithm

further Y'-projected trajectory databases from frequent trajectories Y' found in projections. The pseudocode of the algorithm is presented in Figure 2. The initial call is $TrajPrefixSpan(<>, 0, D_T)$.

4.2 FP-Tree

The physical indexing structure used in our algorithm is a slightly modified *FP-Tree* [9]. The main change consists in storing sequences of elements (as opposed to sets of elements), and allowing a bi-directional traversal of the tree. *FP-Tree* is an undirected acyclic graph with a single root node and several labeled internal nodes. The root of the tree is labeled *null*, and internal nodes are labeled with edge numbers they represent. Each internal node of the tree has a label, a counter representing the support of a sequence from the root to the given node, and a pointer to the next node with the same label (or a *null* pointer if no such node exists). In addition, the index contains a header table with edges ordered by their support and pointers to the first occurrence of an edge within the *FP-Tree*. The tree is constructed during the execution of the *Traj-PrefixSpan* algorithm by pattern growth. Each frequent trajectory discovered by the *Traj-PrefixSpan* algorithm is inserted into *FP-Tree* index for fast lookup. After the frequent trajectory discovery process finishes, the *FP-Tree* contains all frequent trajectories discovered in the database. Generation of movement rules is a straightforward task. For each frequent n-trajectory $X = (X_1, X_2, \ldots, X_n)$, $(n-1)$ movement rules can be generated by splitting the trajectory in every possible place, $T_1 \Rightarrow H_1, T_2 \Rightarrow H_2, \ldots, T_{n-1} \Rightarrow H_{n-1}$.

4.3 Matching Strategies

After frequent trajectories have been found and stored in the *FP-Tree*, they can be used to predict the unknown location of a moving object. For each moving object its known trajectory has to be compared with movement rules generated from frequent trajectories. In the next sections we introduce three matching strategies for scoring a partial trajectory of a moving object with the database of movement rules. In all examples let $X = (X_1, X_2, \ldots, X_m)$ be a partial trajectory of a moving object, for which we are seeking its most probable location. For a given partial trajectory X the set of all matched movement rules is denoted by L_X.

Whole Matcher. The *Whole Matcher* strategy consists in finding all movement rules $T_i \Rightarrow H_i$ such, that $X = T_i$ (i.e., the tail of the rule entirely covers the partial trajectory X). The head H_i can be used as a prediction of a possible location of a moving object. The probability that a moving object follows H_i is given by *confidence* $(T_i \Rightarrow H_i)$. The Whole Matcher strategy yields accurate results, but disallows any deviations of matched rules from the partial trajectory X. Furthermore, in case of long partial trajectories, the Whole Matcher strategy may fail to find a matching movement rule.

Last Matcher. The *Last Matcher* strategy discards all information from the partial trajectory X except for the last traversed edge X_m. The strategy finds all movement rules $T_i \Rightarrow H_i$ such, that $X_m = T_i$. The result of the strategy is the list of edges (movement rule heads H_i) ordered by descending values of *confidence* $(T_i \Rightarrow H_i)$. The Last Matcher strategy finds matching movement rules even for very short partial trajectories, but the predictions in L_X are less reliable, because they ignore the movement history of a moving object.

Longest Last Matcher. The *Longest Last Matcher* strategy is a compromise between the two aforementioned strategies. For a given partial trajectory X it finds all movement rules $T_i \Rightarrow H_i$ such, that T_i covers a part of the partial trajectory X, i.e., there exists $j, 1 \leq j < m$ such, that $T_i = head(X, j)$. The strategy outputs, as the result, the movement rule heads H_i weighted by the relative coverage of the partial trajectory X. For a given movement rule $T_i \Rightarrow H_i$ the strength of the prediction is defined as *confidence* $(T_i \Rightarrow H_i) * \frac{length(T_i)}{length(X)}$. Edges contained in L_X are ordered according to the descending value of the prediction strength.

5   Experiments

In this section we report on the results of the experimental evaluation of the proposed approach. All experiments were conducted on a PC equipped with AMD Athlon XP 2500+ CPU, 521 MB RAM, and a SATA hard drive running under Windows XP SP2 Home Edition. Algorithms and the front-end application were implemented in C# and run within Microsoft .NET 2.0 platform. Synthetic datasets were generated using Network-based Generator of Moving Objects by T.Brinkhoff [4]. Experiments were conducted using the map of Oldenburg. The number of moving objects varied from 1 000 to 10 000, the number of classes of moving objects was set to 10, and the number of time units in each experiment was 200. We set the maximum velocity of moving objects to 50, locations of objects were registered using *PositionReporter* method. All results reported in this section are averaged over 30 different instances of datasets. The experiments measure: the time of mining frequent trajectories, the number of discovered frequent trajectories, the time of matching a partial trajectory with the database of moving rules, and the quality of location prediction.

Figure 3 shows the number of frequent trajectories (depicted on the left-hand side axis of ordinates) and the time of mining frequent trajectories (depicted on the right-hand side axis of ordinates) with respect to the varying value of the *minsup* threshold. Both measured values decrease with the increase of the *minsup* threshold. As can be

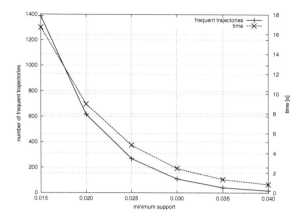

Fig. 3. Minimum support

clearly seen, the correlation between the number of frequent trajectories and the time it takes to mine them is evident. We are pleased to notice that even for low values of *minsup* threshold the algorithm requires less than 20 seconds to complete computations and the number of discovered frequent trajectories remains manageable.

Figure 4 presents the number of frequent trajectories (depicted on the left-hand side axis of ordinates) and the time of mining frequent trajectories (depicted on the right-hand side axis of ordinates) with respect to the varying number of moving objects for a set value of $minsup = 0.025$. Firstly, we notice that the time of mining frequent trajectories is linear w.r.t. the number of moving objects, which is a desirable property of our algorithm. Secondly, we observe a slight decrease in the number of discovered movement rules as the number of moving objects grows (a fivefold increase in the number of moving objects results in a 20% decrease of the number of discovered movement rules). This phenomenon is caused by the fact that a greater number of moving objects is spread more or less uniformly over the movement area, and the *minsup* threshold is expressed as the percentage of the number of all moving objects. Thus, less edges become frequent. For a smaller number of moving objects edges in the center of the city tend to attract more moving objects, and less restrictive *minsup* threshold makes more of these edges frequent, resulting in more movement rules.

Figure 5 shows the number of frequent trajectories (depicted on the left-hand side axis of ordinates) and the time of mining frequent trajectories (depicted on the right-hand side axis of ordinates) with respect to the varying size of an edge cell. The size of a cell is expressed in artificial units. The time of mining steadily decreases with the growth of the cell size. This result is obvious, because larger cells result in less frequent trajectories. On the other hand, the decrease is not linear. For larger cell sizes the number of discovered frequent trajectories is indeed lower. However, discovered frequent trajectories have higher support and tend to be longer, contributing to the overall computation time. The interpretation of the second curve, the number of discovered frequent trajectories, is more tricky. One can notice atypical deviations for cell sizes of 400 and 600 units. These random effects are probably caused by accidental structural influence

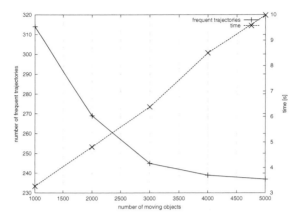

Fig. 4. Number of moving objects

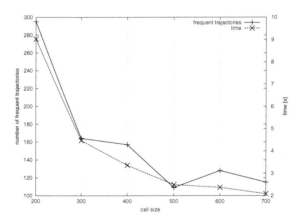

Fig. 5. Cell size

of larger and smaller cell sizes on areas of intensified traffic. The results presented in Figure 5 emphasize the importance of correct setting of the cell size parameter (e.g., the difference in the number of discovered frequent trajectories is 10 when changing the cell size from 300 to 400 units, and it grows to 40 when changing the cell size from 400 to 500 units). Unfortunately, our model does not permit to choose the optimal value of the cell size parameter other than experimentally.

The next two figures present the results of experiments evaluating the accuracy of prediction of the location using movement rules. These experiments were conducted as follows. First, a database of moving objects was generated using a set of fixed parameters. Then, 50 trajectories were randomly drawn from each database. Each test trajectory was then split into a tail and a head. The tail was used as a partial trajectory, for which future location of an object was to be predicted. Finally, the prediction returned from each matching strategy was compared to the known head of the test trajectory and the quality of prediction was computed. Let $X = (X_1, X_2, \ldots, X_m)$ be a randomly selected trajectory of a moving object, divided into $tail(X, k)$ and $head(X, k)$. The tail is used as a partial trajectory for matching. If the next traversed edge, which is X_{k+1} is not contained in the set of matching strategy answers L_X, then the quality of location prediction $Quality(X, L_X) = 0$. Otherwise, the quality of matching is computed as the probability of traversing X_{k+1} diminished by weighted incorrect predictions from L_X that had prediction strength greater than X_{k+1}, i.e.,

$$Quality(X, L_X) = P(X_{k+1}) * (1 - \sum_{j \leq k: X_j \in L_X} \frac{P(X_j) - P(X_{k+1})}{k + 1}) \tag{1}$$

In the above formula we assume that L_X is ordered by the decreasing prediction strength, so stronger predictions have lower indices.

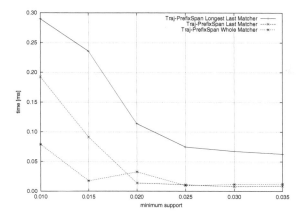

Fig. 6. Prediction time

Figure 6 presents the average time required to match a partial trajectory with the database of movement rules with respect to the varying *minsup* threshold (and, consequently, to the number of discovered movement rules). The *Whole Matcher* and *Last Matcher* strategies perform almost identically, because both strategies can fully utilize the *FP-Tree* index structure. The *Longest Last Matcher* strategy performs slower, because it must traverse a larger part of the *FP-Tree*. Nevertheless, in case of all strategies the matching time is very fast and never exceeds 0.3 ms. We are particularly satisfied with this result, because it supports our thesis that data mining methods can be employed for real-time location prediction.

Figure 7 depicts the average quality of prediction as computed by Equation 1. The prediction quality of the *Whole Matcher* and *Last Matcher* strategies reaches even 95% of accuracy for high *minsup* threshold values. For general settings of the *minsup* threshold the accuracy of both methods remains satisfactory between 75% and 85%. It is worth mentioning that the results depicted in the figure are computed according to our formula, which might be too penalizing for the *Longest Last Matcher* strategy, so the presented numbers are somehow biased towards simple matching strategies. The quality achieved by the *Longest Last Matcher* strategy varies from 35% to over 60%. Surprisingly, the quality of prediction increases with the decrease of the *minsup* threshold. This

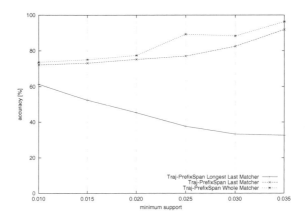

Fig. 7. Quality of prediction

can be explained by the fact that low values of the *minsup* threshold produce more frequent trajectories and more often the correct prediction is placed high in the resulting set L_X. Nevertheless, from the experimental evaluation we conclude that the *Longest Last Matcher* strategy is inferior to the *Whole Matcher* and *Last Matcher* strategies under all conditions.

6   Conclusions

In this paper we have introduced a new data mining model aiming at the efficient prediction of unknown location of moving objects based on movement patterns discovered from raw data. The model represents frequent trajectories of moving objects as movement rules. Movement rules provide a simplification and generalization of a large set of moving objects by transforming original continuous domain of moving object positions into a discretized domain of edges of a superimposed grid. The main thesis of the paper, well proved by conducted experiments, is that data mining techniques can be successfully employed for real-time location prediction in mobile environments. Indeed, while most expensive and burdensome computations (e.g. the discovery of frequent trajectories) can be performed offline and periodically, the online matching of partial trajectories with the database of movement rules is executed very fast. The quality of location prediction is satisfying, but we aim at developing more efficient matching strategies for even better accuracy.

Our future work agenda includes:

– replacing uniform grid cells with differently sized areas that adaptively divide the area of movement based on the density and congestion of moving objects,
– developing new matching strategies,
– including temporal aspects in discovered movement rules,
– including spatial information in movement rules,
– providing more informed decisions to location-based services based on discovered movement rules.

Acknowledgments

The author wishes to express his gratitude to one of his students, Lukasz Rosikiewicz, who greatly contributed to the implementation of all experiments, whose results are presented above.

References

1. Agrawal, R., Srikant, R.: Mining sequential patterns. In: ICDE'95, Taipei, Taiwan, March 6-10, pp. 3–14. IEEE Computer Society, Los Alamitos (1995)
2. Brakatsoulas, S., Pfoser, D., Tryfona, N.: Modeling, storing and mining moving object databases. In: Proceedings of the 8th International Database Engineering and Applications Symposium IDEAS'2004, pp. 68–77 (2004)
3. Brakatsoulas, S., Pfoser, D., Tryfona, N.: Practical data management techniques for vehicle tracking data. In: Proceedings of the 21st International Conference on Data Engineering ICDE'2005, pp. 324–325 (2005)
4. Brinkhoff, T.: A framework for generating network-based moving objects. GeoInformatica 6(2), 153–180 (2002)
5. Ester, M., Frommelt, A., Kriegel, H.-P., Sander, J.: Spatial data mining: Database primitives, algorithms and efficient dbms support. Data Mininig and Knowledge Discovery 4(2/3), 193–216 (2000)
6. Ester, M., Kriegel, H.-P., Sander, J.: Knowledge discovery in spatial databases. In: Proceedings of the 23rd German Conference on Artificial Intelligence, KI'99, pp. 61–74 (1999)
7. Gidofalvi, G., Pedersen, T B: Spatio-temporal rule mining: issues and techniques. In: Proceedings of the 7th International Conference on Data Warehousing and Knowledge Discovery DaWaK'2005, pp. 275–284 (2005)
8. Han, J., Dong, G., Yin, Y.: Efficient mining of partial periodic patterns in time series database. In: ICDE'99, Sydney, Austrialia, 23-26 March, 1999, pp. 106–115. IEEE Computer Society, Los Alamitos (1999)
9. Han, J., Pei, J., Yin, Y.: Mining frequent patterns without candidate generation. In: SIGMOD '00: Proceedings of the 2000 ACM SIGMOD international conference on Management of data, pp. 1–12. ACM Press, New York, NY, USA (2000), doi:10.1145/342009.335372
10. Karimi, H.A., Liu, X.: A predictive location model for location-based services. In: ACM GIS'03, New Orleans, Louisiana, USA, November 7-8, 2003, pp. 126–133. ACM, New York (2003)
11. Koperski, K., Han, J.: Discovery of spatial association rules in geographic databases. In: SSD'95, Portland, Maine, August 6-9, pp. 47–66. Springer, Heidelberg (1995)
12. Li, Y., Han, J., Yang, J.: Clustering moving objects. In: ACM SIGKDD'04, Seattle, Washington, USA, August 22-25, 2004, pp. 617–622. ACM, New York (2004)
13. Mamoulis, N., Cao, H., Kollios, G., Hadjieleftheriou, M., Tao, Y., Cheung, D.W.: Mining, indexing, and querying historical spatiotemporal data. In: ACM SIGKDD'04, Seattle, Washington, USA, August 22-25, 2004, pp. 236–245. ACM, New York (2004)
14. Mokbel, M.F., Ghanem, T.M., Aref, W.G.: Spatio-temporal access methods. IEEE Data Engineering Bulletin 26(2), 40–49 (2003)
15. Morzy, M.: Prediction of moving object location based on frequent trajectories. In: Levi, A., Savaş, E., Yenigün, H., Balcısoy, S., Saygın, Y. (eds.) ISCIS 2006. LNCS, vol. 4263, pp. 583–592. Springer, Heidelberg (2006)
16. Pei, J., Han, J., Mortazavi-Asl, B., Pinto, H., Chen, Q., Dayal, U., Hsu, M.: Prefixspan: Mining sequential patterns by prefix-projected growth. In: ICDE'01, Heidelberg, Germany, April 2-6, 2001, pp. 215–224. IEEE Computer Society, Los Alamitos (2001)

17. Pfoser, D., Tryfona, N.: Capturing fuzziness and uncertainty of spatiotemporal objects. In: Caplinskas, A., Eder, J. (eds.) ADBIS 2001. LNCS, vol. 2151, pp. 112–126. Springer, Heidelberg (2001)

18. Tao, Y., Faloutsos, C., Papadias, D., Liu, B.: Prediction and indexing of moving objects with unknown motion patterns. In: ACM SIGMOD'04, Paris, France, June 13-18, 2004, pp. 611–622. ACM, New York (2004)

19. Trajcevski, G., Wolfson, O., Xu, B., Nelson, P.: Real-time traffic updates in moving objects databases. In: DEXA 2002, pp. 698–704. IEEE Computer Society, Los Alamitos (2002)

20. Wolfson, O., Yin, H.: Accuracy and resource concumption in tracking and location prediction. In: Hadzilacos, T., Manolopoulos, Y., Roddick, J.F., Theodoridis, Y. (eds.) SSTD 2003. LNCS, vol. 2750, pp. 325–343. Springer, Heidelberg (2003)

21. Xu, B., Wolfson, O.: Time-series prediction with applications to traffic and moving objects databases. In: MobiDE 2003, San Diego, California, USA, September 19, 2003, pp. 56–60. ACM, New York (2003)

22. Yang, J., Hu, M.: Trajpattern: Mining sequential patterns from imprecise trajectories of mobile objects. In: Ioannidis, Y., Scholl, M.H., Schmidt, J.W., Matthes, F., Hatzopoulos, M., Boehm, K., Kemper, A., Grust, T., Boehm, C. (eds.) EDBT 2006. LNCS, vol. 3896, pp. 664–681. Springer, Heidelberg (2006)

Categorizing Evolved CoreWar Warriors Using EM and Attribute Evaluation

Doni Pracner, Nenad Tomašev, Miloš Radovanović, and Mirjana Ivanović

University of Novi Sad
Faculty of Science, Department of Mathematics and Informatics
Trg D. Obradovića 4, 21000 Novi Sad
Serbia
doni@neobee.net, tomasev@nspoint.net, {radacha,mira}@im.ns.ac.yu

Abstract. CoreWar is a computer simulation where two programs written in an assembly language called redcode compete in a virtual memory array. These programs are referred to as *warriors*. Over more than twenty years of development a number of different battle strategies have emerged, making it possible to identify different warrior types. Systems for automatic warrior creation appeared more recently, evolvers being the dominant kind. This paper describes an attempt to analyze the output of the CCAI evolver, and explores the possibilities for performing automatic categorization by warrior type using representations based on redcode source, as opposed to instruction execution frequency. Analysis was performed using EM clustering, as well as information gain and gain ratio attribute evaluators, and revealed which mainly brute-force types of warriors were being generated. This, along with the observed correlation between clustering and the workings of the evolutionary algorithm justifies our approach and calls for more extensive experiments based on annotated warrior benchmark collections.

1 Introduction

Among the many approaches to creating artificial intelligence and life, one is concerned with constructing computer programs which run in virtual environments. Many aspects of these environments may be inspired by the real world, with the overall objective to determine how well the programs adapt. In some cases different programs compete for resources and try to eliminate the opposition.

One of the oldest and most popular venues for the development and research of programs executing in a simulated environment is CoreWar, in which programs (referred to as *warriors*) attempt to survive in a looping memory array. The system was introduced in 1984 by A. K. Dewdney in an article in the Scientific American [1]. Basically, two programs are placed in the array end executed until one is completely eliminated from the process queue. The winner is determined through repeated execution of such "battles" with different initial positioning of warriors in the memory. Online competitions are held on a regular basis, with the game being kept alive by the efforts of a small, but devoted community.

Over the course of more than twenty years of development, a number of different battle strategies have emerged, often combining more than one method for eliminating opponents. These strategies closely reflect programmers' ideas about how a warrior

P. Perner (Ed.): MLDM 2007, LNAI 4571, pp. 681–693, 2007.

should go about winning a battle. However, several attempts have been made recently to automatically create new and better warriors, by processes of optimization and evolution. Optimized warriors are essentially human-coded, with only a choice of instruction parameters being automatically calculated to ensure better performance. On the other hand, evolved warriors are completely machine generated through the use of evolutionary algorithms.

In order to evaluate the performance of optimized and evolved warriors, the most common method is to put them against a benchmark set of manually prepared test programs. To get reliable and stable results against every warrior from the benchmark in the usual setting, at least 250 battles are needed, each taking a few seconds to execute. Evolving new warriors from a set of a few thousand programs and iteratively testing them against the benchmark is then clearly a very time demanding process.

The goal of the research presented in this paper is to examine the diversity of warrior pools created by one particular evolver and to test the possibilities of automatic categorization by warrior type (employed strategies), given the information obtained by syntax analysis of warrior source code. The amount of data created by evolver runs usually surpasses the capabilities of human experts to examine and classify the warrior pools. Automated categorization would, therefore, be extremely helpful in the control of diversity levels, and dynamic modification of mutation rates for sustaining the desirable diversity within generations. It would also significantly contribute to our understanding of the nature of the output of evolutionary algorithms, in this case the battle strategies of evolved warriors. Although one may be familiar with every detail of how a particular evolutionary algorithm works, its output is still very much dependent on the performance of warriors against the benchmark, leaving room for many surprises.

There were some attempts in the past to perform automatic categorization of warriors, but these were based on the analysis of execution frequencies of particular instruction types during simulation, which requires the simulation to be run for a certain amount of time [2]. If the source-based approach proved fruitful, it would be possible to come to similar conclusions much quicker, which could, in turn, speed up the whole process of warrior evolution. To the best of our knowledge, this paper presents the first attempt to categorize warriors using *static* (source-based) instead of *dynamic* (execution-based) methods.

The rest of the paper is organized as follows. Section 2 explains the essentials of CoreWar and some basic strategies of human-coded warriors, while Section 3 outlines the principles of the EM clustering algorithm. Section 4 describes the dataset of evolved warriors and how it was processed into the representation suitable for analysis. The analysis, which relies on clustering and attribute evaluation techniques, is the subject of Section 5. The last section provides a summary of the conclusions together with plans for future work.

2   CoreWar

CoreWar is a computer simulation where programs written in an assembly language called corewars (by the 1988 ICWS standard) or redcode (by the 1994 ICWS standard) compete in a virtual memory array. Those programs are referred to as *warriors*. The simulated memory array is called *the core*. It is wrapped around, so that the first memory location in the address space comes right after the last one. The basic unit of memory in the core is one instruction, instead of one bit. The memory array redcode simulator (MARS) controls the execution of the instructions in the core. The execution of instructions is consecutive, apart from the situations arising after executing jump instructions. All arithmetic is modular, depending on the size of the core. All addressing modes are relative.

The goal of a warrior is to take complete control over the core by making the opponent eliminate its own thread of execution from the process queue. There are many ways to achieve this effect, and various different strategies of attack have emerged over time. CoreWar warriors can copy the memory content, read from the core, perform various calculations, mutate and change their behavior, make copies of themselves, place decoys, search for their opponents etc. The starting placement of warriors in the core is done at random, and a predetermined number of fights are staged to decide the winner (3 points are awarded for a win, 1 for a draw, 0 for a loss). Between rounds, the result of the previous fight is stored in a separate memory array called P-space. In some competitions warriors are allowed to access this memory and change their strategy, if necessary, to ensure better performance in future rounds.

CoreWar was introduced by A. K. Dewdney in 1984, in an article published in the Scientific American [1]. Today, CoreWar exists as a programming game with ongoing online competitions on several servers, among which are www.koth.org/ and sal.math.ualberta.ca/. There are many competition leagues, depending on battle parameters, and each of these is called a hill. The warrior currently holding the first place is appropriately called the *king of the hill* (KOTH).

Although the competitions were originally meant as a challenge for testing human skill in making successful CoreWar programs, there were also those who chose to create software capable of autonomously generating or evolving and later evaluating competitive CoreWar programs. On several occasions such warriors were able to outperform warriors coded by humans. This is usually done via the implementation of evolutionary algorithms.

2.1   The Redcode Language

Redcode is a language that is being used as a standard for making CoreWar warriors since 1994. It consists of 19 instructions, 7 instruction modifiers and 8 addressing modes. The warrior files are stored on the disk as *WarriorName.RED*.

The redcode instruction set, although not huge, allows for much creativity and diversity. Each command consists of an instruction name, instruction modifier, A-field addressing mode, A-field value, B-field addressing mode, and the B-field value. The source address is stored in the A-field and the destination address in the B-field. Table 1 summarizes the more important redcode instructions, while Tables 2 and 3 describe all redcode modifiers and addressing modes, respectively. Figure 1(a) depicts the source of an example warrior.

Table 1. Overview of some redcode instructions

Instruction	Description
DAT	Removes the process that executes it from the process queue. It is used to store data. The instruction modifiers play no role here.
MOV	Copies the source to the destination.
ADD	Adds the number in the source field to the number in the destination field. Two additions can be done in parallel if the .F or .X modifier is used.
SUB	Performs subtraction. The functionality is the same as in ADD.
MUL	Performs multiplication. It is not used as frequently as ADD or SUB, however.
DIV	Performs integer division. In case of division by zero, the process demanding the execution of the instruction is removed from the process queue. This is another way of removing enemy processes.
MOD	Gives the remainder of the integer division.
JMP	The unconditional jump instruction, redirecting the execution to the location pointed at by its A-field. The B-field does not affect the jump, so it can be used either to store data, or to modify some other values via the use of incremental/decremental addressing modes.
JMZ	Performs the jump, if the tested value is zero. If the modifier is .F or .X, the jump fails if either of the fields is nonzero. As in the jump instruction, the A-field points to the jump location. The B-field points to the test location. If the jump fails, the instruction following the JMZ will be the next instruction to be executed by this process.
JMN	Performs the jump if the tested value is nonzero. Otherwise functions like JMZ.
DJN	Decreases the destination and jumps if the value is nonzero. The functionality is otherwise the same as in JMZ and JMN.
SPL	Creates a new process and directs its execution to the source value. The old process, being the one that executed the SPL is moved to the next memory location. The new process is executed right after the old process.

Table 2. Overview of redcode instruction modifiers

Modifier	Description
.I	This modifier states that the action is conducted on the whole instruction, and used only when copying an instruction or comparing the content of two memory locations.
.F	Copying, or comparing two fields at the same time.
.X	Copying, or comparing two fields at the same time, A-field of the source to the B-field of the destination, and B-field of the source to the A-field of the destination.
.A	Moving, or comparing, the A field of the source to the A-field of the destination.
.B	Moving, or comparing, the B field of the source to the B-field of the destination.
.AB	Moving, or comparing, the A field of the source to the B-field of the destination.
.BA	Moving, or comparing, the B field of the source to the A-field of the destination.

Table 3. Overview of redcode addressing modes

	Addressing Mode	Description
\$	direct	Points to the instruction x locations away, where x is the respective field value in the executed instruction. It can be omitted.
#	immediate	Points to the current instruction, regardless of the field value.
*	A-field indirect	Points to the instruction $x + y$ locations away, where x is the respective field value and y is the value in the A-field of the instruction x locations away.
@	B-field indirect	Analogous to A-field indirect.
{	A-field predecrement	Indirect mode, also decreasing the A-field value of the instruction pointed to by the respective field in the executed instruction. The decrement is done before calculating the source value of the current instruction.
}	A-field postincrement	Indirect mode, also increasing the A-field value of the instruction pointed to by the respective field in the executed instruction. The increment is done after calculating the source value of the current instruction.
<	B-field predecrement	Analogous to A-field predecrement.
>	B-field postincrement	Analogous to A-field postincrement.

2.2 Warrior Types

As mentioned before, over twenty years of CoreWar competitions had lead to a great increase in diversity of warrior types. Some of the most important warrior categories are given below.

Imps are the simplest kind of warriors which just copy themselves to another memory location in each execution cycle, that way "running around" the core. Imps barely have any offensive capabilities, and are seldom used on their own.

Coreclears attempt to rewrite the whole core with process-killing instructions, that way ensuring a win, in a sense of being positive that the opponent is destroyed.

Stones simply copy DAT instructions over the core, trying to overwrite a part of the enemy code. Up to this moment, many alternate approaches were devised, resulting in warriors copying other instructions as well, not only DATs.

Replicators (papers) follow the logic that in order for the warrior to survive, it should create many processes and let them operate on many copies of the main warrior body, therefore ensuring that some of those copies will survive an enemy attack, since it takes a lot of time to destroy them all. In the meantime, the warrior tries to destroy the enemy process. The warrior in Fig. 1(a) is, in fact, a replicator, referred to as the "black chamber paper."

Scanners (scissors) try to discover the location of enemy code and then start an attack at that location. Since the scanner attack has a greater probability of succeeding, due to the intelligent choice of target location, such a warrior is usually able to invest more time in the attack against that location.

Hybrid warriors combine two or more warrior types in their code, and are nowadays most frequently used in CoreWar tournaments.

Generally, each non-hybrid type of CoreWar warrior is effective over one other warrior type, and is at the same time especially vulnerable to another, with the relationships between types being in line with the rock-paper-scissor metaphor (hence the naming of some warrior types). For more information about the redcode language and warrior types, see [3].

3   Expectation Maximization

The research described in this paper utilizes the *expectation maximization* (EM) clustering algorithm [4] (p. 265), implemented in the WEKA machine learning workbench. This algorithm is probabilistic by nature, and takes the view that while every instance belongs to only one cluster, it is almost impossible to know for certain to which one. Thus, the basic idea is to approximate every attribute with a statistical *finite mixture*. A mixture is a combination of k probability distributions that represent k clusters, that is, the values that are most likely for the cluster members. The simplest mixture is when it is assumed that every distribution is Gaussian (normal), but with different means and variances. Then the clustering problem is to deduce these parameters for each cluster based on the input data. The EM algorithm provides a solution to this problem.

In short, a procedure similar to that of *k-means* clustering ([4], pp. 137–138) is used. At the start, the parameters are guessed and the cluster probabilities calculated. These probabilities are used to re-estimate the parameters, and the process is continued until the difference between the overall log-likelihood at consecutive steps is small enough. The first part of the process is "expectation," i.e. the calculation of cluster probabilities, and the second part – calculating the values of parameters – is the "maximization" of the overall log-likelihood of the distributions given the data.

WEKA's implementation of EM provides an option to automatically determine the number of clusters k using 10-fold cross-validation. This is done by starting with $k = 1$, executing the EM algorithm independently on every fold and calculating the average log-likelihood over the folds. As k is incremented the process is repeated until the average log-likelihood stops increasing.

4   The Dataset

The analyzed data represents a subset of warriors generated by the CCAI evolver [5], which was written by Barkley Vowk from the University of Alberta in summer 2003. The evolutionary approach used in this evolver was the island model [6].

The dataset consists of 26795 warrior files, and is summarized in Table 4. The data was divided into four smaller parts in chronological order. The respective sizes of the parts are 10544, 6889, 4973, 4389, and will be referenced in the text as "generation 1," "generation 2" etc. The first pool was randomly generated, and the others represent the consecutive generations in evolving. One of the reasons why each group is smaller than the previous one is that evolvers reduce diversity in each step, and duplicates are removed before proceeding to the next generation. The benchmark used for this evolution was Optimax [7].

4.1   Selecting the Representation

Inspired by the classical bag-of-words representation for text documents, and the fact that it works for many types of data mining and machine learning problems, we opted for an analogous "bag-of-instructions" representation for CoreWar warriors. Since each instruction may be accompanied by an instruction modifier, two addressing modes and two field values, there are plenty of choices for deriving attributes, possibly leading to a high dimensionality of the representation.

In the end, the decision was made to use a vector with just the bare instruction counts from the warrior source code. The resulting vector has 16 coordinates (attributes), one for each of the command types. The name of the warrior was also added as an attribute. To transform the data into vector form a Java command line application was written, details of which are presented in [3].

Some modifications were introduced to make the information more specific to redcode, and the first alteration was to treat ADD and SUB as the same instruction, being that they can perform the same operation by simply toggling the minus sign in the address field.

The next alteration was done in order to add more information about the structure of the warriors to the representation. For many types of warriors there are specific pairs of commands that appear one after the other. Based on our previous experience with warrior types and coding practices, eight pairs of these two-command combos were added to the representation, namely SPLMOV, MOVJMP, MOVDJN, MOVADD, MOVSUB, SEQSNE, SNEJMP and SEQSLT.

Finally, there are sets of commands specific to some types of *imps*, so a true/false field named "Imp spec" was introduced. Examples of such commands are MOV.I 0,1 and MOV.I #x,1. The presence of any of the commands suggests that an imp structure could be embedded within a warrior.

Figure 1 shows the representation of an example warrior (a) as a vector of attributes (b) described above.

4.2   Removing Duplicates

Besides choosing an appropriate representation, a method for speeding up calculations, as well as improving results, is to remove "too similar" warriors. When clustering the data, warriors which are close to each other in terms of distance between the appropriate vectors in the state-space (containing all the vectors), could easily gravitate smaller groups toward them, thus creating a larger cluster than it should be.

A decision was made to ignore the *address fields*, and therefore duplicates would be any two warriors that have the same sequence of instructions with identical instruction modifiers and address modifiers.

Table 4 summarizes the results of the duplicates search. Most duplicates were removed from generations 1 and 2, 12% and 8% respectively. From generation 3 only about 1% of the files were removed as duplicates, and in set 4 about 4%. In summary, a total of 2153 duplicates were found, which is about 8% of the initial 26795 warriors.

```
boot      SPL.B $1, $0                    DAT:       0
          SPL.B $1, $0                    MOV:       5
          SPL.B $1, $0                    ADD/SUB:   0
          MOV.I {p1, {divide              MUL:       0
divide    SPL.B (p3+1+4000), }c           DIV:       0
p1        SPL.B @(p3+1), }ps1             MOD:       0
          MOV.I }p1, >p1                  JMP:       0
p2        SPL.B @0, }ps2                  JMZ:       1
          MOV.I }p2, >p2                  JMN:       0
          MOV.I #bs2, <1                  DJN:       0
          SPL.B @0, {bs1                  SPL:       7
          MOV.I {p2, {p3                  SEQ:       0
p3        JMZ.A $ps3, *0                  SNE:       0
                                          SLT:       0
                                          NOP:       0
                                          SPLMOV:    4
                                          MOVJMP:    0
                                          MOVDJN:    0
                                          MOVADD:    0
                                          MOVSUB:    0
                                          SEQSNE:    0
                                          SNEJMP:    0
                                          SEQSLT:    0
                                          Imp spec:  false
```

            (a)   (b)

Fig. 1. Example code of a warrior (a), and its attribute vector representation (b)

Table 4. Summary of datasets and results of duplicate detection

Dataset	Files	Duplicates	Reduction
Generation 1	10544	1345	12%
Generation 2	6889	559	8%
Generation 3	4973	56	1%
Generation 4	4389	193	4%
Complete	26795	2153	8%

5 Analysis of Evolved Warriors

5.1 Clustering

First, clustering was performed independently on all warrior generations (and also on the complete set) using the implementation of EM from the WEKA workbench. The number of clusters was automatically determined by cross-validation (see Section 3).

The number of discovered clusters per warrior set and the number of instances per cluster are given in Table 5. In generation 1, only two clusters were found. After examining a portion of the warriors in this set, it appeared that the two clusters that were found consist mostly of various kinds of replicators and some coreclears. This was determined by taking a random sample of 50 warriors from each of the clusters. The only way to achieve absolute confirmation is to manually examine all warriors, which we considered infeasible. However, some insights provided by attribute evaluation (Section 5.2) give additional support to the finding.

Table 5. Clusters per generation and number of warriors per cluster

Dataset	Clusters	Cluster sizes										
Generation 1	2	8081	1146									
Generation 2	4	3456	1857	572	468							
Generation 3	12	88	2112	644	543	526	47	671	36	47	103	38	80
Generation 4	5	2364	1197	357	94	184							
Complete	3	6571	2671	15469									

Compared to generation 1, the number of clusters increases in generations 2 and 3, more precisely 4 and 12 respectively, but this was expected. The warriors in each set were evolved from the previous, and new strategies that had good results were preserved. This means that new groups of warriors with similar strategies should appear in generations 2 and 3, and the clustering algorithm did notice this.

In the last generation, the fourth, the number of clusters decreased to 5. This is most probably due to the reduction of diversity in the warriors that takes place at the end of the process of evolution.

The clustering of the whole dataset resulted in 3 clusters. The reduction of the number may be a consequence of the island model – the larger clusters most likely "absorbing" the smaller ones.

5.2   Attribute Evaluation

To analyze the effects of different attributes on cluster selection, *information gain* (IG) and *gain ratio* (GR) attribute evaluators were used [4]. Because of known shortcomings of both evaluation methods[1], the approach that was utilized was to choose the attributes with the highest gain ratio, but only if their information gain is larger that the average information gain for all attributes ([4], p. 105).

Since attributes in the warrior representation mostly correspond to instructions and instruction pairs, we expected their (in)significance with regards to the clustering to give us some idea about the types of warriors that were grouped together, and also to shed some light on the process of warrior evolution.

Table 6 summarizes the results of attribute evaluation on the complete dataset. It shows that the most informative feature is 'DJN,' being that others with higher GR values have very low information gain. It is interesting to note that the second best is 'MOVDJN' and that these two are also the best two in IG values. However, the rest of the information gain list does not follow in the same order. The 'SPLMOV' attribute

---

[1] IG favors attributes with many distinct values, while GR may give unrealistically high scores to attributes with a low value count.

Table 6. Gain ratio and information gain for the complete dataset

Gain Ratio		Information Gain	
0.40350	MUL	**0.49217**	**DJN**
0.38200	SLT	**0.43917**	**MOVDJN**
0.37550	SEQSNE	0.31378	MOV
0.34570	MOVSUB	0.22883	SPLMOV
0.33690	MOD	0.20510	SPL
0.31520	SNEJMP	0.17605	Imp spec
0.26310	**DJN**	0.17025	DAT
0.24970	SEQSLT	0.15134	ADD/SUB
0.24820	JMZ	0.14137	MOVJMP
0.24070	**MOVDJN**	0.09816	SEQ
0.18620	NOP	0.09345	MOVSUB
0.18270	ADD/SUB	0.07179	JMP
0.18040	Imp spec	0.06768	MOVADD
0.17430	DIV	0.06380	SNE
0.15190	MOVADD	0.06001	JMZ
0.11840	JMN	0.03642	MUL
0.11630	SNE	0.02582	SLT
0.10140	SEQ	0.02370	NOP
0.09350	MOVJMP	0.02184	SEQSNE
0.08930	MOV	0.02126	JMN
0.08240	SPLMOV	0.01484	DIV
0.05780	SPL	0.01138	MOD
0.05420	JMP	0.00730	SNEJMP
0.05110	DAT	0.00117	SEQSLT
0.192436	**AVERAGE**	**0.1191416**	**AVERAGE**

also has high information gain, but shows less in terms of gain ratio. Looking from the CoreWar perspective, 'SPLMOV' and 'MOVDJN' are instruction pairs appearing frequently in both coreclears and replicators, so this result is not surprising. It also suggests that the results might have been significantly different if these attributes had not been used, and an ordinary bag-of-instructions was employed instead.

After the analysis of the complete dataset, an attempt was made to get more information on the actual evolution process by examining the individual generations, keeping in mind that the first generation is (in big part) random.

Table 7 lists the most informative attributes for generations 1–4, with their GR and IG scores. On generation 1, analysis showed that the most informative attribute is 'JMN,' being the first selection or both information gain and gain ratio. 'SLT' is also close, followed by 'MOVSUB' and 'MOVADD' after a large gap.

An interesting observation is that the best attributes for the complete dataset, 'DJN' and 'SPLDJN,' are at the very bottom of the list in generation 1. An interpretation for this is that there was greater variety in the original pool, which did not affect the global dataset at a greater measure, especially when considering the fact that subsequent generations increasingly resemble the complete dataset, as demonstrated below.

Table 7. Most informative attributes for generations 1–4, together with GR and IG scores

Generation 1				Generation 2		
Attribute	**GR**	**IG**		**Attribute**	**GR**	**IG**
JMN	0.47461	0.11450		MOVADD	0.48960	0.34157
SLT	0.41082	0.04551		ADD/SUB	0.37800	0.34566
MOVSUB	0.13870	0.05287		MOVDJN	0.22310	0.43554
MOVADD	0.12117	0.04122		SPLMOV	0.22280	0.56428
				DJN	0.20880	0.43173
				SPL	0.20350	0.54848

Generation 3				Generation 4		
Attribute	**GR**	**IG**		**Attribute**	**GR**	**IG**
SPLMOV	0.36100	0.47913		SPLMOV	0.52450	0.56816
MOVDJN	0.33800	0.31149		MOVJMP	0.43380	0.44600
DJN	0.31500	0.32649		MOVDJN	0.42800	0.33516
Imp spec	0.26400	0.18135		DJN	0.41400	0.33717
SPL	0.22700	0.49328				

In the second warrior set the situation was significantly different compared to generation 1, with 'JMZ' and 'DIV' leading the GR scores, but with low IG. After filtering with the average IG, the list is as follows: 'MOVADD,' 'ADD/SUB,' 'MOVDJN,' 'SPLMOV,' 'DJN,' and 'SPL'. Here 'MOVDJN' and 'DJN,' which were important for the complete dataset, do appear in the list. Also, most of the best attributes from generation 1 do not show, or are a lot lower in the list, except 'MOVADD' and 'ADD/SUB'. This all indicates that much code from generation 1 was discarded during evolution. This is also evident in the reduction of size by 40% between generations 1 and 2.

In the third group, analysis shows that 'SPLMOV,' 'MOVDJN,' 'DJN,' 'ImpSpec,' and 'SPL' had most impact on the clustering process. Compared to the second generation, 'ADD/SUB' and 'MOVADD' which were "inherited" from generation 1 are now gone, leaving a result much closer to the complete set.

In generation 4, 'SPLMOV,' 'MOVJMP,' 'MOVDJN,' and 'DJN' were the most significant attributes with regards to clustering. The only big difference between this and generation 3 is the "climbing" of 'MOVJMP'. This lack of differences is also consistent with the earlier explained way the CCAI evolver works, in the sense that when there are no great improvements to the warriors in the next generation the process is stopped.

6   Conclusions and Future Work

Exploration and generation of CoreWar warriors, assisted by computers, has become increasingly popular in the recent years. Majority of work, however, has concentrated on warrior parameter optimization [7] and the evolution of competitive warriors [8,5,9]. Exploratory analysis (albeit motivated by warrior evolution), by means of automatic categorization based on the analysis of execution frequencies of certain instruction types during simulation, was performed, with some results available in [2], but with no published findings. In the research described in this paper, on the other hand, we attempted to utilize a *static* (source-based) instead of a *dynamic* (execution-based) approach to the analysis and categorization of a set of warriors. The used dataset was the result of warrior evolution conducted by the CCAI evolver [5].

The clustering of the CCAI evolver output was done using the EM algorithm incorporated in the WEKA workbench. Three clusters were detected in the complete dataset. This indicates that the overall diversity of the complete dataset was rather low, which can be explained by the fact that it is difficult for evolutionary algorithms to generate complex structures within the warriors in the evolved population, because small changes and mutations usually render good complex warriors useless, and there is a huge gap between different warrior strategies. Therefore, the most mutation resistant forms prevailed, namely replicators and coreclears.

The complete dataset was divided into 4 subsets, in chronological generational order. After processing, 2, 4, 12 and 5 clusters had been found in generations 1, 2, 3 and 4, respectively (see Table 5). The general tendency of this result was expected, because of varying mutation rates which were decreased at the end of the evolution process, producing a general decrease of diversity in the evolved population.

Information gain and gain ratio analysis showed that 'DJN' and 'MOVDJN' were the most significant attributes in the clustering of the whole dataset (see Table 6). 'SPLMOV' and 'MOVJMP' were also important in clustering of some of the subgroups. This can be explained by the fact that most of the warriors in the dataset were either replicators or coreclears, and these instructions and instruction pairs are seen quite frequently in such warriors.

Attribute analysis generation by generation also showed consistency with the way the evolver works. The greatest changes were exhibited between the original pool and the next generation, and attribute evaluation did register large differences in the informativeness of attributes.

It is also possible to cluster warrior sets according to the scores of evolved warriors against a predetermined benchmark. A diverse benchmark of human-coded warriors manually annotated with their types was created for this purpose, and the score tables have already been generated. The clustering according to the score tables will be conducted and the results compared to those obtained via source-based clustering described in this paper.

An issue with the static source-based warrior representation used in the presented work may be the "garbage" often left over in the source code of evolved warriors – instructions which never actually execute, but effectively introduce noise to the representation. Warriors written by humans, on the other hand, are usually "clean" in this sense. Dynamic representations based on counts of instruction execution are able to deal with this kind of noise, but at the expense of a considerable increase in warrior preprocessing time.

The noted correspondences between the workings of the evolutionary algorithm and clustering indicate that our choice of static warrior representation was to some extent appropriate. However, in order to determine exactly to what extent, and whether the syntax analysis can produce good categorization of evolved warriors, precise measurements are necessary. This may be done through comparison of source-based and score-table-based clusterings, and additionally by training classifiers and comparing classification results with the clusters.

The warrior population evolved by the CCAI evolver was not as diverse in a strategic sense as any human coded warrior group. To see how well clustering and classification algorithms can cope with more diverse datasets, and also to see if the data representation chosen in this project does well in such situations, the whole process will be repeated on some human coded warrior set. Being that human coded warriors often mix several strategies, it would be especially interesting to use probabilistic methods to gain insight into the probabilities of a warrior belonging to classes which were previously identified and annotated. A comparison of static and dynamic representations, on both human-coded and evolved warrior datasets, should then give more definitive answers concerning the feasibility and applicability of automatic warrior categorization.

References

1. Dewdney, A.K.: Computer recreations: In the game called core war hostile programs engage in a battle of bits. Scientific American 250(5), 14–22 (1984)
2. 'Varfar', W.: Wilfiz scores of warriors on the 94nop
   http://redcoder.sourceforge.net/?p=kepler-wilfiz
3. Tomašev, N., Pracner, D.: Categorizing corewar warriors. Seminar paper, Department of Mathematics and Informatics, Faculty of Science, University of Novi Sad (2006)
4. Witten, I.H., Frank, E.: Data Mining: Practical Machine Learning Tools and Techniques, 2nd edn. Morgan Kaufmann Publishers, San Francisco (2005)
5. Vowk, B.: CCAI http://www.math.ualberta.ca/~bvowk/corewar.html
6. Whitley, D., Rana, S., Heckendorn, R.B.: Island model genetic algorithms and linearly separable problems. In: Corne, D.W. (ed.) Evolutionary Computing. LNCS, vol. 1305, Springer, Heidelberg (1997)
7. Zap, S.: Optimax http://www.corewar.info/optimax/
8. Corno, F., Sanchez, E., Squillero, G.: Exploiting co-evolution and a modified island model to climb the core war hill. In: Proceedings of CEC03 Congress on Evolutionary Computation, pp. 2222–2229 (2003)
9. Hillis, D.: Redrace: Evolving core wars page
   http://users.erols.com/dbhillis/

Restricted Sequential Floating Search Applied to Object Selection

J. Arturo Olvera-López, J. Francisco Martínez-Trinidad, and J. Ariel Carrasco-Ochoa

Computer Science Department
National Institute of Astrophysics, Optics and Electronics
Luis Enrique Erro No. 1, Sta. María Tonantzintla, Puebla, CP: 72840, Mexico
{aolvera, fmartine, ariel}@ccc.inaoep.mx

Abstract. The object selection is an important task for instance-based classifiers since through this process the size of a training set could be reduced and then the runtimes in both classification and training steps would be reduced. Several methods for object selection have been proposed but some methods discard relevant objects for the classification step. In this paper, we propose an object selection method which is based on the idea of sequential floating search. This method reconsiders the inclusion of relevant objects previously discarded. Some experimental results obtained by our method are shown and compared against some other object selection methods.

1 Introduction

In supervised classification, a training or sample set (denoted in this paper as T) containing objects (previously assessed) described by a set of values (features) is used for classifying new objects. Commonly T contains objects with non relevant information for classifiers, therefore it is necessary to apply an object selection method over T in order to detect and retain those relevant objects for classification.

Object selection is important for instance-based classifiers because for this kind of classifiers the runtime in training and classification steps depends on the size of the training set. Thus, through the object selection, the runtimes in both training and classification steps could be reduced since these steps are applied over an object subset S ($S \subset T$) instead of using the whole set T.

Sequential search is a method used for finding a sub-optimal solution of a selection problem. This kind of search for selecting consists in evaluating at each step the relevance of each possibility in the partial solution set. This search can be done in the forward or backward direction, the forward search starts with an empty solution set and at each step it evaluates all options and includes the best one. The backward search starts with the whole set and at each step it excludes the worst element. These sequential methods analyze at each step all possibilities for including/excluding one of them but they cannot exclude/include solutions previously included/excluded, this is possible in the sequential floating methods [5] which include/exclude solutions (previously excluded/included) after each inclusion/exclusion.

P. Perner (Ed.): MLDM 2007, LNAI 4571, pp. 694–702, 2007.

Sequential search has been used for the feature selection problem [5, 6] and extended for the object selection problem in [3].

In this paper, we propose a sequential method for object selection. Our method re-considers the inclusion to S of relevant objects previously discarded in the selection process, so that S would include those objects that contribute for improving the quality in S.

This paper has been structured as follows: in section 2 we describe some relevant object selection methods. In section 3 we introduce our object selection method, in section 4 we report comparative results obtained by our method and other object selection methods. Finally, in section 5 conclusions and future work are given.

2 Related Work

One of the first proposed methods for object selection is the *ENN* (*Edited Nearest Neighbor*) [1]. This method is commonly used as noise filter because it deletes noisy objects, that is, objects with a different class in a neighborhood. The *ENN* rule consists in discarding from T those objects that do not belong to their k nearest neighbors' class.

In [2] the *DROP* (*Decremental Reduction Optimization Procedure*) methods were proposed. The selection criterion in *DROP* methods is based on the concept of *associate*. The *associates* of an object O are those objects such that O is one of their k nearest neighbors. *DROP1* starts with $S=T$ and discards the object O if its associates in S can be classified correctly without O. *DROP2* considers the effect of the removal of an object on T, *DROP2* discards O if its associates in T can be classified correctly without O. *DROP3* and *DROP4* apply a noise filter (similar to *ENN*) before starting the selection process. Finally, *DROP5* modifies *DROP2* so that the selection process starts with the nearest enemies (nearest objects with different class).

The sequential search has been used for selecting objects. In [3] the *BSE* (*Backward Sequential Edition*) method was proposed. *BSE* applies the backward sequential search to the object selection problem. This method sequentially analyzes the relevance of each object in the partial object subset and at each step *BSE* discards the object that its deletion maximizes the classification accuracy. This selection process is repeated until the accuracy decreases. *BSE* is an expensive method since at each step it analyzes the impact of excluding each object in the sample.

In [4] the edition schemes *ENN+BSE* and *DROP+BSE* were proposed. These schemes apply a pre-processing step before the selection process using *BSE* so that *BSE* is used over previously reduced object sets. *ENN* and *DROP3,...,DROP5* methods are used by *ENN+BSE* and *DROP+BSE* respectively in the pre-processing step.

3 Proposed Method

Our object selection method is based on the idea of the Sequential Floating Selection (*SFS*) [5], which reconsiders the inclusion/exclusion (in the partial subset) of objects previously discarded/included. *SFS* consists in applying conditional inclusion/exclusion steps after each exclusion/inclusion in the set. This kind of search (as sequential search) can be done in the backward and forward directions.

The backward *SFS* consists in applying after each exclusion step a number of in-clusion steps as long as the classification results are better than the previously evalu-ated ones. The forward *SFS* is the counterpart of backward *SFS*. These floating searches are very expensive therefore we propose an object selection method based on the backward *SFS* but in a restricted way.

Our method named *Restricted Floating Object Selection* (*RFOS*) applies an exclu-sion process followed by the conditional inclusion of discarded objects. The *RFOS* method is shown in figure 1.

RFOS (Training sample *T*)
Let *S*= subset obtained after applying *ENN* or *DROPs* over *T*
Best_val =Classif(S)
Repeat   //exclusion process
    Worst=null
    For each object *O* in *S*
    S'=S-{O}
    Eval = Classif(S')
    If *Eval ≥ Best_val*
      Worst=O
      Best_val=Eval
    If *Worst ≠ null*
      S=S-{Worst}
Until *Worst==null* or $|S|==1$
D=T-S
For each object O_i in *D*   //conditional inclusion
    S'' = S $\cup$ {O_i}
    Eval = Classif(S'')
    If *Eval >Best_val*
      Best_val = Eval
      S=S $\cup$ {O_i}
Return *S*

Fig. 1. *RFOS* method for object selection

RFOS starts applying a pre-processing step followed by the exclusion process and finally the conditional inclusion is applied over the object set previously selected (*S*, *S⊂T*). The exclusion step sequentially discards objects in the partial set. This step analyzes the classification contribution of each object and at each step it excludes the object (*Worst*) with the smallest contribution for the subset quality, in terms of the accuracy of a classifier, which is calculated by the *Classif* function.

The selection process in *RFOS* consists in analyzing (conditional inclusion) the ob-jects discarded from *T* (objects in the set *D=T-S*) for including in *S* those objects that their inclusion improves the classification, that is, an object *O∈ D* is included in *S* only if the classification obtained using *S $\cup$ {O}* is better than the obtained using *S*.

To know whether the classification after the inclusion is better or not, *RFOS* uses a classifier (*Classif* function in figure 1) to evaluate the quality of the sets.

In this work we use *ENN* or *DROP* methods for the pre-processing step but any other object selection method can be used for that step.

The *RFOS* is a restricted floating search method because first it applies only an exclusion process followed by the conditional inclusion. This restricted floating method can be done in the inverse direction (*RFOS-Inv*), that is, first applying an inclusion process followed by the conditional exclusion. The *RFOS-Inv* method is shown in figure 2.

RFOS-Inv (Training sample *T*)
Let *S*= subset obtained after applying *ENN* or *DROPs* over *T*
Best_val = Classif(S)
D=T-S
For each object *O* in *D*   // inclusion process
   S' = S $\cup$ *{O}*
   Eval = Classif(S')
   If *Eval >Best_val*
      Best_val = Eval
      S=S $\cup$ *{O}*
Best_val = Classif(S)   //conditional exclusion
Repeat
     Worst=null
     For each object *O* in *S*
      S''=S-{O}
      Eval = Classif(S'')
      If *Eval* $\geq$ *Best_val*
        Worst=O
        Best_val=Eval
     If *Worst* $\neq$ *null*
      S=S-{Worst}
Until *Worst==null* or |*S*|==1
Return *S*

Fig. 2. *RFOS-Inv* method for object selection

4 Experimental Results

In this section, we show the results obtained by *RFOS* and *RFOS-Inv* over nine datasets obtained from the UCI repository [7] and compare them against *ENN+BSE* and *DROP+BSE* methods.

In all the tables shown in this section, for each method, we show the classification accuracy (*Acc.*) and the percentage of the original training set that was retained by each method (*Str.*), that is 100|*S*|/|*T*|. In addition, we show the classification obtained using the original training set (*Orig.*) and the average results over the nine datasets at the bottom. Also we show the average accuracy difference (*Average diff*) with respect to the original accuracy. In all the experiments 10 fold cross validation was used.

The results obtained by *ENN+BSE* and *DROP+BSE* methods over the datasets are shown in table 1. In table 2 we report the results obtained by *RFOS* using *ENN* and *DROP* methods in the pre-processing step. In table 2, *RFOS(ENN)* is the *RFOS* method using *ENN* for the pre-processing step and by analogy for *RFOS(DROP3),...,*

RFOS(DROP5), the *DROP3,...DROP5* methods were respectively used. Table 3 shows the results obtained applying *RFOS-Inv* method. In tables 1-3 we used as distance function the *Heterogeneous Value Difference Metric* (*HVDM*) [2].

Table 1. Classification (*Acc.*) and retention (*Str.*) results obtained by: original sample (*Orig.*), *ENN+BSE* and *DROP3+BSE...DROP5+BSE* methods

Dataset	Orig.		ENN+BSE		DROP3+BSE		DROP4+BSE		DROP5+BSE	
	Acc.	Str.	Acc.	Str.	Acc.	Str.	Acc.	Str.	Acc.	Str.
Bridges	37.91	100	30.27	51.27	35.45	8.42	36.72	12.02	35.81	14.79
Glass	71.42	100	69.41	21.81	59.78	14.95	59.78	17.18	54.24	15.21
Iris	93.33	100	93.00	8.00	88.00	6.42	88.00	6.64	89.33	6.39
Liver	65.22	100	57.67	26.69	59.77	10.91	61.21	12.36	54.95	11.75
Sonar	86.19	100	71.19	27.24	81.42	12.60	84.83	14.79	84.30	15.17
Tae	51.08	100	46.66	43.85	47.70	14.93	50.00	18.17	46.66	20.08
Thyroid	95.45	100	93.09	5.63	91.19	4.28	91.16	4.39	88.29	3.51
Wine	94.44	100	92.74	8.17	96.07	5.05	96.07	5.05	96.07	4.43
Zoo	91.33	100	91.11	12.59	77.77	11.72	77.77	11.97	83.33	7.76
Average	76.26	100	71.68	22.81	70.79	9.92	71.73	11.40	70.33	11.01
Average diff			-4.58		-5.47		-4.54		-5.93	

Table 2. Classification (*Acc.*) and retention (*Str.*) results obtained by *RFOS*

Dataset	Orig.		RFOS(ENN)		RFOS(DROP3)		RFOS(DROP4)		RFOS(DROP5)	
	Acc.	Str.	Acc.	Str.	Acc.	Str.	Acc.	Str.	Acc.	Str.
Bridges	37.91	100	32.00	58.33	36.45	15.61	35.45	18.29	35.81	24.33
Glass	71.42	100	69.43	29.34	64.48	25.75	65.41	27.46	67.74	26.11
Iris	93.33	100	93.33	10.07	93.00	9.92	93.33	10.29	93.33	10.00
Liver	65.22	100	59.98	33.68	61.70	16.94	65.00	18.39	60.03	19.64
Sonar	86.19	100	72.57	32.27	84.64	21.58	83.52	20.88	83.73	22.59
Tae	51.08	100	50.70	48.88	47.70	25.45	50.00	27.29	53.33	31.34
Thyroid	95.45	100	94.04	7.02	93.98	6.25	94.45	6.77	90.47	5.94
Wine	94.44	100	93.63	10.23	94.44	8.17	94.44	8.17	93.85	8.30
Zoo	91.33	100	91.33	71.14	91.33	14.81	91.33	14.69	91.11	14.93
Average	76.26	100	73.00	33.44	74.19	16.05	74.77	16.91	74.38	18.13
Average diff			-3.26		-2.07		-1.49		-1.89	

Table 3. Classification (*Acc.*) and retention (*Str.*) results obtained by *RFOS-Inv*

Dataset	Orig.		RFOS-Inv(ENN)		RFOS-Inv(DROP3)		RFOS-Inv(DROP4)		RFOS-Inv(DROP5)	
	Acc.	Str.	Acc.	Str.	Acc.	Str.	Acc.	Str.	Acc.	Str.
Bridges	37.91	100	30.54	25.06	35.54	13.23	35.09	16.12	35.54	18.08
Glass	71.42	100	58.35	19.47	43.50	23.36	54.95	20.77	55.49	16.45
Iris	93.33	100	80.66	5.33	92.66	7.18	92.00	7.25	86.00	6.29
Liver	65.22	100	58.84	21.80	59.75	19.25	61.20	20.58	60.30	19.54
Sonar	86.19	100	71.14	16.77	68.26	18.58	70.66	21.20	68.76	22.49
Tae	51.08	100	50.97	14.27	46.66	20.27	45.54	31.86	54.20	30.17
Thyroid	95.45	100	88.83	3.82	94.55	4.85	93.03	5.27	86.96	4.18
Wine	94.44	100	90.00	4.36	88.23	5.18	89.44	5.36	91.04	4.43
Zoo	91.33	100	91.11	15.92	90.00	13.58	78.88	13.45	80.00	14.19
Average	76.26	100	68.94	14.09	68.79	13.94	68.98	15.76	68.70	15.09
Average diff			-7.33		-7.47		-7.29		-7.56	

The runtimes of the experiments reported in tables 1-3 are shown in table 4. Based on the average results, we can observe that because of the inclusion/exclusion steps in *RFOS* and *RFOS-Inv*, their runtimes are higher than the *ENN+BSE* and *DROPs+BSE*.

Table 4. Runtimes (in seconds) spent by the methods shown in tables 1-3

Dataset	ENN+BSE	DROP3+BSE	DROP4+BSE	DROP5+BSE	RFOS (ENN)	RFOS (DROP3)	RFOS (DROP4)	RFOS (DROP5)	RFOS-Inv (ENN)	RFOS-Inv (DROP3)	RFOS-Inv (DROP4)	RFOS-Inv (DROP5)
Bridges	595.3	8.3	13.5	8.6	608.4	31.4	37.1	30.9	379.8	35.4	69.8	38.7
Glass	540.0	14.7	28.0	14.2	545.3	27.0	35.8	25.7	215.6	59.8	46.1	33.7
Iris	420.1	4.5	3.8	2.1	426.9	9.9	9.1	6.3	482.1	7.5	7.9	8.0
Liver	1203.8	68.3	48.9	63.8	1214.0	95.8	74.0	91.5	1211.9	87.6	124.9	120.6
Sonar	1496.8	64.1	65.5	60.3	1509.6	109.6	85.3	83.6	1512.0	142.1	140.7	183.7
Tae	49.6	7.9	13.4	12.8	50.3	11.4	16.5	17.9	55.7	15.3	22.3	26.5
Thyroid	1381.6	2.8	2.2	2.5	1393.1	12.7	11.3	12.8	1140.2	11.4	12.9	12.2
Wine	960.4	5.0	5.3	3.8	969.7	13.7	14.1	12.4	905.7	12.6	14.1	13.5
Zoo	1380.6	6.3	6.5	6.0	1402.9	18.9	15.5	20.4	1020.6	13.1	16.3	17.3
Average	892.02	20.21	20.79	19.34	902.24	36.71	33.19	33.50	769.29	42.76	50.56	50.47

The classifier used in the results shown in tables 1-3 was *k-NN* (*k*=3). The average results reported in tables 1-3 are depicted in figure 3, which shows a scatter graphic of retention (vertical axis) versus accuracy (horizontal axis). On this graphic, the most located at right the best classification accuracy and the most located at bottom the best retention percentage.

Based on results shown in tables 1-3 and figure 3, we can observe that *RFOS* outperformed to *RFOS-Inv* because this method discards relevant objects in the final exclusion step. In addition, the accuracy obtained by *RFOS* is better than the obtained by *ENN+BSE* and *DROP+BSE* schemes; this is because *RFOS* includes relevant objects discarded in the exclusion steps. As a consequence of the final inclusion step, the object sets obtained by *RFOS* are slightly bigger than those obtained by *RFOS-Inv*, *ENN+BSE* and *DROP+BSE*. In this experiment the best accuracy was obtained by *RFOS(DROP4)* in the average case (figure 3).

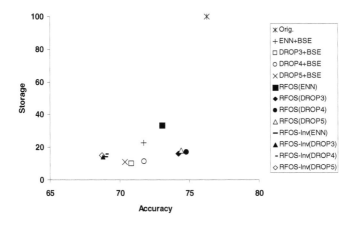

Fig. 3. Scatter graphic from results obtained in tables 1, 2 and 3

In the above results the classifier used was *k-NN*, but it is important to know the performance of the proposed object selection methods using other classifiers. Therefore we applied *RFOS* (the best restricted floating method in above experiments) and *ENN+BSE, DROP+BSE* using *LWR (Locally Weighted Regression)* and *SVM (Support Vector Machines)* classifiers during the selection process (notice that *RFOS, ENN+BSE* and *DROP+BSE* allow us to use any classifier different from *k-NN* in the selection process). In this experiment we have tested only numeric datasets because the classifiers are restricted to this kind of data.

Table 5. Classification *(Acc.)* and retention *(Str.)* results obtained by: original sample *(Orig.)*, *ENN+BSE* and *DROP3+BSE...DROP5+BSE* methods using *LWR*

Dataset	Orig.		ENN+BSE		DROP3+BSE		DROP4+BSE		DROP5+BSE	
	Acc.	Str.	Acc.	Str.	Acc.	Str.	Acc.	Str.	Acc.	Str.
Glass	57.85	100	56.84	50.26	50.71	20.83	55.18	25.54	53.72	21.97
Iris	98.00	100	96.66	20.74	88.00	10.88	88.66	11.18	88.66	8.14
Liver	70.12	100	66.33	31.51	70.99	17.13	68.08	19.00	68.68	16.58
Sonar	64.40	100	65.36	73.29	63.98	21.37	69.26	28.26	63.88	25.21
Thyroid	91.16	100	57.84	51.06	86.10	19.22	87.03	23.66	89.78	18.04
Wine	92.15	100	88.88	57.50	90.96	14.10	88.20	14.10	88.28	9.36
Average	**78.95**	**100**	**71.99**	**47.39**	**75.12**	**17.26**	**76.07**	**20.29**	**75.50**	**16.55**
Average diff			**-6.96**		**-3.82**		**-2.88**		**-3.45**	

Table 6. Classification *(Acc.)* and retention *(Str.)* results obtained by *RFOS* using *LWR*

Dataset	Orig.		RFOS(ENN)		RFOS(DROP3)		RFOS(DROP4)		RFOS(DROP5)	
	Acc.	Str.	Acc.	Str.	Acc.	Str.	Acc.	Str.	Acc.	Str.
Glass	57.85	100	57.79	52.18	53.30	25.13	58.33	26.26	54.54	27.09
Iris	98.00	100	97.33	22.00	96.00	13.40	95.33	13.77	95.33	10.00
Liver	70.12	100	66.34	37.61	73.31	18.64	71.27	21.22	68.78	18.77
Sonar	64.40	100	66.81	74.36	71.00	31.94	65.35	25.96	68.78	30.12
Thyroid	91.16	100	58.66	51.99	91.21	22.58	91.62	25.93	91.19	19.90
Wine	92.15	100	90.62	58.75	90.98	16.22	90.58	16.22	90.73	16.15
Average	**78.95**	**100**	**72.93**	**49.48**	**79.30**	**21.32**	**78.75**	**21.56**	**78.65**	**20.34**
			-6.02		**0.35**		**-0.20**		**-0.30**	

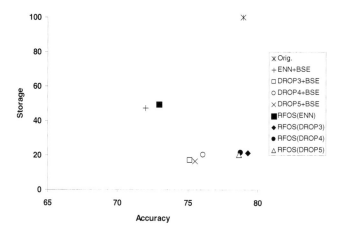

Fig. 4. Scatter graphic from results obtained using *LWR*

In tables 5 and 6 we show the accuracy and retention results obtained using the *LWR* classifier and the average scatter graphic from these results is depicted in figure 4.

Based on tables 5 and 6 we can observe that in all cases *RFOS* outperformed *ENN+BSE* and *DROP+BSE* methods. Figure 4 shows that in the average case, the best method using *LWR* was *RFOS(DROP3)* and the accuracy obtained by the other *RFOS(DROP)* methods was slightly lower than the obtained by the original set.

Also the *SVM* classifier was used for testing *RFOS*, *ENN+BSE* and *DROP+BSE* methods. These results are shown in tables 7-8 and the average results are depicted in figure 5.

Table 7. Classification (*Acc.*) and retention (*Str.*) results obtained by: original sample (*Orig.*), *ENN+BSE* and *DROP3+BSE...DROP5+BSE* methods using *SVM*

Dataset	Orig.		ENN+BSE		DROP3+BSE		DROP4+BSE		DROP5+BSE	
	Acc.	Str.	Acc.	Str.	Acc.	Str.	Acc.	Str.	Acc.	Str.
Glass	65.34	100	66.82	40.82	61.31	17.29	64.87	23.73	61.90	15.42
Iris	96.00	100	96.00	8.89	93.33	3.70	94.00	4.07	94.67	3.33
Liver	69.91	100	69.88	35.31	62.07	17.61	65.84	14.90	63.80	20.32
Sonar	79.38	100	78.60	58.07	72.83	13.68	74.48	14.42	71.57	15.12
Thyroid	72.61	100	72.61	7.23	68.34	3.20	68.20	3.36	67.27	3.31
Wine	97.18	100	96.63	21.68	93.89	3.62	94.97	3.87	92.09	2.75
Average	80.07	100	80.09	28.67	75.30	9.85	77.06	10.73	75.22	10.04
Average diff			0.02		-4.78		-3.01		-4.85	

Table 8. Classification (*Acc.*) and retention (*Str.*) results obtained by *RFOS* using *SVM*

Dataset	Orig.		RFOS(ENN)		RFOS(DROP3)		RFOS(DROP4)		RFOS(DROP5)	
	Acc.	Str.	Acc.	Str.	Acc.	Str.	Acc.	Str.	Acc.	Str.
Glass	65.34	100	69.18	43.24	62.26	20.16	64.95	25.50	63.87	20.14
Iris	96.00	100	96.00	9.78	93.33	4.14	94.00	4.14	94.67	3.70
Liver	69.91	100	69.83	47.68	62.95	20.52	67.03	19.62	67.83	21.71
Sonar	79.38	100	78.90	58.19	74.42	15.48	74.48	14.90	73.16	16.86
Thyroid	72.61	100	72.61	8.16	69.07	5.42	69.23	3.77	69.59	5.78
Wine	97.18	100	96.75	22.75	95.55	5.80	95.55	5.86	92.64	4.80
Average	80.07	100	80.55	31.63	76.26	11.92	77.54	12.30	76.96	12.17
Average diff			0.47		-3.81		-2.53		-3.11	

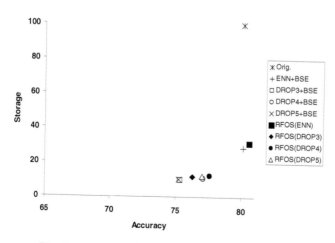

Fig. 5. Scatter graphic from results obtained using *SVM*

According to the results reported in tables 7 and 8, we can notice that using the *SVM* classifier, in all the experiments, again *RFOS* outperformed the *ENN+BSE* and *DROP+BSE* schemes. Figure 5 shows that the best accuracy results using this classifier were obtained by *RFOS(ENN)*.

Based on the results shown in this section, we can observe that the proposed method obtain smaller subsets (with respect to the original size set) without a significantly accuracy reduction. The main benefit of using the subsets obtained is the reduction in training and classification stages for instance-based classifiers.

5   Conclusions

Object selection is an important task for instance-based classifiers since through this process the training set is reduced and also the runtimes in classification and training steps.

Several object selection methods which sequentially discard objects have been proposed, for example, the edition schemes *ENN+BSE* and *DROP+BSE*. It is possible that during the selection process, these methods remove relevant objects for the classification accuracy. In this work, we proposed the *RFOS* method which is an object selection method that includes those relevant objects discarded by the edition schemes.

The experiments show that *RFOS* outperforms *RFOS-Inv*, *ENN+BSE* and *DROP+BSE* (not only for *k-NN* but also for *LWR* and *SVM*), that is, the inclusion of some discarded objects helps to improve the classification.

RFOS is a restricted floating sequential method because it applies only an exclusion process followed by the conditional inclusion, therefore, as future work we will adapt a full floating sequential search for solving the object selection problem.

References

1. Wilson, D.L.: Asymptotic Properties of Nearest Neighbor Rules Using Edited Data. IEEE Transactions on Systems, Man, and Cybernetics 2(3), 408–421 (1972)
2. Randall, W.D., Martínez, T.R.: Reduction Techniques for Instance-Based Learning Algorithms. Machine Learning 38, 257–286 (2000)
3. Olvera-López, J.A., Carrasco-Ochoa, J.A., Martínez-Trinidad, J.F.: Sequential Search for Decremental Edition. In: Gallagher, M., Hogan, J.P., Maire, F. (eds.) IDEAL 2005. LNCS, vol. 3578, pp. 280–285. Springer, Heidelberg (2005)
4. Olvera-López, J.A., Martínez-Trinidad, J.F., Carrasco-Ochoa, J.A.: Edition Schemes based on BSE. In: Sanfeliu, A., Cortés, M.L. (eds.) CIARP 2005. LNCS, vol. 3773, pp. 360–367. Springer, Heidelberg (2005)
5. Pudil, P., Ferri, F.J., Novovičová, J., Kittler, J.: Floating Search Methods for Feature Selection with Nonmonotonic Criterion Functions. In: Proceedings of the 12th International Conference on Pattern Recognition, pp. 279–283. IEEE Computer Society Press, Los Alamitos (1994)
6. Blum, A.L., Langley, P.: Selection of relevant features and examples in machine learning. Artificial Intelligence 97, 245–271 (1997)
7. Blake, C., Keogh, E., Merz, C.J.: UCI repository of machine learning databases. In: Department of Information and Computer Science, University of California, Irvine, CA (1998) http://www.ics.uci.edu/ mlearn/MLRepository.html

Color Reduction Using the Combination of the Kohonen Self-Organized Feature Map and the Gustafson-Kessel Fuzzy Algorithm

Konstantinos Zagoris, Nikos Papamarkos[1], and Ioannis Koustoudis

[1] Image Processing and Multimedia Laboratory
Department of Electrical & Computer Engineering
Democritus University of Thrace
67100 Xanthi, Greece
papamark@ee.duth.gr

Abstract. The color of the digital images is one of the most important components of the image processing research area. In many applications such as image segmentation, analysis, compression and transition, it is preferable to reduce the colors as much as possible. In this paper, a color clustering technique which is the combination of a neural network and a fuzzy algorithm is proposed. Initially, the Kohonen Self Organized Featured Map (KSOFM) is applied to the original image. Then, the KSOFM results are fed to the Gustafson-Kessel (GK) fuzzy clustering algorithm as starting values. Finally, the output classes of GK algorithm define the numbers of colors of which the image will be reduced.

Keywords: Color Reduction, Color Clustering, Neural Networks, Fuzzy Clustering.

1 Introduction

Nowadays the color of the digital images is one of the most widely used information for the image processing researchers. Digital images are usually described by a set of pixels uniformly distributed in a two-dimensional grid. On the one hand, in gray-scale images, the value of each pixel is described by a scalar value. On the other hand, in color images each pixel is expressed by a vector containing the values of three color components. True-type color images consist of more than 16 million different colors, in a 24-bit RGB color space. However, in many applications, such as image segmentation, analysis, compression and transition it is preferable to reduce the colors as much as possible.

The objective of color reduction is to divide a color set of an image into c uniform color clusters. Several techniques have been proposed in the literature for the color reduction of the image. Firstly, there is a group of techniques that repeatedly divide the color histogram in disjoint regions [1]. The methods of octree [2, 3], median-cut (MC) [4] and variance-based algorithm [5] are some of those splitting algorithms.

P. Perner (Ed.): MLDM 2007, LNAI 4571, pp. 703–715, 2007.

The second major class of algorithms is based on cluster analysis of the color space. Techniques in this category attempt to find the optimal palette using vector classifiers like the Growing Neural Gas (GNG) [6], Adaptive Color Reduction [7], FOSART [8-11], Fuzzy ART [12-13] and FCM [14].

Techniques in the third category are general color segmentation techniques, which can be considered as color reduction algorithms. For instance the mean-shift-based procedures for feature space analysis employ the former approach [15-16].

In present paper a color clustering technique which is the combination of a neural network and a fuzzy algorithm is proposed. Initially, the Kohonen Self Organized Featured Map (KSOFM) is applied to the original image. Then, the KSOFM results are fed to the Gustafson-Kessel (GK) fuzzy clustering algorithm as starting values. The resulting classes define the colors of the final image.

The next two sections describe the KSOFM neural network and the GK algorithm used in this work. Section 4 describes the color reduction method through the combine efforts of the above clustering methods. Section 5 presents some experimental results and describes some deductions that derive from them. Finally, in Section 6 some conclusions are drawn and the future directions are defined.

2   Kohonen Self Organized Featured Map (KSOFM)

A major category of neural nets is the self-organized neural nets which do not need supervising during their training phase (unsupervised neural nets). Their goal is to distinguish patterns in their training data and separate them in groups. The KSOFM [17-18] invented by the Prof. Teuvo Kohonen, is such a self-organized neural net. It is a Single Layer Feedforward Network but it differs in how it is trained and in how it revokes a pattern.

Analytically, the Kohonen network consists of two layers, the input and the competitive layer. In the latter layer the output units are arranged in one or two dimensional grids. As depicted in the architecture of the KSOFM in Fig 1, each input class has a feed-forward connection to each output class. So, the neural network maps a set of input vectors into a set of output vectors without supervision. The knowledge the network acquires by the training data is stored in the weights w_{jk}. These weights express the extent to which the connection of the data x_k with the output neuron y_j is important.

In the proposed method the input features are the three components of the RGB color space and the output units are the centers of the classes which depict RGB colors. The number of the output units is equal to the desirable amount of colors from which the image will be reduced.

The training algorithm of the KSOFM is based on competitive learning. Each time an input vector x_k is presented, a winner output neuron y_j is calculated based on the Euclidean Distance.

$$y_j = \arg\min \left\| x_k - w_{jk} \right\| . \tag{1}$$

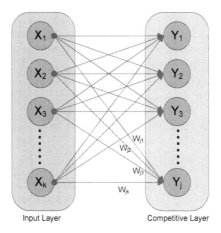

Fig. 1. The Architecture of the Kohonen Self-Organizing Map

Fig. 2. The sub sampling of the image based on the Hilbert's space filling curve

The winner output neuron changes its connections weights w_{jk} as follows:

$$\Delta w_{jk} = n\left(x_k - w_{jk}\right). \qquad (2)$$

The n is a variable that defines the learning rate of the training process and it is not constant but changes each time a new input vector is presented. In this work, the learning rate had the following values: $n_{initially} = 10^{-2}$, $n_{final} = 10^{-4}$, $n_{step} = 10^{-5}$.

One of the most important parts of the training process is the selection of the samples data. It is preferable to have as training samples data a sub-sampling version of the original image instead of the whole image in order to achieve reduction of the computational time. In the proposed color reduction technique, the training samples are selected from the peaks of the well-known Hilbert's space filling curve [19]. As the Fig. 2 depicts, the Hilbert's space filling curve is one dimensional curve which visits every possible pixel within a two dimensional space. This fractal scanning technique is much more efficient in capturing features in a digital image than the commonly used techniques of raster scanning because the neighborhood relation between the pixels is retained.

After the end of the training phase, the KSOFM initially is fed with unknown samples (all the pixels of the image) and then it correlates the samples to the centers of the classes that the neural net converged at the training phase.

3   Gustafson – Kessel Fuzzy Algorithm

3.1   Overview

One major problem of the standard fuzzy c-mean algorithm is that produces spherical classes. For example, if the sets of points presented at Fig 3a, pass through the fuzzy c-mean algorithm for partition into four classes, the result will not be the optimal (Fig 3b).

The Gustafson – Kessel [20] is an extension of the fuzzy c-mean algorithm that deals with this problem by using a covariance matrix in order to detect ellipsoidal classes. That is, each cluster is characterized by its center and its own-inducing matrix A_i that forms the following distance for each class:

$$d_{ik}^2 = \left(x_k - v_i\right)^T A_i \left(x_k - v_i\right) . \tag{3}$$

The $x_k, k \in [1,n]$ are the data vectors and the $v_i, i \in [1,c]$ are the centers of the classes. The A_i is a positive-define matrix adapted according to the topological structure of the data inside a cluster. The following equation (eq. 4) indicates that the objective function of the Gustanfson – Kessel algorithm is linear by A_i and cannot be directly minimized with respect to it.

$$J(U,V) = \sum_{i=1}^{c} \sum_{k=1}^{n} u_{ik}^m d_{ik}^2 . \tag{4}$$

So the A_i must be constrained. This is accomplished by constraining its determinant:

$$|A_i| = \rho_i, \quad \rho_i > 0 \; \forall i . \tag{5}$$

Without any prior knowledge, the ρ_i is fixed at 1 for each cluster. Finaly, the A_i is calculated from the following equation:

$$A_i = \sqrt[h]{\rho_i \det(F_i)} F_i^{-1}, \quad i \in [1,c] . \tag{6}$$

where h represent the number of dimensions of the space that the data reside. Because the RGB color space has three dimensions, the value of h is equal to 3 in this work. F_i is the covariance matrix which shows how the samples are scattered inside a class:

$$F_i = \frac{\sum_{k=1}^{n} (u_{ik})^m (x_k - v_i)(x_k - v_i)^T}{\sum_{k=1}^{n}(u_{ik})^m}, \quad i \in [1,c] .$$

(7)

The weighting parameter m, $m \in (1,\infty)$ influence the crispness or the fuzziness of the resulting partition between the classes. Worth noticing that if the equations 6 and 7 substituted into the equation 3, the outcome will be a squared Mahalanobis distance norm.

Finaly, the $U = [u_{ik}]$ is called partition matrix and is defined as the grade of membership of x_k to the cluster i and it must satisfy the following constraints:

$$0 \le u_{ik} \le 1, \quad i \in [1,c] \text{ and } k \in [1,n] .$$

(8)

$$\sum_{i=1}^{c} u_{ik} = 1, \quad k \in [1,n] .$$

(9)

$$0 < \sum_{k=1}^{n} u_{ik} < n, \quad i \in [1,c] .$$

(10)

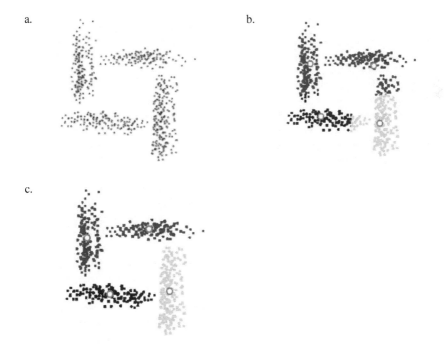

a.

b.

c.

Fig. 3. (a) The points in the 2D space which must separate to four classes. (b) The clustering of the points to four classes through the fuzzy c-mean algorithm. (c) The clustering of the points to four classes through the Gustafson – Kessel algorithm.

3.2 The Algorithm

The Gustafson - Kessel algorithm consists of the following steps:

Step 1: Define the number of the classes c, the weighting parameter m and the cluster volumes ρ_i.

Step 2: Define the termination tolerance $\varepsilon > 0$ and the number of iterations λ. Set a counter α equal to one ($\alpha = 1$).

Step 3: Initialize randomly the partition matrix $U = [u_{ik}]$. In this work, the partition matrix is initialized not randomly but from the connections weights w_{jk} of the KSOFM for each output class.

Step 4: Compute the centers of the classes v_i according to the following equation:

$$v_i = \frac{\sum_{k=1}^{n}(u_{ik})^m x_k}{\sum_{k=1}^{n}(u_{ik})^m}, \quad i \in [1, c] \text{ and } k \in [1, n] . \tag{11}$$

Step 5: Compute the covariance matrix F_i for each class according to the equation (7).

Step 6: Compute the matrix A_i for each class according to the equation 6.

Step 7: Compute the distance d_{ik} of every sample x_k from the center of each class v_i according to the equation 3.

Step 8: Update the partition matrix $U = [u_{ik}]$ for each sample x_k according to the following equation:

$$u_{ik} = \frac{1}{\sum_{j=1}^{c}\left(\dfrac{d_{ik}}{d_{ij}}\right)^{\frac{2}{m-1}}}, \quad i \in [1, c] \text{ and } k \in [1, n] . \tag{12}$$

But when $d_{ik} = 0$ for some x_k and one or more center of classes v_i (very rare case), the partition function u_{ik} cannot be computed. In this case the 0 is assigned to each u_{ik} for which $d_{ik} > 0$ and the membership is distributed arbitrarily among the rest u_{ik} (for which $d_{ik} = 0$) but underlined to the constraint shown at the equation (9).

Step 9: if $\max\left|U^{(\alpha)} - U^{(\alpha-1)}\right| < \varepsilon$ or $\alpha \geq \lambda$ stop, else set $\alpha = \alpha + 1$ and go to step 4.

4 Overview of the Proposed Method

Figure 4 depicts the process of the proposed method. It is a combination of the KSOFM and the Gustafson-Kessel fuzzy algorithm, appropriate for reducing the colors of the image at a preprocessing stage for a segmentation technique. The proposed method consists of the following steps:

Step 1: Assemble the training samples data from the sub-sampling of the original image from the peaks of the Hilbert's space filling curve.

Step 2: Define the number of colors of which the image will be reduced. This number defines also the output classes of the KSOFM and GK algorithm.

Step 3: Feed the KSOFM with the training samples to train it.

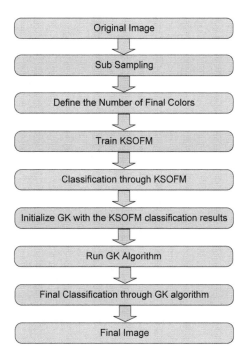

Fig. 4. The process of the proposed image color reduction method

Step 4: Feed each pixel of the original image into the trained KSOFM. This maps (classify) each pixel to one of the output classes and ultimately to one of the reduced colors.

Step 5: Initialize Gustafson – Kessel with the classification results of the KSOFM. That is, populate the partition matrix $U = [u_{ik}]$ from the KSOFM connections weights w_{jk} of each output class.

Step 6: Run the Gustafson – Kessel algorithm.

Step 7: Reduce the colors of the image based on the results of the Gustafson – Kessel fuzzy algorithm.

The result from the above steps is an image which has reduced number of colors as defined in Step 2.

5   Experimental Results

The method proposed in this paper is implemented with the help of a Visual Programming Environment (Borland Delphi). The program can be downloaded at the web address: http://orpheus.ee.duth.gr/download/pythagoras.zip. The proposed technique is tested on several images with satisfactory results.

As the Figures 5, 6, 7, 8 and 9 depicts, the proposed method is compared with other two popular color reduction techniques based on cluster analysis in the color space: the KSOFM and the FCM algorithm. Table 1 presents the parameters of the algorithms during the testing.

The experimental results have shown that the proposed technique has the ability to retain the dominant colors even if the final image consists of a very small number of unique colors. Also, it can merge areas of the image having similar colors. In this point of view, it can be considered as a powerful color image segmentation procedure.

Table 1. The parameters of the algorithms during the testing

KSOFM	Fuzzy C-Mean	KSOFM - GK
Initially Learning Rate: $n_{initially} = 10^{-2}$	$m = 1.2$	Initially Learning Rate: $n_{initially} = 10^{-2}$
Final Learning Rate: $n_{final} = 10^{-4}$	Epochs = 2000	Final Learning Rate: $n_{final} = 10^{-4}$
Step of the Learning Rate: $n_{step} = 10^{-5}$	Termination Tolerance: $\varepsilon = 5 \cdot 10^{-5}$	Step of the Learning Rate: $n_{step} = 10^{-5}$
		KSOFM Termination Tolerance: $\varepsilon = 5 \cdot 10^{-5}$
		$m = 1.2$
		GK termination Tolerance: $\varepsilon = 5 \cdot 10^{-4}$
		Iterations: $\lambda = 100$

A significant disadvantage of the proposed technique is its high computational cost which comes from the determination of the Mahalanobis distance. For an AMD Athlon 64 3000+ (2GHz) based PC with 1GByte RAM, the processing time for a 512x384 image with 119143 colors for all the algorithms is presented at Table 2. The number of colors of which the above image is reduced is six (6).

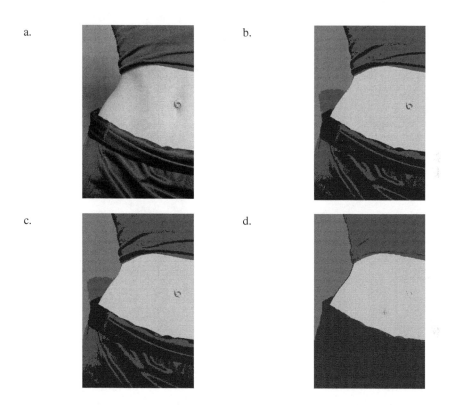

Fig. 5. (a) The original image is constituted of *22410 colors*. (b) The output image, through the KSOFM is constituted of *4 colors*. (c) The output image, through the FCM is constituted of *4 colors*. (d) The output image, through the proposed method (KSOFM-GK) is constituted of *4 colors*.

Table 2. Computational cost for each algorithm processing a 512x384 image with 119143 colors. The final reduced colors are 6.

KSOFM	Fuzzy C-Mean	KSOFM - GK
2.43 seconds	8.32 seconds	43.27 seconds

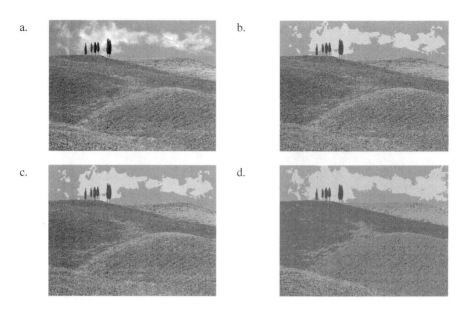

Fig. 6. (a) The original image is constituted of *99760 colors*. (b) The output image, through the KSOFM is constituted of *7 colors*. (c) The output image, through the FCM is constituted of *7 colors*. (d) The output image, through the proposed method (KSOFM-GK) is constituted of *7 colors*.

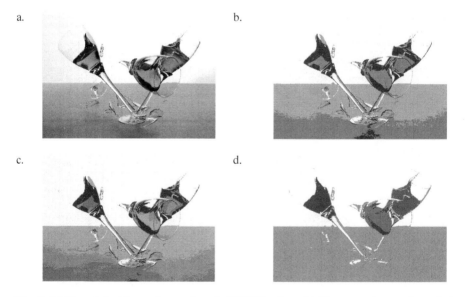

Fig. 7. (a) The original image is constituted of *33784 colors*. (b) The output image, through the KSOFM is constituted of *5 colors*. (c) The output image, through the FCM is constituted of *5 colors*. (d) The output image, through the proposed method (KSOFM-GK) is constituted of *5 colors*.

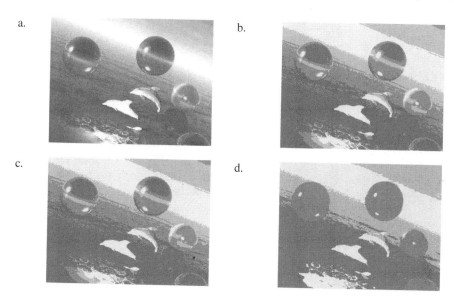

Fig. 8. (a) The original image is constituted of *31655 colors*. (b) The output image, through the KSOFM is constituted of *4 colors*. (c) The output image, through the FCM is constituted of *4 colors*. (d) The output image, through the proposed method (KSOFM-GK) is constituted of *4 colors*.

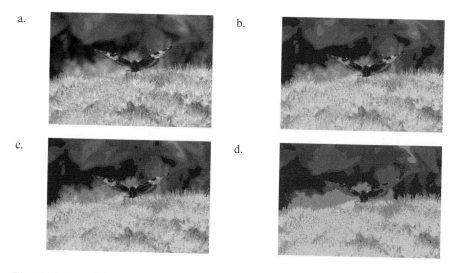

Fig. 9. (a) The original image is constituted of *69656 colors*. (b) The output image, through the KSOFM is constituted of *8 colors*. (c) The output image, through the FCM is constituted of *8 colors*. (d) The output image, through the proposed method (KSOFM-GK) is constituted of *8 colors*.

6 Conclusion

In this paper a color clustering technique is proposed which is based on a combination of a KSOFM neural network and the Gustafson-Kessel fuzzy algorithm. Initially, the KSOFM is applied to the original image and produce a predefined number of color classes. Then, the final color reduction is performed by the Gustafson-Kessel fuzzy clustering algorithm considering the KSOFM clustering results as initial values. Thus, the entire clustering procedure can be considered as an hybrid neuro-fuzzy technique.

The experimental results have shown the ability to retain the image's dominant colors. Also, it can merge areas of the image with similar colors and therefore can be used as a color segmentation procedure. Future directions should include the ability to detect the optimal number of final colors and reduce the high computational cost.

Acknowledgment

This work is co-funded by the project PYTHAGORAS 1249-6.

References

1. Scheunders, P.: A comparison of clustering algorithms applied to color image quantization. Pattern Recognit. Lett. 18, 1379–1384 (1997)
2. Ashdown, I.: Octree color quantization in Radiosity. Wiley, Chichester (1994)
3. Gervautz, M., Purgathofer, W.: A simple method for color quantization: Octree quantization. In: Glassner, A.S. (ed.) Graphics Gems, pp. 287–293. Academic New York (1990)
4. Heckbert, P.: Color image quantization for frame buffer display. Comput. Graph. 16, 297–307 (1982)
5. Wan, S.J., Prusinkiewicz, P., Wong, S.K.M.: Variance based color image quantization for frame buffer display. Color Res. Applicat. 15(1), 52–58 (1990)
6. Fritzke, B.: A growing neural gas network learns topologies. In: Tesauro, G., Touretzky, D.S., Leen, T.K. (eds.) Advances in Neural Information Processing Systems, vol. 7, pp. 625–632. MIT Press, Cambridge, MA (1995)
7. Papamarkos, N., Atsalakis, A., Strouthopoulos, C.: Adaptive color reduction. IEEE Transactions on Systems. Man and Cybernetics Part B: Cybernetics 32 (2002)
8. Baraldi, A., Blonda, P.: A survey of fuzzy clustering algorithms for pattern recognition— part I. IEEE Transactions on Systems, Man, and Cybernetics-Part B: Cybernetics 29(6), 778–785 (1999)
9. Baraldi, A., Blonda, P.: A survey of fuzzy clustering algorithms for pattern recognition-part II. IEEE Transactions on Systems, Man, and Cybernetics—Part B: Cybernetics 29(6), 786–801 (1999)
10. Baraldi, A., Parmiggiani, F.: Novel neural network model combining radial basis function, competitive Hebbian learning rule, and fuzzy simplified adaptive resonance theory. In: Proceedings of the SPIE's Optical Science, Engineering and Instrumentation 1997: Applications of Fuzzy Logic Technology IV, San Diego CA, vol. 3165, pp. 98–112 (1997)
11. Baraldi, A., Parmiggiani, F.: A fuzzy neural network model capable of generating/ removing neurons and synaptic links dynamically. In: Blonda, P., Castellano, M., Petrosino, A. (eds.) Proceedings of the WILF 1997-II Italian Workshop on Fuzzy Logic, pp. 247–259. World Scientific, Singapore (1998)

12. Carpenter, G., Grossberg, S., Rosen, D.B.: Fuzzy ART: fast stable learning and categorization of analog patterns by an adaptive resonance system. Neural Networks 4, 759–771 (1991)
13. Carpenter, G., Grossberg, S., Maukuzon, N., Reynolds, J., Rosen, D.B.: Fuzzy ARTMAP: a neural network architecture for incremental supervised learning of analog multidimensional maps. IEEE Transactions on Neural Networks 3(5), 698–713 (1992)
14. Bezdek, J.C.: Pattern Recognition with Fuzzy Objective Function Algorithms. Plenum Press, New York (1981)
15. Comaniciu, D., Meer, P.: Mean shift: a robust approach toward feature space analysis. IEEE Transactions on Pattern Analysis and Machine Intelligence 24(5), 603–619 (2002)
16. Nikolaou, N., Papamarkos, N.: Color segmentation of complex document images. In: International Conference on Computer Vision Theory and Applications. Setúbal, Portugal, pp. 220–227 (2006)
17. Kohonen, T.: The self-organizing map. Proceedings of IEEE 78(9), 1464–1480 (1990)
18. Kohonen, T.: Self-Organizing Maps, 2nd edn. Springer, Berlin (1997)
19. Sagan, H.: Space-Filling Curves. Springer, New York (1994)
20. Gustafson, E.E., Kessel, W.C.: Fuzzy Clustering with a Fuzzy Covariance Matrix. In: Proc. 18th IEEE Conference on Decision and Control (IEEE CDC, San Diego, CA). Piscataway, NJ, USA, pp. 761–766 (1979)

A Hybrid Algorithm Based on Evolution Strategies and Instance-Based Learning, Used in Two-Dimensional Fitting of Brightness Profiles in Galaxy Images

Juan Carlos Gomez[1] and Olac Fuentes[2]

[1] INAOE, Computer Science Department, Luis Enrique Erro No. 1, Tonantzintla,
Puebla 72000, Mexico
jcgc@inaoep.mx
[2] University of Texas at El Paso, Computer Science Department, 500 West University
Avenue, El Paso 79968, Texas, USA
ofuentes@utep.edu

Abstract. The hybridization of optimization techniques can exploit the strengths of different approaches and avoid their weaknesses. In this work we present a hybrid optimization algorithm based on the combination of Evolution Strategies (ES) and Locally Weighted Linear Regression (LWLR). In this hybrid a local algorithm (LWLR) proposes a new solution that is used by a global algorithm (ES) to produce new better solutions. This new hybrid is applied in solving an interesting and difficult problem in astronomy, the two-dimensional fitting of brightness profiles in galaxy images.

The use of standardized fitting functions is arguably the most powerful method for measuring the large-scale features (e.g. brightness distribution) and structure of galaxies, specifying parameters that can provide insight into the formation and evolution of galaxies. Here we employ the hybrid algorithm ES+LWLR to find models that describe the bi-dimensional brightness profiles for a set of optical galactic images. Models are created using two functions: de Vaucoleurs and exponential, which produce models that are expressed as sets of concentric generalized ellipses that represent the brightness profiles of the images.

The problem can be seen as an optimization problem because we need to minimize the difference between the flux from the model and the flux from the original optical image, following a normalized Euclidean distance. We solved this optimization problem using our hybrid algorithm ES+LWLR. We have obtained results for a set of 100 galaxies, showing that hybrid algorithm is very well suited to solve this problem.

1 Introduction

Galaxies encompass an enormous set of phenomena in the universe, from star formation to cosmology subjects. Thus, study of galaxies is essential to understand many basic questions about the cosmos. Also, there is a huge amount of

P. Perner (Ed.): MLDM 2007, LNAI 4571, pp. 716–726, 2007.
© Springer-Verlag Berlin Heidelberg 2007

astronomical data in images and spectra in surveys (SDSS, 2MASS, etc.) obtained from modern observatories, and it is important to automatically analyze such information in order to extract important physical knowledge.

A very useful way to quantify galaxies and extract knowledge from data is to fit images or spectra with parametric functions [9][11]. The use of standardized fitting functions is arguably the most powerful method for measuring the large-scale features (e.g. brightness profiles) and structure of galaxies (e.g. morphologies), specifying parameters that can provide insight into the formation and evolution of galaxies, since the functions yield a variety of parameters that can be easily compared with the results of theoretical models [4].

Galaxies are composed of distinct elements: stars, gas, dust, planets and dark matter. Old stars are normally present in the central part of a galaxy (also called bulge), while young star, gas and dust are usually in the outer parts (called disk) and dark matter is normally surrounded the galaxy (called halo). Each element contributes in a different way to the light that one galaxy emits; stars producing the light and gas, dust and dark matter dispersing or diffracting it. Galaxy brightness profile describes how this light is distributed over the surface of a galaxy [7]. Thus, studying the brightness profile will lead to understand many subjects about the formation, composition and evolution of galaxies [11]. For example, elliptical galaxies are normally composed only by a bulge and a dark matter halo, which means that ellipticals are old, because they only contains old stars in the central part and they produces a very intense brightness in this part. On the other hand spiral galaxies are usually composed by a bulge, a disk and a dark matter halo, which means spirals are younger than ellipticals because spirals still contain young stars, and gas and dust to produce new stars, where these elements are normally present in the spiral arms. In this case, galaxies present a brilliant central part and a less brilliant disc (with spiral arms) surrounding the bulge.

Thus, fitting of galaxy brightness profiles provides a reasonably detailed description of the radial light distribution with a small number of parameters. Nevertheless, ideally fitting functions would be based upon the physics of the formation and evolutionary processes in galaxies. Unfortunately, these processes are neither simple nor well understood, so the most commonly used functions are derived empirically.

Here we propose a machine learning [8] hybrid algorithm to automatically find models for galaxy brightness profiles; exploring the search space of possible solutions in order to find the best set of parameters that produces a model able to describe the brightness distribution in a galaxy.

Hybridization is referred to a merge or mixture among two or more algorithms, implemented or developed trying to exploit the advantages and avoid the weakness of each particular algorithm. Using such schema, for example in a non-simple search space in an optimization problem, we might initially employ a global algorithm to identify regions of high potential, and then switch, using appropriate switching metrics, to local techniques to rapidly converge on the local minimum. Through hybridization, the optimization strategy can be fitted to suit the specific characteristics of a problem, thereby enhancing the overall robustness and efficiency of the optimization process.

Here we use a different way to perform hybridization: a local algorithm produces one proposed solution that is evaluated and used to produce new solutions by a global algorithm; in this manner an exchange of possible solutions between algorithms occurs. Hybridization is done in each iteration of the global algorithm Evolution Strategies (ES) [10], adding a new solution approximated by the instance-based algorithm Locally Weighted Linear Regression (LWLR) [1] to the set of solutions in ES.

We have obtained fitting for a set of 100 galaxy images, from spiral and elliptical galaxies, using hybrid algorithm, showing that ES+LWLR is a well suited method to solve this problem.

The rest of the paper is structured as follows: in Section 2 a brief description of theory for brightness profile and description of the problem is presented, the hybrid algorithm is shown and explained in Section 3, Section 4 includes the general description of the optimization process, results are presented in Section 5 and Section 6 includes conclusions and future work.

2   Brightness Profile

Surface brightness in a galaxy is literally defined as how much light emits the galaxy [7], and luminosity is defined as the total energy received by unit of area by unit of time. Then, the surface brightness of an astronomical source is the ratio of the source's luminosity F and the solid angle (Ω) subtended by the source.

$$B = \frac{F}{\Omega} \tag{1}$$

The surface brightness distribution in elliptical galaxies depends essentially only on the distance from the centre and the orientation of the major and minor axis. If we consider that elliptical galaxies are only composed by a bulge, and if r is the radius along with the major axis, the surface brightness $I(r)$ is well described by de Vaucoleurs' law $r^{1/4}$ [6]:

$$I_b = I_e \exp\left[-3.3\left[\left(\frac{r^{1/4}}{r_e}\right) - 1\right]\right] \tag{2}$$

where r is the distance from the galactic center, r_e is the mean ratio of the galaxy brightness (the radius where half of the total brightness lies), and I_e is the surface brightness for $r = r_e$.

Although de Vaucoleurs' law is a purely empirical relation, it still gives a remarkably good representation of observed light distribution in bulges. However, in the outer regions of elliptical galaxies changes in light distribution may often occur. Different galaxies differ widely in this respect indicating that the structure of ellipticals is not as simple as it might appear.

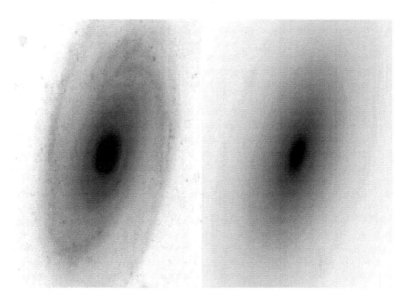

Fig. 1. Example of an observed galaxy image and its modelled brightness profile

Spiral galaxies, and according with observation some ellipticals, are also composed by a disc. The surface brightness profile for a disc galaxy has an exponential distribution:

$$I_d = I_0 \exp\left(-\frac{r}{r_0}\right) \tag{3}$$

where I_0 is the central surface brightness and r_0 is the radial scale length.

Finally, surface brightness distribution in elliptical and spiral galaxies can be described in a more general way as the sum of equations 2 and 3, which is an approximation of the profile using concentric ellipses [7].

$$I = I_b + I_d \tag{4}$$

In fact, it is not expected that equations 2 and 3 fit all the profile measured in the radial range of the galaxy, because sometimes sky subtraction errors in external regions of galaxy can distort the profile. Also the fitting process does not allow the presence of structures, it fits across arms and bars as if they were noise in the data, since models are based only in concentric ellipses.

An example of a galaxy image and its corresponding generated brightness profile using the previous equations is shown in Figure 1.

2.1   Fitting

In order to fit a galaxy image, it is necessary to define a model of its brightness profile that matches the brightness distribution of the original image.

Thus, let $\mathbf{q} = [r_e, I_e, I_0, r_0, i_1, i_2]$ be a vector of brightness parameters, a model for a brightness profile is constructed as follows:

$$\mathbf{m} = I_b(\mathbf{q}) + I_d(\mathbf{q}) \tag{5}$$

where $I_b(\mathbf{q})$ and $I_d(\mathbf{q})$ are the equations 2 and 3 applied with the parameters in $\mathbf{q}$. This model produces an artificial image $\mathbf{m}$ of size $n \times m$ that represent certain brightness distribution.

Then, the final goal in this fitting task is to approximate, efficiently and automatically, the best combination for the following parameters: r_e, mean ratio of the galaxy brightness; I_e, surface brightness in $r = r_e$; I_0, central surface brightness; r_0, radial scale length and two angels i_1 and i_2 which are the rotation angles in x and z axis, that produce a model that matches the brightness distribution of the observed image.

3   Optimization

The process starts with an observed image $\mathbf{o}$, which first is resized to a 256×256 pixel size. This process is done to simplify the fitting task and it is necessary to standardize all the data (the observed and the simulated).

Thus, let $\mathbf{o}$ be the observed image variable of size 256×256, let $\mathbf{m}$ be the simulated image with the same dimensionality as $\mathbf{o}$. The goal of the optimization process is to obtain a model $\mathbf{m}$ that maximizes the following function:

$$f(\mathbf{m}) = 1 - \frac{1}{max} \sqrt{\sum_{i=1}^{256} \sum_{j=1}^{256} (m_{i,j} - o_{i,j})^2} \tag{6}$$

where max is the maximum difference that can exist between images. This is the fitness function used by our hybrid algorithm. The fitness function represents the similarity between both images, and its values range is [0-1], with 1 as a perfect match and 0 as totally different images. At the end, the simulated image that maximizes such equation is the one that was produced by the set of brightness parameters we were looking for.

3.1   Hybrid Algorithm

Our hybrid algorithm, that we called ES+LWLR, is based in the main idea of using the individuals in each generation of ES as the training set for the learning algorithm LWLR [1]. Such algorithm will be used to predict a new individual, which is expected to be closer to the global solution.

LWLR uses implicit knowledge in the ES population about the target function to predict a new individual that is potentially better than the ones in ES. LWLR creates local linear models of the target function around a query point $\mathbf{h}_q$ (original image), and local models achieve an approximation finer of the target function than those based on global models, reaching a more accurate prediction.

Literally LWLR is exploiting the current population configuration for predicting directly a new solution.

In this work we implemented a modified version of $(\mu + \lambda)$ ES [2][3] that includes some changes to the canonical version. We create $\mu = 7$ parent individuals and $\lambda = 14$ children individuals, we use discrete recombination and traditional mutation, but we also include a new way to create offspring with the *average operator* and a dynamical mutation [5] for strategy parameter vectors based on a simple, but effective and easy to understand, multiplication by constant factors.

We employed LWLR in the following way: given a query point $\mathbf{h}_q$, to predict its output parameters $\mathbf{y}_q$, we find the k closest examples in the training, and assign to each of them a weight given by the inverse of its distance to the query point:

$$w_i = \frac{1}{|\mathbf{h}_q - \mathbf{h}_i|} \quad i = 1, \ldots, k$$

where $\mathbf{h}_i$ is the i-st closest example. This measure is called "relevance".

Let W, the weight matrix, be a diagonal matrix with entries $w_1, \ldots, w_k$. Let H be a matrix whose rows are the vectors $\mathbf{h}_1, \ldots, \mathbf{h}_k$, the input parameters of the examples in the training set that are closest to $\mathbf{h}_q$, with the addition of a "1" in the last column. Let Y be a matrix whose rows are the vectors $\mathbf{y}_1, \ldots, \mathbf{y}_k$, the output parameters of these examples. Then the weighted training data are given by $Z = WH$ and the weighted target function is $V = WY$. Then we use the estimator for the target function $\mathbf{y}_q = \mathbf{h}_q^T Z^* V$, where Z^* is the pseudoinverse of Z.

Merging both ideas of ES and LWLR we obtain the ES+LWLR algorithm, where we have: first, the initial population of ES $\mathbf{x}_i$ $i = 1, \ldots, \mu$, where each $\mathbf{x}_i = [r_{e,i}, I_{e,i}, I_{0,i}, r_{0,i}, i_{1,i}, i_{2,i}]$, and their corresponding strategy parameters vectors $\boldsymbol{\sigma}_i$ $i = 1, \ldots, \mu$ are formed by randomly generated values. Then, each $\mathbf{x}_i$ is passed to a module that following equation 5 creates the simulated galaxy image $\mathbf{m}_i$. Afterwards, each $\mathbf{m}_i$ is evaluated using the fitness function 6. The next step consists of an iterative process: if some model has obtained a good match with the original image we stop the process, otherwise we create λ new individuals $\mathbf{x}'_j$ $j = 1, \ldots, \lambda$ using recombination, mutation and average operators. Then we create their corresponding models $\mathbf{m}'_j$ and evaluate them using the fitness function. The next step consist in the hybridization, here we pass the $(\mu + \lambda)$ population, the $\mathbf{m}_k$ $k = 1, \ldots, \mu + \lambda$ models and the original image $\mathbf{o}$ to the LWLR module, where LWLR takes the 7 closest models to the original images an predict the output vector of parameters $\mathbf{y}_q$. This vector is evaluated using the fitness function and returned to ES, replacing the least fit individual of the $(\mu + \lambda)$ population. Afterward we select the best μ individuals from the total population and return to compare the new models with the observed image.

The pseudocode of the ES+LWLR algorithm for this problem is the following:

1. Create $\mu = 7$ parent vectors $\mathbf{x}_i$ $i = 1, \ldots, \mu$, where each vector contains 6 parameters $\mathbf{x}_i = [r_{e,i}, I_{e,i}, I_{0,i}, r_{0,i}, i_{1,i}, i_{2,i}]$, and their corresponding strategy parameter vectors $\boldsymbol{\sigma}_i$ $i = 1, \ldots, \mu$ with the same dimensionality as $\mathbf{x}_i$. Each parameter is chosen through a random process and satisfying the constraints of the problem.

2. For each vector $\mathbf{x}_i$ produce a simulated galaxy image $\mathbf{m}_i$

3. For each $\mathbf{m}_i$ compute the fitness function
$$f(\mathbf{m}_i) = 1 - \frac{1}{max} \sqrt{\sum_{j=1}^{256} \sum_{l=1}^{256} (m_{i,j,l} - o_{j,l})^2}$$

4. If some model $\mathbf{m}_i$ fits good the observed galaxy brightness profile terminate, otherwise continue next step

5. Create new $\lambda = 14$ child individuals in the following way:

   - Create 10% of λ population using discrete recombination, from two parents $\mathbf{x}_a$ and $\mathbf{x}_b$:
$$x'_{k,j} = x_{a,j} \text{ or } x_{b,j}$$
$$\sigma'_{k,j} = \sigma_{a,j} \text{ or } \sigma_{b,j}$$

   - Create 10% of λ population using average operator, from two parents $\mathbf{x}_a$ and $\mathbf{x}_b$, and a random number d between 0-1:
$$\mathbf{x}'_k = d\mathbf{x}_a + (1-d)\mathbf{x}_b$$
$$\boldsymbol{\sigma}'_k = d\boldsymbol{\sigma}_a + (1-d)\boldsymbol{\sigma}_b$$

   - Create 80% of λ population using mutation:
$$\mathbf{x}'_k = \mathbf{x}_a + N(0, \boldsymbol{\sigma}_a)$$

6. For each vector $\mathbf{x}'_k$ produce a simulated galaxy image $\mathbf{m}'_k$ $k = 1, \ldots, \lambda$

7. For each $\mathbf{m}'_k$ compute the fitness function
$$f(\mathbf{m}'_k) = 1 - \frac{1}{max} \sqrt{\sum_{j=1}^{256} \sum_{l=1}^{256} (m'_{k,j,l} - o_{j,l})^2}$$

8. Merge μ and λ populations to obtain a $(\mu + \lambda)$ population

9. Sort the merged population by fitness function, in descending order

10. Pass entire population $\mathbf{x}_k$, its corresponding models $\mathbf{m}_k$ $k = 1, \ldots, \mu + \lambda$ and original observed image $\mathbf{o}$ to LWLR module where:

    (a) Calculate relevance for all the models $w_i = \frac{1}{|\mathbf{o} - \mathbf{m}_i|}$ $i = 1, \ldots, \mu + \lambda$

    (b) Select the $k = 7$ closest examples to form matrix W

    (c) Transform each simulated image $\mathbf{m}_j$ $j = 1, \ldots, k$ in a row for matrix H

    (d) Transform $\mathbf{o}$ in a row

(e) Transform each individual $\mathbf{x}_j$ $j = 1, \ldots, k$ in a row for matrix Y

(f) Do $Z = WH$

(g) Do $V = WY$

(h) Do $\mathbf{y} = \mathbf{o}^T Z^* V$

(i) Return $\mathbf{y}$

11. Do $\mathbf{x}_\mu = \mathbf{y}$
12. Select the best μ individuals from the sorted population

13. Mutate strategy parameter vectors:

$$\sigma'_i = \begin{cases} 0.3\sigma_i & \text{if } \mathbf{x}_i \text{ is a child} \\ 3.3\sigma_i & \text{if } \mathbf{x}_i \text{ is a parent} \end{cases}$$

where these values have been selected experimentally.

14. If some model $\mathbf{m}_i$ fits good the observed galaxy brightness profile or the maximum number of generations is reached terminate, otherwise return to 5

4   Results

In order to evaluate the performance of the hybrid algorithm ES+LWLR we have obtained results for a set of 100 galaxies. Also, we made a comparison with ES algorithm alone to have a best scenario about the improvements reached with the hybrid algorithm.

The fitting of one galaxy image takes on average 12 minutes on a PC with a PIV 3 Ghz processor and 512 MB of RAM, using MatLab. Computational time can be improved if we employ a compiler language such C, C++ or FORTRAN rather than an interpreter one.

The fitness function describes the fitness of each individual by measuring the normalized Euclidean distance with respect to the observed image $\mathbf{o}$, the function is ranged from 0 to 1, with 1 as a perfect match and 0 as totally different images. We say a model $\mathbf{m}*$ matches perfectly the observed image when $f(\mathbf{m}*) = 0$, but since we are matching models with observational data, we do not expect to reach the real maximum. Rather, we are interested in obtaining good approximations to the observed flux distribution.

After a set of experiments we determined that a value of 0.96 for the fitness function is good enough and the model can be acceptable, a bigger value for this threshold would lead to a better fit, albeit at an increase in computation time.

We illustrate the fitting of brightness profiles with a sample of 5 examples, presented in Table 1 that shows 5 galaxy images and their corresponding models approximated by ES+LWLR or ES. The first column indicates the name of the

Table 1. Comparison between ES and ES+LWLR for a set of 5 galaxy images (F.E. means Function Evaluations)

Galaxy	Original Image	Best Model	Difference	Algorithm	F.E.	$f(\mathbf{m})$
NGC2768				ES	5140	0.9824
				ES+LWLR	**4600**	0.9731
NGC2903				ES	**3700**	0.9719
				ES+LWLR	4180	0.9717
NGC3031				ES	5740	0.9616
				ES+LWLR	**730**	0.9575
NGC3344				ES	2430	0.9514
				ES+LWLR	**2020**	0.9513
NGC4564				ES	3670	0.9917
				ES+LWLR	**3160**	0.9615

Table 2. Comparison between ES and ES+LWLR for a set of 100 galaxy images

Algorithm	Function Evaluations (average)	$f(\mathbf{m})$ (average)	Standard deviation	Success (%)
ES	2239	0.97738	0.00563	85
ES+LWLR	**1846**	0.97733	0.00607	86

galaxy, the second, third and fourth columns show the original, the model and the difference images respectively, the fifth one indicates the algorithm (ES or ES+LWLR), the sixth presents the total number of function calls needed by the algorithm to reach convergence, and the last one shows the value for the cost function for the maximum found for each algorithm. We can observe that, as stated before, structure features (such as bridges, tails or spiral arms) of the galaxies are not fitted by the models, but the general distribution of brightness and angles are approximated very closely. Better fit is obtained in the central part, because basically in all the galaxies the centre is formed by a bulge, which can be fitted very well using the de Vaucoleurs' law; outer parts of galaxies are less fitted because the models are known not to be as accurate to describe details about structures.

In Table 2 we present the summarized results for a sample of 100 galaxy images, comparing behaviors of ES and ES+LWLR algorithms. The employed set in this case was formed using 85 spiral galaxies and 15 elliptical galaxies. We can see in the table that both algorithms have similar behaviors in terms of accuracy and average value for fitness function, because both of them present very similar values: 85% and 0.97738 for ES and 86% and 0.97733 for ES+LWLR. Nevertheless, we also observe that ES+LWLR has a better performance since it presents a less number of function evaluations than ES.

5   Conclusions

In this work we have solved the problem of two-dimensional fitting of brightness profiles for spiral and elliptical galaxies using a hybrid algorithm, based on Evolution Strategies and Locally Weighted Linear Regression, an instance based method, this new algorithm is called ES+LWLR. The hybrid algorithm achieved very good results, because was able to find an acceptable solution for almost all the cases in the galaxy images set. The ES+LWLR algorithm shows that knowledge generated by ES, in form of proposed solutions or individuals within a population can be employed to breed a new solution or individual that could be potentially better than those in the present population. This improved solution inserted in the set of current individuals helps to improve the global fitness of the population. Literally, LWLR is exploiting current population configuration for predicting directly a new solution.

Next step with this algorithm is a version where more than one individual is produced by LWLR, taking various training sets extracted from the current population in a random way.

Acknowledgments

This work was partially supported by CONACYT, Mexico, under grants C02-45258/A-1 and 144198.

References

[1] Atkenson, C.G., Moore, A.W., Schaal, S.: Locally Weighted Learning. Artificial Intelligence Review 11, 11–73 (1997)
[2] Back, T., Hammel, U., Schwefel, H.P.: Evolutionary Computation: Comments on the History and Current State. IEEE Transactions on Evolutionary Computation 1, 3–17 (1997)
[3] Back, T., Fogel, D., Michalewicz, Z.: Handbook of Evolutionary Computation. Oxford University Press, London (1997)
[4] Baggett, W.E., Baggett, S.M., Anderson, K.S.J.: Bulge-Disk Decomposition of 659 Spiral and Lenticular Galaxy Brightness Profiles. Astronomical Journal 116, 1626–1642 (1998)
[5] Beyer, H.G.: Toward a Theory of Evolution Strategies: Self-Adaptation. Evolutionary Computation 3, 311–347 (1996)
[6] de Vaucouleurs, G.: Recherches sur les Nebuleuses Extraglactiques. Ann. d'Astrophysics 11, 247 (1948)
[7] Karttunen, H., Kroger, P., Oja, H., Poutanen, M., Donner, K.J.: Fundamental Astronomy. Springer, Heidelberg (2000)
[8] Mitchell, M.: Machine Learning. McGraw-Hill, New York (1997)
[9] Peng, C.Y., Ho, L.C., Impey, C.D., Rix, H.W.: Detailed Structural Decomposition of Galaxy Images. Astrophysical Journal 124, 266–293 (2002)
[10] Rechenberg, I.: Evolutionsstrategie: Optimierung technischer Systeme nach Prinzipien der biologischen Evolution. Frommann-Holzboog, Stuttgart (1973)
[11] Wadadekar, Y., Robbason, B., Kembhavi, A.: Two-dimensional Galaxy Image Decomposition. Astronomical Journal 117, 1219–1228 (1999)

Gait Recognition by Applying Multiple Projections and Kernel PCA

Murat Ekinci, Murat Aykut, and Eyup Gedikli

Computer Vision Lab.
Department of Computer Engineering,
Karadeniz Technical University, Trabzon, Turkey
ekinci@ktu.edu.tr

Abstract. Recognizing people by gait has a unique advantage over other biometrics: it has potential for use at a distance when other biometrics might be at too low a resolution, or might be obscured. In this paper, an improved method for gait recognition is proposed. The proposed work introduces a nonlinear machine learning method, kernel Principal Component Analysis (KPCA), to extract gait features from silhouettes for individual recognition. Binarized silhouette of a motion object is first represented by four 1-D signals which are the basic image features called the distance vectors. The distance vectors are differences between the bounding box and silhouette, and extracted using four projections to silhouette. Classic linear feature extraction approaches, such as PCA, LDA, and FLDA, only take the 2-order statistics among gait patterns into account, and are not sensitive to higher order statistics of data. Therefore, KPCA is used to extract higher order relations among gait patterns for future recognition. Fast Fourier Transform (FFT) is employed as a preprocessing step to achieve translation invariant on the gait patterns accumulated from silhouette sequences which are extracted from the subjects walk in different speed and/or different time. The experiments are carried out on the CMU and the USF gait databases and presented based on the different training gait cycles. Finally, the performance of the proposed algorithm is comparatively illustrated to take into consideration the published gait recognition approaches.

1   Introduction

The image-based individual human identification methods, such as face, fingerprints, palmprints, generally require a cooperative subject, views from certain aspects, and physical contact or close proximity. These methods cannot reliably recognize non-cooperating individuals at a distance in the real world under changing environmental conditions. Gait, which concerns recognizing individuals by the way they walk, is a relatively new biometric without these disadvantages [1]-[6][8]. In other words, a unique advantage of gait as a biometric is that it offers potential for recognition at a distance or at low resolution when the human subject occupies too few image pixels for other biometrics to be perceivable.

P. Perner (Ed.): MLDM 2007, LNAI 4571, pp. 727–741, 2007.
© Springer-Verlag Berlin Heidelberg 2007

Various gait recognition techniques have been proposed and can be broadly divided as model-based and model-free approaches. Model based approaches [13][21] aim to derive the movement of the torso and/or the legs. They usually recover explicit features describing gait dynamics, such as stride dimensions and the kinematics, of joint angles.

Model-free approaches are mainly silhouette-based approaches. The silhouette approach[8][14][9][12][3][2][6] characterizes body movement by the statistics of the patterns produced by walking. These patterns capture both the static and dynamic properties of body shape. A hidden Markov models based framework for individual recognition by gait is presented in [9]. The approach in [14] first extracts key frames from a sequence and then the similarity between two sequences is computed using the normalized correlation. The template matching method in [5] is extended to gait recognition by combining transformation based on canonical analysis and used eigenspace transformation for feature selection. In the work in [8], the similarity between the gallery sequence and the probe sequence is directly measured by computing the correlation corresponding time-normalized frame pairs. The approach in [3] presents self similarity and structural stride parameters (stride and cadence) used PCA applied to self-similarity plots that are derived by differencing. In [2], eigenspace transformation based on PCA is first applied to the distance signals derived from a sequence of silhouette images, then classification is performed on gait patterns produced from the distance vectors. Han *et. al.* [6] used the Gait Energy Image formed by averaging silhouettes and then deployed PCA and multiple discriminant analysis to learn features for fusion.

In this paper, we presents an improved silhouette-based (model-free) approach and kernel PCA is applied to extract the gait features. The main purpose and contributions of this paper:

- An improved spatio-temporal gait representation, we called gait pattern, is first proposed to characterize human walking properties for individual recognition by gait. The gait pattern is created by the distance vectors. The distance vectors are differences between the bounding box and silhouette, and are extracted by using four projections of silhouette.
- A Kernel Principal Component Analysis (KPCA) based method is then applied for feature extraction. KPCA is a state-of-the art nonlinear machine learning method. Experimental results achieved by PCA and KPCA based methods are comparatively presented.
- FFT is employed to achieve translation invariant on the gait patterns which are especially accumulated from silhouette sequences extracted from the subjects walk in different speed and/or different time. Consequently, FFT+KPCA based method is developed to achieve higher recognition for individuals in the database includes training and testing sets do not correspond to the same walking styles.
- A large number of papers in literature reported their performance without using different training numbers. Here, we provide some quantitative comparative experiments to examine the performance of the proposed gait recognition algorithm with different number of training gait cycles of each person.

2   Gait Pattern Representation

In this paper, we only consider individual recognition by activity-specific human motion, i.e., regular human walking, which is used in most current approaches of individual recognition by gait. We first represent the spatio-temporal information in a single 2D gait template (pattern) by using multi-projections of silhouette. We assume that silhouettes have been extracted from original human walking sequences. A silhouette preprocessing procedure [8][17] is then applied on the extracted silhouette sequences. It includes size normalization (proportionally resizing each silhouette image so that all silhouettes have the same height) and horizontal alignment (centering the upper half silhouette part with respect to its horizontal centroid). In a processed silhouette sequence, the process of period analysis of each gait sequence is performed as follows: once the person (silhouette) has been tracked for a certain number of frames, then we take the projections and find the correlation between consecutive frames, and do normalization by subtracting its mean and dividing by its standard deviation, and then smooth it with a symmetric average filter. Further we compute its autocorrelation to find peaks indicate the gait frequency (cycle) information. Hence, we estimate the real period as the average distance between each pair of consecutive major peaks [20][2].

2.1   Representation Construction

Gait pattern is produced from the projections of silhouettes which are generated from a sequence of binary silhouette images, $B_t(x, y)$, indexed spatially by pixel location (x, y) and temporally by time t. An example silhouette and the distance vectors corresponding to four projections are shown in Figure 1. The distance vectors (projections) are the differences between the bounding box and the outer contour of silhouette. There are 4 different image features called the distance vectors; top-, bottom-, left- and right-projections. The size of 1D signals for left- and right-projections is the height of the bounding box. The values in the both signals are the number of columns between bounding box and silhouette at each row. The size of the 1D signals for both top- and bottom-distance vectors is the width of the bounding box, and the values of the signals are the number of rows between the box and silhouette at each column.

Thus, each gait pattern can separately be formed as a new 2D image. For instance, gait pattern image for top-projection is formulated as $P^T(x, t) = \sum_y \overline{B_t}(x, y)$ where each column (indexed by time t) is the top-projections (row sum) of silhouette image $B_t(x, y)$, as shown in Figure 1 (Middle-Top). The meaning of $\overline{B_t}(x, y)$ is complement of silhouette shape, that is empty pixels in the bounding box. Each value $P^T(x, t)$ is then a count of the number of rows empty pixels between the top side of the bounding box and the outer contours in that columns x of silhouette image $B_t(x, y)$. The result is a 2D pattern, formed by stacking row projections (from top of the bounding box to silhouette) together to form a spatio-temporal pattern. A second pattern which represents the bottom-projection $P^B(x, t) = \sum_{-y} \overline{B_t}(x, y)$ can be constructed by stacking row

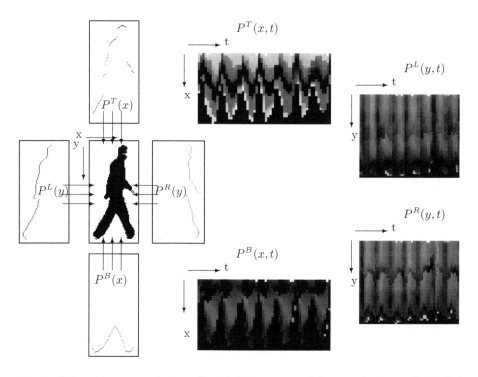

Fig. 1. Silhouette representation. **(Left)** Silhouette and four projections, **(Middle)** Gait patterns produced from top and bottom projections, **(Right)** Gait patterns obtained from left and right projections.

projections (from bottom to silhouette), as shown in Figure 1 (Middle-Bottom). The third pattern $P^L(y,t) = \sum_x \overline{B_t}(x,y)$ is then constructed by stacking columns projections (from left of the bounding box to silhouette) and the last pattern $P^R(y,t) = \sum_{-x} \overline{B_t}(x,y)$ is also finally constructed by stacking columns projections (from right to silhouette), as shown in Figure 1 (Right), respectively. For simplicity of notation, we write $\sum_y$, $\sum_{-y}$, $\sum_x$, and $\sum_{-x}$ as shorthand for $\sum_{y=Top-of-the-box}^{Contour-of-silhouette}$, $\sum_{y=Bottom-of-the-box}^{Contour-of-silhouette}$, $\sum_{x=Left-side-of-the-box}^{Contour-of-silhouette}$, and $\sum_{x=Right-side-of-the-box}^{Contour-of-silhouette}$, respectively.

The variation of each component of the distance vectors can be regarded as gait signature of that object. From the temporal distance vector plots, it is clear that the distance vector is roughly periodic and gives the extent of movement of different part of the subject. The brighter a pixel in 2D patterns in Figure 1 (Middle and Right), the larger value is the value of the distance vector in that position.

3   Human Recognition Using Gait Patterns

In this section, we describe the proposed approach for gait-based human recognition. Binarized silhouettes are produced by using motion segmentation which is

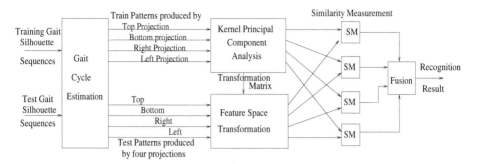

Fig. 2. System diagram of human recognition using the proposed approach

achieved via background modeling using a dynamic background frame estimated and updated in time, for details see to [7]. In the training procedure, each training silhouette sequence is divided into cycles by gait cycle estimation. Training gait patterns are then computed from each cycle. To be achieve translation invariant for the situation that training and test sequences are obtained from the subjects walk different speed and/or different time, the 2D gait pattern is transformed to spectral domain by using frequency transform (FFT). Next, features useful for distinguishing between different persons are extracted by kernel PCA-based nonlinear feature extraction method from the normalized gait pattern. As a result, training gait transformation matrices and training gait features that form feature databases are obtained. This is independently repeated for each gait patterns produced from the projections. In the recognition procedure, each test gait silhouette sequence is processed to generate test gait patterns. These patterns are then transformed by transformation matrices to obtain gait pattern features. Test gait pattern features are compared with training gait pattern features in the database. This is separately performed for each gait pattern features constructed by each projections. Finally a feature fusion strategy is applied to combine gait pattern features at the decision level to improve recognition performance. The system diagram is shown in Figure 2.

3.1 Kernel PCA

The kernel PCA (KPCA) is a technique for nonlinear dimension reduction of data with an underlying nonlinear spatial structure. A key insight behind KPCA is to transform the input data into a higher-dimensional **feature space** [15]. The feature space is constructed such that a nonlinear operation can be applied in the input space by applying a linear operation in the feature space. Consequently, standard PCA can be applied in feature space to perform nonlinear PCA in the input space.

Given k class for training, and each class represents a sequence of the distance vector signals of a person. Multiple sequences of each subject can be added for training, but we have used a sequence includes one gait cycle. Let $P_{i,j}^w$ be the jth distance vector signal in the ith class for w projection to silhouette and M_i

the number of such distance vector signals in the ith class. The total number of training samples is $M_t^w = M_1^w + M_2^w + ... + M_k^w$, as the whole training set can be represented by $[P_{1,1}^w, P_{1,2}^w, .., P_{1,M_1}^w, P_{2,1}^w, ..., P_{k,M_k}^w]$. For ease of understanding, we denote the training samples, $P_{i,j}^w$, as $\chi_i \in \Re^N, i = 1, .., M$, where M is total number of samples.

Thus, given a set of examples $\chi_i \in \Re^N, i = 1, ...M$, which are centered, $\sum_{i=1}^{M} \chi_i = 0$, PCA finds the principal axis by diagonalizing the covariance matrix:

$$C = \frac{1}{M} \sum_{i=1}^{M} \chi_i \chi_j^T \tag{1}$$

Eigenvalue equation, $\lambda v = Cv$ is solved where v is eigenvector matrix. First few eigenvectors are used as the basic vectors of the lower dimensional subspace. Eigen features are then derived by projecting the examples onto these basic vectors [16].

In kernel PCA, the data, χ from input space is first mapped to a higher dimensional feature space by using a map $\Phi : \Re^N \rightarrow F$, and then performing a linear PCA in F. The covariance matrix in this new space F is,

$$\overline{C} = \frac{1}{M} \sum_{i=1}^{M} \Phi(\chi_i) \Phi(\chi_i)^T \tag{2}$$

Now the eigenvalue problem becomes $\lambda V = \overline{C}V$. As mentioned previously we do not have to explicitly compute the nonlinear map Φ. The same goal can be achieved by using the kernel function $k(\chi_i, \chi_j) = (\Phi(\chi_i) \cdot \Phi(\chi_j))$, which implicitly computes the dot product of vector χ_i and χ_j in the higher dimensional space [15]. The most often used kernel functions are Gaussian kernel, polynomial kernels, and sigmoid kernels [15]. Gaussian kernel was used for the experimentation in this work, and it is defined as,

$$k(\chi_i, \chi_j) = exp\left(-\frac{\|\chi_i - \chi_j\|^2}{2\sigma^2}\right), \tag{3}$$

Pairwise similarity between input examples are captured in a matrix K which is also called Gram matrix. Each entry $K_{i,j}$ of this matrix is calculated using kernel function $k(\chi_i, \chi_j)$. Eigenvalue equation in terms of Gram matrix written as (see[15]),

$$M\mathcal{A}\Lambda = K\mathcal{A}, \tag{4}$$

with $\mathcal{A} = (\alpha_1, ..., \alpha_M)$ and $\Lambda = diag(\lambda_1, ..., \lambda_M)$. $\mathcal{A}$ is a M x M orthogonal eigenvector matrix and Λ is a diagonal eigenvalue matrix with diagonal elements in decreasing order. Since the eigenvalue equation is solved for $\mathcal{A}$'s instead of eigenvectors V_i of Kernel PCA, we will have to normalize $\mathcal{A}$ to ensure that eigenvalues of Kernel PCA have unit norm in the feature space, therefore $\alpha_j = \alpha_j/\sqrt{\lambda_j}$. After normalization the eigenvector matrix, V, of Kernel PCA is computed as follows,

$$V = \mathcal{D}\mathcal{A} \tag{5}$$

where $\mathcal{D} = [\Phi(\chi_i)\Phi(\chi_2)\cdots\Phi(\chi_M)]$ is the data matrix in feature space. Now let χ be a test example whose map in the higher dimensional feature space is $\Phi(\chi)$. The Kernel PCA features for this example are derived as follows:

$$F = V^T\Phi(\chi) = \mathcal{A}^T\mathcal{B}, \tag{6}$$

where $\mathcal{B} = [\Phi(\chi_1)\cdot\Phi(\chi)\Phi(\chi_2)\cdot\Phi(\chi)\cdots\Phi(\chi_M)\cdot\Phi(\chi)]^T$.

3.2 Similarity Measurement

Weighted Euclidean Distance (WED) measuring has initially been selected for classification [23], and is defined as follow:

$$WED : d_k = \sum_{i=1}^{N} \frac{(f(i) - f_k(i))^2}{(s_k)^2} \tag{7}$$

where f is the feature vector of the unknown gait pattern, f_k and s_k denote the kth feature vector and its standard deviation, and N is the feature length. In order to increase the recognition performance, a fusion task is developed for the classification results given by each projections.

3.3 Fusion

Two different strategies were developed. In **strategy 1**, each projection is separately treated. Then the strategy is to combine the distances of each projection at the end by assigning equal weight. The final similarity using strategy 1 is calculated as follows:

$$D_i = \sum_{j=1}^{4} w_j * d_{ji} \tag{8}$$

where D_i is the fused distance similarity value, j is the algorithm's index for projection, w its normalized weight, d_i its single projection distance similarity value, and 4 is the number of projections (left, right, top, bottom). In conclusion, if any 2 of the distance similarity values in the 4 projections give maximum similarities for the same person, then the identification is determined as to be positive. Therefore, fusion strategy 1 has rapidly increased the recognition performance in the experiments.

In the experimental studies, it has been seen that some projections have given more robust results than others. For example, while a human moves in the lateral view, with respect to image plane, the back side of the human gives more individual characteristics of gait. The projection corresponding to that side can give more reliable results, and in such case, is called the dominant feature. As a result, **strategy 2** has also been developed to further increase recognition performance. In the strategy 2, if the dominant projection, or at least 2 projections of others, are positive for an individual, then the final identification decision is positive. The dominant feature in this work is automatically assigned by estimating the direction of motion objects under tracking [17].

4   Experiments and Results

We evaluate the performance of the method on CMU's MoBo database[18], and USF database [8].

4.1   CMU Database

This database has 25 subjects (23 males, 2 females) walking on a treadmill. Each subject is recorded performing four different types of walking: slow walk, fast walk, inclined walk, and slow walk holding ball. There are about 8 cycles in each sequence, and each sequences is recorded at 30 frames per second. It also contains six simultaneous motion sequences of 25 subjects, as shown in figure 3.

We did mainly different two type experiments on this database: In type I, all subjects in train set and test set walk on the treadmill at the same walking type. In type II, all subjects walk on the treadmill at different two walking types, and it is called that fast walk and slow walk. We did two kinds of experiment for each type investigation. They are: **I.1**) train on fast walk and test on fast walk, **I.2**) train on slow walk and test on slow walk. Type II: **II.1**) train on slow walk and test on fast walk; **II.2**) train on fast walk and test on slow walk.

First, we use six gait cycles of each person are selected to form a training set, and the rest is used to test. PCA-based method was employed to extract the features from gait patterns, and then the WED based NN is used for classification. The fusion was finally performed to achieve the final decision. We first tested the performance of this algorithm for Type **I**, and it is summarized in Table 1. It can be seen from Table 1 that the right person in the top one match 100% of the times for the cases where testing and training sets correspond to the same walking styles for all viewpoints.

Second, seven kinds of experiment tests were designed: one (two, three, four, five, six, or seven) gait cycle(s) of each person was randomly selected for training, and the other seven gait cycles were used for authentication, respectively. During the experiments, the features are extracted by using the eigenspace method given above. Based on these tests, the matching is separately conducted and the results for Type **I** experiment are given in Figures 4 and 5. The results illustrated in Figures 4 and 5 are obtained from the experiments: train on fast walk and test on fast walk; train slow walk and test on slow walk, respectively. The experimental

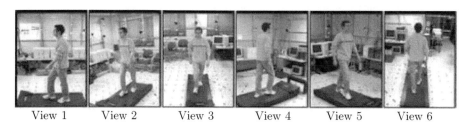

| View 1 | View 2 | View 3 | View 4 | View 5 | View 6 |

Fig. 3. The six CMU database viewpoints

Table 1. Gait Recognition across different views (CMU Data)

Test – Train	CMU Gait Database View Points				
	View 1	View 3	View 4	View 5	View 6
Fast – Fast	100	100	100	100	100
Slow – Slow	100	100	100	100	100

results show that the recognition rate is increased when the more gait cycles are used as training test. We did not need to apply kernel PCA-based feature extraction on the gait patterns, because PCA-based method had achieved the high recognition rates (100%) in this type of the experiments.

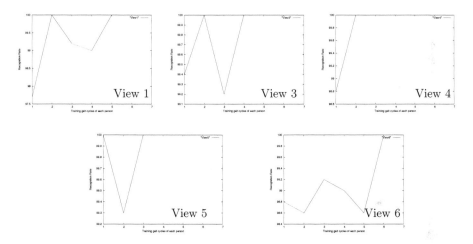

Fig. 4. Illustration of the recognition performance variation with different training gait cycles of each person. Train on fast walk, test on fast walk.

The third experiment, we called Type II, was also done on the gait sequences extracted from the subjects walk on the treadmill with different speed. It is called as slow walk and fast walk. For the case of training with fast walk and testing on slow walk, and vice versa, the dip in performance is caused due to the fact that for some individual as biometrics suggests, there is a considerable change in body dynamics and stride length as a person changes his speed. The results for Type **II** experiments are also summarized in Table 2. Table 2 shows experimental results obtained by different feature extraction methods presented in this paper. In this table, rank1 performance means the percentage of the correct subjects appearing in the first place of the retrieved rank list and rank5 means the percentage of the correct subjects appearing in any of the first five places of the retrieved rank list. The performance in this table is the recognition rate under these two definitions.

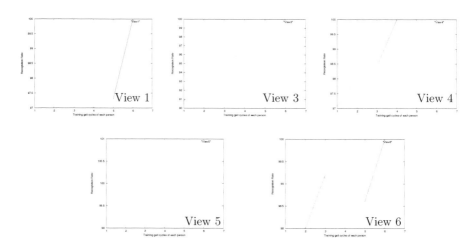

Fig. 5. Illustration of the recognition performance variation with different training gait cycles of each person. Train on slow walk and test on slow walk.

There are 8 gait cycles at the slow walking and fast walking data sets for each view. The 8 cycles in one walking type are used as train set, the 8 cycles in other walking type are used as test set. The gait patterns are produced as explained in section 2.1. The features in the gait patterns are extracted by using four different features extraction methods given in Table 2. When it is considered, it seen that kernel PCA-based feature extraction gives better performance than PCA-based method. There is quite possible translation variant problem between two gait patterns extracted from the subjects walk with different walking styles and/or different times. To achieve translation invariant for the proposed method, the gait pattern in the spatial domain is first transformed to the spectral domain by using one dimensional (1-D) FFT. 1-D FFT process is independently performed in horizontal or vertical directions for the gait patterns produced from both

Table 2. Experiments for two different walking styles with different view points. Each walking styles includes 8 gait cycles.

Train		View 1		View 3		View 4		View 5		View 6	
Test	Method	Rank: 1	5	Rank: 1	5	Rank: 1	5	Rank: 1	5	Rank: 1	5
Slow	PCA	31.5	46	44	64.5	27	58.5	29	44	46	64.5
	KPCA	33	54	46.5	68.5	34.5	60.5	35	54	48	63.5
Fast	FFT+PCA	65	89	80	91.5	63	91	64.5	87	67	87.5
	FFT+KPCA	73	89	76.5	92.5	71.5	94	64	89	76	91.5
Fast	PCA	27	50.5	52	68.5	28	67.5	26	47.5	49	65
	KPCA	39.5	62	53.5	69	31.5	59	24.5	51	49	65
Slow	FFT+PCA	61.5	85	74.5	88	62.5	90.5	64	85	73.5	88
	FFT+KPCA	66.5	89.5	79.5	91.5	61	89.5	67	90	74	88.5

the left and right-projections or for the gait patterns produced from both the top- and bottom-projections, respectively. Then PCA- and kernel PCA-based feature extraction methods are employed to achieve higher recognition rates, as illustrated in Table 2. Consequently, highest recognition rates for most view points were achieved by using FFT+KPCA based feature extraction method.

Table 3 compares the recognition performance of different published approaches on MoBo database. Several papers have published results on this data set, hence, it is a good experiment data set to benchmark the performance of the proposed algorithm. Table 3 lists the reported identification rates for eight algorithms on eight commonly reported experiments. The first row lists the performance of the proposed method. For seven experiments the performance of the proposed algorithm is always highest score. The numbers for given in Table 3 are as read from graphs and tables in the cited papers. The number of the subjects in the training set and test set is 25. In the test experiments for train on fast walk and test on slow walk, or vice versa, 200 gait patterns (25 persons X 8 gait cycles) for each experiment were used to present the performance of the proposed method.

Table 3. Comparison of several algorithm on MoBo dataset

Train Test Viewpoint	Slow Slow View 1	View 3	Fast Fast View 1	View 3	Slow Fast View 1	View 3	Fast Slow View 1	View 3
Proposed method	100	100	100	100	73	76.5	66.5	79.5
BenAbdelkader *et.al.*[3]	100	96	100	100	54	43	32	33
UMD [9][10][11]	72	-	70	-	32	-	58	-
UMD [13]	72	-	76	-	12	-	12	-
CMU [14]	100	-	-	-	76	-	-	-
Baseline [8]	92	-	-	-	72	-	-	-
MIT[19]	100	-	-	-	64	-	-	-

4.2   USF Database

The USF database [8] is finally considered. This database consists of persons walking in elliptical paths in front of the camera. Some samples are shown in Figure 6. For each person, there are up to five covariates: viewpoints (left/right), two different shoe types, surface types (grass/concrete), carrying conditions (with/without a briefcase), and time and clothing. Eight experiments are designed for individual recognition as shown in Table 4. Sarkar et. al. [8] propose a baseline approach to extract human silhouette and recognize an individual in this database. The experiments in this section begin with these extracted binary silhouette data. These data are noisy, e.g., missing of body parts, small holes inside the objects, severe shadow around feet, and missing and adding some parts around the border of silhouettes due to background characteristics. In Table 4, G and C indicate grass and concrete surfaces, A and B indicate shoe types, and L and R indicate left and

Fig. 6. Some sample images in the database described in [22][8]

right cameras, respectively. The number of subjects in each subset is also given in square bracket. Each one also includes 4-5 gait cycle sequence.

The experimental results on the standard USF HumanID Gait database version 1.7 are summarized in Table 4. In this table, the performance of PCA- and KPCA-based feature extraction methods are comparatively illustrated. The matching is also conducted independently based on weighted Euclidean distance classifier. The decision results based on the fusion strategies, explained in section 3.3, are additionally given in Table 4. Fusion 1 and Fusion 2 indicate that the results are produced by using the strategy I and the strategy II, respectively. It is observed from the experiments that, the recognition performance is increased when the strategy II is used in the fusion process.

Table 4. Classification performance for the USF data set, version 1.7

	PCA		KPCA	
Experiment	Fusion 1	Fusion 2	Fusion 1	Fusion 2
CAL[71]	78.8	85.9	84.5	90.1
CAR[71]	85.9	88.7	85.9	87.3
CBL[43]	74.4	86.04	81.3	90.6
CBR[43]	83.7	93.02	79.06	88.3
GAL[68]	86.7	92.6	88.2	92.6
GAR[68]	79.4	82.3	80.8	85.2
GBL[44]	90.9	93.1	93.1	95.4
GBR[44]	77.2	86.3	86.3	90.9

To analyze the relationship between the performance of the proposed method and number of training gait cycles of each person, four kinds of experiment types were designed: one (two, three, or four) training gait cycle(s) of each person was randomly selected for training, and the other gait cycles were used for authentication, respectively. These experimental results are given in Figure 7. KPCA- and PCA-based features extraction methods are comparatively illustrated, as well. In the Figure 7, y-axis indicates recognition rate, and x-axis indicates the number of training gait cycles of each person. When the plotted results in Figure 7 are considered, it can be seen that kernel PCA-based feature extraction approach achieves better performance than PCA-based approach. From the results we can

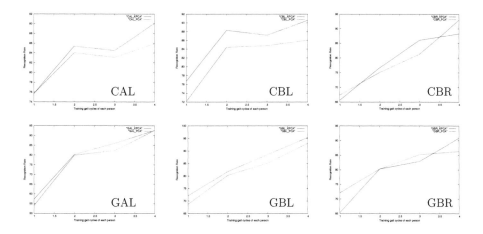

Fig. 7. Illustration of the recognition performance variation with different training gait cycles of each person

report that the accuracy can be greatly improved with the growth of the training gait cycles. For instance, when the proposed algorithm is trained using 1 gait cycle in the experiment **GBL**, an accuracy of 72.1% is achieved. When 4 gait cycles are used for training, a higher accuracy of 95.4% can be gotten. *It is evident that training gait cycle number can play an important role in the matching process. More training gait cycles lead to a high recognition rate.*

Table 5 compares the recognition performance of different published approaches on the USF silhouette version 1.7. The performance of the proposed algorithm is better than other approaches in GBR, GBL, CAR, CBR, CAL, and CBL, and slightly worse in GAL.

Table 5. Comparison of recognition performance using different approaches on USF silhouette sequence version 1.7

Exp.	The method	Baseline[22]	NLPR[2]	UMD-Indirect[9]	UMD-Direct[9]	GEI [6]
GAL	92.6	79	70.42	91	99	100
GBR	90.9	66	58.54	76	89	90
GBL	95.4	56	51.22	65	78	85
CAR	87.3	29	34.33	25	35	47
CBR	88.3	24	21.43	29	29	57
CAL	90.1	30	27.27	24	18	32
CBL	90.6	10	14.29	15	24	31

5   Conclusion

In this paper, we first propose to improve the spatio-temporal gait representation, which is multi-projections of silhouettes developed by our previous work

[20], for individual recognition by gait. As the others contributions and novelties in this paper, **1)** Kernel PCA based features extraction approach for gait recognition is then presented, **2)** FFT-based pre-processing is also proposed to achieve translation invariant for the gait patterns which are produced from silhouette sequences extracted from the subjects walk in different walking styles. **3)** The experimental results were finally submitted to examine the performance of the proposed algorithm with different training gait cycles. The proposed approach achieves highly competitive performance with respect to the published major gait recognition approaches.

Acknowledgments. This research was partially supported by The Research Foundation of Karadeniz Technical University (Grant No: KTU-2004.112.009.001). The authors would like to thank to Dr. R.T. Collins from Carnegie Mellon University, U.S.A., for providing us with the CMU database, Dr. S. Sarkar from the University of South Florida, U.S.A., for providing us with the USF database.

References

1. Nixon, M.S., Carter, J.N.: Automatic Recognition by Gait. Proceeding of the IEEE 94(11), 2013–2023 (2006)
2. Wang, L., Tan, T., Ning, H., Hu, W.: Silhouette Analysis-Based Gait Recognition for Human Identification. IEEE Trans. on PAMI 25(12), 1505–1518 (2003)
3. BenAbdelkader, C., Cutler, R.G., Davis, L.S.: Gait Recognition Using Image Self-Similarity. EURASIP Journal of Applied Signal Processing 4, 1–14 (2004)
4. Veres, G.V., et al.: What image information is important in silhouette-based gait recognition?. In: Proc. IEEE Conference on Computer Vision and Pattern Recognition vol. 2, pp. 776–782 (2004)
5. Huang, P., Harris, C., Nixon, M.S.: Human Gait Recognition in Canonical Space Using Temporal Templates. IEE Vision Image and Signal Processing 146, 93–100 (1999)
6. Han, J., Bhanu, B.: Individual Recognition Using Gait Image Energy. IEEE Trans. on Pattern Analysis and Machine Intelligence 28(2), 316–322 (2006)
7. Ekinci, M., Gedikli, E.: Background Estimation Based People Detection and Tracking for Video Surveillance. In: Yazıcı, A., Şener, C. (eds.) ISCIS 2003. LNCS, vol. 2869, pp. 421–429. Springer, Heidelberg (2003)
8. Sarkar, S., et al.: The HumanID Gait Challenge Problem: Data Sets, Performance, and Analysis. IEEE Transactions on Pattern Analaysis and Machine Intelligence 27(2), 162–177 (2005)
9. Kale, A., et al.: Identification of Humans Using Gait. IEEE Transactions on Image Processing 13(9), 1163–1173 (2004)
10. Kale, A., Cuntoor, N., Chellappa, R.: A Framework for Activity Specific Human Identification. In: IEEE International Conference on Acoustics, Speech, and Signal Processing vol. 4, pp. 3660–3663 (2002)
11. Kale, A., Rajagopalan, A., Cuntoor, N., Krugger, V.: Gait-Based Recognition of Humans Using Continuous HMMMs. In: IEEE International Conference on Automatic Face and Gesture Recognition, pp. 321–326 (2002)
12. Yanxi, L., Collins, R.T., Tsin, T.: Gait Sequence Analysis using Frieze Patterns. In: Heyden, A., Sparr, G., Nielsen, M., Johansen, P. (eds.) ECCV 2002. LNCS, vol. 2351, pp. 657–671. Springer, Heidelberg (2002)

13. BenAbdelkader, C., Cutler, R., Davis, L.: Motion-Based Recognition of People in Eigengait Space. In: IEEE International Conference on Automatic Face and Gesture Recognition, pp. 254–259 (2002)
14. Collins, R., Gross, R.J., Shi, J.: Silhouette-Based Human Identification from Body Shape and Gait. In: IEEE International Conference on Automatic Face and Gesture Recognition, pp. 351–356 (2002)
15. Scholkopf, B., Somala, A.J.: Learning with Kernel: Support Vector Machine, Regularization, Optimization and Beyond. MIT Press, Cambridge (2002)
16. Ali, S., Shah, M.: A Supervised Learning Framework for Generic Object Detection. In: IEEE International Conference on Computer Vision, vol. 2, pp. 1347–1354 (2005)
17. Ekinci, M., Gedikli, E.: Silhouette-Based Human Motion Detection and Analysis for Real-Time Automated Video Surveillance. Turkish Journal of Electrical Engineering and Computer Sciences 13(2), 199–230 (2005)
18. Gross, R., Shi, J.: The CMU motion of body (MOBO) database. Tech. Rep. CMU-RI-TR-01-18, Robotics Institute, Carnegie Mellon University (2001)
19. Lee, L., Grimson, W.: Gait Analysis for Recognition and Classification. In: IEEE, Proc. Int. Conference on Automatic Face and Gesture Recognition, pp. 155–162 (2002)
20. Murat, E.: A New Attempt to Silhouette-Based Gait Recognition for Human Identification. In: Lamontagne, L., Marchand, M. (eds.) Canadian AI 2006. LNCS (LNAI), vol. 4013, pp. 443–454. Springer, Heidelberg (2006)
21. Bazin, A.I., Nixon, M.S.: Gait Verification Using Probabilistic Methods. In: IEEE Workshop on Applications of Computer Vision, pp. 60–65 (2005)
22. Phillips, P., et al.: Baseline Results for the Challenge Problem of Human ID using Gait Analysis. In: IEEE International Conference on Automatic Face and Gesture Recognition, pp. 130–135 (2002)
23. Zhu, Y., Tan, T.: Biometric Personal Identification Based on Handwriting. Pattern Recognition (2), 797–800 (2000)

A Machine Learning Approach
to Test Data Generation:
A Case Study in Evaluation of Gene Finders

Henning Christiansen[1] and Christina Mackeprang Dahmcke[2]

[1] Research group PLIS: Programming, Logic and Intelligent Systems
Department of Communication, Business and Information Technologies
Roskilde University, P.O. Box 260, DK-4000 Roskilde, Denmark
`henning@ruc.dk`
[2] Department of Science, Systems and Models
Roskilde University, P.O. Box 260,
DK-4000 Roskilde, Denmark
`chmada@ruc.dk`

Abstract. Programs for gene prediction in computational biology are examples of systems for which the acquisition of authentic test data is difficult as these require years of extensive research. This has lead to test methods based on semiartificially produced test data, often produced by *ad hoc* techniques complemented by statistical models such as Hidden Markov Models (HMM). The quality of such a test method depends on how well the test data reflect the regularities in known data and how well they generalize these regularities. So far only very simplified and generalized, artificial data sets have been tested, and a more thorough statistical foundation is required.

We propose to use logic-statistical modelling methods for machine-learning for analyzing existing and manually marked up data, integrated with the generation of new, artificial data. More specifically, we suggest to use the PRISM system developed by Sato and Kameya. Based on logic programming extended with random variables and parameter learning, PRISM appears as a powerful modelling environment, which subsumes HMMs and a wide range of other methods, all embedded in a declarative language. We illustrate these principles here, showing parts of a model under development for genetic sequences and indicate first initial experiments producing test data for evaluation of existing gene finders, exemplified by GENSCAN, HMMGene and genemark.hmm.

1   Introduction

A computer program calculating a well-defined mathematical function is either correct or incorrect, and testing is a systematic process aiming to prove the software incorrect. One sort of test is black-box (or external) testing, which consists of running selected data through the program and comparing observed and expected results; established methods exist for designing test data suites

P. Perner (Ed.): MLDM 2007, LNAI 4571, pp. 742–755, 2007.

that increase the chance of finding errors [1]. Systems for information retrieval and extraction, on the other hand, have no such simple correctness criteria and are evaluated by relative measurements such as precision and recall in manually marked-up test data, e.g., the text corpora provided by the TREC and MUC conferences [2,3].

Gene finding programs, whose job is related to information extraction, aim to predict the total number and locations of genes in sequences of up to 3 billion letters. Here the situation is more subtle as a the production of marked-up test sequences may require years of research. In addition to this, it may be a research project in itself, to verify that a new gene suggested by a gene finder is correct.

Following the completion and release of the human genome sequence, a wide range of gene finder programs have been published. The tools differ in which kind of knowledge they integrate in gene modelling; from the fundamental *ab initio* approach, like GENSCAN [4], where generalized knowledge of genes is used, to the more opportunistic models, like GeneWise [5] and GenomeScan [6], where already known sequences of genes, EST's and proteins are used for finding genes by sequence similarity. One major problem with the existing gene prediction tools seems to be the lack of appropriate methods to evaluate their accuracy. The two major groups of human genome sequencing, Celera and Ensemble both predicted about 30,000 genes in the human genome [7,8], but a comparison of the predicted genes [9], revealed a very little overlap. This also seems to be the problem with other gene prediction tools. Which tools are more correct, is not easy to conclude, since most of the new predicted genes are not possible to classify as either correct genes that have not been found yet, or false predicted genes. This also applies to the underlying layer, of telling wrong exons (false positives) from those that have not been found yet. Another problem with the currently used training sets is that they usually consists of relatively simple genes, like the leading Burset/Guigó training set [10], with generalized features like containing few exons, all having tataboxes, having no overlapping genes and so on. Therefore most gene finders get very good at finding simple structured genes, and poor at finding the complex ones. The evaluation of gene finders does also have a problem, with sometimes large overlaps between the data sets used for training and the data sets used for estimating the accuracy of the gene finders [11].

To partly overcome these problems, test methods have been proposes based on semiartificially produced test data, typically by a combination of *ad hoc* techniques and statistical models such as Hidden Markov Models (HMM). The quality of such a test method depends on how well the test data reflect the regularities of known data and how well they generalize these regularities. So far only very simplified, artificial data sets have been tested, and a more thorough statistical foundation is required. A semiartificial data set [11] was generated to overcome these problems, in which a large set of annotated genes were placed randomly in an artificial, randomly generated intergenic background. This background sequence, however, is constructed in rather simplified, using a Markov Model to generate GC content of 38%. The present work is intended as a further step in this direction, demonstrating how the logic-statistical machine-learning system PRISM, introduced by other authors [12,13], can be used for the development of more sophisticated and reliable models. Based on logic programming extended with random variables and parameter learning, PRISM appears as a powerful modelling environment, which subsumes HMMs and a wide range of other methods, all embedded in a declarative language.

We illustrate these principles here, showing parts of a model under development in PRISM for genetic sequences and indicate first experiments producing test data for evaluation of existing gene finders, exemplified by GENSCAN [4], HMMGene [14] and genemark.hmm [15]. The advantage of the approach, that we can demonstrate, is the relative ease and flexibility with which these probabilistic models can be developed, while a claim that the approach may lead to biologically more well-founded models needs to be supported by more extensive testing.

PRISM embeds a both powerful and very flexible modelling language, which makes it possible to embed biological knowledge about genome sequences. A model in PRISM is parameterized by probabilities for random variables, and training the model with known data generates estimates for these probabilities. Using the same model, these probabilities can be used for generating artificial data that mimic in a faithful way sufficiently many properties of the authentic data. The generated data are marked up with information of where the model decided to put in genes, and a given gene finder can be evaluated in a precise way, by comparing its proposals for genes with those of the PRISM model (which, here, would play the role of "true" ones). We describe an experiment testing the three gene finders.

Section 2 gives a brief introduction to PRISM, and section 3 presents fragments a PRISM model for genomic sequences, which is still under development. Section 4 compares briefly with related work, before we show how our model is used for testing three selected gene finders. Sections 5 and 6 describe and evaluate the tests and compare the different gene finders. Section 7 discusses the quality of the test data generated by our model, and section 8 gives perspective and outline plans for future work.

2   Logic-Statistical Modelling in PRISM

By a modeling paradigm based on logic programs, we take a step upwards in formal expressibility compared with those models traditionally used in sequence analysis. Where HMMs are based on regular languages, SCFGs (Stochastic CFGs) as their name indicates on context-free language, PRISM can specify any Turing computable language. In practice, and to stay within computationally tractable cases, we need to be more modest and employ the flexibility of a general programming language to combine existing models and to introduce auxiliary data structures whenever convenient.

PRISM [12,13] represents a logic-statistical modelling system that combines the logic programming language Prolog with probabilistic choice and machine learning, and is implemented as an extension to the B-Prolog language [16]. It includes discrete random variables called *multi-valued random switches*, abbreviated msw's. The system is sound with respect to a probabilistic least Herbrand model semantics, provided that the different msw's are independent (see the references above for details). As an example, we can declare a class of msw's for selecting one out of a four different letters at random as follows.

```
values( nextLetter(_), "ACGT").
```

Recall that a string such as `"ACGT"` is a list of character codes, and in general the `values` directive describes a finite list of possible values. The term `nextLetter(_)` includes a logical variable which means that, for any possible value substituted for it, there exists an msw; e.g., `nextLetter(t1)`, `nextLetter(t2)`. The following fragment shows how an msw typically is used within a rule; the arrow-semicolon is Prolog's generalized if-then-else construction; the notation `0'`*char* indicates the relevant character code.

```
msw( nextLetter(t1), Letter),
(Letter = 0'A -> ...
 ; Letter = 0'C -> ...
 ; Letter = 0'G -> ...
 ; Letter = 0'T -> ... )
```

The dots indicate the different actions taken for the different outcomes. In this way, it is straightforward to write HMM's as PRISM programs (additional msw's may govern state transitions). Other models such as discrete Baysian networks, Stochastic Context-Free Grammars [17], and Hierarchical HMM's [18] can also be described in straightforward ways, and PRISM can be seen as a high-level tool for defining advanced and directly executable probabilistic models, as we do in the present paper. Conditional probabilities can be represented using generic msw names. For example, $P(a = x|b = y)$ is indicated by the code `msw(b,Y)`, `msw(a(Y),X)`.

We show an illustrative example of a model in PRISM, which also illustrates the modular structure we apply in our sequence model. We imagine sequences comprising three types of subsequences `t1`, `t2`, `t3`, put together in random order. Each type represent sequences of the letters `ACGT` but with different relative frequencies for each type. An annotated sequence can be described by a goal as follows, having the annotations in a separate argument, and numbering the letters starting from 1.

```
sequence("AAAACGCGCG",[t1(1,4),t2(5,10)])
```

The annotation is a sequence of descriptors for each subsequence. The following msw's govern the composition of subsequence of different types.

```
values(nextSubstringType(_),[t1,t2,t3]).
values(continueSeq,[yes,no]).
```

The argument to the first one describes the type of a previous subsequence; the last one determines number of subsequence, the higher probability for yes, the longer sequence.[1] The following rule describes the composition at subsequence level. An arbitrary "previous type" called start is assumed for the first subsequence, and an extra argument which keeps track of the position in the string is added; evidently this is a HMM with 5 states, the last one being the implicit final state.

```
seq(Seq,D):-seq(start,Seq,D,1).
```

```
seq(PrevType,Seq,[D1|DMore],N):-
   msw(nextSubstringType(PrevType), SubT),
   subSeq(SubT,D1,Seq,SeqMore,N,M),
   msw(continueSeq, YesNo),
   (YesNo=no -> SeqMore=[], DMore = []
   ; Mplus1 is M+1,
      seq(SubT,SeqMore,DMore,Mplus1)).
```

For each type T, we need to add an appropriate clause subSeq(T,···):- ····. To finish this example, we assume identical definition (but individual probabilities) for each type.

```
subSeq(Type,[L|SeqMore],SeqRest,N,M):-
   msw(nextLetter(Type),L),
   msw(continueSub(Type),YesNo),
   (YesNo=no -> SeqMore=SeqRest, N=M
   ; Nplus1 is N+1,
      subSeq(Type,SeqMore,SeqRest,Nplus1,M)).
```

Notice that we assumed an additional class of msw's to govern the length of each subtype (so this model resembles a 2-level Hierarchical HMM). Such a program can be used in two ways in the PRISM system. In case the probabilities are given in some way or another, the program can generate samples of annotated sequences by a query ?-sample(sequence(S,A)) where PRISM executes the msws by using a random number generator adjusted to the given probabilities.

PRISM can also run in *learner mode*, which means that the program is presented with a number of observed goals, from which it calculates the distribution of probabilities that provides the highest likelihood for explaining the observed data. PRISM uses some quite advanced algorithms and data structures in order to do this in an efficient way; these topics are outside the scope of this paper and we refer to [12,13]. Training with a single observation (somewhat artificial) being the short annotated sequence above, we get for t1 prob. 1.0 for A and for t2 prob. 0.5 for each of C and G. After the training phase, sampling can be used to create *similar* sequences, where the applied notion of similarity is given by the program, i.e., the model for sequence structure that we designed above. PRISM runs also on 64-bit architectures so it can address quite large amounts of storage. The version used for out tests (1.9, 2006) can, with programs as the one shown, easily handle data sets of around 10 sequences of 100,000 letters. For eukaryote

---

[1] As is well-known, this gives a geometric distribution of lengths.

sequence analysis in general, this is a rather small number, and current research aim at increasing this number drastically.[2]

3   A PRISM Model of Genomic Sequences

The example above illustrates the overall structure of our approach to test data generation for gene finders. A model of full genome sequences is under consideration, but for the present task of testing gene finders we have limited to a model of the intergenic regions. We argue that a detailed and biologically faithful model is essential also for these regions, as they contain patterns and fragments that may descend from obsolete functional regions, and thus may confuse a gene finder; this is, in fact, what our experiments reported below seem to indicate.

A sequence is characterized in our model by two intertwined structures. The first level concerns the distribution of *GC islands* which are long stretches with significantly higher frequencies for the two letters; GC islands are important as their presence often indicates subsequences containing genes; remaining regions of the sequences are called *GC sparse*. At a second level, the sequence is considered to consist of mostly arbitrary letters (coloured noise) interspersed with so-called repeat strings, which are either subsequences of a known catalogue of named strings, simple repeats (e.g., (ACCT)n meaning a repetition of the indicated pattern), and low-complexity strings such as CT-rich which is a combination of these letters. A marked up sequence is represented by structure of the following form.

sequence(*sequence-of-ACGT*, *GC-islands*, *repeats*)

The first component is the bare sequence of letters, the second indicates positions of GC islands and GC sparse, and the third one describes the repeats in the sequence. The relationship between these two levels is somewhat tricky as a given repeat substrings can appear in a GC island, in a GC sparse region, or overlap both; however, there is a dependency in the sense that a repeat (or section thereof) with many G's and C's tend to appear more often in GC islands. Since GC islands tend to be much larger that repeats, they are considered the top level structure. The lengths of GC islands and GC sparse regions are determined by empirically determined minimum lengths combined with random variables that decide whether to continue or stop, thus giving geometric distributions upon the minimum lengths. Other random variables are given in two versions, one for GC island and one for GC sparse.

This two-level structure is implemented by hiding the management of GC islands in an abstract data type, which appears as an alternative version of PRISM's msw predicate with the following parameters.

msw(*random-var*, *value*, *GC-islands*, *position*)

---

[2] E.g., using an array representation for sequences instead of Prolog lists as in the current version, may reduce storage for each letter from 24 bytes (on a 64 bit architecture) to 2 bits!.

If a variable, say x(a) is referred to at position 5000 in a sequence, we call msw(x(a), V, ⋯, 5000) and the implementation choses, depending on how position 5000 is classified, the relevant of msw(x(gcIsland,a),V) and msw(x(gcSparse,a),V); in addition, the extended version of msw includes the machinery that describes GC islands in terms of other random variables as explained. Abstracting away a few implementation details, the overall composition of intergenic sequences can be presented as follows; notice that the choice of repeat depends on the previous one, as biologists indicate that certain repeats tend to come in groups; the letter sequences between repeats depend on those as well since some repeats have a tendency to come close to each other. These letter sequences are a kind of coloured noise, where the colours indicate the different letter probabilities in GC islands vs. GC sparse regions.

```
seq(Seq,GCs,Reps):-
    seq(dummy,Seq,GCs,Reps,1).

seq(Prev,Seq1,GCs,Reps,N1):-
    msw(nextRepType(Prev),RepType,GCs,N1),
    letters(Previous,RepType,
                Seq1,Seq2,GCs,N1,N2),
    N2i is N2+1,
    (RepType=noMore -> Seq=[], Reps=[]
    ; RepType=namedRepeat ->
        namedRep(Details,Seq2,Seq3,GCs,N2i,N3),
        Reps=[named(Details)|Reps1],
        seq(RepType,Seq3,GCs,Reps1,N3)
    ; RepType=simpleRepeat ->
        simpleRep(⋯), Reps=⋯, seq(⋯)
    ; RepType=lowComplexRep ->
        lowCmplxRep(⋯), Reps=⋯, seq(⋯)).
```

Predicates namedRep, simpleRep, lowCmplxRep describe the detailed structure of the different sort of repeats. They apply their own random variables to govern which substring is repeated; namedRep has access to an 11Mbyte catalogue of known repeat string strings and employs random variables to determine which portion of the named string is used and its direction: it may appear as a plain copy, reversed, complement (i.e., with letters interchanged A↔T, C↔G), or reverse complement. These predicates are defined in relatively straightforward ways, although a few technicalities are needed in order to handle mutations, which here means that a few "noise letters" have been inserted or deleted in the sequences. Again, we used random variables to govern this and to control which noise letters are chosen.

Named repeats were described in a simplified version in the marked up sequences we had available, as the exact portion used as well as its direction were not given, only its name. To cope with this, we used a preprocess to determine one unique best match of portion and direction and added it to the annotations, where "best" is defined by a few heuristics and a lowest penalty scheme.

In addition, our best match algorithm adds best proposals for where and how mutations have occurred. The use of preprocessing to provide unique best proposals for lacking information in annotations appears to be essential for this form of sequence analysis, as nondeterminism in a PRISM model can introduce backtracking which may be very problematic due to the size of the data being analyzed.

The model is now ready to be trained from a set of marked up sequences; the actual training set is described below. PRISMs learning algorithm produces a set of probabilities for the random variables that gives the highest likelihood for the training data. With these probabilities, the model can thus generate artificial, marked up sequences that are faithful to those properties of the training data that are expressible in the model.

4   Previous Work

The first attempt to overcome the problem of not having large, precisely marked data sets, for evaluation of gene finders was made by embedding random genes in a random sequence approximating intergenic sequence [11]; 42 sequences with accurate gene annotation were used, having an average length of 177160, containing 4.1 genes with an average of 21 exons and 40% GC content. Knowledge about repeated sequences, the existence of GC islands and pseudogenes were not taken into account, and only a simple average GC frequency of 38% was generated by using a fifth order HMM. Not surprisingly, the test of GENSCAN on these sequences did show a lower rate of correct predictions. This is obvious since *ab initio* programs like GENSCAN, recognizes genes by small sequences like tata boxes, splicesites etc. and therefore are dependent on possible sequence variations that will inform it about possibly being close to something relevant; e.g., GC islands. Our approach is to concentrate on including more knowledge about intergenic regions in the model, incorporating GC content variations and intergenic repeats.

5   Results: Evaluation of Three Different Gene Finders

In this section we present results of testing three gene finders; GENSCAN, HMMGene and genemark.hmm, using the data generated by our model, as described in a previous section. A significant drop in accuracy was observed [11] when testing GENSCAN on the semi artificial training set described above. This could be the first step in giving an accurate measure of gene finders, but due to the simplicity of the data set, some degree of uncertainty is evident. Here we have demonstrated a test of the three gene finders using a data set founded on a statistically better ground, and showing a relative high error rate. The model was trained on a data set of 12 intergenic sequences from the human chromosome 17 (NCBI refseq NT_0101718.15), together with a list of specified positions of

repeated sequences and GC islands.[3] We chose to use Repeatmasker [19] to find
repeated sequences in these sequences, since this tool appears well established.
We used standard masking options for this. RepBase [20] was used for generating
a repeater catalogue, in order to train our model on actual repeater sequences.
Furthermore positions of GC island were detected by using CpGplot [21], also
using standard masking options.

After training our model, the learned probabilities were used to generate 12
sequences, together with annotations of inserted repeated elements and GC islands. From these sequences test sets were assembled, with real genes inserted
between the generated sequences. Three test sets were created, each consisting of four generated sequences with three real genes in between. The same
three genes are used in all three test sets, making it possible to compare the
artificial intergenic sequences alone. The sequences of the genes comprise the
genomic region of the gene, from 200 bp before to 200 bp after the CDS.[4] The
length of the generated intergenic regions ranges from 2.2 kb to 30 kb. The three
test sets have varying lengths of; test set A (193,721 bp), test set B (170,545
bp), and test set C (202,368 bp). We report the outcome of predicted exons
in the intergenetic regions, as this is the main area of interest in this paper.
Therefore only false positives are reported. For this we use the evaluation term;
specificity, which describes the relationship between wrong exons and the total number of predicted exons, as described by [10]. The table below shows the
results of the tests performed on the three gene finders, using the three test
sets.

Table 1. Results from the tests on gene finders; GENSCAN, HMMGene, and gene-
mark.hmm. Specificity describes the relationship between wrong hits and total hits,
where 0 is a perfect result and 1 means all predicted exons are wrong.

	Genscan	Hmmgene	Genemark.hmm
Predicted genes in testset A	7	12	10
Predicted genes in testset B	10	11	11
Predicted genes in testset C	11	11	10
Average number	9.33	11.33	10.33
Gene Specificity (Wrong genes / predicted genes)	0.68	0.74	0.71
Average number of predicted exons	54.99	83	41.66
Wrong exons in testset A	50	79	24
Wrong exons in testset B	29	55	34
Wrong exons in testset C	33	73	28
Average number of wrong exons	37.33	69	28.67
Exon Specificity (Wrong exons / predicted exons)	0.68	0.83	0.69

---

[3] One of the reasons for choosing chromosome 17 is due to the fact that it is a rather
well studied chromosome, and the relative gene density is high. This could mean
that most genes are accounted for, and the intergenic sequence therefore is more
accurate to use.

[4] The genomic sequence comprising all exons and introns of the gene.

All the results show a large error rate, both regarding the number of genes and exons predicted. All gene finders have a gene specificity of about 0.7, which is rather high. This result is probably due to the training of the gene finders on single exon-genes, making them more likely to predict many and short genes. As to the results showing the exon specificity, the same high error rate of approximately 0.7 is seen in GENSCAN and genemark predictions. Here HMMGene has an even higher error specificity of 0.83. These results seem to indicate that all the tested gene finders overpredict, when faced with intergenic sequences containing repeated sequences and GC islands.

6   Comparison of the Tested Gene Finders

To determine if the quite similar high error rate is caused by the gene finders finding the same exons, we made a comparison of the predicted set exons. The figure below illustrates the combination of the three predictions, where the predicted exons are considered the same if they match in either end of the other exon.

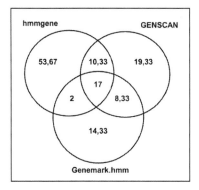

Fig. 1. The predictions of the three tests were joined to find exons that are the same. The given numbers are an average value of the three tests performed.

The result of this comparison shows that very few of the predicted exons are in fact the same. HMMGene has the highest number of predicted exons appearing only within its own prediction. The average number of 17 exons being present in all three predictions, and 37.66 exons being present in at least two predictions, is also not a very high number compared to a total number of 125 predicted exons. These results states that the same error rate between the three gene finders is not a consequence of the predicted exons being the same. In contrary, these differing results are probably due to differences in the underlying models of the gene finders. Therefore the results of table 1 could indicate that the precision of the tested gene finders are much lower than anticipated.

7   Appropriateness of the Generated Data

If the test we are performing is to be used to evaluate a gene finder, the testdata must reflect the essential properties of natural data, and appear as authentic copies. Including most possible dependencies and variables in the data is not possible, but all important features that could affect the outcome of the test must be accounted for. In our model we believe all important features are accounted for, but to examine if this is really the case, an analysis of the three test sets were performed. The length of the intergenic sequences, as well as GC islands and repeated elements did in fact resemble those of the training set. The tool used for finding CG islands in the training material, was also used on the artificial data, to observe if the overall pattern resembled the one of real sequences. The figure below illustrates the GC frequency in data set A, and a section of real sequence from the region the training material was taken from.

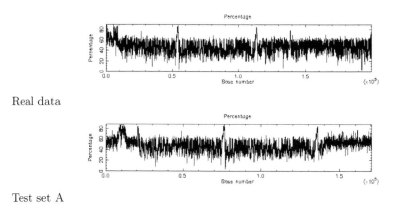

Real data

Test set A

Fig. 2. GC frequencies in real vs. model generated data

Figure 2 seems to indicate that the pattern of the GC percentage in the artificial data resembles the real data. However, when comparing the annotated GC islands and the ones found by CpGPlot, there is a slightly higher amount of GC islands in the CpGPlot. This is probably due to the fact that our model operates on absolute frequencies and the CpGPlot presumably uses a more complex model with relative frequencies. This difference in GC island reporting could help explaining why the three tested gene finders predict too many exons, but it does not explain all of them since the results from the three gene finders, as explained, differed greatly. These facts may also indicate the weakness of our model that it uses a first-order HMM to describe GC islands; a higher order, say 3 or 4, may seem more appropriate.

Furthermore an analysis of the dispersed repeated elements was carried out. Using Repeatmasker to find repeated sequences in the artificial data, and comparing them to the annotated elements, showed a very good match between these. Nearly all of the repeated sequences were found. The alignments of the detected repeaters returned by Repeatmasker showed a similar pattern of mutations in the artificial repeater sequences, compared to those in the training set, which thereby verifies our method for generating mutations. From this we conclude that the model generates fairly realistic artificial intergenic sequences, though this training set, as mentioned, is to be considered as an example of what the model could be trained on. Therefore the evaluation of GENSCAN, HMMGene and genemark will not be an absolute assessment. For this a larger training set is needed, making the probabilities of the model more thoroughly based. We believe however, that this new method of evaluating gene finders, is a new step in the way of finding a more accurate measure for the precision of gene finders.

8   Perspectives and Future Work

A further development of the model presented in this paper, will be a more comprehensive model, which takes more specialized features into account, together with the inclusion of genes. The major issue for a successful training of the model is a larger and better marked up set of sequences; At this point perfectly marked up genomic sequences, which include both genetic and intergenic regions remain sparse and difficult to access.

The use of artificial data sets is a subject under debate. As discussed by [22] the use of artificial data sets have not been used much, since it is very difficult to reflect the whole complexity of real data sets. But how many details must be included in a model for artificial data generation? As [22] state, the goal must be to include those characteristics of real data which may affect the performance of real data sets. Multivariable dependencies are also very important in terms of reflecting real data, and can often be very difficult to reproduce. PRISM however, seems to be a rather powerful modelling tool, which also are capable of reflecting multivariable dependencies, by the inclusion of one msw in another (see section 2). Even if these characteristics are accounted for, the ideal data mining method for one data set might not be ideal for another set. Furthermore, undiscovered correlations could influence the outcome of a given model. Our model only accounts for those characteristics verified by a biological expert, and therefore undiscovered multivariable dependencies are naturally not included. Another important issue, as mentioned, is the balance between including enough information in the model to make it realistic, and including too much, which results in overfitting. This will also be very important to have in mind, when expanding the model. Another important issue is the balance between overfitting and generalization, and the granularity of the model; while a fine-grained model is well suited when large training sets are available but likely subject of over-fitting in case a few training data, it is the other way round for a coarse model which may the the best out of few training date data and and too little out of large sets. Only testing and analyze indicate the right level.

As noticed above, we believe that the specific model described in this paper can be improved concerning the modeling of GC island by using a higher order

HMM, and we may also refer to [23] who discusses consequences of different orders, and proposes an extension to HMM which may be interesting to apply in PRISM-based models.

Our experiences indicates that the elegance of logic programming is not incompatible with the processing of large data sets. With the present technology we can handle around one million letters, which is clearly to little, although with reasonable execution times are reasonable, and plans for extensions of PRISM includes specialized representations for sequences (as opposed to list structures!) as well as adaptation to computer clusters. We hope that the continued refinement of this sort of models of genome sequences, based on a very flexible, powerful and theoretically well-founded paradigm, may lead to models that can be used in future gene finders that provide a higher accuracy than what we can observe at present.

The overall approach of using artificially test data produced by sophisticated, statistical models and machine learning was here motivated by the conditions for sequence analysis, that authentic test data are difficult to require. Similar methods may also have their relevance for embedded systems, with the statistical model simulating a world, and context-aware systems.

Acknowledgement. This work is supported by the CONTROL project, funded by Danish Natural Science Research Council.

References

1. Myers, G.J., Sandler, C., Badgett, T., Thomas, T.M. (Revised by): The Art of Software Testing, 2nd edn. Wiley (2004)
2. TREC: Text REtrieval Conference http://trec.nist.gov/
3. MUC: Message Understanding Conferences
   http://www-nlpir.nist.gov/related_projects/muc/
4. Burge, C., Karlin, S.: Prediction of complete gene structures in human genomic DNA. Journal of Molecular Biology 268, 78–94 (1997)
5. Birney, E., Clamp, M., Durbin, R.: GeneWise and Genomewise. Genome Res. 14(5), 988–995 (2004)
6. Yeh, R.-F., Lim, L.P., Burge, C.B.: Computational Inference of Homologous Gene Structures in the Human Genome. Genome Res. 11(5), 803–816 (2001)
7. Venter, J.C., et al.(>300 authors): The Sequence of the Human Genome. Science 291(5507), 1304–1351 (2001)
8. Lander, E.S., et al.(> 300 authors): Initial sequencing and analysis of the human genome. Nature 409, 860–892 (2001)
9. Hogenesch, J.B., Ching, K.A., Batalov, S., Su, A.I., Walker, J.R., Zhou, Y., Kay, S.A., Schultz, P.G., Cooke, M.P.: A comparison of the Celera and Ensembl predicted gene sets reveals little overlap in novel genes. Cell 106(4), 413–415 (2001)
10. Burset, M., Guigó, R.: Evaluation of Gene Structure Prediction Programs. Genomics 34(3), 353–367 (1996)
11. Guigó, R., Agarwal, P., Abril, J.F., Burset, M., Fickett, J.W.: An Assessment of Gene Prediction Accuracy in Large DNA Sequences. Genome Res. 10(10), 1631–1642 (2000)

12. Sato, T., Kameya, Y.: Parameter learning of logic programs for symbolic-statistical modeling. J. Artif. Intell. Res. (JAIR) 15, 391–454 (2001)
13. Sato, T., Kameya, Y.: Statistical abduction with tabulation. In: Kakas, A.C., Sadri, F. (eds.) Computational Logic: Logic Programming and Beyond. LNCS (LNAI), vol. 2408, pp. 567–587. Springer, Heidelberg (2002)
14. Krogh, A.: Using database matches with for HMMGene for automated gene detection in Drosophila. Genome Research 10(4), 523–528 (2000)
15. Lukashin, A., Borodovsky, M.: Genemark.hmm: new solutions for gene finding. Nucleic Acids Research 26(4), 1107–1115 (1998)
16. Zhou, N.F.: B-Prolog web site (1994–2006) http://www.probp.com/
17. Charniak, E.: Statistical Language Learning. The MIT Press, Cambridge (1993)
18. Fine, S., Singer, Y., Tishby, N.: The hierarchical hidden markov model: Analysis and applications. Machine Learning 32(1), 41–62 (1998)
19. Smit, A., Hubley, R., Green, P.: Repeatmasker web site (2003) http://repeatmasker.org
20. Jurka, J., Kapitonov, V., Pavlicek, A., Klonowski, P., Kohany, O., Walichiewicz, J.: Repbase Update, a database of eukaryotic repetitive elements. Cytogenetic and Genome Research 110(1-4), 462–467 (2005)
21. EMBL-EBI: CpGplot http://www.ebi.ac.uk/emboss/cpgplot/
22. Scott, P.D., Wilkins, E.: Evaluating data mining procedures: techniques for generating artificial data sets. Information & Software Technology 41(9), 579–587 (1999)
23. Wang, J., Hannenhalli, S.: Generalizations of Markov model to characterize biological sequences. BMC Bioinformatics 6 (2005)

Discovering Plausible Explanations of Carcinogenecity in Chemical Compounds

Eva Armengol

IIIA - Artificial Intelligence Research Institute,
CSIC - Spanish Council for Scientific Research,
Campus UAB, 08193 Bellaterra, Catalonia (Spain)
eva@iiia.csic.es

Abstract. The goal of predictive toxicology is the automatic construction of carcinogenecity models. Most common artificial intelligence techniques used to construct these models are inductive learning methods. In a previous work we presented an approach that uses lazy learning methods for solving the problem of predicting carcinogenecity. Lazy learning methods solve new problems based on their similarity to already solved problems. Nevertheless, a weakness of these kind of methods is that sometimes the result is not completely understandable by the user. In this paper we propose an explanation scheme for a concrete lazy learning method. This scheme is particularly interesting to justify the predictions about the carcinogenesis of chemical compounds.

1 Introduction

During the seventies Europe and the United States respectively started long term programs with the aim of developing toxicology chemical databases. The idea was to establish standardized experimental protocols allowing to determine the carcinogenecity of chemical compounds. In particular, the American National Toxicology Program (NTP) established two protocols to be performed on rodents: a short-term protocol (90 days) and a long-term protocol (2 years). To develop both protocols is necessary to sacrify a lot of animals and sometimes the results are not clearly conclusive concerning to carcinogenecity. Moreover, even in the situation of clear carcinogenic activity of a chemical compounds on rodents, there is no certainty that the results may be extrapolable to humans.

The use of computational models applied to toxicology could contribute to reduce the cost of experimental procedures. In particular, artificial intelligence methods such as knowledge discovery and machine learning can be used for building models of carcinogenecity (see [13]). The construction of such models is called *predictive toxicology*. From the machine learning point of view, the goal of the predictive toxicology is a classification task, i.e. toxic compounds are classified as belonging to the positive class and non-toxic compounds are classified as belonging to the negative class.

Most of machine learning approaches use representations of chemical compoudns based on *structure-activity relationship (SAR)* descriptions since there

P. Perner (Ed.): MLDM 2007, LNAI 4571, pp. 756–769, 2007.
© Springer-Verlag Berlin Heidelberg 2007

are easily obtained from commercial drug design tools ([14], www.accelrys.com/ chem/, www.disat.inimib.it/chm/Dragon.htm). Concerning the classification of chemical compounds, a widely used technique to build carcinogenecity models is *inductive logic programming (ILP)*. The main idea of ILP is to induce general descriptions satisfied by a set of examples represented using logical predicates. In these approaches (for instance see [9]), compounds are represented as sets of predicates relating the atoms of the molecule and they also include information about the chemical compunds (such as molecular weight, charge, etc). Nevertheless, due to the wide variety of chemical compounds, the use of inductive learning methods for building a general model of carcinogenesis is very difficult.

In [6] we proposed the use of lazy learning methods, instead of inductive learning methods, to classify chemical compounds. The main difference among both kinds of approaches is that inductive learning methods build a model and then they use it to classify new chemical compounds. Instead, lazy approaches do not build any model, but given a new problem they try to classify it based on both its features and the similarity of that problem with already known problems. This represents an advantage because lazy learning methods are not aware of the variability of the problems but they only focus on the features of the new problem. Concerning to the toxicology domain, since chemical compounds have high variability, inductive learning methods produce models with rules that are too general. Instead, a lazy learning method only focuses on the features of the new chemical compound to assess the similarity of that compound with others compounds with known carcinogenic activity.

A weakness of lazy learning methods is the way they are able to explain the result to the user. The most common way used by case-based reasoning systems to explain the result is to show the user all the cases that support the classification of a new problem. This kind of explanation seems appropriate when domain objects are not too complicated, however when domain objects have a complicated structure the user is not able to detect similarities among the cases. McSherry [18] argues that the most similar case could be a good explanation but it also may have features that could act as arguments against that case and, therefore, against the classification that it proposes. For this reason, McSherry proposes that the explanation of a case-based reasoning system has to explicitly distinguish between the case features supporting a classification and the case features against it. In that way, the user could decide about the final solution of the problem. A related idea proposed in [17] is to use the differences among cases to support the user in understanding why some cases do not satisfy some requirements.

Our approach is based on generating an explanation scheme from the similarities among a problem and a set of cases. As the approaches of McSherry and McCarthy et al. [17], the explanation scheme of our approach is also oriented to the user. The difference of our approach with that of McSherry is that we explain the result using a set of similar cases whereas McSherry explains it using both similarities and differences among the most similar case compared to the problem at hand. An interesting part of the explanation scheme we propose is

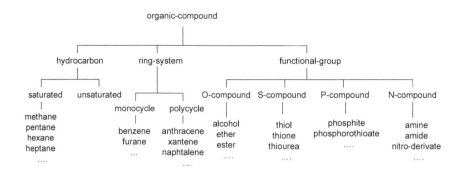

Fig. 1. Partial view of the chemical ontology

that it allows the user to focus on relevant aspects that differentiate carcinogen compounds from non carcinogen compounds.

The structure of the paper is the following. In the next section we briefly describe the formalism of feature terms, the representation we use to describe chemical compounds. Then in Section 3 we introduce LID, the lazy learning method we use to classify chemical compounds and that handles objects represented as feature terms. In Section 4 we introduce the anti-unification concept in which is based the explanation scheme described in Section 5. We end up with some related works and conclusions.

2   Representation of the Chemical Compounds Using Feature Terms

Current approaches using artificial intelligence techniques applied to chemistry use representations inherited from existing tools. These tools describe chemical compounds with a set of structure-activity relationship (SAR) descriptors because they were developed mainly for the task of drug design. In [6] we proposed the use of a representation of chemical compounds based on the *chemical ontology* given by the IUPAC nomenclature (www.chem.qmul.ac.uk/ iupac/). The IUPAC chemical nomenclature is a standard form to describe the (organic and inorganic) molecules from their chemical structure. From our point of view, a formal representation using the IUPAC nomenclature could be very useful since it allows a direct description of the chemical structure, in a way very familiar to the chemist. Our point is that, using the standard nomenclature, the name of a molecule provides enough information to graphically represent its structure. Actually, we represent a compound as a structure with substructures using the chemical ontology that is implicit in the nomenclature of the compounds. Figure 1 shows part of the chemical ontology we have used to represent the compounds in the Toxicology data set.

The implementation of our approach has been done using the *feature terms* formalism [2]. Feature terms is a kind of *relational representation*, i.e. an object

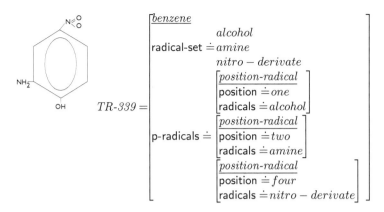

Fig. 2. Representation of TR-339, *2-amino-4-nitrophenol*, with feature terms

is described by its parts and the relationhs among these parts. The intuition behind a feature term is that it can be described as a labelled graph where nodes are objects and links are the features describing the objects. An object, as well as the values of the features of that object belong to a *sort*. A *sort* is described by a set of features, where each feature represents a relation of this sort with another sort. Sorts are related among them by partial order $\preceq$ (see 4.1) that induces a hierarchy of sorts/subsorts relating the concepts of a domain. Thus, the chemical ontology shown in Fig. 1 can be viewed as a sort/subsort hierarchy relating the chemical concepts describing the molecular structure of a chemical compound.

Let us illustrate with an exemple how chemical compounds are represented using feature terms. Figure 2 shows the molecular structure of the chemical compound TR-339, called *2-amino-4-nitrophenol*, and its representation using feature terms. Chemical compound TR-339 is represented by a feature term with *root* TR-339 of sort *benzene* described by two features: radical-set and p-radicals. The value of the feature radical-set is the set {*alcohol, amine, nitro-derivate*}. The value of the feature p-radicals is a set whose elements are of sort *position-radical*. In turn, the sort *position-radical* is described using two features: radicals and position. Values of radicals are those of the feature radical-set meaning the position where the radical is placed. TR-339 has the radical *alcohol* placed in position *one*, the radical *amine* in position *two* and the radical *nitro-derivate* in position *four*. Note that this information has been directly extracted from the chemical name of the compound following the nomenclature rules.

A *leaf* of a feature term is defined as a feature whose value is a (set of) feature term without features. For instance, leaf features of TR-339 are the following: {radical-set, position, radicals, position, radicals, position, radicals}. Notice that there is a leaf position and also a leaf radicals for each value of p-radicals.

Function LID (p, S_D, D, C)
     $S_D :=$ Discriminatory-set (D)
     if stopping-condition(S_D)
         then return $class(S_D)$
         else $f_d :=$ Select-leaf (p, S_D, C)
              $D' :=$ Add-path$(\pi(root(p), f_d), D)$
              $S_{D'} :=$ Discriminatory-set (D', S_D)
              LID $(S_{D'}, p, D', C)$
   end-if
end-function

Fig. 3. The LID algorithm. p is the problem to be solved, D is the similitude term, S_D is the discriminatory set of D, C is the set of solution classes, $class(S_D)$ is the class $C_i \in C$ to which all elements in S_D belong.

A *path* $\Pi(root, f)$ is the sequence of features leading from the root of the feature term to the feature f. Paths of TR-339 from the root to the leaves are the following:

- TR-339.radical-set with value the set $\{alcohol, amine, nitro\text{-}derivate\}$
- TR-339.p-radicals.position with value *one*
- TR-339.p-radicals.radicals with value *alcohol*
- TR-339.p-radicals.position with value *two*
- TR-339.p-radicals.radicals with value *amine*
- TR-339.p-radicals.position with value *four*
- TR-339.p-radicals.radicals with value *nitro-derivate*

3   Lazy Induction of Descriptions

Lazy Induction of Descriptions (LID) is a lazy learning method for classification tasks. LID determines which are the most relevant features of a problem and searches in a case base for cases sharing these relevant features. The problem is classified when LID finds a set of relevant features shared by a subset of cases all o them belonging to a same class. We call *similitude term* the feature term formed by these relevant features and *discriminatory set* the set of cases satisfying the similitude term. A first version of LID was introduced in [3] to assess the risk of complications in diabetic patients. In order to assess the carcinogenecity of chemical compounds, the LID algorithm has been modified to cope with some situation that, although general, they do not occur in the diabetes domain.

Given a problem p, the LID algorithm (Fig. 3) initializes D as a feature term such that $sort(D) = sort(p)$, with no features and with the discriminatory set S_D initialized to the set of cases satisfying D. For the Toxicology domain we set $C = \{positive, negative\}$.

Let D be the current similitude term, the first step is to form the set S_D with all the cases satisfying the similitude term D. When the stopping condition of LID is not satisfied, the next step is to select a leaf for specializing D.

The specialization of a similitude term D is achieved by adding features to it. Given a set F_l of features candidate to specialize D, the next step of LID is the selection of a leaf feature $f_d \in F_l$ to specialize the similitude term D [1]. Selecting the most discriminatory leaf feature in the set F_l is heuristically done using the López de Mántaras' distance (LM) [16] over the features in F_l. LM measures the distance among two partitions and LID uses it to compare each partition P_j induced by a feature f_j with the correct partition P_c. The *correct partition* has two sets: one with the examples belonging to a solution class C_i and the other containing the cases not in C_i. Each feature $f_j \in F_l$ induces in the discriminatory set a partition P_j with two sets, one with the cases where f_j takes the same value than p and the other with the rest. Given two features f_i and f_j inducing respectively partitions P_i and P_j, we say that f_i is *more discriminatory* than f_j iff $LM(P_i, P_c) < LM(P_j, P_c)$. This means that the partition P_i induced by f_i is closer to the correct partition than the partition P_j induced by f_j. LID selects the most discriminatory feature to specialize D.

Let us call f_d the most discriminatory feature in F_l. The specialization of D defines a new similitude term D' by adding to D the sequence of features appearing in the path $\Pi(root(p), f_d)$. After adding the path Π to D, the new similitude term $D' = D + \Pi$ subsumes a subset of cases in S_D, namely $S_{D'}$.

Next, LID is recursively called with the discriminatory set $S_{D'}$ and the similitude term D'. The recursive call of LID has $S_{D'}$ instead of S_D because the cases that are not subsumed by D' will not be subsumed by any further specialization. The process of specialization reduces the discriminatory set at each step, therefore we get a sequence $S_D^n \subseteq S_D^{n-1} \subseteq \ldots \subseteq S_D^0$.

LID has three possible stopping situations: 1) all the cases in the discriminatory set belong to the same solution class, 2) there is no feature allowing to specialize the similitude term, and 3) there are no cases subsumed by the similitude term.

In a previous version of LID [3] there only the stopping conditions 1) and 2) were considered. Now, in the Toxicology domain we have introduced a third stopping condition: the similitude term does not subsumes any case. Let us explain condition 3 in more detail. The similitude term is a feature term of the same sort than p, and the sort of p is any sort of the ontology of organic compounds (Fig. 1). Nevertheless, it is possible than there is no chemical compound of the same sort of p. For instance, let us suppose that cp is the compound TR-496 of sort *eicosane*, then the similitude term is a feature term of sort *eicosane*. This means that LID searches in the case base for chemical compounds of sort *eicosane* but there is not any other chemical compound of that sort, therefore $S_D = \emptyset$. In that situation, LID finishes without giving a solution for p.

When the stopping condition 1) is satisfied because all the cases in S_D belong to a same solution class C_i, then p is classified as belonging to C_i. When $S_D = \emptyset$ then LID gives no classification for p, and finally when the discriminatory set contains cases from several classes, then the *majority criteria* is applied, i.e. p is classified as belonging to the class of the majority of cases in S_D.

---

[1] In fact, the selection of a leaf implies the selection of the path from the root to the leaf.

TR-089

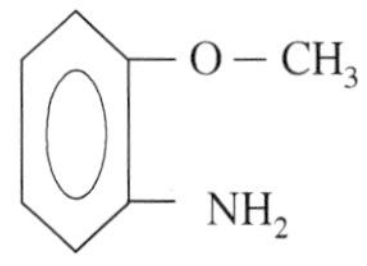

Π_1 = benzene.radical-set   (value = amine)
Π_2 = benzene.radical-set.radical-set   (value = methane)
Π_3 = benzene.radical-set.p-radicals.radicals (value = methane)
Π_4 = benzene.positions.radicals   (value = {methane, ether})
Π_5 = benzene.positions.distance   (value = 1)

Fig. 4. Molecular structure, feature term representation and paths of the chemical compound *TR-089 (resorcinol)*

3.1   Example

In this section we explain the LID algorithm by illustrating the process with the classificacion of the chemical compound *TR-089* (Fig. 4 shows the molecular structure and paths of *TR-089*). The first step is to select a relevant feature, therefore, it is necessary to induce the partitions associated to each feature of *TR-089* and then to compute the distance to the correct partition. Using the LM distance, LID takes the feature radical-set with value *methane*. In such a situation, $D_0 = \Pi_1$ (Fig. 4) and $S_{D_0} = \{$ *TR-491, TR-416, TR-414, TR-372, TR-351, TR-223, TR-171, TR-142, TR-128, TR-127, TR-124, TR-120, TR-114, TR-105, TR-084* $\}$ where some compounds are *positive* and some others are *negative* for carcinogenesis. Therefore LID has to specialize D_0 by selecting a relevant feature to be added to it. The next most relevant feature is radical-set with value *amine*. Now the discriminatory set associated to $D_1 = D_0 + \Pi_2$ is $S_{D_1} = \{$ *TR-084, TR-105, TR-127, TR-142, TR-171, TR-351, TR-372* $\}$ that still contains both *positive* and *negative* compounds. Therefore a new relevant feature has to be selected. Now the selected feature is distance with value 1. The new similitude term is $D_2 = D_1 + \Pi_5$ and the discriminatory set is $S_{D_2} = \{$ *TR-084, TR-127, TR-142, TR-171, TR-372* $\}$ where *TR-171* is the only compound with negative carcinogenecity for male rats. Because LID cannot further specialize D_2, since there are no features able to discriminate the compound *TR-171* from the others, it uses the majority criterion to classify *TR-089* as *positive*.

Notice that in the situation above, the result given by LID after the application of the majority rule seems clear, i.e. TR-089 is positive because *all* the cases assessed as the most similar (except TR-171) are positive. Nevertheless, sometimes such a majority is not so clear. for instance, in the current situation, the user could note that the molecular structures of all the compounds in S_{D_2} are very similar (see Fig. 6) so the question is: why TR-171 is *negative*? In the next section we propose an explanation scheme in order to justify to the user the classifications given by LID.

4   How Results of a Lazy Learning Method Can Be Explained?

Case-based reasoning (CBR) systems predict the solution of a problem based on the similarity between this problem (the *current case*) and already solved problems (cases). Clearly, the key point is the measure used to assess the similarity among the cases. Since the resulting similarity value is difficult to explain, CBR systems often show the retrieved cases (the set of cases that have been assessed as the most similar to the new problem) to the user as an explanation of the prediction: the solution is predicted *because* the problem was similar to the cases shown. Nevertheless, when the cases have a complex structure, simply showing the most similar cases to the user may not be enough. Our proposal, similar to that introduced in [4] is to show the user a symbolic description (the final similitude term given by LID) that makes explicit what the new problem has in common with the retrieved cases.

As we already mentioned, LID has three stopping situations for the classification process of a problem p. For the first one, when all the cases in S_D belong to a same solution class, the similitude term is a good explanation since makes explicit the relevant features shared by p and a subset of classes belonging to a class. However, when the second stopping condition holds, p shares relevant features with cases from different solution classes, therefore the similitude term, by its own, is not a good justification of the result. For this resason, we take the *explanation scheme* introduced in [4] to explain results obtained by LID using the majority rule. This scheme is based on the anti-unification concept.

4.1 The Anti-unification Concept

The explanation scheme we propose is based on the concept of *least general generalization (lgg)*, commonly used in Machine Learning. The partial order $\preceq$ among sorts mentioned in Section 2 gives an informational order among sorts since $s_1 \preceq s_2$ (s_2 is a subsort of s_1) means that s_1 provides *less* information than s_2. Using the partial order $\preceq$ we can define the *least upper bound (lub)* of two sorts $lub(s_1, s_2)$ as the most specific super-sort common to both sorts. For instance, Fig. 1 shows the sort hierarchy representing this chemical ontology. The most general sort is *organic-compound* and most specific sorts are the leafs of this hierarchy (e.g. *pentane, hexane, benzene, furane*, etc). Thus, the super-sort of any two sorts of that hierarchy (for instance *benzene* and *furane*) is always *organic-compound*. The anti-unification concept concerns to the most specific sort of two sort, therefore the *lub(benzene, furane)* (Fig. 1) is the sort *monocycle*. Similarly, *lub(benzene, xantene)=ring-system*, and *lub(methane, O-compound) = organic-compound*.

Now, we can define the *least general generalization* or *anti-unification* of a collection of descriptions represented as feature terms (either generalizations or cases) using the relation *more general than* ($\geq_g$) as follows:

- $AU(d_1, ..., d_k) = g$ such that $(g \geq_g d_1) \wedge ... \wedge (g \geq_g d_k)$ and not exists $(g' \geq_g d_1) \wedge ... \wedge (g' \geq_g d_k)$ such that $g >_g g'$

That is to say, g is the most specific generalization of all those generalizations that cover all the descriptions $d_1, ..., d_k$. $AU(d_1, ..., d_n)$ is a feature term described by all the features common to (or shared by) $d_1, ..., d_2$, i.e it describes *all* aspects in which two ore more descriptions are similar.

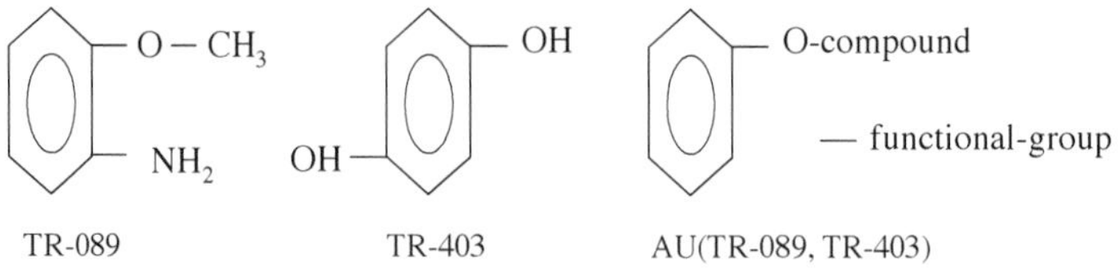

Fig. 5. Graphical representation of both the chemical compound *TR-403* and the anti-unification of *TR-089* and *TR-403*

The anti-unification of the chemical compounds *TR-089* (Fig. 4) and *TR-403* (Fig. 5) is the feature term *AU(TR-089, TR-403)*, shown in Fig. 5. *AU(C-089, C-403)* represents a chemical compound that is a *benzene* with a radical of sort *O-compound* and another radical in a non specified position. See [2] for a more detailed account on feature terms and their anti-unification. In the next section we detail the explanation scheme used to justify the classification of LID.

5   The Explanation Scheme

This section presents the way in which descriptions resulting from the anti-unification of a collection of cases can be used to explain the classification of a new problem in CBR systems. Let S_D the discriminatory set containing cases satisfying the similitude term D given by LID as a result of the classification of a problem p. There are two possible situations: 1) cases in S_D belong to only one class C_i, and 2) cases in S_D belong to several classes.

Concerning the first situation, the similitude term D is a good explanation of why the cases in S_D are similar to p, since it is a description of all that is shared among a subset of cases belonging to a some class C_i and the new problem. Let us to concentrate on the second situation.

Assuming two solution classes C_1 and C_2, let $S_D^1 \subseteq S_D$ be the set of retrieved cases hat belong to a class C_1, and $S_D^2 \subseteq S_D$ the subset of retrieved cases that belong to C_2 ($S_D = S_D^1 \cup S_D^2$). The explanation scheme we proposed in [4] is composed of three descriptions:

- AU^*: the anti-unification of p with all the cases in S_D. This description shows what aspects of the problem are shared by all the retrieved cases, i.e. the k retrieved cases are similar to p because they have in common what is described in AU^*.
- AU^1: the anti-unification of p with the cases in C^1. This description shows what has p in common with the cases in C^1.
- AU^2: the anti-unification of p with the cases in C^2. This description shows what has p in common with the cases in C^2.

This explanation scheme supports the user in the understanding of the classification of a problem p. With the explanation scheme we propose, the similarities

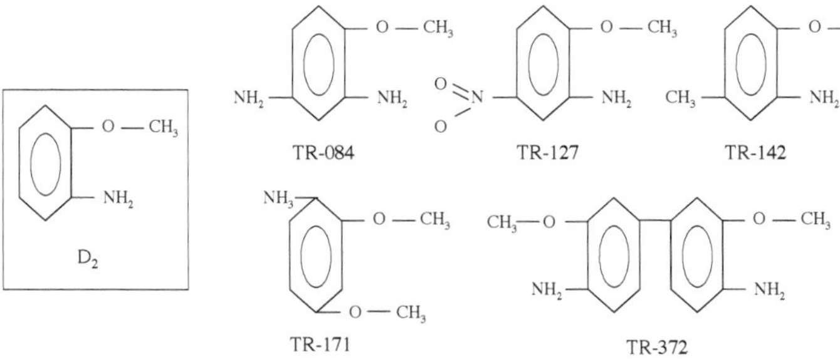

Fig. 6. Graphical representation of the similitude term D_2 and the chemical compounds contained in the discriminatory set S_{D_2} when classifying the compound *TR-089*

among p and the cases of each class are explicitly given to the user, who can decide the final classification of p. This scheme can also be used in situations where more than two classes are present in S_D, our explanation scheme is simply to build one anti-unification description for each one of them. For instance, if cases in the retrieval set belong to 4 classes the explanation scheme consists on the following symbolic descriptions: AU^1, AU^2, AU^3, and AU^4.

When the similitude term AU^* is too general (e.g. most of the features hold the most general sort as value), the meaning is that the cases have low similarity. Conversely, when AU^* is a description with some features holding some specific value, this means that the cases share something more than only the general structure. In this paper instead of AU^* we propose the use of the final similitude term D given by LID. The main difference between AU^* and D is that AU^* shows all the aspects shared by all the retrieved cases whereas D shows the *important* aspects shared by the problem and the retrieved cases, i.e. those aspects considered important to classify the problem.

AU^1 shows the commonalities among the problem p and the retrieved cases belonging to C_1. This allows the user to focus on those aspects that could be relevant to classify p as belonging to C_1. As before, the more specific AU^1 is, the more information it gives for classifying p. Notice that AU^1 could be as general as D; in fact, it is possible that both feature terms are equal. This situation means that p has not too many similar aspects with the cases of C^1. A similar situation may occur with AU^2.

In [7] an example that follows this scheme can be found. Here we illustrate the explanation scheme with the example of the classification of the chemical compound *TR-089* developed in section 3.1. This is an interesting case where the explanation scheme can support the user the search of unclear aspects of the classification of compounds. Figure 6 shows the similitude term D_2 and the discriminatory set $S_{D_2} = \{$ *TR-084, TR-127, TR-142, TR-171, TR-372* $\}$ given by LID when classifying the chemical compound *TR-089*. Concerning the carcinogenesis on male rats, S_{D_2} can be partitioned in the following two subsets:

$S^1_{D_2} = \{ TR\text{-}084,\ TR\text{-}127,\ TR\text{-}142,\ TR\text{-}372 \}$ and $S^1_{D_2} = \{ TR\text{-}171 \}$, where compounds in $S^1_{D_2}$ are *positive* and the compound in $S^2_{D_2}$ is *negative*. The explanation scheme for chemical compound *TR-089* is the following:

- The similitude term D_2 shows that *TR-089* and the compounds in S_{D_2} have in common that they all have a benzene structure with two radicals at distance 1 among them. One of these radicals is an *ether* that in turn has a radical *methane*. The other radical is an *amine*.
- The description AU^1 is the anti-unification of *TR-089* and the chemical compounds considered as *positive* for male rats. In fact, $AU^1 = D_2$, since all positive compounds share, as before, that they are benzenes with two radicals (an *ether* with a radical *methane* and an *amine.*) with distance 1 among them.
- The description AU^2 is the anti-unification of *TR-089* and *TR-171* that is the unique compound in S_{D_2}, i.e. negative for carcinogenesis. Note that also in that case, $AU^2 = D_2$

From the descriptions AU^1 and AU^2 the user can easily observe the similarities and differences among the compounds in C_1 and those in C_2. In the current example, D_2, AU^1 and AU^2 give the same feature term as explanation, which is quite specific since common radicals have specific sorts (*benzene, ether, methane, amine*), therefore the user can conclude that all the compounds are really very similar. So, the question could be why *TR-171* is negative for carcinogenesis. All compounds in S_D (included *TR-171*) are *aromatic amines* which are highly correlated with carcinogenecity [19,1], therefore, in principle *TR-171* should also be carcinogenic. Because the *TR-171* (*2,4 - Dimethoxyaniniline hydrochloride*) is an *aniline* we performed a search on Internet asking for information about experimental results on anilines. We found from the page of the *International Agency for Research on Cancer (IARC)* that there are defined four categories of chemical compounds according to their potential carcinogenic power on humans. In particular, *anilines* are classified on category 3 corresponding to chemical compounds with inadequate evidence of carcinogenecity in humans or those compounds whose experimental evidence on animals is either inadequate or limited. In fact, the NTP report of experimental results of *TR-171* on rodents (see *long term NTP Study Reports* from web page ntp.niehs.nih.gov/) states that studies began when *2,4 - Dimethoxyaniniline hydrochloride* was suspicious to be the cause of the increment of incidence of bladder cancer among dye manufacturing industry workers. Nevertheless the experimental results on rodents did not provide a convincing evidence of the carcinogenic power of the *2,4 - Dimethoxyaniniline hydrochloride*. This means that for chemical experts, *TR-171* was at first sight a potential carcinogen and despite the experimental results on rodents show no evidence of carcinogenecity, toxic activity on other species could not be discarded.

6   Related Work

Concerning to the Predictive Toxicology domain, we have proposed 1) a new approach to represent chemical compounds and 2) a lazy approach for solving the classification task. The most common representation of chemical compounds is using SAR descriptors which represent the compounds from several points of view (structural, physical properties, etc) and they are the basis to build equational models that relate the structure of a chemical compound with its physical-chemical properties. The main difference between the representations based on SAR and our ontological approach is that the former describe the molecular structure of the chemical compounds in an exhaustive way. Instead the representation we propose is more conceptual than SAR in the sense that it directly uses the concepts understood by the chemists.

Some authors use approaches that are not centered on the representation of specific atoms but on molecular structures. For instance, González et al [12] and Deshpande and Karypis [11] represent chemical compounds as labeled graphs, using graph techniques to detect the set of molecular substructures (subgraphs) more frequently occurring in the chemical compounds of the data set. Conceptually, these two approaches are related to ours in that we describe a chemical compound in terms of its radicals (i.e. substructures of the main group).

Concerning the explanation of the solution proposed by a CBR system, there are a lot of possible approaches depending on the kind of explanation we are looking for. Sørmo et al. [20] performed a deep analysis of the different perspectives from whose an explanation can be taken. Related to problem solving tasks, there are two main kinds of explanations that are specially useful: 1) an explanation of how a solution has been reached, and 2) an explanation justifying the result. In this sense, the explanation proposed in this paper justifies the solution proposed by the system. Nevertheless, because part of this explanation scheme (the similitude term) contains the important features that LID used to classify a new problem, the explanation also gives some clues of *how* the system reached the solution.

Most of explanations given by CBR systems are oriented to the user. Sørmo et al. [20], Leake [15] and Cassens [10] also consider that the form of the explanation depends on the user goals. This statement has been proved in the application presented by Bélanger and Martel [8] where the explanations for expert and novice users are completely different. Leake [15] see the process of explanation construction as a form of goal-driven learning where the goals are those facts that need to be explained and the process to achieve them gives the explanation as result. Cassens [10] uses the Activity Theory to systematically analyze how a user evolves in the utilization a system, i.e. how the user model is changing. The idea is that in using a system, the user can change his expectations about it and, in consequence, the explanation of the results would also have to change. In our approach we are considering classification tasks, therefore the user goals are always the same: to classify a new problem. This means that the explanation has to be convincing enough to justify the classification and we assume that the kind of explanation has always the same form, i.e. it does not change along the time. The explanation scheme we have introduced is also oriented to explain the result to the user. Nevertheless, these explanations could also be reused by the system as general domain knowledge as we proposed in [5].

7 Conclusions and Future Work

Lazy learning methods seem to be specially useful on domains such as toxicology, in which object domains are highly variable. Nevertheless, one of the main weakness of the lazy learning methods is how they justify the results to the user. In this paper we have proposed an explanation scheme that supports the user in comparing molecular structures of positive and negative compounds.

The application of that explanation scheme to explain the results of a lazy learning approach to predictive toxicology can be of high utility. Unlike induced leaning methods, a lazy learning method does not build a carcinogenecity model, therefore there is not a clear justification of the result. On the other hand, a chemist needs to focus on both similarities and differences among the molecular structure of chemical compounds. Using our approach, even if it focuses on similarities, the user can easily see the differences among carcinogenic and non carcinogenic compounds. Due to this fact, and because small differences on the molecular structure of compounds may give different carcinogenic activity, the user can revise literature supporting the classification given by the lazy learning method.

As future work we plan to assess the confidence degree of an explanation. This confidence could be assessed taking into account the entropy of the discriminatory set associated to a similitude term. In other words, since LID can finish with a similitude term D satisfied by cases of several classes, a discriminatory set S_D with high entropy means that D is too general, therefore the features included in D, even if considered as relevant, are not actually discriminant. This could be interpreted as a low confidence in the explanation of the classification. Conversely, a discriminatory set with low entropy means that the similitude term D is accurate, therefore the confidence on the classification would be high. This same criteria could be applied to assess the confidence of the parts AU^1 and AU^2 of the explanation scheme.

Acknowledgements. This work has been supported by the MCYT-FEDER Project MID-CBR (TIN2006-15140-C03-01). The author thanks Dr. Lluis Godo for his assistance in the improvement of this paper.

References

1. Ambs, S., Neumann, H.G.: Acute and chronic toxicity of aromatic amines studied in the isolated perfused rat liver. Toxicol. Applied Pharmacol. 139, 186–194 (1996)
2. Armengol, E., Plaza, E.: Bottom-up induction of feature terms. Machine Learning 41(1), 259–294 (2000)
3. Armengol, E., Plaza, E.: Lazy induction of descriptions for relational case-based learning. In: De Reaedt, L., Flach, P. (eds.) ECML-2001. LNCS (LNAI), vol. 2167, pp. 13–24. Springer, Heidelberg (2001)
4. Armengol, E., Plaza, E.: Relational case-based reasoning for carcinogenic activity prediction. Artificial Intelligence Review 20(1–2), 121–141 (2003)

5. Armengol, E., Plaza, E.: Remembering similitude terms in case-based reasoning. In: 3rd Int. Conf. on Machine Learning and Data Mining MLDM-03. LNCS (LNAI), vol. 2734, pp. 121–130. Springer, Heidelberg (2003)

6. Armengol, E., Plaza, E.: An ontological approach to represent molecular structure information. In: Oliveira, J.L., Maojo, V., Martín-Sánchez, F., Pereira, A.S. (eds.) ISBMDA 2005. LNCS (LNBI), vol. 3745, pp. 294–304. Springer, Heidelberg (2005)

7. Armengol, E., Plaza, E.: Using symbolic descriptions to explain similarity on case-based reasoning. In: Radeva, P., Lopez, B., Melendez, J., Vitria, J. (eds.) Artificial Intelligence Research and Develoment, IOS Press, Octubre (2005)

8. Bélanger, M., Martel, J.M.: An automated explanation approach for a decision support system based on mcda. In: Roth-Berghofer, T., Schulz, S. (eds.) Explanation-Aware Computing: Papers from the 2005 Fall Symposium. Technical Report FS-05-04, Menlo Park, California. American Association for Artificial Intelligence, pp. 21–34 (2005)

9. Blockeel, H., Driessens, K., Jacobs, N., Kosala, R., Raeymaekers, S., Ramon, J., Struyf, J., Van Laer, W., Verbaeten, S.: First order models for the predictive toxicology challenge 2001. In: Procs of the Predictive Toxicology Challenge Workshop (2001)

10. Cassens, J.: Knowing what to explain and when. In: Proceedings of the ECCBR 2004 Workshops. Technical Report 142-04. Departamento de Sistemas Informáticos y Programación, Universidad Complutense de Madrid, Madrid, Spain, pp. 97–104 (2004)

11. Deshpande, M., Karypis, G.: Automated approaches for classifying structures. In: Proc. of the 2nd Workshop on Data Mining in Bioinformatics (2002)

12. Gonzalez, J., Holder, L., Cook, D.: Application of graph-based concept learning to the predictive toxicology domain. In: Procs of the Predictive Toxicology Challenge Workshop, Freiburg, Germany (2001)

13. Helma, C., Kramer, S.: A survey of the predictive toxicology challenge 2000-2001. Bioinformatics, pp. 1179–1200 (2003)

14. Katritzky, A.R, Petrukhin, R., Yang, H., Karelson, M.: CODESSA PRO. User's Manual. University of Florida (2002)

15. Leake, D.B.: Issues in goal-driven explanation. In: Proceedings of the AAAI Spring symposium on goal-driven learning, pp. 72–79 (1994)

16. de Mántaras, R.L.: A distance-based attribute selection measure for decision tree induction. Machine Learning 6, 81–92 (1991)

17. McCarthy, K., Reilly, J., McGinty, L., Smyth, B.: Thinking positively - explanatory feedback for conversational recommender systems. In: Procs of the ECCBR 2004 Workshops. TR 142-04. Departamento de Sistemas Informáticos y Programación, Universidad Complutense de Madrid, pp. 115–124 (2004)

18. McSherry, D.: Explanation in recommendation systems. In: Procs of the ECCBR 2004 Workshops. TR 142-04. Departamento de Sistemas Informáticos y Programación, Universidad Complutense de Madrid, pp. 125–134 (2004)

19. Sorensen, R.U.: Allergenicity and toxicity of amines in foods. In: Proceedings of the IFT 2001 Annual Meeting, New Orleans, Louisiana (2001)

20. Sørmo, F., Cassens, J., Aamodt, A.: Explanation in Case-Based Reasoning – Perspectives and Goals. Artificial Intelligence Review 24(2), 109–143 (2005)

One Lead ECG Based Personal Identification with Feature Subspace Ensembles

Hugo Silva[1], Hugo Gamboa[2], and Ana Fred[3]

[1] Instituto de Telecomunicações, Lisbon, Portugal
hugo.silva@lx.it.pt
[2] Escola Superior de Tecnologia de Setúbal, Campus do IPS,
Setúbal, Portugal
hgamboa@est.ips.pt
[3] Instituto de Telecomunicações,
Instituto Superior Técnico, Lisbon, Portugal
afred@lx.it.pt

Abstract. In this paper we present results on real data, focusing on personal identification based on one lead ECG, using a reduced number of heartbeat waveforms. A wide range of features can be used to characterize the ECG signal trace with application to personal identification. We apply feature selection (FS) to the problem with the dual purpose of improving the recognition rate and reducing data dimensionality. A feature subspace ensemble method (FSE) is described which uses an association between FS and parallel classifier combination techniques to overcome some FS difficulties. With this approach, the discriminative information provided by multiple feature subspaces, determined by means of FS, contributes to the global classification system decision leading to improved classification performance. Furthermore, by considering more than one heartbeat waveform in the decision process through sequential classifier combination, higher recognition rates were obtained.

1 Introduction

Fiducial points of the electrocardiographic (ECG) signal, are typically used in clinical applications for diagnostics and evaluation of the cardiac system function [1][2][3]. These points have well characterized reference values, and deviations from those may express multiple anomalies.

The ECG provides a visualization of the electrical activity of the cardiac muscle fibres; as measured from the body surface, the ECG signal is directly related to the physiology of each individual. These measurements are influenced by physiologic factors which include: skin conductivity, genetic singularities, position, shape and size of the heart. Regardless of what factors originate differences in the measurement, the fact that the ECG contains physiologic dependant singularities potentiates its application to personal identification.

Recent research work has been devoted to the characterization of ECG features, unique to an individual, with clear evidence that accurate ECG based

P. Perner (Ed.): MLDM 2007, LNAI 4571, pp. 770–783, 2007.

personal identification is possible [4][5][6]. As a behavioral biometric technique the ECG is very appealing: *it is a non-invasive technique*; *it is not easily replicated or circumvented*; and *it requires the subject to be physiologically active*, among other characteristics.

A wide range of features can be used to characterize the ECG signal trace with application to personal identification [1][7][8][9][3], and a question arises: *for a given feature set, which features are truly relevant for the decision process, and which can be discarded.* The reasons why addressing this question is of paramount importance include: (a) *the curse of dimensionality problem* [10]; and (b) *the fact that some features may misguide the decision process* [11][12].

In pattern recognition, this can be addressed through *feature selection* (FS). Considering a d-dimensional feature representation space (FRS), $F=\{f_1, \cdots, f_d\}$, feature selection consists of determining which subspace $F^* \subset F$, if any, contains the features $f_j \in F$ with most relevant discriminative information [13]. For this purpose, a variety of methods has been proposed [14][15][16].

This paper presents results on real data, for the application of one lead ECG data to personal identification. Previous approaches to the problem [4][5][6], also using real data, have shown the potential of ECG data for subject identification through contingency matrix analysis. In our approach, we study the potential of subject identification using a reduced number of heartbeat waveforms, with the purpose of real-time analysis. We focus on studying the classification performance provided on one hand by a single heartbeat waveform, and on the other hand by multiple heartbeat waveforms. FS and classifier combination techniques are applied to the problem to improve the recognition rates, with positive results when compared to the cases where no FS is performed.

An overview of our *feature subspace ensemble* (FSE) approach is presented: a parallel classifier combination method, in which a global decision is produced by combination of the individual decisions of multiple classifiers, designed using subspaces of the original feature representation space F, obtained by means of FS [17]. Each considered feature subspace contributes to the global decision as a result of the classifier combination process. This allows us to overcome one of the difficulties associated with FS: *retrieval of relevant discriminative information contained in discarded features.* FSE was applied to the problem, and proved to be more effective than a single classifier trained on a single FRS, both for the cases where the original space F, and FS determined subspaces were used.

We evaluate the recognition rate of a single heartbeat waveform for different sizes of the training and validation data, in order to determine the minimum number of patterns necessary to achieve maximum recognition rates. With the same purpose, sequential classifier combination is also employed, to determine how the recognition rate evolves by using a reduced number of heartbeat waveforms for personal identification instead of a single one.

The rest of the paper is organized as follows: Section 2 describes the feature subspace ensemble parallel classifier combination approach. Section 3 details our one lead ECG based personal identification setup and evaluation conditions.

Section 4 presents results for the one lead ECG based personal identification problem. Finally, section 5 summarizes results and presents the main conclusions.

2   Feature Subspace Ensembles

Feature selection is an important tool in classification system design. The classification process is essentially a mapping $F \rightarrow W$, of the original FRS, F, into a set $W = \{w_1, \cdots, w_c\}$ of c categories. FS consists on determining a subspace $F^* \subset F$, containing only the features $f_j \in F$ with the most relevant discriminative information, with the threefold aim of: (a) *improving the discriminative capacitive*; (b) *reducing computational demands*; and (c) *removing redundant or superfluous information* [13]. For this purpose, numerous methods and frameworks have been suggested [18][19][20][14]. In this section, we overview FS and some of the difficulties arising from its usage, and describe a feature subspace ensemble (FSE) method, designed to overcome some of those problems.

Typically, FS methods fall into one of three generic classes: *filter methods*, which are based on the discriminative information provided by individual or groups of features from the original FRS; *wrapper methods*, which are based on the performance of a learning machine; and *embedded methods*, in which the feature subspaces are a consequence of the classifier training process. In general, FS methods are based on the optimization of a feature subspace evaluation criteria, which measures the relevance of F^* in terms of discriminative potential, and usually only suboptimal solutions are guaranteed.

Let $S(A, J, X)$ denote a *feature selection context* (FSC), defined as the FS parameters comprehended by the feature selection algorithm A, the feature subspace evaluation criteria J, and the training data X, through which a given F^* is determined.

As a result of FS, some features from the original FRS are discarded during the process and not incorporated in F^*. Although interesting results are achieved through FS [21][14][15], some difficulties often arise: (a) *solution overfitting to a particular feature selection context* (FSC); (b) *suboptimality of the obtained solutions*; (c) *solution diversity with respect to the FSC*; and (d) *loss of relevant discriminative information contained in features $f_j \in F \setminus F^*$*.

Thus, we devised a more effective method which uses parallel classifier combination rules [12][22], to combine the decisions of multiple, individual classifiers C_r; each designed using its own subspace $F_r^* \subset F$, obtained by means of feature selection in different FSCs. A related approach proposed in [23], uses the combined decision of classifiers constructed on sequentially selected features sets, forcing the full coverage of the original FRS, F.

Let $\mathcal{S} = \{S_1, \cdots, S_p\}$ be a set of p features selection contexts, differing in any combination of the parameters A_r, J_r, or X_r, $(0 < r \leq p)$. In our *feature subspace ensemble* (FSE) approach [17], a set of p feature subspaces $\mathcal{F}^* = \{F_1^*, \cdots, F_p^*\}$ is determined using each FSC, $S_r \in \mathcal{S}$ (thus the term feature subspace ensemble). Using each feature subspace $F_r^* \in \mathcal{F}$, a classifier C_r is designed, forming a set $\mathcal{C} = \{C_1, \cdots, C_p\}$ of p classifiers. For the classification of a given pattern x_i, each individual classifier $C_r \in \mathcal{C}$ produces a decisions $\hat{w}_{C_r}$, and in the end all decisions are combined by a classifier combination strategy [12][24][22][25][26], in order to produce a global decision $\hat{w}_{x_i}$. Figure 1 illustrates the described approach.

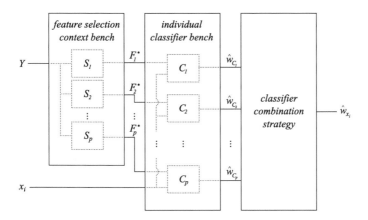

Fig. 1. Feature subspace ensemble (FSE) system. A set $\mathcal{C} = \{C_1, \cdots, C_p\}$ of p classifiers is trained using individual feature subspaces F_r^* obtained for some variation S_r of the FSC. Each classifier $C_r \in \mathcal{C}$ produces an individual decision $\hat{w}_{C_r}$. All individual decisions are combined using a classifier combination strategy to produce a global decision $\hat{w}_{x_i}$.

Figure 2 condenses the results of 50 FS runs on the SAT benchmark data from the UCI machine learning repository [27]; in a given run r, the feature selection context S_r, $(0 < r \leq 50)$, is composed by fixed A_r and J_r (that is, the same type in all runs), and randomly selecting 50% of the available patterns in each run to create X_r. A_r is a sequential forward search (SFS) wrapper framework (later described in section 3.3); J_r is the classification performance of a 1-NN decision rule using X_r as training data to classify the remaining 50% of the available patterns (used as validation set).

In the context of figure 2, in a FRS of dimension $d = 36$ features, the mean feature subspace size was approximately 23 features; an horizontal line indicates the histogram mean. As shown, only a few features are consistently selected in most feature subspaces over all runs, and there is full coverage of the original FRS. This means that there is a great diversity of subspaces with relevant discriminative information, and in a single FS run some of the discarded features may still contain useful information.

Through parallel classifier combination we incorporate in the global decision relevant discriminative information contained in each particular feature subspace, eventually recovering relevant features discarded as a result of a single FS run (e.g., due to a particularly inadequate or misleading FSC). This way, the classification system becomes less sensitive to misleading feature subspaces; the combined decisions of individual classifiers is capable of overcoming inaccurate

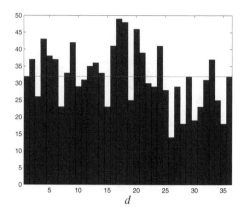

Fig. 2. Illustrative histogram of 50 FS runs on the SAT benchmark data from the UCI machine learning repository. The FSC context S_r of a given run r was established as follows: A_r and J_r are fixed for all runs, A_r being a wrapper sequential forward search framework, and J_r being the classification performance of a 1-NN decision rule trained on X_r to classify a validation set; X_r is randomly selected from the available set of patterns in each run. The horizontal axis corresponds to each of the dimensions of the FRS; the vertical axis corresponds to the number of times a given dimension d was selected. An horizontal line indicates the histogram mean.

decisions resulting from low quality feature subspaces, provided that a sufficient number of feature subspaces exists, that leads to accurate decisions.

In section 3.3 we present a FSE implementation, which we have applied to the ECG based personal identification problem. Comparative results show that, a single classifier designed using a single feature subspace obtained by means of FS, outperforms the case where the original feature representation space F is used, that is, when no feature selection is performed. With feature subspace ensembles further improvements were obtained, outperforming the classification performance of both cases.

3   One Lead ECG Based Personal Identification

3.1   Data Acquisition

Unlike previous work, where ECG recordings were performed at rest [28][6], and in stress potentiating tasks [4], we present preliminary results on real data acquired during a cognitive activity. Twenty six subjects, 18 males and 8 females, between the ages of 18 and 31 years, willingly participated in individual sessions (one per subject), during the course of which their ECG signal was recorded.

In each individual session the subject was asked to complete a concentration task on a computer, designed for an average completion time of 10 minutes. The subject interacted with the computer in a sitting position, using only the mouse as input device. No posture or motion restrictions during the activity were imposed, however, the ECG acquisition was part of a wider multi-modal physiological signal acquisition experiment; therefore due to the placement of other measurement apparatus in the subjects passive hand[1] it was suggested to the subject to reduce the movements of the passive hand to the indispensable minimum.

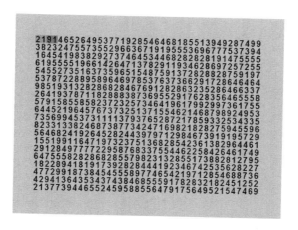

Fig. 3. Illustration of one grid of digits from the concentration task that each subject was asked to complete, and during which the ECG signal acquisition was performed

The task consisted on the presentation of two grids with 800 digits, similar to the one illustrated in figure 3, with the goal of identifying every consecutive pair of digits that added 10. Each grid was traversed in a line wise manner, from the top left to right bottom corner. The task was designed to induce saturation, having the following constraints: in order to be able to move from a current line to the next, the current line would have to be fully traversed; once a new line was moved into, the previously traversed ones could not be accessed. An horizontal bar and a cursor followed the mouse movement along the horizontal axis; the horizontal bar informed the subject of the point until which the current line had been traversed, and the cursor highlighted the pair of consecutive numbers over which the mouse was hovering at a given point. Whenever the user identified a consecutive pair of numbers matching the goal and highlighted by the cursor, he would mark it with a mouse click, and although it was not possible to return to previously visited lines, within the same line the markings could be revised.

A one lead surface mount ECG placement on the V_2 precordial derivation [1][3] was used. Facing the subject, the V_2 derivation is located on the fourth intercostal space over the mid clavicular line, at the right of the sternum. Prior to sensor placement, the area was prepared with abrasive gel and conductive paste was used on the electrodes to improve conductivity.

---

[1] we define active hand as the one used to control the input device; passive hand as the free hand.

3.2   Signal Processing and Feature Extraction

The acquired ECG signals were band-pass filtered in the passing band $2-30Hz$ with a zero-phase forward and reverse scheme [29], to remove high frequency powerline noise and low frequency baseline wander artifacts from the signal.

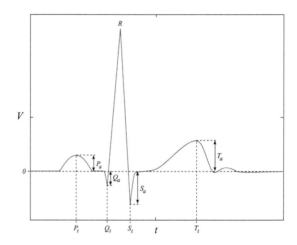

Fig. 4. Measured features from the ECG heartbeat waveform

Each heartbeat waveform was sequentially segmented from the full recording, and after this all waveforms were aligned by their R peaks. From the resulting collection of ECG heartbeat waveforms, the mean wave for groups of 10 heartbeat waveforms (without overlapping), was computed to minimize the effect of outliers. A labeled database was compiled, in which each pattern corresponds to a mean wave.

For each mean waveform, 8 latency and amplitude features were extracted, along with a sub-sampling of the waveform itself. This resulted in a feature representation space F of dimension $d = 53$, with 4 latency features, 4 amplitude features measured at selected points (figure 4), and 45 amplitude values measured at the sub-sampled points. No time limit was imposed to complete the task, and therefore the heartbeat wave form collection of each subject in the database was truncated at approximately 6 minutes[2] to ensure uniform class distribution.

3.3   Feature Selection and Classification

The ECG mean wave database is used for evaluation purposes; 50 data selection runs were performed, where in each run r three mutually exclusive sets X_r, Y_r and Z_r, of randomly selected patterns from the full recording are created. Also a feature subspace F_r^* is determined using the individual feature subspace

---

[2] Corresponding to the fastest completion time over all subjects.

selection framework described next. As a result, 50 feature subspaces will be available as a result of the performed data selection runs. X_r is created with 22.5% of the available patterns and used as training set both for FS and classifier design; Y_r is created with another 22.5% of the available patterns and used as validation set in FS; and the remaining 55% of the available patterns were used to create Z_r, which served as testing set for classification performance assessment.

For our experiments we have employed a wrapper FS framework [16], with a heuristic sequential forward search (SFS) method [30]. SFS is a state space search method, which starts from an initial state $F_{t=0}^* = \emptyset$ and iteratively evolves by constructing at each step all possible super-spaces $F_{t+1} = F_t^* \cup \{f_j \in F \setminus F_t^*\}$, adding each of the features $f_j \in F \setminus F_t^*$ to the optimal subspace F_t^* obtained at the previous step. J is used to evaluate each of the resulting super-spaces F_{t+1}, and F_{t+1}^* is selected as the set which optimizes J. If $J(F_{t+1}^*) < J(F_t^*)$ the search is stopped[3], and $F_r^* = F_t^*$ is considered to be the feature subspace with most relevant discriminative information for a given FSC r. Although conceptually simple, wrapper SFS feature selection has proven to hold comparable results in benchmark data when compared to other (more complex) methods [31][32].

The feature subspace evaluation criteria J in wrapper methods is the optimization of the classification performance of a learning machine. In our implementation, J is trained with X_r, and the recognition error over Y_r is used for feature subspace evaluation; therefore F_r^* is determined as the feature subspace that provides higher recognition rate over the validation set Y_r. Using all feature subspaces computed through SFS during the 50 data selection runs, a feature subspace ensemble $\mathcal{F} = \{F_1^*, \cdots, F_{50}^*\}$ was created, and used for classification performance evaluation of the FSE method.

For classification, we use the k-NN decision rule with an Euclidean neighborhood metric [12]. A 1-NN neighborhood was adopted, since it is a particular case of the k-NN rule where $\hat{w}_{x_i}$ for a given pattern x_i is assigned as the category of the closest pattern from the training set X_r. The same type of classifier is used for feature subspace evaluation criteria J, and for classification performance assessment.

Two types of classification performance analysis were performed. On one hand, we evaluated the recognition rate of a single heartbeat waveform for different sizes of the training and validation data, in order to determine the minimum number of patterns necessary to achieve maximum recognition rates. On the other hand, to determine how the recognition rate evolves by using more than one heartbeat waveforms in personal identification instead of a single one, we evaluated the classification performance achieved by combination of the individual decisions of a reduced set of heartbeat waveforms.

Therefore, additionally to the FSE parallel classifier combination method, sequential classifier combination was also employed. A simple majority voting strategy was adopted as classifier combination rule in both cases [33][24][34].

---

[3] $J(F)$ denotes the usage of J in the evaluation of a given feature subspace F.

4   Results

In this section we present results for the one lead ECG based personal identification. We evaluate the classification performance of a single classifier designed using a single feature subspace both for the cases where no feature selection is performed, and for FS selected feature subspaces. Our feature subspace ensemble method, described in section 2 is also applied to the problem.

Figure 5 illustrates the evolution of the mean recognition error of a single heartbeat waveform (figure 5(a)), and feature subspace size (figure 5(b)), computed over 50 runs according to the methodology described in section 3.3. To determine the minimum number n of patterns necessary to achieve the maximum recognition rate, we experimented training and validation sets (X_r, and Y_r respectively) of different sizes, ranging from a single training and validation pattern $n = 1$ (1 mean heartbeat waveform), to the full set of $n = 9$ patterns (which as described in section 3.3, corresponds to 22.5% of the available patterns in each run). As we can observe the error rate is fairly similar with (curve $iSFS$) and without SFS feature selection (curve $iALL$), although feature selection leads to more compact feature spaces, as illustrated in figure 5(b). An improved recognition rate was achieved with the application of FSE to the problem (curve $eSFS$).

We can observe that even using a single pattern per subject in the training and validation sets, the average recognition error rate is approximately 19.65% using all features, and 19.66% with SFS selected feature subspaces. In this case, the feature subspace ensemble method reduced the recognition error rate to approximately 11.86%. By increasing the number n of patterns in the training and validation sets, the recognition error rate is highly decreased. The minimal values are reached when the whole set of training and validation data is used, with a recognition error rate of 2.80% using all features and 2.58% with SFS selected feature subspaces. In this case, the FSE method further improved the average recognition error rate to 1.91%.

Figure 6 illustrates the feature histogram for the SFS selected feature subspaces over all runs, when the full set of training and validation patterns is used. The mean subspace size is 19.62 features; as we can observe, there is a high feature subspace diversity, and there are several relevant features that not all FSCs lead to. This explains why feature subspace ensembles consistently improved the recognition error rate. We can also observe the presence of irrelevant features, which FS discards or are rarely selected. From figure 5(a) we can see that these, although irrelevant are not misleading the classifier designed using the original feature representation space F, since the recognition error rate is only marginally superior to the results obtained for the classifier design using a single SFS selected feature subspace.

With FSE, a single mean heartbeat waveform, which in our case corresponds to approximately 7 seconds of signal acquisition[4] (since each pattern corresponds

---

[4] this calculation was performed taking as a reference an average normal resting heart rate of 70 beats per minute [3].

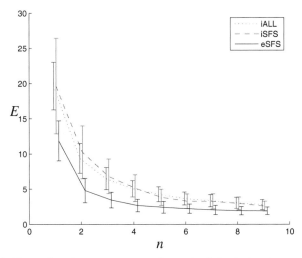

(a) Mean classification error and standard deviation intervals

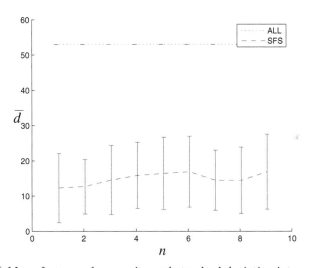

(b) Mean feature subspace size and standard deviation intervals

Fig. 5. Mean recognition error of a single ECG heartbeat waveform (figure 5(a)), and feature subspace size (figure 5(b)). n: number of patterns used for the training and validation sets (X_r and Y_r); E: mean classification error; $\bar{d}$: mean subspace size; *all*: no feature selection; *sfs*: wrapper sequential forward search; the i prefix denotes the curves for individual classifier and subspace cases, and the e prefix denotes the curves for the feature subspace ensemble method.

to the mean wave of a group of 10 heartbeat waveforms), provides 98.09% recognition accuracy, using a training set of 9 patterns (that is, 63 seconds).

Maintaining the methodology described in section 3.3, we also evaluated the recognition rate of personal identification using more than one heartbeat

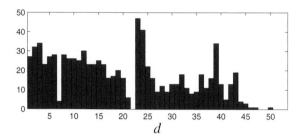

Fig. 6. Histogram for the SFS selected feature subspaces over all runs when the full set of training and validation patterns is used. The horizontal axis corresponds to each dimension of the FRS; the vertical axis corresponds to the number of times a given dimension d was selected. The horizontal line indicates the histogram mean.

waveform. The classification performance obtained for reduced sets of $h = 3, \cdots, 8$ heartbeat waveforms was evaluated, and sequential classifier combination through majority voting was used as classifier combination strategy. Figure 7 illustrates these results. It is important to recall that the FS step was performed to optimize the recognition rate of a single heartbeat waveform (as described in section 3.3). Nonetheless, as we can observe, considering a reduced set of heartbeat waveforms greatly improves the recognition accuracy. The highest recognition rate (99.97%), was obtained by majority voting the individual FSE

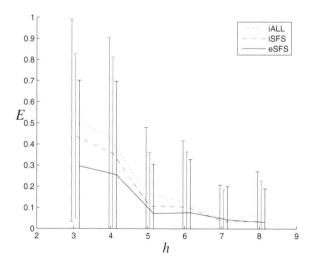

Fig. 7. Mean recognition error and standard deviation intervals for subject identification from sets with a reduced number h of ECG heartbeat waveforms. h: size of the set of heartbeats waveforms; E: mean classification error; all: no feature selection; sfs: wrapper sequential forward search; the i prefix denotes the curves for individual classifier and subspace cases, and the e prefix denotes the curves for the feature subspace ensemble method.

decisions for a group of 8 heartbeat waveforms (the equivalent to 56 seconds of signal acquisition according to the adopted methodology).

5   Conclusions

In this paper we addressed a real data problem of ECG based personal identification using a single, and reduced sets of heartbeat waveforms described in a feature representation space of dimension $d = 53$ measured features. We evaluated the classification performance of a single classifier using the original FRS, and recurred to feature selection to improve the recognition rate and reduce data dimensionality.

We introduced the concept of feature selection context (FSC): the conditions under which a given feature subspace is obtained; and described the generic feature subspace ensemble (FSE) approach: a parallel classifier combination method which uses an association between FS and classifier combination techniques [17]. FSE was designed to overcome some of the difficulties resulting from FS, namely: *FSC overfitting*; *suboptimality of FS methods*; and *recovery of relevant discriminative information contained in features discarded by FS*.

An instantiation of the FSE method using a wrapper heuristic sequential forward search (SFS) framework, 1-NN classifier and the majority voting classifier combination rule, was applied to the ECG based personal identification problem providing higher recognition rates than the single classifier designed using a single FRS cases (both the original FRS, and FS selected through feature subspaces).

Preliminary results have shown that the ECG can be used to identify individuals, particularly useful as a behavioral biometric technique. High recognition rates were achieved using a single heartbeat waveform, and we were able to further improve the results by using sequential classifier combination techniques to combine the individual decisions of a reduced set of heartbeat waveforms. It is important to enhance that, in each evaluation run, a random selection of the patterns was performed from the full recording. This indicates robustness of the ECG signal, since the task during which the signal was acquired was designed to induce saturation.

Through FSE, using a set of 9 training patterns we achieved a personal identification rate of 98.09% from a single heartbeat waveform pattern (which according to the adopted methodology corresponds to 7 seconds of signal acquisition). Using sequential classifier combination in conjunction with FSE, combining the individual decisions from FSE over a reduced set of heartbeat waveforms to produce a global decision, further improved the recognition rates. We were able to achieve a 99.97% subject recognition rate by combining the individual decisions of 8 heartbeat waveforms, which according to the adopted methodology corresponds to 56 seconds of signal acquisition.

FS targets dimensionality reduction and better discriminative ability, by selecting from the original FRS only the features with relevant discriminative information for a given FSC. Classifier combination strategies target the decision refinement, by taking into account multiple individual decisions in order to provide for a global decision. FSE has the potential to combine the advantages of both FS and classifier combination, since through FS reduced dimensionality is achieved; and through classifier combination, the classification system becomes less sensitive to misleading feature subspaces due to particularly inadequate FSCs.

Ongoing and future work includes further validation of the obtained results by including longer databases and a higher number of recordings per individual.

Acknowledgments

This work was partially supported by the Portuguese Foundation for Science and Technology (FCT), Portuguese Ministry of Science and Technology, and FEDER, under grant POSI/EEA-SRI/61924/2004, and by the Institute for Systems and Technologies of Information, Control and Communication (INSTICC) and the Instituto de Telecomunicações (IT), Pólo de Lisboa, under grant P260.

References

1. Chung, E.: Pocketguide to ECG Diagnosis. Blackwell Publishing Professional, Malden (December 2000)
2. Dubin, D.: Rapid Interpretation of EKG's, 6th edn. Cover Publishing Company, Cover (2000)
3. Lipman, B.: Ecg Assessment and Interpretation. F.A. Davis (February 1994)
4. Israel, S., Irvine, J., Cheng, A., Wiederhold, M., Wiederhold, B.: Ecg to identify individuals. Pattern Recognition 38(1), 133–142 (2005)
5. Shen, T., Tompkins, W., Hu, Y.: One-lead ecg for identity verification. In: Proceedings of the Second Joint EMBS/BMES Conference, pp. 62–63 (2002)
6. Biel, L., Petterson, O., Phillipson, L., Wide, P.: Ecg analysis: A new approach in human identification. IEEE Transactions on Instrumentation and Measurement 50(3), 808–812 (2001)
7. Liang, H.: Ecg feature elements identification for cardiologist expert diagnosis. In: ICPR 2004. Proceedings of the 27th Annual International Conference of the IEEE Engineering in Medicine and Biology Society, vol. 1-7, pp. 3845–3848. IEEE Computer Society Press, Los Alamitos (2005)
8. Kunzmann, U., von Wagner, G., Schochlin, J.A.B.: Parameter extraction of ecg signals in real-time. Biomedizinische Technik. Biomedical engineering 47, 875–878 (2002)
9. Duskalov, I., Dotsinsky, I., Christov, I.: Developments in ecg acquisition, preprocessing, parameter measurement, and recording. IEEE Engineering in Medicine and Biology Magazine 17(2), 50–58 (1998)
10. Bellman, R.: Adaptive Control Processes. Princeton University Press, Princeton (1961)
11. Heijden, V., Duin, R., Ridder, D., Tax, D.: Classification, parameter estimation and state estimation - an engineering approach using MATLAB. John Wiley & Sons, Chichester (2004)
12. Duda, R., Hart, P., Stork, D.: Pattern classification, 2nd edn. John Wiley & Sons, Chichester (2001)
13. Guyon, I., Elisseeff, A.: An introduction to variable and feature selection. Journal of Machine Learning Research 3, 1157–1182 (2003)

14. Kudo, M., Sklansky, J.: Comparison of algorithms that select features for pattern classifiers. Pattern Recognition 33, 25–41 (2000)
15. Jain, A., Zongker, D.: Feature selection: Evaluation, application, and small sample performance. IEEE Transactions on Pattern Analysis and Machine Intelligence 19(2), 153–158 (1997)
16. Kohavi, R., John, G.: Wrappers for feature subset selection. Artificial Intelligence 97(1-2), 273–324 (1997)
17. Silva, H., Fred, A.: Feature subspace ensembles: A parallel classifier combination scheme using feature selection. In: MCS 2007. LNCS, vol. 4472, Springer, Heidelberg (to appear, 2007)
18. Forman, G.: A pitfall and solution in multi-class feature selection for text classification. In: ICML '04. Proceedings of the 21st International Conference on Machine Learning, pp. 38–46. ACM Press, New York (2004)
19. Molina, L., Belanche, L., Nebot, A.: Feature selection algorithms: A survey and experimental evaluation. lsi technical report lsi-02-62-r (2002)
20. Kittler, J., Pudil, P., Somol, P.: Advances in statistical feature selection. In: Singh, S., Murshed, N., Kropatsch, W.G. (eds.) ICAPR 2001. LNCS, vol. 2013, pp. 425–434. Springer, Heidelberg (2001)
21. Molina, L., Belanche, L., Nebot, A.: Feature selection algorithms: a survey and experimental evaluation. In: ICDM 2002. Proceedings. IEEE International Conference on Data Mining, pp. 306–313. IEEE Computer Society Press, Los Alamitos (2002)
22. Duin, R., Tax, D.: Experiments with classifier combining rules. In: Kittler, J., Roli, F. (eds.) MCS 2000. LNCS, vol. 1857, Springer, Heidelberg (2000)
23. Skurichina, M., Duin, R.: Combining feature subsets in feature selection. In: Oza, N.C., Polikar, R., Kittler, J., Roli, F. (eds.) MCS 2005. LNCS, vol. 3541, pp. 165–175. Springer, Heidelberg (2005)
24. Fred, A.: Finding consistent clusters in data partitions. In: Kittler, J., Roli, F. (eds.) MCS 2001. LNCS, vol. 2096, pp. 309–318. Springer, Heidelberg (2001)
25. Lam, L.: Classifier combinations: Implementation and theoretical issues. In: Kittler, J., Roli, F. (eds.) MCS 2000. LNCS, vol. 1857, pp. 78–86. Springer, Heidelberg (2000)
26. Kittler, J., Hatef, M., Duin, R., Matas, J.: On combining classifiers. IEEE Transactions on Pattern Analysis and Machine Intelligence 20, 226–239 (1998)
27. Newman, D., Hettich, D., Blake, C., Merz, C.: UCI repository of machine learning databases (1998)
28. Shen, T., Tompkins, W.: Biometric statistical study of one-lead ecg features and body mass index (bmi). In: Proceedings of 27th Annual International Conference of the IEEE Engineering in Medicine and Biology Society, pp. 1162–1165. IEEE Computer Society Press, Los Alamitos (2005)
29. Gustafsson, F.: Determining the initial states in forward-backward filtering. IEEE Transactions on Signal Processing 44(4), 988–992 (1996)
30. Russell, S., Norvig, P.: Artificial Intelligence: A Modern Approach, 2nd edn. Prentice-Hall, Englewood Cliffs (2002)
31. Silva, H.: Feature selection in pattern recognition systems. Master's thesis, Universidade Técnica de Lisboa, Instituto Superior Técnico (2007)
32. Reunanen, J.: Overfitting in making comparisons between variable selection methods. Journal of Machine Learning Research 3, 1371–1382 (2003)
33. Tax, D., Duin, R.: Using two-class classifiers for multiclass classification. In: International Conference on Pattern Recognition, Quebec, Canada (2002)
34. Lam, L., Suen, S.: Application of majority voting to pattern recognition: An analysis of its behavior and performance. IEEE Transactions on Systems, Man, and Cybernetics 27, 553–568 (1997)

Classification of Breast Masses in Mammogram Images Using Ripley's K Function and Support Vector Machine

Leonardo de Oliveira Martins[1], Erick Corrêa da Silva[1],
Aristófanes Corrêa Silva[1], Anselmo Cardoso de Paiva[2], and Marcelo Gattass[3]

[1] Federal University of Maranhão - UFMA, Department of Electrical Engineering
Av. dos Portugueses, SN, Campus do Bacanga, Bacanga
65085-580, São Luís, MA, Brazil
leomartins82@gmail.com, ari@dee.ufma.br, erick_correa@hotmail.com
[2] Federal University of Maranhão - UFMA, Department of Computer Science
Av. dos Portugueses, SN, Campus do Bacanga, Bacanga
65085-580, São Luís, MA, Brazil
paiva@deinf.ufma.br
[3] Pontifical Catholic University of Rio de Janeiro,
Technical Scientific Center, Departament of Informatics
Rua Marquês de São Vicente, 225, Gávea
22453-900 - Rio de Janeiro, RJ - Brasil
mgattass@inf.puc-rio.br

Abstract. Female breast cancer is a major cause of death in western countries. Several computer techniques have been developed to aid radiologists to improve their performance in the detection and diagnosis of breast abnormalities. In Point Pattern Analysis, there is a statistic known as Ripley's K function that is frequently applied to Spatial Analysis in Ecology, like mapping specimens of plants. This paper proposes a new way in applying Ripley's K function to classify breast masses from mammogram images. The features of each nodule image are obtained through the calculate of that function. Then, the samples gotten are classified through a Support Vector Machine (SVM) as benign or malignant masses. SVM is a machine-learning method, based on the principle of structural risk minimization, which performs well when applied to data outside the training set. The best result achieved was 94.94% of accuracy, 92.86% of sensitvity and 93.33% of specificity.

Keywords: Mammogram, Breast Cancer Diagnosis, Ripley's K Function, Texture Analysis, SVM.

1 Introduction

Breast cancer is the most common form of cancer among women in Western countries, and a major cause of death by cancer in the female population. It is well know that the best prevention method is the precocious diagnosis, what lessens the mortality and enhance the treatment [1]. American National Cancer

P. Perner (Ed.): MLDM 2007, LNAI 4571, pp. 784–794, 2007.
© Springer-Verlag Berlin Heidelberg 2007

Institute [2] estimates that every three minutes, a woman is diagnosed with breast cancer, and every 13 minutes, a woman dies from the disease.

Mammography is currently the best technique for reliable detection of early, non-palpable, potentially curable breast cancer [1]. In 1995, the mortality rate from this disease decreased for the first time, due in part to the increasing use of screening mammography [1]. However, the image interpretation is a repetitive task that requires much attention to minute detail, and radiologists vary in their interpretation of mammograms.

Digital mammography represents an enormous advance in detection and diagnosis of breast abnormalities. Through image processing techniques, it is possible to enhance the contrast, color, and sharpness of a digital mammogram. Thus, several possible breast abnormalities may become visible for human beings.

Therefore, in the past decade there has been tremendous interest in the use of image processing and analysis techniques for Computer Aided Detection (CAD)/ Diagnostics (CADx) in digital mammograms. The goal has been to increase diagnostic accuracy as well as the reproducibility of mammographic interpretation. CAD/CADx systems can aid radiologists by providing a second opinion and may be used in the first stage of examination in the near future, allowing to reduce the variability among radiologists in the mammograms interpretation.

Automatic detection and diagnostic of breast lesions has been a highly challenging research area. In [3], a neural-genetic algorithm for feature selection in conjunction with neural and statistical classifiers has obtained a classification rate of 85.0% for testing set. A computer aided neural network classification of suspicious regions on digitized mammograms is also presented in [4]. They use a Radial Basis Function Neural Network (RBFNN) to accomplish the classification, fed by features selected through independent component analysis. That experiments presented a recognition accuracy of 88.23% in the detection of all kinds of abnormalities and 79.31% in the task of distinguishing between benign and malignant regions, outperforming in both cases standard textural features, widely used for cancer detection in mammograms. In [5], the authors proposed a method for discrimination and classification of mammograms with benign, malignant and normal tissues using independent component analysis and multilayer neural networks. The best performance was obtained with probabilistic neural networks, resulting in 97.3% success rate, 100% of specificity and 96% of sensitivity.

Traditionally, texture analysis is accomplished with classical image processing measures, like histogram, Spatial Gray Level Dependence Method, Gray Level Difference Method and others. Ripley's K function is frequently applied to Spatial Analysis in Ecology, like mapping plants specimens. In this paper we intend to investigate the effectivity of a classification methodology that uses Ripley's K function to calculate input measures for a Support Vector Machine, with the purpose to classify masses in mammographic images into two types, benign or malignant.

The main contribution of this work is the application of Ripley's K function to breast nodule characterization, using a Support Vector Machine. For the best of our knowledge, there is no work published applying the Ripley's K function for characterization of textures on medical images, even though this is a classical method in the ecology area.

This work is organized as follows. In Section 2, we present the techniques for feature extraction, and mass diagnosis. Next, in Section 3, the results are shown and we discuss about the application of the techniques under study. Finally, Section 4 presents some concluding remarks.

2   Material and Methods

The methodology proposed in this work to determine breast masses nature is based on three steps. The first one is the image acquisition, that is done obtaining mammograms and selecting manually regions that corresponds to benign and malignant masses.

The second step is the application of some measure function to the mass as a way to define some of its determinant aspects. In this work, we propose the use of Ripley's K function to do this characterization.

The last step is the selection of the most important measures from texture characterization to classify the samples gotten and their use for a SVM based classification into two classes, benign or malignant.

2.1   Image Acquisition

For the development and evaluation of the proposed methodology, we used a public available database of digitized screen-film mammograms: the Digital Database for Screening Mammography DDSM [6].

The DDSM database contains 2620 cases acquired from Massachusetts General Hospital, Wake Forest University, and Washington University in St. Louis School of Medicine. The data are comprised of studies of patients from different ethnic and racial backgrounds. The DDSM contains descriptions of mammographic lesions in terms of the American College of Radiology breast imaging lexicon called the Breast Imaging Reporting and Data System (BI-RADS) [6]. Mammograms in the DDSM database were digitized by different scanners depending on the institutional source of the data. A subset DDSM cases was selected for this study. Cases with mass lesions were chosen by selecting reports that only included the BI-RADS descriptors for mass margin and mass shape. From 2620 cases, 390 images were selected based on this criteria. From this subset, 394 regions of interest were selected manually, 187 represent benign mass and 207 represent malignant mass regions.

2.2   Texture Analysis

Texture can be understood as tonal variations in the spatial domain and determines the overall visual smoothness or coarseness of image features. It reveals important information about the structural arrangements of the objects in the image and their relationship to the environment. Consequently, texture analysis provides important discriminatory characteristics related to variability patterns of digital classifications.

Texture processing algorithms are usually divided into three major categories: structural, spectral and statistical [7]. Structural methods consider textures as repetitions of basic primitive patterns with a certain placement rule [8]. Spectral methods are based on the Fourier transform, analyzing the power spectrum [8]. The third and most important group in texture analysis is represented by statistical methods, which are mainly based on statistical parameters such as the Spatial Gray Level Dependence Method-SGLDM, Gray Level Difference Method-GLDM, Gray Level Run Length Matrices-GLRLM [9], [10], [11].

In practice, some of the most usual terms used by interpreters to describe textures, such as smoothness or coarseness, bear a strong degree of subjectivity and do not always have a precise physical meaning. Analysts are capable of visually extracting textural information from images, but it is not easy for them to establish an objective model to describe this intuitive concept. For this reason, it has been necessary to develop quantitative approaches to obtain texture descriptors. Thus, in a statistical context, textures can be described in terms of an important conceptual component associated to pixels (or other units), their spatial association. This component is frequently analyzed at the global level by quantifying the aggregation or dispersion of the element in study [12].

In this work, the texture analysis is done by quantifying the spatial association between individual pixel values from the nodule image by applying the local form of the Ripley's K function - which will be discussed in a following subsection.

Ripley's K Function

Patterns of point based objects in two or three dimensions or on the surface of the terrestrial or celestial spheres are commonplace; some examples are towns in a region, trees in a forest and galaxies in space. Other spatial patterns such as a sheet of biological cells can be reduced to a pattern of points [13].

Most systems in the natural world are not spatially homogeneous but exhibit some kind of spatial structure. As the name suggests, point pattern analysis comprises a set of tools for looking at the distribution of discrete points [14], for example individual pixels in an image that have been mapped to Cartesian coordinates (x, y).

Point pattern has a long history in statistics and the great majority of them focus on a single distance measurement. There are a lot of indices - most of them use the Poisson distribution [15] as the underlying model for inferences about pattern - used to quantify the intensity of pattern at multiple scales.

Point patterns can be studied by first-order and second-order analysis. The first-order approach uses point-to-point mean distance or derives a mean area per point, and then inverts this to estimate a mean point density from which the test statistics about the expected point density are derived [14]. Second-order analysis looks at a larger number of neighbors beyond the nearest neighbor. This group of methods is used to analyze the mapped positions of objects in the plane or space, such as the stems of trees and assumes a complete census of the objects of interest in the zone (area or volume) under study [14]. One of the most commonly used second-order methods is the Ripley's K function.

Ripley's K function is a tool to make analysis of completely mapped spatial point process data, i.e. data on the locations of events. These are usually recorded in two dimensions, but they may be locations along a line or in 3D space. Completely mapped data include the locations of all events in a predefined study area. Ripley's K function can be used to summarize a point pattern, test hypotheses about the pattern, estimate parameters and fit models [13].

Ripley's K method is based on the number of points tallied within a given distance or distance class. Its typically definition for a given radius, t, is:

$$K(t) = \frac{A}{n^2} \sum_i \sum_j \delta(d_{ij}) \tag{1}$$

for $i \neq j$, where A is the area sampled, n is the total number of points and δ is an indicator function that is 1 if the distance d_{ij} between the points on locations i and j is lower than the radius t, else it takes on 0. In other words, this method counts the number of points within a circle of radius t of each point, as Figure 1 shows.

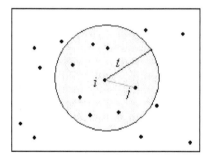

Fig. 1. Schematic illustration measurement of Ripley's K function

It is usual to assume isotropy, i.e. that one unit of distance in the vertical direction has the same effect as one unit of distance in the horizontal direction. Although it is usual to assume stationarity, that means the minimal assumption under which inference is possible from a single observed pattern, $K(t)$ is interpretable for nonstationary processes because $K(t)$ is defined in terms of a randomly chosen event.

As every point in the sample is taken one time to center a plot circle, Ripley's K function provides an inference at global level of the element in study. However, this measure can be also considered in a local form for the ith point [16]:

$$K_i(t) = \frac{A}{n} \sum_{i \neq j} \delta(d_{ij}) \tag{2}$$

2.3   Selection of Most Significant Features

The choice of the minimum set of features that has the power to discriminate the samples is very important to simplify the model and increase its generalization power.

Stepwise selection [17] begins with no variables in the model. At each step, the model is examined. The variable that has the least contribution to the model discriminatory power, as measured by Wilks lambda, is removed if it fails to meet the criterion to stay. Otherwise, the variable not in the model that contributes most to the discriminatory power of the model is included. When all variables in the model meet the criterion to stay and none of the other variables meet the criterion to enter, the stepwise selection process stops.

2.4   Support Vector Machine

The Support Vector Machine (SVM) introduced by V. Vapnik in 1995 is a method to estimate the function classifying the data into two classes [18], [19]. The basic idea of SVM is to construct a hyperplane as the decision surface in such a way that the margin of separation between positive and negative examples is maximized. The SVM term come from the fact that the points in the training set which are closest to the decision surface are called support vectors. SVM achieves this by the structural risk minimization principle that is based on the fact that the error rate of a learning machine on the test data is bounded by the sum of the training-error rate and a term that depends on the Vapnik-Chervonenkis (VC) dimension.

The process starts with a training set of points $x_i \in \Re^n, i = 1, 2, \cdots, l$ where each point x_i belongs to one of two classes identified by the label $y_i \in \{-1, 1\}$. The goal of maximum margin classification is to separate the two classes by a hyperplane such that the distance to the support vectors is maximized. The construction can be thinked as follow: each point x in the input space is mapped to a point $z = \Phi(x)$ of a higher dimensional space, called the feature space, where the data are linearly separated by a hyperplane. The nature of data determines how the method proceeds. There is data that are linearly separable, nonlinearly separable and with impossible separation. This last case be still tracted by the SVM. The key property in this construction is that we can write our decision function using a kernel function $K(x, y)$ which is given by the function $\Phi(x)$ that map the input space into the feature space. Such decision surface has the equation:

$$f(x) = \sum_{i=1}^{l} \alpha_i y_i K(x, x_i) + b \qquad (3)$$

where $K(x, x_i) = \Phi(x).\Phi(x_i)$, and the coefficients α_i and the b are the solutions of a convex quadratic programming problem [18], namely

$$\min_{w, b, \xi} \tfrac{1}{2} w^T \cdot w + C \sum_{i=1}^{l} \xi_i$$
$$\text{subject to } y_i \left[w^T \cdot \phi(x_i) + b \right] \geq 1 - \xi_i \qquad (4)$$
$$\xi_i \geq 0.$$

where $C > 0$ is a parameter to be chosen by the user, which corresponds to the strength of the penality errors and the ξ_i's are slack variables that penalize training errors.

Classification of a new data point x is performed by computing the sign of the right side of Equation 3. An important family of kernel functions is the Radial Basis Function, more commonly used for pattern recognition problems, which has been used in this paper, and is defined by:

$$K(x, y) = e^{-\gamma \|x-y\|^2} \tag{5}$$

where $\gamma > 0$ is a parameter that also is defined by the user.

2.5 Validation of the Classification Methods

To evaluate the performance of a generic classifier, three quantities are usually used. These are the sensitivity (SE), the specificity (SP) and the accuracy rate (A), respectively defined as: $SE = TP/(TP+FN)$, $SP = TN/(TN+FP)$. and $A = (TP+TN)/(TP+TN+FP+FN)$. TP is the number of true positives, i.e. the positive examples correctly classified as positives, TN is the number of true negatives, FP is the number of false positives, i.e. the negative examples incorrectly classified as positives, and FN are the false negatives. Our aim is to obtain a high sensitivity, in order to detect all the positive examples without a significant loss in specificity; from a medical point of view it is indeed crucial to detect all the positive examples, but at the same time we need to significantly reduce the number of false positives.

3 Results and Discussion

The regions of interest (ROI) were manually extracted from each image based on the information provided by the database DDSM. The ROIs are square region sub-images defined to completely enclose the DDSM described abnormality, as shown in Figure 2. To perform the experiments, we take 187 ROIs representing benign masses and 207 representing malignant ones.

In order to find maximum possible information about the masses, we used original images and we also quantized them to 3, 4, 5, 6, and 7 bits (or 8, 16, 32, 64 and 128 gray levels, respectively). For each quantization level we applied Equation 2 to each individual pixel value, being the area $A = \pi \times t^2$. For example, for a nodule quantized to 8 bits, we obtained the 256 function values of $K_i(t)$; first it was obtained $K_i(t)$ for the pixels with density equals to 1, after for the pixels with density equals to 2 and so on, until to be obtained $K_i(t)$ for the pixels with density equals to 255.

For the purpose of carry out the analysis along the entire nodule, we performed the analysis using six different values for the radius t. So, in order to find maximum radius value, we took the ROI's central pixel and then we found out

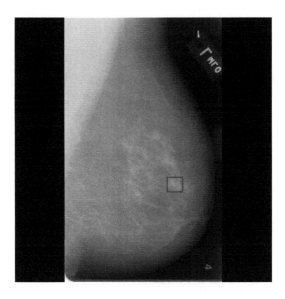

Fig. 2. Illustration of a malignant ROI example (DDSM database reference: B-3086-1)

the farthest one from it (Figure 3). Each circle radius t_i may take values $\frac{1}{6} \times d$, $\frac{1}{3} \times d$, $\frac{1}{2} \times d$, $\frac{2}{3} \times d$, $\frac{5}{6} \times d$ and d, where d is the distance from the central pixel i to the farthest one P.

Figure 4 shows this scheme to compute Ripley's function. This approach makes possible to observe the spatial association among individual pixel values at different locations, from central to peripheral zones of the ROIs.

Thus, we obtained a set of 3024 (equals to $8 + 16 + 32 + 64 + 128 + 256$ gray levels $\times$ 6 concentric circles) different values of $K_i(t)$, for each sample. To make feasible the computation we need to select from all the obtained measures which were the minimum set that has the power to discriminate benign from malignant masses. To do it, we used the stepwise technique that reduced the number of 3024 variables to 83. There are 26 variables for $t = \frac{1}{6} \times d$, 11 variable for $t = \frac{1}{3} \times d$, 8 variables for $t = \frac{1}{2} \times d$, 3 variables for $t = \frac{2}{3} \times d$, 14 for $t = \frac{5}{6} \times d$ and 21 for $t = d$. Analyzing only the way variables are distributed in relation to the gray levels, there are 0 variables for 8 gray levels, 6 for 16 gray levels, 4 for 32 gray levels, 10 for 64 gray levels, 16 for 128 gray levels and 47 for 256 gray levels. In fact, we may see that the more we quantize the images, the more relevant informations are lost.

The next step was the classification of each sample, using a SVM classifier. A library for Support Vector Machines, called LIBSVM [20], was used for training and testing the SVM classifier. We used Radial Basis Function as kernel and the value parameters used were $C = 8192$ and $\gamma = 0.001953125$.

The data set is commonly split in training and test sets, so we generated several pairs of subsets with 315 samples for training and 79 samples for tests. In order to show the best performance results, we select only five pair of subsets

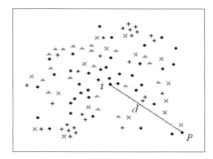

Fig. 3. The distance from the central pixel i to the farthest one P from it

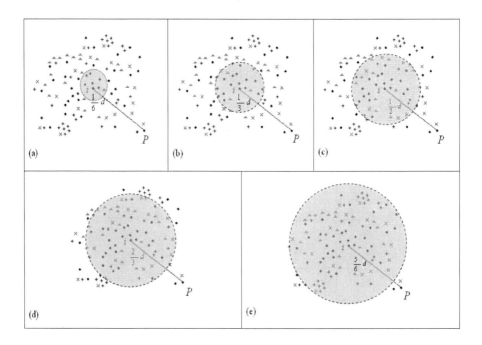

Fig. 4. Schematic illustration of computing the local form of Ripley's K function for different values of t

that present best test accuracy, i.e, that ones that produces less generalization error. Table 1 shows performances measures for each experiment.

Analyzing only test results, we may observe that the best result was 94.94% of accuracy, 92.86% of sensitvity and 93.33% of specificity. A more detailed analysis of test results shows that the methodology presents an average sensibility of 95.29%, an average specificity of 91.55% and an average accuracy of 93.92%. The global accuracy is the sum of all true positives and true negatives detected by the methodology divided by the total of samples, and its average value is equal to 95.94%.

Table 1. Results from running SVM with Ripley's K function

Experiments	Specificity (%)		Sensitivity (%)		Accuracy (%)		Global Accuracy (%)
	Train	Test	Train	Test	Train	Test	
1	93.33	93.33	98.18	92.86	95.87	94.94	95.69
2	93.96	92.11	98.80	95.12	96.51	93.67	95.94
3	94.08	91.43	98.16	95.45	96.19	93.67	95.69
4	95.39	91.43	98.16	95.45	96.83	93.67	96.19
5	95.30	89.47	98.19	97.56	96.83	93.67	96.19

Thus, we verify that the proposed methodology provides a good support for breast masses characterization and classification, resulting in a good generalization power from training to test data. Its success is very encouraging for further investigation and utilization of more complicated databases, with larger number of samples.

4 Conclusion

This paper has presented a new application of a function used traditionally in point pattern analysis, with the purpose of characterizing breast masses as benign or malignant. The measures extracted from Ripley's K function were analyzed and had great discriminatory power, through Support Vector Machine classification. The best test result was 94.94% of accuracy, 93.33% of specificity and 92.86% of specificity, with means that the proposed methodology has also a good generalization power.

The results presented are very encouraging, and they constitute strong evidence that Ripley's K function is an important measure to incorporate into a CAD software in order to distinguish benign and malignant lesions. Nevertheless, there is the need to perform tests with other databases, with more complex cases in order to obtain a more precise behavior pattern.

As future works, we propose the use of other textural measures to be used jointly with Ripley's K funcion, in order to try to reduce the number of false negatives to zero and to find others possible patterns of malignant and benign lesions. We also suggest the application of the function to provide features for automatic detection of masses. Finally, we suggest to apply the proposed mehodology for calcification detection and diagnosis problem, using SVM classification.

Acknowledgements

Leonardo de Oliveira Martins acknowledges receiving scholarships from CAPES (Coordination for the Development of Graduate and Academic Research). Erick Corrêa da Silva acknowledges receiving scholarships from the CNPq (National Research Council of Brazil). The authors also acknowledge CAPES for financial support (process number 0044/05-9).

References

1. (AMS), A.C.S.: Learn about breast cancer (2006) Available at
   http://www.cancer.org
2. (NCI), N.C.I.: Cancer stat fact sheets: Cancer of the breast (2006) Available at
   http://seer.cancer.gov/statfacts/html/breast.html
3. Zhang, P., Verma, B., Kumar, K.: Neural vs. statistical classifier in conjunction
   with genetic algorithm based feature selection. Pattern Recognition Letters 26,
   909–919 (2005)
4. Christoyianni, I., Koutras, A., Dermatas, E., Kokkinakis, G.: Computer aided diagnosis of breast cancer in digitized mammograms. Comput. Med. Imaging Graph 26,
   309–319 (2002)
5. Campos, L., Silva, A., Barros, A.: Diagnosis of breast cancer in digital mammograms using independent component analysis and neural networks. In: Sanfeliu,
   A., Cortés, M.L. (eds.) CIARP 2005. LNCS, vol. 3773, pp. 460–469. Springer, Heidelberg (2005)
6. Heath, M., Bowyer, K., Kopans, D.: Current status of the digital database for
   screening mammography. Digital Mammography, pp. 457–460. Kluwer Academic
   Publishers, Dordrecht (1998)
7. Gonzalez, R.C., Woods, R.E.: Digital Image Processing, 3rd edn. Addison-Wesley,
   Reading, MA (1992)
8. Meyer-Baese, A.: Pattern Recognition for Medical Imaging. Elsevier, Amsterdam
   (2003)
9. Kovalev, V.A., Kruggel, F., Gertz, H.-J., Cramon, D.Y.V.: Three-dimensional texture analysis of MRI brain datasets. IEEE Transactions on Medical Imaging 20,
   424–433 (2001)
10. Li, X.: Texture analysis for optical coherence tomography image. Master's thesis,
    The University of Arizona (2001)
11. Mudigonda, N.R., Rangayyan, R.M., Desautels, J.E.L.: Gradient and texture anal-
    ysis for the classification of mammographic masses. IEEE Transactions on Medical
    Imaging 19, 1032–1043 (2000)
12. Scheuerell, M.D.: Quantifying aggregation and association in three dimensional
    landscapes. Ecology 85, 2332–2340 (2004)
13. Ripley, B.D.: Modelling spatial patterns. J. Roy. Statist. Soc. B 39, 172–212 (1977)
14. Urban, D.L.: Spatial analysis in ecology - point pattern analysis (2003), Available
    at: http://www.nicholas.duke.edu/lel/env352/ripley.pdf
15. Papoulis, A., Pillai, S.U.: Probability, Random Variables and Stochastic Processes,
    4th edn. McGraw-Hill, New York (2002)
16. Dale, M.R.T., Dixon, P., Fortin, M.J., Legendre, P., Myers, D.E., Rosenberg, M.S.:
    Conceptual and mathematical relationships among methods for spatial analysis.
    Ecography 25, 558–577 (2002)
17. Duda, R.O., Hart, P.E.: Pattern Classification and Scene Analysis. Wiley-
    Interscience Publication, New York (1973)
18. Haykin, S.: Redes Neurais: Princípios e Prática, 2nd edn. Bookman, Porto Alegre
    (2001)
19. Burges, C.J.C.: A Tutorial on Support Vector Machines for Pattern Recognition.
    Kluwer Academic Publishers, Dordrecht (1998)
20. Chang, C.C., Lin, C.J.: LIBSVM – a library for support vector machines (2003),
    Available at http://www.csie.ntu.edu.tw/~cjlin/libsvm/

Selection of Experts for the Design of Multiple Biometric Systems

Roberto Tronci, Giorgio Giacinto, and Fabio Roli

Department of Electric and Electronic Engineering, University of Cagliari,
Piazza D'Armi, I-09123 Cagliari, Italy
{roberto.tronci,giacinto,roli}@diee.unica.it

Abstract. In the biometric field, different experts are combined to improve the system reliability, as in many application the performance attained by individual experts (i.e., different sensors, or processing algorithms) does not provide the required reliability. However, there is no guarantee that the combination of any ensemble of experts provides superior performance than those of individual experts. Thus, an open problem in multiple biometric system is the selection of experts to combine, provided that a bag of experts for the problem at hand are available. In this paper we present an extensive experimental evaluation of four combination methods, i.e. the Mean rule, the Product rule, the Dynamic Score Selection technique, and a linear combination based on the Linear Discriminant Analysis. The performance of combination have been evaluated by the Area Under the Curve (AUC), and the Equal Error Rate (EER). Then, four measures have been used to characterise the performance of the individual experts included in each ensemble, namely the AUC, the EER, and two measures of class separability, i.e., the d' and an integral separability measure. The experimental results clearly pointed out that the larger the d' of individual experts, the higher the performance that can be attained by the combination of experts.

1 Introduction

Approaches based on ensemble of experts are widely used in many applications as they avoid the choice of the "best" expert, and typically provide better performance than those provided by individual experts [1]. Ensemble approaches also allow "fusing" experts based on different input sources, so that complementary information can be exploited, and the resulting expert is robust with respect to noise [1]. For this reason, they are widely used in security applications, such as biometric authentication systems, where the goal is to authorise the access to a protected resource by using one or more biometric traits to validate the identity of the person. At present, there is an increasing interest in multi-biometrics, i.e. the combined use of different biometric traits and/or processing algorithms, as in many application the performance attained by individual sensors or processing algorithms does not provide the required reliability [2].

P. Perner (Ed.): MLDM 2007, LNAI 4571, pp. 795–809, 2007.

When combination is performed it would be useful to have some measures allowing to selects the experts to combine from a bag of available experts designed for the task at hand. In the biometric field it is easy to create such a bag of experts as, for example, for a given sensor, a number of processing algorithms and matching techniques are available. The number of available matchers increases in those applications where two or more sensors are used to deploy user authentication mechanisms.

In the case of pattern classifiers, a number of diversity and accuracy measures have been proposed to design a multiple classifier system [1]. The vast majority of these measures are based on the classification errors made by different classifiers. However, these measures are not suited in the biometric field. Biometric experts perform user authentication by the so-called *matchers*, i.e. algorithms that compare the acquired biometry to those stored during the enrolment phase. The output of a matcher is a *matching score*, i.e. a measure stating how much the acquired biometry is likely to be the stored biometry associated to the claimed identity. In order to perform user authentication, a threshold is set so that users with a matching score larger than the threshold are accepted (i.e. assigned to the so-called *genuine* class), otherwise they are rejected (i.e. assigned to the so-called *impostor* class). However, in order to evaluate matchers, typically the ROC curve is used, which represents the relationship between the false acceptance rate and the true acceptance rate for different values of the decision threshold. It is easy to see that the Area Under the Curve (AUC) represents a measure of the performance of the systems. However, as the AUC provides information on the "average" performance of a biometric systems, typically in the biometric field the performance are compared in terms of the Equal Error Rate (EER), i.e. the point of the ROC where the two errors, i.e. the false acceptance rate and the false rejection rate, are equal.

As the combination of experts in a multi-biometric system is usually performed at the score level (i.e., only the matching scores are available), there is no information available about classification "errors", because they are related to the choice of the acceptance threshold. Thus, to design a multiple biometric system we cannot use the diversity and accuracy measures proposed to design a multiple classifier system as they are typically based on the classification errors made by different classifiers. Consequently, we must resort to some performance measure that is not "accuracy based".

To this end, we propose to characterise the individual experts by four measures, i.e., two measures of performance (namely, the AUC, and the EER), and two measures of class separability (namely, the d', and an integral measure). Then, we performed an extensive experimental evaluation on a large bag of biometric experts to seek correlations between the above measures and the performance attained after combination. Four combination rules have been considered: the Mean rule, the Product rule, the Dynamic Score Selection technique [3], and a linear combination based on Linear Discriminant Analysis. It is worth noting that other authors performed extensive experimental evaluation of the performance of the combination of multiple experts [4]. However, while previous works aimed at assessing the improvement in performance that can be attained by combining multiple experts, this paper aims at providing some guidelines to select experts so that their combination allows attaining high performance improvements. In addition, reported experiments will also show which of the four considered combination methods is more suited for combining multiple biometric experts.

The rest of the paper is organised as follows: Sections 2 and 3 present respectively the measures of performance we used to characterise the individual experts and to evaluate the results after combination, and the rules we considered for combining biometric experts. The experimental results are presented in Section 4 and our conclusions are outlined in Section 5.

2   Performance Measures

In biometric systems, performance are assessed by measuring the errors made by rejecting genuine users, and those made by accepting impostor users, for a given value of the acceptance threshold. Let us denote with th an acceptance threshold so that users whose score is larger than th are assigned to the *genuine* class, while users whose score is smaller than th are assigned to the *impostor* class. The two errors, respectively the False Rejection Rate (FRR), and the False Acceptance Rate (FAR) are computed as follows:

$$FRR_j(th) = \int_{-\infty}^{th} p(s_j|s_j \in \text{genuine})\mathrm{d}s_j = P(s_j \leq th|s_j \in \text{genuine}) \quad (1)$$

$$FAR_j(th) = \int_{th}^{\infty} p(s_j|s_j \in \text{impostor})\mathrm{d}s_j = P(s_j > th|s_j \in \text{impostor}) \quad (2)$$

The most widely accepted method used to evaluate the performance of a biometric system is the Receiver Operating Characteristic (ROC) curve. In this curve the value of *1 - FRR* (i.e., the true acceptance rate) is plotted against the value of *FAR* for all possible values of *th*. As the ROC is a graphical measure of performance, to compare two or more biometric experts it is useful to use numerical performance measures. Such a measure can be a summary index related to the ROC, or an index related to the degree of overlapping of the distributions of genuine and impostor scores. In addition, performance can be assessed at a specific point of the ROC that corresponds to a particular working condition. According to the literature on ROC analysis [5], and on biometric system evaluation [6] we selected four measures of performance: the Area Under the ROC Curve (AUC), the Equal Error Rate (EER), the d', and an integral class separation measure.

2.1   Area Under the ROC Curve

In ROC analysis the *Area Under the Curve* (AUC) is the most widely used measure for assessing the performance of a two-class system because it is a more discriminating measure than the accuracy [5]. The AUC can be computed by the numerical integration of the ROC curve, or by the Wilcoxon-Mann-Whitney (WMW) statistic [7]. We decided to use the WMW statistic to estimate the AUC as it is equivalent to the value computed by integrating the ROC, but the resulting estimation is more reliable, as the value of the integral depends on the numerical technique employed. In addition, the AUC can be interpreted as follows: given two randomly chosen users, one from the set of genuine users, and one from the set of impostor users, the AUC is the probability $P(x_p > y_q)$, i.e. the probability of correct pair-wise ranking [8].

According to the WMW statistic, the AUC can be computed as follows. Let us divide into two sets all the scores $\{s_{ij}\}$ produced by a matcher M_j for all the u_i users: $\{x_p\}$, i.e. the set made up of the scores produced by genuine users, and $\{y_q\}$, i.e. the set made up of the scores produced by impostor users.

$$AUC = \frac{\sum_{p=1}^{n_+} \sum_{q=1}^{n_-} I(x_p, y_q)}{n_+ \cdot n_-} \tag{3}$$

where n_+ is the number of genuine users and n_- is the number of impostors, and the function $I(x_p, y_q)$ is[1]:

$$I(x_p, y_q) = \begin{cases} 1 & x_p > y_q \\ 0 & x_p < y_q \end{cases} \tag{4}$$

2.2   Equal Error Rate

The *Equal Error Rate* (EER) is the point of the ROC curve where the two errors, i.e. the FAR and the FRR, are equal. This performance measure is widely used in the biometric field to assess the performance of biometric systems [6].

2.3   d'

The d-prime (d') is a measure of discriminability proposed within the Signal Detection Theory [9]. Given the distributions of the scores produced respectively by genuine and impostor users, the d' is defined as

$$d' = \frac{|\mu_{Gen} - \mu_{Imp}|}{\sqrt{\frac{\sigma_{Gen}^2}{2} + \frac{\sigma_{Imp}^2}{2}}}$$

where μ_{Gen} and μ_{Imp} are the means of the two distributions, while σ_{Gen} and σ_{Imp} are the related standard deviations. It is easy to see that the larger the d', the better the performance.

---

[1] For discrete values $I(x_p, y_q) = 0.5$ if $x_p = y_q$.

2.4   Class Separability

The values of FAR and FRR, for a given value of the threshold, depend on the degree of separation of the distribution of the scores of impostor and genuine users, as it can be easily seen from Equations (1) and (2). Thus it follows that to minimise the errors in different working conditions, it is desirable that the distributions of *genuine* and *impostor* users are separated as much as possible. A measure of the separability between the two distribution is given by:

$$\frac{1}{2} \int_0^1 |p\,(s|\text{genuine}) - p\,(s|\text{impostor})|\,ds \tag{5}$$

where a value of 1 is associated to perfectly separable distributions, while the value of 0 is associated to completely coincident distributions.

3   Combination Rules

Combination of multiple biometric systems can be performed at different representation levels, i.e, the raw data level, and the feature level, the score level, and the decision level [10]. The combination at the raw data level and at the feature level are performed before the matching phase, while the combination at the score and decision levels are performed after the matching phase. The combination at the raw data level can be used only if the raw data comes from the same biometry, and the sensors used are compatible. The combination at the feature level can be performed only when feature sets extracted from different biometric sources are compatible. The combination at the score level allows combining different matching algorithm, different sensors and/or different biometric traits. The combination at the decision level aims at combining authentication outputs (accepted/rejected) from different biometric systems.

This work is focused on the combination at the score level, as it is the most widely used and flexible combination level. In particular, we chose to investigate the performance of four combination methods: the Mean rule, the Product rule, a linear combination whose weights are computed through the Linear Discriminant Analysis (LDA), and a novel technique called Dynamic Score Selection (DSS) [3]. It is worth noting that the combination by LDA and DSS require a training phase in order to estimate the parameters needed to perform the combination.

In the following we will briefly recall the definition of the four combination methods. Let $M = \{M_1 \ldots M_j \ldots M_N\}$ be a set of N matchers and $U = \{u_i\}$ be the set of users. Let also $f_j(\cdot)$ be the function associated to matcher M_j that produces a score s_{ij} for each user u_i, $s_{ij} = f_j(u_i)$.

3.1   Mean Rule

The mean rule is applied directly to the matching scores produced by the set of N matchers, and the resulting score is computed as follows:

$$s_{i,mean} = \frac{1}{N} \sum_{j=1}^{N} s_{ij}$$

3.2   Product Rule

Similarly to the *mean* rule, this fusion rule is applied directly to the matching scores produced by the set of N matchers:

$$s_{i,prod} = \frac{1}{N} \prod_{j=1}^{N} s_{ij}$$

3.3   Linear Combination by Linear Discriminant Analysis

LDA can be used to compute the weights of a linear combination of the scores [9]. The goal of this fusion rule is to attain a fused score such that the within-class variations are minimised, and the between-class variations are maximised. The fused score is computed as follows:

$$s_{i,LDA} = W^t \cdot \mathbf{s}_i$$

where W^t is the transformation vector that takes into account the within and between class variations as

$$W = S_w^{-1}(\mu_{gen} - \mu_{imp})$$

where μ_{gen} is the mean of the genuine distribution, and μ_{imp} is the mean of the impostor distribution, and S_w is the within-class scatter matrix.

3.4   Dynamic Score Selection

The (DSS) is a score selection technique based on the ideal selector defined in [3], where the ideal score selector selects the maximum score for genuine users, and the minimum score for impostor users:

$$s_{i,*} = \begin{cases} \max\{s_{ij}\} \text{ if } u_i \text{ is a genuine user} \\ \min\{s_{ij}\} \text{ if } u_i \text{ is an impostor user} \end{cases}$$

The state of nature for s_{ij} (i.e., if the score is likely to be produced by a "genuine" or an "impostor" user) is estimated using the *Relative Minimum Error* (RME) measure. The RME takes into account two terms: the error committed accepting an impostor, through the difference $FAR_j(-\infty) - FAR_j(s_{ij})$ (i.e., a measure of how likely u_i is a genuine user), and the error committed when a genuine is rejected, through the difference $FRR_j(\infty) - FRR_j(s_{ij})$ (i.e., a measure of how likely u_i is an impostor). These quantities are estimated from a training set. In detail, the *Relative Minimum Error* is computed as follows:

$$RME_{ij} = \frac{[FAR_j(-\infty) - FAR_j(s_{ij})] - [FRR_j(\infty) - FRR_j(s_{ij})]}{|[FAR_j(-\infty) - FAR_j(s_{ij})] + [FRR_j(\infty) - FRR_j(s_{ij})]|} =$$

$$= \frac{FRR_j(s_{ij}) - FAR_j(s_{ij})}{FRR_j(s_{ij}) + FAR_j(s_{ij})}$$

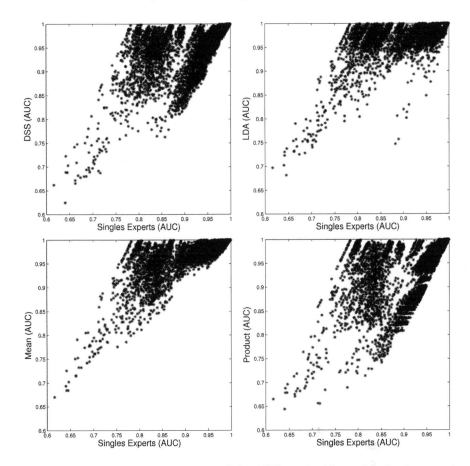

Fig. 1. In these figures the mean value of the AUC attained by each pair of experts is plotted against the AUC of the correspondent combination, using the four combination methods

Summing up, the algorithm of *DSS* is made up of the following steps:

1. Compute for each matcher M_j the value of RME_{ij} for the user u_i
2. Estimate the most reliable state of nature for u_i by selecting the maximum value of $|RME_{ij}|$. Let $k = \text{argmax}_j(|RME_{ij}|)$
3. Select the score s_{sel} based on RME_{ik} as

$$s_{sel} = \begin{cases} \max_j(s_{ij}) & \text{if} \quad RME_{ik} > 0 \\ \min_j(s_{ij}) & \text{if} \quad RME_{ik} < 0 \end{cases}$$

4   Experimental Results

Experiments have been performed using the scores produced by a large number of matchers during the third Fingerprint Verification Competition (FVC2004)

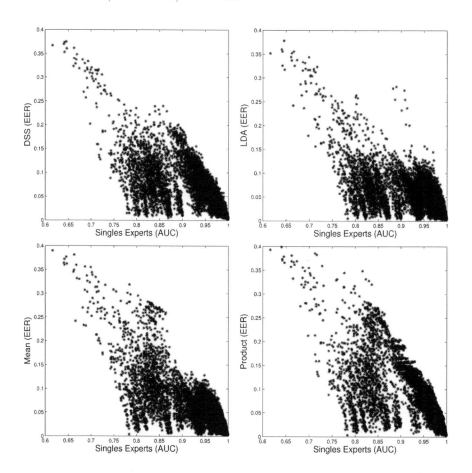

Fig. 2. In these figures the mean value of the AUC attained by each pair of experts is plotted against the EER of the correspondent combination, using the four combination methods

[11] [12]. For our experiments we used 40 experts from the *Open* category. The fingerprint images consists of four different databases, three acquired with different sensors and one created with a synthetic fingerprint generator. For each sensor and for each expert, a set of scores is available, where the scores have been generated by authentication attempts by genuine users, as well as authentication attempts by impostors. For the details on how the scores where obtained and normalised, the reader is referred to [11]. This database is not freely available, so the experiments were ran at the Biometric Systems Lab (University of Bologna, Italy) which organises the competition.

For each sensor, 7750 matching scores are available, 2800 of them belonging to "genuine" users, and 4950 belonging to "impostor" users. In order to create a training set for the LDA fusion rule, and the DSS algorithm, we randomly divided the set of users into four subsets of the same size, each subset made up

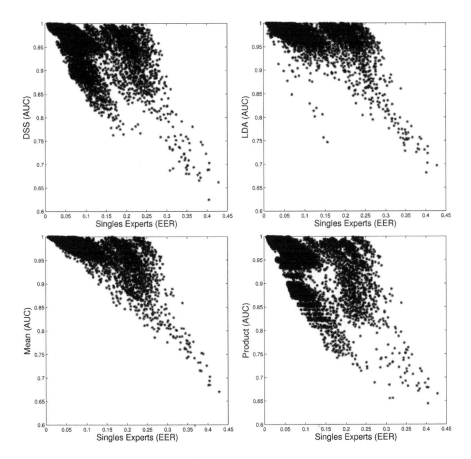

Fig. 3. In these figures the mean value of the EER attained by each pair of experts is plotted against the AUC of the correspondent combination, using the four combination methods

of 700 "genuines" and 1238 "impostors". Each of the four subsets has been used for training, while the remaining three subsets have been used for testing. Using this partitioning of the dataset, we performed an exhaustive multi-algorithmic combination experiment: for each of the four partitioning, and for each sensor, we considered all the possible pairs of experts. Thus, the reported experiments are related to the combination of 13,366 pairs of experts.

For each pair of experts, we computed the mean value of the AUC, the EER, the d' and the separability index. Then, after combining the experts using the four combination rules described in Section 3, we computed the related values of AUC and EER, as they better represent the performance of the resulting systems. It worth remarking that for the AUC, the d' and the separability index the larger the value the better the performance, while the reverse holds for the EER.

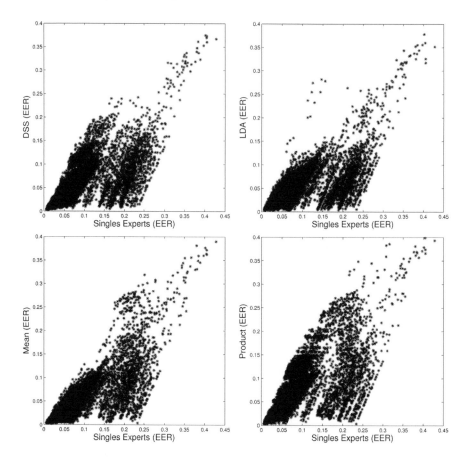

Fig. 4. In these figures the mean value of the EER attained by each pair of experts is plotted against the EER of the correspondent combination, using the four combination methods

In order to evaluate the relationship between the mean performance of the pair experts, and the performance of their combination, we report a graphical representation of the results of the experiments. On the X axis we represent the mean performance of the pair of experts, while on the Y axis we report the performance of their combination. As a result, 32 graphics are reported in Figures (1 - 8).

In Figures (1) and (2) the mean value of the AUC of any pair of experts is plotted against the AUC and the EER, respectively, of all the considered combination methods. The inspection of Figure (1) allows us to conclude that the mean AUC value of a pair of experts is not an useful measure to select the pair of experts whose combination may provide performance improvements. In fact for all the combination rules but the Mean rule, there is no clear relationship between the mean AUC of the pair of experts and the AUC of their combination. In the case of the Mean rule, the AUC of the combination is always greater than

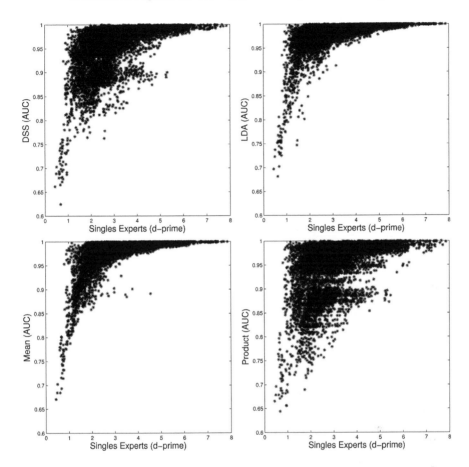

Fig. 5. In these figures the mean value of the d' attained by each pair of experts is plotted against the AUC of the correspondent combination, using the four combination methods

or equal to the mean AUC of the pair of experts. On the other hand, if we evaluate the performance of combination in terms of the EER (Figure (2)), it is clear that the mean AUC of the pair of experts is uncorrelated with the EER attained by the combination. In addition, for all the combined methods, when the mean AUC of the experts is in the range between 0.75 and 0.8, the EER of the combination spans over a wide range of values. Thus, we cannot predict the performance of the combination in terms of EER by taking into account the mean AUC of the individual experts.

In Figures (3) and (4) the mean value of the EER of any pair of experts is plotted against the AUC and the EER attained by the considered combination methods. The graphics plotted in Figure (3) exhibit a behaviour similar to those in Figure (1), as there is no clear relationship between the mean EER of the pair

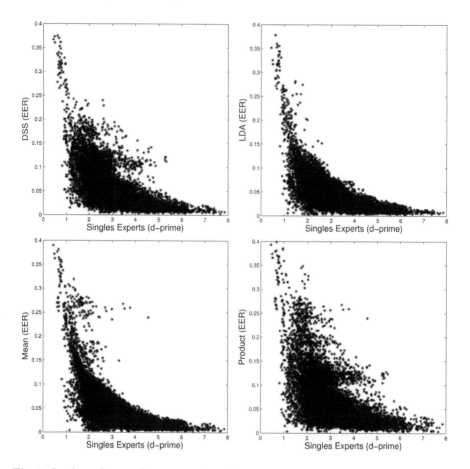

Fig. 6. In these figures the mean value of the d' attained by each pair of experts is plotted against the EER of the correspondent combination, using the four combination methods

of experts and the AUC of their combination. In this case too, the Mean Rule performs better than the other combination methods.

The analysis of Figure (4) shows that there is no correlation between the mean EER of the pair of experts and the EER attained by the combination methods. By comparing Figure (4) with Figure (2) it is easy to see that the graphics plotted have similar behaviour. Therefore, despite the fact that the AUC and the EER are widely used as performance measure to evaluate biometric systems, they are not suited as a measure to select the experts to combine.

Figures (5) and (6) show the mean value of the d' of any pair of experts against the AUC and the EER, respectively, of their combinations. In Figure (5) the larger the d' the larger the AUC, and in Figure (6) the larger the d' the smaller the EER. In addition it is easy to see that for all combination methods but the Product rule, large values of d' guarantee small variance of the performance of

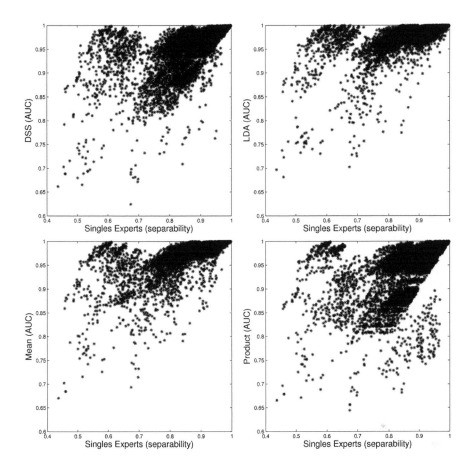

Fig. 7. In these figures the mean value of the Separability attained by each pair of experts is plotted against the AUC of the correspondent combination, using the four combination methods

the combination. Thus, according to these graphics it is clear that the d' is a good measure to choose the experts to combine. Finally, we can say that the d' is more useful to choose the experts to combine than the AUC and the EER.

Figures (7) and (8) show the mean value of the separability of any pair of experts against the AUC and the EER, respectively, of their combinations. These figures point out that the this measure of separability of the pair of experts are not related to the performance of the combinations in terms of the AUC and the EER. In fact, large mean values of the separability of the pair of experts don't always correspond to large values of the AUC or to a small values of the EER of their combinations. By comparing these results with those obtained using the other measures, it is clear that this separability measure is useless if we have to choose the experts to combine.

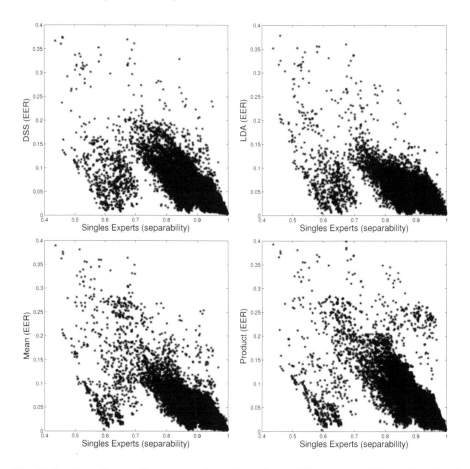

Fig. 8. In these figures the mean value of the Separability attained by each pair of experts is plotted against the EER of the correspondent combination, using the four combination methods

As far as the evaluation of the considered combination methods is concerned Figures (1 - 8) allows to conclude that the Product rule exhibits the worst performance. On the other hand the linear score combination performed through the Mean rule and the LDA provided the highest performance.

5   Conclusions

In this paper, an experimental analysis to evaluate the correlation between the combination methods and performance measures was presented. In particular the aim of the paper was to find which performance measure is the more appropriate to use when we must choose the biometric experts to combine. Moreover we showed the behaviour of the combination methods in terms of different measures.

Reported experiments clearly point out that the d' is the most suitable measure to choose the experts to combine, while the "class separability" is the less appropriate measure to choose the experts. Moreover, despite the fact that the AUC and the EER are typically used to asses the performance of biometric experts, they are not suited to choose the experts to combine as the d' is.

The analysis of the performance of the combination methods clearly show that the worst performance were attained by the Product rule, while the best performance were obtained by the linear score combination methods: the Mean rule and the LDA. The performance of DSS lies between those of the linear combination methods and those of the Product rule. This behaviour can be explained by the fact that the errors made by selection mechanism are heavier than those of fusion techniques.

References

1. Kuncheva, L.I.: Combining Pattern Classifiers: Methods and Algorithms. John Wiley & Sons, Chichester (2004)
2. Jain, A.K., Ross, A.: Multibiometric Systems. Communications of the ACM, Special Issue on Multimodal Interfaces 47, 34–40 (2004)
3. Giacinto, G., Roli, F., Tronci, R.: Score Selection Techniques for Fingerprint Multimodal Biometric Authentication. In: Proc. of 13th ICIAP, pp. 1018–1025 (2005)
4. Fierrez-Aguilar, J., Nanni, L., Ortega-Garcia, J., Cappelli, R., Maltoni, D.: Combining Multiple Matchers for Fingerprint Verification: A Case Study in FVC2004. In: Proc. of 13th ICIAP, pp. 1035–1042 (2005)
5. Huang, J., Ling, C.X.: Using AUC and Accuracy in Evaluating Learning Algorithms. IEEE Transactions on Knowledge and Data Engineering 17, 299–310 (2005)
6. Jain, A., Bolle, R., Pankanti, S.: BIOMETRICS: Personal Identification in Networked society. Kluwer Academic Publishers, Dordrecht (1999)
7. Mann, H.B., Whitney, D.R.: On a test whether one or two random variable is stochastically larger than the other. Ann. Math. Statistic 18, 50–60 (1947)
8. Hanley, J.A., McNeil, B.J.: The meaning and the use of the area under a receiver operanting charateristic curve. Radiology 143, 29–36 (1982)
9. Duda, R.O., Hart, P.E., Stork, D.G.: Pattern Classification. John Wiley & Sons, Chichester (2001)
10. Ross, A., Nandakumar, K., Jain, A.: Handbook of Multibiometrics. LNCS. Springer, Heidelberg (2006)
11. Maio, D., Maltoni, D., Cappelli, R., Wayman, J.L., Jain, A.K.: FVC 2004: Third Fingerprint Verification Competition. In: Proc. ICBA, Hong Kong, pp. 1–7 (2004)
12. FVC2004 Website: http://bias.csr.unibo.it/fvc2004/

Multi-agent System Approach to React to Sudden Environmental Changes

Sarunas Raudys[1,2] and Antanas Mitasiunas[2]

[1] Institute of Mathematics and Informatics and [2] Vilnius University
Akademijos 4, Vilnius-08663, Lithuania
raudys@ktl.mii.lt, antanas.mitasiunas@maf.vu.lt

Abstract. Many processes experience abrupt changes in their dynamics. This causes problems for some prediction algorithms which assume that the dynamics of the sequence to be predicted are constant, or at least only change slowly over time. In this paper the problem of predicting sequences with sudden changes in dynamics is considered. For a model of multivariate Gaussian data we derive expected generalization error of standard linear Fisher classifier in situation where after unexpected task change, the classification algorithm learns on a mixture of old and new data. We show both analytically and by an experiment that optimal length of learning sequence depends on complexity of the task, input dimensionality, on the power and periodicity of the changes. The proposed solution is to consider a collection of agents, in this case non-linear single layer perceptrons (agents), trained by a memetic like learning algorithm. The most successful agents are voting for predictions. A grouped structure of the agent population assists in obtaining favorable diversity in the agent population. Efficiency of socially organized evolving multi-agent system is demonstrated on an artificial problem.

Keywords: Generalization, Dimensionality, Evolution, Learning, Multi-agent systems, Neural networks, Sample size.

1 Introduction

A characteristic feature of current research for the development of intelligent machines is a requirement for intelligent agents and robots to operate in new, unknown environments, to be able adapt to sudden situational changes [1], [2], [3], [4], [5]. A large group of problems from the domain of technology (e.g. the converter-oxygen process of steelmaking, the change of catalyst properties in an oil refinery or, in the process of carbon dioxide conversion) is described by non-stationary probability densities that vary in time gradually [5]. A powerful stream of research of fast adaptation in changing environments had been carried out in analysis of biology, physics, economy and financial time series (see e.g. [6], [7], [8], [9]).

Much work has been done in automatic control, machine learning and pattern recognition communities. In statistical estimation, different models of changes had been utilized in order to develop optimal or close to optimal parameter estimation

P. Perner (Ed.): MLDM 2007, LNAI 4571, pp. 810–823, 2007.

an/or decision making rules [1], [5], [10], [11], [12], [13]. In other part of investigations, fixed length of historic learning data is used to train classification or prediction rules [4], [7], [9]. It was shown theoretically and experimentally that optimal length of training data depends on a power of environmental changes [9]. In situations where environmental changes are taking place unexpectedly, single learning algorithm cannot react adequately to all possible changes. It is difficult to determine optimal length of training sequence and ensure good accuracy. Following a strategy realized in by the Nature in human and artificial immune systems [6, 14] numerous of forecasting or classification algorithms differing in the length of training sequences, input features and complexity of training algorithm started to be used. Like in multiple classifier system approach, a final decision making was allocated to a fusion rule of individual forecasting/classification algorithms [4, 9]. In order to ensure good performance of such collective decision making system, optimal diversity of individual predictors/classifiers should be obtained. In human and natural immune systems the diversity is achieved by suppression of similar classifiers (antibodies) [14]. Another possibility is to split individual predictors/classifiers into groups and let the system to evolve by genetic combination of evolutionary algorithm combined with local search in lengthy series of environmental changes. Under this scenario, learning system inherits parameters of "learning style" genetically [3]. Contrary to standard memetic algorithms [15], a fitness function characterizes learning rapidity after environmental change [3], [16]. To ensure faster adaptation to environmental changes, training with corrupted learning directives started to be utilized [3], [16], [17].

An objective of present paper is to solve methodological problems. In majority of research papers gradual environmental changes had been considered. We examine sudden data changes and consider the Nature inspired prototype schema of collective decision making in changing environments. The paper is organized as follows. Section 2 introduces main terms and notations, considers generalization error of standard Fisher linear discriminant function in stationary situation. This type of classification rule is chosen due to *a possibility to perform theoretical examination* by means of multivariate statistical analysis and by a fact that *the Fisher classifier could be obtained while training linear and non-linear perceptrons* [18], [19], [20], [21]. Section 3 investigates a situation where after the classification task change, the algorithm is trained on a mixture of old and new, the changed, data. Simple, however, rather exact equation to calculate generalization error in non-stationary case is derived. In Section 4 adaptive multi-agent system (MAS) to tackle non-stationary time series is considered. The system is composed of an assortment of adaptive agents that ought to learn: 1) the changing pattern classification tasks by means of iterative percepton training procedure and 2) "learning style" parameters by means of inheritance and mutations. Section 5 discuses possibilities of application of new MAS in analysis or real world time series, considers directions for future research work.

2   Generalization Error in Stationary Case

Standard Fisher linear discriminant function (DF) probably is the most frequently used pattern recognition rule. Discovered seventy years ago it undergone a great variety of modifications and generalizations [21], [22]. In two category and equal prior probability case, classification is performed according to a sign of DF

$$g(x) = (x - \tfrac{1}{2}(\overline{\mathbf{X}}_1 + \overline{\mathbf{X}}_2))^T \mathbf{S}^{-1}(\overline{\mathbf{X}}_1 - \overline{\mathbf{X}}_2), \tag{1}$$

where $x = [x_1, x_2, ..., x_p]^T$ is p-dimensional vector, to be allocated to one of two pattern classes, $\overline{\mathbf{X}}_1$, $\overline{\mathbf{X}}_2$, and $\mathbf{S}$ are sample estimates of mean vectors, μ_1, μ_2, and covariance matrix, Σ, respectively; superscript "T" indicates transposition operation.

Fisher disciminant function may be obtained by training standard sum of squares cost function [18]. Discriminant function (1), however, has more attractive features. While training non-linear single layer perceptron (SLP) in two-category case, we minimize a sum of squares cost function

$$cost = \frac{1}{N_1 + N_2} \sum_{i=1}^{2} \sum_{j=1}^{N_i} (f(\mathbf{x}_{ij}^T \mathbf{w} + w_0) - t_j^{(i)})^2, \tag{2}$$

where $f(arg)$ is activation function, w is p-dimensional weight vector and w_0 is a bias term. In Eq. (2), $t_j^{(i)}$ stands for a desired output, and N_i is a number of training vectors of class Π_i. For standard sigmoid activation function, one can choose: $t_j^{(1)} = 1$, $t_j^{(2)} = 0$.

Profound analysis shows that while minimizing cost (2) one may obtain seven different types of the classifiers. If training sample sizes in two pattern classes $N_2 = N_1 = n/2$, the mean vector of training set is moved to a centre of coordinates and we start total gradient training from the weight vector with zero components, then after the first iteration we obtain Euclidean distance classifier (EDC) based on the mean vectors of the pattern classes. Afterwards, we move towards linear regularized discriminant analysis, standard linear Fisher classifier or the Fisher classifier with pseudo-inverse of the covariance matrix (for an introduction into statistical pattern recognition see e.g. [21], [22]). With a progress of iterative adaptation procedure one has robust discriminant analysis. At the end, when the weights become large, one may approach the minimum empirical error or support vector classifiers [20], [21]. Evolution is a superb peculiarity of total gradient SLP training procedure enabling us to obtain a sequence of diverse classifiers of increasing complexity.

One of most important characteristics of any classification rule is its *generalization error*, a probability of misclassification in terms of discriminant analysis. Generalization error, P_n, of sample based DF (1) is conditioned by two random vectors, $\overline{\mathbf{X}}_1$ and $\overline{\mathbf{X}}_2$, and $p{\times}p$ random covariance matrix, $\mathbf{S}$. In case of multivariate Gaussian data characterized by mean vectors of the pattern classes, μ_1, μ_2, common for both classes covariance matrix Σ and equal prior probabilities of the classes, the generalization error

$$P_n = \frac{1}{2} \sum_{i=1}^{2} \Phi \left\{ \frac{((-1)^{3-i}\mu_i - \frac{1}{2}(\overline{\mathbf{X}}_1 + \overline{\mathbf{X}}_2))^T \mathbf{S}^{-1}(\overline{\mathbf{X}}_1 - \overline{\mathbf{X}}_2)}{\sqrt{(\overline{\mathbf{X}}_1 - \overline{\mathbf{X}}_2)^T \mathbf{S}^{-1}\Sigma \mathbf{S}^{-1}(\overline{\mathbf{X}}_1 - \overline{\mathbf{X}}_2)}} \right\}, \tag{3}$$

where $\Phi(t)$ stands for standard Gaussian cumulative distribution function.

In stationary case the data does not change: $\overline{\mathbf{X}}_i \sim N(\mu_i, 1/N_i \Sigma)$, $\mathbf{S} \sim W(\Sigma_p, n\text{-}2)$, where $N(\mu_i, \Sigma)$ symbolizes a multivariate Gaussian distribution with mean μ_i and covariance Σ. Notation $W(\Sigma_p, n\text{-}2)$ symbolizes a Wishart distribution with $p \times p$ matrix Σ_p and $n\text{-}2 = N_1+N_2-2$ degrees of freedom. Thus, conditional generalization error, P_n, could be considered as a random variable whose distribution density function depends on sample sizes, N_1, N_2, input vector dimensionality, p, and Mahalanobis distance $\delta = ((\mu_1 - \mu_2)^T \Sigma^{-1}(\mu_1 - \mu_2))^{1/2}$. An expectation of conditional probability of misclassification, EP_n, is called an expected classification error or mean generalization error.

Expected classification error is relatively easy to obtain for above multivariate Gaussian data model characterized by common covariance matrices (GCCM), equal prior probabilities of the classes and equal sample sizes, $N_1 = N_2 = n/2$, if one assumes that sample size, N, and dimensionality, p, are high. From above assumption it follows that asymptotically as $N \to \infty$ and $p \to \infty$, distribution density of discriminant function (1) tend to Gaussian one. After calculating its mean and variance, following simple asymptotic formula for generalization error had been derived [23].

$$EP_n = \Phi\{-\frac{\delta}{2\sqrt{(1+\frac{4p}{n\delta^2})(1+\frac{p}{n-p})}}\}, \qquad (4)$$

This formula is valid for standard Fisher linear classifier, linear and non-liner single layer perceptrons trained in a special way. Unfortunately, *we cannot benefit from this formula if classification task has changed abruptly and for learning we use the both, the old and the changed data.*

3 Analysis of Generalization Error in Changing Environments

In this section we will analyze generalization error in situations where the classification task has changed abruptly, however, the researcher does not known this fact and continues to train the classifier both with the old and new data sets. Let us denote by μ_1, μ_2 the mean vectors of the pattern classes *before* the task change, and by Σ a common covariance matrix that does not change. Denote by μ'_1, μ'_2 the mean vectors of the pattern classes *after* the task change. Let learning set size is N vectors of each pattern class and the task has changed N_{new} time steps before. In our analysis we assume, that during each time step we acquire one training vector from each class. Consequently, we use for training $N_{old} = N - N_{new}$ vectors from each of populations $N(\mu_1, \Sigma)$, $N(\mu_2, \Sigma)$ and N_{new} vectors from populations $N(\mu'_1, \Sigma)$, $N(\mu'_2, \Sigma)$.

Denote $\Delta = \mu'_1 - \mu_1 = \mu'_2 - \mu_2$, a common shift vector that characterizes the data change, $\mu = \mu_1 - \mu_2$, and $\alpha = N_{old}/N$, a proportion of old data in training set. A derivation of asymptotic formula for expected generalization error in non-stationary case, when a mixture of old and new task data is used to train the classifier, is rather tedious work. To save the paper's space, below we are presenting a simplified sketch of the derivation. Like in standard linear discriminant analysis (see e.g. [21], [22]), without loss of generality one may assume that $\Sigma = I_p$, $p{\times}p$ dimensional identity matrix, $\mu_1 + \mu_2 = 0$, $\mu = (\delta, 0, 0, \ldots, 0)^T$. Then $(\mu'_1 + \mu'_2)/2 = \Delta$, the estimate of the mean $\frac{1}{2}(\overline{X}_1 + \overline{X}_2))$ will be biased by $\alpha\Delta$, an estimate of covariance matrix will be biased by $\alpha\Delta\Delta^T$. Then Equation (3) can be expressed as

$$P_n = \frac{1}{2}\sum_{i=1}^{2}\Phi\left\{ -\frac{(\frac{1}{2}\mu + (\overline{Y}_1 + \alpha\Delta)(-1)^{3-i})^T (T + \alpha\Delta\Delta^T)^{-1}(\mu + \overline{Y}_2)}{\sqrt{(\mu + \overline{Y}_2)^T (T + \alpha\Delta\Delta^T)^{-2}(\mu + \overline{Y}_2)}} \right\}, \quad (5)$$

where $\overline{Y}_1 \sim N(0, \frac{1}{2N}I_p)$, $\overline{Y}_2 \sim N(0, \frac{2}{N}I_p)$, $T \sim W(I_p, n\text{-}2)$, and $\overline{Y}_1$, $\overline{Y}_2$ and T are statistically independent and $n = 2N$.

For further analysis we will use Bartlett equality to express an inverse of biased covariance matrix as $(T + \alpha\Delta\Delta^T)^{-1} = T^{-1} - \dfrac{T^{-1}\Delta\Delta^T T^{-1}}{1 + \Delta^T T^{-1}\Delta}$. Then

$$P_n = \frac{1}{2}\sum_{i=1}^{2}\Phi\left\{ -\frac{(\frac{1}{2}\mu + (\overline{Y}_1 + \alpha\Delta)(-1)^{3-i})^T (T^{-1} - \dfrac{T^{-1}\Delta\Delta^T T^{-1}}{1 + \Delta^T T^{-1}\Delta})(\mu + \overline{Y}_2)}{\sqrt{(\mu + \overline{Y}_2)^T (T^{-1} - \dfrac{T^{-1}\Delta\Delta^T T^{-1}}{1 + \Delta^T T^{-1}\Delta})^2(\mu + \overline{Y}_2)}} \right\}, \quad (6)$$

To obtain equation for expected generalization error we will use expectations of inverse elements of sample covariance matrix $T^{-1} = ((t^{ij}))$ derived in Appendix of the author's paper [23] using special elements of multivariate statistical analysis. In large dimensionality and large sample size situation, we may write

$$E\, t^{ij} = \begin{cases} n(n-p-1)^{-1} \approx n(n-p)^{-1} & \text{if } j = i \quad \text{and both } p \text{ and } n \text{ are large,} \\ 0 & \text{if } j \neq i, \end{cases}$$

$$E(t^{ij})^2 = \begin{cases} n^2((n-p-1)(n-p-3))^{-1} \approx n^2(n-p)^{-2} & \text{if } j = i \text{ and } p, n \text{ are large,} \\ n^2((n-p)(n-p-1)(n-p-3))^{-1} \approx n^2(n-p)^{-3} & \text{if } j \neq i \text{ and } p, n \text{ are large.} \end{cases}$$

In a similar way, for large p and n using simple, however, tedious algebra we may derive higher moments

$$E\,(\,t^{ij}\,)^3 \approx n^3(n-p)^{-3} \quad \text{if} \quad j=i,$$

$$E\sum_{j=1}^{p}(t^{1j})^2 \approx n^2(n\text{-}p)^{-2} + (p\text{-}1)n^2(n\text{-}p)^{-3} = n^3(n\text{-}p)^{-3}, \quad E\,t^{11}\sum_{j=1}^{p}(t^{1j})^2 \approx n^4(n\text{-}p)^{-4} \text{ and}$$

$$E\,(t^{11})^2\sum_{j=1}^{p}(t^{1j})^2 \approx n^5(n\text{-}p)^{-5}.$$

Equation (6) may be simplified dramatically if we take into account that expectations $E\overline{\mathbf{Y}}_1^T\,\overline{\mathbf{Y}}_2 = 0$, $E\overline{\mathbf{Y}}_2^T\,\overline{\mathbf{Y}}_2 = \delta^2 + 4p/N$, ignore variances of random variables $(t^{ij})^2$ and consider a situation where in the PR task change, the data shift is performed along a line connecting the mean vectors μ_1 and μ_2, i.e., $\delta = (\beta\delta, 0, 0, \ldots, 0)^T$:

$$P_n = \frac{1}{2}\sum_{i=1}^{2}\Phi\left\{-\frac{\delta(1+(1+2(-1)^{3-i}\alpha\beta)}{2\sqrt{(1+4p/(N\delta^2))(n/(n-p))}}\right\}. \tag{7}$$

In case on zero data shift ($\Delta=0$), we obtain simple Equation (4). According to Monte Carlo simulations performed for 104 GCCM populations with different parameters, Eq. (4) outperformed in accuracy other six asymptotic formulae [24]. Speaking in general, *asymptotic analysis of statistical pattern classification algorithms where both learning set size, n, and dimensionality, p, are large, is very fruitful* (see e.g. recent review [25]).

In Figure 1 we present learning curves characterizing expected generalization error of linear Fisher classifier versus leaning set size n calculated according to Eq. (7) for the GCCM data model, where $p=20$, $\delta=3.76$ (asymptotic classification error, $P_\infty=0.03$) and the classification task has changed $2N_{\text{new}}=200$ time steps before the present time moment. Red graph (a) corresponds to $\beta=0.4$, blue graph (b) - to $\beta=0.2$ and black graph (c) – to situation where no task change occurred. By black squares, circles and dots we marked averages of 25 Monte Carlo experiments performed with different artificially generated GCCM data sets. The experiments show a good agreement between Eq. (7) and empirical evaluations even in a case when relatively low dimensionality ($p=20$) and finite learning set size were chosen for validation of the theory.

Most important is the fact that *in a case of classification task change there exist optimal learning set size*, which value depends on:

1) the strength of the pattern recognition task change (value of parameter β) and
2) the time that has passed after the task change, N_{new}.

The place of the minimum is greatly affected by a time passed after the task change. If the change is powerful, the place of the minimum only negligibly exceeds $2N_{\text{new}}$.

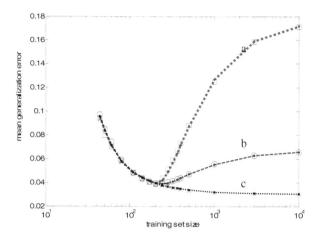

Fig. 1. Generalization error of standard linear Fisher classifier as a function of training set size in situations when the classification tasks changes suddenly: a) large change (red), b) – medium change (blue), c – no change (black). Theoretical and experimental results.

4   Multi-agent System to Tackle Changeability Problem

Calculations performed according to Eq. (7) advocate that at each time moment elapsed after the task change, generalization errors of the classification rules trained on learning sequence of diverse length will be different. In real world situations, the intervals between the environmental changes and their types and strengths are varying unpredictably. Therefore, for each single classification (or prediction) algorithms it is difficult to adapt speedily to all the changes. One may hope that a multi-agent system composed of a great diversity of classifiers (adaptive intelligent agents) would react to the changes more rapidly. As a first attempt to develop such rapidly reacting system we considered a decision making schema composed of m adaptive agents (non-linear single layer perceptrons, SLP). The SLP was selected to model the intelligent agents since the nonlinear SLP has many traits of universality [26]. We believe that such model would provide opportunity to formulate various general statements.

In neural network training, the agents' performance depends also on training conditions, first of all, on initial weights (a position of starting hyperplane in multidimensional feature space and magnitudes of the weights) and differences between target values. If initial weighs are selected correctly, the perceptron learns quickly. Its generalization error could be small if stopping moment would be chosen properly [27].

Large differences between the targets values of the perceptron could lead to very fast training of the perceptron at the very beginning. Later components of the weight vector are becoming large and training process slows down later [28]. In order to ensure fast training all the time, special regularization means have to be used. We used corrupted targets to prevent the weight to grow too large. Corrupted targets (a noise injection) appeared to be useful strategy to increase the SLP based adaptive agent ability to overcome sudden changes of classification tasks [3], [16], [17], [28].

Below we describe a procedure used to create the MAS approach based adaptive pattern recognition system. We consider a long sequence of environmental changes which were mimicked by altering two pattern recognition tasks, Task 1 and Task 2, which artificial agents have to solve. Both classification tasks considered were two-class ten-dimensional Gaussian classes, $N(\mu_1, \Sigma)$, $N(\mu_2, \Sigma)$. We alternated the mean vectors, μ_1, μ_2, and matrix Σ: $\mu'_i = T\mu_i$, $\Sigma = T\begin{bmatrix} 1 & 0.98 \\ 0.98 & 1 \end{bmatrix} T$, with $T = \begin{bmatrix} \rho & 0 \\ 0 & B(t) \end{bmatrix}$ and $B(s) = h^{(-1)^s}$, $s = 1, 2, \ldots, s_{max}$, where parameter h characterizes the strength of the environmental change. Thus, in turn the data were rotated counter clockwise and then clockwise. Other eight features were non-informative Gaussian zero mean vectors. Generalization errors of the trained classifiers were calculated analytically since decision boundary was a hyperplane, and the parameters of GCCM data were known.

In Figure 2 we present two pattern classification tasks in a space of the first two features. On the left and center, we see two tasks corresponding to $B(s) = 1.8$ and $B(s+1) = 1/1.8$, $\mu_1 = (-0.15, 0.15, 0, \ldots, 0)^T$, $\mu_2 = -\mu_1$. Hypothetically, the final weight vector obtained while training with data of Task 1 is used as the initial weight vector while training with the data of Task 2 (dotted line). On the very right, we present the most frequent situation where after the recent task change, the classifier is trained by a mixture of the old and new data. In Fig. 2 we see a situation where learning sets of each of two classes are composed of 20 vectors of Task 1 and 16 vectors of Task 2.

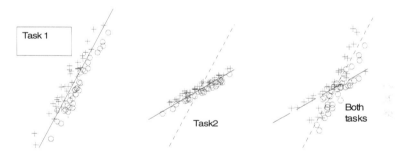

Fig. 2. Changes of the pattern recognition tasks and sets of training data

In each of the experiments, the strengths of the task changes (parameter h) do not vary during $s_{max} = 350$ of the PR task changes. A time interval between two subsequent PR tasks, however, was varying. The time was measured by a number of training epochs of the SLP based classifier. In Figure 3a we have a graph that shows the alteration of the interval. The interval is evaluated as a number of training sessions composed of 10 total gradient descent iterations. So, at the very beginning, the interval between two changes is 240 iterations. Fig 3a shows that the interval diminishes until 80 iterations between the 80th and 200th PR task changes.

To have the classifiers for prediction diverse each of them possessed its individual length of learning sequence (a number of training vectors ($n=2N$) and a fraction, ν, of corrupted targets. Two strategies to generate diverse agents were studied.

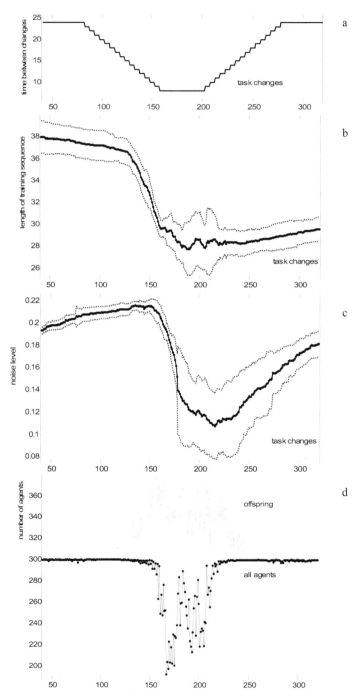

Fig. 3. Dynamics of (a) - a time between the task changes, (b) - a length of training sequence, (c) - a fraction of corrupted (incorrect) training directives and (d) - a total number of agents (black) and that of offspring (green) during the 280 pattern recognition task changes

The strategy with fixed parameters. In this experiment we generated $m=300$ agents with randomly generated values of "learning style" parameters, n and v. The parameters *did not change* during sequences of 350 PR task changes. The non-linear SLP based classifiers were trained by standard total gradient training algorithm with learning step $\eta=1.5$. Typically, after each task change, classification error jumped up suddenly. After 20 – 50 batch iterations, classification error dropped down notably.

The strategy with evolvable parameters. In second experiment, we also generated $m=300$ agents with randomly determined fixed values of n and v which *did change* during the sequence of 350 environmental changes. Variation of parameters, n and v, were governed by a memetic like learning algorithm. The agents which did not succeed adapt rapidly (satisfy requirement $P_{error} < P_{goal} = 0.12$) were replaced by offspring. The offspring started to train from initial weight vector with components equal to zero.

Each offspring, however, inherited the learning sequence length (parameter n) and the fraction, v, of corrupted training signal from a randomly chosen successful agent that satisfied the reproduction requirement, $P_{error} < 0.8 \times P_{goal}$. In addition, small mutations of parameters n and v were introduced. Such strategy resulted that parameters n and v "followed" frequency of the classification task changes (see Figures 3b and 3c where we have a mean and "a mean ± standard deviation" at each time moment). Genetic adaptation causes that during each time moment the agent population is composed of agents with diverse values of n and v. We pay readers attention that the agent population size and that of offspring reacted to environmental changes most rapidly (Figure 3d). Reaction of parameters n and v was much more sluggish. The graphs in Figure 3 are *averages of ten experiments* with evolving populations of 300 agents. In indicates that in Fig. 3bc, the mean values are calculated of 2000 - 3000 existing agents.

Final decision making was performed by "the best" single agent selected after preceding training cycle (ten batch training iterations) in the experiments described below. Performance of decision making system was measured by "*adjustment times*", - a number of training cycles (one cycle = 10 batch iterations) required to decline classification error up to threshold P_{goal} after each PR task change. It means that in our algorithm, *learning rapidity served as a fitness function.*

In Figure 4a we have a histogram of distribution of 300 mean (average) adjustment times (measured in training cycles) calculated in an interval between the 41st – 320th task changes for 300 adaptive agents with different *a priori fixed* learning style parameters. We see, 25 to 60 batch iterations were required to react to the task change. The best from 300 randomly generated individual agents required on average 25.9 training cycles in order to diminish its classification error until the survivability threshold, $P_{goal} = 0.12$. It is the average of the 280 task changes. A distribution of the 280 time intervals is portrayed by a histogram in Figure 3b. The MAS composed of 300 adaptive agents with *a priori fixed* learning style parameters, however, did not outperformed the best individual agent: an average reaction time of such system was actually the same, 26 training cycles.

In Figure 4c we have a histogram of distribution of 300 adjustment times (in tens of batch iterations) of 300 *evolvable agents* registered after the 280th task change. Markedly less, 13 to 39, training cycles were required to react to the task changes. On average, 22.2 training cycles were necessary in the case of evolving MAS. Figure 4c suggests that the distribution of the reaction times is multimodal. It means that during 280 task changes the adaptive agents turned out to be clustered into the groups.

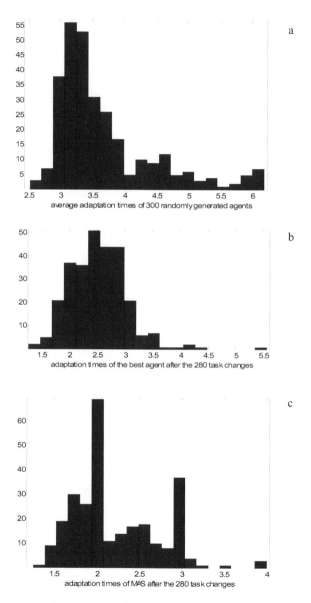

Fig. 4. The histograms of distribution of the number of learning cycles required to decline classification error up to threshold P_{goal} after each task change: (a) - 300 randomly formed individualistic adaptive agents, (b) – the best individualistic agent, (c) – MAS composed of 300 agents with inheritance

It is worth noting that the MAS described above was only the first and simplest attempt developed without an inclusion of various "tricks" that help the population of intelligent adaptive agents to overcome severe environmental changes. Our previous research has shown that organizing the agents into the groups where they help each other, and does not allow other groups to pass away during "hard times" (an "altruism"), inclusion of genetically controlled difference between the target values, $t_j^{(1)}$ and $t_j^{(2)}$ (a stimulation), synthetic emotions (self-stimulation) is also useful (see [16] and references therein).

We performed several experiments where the power of environmental changes (parameter h in expression $B(s)=h^{(-1)^s}$) varied in time. The agents in the population were split into the groups where successful agents transmitted the learning sequence length parameter, n, to offspring of its own group. Preliminary experiments showed that the grouped structure of the agent population assists in obtaining higher differentiation in the values of learning style parameters, n and v. In such cases, we observed certain diversity of the agent population that makes the agent population more resistible to most powerful environmental changes. In such scenario we have a complex interplay between a variety of the parameters that define learning style of the agents and their groups. Results of the research will be published elsewhere later.

In standard evolutionary algorithms, fitness function is related with classification performance. In our multi-agent system, the main criteria used to govern evolution of the agent populations were learning rapidity and survivability of the population. One may hope that further tailoring of the evolution criteria to the requirements of speeding up the reaction time, would make the evolving MAS even more viable.

5 Discussion and Suggestions for Future Research

The analysis has shown that in suddenly changing environments, the decision making algorithms should adjust their learning characteristics to periodicity and magnitude of the changes. First of it, is the length of learning sequence. We derived the equation to calculate generalization error of standard Fisher classifier in dependence on learning set size, a time passed after the classification task alteration and the power of the pattern recognition task change. The equation obtained allows calculating optimal length of learning sequence with respect to minimum generalization error if the data parameters and that of the changes are known.

Unfortunately, in real world problems, the data ant their parameters are unknown and changing in time unexpectedly. Therefore, a prototype of evolving multi-agent pattern recognition system was suggested and investigated experimentally with artificially generated non-stationary multivariate time series. In this system we learned genetically the values of the length of training sequence and the fraction of corrupted training directives. Simulation studies showed that even utilization of simple evolving MAS based on the strategy of "the most rapidly learning agent" outperformed the best individual adaptive agents in final decision making.

Certain research efforts were made to test new collective decision making strategy to predict commodity prices by utilizing real world data [29], [30]. We formulated financial time series forecasting task as pattern classification problem defining the classes as *notable increase, notable decrease* and *insignificant change* of a chosen financial variable at time $t+1$ as compared to time t. A preceding part of the historical data was used to adjust parameters of evolving MAS based forecasting system, and the rest of the data was used to test the MAS like in the experiments described above. The experiments have shown that utilization of evolving multi-classifier forecasting system with fusion of the best agent decisions outperforms decisions of individually acting adaptive pattern classifiers. Results of our research will be published elsewhere later.

The evolving MASs tested in this paper were only one of the first attempts. Many possibilities exist to improve evolution process and collective decision making. At first, the larger number of the parameters that govern learning style of individual agents should be included into the MAS evolution procedure. Second, more sophisticated survival criteria to be used in order to obtain an "optimal diversity" of the agents in the MAS should be developed. Finally, instead of the best instantaneous agent strategy used to make final classification in the present paper, usefulness of more complex fusion rules should be explored (for recent review of such rules see e.g. [31] and references therein).

Acknowledgments

The author thanks Dr. Aistis Raudys and Indrė Žliobaitė for useful discussions and aid in performing experiments with real world financial time series.

References

[1] Oommen, B.J., Masum, H.: Switching models for non-stationary random environments. IEEE Trans. on Systems, Man and Cybernetics 25, 1334–1339 (1995)

[2] Weng, J., McClelland, J., Pentland, A., Sporns, O., Stockman, I., Sur, M., Thelen, E.: Autonomous mental development by robots and animals. Science 291(5504), 599 (2001)

[3] Raudys, S.: Survival of intelligent agents in changing environments. In: Rutkowski, L., Siekmann, J.H., Tadeusiewicz, R., Zadeh, L.A. (eds.) ICAISC 2004. LNCS (LNAI), vol. 3070, pp. 109–117. Springer, Heidelberg (2004)

[4] Kuncheva, L.: Classifier ensembles for changing environments. In: Roli, F., Kittler, J., Windeatt, T. (eds.) MCS 2004. LNCS, vol. 3077, pp. 1–15. Springer, Heidelberg (2004)

[5] Rutkowski, L.: Adaptive probabilistic neural networks for pattern classification in time-varying environment. IEEE Trans. on Neural Networks 15, 811–827 (2004)

[6] Farmer, J.D., Packard, N.H., Perelson, A.S.: The immune system, adaptation, and machine learning. Physica 22D, 187–204 (1986)

[7] Moody, J.: Economic forecasting: challenges and neural network solutions. In: Proc. of the Int. Symposium on Artificial Neural Networks, Hsinchu, Taiwan (1995)

[8] Huang, W., Lai, K.K., Nakamori, Y., Wang, S.: Forecasting foreign exchange rates with artificial neural networks: A review. Int. J. of Information Technology and Decision Making 3(1), 145–165 (2004)

[9] Raudys, S., Zliobaite, I.: Prediction of commodity prices in rapidly changing environments. In: Singh, S., Singh, M., Apte, C., Perner, P. (eds.) ICAPR 2005. LNCS, vol. 3686, pp. 154–163. Springer, Heidelberg (2005)

[10] Figueiredo, R.J.P.: Convergent algorithms for pattern recognition in nonlinearly evolving nonstationary environment. Proceedings of IEEE 56, 188–189 (1968)

[11] Tzypkin, J.Z.: Introduction to the Self-Learning Systems Theory. Nauka, Moscow (1970)

[12] Bartlett, P.L., Ben-David, S., Kulkarni, S.R.: Learning changing concepts by exploiting the structure of change. Machine Learning 41, 153–174 (2000)

[13] Oommen, B.J., Rueda, L.: Stochastic learning-based weak estimation of multinomial random variables and its applications to pattern recognition in non-stationary environments. Pattern Recognition 39, 328–341 (2006)

[14] De Castro, L.N., Timmis, J.: Artificial Immune Systems: A new computational intelligence approach. Springer, London (2001)

[15] Krasnogor, N., Smith, J.: A tutorial for competent memetic algorithms: model, taxonomy, and design issues. IEEE Trans. on Evolutionary Computation 9, 474–488 (2005)

[16] Raudys, S.: Social organization of evolving multiple classifier system functioning in changing environments. Adaptive and Natural Computing Algorithms. In: ICANNGA 2007. LNCS, vol. 4431, 4432, Springer, Heidelberg (2007)

[17] Oommen, B.J., Raghunath, G., Kuipers, B.: Parameter learning from stochastic teachers and stochastic compulsive liars. IEEE Trans. on Systems, Man and Cybernetics 25(B), 820–834 (2006)

[18] Widrow, B., Hoff, M.E.: Adaptive switching circuits. WESCON Convention Record 4, 96–104 (1960)

[19] Haykin, S.: Neural Networks: A comprehensive foundation, 2nd edn. Prentice-Hall, Englewood Cliffs (1999)

[20] Raudys, S.: Evolution and generalization of a single neurone. I. SLP as seven statistical classifiers. Neural Networks 11, 283–296 (1998)

[21] Raudys, S.: Statistical and Neural Classifiers: An integrated approach to design. LNCS. Springer, Heidelberg (2001)

[22] Fukunaga, K.: Introduction to Statistical Pattern Recognition. Academic Press, London (1990)

[23] Raudys, S.: On the amount of a priori information in designing the classification algorithm (in Russian). Engineering Cybernetics N4, 168–174 (1972), see also: http://www.science.mii.lt/mii/raudys/abstracts.html

[24] Wyman, F., Young, D., Turner, D.: A comparison of asymptotic error rate expansions for the sample linear discriminant function. Pattern Recognition 23, 775–783 (1990)

[25] Raudys, S., Young, D.M.: Results in statistical discriminant analysis: a review of the former Soviet Union literature. J. of Multivariate Analysis 89, 1–35 (2004)

[26] Raudys, S.: On the universality of the single-layer perceptron model. In: Rutkowski, L. (ed.) Neural Networks and Soft Computing Physica-Verlag, pp. 79–86. Springer, Heidelberg (2002)

[27] Raudys, S., Amari, S.: Effect of initial values in simple perception. In: IJCNN'98. Proc. 1998 IEEE World Congress on Computational Intelligence, pp. 1530–1535. IEEE Press, Los Alamitos (1998)

[28] Raudys, S.: An adaptation model for simulation of aging process. Int. J. of Modern Physics C. 13, 1075–1086 (2002)

[29] Raudys, S., Zliobaite, I.: The multi-agent system for prediction of financial time series. In: Rutkowski, L., Tadeusiewicz, R., Zadeh, L.A., Zurada, J.M. (eds.) ICAISC 2006. LNCS (LNAI), vol. 4029, pp. 653–662. Springer, Heidelberg (2006)

[30] Raudys, S., Zliobaitė, I.: Evolving multi-agent system for prediction of slow trends and random fluctuations in the oil prices (in preparation 2007)

[31] Raudys, S.: Trainable Fusion Rules. I. Large sample size case. Neural Networks 19, 1506–1516 (2006)

Equivalence Learning in Protein Classification

Attila Kertész-Farkas[1,2], András Kocsor[1,3], and Sándor Pongor[4,5]

[1] Research Group on Artificial Intelligence of the Hungarian Academy of Sciences
and University of Szeged, Aradi vértanúk tere 1, H-6720, Szeged, Hungary
{kfa, kocsor}@inf.u-szeged.hu
[2] Erasmus Program, Technische Universität Dresden, Germany
[3] Applied Intelligence Laboratory, Petofi S. sgt. 43, H-6725, Szeged, Hungary
[4] Bioinformatics Group, International Centre for Genetic Engineering and
Biotechnology, Padriciano 99, I-34012 Trieste, Italy
pongor@icgeb.org
[5] Bioinformatics Group, Biological Research Centre, Hungarian Academy of
Sciences, Temesvári krt. 62, H-6701 Szeged, Hungary

Abstract. We present a method, called equivalence learning, which applies a two-class classification approach to object-pairs defined within a multi-class scenario. The underlying idea is that instead of classifying objects into their respective classes, we classify object pairs either as equivalent (belonging to the same class) or non-equivalent (belonging to different classes). The method is based on a vectorisation of the similarity between the objects and the application of a machine learning algorithm (SVM, ANN, LogReg, Random Forests) to learn the differences between equivalent and non-equivalent object pairs, and define a unique kernel function that can be obtained via equivalence learning. Using a small dataset of archaeal, bacterial and eukaryotic 3-phosphoglycerate-kinase sequences we found that the classification performance of equivalence learning slightly exceeds those of several simple machine learning algorithms at the price of a minimal increase in time and space requirements.

1   Introduction

The classification of proteins is a fundamental task in genome research. In a typical application, a protein sequence object (a string of several tens to several hundred characters) has to be classified into one of the several thousand known classes, based on a string similarity measure. Sequence similarity is thus a key concept since it can imply evolutionary, structural or functional similarity between proteins.

Early methods of protein classification relied on the pairwise comparison of sequences, based on the alignment of sequences using exhaustive dynamic programming methods [1] [2] or faster, heuristic algorithms [3][4]. Pairwise comparison yielded a similarity measure that could be used to classify proteins on an empirical basis. The next generation of methods then used generative models for the protein classes and the similarity of a sequence to a class was assessed by a score computed between the model and the sequence. Hidden Markov Models (HMMs) are now routinely used in protein classification [5], but there are many

P. Perner (Ed.): MLDM 2007, LNAI 4571, pp. 824–837, 2007.

other, simpler types of description in use (for a review, see: [6]). Discriminative models (such as artificial neural networks and support vector machines etc.) are used in a third generation of protein classification methods where the goal is to learn the distinction between class members and non-members. Roughly speaking, 80-90% of new protein sequence data can be classified by simple pairwise comparison. The other, more sophisticated techniques are used mostly to verify whether a new sequence is a novel example of an existing class or it represents a truly new class in itself. As the latter decisions refer to the biological novelty of the data, there is a considerable interest in new, improved classification methods.

Kernel methods represent a subclass of discriminative models in which a pairwise similarity measure calculated between objects is used to learn the decision surface that separates a class from the rest of the database. The kernel function can be regarded as a similarity function which has the additional property of always being positive semi-definite, which allows many novel applications to non-linear problems [7]. In the context of protein classification, kernel methods have an important practical advantage: the similarity measures developed in protein classification can be used to construct kernel functions, and so decision making can directly capitalize on the considerable empirical knowledge accumulated in various fields of protein classification. Over the past decade, many kernels have been developed for sequences such as the String kernel [8], Mismatch kernel [9], Spectrum kernel [10], Local Alignment Kernel [11] and the Fisher kernel [12]. For a good review of these applications, see [7].

This work aims to use a conceptual approach that is slightly different from the mainstream use of kernel functions. Let us first consider a database of objects and a similarity measure computed between each pair of objects. This setup can be visualized as a weighted graph (network) of similarities where the nodes are the proteins and the weighted edges represent the similarities between them. The network can also be represented as a symmetrical matrix in which the cells represent the pairwise comparison measures between the objects. Figure 1a shows a hypothetical database of 8 objects. We can vaguely recognize two groups in which the members are more similar to each other than to the objects outside the groups. Let us now suppose that an expert looks at the similarity data and decides that the two groups represent two classes, A and B, and there is another object that is not a member of either of these. Figure 1b illustrates this new situation. The members of the groups are now connected by an equivalence relation that exists only between the members of a given group. As a result, the similarity matrix becomes a simpler *equivalence matrix* in which only the elements between the members of the same class are non-zero. The aim of this study here is to use a similarity matrix given between a set of objects, and use it to learn the equivalence matrix defined by the classification task, as shown in the bottom part of Figure 1. We term this approach *equivalence learning*, which is characterized as follows. Protein classification methods seek to classify the objects, i.e. the nodes of the similarity network, which is a multi-class problem. In contrast, here we try to classify the edges of the similarity network into just two classes, one signifying equivalence the other the lack of it.

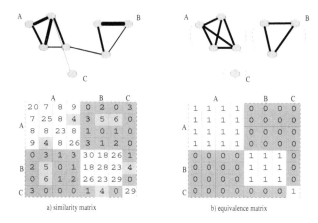

		A				B		C
	20	7	8	9	0	2	0	3
	7	25	8	4	3	5	6	0
A	8	8	23	8	1	0	1	0
	9	4	8	26	3	1	2	0
	0	3	1	3	30	18	26	1
B	2	5	0	1	18	28	23	4
	0	6	1	2	26	23	29	0
C	3	0	0	0	1	4	0	29

a) similarity matrix

		A				B		C
	1	1	1	1	0	0	0	0
	1	1	1	1	0	0	0	0
A	1	1	1	1	0	0	0	0
	1	1	1	1	0	0	0	0
	0	0	0	0	1	1	1	0
B	0	0	0	0	1	1	1	0
	0	0	0	0	1	1	1	0
C	0	0	0	0	0	0	0	1

b) equivalence matrix

Fig. 1. Principle of Equivalence Learning (EL) I. A similarity matrix (left) can be determined by an all vs. all comparison of a database using algorithms such as BLAST or Smith-Waterman. The equivalence matrix (right) is a representation of groups (A,B) and an outlier identified by human experts. EL tries to learn the equivalence.

In more detail, a sequence pair is called equivalent when both of them belong to the same sequence group and is called non-equivalent or distinct when they do not. We define the *equivalence function* that returns one for equivalent sequences and zero for non-equivalent sequences taken from a different group - where the group could be a protein family, superfamily, fold or class. This function can be formulated as follows

$$\delta(s,t) = \begin{cases} 1 & s \text{ and } t \text{ belong to the same sequence group,} \\ 0 & \text{otherwise.} \end{cases}$$

Since the equivalence function defined over the sequence pairs gives a partition (equivalent/non-equivalent sequence pairs), learning this function essentially becomes a two-class classification problem that can be solved by one of the existing classification schemes. For clarity, the process is shown in Figure 2. A key issue here is to decide how we should represent the edges of the similarity network so that we can efficiently predict equivalence. A simple numerical value is not sufficient since, as shown in the example of Figure 1, a value 9 is sufficient for class membership in class A, but not in class B. As a solution, we will use a vectorial representation for the edges, hence we will need projection methods in order to represent a sequence pair in one vector. Now let $P : \mathcal{S} \times \mathcal{S} \to \mathbb{R}^n$ be such a projection function and let the corresponding database be

$$L = \{(x_i, y_i) \mid x_i = P(s_i, t_i), \; y_i = \delta(s_i, t_i), \; s_i, t_i \in \mathcal{S}\}, \tag{1}$$

where $\mathcal{S}$ denotes the set of sequences.

The construction of an equivalence function has been proposed by Cristianini et al who defined the concept as the ideal kernel, along with a kernel alignment

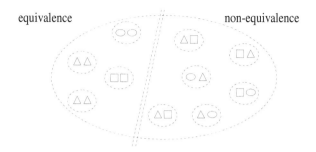

equivalence   non-equivalence

Fig. 2. Principle of equivalence learning II. Equivalence learning is a two-class problem defined on object pairs.

measure that quantifies the difference between a kernel and the ideal one [17]. Semidefinite programming techniques were then used to learn the kernel matrix from data and it was shown that using the labeled part of the data one can learn an embedding too for the unlabeled part [18]. Tsang and Kwok once suggested a feature weighting technique that allows one to approximate the ideal kernel [19].

In this paper we will examine how an equivalence function can be learnt with a standard machine learning method such as Artificial Neural Networks (ANNs) [13], Support Vector Machines (SVMs) [14] Logistic Regression (LogReg) [15] or Random Forest (RF) [16]. These methods learn a function f and for any sequence pair (s, t) return a score $f(z)$, where $z = P(s, t)$ and P is a projection function. If this score is greater than a certain threshold then the sequence pair classified is an equivalent pair, otherwise it is a non-equivalent pair. Unlike a class label predicted for a single protein, the score $f(z)$ can be considered as a similarity measure for a sequence pair (s, t) and thus the learned machine can be regarded as a fast similarity function for a sequence pair. In Section 2.3 we offer a way of constructing a kernel function from a decision boundary $f(z)$ obtained by SVM. Section 3 describes experiments that illustrate how these method can learn the equivalence function and how we can evaluate them in a protein classification context. Section 4 then gives a brief summary along with some conclusions.

2   Methods

2.1   Vectorization Step

Now we will present a way of defining projection functions which map any sequence pair into a fixed-length vector of real numbers, that is $P : \mathcal{S} \times \mathcal{S} \to \mathbb{R}^n$.

First, we define a method to vectorize a single sequence into a vector. The essential idea behind this is that a chosen protein sequence can be effectively captured by asking how similar a protein is to a large collection of other proteins [20][21]. Let us view a fixed set of sequences $\{f_1, f_2, \ldots, f_n\} \subseteq \mathcal{S}$ as a feature set and an arbitrary similarity function D. For a sequence $s \in \mathcal{S}$ let the corresponding vector w whose components are indexed by f_i and its corresponding value w_{f_i} be the similarity score between f_i and the target sequence s, that is $\phi_D : \mathcal{S} \to \mathbb{R}^n$ such that $\phi_D(s)_{f_i} = D(s, f_i)$, where v_i denotes the component of vector v indexed by i. A mapping of this type is also known as an empirical feature map [22].

The next step is to form a vector from two sequence objects. Table 1 summarizes a few simple methods which were used in our experiments. For ease of notation the operators $a \cdot b$, $\sqrt{a}$ and a^n on $\mathbb{R}$ were extended to vectors and they were defined on vectors in a coordinate-wise manner, i.e. for any vector $u, v \in \mathbb{R}^n$ $(u \cdot v)_i = u_i v_i$, $(\sqrt{v})_i = \sqrt{v_i}$ and $(v^n)_i = (v_i)^n$. We should note that unlike other operators the concatenation operator maps to $\mathbb{R}^{2n}$ instead of $\mathbb{R}^n$.

Table 1. Summary of the used vector composition method

Name	Formula	Description
Concatenation	$C_C(u, v) = (u, v)$	gives the concatenation of two vector
Sum	$C_+(u, v) = u + v$	summarizes the vector components
Product	$C_{\bullet}(u, v) = u \cdot v$	product of the vector components
Quadratic	$C_Q(u, v) = (u - v)^2$	quadratic distance between the components
Hellinger	$C_H(u, v) = (\sqrt{u} - \sqrt{v})^2$	a normalized form of quadratic distance

We use the notation $P_V^C : \mathcal{S} \times \mathcal{S} \to \mathbb{R}_+^n$ for the projection function which maps any sequence pair into an n-dimensional vector space. The subscript V of this function denotes the vectorization method for each sequence and the superscript C defines the vector composition method. For example, if the Smith-Waterman (SW) similarity function is used to vectorize a sequence, and the product function $C_{\bullet}$ is used to construct one vector from two, then the projection function we get will be denoted by $P_{\bullet}^{SW}(x, y) = C_{\bullet}(\phi_{SW}(x), \phi_{SW}(y))$.

2.2   Classifier Algorithms

Now the machine learning methods on the set L defined by Eq. 1 should be able to learn an equivalence function. In the following section we will give a short summary of classification methods used in our experiments.

Artificial Neural Networks (ANN) are good at fitting functions and recognizing patterns. In fact there is a proof that a fairly simple neural network can fit any practical function $f(x) = y$. An ANN is composed of interconnected simple elements called neurons and organized in levels. In our study the network structure consisted of one hidden layer with 40 neurons and the output layer consisted of one neuron. In each neuron the log-sigmoid function was used as the transfer function and the Scaled Conjugate Gradient (SCG) algorithm was used for training. The package we applied was the Neural Network Toolbox 5.0 version part of Matlab.

The Support Vector Machine (SVM) gives a decision boundary $f(z) = \langle z, w \rangle + b$ (also called a hyperplane) with the largest margin between the positive and negative classes. Replacing the inner product by a kernel function leads to a nonlinear decision boundary in the feature space. In our experiments the Radial Basis Function kernel was used and its width parameter σ was the median Euclidean distance from any positive training example to the nearest negative example. During the training the class labels 0 for the miscellaneous pair were replaced by -1. Here the SVM used was the LibSVM [23]. The One-Class SVM also also evaluated to learn equivalence members.

The Logistic Regression (LogReg) is one of the generalized linear models which is used when the response variable is a dichotomous variable (i.e. it can take one of two possible values) and the input variables are continuous. Unlike linear regression, logistic regression does not assume a linear relationship between the input and the output, but it does exploit the advantages of the linear methods. To do this, it uses an $ln\left(\frac{p}{1-p}\right) = \langle w, x \rangle + b$ function, called a link function, and thus it leads to a non-linear relationship, where the p is the probability that $y_i = 1$. In our study the LogReg was part of Weka version 3-4.

The Random Forest (RF) technique is a combination of decision trees such that each tree is grown on a bootstrap sample of the training set. For each node the split is chosen from $m \ll M$ variables (M being the number of dimensions) selected from an independent, identically distributed random variable taken from the feature set. In our experiments 50 trees were used and the number of features m was set to $\log l + 1$, where l is the number of input patterns. The RF was part of Weka Version 3-4.

2.3 Learned Kernel Functions

Here we present a way of constructing a kernel function from the decision boundary $f(z) = \langle z, w \rangle - \rho$ obtained by One-Class SVM. After training a One-Class SVM on the set of equivalence members $L = \{x_i \mid x_i = P(s_i, t_i), 1 = \delta(s_i, t_i)\}$ the w parameter of the decision boundary can be expressed as a weighted linear combination of support vectors, that is $w = \sum_i \alpha_i x_i$, where $\alpha_i > 0$ are the corresponding Lagrangian multiplier. Here the support vectors are equivalence sequence pairs which belong to L. Thus the decision function

$$f(P(s,t)) = \sum_i \alpha_i \langle P(s,t), x_i \rangle - \rho \qquad (2)$$

can be regarded as a similarity function over sequence pairs. Moreover, omitting the ρ additive constant from f does not change the essence of similarity. In the following lemma we shall examine what kind of projection function would make a kernel function.

Lemma 1. *Let $P_\bullet^\phi, P_+^\phi, P_Q^\phi, P_H^\phi : \mathcal{S} \times \mathcal{S} \to \mathbb{R}_+^n$ be a symmetric projection function, where $\phi : \mathcal{S} \to \mathbb{R}_+^n$ is an arbitrary positive feature mapping. Using support vectors x_i and their corresponding Lagrangian multiplier $\alpha_i > 0$ the functions*

$$SVK_\bullet(s,t) = \sum_i \alpha_i \exp(\sigma\langle P_\bullet^\phi(s,t), x_i\rangle) \tag{3}$$

$$SVK_+(s,t) = \sum_i \alpha_i \exp(\sigma\langle P_+^\phi(s,t), x_i\rangle) \tag{4}$$

$$SVK_Q(s,t) = \sum_i \alpha_i \exp(-\sigma\langle P_Q^\phi(s,t), x_i\rangle) \tag{5}$$

$$SVK_H(s,t) = \sum_i \alpha_i \exp(-\sigma\langle P_H^\phi(s,t), x_i\rangle) \tag{6}$$

are kernel functions over $\mathcal{S} \times \mathcal{S}$, where $\sigma > 0$. The class of such kernel functions is called the Support Vector Kernel (SVK).

For more detail about kernel functions and their properties, the reader should peruse [25].

Proof. The class of kernel function is closed under direct sum and positive scalar multiplication. Here, it is sufficient to prove that the exponential expressions are kernels. First, let θ a positive valued vector. Then $\langle P_\bullet^\phi(s,t), \theta\rangle$ is a kernel function because it is a weighted inner product. This follows from the fact that

$$\langle P_\bullet^\phi(s,t), \theta\rangle = \langle \phi(s) \cdot \phi(t), \theta\rangle = \phi(s)\Theta\phi(t)$$

where Θ is a diagonal matrix whose diagonal elements are taken from θ. Thus $\exp(\sigma\langle P_\bullet^\phi(s,t), \theta\rangle)$ is a kernel too.

The statement for Eq. 4 follows immediately from

$$\exp(\langle P_+^\phi(s,t), \theta\rangle) = \exp(\langle \phi(s) + \phi(t), \theta\rangle) = \exp(\langle \phi(s), \theta\rangle + \langle \phi(t), \theta\rangle) =$$
$$= \exp(\langle \phi(s), \theta\rangle)\exp(\langle \phi(t), \theta\rangle)$$

because a function of the form $k(x, y) = f(x)f(y)$ is always a kernel function. And $\langle P_Q^\phi(s,t), \theta\rangle$ is a quadratic Euclidian metric weighted by θ, that is

$$\langle P_Q^\phi(s,t), \theta\rangle = \langle \phi^2(s) + \phi^2(t) - 2\phi(s)\phi(t), \theta\rangle$$
$$= \langle \phi^2(s), \theta\rangle + \langle \phi^2(t), \theta\rangle - 2\langle \phi^2(s)\phi^2(t), \theta\rangle$$
$$= \phi(s)\Theta\phi(s) + \phi(t)\Theta\phi(t) - 2\phi(s)\Theta\phi(t).$$

Thus $\exp(-\sigma\langle P_Q^\phi(s,t), \theta\rangle)$ is a kernel. The assertion for $\exp(-\sigma\langle P_H^\phi(s,t), \theta\rangle)$ can be proved in a similar way.

The training points in L are also vectorized sequence pairs represented by a projection function. Hence the training points are a positive-valued vector, and the support vectors obtained are also positive-valued vectors. This proves our statement.   □

3   Experiments

Dataset and performance evaluation. 3PGK is a set of 131 sequences representing the essentially ubiquitous glycolytic enzyme, 3-phosphoglycerate kinase (3PGK,

358 to 505 residues in length) - obtained from 15 archaean, 83 bacterial and 33 eukaryotic species. This dataset was designed to show how an algorithm will generalize to novel, distantly related subtypes of the known protein classes [26]. The dataset is freely available at [27].

The sequences were represented by the so-called pairwise method, where the feature set was the whole train set sequences containing both positive and negative sequences. For the underlying similarity measure we chose two alignment-based sequence comparisons methods, namely the BLAST and Smith-Waterman algorithms. We used version 2.2.4 of the BLAST program with a cutoff score of 25, the Smith-Waterman algorithm was used as implemented in MATLAB. The BLOSUM 62 matrix [28] was used in both cases.

The performance evaluation was carried out by standard receiver operator characteristic (ROC) analysis, which is based on the ranking of the objects to be classified [29]. The analysis was performed by plotting sensitivity vs. 1-specificity at various threshold values, and the resulting curve was integrated to give an "area under the curve" (AUC) value. We should remark here that for a perfect ranking AUC = 1.0 while for a random ranking AUC = 0.5. In our experiments the ROC score for the 3PGK dataset was AUC averaged over 8 tasks.

Equivalence learning. For each classification task, the set of training pairs L defined in Eq. 1 consists of pairs made up from training sequences. The equivalence function δ was calculated via class labels. Here a sequence pair (s, t) is treated as equivalent if s and t belong to the same species and their equivalence score is 1. If they belong to two different species, then their equivalence score is 0. This δ may be regarded as a similarity function and be denoted by D_M where the subscript M stands for a given machine. Thus for an ANN the similarity function obtained is denoted by D_{ANN}. δ was learned on L by several classification methods.

During the evaluation, a test sequence u was paired with each of the train sequences s to check whether they belong to the same group. For the ROC analysis these sequence pairs were ranked by their score obtained via $D_M(u, s)$. As a comparison the sequence pairs were also ranked by their Smith-Waterman, and their BLAST score. The corresponding ROC score for these functions is given in Table 2.

The train pairs and test pairs can be arranged in a matrix M^D whose columns are indexed by train sequences and whose rows are indexed by test sequences. Then an element of $(M^D)_{s,t}$ is a similarity score of (s, t) obtained by a measure D. A heat map representation of a train and test matrix of one of 8 classification tasks is shown in Figure 3 below.

Train set construction. To learn an equivalence function, we randomly selected a small part of the positive and negative pairs from the train sequences. This step is necessary in order to avoid overlearning, to speed up the training and to reduce the training set to a computationally manageable size. In order to select the best number of training pairs we calculated the learned similarity function by varying the number of datasets and repeated the procedure 10 times (Figure 5). We may

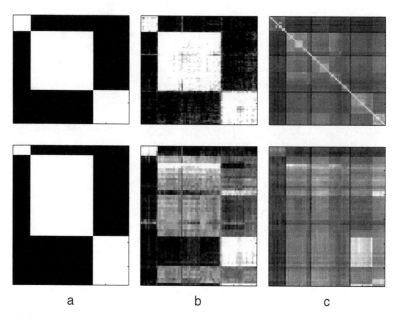

a   b   c

Fig. 3. A heat map representation of train and test matrices from the 3PGK database. The train matrix (above) is an equivalence matrix over train sequences and the test matrix (below) is an equivalence matrix between train and test sequences. The three sequence groups are easily recognizable in the equivalence matrix (a). The equivalence was learnt by RF where the sequence pair was projected by P_+^{SW}. The obtained train and the test matrices are displayed in (b). The Smith-Waterman similarity matrices (c) are also shown for comparison.

Table 2. Evaluation of the equivalence function learning using different vectorization methods.

		D_{SVM}	D_{SVK}	D_{ANN}	D_{LogReg}	D_{RF}
	C_C	0.9146	n.a	0.8600	0.6856	0.8311
SW	C_+	0.9107	0.6308	0.8948	0.7110	0.8446
0.7302[1]	$C_\bullet$	0.9022	0.6508	0.8870	0.6452	0.8516
	C_Q	0.8473	0.7901	0.8135	0.7448	0.8383
	C_H	0.8586	0.7919	0.8429	0.7571	0.8250
	C_C	0.9193	n.a	0.8800	0.6879	0.8605
BLAST	C_+	0.9184	0.6339	0.8906	0.7189	0.8649
0.7372[1]	$C_\bullet$	0.9085	0.6565	0.8839	0.6517	0.8703
	C_Q	0.8561	0.7966	0.8068	0.7530	0.8209
	C_H	0.8587	0.8037	0.8486	0.7617	0.8548

[1] The corresponding ROC score for the similarity method to measure how it can express the equivalence.

conclude here that the standard deviation is generally small and increasing the training points only makes it smaller. We should mention here that a reasonable choice of number of training points depends on the variability of the training

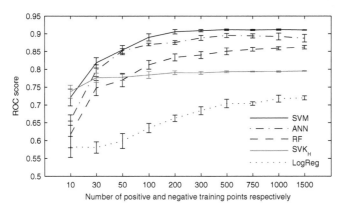

Fig. 4. Dependence of the equivalence learning results on the train set size. Except for the SVK_H case the P_+^{SW} projection method was applied.

set; protein groups in real-life databases are known to be vastly different in the number of members, in average protein size, similarity within group and so on. In our experiments 500 positive and negative training pairs were used for learning, respectively.

In the experiments we dealt only with the situation where the class labels are known for the negative sequences. If the class labels are unavailable, (i.e. it cannot be decided if two negative sequences belong to the same group), there are two potential solutions. First, any (x, y) pair where either or both of them are negative sequences, can be treated as a non-equivalent pair. In this case, the learner learns that a sequence pair belong to the particular species. The second possibility is that only the positive-negative sequence pairs are considered non-equivalent, and negative-negative pairs are removed from the training set. In our study, only the first method gave better results than the second (data not shown).

Iteration. The functions obtained by the machine learning algorithms can be used as an underlying similarity function in the pairwise vectorization approach. This step can be repeated in an iterative fashion. Here, as shown in Figure 5, the results become stable after 3-4 steps. Our empirical test told us that the trained similarity matrices (Figure 5, top) really converge to the ideal equivalence matrix but the test similarity matrix (bottom) kept some of its original mistakes, and during the iteration process these errors became more pronounced.

3.1   Classification Results Via Learned Similarity Functions

These learned similarity functions were evaluated in a protein classification context. A sequence was vectorized by the pairwise vectorization method and the feature set was the whole train sequence set. For the underlying similarity functions D_{ANN}, D_{SVM}, D_{LogReg}, D_{RF} and the original Smith-Waterman were used, while the classification method employed was SVM. The results obtained are listed in the table below.

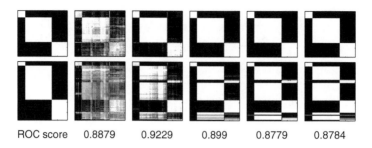

ROC score   0.8879   0.9229   0.899   0.8779   0.8784

Fig. 5. An iteration of ideal similarity learning by RF with P_+^{SW}. The leftmost heat map pair is the target equivalence matrix which was calculated by using class labels.

Table 3. Evaluation of the sequence classification via SVM

		D_{SVM}	D_{ANN}	D_{LogReg}	D_{RF}
	C_C	0.9694	0.9778	0.7709	0.9545
	C_+	0.9749	0.9866	0.7802	0.9700
SVM	$C_\bullet$	0.9759	0.9691	0.8730	0.9712
0.9651[1]	C_Q	0.9641	0.8823	0.9360	0.9434
	C_H	0.9614	0.9242	0.9499	0.9460

[1]The ROC score obtained by SVM classification with SW feature extraction.

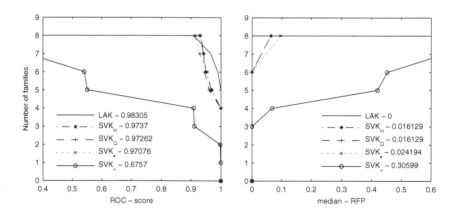

Fig. 6. SVM classification results with kernel functions. Here the BLAST algorithm was used for feature extraction.

Learned kernel results. Experiments for the support vector kernel. The results we got are summarized in Table 3. For a comparison, the Local Alignment Kernel was also evaluated. The program was used with the parameter values suggested by the authors [11].

4   Conclusions

We described a method termed equivalence learning, which applies a two-class classification approach to object-pairs defined within a multi-class scenario. The underlying idea is that instead of classifying objects into their respective classes, we classify object pairs either as equivalent (belonging to the same class) or non-equivalent (belonging to different classes). The method is based on a vectorisation of similarity between the objects. We should note that this is one of the most important steps and the results are sensitive to small changes. We think further methods should be developed to characterize similarity by several values, represented in vector form rather than by just a single value. The application of kernel methods to routine problems in protein classification is apparently hampered by the high dimensionality of vector representations. Equivalence learning is plagued by the same problem, so the reduction of dimensionality may be an important step if equivalence learning is to be applied to real-life protein databases.

The method is more complex than simple machine learning algorithms but its time and storage space requirements are not exceedingly high as compared to these methods, and in some cases it provides a better performance.

Finally we should mention that even though the expression *equivalence learning* may be novel in machine learning, it has already been used in cognitive studies [30]. We hope that this concept will become more popular in machine learning.

Acknowledgements

The work at ICGEB was supported in part by grants from the Ministero dell'Universita e della Ricerca (D.D. 2187, FIRB 2003 (art. 8), "Laboratorio Internazionale di Bioinformatica"). A. Kocsor was supported by the Janos Bolyai fellowship of the Hungarian Academy of Sciences.

References

1. Needleman, S.B., Wunsch, C.D.: A general method applicable to the search for similarities in the amino acid sequence of two proteins. J. Mol. Biol. 48, 443–453 (1970)
2. Smith, T.F., Waterman, M.S.: Identification of common molecular subsequences. J. Mol. Biol. 147, 195–197 (1981)
3. Pearson, W.R.: Rapid and sensitive sequence comparison with FASTP and FASTA. Methods Enzymol. 183, 63–98 (1985)
4. Altschul, S.F., Gish, W., Miller, W., Myers, E.W., Lipman, D.J.: Basic local alignment search tool. J. Mol. Biol. 215, 403–410 (1990)
5. Eddy, S.: HMMER Biological sequence analysis using profile hidden Markov models, Version 2.3.2 (2003) http://hmmer.janelia.org/
6. Mount, D.W.: Bioinformatics: Sequence and Genome Analysis. 2nd edn. Cold Spring Harbor Laboratory Press (2004)

7. Shawe-Taylor, J., Cristianini, N.: Kernel Methods for Pattern Analysis. Cambridge University Press, Cambridge (2004)
8. Lodhi, H., Saunders, C., Cristianini, N., Watkins, C., Shawe-Taylor, J.: String Matching Kernels for Text Classification. Journal of Machine Learning Research 2, 419–444 (2002)
9. Leslie, C., Eskin, E., Weston, J., Noble, W.S.: Mismatch string kernels for SVM protein classification. In: Advances in Neural Information Processing Systems, vol. 15, MIT Press, Cambridge (2003)
10. Leslie, C., Eskin, E., Noble, W.S.: The spectrum kernel: A string kernel for SVM protein classification. In: Proceedings of the Pacific Symposium on Biocomputing (PSB-2002), World Scientific Publishing, Singapore (2002)
11. Vert, J-P., Saigo, H., Akatsu, T.: Local alignment kernels for biological sequences. In: Schölkopf, B., Tsuda, K., Vert, J.-P. (eds.) Kernel methods in Computational Biology, pp. 131–154. MIT Press, Cambridge (2004)
12. Jaakkola, T., Diekhans, M., Haussler, D.: Using the Fisher kernel method to detect remote protein homologies. In: Proc Int. Conf. Intell. Syst. Mol. Biol. pp. 149–158 (1999)
13. Bishop, D.M.: Neural Networks for Pattern Recognition. Clarendon Press, Oxford (1995)
14. Vapnik, V.N.: Statistical Learning Theory. John Wiley & Sons, New York (1998)
15. Rice, J.C.: Logistic regression: An introduction. In: Thompson, B. (ed.) Advances in social science methodology, vol. 3, pp. 191–245. JAI Press, Greenwich, CT (1994)
16. Breiman, L.: Random forests. Machine Learning 45, 5–32 (2001)
17. Cristianini, N., Kandola, J., Elisseeff, A., Shawe-Taylor, J.: On Kernel Target Alignment. In: Advances in Neural Information Processing Systems, vol. 14, pp. 367–373 (2001)
18. Lanckriet, G.R.G., Cristianini, N., Bartlett, P., Ghaoui, L.E., Jordan, M.I.: Learning the Kernel Matrix with Semidefinite Programming. Journal of Machine Learning Research 5, 27–72 (2004)
19. Kwok, T.J., Tsang, I.W.: Learning with idealized Kernels. In: Proc. of the 28. International Confernece on Machine Learning, Washington DC (2003)
20. Liao, L., Noble, W.S.: Combining pairwise sequence similarity and support vector machines for detecting remote protein evolutionary and structural relationships. J. Comput. Biol. 10, 857–868 (2003)
21. Vlahovicek, K., Kajan, L., Agoston, V., Pongor, S.: The SBASE domain sequence resource, release 12: prediction of protein domain-architecture using support vector machines. Nucleic Acids Res. 33, 223–225 (2005)
22. Tsuda, K.: Support vector classification with asymmetric kernel function. Pros. ESANN, 183–188 (1999)
23. Chang, C.-C., Lin, C.-J.: LIBSVM: a library for support vector machines (2001) Software available at http://www.csie.ntu.edu.tw/~cjlin/libsvm
24. Witten, I.H., Frank, E.: Data Mining: Practical Machine Learning tools and Techniques with JAVA implementations. Morgan Kaufman, Seattle, Washington (1999)
25. Berg, C., Christensen, J.P.R., Ressel, P.: Harmonic Analysis on Semigroups: Theory of Positive Definite and Related Functions. Springer, Heidelberg (1984)
26. Kertesz-Farkas, A., Dhir, S., Sonego, P., Pacurar, M., Netoteia, S., Nijveen, H., Leunissen, J., Kocsor, A., Pongor, S.: A comparison of random and supervised cross-validation strategies and benchmark datasets for protein classification (submitted for publication 2007)

27. Sonego, P., Pacurar, M., Dhir, D., Kertész-Farkas, A., Kocsor, A., Gáspári, Z., Leunissen, J.A.M., Pongor, S.: A Protein Classification Benchmark collection for Machine Learning. Nucleid Acids Research
28. Henikoff, S., Henikoff, J.G., Pietrokovski, S.: Blocks+: a non-redundant database of protein alignment blocks derived from multiple compilations. Bioin-formatics 15, 471–479 (1999)
29. Gribskov, M., Robinson, N.L.: Use of Receiver Operating Characteristic (ROC) analysis to evaluate sequence matching. Comput. Chem. 20, 25–33 (1996)
30. Johns, K.W., Williams, D.A.: Acquired equivalence learning with antecedent and consequent unconditioned stimuli. J. Exp. Psychol. Anim. Behav. Process 24m, 3–14 (1998)

Statistical Identification of Key Phrases
for Text Classification

Frans Coenen, Paul Leng, Robert Sanderson, and Yanbo J. Wang

Department of Computer Science, The University of Liverpool,
Ashton Building, Ashton Street, Liverpool L69 3BX, United Kingdom
{frans,phl,azaroth,jwang}@csc.liv.ac.uk

Abstract. Algorithms for text classification generally involve two stages, the first of which aims to identify textual elements (words and/or phrases) that may be relevant to the classification process. This stage often involves an analysis of the text that is both language-specific and possibly domain-specific, and may also be computationally costly. In this paper we examine a number of alternative keyword-generation methods and phrase-construction strategies that identify key words and phrases by simple, language-independent statistical properties. We present results that demonstrate that these methods can produce good classification accuracy, with the best results being obtained using a phrase-based approach.

Keywords: Text Classification, Text Preprocessing.

1 Introduction

The increasing volume and availability of electronic documents, especially those available on-line, has stimulated interest in methods of *text classification* (TC). TC algorithms typically make use of a classifier developed from analysis of a training set of documents that have been previously classified manually. The training process usually involves two stages: first, a *preprocessing* stage to identify relevant textual characteristics in the documents of the training set, and second, a learning stage in which these characteristics are associated with class labels. We are in this paper especially concerned with TC methods that use this second stage to develop *Classification Rules* by a process of Classification Association Rule Mining (CARM).

CARM methods, and other related rule-based classification systems, require the initial preprocessing stage to identify textual components (words or phrases) that can be used in the construction of classification rules of the form $A \rightarrow c$, where A is the conjunction of a set of these components and c is a class label. Much current work on document preprocessing makes use of techniques tailored to either the language in which the documents to be classified are written (e.g. English, Spanish, Chinese, etc.) or the particular domain that the documents describe (e.g. *medline* abstracts, Biological texts, etc.). Knowledge of the language used allows the application of techniques such as natural language parsing

P. Perner (Ed.): MLDM 2007, LNAI 4571, pp. 838–853, 2007.

and stemming, and the use of stop and synonym lists. Knowledge of the domain allows the application of specialised dictionaries and lexicons or the use of sophisticated ontology structures. These approaches can produce very accurate classifiers, but are costly to implement, in terms of human resources, as they are not generally applicable, and the techniques involved may also be relatively costly in computational terms. These reasons motivate a search for methods that will identify relevant words and phrases by statistical techniques, without the need for deep linguistic analysis or domain-specific knowledge.

In this paper we examine a number of such methods for identifying key phrases (and words) in document sets to support TC. The methods all begin by using language-independent statistical methods to identify *significant* words in the document set: i.e. words that are likely to be relevant to the classification task. We investigate a number of strategies for constructing phrases, all of which make use only of simple textual analysis using significant words derived in this way. Eight different methods of generating the significant words are considered, coupled with four phrase formulation algorithms. We compare the phrase-generation methods with results obtained from simpler "bag of words" approaches. Our results demonstrate that the shallow linguistic analysis employed in our preprocessing is nevertheless sufficient to produce good classification accuracy, and that even simple phrase-construction approaches can improve on single-word methods.

The rest of this paper is organised as follows. In section 2 we describe the background and some related work relevant to this study. Section 3 outlines the CARM algorithm and the data sets that we have used to evaluate the various preprocessing strategies. Section 4 describes the methods we use for identification of significant words, and section 5 the phrase-construction algorithms. In section 6 we present experimental results, and in section 7 our conclusions.

2   Previous Work

Text for TC purposes is usually represented using the vector space model, where each document is represented as a single numeric vector d, and d is a subset of some vocabulary V. The vocabulary V is a representation of the set of textual components that are used to characterise documents. Two broad approaches are used to define this: the *bag of words* and the *bag of phrases* approaches.

In the bag of words approach each document is represented by the set of words that is used in the document. Information on the ordering of words within documents as well as the structure of the documents is lost. The vectors representing documents may comprise either (a) Word identification numbers (the *binary representation*), or (b) Words weighted according to the frequency with which they appear in the document (the *term-weighted representation*). The problem with the approach is how to select a limited, computationally manageable, subset of words from the entire set represented in the document base. Methods include the use of stop and synonym lists and stemming, or the use of a domain-dependent set of key words or named entities. These are all options that make use of knowledge of the language in which the documents in the document set are written, an approach which, for reasons discussed above, we wish to avoid.

In the bag of phrases approach each element in a document vector represents a phrase describing an ordered combination of words appearing in sequence, either contiguously or with some maximum word gap. A variety of techniques exist for identifying phrases in documents, most of which again make use of some kind of meta knowledge (either the application domain or the language used in the document set). For example Sharma and Raman in [10] propose a phrase-based text representation for web document management using rule-based Natural Language Processing (NLP) and a Context Free Grammar (CFG). In [4] Katrenko makes an evaluation of the phrase-based representation.

In [6] and [8] a sequence of experiments is described comparing the bag of keywords approach with the bag of phrases approach in the context of text categorisation. The expectation was that the phrase based approach would work better than the keyword approach, because a phrase carries more information; however the reverse was discovered. In [9] a number of reasons for this are given:

1. Phrases have inferior statistical properties.
2. Phrases have lower frequency of occurrence than keywords.
3. The bag of phrases includes many redundant and/or noise phrases.

We hypothesise that these drawbacks can be overcome by the use of appropriate classification algorithms. It is clear that phrases will be found in fewer documents than corresponding key words, but conversely we expect them to have a greater discriminating power. To take advantage of this, we require algorithms that will identify classification rules with relatively low applicability as well as very common ones. To avoid problems of noise, conversely, we require the ability to discard rules that fall below defined thresholds of validity. These requirements point us to the use of CARM algorithms to construct classification rules using the identified words and/or phrases. CARM approaches are based on methods of Association Rule Mining that rely on the examination of large data sets to identify even scarce rules without overfitting. A number of studies (e.g. [1], [7], etc.) have demonstrated that, for many classification problems, CARM approaches can lead to better classification accuracy than other methods. Earlier work by the authors [2] [3], employing a CARM algorithm, TFPC, showed that appropriate selection of thresholds led to high classification accuracy in a wide range of cases. In the present work we seek to apply this algorithm to the TC problem, and to identify parameter values to optimise its accuracy.

3   Experimental Organisation

All experiments described in this paper were undertaken using the authors' TFPC algorithm [2] [3]. TFPC (Total From Partial Classification) is a CARM algorithm that constructs a classifier by identifying Classification Association Rules (CARs) from a set of previously-classified cases. A CAR is a special case of an Association Rule for which the consequent is a class-label. As is the case for association rules in general, CARs can be characterised by their *support* (the relative frequency with which the rule is found to apply), and *confidence* (the ratio of their support to the frequency of the antecedent of the rule). An appropriate selection of threshold values for support and confidence is used to define a set of rules from which the classifier is constructed. The unusual feature of TFPC is that, when the algorithm finds a general rule that meets its threshold requirements, it does not search for any more specific rule whose antecedent is a superset of this. This heuristic makes TFPC less prone to overfitting than other CARM methods that follow an "overfit and prune" strategy, while still enabling the identification of low-support rules. These characteristics make TFPC a realistic choice for TC in potentially noisy environments.

The experimental analysis was undertaken using a subset of the Usenet collection, a set of documents compiled by Lang [5] from 20 different newsgroups, often referred to as the "20 Newsgroup" collection. There are exactly 1,000 documents per group (class) with the exception of one class that contains only 997. For our experiments the collection was split into two data sets covering 10 classes each: NGA.D10000.C10 and NGB.D9997.C10, and the analysis was undertaken using NGA.D10000.C10.

4   Phrase Identification

The phrase identification approach we employed proceeds as follows, for each document in the training set:

1. Remove *common* words, i.e. words that are unlikely to contribute to a characterisation of the document.
2. Remove *rare* words, i.e. words that are unlikely to lead to generally applicable classification rules.
3. From the remaining words select those *significant* words that serve to differentiate between classes.
4. Generate significant phrases from the identified significant words and associated words.

4.1   Noise Word Identification

We define words as continuous sequences of alphabetic characters delimited by non-alphabetic characters, e.g. punctuation marks, white space and numbers. Some non-alphabetic characters (',', '.', ':', ';', '!' and '?'), referred to as *stop marks*, play a role in the identification of phrases (more on this later). All other non-alphabetic characters are ignored.

Common and rare words are collectively considered to be *noise* words. These can be identified by their *support* value, i.e. the percentage of documents in the training set in which the word appears. Common words are words with a support value above a user defined Upper Noise Threshold (UNT), which we refer to as Upper Noise Words (UNW). Rare words are those with a support value below a user defined Lower Noise Threshold (LNT), and are thus referred to as Lower Noise Words (LNW).

The UNT must of course exceed the LNT value, and the distance between the two values determines the number of identified non-noise words and consequently, if indirectly, the number of identified phrases. A phrase, in the context of the TFPC algorithm, represents a possible attribute of a document which may be a component of the antecedent of rules. Some statistics for the NGA.D10000.C10 set, using LNT = 1% and UNT = 50% are presented in Table 1. It can be seen that the majority of words occur in less than 1% of documents, so LNT must be set at a low value so as not to miss any potential significant words. Relatively few words are common, appearing in over 50% of the documents.

Table 1. Statistics for 20 Newsgroup data sets A and B using LNT = 1% and UNT = 50%

Data Set	# words	# LNW	# UNW	% LNW	% UNW
NGA.D10000.C10	49,605	47,981	21	96.73	0.04
NGB.D9997.C10	47,973	46,223	22	96.35	0.05

Tables 2 and 3 list the most common words (support greater than 40%) in the two 20 Newsgroup sets. Figures in parentheses indicate the number of documents where the word appears; recall that there are 10,000 and 9,997 documents in the two sets respectively. Note that NGB.D9997.C10 set contains the additional common word "but".

Table 2. Number of common words (UNT = 40%) in NGA.D10000.C10

a (7,666)	and (7,330)	are (4,519)	be (4,741)	for (6,367)	have (5,135)
i (6,803)	in (7,369)	is (6,677)	it (5,861)	not (4,565)	of (7,234)
on (5,075)	re (5,848)	that (6,012)	the (8,203)	this (5,045)	to (7,682)
with (4,911)	writes (4,581)	you (5,015)			

Table 3. Number of common words (UNT = 40%) in NGB.D9997.C10

a (7,837)	and (7,409)	are (4,807)	be (5,258)	but (4,633)	for (6,401)
have (5,366)	i (6,854)	in (7,579)	is (6,860)	it (6,169)	not (4,849)
of (7,546)	on (5,508)	re (6,267)	that (6,515)	the (8,427)	this (5,333)
to (7,905)	with (4,873)	writes (4,704)	you (5,013)		

4.2   Significant Word Identification

The desired set of *significant* words is drawn from an ordered list of *potential significant* words. A potential significant word is a non-noise word whose *contribution* value exceeds some user specified threshold G. The contribution value of a word is a measure of the extent to which the word serves to differentiate between classes and can be calculated in a number of ways. For the study presented here we considered two methods: (a) Using support counts only, and (b) Term weighting.

Contribution from support counts only is obtained using the following identify:

Contribution G_{wi} of word w with respect to class $i = \frac{S_{wi} \times D}{S_w \times S_i}$

Where D is the total number of documents in the training set, S_i is the number of documents that are labelled as class i, S_{wi} is the number of documents in class i that contain word w, and S_w is the total number of documents that contain word w. The ratio $\frac{S_w}{D}$ describes the overall frequency of occurrence of word w in the document set. If the ratio $\frac{S_{wi}}{S_i}$ is greater than this, then the contribution value G_{wi} will be greater than 1, indicating that w may be a significant word for class i. In practice, of course, even words with little significance may have contribution values slightly greater than 1, so to indicate a significant contribution we require G_{wi} to exceed some threshold value $G > 1$. The maximum value of the contribution can reach is $\frac{D}{S_i}$, obtained when $\frac{S_{wi}}{S_w} = 1$, indicating that w occurs only in class i. In the case of the NGA.D10000.C10 set, we have ten classes of exactly 1,000 documents each, so the maximum contribution value is 10. The algorithm for calculating contribution values using support counts is given in Table 4.

Table 4. Algorithm for calculating contribution using support counts

$G \leftarrow$ significance threshold
$w \leftarrow$ the given word
$C \leftarrow$ set of available classes
$D \leftarrow$ total number of documents
$S_w \leftarrow$ number of documents that contain w
for each c_i in C from $i = 1$ to $|C|$ {
    $S_i \leftarrow$ number of documents labelled as in class c_i
    $S_{wi} \leftarrow$ number of documents in c_i that contain w
    $S_{Li} \leftarrow \frac{S_{wi}}{S_i}$
    contribution $\leftarrow \frac{S_{Li} \times D}{S_w}$
    if (contribution $> G$) then w is a significant word
}

We apply a similar approach when term weighting is used. TF-IDF (Term Frequency - Inverse Document Frequency) [11] is a well established term weighting technique. Our variation of this is defined as follows:

Contribution G_{wi} of word w with respect to class $i = \frac{TF_{wi} \times N}{TF_w \times N_i}$

Where TF_{wi} is the total number of occurrences of w in documents in class i, N is the total number of words in the document set, N_i is the total number of words contained in documents labeled as class i, and TF_w is the total number of occurrences of the word w in the document set. The ratio $\frac{TF_w}{N}$ defines the overall term frequency of w in the document set; if the corresponding ratio $\frac{TF_{wi}}{N_i}$ is significantly greater than this, then a contribution value G_{wi} greater than 1 will indicate a potential significant word. The algorithm for calculating contribution values using term weighting is given in Table 5.

Table 5. Algorithm for calculating contribution using term weighting

---

$G \leftarrow$ significance threshold
$w \leftarrow$ the given word
$C \leftarrow$ set of available classes
$N \leftarrow$ total number of words in the document base
$T_w \leftarrow$ total number of occurrences of word w
for each c_i in C from $i = 1$ to $|C|$ {
$\quad T_{wi} \leftarrow$ total number of occurrences of word w in c_i
$\quad N_i \leftarrow$ total number of words in c_i
$\quad$ contribution $\leftarrow \frac{T_{wi} \times N}{T_w \times N_i}$
$\quad$ if (contribution $> G$) then w is a significant word
}

Thus we have two options for calculating the contribution of a word, using support counts or using term weightings. We place those whose contribution exceeds the threshold G into a potential significant words list ordered according to contribution value. This list may include words that are significant for more than one class, or we may choose to include only those non-noise words with contribution greater than G with respect to one class only (i.e. *uniques*).

From the potential significant words list we select the final list of significant words from which we generate phrases. We have examined two strategies for doing this. The first method, which simply selects the first (most significant) K words from the ordered list, may result in an unequal distribution between classes. In the second approach we select the top $\frac{K}{|C|}$ for each class (where $|C|$ is the number of available classes), so as to include an equal number of significant words for each class. Thus, in summary, we have:

- Two possible contribution selection strategies (support count and term weighting).
- Two potential significant word list construction strategies (include all words with appropriate level of contribution, or alternatively only unique words).
- Two significant word selection strategies (top K or top $\frac{K}{|C|}$ for each class).

These possibilities define eight different methods for the identification of significant words. Tables 6 and 7 illustrate some consequences of these options. Table 6 gives the distribution of significant words per class for the NGA.D10000.C10 set using the "support count, all words and top K strategy" with UNT = 7%, LNT = 0.2%, $G = 3$. Note that the number of significant words per class is not balanced, with the general "forsale" class having the least number of significant words and the more specific "mideast" class the most. Table 7 shows the 10 most significant words for each class using the same strategy and thresholds. The value shown in parentheses is the contribution of the word to that class in each case. Recall that using the support count strategy the highest possible contribution value for the NGA.D10000.C10 set is 10, obtained when the word is unique to a certain class. In the "forsale" category quite poor contribution values are found, while the "mideast" category has many high contribution words.

Table 6. Number of significant words in NGA.D10000.C10 using the "support count, all words and top K strategy" with UNT = 7.0%, LNT = 0.2%, G = 3

Class Label	# Sig. Words	Class Label	# Sig. Words
comp.windows.x	384	rec.motorcycles	247
talk.religion.misc	357	sci.electronics	219
alt.atheism	346	misc.forsale	127
sci.med	381	talk.politics.mideast	1,091
comp.sys.ibm.pc.hardware	175	rec.sport.baseball	360

Table 7. Top 10 significant words per class for NGA.D10000.C10 using the "support count, all words and top K strategy" with UNT = 7.0%, LNT = 0.2%, G = 3

windows.x	motorcycles	religion	electronics	atheism
colormap(10)	behanna(10)	ceccarelli(10)	circuits(9.8)	inimitable(10)
contrib(10)	biker(10)	kendig(10)	detectors(9.6)	mozumder(10)
imake(10)	bikers(10)	rosicrucian(10)	surges(9.5)	tammy(10)
makefile(10)	bikes(10)	atf(9.5)	ic(9.3)	wingate(10)
mehl(10)	cages(10)	mormons(9.5)	volt(9.3)	rushdie(9.8)
mwm(10)	countersteering(10)	batf(9.3)	volts(9.2)	beauchaine(9.7)
olwn(10)	ducati(10)	davidians(9.2)	ir(9.2)	benedikt(9.4)
openlook(10)	fxwg(10)	abortions(9.0)	voltage(9.2)	queens(9.4)
openwindows(10)	glide(10)	feds(8.9)	circuit(8.9)	atheists(9.3)
pixmap(10)	harley(10)	fbi(8.8)	detector(8.9)	sank(9.1)
forsale	med	mideast	hardware	baseball
cod(10)	albicans(10)	aggression(10)	nanao(10)	alomar(10)
forsale(9.8)	antibiotic(10)	anatolia(10)	dma(9.4)	astros(10)
comics(9.5)	antibiotics(10)	andi(10)	vlb(9.4)	baerga(10)
obo(9.0)	candida(10)	ankara(10)	irq(9.3)	baseman(10)
sale(8.8)	diagnosed(10)	apartheid(10)	soundblaster(9.0)	batter(10)
postage(8.6)	dyer(10)	appressian(10)	eisa(8.8)	batters(10)
shipping(8.6)	fda(10)	arabs(10)	isa(8.8)	batting(10)
mint(8.4)	homeopathy(10)	argic(10)	bios(8.7)	bullpen(10)
cassette(8.2)	infections(10)	armenia(10)	jumpers(8.7)	cardinals(10)
panasonic(7.6)	inflammation(10)	armenian(10)	adaptec(8.7)	catcher(10)

5 Phrase Identification

Whichever of the methods described above is selected, we define four different categories of word:

1. **Upper Noise Words (UNW):** Words whose support is above a user defined Upper Noise Threshold (UNT).
2. **Lower Noise Words (LNW):** Words whose support is below a user defined Lower Noise Threshold (LNT).

Table 8. Phrase generation strategies

Delimiters	Contents	Label
Stop marks and noise words	Sequence of one or more significant words and ordinary words	DelSNcontGO
	Sequence of one or more significant words and ordinary words replaced by "wild cards"	DelSNcontGW
Stop marks and ordinary words	Sequence of one or more significant words and noise words	DelSOcontGN
	Sequence of one or more significant words and noise words replaced by "wild cards" .	DelSOcontGW

Table 9. Example of significant word identification process using a document from the NGA.D10000.C10 data set

> *@Class rec.motorcycles*
> *paint jobs in the uk*
> *can anyone recommend a good place for reasonably*
> *priced bike paint jobs, preferably but not*
> *essentially in the london area.*
> *thanks*
> *john somename.*
> *–*
> *acme technologies ltd xy house,*
> *147 somewherex road*

(a) *Example document from NGA.D10000C10 data set in its unprocessed form*

> *paint jobs in the uk can anyone recommend a good place for reasonably*
> *priced bike paint jobs # preferably but not essentially in the london*
> *area # thanks john somename # acme technologies ltd xy house #*
> *somewherex road*

(b) *Document with stop marks indicated and non-alphabetic characters removed*

> *paint jobs in the uk can anyone recommend a good place for reasonably*
> *priced bike paint jobs # preferably but not essentially in the london*
> *area # thanks john somename # acme technologies ltd xy house #*
> *somewherex road*

(c) *Document with lower, upper and significant words marked (all other words are ordinary words)*

3. **Significant Words (SW):** Selected key words that serve to distinguish between classes.
4. **Ordinary Words (OW):** Other non-noise words that were not selected as significant words.

We also identify two groups of categories of words:

1. **Non-Noise Words (NNW):** The union of significant and ordinary words.
2. **Noise Words (NW):** The union of upper and lower noise words.

These categories are all used to describe the construction of phrases. We have investigated four different simple schemes for creating phrases, defined in terms of rules describing the content of phrases and the way in which a phrase is delimited. In all cases, we require a phrase to include at least one significant word. In addition to this, Table 8 shows the four different algorithms used for the experiments described here.

An example illustrates the consequences of each method. In Table 9a we show a document taken from the NGA.D10000.C10 data set (with some proper names changed for ethical reasons). Note that the first line is the class label and plays no part in the phrase generation process. The first stage in preprocessing replaces all stop marks by a # character and removes all other non-alphabetic characters (Table 9b). In Table 9c the document is shown "marked up" after the significant word identification has been completed. Significant words are shown using "wide-tilde" ($\widetilde{abc}...$), upper noise words use "wide-hat" ($\widehat{abc}...$), and lower noise words use "over-line" ($\overline{abc}...$).

In Table 10 we show the phrases used to represent the example document from Table 9 using each of the four different phrase identification algorithms. Where appropriate "wild card" words are indicated by a '?' symbol. Note that a phrase can comprise any number of words, unlike *word-gram* approaches where words are a fixed length. The phrase identified in a document become its *attributes* in the classification process.

Table 10. Example phrases (attributes) generated for example document given in Table 9 using the four advocated phrase identification strategies

Phrase Identification Algorithm	Example of Phrase Representation (Attributes)
DelSNcontGO	$\{\{road\}, \{preferably\}, \{reasonably\ priced\ bike$ $paint\ jobs\}, \{acme\ technologies\ ltd\}\}$
DelSNcontGW	$\{\{road\}, \{preferably\}, \{?\ ?\ bike\ ?\ ?\},$ $\{?\ technologies\ ?\}\}$
DelSOcontGN	$\{\{somewherex\ road\}, \{preferably$ $but\ not\}, \{bike\}, \{technologies\}\}$
DelSOcontGW	$\{\{?\ road\}, \{preferably\ ?\ ?\}, \{bike\},$ $\{technologies\}\}$

6   Experimental Results

Experiments conducted using the NGA.D10000.C10 data set investigated all combinations of the eight different proposed significant word generation strategies with the four proposed different phrase generation approaches. We also investigated the effect of using the generated significant words on their own as a "bag of keywords" representation. The suite of experiments described in this section used the first 9/10th (9,000 documents) as the training set, and the last

1/10th (1,000 documents) as the test set. We used the TFPC algorithm to carry out the classification process. For all the results presented here, the following thresholds were used: support = 0.1%, confidence = 35.0%, UNT = 7.0%, LNT = 0.2%, $G = 3$, and maximum number of significant words threshold of 1,500. These parameters produced a word distribution that is shown in Table 11. As would be expected the number of potential significant words is less when only unique words (unique to a single class) are selected. Note also that using word frequency to calculate the contribution of words leads to fewer significant words being generated than is the case when using the "word support calculation" which considers only the number of documents in which a word is encountered.

Table 11. Number of potential significant words calculated per strategy (NGA.D10000.C10)

Number of Noise Words above UNT	208			
Number of Noise Words below LNT	43,681			
Number of Ordinary Words	4,207			
Number of Significant Words	1,500			
Number of Words	49,596			
	Word Frequency		**Word Support**	
	Unique	All	Unique	All
Number of Potential Significant Words	2,911	3,609	3,188	3,687

Table 12. Number of attributes (phrases) generated (NGA.D10000.C10)

	Word Frequency				**Word Support**			
	Unique		**All**		**Unique**		**All**	
	Dist	Top K	Dist	Top K	Dist	Top K	Dist	Top K
DelSNcontGO	27,551	27,903	26,973	27,020	26,658	25,834	26,335	25,507
DelSNcontGW	11,888	12,474	12,118	13,657	11,970	11,876	11,819	11,591
DelSOcontGN	64,474	63,134	60,561	61,162	59,453	58,083	59,017	57,224
DelSOcontGW	32,913	34,079	32,549	35,090	32,000	32,360	31,542	31,629
Keywords	1,510	1,510	1,510	1,510	1,510	1,510	1,510	1,510

Table 12 shows the number of attributes generated using all the different combinations of the proposed significant word generation and phrase generation strategies, including the case where the significant words alone were used as attributes (the "keyword" strategy). In all cases, the algorithms use as attributes the selected words or phrases, and the ten target classes. Thus, for the keyword strategy the number of attributes is the maximum number of significant words (1,500) plus the number of classes (10). In other experiments, we examined the effect on the keyword strategy of removing the upper limit, allowing up to 4,000 significant words to be used as attributes, but this led to reduced accuracy, suggesting that a limit on the number of words used is necessary to avoid including words whose contribution may be spurious.

Table 13. Classification accuracy (NGA.D10000.C10)

	Word Frequency				Word Support			
	Unique		All		Unique		All	
	Dist	Top K	Dist	Top K	Dist	Top K	Dist	Top K
DelSNcontGO	75.9	73.6	**77.3**	72.4	76.4	73.2	**77.4**	74.5
DelSNcontGW	75.1	71.6	**76.2**	68.5	74.9	71.3	**75.8**	72.3
DelSOcontGN								
DelSOcontGW			**70.9**		70.4	66.0	**71.2**	68.9
Keywords	75.1	73.9	**75.8**	71.2	74.4	72.2	**75.6**	73.7

In the DelSNcontGO and DelSNcontGW algorithms, stop and noise words are used as delimiters. As the results demonstrate, this leads to many fewer phrases being identified than is the case for the other two phrase generation strategies, which use stop words and ordinary words as delimiters. For DelSOcontGN (and to a lesser extent DelSOcontGW) the number of attributes generated usually exceeded the TFPC maximum of 2^{15} (32,767) attributes. This was because these algorithms allow the inclusion of noise words in phrases. Because there are many more noise words (43,889) than ordinary words (4,207), the number of possible combinations for phrases far exceeds the number obtained using the two DelSN strategies. Further experiments which attempted to reduce the number of phrases produced by adjusting the LNT, UNT and G thresholds did not lead to good results, and led us to abandon the DelSOcontGN and DelSOcontGW strategies.

Variations within the DelSN strategies were less extreme. DelSNcontGW produces fewer attributes than DelSNcontGO because phrases that are distinct in DelSNcontGO are collapsed into a single phrase in DelSNcontGW. Intuitively it might seem that identifying more attributes (phrases) would improve the quality of representation and lead to better classification accuracy. In other experiments we increased the number of attributes produced by the DelSNcontGO and Del-SNcontGW strategies by increasing the limit on the number of significant words generated. However, as was the case with the keywords strategy, this did not lead to any better accuracies, presumably because the additional significant words included some that are unhelpful or spurious.

Table 13 shows the percentage classification accuracy results obtained using the different strategies. Because too many phrases were generated using DelSO-contGN and, in some cases, DelSOcontGW for the TFPC algorithm to operate, the results were incomplete for these algorithms, but, as can be seen, results obtained for DelSOcontGW were invariably poorer than for other strategies. In the other cases, it is apparent that better results were always obtained when significant words were distributed equally between classes (columns headed "Dist", noted as "top $\frac{K}{|C|}$" in section 4.2) rather than selecting only the K (1,500) most significant words. Best results were obtained with this policy using a potential significant word list made up all words with a contribution above the G threshold (columns headed "All"), rather than when using only those that were unique to one class. Overall, DelSNcontGO performed slightly better than DelSNcontGW,

Table 14. Number of empty documents in the training set (NGA.D10000.C10)

	Word Frequency				Word Support			
	Unique		All		Unique		All	
	Dist	Top K	Dist	Top K	Dist	Top K	Dist	Top K
DelSNcontGO	190	258	251	299	229	238	224	370
DelSNcontGW	190	226	251	299	229	147	224	370
DelSOcontGN								
DelSOcontGW			251		229	411	224	370
Keywords	190	226	251	299	229	411	224	370

and both phrase-generation strategies outperformed the Keywords-only algorithm. The contribution calculation mechanism used did not appear to make a significant difference to these results.

Table 14 shows the number of "empty" training set documents found in the different cases: that is, documents in which no significant attributes were identified. These represent between 2% and 5% of the total training set. Perhaps more importantly, any such documents in the test set will necessarily be assigned to the default classification. Although no obvious relationship between the frequency of empty documents and classification accuracy is apparent from these results, further investigation of this group of documents may provide further insight into the operation of the proposed strategies.

Table 15 shows execution times in seconds for the various algorithms, including both time to generate rules and time to classify the test set. The key words only approach is faster than DelSNcontGO because many fewer attributes are considered, so TFPC generates fewer frequent sets and rules. However, Del-SNcontGW is fastest as the use of wild card leads to faster phrase matching.

Table 15. Execution times (NGA.D10000.C10)

	Word Frequency				Word Support			
	Unique		All		Unique		All	
	Dist	Top K	Dist	Top K	Dist	Top K	Dist	Top K
DelSNcontGO	244	250	253	242	250	248	328	235
DelSNcontGW	155	148	145	158	157	194	145	224
DelSOcontGN								
DelSOcontGW			370		326	281	278	314
Keywords	183	176	282	287	261	262	235	220

A further set of experiments were conducted to investigate the effects of adjusting the various thresholds. The first of these analysed the effect of changing G. The G value (contribution or siGnificance threshold) defines the minimum contribution that a potential significant word must have. The size of the potential significant word list thus increases with a corresponding decrease in G; conversely, we expect the quality of the words in the list to increase with G.

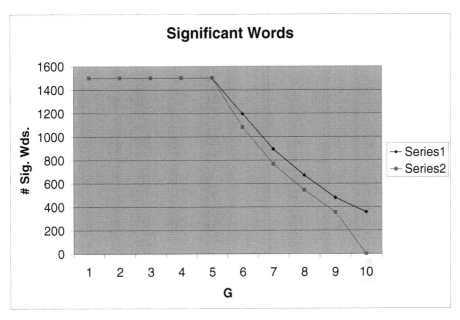

Fig. 1. Relationship between G value and number of significant words identified for NGA.D10000.C10, UNT = 7.0%, LNT = 0.2%, and K = 1,500. Series 1 = word frequency contribution calculation, Series 2 = word support contribution calculation

Figure 1 shows the effect on the number of selected significant words with changes in G, when UNT = 7.0%, LNT = 0.2%, and $K = 1,500$. The figure shows that there is little effect until the value of G reaches a point at which the size of the potential significant words list drops below K, when the number of selected significant words falls rapidly and a corresponding fall in accuracy is also experienced. The drop is slightly less severe using word frequency contribution calculation compared with support count contribution calculation.

In other experiments, varying the support and confidence thresholds had similar effects to those experienced generally in Association Rule Mining. Relatively low support and confidence thresholds are required because of the high variability of text documents, so as not to miss any significant frequent item sets or useful if imprecise rules. Generally we found that a support threshold corresponding to 10 documents produced best results, with a confidence threshold of 35.0%. We also undertook a number of experiments with the LNT and UNT thresholds. Best results were obtained using low values for both (such as those used in the above experiments).

7 Conclusion

In this paper we have described a number of different strategies for identifying phrases in document sets to be used in a "bag of phrases" representation for text classification. Phrases are generated using four different schemes to combine noise, ordinary and significant words. In all eight methods were used to identify significant words, leading overall to 32 different phrase generation strategies that were investigated, as well as 8 keyword only identification strategies.

The main findings of the experiments were:

1. Best results were obtained from a strategy that made use of words that were significant in one or more classes, rather than only those that were unique to one class, coupled with a selection strategy that produced an equal distribution between classes.
2. The most successful phrase based strategy outperformed classification using only keywords.

From the experiments described above we observe that a small subset of the documents to be classified were represented by an empty vector, i.e. they were not represented by any phrases/key words. This suggests that there remain possibilities to improve the strategies considered, which will be the subject of further investigation planned by the authors.

References

1. Ali, K., Manganaris, S., Srikant, R.: Partial classification using association rules. In: Heckerman, D., Mannila, H., Pregibon, D., Uthurusamy, R. (eds.) Proceedings of the Third International Conference on Knowledge Discovery and Data Mining (KDD-97), pp. 115–118. AAAI press, Stanford (1997)
2. Coenen, F., Leng, P., Zhang, L.: Threshold tuning for improved classification association rule mining. In: Ho, T.-B., Cheung, D., Liu, H. (eds.) PAKDD 2005. LNCS (LNAI), vol. 3518, pp. 216–225. Springer, Heidelberg (2005)
3. Coenen, F., Leng, P.: The effect of threshold values on association rule based classification accuracy. Journal of Data and Knowledge Engineering 60(2), 345–360 (2007)
4. Katrenko, S.: Textual data categorization: Back to the phrase-based representation. In: Proceedings of the Second IEEE International Conference on Intelligent Systems, vol. 3, pp. 64–67 (2004)
5. Lang, K.: Newsweeder: Learning to filter netnews. In: Proceedings of the 12th International Conference on Machine Learning (ICML-95), pp. 331–339. Morgan Kaufmann, San Francisco (1995)
6. Lewis, D.D.: An evaluation of phrasal and clustered representations on a text categorization task. In: Belkin, N.J., Ingwersen, P., Pejtersen, A.M. (eds.) Proceedings of the 15th Annual International ACM SIGIR Conference on Research and Development in Information Retrieval (SIGIR-92), pp. 37–50. ACM Press, New York (1992)
7. Li, W., Han, J., Pei, J.: CMAR: Accurate and efficient classification based on multiple class-association rules. In: Cercone, N., Lin, T.Y., Wu, X. (eds.) Proceedings of the 2001 IEEE International Conference on Data Mining (ICDM-01), pp. 369–376. IEEE Computer Society Press, Los Alamitos (2001)

8. Scott, S., Matwin, S.: Feature engineering for text classification. In: Bratko, I., Dzeroski, S. (eds.) Proceedings of the 16th International Conference on Machine Learning (ICML-99), pp. 279–388 (1999)

9. Sebastiani, F.: Machine learning in automated text categorization. ACM Computer Surveys 34(1), 1–47 (2002)

10. Sharma, R., Raman, S.: Phrase-based text representation for managing the web documents. In: Proceedings of the 2003 International Symposium on Information Technology (ITCC-03), pp. 165–169 (2003)

11. Sparck Jones, K.: Exhaustivity and specificity. Journal of Documentation 28, 11–21 (1972) (reprinted in 2004, 60, pp. 493–502)

Probabilistic Model for Structured Document Mapping
Application to Automatic HTML to XML Conversion

Guillaume Wisniewski, Francis Maes, Ludovic Denoyer, and Patrick Gallinari

LIP6 — University of Paris 6
104 avenue du prsident Kennedy
75015 Paris
`name.surname@lip6.fr`

Abstract. We address the problem of learning automatically to map heterogeneous semi-structured documents onto a mediated target XML schema. We adopt a machine learning approach where the mapping between input and target documents is learned from a training corpus of documents. We first introduce a general stochastic model of semi structured documents generation and transformation. This model relies on the concept of meta-document which is a latent variable providing a link between input and target documents. It allows us to learn the correspondences when the input documents are expressed in a large variety of schemas. We then detail an instance of the general model for the particular task of HTML to XML conversion. This instance is tested on three different corpora using two different inference methods: a dynamic programming method and an approximate LaSO-based method.

1 Introduction

With the development and growth of numerical resources, semantically rich data tend to be encoded using semi-structured formats. In these formats, content elements are organized according to some structure, that reflects logical, syntactic or semantic relations between them. For instance, XML and, to a lesser extent, HTML allow us to identify elements in a document (like its title or links to other documents) and to describe relations between those elements (e.g. we can identify the author of a specific part of the text). Additional information such as meta data, annotations, etc., is often added to the content description resulting in richer descriptions.

For many applications, a key problem associated with the widespread of semi-structured resources is heterogeneity: as documents come from different sources, they will have different structures. For instance, in XML document collection focused on a specific domain (like scientific articles), document will come from different sources (e.g. each source corresponds to a journal) and will, therefore, follow different schemas. The schema itself may unknown. For managing or accessing this collection, a correspondence between the different schemas has to

P. Perner (Ed.): MLDM 2007, LNAI 4571, pp. 854–867, 2007.

be established. The same goes for HTML data on the Web where each site will develop its own presentation. If one wants, for example, to develop a movie database, information has to be extracted from each site so that heterogeneous structures may be mapped onto a predefined mediator schema.

```
<table>                                           <cast>
    <tr>                                              <character>
        <td> Korben Dallas </td>                          <actor> Bruce Willis </actor>
        <td> ... </td>                                    <name> Korben Dallas </name>
        <td> <a> Bruce Willis </a> </td>              </character>
    </tr>                                             <character>
    <tr>                                                  <actor> Milla Jovovich </actor>
        <td> Leelo </td>                                  <name> Leelo </name>
        <td> ... </td>                                </character>
        <td> <a> Milla Jovovich </a> </td>        </cast>
    </tr>
</table>
```

Fig. 1. Heterogeneity example: two documents describing the same information coming from two different sources. Both the organization, partitioning and element order differ.

Manual correspondence between heterogeneous schemas or toward a mediated schema is usually performed via document transformation languages, like XSLT. However the multiplicity and the rapid growth of information sources have motivated researchers to work out ways to automate these transformations [1,2]. This heterogeneity problem has been addressed only recently from a content centric perspective for applications in information retrieval [3], legacy document conversion [4], and ontology matching [5]. Depending on the targeted application and on the document sources considered, this semi-structured document mapping problem will take different forms. With heterogeneous XML sources, the correspondence between the different structures will have to handle both the structural and content information. The mapping will provide new structures for the input sources, this is an annotated tree conversion problem which involves tag renaming and document elements reorganization and annotation. For the HTML to XML conversion problem, the context is different. HTML documents are only weakly structured and their format is presentation-oriented. The problem here will be to map this weakly structured visualization oriented format onto a valid XML tree.

In this article, we consider the problem of automatically learning transformations from heterogeneous semi-structured documents onto an XML predefined schema. We adopt a machine learning approach where the transformation is learned directly from examples. We propose a general framework for learning such transformations and focus then on the special case of HTML to XML conversion. The article is organized as follows. The general framework is introduced in Section 2. Section 3 details the HTML to a predefined XML schema conversion problem. Experiments performed on four different corpora are described in Section 4 and related work is reviewed in Section 5.

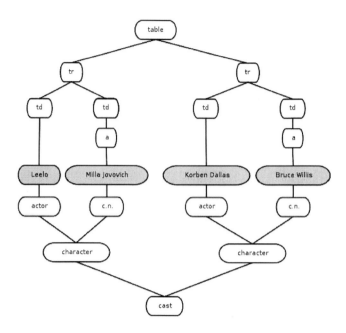

Fig. 2. Toy example of a structured document transformation from HTML data to a predefined schema describing the casting of a movie

2   A Model for Document Structure Mapping

2.1   General Framework

We consider semi-structured documents where content information (text, video, pictures, etc.) is organized according to some schema. In the following, the terms *semi-structured* and *schema* are used in a general sense and are not restricted to XML. The former includes different formats like HTML, XML or PDF and the latter denotes the document organization. We study the problem of learning mappings from a set of heterogeneous documents onto a predefined mediated target XML schema denoted s_T (T holds for Target). The set of possible input schema is denoted $S = \{s_1, ..., s_{|S|}\}$. No assumption is made on the structure of the input documents for the general model. These documents may either follow a well-defined DTD, or may be HTML documents or even plain — unstructured — text documents.

A straightforward approach for learning to map heterogeneous documents onto a target schema is to learn the correspondence for each input schema. This raises different problems: for example representative data have to be collected for each input schema and schemas not represented in the training set cannot be transformed. In order to bypass these limitations, we will introduce an abstract representation of a document, called *meta document*, which will be used as an intermediate representation in our document mapping framework. This abstract representation supposedly contains the information needed for an individual to create the different views of a document corresponding to different schemas. This *meta document* will provide a link between the different representations, it is a variable of our model and its very definition will depend on the precise task we are dealing with. In order to fix the ideas, let us consider an illustration of this concept. In Figure 3, the meta document is represented as a set of relations and content elements which may be stored into a relational database. It may then be used for producing different projections onto different schemas. It may also be transformed into a HTML document for an intranet, into a PDF document or into an XML document following a specific DTD. We denote d_{s_i} the projection of the meta document d onto schema s_i.

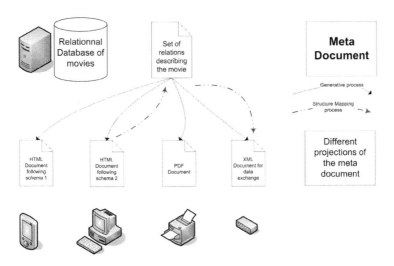

Fig. 3. In this example, a company uses a Database Server to generate different views of a same piece of information of the whole database. Each piece of database describing a particular movie is the *meta document* of the movie.

The meta document d is not necessarily known — in the example of Figure 3, one does not usually have access to the database used to generate the different documents. Different meta documents can produce the same projection onto a schema s_i. For example, different databases can be used to generate the same HTML document. In the proposed model, we will consider that d is a hidden random variable. For the HTML to XML problem dealt with in Section 3, we will propose a specific instance of d.

Our stochastic model of document view generation is described in Figure 4 using a Bayesian network formalism. The meta document d is a latent variable which provides a link between different document representations. a_i is a discrete random variable that represents the author of the projection of d onto d_{s_i} — it identifies the specific process by which d_{s_i} is produced from d. In this model d_{s_i} is fully defined by d and a_i. In practice a_i will simply identify a source. a_T is not represented in this model since the target schema is unique and predefined.

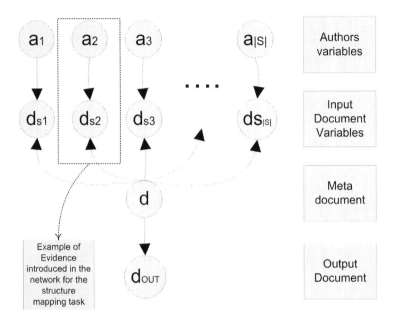

Fig. 4. The belief network representing the generative process of the different views of a meta document

This generative model of document views will serve as a basis for introducing the transformation model.

2.2   Structure Mapping Formalism

We will denote $s_{in(d)}$ the input schema and $d_{s_{in(d)}}$ the projection of d onto this schema. The author of the transformation of d into $s_{in(d)}$ denoted $a_{in(d)}$ may be known or unknown depending on the structure mapping problem. For example, if we want to transform different known Websites into XML documents, the author (the Website) is a known information. Usually there is no information available about the *meta document d* and only a projection of d will be available.

Within this framework, we formalize the mapping problem as follows: given a document $d_{s_{in(d)}}$ and an author $a_{in(d)}$, find the mapping which maximizes the probability of generating a document in the target schema. Formally one could write:

$$d_{s_T} = \underset{d' \in s_T}{\operatorname{argmax}} P(d'|d_{s_{in(d)}}, a_{in(d)}) \qquad (1)$$

In order to solve equation 1 we use the document view model of Figure 4. Let us write the joint probability for the whole Bayesian Network:

$$P(d, d_{s_1}, ..., d_{s_{|S|}}, d_{s_T}, a_1, ..., a_{|S|}) = P(d) \prod_{i=1}^{|S|} P(a_i) \prod_{i=1}^{|S|} P(d_{s_i}|d, a_i) P(d_{s_T}|d)$$

$$(2)$$

Since we only have access to a projection of the document d onto schema $s_{in(d)}$, we will integrate out all unknown variables, leading to:

$$P(d_{s_T}, d_{s_{in(d)}}, a_{in(d)}) = \sum_{\substack{d \\ \{a_k\}_{k \neq in(d)} \\ \{d_{s_j}\}_{j \neq in(d)}}} P(d) \prod_{i=1}^{|S|} P(a_i) \prod_{i=1}^{|S|} P(d_{s_i}|d, a_i) P(d_{s_T}|d) \tag{3}$$

Here the summation over d consists in integrating over all possible instances of the hidden variable d. From this expression, we obtain the final expression for the right term of Equation 1:

$$P(d_{s_T}|d_{s_{in(d)}}, a_{in(d)}) \propto \sum_d P(d)P(d_{s_T}|d)P(d_{s_{in(d)}}|d, a_{in(d)}) \tag{4}$$

The structure mapping problem consists in solving the following equation:

$$d_{s_T} = \underset{d' \in s_T}{\operatorname{argmax}} \sum_d P(d)P(d'|d)P(d_{s_{in(d)}}|d, a_{in(d)}) \tag{5}$$

Here $P(d'|d)$ corresponds to the probability of generating a document into the target schema using the meta document d and $P(d_{s_{in(d)}}|d, a_{in(d)})$ is the probability of generating document $d_{s_{in(d)}}$ according to $a_{in(d)}$. Note that the meta document variable trick allows us to model the processing of heterogeneous databases without having to learn one distinct classifier for each input schema.

Solving equation 5 involves summing over all possible meta-documents d and scoring each possible output d'. In order to efficiently compute the target document probability, we will have to make different simplifying assumptions about the stochastic generation processes corresponding to $P(d_{s_{in(d)}}|d, a_{in(d)})$ and $P(d_{s_T}|d)$. These assumptions will depend on the task and on the type of structure mapping problem. In the following, we will detail these assumptions and the model instance for the HTML to XML conversion task.

3   Model Instance for HTML to XML Conversion

We now focus on learning mappings from heterogeneous HTML sources to a pre-defined XML schema. In this specific HTML to XML conversion task, we consider one possible input schema (HTML) denoted s_{IN} and different possible authors (for example, "IMDB" and "Allocine" fir the movie corpus - see part Experiments). We will make two assumptions: the first one concerning $P(d_{s_{IN}}|d, a_{d,IN})$ and the second one concerning $P(d'|d)$.

3.1   Meta Document Assumption

Tags in HTML documents are mainly used for the rendering of the document and as such do not provide useful information for the transformation. The latter will be essentially based on the content elements of the HTML document. Since tag names and attributes only bring few relevant information in the case of HTML, in the following, the input for the transformation will be the sequence of the document content elements. This assumption models a deterministic process where a meta document d is built from d_{IN} only keeping the sequence of text segments of the input document.

Formally, for the model described in Section 2, a meta document d will be a sequence of text segments denoted $d = (d^1, ..., d^{|d|})$. Let $(d_{IN}^1, ..., d_{IN}^{|d_{IN}|})$ denote the sequence of segment extracted from d_{IN}, the probability $P(d_{s_{IN}}|d, a_{d,IN})$ is defined as follow:

$$P(d_{s_{IN}}|d, a_{d,IN}) = \begin{cases} 0 \text{ if } (d^1, ..., d^{|d|}) \neq (d_{IN}^1, ..., d_{IN}^{|d|}) \\ 1 \text{ elsewhere} \end{cases} \tag{6}$$

3.2 Target Document Model

We now introduce a target document model which will be used for mapping a meta document representation onto a target schema. Under the above hypothesis, this amounts at inferring the probability of XML trees from a sequence of text segments. This model extends a series of document models already proposed for the classification and clustering of XML documents ([6], [7]).

Let $N_{d_T} = (n_1, ..., n_{|N_{d_T}|})$ denote the set of labeled nodes for an XML document d_T and c_i denote the content of node n_i. If n_i is a leaf node of d_T then c_i will be the content of the leaf, if n_i is an internal node of d_T, c_i will be the content of all the leaf nodes descendant of n_i. Let L_{d_T} denote the set of leaves of d_T, and let $d = (d^1, ..., d^{|d|})$ be a meta document, we have $P(d_T|d) = P(d_T|d^1, ..., d^{|d|})$.

Modeling all structural relations from the target tree would involve a very large probabilistic space for random variable d_T. In our model, simplifying assumptions are made so that structure and content information is represented using the local context of each node of the document. These assumptions have already been successfully tested on the categorization and clustering tasks. We will assume that the label of a node only depends on its content, its left sibling (if any) and its father (if any). With these assumptions, we can write[1] (see Figure 5 for the corresponding belief network):

$$P(d_T|d^1, ..., d^{|d|}) = \prod_{n_i \in L_{d_T}} P(c_i|d^i) \prod_{n_i \in N_{d_T}} P(n_i|c_i, sib(n_i), father(n_i)) \tag{7}$$

where n_i is the label of node i (the XML tag), $father(n_i)$ and $sib(n_i)$ correspond to the label of the father node and the label of the left sibling node of n_i. Remind that c_i is the union of all the content information of children of n_i.

---

[1] In order to simplify the equation, we don't write $P(c_i|c_j, c_k, ...)$ for the internal nodes. The content of internal nodes are built by a deterministic process so the probability $P(c_i|c_j, c_k, ...)$ is considered to be equal to 1.

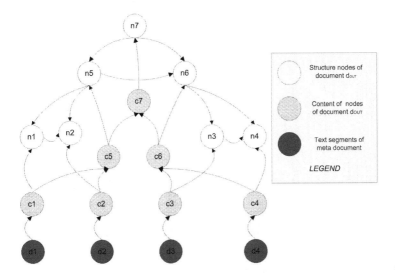

Fig. 5. The belief network representing the dependencies between d_T and the sequence of text segments of the meta-document d. The stochastic process modeled here considers that the input content elements d^i generate the content leaf nodes c_i. The label n_i of node i depends on its left sibling, its father and its content.

We make the additional assumption that the leaf content in d_T is exactly the sequence of elements in d (i.e $P(c_i|d^i) = 0$ if $c_i \neq d^i$)[2] which leads to:

$$P(d_T|d^1, ..., d^{|d|}) = \begin{cases} 0 \text{ if } (d^1, ..., d^{|d|}) \neq (c_1, ..., c_{|d|}) \\ \prod_{n_i \in N_{d_T}} P(n_i|c_i, sib(n_i), father(n_i)) \text{ otherwise} \end{cases} \quad (8)$$

Learning the model: In order to learn the probabilities $P(n_i|c_i, sib(ni), father(n_i))$, we have used a maximum entropy framework [8]. The label for each node is chosen by estimating the probability:

$$P(n_i|c_i, sib(n_i), father(n_i)) = \frac{\exp\left(\langle W_{n_i}, F_{c_i,sib(n_i),father(n_i)} \rangle\right)}{Z_{c_i,sib(n_i),father(n_i)}} \quad (9)$$

where $Z_{c_i,sib(n_i),father(n_i)}$ is a normalizing factor, $F_{c_i,sib(n_i),father(n_i)}$ is a vector representation of the context of node n_i, W_α is the vector of parameters to be learned and $\langle \cdot, \cdot \rangle$ is the scalar product. In this model, we will learn one set of parameters W_α for each possible node label α using a Maximum Entropy method. For the iterative parameter estimation of the Maximum Entropy exponential models, we use one of the quasi Newton methods, namely the Limited Memory BFGS method, which is observed to be more effective than the Generalized Iterative Scaling (GIS) and Improved Iterative Scaling (IIS) for NLP and IE tasks [9].

---

[2] This assumption corresponds to the idea that we don't want to modify the content of the source document in order to generate the target document.

3.3   Final HTML to XML Model

Once the W_α are learned from a training set, the mapping onto the target schema is finally obtained by solving:

$$d_{s_{FINAL}} = \underset{\substack{d_T \text{ such as} \\ (d^1,...,d^{|d|}) = (c_1,...,c_{|d|})}}{\text{argmax}} \prod_{n_i \in N_{d_T}} \frac{\exp\left(\langle W_{n_i}, F_{c_i, sib(n_i), father(n_i)} \rangle\right)}{Z_{c_i, sib(n_i), father(n_i)}} \quad (10)$$

In order to solve this equation, we have used two methods:

1. The first one is based on dynamic programming (DP) (see [10], [11]) and provides an exact solution to Equation 1. Its complexity is $O(n^3.V)$ (see [10] for more details) — where n is the sequence size of d and V is the number of possible internal node labels - which may be prohibitive for large documents.
2. The second one is based on the LaSO algorithm described in Section 4.1) [12]. It allows us to compute an approximation of the maximum in a complexity of $O(|N_{d_{s_T}}|.V.n)$ where $|N_{d_{s_T}}|$ is the number of node of $|d_{s_T}|$.

4   Experiments

4.1   LaSO-Based Model

LaSO is an original method proposed by [12] that describes a general way to make approximate inference in structure mapping problems. This method is especially useful in cases, like ours, in which dynamic programming is too time-consuming. It relies on the observation that inference can be described as a search process and that it is possible to make it faster by learning an adapted heuristic function and using it in a heuristic search algorithm: this method proposed to consider the learning problem and the decoding problem in an integrated manner.

As we show in the next part, LaSo can be applied very easily to our model and allows us to obtain reasonably good results with a lower inference time but a larger training time.

4.2   Corpora

The HTML to XML structure mapping model has been tested on four different collections. One is the INEX'03 corpus [13], which includes XML articles from 20 different journals and proceedings of the IEEE Computer Society. It is made of about 12,000 documents that represent more than 7,000,000 XML elements. The documents have an average length of 500 leaf nodes and 200 internal nodes. There are 139 tags. This is a medium size collection according to the IR criteria, but it is quite a large corpus for the complex structure mapping task. Each INEX document has a corresponding HTML page extracted from the INEX Website which is the input document.

The second collection includes 10,000 movie descriptions extracted from the IMDb Website[3]. Each movie was represented in both, an XML document created from the relational database and a HTML document extracted from the site. The target XML documents have an average length of 100 leaf nodes and 35 internal nodes labeled with 28 possible tags. The documents have a rather regular structure compared to INEX ones: they have fewer tags and share more structural regularities.

The third collection is a set of 39 Shakespearean plays in XML format[4] converted manually to a simple HTML document. There are only a few documents in this collection, however their average length is huge: 4100 leaf nodes and 850 internal nodes. There are 21 different tags. The main challenge of this collection is related to the length of its documents.

The fourth collection, called Mini-Shakespeare, is the smallest one. As in [10], we have randomly selected 60 Shakespearean scenes from the third collection. These scenes have an average length of 85 leaf nodes and 20 internal nodes over 7 distinct tags.

Each collection was randomly split in two equal parts, one for learning and the other for testing. Due to its complexity, dynamic programming was performed only on documents containing less than 150 leafs – this corresponds to 2200 INEX documents, 4000 IMDb documents –, DP was not applicable at all on the third collection.

4.3 Features and Evaluation Measures

The model uses a sparse vector representation of the context of nodes n_i (F_{n_i} in part 3.3). This vector includes structure and content information. Structure is coded through a set of Boolean variables indicating the presence or absence of a particular $(sib(n_i), father(n_i))$ pair. Content is represented by Boolean and real variables. The former encode layout, punctuation and word presence, while the latter represent the size of the content information (in words) and the different word lengths. This sparse representation generates a large vector space: depending on the corpus, there are often more than one million distinct (structure and content) features.

Our first evaluation measure, *Micro*, is the percentage of correctly annotated leaf nodes. It is similar to the word error ratio used in natural language. Since we are dealing with unbalanced classes (*e.g.* INEX documents are essentially made of paragraphs, so this tag is by far the most frequent), we also use a *Macro* score for leaf nodes: the unweighted average of F1 classification scores of each tag. Internal nodes mapping is measured with the *Internal* metric: it is the F1 score of correctly annotated sub-trees, where a sub-tree is correctly annotated when its tag and its content are both correct[5]. The internal metric is similar to the *non-terminal error ratio* used in [10]. The *Full* metric is a *F1* score on all

---

[3] http://www.imdb.com

[4] http://metalab.unc.edu/bosak/xml/eg/shaks200.zip

[5] A sub-tree is correctly annotated if its root node has the right label and if its content is **exactly** the target content. This measure is sometimes called *coverage*.

Table 1. Structure mapping results on four XML collections. Four evaluation measures are used (Experiments performed on a standard 3.2Ghz Computer.)

Collection	Method	Micro	Macro	Internal	Full	Learning time	Testing time
INEX	DP	**79.6%**	**47.5%**	51.5%	**70.5%**	30 min	$\simeq$ 4 days
	LaSO	75.8%	42.9%	**53.1%**	67.5%	> 1 week	**3h20min**
Movie	DP	**95.3%**	**91.2%**	77.1%	**90.4%**	20 min	$\simeq$ 2 days
	LaSO	90.5%	88.6%	**86.8%**	89.6%	> 1 week	**1h15min**
Shakespeare	LaSO	95.3%	78.0%	77.0%	92.2%	$\simeq$ 5 days	**30 min**
Mini-shakespeare	DP	**98.7%**	**95.7%**	**94.7%**	**97.9%**	2 min	$\simeq$ 1 hour
	LaSO	89.4%	83.9%	63.2%	84.4%	20 min	**1 min**

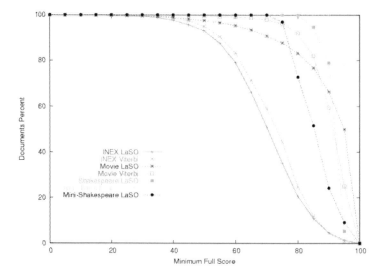

Fig. 6. Percent of documents with more than x% Full score for different values x. We can for example see that the DP method maps correctly more than 80% of the Mini-Shakespeare with a full score included in range [95%, 100%].

built tree components. This is a common measure in the natural language field (under the name of *F1 parsing score*). As a document typically contains more leaf nodes than internal nodes, this measure advantages the leaf score and does not fully inform about the quality of the built tree. These results are shown on Table 1. We also provide the percentage of documents from the test corpus with a Full score greater than than x% (see Figure 6).

4.4   Results

The DP method shows higher scores for leaf nodes classifications than the approximated method based on the LaSO algorithm. For example, with the Movie collection, DP achieves a Micro score of 95.3% whereas LaSO is limited to a score of 90.5%. However, this performance increase has a cost: testing with exact DP inference has a high complexity and may take several days for a collection like INEX, which is unrealistic in practice. It is then limited to short documents. LaSO makes inference fast and practicable for large documents. However, learning is time-consuming. Convergence was not achieved after one week learning on the two real size collections (Movie and INEX). Due to the small number of examples, the huge quantity of features, and the lack of regularization techniques, LaSO also suffers from over-fitting when applied to the Mini-Shakespeare collection.

Best internal scores are achieved by LaSO. This is because LaSO is a top-down parsing method, whereas DP is a bottom-up one. Intuitively, top-down methods may work better on top elements of the trees whereas bottom-up methods are best on bottom elements (leaf nodes).

5   Related Work

In the database community automatic or semi-automatic data integration — known as *schema matching* — has been a major concern for many years. A recent taxonomy and review of these approaches can be found in [5]. [14] describes one of the most complete approach which can handle both ontologies, SQL and XML data.

The matching task is formulated as a supervised multi-label classification problem. While many ideas of the database community can be helpful, their corpora are completely different from the textual corpora used in the IR community: all documents — even XML ones — keep an attribute-value structure like for relational database and are thus much smaller and more regular than for textual documents; textual data hardly appears in those corpora. With database corpora, finding the label of a piece of information is enough to build the corresponding tree because each element usually appears once in the tree structure.

Document structure mapping, also shares similarities with the information extraction task, which aims at automatically extracting instances of specified classes and/or relations from raw text and more recently from HTML pages. Recent works in this field [15] have also highlighted the need to consider structure information and relations between extracted fields.

The document model proposed here is related to other ML models of the literature. Different authors ([16], [10]) have proposed to use natural language formalisms like probabilistic context free grammars (PCFG) to describe the internal structure of documents. Early experiments [11] showed that the complexity of tree building algorithms is so high that they cannot be used on large corpora like INEX. Our specific XML model makes the same kind of independence assumptions as Hierarchical HMMs ([17]) do. The work closest to ours is [10]. They address the HTML to XML document conversion problem. They make use of PCFGs for parsing text segments sequences of and of a maximum entropy classifier for assigning tags to segments.

6 Conclusion

We have proposed a general framework for the structure mapping task on heterogeneous corpora. Our model uses a *meta document* abstraction in order to generate different views of the same document on different schemas and formats. We have then detailed a specific application of this model for the mapping of HTML document onto a mediated XML schema. From our knowledge, this model is today the only one able to handle large amount of documents for the HTML decoding task. For this problem, the *meta document* is a sequence of text segments and the model will find the best XML tree in the target schema. This model has been implemented using two inference methods: a DP exact method and an approximate LaSO algorithm. The results show that, for both methods, the model is able to cope with large corpora of documents. LaSO is faster than DP and this type of method should be investigated further for the transformation problem.

Acknowledgments

This work was supported in part by the IST Programme of the European Community, under the PASCAL Network of Excellence, IST-2002-506778. This publication only reflects the authors' views.

References

1. Chung, C.Y., Gertz, M., Sundaresan, N.: Reverse engineering for web data: From visual to semantic structures. In: ICDE (2002)
2. Zhang, S., Dyreson, C.: Polymorphic xml restructuring. In: IIWeb'06: Workshop on Information Integration on the Web (2006)
3. Wisniewski, G., Gallinari, P.: From layout to semantic: a reranking model for mapping web documents to mediated xml representations. In: Proceedings of the 8th RIAO International Conference on Large-Scale Semantic Access to Content (2007)
4. Chidlovskii, B., Fuselier, J.: Supervised learning for the legacy document conversion. In: DocEng '04: Proceedings of the 2004 ACM symposium on Document engineering, New York, NY, USA, pp. 220–228. ACM Press, New York (2004)
5. Doan, A., Halevy, A.: Semantic integration research in the database community: A brief survey. AI Magazine, Special Issue on Semantic Integration (2005)
6. Denoyer, L., Gallinari, P.: Bayesian network model for semi-structured document classification. Information Processing and Management (2004)
7. Denoyer, L.: Xml document mining challenge. Technical report, LIP6 (2005)
8. Berger, A.L., Pietra, V.J.D., Pietra, S.A.D.: A maximum entropy approach to natural language processing. Comput. Linguist. 22, 39–71 (1996)
9. Malouf, R.: A comparison of algorithms for maximum entropy parameter estimation. In: COLING-02. proceeding of the 6th conference on Natural language learning, Morristown, NJ, USA, pp. 1–7. Association for Computational Linguistics (2002)
10. Chidlovskii, B., Fuselier, J.: A Probabilistic Learning Method for XML Annotation of Documents. In: IJCAI (2005)

11. Denoyer, L., Wisniewski, G., Gallinari, P.: Document structure matching for heterogeneous corpora. In: Workshop SIGIR 2004. Workshop on IR and XML, Sheffield (2004)
12. Daumé III, H., Marcu, D.: Learning as search optimization: approximate large margin methods for structured prediction. In: ICML '05. Proceedings of the 22nd international conference on Machine learning, New York, NY, USA, pp. 169–176. ACM Press, New York (2005)
13. Fuhr, N., Govert, N., Kazai, G., Lalmas, M.: Inex: Initiative for the evaluation of xml retrieval. In: SIGIR'02 Workshop on XML and Information Retrieval (2002)
14. Doan, A., Domingos, P., Halevy, A.: Learning to match the schemas of data sources: A multistrategy approach. Mach. Learn. 50, 279–301 (2003)
15. McCallum, A.: Information extraction: distilling structured data from unstructured text. Queue 3, 48–57 (2005)
16. Young-Lai, M., Tompa, F.W.: Stochastic grammatical inference of text database structure. Machine Learning (2000)
17. Fine, S., Singer, Y., Tishby, N.: The hierarchical hidden markov model: Analysis and applications. Machine Learning 32, 41–62 (1998)

Application of Fractal Theory
for On-Line and Off-Line Farsi Digit Recognition

Saeed Mozaffari[1], Karim Faez[1], and Volker Märgner[2]

[1] Pattern Recognition and Image Processing Laboratory, Electrical Engineering
Department, Amirkabir University of Technology, Tehran, Iran, 15914
{s_mozaffari,kfaez}@aut.ac.ir
[2] Institute of Communications Technology (IfN)
Technical University of Braunschweig, Braunschweig, Germany, 38092
v.maergner@tu-bs.de

Abstract. Fractal theory has been used for computer graphics, image
compression and different fields of pattern recognition. In this paper, a
fractal based method for recognition of both on-line and off-line Farsi/
Arabic handwritten digits is proposed. Our main goal is to verify whether
fractal theory is able to capture discriminatory information from digits
for pattern recognition task. Digit classification problem (on-line and off-
line) deals with patterns which do not have complex structure. So, a general purpose fractal coder, introduced for image compression, is simplified
to be utilized for this application. In order to do that, during the coding
process, contrast and luminosity information of each point in the input
pattern are ignored. Therefore, this approach can deal with on-line data
and binary images of handwritten Farsi digits. In fact, our system represents the shape of the input pattern by searching for a set of geometrical
relationship between parts of it. Some fractal-based features are directly
extracted by the fractal coder. We show that the resulting features have
invariant properties which can be used for object recognition.

Keywords: Fractal theory, Iterated function system, on-line and off-line
Farsi/Arabic handwritten digit recognition.

1 Introduction

The recognition of handwritten alphanumeric is a challenging problem in pattern
recognition. This is due to the large diversity of writing styles and image quality.
English, Chinese and Kanji isolated handwritten character recognition has long
been a focus of study with a high recognition rates reports. But few researches
have been done on Farsi and Arabic. Recognition of Farsi/Arabic script has
progressed slowly mainly due to the special characteristics of these languages,
and the lack of communication among researchers in this field. The reader is
referred to [1][2] for more details on the state of the art of Arabic character
recognition.

Fractal theory of iterated function systems has been extensively investigated
in computer graphics [3] and image compression [4]. The fundamental principle

P. Perner (Ed.): MLDM 2007, LNAI 4571, pp. 868–882, 2007.
© Springer-Verlag Berlin Heidelberg 2007

of fractal coding consists of the representation of any image by a contractive transform of which the fixed point is too close to the original image. Recently its potential in different fields of pattern recognition such as face recognition [5], character and digit recognition [6], texture recognition [7], fingerprint analysis [8] and writer identification and authentication [9] has been explored.

This paper presents a method to recognize on-line and off-line handwritten Farsi digits. The method simplifies a general fractal image coder via discarding contrast and luminosity information during the coding process. The main difference between our work and the one which was reported by Baldoni et al. [6] is that our fractal coder does not use gray level values. Therefore, it can be used for on-line and binary off-line digit recognition.

In section 2, an overview of fractal theory and Iterated Function Systems is presented. Section 3 describes the off-line recognition approach including the proposed fractal coding and decoding algorithms. One dimensional fractal coder for on-line digit recognition is detailed in section 4. Section 5 outlines some feature extraction approaches in the fractal domain. Finally, classification set-up and experimental results are presented in sections 6 and 7 respectively. Conclusion remarks are given in section 8.

2   Overview of Fractal Theory and Coding

The goal of this section is to explain the general concepts of fractal theory in very simple terms, with as little mathematics as possible. Fractal theory is based on the concepts and mathematical results of Iterated Function Systems (IFS). The fundamental principle of fractal coding involves the representation of any image, I, by a contractive transformation T, in which the fixed point (image) is too close to the original image, $I = \lim_{n \to \infty} T(I_n)$.

Banach's fixed point theorem guarantees that the fixed point of such a transformation can be obtained by an iterated application $I_{n+1} = T(I_n)$. The obtained image from IFS is made of modified copies of itself, generated by the elementary transforms. Let us assume an IFS that reduces the input image size and reproduces it three times in a triangle shape. The above IFS consists of three transformations, a reduction, followed by a translation, and repositioning in a triangle shape. If a circle image is given to this system, the fixed point of this IFS is the Sierpinski triangle (Fig 1).

With the introduction of the partitioned IFS (PIFS) by Jacquin [3], Fractal theory became a practical reality for natural image compression. In PIFS, each of the individual mappings operates on a subset of the image, rather than on the entire image as in IFS. A PIFS defines a transform T as the union of affine contractive transforms defined on domains, I_i, included in the image:

$$T(I) = T_1(I_1) \cup T_2(I_2) \cup ... \cup T_n(I_n) \tag{1}$$

The set of all images obtained from all the transformations of sub-images allows to partition the spatial domain of I. So, if the right PIFS is built, the

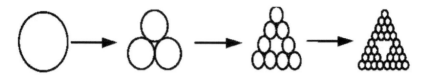

Fig. 1. Generation of fractal image by IFS

initial image would be the attractor of the IFS and could be derived from any initial image. For more details on fractal theory see [3][4].

The aim of fractal compression is to find a set of transforms in a PIFS which generate the initial image as the fixed point (equation 1). In order to construct this PIFS, the image to be encoded is partitioned into non-overlapping *range blocks* (R blocks) and overlapping *domain blocks* (D blocks) as depicted in Fig 2. To apply fixed point theorem, the D blocks are assumed two times larger than R blocks to make sure that the transforms in PIFS are contractive. The task of the fractal coder is to find a D block for each R block in the same image such that the transformation of this block, $W(R)$, minimizes the collage error in equation 2.

$$CollageError = \min \| D - W(R) \|^2 \tag{2}$$

In the above equation, the transformation W, which maps each D block into its corresponding R block, can be any transformation. However, affine transforms are preferred in practice. An affine transformation preserves co-linearity and ratios of distances. It does not necessarily preserve angles or lengths. The transformation W is a combination of geometrical and luminance transformations. According to equation 3, a point at coordinate (x, y) with gray level z is rotated and scaled by the geometrical parameters a, b, c, and d and is offset by e and f.

$$W_i \begin{bmatrix} x \\ y \\ z \end{bmatrix} = \begin{bmatrix} a_i & b_i & 0 \\ c_i & d_i & 0 \\ 0 & 0 & s_i \end{bmatrix} \begin{bmatrix} x \\ y \\ z \end{bmatrix} + \begin{bmatrix} e_i \\ f_i \\ o_i \end{bmatrix} \tag{3}$$

Minimizing equation 2 has two meanings: first, finding a good choice for D_i and second, selecting a good contrast and brightness setting for W_i in equation 3. A choice of D_i, along with a corresponding s_i and o_i, determines a map W_i. It has been proven that parameter s has to be less than 1, in order to obtain a set of contractive transforms [4]. A more complete introduction to fractal image coding are given in [3] and [4].

3   Off-Line Digit Recognition Approach

Off-line recognition involves the automatic conversion of text in an image into letter codes which can be used within computer and text-processing applications. This technique is applied to previously written text, such as any images digitized by an optical scanner. Each character is then located and segmented and the resulting matrix is fed into a preprocessor for smoothing, noise reduction and size normalization. Off-line recognition can be considered as the most general case in

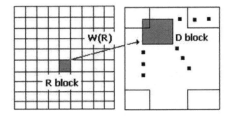

Fig. 2. One of the range to domain block mapping in PIFS representation

which no special device is required for writing. This section focuses on the recognition of off-line handwritten digits and the proposed fractal coder is introduced.

3.1   The Proposed Fractal Image Coder

Unlike many other object recognition systems, such as face and texture recognition, the overall shape of the character is more important than the pixel gray levels in optical character recognition. As a result, many OCR systems work with binary images rather than gray level images and several feature extraction approaches have been proposed for them.

   This section presents a simplified fractal coder for binary image coding. This method concentrates on the geometrical relationship between the range block and its best domain block instead of pixel gray levels distribution. There are two spatial dimensions in equation 3 and the gray level adds a third dimension to it. Therefore, we can easily simplify equation 3 by omitting the third row which includes parameters of luminance transformation to obtain equation 4.

$$v_i \begin{bmatrix} x \\ y \end{bmatrix} = \begin{bmatrix} a_i & b_i \\ c_i & d_i \end{bmatrix} \begin{bmatrix} x \\ y \end{bmatrix} + \begin{bmatrix} e_i \\ f_i \end{bmatrix} \tag{4}$$

   In the above equation, v_i determines how the partitioned ranges are mapped to their domains. As we are dealing with simple patterns (digit images) and range to domain block matching is a very time consuming process, the transformation set is restricted to *isometric* affine transformations in this research. A transformation f is called isometric if it keeps the distance function, d, invariant.

$$d(x, y) = d(f(x), f(y)) \tag{5}$$

   The only isometric affine transformation is the rotation, possibly composed with the flip. Among all rotations, four of them preserve the orientation of a square, namely, the identity, the 90° rotation, the 180° rotation, and the 270° rotation. Composing these rotations with the flip, the following eight deformation matrices are obtained in Table 1. Let us assume a block with 4 pixels like $\begin{bmatrix} 1 & 2 \\ 3 & 4 \end{bmatrix}$. Table 1 shows this block after each transformation. It is supposed that we are dealing with a $M \times M$ binary image $I(x, y)$. The image is partitioned into range and domain blocks. In our experiments, D blocks are twice the size of R blocks. If the R block consists of $N \times N$ pixels, the number of non-overlapping R blocks would be $n_r = \lceil \frac{M}{N} \rceil \times \lceil \frac{M}{N} \rceil$. Since size of D blocks is assumed $2N \times 2N$, the collection D contains $n_d = [M - 2N + 1] \times [M - 2N + 1]$ overlapped squares. Each R/D block is characterized by the number of its pixels and its starting point, which always points to the top-left pixel in the corresponding block. As mentioned before, a D block has 4 times as many pixels as an R block, so we must average the 2×2 sub-squares corresponding to each pixel of R block when minimizing equation 2. This down-sampling method can be done by an averaging transformation. The encoding algorithm can be summarized as follows:

Encoding Algorithm
1- Input the original binary image.
2- Partition the input image into R blocks according to fixed size square blocks partitioning scheme (Fig 2).
3- Create a list of all D blocks.
4- Scan the image from top-to-bottom and from left-to-right.
5- Search for a fractal match. Given a R_i region, loop over all possible D blocks to find the best match using a given metric (equation 2).
6- After finding the best match, select fractal elements. The fractal code of the input image is defined as a set of all n_r range to domain affine transformations. Transformation $f(k)$ consists of five real numbers:

- Starting point of the R block, $rs_k(x, y)$.
- Starting point of the corresponding D block, $ds_k(x, y)$.
- The index of D_k to R_k transformation, T_k.

The transformation Index is a number between 1 and 8 according to Table 1.

Table 1. Eight isometric affine transforms used in this paper

Index	Isometry	Matrix	Pixels order in the transformed block
1	identity	$\begin{bmatrix} 1 & 0 \\ 0 & 1 \end{bmatrix}$	$\begin{bmatrix} 1 & 2 \\ 3 & 4 \end{bmatrix}$
2	x flip	$\begin{bmatrix} -1 & 0 \\ 0 & 1 \end{bmatrix}$	$\begin{bmatrix} 3 & 4 \\ 1 & 2 \end{bmatrix}$
3	y flip	$\begin{bmatrix} 1 & 0 \\ 0 & -1 \end{bmatrix}$	$\begin{bmatrix} 2 & 1 \\ 4 & 3 \end{bmatrix}$
4	$180°$ rotation	$\begin{bmatrix} -1 & 0 \\ 0 & -1 \end{bmatrix}$	$\begin{bmatrix} 4 & 3 \\ 2 & 1 \end{bmatrix}$
5	$x = y$ flip	$\begin{bmatrix} 0 & 1 \\ 1 & 0 \end{bmatrix}$	$\begin{bmatrix} 4 & 2 \\ 3 & 1 \end{bmatrix}$
6	$270°$ rotation	$\begin{bmatrix} 0 & -1 \\ 1 & 0 \end{bmatrix}$	$\begin{bmatrix} 2 & 4 \\ 1 & 3 \end{bmatrix}$
7	$90°$ rotation	$\begin{bmatrix} 0 & 1 \\ -1 & 0 \end{bmatrix}$	$\begin{bmatrix} 3 & 1 \\ 4 & 2 \end{bmatrix}$
8	$x = -y$ flip	$\begin{bmatrix} 0 & -1 \\ -1 & 0 \end{bmatrix}$	$\begin{bmatrix} 1 & 3 \\ 2 & 4 \end{bmatrix}$

3.2   Fractal Image Decoder

The reverse process of generating an image from a fractal model is called decoding or decomposition. The decoding process starts with an $M \times M$ arbitrary initial image. For each fractal transformation $f(k)$, the $N \times N$ domain block d_k is constructed from the initial image, given its start point ds_k stored in the fractal code. Then its corresponding stored affine transformation T_k is applied on it. After down-sampling process, the $N \times N$ obtained block is translated to the corresponding R block located at rs_k. After doing the above process one iteration is completed. The decoding algorithm is usually iterated about 6 to 16 times until the difference between two successive output images is small enough and the fixed point image is obtained. Depending on the image which was coded, the decoding process can be applied to black and white images as well as to gray level images. The results for a gray level image fractal decoding are shown in Fig.3 with different iterations and different R block sizes (N) [14].

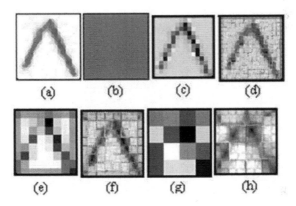

Fig. 3. Fractal image decoding results. (a) Original image. (b) arbitrary initial image. (c) decoded image after 1 iteration for N=4. (d) decoded image after 10 iteration for N=4. (e) decoded image after 1 iteration for N=8. (f) decoded image after 10 iteration for N=8. (g) decoded image after 1 iteration for N=16. (h) decoded image after 10 iteration for N=16.

4   On-Line Digit Recognition Approach

In on-line character recognition systems, the computer recognizes the symbols as they are drawn. The most common writing surface is a digitizing tablet and a special pen which emits the temporal sequence of the pen position points. In on-line recognition, writing line has no width. Additionally, temporal information like velocity and pen lifts are available and can be useful in recognition process. The on-line problem is usually easier than the off-line problem since more information is available.

4.1   One Dimensional Fractal Coder

In on-line recognition, the process is performed on one-dimensional data rather than two-dimensional images in off-line recognition. In this case, we are dealing with a time ordered sequence of points based on the pen positions. Therefore, gray levels, contrast, and luminosity information have no role in on-line digit recognition. Since on-line data is modeled as a set of (x,y) coordinates, we can easily use the proposed fractal image coding presented in section 3.1 for off-line recognition.

We used a re-sampling as preprocessing step to realize uniform a uniform sampled input curve with M sample points. Similar to the block image coding scheme, the digit locus is divided into range and domain *segments* with the length of N and $2N$, respectively. Therefore the input locus can be grouped into $n_r = \lceil \frac{M}{N} \rceil$ range segments and $n_d = [M - 2N + 1]$ domain segments. Each of the R segments shares a common point with its adjacent segment such that the end point of the last segment is the starting point of the next one.

Each of the R and D segments is characterized by the number of their points (N and 2N) and their starting point (rs_k and ds_k). For each range segment, a corresponding domain segment is searched within the digit locus to minimize equation 2 under an appropriate affine transformation. Since each D segment has two times more points than an R segment, during the search for the best D segment a down sampling process is needed. So when we minimize equation 2, the average of each two consecutive points in the D segment is assigned to corresponding R segment point. Similar to 2D fractal codes, each fractal code, $f(k)$, consists of five real numbers (see section 3.1).

4.2   One Dimensional Fractal Decoder

After creating an arbitrary initial locus S with M points, for each fractal code $f(k)$, a D segment with $2N$ points is constructed at the staring point ds_k). After

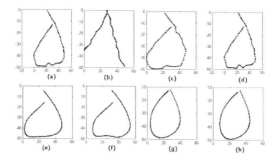

Fig. 4. On-line decoding algorithm's results. (a) original locus. (b) arbitrary initial locus. (c) decoded locus after 1 iteration for N=2. (d) decoded locus after 5 iterations for N=2. (e) decoded locus after 1 iteration for N=4. (f) decoded locus after 5 iterations for N=4. (g) decoded locus after 1 iteration for N=8. (h) decoded locus after 5 iterations for N=8.

down-sampling to N points, the stored affine transformation T_k is performed on the D segment. Then its starting point is shifted to the origin of R segment rs_k. Fig 4 shows the effects of range segment length, N , and number of iterations on the decoded locus.

5   Fractal Based Feature Extraction

The fractal coder attempts to capture some self-similarity information from the given pattern which can be used for pattern recognition. From now onward, we will call the PIFS based features fractal feature. This section presents some fractal features which can be used for on-line and off-line digits recognition.

5.1   Fractal Codes

The Fractal codes, extracted directly from the fractal coder, can be used as a feature vector [5][10]. The fractal coder used the fixed size block/segment partitioning method to divide the input pattern into R blocks/segments. For off-line patterns, the digit image was scanned from top-to-bottom and from left-to-right. The same approach was used for on-line digits in which two adjacent segments share one common point and the input locus is traced from the first segment to the last segment, one after another. Therefore, starting point coordination of the R block/segment is redundant and can be deduced easily. According to sections 3.1 and 4.1, each fractal code contains five real numbers. By omitting (x,y) coordinates of the starting point of the R block/segment (rs_k), a feature vector with the length of $3 \times n_r$ obtained for each input pattern.

5.2   Mapping Vector Accumulator (MVA)

In addition to fractal codes, a secondary subset of the fractal transformation parameters can be used for classification. This feature is based on the relative locations of a range block/segment and its corresponding domain block/segment, called Mapping Vector Accumulator (MVA)[15]. MVA records the angle and magnitude of domain-range mapping vector. The matrix itself is an accumulator, where the angle and magnitude are first quantized and then the appropriate element of the accumulator is incremented. For example, assuming the mapping shown in Fig 5 has an angle $\theta = 78$ and a magnitude $m = 5$. In this case the appropriate element highlighted in the matrix is incremented. Although MVA feature records the relative locations of the range and corresponding domain blocks/segments, it does not show how these blocks/segments are mapped to each other. To add the transform information to the MVA feature, we proposed *Multiple* Mapping Vector Accumulators (MMVA) feature in which MVA matrixes are built for each of the transforms, $M_1, M_2, ..., M_8$, in Table 1. For example, if the range and domain blocks in Fig 5 correspond to each other under the third affine transformation, then the appropriate element in M_3 is increased. After reshaping each of $M_1, M_2, ..., M_8$ matrixes into a vector form, the obtained vectors are concatenated to each other to make the MMVA feature vector, $M = [M_1, M_2, ..., M_8]$.

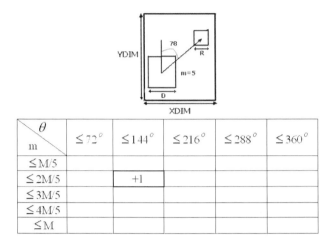

Fig. 5. Mapping Vector Accumulator [15]

5.3   Domain-Range Co-Location Matrix

The Domain-Range Co-Location Matrix (DRCLM) encapsulates information from the relative location of the domain block/segment and the associated range block/segment [15]. So it measures levels of self similarity in different parts of the image. In this method, the image is divided into four equal sized non-overlapping segments. So the matrix is a 4×4 table of numbers illustrating when a mapping occurs from one segment to the other. For example, if a range block in segment 1 is mapped from a domain block in segment 3, then the entry at cell c_{13} in the matrix will be incremented. This is then repeated for all range blocks in the image. Similar to MVA feature, we can have a DRCLM for each affine transformation to compute the Multiple Domain-Range Co-Location Matrix (MDRCLM) for the input pattern. As it was proposed by Linnell and Deravi (2003)[15], two further features can be extracted from MDRCLM matrices. Firstly the difference moment of order 2 and secondly, the entropy of the matrix as:

$$Moment = \sum_i \sum_j (i - j)^2 c_{ij} \tag{6}$$

$$Entropy = \sum_i \sum_j c_{ij} \log c_{ij} \tag{7}$$

These features are similar to those used for texture classification when applied on Co-occurrence matrices. They have been chosen to characterize the self-similarity in the pattern. For example, the difference moment of order 2 is used as a measure of self-similarity in the same segment, as it gives a higher weight to elements on the main diagonal of the matrix. Similarly, the entropy measures the randomness of the entries in the matrix, so it is a good measure of evenly distributed self-similarity across all areas in the pattern [15]. For each of eight

MDRCLM matrices, difference moment and entropy are calculated separately, resulting in a 16 element feature vector.

6   Classification Set-Up

This section describes the two test sets on which we worked, off-line and on-line handwritten digits, and explains different classification methods that we used in this research.

6.1   Test Images

Farsi language has ten digits. Most of the Farsi recognition systems were tested on different databases created only by a few people and no common database is available to compare the results. In this paper off-line and on-line datasets were gathered separately from different people with different educational backgrounds.

Our off-line database contains 480 samples per digit, written by 48 persons. We used 280 samples of each digit for training and the rest for testing. For a given image of a single numeral, two pre-processing tasks have been done to make the system invariant to scale and frame size changes. To remove any difference to the location of a numeral within the image, the bounding rectangle box of each digit was calculated. This bounding box was scaled to 64×64 pixel image for normalization purpose.

For training and testing the proposed system, a data set consisting of 1000 on-line digits written by 100 different persons was collected (100 per digit). 60 samples of each digit were used to train the classifier and the remaining data was utilized in the classification step. The numbers of points in an on-line digit depends on the sampling rate of the digitizing tablet and also on the speed of writing. Therefore, a preprocessing phase was needed to smooth and re-sample the input curve into M spatially uniform points. By interpolating the given sequence, a curve is fitted to it. Then M points were uniformly re-sampled from the obtained curve.

6.2   Classification Methods

In the preceding section, we presented different feature extraction methods. The next step is to evaluate the discriminatory power of these features by means of a classifier. While there are many classification approaches for pattern recognition, we used two different classifiers for on-line and off-line digit recognition.

First a Multilayer Perceptron (MLP) Neural Network with 300 hidden neurons was used. This Neural Network was trained with backpropagation algorithm after 1000 iterations and learning rate of 0.01 [16]. The same structure, learning rate and number of iterations were used for all experiments with Neural Networks in this research.

Fractal transformation is used as the second classifier. The inherent property of fractal theory, based on the IFS fixed point theorem, has also been exploited by some researchers, called fractal transformation. In this method the distortion between an input pattern and the pattern after *one* decoding iteration was used as the basic idea for classification. For more details on fractal transformation classifier see [11][17].

7   Experimental Results

Unfortunately, there is not any standard database for Farsi/Arabic digits recognition to be considered as a benchmark (such as NIST data set for English Digits). Every research group has implemented their system on sets of data gathered by themselves and reported different recognition rates. Among them, the proposed method by Soltanzadeh and Rahmati reached the recognition rate of 99.57% [18] while this was less than 95% for others [19]. To validate the effectiveness of the proposed fractal based features, it is necessary to compare them with other approaches. Such comparison is possible by implementing the concurrent approaches and then applying them with the proposed method on the same database. Therefore, it is very difficult to give comparative results for the proposed methods. We compared the performance of the proposed system with the results of our previous research, which have already been presented separately.

For the off-line digit recognition, first we extracted the fractal features from gray level and binary images with the help of the fractal image coder, described in section 3.1. The aim of this experiment was to verify the effect of luminescence information on fractal based off-line digit classification. The results showed that by the use of the fractal coder, the recognition rates for binary digits were higher than the gray level ones. This can be true because unlike other pattern recognition domains, such as face and texture recognition, the structure and overall shape of a digit are more important than gray level information for classification process. On the other hand, these results should be expected since, the aim of this research was pattern recognition rather than image compression and we had made some simple assumptions about luminescence information in the input image (calculating the average of gray levels in each R/D block during fractal coding and decoding phases). Therefore, the recognition rate improved from 85% for gray level images to 89% for binary images with the help of fractal codes (N=4) and MLP Neural Network classifier. Due to this result, we used binary images instead of gray level digits for the next experiments.

Afterwards, we studied the performance of fractal transformation approach, a classifier based on inherent property of fractal theory, and compared it with MLP Neural Network classifier. According to the comparison made by Tan and Yan [11], this classifier outperformed the others (HMM, PDBNN) in the terms of error rate in the application of face recognition. One drawback of this method is that the complexity of the recognition system is linear to the size of the database, which is not as much the case for Neural Networks. Unlike Neural Networks, fractal transformation does not need re-training process when some classes were added to or removed from the database. In this method, discrimination criterion,

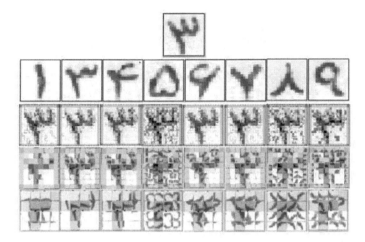

Fig. 6. First row: input image. Second row: some samples in the training set. Third row: results of applying fractal codes of input image with N=4 on second row images after one iteration. Fourth row: results of applying fractal codes of input image with N=8 on second row images after one iteration. Fifth row: results of applying fractal codes of input image with N=16 on second row images after one iteration.

distortion between input and decoded images after one iteration, can highly be affected by the size of Range blocks (N). According to Fig 6, when we used R blocks with the size of 4×4 and 8×8, after one decoding iteration, decoded images were not very different from the input image. By increasing the size of the R blocks in the fractal coding step to 16×16 , the discrimination criterion can be more obvious in fractal transformation classifier. The optimum size of the Range block depends on the size of input image and type of the application. This method can easily be used for on-line digit recognition.

As it was mentioned before, a secondary subset of the fractal features (MVA and DRCLM) can be extracted for digit recognition after the coding phase. By adding transforms information to the MVA and DRCLM features (see sections 5.2 and 5.3), their performance improved from 82% to 85.4% for MVA feature and from 80.2% to 82.6% for DRCLM feature, both with the use of R block size 8×8 and an MLP Neural Network.

Dividing the input image into R and D blocks was another aspect which was studied. Various partitioning methods were proposed for fractal image compression [4]. Quad-tree partitioning technique was also implemented for finding range blocks [12]. A quad-tree partition is a representation of an image as a tree in which each node, corresponding to a square portion of the image, contains four sub-nodes, each corresponding to the four quadrants of the square. The root of the tree is the initial image. Since digit images do not have complicated structures, quad-tree partition takes more time than fixed size square block approach (FSSB) for feature extraction with lower performance (86.2% against 91% recognition rate obtained by FSSB with fractal transformation classifier). For each of

Table 2. Comparison results for off-line digit recognition

Classifier	Feature	Recognition Rate
Fractal Transformation	Fractal N=16	91%
MLP NN	Fractal N=4	89%
MLP NN	MVA N=8	82%
MLP NN	MMVA N=8	85.4%
MLP NN	DRCLM N=8	80.2%
MLP NN	MDRCLM N=8	82.6%

the fractal based methods the best recognition rate (depending on the optimum range block size) is reported. The results of this series of experiments on off-line digit recognition are reported in Table 2.

Although, in the case of on-line digit recognition, additional information such as velocity and pen lifts are available, we only used (x,y) sequence of the plotted points. The experimental results obtained for off-line digit recognition were almost the same for on-line digit recognition. Fractal transformation with R segment length of 8 had the best recognition rate among other fractal based approaches. The results of on-line digit recognition are presented in Table 3.

Table 3. Recognition rates of fractal features for on-line digit recognition

Classifier	Feature	Recognition Rate
Fractal Transformation	Fractal N=8	93.6%
MLP NN	Fractal N=8	91%
MLP NN	MVA N=8	86.5%
MLP NN	MMVA N=8	88.5%
MLP NN	DRCLM N=8	86%
MLP NN	MDRCLM N=8	88.4%

8   Conclusions

The main goal of this paper is to verify whether fractal theory is able to capture discriminatory information from digits for pattern recognition task. In order to be utilized for pattern recognition, a general purpose fractal coder which has previously been used for image compression was simplified. This was achieved by obviating the need to determine the brightness and contrast parameters. In this way we proposed a fractal based approach that can be used for *both* on-line and off-line binary digit recognition. The main difference between the proposed method and the previous works on fractal based object recognition is that we focus on the ability of fractal theory for simple structure one-dimensional pattern recognition. Another aspect that makes our work different from the others is recognizing binary images instead of gray level ones, which is a common task in off-line digit recognition.

In our research, some feature vectors were directly extracted from the appropriate fractal coder and fed to different classifiers. Fractal transformation achieved a high recognition rate but its recognition time was also relatively high and consequently its performance degrades for large databases. Experiments showed recognition rates more than 91% on both on-line and off-line test samples. We believe that the obtained results are adequate to motivate other researchers to study fractal based feature extraction methods for object recognition. Transformation selection for IFS is still an open problem which depends on the application type and needs more consideration.

References

1. Amin, A.: Off-line Arabic characters Recognition: The State Of the Art. Pattern Recognition 31(5), 517–530 (1998)
2. Lorigo, L., Govindaraju, V.: Offline Arabic handwriting recognition: a survey. IEEE Transaction on Pattern Analysis and Machine Intelligence 28(5), 712–724 (2006)
3. Jacquin, E.: Image coding based on a fractal theory of iterated contractive image transformations. IEEE Trans. on Image Processing 1(1), 18–30 (1992)
4. Fisher, Y.: Fractal Image Compression, Theory and Application. LNCS. Springer, Heidelberg (1995)
5. Temdee, P., Khawparisuth, D., Chamnongthai, K.: Face Recognition by using Fractal Encoding and Backpropaga-tion Neural Network. In: International Symposium on Signal Processing and its Application (ISSPA), pp. 159–161 (1999)
6. Baldoni, M., Baroglio, C., Cavagnino, D.: Use of IFS codes for learning 2D isolated-objects classification systems. Computer Vision and Image Understanding 77, 371–387 (2000)
7. Potlapalli, H., Luo, R.C.: Fractal-based classification of natural textures. IEEE Transactions on Industrial Electronics 45(1), 142–150 (1998)
8. Polikarpova, N.: On the fractal features in fingerprint analysis. International Conference on Patter Recognition, pp. 591–595 (1996)
9. Seropian, A., Grimaldi, M., Vincent, V.: Writer identification based on the fractal construction of a reference base. International Conference on Document Analysis and Recognition 2, 1163–1167 (2003)
10. Mozaffari, S., Faez, K., Rashidy Kanan, H.: Feature Comparison between Fractal codes and Wavelet Transform in Handwritten Alphanumeric Recognition Using SVM Classifier. International Conference on Patter Recognition, pp. 331–334 (2004)
11. Tan, T., Yan, H.: Face Recognition by Fractal Transformations. In: International Conference on Acoustics, Speech and Signal Processing (ICASSP), vol. 6, pp. 3537–3540 (1999)
12. Mozaffari, S., Faez, K., Ziaratban, M.: Character Representation and Recognition Using Quadtree-based Fractal Encoding Scheme. International Conference on Document Analysis and Recognition , 819–823 (2005)
13. Ebrahimpour, H., Chandran, V., Sridharan, S.: Face Recognition Using Fractal Codes, pp. 58–61. IEEE Computer Society Press, Los Alamitos (2001)
14. Mozaffari, S., Faez, K., Ziaratban, M.: A Hybrid Structural/Statistical Classifier for Handwritten Farsi/Arabic Numeral Recognition. In: IAPR Conference on Machine VIsion Applications, Japan, May 16-18, 2005, pp. 218–211 (2005)

15. Linnell, T.A., Deravi, F.: Novel Fractal Domain Features for Image Classification. VIE Conference, pp. 33–36 (2003)
16. Fausett, L.V.: Fundamentals of Neural Networks. Prentice-Hall, Englewood Cliffs (1994)
17. Tan, T., Yan, H.: Fractal neighbor distance measure. Pattern Recognition 35, 1371–1387 (2002)
18. Soltanzadeh, H., Rahmati, M.: Recognition of Persian handwritten digits using image profiles of multiple orientations. Pattern Recognition Letters 25, 1569–1576 (2004)
19. Sadri, J., Suen, C.Y, Bui, T.D.: Application of support vector machines for recognition of handwritten Arabic/Persian digits. In: The Proceedings of Second Iranian Conference on Machine Vision and Image Processing, vol. 1, pp. 300–307 (2003)

Hybrid Learning of Ontology Classes

Jens Lehmann

University of Leipzig, Computer Science Department,
Johannisgasse 26, D-04103 Leipzig, Germany
lehmann@informatik.uni-leipzig.de

Abstract. Description logics have emerged as one of the most successful formalisms for knowledge representation and reasoning. They are now widely used as a basis for ontologies in the Semantic Web. To extend and analyse ontologies, automated methods for knowledge acquisition and mining are being sought for. Despite its importance for knowledge engineers, the learning problem in description logics has not been investigated as deeply as its counterpart for logic programs.

We propose the novel idea of applying evolutionary inspired methods to solve this task. In particular, we show how Genetic Programming can be applied to the learning problem in description logics and combine it with techniques from Inductive Logic Programming. We base our algorithm on thorough theoretical foundations and present a preliminary evaluation.

1 Introduction

Ontologies based on *Semantic Web* technologies are now amongst the most prominent paradigms for knowledge representation. The single most popular ontology language in this context is OWL[1]. However, there is still a lack of available ontologies and tools for creating, extending and analysing ontologies are most demanded. Machine Learning methods for the automated learning of classes from instance data can help to overcome these problems.

While the learning, also referred to as *induction*, of logic programs has been studied extensively in the area of *Inductive Logic Programming* (ILP), the analogous problem for description logics has been investigated to a lesser extend, despite recent efforts [9,11]. This is mainly due to the fact that description logics have only recently become a popular knowledge representation paradigm. The rise of the Semantic Web has increased interest in methods for solving the learning problem in description logics.

Genetic Programming (GP) has been shown to deliver human-competitive machine intelligence in many applications [12]. We show how it can be applied to the learning problem in description logics. Further, we discuss the advantages and drawbacks of this approach and propose a framework for hybrid algorithms combining GP and *refinement operators*.

Refinement operators are central in ILP and we can base our methods on a thorough theoretical analysis of their potential and limitations. We will design

---

[1] http://www.w3.org/2004/OWL

P. Perner (Ed.): MLDM 2007, LNAI 4571, pp. 883–898, 2007.

concrete refinement operators and show that they have the desired properties. For the novel algorithm we obtain, a preliminary evaluation is performed, comparing it with standard GP and an other non-evolutionary algorithm for learning in description logics.

The paper is structured as follows. In Section 2 we briefly introduce description logics, Genetic Programming, the learning problem, and refinement operators. Section 3 shows how to apply GP to the learning problem in description logics. The main section is Section 4, in which we show how refinement operators fit in the GP framework. In 5 we report on our prototype implementation. We discuss related work in 6 and draw conclusions in Section 7.

2   Preliminaries

Description Logics. Description logics represent knowledge in terms of *objects*, *concepts*, and *roles*. Objects correspond to constants, concepts to unary predicates, and roles to binary predicates in first order logic. In description logic systems information is stored in a *knowledge base*, which is a set of axioms. It is divided in *TBox*, containing *terminology* axioms, and *ABox*, containing *assertional* axioms.

Table 1. $\mathcal{ALC}$ syntax and semantics

construct	syntax	semantics
concept	A	$A^{\mathcal{I}} \subseteq \Delta^{\mathcal{I}}$
role	r	$r^{\mathcal{I}} \subseteq \Delta^{\mathcal{I}} \times \Delta^{\mathcal{I}}$
top	$\top$	$\Delta^{\mathcal{I}}$
bottom	$\bot$	$\emptyset$
conjunction	$C \sqcap D$	$(C \sqcap D)^{\mathcal{I}} = C^{\mathcal{I}} \cap D^{\mathcal{I}}$
disjunction	$C \sqcup D$	$(C \sqcup D)^{\mathcal{I}} = C^{\mathcal{I}} \cup D^{\mathcal{I}}$
negation	$\neg C$	$(\neg C)^{\mathcal{I}} = \Delta^{\mathcal{I}} \setminus C^{\mathcal{I}}$
existential	$\exists r.C$	$(\exists r.C)^{\mathcal{I}} = \{a \mid \exists b.(a,b) \in r^{\mathcal{I}} \text{ and } b \in C^{\mathcal{I}}\}$
universal	$\forall r.C$	$(\forall r.C)^{\mathcal{I}} = \{a \mid \forall b.(a,b) \in r^{\mathcal{I}} \text{ implies } b \in C^{\mathcal{I}}\}$

We briefly introduce the $\mathcal{ALC}$ description logic, which is the target language of our learning algorithm and refer to [3] for further background on description logics. Let N_I denote the set of objects, N_C denote the set of atomic concepts, and N_R denote the set of roles. As usual in logics, interpretations are used to assign a meaning to syntactic constructs. An *interpretation* $\mathcal{I}$ consists of a non-empty *interpretation domain* $\Delta^{\mathcal{I}}$ and an *interpretation function* $\cdot^{\mathcal{I}}$, which assigns to each object $a \in N_I$ an element of $\Delta^{\mathcal{I}}$, to each concept $A \in N_C$ a set $A^{\mathcal{I}} \subseteq \Delta^{\mathcal{I}}$, and to each role $r \in N_R$ a binary relation $r^{\mathcal{I}} \subseteq \Delta^{\mathcal{I}} \times \Delta^{\mathcal{I}}$. Interpretations are extended to elements as shown in Table 1, and to other elements of a knowledge base in a straightforward way. An interpretation, which satisfies an axiom (set of axioms) is called a model of this axiom (set of axioms). An $\mathcal{ALC}$ concept is in *negation normal form* if negation only occurs in front of concept names.

It is the aim of *inference algorithms* to extract implicit knowledge from a given knowledge base. Standard reasoning tasks include *instance checks*, *retrieval* and *subsumption*. We will only explicitly define the latter. Let C, D be concepts and $\mathcal{T}$ a TBox. *C is subsumed by D*, denoted by $C \sqsubseteq D$, iff for any interpretation $\mathcal{I}$ we have $C^{\mathcal{I}} \subseteq D^{\mathcal{I}}$. *$C$ is subsumed by D with respect to $\mathcal{T}$* (denoted by $C \sqsubseteq_{\mathcal{T}} D$) iff for any model $\mathcal{I}$ of $\mathcal{T}$ we have $C^{\mathcal{I}} \subseteq D^{\mathcal{I}}$. *$C$ is equivalent to D (with respect to $\mathcal{T}$)*, denoted by $C \equiv D$ ($C \equiv_{\mathcal{T}} D$), iff $C \sqsubseteq D$ ($C \sqsubseteq_{\mathcal{T}} D$) and $D \sqsubseteq C$ ($D \sqsubseteq_{\mathcal{T}} C$). *$C$ is strictly subsumed by D (with respect to $\mathcal{T}$)*, denoted by $C \sqsubset D$ ($C \sqsubset_{\mathcal{T}} D$), iff $C \sqsubseteq D$ ($C \sqsubseteq_{\mathcal{T}} D$) and not $C \equiv D$ ($C \equiv_{\mathcal{T}} D$).

Genetic Programming. Genetic Programming is one way to automatically solve problems. It is a systematic method to evolve individuals and has been shown to deliver human-competitive machine intelligence in many applications. The distinctive feature of GP within the area of Evolutionary Computing is to represent individuals (not to be confused with individuals in description logics) as variable length programs. In this article, we consider the case that individuals are represented as trees. Inspired by the evolution in the real world, fit individuals are selected from a population by means of different selection methods. New individuals are created from them using genetic operators like crossover and mutation. We do not introduce GP in detail, but instead refer to [12] for more information.

The Concept Learning Problem in Description Logics. In this section, we introduce the learning problem in Description Logics. In a very general setting learning means that we have a logical formulation of background knowledge and some observations. We are then looking for ways to extend the background knowledge such that we can explain the observations, i.e. they can be deduced from the modified knowledge.

More formally, we are given background knowledge B, positive examples E^+, negative examples E^- and want to find a hypothesis H such that from H together with B the positive examples follow and the negative examples do not follow. It is not required that the same logical formalism is used for background knowledge, examples, and hypothesis. This means, that although we consider learning $\mathcal{ALC}$ concepts in this article, the background knowledge can be a more expressive description logic.

So let a concept name Target, a knowledge base $\mathcal{K}$, and sets E^+ and E^- with elements of the form Target(a) ($a \in N_I$) be given. The learning problem is to find a concept C such that Target $\equiv C$ is an acyclic definition and for $\mathcal{K}' = \mathcal{K} \cup \{\text{Target} \equiv C\}$ we have $\mathcal{K}' \models E^+$ and $\mathcal{K}' \not\models E^-$.

For different solutions of the learning problem the simplest ones are to be preferred by the well-known Occam's razor principle [5]. According to this principle, simpler concepts usually have a higher predictive quality. We measure simplicity as the *length* of a concept, which is defined in a straightforward way, namely as the sum of the number of concept, role, quantifier, and connective symbols occurring in the concept.

Refinement Operators. Learning can be seen as a search process in the space of concepts. A natural idea is to impose an ordering on this search space and use operators to traverse it. This idea is prominent in Inductive Logic Programming [18], where refinement operators are used to traverse ordered spaces. Downward (upward) refinement operators construct specialisations (generalisations) of hypotheses.

A *quasi-ordering* is a reflexive and transitive relation. In a quasi-ordered space $(S, \preceq)$ a *downward (upward) refinement operator* ρ is a mapping from S to 2^S, such that for any $C \in S$ we have that $C' \in \rho(C)$ implies $C' \preceq C$ ($C \preceq C'$). C' is called a *specialisation (generalisation)* of C.

Quasi-orderings can be used for searching in the space of concepts. One such quasi-order is subsumption. If a concept C subsumes a concept D ($D \sqsubseteq C$), then C will cover all examples, which are covered by D. This makes subsumption a suitable order for solving the learning problem.

Definition 1. *A refinement operator in the quasi-ordered space* $(\mathcal{ALC}, \sqsubseteq_{\mathcal{T}})$ *is called an* $\mathcal{ALC}$ *refinement operator.*

We need to introduce some notions for refinement operators. A *refinement chain* of an $\mathcal{ALC}$ refinement operator ρ of length n from a concept C to a concept D is a finite sequence $C_0, C_1, \ldots, C_n$ of concepts, such that $C = C_0, C_1 \in \rho(C_0), C_2 \in \rho(C_1), \ldots, C_n \in \rho(C_{n-1}), D = C_n$. This refinement chain *goes through* E iff there is an i ($1 \leq i \leq n$) such that $E = C_i$. We say that D can be reached from C by ρ if there exists a refinement chain from C to D. $\rho^*(C)$ denotes the set of all concepts, which can be reached from C by ρ. $\rho^m(C)$ denotes the set of all concepts, which can be reached from C by a refinement chain of ρ of length m. If we look at refinements of an operator ρ we will often write $C \leadsto_\rho D$ instead of $D \in \rho(C)$. If the used operator is clear from the context it is usually omitted, i.e. we write $C \leadsto D$.

Refinement operators can have certain properties, which can be used to evaluate their usefulness.

Definition 2. *An* $\mathcal{ALC}$ *refinement operator* ρ *is called*

- *(locally) finite iff* $\rho(C)$ *is finite for any concept* C.
- *(syntactically) redundant iff there exists a refinement chain from a concept* C *to a concept* D, *which does not go through some concept* E *and a refinement chain from* C *to a concept weakly equal to* D, *which does go through* E.
- *proper iff for all concepts* C *and* D, $D \in \rho(C)$ *implies* $C \not\equiv D$.
- *ideal iff it is finite, complete (see below), and proper.*

An $\mathcal{ALC}$ *downward refinement operator* ρ *is called*

- *complete iff for all concepts* C, D *with* $C \sqsubseteq_{\mathcal{T}} D$ *we can reach a concept* E *with* $E \equiv C$ *from* D *by* ρ.
- *weakly complete iff for all concepts* $C \sqsubseteq_{\mathcal{T}} \top$ *we can reach a concept* E *with* $E \equiv C$ *from* $\top$ *by* ρ.

The corresponding notions for upward refinement operators are defined dually.

3   Concept Learning Using Standard GP

To apply GP to the learning problem, we need to be able to represent $\mathcal{ALC}$ concepts as trees. We do this by defining the alphabet $T = N_C \cup \{\top, \bot\}$ and $F = \{\sqcup, \sqcap, \neg\} \cup \{\forall r \mid r \in N_R\} \cup \{\exists r \mid r \in N_R\}$, where T is the set of terminal symbols and F the set of function symbols.

Example 1. The $\mathcal{ALC}$ concept Male $\sqcup$ $\exists$hasChild.Female can be represented as the following tree:

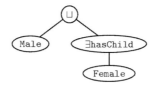

We say that an alphabet has the *closure* property if any function symbol can handle as an argument any data type and value returned by an alphabet symbol. Using the presented encoding the closure property is satisfied, because the way trees are build corresponds exactly to the inductive definition of $\mathcal{ALC}$ concepts. This ensures that tree operations like crossover and mutation can be performed safely, i.e. the obtained trees also represent $\mathcal{ALC}$ concepts.

Fitness Measurement. To be able to apply GP to the learning problem we need to define a fitness measure. To do this, we introduce some notions.

Definition 3 (covered examples). *Let Target be the target concept, $\mathcal{K}$ a knowledge base, and C an arbitrary $\mathcal{ALC}$ concept. The set of positive examples covered by C, denoted by $pos_{\mathcal{K}}(C)$, is defined as:*

$$pos_{\mathcal{K}}(C) = \{Target(a) \mid a \in N_I, K \cup \{Target \equiv C\} \models Target(a)\} \cap E^+$$

Analogously, the set of negative examples covered by C, denoted by $neg_{\mathcal{K}}(C)$, is defined as:

$$neg_{\mathcal{K}}(C) = \{Target(a) \mid a \in N_I, K \cup \{Target \equiv C\} \models Target(a)\} \cap E^-$$

Of course, the fitness measurement should give credit to covered positive examples and penalize covered negative examples. In addition to these classification criteria, it is also useful to bias the GP algorithm towards shorter solutions. A possible fitness functions is:

$$f_{\mathcal{K}}(C) = -\frac{|E^+ \setminus pos_{\mathcal{K}}(C)| + |neg_{\mathcal{K}}(C)|}{|E^+| + |E^-|} - a \cdot |C| \quad (0 < a < 1)$$

The parameter a is the decline in classification accuracy one is willing to accept for a concept, which is shorter by one length unit. Being able to represent solutions and measuring their fitness is already sufficient to apply Genetic Programming to a problem. We discuss some advantages and problems of doing this.

Advantages. GP is a very flexible learning method. It is not only able to learn in $\mathcal{ALC}$, but can also handle other description languages (languages with role constructors can be handled using the framework of Strongly Typed GP). GP has been shown to deliver good results in practice and is especially suited in situations, where approximate solutions are acceptable [13]. An additional advantage is that GP algorithms are parallelizable and can make use of computational resources, i.e. if more resources (time and memory) exist its parameters can be changed to increase the probability of finding good solutions. This may seem obvious, but in fact this does not hold for many (deterministic) solution methods. GP also allows for a variety of extensions and is able to handle noise naturally (the parameter a in the introduced fitness function is one way to handle noise).

Problems of the Standard Approach. Despite the described advantages of GP, there are some notable drawbacks. One problem is that the crossover operator is too destructive. For GP to work well, it should be the case that high fitness individuals are likely to produce high fitness offspring. (This is the reason why selection methods are used instead of random selection.) For crossover on $\mathcal{ALC}$ concepts, small changes in a concept can drastically change its semantics, so it is not very likely that the offspring of high fitness individuals also has a high fitness. Similar problems arise when using GP in ILP and indeed a lot of systems use non-standard operators [7].

Another problem of the standard approach is that we do not use all knowledge we have. An essential insight in Machine Learning [15] is that the approaches, which use most knowledge about the learning problem they want to solve, usually perform best. The standard GP algorithm does not make use of subsumption as quasi-order on concepts. Thus, a natural idea is to enhance the standard GP algorithm by operators, which exploit the subsumption order.

4   Refinement Operators in Genetic Programming

4.1   Transforming Refinement Operators to Genetic Refinement Operators

As argued before, it is useful to modify the standard GP approach to make learning more efficient. The idea we propose is to integrate refinement operators in GP. This aims to resolve the two problems we have outlined above: Well-designed refinement operators are usually less destructive, because applying such an operator to a concept means that only a small change to the concept is performed – syntactically and semantically. Moreover, refinement operators can make use of the subsumption order and, thus, use more available knowledge than the standard GP algorithm. We show how refinement operators and GP can be combined in general and then present a concrete operator.

Some steps need to be done in order to be able to use refinement operators as genetic operators. The first problem is that a refinement operator is a mapping from one concept to an arbitrary number of concepts. Naturally, the idea is to select one of the possible refinements. In order to be able to do this efficiently, we assume that the refinement operators we are looking at are finite.

The second problem when applying refinement operators to GP is that a concrete refinement operator only performs either specialisation or generalisation, but not both. However, in GP we are likely to find too strong as well as too weak concepts, so there is a need for upward and downward refinement. A simple approach is to use two genetic operators: an adapted upward and an adapted downward refinement operator.

Another way to solve the problem is to use one genetic operator, which stochastically chooses whether downward or upward refinement is used. This allows to adjust the probabilities of upward or downward refinement being selected to the classification of the concept we are looking at. For instance consider an overly general concept, i.e. it covers all positive examples, but does also cover some negative examples. In this case we always want to specialize, so the probability for using downward refinement should be 1. In the opposite case for an overly specific concept, i.e. none of the negatives is covered, but some positives, the probability of downward refinement should be 0. How do we assign probabilities to concepts, which are neither overly specific nor overly general? Our approach is as follows:

1. The probability of downward refinement, denoted by $p_\downarrow$, should depend on the percentage of covered negative examples. Using α as variable factor we get:
$$p_\downarrow(\mathcal{K}, C) = \alpha \cdot \frac{|neg_\mathcal{K}(C)|}{|E^-|}$$

   In particular for $|neg_\mathcal{K}(C)| = 0$ (consistent concept) we get $p_\downarrow(\mathcal{K}, C) = 0$.

2. The probability of upward refinement, denoted by $p_\uparrow$, should depend on the percentage of covered positive examples. We use the same factor as in the first case:
$$p_\uparrow(\mathcal{K}, C) = \alpha \cdot \left(1 - \frac{|pos_\mathcal{K}(C)|}{|E^+|}\right)$$

   In particular for $|pos_\mathcal{K}(C)| = |E^+|$ (complete concept) we get $p_\uparrow(\mathcal{K}, C) = 0$.

3. For any concept $p_\downarrow(\mathcal{K}, C) + p_\uparrow(\mathcal{K}, C) = 1$.

From this, we can derive the following formulae for the probabilities of upward and downward refinement:

$$p_\downarrow(\mathcal{K}, C) = \frac{\frac{|neg_\mathcal{K}(C)|}{|E^-|}}{1 + \frac{|neg_\mathcal{K}(C)|}{|E^-|} - \frac{|pos_\mathcal{K}(C)|}{|E^+|}} \qquad p_\uparrow(\mathcal{K}, C) = \frac{1 - \frac{|pos_\mathcal{K}(C)|}{|E^+|}}{1 + \frac{|neg_\mathcal{K}(C)|}{|E^-|} - \frac{|pos_\mathcal{K}(C)|}{|E^+|}}$$

Note that the return value of the formula is undefined, due to division by zero, for complete and consistent concepts. However, in this case C is a learning problem solution and we can stop the algorithm – or continue it to find smaller solutions by just randomly selecting whether upward or downward refinement is used.

This way we have given a possible solution to both problems: transforming the refinement operator to a mapping from a concept to exactly one concept and managing specialisation and generalisation. Overall, for a given finite upward refinement operator $\phi_\uparrow$ and a finite downward refinement operator $\phi_\downarrow$ we can construct a genetic operator ϕ, which is defined as follows (rand selects an element of a given set uniformly at random):

$$\phi_K(C) = \begin{cases} \mathrm{rand}(\phi_\downarrow(C)) & \text{with probability } \dfrac{\frac{|neg_K(C)|}{|E^-|}}{1+\frac{|neg_K(C)|}{|E^-|}-\frac{|pos_K(C)|}{|E^+|}} \\ \mathrm{rand}(\phi_\uparrow(C)) & \text{with probability } \dfrac{1-\frac{|pos_K(C)|}{|E^+|}}{1+\frac{|neg_K(C)|}{|E^-|}-\frac{|pos_K(C)|}{|E^+|}} \end{cases} \quad (1)$$

In the sequel, we will call genetic operators, which are created from upward and downward refinement operators this way, *genetic refinement operators*.

4.2   A Genetic Refinement Operator

To design a suitable refinement operator for learning $\mathcal{ALC}$ concepts, we first look at theoretical limitations. The following theorem [14] is a full analysis of the properties of $\mathcal{ALC}$ refinement operators:

Theorem 1. *Considering the properties completeness, weak completeness, properness, finiteness, and non-redundancy the following are maximal sets of properties (in the sense that no other of the mentioned properties can be added) of $\mathcal{ALC}$ refinement operators:*

1. *{weakly complete, complete, finite}*
2. *{weakly complete, complete, proper}*
3. *{weakly complete, non-redundant, finite}*
4. *{weakly complete, non-redundant, proper}*
5. *{non-redundant, finite, proper}*

We prefer complete operators, because this guarantees that, by applying the operator, we always have the possibility to find a solution of the learning problem. As argued before, we also need a finite operator. This means that a complete and finite operator is the best we can hope for. We will define such an operator in the sequel.

For $A \in N_C$ and background knowledge $K = (\mathcal{T}, \mathcal{A})$, we define $\mathrm{nb}_\downarrow(A) = \{A' \mid A' \in N_C$, there is no $A'' \in N_C$ with $A' \sqsubset_\mathcal{T} A'' \sqsubset_\mathcal{T} A\}$. $\mathrm{nb}_\uparrow(A)$ is defined analogously. Furthermore, we define the operator $\phi_\downarrow$ as shown in Figure 1. It works on concepts in negation normal form, so an input concept has to be converted if necessary.

Proposition 1. $\phi_\downarrow$ *is an $\mathcal{ALC}$ downward refinement operator.*

Proof. We show that $D \in \phi_\downarrow(C)$ implies $D \sqsubseteq_\mathcal{T} C$ by structural induction over $\mathcal{ALC}$ concepts in negation normal form. We can ignore refinements of the form

$$\phi_\downarrow(C) = \begin{cases} \emptyset & \text{if } C = \bot \\ \{\forall r.\top \mid r \in N_R\} \cup \{\exists r.\top \mid r \in N_R\} \cup \{\top \sqcup \top\} & \text{if } C = \top \\ \quad \cup \{A \mid \mathrm{nb}_\uparrow(A) = \emptyset\} \cup \{\neg A \mid \mathrm{nb}_\downarrow(A) = \emptyset\} \\ \{A' \mid A' \in \mathrm{nb}_\downarrow(A)\} \cup \{\bot \mid \mathrm{nb}_\downarrow(A) = \emptyset\} \cup \{A \sqcap \top\} & \text{if } C = A \ (A \in N_C) \\ \{\neg A' \mid A' \in \mathrm{nb}_\uparrow(A)\} \cup \{\bot \mid \mathrm{nb}_\uparrow(A) = \emptyset\} \cup \{\neg A \sqcap \top\} & \text{if } C = \neg A \ (A \in N_C) \\ \{\exists r.E \mid E \in \phi_\downarrow(D)\} \cup \{\exists r.D \sqcap \top\} \cup \{\bot \mid D = \bot\} & \text{if } C = \exists r.D \\ \{\forall r.E \mid E \in \phi_\downarrow(D)\} \cup \{\forall r.D \sqcap \top\} \cup \{\bot \mid D = \bot\} & \text{if } C = \forall r.D \\ \{C_1 \sqcap \cdots \sqcap C_{i-1} \sqcap D \sqcap C_{i+1} \sqcap \cdots \sqcap C_n & \text{if } C = C_1 \sqcap \cdots \sqcap C_n \\ \quad \mid D \in \phi_\downarrow(C_i), 1 \leq i \leq n\} & (n \geq 2) \\ \quad \cup \{C_1 \sqcap \cdots \sqcap C_n \sqcap \top\} \\ \{C_1 \sqcup \cdots \sqcup C_{i-1} \sqcup D \sqcup C_{i+1} \sqcup \cdots \sqcup C_n & \text{if } C = C_1 \sqcup \cdots \sqcup C_n \\ \quad \mid D \in \phi_\downarrow(C_i), 1 \leq i \leq n\} & (n \geq 2) \\ \quad \cup \{C_1 \sqcup \cdots \sqcup C_{i-1} \sqcup C_{i+1} \sqcup \cdots \sqcup C_n \mid 1 \leq i \leq n\} \\ \quad \cup \{(C_1 \sqcup \cdots \sqcup C_n) \sqcap \top\} \end{cases}$$

Fig. 1. Definition of $\phi_\downarrow$

$C \rightsquigarrow C \sqcap \top$, because obviously $C \sqsubseteq_\mathcal{T} C \sqcap \top$ ($C \equiv_\mathcal{T} C \sqcap \top$). We also ignore refinements of the form $C \rightsquigarrow \bot$, for which the claim is also true. All other cases are shown below.

- $C = \bot$: $D \in \phi_\downarrow(C)$ is impossible, because $\phi_\downarrow(\bot) = \emptyset$.
- $C = \top$: $D \sqsubseteq_\mathcal{T} C$ is trivially true for each concept D (and hence also for all refinements).
- $C = A \ (A \in N_C)$: $D \in \phi_\downarrow(C)$ implies that D is also an atomic concept or the bottom concept and $D \sqsubseteq C$.
- $C = \neg A$: $D \in \phi_\downarrow(C)$ implies that D is of the form $\neg A'$ with $A \sqsubseteq_\mathcal{T} A'$. $A \sqsubseteq_\mathcal{T} A'$ implies $\neg A' \sqsubseteq_\mathcal{T} \neg A$ by the semantics of negation.
- $C = \exists r.C'$: $D \in \phi_\downarrow(C)$ implies that D is of the form $\exists r.D'$. We have $D' \sqsubseteq_\mathcal{T} C'$ by induction. For existential restrictions $\exists r.E \sqsubseteq_\mathcal{T} \exists r.E'$ if $E \sqsubseteq_\mathcal{T} E'$ holds in general [4]. Thus we also have $\exists r.D' \sqsubseteq \exists r.C'$.
- $C = \forall r.C'$: This case is analogous to the previous one. For universal restrictions $\forall r.E \sqsubseteq_\mathcal{T} \forall r.E'$ if $E \sqsubseteq_\mathcal{T} E'$ holds in general [4].
- $C = C_1 \sqcap \cdots \sqcap C_n$: In this case one element of the conjunction is refined, so $D \sqsubseteq_\mathcal{T} C$ follows by induction.
- $C = C_1 \sqcup \cdots \sqcup C_n$: One possible refinement is to apply $\phi_\downarrow$ to one element of the disjunction, so $D \sqsubseteq_\mathcal{T} C$ follows by induction. Another possible refinement is to drop an element of the disjunction, when $D \sqsubseteq_\mathcal{T} C$ obviously also holds.

Proposition 2. $\phi_\downarrow$ *is complete.*

Proof. We will first show weak completeness of $\phi_\downarrow$. We do this by structural induction over $\mathcal{ALC}$ concepts in negation normal form, i.e. we show that every concept in negation normal form can be reached by $\phi_\downarrow$ from $\top$.

- Induction Base:
  - $\top$: $\top$ can trivially be reached from $\top$.
  - $\bot$: $\top \leadsto A_1 \leadsto \ldots \leadsto A_n \leadsto \bot$ (descending the subsumption hierarchy)
  - $A \in N_C$: $\top \leadsto A_1 \leadsto \ldots \leadsto A_n \leadsto A$ (descending the subsumption hierarchy until A is reached)
  - $\neg A(A \in N_C)$: $\top \leadsto \neg A_1 \leadsto \ldots \leadsto \neg A_n \leadsto \neg A$ (ascending the subsumption hierarchy of atomic concepts within the scope of a negation symbol)
- Induction Step:
  - $\exists r.C$: $\top \leadsto \exists r.\top \leadsto^* \exists r.C$ (last step by induction)
  - $\forall r.C$: $\top \leadsto \forall r.\top \leadsto^* \forall r.C$ (last step by induction)
  - $C_1 \sqcap \cdots \sqcap C_n$: $\top \leadsto^* C_1$ (by induction) $\leadsto C_1 \sqcap \top \leadsto^* C_1 \sqcap C_2 \leadsto^* C_1 \sqcap \cdots \sqcap C_n$
  - $C_1 \sqcup \cdots \sqcup C_n$: $\top \leadsto \top \sqcup \top \leadsto^* C_1 \sqcup \top$ (by induction) $\leadsto C_1 \sqcup \top \sqcup \top \leadsto^* C_1 \sqcup C_2 \sqcup \top \leadsto^* C_1 \sqcup \cdots \sqcup C_n$

We have shown that $\phi_\downarrow$ is weakly complete. If we have two $\mathcal{ALC}$ concepts C and D in negation normal form with $C \sqsubseteq_{\mathcal{T}} D$, then for a concept $E = D \sqcap C$ we have $E \equiv_{\mathcal{T}} C$. E can be reached by the following refinement chain from D:

$$D \leadsto D \sqcap \top \leadsto^* D \sqcap C \text{ (by weak completeness of } \phi_\downarrow)$$

Thus, we have shown that we can reach a concept equivalent to C, which proves the completeness of $\phi_\downarrow$.

Proposition 3. $\phi_\downarrow$ *is finite.*

Proof. Some rules in the definition of $\phi_\downarrow$ apply $\phi_\downarrow$ recursively, e.g. specialising an element of a conjunction. Since such applications are only performed on inner structures of an input concept, only finitely many recursions are necessary to compute all refinements. This means that it is sufficient to show that every single application of $\phi_\downarrow$ produces finitely many refinements under the assumption that each recursive application of $\phi_\downarrow$ on an inner structure represents a finite set. Since N_R and N_C are finite, this can be verified easily by analysing all cases in Figure 1.

We have shown that $\phi_\downarrow$ is complete and finite, which makes it suitable to be used in a genetic refinement operator. We defined a dual upward refinement operator $\phi_\uparrow$ and showed its completeness and finiteness. The definition of the operator and the proofs are omitted, because they are analogous to what we have shown for $\phi_\downarrow$. From $\phi_\downarrow$ and $\phi_\uparrow$ we can construct a genetic refinement operator as described in Equation 1. This new operator is ready to be used within the GP framework and combines classical induction with evolutionary approaches.

What are the differences between classical refinement operator based approaches and our evolutionary approach? Usually, in a classical algorithm a refinement operator spans a search tree and a search heuristic guides the direction of search. The heuristic corresponds to the fitness function in a GP and usually both bias the search towards small concepts with high classification accuracy.

The search space in a classical algorithm is traversed in a well-structured and often deterministic manner. However, such an algorithm has to maintain (parts of) a search tree, which is usually continuously growing. In a GP, the population has, in most cases, a constant size. This means that a GP can run for a long time without consuming more space (assuming that the individuals themselves are not constantly growing, which is not the case for our genetic refinement operator). In this sense, a GP can be seen as a less structured search with individuals moving stochastically in the search space. Another difference is that classical algorithms often traverse the search space only in one direction (bottom-up or top-down approach), whereas genetic refinement operators use both directions and can start from random points in the search space. In general, it is not clear whether a classical or hybrid approach is to be preferred and the choice – as usual in Machine Learning – depends on the specific problem at hand.

5   Preliminary Evaluation

To perform a preliminary evaluation, we have chosen the FORTE [20] family data set. We transformed it into an OWL ontology about family relationships and defined a new learning task. In our case, the ontology contains two disjoint concepts Male and Female, the roles parent, sibling (symmetric), and married (symmetric and functional). The family tree is described by 337 assertional axioms. As learning target, we have chosen the concept of an uncle. A possible definition of this concept is:

$$\texttt{Male} \sqcap (\exists\, \texttt{sibling}.\exists\, \texttt{parent}.\top \sqcup \exists\, \texttt{married}.\exists\, \texttt{sibling}.\exists\, \texttt{parent}.\top)$$

86 examples, 23 positive and 63 negative, are provided. This learning task can be considered challenging, since the smallest possible solution is long (length 13) and there are no restrictions on the search space. For our experiments, we have chosen to let the GP algorithm run a fixed number of 50 generations. We used a generational algorithm and initialised it using the ramped-half-and-half method with maximum depth 6 for initialisation. The fitness measure in Section 3 with a non-optimised value of $a = 0.0025$ was used, i.e. a length unit is worth a accuracy decline of 0.25% . As selection method, we have chosen rank selection defined in such a way that the highest ranked individual has a ten times higher probability of being selected than the lowest ranked individual.

Since our main contribution is the provision of a new operator, we have tested three sets of operator probabilities. All sets have a 2% probability for mutation. The standard GP set has an 80% probability of crossover (the remaining 18% are used for reproduction). The mixed set uses 40% crossover and 40% genetic refinement operator and the refinement set uses only 5% crossover and 85% genetic refinement operator. We have varied the population size from 100 to 700 and averaged all results over 10 runs. Figure 2 depicts the results we obtained with respect to classification accuracy of the examples (defined as in the first part of Equation 3). Under the assumptions of a t test, the difference in accuracy for

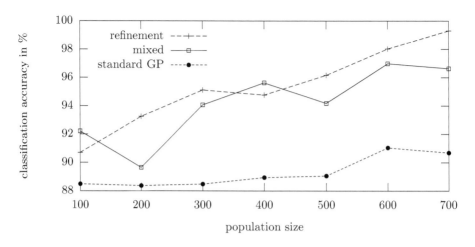

Fig. 2. classification accuracy on family data set

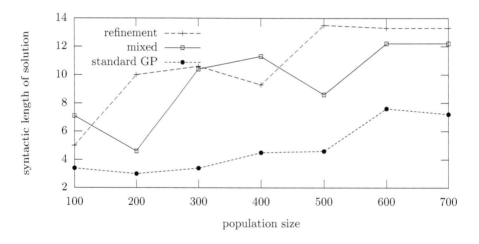

Fig. 3. length of learned concepts on family data set

standard GP compared to one of the two others is statistically significant with a confidence interval of 95% for population sizes higher than 200.

Apart from the classification accuracy, we also measured the length of the concepts, which were returned as solutions by the algorithm. The results are shown in Figure 3. All algorithms were always able to find at least a concept of length 3 with a classification accuracy of 88%. Since the number of $\mathcal{ALC}$ concepts grows exponentially with their length, it is much harder for a learning algorithm to find promising long concepts. In most cases, standard GP failed to do so, whereas the other algorithms had high success rates for high population sizes.

For the experiments we used the reasoner Pellet[2] (version 1.4RC1), which was connected to the learning program using the DIG 1.1 interface[3]. Most of the runtime of the algorithm (98%) was spent for reasoner requests. Since the genetic refinement operator is not proper and performs only small changes on concepts, we built up a caching mechanism for it. Before saving a concept in the cache, we normalized it. First, we defined a linear order over $\mathcal{ALC}$ concepts and ordered elements in conjunctions and disjunctions according to this order. Additionally, we converted the concept to negation normal form and applied equivalence preserving rewriting rules e.g. $C \sqcap \top \rightarrow C$. This techniques allowed us to use the cache for approximately 65% of the computed refinements of the genetic refinement operator. Due to more cache hits, the performance of the genetic refinement operator is even better than for crossover and mutation. The overall runtime varied from approximately 100 seconds on average for a population size of 100 to 950 seconds on average for a population size of 700 on a 2,2 GHz dual core processor[4] with 2 GB RAM.

The YinYang system [11] has a runtime of 200 seconds for this example, a classification accuracy of 73.5%, and a concept length of 12.2 averaged over 10 runs[5]. We could not use other systems for comparison. The system in [6] is not available anymore and the approach in [4] was not fully implemented.

6   Related Work

To the best of our knowledge, there has been no attempt to use evolutionary approaches for learning concepts in description logics. Hence, there is no closely related work we are aware of. Approaches for concept learning in description logics are described in [6,4,9,11]. Although evolutionary methods have not been considered for learning in description logics before, they have been used for inducing logic programs. A recent article [7] provides a good overview.

Evolutionary ILP systems usually use variants of Genetic Algorithms or Genetic Programming. The goal is to learn a set of clauses for a target predicate. EVIL_1 [19] is a system based on Progol [16], where an individual represents a set of clauses (called the *Pittsburgh approach*) and crossover operators are used. REGAL [17] is a system, which consists of a network of genetic nodes to achieve high parallelism. Each individual encodes a partial solution (called the *Michigan approach*). It uses classic mutation and several crossover operators. GNET is a descendant of REGAL. It also uses a network of genetic nodes, but takes a co-evolutionary approach [1], i.e. two algorithms are used to converge to a solution. DOGMA [10] is a system, which uses a combination of the Pittsburgh and Michigan approach on different levels of abstraction. All these systems use a simple bit string representation. This is possible by requiring a fixed template,

---

[2] http://pellet.owldl.com
[3] http://dl.kr.org/dig/
[4] Our current GP implementation does not efficiently use the second CPU.
[5] 3 out of 86 examples could not be used, because YinYang could not calculate most specific concepts for them, which are needed as input for their algorithm.

which the solution must fit in. We did not consider this approach when learning in description logics due to its restricted flexibility.

The following systems use a high level representation of individuals. SIA01 [2] is a bottom-up approach, which starts with a positive example as seed and grows a population until it reaches a bound (so the population size is not fixed as in the standard approach). ECL [8] is a system using only mutation style operators for finding a solution. In contrast GLPS [23] uses only crossover style operators and a tree (more exactly forest) representation of individuals. In [22] a binary representation of clauses is introduced, which is shown to be processable by genetic operators in a meaningful way. [21] extends this framework by a fast fitness evaluation algorithm.

The systems based on Genetic Programming, i.e. SIA01, ECL, and GLPS are closest to our approach. Similar to our research, they also concluded that standard GP is not sufficient to solve their learning problem. As a consequence, they invented new operators. As far as we know, they did not try to connect refinement operators and GP explicitly as we did. We cannot directly compare the operators, which are used in ILP systems, with the genetic operator we have developed, since the target language (logic programs) is different.

7   Conclusions, Further Work

In the article, we have presented a hybrid approach for learning concepts in DLs, which combines GP and refinement operators. We first presented how to solve the problem using standard GP, outlined difficulties and showed how they can be overcome using refinement operators. To the best of our knowledge, this is the first time a framework for transforming refinement operators to genetic operators has been proposed and the first time that evolutionary techniques have been applied to the concept learning problem in description logics. Based on a full property analysis [14], we developed a concrete genetic refinement operator and provided a preliminary evaluation.

In the future we plan to extend our evaluation, propose benchmark data sets for the learning problem, and analyse the interaction between genetic refinement operators and traditional operators.

References

1. Anglano, C., Giordana, A., Bello, G.L., Saitta, L.: An experimental evaluation of coevolutive concept learning. In: Proc. 15th International Conf. on Machine Learning, pp. 19–27. Morgan Kaufmann, San Francisco (1998)
2. Augier, S., Venturini, G., Kodratoff, Y.: Learning first order logic rules with a genetic algorithm. In: Fayyad, U.M., Uthurusamy, R. (eds.) The First International Conference on Knowledge Discovery and Data Mining, Montreal, Canada, August 20-21, 1995, pp. 20–21. AAAI Press, Stanford (1995)
3. Baader, F., Calvanese, D., McGuinness, D.L., Nardi, D., Patel-Schneider, P.F.: The Description Logic Handbook: Theory, Implementation, and Applications. Cambridge University Press, Cambridge (2003)

4. Badea, L., Nienhuys-Cheng, S.-H.: A refinement operator for description logics. In: Cussens, J., Frisch, A.M. (eds.) ILP 2000. LNCS (LNAI), vol. 1866, pp. 40–59. Springer, Heidelberg (2000)
5. Blumer, A., Ehrenfeucht, A., Haussler, D., Warmuth, M.K.: Occam's razor. In: Shavlik, J.W., Dietterich, T.G. (eds.) Readings in Machine Learning, pp. 201–204. Morgan Kaufmann, San Francisco (1990)
6. Cohen, W.W., Hirsh, H.: Learning the CLASSIC description logic: Theoretical and experimental results. In: Doyle, J., Sandewall, E., Torasso, P. (eds.) Proceedings of the 4th International Conference on Principles of Knowledge Representation and Reasoning (KR94), pp. 121–133. Morgan Kaufmann, San Francisco (1994)
7. Divina, F.: Evolutionary concept learning in first order logic: An overview. AI Commun. 19(1), 13–33 (2006)
8. Divina, F., Marchiori, E.: Evolutionary concept learning. In: Langdon, W.B., Cantú-Paz, E., Mathias, K., Roy, R., Davis, D., Poli, R., Balakrishnan, K., Honavar, V., Rudolph, G., Wegener, J., Bull, L., Potter, M.A., Schultz, A.C., Miller, J.F., Burke, E., Jonoska, N. (eds.) GECCO 2002. Proceedings of the Genetic and Evolutionary Computation Conference, New York, July 9-13, 2002, pp. 343–350. Morgan Kaufmann Publishers, San Francisco (2002)
9. Esposito, F., Fanizzi, N., Iannone, L., Palmisano, I., Semeraro, G.: Knowledge-intensive induction of terminologies from metadata. In: McIlraith, S.A., Plexousakis, D., van Harmelen, F. (eds.) ISWC 2004. LNCS, vol. 3298, pp. 441–455. Springer, Heidelberg (2004)
10. Hekanaho, J.: Background knowledge in GA-based concept learning. In: Proc. 13th International Conference on Machine Learning, pp. 234–242. Morgan Kaufmann, San Francisco (1996)
11. Iannone, L., Palmisano, I.: An algorithm based on counterfactuals for concept learning in the semantic web. In: Ali, M., Esposito, F. (eds.) Innovations in Applied Artificial Intelligence. Proceedings of the 18th International Conference on Industrial and Engineering Applications of Artificial Intelligence and Expert Systems, Bari, Italy, June 2005, pp. 370–379 (2005)
12. Koza, J.R., Keane, M.A., Streeter, M.J., Mydlowec, W., Yu, J., Lanza, G.: Genetic Programming IV: Routine Human-Competitive Machine Intelligence. Kluwer Academic Publishers, Dordrecht (2003)
13. Koza, J.R., Poli, R.: A genetic programming tutorial. In: Burke, E. (ed.) Introductory Tutorials in Optimization, Search and Decision Support (2003)
14. Lehmann, J., Hitzler, P.: Foundations of Refinement Operators for Description Logics. In: Proceedings of the 17th International Conference on Inductive Logic programming, ILP 2007 (2007)
15. Mitchell, T.: Machine Learning. McGraw Hill, New York (1997)
16. Muggleton, S.: Inverse entailment and progol. New Generation Computing 13(3&4), 245–286 (1995)
17. Neri, F., Saitta, L.: Analysis of genetic algorithms evolution under pure selection. In: Eshelman, L. (ed.) Proceedings of the Sixth International Conference on Genetic Algorithms, pp. 32–39. Morgan Kaufmann, San Francisco (1995)
18. Nienhuys-Cheng, S.-H., de Wolf, R. (eds.): Foundations of Inductive Logic Programming. LNCS, vol. 1228. Springer, Heidelberg (1997)
19. Reiser, P.G.K., Riddle, P.J.: Evolution of logic programs: Part-of-speech tagging. In: 1999 Congress on Evolutionary Computation, Piscataway, NJ, pp. 1338–1345. IEEE Service Center, Los Alamitos (1999)
20. Richards, B.L., Mooney, R.J.: Refinement of first-order Horn-clause domain theories. Machine Learning 19(2), 95–131 (1995)

21. Tamaddoni-Nezhad, A., Muggleton, S.: A genetic algorithms approach to ILP. In: Matwin, S., Sammut, C. (eds.) ILP 2002. LNCS (LNAI), vol. 2583, pp. 285–300. Springer, Heidelberg (2003)
22. Tamaddoni-Nezhad, A., Muggleton, S.: Searching the subsumption lattice by a genetic algorithm. In: Cussens, J., Frisch, A.M. (eds.) ILP 2000. LNCS (LNAI), vol. 1866, pp. 243–252. Springer, Heidelberg (2000)
23. Wong, M.L., Leung, K.S.: Inducing logic programs with genetic algorithms: The genetic logic programming system. IEEE Expert 10(5), 68–76 (1995)

Discovering Relations Among Entities from XML Documents

Yangyang Wu[1], Qing Lei[1], Wei Luo[1], and Harou Yokota[2]

[1] Department of Computer Science, Huaqiao University, Quanzhou Fujian, China
[2] Department of Computer Science, Tokyo Institute of Technology, Tokyo, Japan
wuyy@hqu.edu.cn

Abstract. This paper addresses relation information extraction problem and proposes a method of discovering relations among entities which is buried in different nest structures of XML documents. The method first identifies and collects XML fragments that contain all types of entities given by users, then computes similarity between fragments based on semantics of their tags and their structures, and clusters fragments by similarity so that the fragments containing the same relation are clustered together, finally extracts relation instances and patterns of their occurrences from each cluster. The results of experiments show that the method can identify and extract relation information among given types of entities correctly from all kinds of XML documents with meaningful tags.

Keywords: relation information extraction, XML document, cluster, occurrence pattern.

1 Introduction

As a new standard for Web information issue and data interchange, XML was designed to describe data and to focus on what data is. In XML documents, tags are "invented" by the authors of the XML documents to mark the data. In fact, the semantic information of XML data is implied in the tags and structure of the document. It can help to infer the content of XML documents. However, there are still greater differences in tags and structures in different XML documents. Therefore, extracting relation information from different XML documents is still a challenging work.

This paper focuses on discovering relations among given types of entities which are buried in different nest structures utilizing the topology of XML data and semantics of XML tags. Different from other Web IE (Information Extraction) tasks, our research aims to find all kinds of relations among given types of entities. Previous Web IE tasks [1] [2] [3] have kept the notion of specific pieces of information based on some pre-defined templates or given relations. However, we treat relations themselves as variables that can be mined. Our method can come up with new relations.

P. Perner (Ed.): MLDM 2007, LNAI 4571, pp. 899–910, 2007.

We observe that the XML document fragments that contain the same relation usually have similar elements and similar topology structures. Therefore, in the procedure for discovering relations, we collect related XML fragments first, and then cluster XML fragments according to semantics of tags and structures of XML fragments so that the fragments containing the same relation are clustered together, finally, extract relation instances and patterns of their occurrences from each cluster.

The rest of the paper is structured as follows. In section 2, we formalize our relation extraction problem. In section 3, we propose a new method and introduce our solution in detail. In section 4, we present some results of experiments. We introduce some related work in section 5. Finally, we draw a conclusion and put forwards further work in section 6.

2   Problem Definition

This paper focuses on the problem of discovering relation information among entities from XML documents. Here we formalize the relation information that will be found, data source (XML file), and the task of extraction as follows.

2.1   Relation Schema and Instance

Define relation information to be found first.

Definition 1. A relation scheme that describes the relationship among n entities is defined as $R(et_1, ..., et_n)$, where R is a relation name and $et_1, ..., et_n$ are named entity types. An instance of the relationship among n entities is an n-tuple $(I(et_1), ..., I(et_n))$, where $I(et_i)$ $(i = 1,...,n)$ is an instance of named entity type et_i.

2.2   XML Document and Fragment

The element is the basic unit of XML document. All XML documents must have a root element. All other elements must be within this root element. Any element can have subelements (child elements). Subelements must be correctly nested within their parent element. Therefore, an XML document can be represented as a labeled tree. Each node v of the tree corresponds to an XML element (or an attribute) and is labeled with the tag name (or the attribute name).

Definition 2. A labeled tree of an XML document d is a pair (N, E), where (1) N is a node set which is the union of element set and attribute set of the document; (2) E is an arc set, $(u,v) \in E$ iff $u,v \in N$ and v is a subelement or an attribute of u. We say that a part of an XML document p is an XML document fragment of d if the labeled tree of p is a subtree of labeled tree of d with root $x \in N$ (namely, the subtree consists of x and all of descendants of x).

Here we model XML documents as a labeled tree with nodes that are related to our mining task.

Definition 3. Let et be an entity type name, $Ex(et)=(et,et^1,et^2, ..., et^m)$ is an extended-name vector of et, where et^i (i=1,2, ..., m) is a synonym, compound word or abbreviation of et.

Namely, an extended-name vector of an entity type name consists of the type name, synonyms, compound words or abbreviations of the type name.

Definition 4. Let p be an XML document fragment, $T= \{t_1, ..., t_v\}$ be the set of tags and attributes of p, $Ex(et)$ be the extended-name vector of entity type name et, we say that fragment p contains entity type et if there exists a $t_i \in T$ such that t_i matches one of element of $Ex(et)$.

Definition 5. Let P be a set of XML document fragments containing entity type $et_1,..., et_n$, $p \in P$, if $\forall p' \in P$, the labeled tree of p' is not a subtree of the labeled tree of p, we call p the minimal fragment containing entity types $et_1,..., et_n$.

Example 2. Both fragment 1 and fragment 2 are minimal fragments containing entity-type "title" and "author".

```
Book-order document:

<?xml version="1.0" encoding="UTF-8"?>
<!-- edited with XML Explorer v2.0 by Mergesoft -->
<Catalog>
  <book>
    <title>Expert One-on-One Oracle</title>
    <bookinfo>
      <author>Thomas Kyte</author>
      <publisher>Apress</publisher>
      <price>$59.99</price>
    </bookinfo>
  </book>
  <book>
    <title>Professional C++ Programming</title>
    <bookinfo>
      <author>Nicholas A. Solter, etc</author>
      <publisher>Wiley</publisher>
      <price>$26.39</price>
    </bookinfo>
  </book>
</Catalog>

Fragment 1:

<book>
  <title>Expert One-on-One Oracle</title>
  <bookinfo>
    <author>Thomas Kyte</author>
    <publisher>Apress</publisher>
    <price>$59.99</price>
  </bookinfo>
</book>
```

```
Fragment 2:

<book>
  <title>Professional C++ Programming </title>
  <bookinfo>
    <author>Nicholas A. Solter, etc </author>
    <publisher>Wiley </publisher>
    <price>$39.99</price>
  </bookinfo>
</book>
```

2.3 Task Description

Let $et_1,\ldots, et_n$ be n entity types related to relations in which we are interested, let $R=\{R_1,\ldots, R_h\}$ denote all of relations among $et_1,\ldots, et_n$, where $R_i=\{r_{ij} \mid r_{ij}=(I(et_1),\ldots, I(et_n))\}$ (i=1,..., h), (j=1,...,|R_i|), let D be a set of XML documents.

The problem of discovering relations among $et_1,\ldots, et_n$ and their occurrence patterns is:

Find all minimal fragments containing entity-type $et_1,\ldots, et_n$, P= $\{p_1,\ldots, p_m\}$, then divide P into $P_1,\ldots, P_h$ so that each P_i corresponds to a relation R_i. For each R_i, set up a relation schema $R_i(et_1,\ldots, et_n)$ and extract relation instances of R_i and their occurrence patterns from P_i.

3 Method Outline

Our goal is to discover and extract relation information among given entity types. An XML document may consist of some fragments with different contents. Therefore, we should extract fragments that contain given entity types from documents first, discarding unrelated parts of document. In General, fragments containing the same relation will have similar elements and structures. There is high similarity between them. Fragments can be clustered by similarity so that a cluster is corresponding to a relation. Thereby, all kinds of relations hidden in documents are discovered. Our solution includes following steps:

(1) Prompt users input entity type names related to desired relations $et_1, \ldots, et_n$.
(2) Generate extended-name vectors of entity types $Ex(et_1), \ldots, Ex(et_n)$ by using WordNet [4] and a user-defined word library.
(3) Extract minimal fragments containing entity types $et_1, \ldots, et_n$ from given XML documents and set up a fragment set P.
(4) Cluster P by similarity so that a cluster is corresponding to a relation.
(5) Select a cluster of P, display the roots of cluster and prompt users to choose a relation name.
(6) Generate a pattern tree for above relation and extract relation instances that match the pattern tree and their occurrence pattern.

Repeat steps 5-6 to extract all kinds of relation information of entity types $et_1, \ldots, et_n$ from P.

Using above method, we call it FCRD (Fragment Clustering based Relation Discovery), all kinds of relations buried in XML documents can be discovered.

3.1 Computing Similarity Between XML Fragments

Fragment clustering is a key step of our method. Fragments about the same kind of relation should be similar to each other. According to Information-Theoretic Definition of Similarity [5], the similarity between A and B can be measured by the ratio between the amounts of information required to state the commonality of A and B and the information required to fully describe what A and B are. If two objects A and B can be viewed from several independent perspectives, their similarity can be computed separately from each perspective. We believe that XML fragments with the same kind of relation information will have the similar set of tags and similar topology structures. Similarity between XML fragments can be measured from the meaning of tags and the topology structure of the fragments. Based on the definition of element similarity and the method for computing structure similarity proposed by [6], we present the methods to calculate semantic similarity and structure similarity between XML fragments respectively first, and then compute overall similarity between XML fragments by weighted average of semantic similarity and structure similarity.

Semantic similarity measure: XML tags are used to mark data. They carry useful semantic information. Therefore, we set up a semantic feature vector model according to tags for each fragment, and then calculate semantic similarity by the semantic feature vectors [7].

Structure similarity measure: we use an adapted sequential pattern mining algorithm for finding maximal similar paths between XML fragments [6]. Path $a_1a_2...a_k$ in fragment p and $b_1b_2...b_k$ in fragment p' are similar iff an item in $Ex(a_i)$ match or partially match an item in $Ex(b_i)$ (i=1,…, k). We compute structure similarity between 2 fragments by the ratio of similar paths to the nested structure of bigger fragment.

Definition 6. Let p_1 and p_2 be XML fragments, the similarity between p_1 and p_2 is defined as follows:

$$Sim(p_1,p_2)=\lambda_1 SemSim(p_1,p_2)+\lambda_2 StrSim(p_1,p_2) . \qquad (1)$$

λ_1, λ_2: the weights of semantic similarity and structure similarity respectively;
SemSim(p_1,p_2): Semantic similarity, it is computed by following steps:
  (1) Each XML fragment is represented as a semantic feature vector.

$$p_1 = (< Ex(t_1^1), score_1^1 >,< Ex(t_2^1), score_2^1 >,..., < Ex(t_m^1), score_m^1 >),$$

$$p_2 = (< Ex(t_1^2), score_1^2 >,< Ex(t_2^2), score_2^2 >,..., < Ex(t_n^2), score_n^2 >)$$

Where, $Ex(t_j^i)$ is the extended-name vector of the jth tag of p_i (i=1, 2). It consists of the tag name, synonyms, compound words or abbreviations of the tag name. $score_j^i$ is the score of similarity of tag t_j^i. The score criterion is as follows:

  6: When the name of tag t_j^i completely matches a tag name of another fragment.

  5: When the name of tag t_j^i completely matches a term in the extended-name vector of a tag in another fragment.

4: When a term in the extended-name vector of the tag t_j^i completely matches a term in the extended-name vector of a tag in another fragment.

3: When the name of tag t_j^i partially matches a tag name of another fragment.

2: When the name of tag t_j^i partially matches a term in the extended-name vector of a tag in another fragment.

1: When a term in the extended-name vector of tag t_j^i partially matches a term in the extended-name vector of a tag in another fragment.

0: When there is no match.

If more than one value is applicable, the maximum value is selected.

If the score>0, we say that this tag (or element) is a similar tag (or element) between p_1 and p_2.

(2) Compute SemSim(p_1, p_2).

$$SemSim(p_1, p_2) = \frac{\sum_{i=1}^{m} score_i^1 + \sum_{j=1}^{n} score_j^2}{6 \times (m+n)} \tag{2}$$

StrSim(p_1,p_2): Structure similarity, it is computed by the nested structure of the fragment with more elements.

$$StruSim(p_1, p_2) = \frac{1}{N+1}[(\sum_{t=1}^{N} \frac{1}{L(R_t)} \times V(R_t)) + MR] \tag{3}$$

N: the number of level-1 subtrees of the fragment with more elements in (p_1,p_2)

R_t: the root of the tth level-1 subtree of the fragment with more elements in $\{p_1,p_2\}$

L(x): a level function; $L(R_t)$ is the number of levels of the subtree with root R_t;

$$V(R_t) = \begin{cases} F(R_t), & If \ R_t \ is \ a \ leaf. \\ F(R_t) + \frac{1}{N(C_t)} \sum_{e \in C_t} \frac{1}{L(e)} V(e), & If \ R_t \ is \ not \ a \ leaf. \end{cases} \tag{4}$$

$$F(R_t) = \begin{cases} 1, & if \ R_t \ is \ in \ the \ similar \ paths \ between \ p_1 and \ p_2 \\ 0, & else \end{cases} \tag{5}$$

C_t: the set of child-notes of R_t

$N(C_t)$: the number of the nodes in C_t

$$MR = \begin{cases} 1 & If \ root \ of \ p_1 \ is \ similar \ to \ the \ root \ of \ p_2 \\ 0 & else \end{cases} \tag{6}$$

Example 3. In figure 1, the semantic similarity of (a) and (b) is 0.77, and the structure similarity is 0.61. When $\lambda_1=\lambda_2=0.5$, the similarity of (a) and (b) is 0.69.

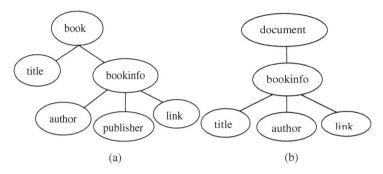

Fig. 1. Two XML Documents

From definition 6, we can prove that Sim(a, b) has following properties.

Proposition 1. The similarity measure Sim(a, b) satisfies:

(1) $\forall$a, b, Sim(a, a) $=1 \geq$ Sim(a, b)
(2) $\forall$a, b, Sim(a, b) = Sim(b, a)
(3) The more similar a and b are, the higher Sim(a, b) is.

3.2 XML Fragment Clustering

We propose a multi-threshold-clustering algorithm to cluster XML fragments so that a cluster is corresponding to a relation. In the procedure of clustering, the algorithm will dynamically choose threshold according to the distribution of similarities. The main steps include: (1) pick up a desired fragment p and compute the similarity between p and all of fragments (include p). In general, there is an obvious difference between the similarities of the fragments that are similar to p and the similarities of the fragments that are dissimilar to p. Therefore, (2) rank similarities in descending order to form a descent curve. (3) Apply the second derivative passing through zero to get inflexions, and then choose the similarity corresponding to the inflexion whose first derivative is the minimum as similarity threshold ε. (4) Select fragments whose similarities are greater than ε to form a candidate cluster. The candidate cluster will become a cluster, if it is large enough, that is greater than the given density in algorithm DBSCAN [8]. The remaining fragments, whose similarities to p are less than ε may contain other relations. Repeating above steps on the rest fragments, more clusters will be obtained.

Algorithm 1. Multi-threshold-clustering algorithm for XML fragments
Input: a set of XML fragments P, density d
Output: a set of XML fragment clusters C
Method:
    (1) Select a desired fragment p_0 from set P.
    (2) For each $p_i \in$ P, compute Sim(p_i, p_0).

(3) Rank all Sim(p_i, p_0) in descending order.

(4) Choose the similarity corresponding to the inflexion whose first derivative is the minimum as threshold ε.

(5) Let S= { p'| Sim(p', p_0)>ε}.

(6) If |S|≥d, then put S into C.

(7) P:=P-S,

Repeat above steps until P=∅, namely, all of the fragments in P have been processed.

3.3 Extracting Relation Instances and Their Occurrence Patterns

Each cluster of XML fragments contains a relation of given types of entities. We first create a pattern tree to represent the relation mining requirement of users, and then apply approximate match to find desired relation instances. Approximate match require only the nodes similar to pattern tree and keep ancestor-descendant relationship. Because the context of a word often determines the meaning of a word in a document, we check the parent node of the root of fragment in original document if necessary while matching. Our tree approximate match is defined as follows.

Definition 7. Let T_p=(V, E_p) be a pattern tree that represents the relation mining requirement, T_f=(W, E_f) be the tree of an XML fragment, we say that there exists an approximate match of T_p in T_f, if there is a function f: V→W, which satisfies:

(1) u=v⇔f(u) =f(v), u, v∈ Domain(f)

(2) name(f(v))∈ Ex(name(v))

(3) u=parent(v)⇔f(u) =ancestor(f(v))

where, condition (1) means that f is one to one; condition (2) means that the node of T_p is similar to corresponding node of T_f , and condition (3) requires f to keep ancestor-descendant relationship.

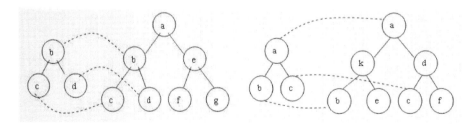

Fig. 2. Two Examples of Approximate Match

Figure 2 shows two examples of approximate match. The first one is a special example for approximate match, in which f keeps parent-child relationship of nodes. It is an example of exactly match.

After finding matches, the next is to extract corresponding elements and attributes from matched subtree, then assemble extracted data in the relation tables and extract occurrence patterns from fragments.

4 Experiment

In order to test our method, we implemented above method on the windows 2000 using Java [10] and chose discovering relations between "title" and "author" as the task of our experiment. Here we focus only on testing ability to discover different relations.

The data set for our experiment came from Wisconsin's XML data bank [11]. The test set D consists of all of XML files in directories bib and club, and some of XML files from directories lindoc and sigrecord. In the experiment, our system extracted 212 minimal fragments with entity types "title" and "author" from D, and set up a set of fragments P.

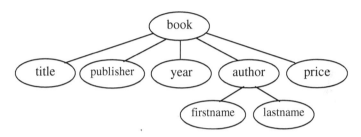

Fig. 3. The tree of bib_0_0.xml

While clustering, the system selected the first fragment bib_0_0.xml (its structure is shown in figure 3) and computed the similarities between bib_0_0.xml and all of fragments of P. Figure 4 (a) shows the similarities between bib_0_0.xml and the fragments extracted from XML files in directory lindoc and sigrecord. The similarities are lower, because bib_0_0.xml and the fragments from directory lindoc and sigrecord contain different relations. Figure 4 (b) lists the similarities between bib_0_0.xml and all of the fragments in P in descending order. It shows that all of similarities between bib_0_0.xml and fragments from directory bib are 1 though they are different not only in their content but also in their structure. For example, bib_0_0.xml contains an author name only, while some fragments contain two or more author names. It is due to their same tag-names and similar paths.

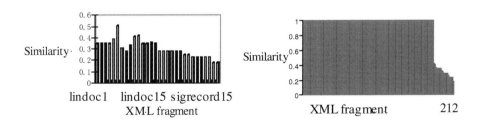

Fig. 4. Similarities between Bib_0_0 and paragraphs of P

Table 1. Some results of experiment 1

Title	Author
Unix Network Programming	Richard Stevens
Crafting a Compiler with C	Charles Fischer
Crafting a Compiler with C	Richard LeBlanc
Problem Solving with C++	Walter Savitch
lex and yacc	John Levine
lex and yacc	Tony Mason
lex and yacc	Doug Brown
Statistics, The Exploration and Analysis of Data	Jay Devore
Statistics, The Exploration and Analysis of Data	Roxy Peck
...	...

Applying the second derivative passing through zero to get inflexions, the result is that 0.6353 was chosen as the similarity threshold ε, which is corresponding to the inflexion whose first derivative is the minimum. With the similarity threshold $\varepsilon=0.6353$ and the density d=10, we got first fragment cluster that contains 182 fragments with book-author relation Book(title, author). All of the fragments in cluster are from directory bib. The system extracted 342 book-author relation instances. Table 1 lists some of them. The occurrence pattern of them is in figure 3. After repeating above process on the remaining fragments, we got other 2 clusters that contain LinuxDocument-author relations Linuxdoc(title, author) and article-author relation Article(title, author) respectively. The experiment extracted all of title-author relation instances from D and partitioned them to 3 relations exactly.

5 Related Work

Extracting entities and relations among them is one of the important tasks for information extraction [9]. There are many researches related to information extraction. As we know, the most similar work to ours is [2] [3]. Sergey Brin [2] and Neel Sundaresan [3] call identifying relation tuples and the patterns of their occurrences the duality problems of the Web, and use similar methods to extract relation tuples and patterns from HTML webpage iteratively. Sergey Brin [2] proposed a technique called DIPRE (Dual Iterative Pattern Relation Expansion) that exploited the duality between sets of patterns and relations to grow the target relation starting from a small sample set. The technique was applied to extract 15257 (author, title) pairs and 346 patterns from 156,000 HTML pages on the Web. Neel Sundaresan and Jeonghee Yi [3] defined and formalized the general duality problem of relations on the web. They solved the problem of identifying acronyms and their expansions through patterns of occurrences of (acronym, expansion) pairs as they occur in web pages. They started off with base sets of (acronym, expansion) pairs, patterns, and acronym formation rules, crawled the web to look for new (acronym, expansion) pairs that conform the patterns in the base set, and then from the set of (acronym, expansion) pairs, new formation rules are extracted. Moreover, new patterns that associate the acronyms were identified. With the extended sets of (acronym, expansion) pairs, patterns, and the rules, they continued crawling the web to discover more of them. Finally their crawler downloaded and analyzed 13,628 web pages, from which 2,694 unique AE-pairs and 948 unique patterns were identified.

Due to the difference of the goals, the method proposed by [2] and [3] cannot be used to our experiment in section 4 directly. In [2], a pattern is a 5-tuple (order, urlprefix, prefix, middle, suffix), where order is a boolean value and the other attributes are strings. In the fragments of our test set D, there are two elements "publisher" and "year" between "title" and "author". The middle strings are different between different title and author. Using the pattern defined in [2], we cannot find new relation instance from our test set. In [3], the authors formalize the duality problem of patterns and relations, but they have not defined general pattern yet. They present a special pattern for problem of finding acronyms and their expansions only. The pattern is a 3-tuple (a_pattern, e_pattern, formation_rule), where the formation_rule is a rule which specifies how an acronym is formed from its expansion. Therefore, their solution of finding acronyms and their expansions cannot be used to our experiment in section 4 directly.

[12] and [13] also consider fragment documents clustering. However, their goal is to integrate XML data sources so that users can conveniently access and acquire more complete information. Their similarity measure just compares the PCDATA values of leaf nodes without considering their semantic similarities. In this paper, semantic information of XML tags is considered when computing similarity between two fragments.

6   Conclusion

XML tags and topology structure of document carry rich semantic information that can be explored for inferring the content of the document. In this paper, we propose a method FCRD to discover and extract relation information among entities from XML documents. According to the types of entities given by users, first, we extract fragments that contain given entity-types from XML documents. Second, we propose a mult-threshold clustering algorithm to cluster extracted fragments by similarity so that the fragments containing different relations are clustered to different groups. Thereby, all kinds of relations are revealed. Finally, we extract all kinds of relations and their occurrence patterns from respective clusters.

Primary experiments show that if extended-entity vectors are generated properly, our method can extract all of relations among given types of entities correctly from XML documents with meaningful tags. We noticed that one of the problems is how to set up a proper dictionary so that system can generate extended-name vector properly for all kinds of application. Next, we will carefully research on this problem. Furthermore, we will improve our similarity measure method and XML fragment clustering algorithm.

Acknowledgments. This work is partially supported by the Natural Science Foundation of Fujian Province of China under Grant No. A0510020, the international Science and Technology Cooperation Project of Fujian Province of China under Grant No. 2004I014.

References

1. Chang, C.-H., et al.: A Survey of Web InformationExtraction Systems. IEEE Transactions on Knowledge And Data Engineering 18(10), 1411–1428 (2006)
2. Brin, S.: Extracting Patterns and Relations from the World Wide Web. In: Schek, H.-J., Saltor, F., Ramos, I., Alonso, G. (eds.) EDBT 1998. LNCS, vol. 1377, pp. 172–183. Springer, Heidelberg (1998)
3. Sundaresan, N., Yi, J.: Mining the Web for Relations. In: Proc. of the 9th international WWW conference on Computer networks, Amsterdam, the Netherlands, pp. 699–711 (2000)
4. WordNet2.0[P] http://www.cogsci.princton.edu./2.0
5. Lin, D.: An information-theoretic definition of similarity. In: Proc. of international conference on machine learning, Madiscon, Wisconsin (1998)
6. Lee, J.-W., Lee, K., Kim, W.: Preparations for Semantics-Based XML mining. In: Proc. of the 2001 IEEE international Conference on Data Mining (ICDM'01), IEEE Computer Society Press, Los Alamitos (2001)
7. Qing, L, Yang-Yang, W: A new Method for Semantics-Based Comparing XML Documents (In Chinese, abstract in English). In: Proc. of National Database Conference 2004. Xiamen China, Computer Science, vol. 31(10s), pp. 433–446 (2004)
8. Han, J., Kamber, M.: Data Mining Concepts and Techniques. Morgan Kaufmann Publishers, San Francisco (2000)
9. Marsh E.N.: MUC-7 Information Extraction Task Definition (version 5.1) In: Proc. of the Seventh Message Understanding Conference, Morgan Kaufman Publishers, Washington DC (In Appendix) (1998)
10. Qing, L., Yang-Yang, W.: Recognizing and Extracting Relations and Patterns from XML Documents. In: Proc. of the 3rd National Symposium of Search Engine and Web Mining, Beijing China, vol. 45(s1), pp. 1757–1761. Journal of Tsinghua University (2005)
11. N. query engine. http://www.cs.wisc.edu/niagara/data.html
12. Liang, W., Yokota, H.: LAX: An Efficient Approximate XML Join Based on Clustered Leaf Nodes for XML Data Integration. In: Jackson, M., Nelson, D., Stirk, S. (eds.) Database: Enterprise, Skills and Innovation. LNCS, vol. 3567, pp. 82–97. Springer, Heidelberg (2005)
13. Liang, W., Yokota, H.: SLAX: An Improved Leaf-Clustering Based Approximate XML Join Algorithm for Integrating XML Data at Subtree Classes. IPSJ Trans. on Databases 47(SIG 8 (TOD 30)), 47–57 (2006)

Author Index

Aburto, Luis   518
Akaho, Shotaro   419
Ao, Fujiang   479
Appice, Annalisa   324
Ariu, Davide   449
Armengol, Eva   756
Aykut, Murat   628, 727

Bacauskiene M.   263
Barile, Nicola   324
Bischl, Bernd   104
Boullé, Marc   548
Bunke, Horst   563

Campedel, Marine   203
Candillier, Laurent   548
Cao, Wenbo   160
Cárdenas, J. Festón   349
Carrasco-Ochoa, J. Ariel   694
Carroll, Douglas   499
Ceci, Michelangelo   324
Chaudhry, Qasim   32
Chen, Haifeng   144
Chen, Huowang   91
Cheng, Haibin   144
Christiansen, Henning   742
Coenen, Frans   339, 838
Corzo, B.   364
Crémilleux, Bruno   533
Cui, Li   616
Czarnowski, Ireneusz   117

da Silva, Erick Corrêa   784
de Oliveira Martins, Leonardo   784
de Paiva, Anselmo Cardoso   784
Denoyer, Ludovic   854
Díaz, F.   364

Ekdahl, Magnus   2
Ekinci, Murat   628, 727

Faez, Karim   868
Fdez-Riverola, F.   364
France, Stephen   499
Fred, Ana   770

Friedrich, Christoph M.   17
Fuentes, Olac   716
Fuller, Robert   464

Gallinari, Patrick   854
Gamboa, Hugo   770
Gattass, Marcelo   784
Géczy, Peter   419
Gedikli, Eyup   727
Geller, James   379
Gelzinis, A.   263
Ghorbani, Ali A.   404
Giacinto, Giorgio   449, 795
Glez-Peña, D.   364
Gomez, Juan Carlos   716
Grim, Jiří   233

Hama, Shiomu   218
Han, Tae Sik   585
Haralick, Robert   160
Hasida, Kôiti   419
Hébert, Céline   533
Holness, Gary F.   601
Hora, Jan   233
Hu, Xuegang   188
Huang, Jian   479
Huang, Kedi   479
Hülsmann, Marco   17

Imiya, Atsushi   218
Ip, Horace H.S.   296
Ivanović, Mirjana   681
Izumi, Noriaki   419

Jain, Anil K.   1
Jędrzejowicz, Piotr   117
Jiang, Guofei   144
Jiang, Jun   296

Kang, Jaewoo   585
Kantardzic, Mehmed   464
Kertész-Farkas, Attila   824
Kilchherr, Vivian   563
Ko, Seung-Kyu   585
Kocsor, András   824

Komazaki, Takuto   218
Koski, Timo   2
Koustoudis, Ioannis   703
Krishnan, Sriram   379
Kudo, Mineichi   490
Kyrgyzov, Ivan O.   203
Kyrgyzov, Olexiy O.   203

Latecki, Longin Jan   61
Lazarevic, Aleksandar   61
Lee, Moonhwi   131
Lehmann, Jens   883
Lei, Qing   899
Leng, Paul   838
León, R. Hernández   349
Li, Chunping   574
Li, Hua   616
Li, Zhoujun   91
Liu, Hongbo   392
Luo, Wei   899

Mackeprang Dahmcke, Christina   742
Maes, Francis   854
Maître, Henri   203
Malerba, Donato   324
Märgner, Volker   868
Marrocco, Claudio   47
Martínez-Trinidad, J. Francisco   694
Méndez, J.R.   364
Meyer, Frank   548
Micarelli, Alessandro   434
Mitasiunas, Antanas   810
Molinara, Mario   47
Morzy, Mikołaj   667
Mozaffari, Saeed   868

Nakamura, Atsuyoshi   490
Neagu, Daniel C.   32
Ng, See-Kiong   76
Ngai, Daniel C.K.   653

Olvera-López, J. Arturo   694

Pagola, José E. Medina   248
Palancar, J. Hernández   349
Papamarkos, Nikos   703
Parikh, Sapankumar   379
Park, Cheong Hee   131
Perdisci, Roberto   449
Pokrajac, Dragoljub   61

Pongor, Sándor   824
Pracner, Doni   681

Radovanović, Miloš   681
Raphael, Benny   174
Raudys, Sarunas   810
Riesen, Kaspar   563
Roli, Fabio   795

Sadoddin, Reza   404
Saitta, Sandro   174
Sakai, Tomoya   218
Sanderson, Robert   838
Sansonetti, Giuseppe   434
Shidara, Yohji   490
Silva, Aristófanes Corrêa   784
Silva, Hugo   770
Smith, Ian F.C.   174
Sprinkhuizen-Kuyper, Ida   310
Suárez, Airel Pérez   248
Sun, Chaofan   286
Szepannek, Gero   104

Tan, Vincent Yan Fu   76
Tomašev, Nenad   681
Tormo, O. Fraxedas   349
Tortorella, Francesco   47
Tronci, Roberto   795
Trundle, Paul R.   32

Valincius, D.   263
van der Maaten, Laurens   310
Vanderlooy, Stijn   310
Verikas, A.   263
Vilalta, Ricardo   286

Wang, Dongbo   188
Wang, Jiaxin   392
Wang, Sheng-De   276
Wang, Tao   91
Wang, Yanbo J.   339, 838
Weber, Richard   518
Weihs, Claus   104
Wisniewski, Guillaume   854
Wong, Hau-San   643
Wu, Kuo-Ping   276
Wu, Xindong   188
Wu, Yangyang   899

Xin, Qin   339

Yan, Yuejin   91, 479
Yang, Jiahai   392
Yokota, Harou   899
Yoshihira, Kenji   144
Yu, Shuang   574

Yu, Zhiwen   643
Yung, Nelson H.C.   653

Zagoris, Konstantinos   703
Zhang, Jiqi   643
Zhang, Shaohong   643
Zhang, Yu   392

Lecture Notes in Artificial Intelligence (LNAI)

Vol. 4597: P. Perner (Ed.), Advances in Data Mining. XI, 353 pages. 2007.

Vol. 4594: R. Bellazzi, A. Abu-Hanna, J. Hunter (Eds.), Artificial Intelligence in Medicine. XVI, 509 pages. 2007.

Vol. 4585: M. Kryszkiewicz, J.F. Peters, H. Rybinski, A. Skowron (Eds.), Rough Sets and Intelligent Systems Paradigms. XIX, 836 pages. 2007.

Vol. 4578: F. Masulli, S. Mitra, G. Pasi (Eds.), Applications of Fuzzy Sets Theory. XVIII, 693 pages. 2007.

Vol. 4573: M. Kauers, M. Kerber, R. Miner, W. Windsteiger (Eds.), Towards Mechanized Mathematical Assistants. XIII, 407 pages. 2007.

Vol. 4571: P. Perner (Ed.), Machine Learning and Data Mining in Pattern Recognition. XIV, 913 pages. 2007.

Vol. 4570: H.G. Okuno, M. Ali (Eds.), New Trends in Applied Artificial Intelligence. XXI, 1194 pages. 2007.

Vol. 4565: D.D. Schmorrow, L.M. Reeves (Eds.), Foundations of Augmented Cognition. XIX, 450 pages. 2007.

Vol. 4562: D. Harris (Ed.), Engineering Psychology and Cognitive Ergonomics. XXIII, 879 pages. 2007.

Vol. 4548: N. Olivetti (Ed.), Automated Reasoning with Analytic Tableaux and Related Methods. X, 245 pages. 2007.

Vol. 4539: N.H. Bshouty, C. Gentile (Eds.), Learning Theory. XII, 634 pages. 2007.

Vol. 4529: P. Melin, O. Castillo, L.T. Aguilar, J. Kacprzyk, W. Pedrycz (Eds.), Foundations of Fuzzy Logic and Soft Computing. XIX, 830 pages. 2007.

Vol. 4511: C. Conati, K. McCoy, G. Paliouras (Eds.), User Modeling 2007. XVI, 487 pages. 2007.

Vol. 4509: Z. Kobti, D. Wu (Eds.), Advances in Artificial Intelligence. XII, 552 pages. 2007.

Vol. 4496: N.T. Nguyen, A. Grzech, R.J. Howlett, L.C. Jain (Eds.), Agent and Multi-Agent Systems: Technologies and Applications. XXI, 1046 pages. 2007.

Vol. 4483: C. Baral, G. Brewka, J. Schlipf (Eds.), Logic Programming and Nonmonotonic Reasoning. IX, 327 pages. 2007.

Vol. 4482: A. An, J. Stefanowski, S. Ramanna, C.J. Butz, W. Pedrycz, G. Wang (Eds.), Rough Sets, Fuzzy Sets, Data Mining and Granular Computing. XIV, 585 pages. 2007.

Vol. 4481: J. Yao, P. Lingras, W.-Z. Wu, M. Szczuka, N.J. Cercone, D. Ślęzak (Eds.), Rough Sets and Knowledge Technology. XIV, 576 pages. 2007.

Vol. 4476: V. Gorodetsky, C. Zhang, V.A. Skormin, L. Cao (Eds.), Autonomous Intelligent Systems: Multi-Agents and Data Mining. XIII, 323 pages. 2007.

Vol. 4452: M. Fasli, O. Shehory (Eds.), Agent-Mediated Electronic Commerce. VIII, 249 pages. 2007.

Vol. 4451: T.S. Huang, A. Nijholt, M. Pantic, A. Pentland (Eds.), Artifical Intelligence for Human Computing. XVI, 359 pages. 2007.

Vol. 4438: L. Maicher, A. Sigel, L.M. Garshol (Eds.), Leveraging the Semantics of Topic Maps. X, 257 pages. 2007.

Vol. 4429: R. Lu, J.H. Siekmann, C. Ullrich (Eds.), Cognitive Systems. X, 161 pages. 2007.

Vol. 4426: Z.-H. Zhou, H. Li, Q. Yang (Eds.), Advances in Knowledge Discovery and Data Mining. XXV, 1161 pages. 2007.

Vol. 4411: R.H. Bordini, M. Dastani, J. Dix, A.E.F. Seghrouchni (Eds.), Programming Multi-Agent Systems. XIV, 249 pages. 2007.

Vol. 4410: A. Branco (Ed.), Anaphora: Analysis, Algorithms and Applications. X, 191 pages. 2007.

Vol. 4399: T. Kovacs, X. Llorà, K. Takadama, P.L. Lanzi, W. Stolzmann, S.W. Wilson (Eds.), Learning Classifier Systems. XII, 345 pages. 2007.

Vol. 4390: S.O. Kuznetsov, S. Schmidt (Eds.), Formal Concept Analysis. X, 329 pages. 2007.

Vol. 4389: D. Weyns, H.V.D. Parunak, F. Michel (Eds.), Environments for Multi-Agent Systems III. X, 273 pages. 2007.

Vol. 4384: T. Washio, K. Satoh, H. Takeda, A. Inokuchi (Eds.), New Frontiers in Artificial Intelligence. IX, 401 pages. 2007.

Vol. 4371: K. Inoue, K. Satoh, F. Toni (Eds.), Computational Logic in Multi-Agent Systems. X, 315 pages. 2007.

Vol. 4369: M. Umeda, A. Wolf, O. Bartenstein, U. Geske, D. Seipel, O. Takata (Eds.), Declarative Programming for Knowledge Management. X, 229 pages. 2006.

Vol. 4342: H. de Swart, E. Orłowska, G. Schmidt, M. Roubens (Eds.), Theory and Applications of Relational Structures as Knowledge Instruments II. X, 373 pages. 2006.

Vol. 4335: S.A. Brueckner, S. Hassas, M. Jelasity, D. Yamins (Eds.), Engineering Self-Organising Systems. XII, 212 pages. 2007.

Vol. 4334: B. Beckert, R. Hähnle, P.H. Schmitt (Eds.), Verification of Object-Oriented Software. XXIX, 658 pages. 2007.

Vol. 4333: U. Reimer, D. Karagiannis (Eds.), Practical Aspects of Knowledge Management. XII, 338 pages. 2006.

Vol. 4327: M. Baldoni, U. Endriss (Eds.), Declarative Agent Languages and Technologies IV. VIII, 257 pages. 2006.

Vol. 4314: C. Freksa, M. Kohlhase, K. Schill (Eds.), KI 2006: Advances in Artificial Intelligence. XII, 458 pages. 2007.

Vol. 4304: A. Sattar, B.-H. Kang (Eds.), AI 2006: Advances in Artificial Intelligence. XXVII, 1303 pages. 2006.

Vol. 4303: A. Hoffmann, B.-H. Kang, D. Richards, S. Tsumoto (Eds.), Advances in Knowledge Acquisition and Management. XI, 259 pages. 2006.

Vol. 4293: A. Gelbukh, C.A. Reyes-Garcia (Eds.), MICAI 2006: Advances in Artificial Intelligence. XXVIII, 1232 pages. 2006.

Vol. 4289: M. Ackermann, B. Berendt, M. Grobelnik, A. Hotho, D. Mladenič, G. Semeraro, M. Spiliopoulou, G. Stumme, V. Svátek, M. van Someren (Eds.), Semantics, Web and Mining. X, 197 pages. 2006.

Vol. 4285: Y. Matsumoto, R.W. Sproat, K.-F. Wong, M. Zhang (Eds.), Computer Processing of Oriental Languages. XVII, 544 pages. 2006.

Vol. 4274: Q. Huo, B. Ma, E.-S. Chng, H. Li (Eds.), Chinese Spoken Language Processing. XXIV, 805 pages. 2006.

Vol. 4265: L. Todorovski, N. Lavrač, K.P. Jantke (Eds.), Discovery Science. XIV, 384 pages. 2006.

Vol. 4264: J.L. Balcázar, P.M. Long, F. Stephan (Eds.), Algorithmic Learning Theory. XIII, 393 pages. 2006.

Vol. 4259: S. Greco, Y. Hata, S. Hirano, M. Inuiguchi, S. Miyamoto, H.S. Nguyen, R. Słowiński (Eds.), Rough Sets and Current Trends in Computing. XXII, 951 pages. 2006.

Vol. 4253: B. Gabrys, R.J. Howlett, L.C. Jain (Eds.), Knowledge-Based Intelligent Information and Engineering Systems, Part III. XXXII, 1301 pages. 2006.

Vol. 4252: B. Gabrys, R.J. Howlett, L.C. Jain (Eds.), Knowledge-Based Intelligent Information and Engineering Systems, Part II. XXXIII, 1335 pages. 2006.

Vol. 4251: B. Gabrys, R.J. Howlett, L.C. Jain (Eds.), Knowledge-Based Intelligent Information and Engineering Systems, Part I. LXVI, 1297 pages. 2006.

Vol. 4248: S. Staab, V. Svátek (Eds.), Managing Knowledge in a World of Networks. XIV, 400 pages. 2006.

Vol. 4246: M. Hermann, A. Voronkov (Eds.), Logic for Programming, Artificial Intelligence, and Reasoning. XIII, 588 pages. 2006.

Vol. 4223: L. Wang, L. Jiao, G. Shi, X. Li, J. Liu (Eds.), Fuzzy Systems and Knowledge Discovery. XXVIII, 1335 pages. 2006.

Vol. 4213: J. Fürnkranz, T. Scheffer, M. Spiliopoulou (Eds.), Knowledge Discovery in Databases: PKDD 2006. XXII, 660 pages. 2006.

Vol. 4212: J. Fürnkranz, T. Scheffer, M. Spiliopoulou (Eds.), Machine Learning: ECML 2006. XXIII, 851 pages. 2006.

Vol. 4211: P. Vogt, Y. Sugita, E. Tuci, C.L. Nehaniv (Eds.), Symbol Grounding and Beyond. VIII, 237 pages. 2006.

Vol. 4203: F. Esposito, Z.W. Raś, D. Malerba, G. Semeraro (Eds.), Foundations of Intelligent Systems. XVIII, 767 pages. 2006.

Vol. 4201: Y. Sakakibara, S. Kobayashi, K. Sato, T. Nishino, E. Tomita (Eds.), Grammatical Inference: Algorithms and Applications. XII, 359 pages. 2006.

Vol. 4200: I.F.C. Smith (Ed.), Intelligent Computing in Engineering and Architecture. XIII, 692 pages. 2006.

Vol. 4198: O. Nasraoui, O. Zaïane, M. Spiliopoulou, B. Mobasher, B. Masand, P.S. Yu (Eds.), Advances in Web Mining and Web Usage Analysis. IX, 177 pages. 2006.

Vol. 4196: K. Fischer, I.J. Timm, E. André, N. Zhong (Eds.), Multiagent System Technologies. X, 185 pages. 2006.

Vol. 4188: P. Sojka, I. Kopeček, K. Pala (Eds.), Text, Speech and Dialogue. XV, 721 pages. 2006.

Vol. 4183: J. Euzenat, J. Domingue (Eds.), Artificial Intelligence: Methodology, Systems, and Applications. XIII, 291 pages. 2006.

Vol. 4180: M. Kohlhase, OMDoc – An Open Markup Format for Mathematical Documents [version 1.2]. XIX, 428 pages. 2006.

Vol. 4177: R. Marín, E. Onaindía, A. Bugarín, J. Santos (Eds.), Current Topics in Artificial Intelligence. XV, 482 pages. 2006.

Vol. 4160: M. Fisher, W. van der Hoek, B. Konev, A. Lisitsa (Eds.), Logics in Artificial Intelligence. XII, 516 pages. 2006.

Vol. 4155: O. Stock, M. Schaerf (Eds.), Reasoning, Action and Interaction in AI Theories and Systems. XVIII, 343 pages. 2006.

Vol. 4149: M. Klusch, M. Rovatsos, T.R. Payne (Eds.), Cooperative Information Agents X. XII, 477 pages. 2006.

Vol. 4140: J.S. Sichman, H. Coelho, S.O. Rezende (Eds.), Advances in Artificial Intelligence - IBERAMIA-SBIA 2006. XXIII, 635 pages. 2006.

Vol. 4139: T. Salakoski, F. Ginter, S. Pyysalo, T. Pahikkala (Eds.), Advances in Natural Language Processing. XVI, 771 pages. 2006.

Vol. 4133: J. Gratch, M. Young, R. Aylett, D. Ballin, P. Olivier (Eds.), Intelligent Virtual Agents. XIV, 472 pages. 2006.

Vol. 4130: U. Furbach, N. Shankar (Eds.), Automated Reasoning. XV, 680 pages. 2006.

Vol. 4120: J. Calmet, T. Ida, D. Wang (Eds.), Artificial Intelligence and Symbolic Computation. XIII, 269 pages. 2006.

Vol. 4118: Z. Despotovic, S. Joseph, C. Sartori (Eds.), Agents and Peer-to-Peer Computing. XIV, 173 pages. 2006.

Vol. 4114: D.-S. Huang, K. Li, G.W. Irwin (Eds.), Computational Intelligence, Part II. XXVII, 1337 pages. 2006.

Vol. 4108: J.M. Borwein, W.M. Farmer (Eds.), Mathematical Knowledge Management. VIII, 295 pages. 2006.

Vol. 4106: T.R. Roth-Berghofer, M.H. Göker, H.A. Güvenir (Eds.), Advances in Case-Based Reasoning. XIV, 566 pages. 2006.